Pronouncing Organism Names (*continued*)

Chlamydomonas klam-i-dō-mō'näs

Chlamydophila pneumoniae kla-mi'dof-i-la nü-mō'nē-ī

C. psittaci sit'a-sē

Chromobacter violaceum krō-mō-bak'-tėr vī-ō-lā'sē-um

Claviceps purpurea kla'vi-seps pür-pü-rē'ä

Clostridium acetobutylicum klôs-tri'dē-um ā-sē-tō-bū-til'i-kum

C. botulinum bot-ū-lī'num

C. difficile dif'fi-sil-ē

C. perfringens pėr-frin'jens

C. tetani te'tän-ē

Coccidioides immitis kok-sid-ē-oi'dēz im'mi-tis

C. posadasii pō-sä-da'sē-ē

Corynebacterium diphtheriae kôr'ē-nē-bak-ti-rē-um dif-thi'rē-ī

Coxiella burnetii käks'ē-el-lä bėr-ne'tē-ē

Cryptococcus neoformans krip'tō-kok-kus nē-ō-fôr'manz

Cryptosporidium coccidi krip'tō-spô-ri-dē-um kok'sid-ē

C. hominis hō'mi-nis

C. parvum pär'vum

Cyclospora cayetanensis sī'klō-spô-rä kī'ē-tan-en-sis

Deinococcus radiodurans dī'nō-kok-kus rā-dē-ō-dür'anz

Desulfovibrio dē'sul-fō-vib-rē-ō

Desulfuromonas dē'sul-für-ō-mō-näs

Echinococcus granulosus ē-kīn-ō-kok'kus gra-nū-lō'sis

Ehrlichia chaffeensis ėr'lik-ē-ä chäf-fen'sis

E. phagocytophila fā-gō-cī-to'fī-lä

Emmonsiella capsulata em'mon-sē-el-lä cap-sül-ä'tä

Entamoeba histolytica en-tä-mē'bä his-tō-li'ti-kä

Enterobacter aerogenes en-te-rō-bak'tėr ā-rä'jen-ēz

E. cloacae klō-ā'ki

Enterobius vermicularis en-te-rō'bē-us ver-mi-kū-lar'is

Enterococcus faecalis en-tė-rō-kok'kus fē-kā'lis

E. faecium fē'sē-um

Epidermophyton ep-ē-der-mō-fī'ton

Erysipelothrix rhusiopathiae ār-ē-sip'e-lō-thriks rü'sī-ō-pa-thē

Escherichia coli esh-ėr-ē'kē-ä kō'lī (or kō'lē)

Euglena ū-glē'nä

Filobasidiella neoformans fī-lo-ba-si-dē-el'lä nē-o-fôr'mäns

Francisella tularensis fran'sis-el-lä tü'lä-ren-sis

Fusobacterium fū-sō-bak-ti'rē-um

Gambierdiscus toxicus gam'bē-ėr-dis-kus toks'i-kus

Gardnerella intestinalis gärd-nė-rel'lä in-tes-ti-nal'is

G. vaginalis va-jin-al'is

Geobacillus stearothermophilus jē-ō-bä-sil'lus ste-är-ō-thėr-mä'fil-us

Giardia lamblia jē-är'dē-ä lam'lē-ä

Gluconobacter glü'kon-ō-bak-tėr

Gonyaulax catanella gon-ē-ō'laks kat-ä-nel'lä

Gymnodinium jim-nō-din'ē-um

Haemophilus ducrcyi hē-mä'fil-us dü-krä'ē

H. influenzae in-flü-en'zī

Halobacterium salinarum ha-lō-bak-ti'rē-um sal-i-när'um

Hartmannella vermiformis hart-mä-nel'lä vêr-mi-fôr'mis

Helicobacter pylori hē'lik-ō-bak-tėr pī'lō-rē

Histoplasma capsulatum his-tō-plaz'mä kap-su-lä'tum

Klebsiella pneumoniae kleb-sē-el'lä nü-mō'ne-ī

Lactobacillus acidophilus lak-tō-bä-sil'lus a-sid-o'fil-us

L. bulgaricus bul-gä'ri-kus

L. caseii kā'sē-ē

L. plantarum plan-tär'um

L. sanfranciscensis san-fran-si-sen'-sis

Lactococcus lactis lak-tō-kok'kus lak'tis

Lagenidium giganteum la-je-ni'dē-um jī-gan'tē-üm

Legionella pneumophila lē-jä-nel'lä nü-mō'fi-lä

Leishmania donovani lish'mä-nē-ä don'ō-vän-ē

L. tropica trop'i-kä

Leptospira interrogans lep-tō-spī'rä in-tėr'rō-ganz

Leuconostoc citrovorum lü-kū-nos'tok sit-rō-vôr'um

L. mesenteroides mes-en-ter-oi'dēz

Listeria monocytogenes lis-te'rē-ä mo-nō-sī-tô'je-nēz

Methanobacterium meth-a-nō-bak-tėr'ē-um

Methanococcus jannaschii meth-a-nō-kok'kus jan-nä'shē-ē

Micrococcus luteus mī-krō-kok'kus lü'tē-us

Micromonospora mī-krō-mō-nos'pōr-ä

(continued on inside back cover)

W9-AUQ-547

Alcamo's FUNDAMENTALS OF
Microbiology

Jones and Bartlett Titles in Biological Science

Alcamo's FUNDAMENTALS OF
Microbiology
Ninth Edition

Jeffrey C. Pommerville

Professor of Biology and Microbiology
Glendale Community College
Glendale, Arizona

JONES AND BARTLETT PUBLISHERS

Sudbury, Massachusetts

BOSTON   TORONTO   LONDON   SINGAPORE

World Headquarters

Jones and Bartlett Publishers
40 Tall Pine Drive
Sudbury, MA 01776
978-443-5000
info@jbpub.com
www.jbpub.com

Jones and Bartlett Publishers Canada
6339 Ormindale Way
Mississauga, Ontario L5V 1J2
Canada

Jones and Bartlett Publishers International
Barb House, Barb Mews
London W6 7PA
United Kingdom

Jones and Bartlett's books and products are available through most bookstores and online booksellers. To contact Jones and Bartlett Publishers directly, call 800-832-0034, fax 978-443-8000, or visit our website www.jbpub.com.

Substantial discounts on bulk quantities of Jones and Bartlett's publications are available to corporations, professional associations, and other qualified organizations. For details and specific discount information, contact the special sales department at Jones and Bartlett via the above contact information or send an email to specialsales@jbpub.com.

Copyright © 2011 by Jones and Bartlett Publishers, LLC

All rights reserved. No part of the material protected by this copyright may be reproduced or utilized in any form, electronic or mechanical, including photocopying, recording, or by any information storage and retrieval system, without written permission from the copyright owner.

Production Credits

Chief Executive Officer: Clayton Jones
Chief Operating Officer: Don W. Jones, Jr.
President, Higher Education and Professional Publishing:
  Robert W. Holland, Jr.
V.P., Sales: William J. Kane
V.P., Design and Production: Anne Spencer
V.P., Manufacturing and Inventory Control: Therese Connell
Publisher, Higher Education: Cathleen Sether
Acquisitions Editor: Molly Steinbach
Senior Editorial Assistant: Jessica Acox

Editorial Assistant: Caroline Perry
Associate Production Editor: Leah Corrigan
Senior Marketing Manager: Andrea DeFronzo
Composition: Shepherd, Inc.
Cover Design: Scott Moden
Senior Photo Researcher and Photographer: Christine Myaskovsky
Cover Image: © Eye of Science/Photo Researchers, Inc.
Printing and Binding: Courier Kendallville
Cover Printing: Courier Kendallville

About the cover: A false-color transmission electron microscope image of the bacterium *Escherichia coli*. These rod-shaped cells are a common inhabitant in the human large intestine where they use the short whisker-like appendages to attach to the intestinal lining.

Library of Congress Cataloging-in-Publication Data

Pommerville, Jeffrey C.
Alcamo's fundamentals of microbiology / Jeffrey C. Pommerville. — 9th ed.
  p. ; cm.
Other title: Fundamentals of microbiology
Includes bibliographical references and index.
ISBN 978-0-7637-6258-2 (alk. paper)
1. Microbiology. 2. Medical microbiology. I. Alcamo, I. Edward. II. Title. III. Title: Fundamentals of microbiology.
[DNLM: 1. Microbiology. QW 4 P787a 2011]
QR41.2.A43 2011
616.9'041—dc22
6048   2010002117

Printed in the United States of America
14 13 12 11 10 10 9 8 7 6 5 4 3 2 1

Brief Contents

Contents

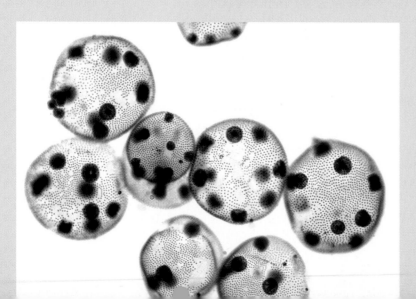

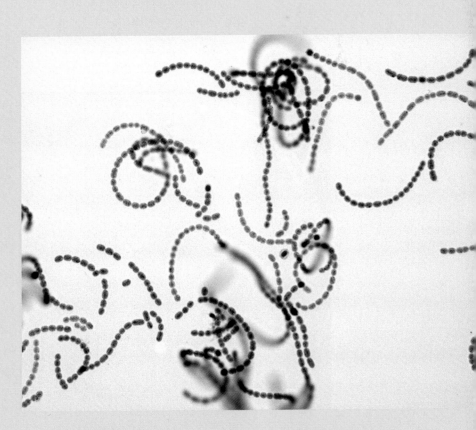

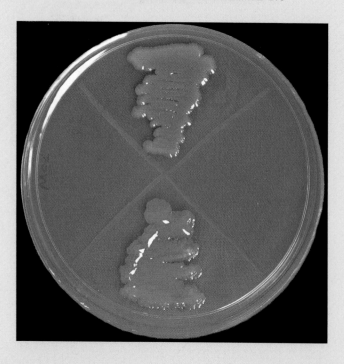

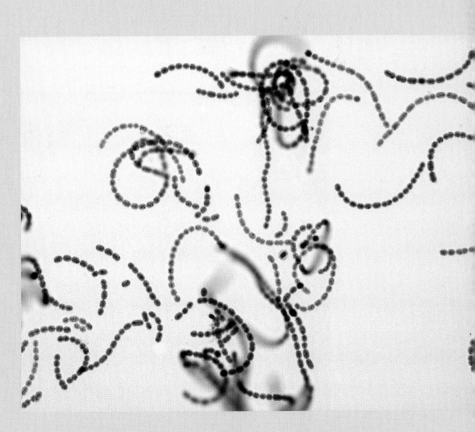

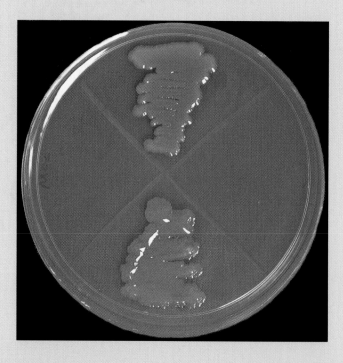

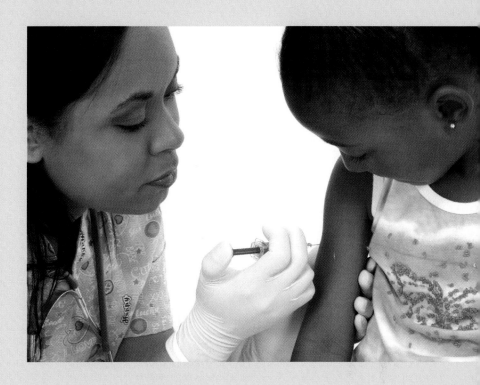

Preface

Do We Need to Know...?

As you embark upon your studies of the microbial world—and a fascinating world it will be—you will wonder how important some topics are that will be covered. You may ask yourself—or your instructor—how important is this topic to my career? If it is nursing, obviously the material on infectious diseases, epidemiology (the scientific and medical study of the causes, transmission, and control of a disease within a population), and immunology (the study of how our bodies fight an infection or disease) are critical. If you are planning on pharmacy school, add antimicrobial drugs to that list.

But what about some of the other "science" topics, like microbial metabolism and genetics? Are these important to a successful career? A few years ago, the forum section on a Web site called allnurses.com ("A Nursing Community for Nurses") asked nurses "What [microbiology] topics keep resurfacing in your nursing classes?" Among the top responses were some obvious ones, such as antibiotics and antibiotic resistance, infectious diseases, and immunity. But also on that list was microbial genetics.

Perhaps more revealing were the responses to a second question: "What do you wish that you had learned better in microbiology that you thought you would never see again [in your nursing classes]?" Among the answers submitted was metabolism.

Another survey published in *Focus on Microbiology Education* in 2006 (Volume 12 No. 2, p. 7–9) asked nurse educators what they thought were important topics for their students to learn in a microbiology class. The top six in importance were:

1. Bacterial structures and their functions;
2. Viral structures and their functions;
3. Epidemiology and public heath issues;
4. Antibiotics;
5. Immunology;
6. Disinfection and antisepsis.

Rounding out the top 10 were:

7. Bacterial metabolism;
8. Fungal structures and their functions;
9. Microbial genetics;
10. Biotechnology (production of vaccines, medicines, and diagnostic techniques).

Notice that in both surveys, the topics of microbial metabolism and microbial genetics are among the top 10 concepts to master and understand. So, make sure you pay attention to what your instructor has to say and what "we need to know" (understand) about these topics. They are important—and they will show up again in your nursing courses!

Besides metabolism and genetics, there is a substantial amount of other information you will need to learn and understand. To facilitate this understanding and coordinate it with class material, I developed a "learning design" format for the textbook (described below) to make reading easier, studying more efficient, and learning uncomplicated. Most importantly, the design allows you to better evaluate your learning and provides you with the tools needed to probe your understanding—that is, chapter learning aids and assessment drills to evaluate your progress. Realize, a prepared student knows her or his mastery before an exam—not as a result of the exam! The "learning design" format facilitates this need.

I am excited that you are using and reading this new, ninth edition of *Alcamo's Fundamentals of Microbiology*. I hope it is very useful in your studies and also enjoyable to read. Always take time to read many of the sidebars (MicroFocus boxes) whether they are assigned or not. They will help in your overall microbiology experience and the realization that microorganisms do rule the world!

Audience

Alcamo's Fundamentals of Microbiology, Ninth Edition, is written for introductory microbiology courses having an emphasis on the biology of human disease. It is geared toward students in health and allied health science curricula such as nursing, dental hygiene, medical assistance, sanitary science, and medical laboratory technology. It also will be an asset to students studying pharmacy, food science, agriculture, environmental science, and health administration. In addition, the text provides a firm foundation for advanced programs in biological sciences, as well as medicine, dentistry, and other health professions.

Organization

Alcamo's Fundamentals of Microbiology, Ninth Edition, is divided into six major areas of concentration. These areas use basic principles as frameworks to provide the unity and diversity of microbiology. Among the principles explored are the variations in structure and growth of microorganisms, the basis for infectious disease and resistance, and the beneficial effects microorganisms have on our lives.

Part 1 deals with the foundations of microbiology. It includes chapters on the origins of microbiology and the universal concepts of growth and metabolism that underpin the science. Part 2 then covers important material on the genetics of microorganisms, including genetic engineering and biotechnology. The discussions carry over to Part 3, where the spectrum of bacterial diseases is surveyed. Part 4 looks at the significance of other microorganisms, including viruses, fungi, and the protozoal and multicellular parasites.

In Part 5 of the text, the emphasis turns to infectious disease and the body's resistance through the immune system. Here, we study the reasons for disease and the means for surviving it. Antibiotics and antibiotic resistance are also covered. Part 6 closes the text with brief discussions of how public health measures interrupt epidemics. Some key insights also are given on the positive effects microorganisms exert through biotechnology.

> ▬ Rems (Roentgen
> Equivalent Man):
> **A measure of radiation dose related to biological effect.**

Marginal Definition

What's New and Important

Besides the continued emphasis on a global perspective on infection, this edition provides detailed updates to microbial structure and function, disease information and statistics, and the immune system.

Chapter Organization

The chapter sequence and number remain the same as in the previous edition, with one exception: the material on physical and chemical control methods has been moved up to Chapter 7, supplying many applications to the more detailed material in the previous chapters.

The "Learning Design" Concept

The text format includes activities designed to encourage student interaction and assessment. These design elements form an integrated study and learning package (learning tools) for student understanding and assessment.

- **Chapter Introductions** provide a stimulating thought or historical perspective to set the tone for the chapter.
- **Key Concepts** present statements identifying the important concepts in the upcoming section and alert you to the significance of that written material.
- **Boldface Terms** highlight important terms and ideas in the text.
- **Marginal Definitions** present succinct definitions of notable terms as they enter the discussion.
- **Marginal Drawings** provide visual images of bacterial shapes and cell arrangements and eukaryotic cells.
- **Marginal Chemical Structures** present structural formulas for many of the antimicrobial drugs described in Chapter 24.
- **Concept and Reasoning Checks** allow you to pause and either summarize the information presented in the previous section or critically reason through a question pertaining to the previous section.
- **MicroFocus Boxes** explore interesting topics concerning microbiology or microorganisms.
- **MicroInquiry Boxes** allow you to investigate (usually interactively) some important aspect of the chapter being studied.

- **Textbook Cases** are embedded in many chapters to help you understand pathogens by presenting contemporary disease outbreaks originally reported by the Centers for Disease Control and Prevention.
- **Figure Questions** further reinforce your understanding of microbiology concepts described in the text.
- **Summary Tables** pull together the similarities and differences of topics discussed in the chapter.
- **Pronouncing Microorganism Names** (inside front and back covers) helps you correctly pronounce those sometimes tongue-twisting microorganism names.
- **Summaries of Key Concepts** condense the major ideas discussed in the chapter. The "learning design" package also includes many useful and important end-of-chapter student assessments.
- **Learning Objectives** outline the important concepts in the chapters through Bloom's Taxonomy, a classification of levels of intellectual skills important in learning.
- **Self-Test questions** (Step A) are multiple-choice questions focusing on concrete "facts" learned in the chapter. Let's face it: there is information that needs to be memorized in order to reason critically.
- **Chapter Review** (Step B) contains questions of a somewhat unconventional type to assist review of the chapter contents.
- **Questions for Thought and Discussion** (Step C) encourage students to use the text to resolve thought-provoking problems with contemporary relevance.
- **Applications** (Step D) are questions requiring students to reason critically through a problem of practical significance.

Being Skeptical

One of the seven types of essay boxes new to this edition is titled "Being Skeptical." A good scientist is a skeptic and skepticism is an important part of science. Skepticism, unlike cynicism, is not unwilling to accept a claim or observation. Skepticism simply says "Prove it!" Science applies scientific reasoning as the method for proof. Thus, a scientist, such as a microbiologist,

MicroFocus Box

must see the evidence and it must be compelling before the observation or statement is provisionally accepted. The claim is still open to further examination and experimentation.

The "Being Skeptical" essays scattered through the textbook present an often-fantastic statement or claim. The essay then examines the claim using reasoning skills and the scientific process, which is sometimes called the scientific method.

Why Pathogens?

Microorganisms perform many useful services for humans when they produce food products, manufacture organic materials in industrial plants, and recycle such elements as carbon and nitrogen. The emphasis of this book, however, is on the tiny, but significant percentage of microorganisms causing human disease, the so-called pathogens. Why do we emphasize pathogens? Here are several reasons:

- Pathogens have regularly altered the course of human history.
- Pathogens are familiar to audiences of microbiology.
- Pathogens add drama to an invisible world of microorganisms.

- Pathogens illustrate ecological relationships between humans and microorganisms.
- Pathogens point up the diversity of microorganisms.

Moreover, the study of pathogens makes basic science relevant and shows how microbiology interfaces with other disciplines such as sociology, economics, history, politics, and geography. Finally, the study of pathogens helps us to understand contemporary newspaper articles, magazine headlines, and stories on the news. And in the end, that makes us better citizens. Indeed, the famous essayist Thomas Mann once wrote, "All interest in disease is only another expression of interest in life."

Additional Resources

Jones and Bartlett offers an array of ancillaries to assist instructors and students in teaching and mastering the concepts in this text. Additional information and review copies of any of the following items are available through your Jones and Bartlett sales representative or by going to www.jbpub.com/biology.

For the Student

Part 6 of this book, "Environmental and Applied Microbiology," is available online with the access code bound into every new copy of this text (in North America). Additional access codes are available for purchase separately.

The Web site we developed exclusively for the ninth edition of this text, http://microbiology.jbpub.com/9e, offers a variety of resources to enhance understanding of microbiology. The site contains eLearning, a free on-line study guide with chapter outlines, chapter essay questions, key term reviews, and short study quizzes.

The *Study Guide* to accompany this textbook contains important information to help you study, take effective class notes, prepare properly for exams, and even to manage your time effectively. The latter is the single most common reason for poor performance in college courses. The *Study Guide* also contains over 3,000 practice exercises and study questions of various types to help you learn and retain the information in the text.

Laboratory Fundamentals of Microbiology, Ninth Edition, is a series of over 30 multipart laboratory exercises providing basic training in the handling of microorganisms and reinforcing ideas and concepts described in the textbook.

Guide to Infectious Diseases by Body Systems is an excellent tool for learning about microbial diseases. Each of the fifteen body systems units presents a brief introduction to the anatomical system and the bacterial, viral, fungal, or parasitic organism infecting the system.

An anthology called *Encounters in Microbiology* (*Volume I, Second Edition,* and *Volume II*) brings together "Vital Signs" articles from *Discover* magazine in which health professionals use their knowledge of microbiology in their medical cases.

For the Instructor

Compatible with Windows® and Macintosh® platforms, the Instructor's Media CD-ROM provides instructors with the following traditional ancillaries:

- The *PowerPoint® Image Bank* provides the illustrations, photographs, and tables (to which Jones and Bartlett Publishers holds the copyright or has permission to reproduce digitally) inserted into PowerPoint slides. You can quickly and easily copy individual images or tables into your existing lecture slides.
- The *PowerPoint Lecture Outline Slides* presentation package, prepared by Jean Revie of South Mountain Community College, provides lecture notes and images for each chapter of *Alcamo's Fundamentals of Microbiology*. Instructors with the Microsoft PowerPoint software can customize the outlines, art, and order of presentation.

The following materials are also available online, at http://www.jbpub.com/catalog/9780763762582.

- The *Instructor's Manual,* provided as a text file, includes chapter summaries and complete chapter lecture outlines and answers to all the end-of-chapter assessments.
- *Chapter Assessments Answers* provide short answers to figure questions, Concept and Reasoning Checks, and all end-of-chapter materials.
- The *Test Bank* is available as straight text files. It has been updated by Cindy Ault of Jamestown College, Jackie Reynolds of Richland College, and Sue Katz of Rogers State University.

Acknowledgments

It is always my pleasure to thank everyone at Jones and Bartlett Publishers who helped put together this new version of the textbook. Cathleen Sether has been a more-than-able publisher—just don't cross international borders with her. Leah Corrigan, the production editor, has been a pleasure to work with, and Lou Bruno continues to display his mastery of the production process; Anne Spencer developed the new design format; Caroline Perry ably assisted everyone; Christine Myaskovsky tracked down many of the great photos that embellish these pages; Deborah Patton read every page and created the index; Shellie Newell was again the "eagle-eye" copy editor; and Elizabeth Morales provided much of the excellent art in this new edition.

The book benefited from the expertise of several fellow microbiologists and biologists. I wish to thank my colleagues Philip Fernandez and Michael McKinley for their input during the writing of this edition. I especially want to thank Brett Miller, a former microbiology student and a GCC biotech major who read every word of the eighth edition, making note of typographic errors, syntax and grammatical errors, and all unclear statements found in the text.

After more than 25 years of university and college instruction, I must thank all my former students who keep me on my toes in the classroom and require me to always be prepared. Their suggestions and evaluations have encouraged me to continually assess my instruction so it can be easily understood. I salute you, and I hope those of you who read this text will let me know what works and what still needs improvement to make your learning efficient and still enjoyable.

Jeff Pommerville
Scottsdale, AZ

About the Author

Today I am a microbiologist, researcher, and science educator. My plans did not start with that intent. While in high school in Santa Barbara, California, I wanted to play professional baseball, study the stars, and own a '66 Corvette. None of these desires would come true—my batting average was miserable (but I was a good defensive fielder), I hated the astronomy correspondence course I took, and I never bought that Corvette.

I found an interest in biology at Santa Barbara City College. After squeaking through college calculus, I transferred to the University of California at Santa Barbara (UCSB) where I received a B.S. in Biology and stayed on to pursue a Ph.D. degree in the lab of Ian Ross studying cell communication and sexual pheromones in a water mold. After receiving my doctorate in Cell and Organismal Biology, my graduation was written up in the local newspaper as a native son who was a fungal sex biologist—an image that was not lost on my three older brothers!

While in graduate school at UCSB, I rescued a secretary in distress from being licked to death by a German Shepherd. Within a year, we were married (the secretary and I). When I finished my doctoral thesis, I spent several years as a postdoctoral fellow at the University of Georgia. I worried that I was involved in too many research projects, but a faculty member told me something I will never forget. He said, "Jeff, it's when you can't think of a project or what to do that you need to worry." Well, I have never had to worry!

I then moved on to Texas A&M University, where I spent eight years in teaching and research—and telling Aggie jokes. Toward the end of this time, after publishing over 30 peer-reviewed papers in national and international research journals, I realized I had a real interest in teaching and education. Leaving the sex biologist nomen behind, I headed farther west to Arizona to join the biology faculty at Glendale Community College, where I continue to teach introductory biology and microbiology.

I have been lucky to be part of several educational research projects and have been honored, with two of my colleagues, with a Team Innovation of the Year Award by the League of Innovation in the Community Colleges. In 2000, I became project director and lead principal investigator for a National Science Foundation grant to improve student outcomes in science through changes in curriculum and pedagogy. I had a fascinating three years coordinating more than 60 science faculty members (who at times were harder to manage than students) in designing and field testing 18 interdisciplinary science units. This culminated with me being honored in 2003 with the Gustav Ohaus Award (College Division) for Innovations in Science Teaching from the National Science Teachers Association.

I am an associate editor for the Journal of Microbiology and Biology Education, the education research journal of the American Society for Microbiology (ASM) and in 2004 was co-chair for the ASM Conference for Undergraduate Educators. From 2006 to 2007, I was the chair of Undergraduate Education Division of ASM. In 2006, I was selected as one of four outstanding instructors at Glendale Community College. The culmination of my teaching career came in 2008 when I was nationally recognized by being awarded the Carski Foundation Distinguished Undergraduate Teaching Award for excellence in teaching microbiology to undergraduate students and encouraging them to subsequent achievement.

I mention all this not to impress but to show how the road of life sometimes offers opportunities in unexpected and unplanned ways. The key though is keeping your "hands on the wheel and your eyes on the prize"; then unlimited opportunities will come your way. With the untimely passing of my friend and professional colleague Ed Alcamo, also the Carski recipient in 2000, I was privileged in 2003 to be offered the opportunity to take over the authorship of Fundamentals of Microbiology. It is an undertaking I continue to relish as I (along with the wonderful folks at Jones and Bartlett) try to evolve a new breed of microbiology textbook reflecting the pedagogy change occurring in science classrooms today. And, hey, who knows—maybe that '66 Corvette could be in my garage yet.

DEDICATION

I dedicate this edition of the book to my late mother and father. Their willingness to let me "explore" the natural world guided me to where I am today. I hope they are pleased with the outcome.

To the Student— Study Smart

Your success in microbiology and any college or university course will depend on your ability to study effectively and efficiently. Therefore, this textbook was designed with you, the student, in mind. The text's organization will help you improve your learning and understanding and, ultimately, your grades. The learning design concept described in the Preface and illustrated below reflects this organization. Study it carefully, and, if you adopt the flow of study shown, you should be a big step ahead in your preparation and understanding of microbiology—and for that matter any subject you are taking.

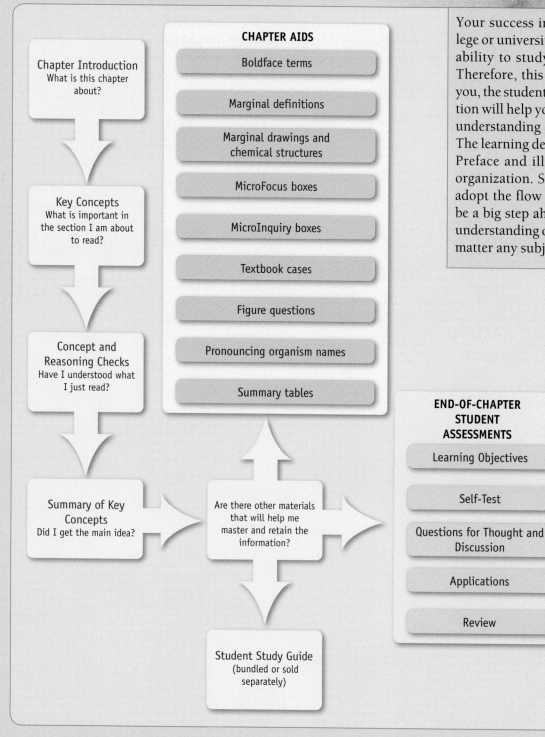

Chapter Introduction
What is this chapter about?

Key Concepts
What is important in the section I am about to read?

Concept and Reasoning Checks
Have I understood what I just read?

Summary of Key Concepts
Did I get the main idea?

Are there other materials that will help me master and retain the information?

Student Study Guide
(bundled or sold separately)

CHAPTER AIDS

Boldface terms

Marginal definitions

Marginal drawings and chemical structures

MicroFocus boxes

MicroInquiry boxes

Textbook cases

Figure questions

Pronouncing organism names

Summary tables

END-OF-CHAPTER STUDENT ASSESSMENTS

Learning Objectives

Self-Test

Questions for Thought and Discussion

Applications

Review

When I was an undergraduate student, I hardly ever read the "To the Student" section (if indeed one existed) in my textbooks because the section rarely contained any information of importance. This one does, so please read on.

In college, I was a mediocre student until my junior year. Why? Mainly because I did not know how to study properly, and, important here, I did not know how to read a textbook effectively. My textbooks were filled with underlined sentences (highlighters hadn't been invented yet!) without any plan on how I would use this "emphasized" information. In fact, most textbooks *assume* you know how to read a textbook properly. I didn't and you might not, either.

Reading a textbook is difficult if you are not properly prepared. So you can take advantage of what I learned as a student and have learned from instructing thousands of students; I have worked hard to make this text user friendly with a reading style that is not threatening or complicated. Still, there is a substantial amount of information to learn and understand, so having the appropriate reading and comprehension skills is critical. Therefore, I encourage you to spend 30 minutes reading this section, as I am going to give you several tips and suggestions for acquiring those skills. Let me show you how to be an active reader. Note: the student *Study Guide* also contains similar information on how to take notes from the text, how to study, how to take class (lecture) notes, how to prepare for and take exams, and perhaps most important for you, how to manage your time effectively. It all is part of this "learning design," my wish to make you a better student.

BE A PREPARED READER

Before you jump into reading a section of a chapter in this text, prepare yourself by finding the place and time and having the tools for study.

Place. Where are you right now as you read these lines? Are you in a quiet library or at home? If at home, are there any distractions, such as loud music, a blaring television, or screaming kids? Is the lighting adequate to read? Are you sitting at a desk or lounging on the living room sofa? Get where I am going? When you read for an educational purpose—that is, to learn and understand something—you need to maximize the environment for reading. Yes, it should be comfortable but not to the point that you will doze off.

Time. All of us have different times during the day when we perform some skill, be it exercising or reading, the best. The last thing you want to do is read when you are tired or simply not "in tune" for the job that needs to be done. You cannot learn and understand the information if you fall asleep or lack a positive attitude. I have kept the chapters in this text to about the same length so you can estimate the time necessary for each and plan your reading accordingly. If you have done your preliminary survey of the chapter or chapter section, you can determine about how much time you will need. If 40 minutes is needed to read—and comprehend (see below)—a section of a chapter, find the place and time that will give you 40 minutes of uninterrupted study. Brain research suggests that most people's brains cannot spend more than 45 minutes in concentrated, technical reading. Therefore, I have avoided lengthy presentations and instead have focused on smaller sections, each with its own heading. These should accommodate shorter reading periods.

Reading Tools. Lastly, as you read this, what study tools do you have at your side? Do you have a highlighter or pen for emphasizing or underlining important words or phrases? Notice, the text has wide margins, which allow you to make notes or to indicate something that needs further clarification. Do you have a pencil or pen handy to make these notes? Or, if you do not want to "deface" the text, make your notes in a notebook. Lastly, some students find having a ruler is useful to prevent your eyes from wandering on the page and to read each line without distraction.

BE AN EXPLORER BEFORE YOU READ

When you sit down to read a section of a chapter, do some preliminary exploring. Look at the section head and subheadings to get an idea of what is discussed. Preview any diagrams, photographs, tables, graphs, or other visuals used. They give you a better idea of what is going to occur. We have used a good deal of space in the text for these features, so use them to your advantage. They will help you learn the written information and comprehend its meaning. Do not try to understand all the visuals, but try to generate a mental "big picture" of what is to come. Familiarize yourself with any symbols or technical jargon that might be used in the visuals.

The end of each chapter contains a **Summary of Key Concepts** for that chapter. It is a good idea to read the summary before delving into the chapter. That way you will have a framework for the chapter before filling in the nitty-gritty information.

BE A DETECTIVE AS YOU READ

Reading a section of a textbook is not the same as reading a novel. With a textbook, you need to uncover the important information (the terms and concepts) from the forest of words on the page. So, the first thing to do is read the complete paragraph. When you have determined the main ideas, highlight or underline them. However, I have seen students highlighting the entire paragraph in yellow, including every *a*, *the*, and *and*. This is an example of highlighting before knowing what is important. So, I have helped you out somewhat. Important terms and concepts are in **bold face** followed by the definition (or the definition might be in the margin). So only highlight or underline with a pen essential ideas and key phrases—not complete sentences, if possible. By the way, the important microbiological terms and major concepts also are in the **Glossary** at the back of the text.

What if a paragraph or section has no bold-faced words? How do you find what is important here? From an English course, you may know that often the most important information is mentioned first in the paragraph. If it is followed by one or more examples, then you can backtrack and know what was important in the paragraph. In addition, I have added section "speed bumps" (called **Concept and Reasoning Checks**) to let you test your learning and understanding before getting too far ahead in the material. These checks also are clues to what was important in the section you just read.

BE A REPETITIOUS STUDENT

Brain research has shown that each individual can only hold so much information in short-term memory. If you try to hold more, then something else needs to be removed—sort of like a full computer disk. So that you do not lose any of this important information, you need to transfer it to long-term memory—to the hard drive if you will. In reading and studying, this means retaining the term or concept; so, write it out in your notebook *using your own words*. Memorizing a term does not mean you have learned the term or understood the concept. By actively writing it out in your own words, you are forced to think and actively interact with the information. This repetition reinforces your learning.

BE A PATIENT STUDENT

In textbooks, you cannot read at the speed that you read your e-mail or a magazine story. There are unfamiliar details to be learned and understood—and this requires being a patient, slower reader. Actually, if you are not a fast reader to begin with, as I am, it may be an advantage in your learning process. Identifying the important information from a textbook chapter requires you to *slow down* your reading speed. Speed-reading is of no value here.

KNOW THE WHAT, WHY, AND HOW

Have you ever read something only to say, "I have no idea what I read!" As I've already mentioned, reading a microbiology text is not the same as reading *Sports Illustrated* or *People* magazine. In these entertainment magazines, you read passively for leisure or perhaps amusement. In *Alcamo's Fundamentals of Microbiology*, you must read actively for learning and understanding—that is, for *comprehension*. This can quickly lead to boredom unless you engage your brain as you read—that is, be an active reader. Do this by knowing the *what*, *why*, and *how* of your reading.

- *What* is the general topic or idea being discussed? This often is easy to determine because the section heading might tell you. If not, then it will appear in the first sentence or beginning part of the paragraph.
- *Why* is this information important? If I have done my job, the text section will tell you why it is important or the examples provided will drive the importance home. These sur-

rounding clues further explain why the main idea was important.

- *How* do I "mine" the information presented? This was discussed under being a detective.

A MARKED UP READING EXAMPLE

So let's put words into action. Below is a passage from the text. I have marked up the passage as if I were a student reading it for the first time. It uses many of the hints and suggestions I have provided. Remember, it is important to read the passage slowly, and concentrate on the main idea (concept) and the special terms that apply.

HAVE A DEBRIEFING STRATEGY

After reading the material, be ready to debrief. Verbally summarize what you have learned. This will start moving the short-term information into the long-term memory storage—that is, *retention*. Any notes you made concerning confusing material should be discussed as soon as possible with your instructor. For microbiology, allow time to draw out diagrams. Again, repetition makes for easier learning and better retention.

In many professions, such as sports or the theater, the name of the game is practice, practice, practice. The hints and suggestions I have given you form a skill that requires practice to

Many Foodborne and Waterborne Diseases Have a Bacterial Cause

KEY CONCEPT
- Bacterial gastrointestinal diseases may arise from intoxications or infections.

We categorize the foodborne and waterborne bacterial diseases as either intoxications or infections. **Intoxications** are illnesses in which bacterial toxins are ingested in food or water. Examples are the toxins causing botulism, staphylococcal food poisoning, and clostridial food poisoning. By contrast, **infections** refer to illnesses in which live bacterial pathogens in food and water are ingested and subsequently grow in the body. Salmonellosis, shigellosis, and cholera are examples. Toxins may be produced, but they are the result of infection.

Determining the etiology of a bacterial disease depends on several factors.

Incubation period. If an individual ingests and swallows a contaminated food or beverage, there is a delay, called the **incubation period**, before the symptoms appear. This period can range from hours to days, depending on the bacterial species and on the infectious dose. During the incubation period, the toxins or microbes pass through the stomach into the intestine where they may directly affect gastrointestinal function or be absorbed into the bloodstream.

Clinical symptoms. The symptoms produced by an intoxication or infection depend

■ **Etiology:** The study of the causes (origins) of disease.

■ **Infectious dose:** The number of organisms consumed to give rise to symptoms of an illness.

perfect and use efficiently. Be patient, things will not happen overnight; perseverance and willingness though will pay off with practice. You might also check with your college or university academic (or learning) resource center. These folks will have more ways to help you to read a textbook better and to study well overall.

CONCEPT MAPS

In science as well as in other subjects you take at the college or university, there often are concepts that appear abstract or simply so complex they are difficult to understand. A **concept map** is one tool to help you enhance your abilities to think and learn. Critical reasoning and the ability to make connections between complex, nonlinear information are essential to your studies and career.

Concept maps are a learning tool designed to represent complex or abstract information visually. Neurobiologists and psychologists tell us that the brain's primary function is to take incoming information and interpret it in a meaningful or practical way. They also have found that the brain has an easier time making sense of information when it is presented in a visual format. Importantly, concept maps not only present the information in "visual sentences" but also take paragraphs of material and present it in an "at-a-glance" format. Therefore, you can use concept maps to

- Communicate and organize complex ideas in a meaningful way
- Aid your learning by seeing connections within or between concepts and knowledge
- Assess your understanding or diagnose misunderstanding

There are many different types of concept maps. The two most used in this textbook are the *process map* or *flow chart* and the *hierarchical map*. The hierarchical map starts with a general concept (the most inclusive word or phrase) at the top of the map and descends downward using more specific, less general words or terms. In several chapters in this textbook process or hierarchical maps are drawn—and you have the opportunity to construct your own hierarchical maps as well.

Concept mapping is the strategy used to produce a concept map. So, let's see how one makes a hierarchical map.

How to Construct a Concept Map

1. Print the central idea (concept or question to be mapped) in a box at the top center of a blank, unlined piece of paper. Use uppercase letters to identify the central idea.
2. Once the concept has been selected, identify the key terms (words or short phrases) that apply to or stem from the concept. Often these may be given to you as a list. If you have read a section of a text, you can extract the terms from that material, as the words are usually boldfaced or italicized.
3. Now, from this list, try to create a hierarchy for the terms you have identified; that is, list them from the most general, most inclusive to the least general, most specific. This ranking may only be approximate and subject to change as you begin mapping.
4. Construct a preliminary concept map. This can be done by writing all of the terms on Post-its®, which can be moved around easily on a large piece of paper. This is necessary as one begins to struggle with the process of building a good hierarchical organization.
5. The concept map connects terms associated with a concept in the following way:
   - The relationship between the concept and the first term(s), and between terms, is connected by an arrow pointing in the direction of the relationship (usually downward or horizontal if connecting related terms).
   - Each arrow should have a label, a very short phrase that explains the relationship with the next term. In the end, each link with a label reads like a sentence.
6. Once you have your map completed, redraw it in a more permanent form. Box in all terms that were on the sticky notes. Remember there may be more than one way to draw a good concept map, and don't be scared off if at first you have some problems mapping; mapping will become more apparent to you after you have practiced this technique a few times using the opportunities given to you in the early chapters of the textbook.

7. Now look at the map and see if it answers the following. Does it:
   - Define clearly the central idea by positioning it in the center of the page?
   - Place all the terms in a logical hierarchy and indicate clearly the relative importance of each term?
   - Allow you to figure out the relationships among the key ideas more easily?
   - Permit you to see all the information visually on one page?
   - Allow you to visualize complex relationships more easily?
   - Make recall and review more efficient?

Example

After reading the section in Chapter 8 on "Protein Synthesis," a student makes a list of the terms used and maps the concept. Using the steps outlined above, the student produces the following hierarchical map. Does it satisfy all the questions asked in (8)?

Practical Uses for Mapping

- **Summarizing textbook readings.** Use mapping to summarize a chapter section or a whole chapter in a textbook. This purpose for mapping is used many times in this text.
- **Summarizing lectures.** Although producing a concept map during the classroom period may not be the best use of the time, making a concept map or maps from the material after class will help you remember the important points and encourage high-level, critical reasoning, which is so important in university and college studies.
- **Reviewing for an exam.** Having concept maps made ahead of time can be a very useful and productive way to study for an exam, particularly if the emphasis of the course is on understanding and applying abstract, theoretical material, rather than on simply reproducing memorized information.
- **Working on an essay.** Mapping also is a powerful tool to use during the early stages of writing a course essay or term paper. Making a

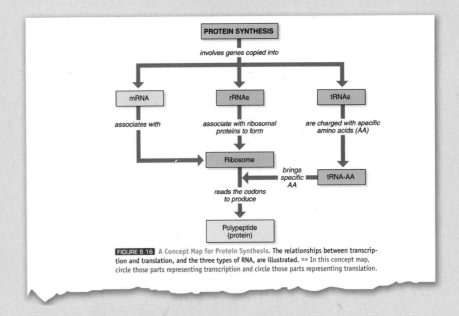

FIGURE 8.16 A Concept Map for Protein Synthesis. The relationships between transcription and translation, and the three types of RNA, are illustrated. »» In this concept map, circle those parts representing transcription and circle those parts representing translation.

concept map before you write the first rough draft can help you see and ensure you have the important points and information you will want to make.

SEND ME A NOTE

In closing, I would like to invite you to write me and let me know what is good about this textbook so I can build on it and what may need improvement so I can revise it. Also, I would be pleased to hear about any news of microbiology in your community, and I'd be happy to help you locate any information not covered in the text. I can be reached at the Department of Biology, Glendale Community College, 6000 W. Olive Avenue, Glendale, Arizona 85302. Feel free to e-mail me at: jeffrey.pommerville@gcmail.maricopa.edu.

I wish you great success in your microbiology course. Welcome! Let's now plunge into the wonderful and sometimes awesome world of microorganisms.

—Dr. P.
Web site: http://gccaz.edu/~jpommerv/

A Tribute to I. Edward Alcamo

Dr. Ignazio Edward Alcamo was a long-time Professor of Microbiology at the State University of New York at Farmingdale and the author of numerous textbooks, lab kits, and educational materials. He was the 2000 recipient of the Carski Foundation Distinguished Undergraduate Teaching Award, the highest honor bestowed upon microbiology educators by the American Society for Microbiology. Dr. Alcamo was educated at Iona College and St. John's University and held a deep belief in the partnership between research scientists and allied health educators. He sought to teach the scientific basis of microbiology in an accessible manner as well as to inspire students with a sense of topical relevance. Michael Vinciguerra, Provost at the SUNY Farmingdale wrote, "In 1970, when I joined the faculty as a chemistry professor, Ed's reputation as an excellent biology educator was already well known."

A prolific author, Dr. Alcamo produced a broad array of publications including several learning guides and textbooks—*Fundamentals of Microbiology*, now in its ninth edition, and the recently published *Microbes and Society, Second Edition*. He also prepared the Encarta encyclopedia entry entitled "Procaryotes," as well as *The Microbiology Coloring Book*, and *Schaum's Outline of Microbiology*. His other books published within the past several years include *AIDS: The Biological Basis, DNA Technology: The Awesome Skill, The Biology Coloring Workbook*, and *Anatomy and Physiology the Easy Way*. In December 2002, after a six-month illness, Dr. Alcamo died of acute myeloid leukemia.

Dr. Alcamo's teaching career was dedicated to the proposition that emphasizing quality in education is central to turning back the tide of fear and uncertainty and enabling doctors to find cures for disease. In the early 1980s, when the early cases of an unknown acquired immunodeficiency syndrome were turning into a mysterious and intractable epidemic, Dr. Alcamo told this to his class:

One afternoon, about 350 years ago, in the countryside near London, a clergyman happened to meet Plague.

"Where are you going?" asked the clergyman.

"To London," responded Plague, "to kill a thousand."

They chatted for another few moments, and each went his separate way.

Some time later, they chanced to meet again. The clergyman said, "I see you decided to show no mercy in London. I heard that 10,000 died there."

"Ah, yes," Plague replied, "but I only killed a thousand. Fear killed the rest."

Foundations of Microbiology

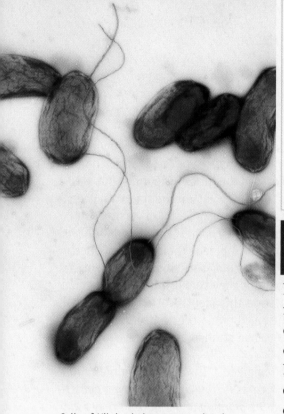

Cells of *Vibrio cholerae,* transmitted to humans in contaminated water and food, are the cause of cholera.

In 1676, a century before the Declaration of Independence, a Dutch merchant named Antony van Leeuwenhoek sent a noteworthy letter to the Royal Society of London. Writing in the vernacular of his home in the United Netherlands, Leeuwenhoek described how he used a simple microscope to observe vast populations of minute, living creatures. His reports opened a chapter of science that would evolve into the study of microscopic organisms and the discipline of microbiology. At that time, few people, including Leeuwenhoek, attached any practical significance to the microorganisms, but during the next three centuries, scientists would discover how profoundly these organisms influence the quality of our lives and the environment around us.

We begin our study of the microorganisms by exploring the grassroot developments that led to the establishment of microbiology as a science. These developments are surveyed in Chapter 1, where we focus on some of the individuals who stood at the forefront of discovery. Today we are in the midst of a third Golden Age of microbiology and our understanding of microorganisms continues to grow even as you read this book. Chapter 1, therefore, is an important introduction to microbiology then and now.

Part 1 also contains a chapter on basic chemistry, inasmuch as microbial growth, metabolism, and diversity are grounded in the molecules and macromolecules these organisms contain and in the biological processes they undergo. The third chapter in Part 1 sets down some basic concepts and describes one of the major tools for studying microorganisms. Much as the alphabet applies to word development, in succeeding chapters we will formulate words into sentences and sentences into ideas as we survey the different groups of microorganisms and concentrate on their importance to public health and human welfare.

Although most microorganisms are harmless—or even beneficial, some cause infectious disease. We will concentrate on the bacterial organisms in Chapter 4, where we survey their structural frameworks. In Chapter 5, we build on these frameworks by examining microbial growth patterns and nutritional requirements. Chapter 6 describes the metabolism of microbial cells, including those chemical reactions that produce energy and use energy. Part 1 concludes by considering the physical and chemical methods used to control microbial growth and metabolism (Chapter 7).

Being a Scientist

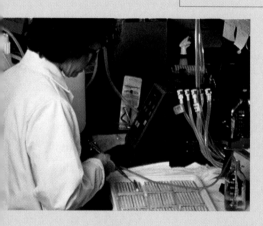

Science may not seem like the most glamorous profession. So, as you read many of the chapters in this text, you might wonder why many scientists have the good fortune to make key discoveries. At times, it might seem like it is the luck of the draw, but actually many scientists have a set of characteristics that put them on the trail to success.

Robert S. Root-Bernstein, a physiology professor at Michigan State University, points out that many prominent scientists like to goof around, play games, and surround themselves with a type of chaos aimed at revealing the unexpected. Their labs may appear to be in disorder, but they know exactly where every tube or bottle belongs. Scientists also identify intimately with the organisms or creatures they study (it is said that Louis Pasteur actually dreamed about microorganisms), and this identification brings on an intuition—a "feeling for the organism." In addition, there is the ability to recognize patterns that might bring a breakthrough. (Pasteur had studied art as a teenager and, therefore, he had an appreciation of patterns.)

The geneticist and Nobel laureate Barbara McClintock once remarked, *"I was just so interested in what I was doing I could hardly wait to get up in the morning and get at it. One of my friends, a geneticist, said I was a child, because only children can't wait to get up in the morning to get at what they want to do."* Clearly, another characteristic of a scientist is having a child-like curiosity for the unknown.

Professor Alcamo once received a letter from a student, asking why he became a microbiologist. *"It was because I enjoyed my undergraduate microbiology course"* he said, *"and when I needed to select a graduate major, microbiology seemed like a good idea. I also think I had some of the characteristics described by Root-Bernstein: I loved to try out different projects; my corner of the world qualified as a disaster area; still I was a nut on organization, insisting that all the square pegs fit into the square holes."*

For this author, science has been an extraordinary opportunity to discover and understand something never before known. Science is fun, yet challenging—and at times arduous, tedious, and frustrating. As with most of us, we will not make the headlines for a breakthrough discovery or find a cure for a disease. However, as scientists we all hope our research will contribute to a better understanding of a biological (or microbiological) phenomenon and will push back the frontiers of knowledge. For me, the opportunity of doing something I love far outweighs the difficulties along the way.

Like any profession, being a scientist is not for everyone. Besides having a bachelor's or higher degree in biology or microbiology, you should be well read in the sciences and capable of working as part of an interdisciplinary team. Of course, you should have good quantitative and communication skills, have an inquisitive mind, and be goal oriented. If all this sounds interesting, then maybe you fit the mold of a scientist. Why not consider pursuing a career in microbiology? Some possibilities are described in other Microbiology Pathways included in this book, but you should also visit with your instructor. Simply stop by the student union, buy two cups of coffee, and you are on your way.

Microbiology: Then and Now

In the field of observation, chance favors only the prepared mind.
—Louis Pasteur (1822–1895)

Space. The final frontier! Really? *The* final frontier? There are an estimated 350 billion large galaxies and more than 10^{22} stars in the visible universe. However, the microbial universe consists of more than 10^{31} microorganisms scattered among an estimated 2 to 3 billion species. So, could understanding these organisms on Earth be as important as studying galaxies in space?

In 1990, microbiologist Stephen Giovannoni of Oregon State University identified in the Sargasso Sea off the southeast United States what is perhaps the most abundant and successful organism on the planet. Called SAR11 (SAR for Sargasso), this bacterial organism, which now goes by the scientific name *Pelagibacter ubique*, has been identified across the oceans of the world. What makes it significant is its population size. Estimated to be 2.4×10^{28} cells, SAR11 alone accounts for 20% of all oceanic bacterial species—and 50% of the bacterial species in the surface waters of temperate oceans in the summer!

SAR11's success story suggests the organism must have a significant impact on the planet. Although such roles remain to be identified and understood, Giovannoni believes SAR11 is responsible for up to 10% of all nutrient recycling on the planet, influencing the cycling of carbon and even affecting climate change.

Also sailing the Sargasso Sea near Bermuda is Craig Venter and his team at the J. Craig Venter Institute. Fresh from his success with the private sector effort to sequence the human genome, Venter's team in 2004 reported the discovery of over 1,800 new microbial species in Sargasso seawater and from them isolated 1.2 million new gene sequences.

Then between 2003 and 2007, Venter's team sailed the world's oceans—á la Charles Darwin—to sample seawater and evaluate the diversity of microorganisms in these waters. For the emerging field of marine molecular

Chapter Preview and Key Concepts

1.1 The Beginnings of Microbiology
1. The discovery of microorganisms was dependent on observations made with the microscope.
2. The emergence of experimental science provided a means to test long-held beliefs and resolve controversies.

MICROINQUIRY 1: Experimentation and Scientific Inquiry

1.2 Microorganisms and Disease Transmission
3. Early epidemiology studies suggested how diseases could be spread and be controlled.
4. Resistance to a disease can come from exposure to and recovery from a mild form of (or a very similar) disease.

1.3 The Classical Golden Age of Microbiology (1854–1914)
5. The germ theory was based on the observations that different microorganisms have distinctive and specific roles in nature.
6. Antisepsis and identification of the cause of animal diseases reinforced the germ theory.
7. Koch's postulates provided a way to identify a specific microorganism as causing a specific infectious disease.
8. Laboratory science and teamwork stimulated the discovery of additional infectious disease agents.
9. Viruses also can cause disease.
10. Many beneficial bacterial species recycle nutrients in the environment.

1.4 Studying Microorganisms
11. The organisms and agents studied in microbiology represent diverse groups.

1.5 The Second Golden Age of Microbiology (1943–1970)
12. Microorganisms and viruses can be used as model systems to study phenomena common to all life.
13. All microorganisms have a characteristic cell structure.
14. Antimicrobial chemicals can be effective in treating infectious diseases.

1.6 The Third Golden Age of Microbiology—Now
15. Infectious disease (natural and intentional) preoccupies much of microbiology.
16. Microbial ecology and evolution are dominant themes in modern microbiology.

Genome:
The complete set of genetic information in a cell, organism, or virus.

microbiology, Venter believes the sequencing of marine microorganisms will provide examples of novel metabolic pathways, identify species that use alternative energy sources, and perhaps help solve critical environmental problems, including climate change.

Giovannoni and Venter are just two of many microbiologists trying to understand the role of microorganisms in the ocean's ecosystems and their dominant role on this planet. But most of all, as "Being a Scientist" identified, Giovannoni's and Venter's primary goal is a voyage of discovery. Since only about 1% of the marine microorganisms have been identified, the microbial universe does represent an inner final frontier!

The science of **microbiology** embraces a biologically diverse group of usually small life forms, encompassing primarily **microorganisms** (bacteria, fungi, algae, and protozoa) and viruses.

Microorganisms (or **microbes** for short) are present in vast numbers in nearly every environment and habitat on Earth, not just the Sargasso Sea. They survive in Antarctica, on top of the tallest mountains, near thermal vents in the deepest parts of the oceans, in the deepest, darkest caves, and even miles down within the crust of the earth. In all, microbes make up more than half of Earth's biomass.

Biomass:
The total weight of living organisms within a defined environment.

The rich diversity of microorganisms is reflected in their profound influence on all aspects of life. Most are harmless or indeed beneficial. For example, they are essential to the recycling of nutrients that form the bodies of all organisms and sustaining all the metabolic cycles of life. They affect our climate and, as a group, produce about 50% of the oxygen gas we breathe and many other organisms use. They have influenced the evolution of life on Earth and actually have outpaced that of the more familiar plants and animals.

Microorganisms survive in, or are purposely put in, many of the foods we eat. Microorganisms and viruses also are in the air we breathe and, at times, in the water we drink. Even closer to home, some 100 billion microorganisms colonize our skin and grow in our mouth, ears, nose, throat, and digestive tract (**FIGURE 1.1A**). Fortunately, the majority of these microbes, called our natural **microbiota**, are actually beneficial in helping us resist disease, and regulating development and nutrition. To be human, we must be "infected."

When most of us hear the word "bacterium" or "virus" though, we think infection or disease. Although such **pathogens** (disease-causing agents) are rare, they periodically have carved out great swatches of humanity as epidemics passed over the land. Some diseases—such as plague, cholera, and smallpox—have become known histori-

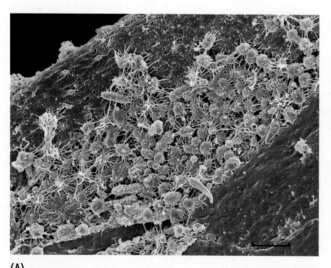

(A)   (B)

FIGURE 1.1 Microbes Are Key to Health and Illness. **(A)** Large numbers of bacteria are found on and in parts of the human body. On the tongue, most are harmless or even beneficial, while a few in our mouth can cause throat infections or lead to tooth decay. (Bar = 5 μm.) **(B)** During the 2009 pandemic of swine flu, people in affected areas such as Mexico wore masks in an attempt to avoid being infected with the virus.

cally as "slate-wipers," a reference to the barren towns they left in their wake (MicroFocus 1.1). Even today, with antibiotics and vaccines to cure and prevent many infectious diseases, pathogens still bring concern and sometimes panic. Just think about the scares that AIDS, severe acquired respiratory syndrome (SARS), and most recently avian and swine influenzas have caused worldwide (FIGURE 1.1B).

Still, if microbes were solely agents of disease, none of us would be here today. Rather "infection" is a way of life—we, all life, and our planet are dependent on microbes!

A major focus of this introductory chapter is to give you an introspective "first look" at microbiology—then and now. We will see how microbes were first discovered and how those that cause infectious disease preoccupied the minds and efforts of so many. Along the way, we continue to see how curiosity and scientific inquiry stimulated the quest for understanding.

Although the study of microorganisms began in earnest with the work of Pasteur and Koch, they were not the first to report microorganisms. To begin our story, we reach back to the 1600s, where we encounter some equally inquisitive individuals.

MICROFOCUS 1.1: History
The Tragedy of Eyam

On the last Sunday in August (Plague Sunday), English pilgrims gather in the English countryside outside the village of Eyam, to pay homage to the townsfolk who in 1665–1666 gave their lives so that others might live. The pilgrims pause, bow their heads, and remember. In 1665, bubonic plague was raging in London. In late August, a traveling tailor arrived in the village of Eyam, about 140 miles north of London. Unknown to him, cloth arriving from London was infected with plague-carrying fleas.

Within a few days, plague began to spread throughout Eyam and villagers debated whether they should flee north. The village rector realized that if the villagers left, they could spread the plague to other towns and villages. So, he made a passionate plea that they stay. After some deep soul-searching, most townsfolk resolved to remain, even though they knew that meant many would die (see figure).

Mompesson's well at Eyam in Derbyshire Peak in Great Britain. This village is best known for being the "plague village" that chose to isolate itself when the plague was discovered there in August 1665, rather than let the infection spread.

The villagers marked off a circle of stones outside the village limits, and people from the adjacent towns brought food and supplies to the barrier, leaving them there for the self-quarantined villagers. Finally, in late 1666, the plague subsided. The rector recorded, *"Now, blessed be God, all our fears are over for none have died of the plague since the eleventh of October and the pest-houses have long been empty.* In the end, 260 of the town's 350 residents succumbed to the plague. Some have suggested this self-sacrificing incident is commemorated in a familiar children's nursery rhyme, one version of which is:

A ring-a-ring of rosies
A pocketful of posies
A tishoo! A tishoo!
We all fall down.

The ring of rosies refers to the rose-shaped splotches on the chest and armpits of plague victims. Posies were tiny flowers the people hoped would sweeten the air and ward off the foul smell associated with the disease. "A tishoo!" refers to the fits of sneezing that accompanied the disease. The last line, the saddest of all, suggests the deaths that befell so many.

1.1   The Beginnings of Microbiology

As the 17th century arrived, an observational revolution was about to begin: Dutch spectacle maker, Zacharias Janssen, was one of several individuals who discovered that if two convex lenses were put together, small objects would appear larger. Many individuals in Holland, England, and Italy further developed the two-lens system. In fact, it was in 1625 that the Italian Francesco Stelluti or Giovanni Faber used the term *microscopio* or "microscope" to refer to this new invention, which Galileo had suggested be called, "the small glass for spying things up close." This combination of lenses, or "compound microscope," would be the forerunner of the modern microscope.

Convex:
Referring to a surface that curves outward.

Microscopy—Discovery of the Very Small

KEY CONCEPT

1. The discovery of microorganisms was dependent on observations made with the microscope.

Robert Hooke, an English natural philosopher (the term *scientist* was not coined until 1833), was one of the most inventive and ingenious minds in the history of science. As the Curator of Experiments for the Royal Society of London, Hooke was the first to take advantage of the magnification abilities of the compound microscope. Although these microscopes only magnified about 25 times (25×), Hooke made detailed studies of many small living objects. In 1665, the Royal Society published his *Micrographia*, which contained descriptions of microscopes and stunning hand-drawn illustrations, including the anatomy of many insects and the structure of cork, where he used the word *cella* to describe the "great many little boxes" he observed and from which today we have the word "*cell*" (FIGURE 1.2A). Importantly, Hooke was the first person to describe and draw a microorganism, a mold he found growing on the sheepskin cover of a book (FIGURE 1.2B).

Micrographia represents one of the most important books in science history because it awakened the learned and general population of Europe to the world of the very small, revolutionized the art of scientific investigation, and showed that the microscope was an important tool for unlocking the secrets of nature.

Antony van Leeuwenhoek, a contemporary of Hooke, was a successful tradesman and dry goods dealer in Delft, Holland. As a cloth merchant, he used hand lenses to inspect the quality of cloth. After seeing Hooke's *Micrographia*, and without much education, Leeuwenhoek became skilled at grinding single pieces of glass into fine lenses, which he placed between two silver or brass plates riveted together (FIGURE 1.2C, D). Using only a single lens, no larger than the head of a pin, his "simple microscope" could magnify objects more than 200×.

The process of "observation" is an important skill for all scientists, including microbiologists— and Leeuwenhoek believed only sound observation and experimentation could be trusted—a requirement that remains a cornerstone of all science inquiry today.

Leeuwenhoek chose to communicate his observations through letters to England's Royal Society. In 1674, one letter described a sample of cloudy surface water from a marshy lake. Placing the sample before his lens, he described hundreds of what he thought were tiny, living animals (probably protozoa and algae), which he called **animalcules.** His curiosity aroused, Leeuwenhoek soon located even smaller animalcules in rainwater, which, reported in his 18th letter in 1676, likely represent the first description of bacteria.

In 1683, he sent his 39th letter to the Royal Society in which he described and illustrated for the first time what almost certainly were swimming bacterial cells taken from dental plaque (FIGURE 1.2E). Leeuwenhoek wrote:

"I then most always saw, with great wonder, that in the said matter there were many very little living animalcules, very prettily a-moving. The biggest sort . . . had a very strong and swift motion, and shot through the water (or spittle) like a pike does through the water. The second sort . . . oft-times spun round like a top . . . and these were far more in number."

Leeuwenhoek's sketches were elegant in detail and clarity. Among the 165 letters sent to the Royal Society, he outlined structural details of protozoa

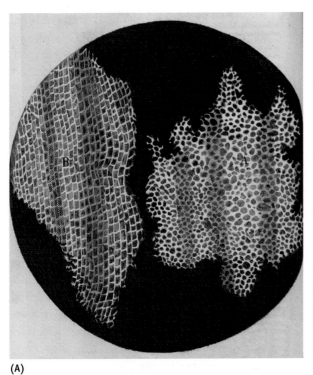

(A)

(B)

(C)

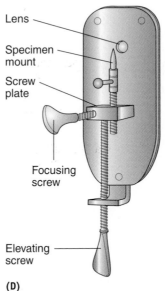

Lens

Specimen mount

Screw plate

Focusing screw

Elevating screw

(D)

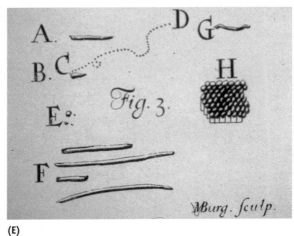

(E)

FIGURE 1.2 **The First Observations and Drawings of Microorganisms.** In his *Micrographia*, Robert Hooke included a drawing of thin shavings of cork that he saw with his microscope (**A**). He also described and drew the structure of a fungal mold (**B**). (**C**) Leeuwenhoek looking through one of his simple microscopes. (**D**) For viewing, he placed an object on the tip of the specimen mount, which was attached to a screw plate. An elevating screw moved the specimen up and down while the focusing screw pushed against the metal plate, moving the specimen toward or away from the lens. (**E**) Leeuwenhoek's drawings of animalcules (bacterial cells) were included in a letter sent to the Royal Society in 1683. He found many of these organisms between his teeth and those of others.

and yeast, and described thread-like fungi and microscopic algae.

Unfortunately, Leeuwenhoek invited no one to work with him, nor did he show anyone how he ground his lenses. Without these lenses, naturalists could not repeat his observations or verify his results, which are key components of scientific inquiry. Still, Leeuwenhoek's observations on the presence and diversity of his "marvelous beasties" and Hooke's *Micrographia* opened the door to a completely new world: the world of the microbe.

CONCEPT AND REASONING CHECKS

1.1 If you were alive in Leeuwenhoek's time, how would you explain the origin for the animalcules he found in materials such as lake water and dental plaque?

Experimentation—Can Life Generate Itself Spontaneously?

KEY CONCEPT

2. The emergence of experimental science provided a means to test long-held beliefs and resolve controversies.

In the early 1600s, most naturalists were "vitalists," individuals who thought life depended on a mysterious "vital force" that pervaded all organisms. This force provided the basis for the doctrine of **spontaneous generation**, which suggested that organisms could arise from non-living matter; that is, where there was putrefaction and decay. Common people embraced the idea, for they too witnessed what appeared to be slime that produced toads and decomposing wheat grains that generated wormlike maggots.

Regarding the latter, Leeuwenhoek suggested that maggots did not arise from wheat grains, but rather from tiny eggs laid in the grain that he could see with his microscope. Such divergent observations concerning spontaneous generation required a new form of investigation—"experimentation"—and a new generation of experimental naturalists arose.

Noting Leeuwenhoek's descriptions, the Italian naturalist Francesco Redi performed one of history's first controlled biological experiments to see if maggots could arise from rotting meat. In 1668, he covered some jars of rotting meat with paper or gauze, thereby preventing the entry of flies, while leaving other jars uncovered. If flies were prevented from entering and landing on pieces of exposed meat, Redi predicted they could not lay their invisible eggs and no maggots would hatch (**FIGURE 1.3**). Indeed, that is exactly what Redi observed and the idea that spontaneous generation could produce larger living creatures soon subsided. However, what about the mysterious and minute animalcules that appeared to straddle the boundary between the nonliving and living world?

In 1748, a British clergyman and naturalist, John Needham, suggested that the spontaneous generation of animalcules resulted from a vital force that reorganized the decaying matter from more complex organisms. To prove this, Needham boiled several tubes of mutton broth and sealed the tubes with corks. After several days, Needham proclaimed that the *"gravy swarm'd with life, with microscopical animals of most dimensions."* He was convinced that putrefaction could generate the vital force needed for spontaneous generation.

Because experiments almost always are subject to varying interpretations, the Italian cleric and naturalist Lazzaro Spallanzani challenged Needham's conclusions and suggested that the duration of heating might not have been long enough. In 1765, he repeated Needham's experiments by boiling the tubes for longer periods. As control experiments, he left some tubes open to the air and stoppered others loosely with corks.

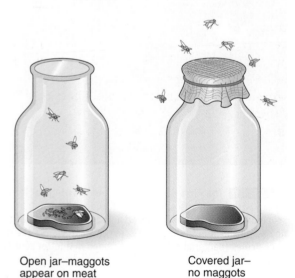

Open jar—maggots   Covered jar—
appear on meat   no maggots

FIGURE 1.3 **Redi's Experiments Refute Spontaneous Generation.** Francesco Redi carried out one of the first biological experiments by placing a piece of meat in an open jar and another in a jar covered with gauze. Maggots arose only in the open jar because flies had access to the meat where they laid their eggs.

After two days, the open tubes were swarming with animalcules, but the stoppered ones had many fewer—and the sealed ones contained none. Spallanzani proclaimed, *"the number of animalcula developed is proportional to the communication with the external air."*

Needham and others countered that Spallanzani's experiments had destroyed the vital force of life with excessive heating and excluded the air necessary for life. The controversy over spontaneous generation of animalcules continued into the mid-1800s. To solve the problem, a new experimental strategy would be needed.

To get at a resolution, the French Academy of Sciences sponsored a contest for the best experiment to prove or disprove spontaneous generation. Louis Pasteur took up the challenge and, through an elegant series of experiments that were a varia- tion of the methods of Needham and Spallanzani, discredited the idea in 1861. MicroInquiry 1 outlines the process of scientific inquiry and Pasteur's winning experiments.

Although Pasteur's experiments generated considerable debate for several years, his exacting and carefully designed experiments marked the end of the long and tenacious clashes over spontaneous generation that had begun two centuries earlier.

However, today there is another form of "spontaneous generation"—this time occurring in the laboratory (MicroFocus 1.2).

CONCEPT AND REASONING CHECKS

1.2 Evaluate the role of experimentation as an important skill to the eventual rejection of spontaneous generation.

1.2 Microorganisms and Disease Transmission

In the 13th century, people knew diseases could be contagious, so quarantines were used to combat disease spread. In 1546, the Italian poet and naturalist Girolamo Fracostoro suggested that transmission could occur by direct human contact, from lifeless objects like clothing and eating utensils, or through the air.

By the mid-1700s, the prevalent belief among naturalists and common people was that disease resulted from an altered chemical quality of the atmosphere or from tiny poisonous particles of decomposed matter in the air, an entity called **miasma** (the word malaria comes from *mala aria,* meaning "bad air"). To protect oneself from the black plague in Europe, for example, plague doctors often wore an elaborate costume they thought would protect them from the plague miasma (FIGURE 1.4).

However, as the 19th century unfolded, more scientists relied on keen observations and experimentation as a way of knowing and explaining divergent observations, including contagion and disease.

Epidemiology—Understanding Disease Transmission

KEY CONCEPT

3. Early epidemiology studies suggested how diseases could be spread and be controlled.

Epidemiology, as applied to infectious diseases, is the scientific study from which the source, cause, and mode of transmission of disease can be identified. The first scientific epidemiological studies, carried out by Ignaz Semmelweis and John Snow, were instrumental in suggesting how diseases were transmitted—and how simple measures could interrupt transmission.

Contagious: Capable of being transmitted between individuals through contact.

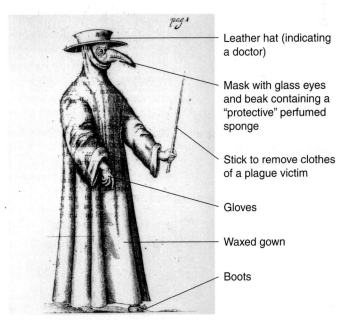

Leather hat (indicating a doctor)

Mask with glass eyes and beak containing a "protective" perfumed sponge

Stick to remove clothes of a plague victim

Gloves

Waxed gown

Boots

FIGURE 1.4 **Dressed for Protection.** This dress was thought to protect a plague doctor from the air (miasma) that caused the plague.

MICROINQUIRY 1

Experimentation and Scientific Inquiry

Science certainly is a body of knowledge as you can see from the thickness of this textbook! However, science also is a process—a way of learning. Often we accept and integrate into our understanding new information because it appears consistent with what we believe is true. But, are we confident our beliefs are always in line with what is actually true? To test or challenge current beliefs, scientists must present logical arguments supported by well-designed and carefully executed **experiments**.

The Components of Scientific Inquiry

There are many ways of finding out the answer to a problem. In science, **scientific inquiry**—or what has been called the "scientific method"—is the way problems are investigated. Let's understand how scientific inquiry works by following the logic of the experiments Louis Pasteur published in 1861 to refute the idea of spontaneous generation.

When studying a problem, the inquiry process usually begins with **observations.** For spontaneous generation, Pasteur's earlier observations suggested that organisms do not appear from nonliving matter (see text discussion of the early observations supporting spontaneous generation).

Next comes the **question**, which can be asked in many ways but usually as a "what," "why," or "how" question. For example, "What accounts for the generation of microorganisms in the beef broth?"

From the question, various hypotheses are proposed that might answer the question. A **hypothesis** is a provisional but testable explanation for an observed phenomenon. In almost any scientific question, several hypotheses can be proposed to account for the same observation. However, previous work or observations usually bias which hypothesis looks most

promising, and scientists then put their "pet hypothesis" to the test first.

Pasteur's previous work suggested that the purported examples of life arising spontaneously in meat or vegetable broths were simply cases of airborne microorganisms landing on a suitable substance and then multiplying in such profusion that they could be seen as a cloudy liquid.

Pasteur's Experiments

Pasteur set up a series of experiments to test the hypothesis that "Life only arises from other life" (see facing page).

Experiment 1A and 1B: Pasteur sterilized a meat broth in glass flasks by heating. He then either left the neck open to the air (A) or sealed the glass neck (B). Organisms only appeared (turned the broth cloudy) in the open flask.

Experiment 2A and 2B: Pasteur sterilized a meat broth in swan-neck flasks (A), so named because their S-shaped necks resembled a swan's neck. No organisms appeared, even after many days. However, if the neck was snapped off or the broth tipped to come in contact with the neck (B), organisms (cloudy broth) soon appeared.

Analysis of Pasteur's Experiments

Let's analyze the experiments. Pasteur had a preconceived notion of the truth and designed experiments to test his hypothesis. In his experiments, only one **variable** (an adjustable condition) changed. In experiment 1, the flask was open or sealed; in experiment 2, the neck was left intact, broken, or allowed to come in contact with the sterile broth. Pasteur kept all other factors the same; that is, the broth was the same in each experiment; it was heated the same length of time; and similar flasks were used. Thus, the experiments had rigorous

controls (the comparative condition). For example, in experiment 1, the control was the flask left open. Such controls are pivotal when explaining an experimental result. Pasteur's finding that no life appeared in the sealed flask (experiment 2A) is interesting, but tells us very little by itself. We only learn something by comparing this to the broken neck (or tipped flask) where life quickly appeared.

Also note that the idea of spontaneous generation could not be dismissed by just one experiment (see "His critics" on facing page). Pasteur's experiments required the accumulation of many experiments, all of which pointed to the same conclusion.

Hypothesis and Theory

When does a hypothesis become a theory? The answer is that there is no set time or amount of evidence that specifies the change from hypothesis to theory. A **theory** is defined as a hypothesis that has been tested and shown to be correct every time by many separate investigators. So, at some point, sufficient evidence exists to say a hypothesis is now a theory. However, theories are not written in stone. They are open to further experimentation and so can be refuted.

As a side note, today a theory often is used incorrectly in everyday speech and in the news media. In these cases, a theory is equated incorrectly with a hunch or belief—whether or not there is evidence to support it. In science, a theory is a general set of principles supported by large amounts of experimental evidence.

Discussion Point

Based on Pasteur's experiments, could one still argue that spontaneous generation could occur? Explain.

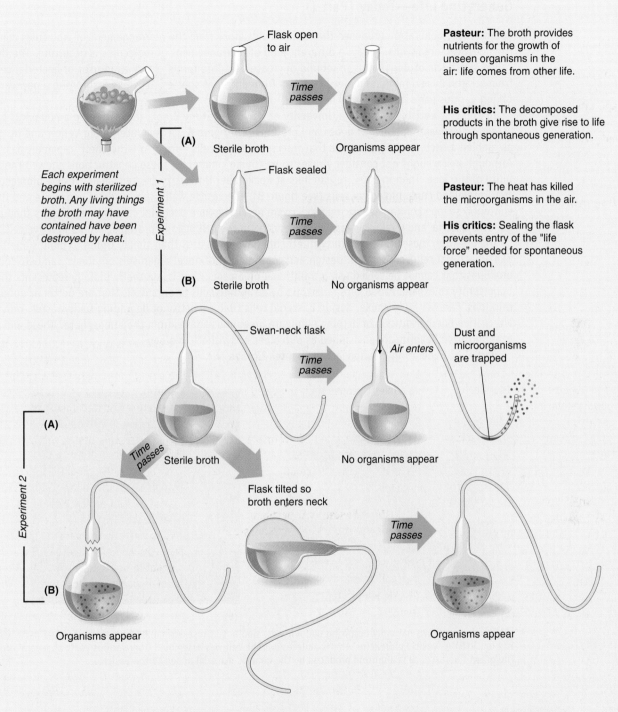

Pasteur: The broth provides nutrients for the growth of unseen organisms in the air: life comes from other life.

His critics: The decomposed products in the broth give rise to life through spontaneous generation.

Pasteur: The heat has killed the microorganisms in the air.

His critics: Sealing the flask prevents entry of the "life force" needed for spontaneous generation.

Flask open to air

Time passes

Sterile broth   Organisms appear

Each experiment begins with sterilized broth. Any living things the broth may have contained have been destroyed by heat.

Experiment 1

(A)

Flask sealed

Time passes

(B)   Sterile broth   No organisms appear

Swan-neck flask

Time passes

Air enters

Dust and microorganisms are trapped

(A)

Time passes

Sterile broth

No organisms appear

Flask tilted so broth enters neck

Experiment 2

Time passes

(B)

Organisms appear   Organisms appear

Pasteur and the Spontaneous Generation Controversy.
(1A) When a flask of sterilized broth is left open to the air, organisms appear. **(1B)** When a flask of sterilized broth is boiled and sealed, no living things appear. **(2A)** Broth sterilized in a swan-neck flask is left open to the air. The curvature of the neck traps dust particles and microorganisms, preventing them from reaching the broth. **(2B)** If the neck is snapped off to allow in air or the flask is tipped so broth enters the neck, organisms soon appear.

Pasteur: No life will appear in the flask because microorganisms will not be able to reach the broth.

His critics: If the "life force" has free access to the flask, life will appear, given enough time.

Many days later the intact flask is still free of any life. Pasteur has refuted the doctrine of spontaneous generation.

MICROFOCUS 1.2: Biotechnology
Generating Life—Today (Part I)

Spontaneous generation proposed that animalcules arose from the rearrangement of molecules coming from decayed organisms. Today, a different kind of rearrangement of molecules is occurring. The field, called **synthetic biology**, aims to rebuild or create new "life forms" (such as viruses or bacterial cells) from scratch by recombining molecules taken from different species. It is like fashioning a new car by taking various parts from a Ford, Chevy, and Toyota.

In 2002, scientists at the State University of New York, Stony Brook, reconstructed a poliovirus by assembling separate poliovirus genes and proteins (see figure A). A year later, Craig Venter and his group assembled a bacteriophage—a virus that infects bacterial cells—from "off-the-shelf" biomolecules. Although many might not consider viruses to be "living" microbes, these constructions showed the feasibility of the idea. Then in 2004, researchers at Rockefeller University created small "vesicle bioreactors" that resembled crude biological cells (see figure B). The vesicle walls were made of egg white and the cell contents, stripped of any genetic material, were derived from a bacterial cell. The researchers then added genetic material and viral enzymes, which resulted in the cell making proteins, just as in a live cell.

Importantly, these steps toward synthetic life have more uses than simply trying to build something like a bacterial cell from scratch. Design and construction of novel organisms or viruses can help solve problems that cannot be solved using traditional organisms; that is, synthetic biology represents the opportunity to expand evolution's repertoire by designing cells or organisms that are better at doing certain jobs. Can we, for example, design bacterial cells that are better at degrading toxic wastes, providing alternative energy sources, or helping eliminate greenhouse gases from the atmosphere? These and many other positive benefits are envisioned as outcomes of synthetic biology.

Part II of Generating Life appears in Chapter 2 (page 58).

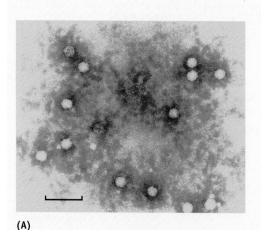
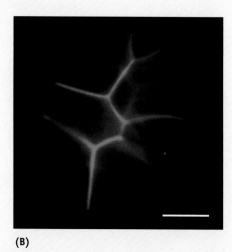

(A)   **(B)**

(A) This image shows naturally occurring polioviruses, similar to those assembled from the individual parts. (Bar = 100 nm.) **(B)** A "vesicle bioreactor" that simulates a crude cell was assembled from various parts of several organisms. The green fluorescence is a protein produced by the genetic material added to the vesicle.

Ignaz Semmelweis was a Hungarian obstetrician who was shocked by the numbers of pregnant women in his hospital who were dying of puerperal fever (a type of blood poisoning also called childbed fever) during labor. He determined the disease was more prevalent in the ward handled by medical students (29% deaths) than in the ward run by midwifery students (3% deaths). This comparative study suggested to Semmelweis that the mode of transmission must involve his medical students. He deduced that the source of contagion must be from cadavers on which the medical students previously

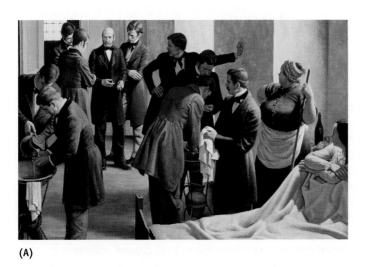

(A)

(B)

FIGURE 1.5 **Blocking Disease Transmission.** (**A**) Semmelweis (background, center-left) believed if hospital staff washed their hands, cases of puerperal fever would be reduced by preventing its spread from staff to patients. (**B**) John Snow (inset) produced a map plotting all the cholera cases in the London Soho district and observed a cluster near to the Broad Street pump (circle).

had been performing autopsies because midwifery students did not work on cadavers. So, in 1847, Semmelweis directed his staff to wash their hands in chlorine water before entering the maternity ward (**FIGURE 1.5A**). Deaths from childbed fever dropped, showing that disease spread could be interrupted. Unfortunately, few physicians initially heeded Semmelweis' recommendations.

In 1854, a cholera epidemic hit London, including the Soho district. With residents dying, English surgeon John Snow set out to discover the reason for cholera's spread. He carried out one of the first thorough epidemiological studies by interviewing sick and healthy Londoners and plotting the location of each cholera case on a district map (**FIGURE 1.5B**). The results indicated most cholera cases clustered to a sewage-contaminated street pump from which local residents obtained their drinking water. Snow then instituted the first known example of a public health measure to interrupt disease transmission—he requested the parish Board of Guardians to remove the street

pump handle! Again, disease spread was broken by a simple procedure.

Snow went on to propose that cholera was not spread by a miasma but rather was waterborne. In fact, he asserted that "organized particles" caused cholera—an educated guess that proved to be correct even though the causative agent would not be identified for another 29 years.

It is important to realize that although the miasma premise was incorrect, the fact that disease was associated with bad air and filth led to new hygiene measures, such as cleaning streets, laying new sewer lines in cities, and improving working conditions. These changes helped usher in the Sanitary Movement and create the infrastructure for the public health systems we have today (MicroFocus 1.3).

CONCEPT AND REASONING CHECKS

1.3 Contrast the importance of the observations and studies by Semmelweis and Snow toward providing a better understanding of disease transmission.

MICROFOCUS 1.3: Public Health
Epidemiology Today

Today, we have a good grasp of disease transmission mechanisms, as we will discuss in Chapter 19. However, even with the advances in sanitation and public health, cholera remains a public health threat in parts of the developing world. In addition, almost 160 years after Semmelweis' suggestions, a lack of hand washing by hospital staff, even in developed nations, remains a major mechanism for disease transmission (see figure). The simple process of washing one's hands still could reduce substantially disease transmission among the public and in hospitals.

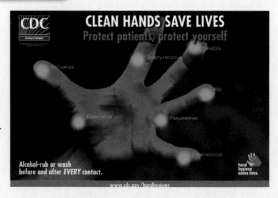

Two of the most important epidemiological organizations today are the Centers for Disease Control and Prevention (CDC) in Atlanta, Georgia, and, on a global perspective, the World Health Organization (WHO) in Geneva, Switzerland. Both employ numerous epidemiologists, popularly called "disease detectives," who, like Snow (but with more expertise), systematically gather information about disease outbreaks in an effort to discover how the disease agent is introduced, how it is spread in a community or population, and how the spread can be stopped. For example, in 2008 more than 1,400 people in 43 states developed similar gastrointestinal symptoms, which CDC investigators traced to bacterial contamination in imported jalapeño peppers. Warnings to not purchase or eat these peppers halted the outbreak and prevented further transmission. And to think, it all started with the seminal work of Semmelweis and Snow.

Variolation and Vaccination—Prevention of Infectious Disease

KEY CONCEPT

4. Resistance to a disease can come from exposure to and recovery from a mild form of (or a very similar) disease.

Besides the controversies over the mechanism of disease transmission, ways to prevent disease from occurring were being considered. In the 1700s, smallpox was prevalent throughout Europe. In England, for example, smallpox epidemics were so severe that one third of the children died before reaching the age of three. Many victims who recovered were blinded from corneal infections and most were left pockmarked.

Significantly, survivors were protected from suffering the disease a second time. These observations suggested that if one contracted a weakened or mild form of the disease, perhaps such individuals would have lifelong resistance.

In the 14th century, the Chinese knew that smallpox survivors would not get re-infected. Spreading from China to India and Africa, the practice of **variolation** developed, which involved blowing a ground smallpox powder into the individual's nose. Europeans followed by inoculating dried smallpox scabs under the skin. Although some individuals did get smallpox, most contracted a mild form of the disease and, upon recovery, were resistant to future smallpox infections.

As an English country surgeon, Edward Jenner learned that milkmaids who occasionally contracted cowpox, a disease of the udders of cows, would subsequently be protected from deadly smallpox. Jenner wondered if intentionally giving cowpox to people would protect them against smallpox and be an effective alternative to variolation. In 1796, he put the matter to the test.

A dairy maid named Sarah Nelmes came to his office, the lesions of cowpox evident on her hand. Jenner took material from the lesions and scratched it into the skin of a boy named James Phipps (**FIGURE 1.6**). The boy soon developed a slight fever, but recovered. Six weeks later Jenner inoculated young Phipps with material from a smallpox lesion. Within days, the boy developed a reaction at the site but failed to show any sign of smallpox.

Jenner repeated his experiments with other children, including his own son. His therapeutic technique of **vaccination** (*vacca* = "cow") worked in all cases and eliminated the risks associated with variolation. In 1798, he published a pamphlet on his work that generated considerable interest.

FIGURE 1.6 The First Vaccination against Smallpox. Edward Jenner performed the first vaccination against smallpox. On May 14, 1796, material from a cowpox lesion was scratched into the arm of eight-year-old James Phipps. The vaccination protected him from smallpox.

Prominent physicians confirmed his findings, and within a few years, Jenner's method of vaccination spread through Europe and abroad. By 1801, some 100,000 people in England had been vaccinated. President Thomas Jefferson wrote to Jenner, *"You have erased from the calendar of human afflictions one of its greatest. Yours is the comfortable reflection that mankind can never forget that you have lived."*

A hundred years would pass before scientists discovered the milder cowpox virus was triggering a defensive mechanism by the body's immune system against the deadlier smallpox virus. It is remarkable that without any knowledge of viruses or disease causation, Jenner accomplished what he did. Again, hallmarks of a scientist—keen observational skills and insight—led to a therapeutic intervention against disease.

CONCEPT AND REASONING CHECKS

1.4 Evaluate the effectiveness of variolation and vaccination as ways to produce disease resistance.

The Stage Is Set

During the early years of the 1800s, several events occurred that helped set the stage for the coming "germ revolution." In the 1830s, advances were made in microscope optics that allowed better resolution of objects. This resulted in improved and more widespread observations of tiny living organisms, many of which resembled short sticks. In fact, in 1838 the German biologist Christian Ehrenberg suggested these "rod-like" looking organisms be called **bacteria** (*bakterion* = "little rod").

The Swiss physician Jacob Henle reported in 1840 that living organisms could cause disease. This was strengthened in 1854 by Filippo Pacini's discovery of rod-shaped cholera bacteria in stool samples from cholera patients. Still, scientists debated whether bacterial organisms could cause disease because such living organisms sometimes were found in healthy people. Therefore, how could these bacterial cells possibly cause disease?

To understand clearly the nature of infectious disease, a new conception of disease had to emerge. In doing so, it would be necessary to demonstrate that a specific bacterial species was associated with a specific infectious disease. This would require some very insightful work, guided by Louis Pasteur in France and Robert Koch in Germany.

1.3 The Classical Golden Age of Microbiology (1854–1914)

Beginning around 1854, microbiology blossomed and continued until the advent of World War I. During these 60 years, many branches of microbiology were established, and the foundations were laid for the maturing process that has led to modern microbiology. We refer to this period as the first, or classical, Golden Age of microbiology.

Louis Pasteur Proposes That Germs Cause Infectious Disease

KEY CONCEPT

5. The germ theory was based on the observations that different microorganisms have distinctive and specific roles in nature.

Born in 1822 in Dôle, France, Louis Pasteur studied chemistry at the École Normale Supérieure in Paris and, in 1854, was appointed Professor of Chemistry at the University of Lille in northern France (**FIGURE 1.7A**). Pasteur was among the first scientists who believed that problems in science could be solved in the laboratory with the results having practical applications.

Always one to tackle big problems, Pasteur soon set out to understand the process of fermentation and the other processes that can accompany it. The prevailing theory held that fermentation resulted from the chemical breakdown of grape juice. No living agent seemed to be involved. However, Pasteur's microscope observations

Fermentation: A splitting of sugar molecules into simpler products, including alcohol, acid, and gas (CO_2).

(A)

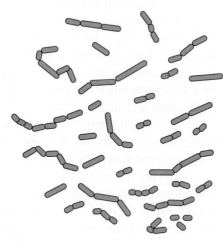

(B)

FIGURE 1.7 Louis Pasteur and Fermentation Bacteria.
(A) Louis Pasteur as a 46-year-old professor of chemistry
at the University of Paris. **(B)** The following is part of
a description of the living bacterial cells he observed.
*"A most beautiful object: vibrios all in motion, advancing
or undulating. They have grown considerably in bulk and
length since the 11th; many of them are joined together in
long sinuous chains . . . "* Pasteur concluded these bacterial cells can live without air or free oxygen; in fact, *"the
presence of gaseous oxygen operates prejudicially against
the movements and activity of those vibrios."*

consistently revealed large numbers of tiny yeast
cells in the juice that were overlooked by other
scientists. When he mixed yeast in a sugar-water
solution, the yeast grew and the quantity of yeast
increased. Yeast must be living organisms and one
of the living "ferments" responsible for the fermentation process.

Pasteur also demonstrated that wines, beers,
and vinegar each contained different and specific
types of microorganisms. For example, in studying a local problem of wine souring, he observed
that only soured wines contained populations of
bacterial cells. These cells must have contaminated
a batch of yeast and produced the acids that caused
the souring. In addition, Pasteur discovered the
process occurred in the absence of oxygen gas
(**FIGURE 1.7B**).

Pasteur recommended a practical solution for
the "wine disease" problem: heat the wine to 55°C
after fermentation but before aging. His controlled
heating technique, known as **pasteurization**, soon
was applied to other products, especially milk.

Pasteur's experiments demonstrated that yeast
and bacterial cells are tiny, living factories in which
important chemical changes take place. Therefore,
if microorganisms represented agents of change,
perhaps human infections could be caused by
those microorganisms that cause disease—**germs**.

In 1857, Pasteur published a short paper on
wine souring by bacterial cells in which he implied
that germs (bacteria) also could be related to
human illness. Five years later, after he disproved
spontaneous generation, he formulated the **germ
theory** of disease, which holds that some microorganisms are responsible for infectious disease.

CONCEPT AND REASONING CHECKS

1.5 Describe how wine fermentation and souring led
Pasteur to propose the germ theory.

Pasteur's Work Stimulates Disease Control and Reinforces Disease Causation

KEY CONCEPT

6. Antisepsis and identification of the cause of animal
diseases reinforced the germ theory.

Pasteur had reasoned that if microorganisms were
acquired from the environment, their spread could
be controlled and the chain of disease transmission broken.

Joseph Lister was Professor of Surgery at
Glasgow Royal Infirmary in Scotland, where
more than half his amputation patients died—
not from the surgery—but rather from postoperative infections. Hearing of Pasteur's germ theory,
Lister argued that these surgical infections resulted

from living organisms in the air. Knowing that carbolic acid had been effective on sewage control, in 1865 he used a carbolic acid spray in surgery and on surgical wounds (**FIGURE 1.8**). The result was spectacular—the wounds healed without infection. His technique would soon not only revolutionize medicine and the practice of surgery, but also lead to the practice of **antisepsis,** the use of chemical methods for disinfection of external living surfaces, such as the skin (Chapter 7). Once again, practical applications from the laboratory triumphed.

In an effort to familiarize himself with biological problems, Pasteur turned his attention to pébrine, a disease of silkworms. By 1870, he identified a protozoan as the infectious agent in silkworms and the mulberry leaves fed to the worms. By separating the healthy silkworms from the diseased silkworms and their food, he managed to quell the spread of disease. The identification of the protozoan was crucial to supporting the germ theory and Pasteur would never again doubt the ability of microorganisms to cause infectious disease. Now infectious disease would be his only interest.

In 1865, cholera engulfed Paris, killing 200 people a day. Pasteur tried to capture the responsible pathogen by filtering the hospital air and trapping the bacterial cells in cotton. Unfortunately, Pasteur could not grow or separate one bacterial species apart from the others because his broth cultures allowed the organisms to mix freely. Although Pasteur demonstrated that bacterial inoculations made animals ill, he could not pinpoint an exact cause.

To completely validate the germ theory, what was missing was the ability to isolate a specific bacterial species from a diseased individual and demonstrate the isolated organism caused the same disease.

CONCEPT AND REASONING CHECKS

1.6 Assess Lister's antisepsis procedures and Pasteur's work on pébrine toward supporting the germ theory.

Robert Koch Formalizes Standards to Identify Germs with Infectious Disease

KEY CONCEPT

7. Koch's postulates provided a way to identify a specific microorganism as causing a specific infectious disease.

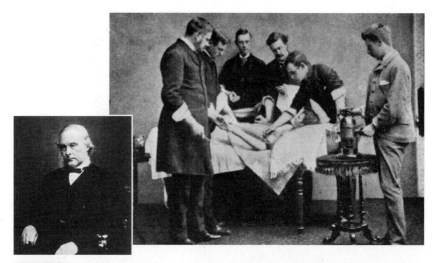

FIGURE 1.8 Lister and Antisepsis. By 1870, Joseph Lister (inset) and his students were using a carbolic acid spray in surgery and on surgical wounds to prevent postoperative infections.

Robert Koch (**FIGURE 1.9A**) was a German country doctor who was well aware of anthrax, a deadly disease that periodically ravaged cattle and sheep, and could cause disease in humans.

In 1875, Koch injected mice with the blood from such diseased sheep and cattle. He then performed meticulous autopsies and noted the same symptoms in the mice that had appeared in the sheep and cattle. Next, he isolated from the blood a few rod-shaped bacterial cells (called **bacilli**) and grew them in the aqueous humor of an ox's eye. With his microscope, Koch watched for hours as the bacilli multiplied, formed tangled threads, and finally reverted to highly resistant spores. He then took several spores on a sliver of wood and injected them into healthy mice. The symptoms of anthrax appeared within hours. Koch autopsied the animals and found their blood swarming with anthrax bacilli. He reisolated the bacilli in fresh aqueous humor. The cycle was now complete. The bacilli definitely were the causative agent of anthrax.

When Koch presented his work, scientists were astonished. Here was the verification of the germ theory that had eluded Pasteur. Koch's procedures became known as **Koch's postulates** and were quickly adopted as the formalized standards for relating a specific organism to a specific disease (**FIGURE 1.9B**).

Broth:
A liquid containing nutrients for microbial growth.

(A)

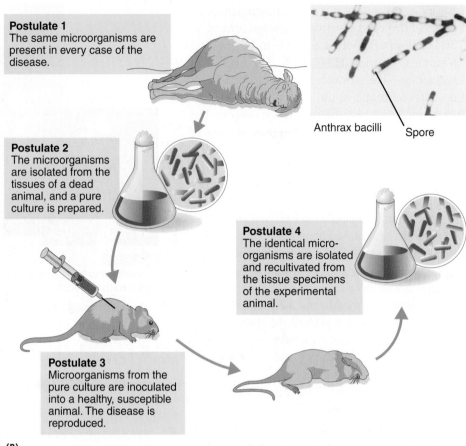

Postulate 1
The same microorganisms are present in every case of the disease.

Postulate 2
The microorganisms are isolated from the tissues of a dead animal, and a pure culture is prepared.

Postulate 3
Microorganisms from the pure culture are inoculated into a healthy, susceptible animal. The disease is reproduced.

Postulate 4
The identical microorganisms are isolated and recultivated from the tissue specimens of the experimental animal.

Anthrax bacilli   Spore

(B)

FIGURE 1.9 **A Demonstration of Koch's Postulates.** Robert Koch (**A**) developed what became known as Koch's postulates (**B**) that were used to relate a single microorganism to a single disease. The insert (in the upper right) is a photo of the rod-shaped anthrax bacteria. Many rods are swollen with spores (white ovals).

Koch Develops Pure Culture Techniques

In 1877, Koch developed methods for staining bacterial cells and preparing permanent visual records. Then, in 1880, Koch accepted an appointment to the Imperial Health Office, and while there, he observed a slice of potato on which small masses of bacterial cells, which he termed **colonies**, were growing and multiplying. So, Koch tried adding gelatin to his broth to prepare a solid culture surface. He then inoculated bacterial cells on the surface and set the dish aside to incubate. Within 24 hours, visible colonies would be growing on the surface. By 1884, agar replaced gelatin as the preferred solidifying agent (MicroFocus 1.4).

Koch now could inoculate laboratory animals with a **pure culture** of bacterial cells and be certain that only one bacterial species was involved.

Agar:
A complex polysaccharide derived from marine algae.

CONCEPT AND REASONING CHECKS

1.7 Why was pure culture crucial to Koch's postulates and the germ theory?

Competition Fuels the Study of Infectious Disease

KEY CONCEPT

8. Laboratory science and teamwork stimulated the discovery of additional infectious disease agents.

The period of the 1860s took a toll on Pasteur. His father and three of his five children died, and a stroke in 1868 left him partially paralyzed in his left arm and leg. However, he soon wrote that the work on silkworms was *"a good preparation for the investigations that we are about to undertake."*

Research studies conducted in a laboratory were becoming the normal method of work.

MICROFOCUS 1.4: History
Jams, Jellies, and Microorganisms

One of the major developments in microbiology was Robert Koch's use of a solid culture surface on which bacterial colonies would grow. He accomplished this by solidifying beef broth with gelatin. When inoculated onto the surface of the nutritious medium, bacterial cells grew vigorously at room temperature and produced discrete, visible colonies.

On occasion, however, Koch was dismayed to find that the gelatin turned to liquid. It appeared that certain bacterial species were producing a chemical substance to digest the gelatin. Moreover, gelatin liquefied at the warm incubator temperatures commonly used to cultivate certain bacterial species.

Walther Hesse, an associate of Koch's, mentioned the problem to his wife and laboratory assistant, Fanny Eilshemius Hesse. She had a possible solution. For years, she had been using a seaweed-derived powder called agar (pronounced ah'gar) to solidify her jams and jellies. Agar was valuable because it mixed easily with most liquids and once gelled, it did not liquefy, even at the warm incubator temperatures.

Fanny Hesse.

In 1880, Hesse was sufficiently impressed to recommend agar to Koch. Soon Koch was using it routinely to grow bacterial species, and in 1884 he first mentioned agar in his paper on the isolation of the bacterial organism responsible for tuberculosis. It is noteworthy that Fanny Hesse may have been among the first Americans (she was originally from New Jersey) to make a significant contribution to microbiology.

Another point of interest: The common petri dish (plate) also was invented about this time (1887) by Julius Petri, another of Koch's assistants.

Pasteur's lab and coworkers now were primarily interested in the mechanism of infection and immunity, and the practical applications that could be derived, while Koch's lab focused on procedural methods such as isolation, cultivation, and identification of specific pathogens. A competition arose that would last into the next century.

The Pasteur Lab. Pasteur continued work with anthrax and found that the bacilli were "filterable." When passed through a filter, only the clear fluid from the broth passed through; it could not trigger disease in rabbits. The anthrax bacteria were trapped on the filter and just a small drop was sufficient to kill the animals. These and other experiments further validated the germ theory.

One of Pasteur's more remarkable discoveries was made in 1881. For months, he and his coworker Charles Chamberland had been working on ways to attenuate the bacterial cells of chicken cholera using heat, different growth conditions,

successive inoculations in animals, and virtually anything that might damage the cells. Finally, they developed a weak strain by suspending the bacterial cells in a mildly acidic medium and allowing the culture to remain undisturbed for a long period. When the bacterial cells were inoculated into chickens and later followed by a dose of lethal pathogen, the animals did not develop cholera. This attenuation principle is the basis for many vaccines today. Pasteur also applied the principle to anthrax in 1881 and, in a public demonstration, found he could protect sheep against this disease as well (FIGURE 1.10).

Pasteur reached the zenith of his career in 1885 when he successfully immunized a young boy against the dreaded disease rabies. Although he never could culture the causative agent of rabies, Pasteur could cultivate it in spinal cord tissue of experimental animals. After his coworker Émile Roux tested the vaccine with success in

Attenuate:
To reduce or weaken.

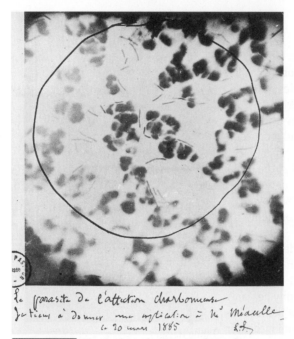

Le parasite de l'affection charbonneuse.
Je tiens à donner une explication à M. Méaulle
le 20 mars 1885   *L.P.*

FIGURE 1.10 The Anthrax Bacillus. A photomicrograph of the anthrax bacillus taken by Louis Pasteur in 1885. Pasteur circled the bacilli (the tiny rods) in tissue and annotated the photograph, "the parasite of Charbonneuse." ("Charbonneuse" is the French equivalent of anthrax.)

dogs—all immunized animals survived a rabies exposure—the ultimate test arrived. A 9-year-old boy, Joseph Meister, had been bitten and mauled by a rabid dog. Pasteur gave the boy the untested (in humans) rabies vaccine (MicroFocus 1.5). The treatment lasted 10 days and the boy recovered and remained healthy. The rabies vaccine was a triumph because it fulfilled his dream of applying the principles of science to practical problems. Such successes helped establish the Pasteur Institute in Paris, one of the world's foremost scientific establishments. Pasteur presided over the Institute until his death in 1895.

The Koch Lab. Koch also reached the height of his influence in the 1880s. In 1882, he identified and grew the bacterium responsible for tuberculosis (TB) in pure culture. In 1883, he interrupted his work on TB to lead a group of German scientists studying cholera in Egypt and India. In both countries, Koch isolated a comma-shaped bacillus and confirmed John Snow's suspicion that water is the key to transmission. In 1891, as director of Berlin's Institute for Infectious Diseases, Koch returned to his work on TB. Unfortunately, his supposed cure was a total failure—the only tarnish

on a spectacular career. Still, his TB studies were significant and ultimately gained him the 1905 Nobel Prize in Physiology or Medicine. He died of a stroke in 1910.

The germ theory set a new course for studying and treating infectious disease. The studies carried out by Pasteur and Koch made the discipline of **bacteriology**, the study of bacterial organisms, a well-respected field of study. In fact, a new generation of international scientists, including several from the Pasteur and Koch labs, stepped in to expand the work on infectious disease (TABLE 1.1).

CONCEPT AND REASONING CHECKS

1.8 Assess the importance of the science laboratory and teamwork to the increasing identification of pathogenic bacteria.

Other Global Pioneers Contribute to New Disciplines in Microbiology

KEY CONCEPTS

9. Viruses also can cause disease.
10. Many beneficial bacterial species recycle nutrients in the environment.

Although the list of identified microbes was growing, the agents responsible for diseases such as measles, mumps, smallpox, and yellow fever continued to elude identification. In 1892, a Russian scientist, Dimitri Ivanowsky, used a filter developed by Pasteur's group to trap what he thought were bacterial cells responsible for tobacco mosaic disease, which produces mottled and stunted tobacco leaves. Surprisingly, Ivanowsky discovered that when he applied the liquid that passed through the filter to healthy tobacco plants, the leaves became mottled and stunted. Ivanowsky assumed bacterial cells somehow had slipped through the filter.

Unaware of Ivanowsky's work, Martinus Beijerinck, a Dutch investigator, did similar experiments in 1899 and suggested tobacco mosaic disease was a "contagious, living liquid" that acted like a poison or virus (*virus* = "poison"). In 1898, the first "filterable virus" responsible for an animal disease—hoof-and-mouth disease—was discovered, and in 1901 American Walter Reed concluded that the agent responsible for yellow fever in humans also was a filterable agent. With these discoveries, the discipline of **virology**, the study of viruses, was launched.

TABLE

1.1 Other International Scientists and Their Accomplishments during the Classical Golden Age of Microbiology

Investigator (Year)	Country	Accomplishment
Otto Obermeier (1868)	Germany	Observed bacterial cells in relapsing fever patients
Ferdinand Cohn (1872)	Germany	Established bacteriology as a science; produced the first bacterial taxonomy scheme
Gerhard Hansen (1873)	Norway	Observed bacterial cells in leprosy patients
Albert Neisser (1879)	Germany	Discovered the bacterium that causes gonorrhea
*Charles Laveran (1880)	France	Discovered that malaria is caused by a protozoan
Hans Christian Gram (1884)	Denmark	Introduced staining system to identify bacterial cells
Pasteur Lab		
Elie Metchnikoff (1884)	Ukraine	Described phagocytosis
Emile Roux and Alexandre Yersin (1888)	France	Identified the diphtheria toxin
Koch Lab		
Friedrich Löeffler (1883)	Germany	Isolated the diphtheria bacillus
Georg Gaffky (1884)	Germany	Cultivated the typhoid bacillus
*Paul Ehrlich (1885)	Germany	Suggested some dyes might control bacterial infections
Shibasaburo Kitasato (1889)	Japan	Isolated the tetanus bacillus
Emil von Behring (1890)	Germany	Developed the diphtheria antitoxin
Theodore Escherich (1885)	Germany	Described the bacterium responsible for infant diarrhea
Daniel E. Salmon (1886)	United States	Developed the first heat-killed vaccine
Richard Pfeiffer (1892)	Germany	Identified a bacterial cause of meningitis
William Welch and George Nuttall (1892)	United States	Isolated the gas gangrene bacillus
Theobald Smith and F. Kilbourne (1893)	United States	Proved that ticks transmit Texas cattle fever
S. Kitasato and A. Yersin (1894)	Japan France	Independently discovered the bacterium causing plague
Emile van Ermengem (1896)	Belgium	Identified the bacterium causing botulism
*Ronald Ross (1898)	Great Britain	Showed mosquitoes transmit malaria to birds
Kiyoshi Shiga (1898)	Japan	Isolated a cause of bacterial dysentery
Walter Reed (1901)	United States	Studied mosquito transmission of yellow fever
David Bruce (1903)	Great Britain	Proved that tsetse flies transmit sleeping sickness
Fritz Schaudinn and Erich Hoffman (1903)	Germany	Discovered the bacterium responsible for syphilis
*Jules Bordet and Octave Gengou (1906)	France	Cultivated the pertussis bacillus
Albert Calmette and Camille Guèrin (1906)	France	Developed immunization process for tuberculosis
Howard Ricketts (1906)	United States	Proved that ticks transmit Rocky Mountain spotted fever
Charles Nicolle (1909)	France	Proved that lice transmit typhus fever
George McCoy and Charles Chapin (1911)	United States	Discovered the bacterial cause of tularemia

*Nobel Prize winners in Physiology or Medicine.

MICROFOCUS 1.5: History
The Private Pasteur

The notebooks of Louis Pasteur had been an enduring mystery of science ever since the scientist himself requested his family not to show them to anyone. But in 1964, Pasteur's last surviving grandson donated the notebooks to the National Library in Paris, and after soul-searching for a decade, the directors made them available to a select group of scholars. Among the group was Gerald Geison of Princeton University. What Geison found stripped away part of the veneration conferred on Pasteur and showed another side to his work.

In 1881, Pasteur conducted a trial of his new anthrax vaccine by inoculating half a flock of animals with the vaccine, then exposing the entire flock to the disease. When the vaccinated half survived, Pasteur was showered with accolades. However, Pasteur's notebooks, according to Geison, reveal that he had prepared the vaccine not by his own method, but by a competitor's. (Coincidentally, the competitor suffered a nervous breakdown and died a month after the experiment ended.)

Pasteur also apparently sidestepped established protocols when he inoculated two boys with a rabies vaccine before it was tested on animals. Fortunately, the two boys survived, possibly because they were not actually infected or because the vaccine was, indeed, safe and effective. Nevertheless, the untested treatment should not have been used, says Geison. His book, *The Private Science of Louis Pasteur* (Princeton University Press, 1995) places the scientist in a more realistic light and shows that today's pressures to succeed in research are little different than they were more than a century ago.

While many scientists were advancing medical microbiology, others devoted their research to the environmental importance of microorganisms. The Russian scientist Sergei Winogradsky discovered bacterial cells that metabolized sulfur and developed the concept of **nitrogen fixation,** where bacterial cells convert nitrogen gas (N_2) into ammonia (NH_3). Beijerinck was the first to obtain pure cultures of microorganisms from the soil and water by enriching the growth conditions. Together with Winogradsky, he developed many of the laboratory materials essential to the study of environmental microbiology, while adding to the understanding of the essential roles microorganisms play in the environment.

Today, along with Giovannoni and Venter, many microbiologists continue to search for and understand the roles of microorganisms. In fact, with less than 2% of all microorganisms on Earth having been identified and many fewer cultured, there is still a lot to be discovered in the microbial world!

CONCEPT AND REASONING CHECKS

1.9 Describe how viruses were discovered as disease-causing agents.

1.10 Judge the significance of the work pioneered by Winogradsky and Beijerinck.

1.4   Studying Microorganisms

Besides bacteriology and virology, other disciplines also were developing at the beginning of the 20th century. This included **mycology,** the study of fungi; **protozoology,** the study of the protozoa; and **phycology,** the study of algae (FIGURE 1.11).

The applications of microbiological knowledge also were important to the development of epidemiology, infection control, and **immunology,** which is the study of bodily defenses against microorganisms and other agents.

The Spectrum of Microorganisms and Viruses Is Diverse

KEY CONCEPT

11. The organisms and agents studied in microbiology represent diverse groups.

By the end of the classical Golden Age of microbiology, the diversity of microbes included more than just bacterial species. Let's briefly survey what we know about these groups today.

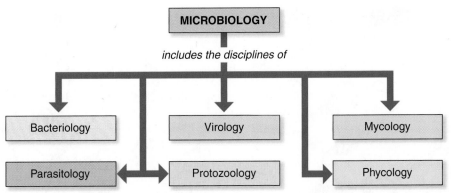

FIGURE 1.11 **Microbiology Disciplines by Organism or Agent Studied.** This simple concept map shows the relationship between microbiology and the organisms or agents that make up the various disciplines. Parasitology is the study of animal parasites. Some of these parasites cause disease in humans, which is why parasitology is included with the other disciplines of microbiology.

Bacteria. It is estimated that there may be more than 10 million bacterial species. Most are very small, single-celled (unicellular) organisms (although some form filaments, and many associate in a bacterial mass called a "biofilm"). The cells may be spherical, spiral, or rod-shaped (**FIGURE 1.12A**), and they lack the cell nucleus and most of the typical cellular compartments typical of other microbes and multicellular organisms. Some bacterial species, like the **cyanobacteria,** carry out photosynthesis (**FIGURE 1.12B**).

Besides the disease-causing members, some are responsible for food spoilage while others are useful in the food industry. Many bacterial species, along with several fungi, are **decomposers,** organisms that recycle nutrients from dead organisms.

Archaea. Based on recent biochemical and molecular studies, many bacterial species have been reassigned into another group, called the *Archaea*. Many archaeal species can be found in environments that are extremely hot (such as the Yellowstone hot springs), extremely salty (such as the Dead Sea), or in areas of extremely low pH (such as acid mine drainage). Adaptations to these environments are partly why they have been collected into their own unique group. Most bacterial and archaeal species absorb their food from the environment.

Viruses. Although not correctly labeled as microorganisms, currently there are more than 3,600 known types of viruses. Viruses are not cellular and cannot be grown in pure culture. They have a core of nucleic acid (DNA or RNA) surrounded by a protein coat. Among the features used to identify viruses are morphology (size, shape), genetic material (RNA, DNA), and biological properties (organism or tissue infected).

Viruses infect organisms for one reason only—to replicate. Viruses in the air or water, for example, cannot replicate because they need the metabolic machinery inside a cell. Of the known viruses, only a small percentage causes disease in humans. Polio, the flu, measles, AIDS, and small-pox are examples (**FIGURE 1.12C**).

The other group of microbes has a cell nucleus and a variety of internal cellular compartments. Many of the organisms are familiar to us.

Fungi. The fungi include the unicellular yeasts and the multicellular mushrooms and molds (**FIGURE 1.12D**). About 100,000 species of fungi have been described; however, there may be as many as 1.5 million species in nature.

Most fungi grow best in warm, moist places and secrete digestive enzymes that break down nutrients into smaller bits that can be absorbed easily. Fungi thus live in their own food supply. If that food supply is a human, disease may result.

Some fungi provide useful products including antibiotics, such as penicillin. Others are used in the food industry to impart distinctive flavors in foods such as Roquefort cheeses. Together with many bacterial species, numerous molds play a major role as decomposers.

Protista. The protista consist of single-celled protozoa and algae. Some are free living while others live in association with plants or animals. Locomotion may be achieved by flagella or cilia, or by a crawling movement.

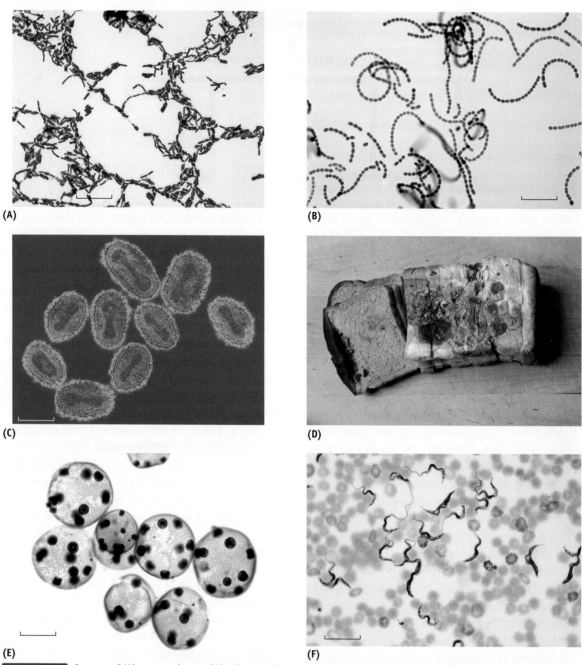

FIGURE 1.12 Groups of Microorganisms. (**A**) A bacterial smear showing the rod shaped cells of *Bacillus cereus* (stained blue), a normal inhabitant of the soil. (Bar = 10 μm.) (**B**) Filamentous strands of *Anabaena*, a cyanobacterium that carries out photosynthesis. (Bar = 100 μm.) (**C**) Smallpox viruses. (Bar = 100 nm.) (**D**) A typical blue-gray *Penicillium* mold growing on a loaf of bread. (**E**) The colonial green alga, *Volvox*. (Bar = 300 μm.) (**F**) The ribbon-like cells of the protozoan *Trypanosoma*, the causative agent of African sleeping sickness. (Bar = 10 μm.)

Different protista obtain nutrients in different ways. Protozoa either absorb nutrients from the surrounding environment or ingest algae and bacterial cells. The unicellular, colonial, or filamentous algae carry out photosynthesis (FIGURE 1.12E). Most protozoa are helpful in that they are important in lower levels of the food chain, providing food for living organisms such as snails, clams,

and sponges. Some protozoa are capable of causing diseases in animals, including humans; these include malaria, several types of diarrhea, and sleeping sickness (FIGURE 1.12F).

CONCEPT AND REASONING CHECKS

1.11 Why have microorganisms been separated into a variety of different groups?

1.5   The Second Golden Age of Microbiology (1943–1970)

The 1940s brought the birth of molecular genetics to biology. Many biologists focused on understanding the genetics of organisms, including the nature of the genetic material and its regulation.

Molecular Biology Relies on Microorganisms

KEY CONCEPT

12. Microorganisms and viruses can be used as model systems to study phenomena common to all life.

In 1943, the Italian-born microbiologist Salvador Luria and the German physicist Max Dulbrück carried out a series of experiments with bacterial cells and viruses that marked the second Golden Age of microbiology. They used a common gut-inhabiting bacterium, *Escherichia coli,* to address a basic question regarding evolutionary biology: Do mutations occur spontaneously or does the environment induce them? Luria and Dulbrück showed that bacterial cells could develop spontaneous mutations that generate resistance to viral infection. Besides the significance of their findings to microbial genetics, the use of *E. coli* as a microbial model system showed to other researchers that microorganisms could be used to study general principles of biology.

Biologists were quick to jump on the "microbial bandwagon." Experiments carried out by Americans George Beadle and Edward Tatum in the 1940s ushered in the field of molecular biology by using the fungus *Neurospora* to show that "one gene codes for one enzyme." Oswald Avery, Colin MacLeod, and Maclyn McCarty, working with the bacterial species *Streptococcus pneumoniae,* suggested in 1944 that deoxyribonucleic acid (DNA) is the genetic material in cells. In 1953, American biochemist Alfred Hershey and geneticist Martha Chase, using a virus that infects bacterial cells, provided irrefutable evidence that DNA is the substance of the genetic material. These experiments and discoveries, which will be discussed in more detail in Chapter 8, placed microbiology in the middle of the molecular biology revolution.

CONCEPT AND REASONING CHECKS

1.12 What roles did microorganisms and viruses play in understanding general principles of biology?

Two Types of Cellular Organization Are Realized

KEY CONCEPT

13. All microorganisms have a characteristic cell structure.

The small size of bacterial cells hindered scientists' abilities to confirm that these cells were "cellular" in organization. In the 1940s and 1950s, a new type of microscope—the electron microscope—was being developed that could magnify objects and cells thousands of times better than typical light microscopes. With the electron microscope, for the first time bacterial cells were seen as being cellular like all other microbes, plants, and animals. However, studies showed that they were organized in a fundamentally different way from other organisms.

It was known that animal and plant cells contained a cell nucleus that houses the genetic instructions in the form of chromosomes and was separated physically from other cell structures by a membrane envelope (**FIGURE 1.13A**). This type of cellular organization is called **eukaryotic** (*eu* = "true"; *karyon* = "nucleus"). Microscope observations of the protista and fungi had revealed that these organisms also have a eukaryotic organization. Thus, not only are all plants and animals eukaryotes, so are the microorganisms that comprise the fungi and protista.

Studies with the electron microscope revealed that bacterial (and archaeal) cells had few of the cellular compartments typical of eukaryotic cells. They lacked a cell nucleus, indicating the bacterial chromosome (DNA) was not surrounded by a membrane envelope (**FIGURE 1.13B**). Therefore, members of the *Bacteria* and *Archaea* have a **prokaryotic** (*pro* = "before") type of cellular organization and represent prokaryotes. (By the way, because viruses lack a cellular organization, they are neither prokaryotes nor eukaryotes.) As we will see in Chapter 4, there are many differences between bacterial and archaeal cells, blurring the use of the term "prokaryote."

Mutations: Permanent alterations in DNA base sequences.

CONCEPT AND REASONING CHECKS

1.13 Distinguish between prokaryotic and eukaryotic cells.

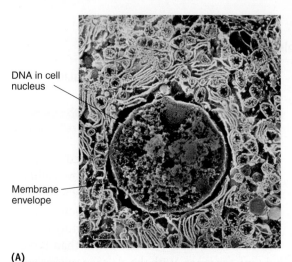

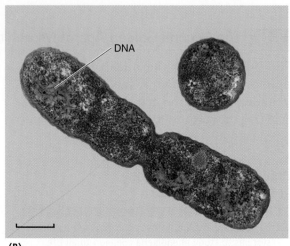

(A)   **(B)**

FIGURE 1.13 **False Color Images of Eukaryotic and Prokaryotic Cells.** (**A**) A scanning electron microscope image of a eukaryotic cell. All eukaryotes, including the protozoa, algae, and fungi, have their DNA (pink) enclosed in a cell nucleus with a membrane envelope. (Bar = 3 μm.) (**B**) A false-color transmission electron microscope image of a dividing *Escherichia coli* cell. The DNA (orange) is not surrounded by a membrane. (Bar = 0.5 μm.)

Antibiotics Are Used to Cure Infectious Disease

KEY CONCEPT

14. Antimicrobial chemicals can be effective in treating infectious diseases.

In 1910, another coworker of Koch's, Paul Ehrlich, synthesized the first "magic bullet"— a chemical that could kill pathogens. Called salvarsan, Ehrlich showed that this arsenic-containing compound cured syphilis, a sexually transmitted disease. Antibacterial **chemotherapy**, the use of antimicrobial chemicals to kill microbes, was born.

In 1929, Alexander Fleming, a Scottish scientist, discovered a mold growing in one of his bacterial cultures (**FIGURE 1.14A, B**). His curiosity aroused, Fleming observed that the mold, a species of *Penicillium,* killed the bacterial cells and colonies that were near the mold. He named the antimicrobial substance penicillin and developed an assay for its production. In 1940, British biochemists Howard Florey and Ernst Chain purified penicillin and carried out clinical trials that showed the antimicrobial potential of the natural drug (MICROFOCUS 1.6).

Additional magic bullets also were being discovered. The German chemist Gerhard Domagk discovered a synthetic chemical dye, called prontosil, which was effective in treating *Streptococcus* infections. Examination of soil bacteria led Selman Waksman to the discovery of actinomycin and streptomycin, the latter being the first effective agent against tuberculosis. He coined the term **antibiotic** to refer to those antimicrobial substances naturally produced by mold and bacterial species that inhibit growth or kill other microorganisms.

The push to market effective antibiotics was stimulated by a need to treat deadly infections in casualties of World War II (**FIGURE 1.14C**). By the 1950s, penicillin and several additional antibiotics were established treatments in medical practice. In fact, the growing arsenal of antibiotics convinced many that the age of infectious disease was waning. In fact, by the mid-1960s, many believed all major infections would soon disappear due to antibiotic chemotherapy.

Partly due to the perceived benefits of antibiotics, interest in microbes was waning by the end of the 1960s as the knowledge gained from bacterial studies was being applied to eukaryotic organisms, especially animals. What was ignored was the mounting evidence that bacterial species were becoming resistant to antibiotics.

CONCEPT AND REASONING CHECKS

1.14 Contrast Ehrlich's salvarsan and Domagk's prontosil from those drugs developed by Fleming, Florey and Chain, and Waksman.

(A)

(C)

FIGURE 1.14 **Fleming and Penicillin.** (**A**) Fleming in his laboratory. (**B**) Fleming's notes on the inhibition of bacterial growth by the fungus *Penicillium*. (**C**) A World War II poster touting the benefits of penicillin and illustrating the great enthusiasm in the United States for treating infectious diseases in war casualties.

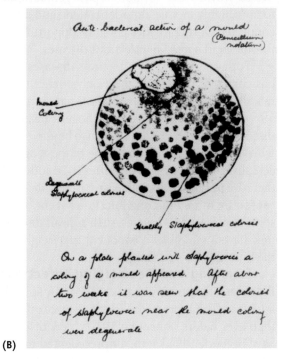

(B)

MICROFOCUS 1.6: History

Hiding a Treasure

Their timing could not have been worse. Howard Florey, Ernest Chain, Norman Heatley, and others of the team had rediscovered penicillin, refined it, and proved it useful in infected patients. But it was 1939, and German bombs were falling on London. This was a dangerous time to be doing research into new drugs and medicines. What would they do if there was a German invasion of England? If the enemy were to learn the secret of penicillin, the team would have to destroy all their work. So, how could they preserve the vital fungus yet keep it from falling into enemy hands?

Heatley made a suggestion. Each team member would rub the mold on the inside lining of his coat. The *Penicillium* mold spores would cling to the rough coat surface where the spores could survive for years (if necessary) in a dormant form. If an invasion did occur, hopefully at least one team member would make it to safety along with his "moldy coat." Then, in a safe country the spores would be used to start new cultures and the research could continue. Of course, a German invasion of England did not occur, but the plan was an ingenious way to hide the treasured organism.

The whole penicillin story is well told in *The Mold in Dr. Flory's Coat* by Eric Lax (Henry Holt Publishers, 2004).

1.6   The Third Golden Age of Microbiology—Now

Microbiology finds itself on the world stage again, in part from the biotechnology advances made in the latter part of the 20th century. **Biotechnology** frequently uses the natural and genetically engineered abilities of microbial agents to carry out biological processes for industrial/commercial/medical applications. It has revolutionized the way microorganisms are genetically manipulated to act as tiny factories producing human proteins, such as insulin, or new synthetic vaccines, such as the hepatitis B vaccine. In the latest Golden Age, microbiology again is making important contributions to the life sciences and humanity.

Microbiology Continues to Face Many Challenges

KEY CONCEPT

15. Infectious disease (natural and intentional) preoccupies much of microbiology.

The third Golden Age of microbiology faces several challenges, many of which still concern the infectious diseases that are responsible for 26% of all deaths globally (FIGURE 1.15).

A New Infectious Disease Paradigm. Infectious disease remains a major concern worldwide. Even in the United States, more than 100,000 people die each year from bacterial infections, making them the fourth leading cause of death. In fact, on a global scale, infectious diseases are spreading geographically faster than at any time in history. It is estimated that more than 2.5 billion people will be traveling by air in 2010, making an outbreak or epidemic in any one part of the world only a few airline hours away from becoming a potentially dangerous threat in another part of the world. It is a sobering thought to realize that since 2002, the World Health Organization (WHO) has verified more than 1,100 epidemic events worldwide. So, unlike past generations, today's highly mobile, interdependent, and interconnected world provides potential opportunities for the rapid spread of infectious diseases.

Today, our view of infectious diseases also has changed. In Pasteur and Koch's time, it was mainly a problem of finding the germ that caused a specific disease. Today, new pathogens are being discovered that were never known to be associated with infectious disease and some of these agents actually cause more than one disease. In addition, there are **polymicrobial diseases**; that is, diseases caused by more than one infectious agent. Even some noninfectious diseases, such as heart disease, may have a microbial component that heightens the illness.

Emerging and Reemerging Infectious Diseases. Infectious diseases are not only spreading faster, they appear to be emerging more quickly than ever before. Since the 1970s, new diseases have been identified at the unprecedented rate of

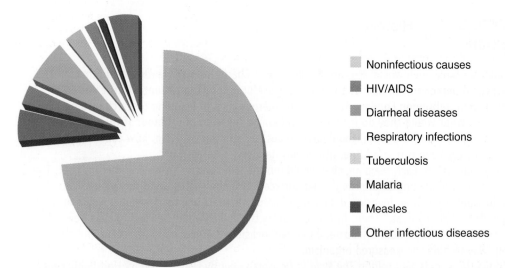

Noninfectious causes

HIV/AIDS

Diarrheal diseases

Respiratory infections

Tuberculosis

Malaria

Measles

Other infectious diseases

FIGURE 1.15 **Global Mortality—All Ages.** On a global scale, infectious diseases account for about 26% of all deaths. Noninfectious causes include chronic diseases, injuries, nutritional deficiencies, and maternal and perinatal conditions. *Source:* World Health Statistics 2008: World Health Organization.

one or more per year. There are now nearly 40 diseases that were unknown a generation ago. For example, the food chain has undergone considerable and rapid changes over the last 50 years, becoming highly sophisticated and international. Although the safety of food has dramatically improved overall, progress is uneven and foodborne outbreaks from microbial contamination are common in many countries. The trading of contaminated food between countries increases the potential that outbreaks will spread.

Emerging infectious diseases are those that have recently surfaced in a population. Among the more newsworthy have been AIDS, hantavirus pulmonary syndrome, Lyme disease, mad cow disease, and most recently, SARS, and swine flu. There is no cure for any of these. **Reemerging infectious diseases** are ones that have existed in the past but are now showing a resurgence in incidence or a spread in geographic range. Among the more prominent reemerging diseases are cholera, tuberculosis, dengue fever, and, for the first time in the Western Hemisphere, West Nile virus disease (FIGURE 1.16A). The cause for the reemergence may be antibiotic resistance or a population of susceptible individuals. Climate change also may become implicated in the upsurge and spread of disease as more moderate temperatures advance to more northern and southern latitudes.

Increased Antibiotic Resistance. Another challenge concerns our increasing inability to fight infectious disease because most pathogens are now resistant to one or more antibiotics and antibiotic resistance is developing faster than new antibiotics are being discovered. Ever since it was recognized that pathogens could mutate into "superbugs" that are resistant to many drugs, a crusade has been waged to restrain the inappropriate use of these drugs by doctors and to educate patients not to demand them in uncalled-for situations.

The challenge facing microbiologists and drug companies is to find new and effective antibiotics to which pathogens will not quickly develop resistance before the current arsenal is completely useless. Unfortunately, the growing threat of antibiotic resistance has been accompanied by a decline in new drug discovery and an increase in the time to develop a drug from discovery to market. Thus, antibiotic resistance is a major challenge for microbiology today. If actions are not taken to

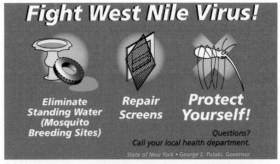

(A)

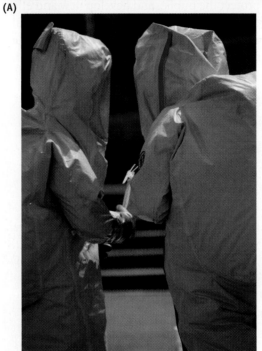

(B)

FIGURE 1.16 **Emerging Disease Threats: Natural and Intentional. (A)** There have been and will continue to be natural disease outbreaks. West Nile virus (WNV) is just one of several agents responsible for emerging or reemerging diseases. Methods have been designed that individuals can use to protect themselves from mosquitoes that spread the WNV. **(B)** Combating the threat of bioterrorism often requires special equipment and protection because many agents seen as possible bioweapons could be spread through the air.

contain and reverse resistance, the world could be faced with previously treatable diseases that have again become untreatable, as in the days before antibiotics were developed.

Bioterrorism. Perhaps it is the potential misuse of microbiology that has brought microbiology to the attention of the life science community and the public. **Bioterrorism** involves the intentional or threatened use of biological agents to cause fear in or actually inflict death or disease upon a large population. Most of the recognized biological agents are microorganisms, viruses, or microbial toxins that are bringing diseases like anthrax, smallpox, and plague back into the human psyche (FIGURE 1.16B). To minimize the use of these agents to inflict mass casualties, the challenge to the scientific community and microbiologists is to improve the ways that bioterror agents are detected, discover effective measures to protect the public, and develop new and effective treatments for individuals or whole populations. If there is anything good to come out of such challenges, it is that we will be better prepared for potential natural emerging infectious disease outbreaks, which initially might be difficult to tell apart from a bioterrorist attack.

Bioremediation:
The use of microorganisms to remove or decontaminate toxic materials in the environment.

CONCEPT AND REASONING CHECKS

1.15 Describe the natural and intentional disease threats challenging microbiology.

Microbial Ecology and Evolution Are Helping to Drive the New Golden Age

KEY CONCEPT

16. Microbial ecology and evolution are dominant themes in modern microbiology.

Since the time of Pasteur, microbiologists have wanted to know how a microbe interacts, survives, and thrives in the environment. Today, microbiology is less concerned with a specific microbe and more concerned with the process and mechanisms that link microbial agents.

Microbial Ecology. Traditional methods of microbial ecology require organisms from an environment be cultivated in the laboratory so that they can be characterized and identified. However, up to 99% of microorganisms do not grow well in the lab (if at all) and therefore could not be studied. Today, many microbiologists, armed with genetic, molecular, and biotechnological tools, can study and characterize these unculturable microbes. Such investigations are producing a new understanding of microbial communities and their influence on the ecology of all organisms. SAR11 and the plans of Craig Venter, mentioned in the chapter's opening piece, are but two examples.

Today we are learning that most microbes do not act as individual entities; rather, in nature they survive in complex communities called a **biofilm** (FIGURE 1.17A). Microbes in biofilms act very differently than individual cells and can be difficult to treat when biofilms cause infectious disease. If you or someone you know has had a middle ear infection, the cause was a bacterial biofilm.

The discovered versatility of many bacterial and archaeal species is being applied to problems that have the potential to benefit humankind. Bioremediation is one example where the understanding of microbial ecology has produced a useful outcome (FIGURE 1.17B). Other microbes hold potential to solve ecological impacts caused by toxic wastes, fertilizers, and pesticides released into the environment.

(A)

(B)

FIGURE 1.17 Microbial Ecology—Biofilms and Bioremediation. **(A)** The slimy, and often smelly, film seen in a flower vase is an example of a biofilm. **(B)** Microbes can be used to clean up toxic spills. A shoreline coated with oil from an oil spill can be sprayed with microorganisms that, along with other measures, help degrade oil.

Microbial Evolution. It was Charles Darwin—another of the scientists in this chapter who combined observation with a "prepared mind"—who first described the principles of evolution, which represents the foundation for all biology and medicine.

Like all life, microorganisms evolve. Because most have relatively short generation times, they represent experimental (model) systems in which evolutionary processes can be observed directly; microbial evolution is an experimental science. That makes it possible today to "replay history" by following the accumulation of unpredictable, chance events that lead to evolutionary novelty. For example, when considering the challenges facing microbiology today, current research is putting together a better understanding of the superfast evolution and spread of antibiotic resistance. It is also helping us better understand the mechanisms and evolution of emerging infectious diseases.

Researchers once thought that they would not be able to work out the evolutionary history of microbes. Today, thanks to the availability of sequenced genomes for groups of related and unrelated microbes, and new analytical approaches, researchers are constructing a family tree that more clearly illustrates evolutionary relationships (Chapter 3). Then, by comparing the genomes of different microorganisms, they can better understand why there are so many strains of certain bacterial species and how these new strains evolved.

Microbial evolution today is providing the means to more accurately understand what happens in real world evolution. Indeed, microbial evolution represents the organization for the biological and microbiological knowledge contained within this text.

CONCEPT AND REASONING CHECKS

1.16 Give some examples of how microbial ecology and evolution are helping drive the new golden age of microbiology.

■ SUMMARY OF KEY CONCEPTS

1.1 The Beginnings of Microbiology

1. The observations with the microscope made by Hooke and especially Leeuwenhoek, who reported the existence of **animalcules** (microorganisms), sparked interest in an unknown world of microscopic life.
2. The controversy over **spontaneous generation** initiated the need for accurate scientific experimentation, which then provided the means to refute the concept.

1.2 Microorganisms and Disease Transmission

3. Semmelweis and Snow believed that infectious disease could be caused by something transmitted from the environment and that the transmission could be interrupted.
4. Edward Jenner determined that disease (smallpox) could be prevented through **vaccination** with a similar but milder disease-causing agent.

1.3 The Classical Golden Age of Microbiology (1854–1914)

5. Pasteur's fermentation experiments indicated that microorganisms could induce chemical changes. He proposed the **germ theory** of disease, which stated that human disease could be due to chemical changes brought about by microorganisms in the body.
6. Lister's use of **antisepsis** techniques and Pasteur's studies of pébrine supported the germ theory and showed how diseases can be controlled.
7. Koch's work with anthrax allowed him to formalize the methods (**Koch's postulates**) for relating a single microorganism to a single disease. These postulates were only valid after he discovered how to make pure cultures of bacterial species.
8. Laboratory science arose as Pasteur and Koch hunted down the microorganisms of infectious disease. Pasteur's lab studied the mechanisms for infection and developed vaccines for chicken cholera, animal anthrax, and human rabies. Koch's lab focused on isolation, cultivation, and identification of pathogens such as those responsible for cholera and tuberculosis.
9. Ivanowsky and Beijerinck provided the first evidence for viruses as infectious agents.
10. Winogradsky and Beijerinck examined the beneficial roles of noninfectious microorganisms in the environment.

1.4 Studying Microorganisms

11. Microbes include the "bacteria" (*Bacteria* and *Archaea*), viruses, fungi (yeasts and molds), and protista (protozoa and algae).

1.5 The Second Golden Age of Microbiology (1943–1970)

12. Many of the advances toward understanding molecular biology and general principles in biology were based on experiments using microbial model systems.
13. With the advent of the electron microscope, microbiologists realized that there were two basic types of cellular organization: **eukaryotic** and **prokaryotic**.
14. Following from the initial work by Ehrlich, antibiotics were developed as "magic bullets" to cure many infectious diseases.

1.6 The Third Golden Age of Microbiology—Now

15. In the 21st century, fighting infectious disease, identifying **emerging** and **reemerging infectious diseases**, combating increasing antibiotic resistance, and countering the **bioterrorism** threat are challenges facing microbiology, health care systems, and society.
16. **Microbial ecology** is providing new clues to the roles of microorganisms in the environment. The understanding of **microbial evolution** has advanced with the use of genomic technologies and has expanded our understanding of microorganism relationships.

LEARNING OBJECTIVES

After understanding the textbook reading, you should be capable of writing a paragraph that includes the appropriate terms and pertinent information to answer the objective.

1. Identify the significant contributions made by Hooke and Leeuwenhoek that foreshadowed the beginnings of microbiology.
2. Discuss **spontaneous generation** and compare the experiments that led to its downfall.
3. Assess the importance of the work carried out by Semmelweis and by Snow that went against the **miasma** idea and established the field of **epidemiology.**
4. Explain how Jenner's work differs from earlier practices for preventing infectious disease.
5. Discuss Pasteur's early studies suggesting that germs could cause disease.
6. Describe how Lister's surgical work and Pasteur's studies of pébrine further strengthen the **germ theory** of disease.
7. Judge the importance of (a) the germ theory of disease and (b) **Koch's postulates** to the identification of microbes as agents of infectious disease.
8. Identify several discoveries made in the laboratories of Pasteur and Koch.
9. Describe how viruses were discovered.
10. Describe the contributions Winogradsky and Beijerinck made to environmental microbiology.
11. Provide several reasons why the "microbial agents" are placed in different groups.
12. Illustrate how microorganisms and viruses make good model systems.
13. Explain why *Bacteria* and *Archaea* are **prokaryotic** cells and all other organisms are **eukaryotic** cells.
14. Define **chemotherapy** and explain why **antibiotics** were referrd to as "magic bullets."
15. Outline the major challenges facing microbiology today.
16. Assess the importance of **microbial ecology** and **microbial evolution** to the current golden age of microbiology.

STEP A: SELF-TEST

Each of the following questions is designed to assess your ability to remember or recall factual or conceptual knowledge related to this chapter. Read each question carefully, then select the **one** answer that best fits the question or statement. Answers to the even-numbered questions can be found in **Appendix C.**

1. Who was the first person to see bacterial cells with the microscope?
   A. Pasteur
   B. Koch
   C. Leeuwenhoek
   D. Hooke
2. What process was studied by Redi and Spallanzani?
   A. Spontaneous generation
   B. Fermentation
   C. Variolation
   D. Antisepsis
3. What is the name for the field of study established by Semmelweis and Snow in the mid 1800s?
   A. Immunology
   B. Bacteriology
   C. Virology
   D. Epidemiology
4. The process of _____ involved the inoculation of dried smallpox scabs under the skin.
   A. vaccination
   B. antisepsis
   C. variolation
   D. immunization
5. The process of controlled heating, called _____, was used to keep wine from spoiling.
   A. curdling
   B. fermentation
   C. pasteurization
   D. variolation

6. What surgical practice was established by Lister?
   A. Antisepsis
   B. Chemotherapy
   C. Variolation
   D. Sterilization
7. Which one of the following statements is NOT part of Koch's postulates?
   A. The microorganism must be isolated from a dead animal and pure cultured.
   B. The microorganism and disease can be identified from a mixed culture.
   C. The pure cultured organism is inoculated into a healthy, susceptible animal.
   D. The same microorganism must be present in every case of the disease.
8. Match the lab with the correct set of identified diseases.
   A. Pasteur: tetanus and tuberculosis
   B. Koch: anthrax and rabies
   C. Koch: cholera and tuberculosis
   D. Pasteur: diphtheria and typhoid
9. What group of microbial agents would eventually be identified from the work of Ivanowsky and Beijerinck?
   A. Viruses
   B. Fungi
   C. Protozoa
   D. Bacteria
10. What microbiological field was established by Winogradsky and Beijerinck?
   A. Virology
   B. Microbial ecology
   C. Bacteriology
   D. Mycology

11. What group of microorganisms has a variety of internal cell compartments and acts as decomposers?
    A. *Bacteria*
    B. Viruses
    C. *Archaea*
    D. Fungi
12. Which one of the following organisms was NOT a model organism related to the birth of molecular genetics?
    A. *Streptococcus*
    B. *Penicillium*
    C. *Escherichia*
    D. *Neurospora*
13. Which group of microbial agents is eukaryotic?
    A. *Bacteria*
    B. Viruses
    C. *Archaea*
    D. Algae

14. The term antibiotic was coined by _____ to refer to antimicrobial substances naturally derived from _____.
    A. Waksman; bacteria and fungi
    B. Domagk; other living organisms
    C. Fleming; fungi and bacteria
    D. Ehrlich; bacteria
15. Which one of the following is NOT considered an emerging infectious disease?
    A. Polio
    B. Hantavirus pulmonary disease
    C. Lyme disease
    D. AIDS
16. A _____ is a mixture of _____ that form as a complex community.
    A. genome; genes
    B. biofilm; microbes
    C. biofilm; chemicals
    D. miasma; microbes

STEP B: REVIEW

The answers to even-numbered quesitons or statements can be found in **Appendix C**.

17. Construct a concept map for **Microbial Agents** using the following terms.

algae	fungi	viruses
Archaea	microorganisms	
cyanobacteria	nucleated cells	
decomposers	protista	
Bacteria	protozoa	

On completing your study of these pages, test your understanding of their contents by deciding whether the following statements are true (T) or false (F). If the statement is false, substitute a word or phrase for the underlined word or phrase to make the statement true.

18. _____ Leeuwenhoek was a vitalist who believed mice could spontaneously generate from putrefaction and decay.

19. _____ Pasteur proposed that "wine disease" was a souring of wine caused by yeast cells.
20. _____ Antisepsis is the use of chemical methods for disinfecting living surfaces.
21. _____ Separate bacterial colonies can be observed in a broth culture.
22. _____ Semmelweis proposed that cholera was a waterborne disease.
23. _____ Some bacterial species can convert nitrogen gas (N$_2$) into ammonia (NH$_3$).
24. _____ Fungi are eukaryotic microorganisms.
25. _____ Robert Koch was a French country doctor.
26. _____ Variolation involved inoculating individuals with smallpox scabs.
27. _____ Mycology is the scientific study of viruses.

STEP C: APPLICATIONS

Answers to the even-numbered questions can be found in **Appendix C**.

28. As a microbiologist in the 1940s, you are interested in discovering new antibiotics that will kill bacterial pathogens. You have been given a liquid sample of a chemical substance to test in order to determine if it kills bacterial cells. Drawing on the culture techniques of Robert Koch, design an experiment that would allow you to determine the killing properties of the sample substance.

29. As an environmental microbiologist, you discover a new species of microbe. How could you determine if it has a prokaryotic or eukaryotic cell structure? Suppose it has a eukaryotic structure. What information would be needed to determine if it is a member of the protista or fungi?
30. On the front page of this chapter there is a quote from Louis Pasteur. How does this quote apply to the work done by (a) Semmelweis, (b) Snow, and (c) Fleming?

STEP D: QUESTIONS FOR THOUGHT AND DISCUSSION

Answers to the even-numbered questions can be found in **Appendix C**.

31. Many people are fond of pinpointing events that alter the course of history. In your mind, which single event described in this chapter had the greatest influence on the development of microbiology? What event would be in second place?

32. One of the foundations of scientific inquiry is proper experimental design involving the use of controls. What is the role of a control in an experiment? For each of the experiments described in the section on spontaneous generation, identify the control(s) and explain how the interpretation of the experimental results would change without such controls.

33. One reason for the rapid advance in knowledge concerning molecular biology during the second Golden Age of microbiology was because many researchers used microorganisms as model systems. Why would bacterial cells be more advantageous to use for research than, say, rats or guinea pigs?

34. When you tell a friend that you are taking microbiology this semester, she asks, "*Exactly what is microbiology?*" How do you answer her?

35. As microbiologists continue to explore the microbial universe, it is becoming more apparent that microbes are "invisible emperors" that rule the world. Now that you have completed Chapter 1, provide examples to support the statement: Microbes Rule!

36. Who would you select as the "father of microbiology?" (a) Leeuwenhoek or (b) Pasteur and Koch. Support your decision.

HTTP://MICROBIOLOGY.JBPUB.COM/9E

The site features eLearning, an online review area that provides quizzes and other tools to help you study for your class. You can also follow useful links for in-depth information, read more MicroFocus stories, or just find out the latest microbiology news.

The Chemical Building Blocks of Life

Chapter Preview and Key Concepts

2.1 The Elements of Life
1. Atoms are composed of charged and uncharged particles.
2. Isotopes and ions are atoms of an element with varying mass numbers or electrical charges.
3. The chemical properties of atoms are strongly governed by the number of electrons in their outermost electron shell.

2.2 Chemical Bonding
4. Unstable atoms can be linked through ionic interactions.
5. Unstable atoms can interact through the sharing of outer shell electron pairs.
6. Hydrogen bonding is a weak electrostatic attraction between atoms.
7. Chemical reactions convert reactants into products.

2.3 Water, pH, and Buffers
8. Water is the solvent of life.
9. The concentration of hydrogen ions is expressed in pH units.
10. Buffers prevent pH shifts.

2.4 Major Organic Compounds of Living Organisms
11. Functional groups represent the set of atoms involved in chemical reactions.
12. Carbohydrates provide energy and structural materials.
13. Lipids store energy and are components of membranes.
14. Nucleic acids store, transport, and control hereditary information.
15. Proteins fold into diverse three-dimensional shapes.
    MICROINQUIRY 2: Is Protein or DNA the Genetic Material?

The significant chemicals in living tissue are rickety and unstable, which is exactly what is needed for life.
—Isaac Asimov (1920–1992)

The origin of life is one of the great unsolved problems of science. We are fairly certain that microbial life established itself on Earth about three and a half billion years ago, although no one can definitely say how or where life originated. In fact, there are many hypotheses. About all that is known is that a common primitive life form gave rise to bacterial and archaeal cells. But even this is not certain. Assuming that life did arise here on Earth, did such life arise but once or was there opportunity for "life" arising more than once—and in a different form?

Paul Davies is a theoretical physicist and astrobiologist and director of BEYOND: Center for Fundamental Concepts in Science at Arizona State University. One of the "big questions" he and others are pursuing is whether "alien" life may be hiding right in front of our noses. His hypothesis is that perhaps life formed several times on planet Earth and still exists here today in a so-called "shadow biosphere." To pursue this controversial idea, scientists have begun searching high and low (literally in the air and deep in the crust of the earth among other places) for evidence of "alien" life-forms—organisms that as the result of a "second genesis" would differ fundamentally from all known microbial life because they arose independently. Most likely, such organisms would be microbial-like. Davies believes what might make them different is an alternate and distinctive chemistry, the topic of this chapter. Microbes and all known life essentially use the same cellular chemistry—be it quite diverse—and work with an almost identical genetic code. Perhaps these undiscovered life forms would look like bacterial or archaeal cells, but their biochemistry might single them out as "alien."

Biosphere:
That part of the earth—including the air, soil, and water—where life occurs.

Some potential markers for detecting this shadow life include: (a) proteins—perhaps there are life forms that make proteins from more than, or in addition to, the 20 amino acids commonly used as building blocks for known life; (b) biomolecules—research suggests that arsenic, although poisonous to life as we know it but present in ocean vents and hot springs, can mimic the behavior of phosphorus and thus could replace phosphorus in building important biomolecules including DNA and membranes; and (c) silicon—many scientists have proposed, and science fiction writers written stories about, using silicon instead of carbon as the backbone of biological molecules.

As mentioned in the opening chapter, microbiologists have identified but a miniscule fraction of the organisms forming the microbial universe here on Earth. Because we know so little about the diversity of these microbes, is it possible there is or are other "alien" forms yet to be discovered? Importantly, that is what **science** is all about—the field of study that based on evidence tries to describe and comprehend the nature of the universe in whole or part, wherever that might take us.

In Chapter 1, you learned that microorganisms are found in most, if not all, habitats on Earth and perhaps "signatures" of other "alien" forms are out there as well. You now know that some microbes can survive high temperatures of a hot spring, the acidic runoff from a mine, or alkaline, hypersaline conditions of some lakes (**FIGURE 2.1**). In these cases, as with all habitats where microbes exist, their survival depends on **cellular chemistry**, the chemical reactions between atoms and molecules that provide for the unique metabolism found in microbial cells.

The basic principles of chemistry also permit microbiologists to understand how pathogens make a living in the human body and how the body responds. For example, studying microbial chemistry means determining:

- How microbes cause disease.
- How the immune system attempts to combat infections.
- How antibiotics and vaccines can eliminate or protect against infections.

This chapter serves as a primer or review of the fundamental concepts of chemistry that form a foundation for the chapters ahead. We will identify the elements making up all known substances and show how these elements combine to form the major groups of organic compounds found in all "known" forms of life. Realize the time invested now to understand or refresh your memory about chemistry will make subsequent chapters easier and prepare you for a rewarding learning experience as you continue your study of microbiology.

(A)

(B)

FIGURE 2.1 Cellular Chemistry Allows Microbes to Colonize and Survive in Earth's Extreme Environments. (**A**) Yellowstone's Grand Prismatic Spring. The gentle flow of heated water spreads out in terraces, with green and red algae thriving in the warm, shallow water toward the edge. (**B**) Microbial communities can thrive in extremely alkaline and salty waters, such as Mono Lake in California. »» What other extreme environments can you identify where microbes might survive?

2.1 The Elements of Life

As far as scientists know, all matter in the physical universe—be it a rock, a tree, or a microbe—is built of substances called chemical elements. **Chemical elements** are the most basic forms of matter and they cannot be broken down into other substances by ordinary chemical means.

Ninety-two naturally occurring elements have been discovered, while additional elements have been made in the laboratory or nuclear reactor. One or two letters, many standing for its English, Latin, or Greek name, designate each element. For example, H is the symbol for hydrogen, O for oxygen, Cl for chlorine, and Mg for magnesium. Some Latin abbreviations include Na (*natrium* = "sodium"), K (*kalium* = "potassium"), and Fe (*ferrum* = "iron").

Only about 25 of the 92 naturally occurring elements are essential to the survival of living organisms. Many of these are major elements needed in relatively large amounts (TABLE 2.1). Note that just six of these elements—carbon, hydrogen, nitrogen, oxygen, phosphorus, and sulfur—make up 98% of the weight in both human and bacterial cells. (The acronym CHNOPS is helpful in remembering these six important elements.)

In addition, there are a number of elements needed in much smaller amounts. These elements vary from organism to organism, but often include such elements as sodium (Na), calcium (Ca), manganese (Mn), iron (Fe), copper (Cu), and zinc (Zn).

Matter Is Composed of Atoms

KEY CONCEPT

1. Atoms are composed of charged and uncharged particles.

An **atom** is the smallest unit of an element having the properties of that element; it cannot be broken down further without losing the quality of the element. Simply stated, carbon consists of carbon atoms, oxygen of oxygen atoms, and so forth. If you split a carbon atom into simpler parts, it no longer has the properties of carbon.

An atom consists of a positively charged core, the **atomic nucleus** (FIGURE 2.2A). The atomic nucleus contains most of the atom's mass and two kinds of tightly packed particles called **protons**

and **neutrons**. Although protons and neutrons have about the same mass, protons bear a positive electrical charge (value = +1), while neutrons have no charge.

The number of protons in an atom defines each element. For example, carbon atoms always have six protons. If there are seven protons, it is no longer carbon but rather the element nitrogen. The number of protons also represents the **atomic number** of the atom. As shown in Table 2.1, carbon with six protons has an atomic number of 6. The **mass number** is the number of protons and neutrons combined. Because carbon atoms have six protons and usually six neutrons in the atomic nucleus, the mass number of carbon is 12.

Surrounding the atomic nucleus is a negatively charged cloud of **electrons** (value = −1). In any uncharged atom, the number of electrons is equal to the number of protons; that is, an atom has no net electrical charge. Although it is impossible to predict at any moment where a particular electron might be located, we can identify

Matter: Anything that occupies space and has mass.

Mass: The quantity of matter in a sample.

TABLE 2.1 Some of the Major Elements of Humans and Bacteria

Human Cell

Element	Symbol	Percentage by Weight	Atomic Number	Mass Number
Oxygen	O	65	8	16
Carbon	C	18	6	12
Hydrogen	H	10	1	1
Nitrogen	N	3	7	14
Phosphorus	P	1	15	31
Sulfur	S	0.9	16	32
Sodium	Na	0.2	11	23
Magnesium	Mg	0.1	12	24
Chlorine	Cl	0.2	17	36
Potassium	K	0.4	19	39

Bacterial Cell

Element	Symbol	Percentage by Weight	Atomic Number	Mass Number
Oxygen	O	72		
Carbon	C	12		
Hydrogen	H	10		
Nitrogen	N	3		
Phosphorus	P	0.6		
Sulfur	S	0.3		

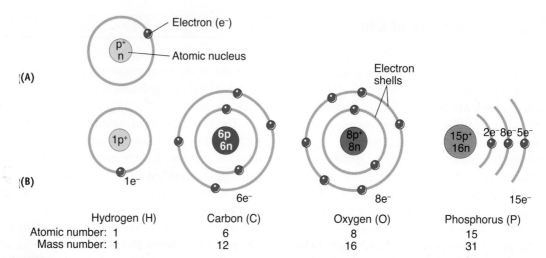

FIGURE 2.2 The Atomic Structure of an Atom and the Electron Configurations for Four Biologically Important Elements. (**A**) The atom is composed of protons and neutrons in the atomic nucleus, and electrons that move about the nucleus in electron shells. (**B**) The atomic structure of four biologically essential elements illustrates that the number of protons equals the number of electrons (though not necessarily equal to the number of neutrons). »» Knowing the mass number for an element, what does that tell you about the mass of an electron?

the spaces within the atom where electrons are usually located. These spaces are called **electron shells**, each shell representing a different energy level. **FIGURE 2.2B** provides a simple diagram of the structures, atomic numbers, and mass numbers of four atoms essential to life.

CONCEPT AND REASONING CHECKS

2.1 How does the atomic number differ from the mass number?

Atoms Can Vary in the Number of Neutrons or Electrons

KEY CONCEPT

2. Isotopes and ions are atoms of an element with varying mass numbers or electrical charges.

Although the number of protons is the same for all atoms in an element, the number of neutrons in an element may vary, altering its mass number. Most carbon atoms, for example, have a mass number of 12, but some carbon atoms have eight neutrons, rather than six, in the atomic nucleus and, hence, a mass number of 14. Atoms of the same element that have different numbers of neutrons are called **isotopes**. Therefore, carbon-12 and carbon-14 (symbolized as ^{12}C and ^{14}C) are isotopes of carbon.

Some isotopes are unstable and give off energy in the form of radiation. Such **radioisotopes** are useful in research and medicine. ^{14}C can be incorporated into an organic substance and isotopes of

other elements can be used as radioactive tracers to follow the fate of the substance, as MicroInquiry 2 demonstrates (see page 59).

Atoms are uncharged when they contain equal numbers of electrons and protons. Should an atom acquire an electrostatic charge, it is called an **ion** (**FIGURE 2.3**). The addition of one or more electrons to an atom means there is/are more negatively charged electrons than positively charged protons. Such a negatively charged ion is called an **anion**. By contrast, the loss of one or more electrons leaves the atom with extra protons and yields a positively charged ion, called a **cation**. As we will see, ion formation is important to some forms of chemical bonding.

CONCEPT AND REASONING CHECKS

2.2 How does an isotope differ from an ion?

Electron Placement Determines Chemical Reactivity

KEY CONCEPT

3. The chemical properties of atoms are strongly governed by the number of electrons in their outermost electron shell.

As shown in Figure 2.2B, each shell can hold a maximum number of electrons. The shell closest to the nucleus can accommodate two electrons, while the second and third shells each can hold eight. Other shells also have maximum numbers but usually no more than 18 are present in those outer shells. Because the 25 essential elements are

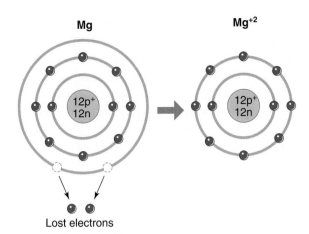

FIGURE 2.3 Formation of an Ion. Ions can be formed by the loss or gain of one or more electrons. Here, a magnesium (Mg) atom has lost two electrons to become a magnesium ion (Mg^{+2}). »» For Mg, what does the superscript denote?

of lower mass number, only the first few shells are of significance to life. Inner shells are filled first and, if there are not enough electrons to completely fill the shell, the outermost shell is left incompletely filled.

Atoms with an unfilled outer electron shell are unstable but can become stable by interacting with another unstable atom. A carbon atom, with six electrons, has two electrons in its first shell and only four in the second (see Figure 2.2B). For this reason, carbon is extremely reactive in "finding" four more electrons and, as we will see, forms innumerable combinations with other elements. Therefore, only atoms with unfilled outer shells will participate in a chemical reaction.

The shells of a few elements normally are filled completely. Each of these elements, called an inert gas, are chemically stable and exist as separate atoms in nature. Helium (atomic number 2) and

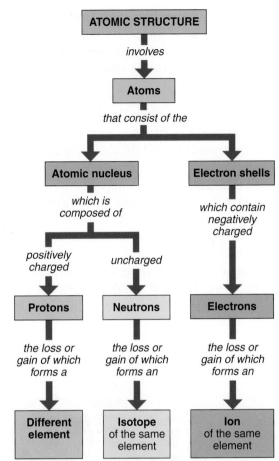

FIGURE 2.4 A Concept Map for Atomic Structure. The map shows the relationships between atoms, elements, isotopes, and ions. »» Give three examples of how a loss or gain of protons gives rise to a different element.

neon (atomic number 10) are examples, as each has its outermost electron shell filled. **FIGURE 2.4** summarizes atomic structure.

CONCEPT AND REASONING CHECKS

2.3 Looking at Figure 2.2B, do these atoms have filled outer electron shells? Explain.

2.2 Chemical Bonding

Isaac Asimov's opening quote that chemicals are "rickety and unstable" applies to how atoms interact. When the electron shells of two unstable atoms come close, the electron shells overlap, an energy exchange takes place, and each of the participating atoms assumes an electron configuration more stable than its original unstable configuration. When two or more atoms are linked together, the force holding them is called

a **chemical bond**. Chemical bonds are the result of these rickety, unstable atoms filling their outer electron shells.

The rearrangement of atoms through chemical bonding can occur in one of two major ways: atoms, as ions, can interact electrostatically; or, each uncharged atom can share electrons with one or more other atoms. In both cases, the result is atoms having full electron shells.

Ionic Bonds Form between Oppositely Charged Ions

KEY CONCEPT

4. Unstable atoms can be linked through ionic interactions.

In the formation of an **ionic bond**, one atom gives up its outermost electrons to another. The reaction between sodium and chlorine is illustrative of how these atoms become ions and then form ionic bonds (**FIGURE 2.5**). Both atoms now have their outermost shell filled. Because opposite electrical charges attract each other, the chloride ions and sodium ions come together to form stable sodium chloride (NaCl).

Salts are typically formed through ionic bonding. Besides sodium and chloride, important salts are formed from other ions, including calcium (Ca^{+2}), potassium (K^{+2}), magnesium (Mg^{+2}), and iron (Fe^{+2} or Fe^{+3}). Although ionic bonds are relatively weak, they play important roles in protein structure and the reactions between antigens and antibodies in the immune response (Chapter 21).

When two or more different elements interact with one another to achieve stability, they form a **compound**. Each compound, like each element, has a definite formula and set of properties that distinguish it from its components. For example, sodium (Na) is an explosive metal and chlorine (Cl) is a poisonous gas, but the compound they form is crystals of edible table salt (NaCl).

CONCEPT AND REASONING CHECKS

2.4 Construct a diagram to show how the salt calcium chloride ($CaCl_2$) is formed.

Covalent Bonds Share Electrons

KEY CONCEPT

5. Unstable atoms can interact through the sharing of outer shell electron pairs.

Atoms also can achieve stability by sharing electrons between atoms, the sharing producing a **covalent bond**. Such strong bonds are very important in biology because the CHNOPS elements of life usually enter into covalent bonds with themselves or one another.

Covalent bonding occurs frequently in carbon because this element has four electrons in its outer shell. The carbon atom is not strong enough to acquire four additional electrons, but it is sufficiently strong to retain the four it has. It therefore enters into a variety of covalent bonds with other atoms or groups of atoms. The vast array of carbon compounds that can be formed is responsible for the chemistry of life.

Many of the microbes residing in the ruminant stomach of a cow produce methane or natural gas (CH_4) as a by-product of cellulose digestion. This gas is a good example to illustrate covalent bonding between carbon and hydrogen (**FIGURE 2.6A**). A carbon atom shares each of its four outer shell electrons with the electron of a hydrogen atom, forming four single covalent bonds.

Scientists often draw chemical structures as **structural formulas**; that is, chemical diagrams showing the order and arrangement of atoms. In Figure 2.6, each line between carbon and hydro-

1. Sodium gives up its outer shell electron to chlorine, resulting in sodium and chloride ions.

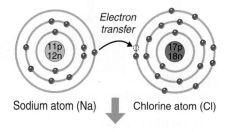

Sodium atom (Na)   Chlorine atom (Cl)

2. The stable outer shells bring about the attraction of opposite electrical charges.

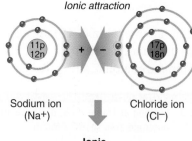

Sodium ion (Na⁺)   Chloride ion (Cl⁻)

3. The attraction of opposite charges is the ionic bond, forming the salt sodium chloride (NaCl).

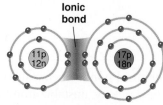

Sodium chloride (NaCl)

FIGURE 2.5 **Ion Formation and Ionic Bonding.** The transfer of an electron from sodium to chlorine generates oppositely charged ions that are attracted to one another by forming an ionic bond. »» Explain why the sodium ion has a net positive charge and the chloride ion has a net negative charge.

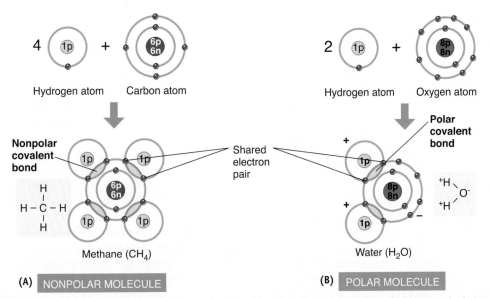

FIGURE 2.6 Chemical Bonding. (**A**) A nonpolar covalent bond involves the equal sharing of electron pairs between atoms, the example shown here being the simple organic compound methane. (**B**) A polar covalent bond involves the unequal sharing of electron pairs between hydrogen and oxygen (or nitrogen) atoms, such as in this molecule of water. »» How does a nonpolar covalent bond differ from a polar covalent bond?

gen (C—H) represents a single covalent bond between a pair of shared electrons. Other molecules, such as carbon dioxide (CO_2), share two pairs of electrons and therefore two lines are used to indicate the **double covalent bond**: O=C=O. However, in all cases, the atoms now are stable because the outer electron shell of each atom is filled through this sharing.

A **molecule** is two or more atoms held together by covalent bonds. Molecules may be composed of only one kind of atom, as in oxygen gas (O_2), or they may consist of different kinds of atoms in substances such as water (H_2O), carbon dioxide (CO_2), and the simple sugar glucose ($C_6H_{12}O_6$). As shown by these examples, the kinds and amounts of atoms (the subscript) in a molecule is called the **molecular formula**. (Note that the presence of one atom is represented without the subscript "1".)

The simplest derivatives of carbon are the **hydrocarbons**, molecules consisting solely of hydrogen and carbon. Methane is the most fundamental hydrocarbon (MICROFOCUS 2.1). Other hydrocarbons consist of chains of carbon atoms and, in some cases, the chains may be closed to form a ring.

When atoms bond together to form a molecule, they establish a geometric relationship determined largely by the electron configuration. Notice in the hydrocarbons drawn that the covalent bonds are distributed equally around each carbon atom. Each of these examples of the "equal sharing" of electron pairs represents a **nonpolar molecule**—there are no electrical charges (poles) and the bonds are called nonpolar covalent bonds.

MICROFOCUS 2.1: Evolution

Earth's Early Microbial Chemistry

In 2005, NASA's Cassini spacecraft made several flybys of Saturn's largest moon, Titan, and deployed a probe that landed on the moon's surface. (see figure). Although the presence of methane gas (CH_4) in the upper atmosphere was no surprise, the detection of several types of complex organic materials was surprising. In many ways, these conditions are similar to those that might have existed on Earth early in its history when life was first getting started.

Many scientists believe that long before oxygen gas (O_2) dominated the atmosphere, methane was a major component, giving the atmosphere a pinkish-orange color similar to Titan's today. If so, then the first organisms to evolve on Earth might have been oxygen-intolerant methane-producers called "methanogens." These microbes sustained the atmosphere for perhaps a billion years before the oxygen-producing microbes, the cyanobacteria, took hold some 2.7 billion years ago (see MicroFocus 5.4).

In the previous billion years, methanogens thrived in many of the very warm environments where hydrogen gas (H_2) dominated and could use H_2 with CO_2 for energy production. Methane gas would be a by-product. The continued accumulation of methane would warm up the planet. But with the increasing concentration of methane, sunlight would link methane molecules together and produce hydrocarbons.

Hydrocarbons would condense as a haze of atmospheric particles. Importantly, the haze would have a cooling effect on the atmosphere and shift life to those methanogens that preferred cooler temperatures. The changing chemistry also may have given oxygen-producing microbes a foothold and, along with the hydrocarbon haze, led to the first ice age about 2.3 billion years ago.

Eventually, methanogens either died out or "retreated" to oxygen-free environments where methane still dominated. Today, methanogens make up about half of all the archaeal species, further supporting their ancestors as being among the first life to evolve.

On Earth, we may never be able to verify any hypothesis concerning the pre-biological chemistry or origins of life; perhaps we can by exploring other worlds, such as Mars or Saturn's moon Titan.

An artist's rendering of Huygens, the probe carried by Cassini and sent through Titan's atmosphere to land on the moon's surface; an event that occurred on January 14, 2005.

Not all molecules are nonpolar. Indeed, one of the most important molecules to life, water, is a **polar molecule**—it has electrically charged poles (**FIGURE 2.6B**). Here the adjacent atoms do not equally share the electron pairs. Rather, oxygen has a stronger "pull" on the electrons and thus has a slight negative charge. The hydrogen atoms are then left with a slight positive charge. The water molecule therefore consists of polar covalent bonds.

CONCEPT AND REASONING CHECKS

2.5 Why do atoms share electron pairs?

Hydrogen Bonds Form between Polar Groups or Molecules

KEY CONCEPT

6. Hydrogen bonding is a weak electrostatic attraction between atoms.

A **hydrogen bond** involves the attraction of a partially positive hydrogen atom that is covalently bonded to one polar molecule toward another polar molecule having either a partially negative oxygen atom or nitrogen atom (H^+-O^-) or nitrogen atom (H^+-N^-). Although hydrogen bonds are much weaker

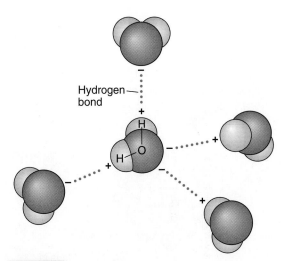

Hydrogen bond

FIGURE 2.7 **Hydrogen Bonding between Water Molecules.** The charged regions of each polar water molecule are attracted to oppositely charged regions of neighboring molecules through hydrogen bond formation. »» What is the origin of the (+) and (-) charges on hydrogen and oxygen in a water molecule?

than covalent bonds, hydrogen bonds provide the "glue" to hold water molecules together (FIGURE 2.7). These bonds also are important to the structure of proteins and nucleic acids, two of the major components of living cells.

TABLE 2.2 summarizes the different types of chemical bonds we have discussed.

CONCEPT AND REASONING CHECKS

2.6 Construct a diagram to show the hydrogen bonding in liquid ammonia (NH_3).

Chemical Reactions Change Bonding Partners

KEY CONCEPT

7. Chemical reactions convert reactants into products.

A **chemical reaction** is a process in which atoms or molecules interact to form new bonds.

Different combinations of atoms or molecules result from the reaction; that is, bonding partners change. However, the total number of interacting atoms remains constant. For chemical reactions, an arrow is used to indicate in which direction the reaction will proceed. By convention, the atoms or molecules drawn to the left of the arrow are the **reactants** and those to the right are the **products** of the reaction.

In biology, many chemical reactions are based on the assembly of larger compounds or the tearing down of larger compounds into smaller ones. In a "synthesis" reaction, smaller reactants are put together into larger products. If water is involved as a product, often it is called a **dehydration synthesis (condensation) reaction**:

$$C_6H_{12}O_6 \quad + \quad C_6H_{12}O_6 \quad \rightarrow \quad C_{12}H_{22}O_{11} \quad + \quad H_2O$$
Glucose · · · · · · · · · Glucose · · · · · · · · · Maltose · · · · · · · · · Water

The reverse is a "decomposition" reaction, where a larger reactant is broken into smaller products. Often in biology, water is one of the reactants used to break a molecule, so it is referred to as a **hydrolysis reaction** (*hydro* = "water"; *lysis* = "break"):

$$C_{12}H_{22}O_{11} \quad + \quad H_2O \quad \rightarrow \quad C_6H_{12}O_6 \quad + \quad C_6H_{12}O_6$$
Maltose · · · · · · · · · Water · · · · · · · · · Glucose · · · · · · · · · Glucose

Most importantly, the new products formed have the same number and types of atoms that were present in the reactants. In forming new products, chemical reactions only involve a change in the bonding partners. No atoms have been gained or lost from any of these reactions.

CONCEPT AND REASONING CHECKS

2.7 In the dehydration synthesis and hydrolysis reactions drawn above, what are the reactants and products in each reaction?

TABLE			
2.2	**Three Types of Chemical Bonds in Living Organisms**		
Type	Chemical Basis	Strength	Example
Ionic	Attraction between oppositely-charged ions	Weak	Sodium chloride; salts
Covalent	Sharing of electron pairs between atoms	Strong	Glucose
Hydrogen	Attraction of a hydrogen nucleus (a proton) to negatively charged oxygen or nitrogen atoms in the same or neighboring molecules	Weak	Water

2.3   Water, pH, and Buffers

All organisms are composed primarily of water. Humans are about 90% water while bacterial and archaeal organisms are about 70% water by weight. No organism can survive and grow without water.

Water Has Several Unique Properties

KEY CONCEPT

8.   Water is the solvent of life.

Liquid water is the medium in which all cellular chemical reactions occur. Being polar, water molecules are attracted to other polar molecules and act as the universal solvent in cells. Take for example what happens when you put a solute like salt in water. The salt is **hydrophilic** because it easily dissolves into separate sodium and chloride ions (FIGURE 2.8) as water molecules break the weak ionic bonds and surround each ion in a sphere of water molecules. An **aqueous solution**, which consists of solutes in water, is essential for chemical reactions to occur (MicroFocus 2.2). Molecules that do not dissolve in water are **hydrophobic**.

Water molecules also are reactants in many chemical reactions. The example of the hydrolysis reaction shown on the previous page involved water in splitting maltose into two molecules of glucose.

As you have learned, the polar nature of water molecules leads to hydrogen bonding. By forming a large number of hydrogen bonds between water molecules, it takes a large amount of heat energy to increase the temperature of water. Likewise, a large amount of heat must be lost before water decreases temperature. So, by being 70% to 90% water, cells are bathed in a solvent that maintains a more consistent temperature even when the environmental temperatures change.

CONCEPT AND REASONING CHECKS

2.8   What characteristic of water gives the molecule its unique properties?

Acids and Bases Affect a Solution's pH

KEY CONCEPT

9.   The concentration of hydrogen ions is expressed in pH units.

Solvent:
The liquid doing the dissolving to form a solution.

Solute:
The substance dissolved in the solvent.

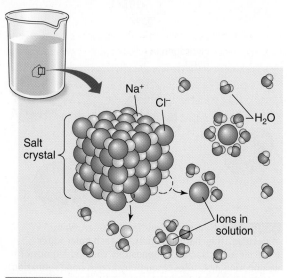

FIGURE 2.8   Solutes Dissolve in Water. Water molecules surround Na^+ and Cl^- ions, facilitating their dissolving into solution. »» When dissolving, why do the H's of water surround Cl^- while the O's surround the Na^+?

In an aqueous solution, most of the water molecules remain intact. However, some can dissociate spontaneously into hydrogen ions (H^+) and hydroxide ions (OH^-) only to rapidly recombine. This can be represented as follows, where the double arrow indicates a reversible reaction:

$$H_2O \leftrightarrow H^+ + OH^-$$

Besides water, other compounds in cells can release H^+ when they dissolve in water. For our purposes, an **acid** is a chemical substance that donates H^+ to a solution.

Acids are distinguished by their sour taste. Some common examples are acetic acid in vinegar, citric acid in citrus fruits, and lactic acid in sour milk products. Strong acids can donate large numbers of hydrogen ions to a solution. Hydrochloric acid (HCl), sulfuric acid (H_2SO_4), and nitric acid (HNO_3) are examples. Weak acids, typified by carbonic acid (H_2CO_3), donate a smaller number of hydrogen ions.

By contrast, a **base** is a substance that combines with H^+ in solution. Bases have a bitter taste. Strong bases take up numerous hydrogen ions from a solution. Potassium hydroxide (KOH), a material used to make soap, is among them.

Acids and bases frequently react with each other because of their opposing chemical charac-

MICROFOCUS 2.2: Tools

The Relationship between Mass Number and Molecular Weight

Often scientists need to make solutions that have a specific concentration of solutes. To make these solutions, one needs to know how much a particular molecule or solute weighs, which is referred to as the **molecular weight**. The calculation of the molecular weight simply consists of adding together the mass number of all the individual atoms in a molecule, such as water, carbon dioxide, or glucose. Thus, the molecular weight of a water molecule is 18 daltons, while the molecular weight of a glucose molecule is 180 daltons. Other molecules can reach astonishing proportions—antibodies of the immune system may have a molecular weight of 150,000 daltons and the bacterial toxin causing botulism is of over 900,000 daltons.

Daltons: Units to measure the weight of atomic particles or molecules; equivalent to atomic mass units used in chemistry (one-twelfth the weight of an atom of ^{12}C).

	Hydrogen (H)	Carbon (C)	Oxygen (O)	Water (H_2O)	Carbon dioxide (CO_2)	Glucose ($C_6H_{12}O_6$)
Mass number:	1 H = 1	1 C = 12	1 O = 16	2 H = 2 1 O = 16	1 C = 12 2 O = 32	6 C = 72 12 H = 12 6 O = 96
Molecular weight: (in daltons)	—	—	—	18	44	180

teristics. An "exchange reaction" involving hydrochloric acid (HCl) and sodium hydroxide (NaOH) is one example:

$$HCl + NaOH \rightarrow NaCl + H_2O$$

To indicate the concentration of H^+ in a solution, the Danish chemist Søren P. L. Sørensen introduced the symbol **pH** (power of hydrogen ions) and the **pH scale**. This numerical scale extends from 0 (extremely acidic; high H^+) to 14 (extremely basic or alkaline; low H^+) and is based on actual calculations of the number of hydrogen ions present when a substance mixes with water. A substance with a pH of 7, such as pure water, is said to be neutral; solutions that gain H^+ are said to be **acidic** and have a pH lower than 7; solutions that lose H^+ are **basic** (or alkaline) and have a pH greater than 7. The pH scale is logarithmic; that is, every time the pH changes by one unit, the [H^+] changes 10 times. For example, lemon juice (pH 2) and black coffee (pH 5) differ a thousandfold (10^3) in H^+ concentration. **FIGURE 2.9** summarizes the

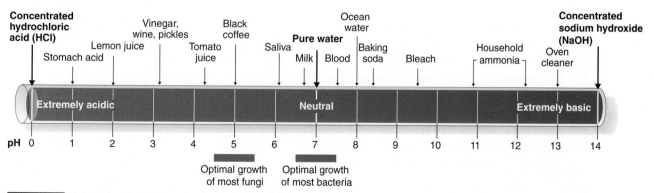

FIGURE 2.9 A Sample of pH Values for Some Common Substances. Most fungi prefer a slightly acidic pH for growth compared to most bacteria. »» On the pH scale, notice that many of the beverages we drink (e.g., wine, tomato juice, coffee) are fairly acidic. However, we would never normally drink alkaline solutions (e.g., commercial bleach, ammonia). Propose an explanation for these observations.

MICROFOCUS 2.3: Environmental Microbiology

Just South of Chicago

All you need is a map, some pH paper, and a few collection vials. When in Chicago, use your map to find the Lake Calumet region just southeast of Chicago. When you arrive, pull out your pH paper and sample some of the groundwater in the region near the Calumet River. You will be shocked to discover the pH is greater than 12—almost as alkaline as oven cleaner! In fact, this might be one of the most extreme pH environments on Earth.

How did the water get this alkaline and could anything possibly live in the groundwater?

The groundwater in the area near Lake Calumet became strongly alkaline as a result of the steel slag that has been dumped into the area for more than 100 years. Used to fill the wetlands and lakes, water and air chemically react with the slag to produce lime [calcium hydroxide, $Ca(OH)_2$]. It is estimated that 10 trillion cubic feet of slag and the resulting lime has pushed the pH to such a high value.

Now use your collection vials to collect some samples of the water. Back in the lab you will be surprised to find that there are bacterial communities present in the water. Hydrogeologists who have collected such samples have discovered some bacterial species that until then had only been found in Greenland and deep gold mines of South Africa. Other identified species appear to use the hydrogen resulting from the corrosion of the iron for energy.

How did these bacterial organisms get there? The hydrogeologists propose that the bacterial species have always been there and have simply adapted to the environment over the last 100 years when slag has been dumped. Otherwise, the microbes must have been imported in some way.

So, once again, provide a specific environment and they will come (or evolve)—the microbes that is.

---

pH values of several common substances. With regard to organisms, fungi prefer a slightly acidic environment compared to the more neutral environment preferred by most bacteria—although there are some spectacular exceptions (MICROFOCUS 2.3).

CONCEPT AND REASONING CHECKS

2.9 If the pH of an aqueous solution changes from 7 to 5, how many times has the [H^+] changed?

Cell Chemistry Is Sensitive to pH Changes

KEY CONCEPT

10. Buffers prevent pH shifts.

As microorganisms—and all organisms—take up or ingest nutrients and undergo metabolism, chemical reactions occur that use up or produce H^+. It is important for all organisms to balance the acids and bases in their cells because chemical reactions and organic compounds are very sensitive to pH shifts. Proteins are especially vulnerable, as we will soon see. If the internal cellular pH is not maintained, these proteins may be destroyed. Likewise, when most microbes grow in a microbiological nutrient medium, the waste products produced may lower the pH of the medium, which could kill the organisms.

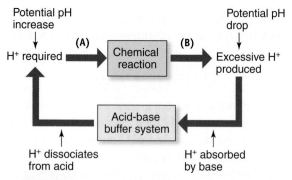

FIGURE 2.10 A Hypothetical Example for pH Shifts. An acid/base buffer system can prevent pH shifts from occuring as a result of a chemical reaction. If the reaction is using up H^+ (**A**), the acid component prevents a pH rise by donating H^+ to offset those used. If the reaction is producing excess H^+ (**B**), the base can prevent a pH drop by "absorbing" them. »» Propose what would happen if a chemical reaction continued to release excessive H^+ for a prolonged period of time.

To prevent pH shifts, cells and the growth media contain **buffers**, which are substances that maintain a specific pH. The buffer does not necessarily maintain a neutral pH, but rather whatever pH is required for that environment.

Most biological buffers consist of a weak acid and a weak base (**FIGURE 2.10**). If an excessive number of H^+ are produced (potential pH drop), the base can absorb them. Alternatively, if there is a decrease in the hydrogen ion concentration (potential pH increase), the weak acid can dissociate, replacing the lost hydrogen ions.

CONCEPT AND REASONING CHECKS

2.10 If the pH does drop in a cell, what does that tell you about the buffer system?

2.4 Major Organic Compounds of Living Organisms

As mentioned in the last section, a typical prokaryotic cell is about 70% water. If all the water is evaporated, the predominant "dry weight" remaining consists of **organic compounds**, which are those compounds related to or having a carbon basis: the carbohydrates, lipids, proteins, and nucleic acids (FIGURE 2.11). Except for the lipids, each class represents a **polymer** (*poly* = many; *mer* = part) built from a very large number of building blocks called **monomers** (*mono* = one).

Functional Groups Define Molecular Behavior

KEY CONCEPT

11. Functional groups represent the set of atoms involved in chemical reactions.

Before we look at the major classes of organic compounds, we need to address one question. The monomers building carbohydrates, nucleic acids, and proteins are essentially stable molecules because their outer shells are filled through covalent bonding. Why then should these molecules take part in chemical reactions to build polymers?

The answer is that these monomers are not completely stable. Projecting from the carbon skeletons or other atoms on these biological molecules are groups of atoms called functional groups. **Functional groups** represent points where further chemical reactions can occur if facilitated by a specific enzyme. The reactions will not happen spontaneously.

There is a small number of functional groups but their differences and placement on compounds makes possible a large variety of chemical reactions. The important functional groups in living organisms are identified in TABLE 2.3 .

Functional groups on monomers can interact to form larger molecules or polymers through dehydration synthesis reactions. In addition, functional groups can be critical for the decomposition of larger polymers into monomers through hydrolysis reactions.

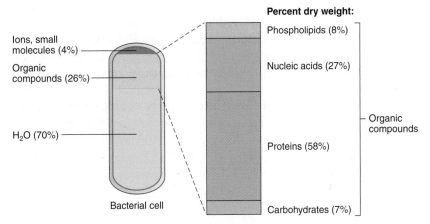

FIGURE 2.11 **Organic Compounds in Bacterial Cells.** Organic compounds are abundant in cells. The approximate composition of these compounds in a bacterial cell is similar to the percentages found in other microbes. »» Propose a reason why proteins make up almost 60% of the dry weight of a bacterial cell.

TABLE

2.3 Common Functional Groups on Organic Compounds

Functional Group	Shorthand	Structural Formula
Hydroxyl	—OH	—O—H
Carboxyl	—COOH	$\overset{\displaystyle O}{\overset{\displaystyle \|}{-C}}-OH$
Carbonyl	—CO—	$\overset{\displaystyle O}{\overset{\displaystyle \|}{-C}}-$
Amino	—NH₂	$\overset{\displaystyle H}{\overset{\displaystyle \|}{-N}}-H$
Sulfhydryl	—SH	—S—H
Phosphate	—H₂PO₄	$-O-\overset{\displaystyle OH}{\underset{\displaystyle OH}{\overset{\displaystyle \|}{\underset{\displaystyle \|}{P}}}}=O$

Enzyme:
A protein that facilitates a specific chemical reaction.

If you are unsure about the role of these functional groups, do not worry. We will see how specific functional groups interact through dehydration synthesis reactions as we now visit each of the four classes of organic compounds.

CONCEPT AND REASONING CHECKS

2.11 Why must dehydration synthesis and hydrolysis reactions be controlled by enzymes?

Carbohydrates Consist of Sugars and Sugar Polymers

KEY CONCEPT

12. Carbohydrates provide energy and structural materials.

Carbohydrates are organic compounds composed of carbon, hydrogen, and oxygen atoms that build sugars and starches. In simple sugars, like glucose ($C_6H_{12}O_6$), the ratio of hydrogen to oxygen is 2 to 1, the same as in water. The carbohydrates therefore are considered "hydrated carbon." However, the atoms are not present as water molecules bound to carbon but rather carbon covalently bonded to hydrogen and hydroxyl groups (H–C–OH).

Carbohydrates function as major fuel sources in cells. They also function as structural molecules in cell walls and nucleic acids. Often the carbohydrates are termed saccharides (*sacchar* = "sugar") and are divided into three groups.

Monosaccharides and **disaccharides** are simple sugars; many represent the monomers for the complex **polysaccharides**. Glucose, a six-carbon sugar, is one of the most widely encountered monosaccharides (**FIGURE 2.12A**). Glucose serves as the basic supply for cellular energy in the world. Estimates vary, but many scientists estimate half the world's carbon exists as glucose. Such sugars are synthesized from water and carbon dioxide through the process of **photosynthesis**. Algae and cyanobacteria are microorganisms that have the chemical machinery for this process, which is described in Chapter 6.

Disaccharides (*di* = two) are composed of two monosaccharides held together by a covalent bond. Sucrose (table sugar) is an example. It is constructed from a glucose and fructose molecule through a dehydration synthesis reaction. Sucrose is a starting point in wine fermentations. Maltose, another disaccharide, is composed of two glucose monomers, occurs in cereal grains,

such as barley, and is fermented by yeasts for energy (**FIGURE 2.12B**). An important by-product of the fermentation is the formation of alcohol in beer. Lactose, a third common disaccharide, is composed of the monosaccharides glucose and galactose. Lactose is known as "milk sugar" because it is the principal sugar in milk. Under controlled industrial conditions, microorganisms digest the lactose for energy; in the process, they produce the acid in yogurt, sour cream, and other sour dairy products.

In the microbial world, the real significance of monosaccharides is as building blocks for polysaccharides (**FIGURE 2.12C**). **Polysaccharides** (*poly* = many) are complex carbohydrates formed by joining together hundreds of thousands of similar monomers. Covalent bonds resulting from the reactions link the units together.

Starch and glycogen are common storage polysaccharides in algal and some bacteria cells, where they function as a stored energy source. Cellulose, a structural polysaccharide, is a component of the cell walls of many algae while chitin, built from chains of another glucose derivative, N-acetylglucosamine, forms the cell walls of fungi. Some bacterial cells also produce dextran that enables the cells to attach to surfaces (MicroFocus 2.4). In most bacterial cells, the cell wall is composed of carbohydrate and protein. The carbohydrate building block is a disaccharide of N-acetylglucosamine and N-acetylmuramic acid linked in long chains. In Chapter 4, we will examine the cell wall in more detail.

CONCEPT AND REASONING CHECKS

2.12 Explain how carbohydrate monomers are assembled into energy storage and cell wall polymers. Give examples.

Lipids Are Water-Insoluble Compounds

KEY CONCEPT

13. Lipids store energy and are components of membranes.

The **lipids** are a broad group of nonpolar organic compounds that are **hydrophobic**; they do not dissolve in water. Like carbohydrates, lipids are composed of carbon, hydrogen, and oxygen, but the proportion of oxygen is much lower. Lipids serve many microorganisms, but not bacterial species, as important stored energy sources.

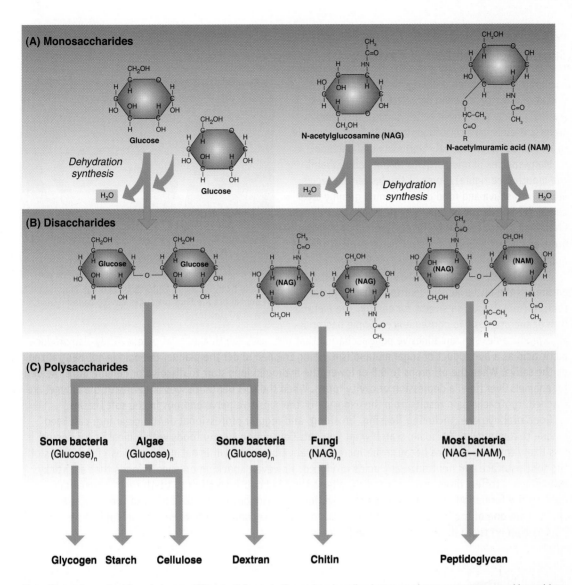

Note: The polysaccharides of glucose [(Glucose)ₙ] vary in the carbon bonding between glucose monomers and branching of the polymer chains.

FIGURE 2.12 Carbohydrate Monomers Are Built into Polymers. (**A**) There are many monosaccharides used by organisms that can be combined into disaccharides (**B**) or assembled into long polymers called polysaccharides (**C**). »» What type of chemical reaction is required to link glucose into a long polymer such as cellulose?

Lipids consist of a three-carbon glycerol molecule and up to three long-chain fatty acids (triglyceride) (**FIGURE 2.13A**). Each fatty acid is a long nonpolar hydrocarbon chain containing between 16 and 18 carbon atoms. Bonding of each fatty acid to the glycerol molecule occurs by a dehydration synthesis reaction between the hydroxyl and carboxyl functional groups.

A fatty acid is considered to be **saturated** if it contains the maximum number of hydrogen atoms extending from the carbon backbone, that is, no

double covalent bonds between carbon atoms. A fatty acid is **unsaturated** if it contains less than the maximum hydrogen atoms; that is, there is one or more double covalent bonds between a few carbon atoms.

Another type of lipid found in cell membranes is the **phospholipids**, which have only two fatty acid tails attached to glycerol (**FIGURE 2.13B**). In place of the third fatty acid there is a phosphate group, representing a functional group that is polar and can actively interact with other polar

MICROFOCUS 2.4: Public Health

Sugars, Acid, and Dental Cavities

Each of us at some time probably has feared the dentist's drill after a cavity has been detected. Dental cavities (caries) usually result from eating too much sugar or sweets. These sugars contribute to cavity formation only in an indirect way. The real culprits are oral bacteria.

Many species of microorganisms normally inhabit the mouth (see figure). Some of the bacterial species, along with saliva and food debris, form a gummy layer called **dental plaque**, a type of biofilm mentioned in Chapter 1. If not removed, plaque accumulates on the grooved chewing surfaces of back molars and at the gum line.

Plaque starts to accumulate within 20 minutes after eating. As the bacterial cells multiply, they digest the sucrose (table sugar) in sweets for energy. The metabolism of sucrose has two consequences. Some bacterial

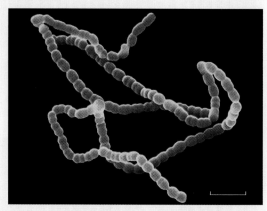

Cells of *Streptococcus mutans*, one of the major agents in dental plaque that produces cavity-causing acid. (Bar = 10 μm.)

cells produce dextran, an adhesive polysaccharide that increases the thickness of plaque. They also produce lactic acid as a by-product of sugar metabolism. Being trapped under the plaque, the acid is not neutralized by the saliva. When the pH drops to 5.5 or lower, the hydrogen ions start to dissolve or demineralize the dental enamel. Over time, a depression or cavity forms. When the soft dental tissues underneath the enamel are reached, toothache pain results from the exposure of the sensitive nerve endings in the soft tissues.

Good oral hygiene, including flossing, brushing, and regular professional dental cleaning, can keep plaque to a minimum. At home, watch what you eat. Consuming sugary foods with a meal or for dessert is less likely to cause cavities because the increased saliva produced while eating helps wash food debris off the tooth surface and neutralize any acids produced. However, snacking on sugary foods that are sticky, like caramel, toffee, dried fruit, or candies, allows the food debris to cling to teeth for a longer time, causing the formation of more plaque and providing a continuous acid attack on your teeth. No wonder cavities are one of the most prevalent infectious diseases, second only to the common cold. More detailed information on tooth decay is provided in Chapter 11.

molecules. We will have more to say about phospholipids and membranes in the next chapter. Some bacterial toxins are a combination of polysaccharide and lipid (TEXTBOOK CASE 2).

Other types of lipids include the waxes and sterols. Waxes are composed of long chains of fatty acids and form part of the cell wall in *Mycobacterium tuberculosis*, the bacterium causing tuberculosis. **Sterols**, such as cholesterol, are very different from lipids and are included with lipids solely because they too are hydrophobic molecules (FIGURE 2.13C). Sterols, composed of several rings of carbon atoms with side chains, stabilize membranes of algae, protozoa, and fungi, and the bacterium *Mycoplasma*. Sterol-like molecules are used in most bacterial cells to control membrane stability and flexibility.

CONCEPT AND REASONING CHECKS

2.13 Why are lipids not considered polymers in the sense that polysaccharides are?

Nucleic Acids Are Large, Information-Containing Polymers

KEY CONCEPT

14. Nucleic acids store, transport, and control hereditary information.

The **nucleic acids**, among the organic compounds found in organisms, are organic compounds composed of carbon, hydrogen, oxygen, nitrogen, and phosphorus atoms. Two types function in all living things: **deoxyribonucleic acid (DNA)** and **ribonucleic acid (RNA)**.

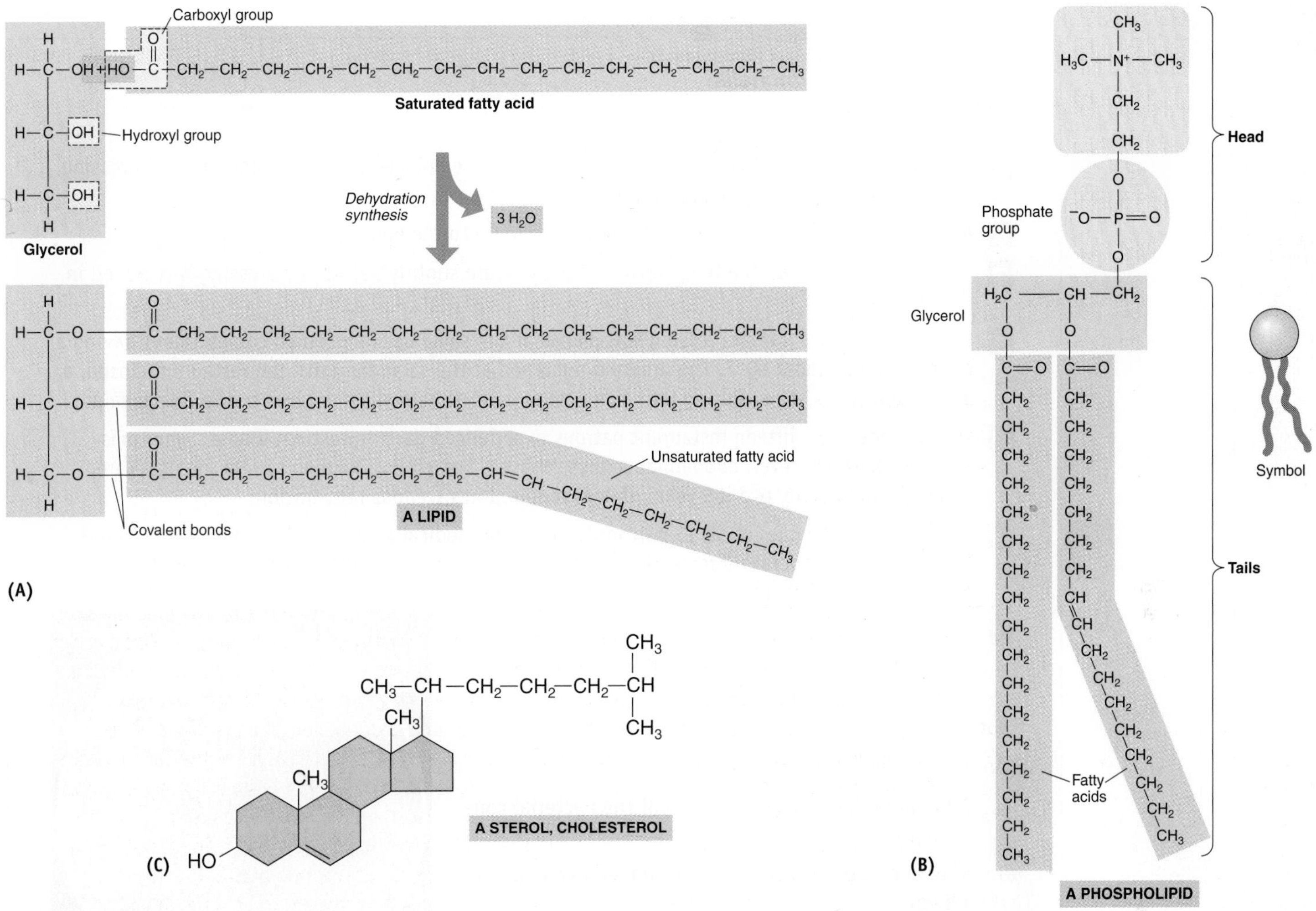

FIGURE 2.13 Lipid and Lipid-Related Compounds. (**A**) A lipid, such as a triglyceride, consists of glycerol and fatty acids. (**B**) A phospholipid consists of glycerol attached to two fatty acids and a phosphate head group. Inset: The symbol for the structure of a phospholipid. (**C**) The sterol cholesterol. »» *Why are all these compounds considered "lipids"?*

Both DNA and RNA are macromolecules composed of repeating monomers called **nucleotides** (**FIGURE 2.14A**). Each nucleotide has three components: a sugar molecule, a phosphate group, and a nitrogenous base. The sugar in DNA is deoxyribose, while in RNA it is ribose. The **nitrogenous bases** are nitrogen-containing compounds. In DNA, the purine bases are adenine (A) and guanine (G), while the pyrimidine bases are cytosine (C) and thymine (T). In RNA, adenine, guanine, and cytosine also are present, but uracil (U) is found instead of thymine.

Nucleotides are covalently joined through dehydration synthesis reactions between the sugar of one nucleotide and the phosphate of the adjacent nucleotide to eventually form a **polynucleotide** (**FIGURE 2.14B**).

DNA. In 1953, James Watson, Francis Crick, Rosalind Franklin, and Maurice Wilkins published papers describing how a complete DNA molecule consists of two polynucleotide strands opposed to each other in a ladder-like arrangement (**FIGURE 2.14C**). Guanine and cytosine line up opposite one another, and thymine and adenine oppose each other in the two strands. The complementary base pairs in the double-stranded DNA molecule are held together by hydrogen bonds. The double strand then twists to form a spiral arrangement called the **DNA double helix**.

DNA is the genetic material in all living organisms. The genetic information exists in discrete units called **genes**, which are sequences of nucleotides that encode information to regulate and synthesize proteins (Chapter 8). In bacterial and archaeal cells, these genes are found on a single circular chromosome, while in most eukaryotic microbes, the genes are located on several linear chromosomes. Genes only

Chromosome: A DNA molecule containing the hereditary information in the form of genes.

Textbook CASE 2

An Outbreak of *Salmonella* Food Poisoning

1 During the morning of October 17, 1991, a restaurant employee prepared a Caesar salad dressing, cracking fresh eggs into a large bowl containing olive oil.

2 Anchovies, garlic, and warm water then were mixed into the eggs and oil.

3 The warm water raised the temperature of the mixture slightly before the dressing was placed in the refrigerator.

4 Later that day, the Caesar dressing was placed at the salad bar in a cooled compartment having a temperature of about 60°F. The dressing remained at the salad bar until the restaurant closed, a period of 8 to 10 hours. During that time, many patrons helped themselves to the Caesar salad.

5 Within three days, fifteen restaurant patrons experienced gastrointestinal illness. Symptoms included diarrhea, fever, abdominal cramps, nausea, and chills. Thirteen sought medical care, and eight (all elderly over 65 years of age) required intravenous rehydration.

6 From the stool samples of all 13 patrons who sought medical attention, bacteria were cultured (see figure) and laboratory tests identified *Salmonella enterica* serotype Enteritidis as the causative agent (see Chapter 11).

7 *S. enterica* serotype Enteritidis produces a lipopolysaccharide toxin that causes the symptoms experienced by all the affected patrons.

Questions:

(Answers can be found in Appendix D.)

A. What might have been the origin of the bacterial contamination?

B. What conditions would have encouraged bacterial growth?

C. How could the outbreak have been prevented?

D. What types of organic compounds form the lipopolysaccharide toxin?

E. Why did so many of the elderly patrons develop a serious illness?

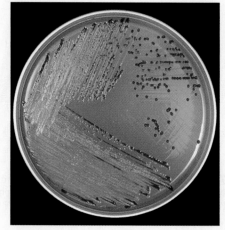

A culture plate of *Salmonella*.

For additional information see http://www.cdc.gov/ncidod/dbmd/diseaseinfo/salment_g.htm.

---

carry the information to regulate or synthesize proteins.

RNA. Besides having uracil as a base and ribose as the sugar, RNA molecules in cells are single-stranded polynucleotides. Biologists once viewed RNAs as the intermediaries, involved in carrying gene information or as structural molecules needed to construct proteins. This certainly is a major role for RNA but not the only role.

In viruses such as the influenza and measles viruses, RNA is the genetic information, not DNA. Other RNA molecules play key roles in regulating gene activity, while several small RNAs control various cellular processes in microbial cells.

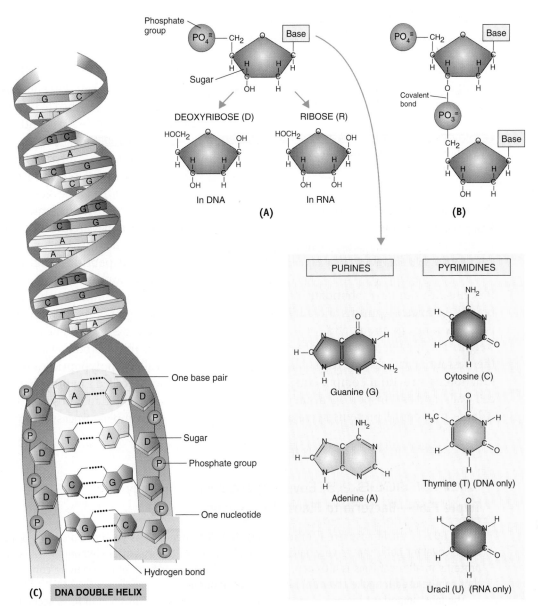

FIGURE 2.14 The Molecular Structures of Nucleotide Components and the Construction of DNA. (**A**) The sugars in nucleotides are ribose and deoxyribose, which are identical except for one additional oxygen atom in RNA. The nitrogenous bases include adenine and guanine, which are large purine molecules, and thymine, cytosine, and uracil, which are smaller pyrimidine molecules. Note the similarities in the structures of these bases and the differences in the side groups. (**B**) Nucleotides are bonded together by dehydration synthesis reactions. (**C**) The two polynucleotides of DNA are held together by hydrogen bonds between adenine (A) and thymine (T) or guanine (G) and cytosine (C) to form a double helix. »» If a segment of one strand of DNA has the bases TTAGGCACG, what would be the sequence of bases in the complementary strand?

The nucleic acids cannot be altered without injuring the organism or killing it. Ultraviolet light damages DNA, and thus it can be used to control microbes on an environmental surface. Chemicals, such as formaldehyde, alter the nucleic acids of viruses and can be used in the preparation of vaccines. Certain antibiotics interfere with DNA or RNA function and thereby kill bacteria.

Chapters 8 and 9 are devoted to the role of nucleic acids in the genetics of microbes and how the genetic information can be manipulated for medical or industrial applications.

It also is important to point out that nucleotides have other roles in cells besides being part of DNA or RNA. Adenine nucleotides with three attached phosphate groups form **adenosine** triphosphate (ATP), which is the cellular energy currency in all cells. Other nucleotides can be part of the structure of some enzymes, while independent, modified nucleotides called cyclic adenosine monophosphate (cAMP) act as chemical signals in many microbes.

CONCEPT AND REASONING CHECKS

2.14 How does the structure of DNA differ from that of RNA?

Proteins Are the Workhorse Polymers in Cells

KEY CONCEPT

15. Proteins fold into diverse three-dimensional shapes.

Proteins are the most abundant organic compounds in microorganisms and all living organisms, making up about 58% of the cell's dry weight. The high percentage of protein indicates their essential and diverse roles. Many proteins function as structural components of cells and cell walls, and as transport agents in membranes. A large number of proteins serve as enzymes.

Proteins are composed of carbon, hydrogen, oxygen, nitrogen, and, usually, sulfur atoms.

Proteins are polymers built from nitrogen-containing monomers called **amino acids** (MicroFocus 2.5). At the center of each amino acid is a carbon atom attached to two functional groups: an amino group ($-NH_2$) and a carboxyl group ($-COOH$) (FIGURE 2.15A). Also attached to the carbon is a side chain, called the **R group**. Each of the 20 amino acids differs only by the atoms composing the R group. These side chains, many being functional groups, are essential in determining the final shape, and therefore function, of the protein.

In protein formation, two amino acids (sometimes called peptides) are joined together by a covalent bond when the amino group of one amino acid is linked to the carboxyl of another amino acid through a dehydration synthesis reaction (Figure 2.15A). Repeating the reaction hundreds of times produces a long chain of amino acids called a **polypeptide** and the covalent bond therefore is called a **peptide bond**. How the amino acids are slotted into position to build a polypeptide is a complex process discussed in Chapter 8.

MICROFOCUS 2.5: Environmental Microbiology

Triple Play—Bacteria to Plants to Humans

About 80% of the atmosphere is nitrogen gas (N_2). Nitrogen gas, as you have discovered, contains a triple covalent bond that is very hard to break. Yet, one of the essential elements in nucleic acids and proteins in all organisms is nitrogen. So, how can the gaseous form of nitrogen be converted into a form that can be used to make essential biological compounds?

The most important biological process to break the triple covalent bond in N_2 is accomplished by a few bacterial species commonly found in root nodules of pea and bean plants (legumes) or in the soil close to the plant roots. These bacterial organisms contain an enzyme, called nitrogenase, which converts N_2 into ammonia that then can be further converted by microbial action into forms used by legumes and other plants. The process is called **nitrogen fixation**:

$$N_2 \rightarrow NH_3 \text{ (ammonia)}$$

In contact with water, the gaseous ammonia is converted to ammonium ions (NH_4^+), which serve as a source of nitrogen for nucleic acid and amino acid synthesis by bacterial cells and plants. We then get our nitrogen for amino acids and nucleotides from eating plants or through exchange reactions of carbohydrate metabolism that convert sugars into nucleotides or amino acids. Chapter 26 describes the nitrogen cycle in more detail.

The important point to remember in all this chemistry is that the initial fixation of nitrogen is dependent on bacterial chemistry. In fact, without nitrogen fixation, life as we know it would not exist.

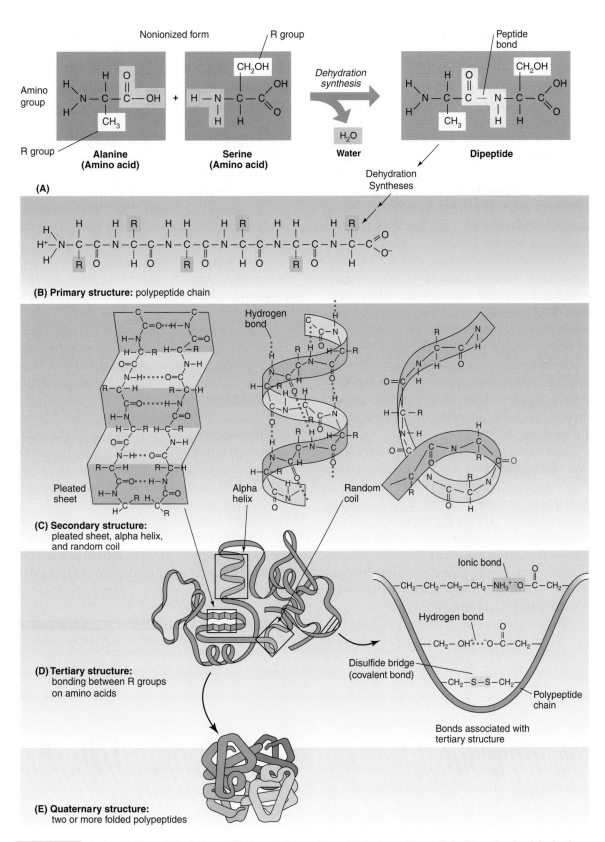

FIGURE 2.15 Amino Acids and Their Assembly into Polypeptides. (**A**) Amino acids are linked together by dehydration synthesis reactions. (**B**) As they get longer, the sequence of amino acids forms the primary structure, which takes on a secondary structure (**C**). (**D**) The whole polypeptide folds into a tertiary structure through bonding between R groups. Some proteins consist of more than one polypeptide, forming a quaternary structure (**E**). »» Explain why each and every protein must have three or four levels of folding.

However, the sequence of amino acids is critical because a single amino acid improperly positioned may change the three-dimensional shape and function of the protein.

Because proteins have tremendously diverse roles, they come in many sizes and shapes. The final shape depends on several factors associated with the amino acids. The sequence of amino acids in the polypeptide represents the **primary structure** (FIGURE 2.15B). Each protein that has a different function will have a different primary structure. However, the sequence of amino acids alone is not sufficient to confer function.

Many polypeptides have regions folded into a corkscrew shape or **alpha helix**. These regions represent part of the protein's **secondary structure** (FIGURE 2.15C). Hydrogen bonds between amino groups (–NH) and carbonyl groups (–C=O) on nearby amino acids maintain this structure. A secondary structure also may form when the hydrogen bonds cause portions of the polypeptide chain to zigzag in a flat plane, forming a **pleated sheet**. Other regions may not interact and remain in a random coil.

Many polypeptides also have a **tertiary structure** (FIGURE 2.15D). Such a three-dimensional (3-D) shape of a polypeptide is folded back on itself much like a spiral telephone cord. Ionic and hydrogen bonds between R groups on amino acids in proximity to each other help form and maintain the polypeptide in its tertiary structure. In addition, covalent bonds, called **disulfide bridges**, between sulfur atoms in R groups are important in stabilizing tertiary structure.

The ionic and hydrogen bonds helping hold a protein in its 3-D shape are relatively weak associations. As such, these interactions in a protein are influenced by environmental conditions. When subjected to heat, pH changes, or certain chemicals, these bonds may break, causing the polypeptide to unfold and lose its biological activity. This loss of 3-D shape is referred to as **denaturation**. For example, the white of a boiled egg is denatured egg protein (albumin) and cottage cheese is denatured milk protein. Should enzymes be denatured, the important chemical reactions they facilitate will be interrupted and death of the organism may result. Viruses also can be destroyed by denaturing the proteins found on and in the virus.

Now you should understand the importance of buffers in cells; by preventing pH shifts, they prevent protein denaturation and maintain protein function.

Many proteins are single polypeptides. However, other proteins contain two or more polypeptides to form the complete and functional protein; this is called the **quaternary structure** (FIGURE 2.15E). Each polypeptide chain is folded into its tertiary structure and the unique association between separate polypeptides produces the quaternary structure. The same types of chemical bonds are involved as in tertiary structure.

The four major classes of organic compounds are summarized in FIGURE 2.16 . MicroFocus 2.6 looks at the origins of the monomers and polymers discussed in this chapter, while MicroInquiry 2 uses the radioactive attributes of two chemical elements to discover whether protein or DNA is the genetic material.

CONCEPT AND REASONING CHECKS

2.15 Why does a denatured protein no longer have biological activity?

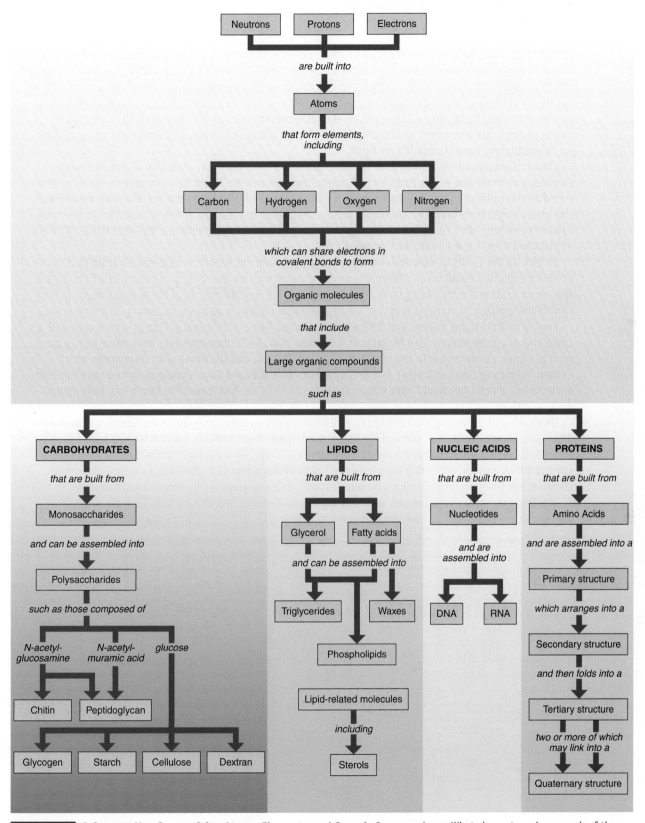

FIGURE 2.16 A Concept Map Summarizing Atoms, Elements, and Organic Compounds. »» What elements make up each of the four large organic compounds?

MICROFOCUS 2.6: Evolution
Generating Life—Today (Part II)

In MicroFocus 1.2, we discussed the idea of creating life artificially—through the discipline called **synthetic biology**. The idea concerns the ability to create "new" life in a test tube. As you discovered, some scientists already have made initial strides in this direction by reconstructing a virus from scratch. Others are trying to design "new organisms" by piecing together specific cellular and genetic parts taken from various microbes. So far, these "bioreactors" do not represent cellular life. However, the potential to build synthetic life also opens up the possibility of assembling, in the lab, a "primordial organism" similar to what might have started life on Earth.

Chemical precursors of life may have involved gases like formaldehyde, methane, water, hydrogen cyanide, and ammonia. As the Earth formed, these "molecular seeds" were brought together and concentrated at the Earth's surface. Through a process of chemical evolution, chemical reactions between these precursors might have produced larger and more complicated compounds. If the products could be stabilized in some way, they could form into simple sugars, nucleotides, and amino acids—the building blocks of carbohydrates, nucleic acids, and proteins.

In fact, in the 1950s Stanley Miller and Harold Urey carried out experiments demonstrating that such chemical evolution could occur. Mixing the primeval gases with an energy source produced within days a few amino acids and nitrogenous bases. A more recent 2008 re-examination of these experiments using conditions similar to a volcanic eruption found additional organic molecules, including 22 amino acids.

Most scientists today believe that RNA was the original "genetic information" in primitive cells and could also act as an enzyme (see MicroFocus 6.2, page 164). Such ribozymes may have arose from sugars, such as ribose, combining with the nitrogenous bases (A, G, C, and U). Along with phosphate, which existed in volcanic areas and other deposits on Earth, the bases and sugars could be linked together into nucleotides, which then would form into long chains of RNA. So, RNA-based life forms may have spread and evolved on Earth for millions of years.

Generating or recreating such life today may be the way to verify the hypothesis. Scientists, such as John Szostak and his team at Harvard Medical School, have mixed RNA nucleotides with clay and then added fatty acids. The spontaneous result was a primordial cell—a lipid (membrane) bubble containing short RNA polynucleotides. A key experiment is to demonstrate that the RNAs can carry out simple chemical reactions such as hydrolysis or dehydration synthesis reactions by acting as a ribozyme.

The researchers have not created synthetic life—yet. However, the experiments do indicate their ideas have promise and may go a long way toward telling scientists how life originated on Earth.

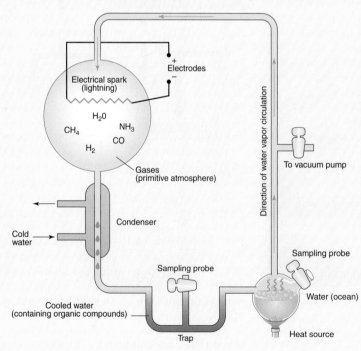

MICROINQUIRY 2

Is Protein or DNA the Genetic Material?

In the early 1950s, there were scientists who still debated whether protein or DNA was the genetic material in cells. To settle the controversy, in 1952 Alfred Hershey and Martha Chase carried out a series of experiments to trace the fates of protein and DNA, and in so doing hopefully settle the debate.

It was known that some viruses that infect bacterial cells were composed of DNA and protein, and that the virus genetic material needed to enter the bacterial cells to direct the production of more viruses. Because the viruses left a viral coat on the surface, what actually entered the cells—protein or DNA? Whichever did must be the genetic material.

Several biologically important elements have isotopes that are radioactive. The table to the right lists a few such elements. Hershey and Chase decided to radioactively label the viruses such that the protein and DNA could be identified by their unique radioactive profiles.

Some Radioactive Isotopes

Element	Common Form	Radioactive Form
Hydrogen	^1H	^3H (tritium)
Carbon	^{12}C	^{14}C
Phosphorus	^{31}P	^{32}P
Sulfur	^{32}S	^{35}S

2a. Which of the radioactive elements would only label protein?

2b. Which of the radioactive elements would only label DNA?

2c. Could ^3H or ^{14}C have been used? Explain.

The Hershey and Chase experiment is outlined in the figure below. From the two experiments, they could then measure the radioactivity in the pellet (bacterial cells) and the fluid (virus coats) and determine which radioactive isotope was associated with the bacterial cells.

2d. If protein is the genetic material, which isotope should be associated with the bacterial cells?

2e. If DNA is the genetic material, which isotope should be associated with the bacterial cells?

2f. So, did the experiments carried out by Hershey and Chase support or refute Avery's work that DNA was the genetic material?

Answers can be found on the student companion Web site in **Appendix D**.

DNA — Protein — Bacterial virus

A. Viruses containing radioactive sulfur (^{35}S) or (^{32}P) are mixed with bacteria so that infection can occur.

B. After infection, the mixture is placed in a blender to separate viruses outside the cells from the bacterial cells.

C. The mixture is centrifuged in a test tube to separate the cells from the viruses.

D. Radioactivity in the pellet and liquid is measured.

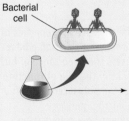

Experiment 1: Viruses are produced that have radioactive sulfur (^{35}S) incorporated into the virus protein.

Bacterial cell

Empty protein shell

Centrifuge

Pellet (bacterial cells and contents)

Experiment 1 results: All radioactivity is ^{35}S and it is detected in the liquid.

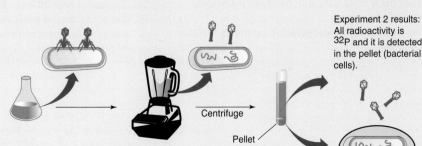

Experiment 2: Viruses are produced that have radioactive phosphorus (^{32}P) incorporated into the virus DNA.

Centrifuge

Pellet

Experiment 2 results: All radioactivity is ^{32}P and it is detected in the pellet (bacterial cells).

Is Protein or DNA the Genetic Material? The Hershey-Chase experiment.

◼ SUMMARY OF KEY CONCEPTS

2.1 The Elements of Life

1. **Atoms** consist of an **atomic nucleus** (with **neutrons** and positively charged **protons**) surrounded by a cloud of negatively charged **electrons**.

2. **Isotopes** of an element have different numbers of neutrons. Some unstable ones, called **radioisotopes**, are useful in research and medicine. If an atom gains or loses electrons, it becomes an electrically charged **ion**. Many ions are important in microbial metabolism.

3. Each electron shell holds a maximum number of electrons. **Chemical bonding** occurs between atoms to fill the outer shells with electrons.

2.2 Chemical Bonding

4. **Ionic bonds** result from the attraction of oppositely charged ions. Compounds called **salts** result.

5. Most atoms achieve stability through a sharing of electrons, forming **covalent bonds**. The equal sharing of electrons produces **nonpolar molecules** (no electrical charge). Atomic interactions between hydrogen and oxygen (or nitrogen) produce unequal sharing of electrons, which generates **polar molecules** (have electrical charges).

6. Separate polar molecules, like water, are electrically attracted to one another and form **hydrogen bonds**, involving positively charged hydrogen atoms and negatively charged oxygen (or nitrogen) atoms.

7. In a chemical reaction, the atoms in the **reactant** change bonding partners in forming one or more **products**. Two common chemical reactions in cells are **dehydration synthesis reactions** and **hydrolysis reactions**. In these reactions, the number of atoms is the same in the reactants and products.

2.3 Water, pH, and Buffers

8. All chemical reactions in organisms occur in liquid water. Being polar, water has unique properties. These include its role as a solvent, as a chemical reactant, and as a factor to maintain a fairly constant temperature.

9. **Acids** donate hydrogen ions (H^+) while **bases** acquire H^+ from a solution. The **pH scale** indicates the number of H^+ in a solution and denotes the relative acidity of a solution.

10. **Buffers** are a mixture of a weak acid and a weak base that maintain acid/base balance in cells. Excess H^+ can be absorbed by the base and too few H^+ can be provided by the acid.

2.4 Major Organic Compounds of Living Organisms

11. The building of large organic compounds depends on the **functional groups** found on the building blocks called **monomers**. Functional groups on monomers interact through dehydration synthesis reactions to form a covalent bond between monomers.

12. **Carbohydrates** include **monosaccharides** such as glucose that are linked into **polysaccharides** that represent energy and structural molecules.

13. **Lipids** serve as energy sources, but their major role is as **phospholipids** in cell membranes. Other lipids include the **sterols**.

14. The genetic instructions for living organisms are composed of two types of **nucleic acids: deoxyribonucleic acid** (**DNA**), which stores and encodes the hereditary information; and **ribonucleic acid** (**RNA**), which transmits the information to make proteins, controls genes, and helps regulate genetic activity.

15. **Proteins** are chains of **amino acids** connected by **peptide bonds**. Proteins are used as enzymes and as structural components of cells. **Primary**, **secondary**, and **tertiary structures** form the functional shape of many proteins, which can unfold by **denaturation**. Many proteins are the result of two or more polypeptides bonding together (**quaternary structure**).

◼ LEARNING OBJECTIVES

After understanding the textbook reading, you should be capable of writing a paragraph that includes the appropriate terms and pertinent information to answer the objective.

1. Contrast the properties of **protons**, **neutrons**, and **electrons**. Assess the importance of these atomic particles to atomic structure.

2. Summarize how elements can form **isotopes** and **ions**.

3. Describe the basis for chemical bonding.

4. Distinguish between ions and **ionic bonds**.

5. Compare and contrast **polar molecules** with **nonpolar molecules**.

6. Explain how **hydrogen bonds** form.

7. Describe the differences between a **dehydration synthesis reaction** and a **hydrolysis reaction**.

8. Identify several properties of water.

9. Contrast an **acid** and a **base**.

10. Assess the importance of **buffers** to chemical reactions.

11. Identify the role of **functional groups** on organic molecules.

12. List the **polysaccharides** found in cells or organisms. Explain their role in microorganisms.

13. Explain how dehydration synthesis forms a **lipid**. Contrast between **saturated** and **unsaturated** fatty acids.

14. Summarize how **DNA** and **RNA** differ in structure and function.

15. Show how **amino acids** link together and name the specific type of bond formed between these amino acids, and compare the four levels of protein structure.

STEP A: SELF-TEST

Each of the following questions is designed to assess your ability to remember or recall factual or conceptual knowledge related to this chapter. Read each question carefully, then select the **one** answer that best fits the question or statement. Answers to even-numbered questions can be found in **Appendix C**.

1. These positively charged particles are found in the atomic nucleus.
   A. Protons only
   B. Electrons only
   C. Protons and neutrons
   D. Neutrons only

2. Atoms of the same element that have different numbers of neutrons are called _____.
   A. isotopes
   B. ions
   C. isomers
   D. inert elements

3. If an element has two electrons in the first shell and seven in the second shell, the element is said to be what?
   A. Unstable
   B. Unreactive
   C. Stable
   D. Inert

4. For _____ bonding, one or more electrons are transferred between atoms.
   A. hydrogen
   B. ionic
   C. peptide
   D. covalent

5. The covalent bonding of atoms forms a/an
   A. molecule.
   B. ion.
   C. element.
   D. isomer.

6. The _____ bond is a weak bond that can exist between poles of adjacent molecules.
   A. hydrogen
   B. ionic
   C. polar covalent
   D. nonpolar covalent

7. In what type of chemical reaction are the products of water removed during the formation of covalent bonds?
   A. Hydrolysis
   B. Ionization
   C. Dehydration synthesis
   D. Decomposition

8. A _____ dissolves in water.
   A. solvent
   B. hydrophobic molecule
   C. solute
   D. nonpolar molecule

9. The pH scale relates the measure of _____ of a chemical substance.
   A. ionization
   B. denaturation
   C. acidity
   D. buffering

10. Which one of the following statements about buffers is *false*?
    A. They work inside cells.
    B. They consist of a weak acid and weak base.
    C. They prevent pH shifts.
    D. They enhance chemical reactions.

11. A functional group designated—COOH is known as a/an
    A. carboxyl.
    B. carbonyl.
    C. amino.
    D. hydroxyl.

12. Which one of the following is NOT a polysaccharide?
    A. Chitin
    B. Glycogen
    C. Cellulose
    D. Lipid

13. How do the lipids differ from the other organic compounds?
    A. They are the largest organic compounds.
    B. They are nonpolar compounds.
    C. They have no biological role.
    D. They are not used for energy storage.

14. Both DNA and RNA are composed of _____.
    A. polynucleotides
    B. genes
    C. polysaccharides
    D. polypeptides

15. The _____ structure of a protein is the sequence of amino acids.
    A. primary
    B. secondary
    C. tertiary
    D. quaternary

STEP B: REVIEW

Answers to even-numbered questions can be found in **Appendix C.**

16. Construct a concept map for **chemical bonds** using the following terms:

hydrocarbon   nonpolar covalent bonds
hydrogen bonds   polar covalent bonds
ionic bonds   salts
methane   water
NaCl

17. Use the following list to identify the structures (i–v) drawn below.
    A. Amino acid
    B. Monosaccharide
    C. Nucleotide
    D. Lipid
    E. Disaccharide
    F. Polysaccharide
    G. Sterol

18. Identify any and all functional groups on each structure (i–v).

(i)

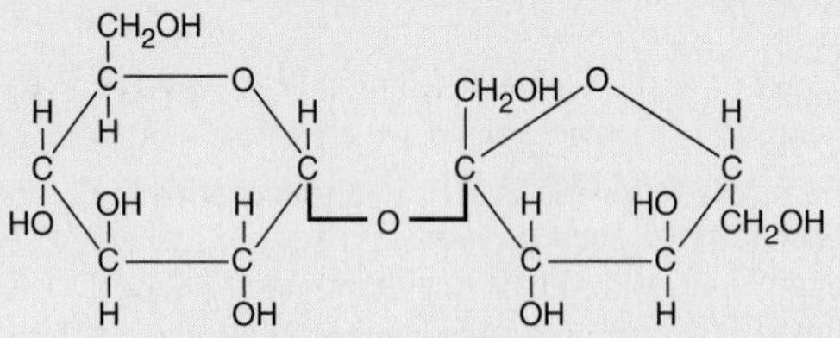

(ii)

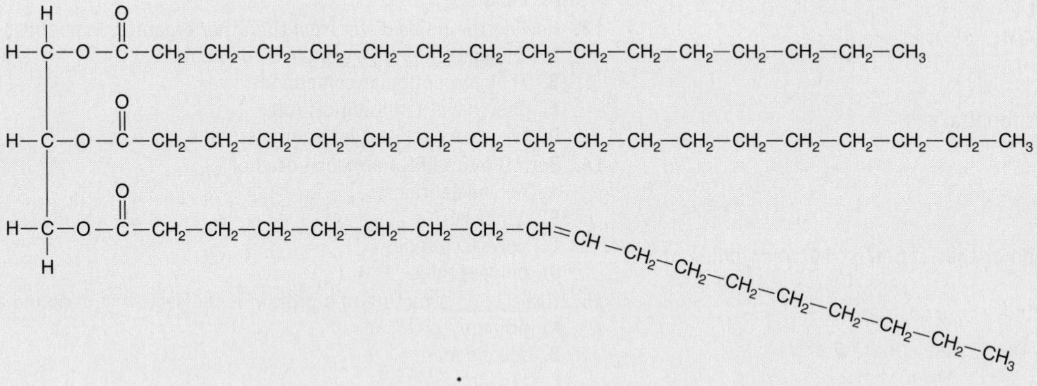

(iii)

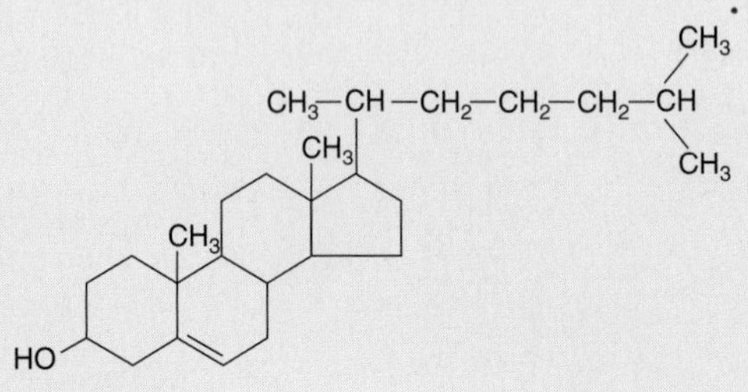

(iv)

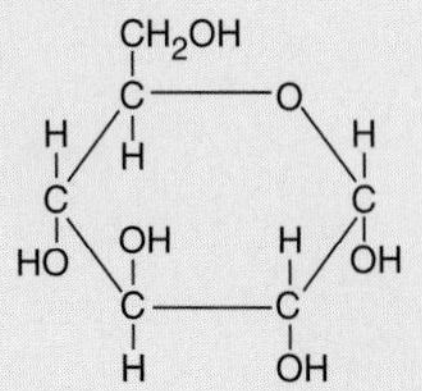

(v)

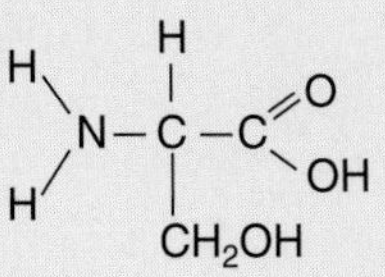

STEP C: APPLICATIONS

Answers to even-numbered questions can be found in **Appendix C**.

19. You want to grow a bacterial species that is acid-loving; that is, it grows best in very acid environments. Would you want to grow it in a culture that has a pH of 2.0, 6.8, or 11.5? Explain.

20. You are given two beakers of a broth growth medium. However, only one of the beakers of broth is buffered. How could you determine which beaker contains the buffered broth solution? Hint: You are provided with a bottle of concentrated HCl and pH papers that indicate a solution's pH.

21. The microbial community in a termite's gut contains the enzyme cellulase. How does this benefit the termite and the termite's microbial community?

STEP D: QUESTIONS FOR THOUGHT AND DISCUSSION

Answers to even-numbered questions can be found in **Appendix C**.

22. Propose a reason why organic molecules tend to be so large.

23. Bacterial cells do not grow on bars of soap even though the soap is wet and covered with bacterial organisms after one has washed. Explain this observation.

24. Suppose you had the choice of destroying one class of organic compounds in bacterial cells to prevent their spread. Which class would you choose? Why?

25. Milk production typically has the bacterium *Lactobacillus* added to the milk before it is delivered to market. This organism produces lactic acid. (a) Why would this organism be added to the milk and (b) why was it chosen?

26. The toxin associated with the foodborne disease botulism is a protein. To avoid botulism, home canners are advised to heat preserved foods to boiling for at least 12 minutes. How does the heat help?

27. Justify Isaac Asimov's quote, "The significant chemicals in living tissue are rickety and unstable, which is exactly what is needed for life," to the atoms, molecules, and organic compounds described in this chapter.

HTTP://MICROBIOLOGY.JBPUB.COM/9E

The site features eLearning, an online review area that provides quizzes and other tools to help you study for your class. You can also follow useful links for in-depth information, read more MicroFocus stories, or just find out the latest microbiology news.

3

Concepts and Tools for Studying Microorganisms

We think we have life down; we think we understand all the conditions of its existence; and then along comes an upstart bacterium, live or fossilized, to tweak our theories or teach us something new.
—Jennifer Ackerman in *Chance in the House of Fate* (2001)

The oceans of the world are a teeming but invisible forest of microorganisms and viruses. For example, one liter of seawater contains more than 25,000 different bacterial species.

A substantial portion of these marine microbes represent the **phytoplankton** (*phyto* = "plant"; *plankto* = "wandering"), which are floating communities of cyanobacteria and eukaryotic algae. Besides forming the foundation for the marine food web, the phytoplankton account for 50% of the photosynthesis on earth and, in so doing, supply about half the oxygen gas we and other organisms breathe.

While sampling ocean water, scientists from MIT's Woods Hole Oceanographic Institution discovered that many of their samples were full of a marine cyanobacterium, which they eventually named *Prochlorococcus*. Inhabiting tropical and subtropical oceans, a typical sample often contained more than 200,000 (2×10^5) cells in one drop of seawater.

Studies with *Prochlorococcus* suggest the organism is responsible for almost 50% of the photosynthesis in the open oceans (**FIGURE 3.1**). This makes *Prochlorococcus* the smallest and most abundant marine photosynthetic organism yet discovered.

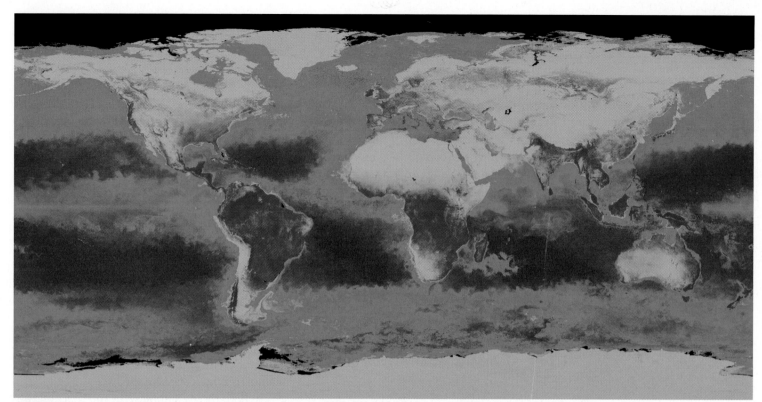

FIGURE 3.1 **Photosynthesis in the World's Oceans.** This global satellite image (false color) shows the distribution of photosynthetic organisms on the planet. In the aquatic environments, red colors indicate high levels of chlorophyll and productivity, yellow and green are moderate levels, and blue and purple areas are the "marine deserts." »» How do the landmasses where photosynthesis is most productive (green) compare in size to photosynthesis in the oceans?

The success of *Prochlorococcus* is due, in part, to the presence of different ecotypes inhabiting different ocean depths. For example, the high sunlight ecotype occurs in the surface waters while the low-light type is found below 50 meters. This latter ecotype compensates for the decreased light by increasing the amount of cellular chlorophyll that can capture the available light.

In terms of nitrogen sources, the high-light ecotype only uses ammonium ions (NH_4^+) (see MicroFocus 2.5). At increasing depth, NH_4^+ is less abundant so the low-light ecotype compensates by using a wider variety of nitrogen sources.

These and other attributes of *Prochlorococcus* illustrate how microbes survive through change. They are of global importance to the functioning of the biosphere and, directly and indirectly, affect our lives on Earth.

Once again, we encounter an interdisciplinary group of scientists studying how microorgan-isms influence our lives and life on this planet. Microbial ecologists study how the phytoplankton communities help in the natural recycling and use of chemical elements such as nitrogen. Evolutionary microbiologists look at these microorganisms to learn more about their taxonomic relationships, while microscopists, biochemists, and geneticists study how *Prochlorococcus* cells compensate for a changing environment of sunlight and nutrients.

This chapter focuses on many of the aspects described above. We examine how microbes maintain a stable internal state and how they can exist in "multicellular", complex communities. Throughout the chapter we are concerned with the relationships between microorganisms and the many attributes they share. Then, we explore the methods used to name and catalog microorganisms. Finally, we discuss the tools and techniques used to observe the microbial world.

Ecotypes:
Subgroups of a species that have special characteristics to survive in their ecological surroundings.

Biosphere:
That part of the earth—including the air, soil, and water—where life occurs.

3.1   The Bacteria/Eukaryote Paradigm

In the news media or even in scientific magazines and textbooks, bacterial and archaeal species often are described as "simple organisms" compared to the "complex organisms" representing multicellular plant and animal species. This view represents a mistaken perception. Despite their microscopic size, bacterial and archaeal organisms exhibit every complex feature, or emerging property, common to all living organisms. These include:

- DNA as the hereditary material controlling structure and function.
- Complex biochemical patterns of growth and energy conversions.
- Complex responses to stimuli.
- Reproduction to produce offspring.
- Adaptation from one generation to the next.

Focusing on the *Bacteria*, what is the evidence for complexity?

Bacterial Complexity: Homeostasis and Biofilm Development

KEY CONCEPT

1. Bacterial cells undergo biological processes as complex as in eukaryotes.

Historically, when one looks at bacterial cells even with an electron microscope, often there is little to see (FIGURE 3.2A). "Cell structure," representing the cell's physical appearance or its components and the "pattern of organization," referring to the configuration of those structures and their relationships to one another, do give the impression of simpler cells.

But what has been overlooked is the "cellular process," the activities all cells carry out for the continued survival of the cell (and organism). At this level, the complexity is just as intricate as in any eukaryotic cell. So, in reality, bacteria cells carry out many of the same cellular processes as eukaryotes—only without the need for an elaborate, visible structural organization.

Homeostasis. All organisms continually battle their external environment, where factors such as temperature, sunlight, or toxic chemicals can have serious consequences. Organisms strive to maintain a stable internal state by making appropriate metabolic or structural adjustments. This ability to adjust yet maintain a relatively steady internal state is called **homeostasis** (*homeo* = "similar"; *stasis* = "state"). Two examples illustrate the concept (FIGURE 3.2B).

The low-light *Prochlorococcus* ecotype mentioned in the chapter introduction lives at depths of below 50 meters. At these depths, transmitted sunlight decreases and any one nitrogen source is less accessible. The ecotype compensates for the light reduction and nitrogen limitation by (1) increasing the amount of cellular **chlorophyll** to capture light and (2) using a wider variety of available nitrogen sources. These adjustments maintain a steady internal state.

For our second example, suppose a patient is given an antibiotic to combat a bacterial infection. In response, the infecting bacterium compensates for the change by breaking the structure of the antibiotic. The adjustment, antibiotic resistance, maintains homeostasis in the bacterial cell.

In both these examples, the internal environment is maintained despite a changing environment. Such, often complex, homeostatic controls are critical to all microbes, including bacterial species.

Biofilm Development. One of the emerging properties of life is that cells must cooperate with one another. This is certainly true in animals and plants, but it is true of most bacterial organisms as well.

The early studies of disease causation done by Pasteur and Koch (see Chapter 1) certainly required pure cultures to associate a specific disease with one specific microbe. However, today it is necessary to abolish the impression that bacteria are self-contained, independent organisms. In nature few species live such a pure and solitary life. In fact, it has been estimated that up to 99% of bacterial species live in communal associations called **biofilms**; that is, in a "multicellular state" where survival requires chemical communication and cooperation between cells.

As a biofilm forms, the cells become embedded in a matrix of excreted polymeric substances produced by the bacterial cells (FIGURE 3.2C .) These sticky substances are composed of charged and neutral polysaccharides that hold the biofilm together and cement it to nonliving or living surfaces, such as metals, plastics, soil particles,

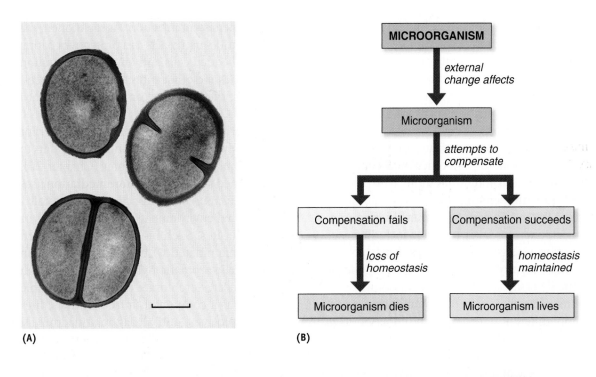

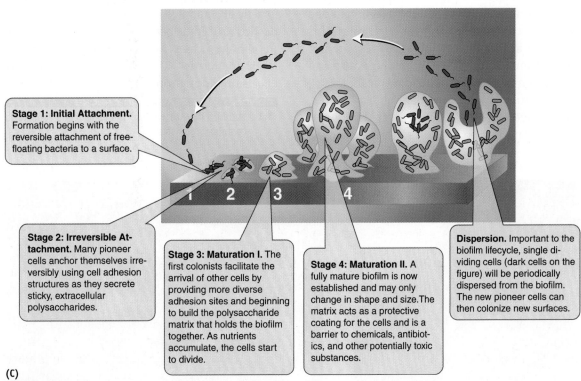

FIGURE 3.2 Simpler, Unicellular Organisms? (**A**) This false-color electron microscope image of *Staphylococcus aureus* gives the impression of simplicity in structure. (Bar = 0.5μm) (**B**) A concept map illustrating how bacterial organisms, like all microorganisms, have to compensate for environmental changes. Survival depends on such homeostatic abilities. (**C**) The formation of a biofilm is an example of intercellular cooperation in the development of a multicellular structure. »» Using the concept map in (**B**), explain how *Prochlorococcus* compensates for low-light conditions in its environment. (**C**) Modified from David G. Davies, Binghamton University, Binghamton NY.

medical indwelling devices, or human tissue. The mature, fully functioning biofilm is like a living tissue with a primitive circulatory system made of water channels to bring in nutrients and eliminate wastes. A biofilm is a complex, metabolically cooperative community made up of peacefully coexisting species.

It is during this colonization that the cells are able to "speak to each other" and cooperate through chemical communication. This process, called **quorum sensing**, involves the ability of bacteria to sense their numbers, and then to communicate and coordinate behavior, including gene expression, via signaling molecules. Thus, biofilms are characterized by structural heterogeneity, genetic diversity, and complex community interactions. The cells within the community are profoundly different in behavior and function from those of their independent, free-living cousins. MicroFocus 3.1 describes a few examples.

Biofilms can also be associated with infections. Development of a fatal lung infection (cystic fibrosis pneumonia), middle ear infections (otitis media), and tooth decay (dental caries) are but a few examples (FIGURE 3.3A). Biofilms also can develop on improperly cleaned medical devices, such as artificial joints, mechanical heart valves, and catheters (FIGURE 3.3B), such that when implanted into the body, the result is a slow devel-

oping but persistent infection. As mentioned, the polysaccharide matrix acts as a protective coating for the embedded cells and impedes penetration by antibiotics and other antimicrobial substances. As a result, the infection can be extremely hard to eradicate.

On the other hand, biofilms can be useful. For example, sewage treatment plants use biofilms to remove contaminants from water (Chapter 26). As mentioned in Chapter 1, **bioremediation** uses microorganisms to remove or clean up chemically-contaminated environments, such as oil spills or toxic waste sites. Such biofilms have been used at sites contaminated with toxic organics, such as "polycyclic aromatic hydrocarbons" that can lead to cancer. Perchlorate (ClO_4^-) is a soluble anion that is a component in rocket fuels, fireworks, explosives, and airbag manufacture. It is toxic to humans and is highly persistent in drinking water, especially in the western United States. Natural subterranean biofilms are being genetically modified so the cells contain the genes needed to degrade perchlorate from groundwater. In both these cases, a concentrated community of microorganisms—a biofilm—can have positive effects on the environment.

CONCEPT AND REASONING CHECKS

3.1 Support the statement "Bacterial cells represent complex organisms."

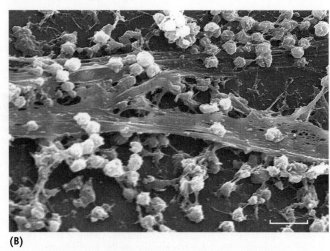

(A)   (B)

FIGURE 3.3 Biofilms in Disease. (**A**) A false-color electron microscope image of a tooth surface showing the plaque biofilm (purple) containing bacteria cells. The red cells are red blood cells. (Bar = 60 µm.) (**B**) An electron microscope image of *Staphylococcus aureus* contamination on a catheter. The fibrous-looking substance is part of the biofilm. (Bar = 3 µm.) »» What is the best way to minimize such biofilms on the teeth?

MICROFOCUS 3.1: Environmental Microbiology
The Power of Quorum Sensing

As the chapter opener stated, the microbial world is truly immense and we are continually surprised by what we find. Take quorum sensing for example. The discovery that bacterial cells can communicate with each other changed our general perception of bacterial species as single, simple organisms inhabiting our world. Here are two examples.

Vibrio fischeri

Vibrio fischeri is a light-emitting, marine bacterial organism found at very low concentrations around the world. At these low concentrations, the cells do not emit any light (see figure). However, juvenile Hawaiian bobtail squids selectively draw up the free-living *V. fischeri* and the bacterial cells take up residence in what will be the squids' functional adult light organ called the photophore. The bacterial cells are maintained in this organ for the entire life of the squid. Why take up these bacterial cells?

The bobtail squid is a nocturnal species that hunts and feeds in shallow marine waters. On moonlit nights, the light casts a moving shadow of the squid on the sandy bottom. Such movements can attract squid predators. The *V. fischeri* cells confined to the photophore grow to high concentrations (about 10^{11} cells/ml). Sensing their high numbers, the *V. fischeri* cells start chemically "chatting" with one another and produce a signaling molecule that triggers the synthesis of the bacterial enzyme luciferase. This enzyme oxidizes bacterial luciferin to oxyluciferin and energy. Now here is the quorum sensing finale: The energy is given off as cold light (bioluminescence)—the squid's photophore shines. The squid modulates the light to match that of the moonlight and directs the bacterial glow toward the bottom of the shallow waters, eliminating the bottom shadows and camouflaging itself from any predators.

 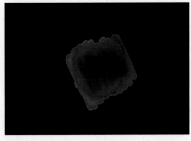

Photographs of *Vibrio fischeri* growing in a culture plate (left) and triggered to bioluminesce (right).

Myxobacteria

One of the first organisms in which quorum sensing was observed was in the myxobacteria, a bacterial group that predominantly lives in the soil. Individual myxobacterial cells are always evaluating both their own nutritional status and that of their community. The myxobacterial cells can move actively by gliding and, on sensing food (bacterial, yeast, or algal cells), typically travel in "swarms" (also known as "wolf packs") that are kept together by intercellular molecular signals. This form of quorum sensing coordinates feeding behavior and provides a high concentration of extracellular enzymes from the "multicellular" swarm needed to digest the prey. Like a lone wolf, a single cell could not effectively carry out this behavior.

Under nutrient starvation, a different behavior occurs—the cells aggregate into **fruiting bodies** that facilitate species survival. During this developmental program, approximately 100,000 cells coordinately construct the macroscopic fruiting body. In *Myxococcus xanthus*, the myxobacterial cells first respond by triggering a quorum-sensing A-signal that helps them assess starvation and induce the first stage of aggregation. Later, the morphogenetic C-signal helps to coordinate fruit body development, as many myxobacterial cells die in forming the stalk while the remaining viable cells differentiate into environmentally resistant and metabolically quiescent myxospores.

Bacteria and Eukaryotes: The Similarities in Organizational Patterns

KEY CONCEPT

2. There are organizational patterns common to all living organisms.

In the 1830s, Matthias Schleiden and Theodor Schwann developed part of the **cell theory** by demonstrating all plants and animals are composed of one or more cells, making the cell the fundamental unit of life. (Note: about 20 years later, Rudolph Virchow added that all cells arise from pre-existing cells.) Although the concept of a microorganism was just in its infancy at the time, the theory suggests that there are certain organizational patterns common to all organisms.

Genetic Organization. All organisms have a similar genetic organization whereby the hereditary material is communicated or expressed (Chapter 9). The organizational pattern for the hereditary material is in the form of one or more **chromosomes**. Structurally, most bacterial cells

Metabolism:
All the chemical reactions occurring in an organism or cell.

have a single, circular DNA molecule without an enclosing membrane (**FIGURE 3.4**.) Eukaryotic cells, however, have multiple, linear chromosomes enclosed by the membrane envelope of the cell nucleus.

Compartmentation. All organisms have an organizational pattern separating the internal compartments from the surrounding environment but allowing for the exchange of solutes and wastes. The pattern for compartmentation is represented by the cell. All cells are surrounded by a **cell membrane** (known as the **plasma membrane** in eukaryotes), where the phospholipids form the impermeable boundary to solutes while membrane proteins are the gates through which the exchange of solutes and wastes occurs, and across which chemical signals are communicated. We have more to say about membranes in the next chapter.

Metabolic Organization. The process of metabolism is a consequence of compartmentation. By being enclosed by a membrane, all cells

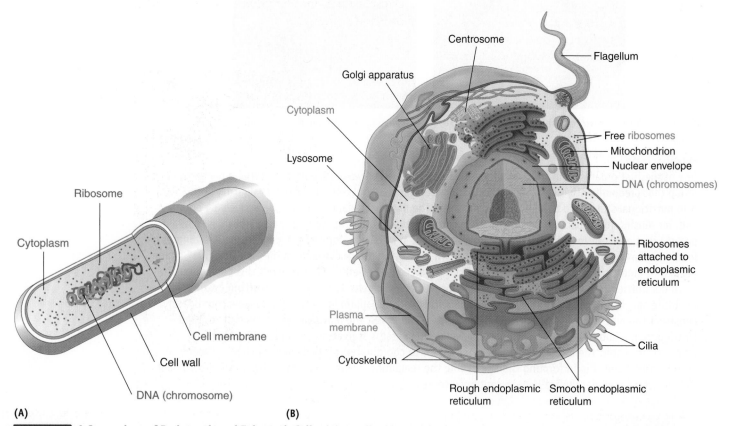

FIGURE 3.4 **A Comparison of Prokaryotic and Eukaryotic Cells.** (**A**) A stylized bacterial cell as an example of a prokaryotic cell. Relatively few visual compartments are present. (**B**) A protozoan cell as a typical eukaryotic cell. Note the variety of cellular subcompartments, many of which are discussed in the text. Universal structures are indicated in red. »» List the ways you could microscopically distinguishing a eukaryotic microbial cell from a bacterial cell.

have an internal environment in which chemical reactions occur. This space, called the **cytoplasm**, represents everything surrounded by the membrane and, in eukaryotic cells, exterior to the cell nucleus. If the cell structures are removed from the cytoplasm, what remains is the **cytosol**, which consists of water, salts, ions, and organic compounds as described in Chapter 2.

Protein Synthesis. All organisms must make proteins, which we learned in Chapter 2 are the workhorses of cells and organisms. The structure common in all cells is the **ribosome**, an RNA-protein machine that cranks out proteins based on the genetic instructions it receives from the DNA (Chapter 8). Although the pattern for protein synthesis is identical, structurally bacterial ribosomes are smaller than their counterparts in eukaryotic cells.

CONCEPT AND REASONING CHECKS

3.2 The cell theory states that the cell is the fundamental unit of life. Summarize those processes all cells have that contribute to this fundamental unit.

Bacteria and Eukaryotes: The Structural Distinctions

KEY CONCEPT

3. Bacteria and eukaryotes have distinct subcellular compartments.

In the cytoplasm, eukaryotic microbes have a variety of structurally discrete, often membrane-enclosed, subcellular compartments called **organelles** to carry out specialized functions (Figure 3.4). Bacterial cells also have subcellular compartments—they just are not readily visible or membrane enclosed.

Protein/Lipid Transport. Eukaryotic microbes have a series of membrane-enclosed organelles that compose the cell's **endomembrane system**, which is designed to transport protein and lipid cargo through and out of the cell. This system includes the **endoplasmic reticulum (ER)**, which consists of flat membranes to which ribosomes are attached (rough ER) and tube-like membranes without ribosomes (smooth ER). These portions of the ER are involved in protein and lipid synthesis and transport, respectively.

The **Golgi apparatus** is a group of independent stacks of flattened membranes and vesicles where the proteins and lipids coming from the ER are processed, sorted, and packaged for trans-

port. **Lysosomes**, somewhat circular, membrane-enclosed sacs containing digestive (hydrolytic) enzymes, are derived from the Golgi apparatus and, in protozoal cells, break down captured food materials.

Bacteria lack an endomembrane system, yet they are capable of manufacturing and modifying proteins and lipids just as their eukaryotic relatives do. However, many bacterial cells contain so-called **microcompartments** surrounded by a protein shell (FIGURE 3.5.) These microcompartments represent a type of organelle since the shell proteins can control transport similar to membrane-enclosed organelles.

Energy Metabolism. Cells and organisms carry out one or two types of energy transformations. Through a process called **cellular respiration**, all cells convert chemical energy into cellular energy for cellular work. In eukaryotic microbes, this occurs in the cytosol and in membrane-enclosed organelles called **mitochondria** (sing., mitochondrion). Bacterial (and archaeal) cells lack mitochondria; they use the cytosol and cell membrane to complete the energy converting process.

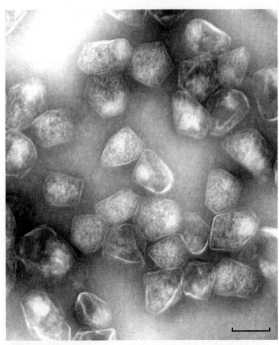

FIGURE 3.5 **Microcompartments.** Purified bacterial microcompartments from *Salmonella enterica* are composed of a complex protein shell that encases metabolic enzymes. (Bar = 100 nm.) »» How do these bacterial microcompartments differ structurally from a eukaryotic organelle?

Vesicles: Membrane-enclosed spheres involved with secretion and storage.

A second energy transformation, **photosynthesis**, involves the conversion of light energy into chemical energy. In algal protists, photosynthesis occurs in membrane-bound **chloroplasts**. Some bacteria, such as the cyanobacteria we have mentioned, also carry out almost identical energy transformations. Again, the cell membrane or elaborations of the membrane represent the chemical workbench for the process.

Cell Structure and Transport. The eukaryotic **cytoskeleton** is organized into an interconnected system of cytoplasmic fibers, threads, and interwoven molecules that give structure to the cell and assist in the transport of materials throughout the cell. The main components of the cytoskeleton are microtubules that originate from the **centrosome** and microfilaments, each assembled from different protein subunits. Bacterial cells to date have no similar physical cytoskeleton, although proteins related to those that construct microtubules and microfilaments aid in determining the shape in some bacterial cells as we will see in Chapter 4.

Cell Motility. Many microbial organisms live in watery or damp environments and use the process of cell motility to move from one place to another. Many algae and protozoa have long, thin protein projections called **flagella** (sing., flagellum) that, covered by the plasma membrane, extend from the cell. By beating back and forth, the flagella provide a mechanical force for motility. Many bacterial cells also exhibit motility; however, the flagella are structurally different and without a cell membrane covering. The pattern of motility also is different, providing a rotational propeller-like force for movement (Chapter 4).

Some protozoa also have other membrane-enveloped appendages called **cilia** (sing., cilium) that are shorter and more numerous than flagella. In some motile protozoa, they wave in synchrony and propel the cell forward. No bacterial cells have cilia.

Water Balance. The aqueous environment in which many microorganisms live presents a situation where the process of diffusion occurs, specifically the movement of water, called **osmosis**, into the cell. Continuing unabated, the cell would eventually swell and burst (cell lysis) because the cell or plasma membrane does not provide the integrity to prevent lysis.

Most bacterial and some eukaryotic cells (fungi, algae) contain a **cell wall** exterior to the cell or plasma membrane. Although the structure and organization of the wall differs between groups (see Chapter 2), all cell walls provide support for the cells, give them shape, and help them resist the pressure exerted by the internal water pressure.

A summary of the bacteria and eukaryote processes and structures is presented in **TABLE 3.1**.

CONCEPT AND REASONING CHECKS

3.3 Explain how variation in cell structure between bacteria and eukaryotes can be compatible with a similarity in cellular processes between these organisms.

Diffusion:
The movement of a substance from where it is in a higher concentration to where it is in a lower concentration.

TABLE 3.1   Comparison of Bacterial and Eukaryotic Cell Structure

Process	Cell Structure or Compartment	
	Bacterial	Eukaryotic
Genetic organization	Circular DNA chromosome	Linear DNA chromosomes
Compartmentation	Cell membrane	Plasma membrane
Metabolic organization	Cytoplasm	Cytoplasm
Protein synthesis	Ribosomes	Ribosomes
Protein/lipid transport	Cytoplasm	Endomembrane system
Energy metabolism	Cytoplasm and cell membrane	Mitochondria and chloroplasts
Cell structure and transport	Proteins in cytoplasm	Protein filaments in cytoplasm
Cell motility	Bacterial flagella	Eukaryotic flagella or cilia
Water balance	Cell wall	Cell wall

If you open any catalog, items are separated by types, styles, or functions. For example, in a fashion catalog, watches are separated from shoes and, within the shoes, men's, women's, and children's styles are separated from one another. Even the brands of shoes or their use (e.g., dress, casual, athletic) may be separated.

With such an immense diversity of organisms on planet Earth, the human drive to catalog these organisms has not been very different from cataloging watches and shoes; both have been based on shared characteristics. In this section, we shall explore the principles on which microorganisms are classified and cataloged.

Classification Attempts to Catalog Organisms

KEY CONCEPT

4. Organisms historically were grouped by shared characteristics.

In the 18th century, Carolus Linnaeus, a Swedish scientist, began identifying living organisms according to similarities in form (resemblances) and placing organisms in one of two "kingdoms"— Vegetalia and Animalia (**FIGURE 3.6**). This system was well accepted until the mid-1860s when a German naturalist, philosopher, and physician, Ernst Haeckel, identified a fundamental problem in the two-kingdom system. The unicellular (microscopic) organisms being identified by Haeckel, Pasteur, Koch, and their associates did not conform to the two-kingdom system of multicellular organisms. Haeckel constructed a third kingdom, the Protista, in which all the known unicellular organisms were placed. The bacterial organisms, which he called "moneres," were near the bottom of the tree, closest to the root of the tree.

With improvements in the design of light microscopes, more observations were made of bacterial and protist organisms. In 1937, a French biologist, Edouard Chatton, proposed that there was a fundamental dichotomy among the Protista. He saw bacteria as having distinctive properties (not articulated in his writings) in "the prokaryotic nature of their cells" and should be separated from all other protists "which have eukaryotic cells." With the development of the electron microscope

in the 1950s, it became apparent that some protists had a membrane-enclosed nucleus and were identified, along with the plants and animals, as being eukaryotes while other protists (the bacteria) lacked this structure and were considered to be prokaryotes (see Chapter 1). Thus, in 1956, Herbert Copland suggested bacteria be placed in a fourth kingdom, the Monera.

But there was still one more problem with the kingdom Protista. Robert H. Whittaker, a botanist at the University of California, saw the fungi as yet another kingdom of organisms. The fungi are the only eukaryotic group that must externally digest their food prior to absorption and, as such, live in the food source. For this and other reasons, Whittaker in 1959 refined the four-kingdom system into five kingdoms, identifying the kingdom *Fungi* as a separate, multicellular, eukaryotic kingdom distinguished by an absorptive mode of nutrition (Chapter 17).

The five kingdom system rested safely for about 15 years. In the late 1970s, Carl Woese, an evolutionary biologist at the University of Illinois, began a molecular analysis of living organisms based on comparisons of nucleotide sequences of genes coding for the small subunit ribosomal RNA (rRNA) found in all organisms. These analyses revealed yet another dichotomy, this time among the prokaryotes. By 1990, it was clear that the kingdom Monera contained two fundamentally unrelated groups, what Woese initially called the *Bacteria* and *Archaebacteria*. These two groups were as different from each other as they were different from the eukaryotes.

CONCEPT AND REASONING CHECKS

3.4 What four events changed the cataloging of microorganisms.

Kingdoms and Domains: Trying to Make Sense of Taxonomic Relationships

KEY CONCEPT

5. The three-domain system shows the taxonomic relationships between living organisms.

What many of these scientists are or were doing is **systematics**; that is, studying the diversity of life and its evolutionary relationships. **Systematic biologists—systematists** for short— identify, describe, name, and classify organisms

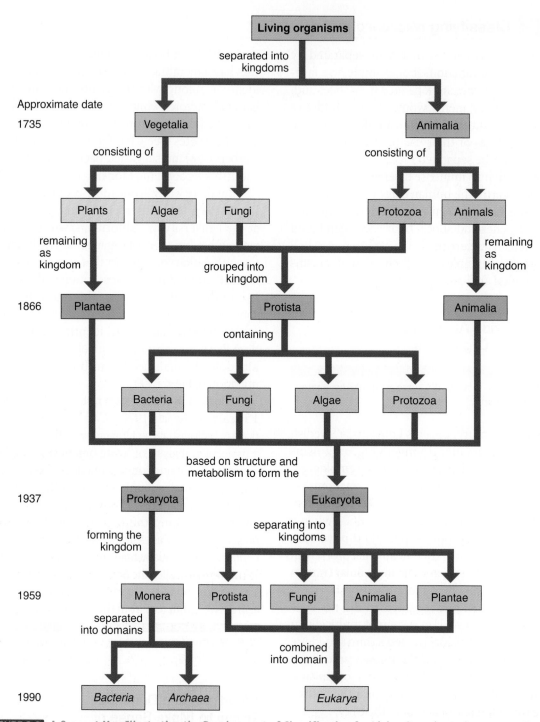

FIGURE 3.6 A Concept Map Illustrating the Development of Classification for Living Organisms. Over some 140 years, new observations and techniques have been used to reclassify and reorganize living organisms. »» Of the plants, algae, fungi, bacteria, protozoa, and animals, which are in each of the three domains? Modified from Schaechter, Ingraham, and Neidhardt. *Microbe*. ASM Press, 2006, Washington, D.C.

(**taxonomy**), and organize their observations within a framework that shows taxonomic relationships.

Often it is difficult to make sense of taxonomic relationships because new information that is more detailed keeps being discovered about organisms. This then motivates taxonomists to figure out how the new information fits into the known classification schemes—or how the schemes need to be modified to fit the new information. This is no clearer than the most recent taxonomic revolution that, as the opening quote states, has come along to "tweak our theories or teach us something new."

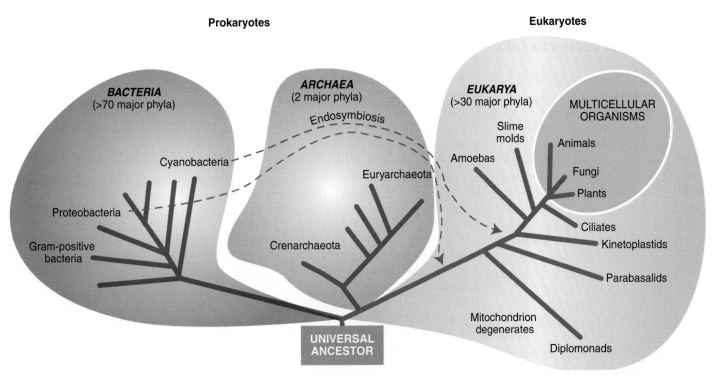

Prokaryotes

Eukaryotes

BACTERIA
(>70 major phyla)

ARCHAEA
(2 major phyla)

EUKARYA
(>30 major phyla)

MULTICELLULAR
ORGANISMS

Endosymbiosis

Cyanobacteria

Euryarchaeota

Slime
molds

Animals

Proteobacteria

Amoebas

Fungi

Plants

Gram-positive
bacteria

Crenarchaeota

Ciliates

Kinetoplastids

Parabasalids

Mitochondrion
degenerates

Diplomonads

**UNIVERSAL
ANCESTOR**

FIGURE 3.7 The Three-Domain System Forms the "Tree of Life". Fundamental differences in genetic endowments are the basis for the three domains of all organisms on Earth. Some 3.5 billion years ago, a universal ancestor arose from which all modern day organisms descended. »» What cellular characteristic was the major factor stimulating the development of the three-domain system?

Carl Woese, along with George Fox and coworkers at the University of Illinois, Urbana-Champaign, proposed a new classification scheme with a new most inclusive taxon, the **domain**. The new scheme initially came from work that compared the DNA nucleotide base sequences for the RNA in ribosomes, those protein manufacturing machines needed by all cells. Woese and Fox's results were especially relevant when comparing those sequences from a group of bacterial organisms formerly called the archaebacteria (*archae* = "ancient"). Many of these bacterial forms are known for their ability to live under extremely harsh environments. Woese discovered that the nucleotide sequences in these archaebacteria were different from those in other bacterial species and in eukaryotes. After finding other differences, including cell wall composition, membrane lipids, and sensitivity to certain antibiotics, the evidence pointed to there being three taxonomic lines to the "tree of life".

One goal of systematics, and the main one of interest here, is to reconstruct the **phylogeny** (*phylo* = "tribe"; *geny* = "production"), the evolutionary history of a species or group of species. Systematists illustrate phylogenies with **phylogenetic trees**, which identify inferred relationships

among species. Because, like all hypotheses, they are revised as scientists gather new data, trees change as our knowledge of diversity increases.

In Woese's **three-domain system**, one branch of the phylogenetic tree includes the former archaebacteria and is called the domain *Archaea* (**FIGURE 3.7**). The second encompasses all the remaining true bacteria and is called the domain *Bacteria*. The third domain, the *Eukarya*, includes the four remaining kingdoms (Protista, Plantae, Fungi, and Animalia).

In 1996, Craig Venter and his coworkers deciphered the DNA base sequence of the archaean *Methanococcus jannaschii* and showed that almost two thirds of its genes are different from those of the *Bacteria*. They also found that proteins replicating the DNA and involved in RNA synthesis have no counterpart in the *Bacteria*. The three-domain system now is on firm ground.

MICROINQUIRY 3 examines a scenario for the evolution of the eukaryotic cell.

Taxon (pl., taxa): Subdivisions used to classify organisms.

CONCEPT AND REASONING CHECKS

3.5 It has been said that Woese "lifted a whole submerged continent out of the ocean." What is the "submerged continent" and why is the term "lifted" used?

MICROINQUIRY 3

The Evolution of Eukaryotic Cells

Biologists and geologists have speculated for decades about the chemical evolution that led to the origins of the first prokaryotic cells on Earth (see Micro-Focus 2.1 and 2.6). Whatever the origin, the first ancestral prokaryotes arose about 3.8 billion years ago and remained the sole inhabitants for some 1.5 billion years.

Scientists also have proposed various scenarios to account for the origins of the first eukaryotic cells. The oldest known fossils thought to be eukaryotic are about 2 billion years old.

A key concern here is figuring out how different membrane compartments arose to evolve into what are found in the eukaryotic cells today. Debate on this long intractable problem continues, so here we present some of the ideas that have fueled such discussions.

At some point around 2 billion years ago, the increasing number of metabolic reactions occurring in presumably larger prokaryotes started to interfere with one another. As cells increased in size, the increasing volume of cell cytoplasm outpaced the ability of the cell surface (membrane) to be an effective "workbench" for servicing the metabolic needs of the whole cell. Complexity would necessitate more extensive workbench surface through compartmentation.

The Endomembrane System May Have Evolved through Invagination

Similar to today's bacterial and archaeal cells, the cell membrane of an ancestral prokaryote may have had specialized regions involved in protein synthesis, lipid synthesis, and nutrient hydrolysis. If the invagination of these regions occurred, the result could have been the internalization of these processes as independent internal membrane systems. For example, the membranes of the endoplasmic reticulum may have originated by multiple invagination events of the cell membrane (**Figure A1**).

Biologists have suggested that the elaboration of the evolving ER surrounded the nuclear region and DNA, creating the nuclear envelope. Surrounded and protected by a double membrane, greater genetic complexity could occur as the primitive eukaryotic cell continued to evolve in size and function. Other internalized membranes could give rise to the Golgi apparatus.

Chloroplasts and Mitochondria Arose from a Symbiotic Union of Engulfed Bacteria

Mitochondria and chloroplasts are not part of the extensive endomembrane system. Therefore, these energy-converting organelles probably originated in a different way.

The structure of modern-day chloroplasts and mitochondria is very similar to a bacterial cell. In fact, mitochondria, chloroplasts, and bacteria share a large number of similarities (see **Table**). In addition, there are bacterial cells alive today that carry out cellular respiration similarly to mitochondria and other bacterial cells (the cyanobacteria) that can carry out photosynthesis similarly to chloroplasts.

These similar functional patterns, along with other chemical and molecular similarities, suggested to Lynn Margulis at the University of Massachusetts, Amherst, that present-day chloroplasts and mitochondria represent modern representatives of what were once, many eons ago, free-living prokaryotes. Margulis, therefore, proposed the **endosymbiont model** for the origin of mitochondria and chloroplasts. The hypothesis suggests, in part, that mitochondria evolved from a prokaryote that carried out cellular respiration and which was "swallowed" (engulfed) by a primitive eukaryotic cell. The bacterial partner then lived within (*endo*) the eukaryotic cell in a mutually beneficial association (*symbiosis*) (**Figure A2**).

Likewise, a photosynthetic prokaryote, perhaps a primitive cyanobacterium, was engulfed and evolved into the chloroplasts present in plants and algae today (**Figure A3**). The theory also would explain why both organelles have two membranes. One was the cell membrane of the engulfed bacterial cell and the other was the plasma membrane resulting from the engulfment process. By engulfing these prokaryotes and not destroying them, the evolving eukaryotic cell gained energy-conversion abilities, while the symbiotic bacterial cells gained a protected home.

If the first ancestral prokaryote appeared about 3.5 billion years ago and the first single-celled eukaryote about 2 billion years ago, then it took some 1.5 billion years of evolution for the events described above to occur (see Figure 8.2). With the appearance of the first eukaryotic cells, a variety of single-celled forms evolved, many of which were the very ancient ancestors of the single-celled eukaryotic organisms that exist today.

Obviously, laboratory studies can only hypothesize at mechanisms to explain how cells evolved and can only suggest—not prove—what might have happened billions of years ago. The description here is a very simplistic view of how the first eukaryotic cells might have evolved. Short of inventing a time machine, we may never know the exact details for the origin of eukaryotic cells and organelles.

Discussion Point

Determine which endosymbiotic event must have come first: the engulfment of the bacterial progenitor of the chloroplast or the engulfment of the bacterial progenitor of the mitochondrion.

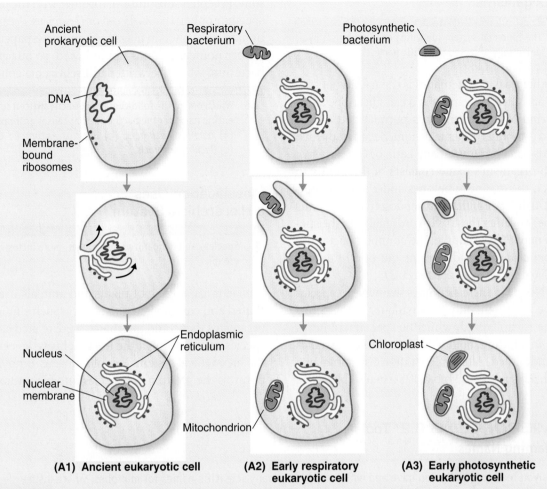

Ancient prokaryotic cell

DNA

Membrane-bound ribosomes

Respiratory bacterium

Photosynthetic bacterium

Nucleus

Nuclear membrane

Endoplasmic reticulum

Mitochondrion

Chloroplast

(A1) Ancient eukaryotic cell

(A2) Early respiratory eukaryotic cell

(A3) Early photosynthetic eukaryotic cell

FIGURE A **Possible Origins of Eukaryotic Cell Compartments.** (A1) Invagination of the cell membrane from an ancient prokaryotic cell may have led to the development of the cell nucleus as well as to the membranes of the endomembrane system, including the endoplasmic reticulum. (A2) The mitochondrion may have resulted from the uptake and survival of a bacterial cell that carried out cellular respiration. (A3) A similar process, involving a bacterial cell that carried out photosynthesis, could have accounted for the origin of the chloroplast.

TABLE

Similarities between Mitochondria, Chloroplasts, Bacteria, and Microbial Eukaryotes

Characteristic	Mitochondria	Chloroplasts	Bacteria	Microbial Eukaryotes
Average size	1–5 μm	1–5 μm	1–5 μm	10–20 μm
Nuclear envelope present	No	No	No	Yes
DNA molecule shape	Circular	Circular	Circular	Linear
Ribosomes	Yes; bacterial-like	Yes; bacterial-like	Yes	Yes; eukaryotic-like
Protein synthesis	Make some of their proteins	Make some of their proteins	Make all of their proteins	Make all of their proteins
Reproduction	Binary fission	Binary fission	Binary fission	Mitosis and cytokinesis

Nomenclature Gives Scientific Names to Organisms

KEY CONCEPT

6. The binomial system identifies each organism by a universally accepted scientific name.

Another goal of systematics is the naming of species and their placement in a classification. In his *Systema Naturae*, Linnaeus popularized a two-word (binomial) scheme of nomenclature, the two words usually derived from Latin or Greek stems. Each organism's name consists of the **genus** to which the organism belongs and a **specific epithet**, a descriptor that further describes the genus name. Together these two words make up the **species** name. For example, the common bacterium *Escherichia coli* resides in the gut of all humans (*Homo sapiens*) (MICROFOCUS 3.2).

Notice in these examples that when a species name is written, only the first letter of the genus name is capitalized, while the specific epithet is not. In addition, both words are printed in italics or underlined. After the first time a species name has been spelled out, biologists usually abbreviate the genus name using only its initial genus letter or some accepted substitution, together with the full specific epithet; that is, *E. coli* or *H. sapiens*. A cautionary note: often in magazines and newspapers, proper nomenclature is not followed, so our gut bacterium would be written as Escherichia coli.

CONCEPT AND REASONING CHECKS

3.6 Which one of the following is a correctly written scientific name for the bacterium that causes anthrax? (a) *bacillus Anthracis*; (b) *Bacillus Anthracis*; or (c) *Bacillus anthracis*.

Classification Uses a Hierarchical System

KEY CONCEPT

7. Species can be organized into higher, more inclusive groups.

Linnaeus' cataloging of plants and animals used shared and common characteristics. Such similar organisms that could interbreed were related as a species, which formed the least inclusive level of the hierarchical system. Part of Linnaeus' innovation was the grouping of species into higher taxa

---

MICROFOCUS 3.2: Tools

Naming Names

As you read this book, you have and will come across many scientific names for microbes, where a species name is a combination of the genus and specific epithet. Not only are many of these names tongue twisting to pronounce (many are listed with their pronunciation inside the front and back covers), but how in the world did the organisms get those names? Here are a few examples.

Genera Named after Individuals

Escherichia coli: named after Theodore Escherich who isolated the bacterial cells from infant feces in 1885. Being in feces, it commonly is found in the colon.

Neisseria gonorrhoeae: named after Albert Neisser who discovered the bacterial organism in 1879. As the specific epithet points out, the disease it causes is gonorrhea.

Genera Named for a Microbe's Shape

Vibrio cholerae: vibrio means "comma-shaped," which describes the shape of the bacterial cells that cause cholera.

Staphylococcus epidermidis: staphylo means "cluster" and *coccus* means "spheres." So, these bacterial cells form clusters of spheres that are found on the skin surface (epidermis).

Genera Named after an Attribute of the Microbe

Saccharomyces cerevisiae: in 1837, Theodor Schwann observed yeast cells and called them *Saccharomyces* (*saccharo* = "sugar"; *myce* = "fungus") because the yeast converted grape juice (sugar) into alcohol; *cerevisiae* (from *cerevisia* = "beer") refers to the use of yeast since ancient times to make beer.

Myxococcus xanthus: myxo means "slime," so these are slime-producing spheres that grow as yellow (*xantho* = "yellow") colonies on agar.

Thiomargarita namibiensis: see MicroFocus 3.5.

TABLE 3.2	Taxonomic Classification of Humans, Brewer's Yeast, and a Common Bacterium		
	Humans	Brewer's Yeast	*Escherichia coli*
Domain	*Eukarya*	*Eukarya*	*Bacteria*
Kingdom	Animalia	Fungi	
Phylum	Chordata	Ascomycota	Proteobacteria
Class	Mammalia	Saccharomycotina	Gammaproteobacteria
Order	Primates	Saccharomycetales	Enterobacteriales
Family	Hominidae	Saccharomycetaceae	Enterobacteriaceae
Genus	*Homo*	*Saccharomyces*	*Escherichia*
Species	*H. sapiens*	*S. cerevisiae*	*E.coli*

that also were based on shared, but more inclusive, similarities.

Today several similar species are grouped together into a genus (pl., genera). A collection of similar genera makes up a **family** and families with similar characteristics make up an **order**. Different orders may be placed together in a **class** and classes are assembled together into a **phylum** (pl., phyla). All phyla would be placed together in a kingdom and/or domain, the most inclusive level of classification. TABLE 3.2 outlines the taxonomic hierarchy for three organisms.

In prokaryotes, an organism may belong to a rank below the species level to indicate a special characteristic exists within a subgroup of the species. Such ranks have practical usefulness in helping to identify an organism. For example, two biotypes of the cholera bacterium, *Vibrio cholerae*, are known: *Vibrio cholerae* classic and *Vibrio cholerae* El Tor. Other designations of ranks include subspecies, serotype, strain, morphotype, and variety.

David Hendricks Bergey devised one of the first systems of classification for the bacterial species in 1923. Today, the proper taxonomic classification for the *Bacteria* and *Archaea* can be found in the second edition of *Bergey's Manual of Systematic Bacteriology*. The first two volumes of this 5-volume compendium have been published. The tremendous changes that have taken place in taxonomy can be seen by the addition of more than 2,200 new species and 390 new genera to the first volume of the new second edition.

CONCEPT AND REASONING CHECKS

3.7 How would you describe an order in the taxonomic classification?

Many Methods Are Available to Identify and Classify Microorganisms

KEY CONCEPT

8. Identification and classification of microorganisms may use different methods.

There are several traditional and more modern criteria that microbiologists can use to identify and classify microorganisms. For example, a medical identification usually uses physical, staining, and biochemical methods (metabolic tests). In fact, *Bergey's Manual of Determinative Bacteriology*, now in its ninth edition, is the primary source for making routine medical identifications of bacterial pathogens. On the other hand, many emerging biotechnologies (Chapter 9) depend on a thorough knowledge of a microorganism's biochemistry, molecular biology, and phylogenetic relatedness. More molecular methods will be required here.

Let's briefly review some of the more determinative methods and a few molecular methods for classification.

Physical Characteristics. These include differential staining reactions to help determine the organism's shape (morphology), and the size and arrangement of cells. Other characteristics can include oxygen, pH, and growth temperature requirements. Spore-forming ability and motility are additional determinants. Unfortunately, there

Biotypes: Populations or groups of individuals having the same genetic constitution (genotype).

are many bacterial and archaeal organisms that have the same physical characteristics, so other distinguishing features are needed.

Biochemical Tests. As microbiologists better understood bacterial physiology, they discovered there were certain metabolic properties that were present only in certain groups.

Today, a large number of biochemical tests exist and often a specific test can be used to eliminate certain groups from the identification process. Among the more common tests are: fermentation of carbohydrates, the use of a specific substrate, and the production of specific products or waste products. But, as with the physical characteristics, often several biochemical tests are needed to differentiate between species.

These identification tests are important clinically, as they can be part of the arsenal available to the clinical lab that is trying to identify a pathogen. Many of these tests use rapid identification methods (MICROFOCUS 3.3) or automated systems (FIGURE 3.8).

Serological Tests. Microorganisms are antigenic, meaning they are capable of triggering the production of antibodies. Solutions of

Antibodies:
Proteins produced by the immune system in response to a specific chemical configuration (antigen).

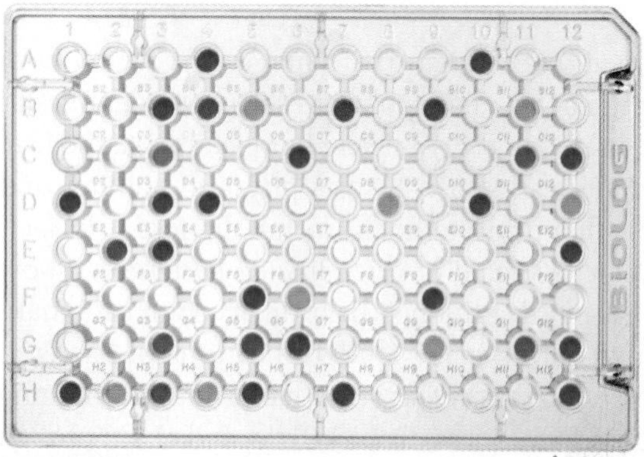

such collected antibodies, called **antisera**, are commercially available for many medically important pathogens. For example, mixing a *Salmonella* antiserum with *Salmonella* cells will cause the cells to clump together or agglutinate. If a foodborne illness occurs, the antiserum may be useful in identifying if *Salmonella* is the pathogen. More information about serological testing will be presented in Chapter 22.

Nucleic Acid Analysis. In 1984, the editors of *Bergey's Manual of Systematic Bacteriology* noted that there is no "official" classification of bacterial species and that the closest approximation to an official classification is the one most widely accepted by the community of microbiologists. The editors stated that a comprehensive classification might one day be possible. Today, the fields of molecular genetics and genomics have advanced the analysis and sequencing of nucleic acids. This has given rise to a new era of molecular taxonomy.

Molecular taxonomy is based on the universal presence of ribosomes in all living organisms. In particular, it is the RNAs in the ribosome, called ribosomal RNA (rRNA), which are of most interest and the primary basis of Woese's construction of the three-domain system. Many scientists today believe the genes for rRNA are the most accurate measure for precise bacterial classification in all taxonomic classes. Other techniques, including the polymerase chain reaction and nucleic acid hybridization, will be mentioned in later chapters.

The vast number of tests and analyses available for bacterial cells can make it difficult to know which are relevant for pathogen identification purposes. One widely used technique in many disciplines is the **dichotomous key**. There are various forms of dichotomous keys, but one very useful construction is a flow chart where a series of positive or negative test procedures are listed down the page. Based on the dichotomous nature of the test (always a positive or negative result), the flow chart immediately leads to the next test result. The result is the identification of a specific organism. A simplified example is shown in MICROFOCUS 3.4.

FIGURE 3.8 A Biolog MicroPlate®. The BIOLOG system is capable of identifying bacteria by assessing the bacterium's ability to use any of 95 different substrates in a 96-well microtiter plate. The use of any substrate results in a reduction of the dye in each well, resulting in purple color development. The intensity of the purple coloration indicates the degree of substrate usage and is read by a computer-linked automated microtiter reader. The first well (upper left) is a negative control with no substrate. »» Of the methods described on this page, which is/are most likely to be used in this more automated system? Explain.

CONCEPT AND REASONING CHECKS

3.8 Why are so many tests often needed to identify a specific bacterial species?

MICROFOCUS 3.3: Tools

Rapid Identification of Enteric Bacteria

In recent years, a number of miniaturized systems have been made available to microbiologists for the rapid identification of enteric bacteria. One such system is the Enterotube II, a self-contained, sterile, compartmentalized plastic tube containing 12 different media and an enclosed inoculating wire. This system permits the inoculation of all media and the performance of 15 standard biochemical tests using a single bacterial colony. The media in the tube indicate by color change whether the organism can carry out the metabolic reaction. After 24 hours of incubation, the *positive* tests are circled and all the circled numbers in each boxed section are added to yield a 5-digit ID for the organism being tested. This 5-digit number is looked up in a reference book or computer software to determine the identity of the bacterium.

(A) An uninoculated tube.

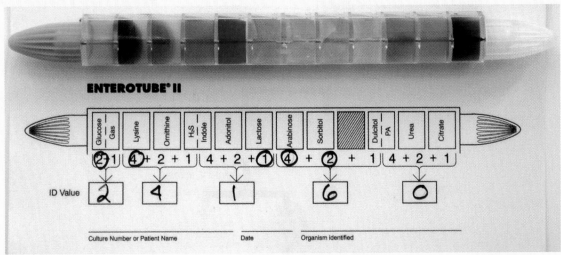

(B) An inoculated tube incubated for 24 hours.

VALUE	IDENTIFICATION	TESTS	TESTS
			MOT VP
24040	*ESCHERICHIA VULNERIS	NONE	+ −
	KLEBSIELLA OZAENAE	ADO−	− −
	HAFNIA ALVEI	ORN−	+ +
			RAF
24060	ESCHERICHIA COLI (AD)	IND−	−
	KLEBSIELLA OZAENAE	ADO−	+
24061	SERRATIA LIQUIFACIENS	ORN−	
24070	ESCHERICHIA COLI (AD)	IND−	
24160	ESCHERICHIA COLI	IND−	
			PUNGENT ODOR
24161	*SERRATIA ODORIFERA BIOGP 2	NONE	+
24163	KLEBSIELLA PNEUMONIAE	ADO−	
24170	ESCHERICHIA COLI	IND−	
24173	KLEBSIELLA PNEUMONIAE	ADO−	
24200	KLEBSIELLA OZAENAE	ARA−	
			RAF
24220	*SERRATIA MARCESCENS BIOGP 1	NONE	−
	KLEBSIELLA OZAENAE	ARA−	+
24221	SERRATIA MARCESCENS	ORN−	

As seen from the reference, 24160 is *Escherichia coli*.

MICROFOCUS 3.4: Tools

Dichotomous Key Flow Chart

A medical version of a taxonomic key (in the form of a dichotomous flow chart) can be used to identify very similar bacterial species based on physical and biochemical characteristics.

In this simplified scenario, an unknown bacterium has been cultured and several tests run. The test results are shown in the box at the top. Using the test results and the flow chart, identify the bacterial species that has been cultured.

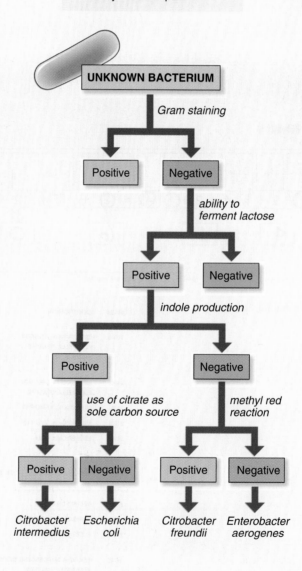

Microbiology Test Results

• Gram stain: gram-negative rods

Biochemical tests:
  • Citrate test: negative
  • Lactose fermentation: positive
  • Indole test: positive
  • Methyl red test: positive

UNKNOWN BACTERIUM

Gram staining

Positive   Negative

ability to ferment lactose

Positive   Negative

indole production

Positive   Negative

use of citrate as sole carbon source   methyl red reaction

Positive   Negative   Positive   Negative

Citrobacter intermedius   Escherichia coli   Citrobacter freundii   Enterobacter aerogenes

3.3 Microscopy

The ability to see small objects all started with the microscopes used by Robert Hooke and Antony van Leeuwenhoek. By now, you should be aware that microorganisms usually are very small. Before we examine the instruments used to "see" these tiny creatures, we need to be familiar with the units of measurement.

Many Microbial Agents Are In the Micrometer Size Range

KEY CONCEPT

9. Metric system units are the standard for measurement.

One physical characteristic used to study microorganisms and viruses is their size. Because they are so small, a convenient system of measurement is used that is the scientific standard around the world. The measurement system is the metric system, where the standard unit of length is the meter and is a little longer than a yard (see Appendix A). To measure microorganisms, we need to use units that are a fraction of a meter. In microbiology, the common unit for measuring length is the **micrometer (μm)**, which is equivalent to a millionth (10^{-6}) of a meter. To appreciate how small a micrometer is, consider this: Comparing a micrometer to an inch is like comparing a housefly to New York City's Empire State Building, 1,472 feet high.

Microbial agents range in size from the relatively large, almost visible protozoa (100 μm) down to the incredibly tiny viruses (0.02 μm) (**FIGURE 3.9**). Most bacterial and archaeal cells are about 1 μm to 5 μm in length, although notable exceptions have been discovered recently

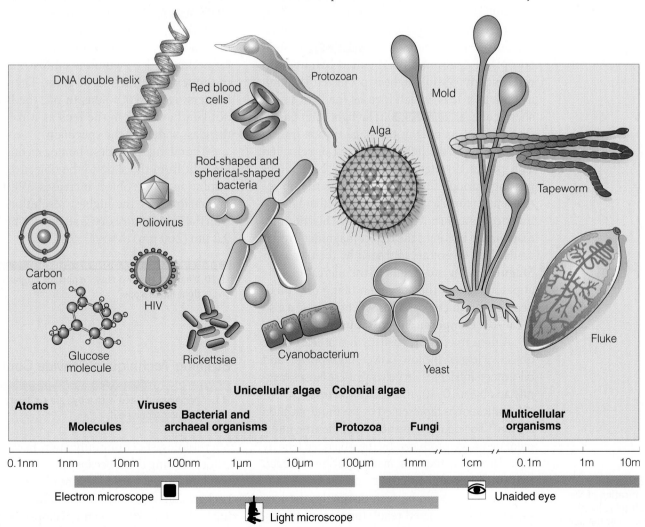

FIGURE 3.9 Size Comparisons Among Various Atoms, Molecules, and Microorganisms (not drawn to scale). Although tapeworms and flukes usually are macroscopic, the diseases these parasites cause are studied by microbiologists. »» Which domain on average has the smallest organisms and which has the largest?

(MICROFOCUS 3.5). Because most viruses are a fraction of one micrometer, their size is expressed in nanometers. A **nanometer (nm)** is equivalent to a billionth (10^{-9}) of a meter; that is, 1/1,000 of a μm. Using nanometers, the size of the poliovirus, among the smaller viruses, measures 20 nm (0.02 μm) in diameter.

CONCEPT AND REASONING CHECKS

3.9 If a bacterial cell is 0.75 μm in length, what is its length in nanometers?

Light Microscopy Is Used to Observe Most Microorganisms

KEY CONCEPT

10. Light microscopy uses visible light to magnify and resolve specimens.

The basic microscope system used in the microbiology laboratory is the **light microscope**, in which visible light passes directly through the lenses and specimen (**FIGURE 3.10A**). Such an optical configuration is called **bright-field microscopy**. Visible light is projected through a condenser lens, which focuses the light into a sharp cone (**FIGURE 3.10B**). The light then passes through the opening in the stage. When hitting the glass slide, the light is reflected or refracted as it passes through the specimen. Next, light passing through the specimen enters the objective lens to form a magnified intermediate image inverted from that of the specimen. This intermediate image becomes the object magnified by the ocular lens (eyepiece) and seen by the observer. **Magnification** thus refers to the increase in the apparent size of the specimen being observed. Because this microscope has several lenses, it also is called a **compound microscope**.

A light microscope usually has at least three objective lenses: the low-power, high-power, and oil-immersion lenses. In general, these lenses magnify an object 10, 40, and 100 times, respectively. (Magnification is represented by the multiplication sign, ×.) The ocular lens then magnifies the intermediate image produced by the objective lens by 10×. Therefore, the total magnification achieved is 100×, 400×, and 1,000×, respectively.

For an object to be seen distinctly, the lens system must have good **resolving power**; that is, it must transmit light without variation and allow closely spaced objects to be clearly distinguished. For example, a car seen in the distance at night may appear to have a single headlight because at that distance the unaided eye lacks resolving power. However, using binoculars, the two headlights can be seen clearly as the resolving power of the eye increases.

When switching from the low-power (10×) or high-power (40×) lens to the oil-immersion lens (100×), one quickly finds that the image has become fuzzy. The object lacks resolution, and the resolving power of the lens system appears to be poor. The poor resolution results from the refraction of light.

Both low-power and high-power objectives are wide enough to capture sufficient light for viewing. The oil-immersion objective, on the other hand, is so narrow that most light bends away and would miss the objective lens **FIGURE 3.10C**. The **index of refraction** (or refractive index) is a measure of the light-bending ability of a medium. Immersion oil has an index of refraction of 1.5, which is almost identical to the index of refraction of glass. Therefore, by immersing the 100× lens in oil, the light does not bend away from the lens as it passes from the glass slide and the specimen.

The oil thus provides a homogeneous pathway for light from the slide to the objective, and the resolution of the object increases. With the oil-immersion lens, the highest resolution possible with the light microscope is attained, which is near 0.2 μm (200 nm) (MICROFOCUS 3.6).

CONCEPT AND REASONING CHECKS

3.10 What are the two most important properties of the light microscope?

Staining Techniques Provide Contrast

KEY CONCEPT

11. Specimens stained with a dye are contrasted against the microscope field.

Microbiologists commonly stain bacterial cells before viewing them because the cytoplasm lacks color, making it hard to see the cells on the bright background of the microscope **field**. Several staining techniques have been developed to provide **contrast** for bright-field microscopy.

Total magnification: The magnification of the ocular multiplied by the magnification of the objective lens being used.

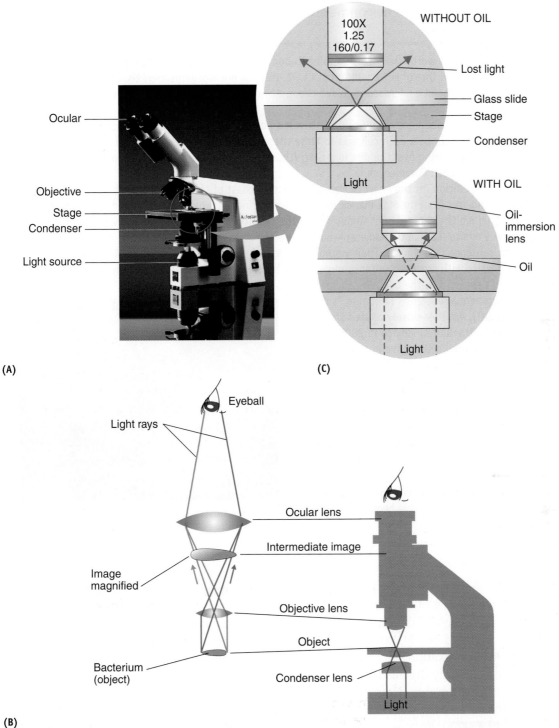

FIGURE 3.10 The Light Microscope. (**A**) The light microscope is used in many instructional and clinical laboratories. Note the important features of the microscope that contribute to the visualization of the object. (**B**) Image formation in the light microscope requires light to pass through the objective lens, forming an intermediate image. This image serves as an object for the ocular lens, which further magnifies the image and forms the final image the eye perceives. (**C**) When using the oil immersion lens (100×), oil must be placed between and continuous with the slide and objective lens. »» Why must oil be used with the 100× oil-immersion lens?

MICROFOCUS 3.5: Environmental Microbiology
Biological Oxymorons

An oxymoron is a pair of words that seem to refer to opposites, such as jumbo shrimp, holy war, old news, and sweet sorrow. One of the characteristics we used for microorganisms is that most are invisible to the naked eye; you need a microscope to see them. Always true? So how about the oxymoron: macroscopic microorganism?

In 1993, researchers at Indiana University discovered near an Australian reef macroscopic bacterial cells in the gut of surgeonfish. Each cell was so large that a microscope was not needed to see it. The spectacular giant, measuring over 0.6 mm in length (that's 600 μm compared to 2 μm for *Escherichia coli*) even dwarfs the protozoan *Paramecium*.

While on an expedition off the coast of Namibia (western coast of southern Africa) in 1997, Heide Schultz and teammates from the Max Planck Institute for Marine Microbiology in Bremen, Germany, found another bacterial monster in sediment samples from the sea floor. These bacterial cells were spherical being about 0.1 mm to 0.3 mm in diameter—but some as large as 0.75 mm—about the diameter of the period in this sentence (see figure). Their volume is about 3 million times greater than that of *E. coli*. The cells, shining white with enclosed sulfur granules, were held together in chains by a mucus sheath looking like a string of pearls. Thus, the bacterial species was named *Thiomargarita namibiensis* (meaning "sulfur pearl of Namibia"). Another closely related strain was discovered in the Gulf of Mexico in 2005.

How does a bacterial cell survive in so large a size? The trick is to keep the cytoplasm as a thin layer plastered against the edge of the cell so materials do not need to travel (diffuse) far to get into or out of the cell. The rest of the cell is a giant "bubble," called a vacuole, in which nitrate and sulfur are stored as potential energy sources. Thus, the actual cytoplasmic layer is microscopic and as close to the surface as possible.

Yes, the vast majority of microorganisms are microscopic, but exceptions have been found in some exotic places.

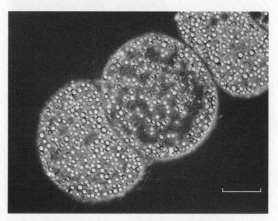

A phase microscopy image showing a chain of *Thiomargarita namibiensis* cells. (Bar = 250 μm.)

To perform the **simple stain technique**, bacterial cells in a droplet of water or broth are smeared on a glass slide and the slide air-dried. Next, the slide is passed briefly through a flame in a process called **heat fixation**, which bonds the cells to the slide, kills any organisms still alive, and increases stain absorption. Now the slide is flooded with a **basic** (cationic) **dye** such as methylene blue (FIGURE 3.11A). Because cationic dyes have a positive charge, the dye is attracted to the cytoplasm and cell wall, which primarily have negative charges. By contrasting the blue cells against the bright background, the staining procedure allows the observer to measure cell size and determine cell shape. It also can provide information about how cells are arranged with respect to one another (Chapter 4).

The **negative stain technique** works in the opposite manner (FIGURE 3.11B). Bacterial cells are mixed on a slide with an **acidic** (anionic) **dye** such as nigrosin (a black stain) or India ink (a black drawing ink). The mixture then is pushed across the face of the slide and allowed to air-dry. Because the anionic dye carries a negative charge, it is repelled from the cell wall and cytoplasm. The stain does not enter the cells and the observer sees clear or white cells on a black or gray background. Because this technique avoids chemical reactions and heat fixation, the cells appear less shriveled and less distorted than in a simple stain. They are closer to their natural condition.

The **Gram stain technique** is an example of a **differential staining procedure**; that is, it allows the observer to differentiate (separate) bacterial

MICROFOCUS 3.6: Tools

Calculating Resolving Power

The resolving power (RP) of a lens system is important in microscopy because it indicates the size of the smallest object that can be seen clearly. The resolving power varies for each objective lens and is calculated using the following formula:

$$RP = \frac{\lambda}{2 \times NA}$$

In this formula, the Greek letter λ (lambda) represents the wavelength of light; for white light, it averages about 550 nm. The symbol NA stands for the numerical aperture of the lens and refers to the size of the cone of light that enters the objective lens after passing through the specimen. This number generally is printed on the side of the objective lens (see Figure 3.10C). For an oil-immersion objective with an NA of 1.25, the resolving power may be calculated as follows:

$$RP = \frac{550\,nm}{2 \times 1.25} = \frac{550}{2.5} = 220 \text{ nm or } 0.22 \text{ }\mu m$$

Because the resolving power for this lens system is 220 nm, any object smaller than 220 nm could not be seen as a clear, distinct object. An object larger than 220 nm would be resolved.

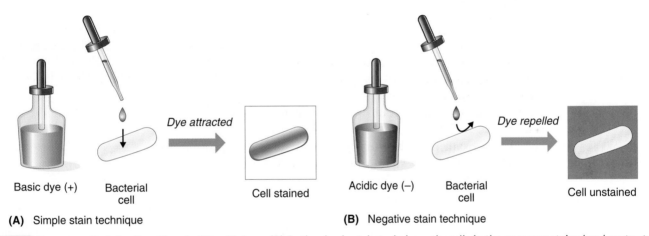

(A) Simple stain technique   **(B)** Negative stain technique

FIGURE 3.11 **Important Staining Reactions in Microbiology.** **(A)** In the simple stain technique, the cells in the smear are stained and contrasted against the light background. **(B)** With the negative stain technique, the cells are unstained and contrasted against a dark background.
»» Explain how the simple and negative staining procedures stain and do not stain cells, respectively.

cells visually into two groups based on staining differences. The Gram stain technique is named for Christian Gram, the Danish physician who first perfected the technique in 1884.

The first two steps of the technique are straightforward (**FIGURE 3.12A**). Air-dried, heat-fixed smears are (1) stained with crystal violet, rinsed, and then (2) a special Gram's iodine solution is added. All bacterial cells would appear blue-purple if the procedure was stopped and the sample viewed with the light microscope. Next, the smear is (3) rinsed with a decolorizer, such as 95% alcohol or an alcohol-acetone mixture.

Observed at this point, certain bacterial cells may lose their color and become transparent. These are the **gram-negative** bacterial cells. Others retain the crystal violet and represent the **gram-positive** bacterial cells. The last step (4) uses safranin, a red cationic dye, to counterstain the gram-negative organisms; that is, give them a orange-red color. So, at the technique's conclusion, gram-positive cells are blue-purple while gram-negative cells are orange-red (**FIGURE 3.12B**). Similar to simple staining, gram staining also allows the observer to determine size, shape, and arrangement of cells.

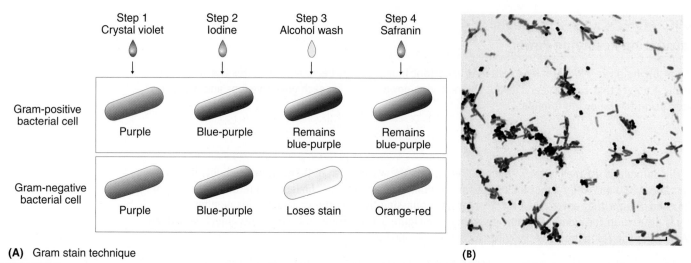

(A) Gram stain technique

(B)

FIGURE 3.12 **Important Staining Reactions in Microbiology.** The Gram stain technique is a differential staining procedure. **(A)** All bacterial cells stain with the crystal violet and iodine, but only gram-negative cells lose the color when alcohol is applied. Subsequently, these bacterial cells stain with the safranin dye. Gram-positive cells remain blue purple. **(B)** This light micrograph demonstrates the staining results of a Gram stain for differentiating between gram-positive and gram-negative cells. (Bar = 10 μm.) »» Besides identifying the Gram reaction, what other characteristics can be determined using the Gram stain procedure?

Knowing whether a bacterial cell is Gram positive or Gram negative is important for microbiologists and clinical technicians who use the results from the Gram stain technique to classify it in *Bergey's Manual* or aid in the identification of an unknown bacterial species (TEXTBOOK CASE 3).

Gram-positive and gram-negative bacterial cells also differ in their susceptibility to chemical substances such as antibiotics (gram-positive cells are more susceptible to penicillin, gram-negative cells to tetracycline). Also, gram-negative cells have more complex cell walls, as described in Chapter 4, and gram-positive and gram-negative bacterial species can produce different types of toxins.

> **Toxins:**
> Chemical substances that are poisonous.

One other differential staining procedure, the **acid-fast technique**, deserves mention. This technique is used to identify members of the genus *Mycobacterium*, one species of which causes tuberculosis. These bacterial organisms are normally difficult to stain with the Gram stain because the cells have very waxy walls that resist the dyes. However, the cell will stain red when treated with carbol-fuchsin (red dye) and heat (or a lipid solubilizer) (**FIGURE 3.13**). The cells then retain their color when washed with a dilute acid-alcohol solution. Other stained genera lose the red color easily during the acid-alcohol wash. The *Mycobacterium* species, therefore, is called acid resistant or "acid fast." Because they stain red and break sharply

when they reproduce, *Mycobacterium* species often are referred to as "red snappers."

CONCEPT AND REASONING CHECKS

3.11 What would happen if a student omitted the alcohol wash step when doing the Gram stain procedure?

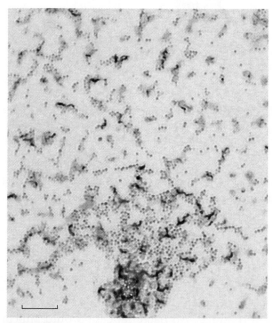

FIGURE 3.13 *Mycobacterium tuberculosis.* The acid-fast technique is used to identify species of *Mycobacterium*. The cells retain the red dye after an acid-alcohol wash. (Bar = 10 μm.) »» Why are cells of *Mycobacterium* resistant to Gram staining?

Textbook CASE 3

Bacterial Meningitis and a Misleading Gram Stain

1 A woman comes to the hospital emergency room complaining of severe headache, nausea, vomiting, and pain in her legs. On examination, cerebral spinal fluid (CFS) was observed leaking from a previous central nervous system (CNS) surgical site.

2 The patient indicates that 6 weeks and 8 weeks ago she had undergone CNS surgery after complaining of migraine headaches and sinusitis. Both surgeries involved a spinal tap. Analysis of cultures prepared from the CFS indicated no bacterial growth.

3 The patient was taken to surgery where a large amount of CFS was removed from underneath the old incision site. The pinkish, hazy fluid indicated bacterial meningitis, so among the laboratory tests ordered was a Gram stain.

4 The patient was placed on antibiotic therapy, consisting of vancomycin and cefotaxime.

5 Laboratory findings from the gram-stained CFS smear showed a few gram-positive, spherical bacterial cells that often appeared in pairs. The results suggested a *Streptococcus pneumoniae* infection.

6 However, upon reexamination of the smear, a few gram-negative spheres were observed.

7 When transferred to a blood agar plate, growth occurred and a prepared smear showed many gram-negative spheres (see figure). Further research indicated that several genera of gram-negative bacteria, including *Acinetobacter,* can appear gram-positive due to under-decolorization during the alcohol wash step.

8 Although complicated by the under-decolorization outcome, the final diagnosis was bacterial meningitis due to *Acinetobacter baumanii*.

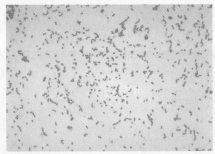

A gram-stained preparation from the blood agar plate.

Questions:

(Answers can be found in Appendix D.)

A. From the gram-stained CSF smear, what color were the gram-positive bacterial spheres?

B. After reexamination of the CFS smear, assess the reliability of the gram-stained smear.

C. What reagent is used for the decolorization step in the Gram stain?

Adapted from: Harrington, B. J. and Plenzler, M., 2004. Misleading gram stain findings on a smear from a cerebrospinal fluid specimen. *Lab. Med.* 35(8): 475–478.
For additional information see http://www.cdc.gov/ncidod/hip/aresist/acin_general.htm.

Light Microscopy Has Other Optical Configurations

KEY CONCEPT

12. Different optical configurations provide detailed views of cells.

Bright-field microscopy provides little contrast (**FIGURE 3.14A**). However, a light microscope can be outfitted with other optical systems to improve contrast of microorganisms without staining.

Three systems commonly employed are mentioned here.

Phase-contrast microscopy uses a special condenser and objective lenses. This condenser lens on the light microscope splits a light beam and throws the light rays slightly out of phase. The separated beams of light then pass through and around the specimen, and small differences in the refractive index within the specimen show up as different degrees of brightness and contrast. With

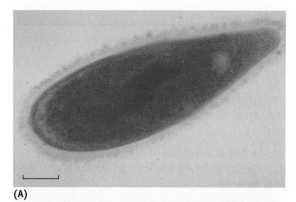

(A)

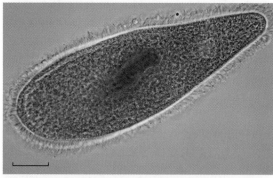

(B)

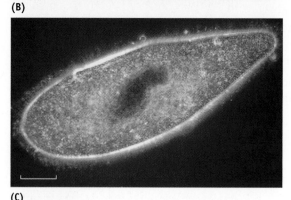

(C)

FIGURE 3.14 **Variations in Light Microscopy.** The same *Paramecium* specimen seen with three different optical configurations: **(A)** bright-field, **(B)** phase-contrast, and **(C)** dark-field. (Bar = 25 μm.) »» What advantage is gained by each of the three microscopy techniques?

phase-contrast microscopy, microbiologists can see organisms alive and unstained (**FIGURE 3.14B**). The structure of yeasts, molds, and protozoa is typically studied with this optical configuration.

Dark-field microscopy also uses a special condenser lens mounted under the stage. The condenser scatters the light and causes it to hit the specimen from the side. Only light bouncing off the specimen and into the objective lens

makes the specimen visible, as the surrounding area appears dark because it lacks background light (**FIGURE 3.14C**). Dark-field microscopy provides good resolution and often illuminates parts of a specimen not seen with bright-field optics. Dark-field microscopy also is the preferred way to study motility of live cells.

Dark-field microscopy helps in the diagnosis of diseases caused by organisms near the limit of resolution of the light microscope. For example, syphilis, caused by the spiral bacterium *Treponema pallidum,* has a diameter of only about 0.15 μm. Therefore, this bacterial species may be observed in scrapings taken from a lesion of a person who has the disease and observed with dark-field microscopy.

Fluorescence microscopy is a major asset to clinical and research laboratories. The technique has been applied to the identification of many microorganisms and is a mainstay of modern microbial ecology and especially clinical microbiology.

For fluorescence microscopy, microorganisms are coated with a fluorescent dye, such as fluorescein, and then illuminated with ultraviolet (UV) light. The energy in UV light excites electrons in fluorescein, and they move to higher energy levels. However, the electrons quickly drop back to their original energy levels and give off the excess energy as visible light. The coated microorganisms thus appear to fluoresce; in the case of fluorescein, they glow a greenish yellow. Other dyes produce other colors (**FIGURE 3.15**).

An important application of fluorescence microscopy is the **fluorescent antibody technique** used to identify an unknown organism. In one variation of this procedure, fluorescein is chemically attached to antibodies, the protein molecules produced by the body's immune system. These "tagged" antibodies are mixed with a sample of the unknown organism. If the antibodies are specific for that organism, they will bind to it and coat the cells with the dye. When subjected to UV light, the organisms will fluoresce. If the organisms fail to fluoresce, the antibodies were not specific to that organism and a different tagged antibody is tried.

More recently, such methods have revolutionized our understanding of the subcellular organi-

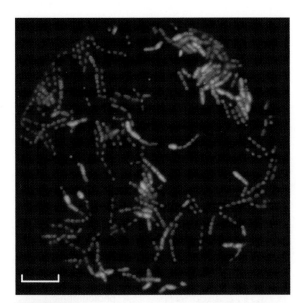

FIGURE 3.15 Fluorescence Microscopy. Fluorescence microscopy of sporulating cells of *Bacillus subtilis*. DNA has been stained with a dye that fluoresces red and a sporulating protein with fluorescein (green). RNA synthesis activity is indicated by a dye that fluoresces blue. (Bar = 15 μm.) »» What advantage is gained by using fluorescence optics over the other light microscope optical configurations?

zation in bacterial cells. We will see the results in the next chapter.

CONCEPT AND REASONING CHECKS

3.12 What optical systems can improve specimen contrast over bright-field microscopy?

Electron Microscopy Provides Detailed Images of Cells, Cell Parts, and Viruses

KEY CONCEPT

13. Electron microscopy uses a beam of electrons to magnify and resolve specimens.

The **electron microscope** grew out of an engineering design made in 1932 by the German physicist Ernst Ruska (winner of the 1986 Nobel Prize in Physics). Ruska showed that electrons will flow in a sealed tube if a vacuum is maintained to prevent electron scattering. Magnets, rather than glass lenses, pinpoint the flow onto an object, where the electrons are absorbed, deflected, or transmitted depending on the density of structures within the object (**FIGURE 3.16**). When projected onto a screen underneath, the electrons form a final image that outlines the structures. As mentioned

in Chapter 1, the early days of electron microscopy produced electron micrographs that showed bacterial cells indeed were cellular but their structure was different from eukaryotic cells.

The power of electron microscopy is the extraordinarily short wavelength of the beam of electrons. Measured at 0.005 nm (compared to 550 nm for visible light), the short wavelength dramatically increases the resolving power of the system and makes possible the visualization of viruses and detailed cellular structures, often called the **ultrastructure** of cells. The practical limit of resolution of biological samples with the electron microscope is about 2 nm, which is 100× better than the resolving power of the light microscope. The drawback of the electron microscope is that the method needed to prepare a specimen kills the cells or organisms.

Two types of electron microscopes are commonly in use. The **transmission electron microscope (TEM)** is used to view and record detailed structures within cells (**FIGURE 3.17A**). Ultrathin sections of the prepared specimen must be cut because the electron beam can penetrate matter only a very short distance. After embedding the specimen in a suitable plastic mounting medium or freezing it, scientists cut the specimen into sections with a diamond knife. In this manner, a single bacterial cell can be sliced, like a loaf of bread, into hundreds of thin sections.

Several of the sections are placed on a small grid and stained with heavy metals such as lead and osmium to provide contrast. The microscopist then inserts the grid into the vacuum chamber of the microscope and focuses a 100,000-volt electron beam on one portion of a section at a time. An image forms on the screen below or can be recorded on film. The electron micrograph may be enlarged with enough resolution to achieve a final magnification approaching 2 million ×.

The **scanning electron microscope (SEM)** was developed in the late 1960s to enable researchers to see the surfaces of objects in the natural state and without sectioning. The specimen is placed in the vacuum chamber and covered with a thin coat of gold. The electron beam then scans across the specimen and knocks loose showers of electrons that are captured by a detector. An image builds line by line, as in a television receiver. Electrons

■ Electron micrographs: Images recorded on electron-sensitive film.

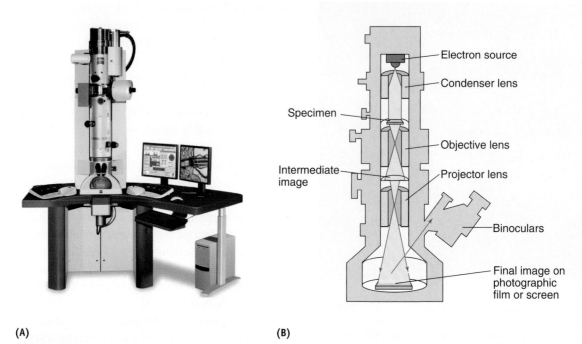

(A)   **(B)**

FIGURE 3.16 **The Electron Microscope.** **(A)** A transmission electron microscope (TEM). **(B)** A schematic of a TEM. A beam of electrons is emitted from the electron source and electromagnets are used to focus the beam on the specimen. The image is magnified by objective and projector lenses. The final image is projected on a screen, television monitor, or photographic film. »» How does the path of the image for the transmission electron microscope compare with that of the light microscope (Figure 3.10)?

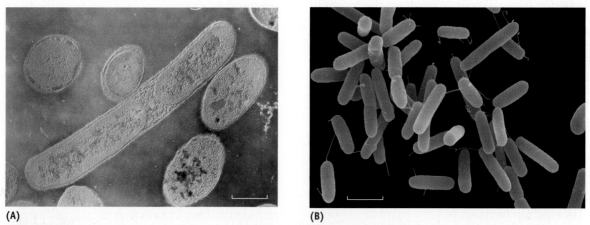

(A)   **(B)**

FIGURE 3.17 **Transmission and Scanning Electron Microscopy Compared.** The bacterium *Pseudomonas aeruginosa* (false-color images) as seen with two types of electron microscopy. **(A)** A view of sectioned cells seen with the transmission electron microscope. (Bar = 1.0 μm.) **(B)** A view of whole cells seen with the scanning electron microscope. (Bar = 3.0 μm.) »» What types of information can be gathered from each of these electron micrographs?

that strike a sloping surface yield fewer electrons, thereby producing a darker contrasting spot and a sense of three dimensions. The resolving power of the conventional SEM is about 10 nm and magnifications with the SEM are limited to about 20,000×. However, the instrument provides vivid

and undistorted views of an organism's surface details (**FIGURE 3.17B**).

The electron microscope has added immeasurably to our understanding of the structure and function of microorganisms by letting us penetrate their innermost secrets. In the chapters ahead, we

TABLE 3.3 Comparison of Various Types of Microscopy				
Type of Microscopy	Special Feature	Appearance of Object	Magnification Range	Objects Observed
Light				
Bright-field	Visible light illuminates object	Stained microorganisms on clear background	100×–1,000×	Arrangement, shape, and size of killed microorganisms (except viruses)
Phase-contrast	Special condenser throws light rays "out of phase"	Unstained microorganisms with contrasted structures	100×–1,000×	Internal structures of live, unstained eukaryotic microorganisms
Dark-field	Special condenser scatters light	Unstained microorganisms on dark background	100×–1,000×	Live, unstained microorganisms; motility of live cells
Fluorescence	UV light illuminates fluorescent-coated objects	Fluorescing microorganisms on dark background	100×–1,000×	Outline of microorganisms coated with fluorescent-tagged antibodies
Electron				
Transmission	Short-wavelength electron beam penetrates sections	Alternating light and dark areas contrasting internal cell structures	100×–2,000,000×	Ultrathin slices of microorganisms and internal components
Scanning	Short-wavelength electron beam knocks loose electron showers	Microbial surfaces	10×–20,000×	Surfaces and textures of microorganisms and cell components

will encounter many of the structures discovered with electron microscopy, and we will better appreciate microbial physiology as it is defined by microbial structures.

The various types of light and electron microscopy are compared in **TABLE 3.3**.

CONCEPT AND REASONING CHECKS

3.13 What type of electron microscope would be used to examine (a) the surface structures on a *Paramecium* cell and (b) the organelles in an algal cell?

SUMMARY OF KEY CONCEPTS

3.1 The Bacteria/Eukaryotic Paradigm

1. All living organisms share the common emergent properties of life, attempt to maintain a stable internal state called **homeostasis**, and interact through a multicellular association (a **biofilm**) involving chemical communication and cooperation between cells.

2. Bacterial and eukaryotic cells share certain organizational patterns, including genetic organization, compartmentation, metabolic organization, and protein synthesis.

3. Although bacterial and eukaryotic cells carry out many similar processes, eukaryotic cells often contain membrane-enclosed compartments (**organelles**) to accomplish the processes.

3.2 Classifying Microorganisms

4. Many systems of classification have been devised to catalog organisms based on shared characteristics.

5. Based on several molecular and biochemical differences, Woese proposed a **three-domain system** where the prokaryotes are separated into two domains, the *Bacteria* and *Archaea*. The kingdoms Protista, Fungi, Plantae, and Animalia are placed in the domain *Eukarya*.

6. Part of an organism's binomial name is the **genus** name; the remaining part is the **specific epithet** that describes the genus name. Thus, a **species** name consists of the genus and specific epithet.

7. Organisms are properly classified using a standardized hierarchical system from species (the least inclusive) to domain (the most inclusive).

8. *Bergey's Manual* is the standard reference to identify and classify bacterial species. Criteria have included traditional characteristics, but modern molecular methods have led to a reconstruction of evolutionary events and organism relationships.

3.3 Microscopy

9. Another criterion of a microorganism is its size, a characteristic that varies among members of different groups. The **micrometer** (**μm**) is used to measure the dimensions of bacterial, protozoal, and fungal cells. The **nanometer** (**nm**) is commonly used to express viral sizes.

10. The instrument most widely used to observe microorganisms is the **light microscope**. Light passes through several lens systems that magnify and resolve the object being observed. Although **magnification** is important, **resolving power** is key. The light microscope can magnify up to 1,000× and resolve objects as small as 0.2 μm.

11. For bacterial cells, staining generally precedes observation. The **simple**, **negative**, **Gram**, **acid-fast**, and other staining techniques can be used to impart contrast and determine structural or physiological properties.

12. Microscopes employing **phase-contrast**, **dark-field**, and **fluorescence** optics have specialized uses in microbiology to contrast cells without staining.

13. To increase resolving power and achieve extremely high magnification, the electron microscope employs a beam of electrons to magnify and resolve specimens. To observe internal details (**ultrastructure**), the **transmission electron microscope** is most often used; to see whole objects or surfaces, the **scanning electron microscope** is useful.

LEARNING OBJECTIVES

After understanding the textbook reading, you should be capable of writing a paragraph that includes the appropriate terms and pertinent information to answer the objective.

1. Assess the importance of **homeostasis** to cell (organismal) survival and contrast bacteria as **unicellular** and **multicellular** organisms.
2. Describe the four organizational patterns common to all organisms.
3. Identify the structural distinctions between bacterial and eukaryotic cells.
4. Explain how knowledge of shared characteristics changed the classification of living organisms from Linnaeus to Woese.
5. Explain the assignment of organisms to the **three-domain system** of **classification**.
6. Write scientific names of organisms using the **binomial system**.
7. Identify the **taxa** used to classify organisms from least to most inclusive taxa.
8. Contrast the determinative methods used to identify bacterial species.
9. Identify how microbial agents are measured using metric system units.
10. Assess the importance of **magnification** and **resolving power** to microscopy.
11. Summarize the **Gram stain** procedure.
12. Identify the optical configurations that provide contrast with **light microscopy**.
13. Compare the uses of the **transmission** and **scanning electron microscopes**.

STEP A: SELF-TEST

Each of the following questions is designed to assess your ability to remember or recall factual or conceptual knowledge related to this chapter. Read each question carefully, then select the *one* answer that best fits the question or statement. Answers to even-numbered questions can be found in **Appendix C**.

1. What is the term that describes the ability of organisms to maintain a stable internal state?
   A. Metabolism
   B. Homeostasis
   C. Biosphere
   D. Ecotype

2. Which one of the following is NOT an organizational pattern common to all organisms?
   A. Genetic organization
   B. Protein synthesis
   C. Compartmentation
   D. Microcompartments

3. Which one of the following is NOT found in bacterial cells?
   A. Ribosomes
   B. DNA
   C. Mitochondria
   D. Cytoplasm

4. Who is considered to be the father of modern taxonomy?
   A. Woese
   B. Whittaker
   C. Haeckel
   D. Linnaeus

5. _____ was first used to catalog organisms into one of three domains.
   A. Photosynthesis
   B. Ribosomal RNA genes
   C. Nuclear DNA genes
   D. Mitochondrial DNA genes

6. Which one of the following is the correct genus name for the bacterial organism that causes syphilis?
   A. pallidum
   B. *Treponema*
   C. *pallidum*
   D. *T. pallidum*

7. Several classes of organisms would be classified into one
   A. order.
   B. genus.
   C. phylum.
   D. family.

8. An important method used in the rapid identification of a pathogen is _____.
   A. rRNA gene sequencing
   B. polymerase chain reaction
   C. molecular taxonomy
   D. biochemical tests

9. Most bacterial cells are measured using what metric system of length?
   A. Millimeters (mm)
   B. Micrometers (μm)
   C. Nanometers (nm)
   D. Centimeters (cm)

10. Resolving power is the ability of a microscope to
    A. estimate cell size.
    B. magnify an image.
    C. see two close objects as separate.
    D. keep objects in focus.

11. Before bacterial cells are simple stained and observed with the light microscope, they must be
    A. smeared on a slide.
    B. heat fixed.
    C. air dried.
    D. All the above (A–C) are correct.

12. If you wanted to study bacterial motility you would most likely use
    A. a transmission electron microscope.
    B. a light microscope with dark-field optics.
    C. a scanning electron microscope.
    D. a light microscope with phase-contrast optics.

13. If you wanted to study the surface of a bacterial cell, you would use
    A. a transmission electron microscope.
    B. a light microscope with phase-contrast optics.
    C. a scanning electron microscope.
    D. a light microscope with dark-field optics.

STEP B: REVIEW

Answers to even-numbered questions or statements can be found in **Appendix C**.

14. Construct a concept map for **Living Organisms** using the following terms (terms can be used more than once).

bacterial cells	Golgi apparatus
cell membrane	lysosomes
chloroplasts	microcompartments
cytoplasm	mitochondria
cytoskeleton	nucleus
cytosol	RER
DNA region	ribosomes
eukaryotic cells	SER
flagella	

15. Construct a concept map for **staining techniques** using the following terms only once.

acid-fast technique	differential stain procedure
acidic dye	gram negative
basic dye	gram positive
blue-purple cells	gram stain technique
cell arrangement	*Mycobacterium*
cell shape	negative stain technique
cell size	orange-red cells
contrast	simple stain technique

Match the statement on the left to the term on the right by placing the letter of the term in the available space.

Statement

16. _____ Major group of organisms whose cells have no nucleus or organelles in the cytoplasm.

17. _____ Bacterial organisms capable of photosynthesis.

18. _____ Type of electron microscope for which cell sectioning is not required.

19. _____ These structures carry out protein synthesis in all cells.

20. _____ The organelle, absent in bacteria, that carries out the conversion of chemical energy to cellular energy in eukaryotes.

21. _____ Domain in which fungi and protista are classified.

22. _____ Staining technique that differentiates bacterial cells into two groups.

23. _____ Category into which two or more genera are grouped.

24. _____ The staining technique employing a single cationic dye.

25. _____ Type of microscopy using UV light to excite a dye-coated specimen.

Term

A. *Bacteria*	J. Homeostasis
B. Chloroplast	K. Mitochondrion
C. Cyanobacteria	L. Negative
D. Dark-field	M. Phase-contrast
E. *Eukarya*	N. Ribosomes
F. Family	O. Scanning
G. Fluorescence	P. Simple
H. Fungi	Q. Taxonomy
I. Gram	R. Transmission

HTTP://MICROBIOLOGY.JBPUB.COM/9E

The site features eLearning, an online review area that provides quizzes and other tools to help you study for your class. You can also follow useful links for in-depth information, read more MicroFocus stories, or just find out the latest microbiology news.

STEP C: APPLICATIONS

Answers to even-numbered questions can be found in **Appendix C**.

26. A student is performing the Gram stain technique on a mixed culture of gram-positive and gram-negative bacterial cells. In reaching for the counterstain in step 4, he inadvertently takes the methylene blue bottle and proceeds with the technique. What will be the colors of gram-positive and gram-negative bacteria at the conclusion of the technique?

27. Would the best resolution with a light microscope be obtained using red light ($\lambda = 680$ nm), green light ($\lambda = 520$ nm), or blue light ($\lambda = 500$ nm)? Explain your answer.

28. Identify the cell structures (a–p) indicated in drawings (A) and (B) below.

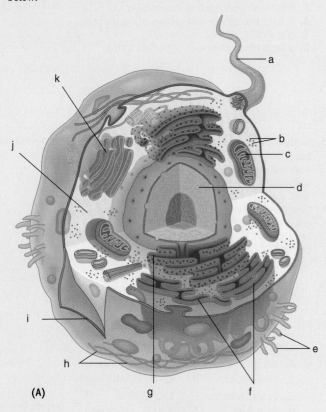

(A)

29. The electron micrograph below shows a group of bacterial cells. The micrograph has been magnified 5,000×. At this magnification, the cells are about 10 mm in length. Calculate the actual length of the bacterial cells in micrometers (μm)?

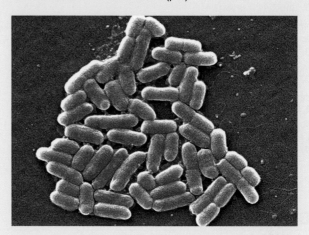

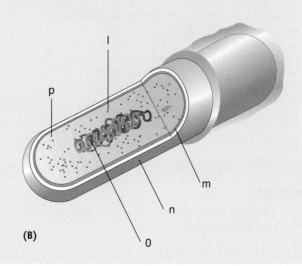

(B)

STEP D: QUESTIONS FOR THOUGHT AND DISCUSSION

Answers to even-numbered questions can be found in **Appendix C**.

30. A local newspaper once contained an article about "the famous bacteria *E. coli*." How many things can you find wrong in this phrase? Rewrite the phrase correctly.

31. Microorganisms have been described as the most chemically diverse, the most adaptable, and the most ubiquitous organisms on Earth. Although your knowledge of microorganisms still may be limited at this point, try to add to this list of "mosts."

32. Bacteria lack the cytoplasmic organelles commonly found in the eukaryotes. Provide a reason for this structural difference.

33. A new bacteriology laboratory is opening in your community. What is one of the first books that the laboratory director will want to purchase? Why is it important to have this book?

34. In a respected science journal, an author wrote, "Linnaeus gave each life form two Latin names, the first denoting its genus and the second its species." A few lines later, the author wrote, "Man was given his own genus and species *Homo sapiens*." What is conceptually and technically wrong with both statements?

35. A student of general biology observes a microbiology student using immersion oil and asks why the oil is used. "To increase the magnification of the microscope" is the reply. Do you agree? Why?

36. Every state has an official animal, flower, or tree, but only Oregon has a bacterial species named in its honor: *Methanohalophilus oregonese*. The specific epithet *oregonese* is obvious, but can you decipher the meaning of the genus name?

Cell Structure and Function in the *Bacteria* and *Archaea*

Our planet has always been in the "Age of Bacteria," ever since the first fossils—bacteria of course—were entombed in rocks more than 3 billion years ago. On any possible, reasonable criterion, bacteria are—and always have been—the dominant forms of life on Earth.
—Paleontologist Stephen J. Gould (1941–2002)

"Double, double toil and trouble; Fire burn, and cauldron bubble"* is the refrain repeated several times by the chanting witches in Shakespeare's *Macbeth* (Act IV, Scene 1). This image of a hot, boiling cauldron actually describes the environment in which many bacterial, and especially archaeal, species happily grow! For example, some species can be isolated from hot springs or the hot, acidic mud pits of volcanic vents (FIGURE 4.1).

When the eminent evolutionary biologist and geologist Stephen J. Gould wrote the opening quote of this chapter, he as well as most microbiologists had no idea that embedded in these "bacteria" was another whole domain of organisms. Thanks to the pioneering studies of Carl Woese and his colleagues, it now is quite evident there are two distinctly different groups of "prokaryotes"—the *Bacteria* and the *Archaea* (see Chapter 3). Many of the organisms Woese and others studied are organisms that would live a happy life in a witch's cauldron because they can grow at high temperatures, produce methane gas, or survive in extremely acidic and hot environments—a real cauldron! Termed **extremophiles**, these members of the domains *Bacteria*

(A)   (B)

FIGURE 4.1 **Life at the Edge.** Bacterial and archaeal extremophiles have been isolated from the edges of natural cauldrons, including (**A**) the Grand Prismatic Spring in Yellowstone National Park, Wyoming, where the water of the hot spring is over 70°C, or (**B**) the mud pools surrounding sulfurous steam vents of the Solfatara Crater in Pozzuoli, Italy, where the mud has a very low pH and a temperature above 90°. »» How do extremophiles survive under these extreme conditions?

and *Archaea* have a unique genetic makeup and have adapted to extreme environmental conditions.

In fact, Gould's "first fossils" may have been archaeal species. Many microbiologists believe the ancestors of today's archaeal species might represent a type of organism that first inhabited planet Earth when it was a young, hot place (see MicroFocus 2.1). These unique characteristics led Woese to propose these organisms be lumped together and called the Archaebacteria (*archae* = "ancient").

Since then, the domain name has been changed to *Archaea* because (1) not all members are extremophiles or related to these possible ancient ancestors and (2) they are not *Bacteria*—they are *Archaea*. Some might also debate using the term prokaryotes when referring to both domains, as organisms in the two domains are as different from each other as they are from the *Eukarya*.

As more microbes have had their complete genomes sequenced, it now is clear that there are unique as well as shared characteristics between species in the domains *Bacteria* and *Archaea*.

In this chapter, we examine briefly some of the organisms in the domains *Bacteria* and *Archaea*. However, because almost all known "prokaryotic" pathogens of humans are in the domain *Bacteria*, we emphasize structure within this domain. As we see in this chapter, a study of the structural features of bacterial cells provides a window to their activities and illustrates how the *Bacteria* relate to other living organisms.

As we examine bacterial and archaeal cell structure, we can assess the dogmatic statement that these cells are characterized by a lack of a cell nucleus and internal membrane-bound organelles. Before you finish this chapter, you will be equipped to revise this view.

4.1   Diversity among the *Bacteria* and *Archaea*

In this section, we discuss bacterial and archaeal diversity using the current classification scheme, which is based in large part on nucleotide sequence data. There are some 7,000 known bacterial and archaeal species and a suspected 10 million species. In this section, we will highlight a few phyla and groups using the phylogenetic tree in **FIGURE 4.2**.

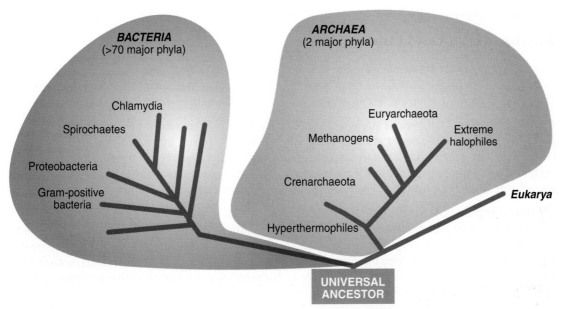

FIGURE 4.2 The Phylogenetic Tree of *Bacteria* and *Archaea*. The tree shows several of the bacterial and archaeal phyla discussed in this chapter. »» What is common to the branch base of both the *Bacteria* and *Archaea*?

The Domain *Bacteria* Contains Some of the Most Studied Microbial Organisms

KEY CONCEPT

1. The *Bacteria* are classified into several major phyla.

There are about 18 phyla of *Bacteria* identified from culturing or nucleotide sequencing. It should come as no shock to you by now to read that the vast majority of these phyla play a positive role in nature (MICROFOCUS 4.1). Although not unique to just the bacterial phyla, they digest sewage into simple chemicals; they extract nitrogen from the air and make it available to plants for protein production; they break down the remains of all that die and recycle the carbon and other elements; and they produce oxygen gas that we and other animals breathe.

Of course, we know from Chapter 1 and personal experience that some bacterial organisms are harmful—many human pathogens are members of the domain *Bacteria*. Certain species multiply within the human body, where they disrupt tissues or produce toxins that result in disease.

The *Bacteria* have adapted to the diverse environments on Earth, inhabiting the air, soil, and water, and they exist in enormous numbers on the surfaces of virtually all plants and animals. They can be isolated from Arctic ice, thermal hot springs, the fringes of space, and the tissues of animals. Bacterial species, along with their archaeal relatives, have so completely colonized every part of the Earth that their mass is estimated to outweigh the mass of all plants and animals combined. Let's look briefly at some of the major phyla and other groups.

Proteobacteria. The **Proteobacteria** (*proteo* = "first") contains the largest and most diverse group of species and includes many familiar gram-negative genera, such as *Escherichia* (FIGURE 4.3A). The phylum also includes some of the most recognized human pathogens, including species of *Shigella*, *Salmonella*, *Neisseria* (responsible for gonorrhea), *Yersinia* (responsible for plague), and *Vibrio* (responsible for cholera). It is likely that the mitochondria of the *Eukarya* were derived through endosymbiosis from an ancestor of the Proteobacteria (see MicroInquiry 3).

The group also includes the rickettsiae (sing., rickettsia), which were first described by Howard Taylor Ricketts in 1909. These tiny bacterial cells can barely be seen with the most powerful light microscope. They are transmitted among humans primarily by arthropods, and are cultivated only in living tissues such as chick embryos. Different species cause a number of important diseases, including Rocky Mountain spotted fever and

Arthropods: Animals having jointed appendages and segmented body (e.g., ticks, lice, fleas, mosquitoes).

MICROFOCUS 4.1

Bacteria in Eight Easy Lessons[1]

Mélanie Hamon, an assistante de recherché at the Institut Pasteur in Paris, says that when she introduces herself as a bacteriologist, she often is asked, "Just what does that mean?" To help explain her discipline, she gives us, in eight letters, what she calls "some demystifying facts about bacteria."

Basic principles: Their average size is 1/25,000th of an inch. In other words, hundreds of thousands of bacteria fit into the period at the end of this sentence. In comparison, human cells are 10 to 100 times larger with a more complex inner structure. While human cells have copious amounts of membrane-contained subcompartments, bacteria more closely resemble pocketless sacs. Despite their simplicity, they are self-contained living beings, unlike viruses, which depend on a host cell to carry out their life cycle.

Astonishing: Bacteria are the root of the evolutionary tree of life, the source of all living organisms. Quite successful evolutionarily speaking, they are ubiquitously distributed in soil, water, and extreme environments such as ice, acidic hot springs or radioactive waste. In the human body, bacteria account for 10% of dry weight, populating mucosal surfaces of the oral cavity, gastrointestinal tract, urogenital tract and surface of the skin. In fact, bacteria are so numerous on earth that scientists estimate their biomass to far surpass that of the rest of all life combined.

Crucial: It is a little known fact that most bacteria in our bodies are harmless and even essential for our survival. Inoffensive skin settlers form a protective barrier against any troublesome invader while approximately 1,000 species of gut colonizers work for our benefit, synthesizing vitamins, breaking down complex nutrients and contributing to gut immunity. Unfortunately for babies (and parents!), we are born with a sterile gut and "colic" our way through bacterial colonization.

Tools: Besides the profitable relationship they maintain with us, bacteria have many other practical and exploitable properties, most notably, perhaps, in the production of cream, yogurt and cheese. Less widely known are their industrial applications as antibiotic factories, insecticides, sewage processors, oil spill degraders, and so forth.

Evil: Unfortunately, not all bacteria are "good," and those that cause disease give them all an often undeserved and unpleasant reputation. If we consider the multitude of mechanisms these "bad" bacteria—pathogens—use to assail their host, it is no wonder that they get a lot of bad press. Indeed, millions of years of coevolution have shaped bacteria into organisms that "know" and "predict" their hosts' responses. Therefore, not only do bacterial toxins know their target, which is never missed, but bacteria can predict their host's immune response and often avoid it.

Resistant: Even more worrisome than their effectiveness at targeting their host is their faculty to withstand antibiotic therapy. For close to 50 years, antibiotics have revolutionized public health in their ability to treat bacterial infections. Unfortunately, overuse and misuse of antibiotics have led to the alarming fact of resistance, which promises to be disastrous for the treatment of such diseases.

Ingenious: The appearance of antibiotic-resistant bacteria is a reflection of how adaptable they are. Thanks to their large populations they are able to mutate their genetic makeup, or even exchange it, to find the appropriate combination that will provide them with resistance. Additionally, bacteria are able to form "biofilms," which are cellular aggregates covered in slime that allow them to tolerate antimicrobial applications that normally eradicate free-floating individual cells.

A long tradition: Although "little animalcules" were first observed in the 17th century, it was not until the 1850s that Louis Pasteur fathered modern microbiology. From this point forward, research on bacteria has developed into the flourishing field it is today. For many years to come, researchers will continue to delve into this intricate world, trying to understand how the good ones can help and how to protect ourselves from the bad ones. It is a great honor to be part of this tradition, working in the very place where it was born.

---

[1]Republished with permission of the author, the Institut Pasteur, and the Pasteur Foundation. The original article appeared in *Pasteur Perspectives* Issue 20 (Spring 2007), the newsletter of the Pasteur Foundation, which may be found at www.pasteurfoundation.org © Pasteur Foundation.

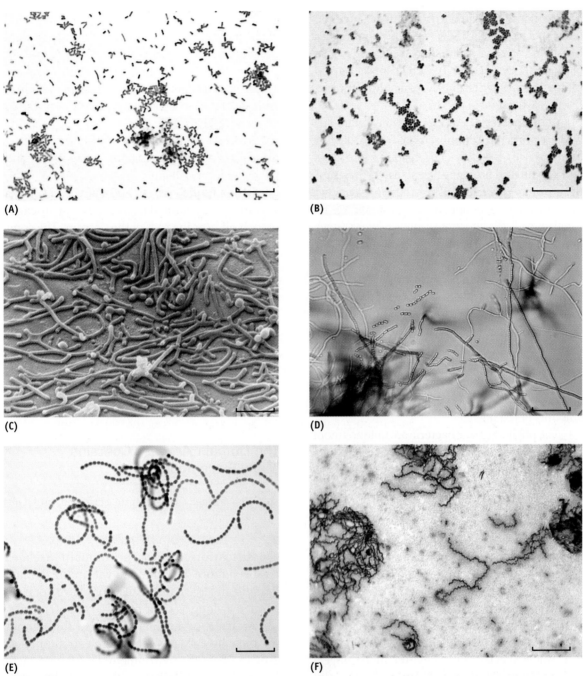

FIGURE 4.3 Members of the Domain *Bacteria*. (**A**) *Escherichia coli* (Bar = 10 μm.), (**B**) *Staphylococcus aureus* (Bar = 10 μm.), (**C**) *Mycoplasma* species (Bar = 2 μm.), (**D**) *Streptomyces* species (Bar = 20 μm.), (**E**) *Anabaena* species (Bar = 100 μm.), and (**F**) *Treponema pallidum* (Bar = 10 μm.). All images are light micrographs except (**C**), a false-color scanning electron micrograph. »» What is the Gram staining result for *E. coli* and *S. aureus*?

typhus fever. Chapter 12 contains a more thorough description of their properties.

Firmicutes. The **Firmicutes** (*firm* = "strong"; *cuti* = "skin") consists of many species that are grampositive. As we will see in this chapter, they share a similar thick "skin," which refers to their cell wall structure. Genera include *Bacillus* and *Clostridium*,

specific species that are responsible for anthrax and botulism, respectively. Species within the genera *Staphylococcus* and *Streptococcus* are responsible for several mild to life-threatening human illnesses (FIGURE 4.3B).

Also within the Firmicutes is the genus *Mycoplasma*, which lacks a cell wall but is otherwise phylogenetically related to the gram-positive bacterial species (**FIGURE 4.3C**). Possibly the smallest free-living bacterial cell, one species causes a form of pneumonia (Chapter 10) while another mycoplasmal illness represents a sexually-transmitted disease (Chapter 13).

Actinobacteria. Another phylum of gram-positive species is the **Actinobacteria**. Often called the actinomycetes, these bacterial organisms form a system of branched filaments that somewhat resemble the growth form of fungi. The genus *Streptomyces* is the source for important antibiotics (**FIGURE 4.3D**). Another medically important genus is *Mycobacterium*, one species of which is responsible for tuberculosis.

Cyanobacteria. In Chapter 3 we discussed the cyanobacteria. They are phylogenetically related to the gram-positive species and can exist as unicellular, filamentous, or colonial forms (**FIGURE 4.3E**). Once known as blue-green algae because of their pigmentation, pigments also may be black, yellow, green, or red. The periodic redness of the Red Sea, for example, is due to blooms of cyanobacteria whose members contain large amounts of red pigment.

The phylum **Cyanobacteria** are unique among bacterial groups because they carry out photosynthesis similar to unicellular algae (Chapter 6) using the light-trapping pigment chlorophyll. Their evolution on Earth was responsible for the "oxygen revolution" that transformed life on the young planet. In addition, chloroplasts probably are derived from the endosymbiotic union with a cyanobacterial ancestor.

Chlamydiae. Roughly half the size of the rickettsiae, members of the phylum **Chlamydiae** are so small that they cannot be seen with the light microscope and are cultivated only within living cells. Most species in the phylum are pathogens and one species causes the gonorrhea-like disease known as chlamydia. Chlamydial diseases are described in Chapter 13.

Spirochaetes. The phylum **Spirochaetes** contains more than 340 gram-negative species that possess a unique cell body that coils into a long helix and moves in a corkscrew pattern. The ecological niches for the spirochetes is diverse: from free-living species found in mud and sediments, to symbiotic species present in the digestive tracts of insects, to the pathogens found in the urogenital tracts of vertebrates. Many spirochetes

Blooms:
Sudden increases in the numbers of cells of an organism in an environment.

are found in the human oral cavity; in fact, some of the first animalcules seen by Leeuwenhoek were probably spirochetes from his teeth scrapings (see Chapter 1). Among the human pathogens are *Treponema pallidum*, the causative agent of syphilis and one of the most common sexually transmitted diseases (**FIGURE 4.3F**; Chapter 13); and specific species of *Borrelia*, which are transmitted by ticks or lice and are responsible for Lyme disease and relapsing fever (Chapter 12).

Other Phyla. There are many other phyla within the domain *Bacteria*. Several lineages branch off near the root of the domain. The common link between these organisms is that they are **hyperthermophiles**; they grow at high temperatures. Examples include *Aquifex* and *Thermotoga*, which typically are found in earthly cauldrons such as hot springs.

CONCEPT AND REASONING CHECKS

4.1 What three unique events occurred within the Proteobacteria and Cyanobacteria that contributed to the evolution of the Eukarya and the oxygen-rich atmosphere on Earth?

The Domain *Archaea* Contains Many Extremophiles

KEY CONCEPT

2. The *Archaea* are currently classified into two major phyla.

Classification within the domain *Archaea* has been more difficult than within the domain *Bacteria*, in large part because they have not been studied as long as their bacterial counterparts.

Archaeal organisms are found throughout the biosphere. Many genera are extremophiles, growing best at environmental extremes, such as very high temperatures, high salt concentrations, or extremes of pH. However, most species exist in very cold environments although there are archaeal genera that thrive under more modest conditions. The archaeal genera can be placed into one of two major phyla.

Euryarchaeota. The **Euryarchaeota** contain organisms with varying physiologies, many being extremophiles. Some groups, such as the **methanogens** (*methano* = "methane"; *gen* = "produce") are killed by oxygen gas and therefore are found in environments devoid of oxygen gas. The pro-

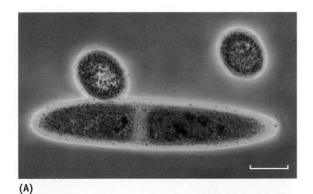

(A)

(B)

(C)

FIGURE 4.4 Members of the Domain *Archaea*. (**A**) A false-color transmission electron micrograph of the methanogen *Methanospirillum hungatei*. (Bar = 0.5 μm.) (**B**) An aerial view above Redwood City, California, of the salt ponds whose color is due to high concentrations of extreme halophiles. (**C**) A false-color scanning electron micrograph of *Sulfolobus*, a hyperthermophile that grows in waters as hot as 90°C. (Bar = 0.5 μm.) »» What advantage is afforded these species that grow in such extreme environments?

duction of methane (natural) gas is important in their energy metabolism (**FIGURE 4.4A**). In fact, these archaeal species release more than 2 billion tons of methane gas into the atmosphere every year. About a third comes from the archaeal species living in the stomach (rumen) of cows (see Chapter 2).

Another group is the **extreme halophiles** (*halo* = "salt"; *phil* = "loving"). They are distinct from the methanogens in that they require oxygen gas for energy metabolism and need high concentrations of salt (NaCl) to grow and reproduce. The fact that they often contain pink pigments makes their identification easy (**FIGURE 4.4B**). In addition, some extreme halophiles have been found in lakes where the pH is greater than 11.

A third group is the **hyperthermophiles** that grow optimally at high temperatures approaching or surpassing 100°C.

Crenarchaeota. The second phylum, the **Crenarchaeota**, are mostly hyperthermophiles growing at temperatures above 80°C. Hot sulfur springs are one environment where these archaeal species also thrive. The temperature is around 75°C but the springs are extremely acidic (pH of 2–3). Volcanic vents are another place where these organisms can survive quite happily (**FIGURE 4.4C**). Other species are dispersed in open oceans, often inhabiting the cold ocean waters (–3°C) of the deep sea environments and polar seas.

TABLE 4.1 summarizes some of the characteristics that are shared or are unique among the three domains.

CONCEPT AND REASONING CHECKS

4.2 Compared to the more moderate environments in which some archaeal species grow, why have others adapted to such extreme environments?

4.1   Some Major Differences between *Bacteria*, *Archaea*, and *Eukarya*

Characteristic	Bacteria	Archaea	Eukarya
Cell nucleus	No	No	Yes
Chromosome form	Single, circular	Single, circular	Multiple, linear
Histone proteins present	No	Yes	Yes
Peptidoglycan cell wall	Yes	No	No
Membrane lipids	Ester-linked	Ether-linked	Ester-linked
Ribosome sedimentation value	70S	70S	80S
Ribosome sensitivity to diphtheria toxin	No	Yes	Yes
First amino acid in a protein	Formylmethionine	Methionine	Methionine
Chlorophyll-based photosynthesis	Yes (cyanobacteria)	No	Yes (algae)
Growth above 80°C	Yes	Yes	No
Growth above 100°C	No	Yes	No

4.2   Cell Shapes and Arrangements

Bacterial and archaeal cells come in a bewildering assortment of sizes, shapes, and arrangements, reflecting the diverse environments in which they grow. As described in Chapter 3, these three characteristics can be studied by viewing stained cells with the light microscope. Such studies show that most, including the clinically significant ones, appear in one of three different shapes: the rod, the sphere, or the spiral.

Variations in Cell Shape and Cell Arrangement Exist

KEY CONCEPT

3.   Many bacterial cells have a rod, spherical, or spiral shape and are organized into a specific cellular arrangement.

A bacterial cell with a rod shape is called a **bacillus** (pl., bacilli). In various species of rod-shaped bacteria, the cylindrical cell may be as long as 20 μm or as short as 0.5 μm. Certain bacilli are slender, such as those of *Salmonella typhi* that cause typhoid fever; others, such as the agent of anthrax (*Bacillus anthracis)*, are rectangular with squared ends; still others, such as the diphtheria bacilli (*Corynebacterium diphtheriae*), are club shaped. Most rods occur singly, in pairs called **diplobacillus**, or arranged into a long chain called **streptobacillus** (*strepto* = "chains") (FIGURE 4.5A). Realize there are two ways to use the word "bacillus": to

denote a rod-shaped bacterial cell, and as a genus name (*Bacillus*).

A spherically shaped bacterial cell is known as a **coccus** (pl., cocci; *kokkos* = "berry") and tends to be quite small, being only 0.5 μm to 1.0 μm in diameter. Although they are usually round, they also may be oval, elongated, or indented on one side. Many bacterial species that are cocci stay together after division and take on cellular arrangements characteristic of the species (FIGURE 4.5B). Cocci remaining in a pair after reproducing represent a **diplococcus**. The organism that causes gonorrhea, *Neisseria gonorrhoeae*, and one type of bacterial meningitis (*N. meningitidis*) are diplococci. Cocci that remain in a chain are called **streptococcus**. Certain species of streptococci are involved in strep throat (*Streptococcus pyogenes*) and tooth decay (*S. mutans*), while other species are harmless enough to be used for producing dairy products such as yogurt (*S. lactis*). Another arrangement of cocci is the **tetrad**, consisting of four cocci forming a square. A cube-like packet of eight cocci is called a **sarcina** (*sarcina* = "bundle"). *Micrococcus luteus*, a common inhabitant of the skin, is one example. Other cocci may divide randomly and form an irregular grape-like cluster of cells called a **staphylococcus** (*staphylo* = "cluster"). A well-known example, *Staphylococcus aureus*, is often a cause of food poisoning, toxic shock syndrome, and several skin infections. The latter are known in the modern vernacular as "staph" infections. Notice again that the words "streptococcus" and "staphy-

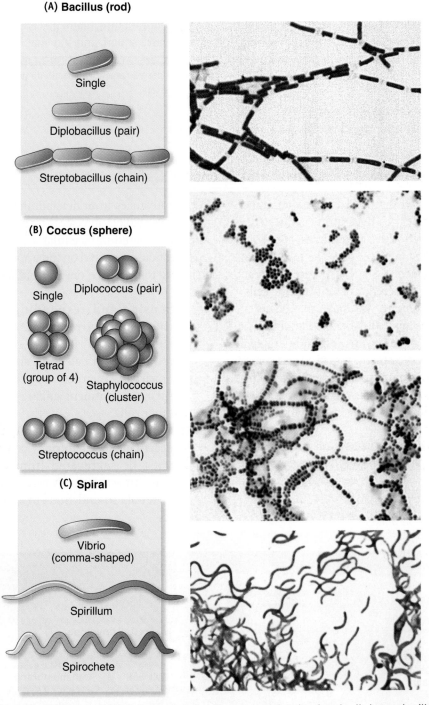

(A) Bacillus (rod)

Single

Diplobacillus (pair)

Streptobacillus (chain)

(B) Coccus (sphere)

Single   Diplococcus (pair)

Tetrad
(group of 4)   Staphylococcus
(cluster)

Streptococcus (chain)

(C) Spiral

Vibrio
(comma-shaped)

Spirillum

Spirochete

FIGURE 4.5 Variation in Shape and Cell Arrangements. Many bacterial and archaeal cells have a bacillus (**A**) or coccus (**B**) shape. Most spiral shaped-cells (**C**) are not organized into a specific arrangement. »» In photomicrograph (C), identify the vibrio and the spirillum forms.

lococcus" can be used to describe cell shape and arrangement, or a bacterial genus (*Streptococcus* and *Staphylococcus*).

The third common shape of bacterial cells is the **spiral**, which can take one of three forms (**FIGURE 4.5C**). The **vibrio** is a curved rod that resembles a comma. The cholera-causing organism *Vibrio cholerae* is typical. Another spiral form called **spirillum** (pl., spirilla) has a helical shape with a thick, rigid cell wall and flagella that assist movement. The spiral-shaped form known as **spirochete** has a thin, flexible cell wall but no flagella in the traditional sense. Movement in these organisms occurs by contractions of endoflagella that run the length of the cell. The organism causing syphilis, *Treponema pallidum,* typifies a spirochete. Spiral-shaped bacterial cells can be from 1 µm to 100 µm in length.

In addition to the bacillus, coccus, and spiral shapes, other variations exist. Some bacterial species have appendaged bacterial cells while others consist of branching filaments; and some archaeal species have square and star shapes.

CONCEPT AND REASONING CHECKS

4.3   Propose a reason why bacilli do not form tetrads or clusters.

4.3   An Overview to Bacterial and Archaeal Cell Structure

In the last chapter we discovered that bacterial and archaeal cells appear to have little visible structure when observed with a light microscope. This, along with their small size, gave the impression they are homogeneous, static structures with an organization very different from eukaryotic cells.

However, the point was made that bacterial and archaeal species still have all the complex processes typical of eukaryotic cells. It is simply a matter that, in most cases, the structure and sometimes pattern to accomplish these processes is different from the membranous organelles typical of eukaryotic species.

Cell Structure Organizes Cell Function

KEY CONCEPT

4.   Bacterial and archaeal cells are organized at the cellular and molecular levels.

Recent advances in understanding bacterial and archaeal cell biology indicate these organisms exhibit a highly ordered intracellular organization. This organization is centered on three specific processes that need to be carried out (**FIGURE 4.6**). These are:

- **Sensing and responding to the surrounding environment.** Because most bacterial and archaeal cells are surrounded by a cell wall, some pattern of "external structures" is necessary to sense their environment and respond to it or other cells.
- **Compartmentation of metabolism.** As described in Chapter 3, cell metabolism must be segregated from the exterior environment and yet be able to transport materials to and from that environment.

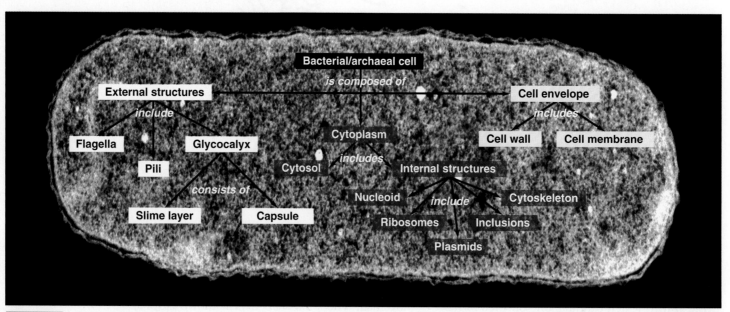

FIGURE 4.6   A Concept Map for Studying Bacterial and Archaeal Cell Structure. Not all cells have all the structures shown here. »» Why can't we see in the TEM image of a bacterial cell all the structures outlined in the concept map?

In addition, protection from osmotic pressure due to water movement into cells must be in place. The "cell envelope" fulfills those roles.

- **Growth and reproduction.** Cell survival demands a complex metabolism that occurs within the aqueous "cytoplasm." These processes and reproduction exist as internal structures or subcompartments localized to specific areas within the cytoplasm.

Our understanding of bacterial and archaeal cell biology is still an emerging field of study. However, there is more to cell structure than previously thought—smallness does not equate with simplicity.

Although this chapter is primarily looking at bacterial cells at the cellular level, it also is important to realize that bacterial and archaeal cells, like their eukaryotic counterparts, are organized on the molecular level as well. Specific cellular proteins can be localized to specific regions of the cell. For example, as the name suggests, *Streptococcus pyogenes* has spherical cells. Yet many of the proteins that confer its pathogenic nature in causing diseases like strep throat are secreted from a specific area of the surface. *Yersinia pestis,* which is the agent responsible for plague, contains a specialized secretion apparatus through which proteins are released. This apparatus only exists on the bacterial surface that is in contact with the target human cells.

So, the cell biology studies are not only important in their own right in understanding cell structure, these studies also may have important significance to clinical microbiology and the fight against infectious disease. As more is discovered about these cells and how they truly differ from eukaryotic cells, the better equipped we will be to develop new antimicrobial agents that will target the subcellular organization of pathogens. In an era when we have fewer effective antibiotics to fight infections, the application of the understanding of cell structure and function may be very important.

On the following pages, we examine some of the common structures found in an idealized bacterial cell, as no single species contains all the structures (**FIGURE 4.7**). Our journey starts by examining the structures on or extending from

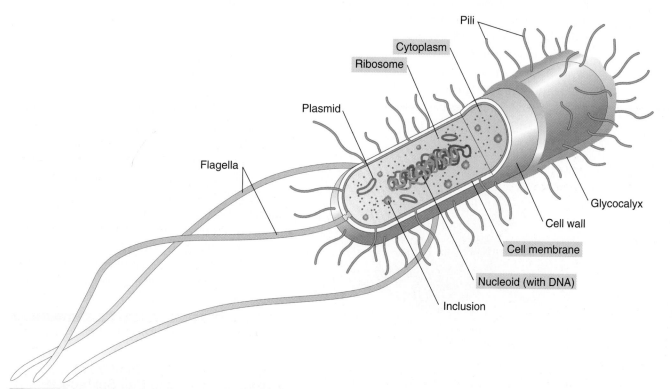

FIGURE 4.7 **Bacterial Cell Structure.** The structural features of a composite, "idealized" bacterial cell. Structures highlighted in blue are found in all bacterial and archaeal species. »» Which structures represent (a) external structures, (b) the cell envelope, and (c) cytoplasmic structures?

the surface of the cell. Then, we examine the cell envelope and spend some time discussing the cell membrane. Our journey then plunges into the cell cytoplasm. All cells must control and coordinate many metabolic processes that need to be separated from one another. We will discover the cytoplasmic subcellular compartmentation that provides this function.

CONCEPT AND REASONING CHECKS

4.4 What is gained by bacterial and archaeal cells being organized into three general sets of structures—external, envelope, and cytoplasmic?

4.4 External Cell Structures

Mucosal:
Referring to the mucous membranes lining many body cavities exposed to the environment.

Virulence factor:
A pathogen-produced molecule or structure that allows the cell to invade or evade the immune system and possibly cause disease.

Bacterial and archaeal cells need to respond to and monitor their external environment. This is made difficult by having a cell wall that "blindfolds" the cell. Many cells have solved this sensing problem by possessing structures that extend from the cell surface into the environment.

Pili Are Protein Fibers Extending from the Cell Surface

KEY CONCEPT

5. Pili allow cells to attach to surfaces or other cells.

Numerous short, thin fibers, called **pili** (sing., pilus; *pilus* = "hair"), protrude from the surface of most gram-negative bacteria (**FIGURE 4.8**). The rigid fibers, composed of protein, act as scaffolding onto which specific adhesive molecules, called **adhesins**, are attached. Therefore, the function of pili is to attach cells to surfaces forming biofilms or, in the case of human pathogens, on human cell and tissue surfaces. This requires that the pili on different bacterial species have specialized adhes-ins to "sense" the appropriate cell. For example, the pili adhesins on *Neisseria gonorrhoeae* cells specifically anchor the cells to the mucosal surface of the urogenital tract whereas the adhesins on *Bordetella pertussis* (causative agent of whooping cough) adhere to cells of the mucosal surface of the upper respiratory tract. In this way, the pili act as a virulence factor by enhancing attachment to host cells, facilitating tissue colonization, and possibly leading to disease development. Without the chemical mooring line lashing the bacterial cells to host cells, it is less likely the cells could infect host tissue (MICROFOCUS 4.2).

Besides these attachment pili, some bacterial species produce flexible **conjugation pili** that establish contact between appropriate cells, facilitating the transfer of genetic material from donor to recipient through a process called conjugation (Chapter 9). Conjugation pili are longer than attachment pili and only one or a few are produced on a cell.

Until recently, attachment pili were thought to be specific to only certain species of gram-negative bacteria. However, research now indicates that extremely thin pili are present on at least some gram-positive bacteria, including the pathogen *Corynebacterium diphtheriae* and *Streptococcus* species. However, very little is known about their function, although they probably play a very similar role to the pili on gram-negative cells.

It should be noted that microbiologists often use the term "pili" interchangeably with "fimbriae" (sing., fimbria; *fimbria* = "fiber").

CONCEPT AND REASONING CHECKS

4.5 What would happen if pili lacked adhesins?

Flagella Are Long Appendages Extending from the Cell Surface

KEY CONCEPT

6. Flagella provide motility.

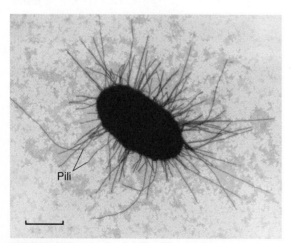

FIGURE 4.8 **Bacterial Pili.** False-color transmission electron micrograph of an *Escherichia coli* cell (blue) with many pili (green). (Bar = 0.5 μm.) »» What function do pili play?

MICROFOCUS 4.2: Public Health
Diarrhea Doozies

They gathered at the clinical research center at Stanford University to do their part for the advancement of science (and earn a few dollars as well). They were the "sensational sixty"—sixty young men and women who would spend three days and nights and earn $300 to help determine whether hair-like structures called pili have a significant place in disease.

A number of nurses and doctors were on hand to help them through their ordeal. The students would drink a fruit-flavored cocktail containing a special diarrhea-causing strain of *Escherichia coli*. Thirty cocktails had *E. coli* with normal pili, while thirty had *E. coli* with pili mutated beyond repair. The hypothesis was that the bacterial cells with normal pili would latch onto intestinal tissue and cause diarrhea, while those with mutated pili would be unable to attach and would be swept away by the rush of intestinal movements and not cause intestinal distress. At least that's what the sensational sixty would either verify or prove false.

On that fateful day in 1997, the experiment began. Neither the students nor the health professionals knew who was drinking the diarrhea cocktail and who was getting the "free pass"; it was a so-called double-blind experiment. Then came the waiting. Some experienced no symptoms, but others felt the bacterial onslaught and clutched at their last remaining vestiges of dignity. For some, it was three days of hell, with nausea, abdominal cramps, and numerous bathroom trips; for others, luck was on their side, and investing in a lottery ticket seemed like a good idea.

When it was all over, the numbers appeared to bear out the hypothesis: The great majority of volunteers who drank the mutated bacterial cells experienced no diarrhea, while the great majority of those who drank the normal bacterial cells had attacks of diarrhea, in some cases real doozies.

All appeared to profit from the experience: The scientists had some real-life evidence that pili contribute to infection; the students made their sacrifice to science and pocketed $300 each; and the local supermarket had a surge of profits from unexpected sales of toilet paper, Pepto-Bismol, and Imodium.

Numerous species in the domains *Bacteria* and *Archaea* are capable of some type of locomotion. This can be in the form of flagellar motility or gliding motility.

Flagellar Motility. Many bacterial and archaeal cells are motile by using remarkable "nanomachines" called **flagella** (sing., flagellum). Depending on the species, one or more flagella may be attached to one or both ends of the cell, or at positions distributed over the cell surface (FIGURE 4.9A).

Flagella range in length from 10 μm to 20 μm and are many times longer than the diameter of the cell. Because they are only about 20 nm thick, they cannot be seen with the light microscope unless stained. However, their existence can be inferred by using dark-field microscopy to watch the live cells dart about.

In the domain *Bacteria*, each flagellum is composed of a helical filament, hook, and basal body (FIGURE 4.9B). The hollow filament is composed of long, rigid strands of protein while the hook attaches the filament to a basal body anchored in the cell membrane and cell wall.

The basal body is an assembly of more than 20 different proteins that form a central rod and set of enclosing rings. Gram-positive bacteria have a pair of rings embedded in the cell membrane and one in the cell wall, while gram-negative bacteria have a pair of rings embedded in the cell membrane and another pair in the cell wall.

In the domain *Archaea*, flagellar protein composition and structure differs from that of the *Bacteria*; motility appears similar though.

The basal body represents a powerful biological motor or rotary engine that generates a propeller-type rotation of the flagellum. The energy for rotation comes from the diffusion of protons (hydrogen ions; H^+) into the cell through proteins associated with the basal body. This energy is sufficient to produce up to 1,500 rpm by the filament, driving the cell forward.

What advantage is gained by cells having flagella? In nature, there are many chemical nutrients in the environment that cells need to survive. Cells will move toward such **attractants** by using their flagella to move up the concentration gradient; that is, toward the attractant. The process is called **chemotaxis**.

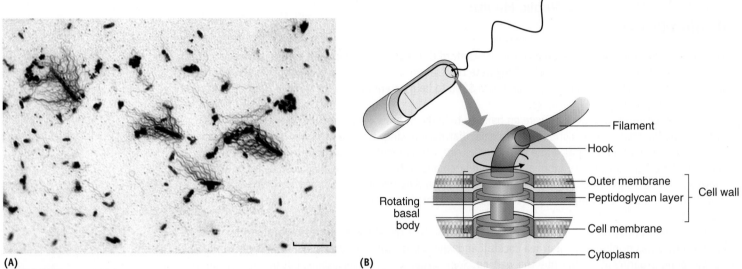

FIGURE 4.9 **Bacterial Flagella. (A)** A light micrograph of *Proteus vulgaris* showing numerous flagella extending from the cell surface. (Bar = 10 μm.) Note that the length of a flagellum is many times the width of the cell. **(B)** The flagellum on a gram-negative bacterial cell is attached to the cell wall and membrane by two pairs of protein rings in the basal body. »» Why is the flagellum referred to as a "nanomachine"?

Temporal sensing:
One that compares the chemical environment and concentration from one moment to the next.

Being so small, the cells sense their chemical surroundings using a temporal sensing system. In the absence of a gradient, the flagella all rotate as a bundle counterclockwise and the cell moves straight ahead in short bursts called "runs" (**FIGURE 4.10A**). These runs can last a few seconds and the cells can move up to 10 body lengths per second (the fastest human can run about 5–6 body lengths per second). A reversal of flagellar rotation (clockwise rotation) causes the cell to "tumble" randomly for a second as the flagella become unbundled. Then, the motor again reverses direction and another run occurs in a new direction.

If an attractant gradient is present, cell behavior changes; cells moving up the gradient now experience longer periods when the motor turns counterclockwise (lengthened runs) and shorter periods when it turns clockwise (shortened tumbles) (**FIGURE 4.10B**). The combined result is a net movement toward the attractant; that is, up the concentration gradient.

Similar types of motile behavior are seen in photosynthetic organisms moving toward light (phototaxis) or other cells moving toward oxygen gas (aerotaxis). MICROFOCUS 4.3 investigates how flagella may have evolved.

One additional type of flagellar organization is found in the spirochetes, a group of gram-negative, coiled bacterial species. The cells are motile by flagella that extend from one or both poles of the cell but fold back along the cell body (**FIGURE 4.11**). Such **endoflagella** and the cell body are surrounded by an outer sheath membrane. Motility results from the torsion generated on the cell by the normal rotation of the flagella. The resulting motility is less regular and more jerky than with flagellar motility.

Gliding Motility. Some bacterial cells can move about without flagella by gliding across a solid surface. The motility occurs along the long axis of bacillus- or filamentous-shaped cells and usually is slower than flagellar motility. The cyanobacteria and myxobacteria (see Chapter 3) are two examples of organisms with gliding motility.

How the cells actually move is not completely understood. It appears that the force for gliding is generated by cytoplasmic proteins (motor proteins) that move along a helical track pushing the cell forward.

CONCEPT AND REASONING CHECKS

4.6 Explain how flagella move cells during a "run."

The Glycocalyx Is an Outer Layer External to the Cell Wall

KEY CONCEPT

7. A glycocalyx protects against desiccation, attaches cells to surfaces, and helps pathogens evade the immune system.

Many bacterial species secrete an adhering layer of polysaccharides, or polysaccharides and small

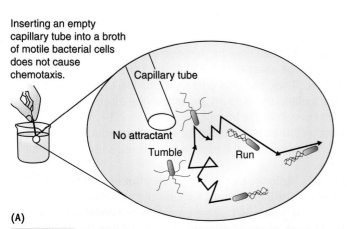

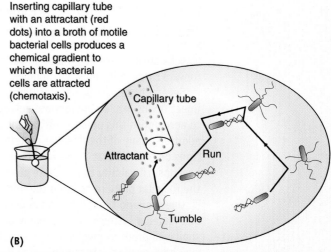

FIGURE 4.10 **Chemotaxis.** Chemotaxis represents a behavioral response to chemicals. (**A**) Rotation of the flagellum counterclockwise causes the bacterial cell to " run," while rotation of the flagellum clockwise causes the bacterial cell to "tumble," as shown. (**B**) During chemotaxis to an attractant, such as sugar, flagellum behavior leads to longer runs and fewer tumbles, which will result in biased movement toward the attractant. »» Predict the behavior of a bacterial cell if it sensed a repellant; that is a potential harmful or lethal chemical.

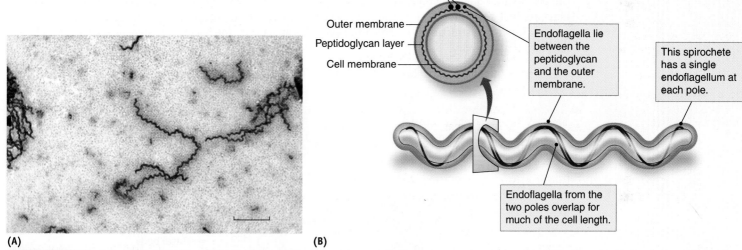

FIGURE 4.11 **The Spirochete Endoflagella.** (**A**) A light micrograph of *Treponema pallidum* shows the corkscrew-shaped spirochete cell. (Bar = 10 μm.) (**B**) Diagram showing the positioning of endoflagella in a spirochete. »» How are endoflagella different from true bacterial flagella?

proteins, called the **glycocalyx** (*glyco* = "sweet"; *calyx* = "coat"). The layer can be thick and covalently bound to the cell, in which case it is known as a **capsule**. A thinner, loosely attached layer is referred to as a **slime layer**. Colonies containing cells with a glycocalyx appear moist and glistening. The actual capsule can be seen by light microscopy when observing cells in a negative stain preparation or by transmission electron microscopy (**FIGURE 4.12**).

The glycocalyx serves as a buffer between the cell and the external environment. Because of its high water content, the glycocalyx can protect cells from desiccation. Another major role of the gly-

cocalyx is to allow the cells to attach to surfaces. The glycocalyx of *V. cholerae*, for example, permits the cells to attach to the intestinal wall of the host. The glycocalyx of pathogens therefore represents another virulence factor.

Other encapsulated pathogens, such as *Streptococcus pneumoniae* (a principal cause of bacterial pneumonia) and *Bacillus anthracis,* evade the immune system because they cannot be easily engulfed by white blood cells during phagocytosis. Scientists believe the repulsion between bacterial cell and phagocyte is due to strong negative charges on the capsule and phagocyte surface.

Encapsulated:
A cell having a capsule.

Phagocytosis:
A process whereby certain white blood cells (phagocytes) engulf foreign matter and often destroy microorganisms.

MICR⊙F⊙CUS 4.3: Evolution

The Origin of the Bacterial Flagellum

Flagella are an assembly of protein parts forming a rotary engine that, like an outboard motor, propel the cell forward through its moist environment. Recent work has shown how such a nanomachine may have evolved.

Several bacterial species, including *Yersinia pestis,* the agent of bubonic plague, contain structures to inject toxins into an appropriate eukaryotic host cell. These bacterial cells have a hollow tube or needle to accomplish this process, just as the bacterial flagellum and filament are hollow (see diagram below). In addition, many of the flagellar proteins are similar to part of the injection proteins. In 2004, investigations discovered that *Y. pestis* cells actually contain all the genes needed for a flagellum—but the cells have lost the ability to use these genes. *Y. pestis* is nonmotile and it appears that the cells use a subset of the flagellar proteins to build the injection device.

One scenario then is that an ancient cell evolved a structure that was the progenitor of the injection and flagellar systems. In fact, many of the proteins in the basal body of flagellar and injection systems are similar to proteins involved in proton (hydrogen ion; H^+) transport. Therefore, a proton transport system may have evolved into the injection device and, through diversification events, evolved into the motility structure present on many bacterial cells today.

The fascinating result of these investigations and proposals is it demonstrates that structures can evolve from other structures with a different function. It is not necessary that evolution "design" a structure from scratch but rather it can modify existing structures for other functions.

Individuals have proposed that the complexity of structures like the bacterial flagellum are just too complex to arise gradually through a step-by-step process. However, the investigations being conducted illustrate that a step-by-step evolution of a specific structure is not required. Rather, there can be cooperation, where one structure is modified to have other functions. The bacterial flagellum almost certainly falls into that category.

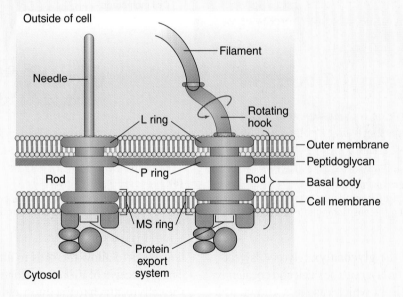

A bacterial injection device (left) compared to a bacterial flagellum (right).
Both have a protein export system in the base of the basal body.

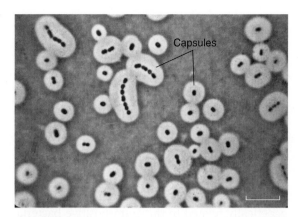

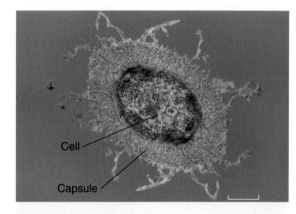

(A)   **(B)**

FIGURE 4.12 The Bacterial Glycocalyx. (**A**) Demonstration of the presence of a capsule in an *Acinetobacter* species by negative staining and observed by phase-contrast microscopy. (Bar = 10 µm.) (**B**) A false-color transmission electron micrograph of *Escherichia coli*. The cell is surrounded by a thick capsule (pink). (Bar = 0.5 µm.) »» How does the capsule provide protection for the bacterial cell?

A slime layer usually contains a mass of tangled fibers of a polysaccharide called dextran (see Chapter 2). The fibers attach the bacterial cell to tissue surfaces. A case in point is *Streptococcus mutans,* an important cause of tooth decay previously discussed in MicroFocus 2.4. This species forms dental plaque, which represents a type of biofilm on the tooth surface. TEXTBOOK CASE 4 (p. 116) details a medical consequence of a biofilm.

CONCEPT AND REASONING CHECKS

4.7 Under what circumstances might it be advantageous to a bacterial cell to have a capsule rather than a slime layer?

4.5 The Cell Envelope

The **cell envelope** is a complex structure that forms the two "wrappers"—the **cell wall** and the **cell membrane**—surrounding the cell cytoplasm. The cell wall is relatively porous to the movement of substances whereas the cell membrane regulates transport of nutrients and metabolic products.

The Bacterial Cell Wall Is a Tough and Protective External Shell

KEY CONCEPT

8. Bacterial cell walls help maintain cell shape and protect the cell membrane from rupture.

The fact that most all bacterial and archaeal cells have a cell wall suggests the critical role this structure must play. By covering the entire cell surface, the cell wall acts as an exoskeleton to protect the cell from injury and damage. It helps, along with the cytoskeleton (see Section 4.6), to maintain the shape of the cell and reinforce the cell envelope against the high intracellular water (osmotic) pressure pushing against the cell membrane. As described in Chapter 3, most microbes live in an environment where there are more dissolved materials inside the cell than outside. This hypertonic condition in the cell means water diffuses inward, accounting for the increased osmotic pressure. Without a cell wall, the cell would rupture or undergo **lysis** (**FIGURE 4.13**). It is similar to blowing so much air into a balloon that the air pressure bursts the balloon.

The bacterial cell wall differs markedly from the walls of archaeal cells and cells of eukaryotic

Hypertonic:
A solution with more dissolved material (solutes) than the surrounding solution.

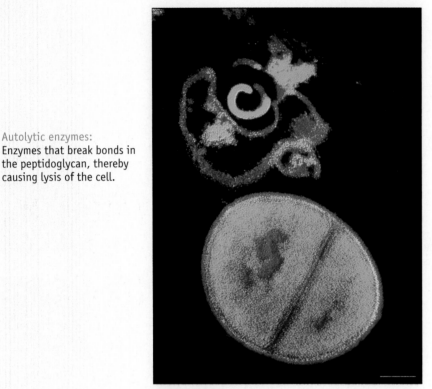

FIGURE 4.13 Cell Rupture (Lysis). A false-color electron micrograph showing the lysis of a *Staphylococcus aureus* cell. The addition of the antibiotic penicillin interferes with the construction of the peptidoglycan in new cells, and they quickly burst (top cell). (Bar = 0.25 μm.) »» Where is the concentration of dissolved substances (solutes) higher, inside the cell or outside? Explain how this leads to cell lysis.

Autolytic enzymes:
Enzymes that break bonds in the peptidoglycan, thereby causing lysis of the cell.

Endotoxin:
A poison that can activate inflammatory responses, leading to high fever, shock, and organ failure.

microorganisms (algae and fungi) in containing **peptidoglycan**, which is a network of disaccharide chains (glycan strands) cross-linked by short peptides (**FIGURE 4.14A**). Each disaccharide in this very large molecule is composed of two monosaccharides, N-acetylglucosamine (NAG) and N-acetylmuramic acid (NAM) (see Chapter 2). The carbohydrate backbone can occur in multiple layers connected by side chains of four amino acids and peptide cross-bridges.

There is more to a bacterial cell wall than just peptidoglycan, so several forms of cell wall architecture exist.

Gram-Positive Walls. Most gram-positive bacterial cells have a very thick, rigid peptidoglycan cell wall (**FIGURE 4.14B**). The abundance and thickness (25 nm) of this material may be one reason why they retain the crystal violet in the Gram stain (see Chapter 3). The multiple layers of glycan strands are cross-linked to one another both in the same layer as well as between layers.

The gram-positive cell wall also contains a sugar-alcohol called **teichoic acid**. Wall teichoic acids, which are bound to the glycan chains, are essential for cell viability—if the genes for teichoic acid synthesis are deleted, cell death occurs. Still, the function of the teichoic acids remains unclear. They may help maintain a surface charge on the cell wall, control the activity of autolytic enzymes acting on the peptidoglycan, and/or maintain permeability of the cell wall layer.

The bacterial genus *Mycobacterium* is phylogenetically related to the gram-positive bacteria. However, these rod-shaped cells have evolved another type of wall architecture to protect the cell membrane from rupture. The cell wall is composed of a waxy lipid called **mycolic acid** that is arranged in two layers that are covalently attached to the underlying peptidoglycan. Such a hydrophobic layer is impervious to the Gram stains, so stain identification of *M. tuberculosis* is carried out using the acid-fast stain procedure (see Chapter 3).

Gram-Negative Walls. The cell wall of gram-negative bacterial cells is structurally quite different from that of the gram-positive wall (**FIGURE 4.14C**). The peptidoglycan layer is two-dimensional; the glycan strands compose just a single layer or two. This is one reason why it loses the crystal violet dye during the Gram stain. Also, there is no teichoic acid present.

The unique feature of the gram-negative cell wall is the presence of an **outer membrane**, which is separated by a gap, called the **periplasm**, from the cell membrane. This gel-like compartment contains digestive enzymes and transport proteins to speed entry of nutrients into the cell. The peptidoglycan layer is located in the periplasm and attached to lipoproteins in the cell membrane.

The inner half of the outer membrane contains phospholipids similar to the cell membrane. However, the outer half is composed primarily of **lipopolysaccharide (LPS)**, which consists of polysaccharide attached to a unique lipid molecule known as **lipid A**. The so-called O polysaccharide is used to identify variants of a species (e.g., strain O157:H7 of *E. coli*). On cell death, lipid A is released and represents an endotoxin that can be toxic if ingested (Chapter 19).

The outer membrane also contains unique proteins called **porins**. These proteins form pores in the outer membrane through which small,

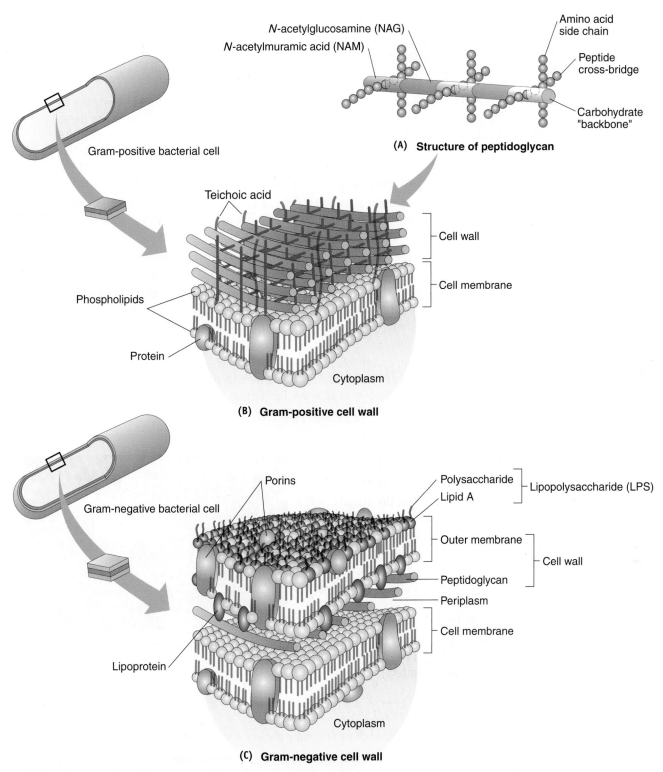

FIGURE 4.14 A Comparison of the Cell Walls of Gram-Positive and Gram-Negative Bacterial Cells. (**A**) The structure of peptidoglycan is shown as units of NAG and NAM joined laterally by amino acid cross-bridges and vertically by side chains of four amino acids. (**B**) The cell wall of a gram-positive bacterial cell is composed of peptidoglycan layers combined with teichoic acid molecules. (**C**) In the gram-negative cell wall, the peptidoglycan layer is much thinner, and there is no teichoic acid. Moreover, an outer membrane overlies the peptidoglycan layer such that both comprise the cell wall. Note the structure of the outer membrane in this figure. It contains porin proteins and the outer half is unique in containing lipopolysaccharide. »» Simply based on cell wall structure, assess the potential of gram-positive and gram-negative cells as pathogens.

Textbook CASE 4

An Outbreak of *Enterobacter cloacae* Associated with a Biofilm

Hemodialysis is a treatment for people with severe chronic kidney disease (kidney failure). The treatment filters the patient's blood to remove wastes and excess water. Before a patient begins hemodialysis, an access site is created on the lower part of one arm. Similar to an intravenous (IV) site, a tiny tube runs from the arm to the dialysis machine. The patient's blood is pumped through the dialysis machine, passed through a filter or artificial kidney called a dialyzer, and the cleaned blood returned to the patient's body at the access site. The complete process can take 3 to 4 hours.

1 During September 1995, a patient at an ambulatory hemodialysis center in Montreal, Canada received treatment on a hemodialysis machine to help relieve the effects of kidney disease. The treatment was performed without incident.

2 The next day, a second patient received treatment on the same hemodialysis machine. His treatment also went normally, and he returned to his usual activities after the session was completed.

3 In the following days, both patients experienced bloodstream infections (BSIs). They had high fever, muscular aches and pains, sore throat, and impaired blood circulation. Because the symptoms were severe, the patients were hospitalized. Both patients had infections of *Enterobacter cloacae*, a gram-negative rod.

4 In the following months, an epidemiological investigation reviewed other hemodialysis patients at that center. In all, seven additional adult patients were identified who had used the same hemodialysis machine. They discovered all seven had similar BSIs.

5 Inspection of the hemodialysis machine used by these nine patients indicated the presence of biofilms containing *Enterobacter cloacae*, which was identical to those samples taken from the patients' bloodstreams (see figure).

6 Further study indicated that the dialysis machine was contaminated with *E. cloacae*, specifically where fluid flows.

7 It was discovered that hospital personnel were disinfecting the machines correctly. The problem was that the valves in the drain line were malfunctioning, allowing a backflow of contaminated material.

8 Health officials began a hospital education program to ensure that further outbreaks of infection would be minimized.

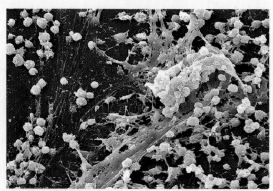

Similar to the description in this textbook case, biofilms consisting of *Staphylococcus* cells can contaminate hemodialysis machines.

Questions:

(Answers can be found in Appendix D.)

A. Suggest how the hemodialysis machine originally became contaminated.

B. Why weren't the other five cases of BSI correlated with the hemodialysis machine until the epidemiological investigation was begun?

C. How could future outbreaks of infection be prevented?

For additional information see http://www.cdc.gov/mmwr/preview/mmwrhtml/00051244.htm.

TABLE 4.2 A Comparison of Gram-Positive and Gram-Negative Cell Walls		
Characteristic	Gram Positive	Gram Negative
Peptidoglycan	Yes, thick layer	Yes, thin layer
Teichoic acids	Yes	No
Outer membrane	No	Yes
Lipopolysaccharides (LPS)	No	Yes
Porin proteins	No	Yes
Periplasm	No	Yes

hydrophilic molecules (sugars, amino acids, some ions) pass into the periplasm. Larger, hydrophobic molecules cannot pass, partly accounting for the resistance of gram-negative cells to many antimicrobial agents, dyes, disinfectants, and lysozyme.

Before leaving the bacterial cell walls, a brief mention should be made of bacterial species that lack a cell wall. The mycoplasmas are a wall-less genus that is again phylogenetically related to the gram-positive bacteria. Taxonomists believe that the mycoplasmas once had a cell wall but lost it because of their parasitic relationship with their host. To help protect the cell membrane from rupture, the mycoplasmas are unusual in containing sterols in the cell membrane (see Chapter 2). TABLE 4.2 summarizes the major differences between the two major types of bacterial cell walls.

CONCEPT AND REASONING CHECKS

4.8. Penicillin and lysozyme primarily affect peptidoglycan synthesis in gram-positive bacterial cells. Why are these agents less effective against gram-negative bacterial cells?

The Archaeal Cell Wall Also Provides Mechanical Strength

KEY CONCEPT

9. Archaeal cell walls have crystalline layers.

Archaeal species vary in the type of wall they possess. None have the peptidoglycan typical of the *Bacteria*. Some species have a **pseudopeptidoglycan** where the NAM is replaced by N-acetyltalosamine uronic acid (NAT). Other archaeal cells have walls made of polysaccharide, protein, or both.

The most common cell wall among archaeal species is a surface layer called the **S-layer**. It consists of hexagonal patterns of protein or glycoprotein that self-assemble into a crystalline lattice 5 nm to 25 nm thick.

Although the walls may be structurally different and the molecules form a different structural pattern, the function is the same as in bacterial species—to provide mechanical support and prevent osmotic lysis.

CONCEPT AND REASONING CHECKS

4.9 Distinguish between peptidoglycan and pseudopeptidoglycan cell walls.

The Cell Membrane Represents the Interface between the Cell Environment and the Cell Cytoplasm

KEY CONCEPT

10. Molecules and ions cross the cell membrane by facilitated diffusion or active transport.

A **cell** (or **plasma**) **membrane** is a universal structure that separates external from internal (cytoplasmic) environments, preventing soluble materials from simply diffusing into and out of the cell. One exception is water, which due to its small size and overall lack of charge can diffuse slowly across the membrane.

The bacterial cell membrane, which is about 7 nm thick, is 40% phospholipid and 60% protein. In illustrations, the cell membrane appears very rigid (FIGURE 4.15). In reality, it is quite fluid, having the consistency of olive oil. This means the mosaic of phospholipids and proteins are not cemented in place, but rather they can move

Hydrophilic:
Pertaining to molecules or parts of molecules that are soluble in water.

Hydrophobic:
Pertaining to molecules or parts of molecules that are not soluble in water.

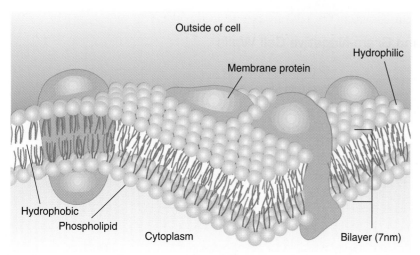

FIGURE 4.15 **The Structure of the Bacterial Cell Membrane.** The cell membrane of a bacterial cell consists of a phospholipid bilayer in which are embedded integral membrane proteins. Other proteins and ions may be associated with the integral proteins or the phospholipid heads. »» Why is the cell membrane referred to as a fluid mosaic structure?

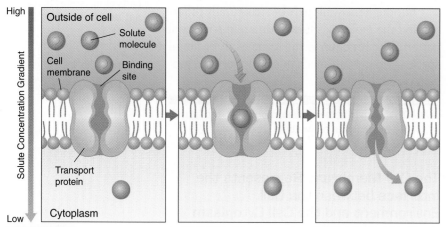

FIGURE 4.16 **Facilitated Transport through a Membrane Protein.** Many transport proteins facilitate the diffusion of nutrients across the lipid bilayer. The transport protein forms a hydrophilic channel through which a specific solute can diffuse. »» Why would a solute move through a membrane protein rather than simply across the lipid bilayer?

pokes holes in the bilayer, while some detergents and alcohols dissolve the bilayer. Such action allows the cytoplasmic contents to leak out of bacterial cells, resulting in death through cell lysis.

A diverse population of membrane proteins populates the phospholipid bilayer. These membrane proteins often have stretches of hydrophobic amino acids that interact with the hydrophobic fatty acid chains in the membrane. These proteins span the width of the bilayer and are referred to as "integral membrane proteins". Other proteins, called "peripheral membrane proteins", are associated with the polar heads of the bilayer.

The membrane proteins carry out numerous important functions. Some represent enzymes needed for cell wall synthesis or for energy metabolism. As mentioned, bacterial and archaeal cells lack mitochondria and part of that organelle's function is carried out by the cell membrane. Other membrane proteins help anchor the DNA to the membrane during replication or act as receptors of chemical information, sensing changes in environment conditions and triggering appropriate responses.

Perhaps the largest group of integral membrane proteins is involved as transporters of charged solutes, such as amino acids, simple sugars, nitrogenous bases, and ions across the lipid bilayer. The transport systems are highly specific though, only transporting a single molecular species or a very similar class of molecules. Therefore, there are many different transport proteins to regulate the diverse molecular traffic that must flow into or out of a cell.

The transport process can be passive or active. In **facilitated diffusion**, integral membrane proteins facilitate the movement of materials down their concentration gradient; that is, from an area of higher concentration to one of lower concentration (**FIGURE 4.16**). By acting as a conduit for diffusion or as a transporter through the hydrophobic bilayer, hydrophilic solutes can enter or leave without the need for cellular energy.

Unlike facilitated diffusion, **active transport** allows different concentrations of solutes to be established outside or inside of the cell against the concentration gradient. These membrane proteins act as "pumps" and, as such, demand an energy input from the cell. Cellular processes such as cell energy production and flagella rotation also depend on active transport.

laterally in the membrane. This dynamic model of membrane structure therefore is called the **fluid mosaic model**.

The phospholipid molecules, typical of most biological membranes, are arranged in two parallel layers (a bilayer) and represent the barrier function of the membrane. The phospholipids contain a charged phosphate head group attached to two hydrophobic fatty acid chains (see Chapter 2). The fatty acid "tails" are the portion that forms the permeability barrier. In contrast, the hydrophilic head groups are exposed to the aqueous external or cytoplasmic environments.

Several antimicrobial substances act on the membrane bilayer. The antibiotic polymyxin

CONCEPT AND REASONING CHECKS

4.10 Justify the necessity for phospholipids and proteins in the cell membrane.

The Archaeal Cell Membrane Differs from Bacterial and Eukaryal Membranes

KEY CONCEPT

11. Archaeal membranes are structurally unique.

Besides the differences in gene sequences for ribosomal RNA in the domain *Archaea*, another major difference used to separate the archaeal organisms into their own domain is the chemical nature of the cell membrane.

The manner in which the hydrophobic lipid tails are attached to the glycerol is different in the *Archaea*. The tails are bound to the glycerol by "ether linkages" rather than the "ester linkages" found in the domains *Bacteria* and *Eukarya* (**FIGURE 4.17A**).

Also, typical fatty acid tails are absent from the membranes; instead, repeating five-carbon units are linked end-to-end to form lipid tails longer than the fatty acid tails. The result is a lipid **monolayer** rather than a bilayer (**FIGURE 4.17B**). This provides an advantage to the hyperthermophiles by preventing a peeling in two of the membrane, which would occur with a typical phospholipid bilayer structure.

CONCEPT AND REASONING CHECKS

4.11 What is unique about archaeal membrane structure?

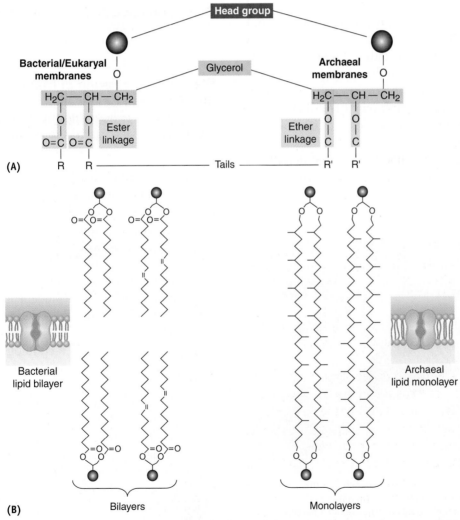

FIGURE 4.17 **Structure of Cell Membranes.** (**A**) Bacterial and eukaryal cell membranes involve an ester linkage joining the glycerol to the fatty acid tails (R) while archaeal membranes have an ether linkage to the isoprenoid tails (R'). (**B**) Bacterial and eukaryal membranes form a bilayer while archaeal membranes are a monolayer. »» What identifies an ester linkage from an ether linkage?

4.6   The Cell Cytoplasm and Internal Structures

The cell membrane encloses the **cytoplasm**, which is the compartment within which most growth and metabolism occurs. The cytoplasm consists of the **cytosol**, a semifluid mass of proteins, amino acids, sugars, nucleotides, salts, vitamins, and ions—all dissolved in water (see Chapter 2)—and several bacterial structures or subcompartments, each with a specific function.

The Nucleoid Represents a Subcompartment Containing the Chromosome

KEY CONCEPT

12. The nucleoid contains the cell's essential genetic information.

The chromosome region in bacterial and archaeal cells appears as a diffuse mass termed the **nucleoid** (FIGURE 4.18). The nucleoid does not contain a covering or membrane; rather, it represents a central subcompartment in the cytoplasm where the DNA aggregates and ribosomes are absent. Usually there is a single chromosome per cell and, with few exceptions, exists as a closed loop of DNA and protein.

The DNA contains the essential hereditary information for cell growth, metabolism, and reproduction. Because most cells only have one chromosome, the cells are genetically haploid. Unlike eukaryotic microorganisms and other eukaryotes, the nucleoid and chromosome do not undergo mitosis and having but the one set of genetic information cannot undergo meiosis.

The complete set of genes in an organism, called the **genome**, varies by species. For example, the genome of *E. coli*, a typical bacterial species in the mid-size range, contains about 4,300 genes. In all cases, these genes determine what proteins and enzymes the cell can make; that is, what metabolic reactions and activities can be carried out. For *E. coli*, this equates to some 2,000 different proteins. Extensive coverage of bacterial DNA is presented in Chapter 8.

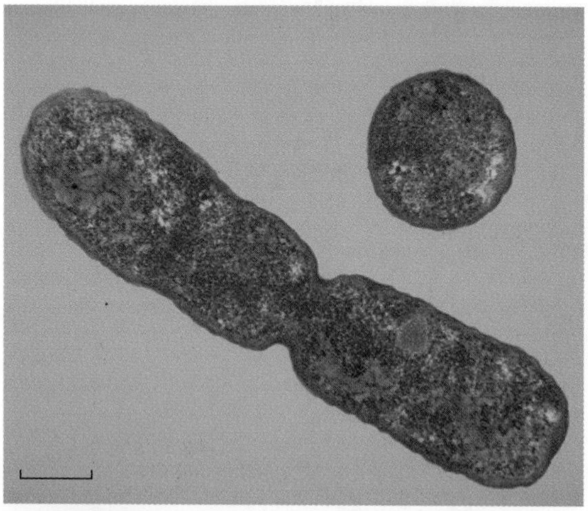

CONCEPT AND REASONING CHECKS

4.12 Why do we say that the bacterial chromosome contains the "essential hereditary information"?

Plasmids Are Found in Many Bacterial and Archaeal Cells

KEY CONCEPT

13. Plasmids contain nonessential genetic information.

Besides a nucleoid, many bacterial and archaeal cells also contain smaller molecules of DNA called **plasmids**. About a tenth the size of the chromosome, these stable, extrachromosomal DNA molecules exist as closed loops containing 5 to 100 genes. There can be one or more plasmids in a cell and these may contain similar or different genes. Plasmids replicate independently of the chromosome and can be transferred between cells during recombination. They also represent important vectors in industrial technologies that use genetic engineering. Both topics are covered in Chapter 9.

Although plasmids may not be essential for cellular growth, they provide a level of genetic flexibility. For example, some plasmids possess genes for disease-causing toxins and many carry genes for chemical or antibiotic resistance. For this latter reason, these genetic elements often are called **R plasmids** (R for resistance).

CONCEPT AND REASONING CHECKS

4.13 What properties distinguish the bacterial chromosome from a plasmid?

Haploid:
Having a single set of genetic information.

Vectors:
Genetic elements capable of incorporating and transferring genetic information.

FIGURE 4.18 **The Bacterial Nucleoid.** In this false-color transmission electron micrograph of *Escherichia coli*, nucleoids (orange) occupy a large area in a bacterial cell. Both longitudinal (center cells) and cross sections (cell upper right) of *E. coli* are visable. (Bar = 0.5 μm.) »» How does a nucleoid differ from the eukaryotic cell nucleus described in the previous chapter?

Other Subcompartments Exist in the Cell Cytoplasm

KEY CONCEPT

14. Ribosomes, microcompartments, and inclusions carry out specific intracellular functions.

For a long time the cytoplasm was looked at as a bag enclosing the genetic machinery and biochemical reactions. As more studies were carried out with the electron microscope and with biochemical techniques, it became evident that the cytoplasm contained "more than meets the eye."

Ribosomes. One of the universal cell structures mentioned in Chapter 3 was the **ribosome**. There are hundreds of thousands of these nearly spherical particles in the cell cytoplasm, which gives it a granular appearance when viewed with the electron microscope (**FIGURE 4.19A**). Their relative size is measured by how fast they settle when spun in a centrifuge. Measured in Svedberg units (S), bacterial and archaeal ribosomes represent 70S particles.

The ribosomes are built from RNA and protein and are composed of a small subunit (30S) and a large subunit (50S) (**FIGURE 4.19B**). For proteins to be synthesized, the two subunits come together to form a 70S functional ribosome (Chapter 8). Some antibiotics, such as streptomycin and tetracycline, prevent bacterial and archaeal ribosomes from carrying out protein synthesis.

Microcompartments. Recently, some bacterial species have been discovered that contain **microcompartments**. The microcompartments appear to be unique to the *Bacteria* and consist of a polyprotein shell 100 to 200 nm in diameter (see Chapter 3). The shell surrounds various types of enzymes and, in the cyanobacteria, microcompartments called "carboxysomes" function to enhance carbon dioxide fixation. In some non-photosynthetic species, microcompartments limit diffusion of volatile or toxic metabolic products.

Inclusions. Cytoplasmic structures, called **inclusions**, can be found in the cytoplasm. Many of these bodies store nutrients or the monomers for cellular structures. For example, some inclusions consist of aggregates or granules of polysaccharides (glycogen), globules of elemental sulfur, or lipid. Other inclusion bodies can serve as important identification characters for bacterial pathogens. One example is the diphtheria bacilli that contain **metachromatic granules**, or **volutin**, which are deposits of polyphosphate (long chains of inorganic phosphate) along with calcium and other ions. These granules stain with dyes such as methylene blue.

Some aquatic and marine forms float on the water surface, which is made possible by the presence of **gas vesicles**, cytoplasmic compartments built from a water-tight protein shell. These vesicles decrease the density of the cell, which generates and regulates their buoyancy.

Centrifuge:
An instrument that spins particles suspended in liquid at high speed.

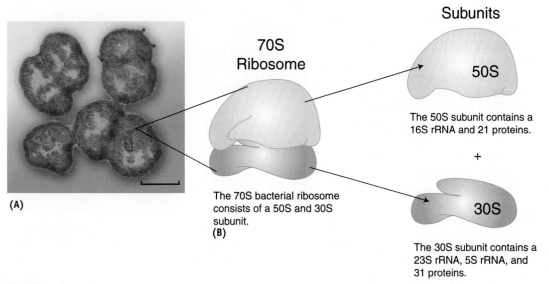

Subunits

70S Ribosome

50S

The 50S subunit contains a 16S rRNA and 21 proteins.

+

The 70S bacterial ribosome consists of a 50S and 30S subunit.

(B)

30S

The 30S subunit contains a 23S rRNA, 5S rRNA, and 31 proteins.

(A)

FIGURE 4.19 The Bacterial Ribosome. (A) A false-color electron microscope image of *Neisseria gonorrhoeae*, showing the nucleoid (yellow) and ribosomes (blue). (Bar = 1μm.) **(B)** The functional 70S ribosome is assembled from a small (30S) and large (50S) subunit. »» How many rRNA molecules and proteins construct a 70S ribosome?

The **magnetosome**, another type of inclusion or subcompartment, is described in MicroFocus 4.4. These bacterial inclusions are invaginations of the cell membrane, which are coordinated and positioned by cytoskeletal filaments similar to eukaryotic microfilaments.

CONCEPT AND REASONING CHECKS

4.14 Provide the roles for the subcompartments found in bacterial cells.

Many Bacterial/Archaeal Cells Have a "Cytoskeleton"

KEY CONCEPT

15. Cytoskeletal proteins regulate cell division and help determine cell shape.

Until recently, the dogma was that bacterial and archaeal cells lacked a cytoskeleton, which is a common feature in eukaryotic cells (see Chapter 3). However, it now appears cytoskeletal

---

MICROFOCUS 4.4: Environmental Microbiology
A "Not So Fatal" Attraction

To get from place to place, humans often require the assistance of maps, GPS systems, or gas station attendants. In the microbial world, life is generally more simple, and traveling is no exception.

In the early 1980s, Richard P. Blakemore and his colleagues at the University of New Hampshire observed mud-dwelling bacterial cells gathering at the north end of water droplets. On further study, they discovered each cell had a chain of aligned magnetic particles acting like a compass directing the organism's movements (magnetotaxis). Additional interdisciplinary investigations by microbiologists and physicists have shown the magnetotactic bacteria contain a linear array of 15–20 membrane-bound vesicles along the cell's long axis (see figure). Each vesicle, called a **magnetosome**, is an invagination of the cell membrane and contains the protein machinery to nucleate and grow a crystal of magnetite (Fe_3O_4) or greigite (Fe_3S_4). The chain of magnetosomes is organized by filaments that are similar to eukaryotic actin. By running parallel to the magnetosome chain, the filaments organize the vesicles into a chain. As each vesicle accumulates the magnetite or greigite crystal to form a magnetosome, magnetostatic interactions between vesicles stabilizes the linear aggregation.

To date, all magnetotactic bacterial cells are motile, gram-negative cells common in aquatic and marine habitats, including sediments where oxygen is absent. This last observation is particularly noteworthy because it explains why these organisms have magnetosomes.

It originally was thought that magnetotaxis was used to guide cells to those regions of the habitat with no oxygen; in other words, they travel downward toward the sediment. More recent studies have shown that some magnetotactic bacteria actually prefer low concentrations of oxygen. So the opinion now is that both magnetotaxis and aerotaxis work together to allow cells to "find" the optimal point within an oxygen gradient. This "not so fatal" attraction permits the bacteria to reach a sort of biological nirvana and settle in for a life of environmental bliss.

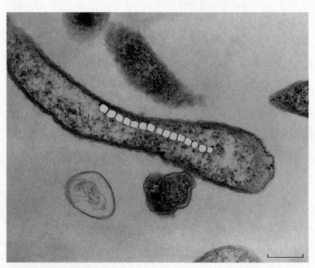

Bacterial magnetosomes (yellow) are seen in this false-color transmission electron micrograph of a magnetotactic marine spirillum. (Bar = 1 μm.)

proteins homologous to those in the eukaryotic cytoskeleton are present.

The first protein discovered was a homolog of the eukaryotic protein tubulin, which forms filaments that assemble into microtubules. The homolog forms filaments similar to those in microtubules but the filaments do not assemble into microtubules. These tubulin-like proteins have been found in all bacterial and archaeal cells examined and appear to function in the regulation of cell division. During this process, the protein localizes around the neck of the dividing cell where it recruits other proteins needed for the deposition of a new cell wall between the dividing cells (MICROFOCUS 4.5).

Protein homologs remarkably similar in three-dimensional structure to eukaryotic microfilaments assemble into filaments that help determine cell shape in *E. coli* and *Bacillus subtilis*. These homologs have been found in most non-spherical cells where they form a helical network beneath the cell membrane to guide the proteins involved in cell wall formation (**FIGURE 4.20A**). The homologs also are involved with chromosome segregation during cell division and magnetosome formation.

Intermediate filaments (IF), another component of the eukaryotic cytoskeleton in some metazoans, have a homolog as well. The protein, called crescentin, helps determine the characteristic crescent shape of *Caulobacter crescentus* cells.

Homolog:
An entity with similar attributes due to shared ancestry.

Metazoans:
Members of the vertebrates, nematodes, and mollusks.

MICROFOCUS 4.5: Public Health
The Wall-less Cytoskeleton

Sometimes the lack of something can speak loudly. Take for example the mycoplasmas such as *Mycoplasma pneumoniae* that causes primary atypical pneumonia (walking pneumonia). This, as well as other *Mycoplasma* species, lack a cell wall. How then can they maintain a defined cell shape (see figure)?

Transmission electron microscopy has revealed that mycoplasmal cells contain a very complex cytoskeleton and further investigations indicate the cytoskeletal proteins are very different from the typical cytoskeletal homologs found in other groups of the Firmicutes. For example, *Spiroplasma citri,* which causes infections in other animals, has a fibril protein cytoskeleton that is laid down as a helical ribbon. This fibril protein has not been found in any other organisms. Because the cells are spiral shaped, the ribbon probably is laid down in such a way to determine cell shape, suggesting that shape does not have to be totally dependent on a cell wall.

In *Mycoplasma genitalium,* which is closely related to *M. pneumoniae* and causes human urethral infections, a eukaryotic-like tubulin homolog has been identified, but none of the other proteins have been identified that it recruits at the division neck for cell division. Surprising? Not really. Important? Immensely! Because mycoplasmas do not have a cell wall, why would they require those proteins that lay down a peptidoglycan cross wall between cells? So the lack of something (wall-forming proteins) tells us what those proteins must do in their gram-positive relatives that do have walls.

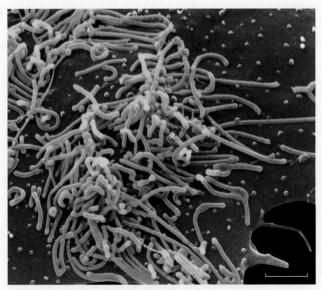

False-color scanning electron micrograph of *Mycoplasma pneumoniae* cells. (Bar = 2.5 µm.)

(A)

FIGURE 4.20 **A Bacterial Cytoskeleton.** Bacterial cells have proteins similar to those that form the eukaryotic cytoskeleton. **(A)** Microfilament-like proteins form helical filaments that curve around the edges of these cells of *Bacillus subtilis*. (Bar = 1.5 μm.) **(B)** Three-dimensional model of a helical *Caulobacter crescentus* cell (green) with a helical cytoskeletal filament of crescentin (pink). »» What would be the shape of these cells without the cytoskeletal proteins?

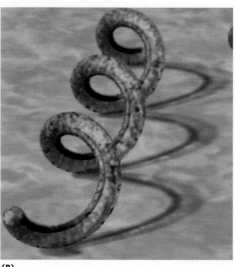

(B)

In older cells that become filamentous, crescentin maintains the helical shape of the cells by aligning with the inner cell curvature beneath the cytoplasmic membrane (**FIGURE 4.20B**).

Even though the evolutionary relationships are quite distant between bacterial/archaeal and eukaryotic cytoskeletal proteins based on protein sequence data, the similarity of their three-dimensional structure and function is strong evidence supporting homologous cytoskeletons.

CONCEPT AND REASONING CHECKS

4.15 Evaluate the relationship between the eukaryotic cytoskeleton and the cytoskeletal protein homologs in bacterial and archaeal cells.

4.7   The Bacteria/Eukaryote Paradigm—Revisited

TABLE 4.3 summarizes the structural features of bacterial and archaeal cells. One of the take-home lessons from the table, and discussions of cell structure and function explored in this chapter (and the initial discussion of the bacteria/eukaryote paradigm in Chapter 3), is the ability of these organisms to carry out the "complex" metabolic and biochemical processes typically associated with eukaryotic cells—usually without the need for elaborate membrane-enclosed subcompartments.

What Is a Prokaryote?

KEY CONCEPT

16. Cellular processes in bacterial cells can be similar to those in eukaryotic cells.

Earlier in this chapter, the intricate subcellular compartmentation was discussed for several cell structures. What about other major cellular processes such as making proteins? This requires two processes, that of transcription and translation (Chapter 8). In eukaryotic cells, these processes are spatially separated into the cell nucleus (transcription) and the cytoplasm (translation).

In bacterial and archaeal cells, there also can be spatial separation between transcription and translation **FIGURE 4.21**. The RNA polymerase molecules needed for transcription are localized to a region separate from the ribosomes and other proteins that perform translation. So, even without a nuclear membrane, these cells can separate the process involved in

TABLE

4.3 **A Summary of the Structural Features of Bacterial and Archaeal Cells**

Structure	Chemical Composition	Function	Comment
External Structures			
Pili	Protein	Attachment to surfaces Genetic transfer	Found primarily in gram-negative bacteria
Flagella	Protein	Motility	Present in many rods and spirilla; few cocci; vary in number and placement
Glycocalyx	Polysaccharides and small proteins	Buffer to environment Attachment to surfaces	Capsule and slime layer Contributes to disease development Found in plaque bacteria and biofilms
Cell Envelope			
Cell wall		Cell protection Shape determination Cell lysis prevention	
Bacterial	Gram positives: thick peptidoglycan and teichoic acid Gram negatives: little peptidoglycan and an outer membrane		Site of activity of penicillin and lysozyme Gram-negative bacteria release endotoxins
Archaeal	Pseudopeptidoglycan Protein		S-layer

Cell membrane
   Bacterial
   Archaeal
	Protein and
	phospholipid
		Cell boundary
		Transport into/out of cell
		Site of enzymatic reactions
			Lipid bilayer
			Lipid monolayer

Internal Structures			
Nucleoid	DNA	Site of essential genes	Exists as single, closed loop chromosome
Plasmids	DNA	Site of nonessential genes	R plasmids
Ribosomes	RNA and protein	Protein synthesis	Inhibited by certain antibiotics
Microcompartments	Various metabolic enzymes	Carbon dioxide fixation Retention of volatile or toxic metabolites	Enzymes are enclosed in a protein shell
Inclusions	Glycogen, sulfur, lipid	Nutrient storage	Used as nutrients during starvation periods
Metachromatic granules	Polyphosphate	Storage of polyphosphate and calcium ions	Found in diphtheria bacilli
Gas vesicles	Protein shells	Buoyancy	Helps cells float
Magnetosome	Magnetite/greigite	Cell orientation	Helps locate preferred habitat
Cytoskeleton	Proteins	Cell division, chromosomal segregation, cell shape	Functionally similar to eukaryotic cytoskeletal proteins

MICRO◉INQUIRY 4
The Prokaryote/Eukaryote Model

"It is now clear that among organisms there are two different organizational patterns of cells, which Chatton . . . called, with singular prescience, the eukaryotic and prokaryotic type. The distinctive property of bacteria and blue-green algae is the prokaryotic nature of their cells. It is on this basis that they can be clearly segregated from all other protists (namely, other algae, protozoa and fungi), which have eukaryotic cells."

Stanier and van Niel (1962)
—*The concept of a bacterium.*

The idea of a tree of life extends back centuries and originates not with scientific thinking, but rather with folklore and culture, and often focused on immortality or fertility (see figure).

The development of the three-domain tree of life, on the other hand, represents the evolutionary relationships between species. Its development has made a profound change in biology. Instead of

Glass mosaic of tree of life on a wall of the 16th century Sim Wat Xiang Thong Luang Prabang UNESCO World Heritage Site, Laos.

two kinds of organisms—prokaryotes and eukaryotes—there are three: *Bacteria*, *Archaea*, and *Eukarya*.

In 2006, Norm Pace, a molecular biologist turned evolutionist at the University of Colorado, Boulder, suggested that the massive data bank of gene sequences identified since 1995 shows just how different archaeal organisms are from bacterial organisms and, in some ways, the archaeal ones are more similar to eukaryotic organisms. Therefore, Pace says, "we need to reassess our understanding of the course of evolution at the most fundamental level." Among items needing reassessment is the prokaryotic/eukaryotic paradigm—the tradition (folklore) if you will—that if an organism is not a eukaryote, it must be a prokaryote.

The quote at the top of the page refers to Edouard Chatton who coined the terms "prokaryotic" and "eukaryotic." Interestingly, neither he, nor Stanier and van Niel, ever really made mention as to

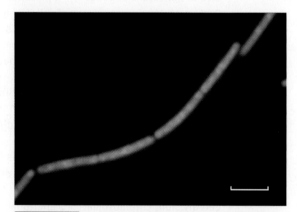

FIGURE 4.21 Spatial Separation of Transcription and Translation. In these cells of *Bacillus subtilis*, fluorescence microscopy was used to identify RNA polymerase (transcription) using a red fluorescent protein and ribosomes (translation) using a green fluorescent protein. Separate subcompartments are evident. (Bar = 3 μm.) »» What does the spatial separation indicate concerning compartmentation?

making cellular proteins, in a manner similar to eukaryotic cells.

Traditionally, a prokaryote was an organism without a cell nucleus; that is, without a membrane surrounding the DNA or chromosome. But is that a fair way to describe all bacterial and archaeal organisms? In this chapter, you have learned that there are many basic differences between bacterial and archaeal cells, yet do we lump them together just because they both lack a cell nucleus? Some scientists say no—the terms "prokaryote" and "prokaryotic" are not appropriate for these two domains of life.

So, to finish this chapter, take a look at MicroInquiry 4, which discusses the prokaryotic/eukaryotic model for living organisms.

CONCEPT AND REASONING CHECKS

4.16 Make a list of the various subcompartments in bacterial cells.

what a prokaryotic cell is. Yet if you look in any introductory biology textbook, prokaryote is defined as a group of organisms that lack a cell nucleus. According to Pace, "the prokaryote/eukaryote model for biological diversity and evolution is invalid." How can all organisms without a cell nucleus be called "prokaryotic," especially because the eukaryotic cell nucleus appears to be descended from as ancient a line of cells as the *Archaea?* Yes, the concept of a nuclear membrane (or not) is important, but no more important than other cellular properties. And the problem is that the word "prokaryote" is so engrained in the culture of biology and in the scientific mind of biologists—and students—that inappropriate inferences about organisms are made using this term. Pace does not buy the argument that the term "prokaryote" can be used to identify organisms that are not eukaryotes because the *Bacteria* are very different from the *Archaea* and, therefore, should not be put under the umbrella of "prokaryote."

Pace believes saying that prokaryotes lack a cell nucleus is a scientifically invalid description; although open to debate, he says no one can define what a prokaryote is—only what it is not (e.g., no nucleus, no mitochondria, no chloroplasts, no endomembrane system, etc.). Therefore, lumping the *Bacteria* and *Archaea* conceptually dismisses the fundamental and important differences between these two kinds of organisms and reinforces an incorrect understanding of biological organization and evolution.

Pace believes it is time to delete the term prokaryote as a term for bacterial and archaeal organisms. Because it has long been used by all biology texts, including this one (although in many cases "bacteria" and "archaea" have replaced the term prokaryote), Pace says he realizes "it is hard to stop using the word, prokaryote."

Discussion Point
There is no doubt that bacterial and archaeal organisms are very different entities. So, if "prokaryote" is to be deleted from the biological vocabulary, what can we call the Bacteria *and* Archaea? *Can you think of some positive characters that would define both bacterial and archaeal cells? If so, then how about inventing a common noun and adjective for both? Or do we simply speak of the bacteria, archaea, and eukaryotes separately?*

■ SUMMARY OF KEY CONCEPTS

4.1 Diversity among the *Bacteria* and *Archaea*

1. The **phylogenetic tree** of life contains many bacterial phyla and groups, including the **Proteobacteria**, **Gram-positive bacteria**, **Cyanobacteria**, **Chlamydiae**, and **Spirochaetes**.
2. Many organisms in the domain *Archaea* live in extreme environments. The **Euryarchaeota** (methanogens, extreme halophiles, and the thermoacidophiles) and the **Crenarchaeota** are the two phyla.

4.2 Cell Shapes and Arrangements

3. **Bacilli** have a cylindrical shape and can remain as single cells or be arranged into **diplobacilli** or chains (**streptobacilli**). **Cocci** are spherical and form a variety of arrangements, including the **diplococcus**, **streptococcus**, and **staphylococcus**. The spiral-shaped bacteria can be curved rods (**vibrios**) or spirals (**spirochetes** and **spirilla**). Spirals generally appear as single cells.

4.3 An Overview to Bacterial and Archaeal Cell Structure

4. Cell organization is centered on three specific processes: sensing and responding to environmental changes, compartmentalizing metabolism, and growing and reproducing.

4.4 External Cell Structures

5. **Pili** are short hair-like appendages found on many gram-negative bacteria to facilitate attachment to a surface. **Conjugation pili** are used for genetic transfer of DNA.
6. One or more **flagella**, found on many rods and spirals, provide for cell motility. Each flagellum consists of a **basal body** attached to the flagellar filament. In nature, flagella propel bacterial cells toward nutrient sources (**chemotaxis**). Spirochetes have **endoflagella**, while other bacterial species undergo gliding motility.
7. The **glycocalyx** is a sticky layer of polysaccharides that protects the cell against desiccation, attaches it to surfaces, and helps evade immune cell attack. The glycocalyx can be thick and tightly bound to the cell (**capsule**) or thinner and loosely bound (**slime layer**).

4.5 The Cell Envelope

8. The **cell wall** provides structure and protects against cell lysis. Gram-positive bacteria have a thick wall of **peptidoglycan** strengthened with **teichoic acids**. Gram-negative cells have a single layer of peptidoglycan and an **outer membrane** containing **lipopolysaccharide** and **porin proteins**.

9. Archaeal cell walls lack peptidoglycan but may have either a **pseudopeptidoglycan** or **S-layer**.
10. The **cell membrane** represents a permeability barrier and the site of transfer for nutrients and metabolites into and out of the cell. The cell membrane reflects the **fluid mosaic model** for membrane structure in that the lipids are fluid and the proteins are a mosaic that can move laterally in the bilayer.
11. The archaeal cell membrane links lipids through an ether linkage and the lipid tails are bonded together into a single **monolayer**.

4.6 The Cell Cytoplasm and Internal Structures

12. The DNA (**bacterial chromosome**), located in the **nucleoid**, is the essential genetic information and represents the organism's **genome**.

13. Bacterial and archaeal cells may contain one or more **plasmids**, circular pieces of nonessential DNA that replicate independently of the chromosome.
14. **Ribosomes** carry out protein synthesis, **microcompartments** carry out species-specific processes, while inclusions store nutrients or structural building blocks.
15. The **cytoskeleton**, containing protein homologs to the cytoskeletal proteins in eukaryotic cells, helps determine cell shape, regulates cell division, and controls chromosomal segregation during cell division.

4.7 The Bacteria/Eukaryote Paradigm—Revisited

16. Cell biology investigations are showing that compartmentation in bacterial cells can occur; it simply does not require the diverse membranous organelles typical of eukaryotic cells.

LEARNING OBJECTIVES

After understanding the textbook reading, you should be capable of writing a paragraph that includes the appropriate terms and pertinent information to answer the objective.

1. Identify the major bacterial phyla described in this chapter and provide characteristics for each group.
2. Explain why many archaeal organisms are considered **extremophiles**.
3. Compare the various shapes and arrangements of bacterial and archaeal cells.
4. Summarize how the processes of sensing and responding to the environment, compartmentation of metabolism, and growth and metabolism are linked to cell structure.
5. Assess the role of **pili** to bacterial colonization and infection.
6. Describe the structure of bacterial flagella and discuss how they function in **chemotaxis**.
7. Differentiate between a **capsule** and **slime layer**. Identify their roles in cell survival.

8. Compare and contrast the structure of a **gram-positive cell wall** with a **gram-negative cell wall**.
9. Summarize the differences between bacterial and archaeal cell walls.
10. Justify the need for a **cell membrane** surrounding all bacterial and archaeal cells.
11. Explain how the structure of archaeal cell membranes differs from bacterial cell membranes.
12. Describe the structure of the **nucleoid**.
13. Judge the usefulness of **plasmids** to cell metabolism and organismal survival.
14. List the typical **inclusions** found in the bacterial cell cytoplasm and identify their contents or roles.
15. Describe three roles that the bacterial **cytoskeleton** plays.
16. Justify the statement, "Bacterial cells are as highly organized subcellularly as are eukaryotic cells."

STEP A: SELF-TEST

Each of the following questions is designed to assess your ability to remember or recall factual or conceptual knowledge related to this chapter. Read each question carefully, then select the *one* answer that best fits the question or statement. Answers to even-numbered questions can be found in **Appendix C**.

1. Which one of the following is NOT a genus within the gram-positive bacteria?
   A. *Staphylococcus*
   B. Methanogens
   C. *Mycoplasma*
   D. *Bacillus* and *Clostridium*
2. The domain *Archaea* includes all the following groups *except* the
   A. mycoplasmas.
   B. extreme halophiles.
   C. Crenarchaeota.
   D. Euryarchaeota.

3. Spherical bacterial cells in chains would be a referred to as a _____ arrangement.
   A. vibrio
   B. streptococcus
   C. staphylococcus
   D. tetrad
4. Intracellular organization in bacterial and archaeal species is centered around
   A. compartmentation of metabolism.
   B. growth and reproduction.
   C. sensing and responding to environment.
   D. All the above (**A–C**) are correct.

5. Which one of the following statements does NOT apply to pili?
   A. Pili are made of protein.
   B. Pili allow for attachment to surfaces.
   C. Pili facilitate nutrient transport.
   D. Pili contain adhesins.

6. Flagella are
   A. made of carbohydrate and lipid.
   B. found on all bacterial cells.
   C. shorter than pili.
   D. important for chemotaxis.

7. Capsules are similar to pili because both
   A. contain DNA.
   B. are made of protein.
   C. contain dextran fibers.
   D. permit attachment to surfaces.

8. Gram-negative bacteria would stain _____ with the Gram stain and have _____ in the wall.
   A. orange-red; teichoic acid
   B. orange-red; lipopolysaccharide
   C. purple; peptidoglycan
   D. purple; teichoic acid

9. The cell membrane of archaeal cells contains
   A. a monolayer.
   B. sterols.
   C. ester linkages.
   D. All the above (A–C) are correct.

10. The movement of glucose into a cell occurs by
    A. facilitated diffusion.
    B. active transport.
    C. simple diffusion.
    D. phospholipid exchange.

11. When comparing bacterial and archaeal cell membranes, only bacterial cell membranes
    A. have three layers of phospholipids.
    B. have a phospholipid bilayer.
    C. are fluid.
    D. have ether linkages.

12. Which one of the following statements about the nucleoid is NOT true?
    A. It contains a DNA chromosome.
    B. It represents a nonmembranous subcompartment.
    C. It represents an area devoid of ribosomes.
    D. It contains nonessential genetic information.

13. Plasmids
    A. replicate with the bacterial chromosome.
    B. contain essential growth information.
    C. may contain antibiotic resistance genes.
    D. are as large as the bacterial chromosome.

14. Which one of the following is NOT a structure or subcompartment found in bacterial cells?
    A. Microcompartments
    B. Volutin
    C. Ribosomes
    D. Mitochondria

15. The bacterial cytoskeleton
    A. transports vesicles.
    B. helps determine cell shape.
    C. is organized identical to its eukaryotic counterpart.
    D. centers the nucleoid.

16. The bacterial cell is capable of
    A. spatial separation of metabolic processes.
    B. carrying out complex metabolic processes.
    C. subcompartmentalizing biochemical processes.
    D. All the above (A–C) are correct.

STEP B: REVIEW

Answers to even-numbered questions or statements can be found in **Appendix C.**

17. Construct a concept map for the **domain *Bacteria*** using the following terms.

Actinobacteria	hyperthermophiles
Bacillus	*Mycoplasma*
blooms	Proteobacteria
Chlamydiae	rickettsiae
Cyanobacteria	Spirochaetes
Escherichia	*Staphylococcus*
Firmicutes	*Streptomyces*
gram-negative species	*Treponema*
gram-positive species	

18. Construct a concept map for the **Cell Envelope** using the following terms.

active transport	membrane proteins
cell membrane	NAG
cell wall	NAM
endotoxin	outer membrane
facilitated transport	peptidoglycan
fluid-mosaic model	periplasm
gram-negative wall	phospholipids
gram-positive wall	polysaccharide
lipid A	porin proteins
lipopolysaccharide (LPS)	teichoic acid

Identify and label the structure on the accompanying bacterial cell from each of the following descriptions. Some separate descriptions may apply to the same structure.

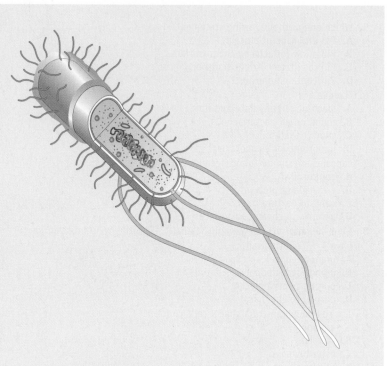

Descriptions

19. An essential structure for chemotaxis, aerotaxis, or phototaxis.
20. Contains nonessential genetic information that provides genetic variability.
21. The structure that synthesizes proteins.
22. The protein structures used for attachment to surfaces.
23. Contains essential genes for metabolism and growth.
24. Prevents cell desiccation.
25. A 70S particle.
26. Contains peptidoglycan.
27. Regulates the passage of substances into and out of the cell.
28. Extrachromosomal loops of DNA.
29. Represents a capsule or slime layer.
30. The semifluid mass of proteins, amino acids, sugars, salts, and ions dissolved in water.

STEP C: APPLICATIONS

Answers to even-numbered questions can be found in **Appendix C**.

31. A bacterium has been isolated from a patient and identified as a gram-positive rod. Knowing that it is a human pathogen, what structures would it most likely have? Explain your reasons for each choice.
32. Another patient has a blood infection caused by a gram-negative bacterium. Why might it be dangerous to prescribe an antibiotic to treat the infection?
33. In the research lab, the gene for the cytoskeletal protein similar to eukaryotic tubulin is transferred into the DNA chromosome of a coccus-shaped bacterium. When this cell undergoes cell division, predict what shape the daughter cells will exhibit. Explain your answer.

STEP D: QUESTIONS FOR THOUGHT AND DISCUSSION

Answers to even-numbered questions can be found in **Appendix C**.

34. In reading a story about a bacterium that causes a human disease, the word "bacillus" is used. How would you know if the article is referring to a bacterial shape or a bacterial genus?
35. Suppose this chapter on the structure of bacterial and archaeal cells had been written in 1940, before the electron microscope became available. Which parts of the chapter would probably be missing?
36. Why has it taken so long for microbiologists to discover microcompartments and a cytoskeleton in bacterial and archaeal cells?
37. Apply the current understanding of the bacteria/eukaryote paradigm to the following statement: "Studying the diversity of life only accentuates life's unity."

HTTP://MICROBIOLOGY.JBPUB.COM/9E

The site features eLearning, an online review area that provides quizzes and other tools to help you study for your class. You can also follow useful links for in-depth information, read more MicroFocus stories, or just find out the latest microbiology news.

Microbial Growth and Nutrition

But who shall dwell in these worlds if they be inhabited? . . . Are we or they Lords of the World? . . .
—Johannes Kepler (quoted in *The Anatomy of Melancholy*)

Books have been written about it; movies have been made; even a radio play in 1938 about it frightened thousands of Americans. What is it? Martian life. In 1877 the Italian astronomer, Giovanni Schiaparelli, saw lines on Mars, which he and others assumed were canals built by intelligent beings. It wasn't until well into the 20th century that this notion was disproved. Still, when we gaze at the red planet, we wonder: Did life ever exist there?

We are not the only ones wondering. Astronomers, geologists, and many other scientists have asked the same question. Today microbiologists have joined their scientific colleagues, wondering if microbial life once existed on the Red Planet or, for that matter, elsewhere in our Solar System. In 1996, NASA scientists reported finding what looked like fossils of microbes inside a meteorite thought to have come from Mars. Although most now believe these "fossils" are not microbial, it only fueled the debate.

Could microbes, as we know them here on Earth, survive on Mars where the temperatures are far below 0°C and—as far as we know—there is little, if any, liquid water? Researchers, using a device to simulate the Martian environment, placed in it microbes known to survive extremely cold environments here on Earth. Their results indicated that members of the *Archaea*, specifically the methanogens, could grow at the cold temperatures and low pressures known to exist on Mars. They concluded that life could have existed on the Red Planet in the past or, as the quote says, "dwell in these worlds [today] if they be inhabited."

TABLE

5.1   Some Microbial Record Holders

Hottest environment (Juan de Fuca ridge)—121°C (250°F): Strain 121 (*Archaea*)

Coldest environment (Antarctica)—5°F (−15°C): Cryptoendoliths (*Bacteria* and lichens)

Highest radiation survival—5MRad, or 5000× what kills humans: *Deinococcus radiodurans* (*Bacteria*)

Deepest—3.2 km underground: Many *Bacterial* and *Archaeal* species

Most acid environment (Iron Mountain, CA)—pH 0.0 (most life is at least a factor of 100,000 less acidic): *Ferroplasma acidarmanus* (*Archaea*)

Most alkaline environment (Lake Calumet, IL)—pH 12.8 (most life is at least a factor of 1000 less basic): Proteobacteria (*Bacteria*)

Longest in space (NASA satellite)–6 years: *Bacillus subtilis* (*Bacteria*)

High pressure environment (Mariana Trench)—1200 times atmospheric pressure: *Moritella, Shewanella* and others (*Bacteria*)

Saltiest environment (Eastern Mediterranean basin)—47% salt, (15 times human blood saltiness): Several *Bacterial* and *Archaeal Species*

Source: http://www.astrobio.net/news/.

So, microbiologists have joined the search for extraterrestrial life. This seems a valid pursuit since the **extremophiles** found here on Earth survive, and even require, living in extreme environments (TABLE 5.1)—some not so different from Mars (FIGURE 5.1). If life did or does exist on Mars, it almost certainly was or is microbial—most likely bacterial or archaeal-like organisms.

In 2004, NASA sent two spacecrafts to Mars to look for indirect signs of past life. Scientists here on Earth monitored instruments on the Mars rovers, Spirit and Opportunity, designed to search for signs suggesting water once existed on the planet. Some findings suggest there are areas where salty seas once washed over the plains of Mars, creating a life-friendly environment.

Opportunity found evidence for ancient shores of a large body of surface water that contained currents, which left their marks in rocks at the bottom of what once was a sea. Scientists reported in 2008 that a more recent spacecraft, the Mars Phoenix Lander, detected water ice near the Martian soil surface.

Did or does life exist on Mars? Perhaps one day when human explorers or more sophisticated spacecraft reach Mars, we will know.

Whether microorganisms are here on Earth in a moderate or extreme environment, or on Mars,

FIGURE 5.1 **The Martian Surface?** This barren-looking landscape is not Mars but the Atacama Desert in Chile. It looks similar to photos taken by the Mars rovers Spirit and Opportunity. »» Does this area look like a habitable place for life, even microbial life?

there are certain physical and chemical requirements they must possess to survive, reproduce, and grow. In this chapter, we explore the process of cell reproduction in bacterial cells as compared to that in eukaryotic microbial cells. We also examine the physical and chemical conditions required for growth of bacterial and archaeal cells, and discover the ways that microbial growth can be measured.

As we have been emphasizing in this text, the domains of organisms may have different structures and patterns, but they carry out many of the same processes.

5.1 Microbial Reproduction

Growth in the microbial world usually refers to an increase in the numbers of individuals; that is, an increase in the population size with each cell carrying the identical genetic instructions of the parent cell. **Asexual reproduction** is a process to maintain genetic constancy while increasing cell numbers. In eukaryotic microbes, an elaborate interaction of microtubules and proteins with pairs of chromosomes in the cell nucleus allows for the precise events of mitosis and cytokinesis. Bacterial and archaeal cells divide without the microtubular involvement.

Most Bacteria Reproduce by Binary Fission

KEY CONCEPT

1. Binary fission produces genetically identical daughter cells.

Most bacterial organisms reproduce by an asexual process called **binary fission**, which usually occurs after a period of growth in which the cell doubles in mass. At this time, the chromosome (DNA) replicates and the two DNA molecules separate (**FIGURE 5.2A**). Chromosome segregation is not well understood. Unlike eukaryotic cells, bacterial cells lack a mitotic spindle to separate replicated chromosomes. The segregation process does involve specialized chromosomal-associated proteins but there is no clear picture describing how most of these proteins work to ensure accurate chromosome segregation.

In any event, cell fission at midcell involves the synthesis of a partition, or septum, that separates the mother cell into two genetically identical daughter cells (**FIGURE 5.2B**). A eukaryotic tubulin homolog found in bacterial cells (see Chapter 4) is part of the fission ring apparatus that organizes the inward growth of the cell envelope.

Cell separation then occurs by dissolution of the material in the septum. Depending on the growth conditions, the septum may dissolve at a slow enough rate for chains of connected cells (streptobacilli) to form.

Reproduction by binary fission seems to confer immortality because there is never a moment at which the first bacterial cell has

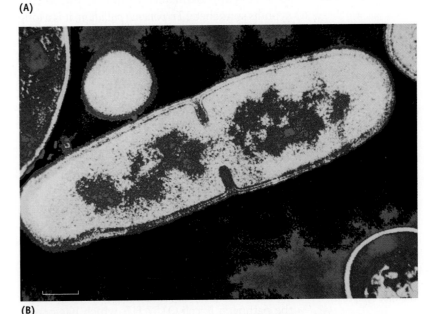

① Cell elongates and DNA is replicated.

Cell wall   Cell membrane

Replicated DNA molecules

② Cell wall and cell membrane begin to invaginate.

Fission ring apparatus

③ Cross-wall forms two distinct cells.

④ Cells separate.

(A)

(B)

FIGURE 5.2 The Process of Binary Fission. (**A**) Binary fission in a rod-shaped cell begins with DNA replication and segregation. Inward growth occurs at the midcell fission ring apparatus of the cell envelope separates the mother cell into two genetically-identical daughter cells. (**B**) A false-color transmission electron micrograph of a cell of *Bacillus licheniformis* undergoing binary fission. The inward growth of the cell wall and membrane is evident at midcell. (Bar = 0.25 µm.) »» How would binary fission differ for a prokaryotic organism having cells arranged in chains and another that forms single cells?

died. Each mother cell undergoes binary fission to become the two young daughter cells. However, the perception of immortality has been challenged by experiments suggesting bacterial cells do age (MicroFocus 5.1).

CONCEPT AND REASONING CHECKS

5.1 Propose an explanation as to how a bacterial cell "knows" when to divide.

Bacterial and Archaeal Cells Reproduce Asexually

KEY CONCEPT

2. Environmental and genetic factors affect a cell's generation time.

The interval of time between successive binary fissions of a cell or population of cells is known as the **generation time** (or doubling time). Under optimal conditions, some species have a very fast generation time; for others, it is much slower. For example, the optimal generation time for *Staphylococcus aureus* is about 30 minutes; for *Mycobacterium tuberculosis*, the agent of tuberculosis, it is approximately 15 hours; and for the syphilis spirochete, *Treponema pallidum*, it is a long 33 hours.

One enterprising mathematician calculated that if *Escherichia coli* binary fissions were to continue at their optimal generation time (15 minutes) for 36 hours, the bacterial cells would cover the surface of the Earth! Thankfully, this will not occur because of the limitation of nutrients and the loss of ideal physical factors required for growth. The majority of the bacterial cells would starve to death or die in their own waste.

The generation time is useful in determining the amount of time that passes before disease symptoms appear in an infected individual; faster division times often mean a shorter incubation period for a disease. For example, suppose you eat an undercooked hamburger contaminated with the pathogen *E. coli* O157:H7, which has one of the shortest generation times—just 20 minutes under optimal conditions (FIGURE 5.3). If you ingested one cell (more likely several hundred at least) at 8:00 PM this evening, two would be present by 8:20, four by 8:40, and eight by 9:00. You would have over 4,000 by midnight. By 3:00 AM, there would be over 2 million. Depending on the response of the immune system, it is quite likely

Incubation period:
The time from entry of a pathogen into the body until the first symptoms appear.

that sometime during the night you would know you have food poisoning.

Bacterial and archaeal organisms are subject to the same controls on growth as all other organisms on Earth. Let's examine the most important growth factors conferring optimal generation times.

CONCEPT AND REASONING CHECKS

5.2 If it takes *E. coli* 7 hours to reach some 2 million cells, how long would it take *T. pallidum* to reach that same number under optimal conditions?

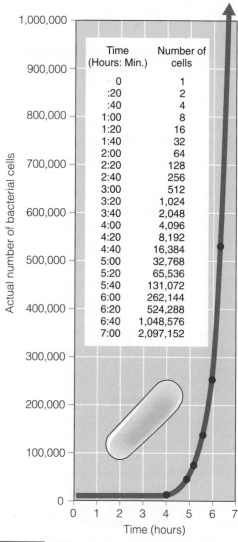

Time (Hours: Min.)	Number of cells
0	1
:20	2
:40	4
1:00	8
1:20	16
1:40	32
2:00	64
2:20	128
2:40	256
3:00	512
3:20	1,024
3:40	2,048
4:00	4,096
4:20	8,192
4:40	16,384
5:00	32,768
5:20	65,536
5:40	131,072
6:00	262,144
6:20	524,288
6:40	1,048,576
7:00	2,097,152

FIGURE 5.3 A Skyrocketing Bacterial Population. The number of *Escherichia coli* cells progresses from 1 cell to 2 million cells in a mere 7 hours. The J-shaped growth curve gets steeper and steeper as the hours pass. Only a depletion of food, buildup of waste, or some other limitation will halt the progress of the curve. »» What is the generation time for the bacterial species in this figure?

It used to be thought that bacteria do not age—they are immortal. This might seem obvious considering a mother cell divides at the mid-point by binary fission to become the two genetically equal daughter cells. However, new research suggests that although the DNA may be identical, after several generations of binary fission, the population consists of cells of different ages and the oldest ones have the longest generation time.

Eric Stewart and his collaborators at INSERM, the National Institute of Health and Medical Research in Paris, filmed *Escherichia coli* cells as they divided into daughter cells on a specially designed microscope slide. A record of every daughter cell (total of 35,000 individual cells) was recorded for nine generations over a period of six hours. Then, a custom-designed computer system analyzed the micrographs.

The group's results suggest that although the cells may divide symmetrically, daughter cells are not morphologically or physiologically symmetrical. Each contains cellular poles of different ages.

When a mother cell divides, each daughter cell inherits one end or pole of the mother cell. The region where the cells split develops into the other pole (see figure). For example, in the first division, the mother cell splits with a new wall (red). When the daughter cells grow in size they contain an old pole (brown) and a new one (red). When each of these cells divides, two have the oldest pole (brown) and youngest pole (green) while the two other cells have a younger pole (red) and a youngest pole (green). So after just two divisions, there are two populations of daughter cells: two have oldest and youngest poles while two have younger poles and youngest poles. According to Stewart's group, the two cells with the oldest/youngest poles grew 2.2 percent slower than the cells with younger/youngest poles.

As more and more binary fissions occur, the difference in age between daughter cells will continue to increase. The bottom line is that cells inheriting older and older poles experience longer generation times, reduced rates of offspring formed, and increased risk of dying compared to cells with younger, newer poles. This loss of fitness is called **senescence**. Note: Stewart's group could not follow any cells to actual death because the cell populations eventually had so many cells, even their computer program could not keep them all independently recorded.

Exactly why the older cells senesce is not understood. However, if the results from Stewart's group are verified by others, it would at least appear that bacterial cells cannot escape the aging process. Even a microbe's life is limited.

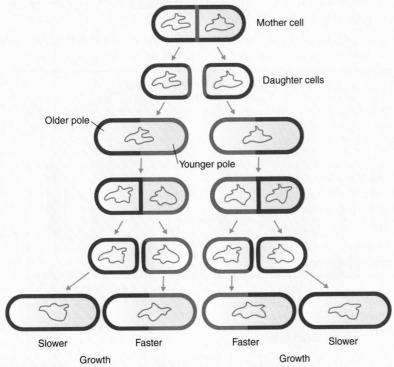

Two successive binary fissions produce daughter cells with various aged poles (brown = oldest; red = younger; green = youngest).

5.2 Microbial Growth

In the previous section, we discovered how fast some bacterial cells can grow under ideal circumstances. Let's look at the dynamics of bacterial growth in a little more detail.

A Bacterial Growth Curve Illustrates the Dynamics of Growth

KEY CONCEPT

3. Bacterial population growth goes through four phases.

A typical **bacterial growth curve** for a population illustrates the events occurring over time (**FIGURE 5.4**). Whether several bacterial cells infect the human respiratory tract or are transferred to a tube of fresh growth medium in the laboratory, four distinct phases of growth occur: the lag phase; the logarithmic phase; the stationary phase; and the decline phase.

1. The Lag Phase. The first portion of the growth curve during which time no cell divisions occur is called the **lag phase**. At this time, bacterial cells are adapting to their new environment. In the respiratory tract, scavenging white blood cells may engulf and destroy some of the cells; in growth media, some cells may die from the shock of transfer or the inability to adapt to the new

environment. The actual length of the lag phase depends on the metabolic activity in the remaining cells. They must grow in size, store nutrients, and synthesize essential enzymes and other cell constituents—all in preparation for binary fission.

2. The Log Phase. The population now enters an active stage of growth called the **logarithmic phase** (or **log phase**). This is the exponential growth described above for *E. coli*. In the log phase, all cells are undergoing binary fission and the generation time is dependent on the species and environmental conditions present. As each generation time passes, the number of cells doubles and the graph rises in a straight line on a logarithmic scale.

During an infection, disease symptoms usually develop during the log phase because the bacterial cells cause tissue damage. If the bacterial cells produce toxins, tissue destruction may become apparent. In a broth tube, the medium becomes cloudy (turbid) due to increasing cell numbers. If plated on solid growth medium, bacterial growth will be so vigorous that visible colonies appear and each colony may consist of millions of cells (**FIGURE 5.5**). Vulnerability to antibiotics is also highest at this active stage of growth because many antibiotics affect actively metabolizing cells.

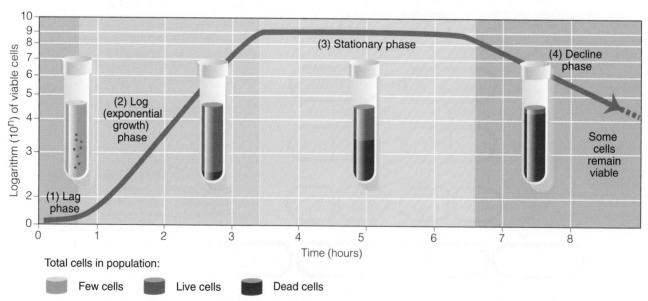

Total cells in population:

Few cells   Live cells   Dead cells

FIGURE 5.4 **The Growth Curve for a Bacterial Population.** (**A**) During the lag phase, the population numbers remain stable as bacterial cells prepare for division. (**B**) During the logarithmic (exponential growth) phase, the numbers double with each generation time. Environmental factors later lead to stationary phase (**C**), which involves a stabilizing population. (**D**) The decline phase is the period during which cell death becomes substantial. »» Why would antibiotics work best to kill or inhibit cells in the log phase?

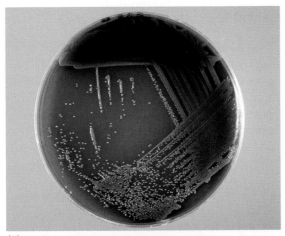

(A)

(B)

FIGURE 5.5 Two Views of Bacterial Colonies.
(**A**) Bacterial colonies cultured on blood agar in a culture dish. Blood agar is a mixture of nutrient agar and blood cells. It is widely used for growing bacterial colonies.
(**B**) Close-up of typhoid bacteria (*Salmonella typhi*) colonies being cultured on a growth medium. »» How did each colony in (**A**) or (**B**) start?

3. The Stationary Phase. After some days (in an infection) or hours (in a culture tube), the vigor of the population changes and, as the reproductive and death rates equalize, the population enters a plateau, called the **stationary phase**. In the respiratory tract, antibodies from the immune system are attacking the bacterial cells, and phagocytosis by white blood cells adds to their destruction. In the culture tube, available nutrients become scarce and waste products accumulate. Factors such as oxygen also may be in short supply. This limitation of nutrients and buildup of waste materials leads to the death of many cells—but not all as MicroFocus 5.2 describes.

4. The Decline Phase. If nutrients in the external environment remain limited or the quantities become exceedingly low, the population enters a **decline phase** (or logarithmic death phase). Now the number of dying cells far exceeds the number of new cells formed. A bacterial glycocalyx may forestall death by acting as a buffer to the environment, and flagella may enable organisms to move to a new location. For many species, though, the history of the population ends with the death of the last cell. When we discuss the progression of human diseases in Chapter 19, we will see a similar curve for the stages of a disease.

For some bacterial species, especially soil bacteria, they can escape cell death by forming endospores. Let's examine these amazing dormancy structures next.

CONCEPT AND REASONING CHECKS

5.3 In a broth tube, describe the status of the bacterial cell population in each phase of the bacterial growth curve.

Endospores Are a Response to Nutrient Limitation

KEY CONCEPT

4. Endospores are dormant structures that can endure times of nutrient stress.

A few gram-positive bacterial species, especially soil bacteria belonging to the genera *Bacillus* and *Clostridium,* produce highly resistant structures called **endospores** or, simply, spores (**FIGURE 5.6**). As described in the previous section, bacterial cells normally grow, mature, and reproduce as vegetative cells. However, when nutrients such as carbon or nitrogen are limiting and the population density reaches a critical mass, species of *Bacillus* and *Clostridium* enter stationary phase and begin spore formation or **sporulation**.

Sporulation begins when the bacterial chromosome replicates and binary fission is characterized by an asymmetric cell division (**FIGURE 5.7**). The smaller cell, the prespore, will become the mature endospore, while the larger mother cell will commit itself to maturation of the endospore before undergoing lysis.

Depending on the exact asymmetry of cell division, the endospore may develop at the end of the cell, near the end, or at the center of the cell (the position is useful for species identification purposes).

Vegetative:
Referring to cells actively metabolizing and obtaining nutrients.

MICROFOCUS 5.2: Evolution/Environmental Microbiology
The Secret of Success Is Constancy to Purpose

The title quote above was said by Benjamin Disraeli, a British statesman and literary figure in the mid-1800s. It not only applies to human success in life but also to success of microbial species. Here's how it applies in the bacterial world.

As bacterial cells grow logarithmically, they eventually start running short of food and nutrients to sustain this log phase growth. Sensing that nutrients are becoming limited, as the population enters stationary phase, a few cells of many bacterial species enter a dormant (non-dividing) state and are referred to as persister (survivor) cells (see figure). Should the population go through a death phase, these persister cells "live on."

In this sense, the secret to evolutionary success for many species is constancy of purpose; always form some persister cells to ensure survival of the population should the species find itself in an unfamiliar or hostile environment. Indeed, such altruistic-like behavior may be responsible for some hard-to-treat infectious diseases, such as tuberculosis (TB), where persister cells can "hide out" in the body as a latent TB infection (Chapter 10). The constancy of purpose—to survive—also means persister cells would be unaffected by antibiotics, which tend to affect growing cells, not the dormant persister cells. The secret to successful survival here is antibiotic resistance, which, in part, would account for the recalcitrant properties exhibited by bacterial communities in a biofilm (see Chapter 3). In the case of the "hibernating" TB cells, persister cells could survive the antibiotic assault and later repopulate the infection. But genetic analysis of persister cells may have a solution. Biologist Kim Lewis at Northeastern University in Boston has discovered a gene in persister cells that codes for a protein that triggers dormancy. If a medical treatment could be devised to delete or deactivate the gene, the secret to success in pathogens could be defeated, possibly saving thousands of lives every year. So, in the end, the constancy of purpose carried out by medical researchers to understand the persister problem perhaps will be the ultimate secret of success.

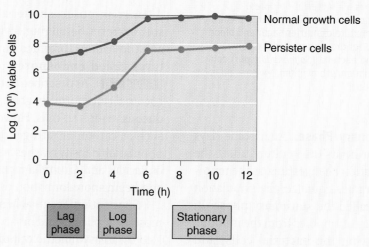

Persister cells increase as the growth curve progresses. Adapted from Lewis, K. 2007. *Nature Reviews Microbiology* 5: 48–56.

The prespore cell contains cytoplasm and DNA, and a large amount of **dipicolinic acid**, a unique organic substance that helps stabilize the proteins and DNA. After the cell is engulfed by the mother cell, thick layers of peptidoglycan form the cortex, followed by a series of protein coats that further protect the contents. The mother cell then disintegrates and the spore is freed. It should be stressed that sporulation is not a reproductive process. Rather, the endospore represents a dormant stage in the life of the bacterial species.

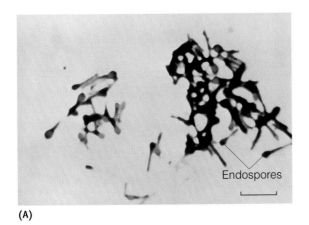

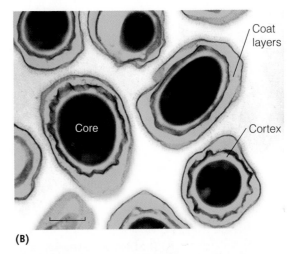

Coat
layers

Core

Cortex

(A)

(B)

FIGURE 5.6 Three Different Views of Bacterial
Endospores. (A) A light microscope image of *Clostridium*
cells showing terminal spore formation. Note the characteristic drumstick appearance of the spores. (Bar =
5.0 μm.) (B) The fine structure of a *Bacillus anthracis*
spore seen using the transmission electron microscope.
The visible spore structures include the core, cortex,
and coat layers. (Bar = 0.5 μm.) (C) A scanning electron
microscope view of a germinating spore (arrow). Note that
the spore coat divides equatorially along the long axis,
and as it separates, the vegetative cell emerges. (Bar =
2.0 μm.) »» If an endospore is resistant to so many environmental conditions, how does a spore "know" conditions are favorable for germination?

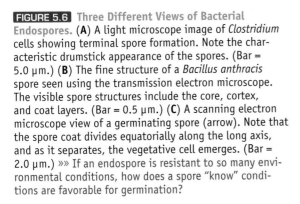

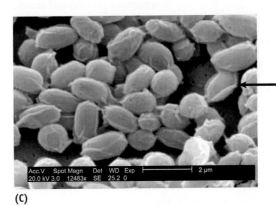

Acc.V Spot Magn Det WD Exp
20.0 kV 3.0 12483x SE 25.2 0   2 μm

(C)

Endospores are probably the most resistant living structures known. Desiccation has little effect
on the spore. By containing little water, endospores
also are heat resistant and undergo very few chemical reactions. These properties make them difficult
to eliminate from contaminated medical materials and food products. For example, endospores
can remain viable in boiling water (100°C) for
2 hours. When placed in 70% ethyl alcohol,
endospores have survived for 20 years. Humans
can barely withstand 500 rems of radiation, but
endospores can survive one million rems. In this
dormant condition, endospores can "survive" for
extremely long periods of time (MICROFOCUS 5.3).

When the environment is favorable for cell
growth, the protective layers break down and each
endospore germinates into a vegetative cell.

A few serious diseases in humans are caused
by spore formers. The most newsworthy has been
Bacillus anthracis, the agent of the 2001 anthrax
bioterror attack through the mail. This potentially
deadly disease, originally studied by Koch and
Pasteur, develops when inhaled spores germinate

in the lower respiratory tract and the resulting vegetative cells secrete two deadly toxins. Botulism,
gas gangrene, and tetanus are diseases caused by
different species of *Clostridium*. Clostridial endospores often are found in soil, as well as in human
and animal intestines. However, the environment
must be free of oxygen for the spores to germinate to vegetative cells. Dead tissue in a wound
provides such an environment for the development of tetanus and gas gangrene, and a vacuum-
sealed can of food is suitable for the development
of botulism.

Killing endospores can be a tough task.
Heating them for many hours under high pressure
will do the trick. If they contaminate machinery,
such as they did in mail sorting equipment in the
2001 anthrax attacks, there are potent but highly
dangerous chemical methods to kill the spores
(Chapter 7). Postal workers who were exposed to
the spores were effectively treated with antibiotics
that can kill any newly germinated endospores
before the vegetative cells can produce and secrete
the deadly toxins.

■ Rems (Roentgen
Equivalent Man):
A measure of radiation dose
related to biological effect.

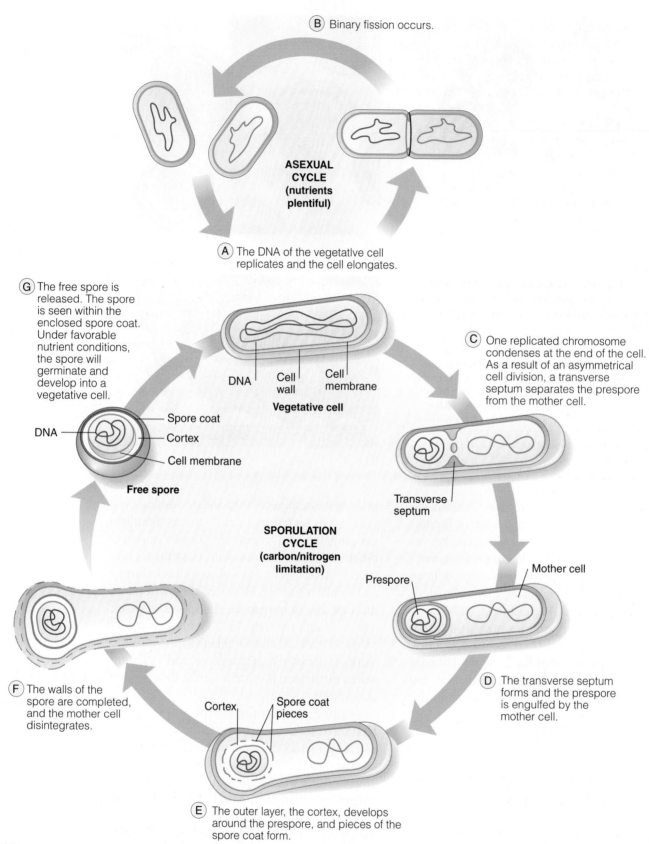

FIGURE 5.7 **The Formation of a Bacterial Spore by *Bacillus subtilis*.** (**A, B**) When nutrient conditions can support growth and reproduction, vegetative cells continue through cycles of binary fission. (**C–G**) When nutrient conditions become limiting (e.g., carbon, nitrogen), endospore formers, such as *B. subtilis*, enter the sporulation cycle shown here. »» Hypothesize how a vegetative cell "knows" nutrient conditions are limiting.

MICROFOCUS 5.3: Being Skeptical
Germination of 25 Million-Year-Old Endospores?

Endospores have been recovered and germinated from various archaeological sites and environments. Living spores have been recovered and germinated from the intestines of Egyptian mummies several thousand years old. In 1983, archaeologists found viable spores in sediment lining Minnesota's Elk Lake. The sediment was over 7,500 years old.

All these reports though pale in comparison to the controversial discovery reported in 1995 by researcher Raul Cano of California Polytechnic State University, San Louis Obispo. Cano found bacterial spores in the stomach of a fossilized bee trapped in amber—a hardened resin—produced from a tree in the Dominican Republic. The fossilized bee was about 25 million-years-old. When the amber was cracked open and the material from the abdomen of the bee extracted and placed in nutrient medium, the equally ancient spores germinated. With microscopy, the cells from a colony were very similar to *Bacillus sphaericus,* which is found today in bees in the Dominican Republic. Is it possible for an endospore to survive for 25 million years—even if it is encased in amber?

Critics were quick to claim the bacterial species may represent a modern-day species that contaminated the amber sample being examined. However, Professor Cano had carried out appropriate and rigorous decontamination procedures and sterilized the amber sample before cracking it open. He also carried out all the procedures in a class II laminar flow hood, which prevents outside contamination from entering the working area. In addition, the hood had never been used for any other bacterial extraction processes. Several other precautions were added to eliminate any chance that the spores were modern-day contaminants from an outside source. Still, many scientists question whether all contamination sources had been identified.

The major question that remains is whether DNA can remain intact and functional after so long a period of dormancy. Does it really have a capability of replication and producing new vegetative growth? Granted, the DNA presumably was protected in a resistant spore, but could DNA remain intact for 25 million years?

Research on bacterial DNA suggests the maximum survival time is about 400,000 to 1.5 million years. If true, then the 25 million-year-old spores could not be viable. But that is based on current predictions and they may be subject to change as more research is carried out with ancient DNA.

The verdict? It seems unlikely that such ancient endospores could germinate after 25 million years. Perhaps new evidence will change that perception.

CONCEPT AND REASONING CHECKS

5.4 Hypothesize why gram-negative and most gram-positive bacterial species cannot produce endospores.

Optimal Microbial Growth Is Dependent on Several Physical Factors

KEY CONCEPT

5. Growth of microbial populations is sensitive to temperature, oxygen gas, and pH.

Now that we have examined reproduction and growth, let's examine the essential physical and chemical factors influencing cell growth.

Temperature. Temperature is one of the most important factors governing growth. Every microbial species has an optimal growth temperature and an approximate 30°C operating range, from minimum to maximum, over which the cells will grow albeit with a slower generation time (**FIGURE 5.8**). In general, most microbes can be assigned to one of three groups—psychrophiles, mesophiles, or thermophiles—based on their optimal growth temperature as well as their minimal and maximal growth temperatures.

Microbes that have their optimal growth rates below 15°C but can still grow at 0°C to 20°C are called **psychrophiles** (*psychro* = "cold"). Because about 70% of the Earth is covered by oceans having deep water temperatures below 5°C, psychrophiles represent a group of bacterial and archaeal extremophiles that make up a large portion of the global microbial community. In fact, many psychrophiles can grow as fast at 4°C as *E. coli* does at 37°C. On the other hand, at these low temperatures, psychrophiles could not be human

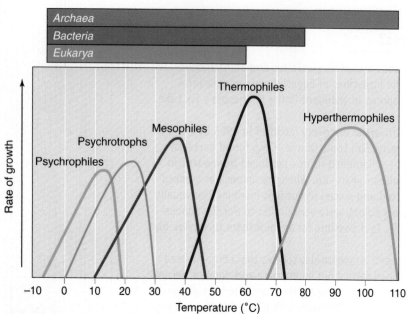

FIGURE 5.8 **Growth Rates for Different Microorganisms in Response to Temperature.** Temperature optima and ranges define the growth rates for *Bacteria*, *Archaea*, and *Eukarya*. Notice that the growth rates decline quite rapidly to either side of the optimal growth temperature. »» Propose what adaptations are needed for prokaryotes to survive at the psychrophilic or thermophilic extremes.

pathogens because they cannot grow at the warmer 37°C body temperature.

Another group of "cold-loving" microbes are the **psychrotrophs** or **psychrotolerant** microorganisms. These species have a higher optimal growth temperature as well as a higher minimal and maximal growth temperature. Psychrotrophs can be found in water and soil in temperate regions of the world but are perhaps most commonly encountered on spoiled refrigerated foods (4°C). Some bacterial and archaeal species are psychrotolerant as are several microbial eukaryotic species, including fungi (molds). When such foods are consumed without heating, the toxins may cause food poisoning. One example is *Campylobacter*, the most frequently identified cause of infective diarrhea (TEXTBOOK CASE 5).

At the opposite extreme are the **thermophiles** (*thermo* = "heat") that multiply best at temperatures around 60°C but still multiply from 40°C to 70°C. Thermophiles are present in compost heaps and hot springs, and are important contaminants in dairy products because they survive pasteurization temperatures. However, thermophiles pose little threat to human health because they do not grow well at the cooler temperature of the body.

Opposite to the psychrophiles, thermophiles have highly saturated fatty acids in their cell membranes to stabilize these structures. They also contain heat-stable proteins and enzymes.

There also are many archaeal species that grow optimally at temperatures that exceed 80°C. These **hyperthermophiles** have been isolated from seawater brought up from hot-water vents along rifts on the floor of the Pacific Ocean. Because the high pressure keeps the water from boiling, some archaeal species can grow at an astonishing 121°C (Table 5.1).

Most of the best-characterized bacterial species are **mesophiles** (*meso* = "middle"), which thrive at the middle temperature range of 10°C to 45°C. This includes the pathogens that grow in warm-blooded animals, including humans, as well as those species found in aquatic and soil environments in temperate and tropical regions of the world. *E. coli* is a typical mesophile.

Oxygen. The growth of many microbes depends on a plentiful supply of oxygen, and in this respect, such **obligate aerobes** must use oxygen gas as a final electron acceptor to make cellular energy (Chapter 6). Other species, such as *Treponema pallidum*, the agent of syphilis, are termed **microaerophiles** because they survive in environments where the concentration of oxygen is relatively low. In the body, certain microaerophiles cause disease of the oral cavity, urinary tract, and gastrointestinal tract. Conditions can be established in the laboratory to study these microbes (FIGURE 5.9A).

The **anaerobes**, by contrast, are microbes that do not or cannot use oxygen. Some are **aerotolerant**, meaning they are insensitive to oxygen. Many bacterial and archaeal species, as well as a few fungal and protozoal species, are **obligate anaerobes**, which are inhibited or killed if oxygen is present. This means they need other ways to make cell energy. Some anaerobic bacterial species, such as *Thiomargarita namibiensis* discussed in MicroFocus 3.5, use sulfur in their metabolic activities instead of oxygen, and therefore they produce hydrogen sulfide (H_2S) rather than water (H_2O) as a waste product of their metabolism. Others we have already encountered, such as the ruminant archaeal organisms that produce methane as the by-product of the energy conversions. In fact, life originated on Earth in an anaerobic

Textbook CASE | **5**

An Outbreak of Campylobacteriosis Caused by *Campylobacter jejuni*

1 On August 15, a cook began his day by cutting up raw chickens to be roasted for dinner.

2 He also cut up lettuce, tomatoes, cucumbers, and other salad ingredients on the same countertop. The countertop surface where he worked was unusually small.

3 For lunch that day, the cook prepared sandwiches on the same countertop. Most were garnished with lettuce.

4 Restaurant patrons enjoyed sandwiches for lunch and roasted chicken for dinner. Many patrons also had a portion of salad with their meal.

5 During the next three days, 14 people experienced stomach cramps, nausea, and vomiting.

Campylobacter jejuni agar culture.

6 Public health officials learned that all the affected patrons had eaten salad with lunch or dinner. *Campylobacter,* a bacterial pathogen of the intestines, was located in their stools.

7 On inspection, microbiologists concluded that the chicken was probably contaminated with *Campylobacter jejuni* (see figure). However, the microbiologists concluded that the cooked chicken was not the cause of the illness.

8 Rather, *C. jejuni* from the raw chicken was the source.

Questions:

(Answers can be found in Appendix D.)

A. Why would the cooked chicken not be the source for the illness?

B. Why was the raw chicken identified as the source?

C. How, in fact, did the patrons become ill?

For additional information see http://www.cdc.gov/mmwr/preview/mmwrhtml/00051427.htm.

environment consisting of methane and other gases (MICROFOCUS 5.4).

Some species of anaerobic bacteria cause disease in humans. For example, the *Clostridium* species that cause tetanus and gas gangrene multiply in the dead, anaerobic tissue of a wound and produce toxins causing tissue damage. Another species of *Clostridium* multiplies in the oxygen-free environment of a vacuum-sealed can of food, where it produces the lethal toxin of botulism.

Among the most widely used methods to establish anaerobic conditions in the laboratory

is the GasPak system, in which hydrogen reacts with oxygen in the presence of a catalyst to form water, thereby creating an oxygen-free atmosphere (**FIGURE 5.9B, C**).

Many microbes are neither aerobic nor anaerobic but **facultative**, meaning they can grow in either the presence or a reduced concentration of oxygen. This group includes many staphylococci and streptococci as well as members of the genus *Bacillus* and a variety of intestinal rods, among them *E. coli*. A facultative aerobe prefers anaerobic conditions (but also grows aerobically), while a

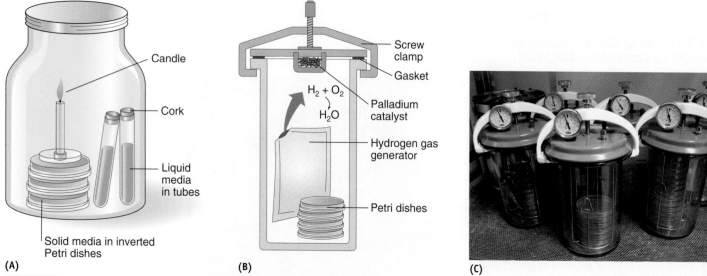

(A)   **(B)**   **(C)**

FIGURE 5.9 **Bacterial Cultivation in Different Gas Environments.** Two types of cultivation methods are shown for bacterial species that grow poorly in an oxygen-rich environment. (**A**) A candle jar, in which microaerophilic bacterial species grow in an atmosphere where the oxygen is reduced by the burning candle. (**B**, **C**) An anaerobic jar, in which hydrogen is released from a generator and then combines with oxygen through a palladium catalyst to form water and create an anaerobic environment. »» In which jar would a microbe grow?

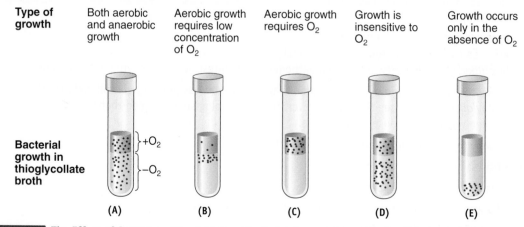

FIGURE 5.10 **The Effect of Oxygen on Microbial Growth.** Each tube contains a thioglycollate broth into which was inoculated a different bacterial species. »» Identify the O_2 requirement in each thioglycollate tube based on the growth density [example: (**A**) represents facultative microbes].

facultative anaerobe prefers oxygen-rich conditions (but also grows anaerobically).

A common way to test an organism's oxygen sensitivity is to use a **thioglycollate broth**, which binds free oxygen so that only fresh oxygen entering at the top of the tube would be available (**FIGURE 5.10**).

Finally, there are bacterial species said to be **capnophilic** (*capno* = "smoke"); they require an atmosphere low in oxygen but rich in carbon dioxide. Members of the genera *Neisseria* and *Streptococcus* are capnophiles.

pH. The cytoplasm of most microorganisms has a pH near 7.0. This means that the majority of species grow optimally at neutral pH (see Chapter 2) and have a growth pH range that covers three pH units. However, some pH-hearty bacterial species, such as *Vibrio cholerae*, can tolerate acidic conditions as low as pH 2.0 and alkaline conditions as high as pH 9.5.

Acid-tolerant bacteria called **acidophiles** are valuable in the food and dairy industries. For example, certain species of *Lactobacillus* and *Streptococcus* produce the acid that converts milk

MICROFOCUS 5.4: Evolution

"It's Not Toxic to Us!"

It's hard to think of oxygen as a poisonous gas, but billions of years ago, oxygen was as toxic as cyanide. One whiff by an organism and a cascade of highly destructive oxidation reactions was set into motion. Death followed quickly.

Difficult to believe? Not if you realize that the ancient *Bacteria* and *Archaea* relied on fermentation and anaerobic chemistry for their energy needs. They took organic materials from the environment and digested them to release the available energy. The atmosphere was full of methane, hydrogen, ammonia, carbon dioxide, and other gases. But no oxygen. And it was that way for hundreds of millions of years.

Then, some 3.5 billion years ago, along came the cyanobacteria with their ability to perform photosynthesis. Chlorophyll and chlorophyll-like pigments evolved, and the bacterial cells could now trap radiant energy from the sun and convert it to chemical energy in carbohydrates. But there was a downside: Oxygen was a waste product of the process—and it was deadly because the oxygen radicals (O_2^-, $OH\bullet$) produced could disrupt cellular metabolism by "tearing away" electrons from other molecules.

As millions of microbial species died off in the toxic oceans and atmosphere, others "escaped" to oxygen-free environments that are still in existence today. A few species survived because they evolved the enzymes to safely tuck away oxygen atoms in a nontoxic form—that form was water.

Also coming into existence were millions of new microbial species, some merely surviving and others thriving in the oxygen-rich environment. One of the modern-day survivors of these first communities are the **stromatolites**. These rock-like looking structures are still found in a few places on Earth, such as Sharks Bay off the western coast of Australia (see figure). These structures formed from ocean sediments and calcium carbonate that became trapped in the microbial community, building a rock-like fortress dead on the inside but alive on the surface. The top few inches in the crown of a stromatolite contain the oxygen-evolving, photosynthetic cyanobacteria, while below these species are other bacterial species that can tolerate oxygen and sunlight. Buried beneath these organisms are other bacterial species that survive the anaerobic, dark niche of the stromatolite where neither oxygen nor sunlight can reach. A couple of billion years would pass before one particularly well-known species of oxygen-breathing creature evolved: *Homo sapiens*.

Stromatolites, Shark Bay, Western Australia.

to buttermilk and cream to sour cream. These species pose no threat to good health even when consumed in large amounts. The "active cultures" in a cup of yogurt are actually acidophilic bacterial species. **Extreme acidophiles** are found among the *Archaea* as we saw in MicroFocus 2.3.

The majority of known bacterial species, however, do not grow well under acidic conditions. Thus, the acidic environment of the stomach helps deter disease, while providing a natural barrier to the organs beyond. In addition, you may have noted certain acidic foods such as lemons, oranges, and other citrus fruits as well as tomatoes and many vegetables are hardly ever contaminated by bacterial growth. However, such damaged produce may be subject to fungal growth because many fungi grow well at a pH of 5 or lower.

Hydrostatic and Osmotic Pressure. Further environmental factors can influence the growth of microbial cells. Psychrophiles in deep ocean waters and sediments are under extremely high hydrostatic pressure. In some deep marine trenches the hydrostatic pressure is tremendous—as high as 16,000 pounds per square inch (psi). Some extremophiles may be the only organisms able to withstand the pressure. Such **barophiles** in fact will die quite quickly at normal atmospheric pressures (14.7 psi).

We have discussed osmotic pressure previously in regard to the pressure water exerts on cells and the necessity for many microbial cells to have cell walls to prevent rupture of the cell or plasma membrane (see Chapter 4). In a reverse scenario, should the environment have more dissolved materials, water would leave the cells and the cells would plasmolyze. This is the principle behind salting meats and other food products, and using sugar as a preservative in jams and jellies. A high salt or sugar concentration will prevent growth and may even kill the cells (Chapter 7). Microbes, like *E. coli*, that are unable to grow in the presence of salt (NaCl) are called **nonhalophiles**.

There are microorganisms though that are salt-loving. These **halophiles** require salt to survive. Marine microoganisms represent halophiles surviving well in 3.5% NaCl. *Staphylococcus* and some other microbial species are considered **halotolerant**; that is, they grow best without NaCl but can tolerate low concentrations of NaCl. The **extreme halophiles** represent groups of the *Archaea* that tolerate salt concentrations of 15 to 30%.

The ability of microbes to withstand some very extreme conditions suggests they could live on other worlds (MicroFocus 5.5). **FIGURE 5.11** summarizes the physical factors influencing microbial growth.

CONCEPT AND REASONING CHECKS

5.5 Identify what would be extremophile conditions for each of the physical factors described in this section.

Hydrostatic pressure:
The pressure exerted by the weight of water.

5.3 Culture Media and Growth Measurements

In this chapter, we have been discussing microbial growth and the physical factors that control growth. To complete our analysis of growth and nutrition, we need to identify the chemical media used to grow and separate specific microorganisms, and consider the measurements used to evaluate growth.

A critical development in the design of culture media and the analysis of cell growth was the introduction of agar by Robert Koch (see Chapter 1). **Agar** is a polysaccharide derived from marine red algae. It contains no essential nutrients and is a unique colloid that remains liquid until cooled to below approximately 36°C. The solidified medium can be used to cultivate many different types of microbes, isolate pure cultures, or accomplish other tasks, such as a medium for measuring population growth.

Culture Media Are of Two Basic Types

KEY CONCEPT

6. Culture media contain the nutrients needed for optimal microbial growth.

Since the time of Pasteur and Koch, microbiologists have been growing bacterial and other microbial species in laboratory cultures; that is, in ways to mimic the natural environment. Today, many of the media used in the medical diagnostic bacteriology laboratory have their origins in the first Golden Age of Microbiology (see Chapter 1). These early media often contain

Colloid:
Aggregates of molecules in a finely divided state dispersed in a solid medium.

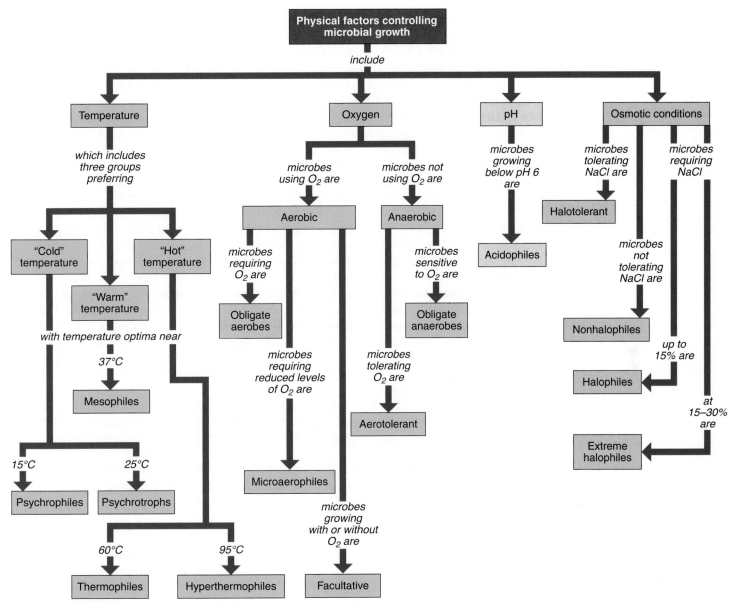

FIGURE 5.11 **Classes of Microbes Based on Physical Factors.** This concept map summarizes the classes of microbes requiring specific physical factors for growth. »» *Escherichia coli* is a mesophilic, facultative, nonhalophile. What specific physical factors does this organism require for optimal growth?

blood or serum to mimic the environment in the human body.

For the isolation and identification of microorganisms, two types of culture media are commonly used. A chemically undefined medium, or **complex medium** contains nutrients in which the exact components or their quantity is not completely known. For example, in nutrient broth or nutrient agar media it is not known precisely what

carbon and energy sources or other growth factors are present because complex media typically contain animal or plant digests (e.g., beef extract, soybean extract) or yeast extracts of an undefined nature (TABLE 5.2A). Complex media are commonly used in the teaching laboratory because the purpose is simply to grow microbes and not be concerned about what specific nutrients are needed to accomplish this action.

MICROFOCUS 5.5: Environmental Biology

War of the Worlds—On Mars

In Steven Spielberg's 2005 film *War of the Worlds,* adapted from H. G. Wells' 1898 novel, what were thought to be falling stars or meteorites turn out to be Martian spaceships fleeing a dying world. When the curious come to examine the crash sites in the countryside, they discover the alien spacecrafts are filled with tentacled Martian invaders and their robotic war machines. Metallic appendages emerge from the crash craters and begin to destroy everything in their path. The war between Mars and Earth has begun. Although this is science fiction, in reality the scenario could happen—only on Mars.

The United States has sent several spacecraft to Mars since the first Viking landers in 1976. Recently, an international team of scientists carried out studies suggesting terrestrial microbes could hitch a ride to Mars on such a craft—and even survive the journey. The team believes most spacecraft that have touched down on Mars were not thoroughly sterilized by heat or radioactivity, so they could be carrying living microbes from Earth. NASA scientists have assumed Mars' thin atmosphere, which allows intense ultraviolet (UV) radiation to reach the planet's surface—triple Earth's intensity—would kill any life inadvertently carried on the spacecraft. In laboratory tests, Martian-level doses of UV radiation destroyed most microbes in just seconds.

The reason the international team has raised the microbe alarm is from the tests they carried out. They tested the endurance of a particularly hardy cyanobacterium that thrives in the dry deserts of Antarctica. The extremophile, called *Chroococcidiopsis,* inhabits porous rocks near the rock surface where temperature and humidity are very low. The team found that most dormant spores of *Chroococcidiopsis* were killed after five minutes of a Martian UV dose. However, a few spores remained alive if they were buried by just 1 mm of soil. So, microbes might survive—and potentially grow—if protected from UV radiation and deposited in an environment with water and nutrients.

Until 2008, American spacecraft had not landed in areas known to have such "habitable" conditions. Then, NASA's Mars lander, the Phoenix mission, landed in the northern arctic region. It dug into the subsurface and detected water ice—areas where earthly microbial aliens could establish a foothold from a contaminated spacecraft.

If true, and Martian life also was present in these regions, is it possible that earthly tentacled (piliated) bacterial or radiation-resistant archaeal invaders might start a war of the worlds—on Mars?

TABLE 5.2   Composition of a Complex and a Synthetic Growth Medium

Ingredient	Nutrient Supplied	Amount
A. Complex Agar Medium		
Peptone	Amino acids, peptides	5.0 g
Beef extract	Vitamins, minerals, other nutrients	3.0 g
Sodium chloride (NaCl)	Sodium and chloride ions	8.0 g
Agar		15.0 g
Water		1.0 liter
B. Synthetic Broth Medium		
Glucose	Simple sugar	5.0 g
Ammonium phosphate $((NH_4)_2HPO_4)$	Nitrogen, phosphate	1.0 g
Sodium chloride (NaCl)	Sodium and chloride ions	5.0 g
Magnesium sulfate $(MgSO_4 \cdot 7H_2O)$	Magnesium ions, sulphur	0.2 g
Potassium phosphate (K_2HPO_4)	Potassium ions, phosphate	1.0 g
Water		1.0 liter

The second type of medium is a chemically defined or **synthetic medium**. In this medium, the precise chemical composition and amount of all components are known (TABLE 5.2B). This medium is used when trying to determine an organism's specific growth requirements.

CONCEPT AND REASONING CHECKS

5.6   Compare and contrast complex and synthetic media.

Culture Media Can Be Devised to Select for or Differentiate between Microbial Species

KEY CONCEPT

7.   Special chemical formulations are used to isolate and identify some bacteria.

In the clinical laboratory, the basic ingredients of the growth media can be modified in one of three ways to provide fast and critical information about the organism causing an infection or disease (TABLE 5.3).

TABLE

5.3   A Comparison of Special Culture Media

Name	Components	Uses	Examples
Selective medium	Growth stimulants Growth inhibitors	Selecting certain microbes out of a mixture	Mannitol salt agar for staphylococci
Differential medium	Dyes Growth stimulants Growth inhibitors	Distinguishing different microbes in a mixture	MacConkey agar for gram-negative bacteria
Enriched medium	Growth stimulants	Cultivating fastidious microbes	Blood agar for streptococci; chocolate agar for *Neisseria* species

A **selective medium** contains ingredients to inhibit the growth of certain microbes in a mixture while allowing the growth of others. The basic growth medium may contain extra salt (NaCl) or a dye to inhibit the growth of intolerant or sensitive organisms but permits the growth of those species or pathogens one wants to isolate (FIGURE 5.12A).

Another modification to a basic growth medium is the addition of one or more compounds that allow one to differentiate between very similar species based on specific biochemical or physiological properties. This **differential medium** contains in the culture plate specific chemicals to indicate which species possess and which lack a particular biochemical process. Such indicators make it easy to distinguish visually colonies of one organism from colonies of other similar organisms on the same culture plate (FIGURE 5.12B). MicroInquiry 5 looks closer at these two approaches to identify or separate similar bacterial species.

Although many microorganisms grow well in nutrient broth and nutrient agar, certain so-called fastidious organisms may require an **enriched medium** containing special nutrients (MicroFocus 5.6).

Fastidious: Having complex nutritional requirements.

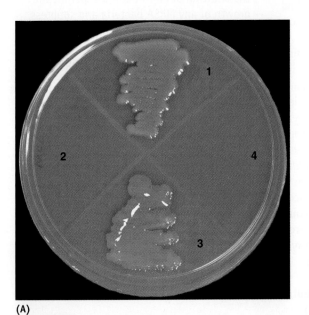

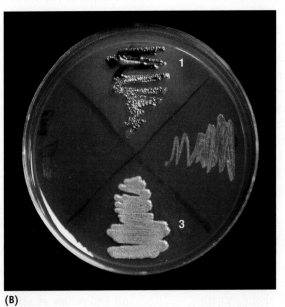

(A)   (B)

FIGURE 5.12 Selective and Differential Media. **(A)** Four different bacterial species (1–4), two gram positive and two gram negative, were streaked onto separate sections of a special plate and allowed to incubate for 48 hours. This medium only supports the growth of gram-negative species. **(B)** Because the two gram-negative species cannot be visually distinguished from one another, they were then streaked into another special medium plate and incubated for 48 hours. This medium allows one to distinguish between human enteric bacteria, where *Escherichia coli* produces a green metallic sheen while species like *Enterobacter aerogenes* produce a pink color. »» Which special medium is selective and which is differential? Explain your reasoning.

MICROINQUIRY 5

Identification of Bacterial Species

It often is necessary to identify a bacterial species or be able to tell the difference between similar-looking species in a mixture. In microbial ecology, it might be necessary to isolate certain naturally growing species from others in a mixture. In the clinical and public health setting, microbes might be pathogens associated with disease or poor sanitation. In addition, some may be resistant to standard antibiotics normally used to treat an infection. In all these cases, identification can be accomplished by modifying the composition of a complex or synthetic growth medium. Let's go through two scenarios.

■ Suppose you are an undergraduate student in a marine microbiology course. On a field trip, you collect some seawater samples and, now back in the lab, you want to grow only photosynthetic marine microbes.

How would you select for photosynthetic microbes? First, you know the photosynthetic organisms manufacture their own food, so their energy source will be sunlight and not the organic compounds typically found in culture media (see Table 5.2). So, you would need to use a synthetic medium but leave out the glucose. Also, knowing the salts typically in ocean waters, you would want to add them to the medium. You would then inoculate a sample of the collected material into a broth tube, place the tube in the light, and incubate for one week at a temperature typical of where the organisms were collected.

5a. What would you expect to find in the broth tube after one week's incubation?

What you have used in this scenario is a **selective medium**; that is, one that will encourage the growth of photosynthetic microbes (light and sea salts) and suppress the growth of non-photosynthetic micro-

organisms (no carbon = no energy source). So, only marine photosynthetic microbes should be present.

■ As an infection disease officer in a local hospital, you routinely swab critical care areas to determine if there are any antibiotic resistant bacteria present. You are especially concerned about methicillin-resistant *Staphylococcus aureus* (MRSA) as it frequently can cause disease outbreaks in a hospital setting. One swab you put in a broth tube showed turbidity after 48 hours.

5b. Knowing that *Staphylococcus* species are halotolerant, how could you devise an agar medium to visually determine if any of the growth is due to *Staphylococcus aureus*?

Again, a selective medium would be used. It would be prepared by adding 7.5% salt to a complex agar medium. A sample from the broth tube would be streaked on the plate and incubated at 37°C for 48 hours.

5c. What would you expect to find on the agar plate after 48 hours?

Your selective medium contained 10 discrete colonies. You do a Gram stain and discover that all the colonies contain clusters of purple spheres; they are gram-positive. However, there are other species of *Staphylococcus* that do not cause disease. One is *S. epidermidis*, a common skin bacterium. A Gram stain therefore is of no use to differentiate *S. aureus* from *S. epidermidis*.

5d. Knowing that only *S. aureus* will produce acid in the presence of the sugar mannitol, how could you design a differential broth medium to determine if any of the colonies are *S. aureus*? (Hint: phenol red is a pH indicator that is red at neutral pH and yellow at acid pH).

You can identify each bacterial species by taking a complex broth medium, such as

nutrient broth, and adding salt and mannitol (mannitol salts broth) and phenol red. Next, you inoculate a sample of each colony into a separate tube. You inoculate the 10 tubes and incubate them for 48 hours at 37°C.

5e. The broth tubes are shown below. What do the results signify? Which tubes contain which species of *Staphylococcus*?

This method is an example of a **differential medium** because it allowed you to visually differentiate or distinguish between two very similar bacterial species.

Knowing which colonies on the original selective medium plate are *S. aureus*, you need to determine which, if any, are resistant to the antibiotic methicillin.

5f. How could you design an agar medium to identify any MRSA colonies?

5g. If the plates are devoid of growth, what can you conclude?

Again, you have used a selective medium; the addition of methicillin will permit the growth of any MRSA bacteria and suppress the growth of staphylococci sensitive to methicillin.

Answers can be found in **Appendix D**.

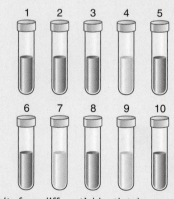

Results from differential broth tubes.

MICROFOCUS 5.6: Public Health
"Enriching" Koch's Postulates

On July 21–23, 1976, some 5,000 Legionnaires attended the Bicentennial Convention of the American Legion in Philadelphia, PA. About 600 of the Legionnaires stayed at the Bellevue Stratford Hotel. As the meeting was ending, several Legionnaires who stayed at the hotel complained of flu-like symptoms. Four days after the convention ended, an Air Force veteran who had stayed at the hotel died. He would be the first of 34 Legionnaires over several weeks to succumb to a lethal pneumonia, which became known as Legionnaires' disease.

As with any new disease, epidemiological studies look for the source of the disease. The Centers for Disease Control and Prevention (CDC) had an easy time tracing the source back to the Bellevue Stratford Hotel. Epidemiological studies also try to identify the causative agent. Using Koch's postulates, CDC staff collected tissues from lung biopsies and sputum samples. However, no microbes could be detected on slides of stained material. By December 1976, they were no closer to identifying the infectious agent.

How can you verify Koch's postulates if you have no infectious agent? It was almost like being back in the times of Pasteur and Koch. Why was this bacterial species so difficult to culture on bacteriological media? Perhaps it was a virus.

After trying 17 different culture media formulations, the infectious agent was finally cultured. It turns out it was a bacterial species, but one with fastidious growth requirements. The initial agar medium contained a beef infusion, amino acids, and starch. When this medium was enriched with 1% hemoglobin and 1% isovitalex, small, barely visible colonies were seen after five days of incubation at 37°C. Investigators then realized the hemoglobin was supplying iron to the bacterium and the isovitalex was a source of the amino acid cysteine. Using these two chemicals in pure form, along with charcoal to absorb bacterial waste, a pH of 6.9, and an atmosphere of 2.5% CO_2, bacterial growth was significantly enhanced (see figure). From these cultures, a gram-negative rod was confirmed and the organism was appropriately named *Legionella pneumophila*.

With an enriched medium to pure culture the organism, susceptible animals (guinea pigs) could be injected as required by Koch's postulates. *L. pneumophila* then was recovered from infected guinea pigs, verifying the organism as the causative agent of Legionnaires' disease.

Today, we know *L. pneumophila* is found in many aquatic environments, both natural and artificial. At the Bellevue Stratford Hotel, epidemiological studies indicated guests were exposed to *L. pneumophila* as a fine aerosol emanating from the air-conditioning system. Through some type of leak, the organism gained access to the system from the water cooling towers.

Koch's postulates are still useful—it's just hard sometimes to satisfy the postulates without an isolated pathogen.

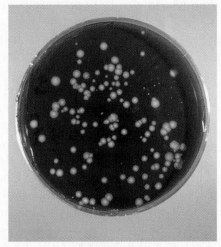

Colonies of *L. pneumophila* on an enriched medium.

Many of the *Bacteria* and *Archaea* are impossible to cultivate in any laboratory culture medium yet devised. In fact, less than 1% of the species in natural water and soil samples can be cultured. So, it is impossible to estimate accurately microbial diversity in an environment based solely on culturability. Such nonculturable organisms are said to be in a **VBNC (viable but non-culturable)** state. Procedures for identifying VBNC organisms include direct microscopic examination and, most commonly, amplification of diagnostic gene sequences or 16S rRNA sequences as mentioned in the introduction to Chapter 1.

Why are these organisms non-culturable? Microbiologists believe that part of the reason may be due to their presence in a "foreign" environment because most species have adapted to their own familiar and specific environment; a complex or synthetic medium is not their typical home. Therefore, these species go into a type of dormancy state and do not divide; that is, they are viable, but not culturable (see MicroFocus 5.2). Studies on VBNC *Bacteria* and *Archaea* present a vast and as yet unexplored field, which is important not only for detection of human pathogens, but also to reveal the diversity of these domains.

CONCEPT AND REASONING CHECKS

5.7 List reasons why many bacterial and archaeal species cannot be cultured in existing complex or synthetic growth media.

Population Measurements Are Made Using Pure Cultures

KEY CONCEPT

8. Two standard methods are available to produce pure cultures.

Microorganisms rarely occur in nature as a single species. Rather, they are mixed with other species, in a so-called mixed culture most often as a **biofilm** (see Chapter 3). Therefore, to study a species, microbiologists and laboratory technologists must use a **pure culture**—that is, a population consisting of only one species. This is particularly important when trying to identify a pathogen, as Pasteur discovered when trying to identify the agent responsible for cholera (see Chapter 1).

Aseptic technique:
The practice of transferring microorganisms to a sterile culture medium without introducing other contaminating organisms.

Subculturing:
The process of transferring bacteria from one tube or plate to another.

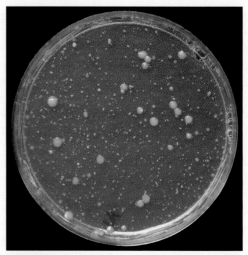

FIGURE 5.13 **A Pour Plate.** The dispersed bacterial cells grow as individual, discrete colonies. »» By looking at this plate, how would you know the original broth culture was a mixture of bacterial species?

If one has a mixed broth culture of bacterial species, how can the organisms be isolated as pure colonies? Two established methods are available. The first is the **pour-plate method**. Here, a small volume of the mixed culture is placed in a sterile culture dish. A molten agar medium is then poured into the dish and allowed to harden. During a 24 to 48 hour incubation, the cells divide to form discrete colonies throughout the agar (**FIGURE 5.13**).

A second, more commonly used technique, called the **streak-plate method**, uses a single plate of nutrient agar (**FIGURE 5.14A–D**). An inoculum from a mixed culture is removed with a sterile loop or needle using aseptic technique, and a series of streaks is made on the surface of one area of the plate. The loop is flamed, touched to the first area, and a second series is made in a second area. Similarly, streaks are made in the third and fourth areas, thereby spreading out the individual cells so they grow as separated colonies. On incubation, each cell will grow exponentially to form a discrete colony on the plate (**FIGURE 5.14E**)

In both methods, the researcher, technologist, or student can select samples of the colonies for further testing and subculturing.

CONCEPT AND REASONING CHECKS

5.8 Explain the difference between the pour-plate and streak-plate methods.

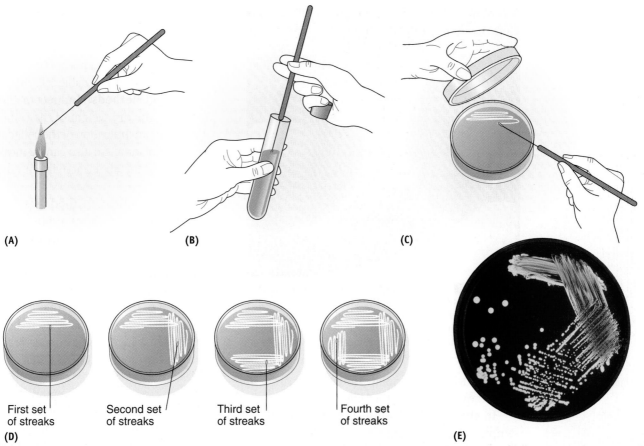

FIGURE 5.14 **The Streak-Plate Method.** (**A**) A loop is sterilized, (**B**) a sample of cells is obtained from a mixed culture, and (**C**) streaked near one edge of the plate of medium. (**D**) Successive streaks are performed, and the plate is incubated. (**E**) Well-isolated and defined colonies illustrate a successful isolation. »» Justify the need to streak a mixed sample over four areas on a culture plate.

Population Growth Can Be Measured in Several Ways

KEY CONCEPT

9. Microbial growth can be measured by direct and indirect methods.

Microbial growth in a medium, can be measured by direct and indirect methods.

Direct Methods. There are a number of ways to directly measure cell numbers. Scientists may wish to perform a **direct microscopic count** using a known sample of the culture on a specially designed counting chamber (**FIGURE 5.15**). However, this procedure will count both live and dead cells.

In the **most probable number test**, microbial samples are added to numerous lactose broth tubes and the presence or absence of gas formed in fermentation gives a rough statistical estimation of the cell number. This technique has been used

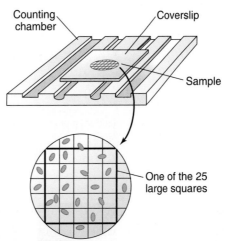

(A) The counting chamber is a specially marked slide containing a grid of 25 large squares of known area. The total volume of liquid held is 0.00002 ml (2×10^{-5} ml).

(B) The counting chamber is placed on the stage of a light microscope. The number of cells are counted in several of the large squares to determine the average number.

FIGURE 5.15 **Direct Microscopic Count.** This procedure can be used to estimate the total number of live and dead cells in a culture sample. »» Suppose the average number of cells per square was 14. Calculate the number of cells in a 10 ml sample.

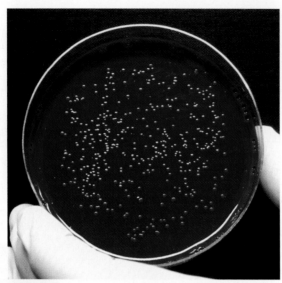

FIGURE 5.16 The Standard Plate Count. Individual bacterial colonies have grown on this blood agar plate. Each colony represents a colony-forming unit (CFU) since it developed from a single bacterial cell. »» If a 0.1 ml sample of a 10^4 dilution contained 250 colonies, how many bacterial cells were in 10 ml of the original broth culture?

cells could clump together on a plate and grow as a single colony, the standard plate count is expressed as the number of **colony-forming units** (**CFUs**). After incubation, the number of CFUs will be used to estimate the number of viable cells originally plated.

Indirect Methods. Indirect methods include measuring the dry weight of the cell population, which gives an indication of the cell mass. Oxygen uptake in metabolism also can be measured as an indication of metabolic activity and therefore cell number.

Another indirect method uses a spectrophotometer to measure the cloudiness, or **turbidity**, of a broth culture. This instrument detects the amount of light scattered by a suspension of cells placed in the spectrophotometer such that the amount of light scatter (optical density, OD) is a function of the cell mass; that is, the more cells present, the more light is scattered or absorbed, resulting in a higher absorbance reading on the spectrophotometer (**FIGURE 5.17**). A standard curve can be generated to serve as a measure of cell numbers. However, because more than 10 million cells are needed to make a reading on the spectrophotometer, turbidity is not a useful way to study the growth of small populations of bacterial cells.

for measuring water quality and is described in MicroInquiry 26 in Chapter 26.

In the **standard plate count procedure**, samples of a broth culture are spread on agar plates mixed in agar using the pour-plate method (**FIGURE 5.16**). The assumption is that each cell will undergo multiple rounds of cell division to produce separate colonies on the plate. Because two or more

CONCEPT AND REASONING CHECKS

5.9 Distinguish between direct and indirect methods to measure population growth.

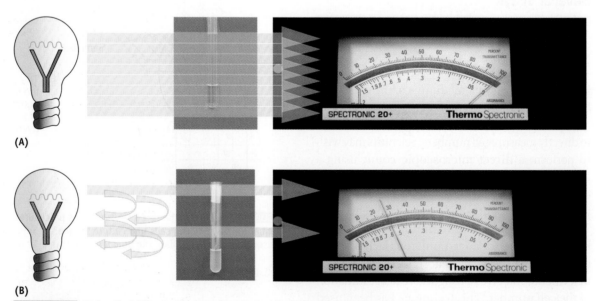

(A)

(B)

FIGURE 5.17 Using Turbidity to Measure Population Growth. (**A**) As light passes through a sterile broth tube in the spectrophotometer, the instrument is standardized at 0 absorbance. (**B**) As a bacterial population in a broth tube grows, the cells will scatter more of the light, which on the spectrophotometer is detected as an increase in absorbance. »» Why do turbidity measurements represent an indirect method to measure population growth?

SUMMARY OF KEY CONCEPTS

5.1 Microbial Reproduction

1. Bacterial reproduction involves DNA replication and **binary fission** to produce genetically identical daughter cells.
2. Binary fissions occur at intervals called the **generation time**, which may be as short as 20 minutes.

5.2 Microbial Growth

3. The dynamics of the **bacterial growth curve** show how a microbial population grows logarithmically, reaches a certain peak and levels off, and then may decline.
4. **Sporulation** is a dormancy response in a few bacterial species to nutrient limitation and high population density. The **endospores** formed are resistant to many harsh environmental conditions.
5. Temperature, oxygen, pH, and hydrostatic/osmotic pressure are physical factors that influence microbial growth. Away from the optimal condition, growth slows within a set range.

5.3 Culture Media and Growth Measurements

6. **Complex** and **synthetic media** contain the nutrients for microbial growth.
7. Complex or synthetic media can be modified to select for a desired microbial species, to differentiate between two similar species, or to enrich for species requiring special nutrients.
8. **Pure cultures** can be produced from a mixed culture by the **pour-plate method** or the **streak-plate method**. In both cases, discrete colonies can be identified that represent only one microbial species.
9. Microbial growth can be measured by **direct microscopic count**, the **most probable number test**, and the **standard plate count** procedure. Indirect methods include dry weight, oxygen uptake, and **turbidity measurements**.

LEARNING OBJECTIVES

After understanding the textbook reading, you should be capable of writing a paragraph that includes the appropriate terms and pertinent information to answer the objective.

1. Describe the process of **binary fission**.
2. Summarize the uses for knowing a microbe's **generation time**.
3. Compare the events of each phase of a **bacterial growth curve**.
4. Contrast the stages of bacterial **sporulation** and assess the importance of the process.
5. Identify the 4 major physical factors governing microbial growth and describe how microorganisms have adapted to these physical environments.
6. Contrast the chemical composition of **complex** and **synthetic media**.
7. Explain how **selective** and **differential** media are each constructed.
8. Explain the procedures used in the **pour-plate** and **streak-plate** methods.
9. Judge the usefulness of **direct** and **indirect methods** to measure microbial growth.

STEP A: SELF-TEST

Each of the following questions is designed to assess your ability to remember or recall factual or conceptual knowledge related to this chapter. Read each question carefully, then select the *one* answer that best fits the question or statement. Answers to even-numbered questions can be found in **Appendix C**.

1. Which one of the following statements does NOT apply to bacterial reproduction?
   A. A fission ring apparatus is present.
   B. Septum formation occurs.
   C. A spindle apparatus is used.
   D. Symmetrical cell division occurs.
2. If a bacterial cell in a broth tube has a generation time of 40 minutes, how many cells will there be after 6 hours of optimal growth?
   A. 18
   B. 64
   C. 128
   D. 512
3. A bacterial species generation time would be determined during the _____ phase.
   A. decline
   B. lag
   C. log
   D. stationary
4. Which one of the following is NOT an event of sporulation?
   A. Symmetrical cell divisions
   B. Mother cell disintegration
   C. DNA replication
   D. Prespore engulfment by the mother cell
5. A microbe that is a microaerophilic mesophile would grow optimally at _____ and _____ .
   A. high O_2; 30°C
   B. low O_2; 20°C
   C. no O_2; 30°C
   D. low O_2; 37°C
6. If the carbon source in a growth medium is beef extract, the medium must be an example of a/an _____ medium.
   A. complex
   B. chemically defined
   C. enriched
   D. synthetic
7. A _____ medium would involve the addition of the antibiotic methicillin to identify methicillin-resistant bacteria.
   A. differential
   B. selective
   C. thioglycollate
   D. VBNC

8. Which one of the following is NOT part of the streak-plate method?
   A. Making four sets of streaks on a plate.
   B. Diluting a mixed culture in molten agar.
   C. Using a mixed culture.
   D. Using a sterilized loop.

9. Direct methods to measure bacterial growth would include all the following except the _____ .
   A. total bacterial count
   B. microscopic count
   C. turbidity measurements
   D. most probable number

STEP B: REVIEW

Answers to even-numbered questions or statements can be found in **Appendix C**.

10. Use the log phase growth curves (1, 2, or 3) below to answer each of the following questions (a–c).

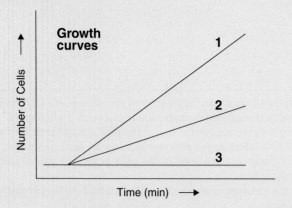

Growth curves

Number of Cells

Time (min) →

_____ a. Which curve (1, 2, or 3) best represents the growth curve for a mesophile incubated at 60°C?

_____ b. Which curve (1, 2, or 3) best represents a non-halophile growing in 5% salt?

_____ c. Which curve (1, 2, or 3) best represents an acidophile growing at pH 4?

11. Construct a concept map for **Growth Measurements** using the following terms.

cell mass
colony-forming units
direct methods
direct microscopic count
dry weight
indirect methods

metabolic activity
most probable number test
oxygen uptake
standard plate count
turbidity

On completing your study of these pages, test your understanding of their contents by deciding whether the following statements are true (T) or false (F). If the statement is false, substitute a word or phrase for the underlined word or phrase to make the statement true.

12. _____ Endospores are produced by some gram-negative bacterial species.

13. _____ Obligate aerobes use oxygen gas as a final electron acceptor in energy production.

14. _____ The most common growth medium used in the teaching laboratory is a complex medium.

15. _____ The majority of bacterial and archaeal organisms that have been discovered can be cultured in growth media.

16. _____ A standard plate count procedure is an example of a direct method to estimate population growth.

17. _____ In attempting to culture a fastidious bacterial pathogen, a differential medium would be used.

18. _____ Acidophiles grow best at pHs greater than 9.

19. _____ Mesophiles have their optimal growth near 37°C.

20. _____ Bacterial and archaeal cells lack a mitotic spindle to separate chromosomes.

21. _____ The fastest doubling time would be found in the lag phase of a bacterial growth curve.

22. _____ If *E. coli* cells are placed in distilled water, they will lyse.

23. _____ Halophiles would dominate in marine environments.

STEP C: APPLICATIONS

Answers to even-numbered questions can be found in **Appendix C**.

24. Consumers are advised to avoid stuffing a turkey the night before cooking, even though the turkey is refrigerated. A homemaker questions this advice and points out that the bacterial species of human disease grow mainly at warm temperatures, not in the refrigerator. What explanation might you offer to counter this argument?

25. Public health officials found that the water in a Midwestern town was contaminated with sewage bacteria. The officials suggested that homeowners boil their water for a couple of minutes before drinking it. (a) Would this treatment sterilize the water? Why? (b) Is it important that the water be sterile? Explain.

STEP D: QUESTIONS FOR THOUGHT AND DISCUSSION

Answers to even-numbered questions can be found in **Appendix C**.

26. To prevent decay by bacterial species and to display the mummified remains of ancient peoples, museum officials place the mummies in glass cases where oxygen has been replaced with nitrogen gas. Why do you think nitrogen is used?

27. Extremophiles are of interest to industrial corporations, who see these organisms as important sources of enzymes that function at temperatures of 100°C and pH levels of 10 (the enzymes have been dubbed "extremozymes"). What practical uses can you foresee for these enzymes?

28. During the filming of the movie *Titanic,* researchers discovered at least 20 different bacterial and archaeal species literally consuming the ship, especially a rather large piece of the midsection. What type of environmental conditions are these bacterial and archaeal species subjected to at the wreck's depth of 12,600 feet?

29. Although thermophilic bacteria are presumably harmless because they do not grow at body temperatures, they may still present a hazard to good health. Can you think of a situation in which this might occur?

HTTP://MICROBIOLOGY.JBPUB.COM/9E

The site features eLearning, an online review area that provides quizzes and other tools to help you study for your class. You can also follow useful links for in-depth information, read more MicroFocus stories, or just find out the latest microbiology news.

6

Metabolism of Microorganisms

Life is like a fire; it begins in smoke and ends in ashes.
—Ancient Arab proverb connecting energy to life

Charlie Swaart had been a social drinker for years. A few beers or drinks with his pals, but no lasting alcoholic consequences. Then, in 1945, he began a nightmare that would make medical history.

One October day, while stationed in Tokyo after World War II, Swaart suddenly became drunk for no apparent reason. He had not had any alcohol for days, but suddenly he felt like he had been partying all night. After sleeping it off, he would be fine the next day.

Unfortunately, this "behavior" returned time and time again. For years thereafter, the episodes continued—bouts of drunkenness and monumental hangovers without drinking so much as a beer! Doctors were puzzled as they could detect alcohol on his breath and in his blood. Was this some type of internal metabolism gone haywire? Was it the result of a bacterial infection? It didn't seem likely. They warned him though not to drink any additional alcohol for fear of damaging his liver. Swaart followed their advice to the letter; still, he experienced periods of drunkenness.

Twenty years passed before Swaart, known as the "drinkless drunk," learned of a similar case in Japan. A Japanese businessman had endured years of social and professional disgrace before doctors discovered a yeast-like fungus in his intestine. Studying this eukaryotic microbe showed that the fungal cells were fermenting carbohydrates to alcohol right there in his intestine. The fungus was identified as *Candida albicans* (**FIGURE 6.1**). Now having *C. albicans* in one's intestine is not uncommon; but, finding fermenting *C. albicans* was historic.

Swaart learned that an antibiotic had worked to kill the yeast cells in the Japanese man's gut. With this knowledge, he approached his doctor. Sure enough, lab tests showed massive colonies of *C. albicans* in Swaart's intestine too. The sugar in a cup of coffee or any carbohydrate in pasta, cake, or candy

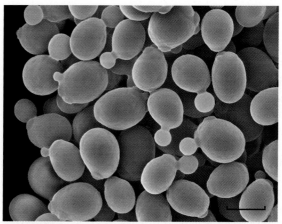

FIGURE 6.1 *Candida albicans* **Cells.** This false-color scanning electron microscope image of *C. albicans* shows daughter cells (yellow) budding from the mother cells. (Bar = 10 μm.) »» To what domain of organisms does *C. albicans* belong?

TABLE

6.1   A Comparison of Two Key Aspects of Cellular Metabolism

Anabolism	Catabolism
Buildup of small molecules	Breakdown of large molecules
Products are large molecules	Products are small molecules
Photosynthesis	Glycolysis, citric acid cycle
Mediated by enzymes	Mediated by enzymes
Energy generally is required (endergonic)	Energy generally is released (exergonic)

could bring on drunkenness. However, to cure his illness, Swaart had to travel back to Japan to get the effective antibiotic.

Researchers believed the atomic blasts of Hiroshima and Nagasaki in 1945 may have caused a normal *C. albicans* to mutate to a fermenting form, which somehow found its way into Swaart's digestive system. One can only wonder if there are many other individuals who have become living fermentation vats for the fungus. For Charlie Swaart, though, the nightmare was finally over.

The process of fermentation described here was in a eukaryotic microbe. However, many other types of fermentation processes also occur in microorganisms and they are but one aspect of the broad topic of microbial metabolism.

Metabolism refers to all the biochemical reactions taking place in an organism. These processes are divided into two general categories. **Anabolism** builds larger organic compounds such as carbohydrates and proteins from simpler monomers, including glucose and amino acids. **Catabolism**

hydrolyzes these polymers into simpler molecules such as carbon dioxide, ammonia, and water (see Chapter 2).

From an energy perspective, anabolic reactions form bonds, which require energy. Such energy-requiring processes are **endergonic** (*end* = "inner"; *ergon* = "work") **reactions**. In contrast, catabolic reactions break bonds, releasing energy. These processes are called **exergonic** (*ex* = "outside of") **reactions**. **TABLE 6.1** compares anabolism and catabolism, which are reactions often taking place simultaneously in cells and organisms. Realize metabolism also includes conversion reactions that transform one molecule into another without any type of catabolic or anabolic event.

In this chapter, we examine the types of metabolism exhibited by microorganisms. Much of the chapter discusses the catabolic reactions involved in energy conversions forming adenosine triphosphate (ATP). Because the chapter emphasizes the role of carbohydrates in the energy conversions, it might be worthwhile to revisit Chapter 2 and refresh yourself with these types of organic compounds.

6.1   Enzymes and Energy in Metabolism

The growth we examined in the previous chapter depends on metabolic processes that occur in the cell cytosol, on the cell (plasma) membrane, in the periplasmic space (gram-negative bacteria), in eukaryotic organelles, and outside the cell. To

carry out these reactions, cells need a large variety of enzymes. Therefore, we begin our study of metabolism with a discussion of these essential proteins, which have been known only since the early 1900s (MicroFocus 6.1).

MICROFOCUS 6.1: History
"Hans, Du Wirst Das Nicht Glauben!"

Louis Pasteur's discovery of the role yeast cells play in fermentation heralded the beginnings of micro-biology because it showed that tiny organisms could bring about important chemical changes. However, it also opened debate on how yeasts accomplished fermentation. Lively controversies ensued among scientists; some suggesting sugars from grape juice entered yeast cells to be fermented; others believing fermentation occurred outside the cells. The question would not be resolved until a fortunate accident happened in the late 1890s.

In 1897, two German chemists, Eduard and Hans Buchner, were preparing yeast as a nutritional supplement for medicinal purposes. They ground yeast cells with sand and collected the cell-free "juice." To preserve the juice, they added a large quantity of sugar (as was commonly done at that time) and set the mixture aside. Several days later Eduard noticed an unusual alcoholic aroma coming from the mixture. Excitedly, he called to his brother, "Hans, Du Wirst Das Nicht Glauben!" ("Hans, you'll never believe this!") One taste confirmed their suspicion: The sugar had fermented to alcohol.

The discovery by the Buchner brothers was momentous because it demonstrated that a chemical substance inside yeast cells brings about fermentation, and that fermentation can occur without living cells. The chemical substance came to be known as an "enzyme," meaning "in yeast."

In 1905, the English chemist Arthur Haden expanded the Buchner study by showing that "enzyme" is really a multitude of chemical compounds and should better be termed "enzymes." Thus, he added to the belief that fermentation is a chemical process. Soon, many chemists became biochemists, and biochemistry gradually emerged as a new scientific discipline.

Enzymes Catalyze All Chemical Reactions in Cells

KEY CONCEPT

1. Enzyme catalysis means enzymes must have specific chemical properties.

Enzymes are proteins (or in a few instances RNA molecules) that increase the probability of chemical reactions while themselves remaining unchanged. They accomplish in fractions of a second what otherwise might take hours, days, or longer to happen spontaneously under normal biological conditions. For example, even though organic molecules like amino acids have functional groups, it is highly unlikely they would randomly bump into one another in the precise way needed for a chemical reaction (dehydration synthesis) to occur and for a new peptide bond to be formed (see Chapter 2). Thus, the reaction rate would be very slow were it not for the activity of enzymes.

Enzymes have several common characteristics.

1. Enzymes are reusable. Once a chemical reaction has occurred, the enzyme is released to participate in another identical reaction. In fact, the same enzyme can catalyze the same reaction 100 to 1,000 times each second.

2. Enzymes are highly specific. An enzyme that functions in one type of chemical reaction usually will not participate in another type of reaction. That means there must be thousands of different enzymes to catalyze the thousands of different chemical reactions of metabolism occurring in a microbial organism.

3. Enzymes have an active site. Each enzyme has a special pocket or cleft called an **active site**, which has a specific three-dimensional shape complementary to a reactant (called a **substrate**). The active site positions the substrate such that it is highly likely a chemical reaction will occur to form one or more **products**.

4. Enzymes are required in minute amounts. Because an enzyme can be used thousands of times to catalyze the same reaction, only minute amounts of a particular enzyme are needed to ensure that a fast and efficient metabolic effect occurs.

Many enzymes can be identified by their names, which often end in "-ase." For example, sucrase is the enzyme that breaks down sucrose and ribonuclease digests ribonucleic acid. In terms of anabolic metabolism, polymerases link together nucleotides and transferases link together the NAG and NAM units to build the bacterial cell wall peptidoglycan.

CONCEPT AND REASONING CHECKS

6.1 List the characteristics of enzymes.

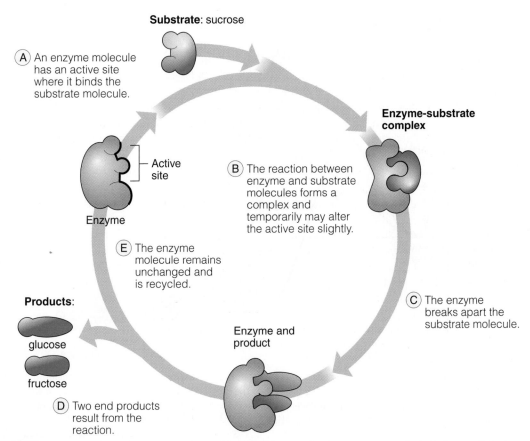

FIGURE 6.2 The Mechanism of Enzyme Action. Although this example shows an enzyme hydrolyzing a substrate (sucrose), enzymes also catalyze dehydration reactions, which, in this case, would combine glucose and fructose into sucrose. »» How do enzymes recognize specific substrates?

Enzymes Act through Enzyme-Substrate Complexes

KEY CONCEPT

2. Cellular chemical reactions occur at the enzyme's active site.

Enzymes function by aligning substrate molecules in such a way that a reaction is highly favorable. In the hydrolysis reaction shown in **FIGURE 6.2**, the three-dimensional shape of the enzyme's active site recognizes and holds the substrate in an **enzyme-substrate complex**. While in the complex, chemical bonds in the substrate are stretched or weakened by the enzyme, causing the bond to break. In a synthesis reaction, by contrast, the electron shells of the substrates in the enzyme-substrate complex are forced to overlap in the spot where the chemical bond will form.

Thus, in a hydrolysis or dehydration reaction, recognition of the substrate(s) is a precisely controlled, nonrandom event.

Looking at sucrose again, the bonds holding glucose and fructose together will not break spontaneously. The reason is the bond between the monosaccharides is stable and there is a substantial energy barrier preventing a reaction (**FIGURE 6.3A**). The job of sucrase is to bind the substrate and lower the energy barrier so that it is much more likely that the reaction will occur. In other words, the bond holding glucose to fructose needs to be destabilized (i.e., stretched, weakened) by the enzyme (**FIGURE 6.3B**). This energy barrier is called the **activation energy**. Enzymes, then, play a key role in metabolism because they provide an alternate reaction pathway of less resistance; that is, with a lower activation energy barrier. They assist in the destabilization of chemical bonds and the formation of new ones by separating or joining atoms in a carefully orchestrated fashion.

Some enzymes are made up entirely of protein. An example is **lysozyme**, the enzyme in human tears and saliva that hydrolyzes the bond between

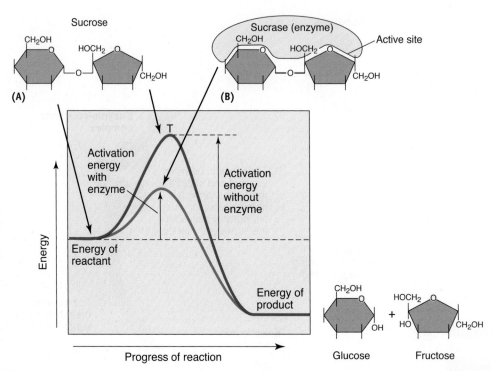

FIGURE 6.3 **Enzymes and Activation Energy.** Enzymes lower the activation energy barrier required for chemical reactions of metabolism. (**A**) The hydrolysis of sucrose is unlikely because of the high activation energy barrier. (**B**) When sucrase is present, the enzyme effectively lowers the activation energy barrier at the transition state (**T**), making the hydrolysis reaction highly favorable. »» How does the enzyme lower the activation energy of a stable substrate?

NAG and NAM in the cell walls of gram-positive bacterial cells. Other enzymes, however, may contain small, nonprotein substances, called **cofactors**, that participate in the catalytic reaction. Inorganic cofactors are metal ions, such as magnesium (Mg^{2+}), iron (Fe^{2+}), and zinc (Zn^{2+}). When the nonprotein cofactor is a small organic molecule, it is referred to as a **coenzyme,** most of which are derived from vitamins. Examples of two important coenzymes are nicotinamide adenine dinucleotide (NAD^+) and flavin adenine dinucleotide (FAD). These coenzymes play a significant role as electron carriers in metabolism, and we shall encounter them in our ensuing study of microbial metabolism.

CONCEPT AND REASONING CHECKS

6.2 Assess the role of the active site in "stimulating" a chemical reaction.

Enzymes Often Team Up in Metabolic Pathways

KEY CONCEPT

3. Metabolism often involves a series of chemical reactions controlled by separate enzymes.

There are many examples, such as the sucrose example, where an enzymatic reaction is a sin-

gle substrate to product reaction. However, cells more often use metabolic pathways. A **metabolic pathway** is a sequence of chemical reactions, each reaction catalyzed by a different enzyme, in which the product (output) of one reaction serves as a substrate (input) for the next reaction (**FIGURE 6.4A**). The pathway starts with the initial substrate and finishes with the final end product. The products of "in-between" stages are referred to as "intermediates."

Metabolic pathways can be anabolic, where larger molecules are synthesized from smaller monomers. In contrast, other pathways are catabolic because they break larger molecules into smaller ones. Such pathways may be linear, branched, or cyclic. We will see many of these pathways in the microbial metabolism sections ahead.

CONCEPT AND REASONING CHECKS

6.3 If a metabolic pathway has eight intermediates, how many different enzymes are involved?

Enzyme Activity Can Be Inhibited

KEY CONCEPT

4. Metabolism can control and be controlled by enzymes.

Both environmental factors by themselves and in concert with metabolic pathways can inhibit enzyme activity permanently or temporarily.

Enzyme Inhibition by Environmental Factors. In the last chapter, we spent some time investigating the physical factors affecting microbial growth. For example, temperature affects an enzyme's reaction rate, which slows down the further temperature deviates from the optimal. Since most enzymes are proteins, they are sensitive to changes in temperature; indeed, high temperature can denature a protein, perhaps bringing metabolism to a sudden halt.

We also discussed pH and how it affects microbial growth. Again, an increase or decrease in protons (H^+) will interfere with an enzyme's reaction rate, extreme change leading to not only enzyme denaturation, but to enzyme and metabolic inhibition.

In addition, chemicals applied "environmentally" may inhibit enzyme action. Alcohols and phenol inactivate enzymes and precipitate proteins, making these chemical agents effective antiseptics or disinfectants (Chapter 7). Other natural chemicals interfere with enzyme action (e.g., penicillin) or with a cell's ability to carry out a critical enzyme reaction (e.g., sulfa drugs), making these agents effective antibiotics (Chapter 24).

Enzyme Inhibition through Pathway Modulation. The same chemical reaction does not occur in a cell all the time, even if the substrate is present. Rather, cells regulate the enzymes so that they are present or active only at the appropriate time during metabolism.

One of the most common ways of modulating enzyme activity is for the final end product of a metabolic pathway to inhibit an enzyme in that pathway (**FIGURE 6.4B**). If the first enzyme in the pathway is inhibited, then no product is available as input for the rest of the pathway. Such **feedback inhibition** is typical of many metabolic pathways in cells. In general, when the final end product or any molecule binds to a non-active site on the enzyme, the shape of the active site changes and can no longer bind substrate. This type of modulation is referred to as **noncompetitive inhibition**.

Another way of modulating an enzyme is by blocking its active site so the normal substrate cannot bind. Such **competitive inhibition** occurs in the following way (**FIGURE 6.4C**). If a molecule resembles the normal substrate, it binds reversibly to the active site, competing with the normal sub-

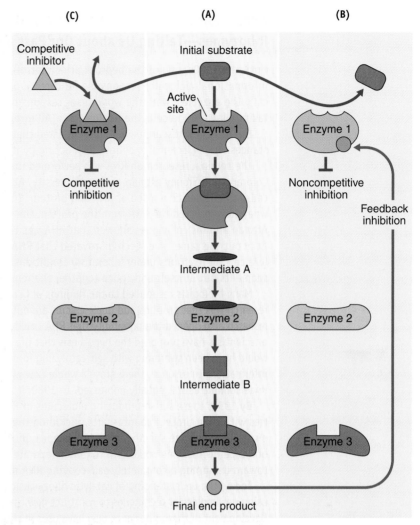

FIGURE 6.4 Metabolic Pathways and Enzyme Inhibition. (A) In a metabolic pathway, a series of enzymes transforms an initial substrate into a final end product. **(B)** If excess final end product accumulates, it "feeds back" on the first enzyme in the pathway and inhibits the enzyme by binding at another site on the enzyme. **(C)** In competitive inhibition, a substrate that resembles the normal substrate competes with the substrate for the enzyme's active site. Competitive inhibition would reduce the productivity of the metabolic pathway by slowing down or stopping the pathway. In both **(B)** and **(C)**, the whole pathway can become temporarily inoperative. »» Hypothesize why most substrates cannot be converted into a final end product in one enzymatic step.

strate for the active site. Sitting in the active site, this competitive inhibitor cannot be converted to product and does not allow the normal substrate to bind.

The dogma in biology used to say that all enzymes were proteins. Although this statement usually is true, there are cases where ribonucleic acid (RNA) can have catalytic effects (MICROFOCUS 6.2).

CONCEPT AND REASONING CHECKS

6.4 Describe how an enzyme could be modulated by competitive and noncompetitive inhibition.

MICROFOCUS 6.2: History/Biotechnology
Ribozymes—Telling Us about Our Past and Helping with Our Future

Until the 1980s, one of the bedrock principles of biology held that nucleic acids (DNA and RNA) were the informational molecules responsible for directing the metabolic reactions in the cell. Proteins, specifically the enzymes, were the workhorses responsible for catalyzing the thousands of chemical reactions taking place in the cell. The dogma was, "All enzymes are proteins."

In 1981, new research evidence suggested that RNA molecules could act as catalysts in certain circumstances. Today, scientists believe RNA acting by itself can trigger specific chemical reactions.

The seminal research on RNA was performed independently by Thomas R. Cech of the University of Colorado and Sidney Altman of Yale University. Altman had found an unusual enzyme in some bacterial cells, an enzyme composed of RNA and protein. Initially, he thought the RNA was a contaminant, but when he separated the RNA from the protein, the bacterial enzyme could not function. After several years, Altman and his colleagues showed that RNA was the enzyme's key component because it could act alone. At about the same time, Cech discovered that RNA molecules from *Tetrahymena*, a protozoan, could catalyze certain reactions under laboratory conditions. He showed that a molecule of RNA could cut internal segments out of itself and splice together the remaining segments.

Many biologists responded to the findings of Cech and Altman with disbelief. The implication of the research was that proteins and nucleic acids are not necessarily interdependent, as had been assumed. The research also opened the possibility that RNA could have evolved on Earth without protein. In fact, a number of scientists have proposed the hypothesis that life may have started in a primeval "RNA world." This world would have been swarming with self-catalyzing forms of RNA having the ability to reproduce and carry genetic information. In essence, there arose a whole new way of imagining how life might have begun. The Nobel Prize committee was equally impressed. In 1989, it awarded the Nobel Prize in Chemistry to Cech and Altman.

By 1990, these self-reproducing molecules of RNA had a name—**ribozymes**. They share many similarities with their protein counterparts, including the presence of binding pockets that, like active sites on enzymes, recognize specific molecular shapes. Biochemists at Massachusetts General Hospital showed that one type of ribozyme could join together separate short nucleotide segments. The research was a step toward designing a completely self-copying RNA molecule.

Today, the understanding of catalytic ribozymes goes beyond the research laboratory. Several companies are using new molecular techniques to construct new catalytic ribozymes in what is termed "directed evolution." Development of these ribozymes may have uses in clinical diagnostics and as therapeutic agents. For example, in diagnostic applications, ribozymes are being developed to identify potential new drugs. Other companies are using ribozymes as biosensors to detect viral contaminants in blood. These catalytic molecules also may be useful in fighting infectious diseases by inactivating RNA molecules in viruses or other pathogens.

So, ribozymes have much to offer in understanding our very distant past as well as providing for a healthier future.

Energy in the Form of ATP Is Required for Metabolism

KEY CONCEPT

5. ATP is the universal energy currency in cells.

In many metabolic reactions, energy is needed, along with enzymes, for the reactions to occur. The cellular "energy currency" is a compound called **adenosine triphosphate (ATP)** (**FIGURE 6.5A**). In bacterial and archaeal cells, the ATP is formed on the cell membrane, while in eukaryotes the reactions occur primarily in the mitochondria. An ATP molecule acts like a por-

table battery. It provides the needed energy for activities such as binary fission, flagellar motion, active transport, and spore formation. On a more chemical level, it fuels protein synthesis and carbohydrate breakdown. It is safe to say that a major share of microbial functions depends on a continual supply of ATP. Should the supply be cut off, the cell dies very quickly, as ATP cannot be stored.

ATP molecules are relatively unstable. In Figure 6.5A, notice that the three phosphate groups all have negative charges on an oxygen atom. Like charges repel, so the phosphate groups in ATP, being tightly packed together, are very

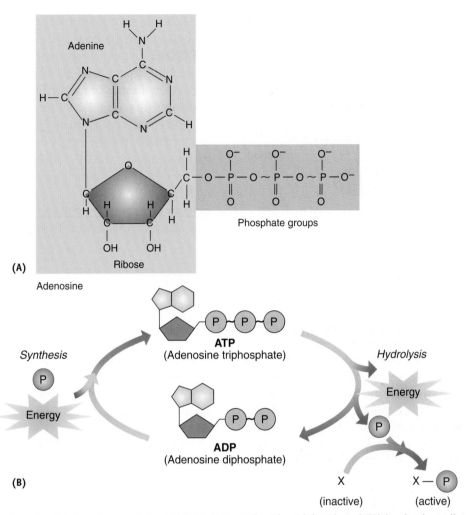

FIGURE 6.5 Adenosine Triphosphate and the ATP/ADP Cycle. Adenosine triphosphate (ATP) is a key immediate energy source for all microbial cells and other living things. (**A**) The ATP molecule is composed of adenine and ribose bonded to one another and to three phosphate groups. (**B**) When the ATP molecule breaks down, it releases a phosphate group and energy, and becomes adenosine diphosphate (ADP). The freed phosphate can activate another chemical reaction through phosphorylation. For the synthesis of ATP, energy and a phosphate group must be supplied to an ADP molecule. »» What genetic molecule closely resembles ATP?

unstable. Breaking the so-called "high-energy bond" holding the last phosphate group on the molecule produces a more stable **adenosine diphosphate** (**ADP**) molecule and a free phosphate group (**FIGURE 6.5B**). ATP hydrolysis is analogous to a spring compacted in a box. Open the box (hydrolyze the phosphate group) and you have a more stable spring (a more stable ADP molecule). The release of the spring (the freeing of a phosphate group) provides the means by which work can be done. Thus, the hydrolysis of the unstable phosphate groups in ATP molecules to a more stable condition is what drives other energy-requiring reactions through the transfer of

phosphate groups (Figure 6.5B). The addition of a phosphate group to another molecule is called **phosphorylation**.

Because ATP molecules are unstable, they cannot be stored. Therefore, microbial cells synthesize large organic compounds like glycogen or lipids for energy storage. As needed, the chemical energy in these molecules can be released in catabolic reactions and used to reform ATP from ADP and phosphate (Figure 6.5B). This **ATP/ADP cycle** occurs continuously in cells. It has been estimated that a typical bacterial cell must reform about 3 million ATP molecules per second from ADP and phosphate to supply its energy needs.

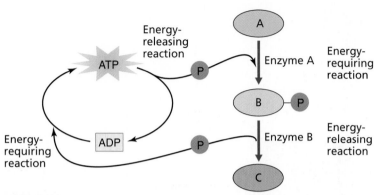

FIGURE 6.6 **A Metabolic Pathway Coupled to the ATP/ADP Cycle.** In this metabolic pathway, enzyme A catalyzes an energy-requiring reaction where the energy comes from ATP hydrolysis. Enzyme B converts the phosphorylated substrate to the end product. Being an energy-releasing reaction, the free phosphate can be coupled to the reformation of ATP. »» What are the terms for energy-requiring and energy-releasing reactions?

It might be a good idea to review what we have covered to this point, which is summarized in **FIGURE 6.6**. Enzymes regulate metabolic reactions by binding an appropriate substrate at its active site. Often a series of metabolic steps (metabolic pathway) are required to form the final product. Some steps in the pathway may require energy (endergonic); this energy is supplied by ATP as it is hydrolyzed to ADP. Other reactions release energy (exergonic), which may be used to reform ATP from ADP.

CONCEPT AND REASONING CHECKS

6.5   Judge the importance of the ATP cycle to microbial metabolism.

6.2   The Catabolism of Glucose

Since the early part of the twentieth century, the chemistry of glucose catabolism has been the subject of intense investigation by biochemists because glucose is a key source of energy for ATP production. Moreover, the process of glucose catabolism is very similar in all organisms, making this "metabolic interlock" one feature that unites all life.

Glucose Contains Stored Energy That Can Be Extracted

KEY CONCEPT

6.   Glucose is a primary source for generating ATP.

Mole:
The molecular weight of a substance expressed in grams.

Calories:
Units of energy defined in the amount of heat required to raise one gram of water 1°C.

A mole of glucose (180 g) contains about 686,000 calories of energy. This fact can be demonstrated in the laboratory by setting fire to a mole of glucose and measuring the energy released. In a cell, however, not all the energy is set free from glucose, nor can the cell trap all that is released. The process accounts for the transfer of about 40% of the glucose energy to ATP energy; that is, chemical energy to cellular energy.

Virtually all cells make ATP by harvesting energy from exergonic metabolic pathways, such as the hydrolysis of food molecules. Such a process is called **cellular respiration**. If cells consume oxygen in making ATP, the process is called **aerobic respiration**. In other instances, cells can make almost equally substantial amounts of ATP without using oxygen, in which case it is called

anaerobic respiration. In these instances, another inorganic molecule will replace oxygen.

A form of "anaerobic metabolism" is **fermentation**, which we will examine later in this chapter.

The catabolism of glucose or another molecule does not take place in one chemical reaction, nor do ATP molecules form all at once. Rather, the energy in glucose is extracted and transferred slowly to ATP through metabolic pathways (**FIGURE 6.7**). It is similar to the proverb quoted at the beginning of this chapter, "*Life is like a fire; it begins in smoke and ends in ashes.*" The catabolism of glucose starts with a little energy being converted to ATP (the smoke), which builds to a point where large amounts of energy are converted to ATP (the fire), and the original glucose molecule has been depleted of its useful energy (the ashes).

To begin our study of cellular respiration, we shall follow the process of aerobic respiration as it occurs in obligate aerobes. The process is represented in the following summary form:

$$C_6H_{12}O_6 + 6\,O_2 + 38\,ADP + 38\,P$$

Glucose   Oxygen

$$\downarrow$$

$$6\,CO_2 + 6\,H_2O + 38\,ATP$$

Carbon   Water
dioxide

The events summarized in the overall reaction are conveniently divided into three stages: glycolysis, the citric acid cycle, and oxidative phosphorylation. Let's examine each of these in sequence. To simplify our discussion of glucose catabolism, we shall follow the fate of one glucose molecule.

CONCEPT AND REASONING CHECKS

6.6 In the summary equation for cellular respiration, identify where in the products each of the substrate atoms ends up.

Glycolysis Is the First Stage of Energy Extraction

KEY CONCEPT

7. Glycolysis is a metabolic pathway yielding ATP and NADH.

The splitting of glucose, called **glycolysis** (*glyco* = "sweet"), occurs in the cytosol of all microorganisms and involves a metabolic pathway that converts an initial 6-carbon substrate, **glucose**, into two 3-carbon molecules called **pyruvate**. Between glucose and pyruvate, there are eight intermediates formed, each catalyzed by a specific enzyme. FIGURE 6.8 illustrates the process. For easy referral, numbers in circles identify each reaction, and it would be helpful to refer to the figure as the discussion proceeds.

The first part of glycolysis is endergonic; one molecule of ATP is hydrolyzed (consumed) in reaction (**1**) and a second in reaction (**3**). In both cases, the phosphate group from ATP attaches to the product. Thus, reaction (**1**) produces glucose-6-phosphate, and reaction (**3**) yields fructose-1,6-bisphosphate (bis means "two separate;" that is, two separate phosphate molecules).

After the splitting of fructose-1,6-bisphosphate into two 3-carbon molecules, each passes through an additional series of conversions that ultimately form pyruvate. During reactions (**7**) and (**10**), ATP is generated. In both exergonic steps, enough energy is released to synthesize an ATP molecule from ADP and phosphate, resulting in a total of four ATP molecules. Because these ATP molecules were the result of the transfer of a phosphate from a substrate to ADP, we say these ATP molecules were the result of **substrate-level phosphorylation**. Considering two ATP molecules were consumed in reactions (**1**) and (**3**), the net gain from glycolysis is two molecules of ATP.

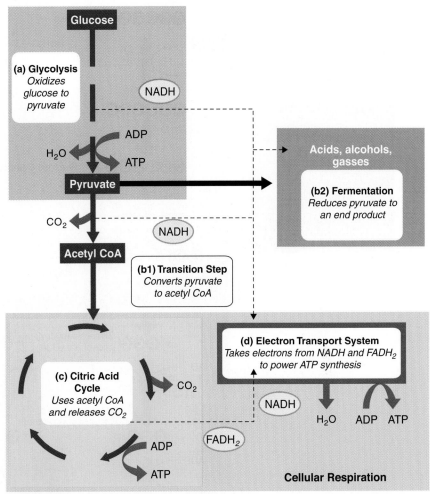

FIGURE 6.7 A Metabolic Map of Aerobic and Anaerobic Pathways for ATP Production. The production of ATP by microorganisms can be achieved through glycolysis (a) following a cellular respiration pathway (b1, c, d) or fermentation pathway (b2). »» Using this map, show how *"Life is like a fire; it begins in smoke and ends in ashes."*

Before we proceed, take note of reaction (**6**). This enzymatic reaction releases two high-energy electrons and two protons (H^+), which are picked up by the coenzyme NAD^+, reducing each to NADH. This and similar events will have great significance shortly as an additional source to generate ATP.

Reducing (reduction): Referring to the process of a substance gaining electron pairs.

CONCEPT AND REASONING CHECKS

6.7 At the end of glycolysis, where is the energy that was originally in glucose?

The Citric Acid Cycle Extracts More Energy from Pyruvate

KEY CONCEPT

8. The citric acid cycle yields additional ATP and NADH, as well as $FADH_2$.

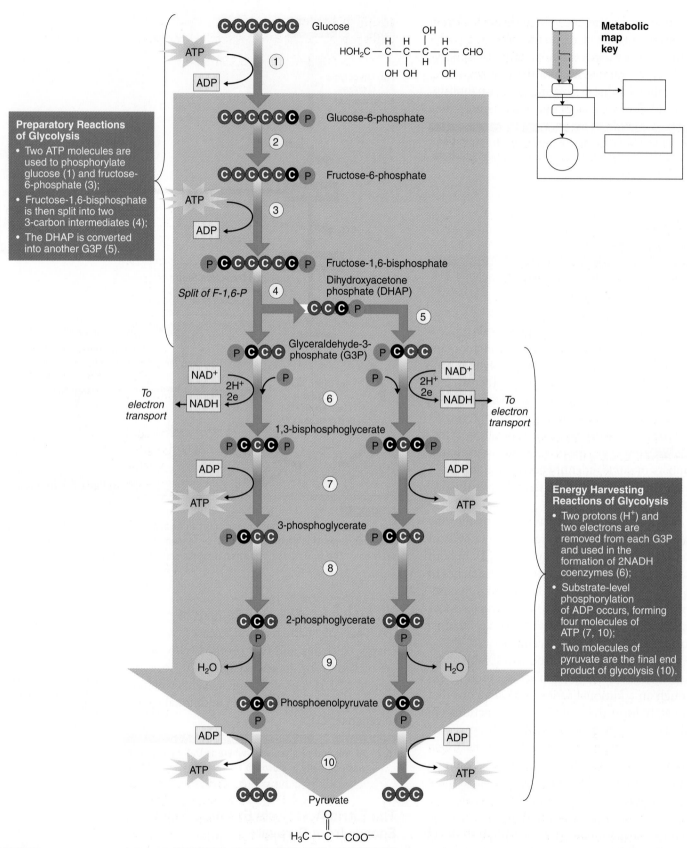

FIGURE 6.8 The Reactions of Glycolysis. Glycolysis is a metabolic pathway that converts glucose, a 6-carbon sugar, into two 3-carbon pyruvate products. In the process, two NADH coenzymes and a net gain of two ATP molecules occur. Carbon atoms are represented by circles. The dark circles represent carbon atoms bonded to phosphate groups. »» How many substrate-level phosphorylation events occurred during glycolysis?

The **citric acid cycle** (also called the **Krebs cycle** in honor of Hans Krebs and colleagues who worked out the pathway) is a series of chemical reactions that are referred to as a cycle because the end product formed is used as one substrate to initiate the pathway. All of the reactions are catalyzed by enzymes, and all take place along the cell membrane of bacterial and archaeal cells. In eukaryotic microbes, including the protozoa, algae, and fungi, the cycle occurs in the mitochondria.

The citric acid cycle is somewhat like a constantly turning wheel. Each time the wheel comes back to the starting point, something must be added to spin it for another rotation. That something is the pyruvate molecule derived from glycolysis. FIGURE 6.9 shows the citric acid cycle. The reactions are identified by capital letters in circles to guide us through the cycle.

Before pyruvate molecules enter the cycle, they undergo oxidation, indicated in reaction (**A**). An enzyme removes a carbon atom from each of the two pyruvate molecules and releases the carbons as two carbon dioxide molecules (2 CO_2). The remaining two carbon atoms of pyruvate are combined with **coenzyme A (CoA)** to form **acetyl-CoA**. Equally important, the lost electrons from pyruvate, along with two protons are transferred to NAD^+ to form NADH.

The two remaining carbons from pyruvate are now ready to enter the citric acid cycle. In reaction (**B**), each acetyl-CoA unites with a 4-carbon oxaloacetate to form citrate, a 6-carbon molecule. (Citrate, or citric acid, may be familiar to you as a component of soft drinks.) Citrate undergoes a series of reactions (**C, D**), forming a 4-carbon succinate. The two last carbons from glucose are released as CO_2. Succinate then undergoes a series of modifications (reactions **E, F**, and **G**) reforming oxaloacetate. The cycle is now complete, and oxaloacetate is ready to unite with another molecule of acetyl-CoA.

Several features of the cycle merit closer scrutiny. First, we shall follow the carbon. Pyruvate, with three carbon atoms, emerges from glycolysis, but after one turn of the cycle, its carbon atoms exist in three molecules of CO_2. There are two molecules of pyruvate from glycolysis, so when the second molecule enters the cycle, its carbon atoms also will form three CO_2 molecules. Remember that we began with a 6-carbon glucose molecule;

six CO_2 molecules now have been produced. This fulfills part of the equation for aerobic respiration:

$$\underline{C_6}H_{12}O_6 + 6\ O_2 + 38\ ADP + 38\ P$$
$$\downarrow$$
$$\underline{6}\ \underline{C}O_2 + 6\ H_2O + 38\ ATP$$

The second feature of the cycle is reaction (**D**). Here a molecule of ATP forms. Because we have two pyruvate molecules entering the cycle (per molecule of glucose), a second ATP molecule will form from GTP when the second pyruvate passes through the cycle.

Last, and perhaps of most importance, are reactions **C, D, E**, and **G**. Reactions **C, D**, and **G**, like reaction 6 in glycolysis and the transition step, are associated with NAD^+ and again are reduced to NADH. Two NADH molecules are produced in each step for a total of six. In addition, reaction **E** accomplishes much of the same result except it is associated with another coenzyme, FAD. It too receives two electrons and two protons from the reaction, being reduced to $FADH_2$. For the two pyruvate molecules starting the process, two $FADH_2$ molecules are formed.

In summary, glycolysis and the citric acid cycle have extracted as much energy as possible from glucose and pyruvate (FIGURE 6.10). This has amounted to a small gain of ATP molecules formed from one glucose molecule. However, the 10 NADH and 2 $FADH_2$ molecules formed are most significant. Let's see how.

Oxidation:
The process of removing electron pairs from a substance.

CONCEPT AND REASONING CHECKS

6.8 Identify the initial substrates and final end products of the citric acid cycle.

Oxidative Phosphorylation Is the Process by Which Most ATP Molecules Form

KEY CONCEPT

9. NADH and $FADH_2$ provide the starting materials for oxidative phosphorylation.

Oxidative phosphorylation refers to a sequence of reactions in which two events happen: Pairs of electrons are passed from one chemical substance to another (electron transport), and the energy released during their passage is used to combine phosphate with ADP to form ATP (ATP synthesis). The adjective "oxidative" is derived from the term

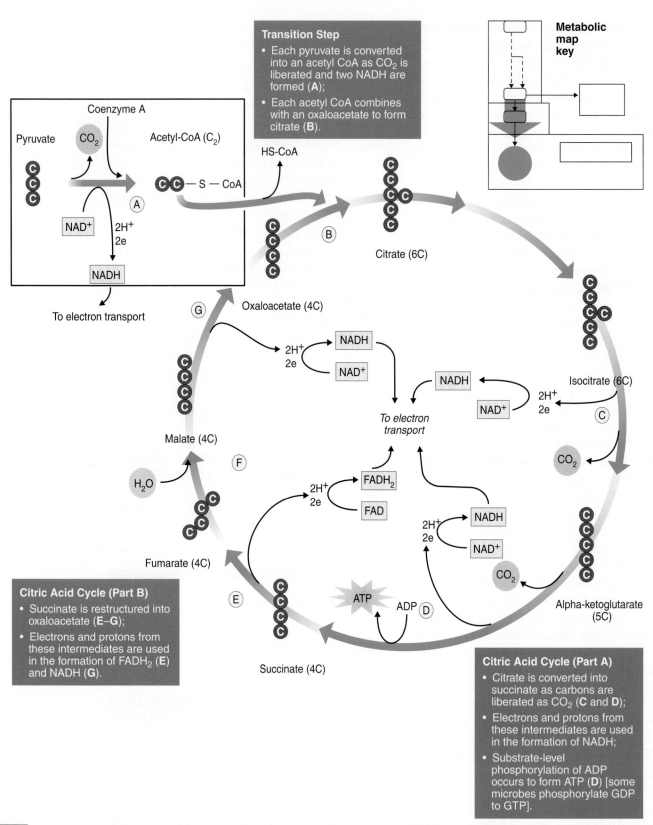

Transition Step
- Each pyruvate is converted into an acetyl CoA as CO_2 is liberated and two NADH are formed (**A**);
- Each acetyl CoA combines with an oxaloacetate to form citrate (**B**).

Metabolic map key

Coenzyme A

Pyruvate CO_2 Acetyl-CoA (C_2)

HS-CoA

— S — CoA

(A)

NAD⁺   2H⁺ 2e

NADH

To electron transport

(B)

Citrate (6C)

(G)   Oxaloacetate (4C)

2H⁺ 2e   NADH

NAD⁺

To electron transport

NADH

NAD⁺   2H⁺ 2e   Isocitrate (6C)

(C)

Malate (4C)

(F) H_2O CO_2

FADH₂

FAD   2H⁺ 2e

Fumarate (4C)

Citric Acid Cycle (Part B)
- Succinate is restructured into oxaloacetate (**E–G**);
- Electrons and protons from these intermediates are used in the formation of FADH₂ (**E**) and NADH (**G**).

(E)   Succinate (4C)

ATP   ADP   (D)

2H⁺ 2e   NADH

NAD⁺

CO_2

Alpha-ketoglutarate (5C)

Citric Acid Cycle (Part A)
- Citrate is converted into succinate as carbons are liberated as CO_2 (**C** and **D**);
- Electrons and protons from these intermediates are used in the formation of NADH;
- Substrate-level phosphorylation of ADP occurs to form ATP (**D**) [some microbes phosphorylate GDP to GTP].

FIGURE 6.9 **The Reactions of the Citric Acid Cycle.** Pyruvate from glycolysis combines with coenzyme A to form acetyl-CoA (transition step). This molecule then joins with oxaloacetate to form citrate (citric acid cycle—Part A). Each turn of the cycle releases CO_2, produces ATP, and forms NADH and FADH₂ coenzymes as oxaloacetate is replaced (citric acid cycle—Part B). »» Why are so many reactions required to extract energy out of pyruvate?

oxidation, which, as defined earlier, refers to the loss of electron pairs from molecules. Its counterpart is **reduction**, which refers to a gain of electron pairs by molecules. Phosphorylation, as we already have seen, implies adding a phosphate to another molecule. So, in oxidative phosphorylation, the loss and transport of electrons will enable ADP to be phosphorylated to ATP. Oxidative phosphorylation takes place at the cell membrane in bacterial and archaeal cells and in the mitochondrial inner membrane of eukaryotic microbes.

Oxidative phosphorylation is responsible for producing 34 molecules of ATP per glucose. The overall sequence involves the NAD$^+$ and FAD coenzymes that underwent reduction to NADH and FADH$_2$ during glycolysis and the citric acid cycle; remember, they gained two electrons in those metabolic pathways. In oxidative phosphorylation, the coenzymes will be reoxidized by transferring those two electrons to a series of electron carriers (FIGURE 6.11). These carriers, called **cytochromes** (*cyto* = "cell"; *chrome* = "color"), are a set of proteins containing iron cofactors that accept and release electron pairs. Together, the cytochrome complexes (I–IV) form an **electron transport chain**. The last link in the chain is oxygen gas.

In the oxidative phosphorylation process shown in Figure 6.11, the two electrons with each NADH and FADH$_2$ are passed to cytochrome complex I in the chain (**A**). The reoxidized coenzymes, NAD$^+$ and FAD, return to the cytosol (**B**) to be used again in glycolysis or the citric acid cycle. Like walking along stepping stones, each electron pair is passed from one cytochrome complex to the next down the chain (**C**) until the electron pair is finally transferred from complex IV to oxygen. Oxygen also acquires two protons (4 H$^+$) from the cytosol (**D**) and becomes water (2 H$_2$O). Oxygen's role is of great significance because if oxygen were not present, there would be no way for cytochromes to unload their electrons and the entire system would soon back up like a jammed conveyer belt and come to a halt. Oxygen's role also is reflected in the equation for aerobic respiration:

$$C_6\underline{H_{12}}O_6 + \underline{6\ O_2} + 38\ ADP + 38\ P$$
$$\downarrow$$
$$6\ CO_2 + \underline{6\ H_2O} + 38\ ATP$$

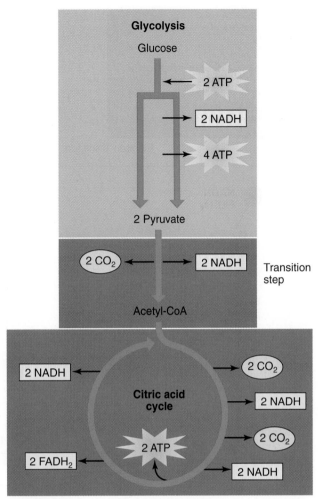

FIGURE 6.10 **Summary of Glycolysis and the Citric Acid Cycle.** Glycolysis and the citric acid cycle are metabolic pathways to extract chemical energy from glucose to generate cellular energy (ATP). »» Every time one glucose molecule is broken down to CO$_2$ and water, how many ATP, NADH, and FADH$_2$ molecules are gained?

So, what is the importance of the electron transport chain because no ATP has been made? The actual mechanism for ATP synthesis comes from the pumping of protons by a process called **chemiosmosis** (*osmos* = "push"). First proposed by Nobel Prize–winner Peter Mitchell, chemiosmosis uses the power of proton movement across a membrane to conserve energy for ATP synthesis.

What happens in chemiosmosis also is shown in Figure 6.11. As the electrons pass between cytochrome complexes, the electrons gradually lose energy. The energy, however, is not lost in the sense that it is gone forever. Instead, the energy is used at three transition points to "pump" protons (H$^+$) across the membrane from the cytosol to

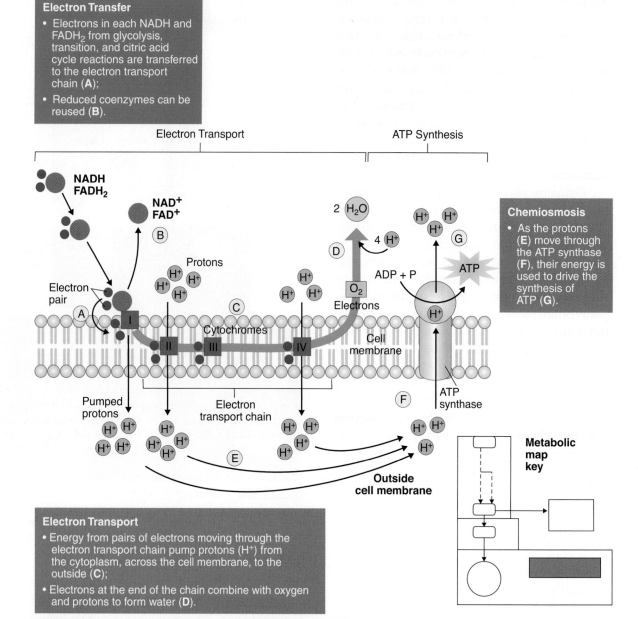

Electron Transfer
- Electrons in each NADH and $FADH_2$ from glycolysis, transition, and citric acid cycle reactions are transferred to the electron transport chain (**A**);
- Reduced coenzymes can be reused (**B**).

Electron Transport   ATP Synthesis

NADH $FADH_2$

NAD^+ FAD^+

Chemiosmosis
- As the protons (**E**) move through the ATP synthase (**F**), their energy is used to drive the synthesis of ATP (**G**).

Protons

Electron pair

Cytochromes

Cell membrane

Pumped protons

Electron transport chain

Outside cell membrane

ATP synthase

Metabolic map key

Electron Transport
- Energy from pairs of electrons moving through the electron transport chain pump protons (H^+) from the cytoplasm, across the cell membrane, to the outside (**C**);
- Electrons at the end of the chain combine with oxygen and protons to form water (**D**).

FIGURE 6.11 **Oxidative Phosphorylation in Bacterial Cells.** (**A**) Originating in glycolysis and the citric acid cycle, coenzymes NADH and $FADH_2$ transport electron pairs to the electron transport chain in the cell membrane, which fuels the transport of protons (H^+) across the cell membrane. Protons then reenter the cytosol through a protein channel in the ATP synthase enzyme. ADP molecules join with phosphates as protons move through the channel, producing ATP. »» What would happen to the oxidative phosphorylation process if this cell were deprived of oxygen?

the area outside of the cell membrane (**E**). Soon a large number of protons have built up outside the membrane, and because they cannot easily reenter the cell, they represent a large concentration of potential energy (much like a boulder at the top of a hill). The protons are positively charged, so there also is a buildup of charges outside the membrane.

Suddenly, a series of channels opens and the proton flow reverses (**F**). Each "channel" is contained within a large, membrane-spanning enzyme complex called **ATP synthase**, which has binding sites for ADP and phosphate. As the protons rush through the channel, they release their energy, and the energy is used to synthesize

ATP molecules from ADP and phosphate ions (**G**), as MICROINQUIRY 6 explains. Three molecules of ATP can be synthesized for each pair of electrons originating from NADH; two molecules of ATP are produced for each pair of electrons from $FADH_2$ because the coenzyme interacts further down the chain. MICROFOCUS 6.3 highlights a novel way of using the bacterial respiratory process to generate electricity.

If the cell membrane is damaged so chemiosmosis cannot take place, the synthesis of ATP ceases and the organism rapidly dies. This is one reason why damage to the bacterial cell membrane, such as with antibiotics or detergent disinfectants, is so harmful.

The ATP yield from aerobic respiration is summarized in **FIGURE 6.12**. The reactions can generate up to 38 ATP molecules from one glucose molecule. It also completes the equation for aerobic respiration:

$$C_6H_{12}O_6 + 6\ O_2 + \textbf{\underline{38 ATP}} + \textbf{\underline{38 P}}$$
$$\downarrow$$
$$6\ CO_2 + 6\ H_2O + \textbf{\underline{38 ATP}}$$

CONCEPT AND REASONING CHECKS

6.9 Why are NADH and $FADH_2$ so critical to cell energy metabolism?

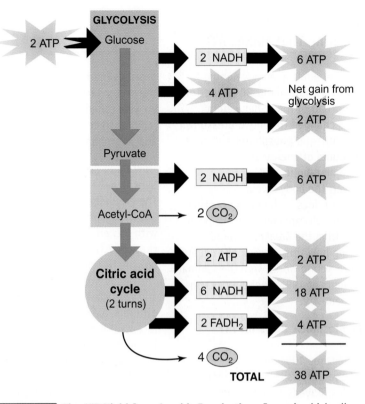

FIGURE 6.12 The ATP Yield from Aerobic Respiration. In a microbial cell, 38 molecules of ATP can result from the metabolism of a molecule of glucose. Each NADH molecule accounts for the formation of three molecules of ATP; each molecule of $FADH_2$ accounts for two ATP molecules. »» From this diagram, what is the single most important reactant for ATP synthesis?

Other Aspects of Catabolism

The catabolism of glucose is a process central to the metabolism of all organisms as it provides a glimpse of how organisms obtain energy for life. In this section, we examine how cells obtain energy from other organic compounds (fats and proteins) by directing those compounds into the process of cellular respiration. We also discover how modifications to cellular respiration and glucose metabolism allow anaerobic organisms to use glucose and generate ATP without having oxygen gas as the final electron acceptor in electron transport.

Other Nutrients Represent Potential Energy Sources

KEY CONCEPT

10. Other carbohydrates as well as fats and proteins can supply chemical energy for ATP production.

A wide variety of monosaccharides, disaccharides, and polysaccharides serve as useful energy sources. All must go through a series of preparatory conversions before they are processed in glycolysis, the citric acid cycle, and oxidative phosphorylation.

In preparation for entry into the scheme of metabolism, different carbohydrates use different pathways (**FIGURE 6.13**). Sucrose, for example, is first digested by the enzyme sucrase into its constituent molecules, glucose and fructose. The glucose molecule enters the glycolysis pathway directly, but the fructose molecule first is converted to fructose-1-phosphate. The latter then undergoes further conversions and a molecular split before it enters the scheme as DHAP [reaction (4)]. Lactose, another disaccharide, is broken in two by the enzyme lactase to glucose and

MICROINQUIRY 6
The Machine That Makes ATP

Every day, an adult human weighing 160 pounds uses up about 80 pounds of ATP (about half of his or her weight). The ATP is changed to its two breakdown products, ADP and phosphate, and enormous amounts of energy are made available to do metabolic work.

However, the body's weight does not go down, nor does it change perceptibly because the cells are constantly regenerating ATP from the breakdown products. Discovering how this is accomplished and how the recycling works were the seminal achievements of 1997's Nobel Prize winners in Chemistry.

The Chemiosmotic Basis for ATP Synthesis

One of the three 1997 winners was Paul D. Boyer at the University of California at Los Angeles. Boyer's work expanded the pioneering work of Peter Mitchell, who developed the concept of chemiosmosis. Chemiosmosis proposes that electron transport between cytochrome complexes provides the energy to "pump" protons (H^+) across the membrane; in the case of bacterial and archaeal cells, this is from the cytosol to the environment. As explained in the text, this proton gradient

provides the force or potential to drive the protons back into the cell through an enzyme called ATP synthase. This flow of rapidly streaming H^+ brings together ADP and phosphate to form ATP.

How Does Proton Flow Cause ATP Synthesis?

The groundbreaking research as to how the ATP synthase works came from studies with *Escherichia coli* cells. Today, we know that the ATP synthase consists of two polypeptide complexes (see **Figure A**). The headpiece (F_1) faces into the cytosol and consists of nine polypeptides of five different types (α, β, γ, ε, and δ) and represents the catalytic complex for converting ADP + P to ATP. The basal unit (F_0) is embedded in the cell membrane and consists of 15 polypeptides of three different types (a, b, and c). The basal unit contains the proton-transporting channel through the membrane. So, an ATP synthase consists of 24 polypeptides—a veritable nanomachine.

Boyer took the three complexes and hypothesized how they could manufacture ATP. His ideas plus newer findings have been merged into the current model (see **Figure B**):

1. The flow of protons through the basal unit c proteins causes the basal unit to spin (somewhat similar to the turning of a waterwheel).
2. The γ and ε polypeptides in the headpiece also spin and, as they spin, δ makes contact with each of the β subunits.
3. Each β subunit changes shape, and like an enzyme's active site, allows the subunit to bind an ADP + P and catalyze the production of an ATP.
4. When a β subunit returns to its original shape, it releases the ATP.

Because there are three β subunits in the headpiece, three ATP molecules are produced each time the basal unit and the γ and ε polypeptides make one complete rotation.

Discussion Point
Ribosomes, flagella, and ATP synthase all represent "nanomachines" to carry out specific functions in cells. Discuss the concept of a bacterial cell as being an assemblage of nanomachines.

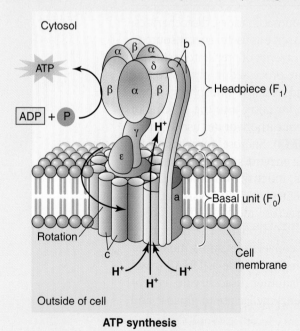

ATP synthesis

FIGURE A   A bacterial ATP synthase enzyme consists of 24 polypeptides in 2 complexes, the basal unit embedded in the cell membrane and the headpiece that projects into the cytosol.

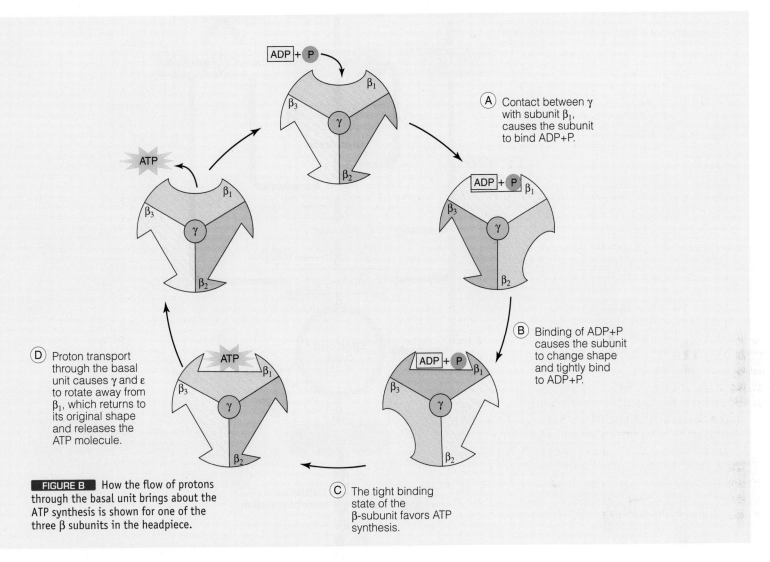

A Contact between γ with subunit β_1, causes the subunit to bind ADP+P.

B Binding of ADP+P causes the subunit to change shape and tightly bind to ADP+P.

C The tight binding state of the β-subunit favors ATP synthesis.

D Proton transport through the basal unit causes γ and ε to rotate away from β_1, which returns to its original shape and releases the ATP molecule.

FIGURE B How the flow of protons through the basal unit brings about the ATP synthesis is shown for one of the three β subunits in the headpiece.

MICROFOCUS 6.3: Biotechnology

Bacteria Not Included

How many toys (child or adult) or electronic devices do you purchase each year where batteries are needed to run the device? And often batteries are not included. Today, a new type of battery is being developed—one that converts sugar not into ATP but rather into electricity. The battery is one packed with bacterial cells.

Realize that cellular respiration involves the generation of minute electrical currents. During cellular respiration, electrons are transferred to cofactors like NAD^+ and then passed along a chain of cytochrome complexes during oxidative phosphorylation. Swades Chaudhuri and Derek Lovely of the University of Massachusetts at Amherst have taken this idea and applied it to developing a new type of fuel cell or battery.

The scientists mixed the bacterial species *Rhodoferax ferrireducens*, which they found in aquifer sediments in Virginia, with a variety of common sugars. When placed in a chamber with a graphite electrode, *R. ferrireducens* metabolized the sugar, stripped off the electrons, and transferred them directly to the electrode. The result: a current was produced. In addition, the bacterial cells continued to grow, so a stable current could be produced with high efficiency.

Although it is still a long way from producing a reliable, long-lasting bacterial battery, researchers believe much of the agricultural or industrial waste produced today could be the "sugar" used in making these bacterial batteries. So, as Sarah Graham reported for *Scientific American.com*, "Perhaps one day electronics will be sold with the caveat 'bacteria not included.' "

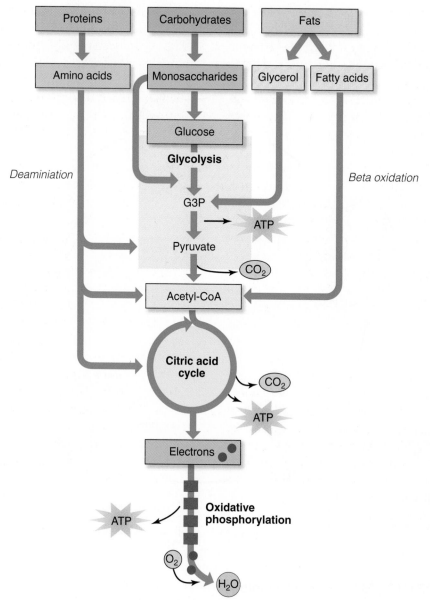

FIGURE 6.13 **Carbohydrate, Protein, and Fat Metabolism.** Besides glucose, other carbohydrates as well as proteins and fats can be sources of energy by providing electrons and protons for cellular respiration. The intermediates enter the pathway at various points. »» Why would more ATP be produced from the products of one fatty acid entering the pathway at acetyl-CoA than from one amino acid entering at acetyl-CoA?

galactose. Galactose undergoes a series of changes before it is ready to enter glycolysis in the form of glucose-6-phosphate [reaction (**2**)].

Stored polysaccharides, such as starch and glycogen, are metabolized by enzymes that remove one glucose unit at a time and convert it to glucose-1-phosphate. An enzyme converts this compound to glucose-6-phosphate, ready for entry into the glycolysis pathway. The point is that carbohydrates other than glucose also are used as chemical energy sources.

The economy of metabolism is demonstrated further when we consider protein and fat catabolism (Figure 6.13). Fats are extremely valuable energy sources because their chemical bonds contain enormous amounts of chemical energy. Although proteins are generally not considered energy sources, cells use them for energy when carbohydrates and fats are in short supply. Both fats and proteins are broken down through glucose catabolism as well as through other pathways. Basically, the proteins and fats undergo a series of enzyme-catalyzed conversions and form components normally occurring in carbohydrate metabolism. These components then continue along the metabolic pathways as if they originated from carbohydrates.

Proteins are broken down to amino acids (see Chapter 2). Enzymes then convert many amino acids to pathway components by removing the amino group and substituting a carbonyl group. This process is called **deamination**. For example, alanine is converted to pyruvate and aspartic acid is converted to oxaloacetate. For certain amino acids, the process is more complex, but the result is the same: The amino acids become pathway intermediates of cellular respiration.

Fats consist of three fatty acids bonded to a glycerol molecule (see Chapter 2). To be useful for energy purposes, the fatty acids are separated from the glycerol by the enzyme lipase. Once this has taken place, the glycerol portion is converted to DHAP. For fatty acids, there is a complex series of conversions called **beta oxidation**, in which each long-chain fatty acid is broken by enzymes into 2-carbon units. Other enzymes then convert each unit to a molecule of acetyl-CoA ready for the citric acid cycle. We previously noted that for each turn of the cycle, 16 molecules of ATP are derived. A quick calculation should illustrate the substantial energy output from a 16-carbon fatty acid (eight 2-carbon units).

In most habitats, microbes can use a large and diverse set of chemical compounds as potential energy sources. When competing for these food resources, some bacterial organisms can "raise a stink" (MicroFocus 6.4).

CONCEPT AND REASONING CHECKS

6.10 Describe how lipids and proteins are prepared for entry into the cellular respiration pathway.

Anaerobic Respiration Produces ATP Using Other Final Electron Acceptors

KEY CONCEPT

11. ATP can be produced through chemiosmosis without oxygen gas.

Nearly all eukaryotic microbes, as well as multicellular animals and plants, carry out aerobic respiration, using oxygen as the final electron acceptor in the electron transport chain. However, many bacterial and archaeal organisms exist in environments where oxygen is scarce, such as in wetland soil and water, and within the human and animal digestive tracts. In these environments, the organisms have evolved a respiratory process called **anaerobic respiration** that relies on terminal, usually inorganic, electron acceptors other than oxygen for ATP production. Considering the immense number of species that live in such anaerobic environments, anaerobic respiration is extremely important ecologically.

The facultative species *Escherichia coli*, for example, uses nitrate (NO_3^-) with which electrons combine to form nitrite (NO_2^-) or another nitrogen product. The obligate anaerobe *Desulfovibrio* uses sulfate (SO_4^{2-}) for anaerobic respiration. The sulfate combines with the electrons from the cytochrome chain and changes to hydrogen sulfide (H_2S). This gas gives a rotten egg smell to the environment (as in a tightly compacted landfill). A final example is exhibited by the archaeal methanogens, *Methanobacterium* and *Methanococcus*. These obligate anaerobes use carbonate (CO_3) as a final electron acceptor and, with hydrogen nuclei, form large amounts of methane gas (CH_4).

In anaerobic respiration, the amount of ATP produced is less than in aerobic respiration. There are several reasons for this. First, only a portion of the citric acid cycle functions in anaerobic respiration, so fewer reduced coenzymes are available to the electron transport chain. Also, not all of the cytochrome complexes function during anaerobic respiration, so the ATP yield will be less. The exact amount of ATP produced therefore will depend on the organism and where in the respiratory pathway intermediates enter.

CONCEPT AND REASONING CHECKS

6.11 Why would obligate anaerobes tend to grow slower than obligate aerobes?

Fermentation Produces ATP Using an Organic Final Electron Acceptor

KEY CONCEPT

12. Fermentation generates ATP in the absence of exogenous electron acceptors.

In environments that are anoxic and without the alternative electron acceptors needed by anaerobes, much of the organic material

Anoxic: Without oxygen gas (O_2).

MICROFOCUS 6.4: Environmental Microbiology
Microbes "Raise a Stink"

We are all familiar with body odor and bad breath. When one does not maintain a level of cleanliness or hygiene, bacterial species on the skin surface or in the mouth can grow out of proportion and, as they metabolize compounds like proteins, they produce noticeably unpleasant, smelly odors.

On a more environmental level, there are the smells that often come from decaying or rotting foods. As decomposers, these microbes also give off foul odors caused by the presence of bacteria colonizing the dead carrion. Competing in the environment with other animal scavengers for food, do these bacterially-produced odors have a useful role in repelling or deterring animal species from consuming important food resources? This is what Mark Hay and collaborators at Georgia Institute of Technology wanted to know, especially with respect to marine ecosystems. Their hypothesis: Decaying food resources make these resources repugnant to larger animal species like crabs or fish.

To test their hypothesis, the research team baited crab traps near Savannah, Georgia, with menhaden, a typical bait-fish for crabs. Some traps contained microbe-laden menhaden carrion that had been allowed to rot for one or two days, while other traps contained freshly thawed carrion having relatively few microbes. When the traps were inspected, those with fresh carrion had more than twice the number of animals per trap than did the traps with microbe-laden carrion. Lab studies with stone crabs showed they too avoided the microbe-laden, rotting food, but readily consumed the freshly thawed menhaden carrion.

To examine the role of bacteria in the avoidance behavior by stone crabs (see figure), some menhaden was allowed to rot in water without the antibiotic chloramphenicol while other samples contained the antibiotic in the water to prevent or inhibit microbial growth. Again, the study observed that the crabs readily ate the antibiotic-incubated menhaden but avoided the menhaden without antibiotic; the bacteria were in some way responsible for the aversion.

Finally, the researchers used organic extracts prepared from the microbe-laden carrion and mixed these chemical substances with freshly thawed menhaden. Again, the crabs were repelled. Exactly what chemical compounds were responsible for the behavior were not evident from the study. In summary, it appears that bacteria not only act as decomposers and pathogens in the environment, but also can compete very successfully with relatively large animal consumers for mutually attractive food sources.

A stone crab.

will be catabolized through fermentation. **Fermentation** is the enzymatic process for producing ATP using endogenous organic compounds as both electron donors and acceptors—exogenous electron acceptors (O_2, NO_3^-, SO_4^{2-}, CO_3) are absent.

The chemical process of fermentation makes a few ATP molecules in the absence of cellular respiration. However, the citric acid cycle and oxidative phosphorylation are shut down, so the products of glycolysis (pyruvate) are shuttled through a pathway that produces other final end products. In these pathways, pyruvate is the intermediary accepting the electrons.

In all cases, no matter what the end product, fermentation ensures a constant supply of NAD$^+$ for glycolysis and the production of two ATP molecules per glucose (**FIGURE 6.14A**).

For example, in the fermentation of glucose by *Streptococcus lactis*, the conversion of pyruvate to lactic acid is a way to reform NAD$^+$ coenzymes so glycolysis can still make two ATP molecules for every glucose molecule consumed (**FIGURE 6.14B**).

The diversity of fermentation chemistry extends to some eukaryotic microbes as well. In yeasts like *Saccharomyces*, when pyruvate is converted to ethyl alcohol (ethanol), NAD$^+$ is reformed.

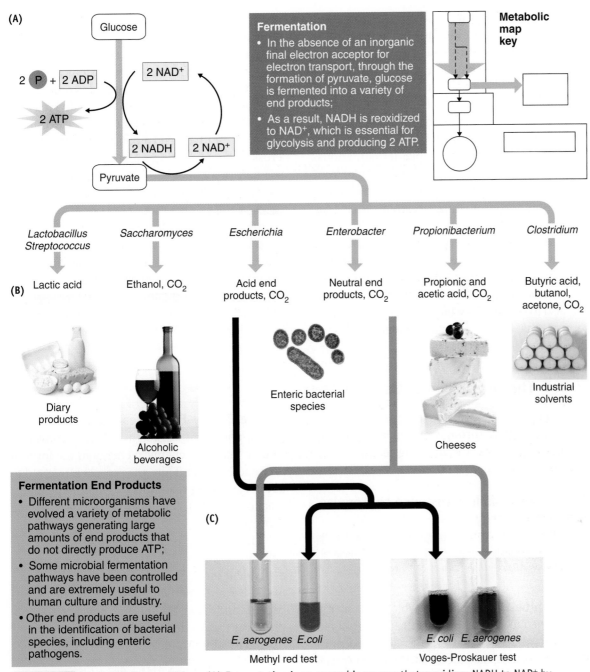

(A)

Glucose

2 P + 2 ADP

2 ATP

2 NAD$^+$

2 NADH   2 NAD$^+$

Pyruvate

Fermentation

- In the absence of an inorganic final electron acceptor for electron transport, through the formation of pyruvate, glucose is fermented into a variety of end products;

- As a result, NADH is reoxidized to NAD$^+$, which is essential for glycolysis and producing 2 ATP.

Metabolic map key

(B)

Lactobacillus *Streptococcus*	*Saccharomyces*	*Escherichia*	*Enterobacter*	*Propionibacterium*	*Clostridium*
Lactic acid	Ethanol, CO_2	Acid end products, CO_2	Neutral end products, CO_2	Propionic and acetic acid, CO_2	Butyric acid, butanol, acetone, CO_2

Diary products

Alcoholic beverages

Enteric bacterial species

Cheeses

Industrial solvents

Fermentation End Products

- Different microorganisms have evolved a variety of metabolic pathways generating large amounts of end products that do not directly produce ATP;

- Some microbial fermentation pathways have been controlled and are extremely useful to human culture and industry.

- Other end products are useful in the identification of bacterial species, including enteric pathogens.

(C)

E. aerogenes   *E.coli*

Methyl red test

E. coli   *E. aerogenes*

Voges-Proskauer test

FIGURE 6.14 **Microbial Fermentation.** (**A**) Fermentation is an anaerobic process that reoxidizes NADH to NAD$^+$ by converting organic materials into fermentation end products (**B**). Fermentation end products also provide a way to help identify bacterial species (**C**). »» How are lactic acid and alcohol fermentation identical in purpose?

The energy benefits to fermentative organisms are far less than in cellular respiration. In fermentation, each glucose passing through glycolysis yields two ATP molecules and the production of fermentation end products. This is in sharp contrast to the 38 molecules from cellular respiration. It is clear that cellular respiration is the better choice for energy conservation, but under anoxic conditions, there may be little alternative if life

for *S. lactis, Saccharomyces,* or any fermentative microorganism is to continue.

Although fermentation end products are waste products to the microbes producing them, industries see many of these products in a very different light (Figure 6.14B).

In a dairy plant, the process of **lactic acid fermentation** by *S. lactis* is carefully controlled so the acid will curdle fresh milk to make buttermilk or yogurt. The liquor industry uses the ethyl alcohol produced in **alcohol fermentation** to make alcoholic beverages such as beer and wine (Chapter 27). Fermentation of carbohydrates to alcohol by *C. albicans* also may take place in the human body, such as that of Charlie Swaart's, as described in the opener of this chapter.

As indicated in Figure 6.14B, other industries also make use of microbial fermentations. For example, Swiss cheese develops its flavor from the propionic acid produced during fermentation and gets its holes from trapped carbon dioxide gas resulting from fermentation.

The chemical industries also have harnessed the power of fermentation in the production of acetone, butanol, and other industrial solvents. Thus, fermentation is useful not only to microorganisms, but also to consumers who enjoy its products and industry that benefits from the products.

The ability of microbes to carry out different fermentation reactions that produce different end products can be very useful in species identification (see Chapter 3). Therefore, specific tests have been developed to detect particular end products. For example, with enteric species the "methyl red test" maintains a red colored solution if a species can ferment glucose to acid end products, while the "Voges-Proskauer test" produces a brownish-red colored solution if a species forms neutral end products from the acids produced through glucose fermentation (FIGURE 6.14C). These and other physiological and biochemical tests are often essential to the clinical lab microbiologist to identify a potential pathogen—and most likely for you in your microbiology lab class to identify an unknown bacterial species.

> **CONCEPT AND REASONING CHECKS**
>
> **6.12** Justify the need for some microbes to produce fermentation end products.

Enteric:
Referring to the intestines.

6.4 The Anabolism of Carbohydrates

Although the anabolism or synthesis of carbohydrates takes place through various mechanisms in microorganisms, the unifying feature is the requirement for energy.

Photosynthesis Is a Process to Acquire Chemical Energy

> **KEY CONCEPT**
>
> **13.** Photosynthesis converts light energy into chemical energy usually in the form of carbohydrates.

Photosynthesis is a process by which light energy is converted to chemical energy that is then stored as carbohydrate or other organic compounds. In the cyanobacteria, the process takes place in special thylakoid membranes, which contain chlorophyll or chlorophyll-like pigments (FIGURE 6.15A). Among eukaryotes, photosynthesis occurs in the chloroplasts of such organisms as diatoms, dinoflagellates, and green algae (FIGURE 6.15B). In all cases, these microbes carry out **oxygenic photosynthesis**; that is, where oxygen gas (O_2) is a by-product of the process.

The phases of photosynthesis are shown in FIGURE 6.16 , where the sequence of stages is labeled by number.

The Energy-Fixing Reactions. In the first part of photosynthesis, light energy is converted into or "fixed" as chemical energy in the form of ATP and NADPH. Thus, the process is referred to as **energy-fixing reactions**, or "light-dependent" reactions, because light energy is required for the process.

Like the reactions of cellular respiration, the energy-fixing reactions of photosynthesis are dependent on electrons and protons. The source of these atomic particles is water. In the cyanobacteria and algae, the splitting of water not only produces the needed atomic particles, it releases oxygen as a by-product:

$$2H_2O \rightarrow 4H^+ + 4e^- + O_2$$

As we have discussed in Chapters 2 and 5, the ability of the ancestors of modern-day cyanobacteria to generate oxygen gas profoundly changed the atmosphere of Earth some 3.5 billion years ago, leading to the evolution of aerobic organisms carrying out aerobic respiration.

Light energy is absorbed by the green pigment **chlorophyll *a***, a magnesium-containing, lipid-soluble compound (Figure 6.16A). Chlorophylls and accessory pigments make up light-receiving

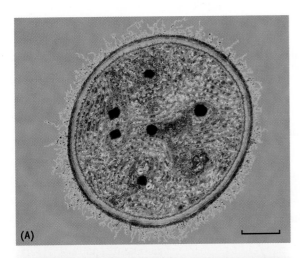

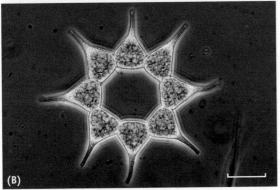

FIGURE 6.15 Photosynthetic Microbes. **(A)** False-color transmission electron micrograph of a cyanobacterium displaying the membranes (green) along which photosynthetic pigments are located. (Bar = 2 μm.) **(B)** A light micrograph of the colonial green alga *Pediastrum*. (Bar = 5 μm). »» What membranes in algae are analagous to the cyanobacterial membranes?

complexes called **photosystems**. The light excites pigment molecules in photosystem II, resulting in the loss of one electron (**1**). These electrons are replaced from the splitting of water.

Excited electrons are immediately accepted by the first of a series of electron carriers (**2**). The electrons are passed along the membrane carriers and cytochrome complexes, and eventually the electrons are taken up by other chlorophyll pigments that form photosystem I.

As the electrons move between cytochromes, energy is made available for proton pumping across the thylakoid membrane of the cyanobacterium, followed by chemiosmosis. As described for oxidative phosphorylation, ATP is formed when protons pass back across the membrane and release their energy. Because light was involved in the formation of ATP, this process is called **photophosporylation**.

The electrons in photosystem I again are excited by light energy (**3**) and are boosted out of the pigment molecules to the first of another set of membrane carriers, and finally to a coenzyme called nicotinamide adenine dinucleotide phosphate ($NADP^+$). The coenzyme functions much like NAD^+ in that $NADP^+$ receives pairs of electrons and protons from water molecules to form NADPH (**4**).

The Carbon-Fixing Reactions. In the second stage of photosynthesis, another cyclic metabolic pathway forms carbohydrates (Figure 6.16B). The process is known as the **carbon-fixing reactions** because the carbon in carbon dioxide is trapped (or fixed) into carbohydrates and other organic compounds. It also is called the "Calvin cycle," named after Melvin Calvin who worked out the sequence of reactions. An enzyme bonds carbon dioxide to a 5-carbon organic substance called ribulose 1,5-bisphosphate (RuBP)(**5**). (The enzyme is called ribulose bisphosphate carboxylase.) The resulting 6-carbon molecule then splits to form two molecules of 3-phosphoglycerate (3PG). In the next step, the products of the energy-fixing reactions, ATP and NADPH, drive the conversion of 3PG to glyceraldehyde-3-phosphate (G3P)(**6**). Two molecules of G3P then condense with each other to form a molecule of glucose (**7**). Thus, the overall formula for photosynthesis may be expressed as:

$$6\ CO_2 + 6\ H_2O + ATP$$
$$\downarrow light$$
$$C_6H_{12}O_6 + 6\ O_2 + ADP + P$$

Notice that this reaction is the reverse of the equation for aerobic respiration. The fundamental difference is that aerobic respiration is a catabolic, energy-yielding process, while photosynthesis is an anabolic, energy-trapping process.

To finish off the cycle, most G3P molecules undergo a complex series of enzyme-catalyzed reactions that require ATP to reform RuBP (**8**). However, some G3P exits the cycle and combines in pairs to form glucose. The sugar then can be used for cell respiration, stored as glycogen, or used for other cellular purposes.

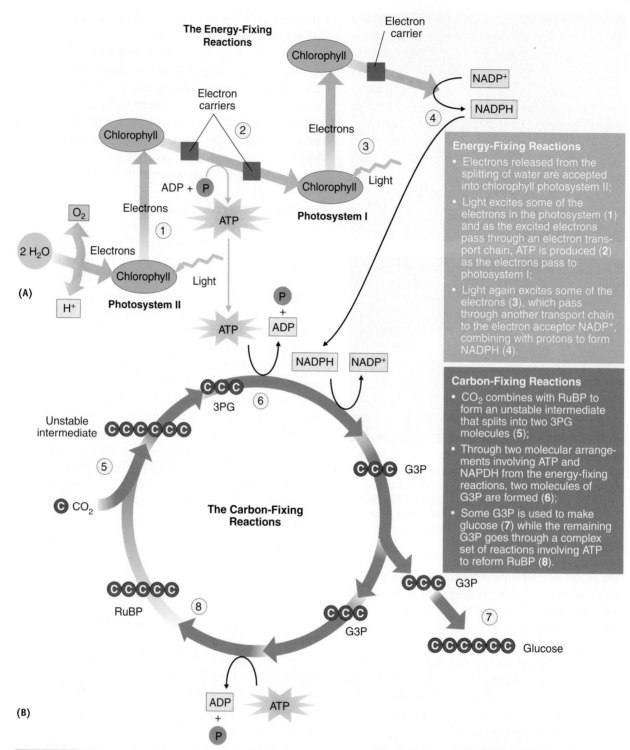

FIGURE 6.16 Photosynthesis in Cyanobacteria and Algae. (**A**) The energy-fixing reactions generate ATP and NADPH. (**B**) The carbon-fixing reactions unite carbon dioxide with ribulose bisphosphate (RuBP) to form an unstable 6-carbon molecule. The latter splits to form two molecules of 3-phosphoglycerate (3PG) then glyceraldehyde-3-phosphate (G3P). ATP and NADPH from the energy-fixing reactions are used in the latter conversion. Condensations of two 3-carbon G3P molecules yields glucose, and the remainder are used to form RuBP to continue the process. »» How are the carbon-fixing reactions of photosynthesis dependent on the energy-fixing reactions?

In addition to the cyanobacteria, a few other groups of bacteria trap energy by photosynthesis. Two such groups are the green bacteria and purple bacteria, so named because of the colors imparted by their pigments. These bacterial organisms have chlorophyll-like pigments known as **bacteriochlorophylls** to distinguish them from other chlorophylls. In the energy-fixing reactions, the organisms do not use water as a source of hydrogen ions and electrons. Consequently, no oxygen is liberated and the process is therefore called **anoxygenic photosynthesis**. Instead of water, a series of inorganic or organic substances, such as hydrogen sulfide gas (H_2S) and fatty acids, are used as a source of electrons and hydrogen ions. Thus, the green and purple bacteria commonly live under anaerobic conditions in environments such as sulfur springs and stagnant ponds.

CONCEPT AND REASONING CHECKS

6.13 Compare and contrast the processes of glucose catabolism (aerobic respiration) with glucose anabolism (photosynthesis).

6.5 Patterns of Metabolism

Microorganisms must meet certain nutritional requirements for growth. Besides water, which is an absolute necessity, they need nutrients that can serve as energy sources and raw materials for the synthesis of cell components. These generally include proteins for structural compounds and enzymes, carbohydrates for energy, and a series of vitamins, minerals, and inorganic salts.

Autotrophs and Heterotrophs Get Their Energy and Carbon in Different Ways

KEY CONCEPT

14. Autotrophs and heterotrophs vary in their energy and carbon sources.

Two different patterns exist for satisfying an organism's metabolic needs. These patterns are called **autotrophy** and **heterotrophy**. They are primarily based on the source of carbon used for making cell components (**FIGURE 6.17**).

Autotrophs. Organisms that synthesize their own foods from simple carbon sources such as CO_2 are referred to as **autotrophs** (*auto* = "self"; *troph* = "nourish"). Those that use light as the energy source, such as the cyanobacteria, are **photoautotrophs**. Microorganisms, including the cyanobacteria and algae, can carry out photosynthesis using water and producing oxygen gas (oxygenic photosynthesis), while other bacterial species carry out **anoxygenic photosynthesis**; they use hydrogen sulfide (H_2S) rather than water and produce sulfur granules (S) rather than oxygen gas (thus anoxygenic).

$$2H_2S + CO_2 \xrightarrow{\text{light}} \text{Carbohydrates} + H_2O + 2S$$

Another group of autotrophs do not use light as an energy source. Instead, they use inorganic compounds and are referred to as **chemoautotrophs**. For example, species of *Nitrosomonas* convert ammonium ions (NH_4^+) into nitrite ions (NO_2^-) under aerobic conditions, thereby obtaining ATP. The genus *Nitrobacter* then converts the nitrite ions into nitrate ions (NO_3^-), also as an ATP-generating mechanism. In addition to providing energy to both bacterial species, these reactions have great significance in the environment as a critical part of the nitrogen cycle (Chapter 26). By preserving nitrogen in the soil in the form of nitrate or ammonia, it can be used by green plants to form amino acids. Another example of chemoautotrophy involving a symbiosis between animal and bacterial cells is described in MicroFocus 6.5.

Nitrogen cycle: A biogeochemical cycle that cycles nitrogen gas into nitrogenous compounds and back again.

Heterotrophs. Many microorganisms are **heterotrophs** (*hetero* = "other"). Such heterotrophic organisms obtain their energy and carbon in one of two ways. The **photoheterotrophs** use light as their energy source and preformed organic compounds such as fatty acids and alcohols as sources of carbon. Photoheterotrophs include certain green nonsulfur and purple nonsulfur bacteria.

The **chemoheterotrophs** use preformed organic compounds for both their energy and carbon sources. Glucose would be one example.

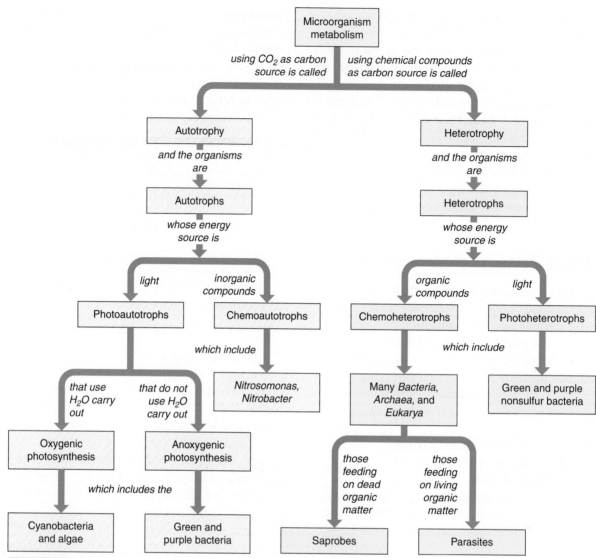

FIGURE 6.17 **Microbial Metabolic Diversity.** This concept map summarizes microbial metabolism based on carbon sources (CO_2 or chemical compounds) and energy sources (light or inorganic/organic compounds). »» Why do the *Bacteria* and *Archaea* show such diversity of metabolic types?

Those chemoheterotrophic microorganisms that feed exclusively on dead organic matter are commonly called **saprobes**. In contrast, chemoheterotrophs that feed on living organic matter, such as human tissues, are commonly known as **parasites**.

The term **pathogen** is used if the parasite causes disease in its host organism. We will certainly see many examples of this in upcoming chapters.

CONCEPT AND REASONING CHECKS

6.14 Why are there no autotrophic parasites or pathogens?

MICROFOCUS 6.5: Environmental Microbiology

Life in the Deep

Deep-sea hydrothermal vents, or black smokers, are truly exotic places, representing volcanically active mid-ocean ridges several thousand meters below the ocean surface where the Earth's crust has cracked as tectonic plates move apart. The vents in these extreme environments spew forth extremely hot water that can be as high as 300°C, compared to the chilling 2°C for the surrounding deep ocean water. And you guessed it—microbes are found here often in symbiotic association with one of the vent residents, the giant tube worm *Riftia pachyptila*. These deep-sea aliens share the cooler parts of the vent ecosystem (12–222°C) with other vent creatures like crabs, lobsters, and octopuses, and live in colonies made up of hundreds of individuals (see photo). *Riftia* lacks a mouth and a gut, yet somehow is able to grow to more than 7 feet in length. To accomplish this growth, they wave their bright red plumes, exposing and absorbing chemicals from the vent fluids, making these chemicals available to the symbiotic bacteria living within the tube worm, and the bacteria, in turn, using these chemicals to provide the tube worm with sustenance.

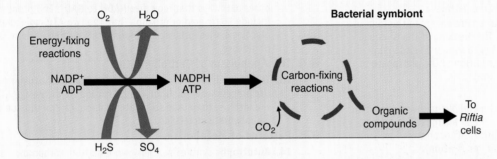

The energy- and carbon-fixing reactions in the bacterial symbiont of *Riftia*.

These communities represent chemoautotrophic proteobacteria (see Chapter 4). The cells receive inorganic compounds [hydrogen sulfide (H_2S) and O_2] from *Riftia's* circulatory system, the chemicals then serving as the energy source, along with carbon dioxide gas as the carbon source, for growth (see drawing).

Specifically, the worms remove hydrogen sulfide and oxygen from the vent seawater and deliver it to the bacteria that have been gathered together in an organ called the trophosome to population densities surpassing 10 billion bacterial cells per gram of worm tissue. The bacterial symbionts oxidize the sulfide and use some of the released energy to fix CO_2 via the carbon-fixing reactions into organic compounds, some of which are translocated back to the tube worm's tissues, supporting its growth. Life in the deep is truly remarkable—and microbes will always be there.

Tube worms on a hydrothermal vent.

SUMMARY OF KEY CONCEPTS

6.1 Enzymes and Energy in Metabolism

The two major themes of microbial metabolism are **catabolism** (the breakdown of organic molecules) and **anabolism** (the synthesis of organic molecules).

1. Microorganisms and all living organisms use **enzymes**, protein molecules that speed up a chemical change, to control cellular reactions.

2. Enzymes bind to substrates at their **active site** (enzyme-substrate complex) where functional groups are destabilized, lowering the **activation energy**.

3. Many metabolic processes occur in **metabolic pathways**, where a sequence of chemical reactions is catalyzed by different enzymes.

4. Enzymes can be inhibited by physical agents, which can denature enzymes and inhibit their action in cells. Enzymes are modulated through **feedback inhibition**. The final end product often inhibits the first enzyme in the pathway.

5. Many metabolic reactions in cells require energy in the form of **adenosine triphosphate (ATP)**. The breaking of the terminal phosphate produces enough energy to supply an endergonic reaction, and often involves the addition of the phosphate to another molecule (**phosphorylation**).

6.2 The Catabolism of Glucose

6. **Cellular respiration** is a series of metabolic pathways in which chemical energy is converted to cellular energy (ATP). It may require oxygen gas (**aerobic respiration**) or another inorganic final electron acceptor (**anaerobic respiration**).

7. **Glycolysis**, the catabolism of glucose to pyruvate, extracts some energy from which two ATP and 2 NADH molecules result.

8. The catabolism of pyruvate into carbon dioxide and water in the **citric acid cycle** extracts more energy as ATP, NADH, and $FADH_2$. Carbon dioxide gas is released.

9. The process of **oxidative phosphorylation** involves the oxidation of NADH and $FADH_2$, the transport of freed electrons along a cytochrome chain, the pumping of protons across the cell membrane, and the synthesis of ATP from a reversed flow of protons. These last two steps are referred to as **chemiosmosis**.

6.3 Other Aspects of Catabolism

10. Other carbohydrates, such as sucrose, lactose, and polysaccharides, represent energy sources that can be metabolized through cellular respiration. Besides carbohydrates, proteins and fats can be metabolized through the cellular respiratory pathways to produce ATP.

11. The **anaerobic respiration** of glucose uses different final electron acceptors in oxidative phosphorylation. Glycolysis and parts of the citric acid cycle still function, and ATP synthesis occurs.

12. In **fermentation**, the catabolism of glucose can continue without a functional citric acid cycle or oxidative phosphorylation process. To maintain a steady supply of NAD^+ for glycolysis and ATP synthesis, pyruvate is redirected into other pathways that reoxidize NADH to NAD^+. End products include lactic acid or ethanol. Only the two ATP molecules of glycolysis are synthesized in fermentation from each molecule of glucose.

6.4 The Anabolism of Carbohydrates

13. The anabolism of carbohydrates can occur by **photosynthesis**, the process whereby light energy is used to synthesize ATP, and the latter is then used to fix atmospheric carbon dioxide into carbohydrate molecules.

6.5 Patterns of Metabolism

14. **Autotrophs** synthesize their own food from carbon dioxide and light energy (**photoautotrophs**) or carbon dioxide and inorganic compounds (**chemoautotrophs**). **Heterotrophs** obtain their carbon from organic compounds and energy from light (**photoheterotrophs**) or from organic compounds (**chemoheterotrophs**).

LEARNING OBJECTIVES

After understanding the textbook reading, you should be capable of writing a paragraph that includes the appropriate terms and pertinent information to answer the objective.

1a. Contrast **anabolism** and **catabolism** as biochemical reactions and as energy processes.

1b. Describe the four properties of **enzymes**.

2. State the role of an **enzyme-substrate complex** to regulating metabolism.

3. Judge the importance of **metabolic pathways** in microbial cells.

4. Compare the mechanisms of **noncompetitive** and **competitive inhibition**.

5. Assess the role of **ATP** and the **ATP/ADP cycle** in cell metabolism.

6. Explain the importance of **glucose** to energy metabolism.

7. Summarize the important steps of **glycolysis**.

8. Identify the importance of the **citric acid cycle** to **cellular respiration**.

9. Construct an **electron transport pathway**, indicating the important steps in the synthesis of ATP.

10. Identify what other compounds, besides glucose, can be used to supply chemical energy for ATP production.

11. Compare and contrast **aerobic** and **anaerobic respiration**.

12. Summarize the steps in **fermentation** and identify the reason why pyruvate is converted into a final end product.

13. Summarize the importance of (a) the **energy-fixing reactions** and (b) the **carbon-fixing reactions** of photosynthesis.

14. Distinguish between the energy and carbon sources for the four nutritional classes of microorganisms.

STEP A: SELF-TEST

Each of the following questions is designed to assess your ability to remember or recall factual or conceptual knowledge related to this chapter. Read each question carefully, then select the **one** answer that best fits the question or statement. Answers to even-numbered questions can be found in **Appendix C**.

1. Enzymes are
   A. inorganic compounds.
   B. destroyed in a reaction.
   C. proteins.
   D. vitamins.

2. Enzymes combine with a _____ at the _____ site to lower the activation energy.
   A. substrate; active
   B. product; noncompetitive
   C. product; active
   D. coenzyme; active

3. Which one of the following is NOT a metabolic pathway?
   A. Citric acid cycle
   B. The carbon-fixing reactions
   C. Glycolysis
   D. Sucrose → glucose + fructose

4. If an enzyme's active site becomes deformed, _____ inhibition was likely responsible.
   A. metabolic
   B. competitive
   C. noncompetitive
   D. cellular

5. Which one of the following is NOT part of an ATP molecule?
   A. Phosphate groups
   B. Cofactor
   C. Ribose
   D. Adenine

6. The use of oxygen gas (O_2) in an exergonic pathway generating ATP is called
   A. anaerobic respiration.
   B. photosynthesis.
   C. aerobic respiration.
   D. fermentation.

7. Which one of the following is NOT produced during glycolysis?
   A. ATP
   B. NADH
   C. Pyruvate
   D. Glucose

8. All the following are produced during the citric acid cycle *except*:
   A. CO_2.
   B. O_2.
   C. ATP.
   D. NADH.

9. The electron transport chain is directly involved with
   A. ATP synthesis.
   B. CO_2 production.
   C. H^+ pumping.
   D. generating oxygen gas.

10. Which one of the following macromolecules would NOT normally be used for microbial energy metabolism?
    A. DNA
    B. Proteins
    C. Carbohydrates
    D. Fats

11. Anaerobic respiration does NOT
    A. use an electron transport system.
    B. use oxygen gas (O_2).
    C. occur in bacterial cells.
    D. generate ATP molecules.

12. In fermentation, the conversion of pyruvate into a final end product is critical for the production of
    A. CO_2.
    B. glucose.
    C. NAD^+.
    D. O_2.

13. Which one of the following is the correct sequence for the flow of electrons in the energy-fixing reactions of photosynthesis?
    A. Water—photosystem I—photosystem II—NADPH
    B. Photosystem I—NADPH—water—photosystem II
    C. Water—photosystem II—photosystem I—NADPH
    D. NADPH—photosystem II—photosystem I—water

14. Microorganisms that use organic compounds as energy and carbon sources are
    A. chemoheterotrophs.
    B. chemoautotrophs.
    C. photoautotrophs.
    D. photoheterotrophs.

STEP B: REVIEW

15. Identify on the metabolic map key where each reactant is used and where each product is produced in the cellular respiration summary equation.

$$C_6H_{12}O_6 + 6O_2 + 38\ ADP + 38\ P$$
$$\downarrow$$
$$6\ CO_2 + 6\ H_2O + 38\ ATP$$

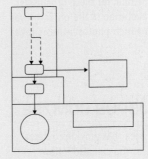

For each choice, circle the word or term that best completes each of the following statements. The answers to even-numbered statements are listed in **Appendix C**.

16. The sum total of all an organism's biochemical reactions is known as (catabolism, metabolism); it includes all the (synthesis, digestion) reactions called anabolism and all the breakdown reactions known as (inactivation, catabolism).

17. Enzymes are a group of (carbohydrate, protein) molecules that generally (slow down, speed up) a chemical reaction by converting the (substrate, active site) to end products.

18. The aerobic respiration of glucose begins with the process of (oxidative phosphorylation, glycolysis) and requires that (amino acids, energy) be supplied by (ATP, NADH) molecules.

19. The process of (fermentation, the citric acid cycle) takes place in the absence of (oxygen, carbon dioxide) and begins with a molecule of (glucose, protein) and ends with molecules of (amino acids, an organic end product).

20. In oxidative phosphorylation, pairs of (protons, electrons) are passed among a series of (chromosomes, cytochromes) with the result that (oxygen, energy) is released for (NAD^+, ATP) synthesis.

21. In the citric acid cycle, (glucose, pyruvate) undergoes a series of changes and releases its (carbon, nitrogen) as (carbon dioxide, nitrous oxide) and its electrons to (NAD^+, ATP).

22. For use as energy compounds, proteins are first digested to (uric, amino) acids, which then lose their (carboxyl, amino) groups in the process of (fermentation, deamination) and become intermediates of cellular respiration.

23. Ribulose 1,5-bisphosphate bonds with (carbon monoxide, carbon dioxide) molecules during (fermentation, photosynthesis), a process that ultimately results in molecules of (pyruvate, glucose).

24. Chemoautotrophs use energy from (light, inorganic compounds) to synthesize (carbohydrates, oxygen gas) and are typified by species of (*Staphylococcus, Nitrosomonas*).

25. Fats are broken down to (fatty acids, coenzymes), which are converted through (beta oxidation, deamination) reactions to (glucose, two-carbon units) and eventually enter (cellular respiration, photosynthesis).

STEP C: APPLICATIONS

Answers to even-numbered questions can be found in **Appendix C**.

26. You have two flasks with broth media. One contains a species of cyanobacteria. The other flask contains *E. coli*. Both flasks are sealed and incubated under optimal growth conditions for two days. Assuming the cell volume and metabolic rate of the bacterial cells is identical in each flask, why would the carbon dioxide concentration be higher in the *E. coli* flask than in the cyanobacteria flask after the two-day incubation?

27. A stagnant pond usually has a putrid odor because hydrogen sulfide has accumulated in the water. A microbiologist recommends that tons of green bacteria be added to remove the smell. What chemical process does the microbiologist have in mind? Do you think it will work?

28. Citrase is the enzyme that converts citrate to α-ketoglutarate in the citric acid cycle. A chemical company has isolated a mutant microorganism that cannot produce this enzyme and proposes to use the microorganism to manufacture a particular product. What do you suppose the product is? How might this product be useful?

STEP D: QUESTIONS FOR THOUGHT AND DISCUSSION

Answers to even-numbered questions can be found in **Appendix C**.

29. A student goes on a college field trip and misses the microbiology exam covering microbial metabolism. Having made prior arrangements with the instructor for a make-up exam, he finds one question on the exam: "Discuss the interrelationships between anabolism and catabolism." How might you have answered this question?

30. If ATP is such an important energy source for microbes, why do you think it is not added routinely to the growth medium for these organisms?

31. One of the most important steps in the evolution of life on Earth was the appearance of certain organisms in which photosynthesis takes place. Why was this critical?

32. A population of a *Bacillus* species is growing in a soil sample. Suppose glycolysis came to a halt in these bacterial cells. Would this mean that the citric acid cycle would also stop? Why?

33. While you are taking microbiology, a friend is enrolled in a general biology course. You both are studying cell energy metabolism. Your friend looks puzzled when you tell him that the citric acid cycle and electron transport occur in the bacterial cytoplasm and cell membrane. Your friend insists these processes occur in the mitochondrion. Who is correct and why?

🌐 HTTP://MICROBIOLOGY.JBPUB.COM/9E

The site features eLearning, an online review area that provides quizzes and other tools to help you study for your class. You can also follow useful links for in-depth information, read more MicroFocus stories, or just find out the latest microbiology news.

Control of Microorganisms: Physical and Chemical Methods

Advances in our understanding of hygiene, sanitation, and pathology that followed the development of the germ theory have done more to extend life expectancy and change the nature of society than any other medical innovation.
—Dr. Harry Burns, Chief Medical Officer for Scotland

For personal hygiene, washing our hands, taking regular showers or baths, brushing our teeth with fluoride toothpaste, and using an underarm deodorant are common practices we use to control microorganisms on our bodies. In our homes, we try to keep microbes in check by cooking and refrigerating foods, cleaning our kitchen counters and bathrooms with disinfectant chemicals, and washing our clothes with detergents.

In our attempt to be hygiene-minded consumers, sometimes we have become excessively "germphobic." The news media regularly report about this scientific study or that survey identifying places in our homes (e.g., toilets, kitchen drains) or environment (e.g., public bathrooms, drinking-water fountains) that represent infectious dangers. Consumer groups distribute pamphlets on "microbial awareness" and numerous companies have responded by manufacturing dozens of household antimicrobial products—some useful, but many unnecessary (FIGURE 7.1A).

Our desire to protect ourselves from microbes also stems from events beyond our doorstep. The news media again report about dangerous disease outbreaks, many of which result from a lack of sanitary controls or a lack of vigilance to maintain those controls. In our communities, we expect our drinking water to be clean. That goes for our hospitals as well. Nowhere

Chapter Preview and Key Concepts

7.1 General Principles of Microbial Control
1. Microbes are kept under control either by eliminating them or reducing their numbers.

7.2 Physical Methods of Control
2. Microbes and viruses are killed at temperatures above their temperature range for growth or replication.
3. Dry heat uses hot air with little, if any, moisture.
   MicroInquiry 7: Exploring Heat as an Effective Control Method
4. Moist heat more easily penetrates objects and materials.
5. Filtration removes microbes from the air or water.
6. UV light can be bactericidal.
7. X rays and gamma rays also are microbicidal.
8. Dehydration and cold temperatures slow microbial growth.

7.3 General Principles of Chemical Control
9. Disinfectants and antiseptics are key to proper sanitation and public health.
10. Disinfectants and antiseptics are defined by their properties.
11. Standards have been established to know the relative effectiveness of a chemical agent.

7.4 Chemical Methods of Control
12. Chlorine and iodine are good disinfectant agents.
13. Many phenolic derivatives are used as disinfectants or antiseptics.
14. Mercury, copper, and silver compounds can be useful disinfectants.
15. Alcohols are widely used skin antiseptics.
16. Cationic detergents are bacteriostatic.
17. Hydrogen peroxide can be used as an antiseptic rinse.
18. Aldehydes and gases can be used for sterilization.

(A)   **(B)**

FIGURE 7.1 **Controlling Microorganisms.** **(A)** Household cleaning products are diverse and formulated for every cleaning need to maintain a sanitary condition. **(B)** This African shantytown has slum houses, open sewage, and littered walkways. It is not surprising that in these unsanitary conditions diseases such as cholera and typhoid are common. »» In these examples, does controlling microorganisms mean sterilization or simply reducing the number of microbes to a safe level? Explain.

is this more important than in the operating rooms and surgical wards. Here, hospital personnel must maintain scrupulous levels of cleanliness and have surgical instruments that are sterile. Yet, **noso-comial** (hospital-acquired) **infections** do occur when hygiene barriers are breached.

Microbial control also is a global endeavor. Government and health agencies in many developing nations often lack the means (financial, medical, social) to maintain proper sanitary conditions. These circumstances can result in outbreaks of diseases such as diphtheria, malaria, measles, meningitis, and cholera. Cholera, as an example, tends to be associated with poverty-stricken areas where overcrowding and inadequate sanitation practices generate contaminated water supplies and food (**FIGURE 7.1B**).

We do need to be hygiene conscious. If the procedures and methods to control pathogens fail or are not monitored properly, serious threats to health and well-being may occur. Just think back to our description of Ignaz Semmelweis and childbed fever or John Snow and the London cholera outbreak (see Chapter 1). Importantly, then, as now, the successful control of microorganisms usually requires simple methods and procedures. These methods are not products of the modern era. As the opening quote reminds us, proper hygiene and sanitation have extended our lives and afforded us the opportunity to contribute more fully to society.

Now that we have a good understanding of microbial growth and metabolism, we can examine a variety of physical and chemical methods used for controlling the growth and spread of microorganisms. Our study begins by outlining some general principles and terminology and then identifies physical methods commonly used today. We also explore chemical methods for microbial control and discuss the spectrum of antiseptics and disinfectants. Whether the methods are physical or chemical, they are integral in public health practices to ensure continued good health and protection from infectious disease. Controlling microbial growth within the body as a result of infection using antimicrobial drugs (e.g., antibiotics) will be discussed in Chapter 24.

7.1   General Principles of Microbial Control

Although your immune system usually does a great job keeping pathogens out of the body, external control measures can further guard against contact with these potential invaders.

The effective control of disease-causing microbes requires an understanding of the procedures and agents available today, including the physical and chemical methods used to limit microbial growth and/or microbial transmission.

First, let's establish some basic vocabulary generally used by the public and by health officials when talking about microbial control.

Sterilization and Sanitization Are Key to Good Public Health

KEY CONCEPT

1. Microbes are kept under control either by eliminating them or reducing their numbers.

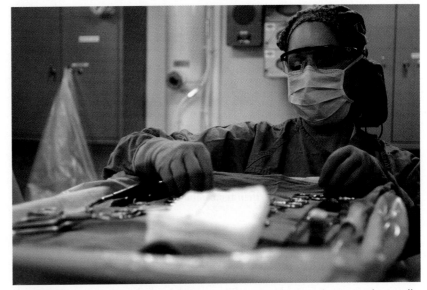

FIGURE 7.2 **Sterile Surgical Instruments.** Here a surgical nurse is unwrapping sterile surgical instruments. »» How would you sterilize these instruments?

Microbiologically speaking, **sterilization** involves the destruction or removal of all living microbes, spores, and viruses on an object or in an area. For example, in a surgical operation, the surgeon uses sterile instruments previously treated in some way to kill any microbes present on them (**FIGURE 7.2**).

Everyday experiences bring us in contact with sterile materials. An unopened can of corn or peas is sterile inside. During the canning process, companies use special sterilization procedures to kill all the microbes on the vegetables and in the tin can. Agents that kill microbes are **microbicidal** (*-cide* = "kill") or more simply called "germicides." If the agent specifically kills bacteria, it is **bactericidal**; if it kills fungi, it is **fungicidal**. Many physical methods and chemical agents are capable of destroying microbes on nonliving materials or on the skin surface. However, once exposed to the air and surroundings, sterile objects will again be contaminated with microbes in the air or the surrounding area.

More often, in our daily experiences we are likely to encounter materials where microbial populations have been reduced or where their growth has been inhibited. **Sanitization** involves those procedures reducing the numbers of patho-

genic microbes or discouraging (inhibiting) their growth. Given enough time, these pathogens will grow and some could possibly cause spoilage or a health problem. So, a tasty wedge of cheese in the refrigerator might look fine today, but in a few weeks it may have a mold growing on it. The toilet bowl "sanitized" with a disinfectant today contains few pathogens. Tomorrow it may again be an area with increased numbers of bacterial species. Many chemical agents are **microbiostatic** (*-static* = "remain in place"); they reduce microbial numbers or inhibit their growth. Again, agents can be **bacteriostatic** or **fungistatic**.

Sanitary measures to control pathogens are very important in areas frequented by the public. City and state sanitation agencies monitor drinking water quality and the preparation of food in public eating establishments to ensure pathogen elimination. Public health depends on good sanitary practices at home and in the workplace.

Contaminated: In microbiology, a once sterile object that is again harboring microorganisms and/or viruses.

CONCEPT AND REASONING CHECKS

7.1 Assess the importance of maintaining sterility or sanitary conditions.

7.2   Physical Methods of Control

With this background, let's now examine some specific physical agents that kill microorganisms or inhibit their growth. Although there are several methods and agents used today that affect microbial survival, they generally include temperature, filtration, radiation, and osmotic pressure.

Heat Is One of the Most Common Physical Control Methods

KEY CONCEPT

2.   Microbes and viruses are killed at temperatures above their temperature range for growth or replication.

The Citadel is a novel by A. J. Cronin that follows the life of a young British physician, beginning in the 1920s. Early in the story, the physician, Andrew Manson, begins his practice in a small coal-mining town in Wales. Almost immediately, he encounters an epidemic of typhoid fever. When his first patient dies of the disease, Manson becomes terribly distraught. However, he realizes the epidemic can be halted, and in the next scene, he is tossing all of the patient's bed sheets, clothing, and personal effects into a huge bonfire.

The killing effect of heat on microorganisms has long been known. Heat is fast, reliable, and relatively inexpensive. Above the growth range temperature for a microbe (see Chapter 5), enzymes and other proteins as well as nucleic acids are denatured (see Chapter 2). Heat also drives off water, and because all organisms depend on water, this loss may be fatal.

The killing rate of heat may be expressed as a function of time and temperature. For example, bacilli of *Mycobacterium tuberculosis* are destroyed in 30 minutes at 58°C, but in only 2 minutes at 65°C, and in a few seconds at 72°C. Each microbial species has a **thermal death time**, the time necessary for killing the population at a given temperature. Each species also has a **thermal death point**, the minimal temperature at which it dies in a given time. These measurements are particularly important in the food industry, where heat is used for preservation (Chapter 25). MICROINQUIRY 7 examines heat and microbial killing.

CONCEPT AND REASONING CHECKS

7.2   How does the thermal death time differ from the thermal death point?

Foot-and-mouth disease: A highly contagious viral disease affecting cattle, sheep, and pigs, in which the animal develops ulcers in the mouth and near the hooves.

Dry Heat Has Useful Applications

KEY CONCEPT

3.   Dry heat uses hot air with little, if any, moisture.

The form in which dry heat is used depends on the nature of the substance to be treated.

Incineration.   Using a direct flame can incinerate microbes very rapidly. For example, the flame of the Bunsen burner is employed for a few seconds to sterilize the bacteriological loop before removing a sample from a culture tube (**FIGURE 7.3**). Flaming the lip of the tube also destroys organisms that happen to contact the lip, while burning away lint and dust.

Disposable hospital gowns and certain plastic apparatus are examples of materials that may be incinerated. In past centuries, the bodies of disease victims were burned to prevent spread of the plague. It still is common practice to incinerate the carcasses of cattle that have died of anthrax and to put the contaminated field to the torch because anthrax spores cannot be destroyed adequately by other means. The 2001 outbreak of foot-and-mouth disease in British cattle required the mass incineration of thousands of cattle as a means to stop the spread of the disease (**FIGURE 7.4**).

Dry Heat.   The hot-air oven uses radiating dry heat for sterilization. This type of energy does not penetrate materials easily, and therefore, long periods of exposure to high temperatures are necessary. For example, at a temperature of 160°C (320°F), a period of two hours is required for the destruction of bacterial spores. Higher temperatures are not recommended because the wrapping paper used for equipment tends to char at 180°C. The hot-air method is useful for sterilizing dry powders and water-free oily substances, as well as for many types of glassware, such as pipettes, flasks, and syringes. Dry heat does not corrode sharp instruments as steam often does, nor does it erode the ground glass surfaces of nondisposable syringes.

The effect of dry heat on microorganisms is equivalent to that of baking. The heat changes microbial proteins by oxidation reactions and creates an arid internal environment, thereby burning microorganisms slowly. It is essential that organic matter such as oil or grease films be removed from the materials, because such substances insulate

MICROINQUIRY 7
Exploring Heat as an Effective Control Method

Is that can of unopened peas in your pantry sterile? Yes, because companies, like General Mills (manufacturer of Green Giant® products) and the food industry in general, have established appropriate procedures for sterilizing commercial foods.

Sterilization depends on several factors. Identifying the type(s) of microbes in a product can determine whether the heating process will sterilize or eliminate only potential disease-causing species. In many foods, the microbes usually are not in water but rather in the food material. Microbes in powders or dry materials will require a different length of time to sterilize the product than microbes in organic matter.

Environmental conditions also influence the sterilization time. Microbes in acidic or alkaline materials decrease sterilization times while microbes in fats and oils, which slow heat penetration, increase sterilization times. It must be remembered that sterilization times are not precise values. However, by knowing these factors, heating the product to temperatures above the maximal range for microbial growth will kill microbes rapidly and effectively. Let's explore the factors of time and temperature. Answers can be found in **Appendix D**.

As we described for the microbial growth curve in Chapter 4, microbial death occurs in an exponential fashion. Look at the **table** to the right. The table records the death of a microbial population by heating. Notice that the cells die at a constant rate. In this generalized example, each minute 90% of the cells die (10% survive). Therefore, if you know the initial number of microorganisms, you can predict the thermal death time (TDT), which is defined as the minimal time, at a specified temperature, required to kill a population of microorganisms.

The food industry depends on knowing a microorganism's heat sensitivity when planning the canning or packaging of many foods as excessive heat can

affect the texture and flavor of the food product. One way the industry determines this sensitivity is by using standard curves that take into account the factors mentioned above. The graph drawn below represents three such curves, each representing a different bacterial species treated at the same temperature (60°C) in the same food material (curve B represents the plotted data from the table).

7a. If you had to sterilize this food product that initially contained 10^6 bacteria, how long would it take for each bacterial species? (Hint: $10^0 = 1$)

Another value of importance is the **decimal reduction time (DRT)** or **D value**, which is the time required at a specific temperature to kill 90% of the viable organisms. These are the values typically used in the canning industry. D values are usually identified by the temperature used for killing. In the graph to the right, the temperature was 60°C, so D is written as D60. Look at the graph again.

7b. Calculate the D60 values for the three bacterial populations depicted in curves A, B, and C.

On another day in the canning plant, you need to sterilize a food product. However, the only information you have is a D70 = 12 minutes for the microorganism in the food product. Assuming the D value is for the volume you have to sterilize and there are 10^8 bacteria in the food product:

7c. At what temperature will you treat the food product?

7d. How long will it take to sterilize the product?

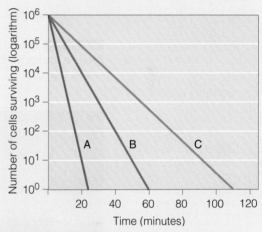

Standard curves for death of three microbial species (A, B, C).

TABLE		
Microbial Death Rate		
Time (minutes)	Number of Cells Surviving	% Killed
0	1,000,000	—
10	100,000	90
20	10,000	90
30	1,000	90
40	100	90
50	10	90
60	1	90

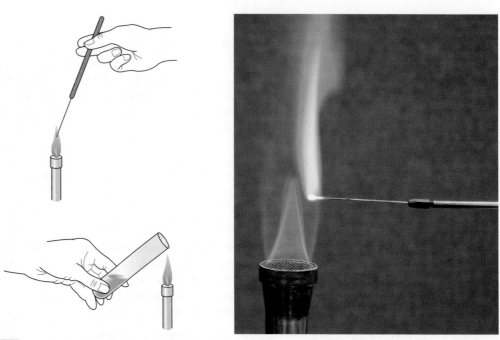

FIGURE 7.3 **Use of the Direct Flame as a Sterilizing Agent.** A few seconds in the flame of a laboratory Bunsen or Fisher burner is usually sufficient to effect sterilization of a culture tube lip or inoculation loop. »» Why is it necessary to flame a culture tube lip?

FIGURE 7.4 **Incineration Is an Extreme Form of Heat.** Nearly four million hoofed animals infected with or exposed to foot-and-mouth disease in England in 2001 were incinerated and buried. »» How is incineration similar to flame sterilization?

against dry heat. Moreover, the time required for heat to reach sterilizing temperatures varies according to the material. This factor must be considered in determining the total exposure time.

CONCEPT AND REASONING CHECKS

7.3 Explain how dry heat is used to "eliminate" microorganisms.

Moist Heat Is More Versatile Than Dry Heat

KEY CONCEPT

4. Moist heat more easily penetrates objects and materials.

There are several ways that moist heat is used to control microbes and sterilize materials.

Moist Heat as Boiling Water. Boiling water is an example of moist heat that penetrates materials much more rapidly than dry heat because water molecules conduct heat better than air. Therefore, moist heat can be used at a lower temperature and shorter exposure time than for dry heat.

Moist heat kills microorganisms by denaturing their proteins. **Denaturation** is a change in the chemical or physical property of a protein. It includes structural alterations due to destruction of the chemical bonds holding proteins in a three-dimensional form. As proteins revert to a two-dimensional structure, they coagulate (denature) and become nonfunctional. Egg protein undergoes a similar transformation when it is boiled. (You might find reviewing the chemical structure of proteins in Chapter 2 helpful in understanding this process.) The coagulation of proteins requires less energy than oxidation, and, therefore, less heat need be applied.

Boiling water is not considered a sterilizing agent because the destruction of bacterial spores and the inactivation of viruses cannot always be assured. Under ordinary circumstances, with microorganisms at concentrations of less than 1 million per milliliter, most species of microorganisms can be killed within 10 minutes. Indeed, the process may require only a few seconds. However, fungal spores, protozoal cysts, and large concentrations of hepatitis A viruses require up to 30 minutes' exposure. Bacterial spores often require two hours or more. Because inadequate information exists on the heat tolerance of many species of microorganisms, boiling water is not reliable for sterilization purposes (FIGURE 7.5).

If boiling water must be used to destroy microorganisms, then materials must be thoroughly cleaned to remove traces of organic matter, such as blood or feces. The minimum exposure period should be 30 minutes, except at high altitudes, where it should be increased to compensate for the lower boiling point of water. All materials should be well covered. Baking soda may be added at a 2% concentration to increase the efficiency of the process.

Sterilization with Pressurized Steam. Moist heat in the form of pressurized steam is regarded as the most dependable method for sterilization, including the destruction of bacterial spores. This method is incorporated into a device called the **autoclave**. When the pressure of a gas increases, the temperature of the gas also increases proportionally. Because steam is a gas, increasing its pressure in a closed system increases its temperature. As the water molecules in steam become more energized, their penetration increases substantially. This principle is used to reduce cooking time in the home pressure cooker and to reduce sterilizing time in the autoclave. During autoclaving, the sterilizing agent is the moist heat, not the pressure.

Autoclaves contain a sterilizing chamber into which articles are placed, and a steam jacket where steam is maintained (FIGURE 7.6). As steam flows from the steam jacket into the sterilizing chamber, cool air is forced out and a special valve increases the pressure to 15 pounds/square inch (lb/in^2) above normal atmospheric pressure. The temperature rises to 121.5°C, and the superheated steam rapidly conducts heat into microorganisms. The time for destruction of the most resistant bacterial species is about 15 minutes. For denser objects or larger volumes, more than 30 minutes of exposure may be required. The conditions must be carefully controlled to assure sterilization has been accomplished (MicroFocus 7.1).

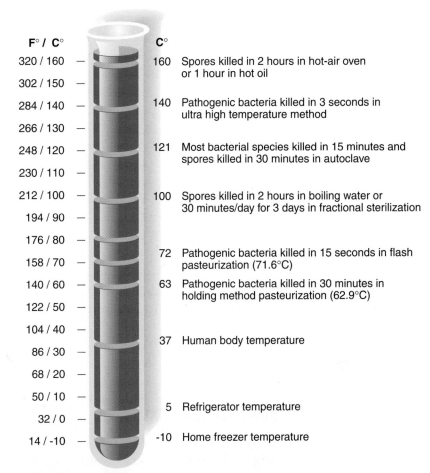

F° / C°		C°	
320 / 160 —		160	Spores killed in 2 hours in hot-air oven or 1 hour in hot oil
302 / 150 —			
284 / 140 —		140	Pathogenic bacteria killed in 3 seconds in ultra high temperature method
266 / 130 —			
248 / 120 —		121	Most bacterial species killed in 15 minutes and spores killed in 30 minutes in autoclave
230 / 110 —			
212 / 100 —		100	Spores killed in 2 hours in boiling water or 30 minutes/day for 3 days in fractional sterilization
194 / 90 —			
176 / 80 —			
158 / 70 —		72	Pathogenic bacteria killed in 15 seconds in flash pasteurization (71.6°C)
140 / 60 —		63	Pathogenic bacteria killed in 30 minutes in holding method pasteurization (62.9°C)
122 / 50 —			
104 / 40 —			
86 / 30 —		37	Human body temperature
68 / 20 —			
50 / 10 —			
32 / 0 —		5	Refrigerator temperature
14 / -10 —		-10	Home freezer temperature

FIGURE 7.5 **Temperature and the Physical Control of Microorganisms.** Notice that materials containing bacterial endospores require longer exposure times and higher temperatures for killing. »» Pure water boils and freezes at what temperatures on the Celsius scale?

The autoclave is used to control microorganisms in both hospitals and laboratories. It is employed for blankets, bedding, utensils, instruments, intravenous solutions, and a broad variety of other objects. The laboratory technician uses it to sterilize bacteriological media and destroy pathogenic cultures. The autoclave is equally valuable for glassware and metalware, and is among the first instruments ordered when a microbiology laboratory is established.

The autoclave has certain limitations. For example, some plasticware melts in the high heat and sharp instruments often become dull. Moreover, many chemicals break down during the sterilization process and oily substances cannot be treated because they do not mix with water.

In recent years a new form of autoclave, called the **prevacuum autoclave**, has been developed for sterilization procedures. This machine draws

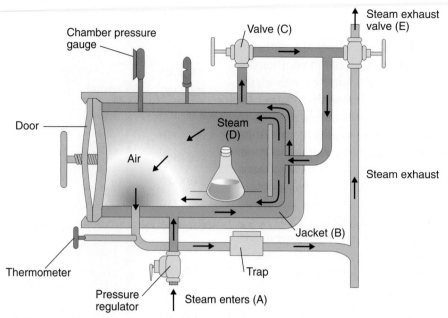

FIGURE 7.6 **Operating an Autoclave.** Steam enters through the port (A) and passes into the jacket (B). After the air has been exhausted through the vent, a valve (C) opens to admit pressurized steam (D), which circulates among and through the materials, thereby sterilizing them. At the conclusion of the cycle, steam is exhausted through the steam exhaust valve (E). »» Is it the steam or the pressure that kills microorganisms in an autoclave? Explain.

MICROFOCUS 7.1: Tools
Autoclave Quality Assurance

A nosocomial outbreak of *Pseudomonas* in a Thailand hospital illustrates the need to carefully monitor the autoclave during use.

The problem began when hospital pharmacists prepared bottles of basal salts solution for use in the hospital operating rooms. To sterilize the solutions, the bottles were placed in the autoclave and left to run on its automatic cycle.

The bottles were then delivered to surgery to be used to irrigate the eyes of patients undergoing cataract surgery. Some bottles were left unused.

Following surgery, three cataract patients developed eye inflammations. The organism isolated from the patients was a pathogenic strain of *Pseudomonas* and the infected patients were treated with antibiotics.

An autoclave can be used to sterilize many dry and liquid materials.

Health investigators tested the unused bottles of salt solution as well as the tubes attached to the now-empty bottles. They found the identical strain of *Pseudomonas*.

Examining the pharmacy records, investigators noted that the autoclave pressure had reached only 10 to 12 lb/in^2, rather than the required 15 lb/in^2. The salts solution apparently was not sterilized.

There are several ways to assure that materials are properly sterilized. Autoclaves have temperature and pressure gauges visible from the outside and most models can produce a paper record of the temperature, time, and pressure.

To gauge the success of sterilization, materials usually are autoclaved with autoclave tape, which turns color if the object inside the material has been autoclaved correctly. Biological indicators also can be used. A strip containing spores of *Geobacillus stearothermophilus* can be included with the objects treated. At the conclusion of the cycle, the strip is placed in a nutrient broth medium and incubated. If the sterilization process has been unsuccessful, the spores will germinate and their metabolism will change the color of a pH indicator in the growth medium.

air out of the sterilizing chamber at the beginning of the cycle. Saturated steam then is used at a temperature of 132°C to 134°C at a pressure of 28 to 30 lb/in^2. The time for sterilization is now reduced to as little as four minutes. A vacuum pump operates at the end of the cycle to remove the steam and dry the load. The major advantages of the prevacuum autoclave are the minimal exposure time for sterilization and the reduced time to complete the cycle.

Sterilization without Pressurized Steam. In the years before the development of the autoclave, liquids and other objects were sterilized by exposure to free-flowing steam at 100°C for 30 minutes on each of three successive days, with incubation periods at room temperature between the steaming. The method was called **fractional sterilization** because a fraction of the sterilization was accomplished on each day. It was also called **tyndallization** after its developer, John Tyndall.

Sterilization by the fractional method is achieved by an interesting series of events. During the first day's exposure, steam kills virtually all organisms except bacterial spores. During overnight incubation, the spores germinate and the viable cells multiply only to be killed on the second day's 100°C exposure. Again, the material is cooled and any remaining spores may germinate, the resulting cells only to be killed on the third day. Although the method usually results in sterilization, occasions arise when several spores fail to germinate. The method also requires that spores be in a suitable medium, such as broth, for germination.

Fractional sterilization has assumed renewed importance in modern microbiology with the development of high-technology instrumentation and new chemical substances. Often, these materials cannot be sterilized at autoclave temperatures, or by long periods of boiling or baking, or with chemicals.

Pasteurization. The final example of moist heat involves the process of **pasteurization**, which reduces the bacterial population of a liquid such as milk and destroys organisms that may cause spoilage and human disease (**FIGURE 7.7**). Spores are not affected by pasteurization.

One method for milk pasteurization, called the **holding (or batch) method**, involves heating at 63°C for 30 minutes. Although any thermophilic

FIGURE 7.7 **The Pasteurization of Milk.** Milk is pasteurized by passing the liquid through a heat exchanger. The flow rate and temperature are monitored carefully. Following heating, the liquid is rapidly cooled. »» Why is the liquid rapidly cooled?

bacteria would thrive at this temperature, they are of little consequence because they cannot grow at body temperature. For decades, pasteurization has been aimed at destroying *Mycobacterium tuberculosis,* long considered the most heat-resistant bacterial species. More recently, however, attention has shifted to destruction of *Coxiella burnetii,* the agent of Q fever (Chapter 10), because this bacterial organism has a higher resistance to heat. Because both organisms are eliminated by pasteurization, dairy microbiologists assume other pathogenic bacteria also are destroyed. Pasteurization also is used to eliminate the *Salmonella* and *Escherichia coli* that can contaminate fruit juices. Two other methods are the **flash pasteurization method** at 71.6°C for 15 seconds and the **ultra high temperature (UHT) method** at 140°C for 3 seconds. The UHT method is the only method that sterilizes the liquid (MICROFOCUS 7.2). These methods are discussed in Chapter 25.

Although heat is a valuable physical agent for controlling microorganisms, sometimes it is impractical to use. For example, no one would suggest removing the microbial population from a tabletop by using a Bunsen burner, nor can heat-sensitive solutions be subjected to an autoclave. In instances such as these and numerous others, a heat-free method must be used.

CONCEPT AND REASONING CHECKS

7.4 Summarize the ways that moist heat controls or sterilizes materials or beverages.

MICROFOCUS 7.2: Being Skeptical
Milk Stays Fresh Longer If It's Organic

If you ever go shopping for milk at the local market, or especially one specializing in "natural and organic foods," you might notice that the "Best by" date expires much sooner on a carton of regular milk than on one that is organic milk. In fact, the date on the organic milk may be some 3 weeks longer than on the regular milk, which typically is 5 to 7 days from the store delivery date. So, being organic, does that ensure a longer shelf life?

The fact that the milk is organic has nothing to do with its longer shelf life. Labeling the milk as "organic" only means that the cows on the dairy farm were not given antibiotics or hormones like bovine growth hormone (BGH), which stimulates a cow's milk production (see figure). The reason it has a longer shelf life is due to the pasteurization process. Organic milk is subjected to the ultra high temperature (UHT) process (ultra-pasteurized) where the milk is heated to 140°C for three sec-

A carton of organic milk.

onds. This kills all the microorganisms that may be in the liquid—it is sterile. Most regular milk today is subjected to the flash pasteurization method where the milk is at about 72°C for 15 seconds. This "high temperature, short duration" process does not kill all microbes that may be in the milk; only the pathogens have been eliminated. Because there are bacterial species that are psychrotolerant, the milk can spoil if left on the refrigerated shelf too long.

Regular milk also could be subjected to UHT. However, it usually is not because it has to travel only a short distance to market; organic products, on the other hand, are not often produced throughout the country, so they have further to travel to reach the consumer. So, UHT preserves the product longer. Although not found commonly within the United States, room-temperature Parmalat milk is a product of UHT and can be found commonly in Europe and other parts of the world.

The verdict? "Organic" is not defined as "longer shelf life." If shelf life is important to you, simply look for products treated by UHT. Also of note: UHT does burn some of the sugars in the milk, so the milk may taste slightly "caramelized," something some people find less tasty.

Filtration Traps Microorganisms

KEY CONCEPT

5. Filtration removes microbes from the air or water.

Filters came into prominent use in microbiology as interest in viruses grew during the 1890s. Previous to that time, filters were used to trap airborne organisms and sterilize bacteriological media, but they became essential for separating viruses from other microorganisms. Among the early pioneers of filter technology was Charles Chamberland, an associate of Pasteur. His porcelain filter was important to early virus research. Another pioneer was Julius Petri (inventor of the Petri dish), who developed a sand filter to separate bacterial cells from the air.

Filtration is a mechanical method that can be used to remove microorganisms from a solution or gas. Several types of filters are used in the microbiology laboratory.

The most common is the **membrane filter**, which consists of a pad of cellulose acetate or polycarbonate mounted in a holding device (**FIGURE 7.8A**). As fluid passes through the filter, organisms are trapped in the pores of the filtering material. The solution dripping through the filter into the receiving container is decontaminated or, in some cases, sterilized. Membrane filters are used to purify such heat-sensitive liquids as beverages, some bacteriological media, toxoids, many pharmaceuticals, and blood solutions.

The membrane filter is particularly valuable because bacterial cells trapped on the filter multiply and form colonies on the filter pad when the pad is placed on a plate of culture medium. Microbiologists then can count the colonies to determine the number of bacteria originally present (**FIGURE 7.8B, C**).

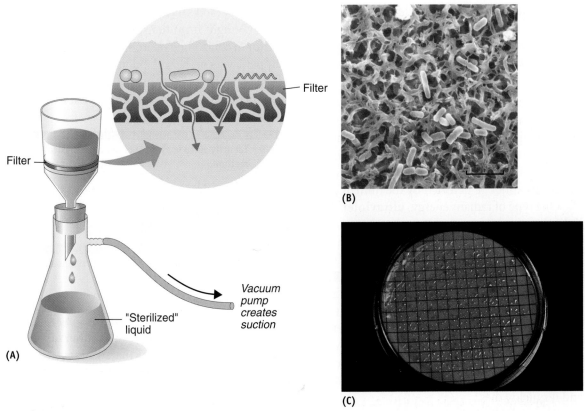

(A)

(B)

(C)

FIGURE 7.8 **The Principle of Filtration.** Filtration is used to remove microorganisms from a liquid. The effectiveness of the filter is proportional to the size of its pores. (**A**) Bacteria-laden liquid is poured into a filter, and a vacuum pump helps pull the liquid through and into the flask below. The bacterial cells are larger than the pores of the filter, and they become trapped. (**B**) A view of *Escherichia coli* cells trapped in the pores of a 0.45-μm nylon membrane filter. (Bar = 5 μm.) (**C**) *E. coli* colonies growing on a membrane filter. »» Why aren't viruses also trapped on the filter?

Air also can be filtered to remove microorganisms. The filter generally used is a **high-efficiency particulate air (HEPA) filter**, which consists of a mat of randomly arranged fibers that trap particles, microorganisms, and spores. As part of a **biological safety cabinet**, HEPA filters can trap over 99% of all particles, including microorganisms and spores with a diameter larger than 0.3 μm (**FIGURE 7.9**). The air entering surgical units and specialized treatment facilities, such as burn units, also are HEPA filtered to exclude microorganisms. In some hospital wards, such as for respiratory diseases and in certain pharmaceutical filling rooms, the air is recirculated through HEPA filters to ensure air purity.

CONCEPT AND REASONING CHECKS

7.5 Determine the uses for filtration in a health care setting.

FIGURE 7.9 **A Biological Safety Cabinet.** The cabinet shown has a metal grid at the top that covers a HEPA filter through which air enters the cabinet. As the filtered air, free of contaminants and microbes, moves into and across the workspace, it exits out the bottom front and rear. A UV light is also positioned at the top rear to decontaminate the metal surfaces maintaining a contaminant-free workspace when the cabinet is not in use. »» Why is air moved out of the cabinet rather than into the cabinet?

Ultraviolet Light Can Be Used to Control Microbial Growth

KEY CONCEPT

6. UV light can be bactericidal.

Visible light is a type of radiant energy detected by the light-sensitive cells of the eye. The wavelength of this energy is between 400 and 800 nanometers (nm). Other types of radiation have wavelengths longer or shorter than that of visible light, and therefore cannot be detected by the human eye.

One type of radiant energy, **ultraviolet (UV) light**, is useful for controlling microorganisms. Ultraviolet light has a wavelength between 100 and 400 nm, with the energy at about 265 nm most destructive to bacterial cells (**FIGURE 7.10**). When microorganisms are subjected to UV light, cellular DNA absorbs the energy, and adjacent thymine molecules (in the same strand) link together, kinking the double helix and disrupting DNA replication (Chapter 8). The damaged organism can no longer produce critical proteins or reproduce, and it quickly dies.

Ultraviolet light effectively reduces the microbial population where direct exposure takes place. It is used to limit airborne or surface contamination in a hospital room, morgue, pharmacy, toilet facility, or food service operation. It is noteworthy that UV light from the sun may be an important factor in controlling microorganisms in the air and upper layers of the soil, but it may not be effective against all bacterial spores. Ultraviolet light does not penetrate liquids or solids, and it can lead to human skin cancer.

CONCEPT AND REASONING CHECKS

7.6 Identify some uses for UV light as a physical control method.

Other Types of Radiation Also Can Sterilize Materials

KEY CONCEPT

7. X rays and gamma rays also are microbicidal.

Looking back at Figure 7.10, there are two other forms of radiation useful for destroying microorganisms. These are **X rays** and **gamma rays**. Both have wavelengths shorter than the wavelength of UV light. As X rays and gamma rays pass through microbial molecules, they force electrons out of their shells, thereby creating ions. For this reason, the radiations are called **ionizing radiations**. The ions quickly combine, mainly with cellular water, and the free radicals generated affect cell metabolism and physiology. Ionizing radiations currently are used to sterilize such heat-sensitive pharmaceuticals as vitamins, hormones, and antibiotics as well as certain plastics and suture materials.

Ionizing radiations also have been approved for controlling microorganisms, and for preserving foods, as noted in MicroFocus 7.3. The approval has generated much controversy, fueled by activists concerned with the safety of factory workers and consumers. First used in 1921 to inactivate *Trichinella spiralis,* the agent of trichinellosis, irradiation now is used as a preservative in more than 40 countries for over 100 food items, including potatoes, onions, cereals, flour, fresh fruit, and poultry (**FIGURE 7.11A**). The U.S. Food and Drug Administration (FDA) approved cobalt-60 and cesium-137 irradiation to preserve or extend the shelf life of several foods. This includes irradiating poultry and red meats such as beef, lamb, and pork. In 2008, the FDA approved the irradiation of fresh and bagged spinach, and iceberg lettuce, to reduce potential foodborne illness. Irradiation has been used to prepare many meals for the U.S. military and the American astronauts (**FIGURE 7.11B**). What is called

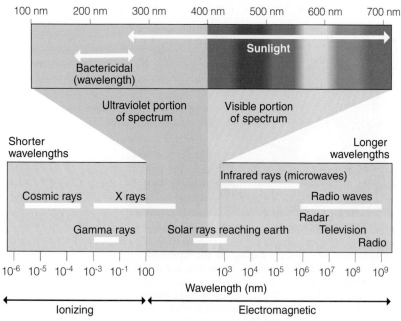

FIGURE 7.10 **The Ionizing and Electromagnetic Spectrum of Energies.** The complete spectrum is presented at the bottom of the chart, and the ultraviolet and visible sections are expanded at the top. Notice how the bactericidal energies overlap with the UV portion of sunlight. This may account for the destruction of microorganisms in the air and in upper layers of soil. »» How does UV light kill bacteria?

MICROFOCUS 7.3: Public Health

"No, the Food Does Not Glow!"

In the United States, there are more than 76 million cases of foodborne disease accounting for more than 300,000 hospitalizations and 5,000 deaths each year. One major source of foodborne disease is agricultural produce contaminated with intestinal pathogens. Another source is from improperly cooked or handled meats, or poultry harboring human intestinal pathogens, such as *Escherichia coli* O157:H7, *Campylobacter*, *Listeria*, and *Salmonella*.

Irradiation has the potential to greatly limit such illnesses. The Centers for Disease Control and Prevention (CDC) have estimated that if just 50% of the meat and poultry sold in the United States was

Irradiated strawberries displaying the "radura" symbol, meaning they have been treated by irradiation.

irradiated, there would be 900,000 fewer cases of foodborne illness and 350 fewer deaths each year.

Yet today manufacturers continue to wrestle with the concept of food irradiation as they constantly confront a leery public, some of who still have visions of Hiroshima and Nagasaki. In the United States, just 10% of the herbs and spices are irradiated and only 0.002% of fruits, vegetables, meats, and poultry are irradiated.

Food irradiation is entirely different from atomic radiation. The irradiation comes from gamma rays produced during the natural decay of cobalt-60 or cesium-137. The most common method involves electron beams (e-beams) not unlike those used in an electron microscope. None of these types of radiations produce radioactivity—the irradiated food does not glow (see figure).

Low doses of irradiation are used for disinfestations and extending the shelf life of packaged foods. As mentioned in the chapter narrative, a pasteurizing dose is used on meats, poultry, and other foods. Such levels do not eliminate all microbes in the food, but, similar to pasteurization, helps to reduce the dangers of pathogen-contaminated or cross-contaminated meats and poultry.

During the irradiation, the gamma rays or electrons penetrate the food, and, just as in cooking, cause molecular changes in any contaminating microorganisms, which ultimately leads to their death.

Irradiation of foods also has its limitations. The irradiation dose will not kill bacterial endospores, inactivate viruses, or neutralize toxins. Therefore, irradiated food still must be treated in a sanitary fashion. Nutritional losses are similar to those occurring in cooking and/or freezing. Otherwise, there are virtually no known changes in the food, and there is no residue.

(A)

BBQ Beef Brisket

PRODUCED AT
NATICK SOLDIER CENTER
COMBAT FEEDING
PROCESSED AT
FOOD TECHNOLOGY SERVICE, INC.

FOR USE BY

NASA NATIONAL AERONAUTICS AND SPACE ADMINISTRATION

TREATED BY   IRRADIATION
FOR SHELF STABILITY

(B)

FIGURE 7.11 Food Irradiation. **(A)** The FDA has approved irradiation as a preservation method for numerous foods, including many fruits and vegetables, as well as poultry and red meats. **(B)** Many otherwise perishable foods eaten by NASA astronauts are prepared by irradiation. »» Does irradiation sterilize the treated product? Explain.

MICROFOCUS 7.4: Public Health

Microwave That Sponge!

Kitchens are often considered to be one of the microbially dirtiest places in the home. Raw fruits, vegetables, and especially meats and poultry can carry substantial microbes that can end up on cutting boards, washcloths, and especially that kitchen sponge. These "hot spots" need to be kept clean with disinfectants and sanitized. But how do you clean that kitchen sponge that harbors millions of bacterial cells picked up in the juices from raw chicken or ground beef? Microwave it!

A study published in the *Journal of Environmental Health* in 2006 suggests that putting that kitchen sponge or plastic scrubbing pad in the microwave oven for two minutes will kill bacterial pathogens

A sponge can be microwaved to kill microbial contaminants.

like *Escherichia coli* and *Campylobacter jejuni*, two species commonly found on raw ground beef and chicken (see figure). Simply washing the sponge and placing it in the dishwasher may clean it, but it will not get rid of bacteria buried deep in the nooks and crannies of the sponge. The sponge needs to be decontaminated.

The University of Florida researchers came to their conclusions after doing experiments with sponges and pads that were soaked in wastewater containing *E. coli* cells, *Bacillus cereus* endospores, and a "cocktail" of other microbes and viruses. The researchers used a standard kitchen microwave oven to heat the sponges and scrub pads for varying lengths of time. They were then squeezed to remove the water from which the researchers determined the microbes remaining. They compared the test sponges and pads with control sponges and pads that had been soaked but not placed in the microwave oven.

Just two minutes of microwaving on full power killed or inactivated more than 99% of all the bacterial cells and viruses in the sponges and pads; four minutes was necessary to kill the *B. cereus* endospores (not surprising for such resistant structures).

Because microwaves work by exciting water molecules and generating frictional heat, rather than by the direct effects of the microwave radiation, it is important to make sure the sponge or scrubbing pad is wet before putting it in the microwave. And how often should you microwave that sponge? The University of Florida researchers suggest every other day should be fine.

a **pasteurizing dose** is used on meats, poultry, and other foods. Such levels are not intended to eliminate all microbes in the food, but, like pasteurization of milk, to eliminate the pathogens. The foods are not necessarily sterile.

Another form of energy, the microwave, has a wavelength longer than that of ultraviolet light and visible light. In a microwave oven, microwaves are absorbed by water molecules, which are set into high-speed motion, and the heat of friction from these excited molecules is transferred to foods. In fact, the microwave can be an excellent way to sterilize your kitchen sponge (MICROFOCUS 7.4).

CONCEPT AND REASONING CHECKS

7.7 Identify some uses for X rays and gamma rays as a physical control method.

Preservation Methods Retard Spoilage by Microorganisms in Foods

KEY CONCEPT

8. Dehydration and cold temperatures slow microbial growth.

Over the course of many centuries, various physical methods have evolved for controlling microorganisms in food. Though valuable for preventing the spread of infectious agents, these procedures are used mainly to retard spoilage and prolong the shelf life of foods, rather than for sterilization. Irradiation is an example of a preservation method.

Drying is useful in the preservation of various meats, fish, cereals, and other foods. Because water is necessary for life, it follows that where there is no water, there is virtually no life. Many nonperishable foods (such as cereals, rice, and sugar) in the kitchen pantry represent such shelf-stable products.

Preservation by salting is based upon the principle of osmotic pressure. When food is salted (usually sodium chloride), water diffuses out of microorganisms toward the higher salt concentration and lower water concentration in the surrounding environment. This flow of water, called **osmosis**, leaves the microorganisms dehydrated, and they die. The same phenomenon occurs in highly sugared foods (usually sucrose) such as syrups, jams, and jellies. However, fungal contamination (molds and yeasts) and growth at the surface may occur because they can tolerate low water and high sugar concentrations.

Low temperatures found in the refrigerator and freezer retard spoilage by lowering the metabolic rate of microorganisms and thereby reducing their rate of growth (see Chapter 5). Spoilage is not totally eliminated in cold foods, however, and many psychrotrophs remain alive, even at freezer temperatures. These organisms multiply rapidly when food thaws, which is why prompt cooking is recommended.

Note in these examples that there are significant differences between killing microorganisms, holding them in check, and reducing their numbers. The preservation methods are described as bacteriostatic because they prevent the further multiplication of food-borne pathogens such as *Salmonella* and *Clostridium*. A more complete discussion of food preservation as it relates to public health is presented in Chapter 25. **TABLE 7.1** and **FIGURE 7.12** summarize the physical agents used for controlling microorganisms.

> Osmotic pressure: The pressure applied to a solution to stop the inward diffusion (osmosis) of a solvent through a semi-permeable membrane.

CONCEPT AND REASONING CHECKS

7.8 Explain how salting foods acts as a preservation method.

TABLE

7.1 A Summary of Physical Agents Used to Control Microorganisms

Physical Method	Conditions	Instrument	Object of Treatment	Examples of Uses	Comment
Incineration	A few seconds	Flame	All microorganisms	Laboratory instruments	Object must be disposable or heat-resistant
Hot air	160°C for 2 hr	Oven	Bacterial spores	Glassware Powders Oily substances	Not useful for fluid materials
Boiling water	100°C for 10 min	—	Vegetative microorganisms	Wide variety of objects	Total immersion and precleaning necessary
	100°C for 2 hr+	—	Bacterial spores		
Pressurized steam	121°C for 15 min at 15 lb/in²	Autoclave	Bacterial spores	Instruments Surgical materials Solutions and media	Broad application in microbiology
Fractional sterilization	30 min/day for 3 successive days	Sterilizer	Bacterial spores	Materials not sterilized by other methods	Long process Sterilization not assured
Pasteurization	Holding method Flash method UHT method	Pasteurizer	Pathogenic microorganisms	Dairy products Beverages	Sterilization achieved with UHT
Filtration	Entrapment in pores	Membrane filter HEPA filter	All microorganisms	Fluids Air	Many adaptations
Ultraviolet light	265 nm energy	Generator	All microorganisms	Surface and air sterilization	Not useful in fluids
X rays Gamma rays	Short wave-length energy	Generator	All microorganisms	Heat-sensitive materials	Extending food shelf life
Dehydration	Osmotic conditions	—	All microorganisms	Salted and sugared foods	Food preservation
Refrigeration/ Freezing	5°C/−10°C	Refrigerator/ Freezer	All microorganisms	Numerous foods	Spoilage/Food preservation

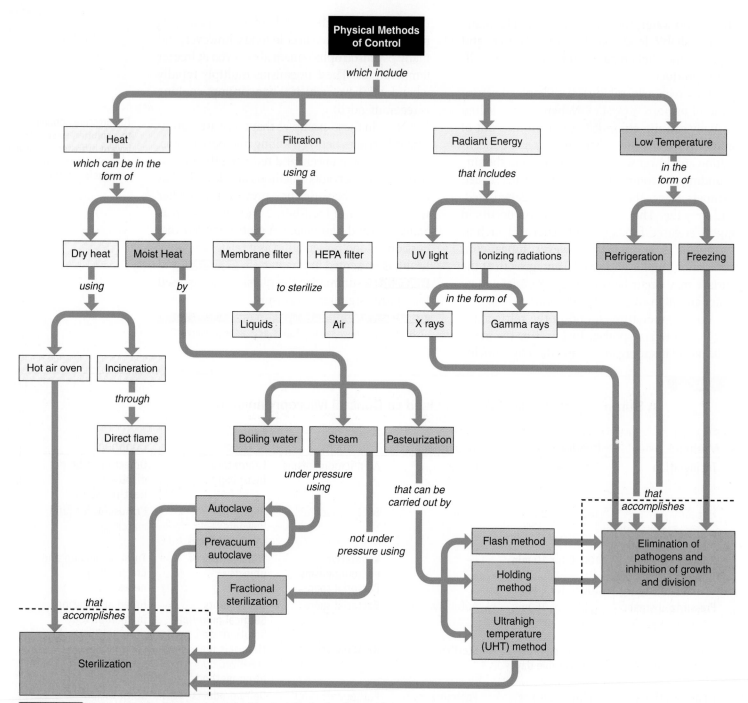

FIGURE 7.12 **A Concept Map Summarizing the Physical Methods of Microbial Control.** Note that some methods sterilize while others tend to inhibit growth and division. »» What is common to most of the sterilization methods?

7.3 General Principles of Chemical Control

Sanitation and disinfection methods are not unique to the modern era. The Bible refers often to cleanliness and prescribes certain dietary laws to prevent consumption of what was believed to be contaminated food. Egyptians used resins and aromatics for embalming, and ancient peoples burned sulfur for deodorizing and sanitary purposes. Arabian physicians first suggested using mercury to treat syphilis. Over the centuries, spices were used as preservatives as well as masks for foul odors, making Marco Polo's trips to Asia for new spices a necessity as well as an adventure.

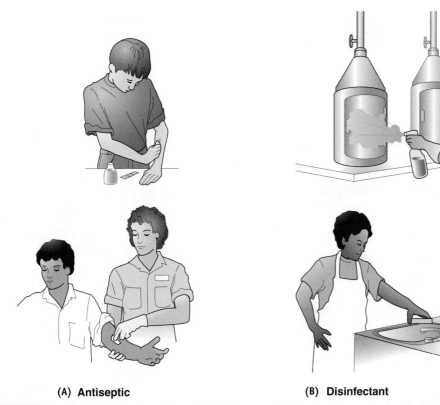

(A) Antiseptic   **(B) Disinfectant**

FIGURE 7.13 Sample Uses of Antiseptics and Disinfectants. (**A**) Antiseptics are used on body tissues, such as on a wound or before piercing the skin to take blood. (**B**) A disinfectant is used on inanimate objects, such as equipment used in an industrial process or tabletops and sinks. »» Why aren't disinfectants normally used as antiseptics?

And fans of Western movies probably have noted that American cowboys practiced disinfection by pouring whiskey onto gunshot wounds.

Chemical Control Methods Are Dependent on the Object to Be Treated

KEY CONCEPT

9. Disinfectants and antiseptics are key to proper sanitation and public health.

As early as 1830, the United States Pharmacopoeia listed tincture of iodine as a valuable antiseptic, and soldiers in the Civil War used it in plentiful amounts. Joseph Lister established the principles of aseptic surgery using carbolic acid (phenol) for treating wounds (see Chapter 1).

As we have discussed, the physical agents for controlling microorganisms generally are intended to achieve sterilization. Chemical agents, by contrast, rarely achieve sterilization. Instead, they are expected only to destroy the pathogenic organisms on or in an object or area. The process of destroying pathogens is called **disinfection** and the object is said to be disinfected. If the object treated is lifeless, such as a tabletop, the chemical agent used is called a **disinfectant**. However, if the object treated is living, such as a tissue of the human body, the chemical agent used is an **antiseptic** (**FIGURE 7.13**). It is important to note that even though a particular chemical may be used as a disinfectant as well as an antiseptic (e.g., iodine), the precise formulations are so different that its ability to kill or inactivate microorganisms differs substantially in the two products.

Antiseptics and disinfectants are usually microbicidal; they inactivate the major enzymes of an organism and interfere with its metabolism so that it dies. A chemical also may be microbiostatic, disrupting minor chemical reactions and slowing the metabolism, which results in a longer time between cell divisions. Although a subtle difference sometimes exists between the two chemical agents, the terms indicate effectiveness in a particular situation.

The word **sepsis** (*seps* = "putrid") refers to a condition in which microbes or their toxins are present in tissues or the blood; thus, we have **septicemia**, meaning "microbial infection of the

Tincture:
A substance dissolved in alcohol.

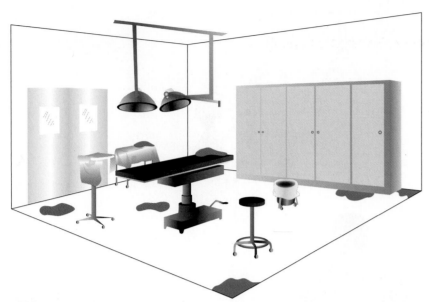

FIGURE 7.14 **A Pathogen-Contaminated Operating Room.** Following a "dirty operation" (for an infectious disease), the operating room (OR) may have become broadly contaminated with the pathogen. The purple areas indicate where the pathogen was cultured from a swab sample. »» Will the OR need to be sterilized, disinfected, or sanitized? Explain.

blood," and antiseptic, "against infection." It also is the origin of the term **asepsis**, meaning "free of disease-causing microbes."

Other expressions are associated with chemical control. To **sanitize** an object is to reduce the microbial population to a safe level as determined by public health standards. For example, in dairy and food-processing plants, the equipment usually is sanitized through the process of sanitization. Commercial establishments, such as restaurants, depend on disinfectants to maintain a sanitary kitchen and work establishment. To **degerm** an object is merely to remove organisms from its surface. Washing with soap and water degerms the skin surface but has little effect on microorganisms deep in the skin pores.

FIGURE 7.14 illustrates the extent of "cleanup" necessary in an operation room following an operation on an infected patient.

CONCEPT AND REASONING CHECKS

7.9 Distinguish between (a) an antiseptic and a disinfectant and (b) disinfection and sanitization.

Chemical Agents Are Important to Laboratory and Hospital Safety

KEY CONCEPT

10. Disinfectants and antiseptics are defined by their properties.

To be useful as an antiseptic or disinfectant, a chemical agent must have a number of properties. The agent should be:

- Able to kill or slow the growth of microorganisms.
- Nontoxic to animals or humans, especially if it is used as an antiseptic.
- Soluble in water and have a substantial shelf life during which its activity is retained.
- Useful in much diluted form and able to perform its job in a relatively short time.

Other characteristics also will contribute to the value of a chemical agent: It should not separate on standing, it should penetrate well, and it should not corrode instruments. The chemical will have a distinct advantage if it does not combine with organic matter such as blood or feces, because organic matter can bind and "use up" the chemical. Of course, the chemical should be easy to obtain and relatively inexpensive.

Because disinfection is essentially a chemical process, several chemical parameters should be considered when selecting an antiseptic or disinfectant.

- **Temperature.** It is important to know at what temperature the disinfection is to take place because a chemical reaction occurring at 37°C (body temperature) may not occur at 25°C (room temperature).
- **pH.** A particular chemical may be effective at a certain pH but not another.
- **Stability.** The chemical reaction may be very rapid with one agent and slower with another. Thus, if long-term disinfection is desired, the second agent may be preferable.

Two other considerations are the type of microorganism to be eliminated and the surface treated. For instance, the removal of bacterial spores requires more vigorous treatment than the removal of vegetative cells. Also, a chemical applied to a laboratory bench is considerably different from one used on a wound or for sterilizing an object.

It therefore is imperative to distinguish the antiseptic or disinfectant nature of a chemical before proceeding with its use. Indeed, chemical agents formulated as disinfectants are regu-

TABLE 7.2	Phenol Coefficients of Some Common Antiseptics and Disinfectants[a]	
Chemical Agent	*Staphylococcus aureus*	*Salmonella typhi*
Phenol	1	1
Chloramine	133	100
Tincture of iodine	6.3	5.8
Lysol	3.5	1.9
Mercurochrome	5.3	2.7
Ethyl alcohol	0.04	0.04
Formalin	0.3	0.7
Cetylpyridinium chloride	337	228

[a]All PC values were determined at 37°C.

lated and registered by the U.S. Environmental Protection Agency (EPA), while chemicals formulated as antiseptics are regulated by the FDA.

CONCEPT AND REASONING CHECKS

7.10 Summarize the properties important in the selection of a disinfectant or antiseptic.

Antiseptics and Disinfectants Can Be Evaluated for Effectiveness

KEY CONCEPT

11. Standards have been established to know the relative effectiveness of a chemical agent.

Currently, there are more than 8,000 disinfectants and antiseptics for hospital use and thousands more for general use. Evaluating these chemical agents is a tedious process because of the broad diversity of conditions under which they are used.

One measure of effectiveness for chemical agents is the **phenol coefficient** (**PC**). This is a number indicating the disinfecting ability of an antiseptic or disinfectant in comparison to phenol under identical conditions (TABLE 7.2). A PC higher than 1 indicates the chemical is more effective than phenol; a number less than 1 indicates poorer disinfecting ability than phenol. For example, antiseptic A may have a PC of 78.5, while antiseptic B has a PC of 0.28. These numbers are used relative to each other rather than to phenol because phenol is allergenic and irritating to tissues and thus is not used in a concentrated form.

The phenol coefficient is determined by a laboratory procedure in which dilutions of phenol and the test chemical are mixed with standardized bacterial species, such as *Staphylococcus aureus*, *Salmonella typhi*, or other species. The laboratory technician then determines which dilutions have killed the organisms after a 10-minute exposure but not after a 5-minute exposure. The test has many drawbacks, especially because it is performed in the laboratory rather than in a real-life situation. Nor does it take into account many of the factors cited above, such as tissue toxicity, activity in the presence of organic matter, or temperature variations.

A more practical way of determining the value of a chemical agent is by an **in-use test**. For example, swab samples from a floor are taken before and after the application of a disinfectant to determine the level of disinfection. Another method is to dry standardized cultures of a bacterial species on small stainless steel cylinders and then expose the cylinders to the test chemical. After an established period of time, the organism is tested for survival rates. These methods of standardization have value under certain circumstances. However, it is conceivable that a universal test may never be developed, in view of the huge variety of chemical agents available and the numerous conditions under which they are used.

CONCEPT AND REASONING CHECKS

7.11 Assess the need to know a chemical agent's effectiveness.

7.4   Chemical Methods of Control

The chemical agents currently in use for controlling microorganisms range from very simple substances, such as halogen ions, to very complex compounds, typified by detergents. Many of these agents in nature have been used for generations, while others represent the latest products of chemical companies. In this section, we shall survey several groups of chemical agents and indicate how they are best applied in the chemical control of microorganisms. MICROFOCUS 7.5 identifies some common but surprising antiseptics.

Halogens Oxidize Proteins

KEY CONCEPT
12. Chlorine and iodine are good disinfectant agents.

The **halogens** are a group of highly reactive elements whose atoms have seven electrons in the outer shell (see Chapter 2). Two halogens, chlorine and iodine, are commonly used for disinfection. In microorganisms, halogens are believed to cause the release of atomic oxygen, which then combines with and inactivates certain cytoplasmic proteins, such as enzymes. Killing almost always occurs within 30 minutes after application.

Chlorine (Cl) is effective against a broad variety of organisms, including most gram-positive and gram-negative bacteria, and many viruses, fungi, and protozoa. However, it is not sporicidal.

Chlorine is available in a gaseous form and as both organic and inorganic compounds (FIGURE 7.15). It is widely used in municipal water supplies and swimming pools, where it keeps microbial populations at low levels. Chlorine combines readily with numerous ions in water; therefore, enough chlorine must be added to ensure a residue remains for antimicrobial activity. In municipal water, the residue of chlorine is usually about 0.2 to 1.0 parts per million (ppm) of free chlorine. One ppm is equivalent to 0.0001 percent, an extremely small amount.

Chlorine also is available as sodium hypochlorite (Clorox®) or calcium hypochlorite. The latter, also known as chlorinated lime, was used by Semmelweis in his studies (see Chapter 1).

Na^+ [OCl]⁻

Sodium hypochlorite

Ca^{2+} [OCl]⁻$_2$

Calcium hypochlorite

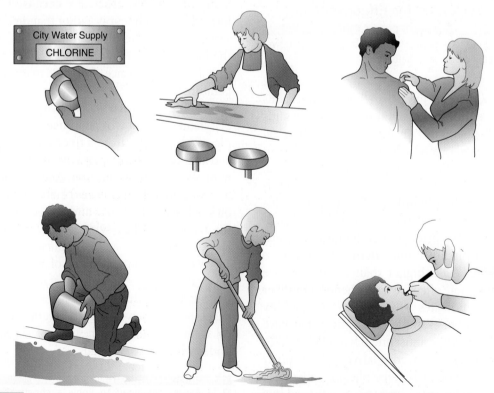

FIGURE 7.15 Some Practical Applications of Disinfection with Chlorine Compounds. Different chlorine compounds have been used as both disinfectants and antiseptics. »» In each of the above illustrations, is the chemical agent being used as a disinfectant or an antiseptic?

Today, we live in an age when alternative and herbal medicine claims are always in the news, and these reports have generated a whole industry of health products that often make unbelievable claims. With regard to "natural products," are there some that have genuine medicinal and antiseptic properties?

Cinnamon

Professor Daniel Y. C. Fung, Professor of Food Science and Food Microbiology at Kansas State University in Manhattan, Kansas, believes cinnamon might be an antiseptic that can control pathogens, at least in fruit beverages. Fung's group added cinnamon to commercially pasteurized apple juice. They then added typical foodborne pathogens (*Salmonella typhimurium, Yersinia enterocolitica,* and *Staphylococcus aureus*) and viruses. After one week of monitoring the juice at refrigerated and room temperatures, the investigators discovered the pathogens were killed more readily in the cinnamon blend than in the cinnamon-free juice. In addition, more bacterial organisms and viruses were killed in the juice at room temperature than when refrigerated.

Garlic

In 1858, Louis Pasteur examined the properties of garlic as an antiseptic. During World War II, when penicillin and sulfa drugs were in short supply, garlic was used as an antiseptic to disinfect open wounds and prevent gangrene. Since then, numerous scientific studies have tried to discover garlic's antiseptic powers.

Many research studies have identified a sulfur compound, allicin, as one key to garlic's antiseptic properties. When a raw garlic clove is crushed or chewed, allicin gives garlic its characteristic taste and smell. Laboratory studies using garlic suggest that this compound is responsible for combating the microbes causing the common cold, flu, sore throat, sinusitis, and bronchitis. The findings indicate that the compound blocks key enzymes that bacterial cells and viruses need to invade and damage host cells.

Honey

For nearly three decades, Professor Peter Molan, associate professor of biochemistry and director of the Waikato Honey Research Unit at the University of Waikato, New Zealand, has been studying the medicinal properties of and uses for honey. Its acidity, between 3.2 and 4.5, is low enough to inhibit many pathogens. Its low water content (15% to 21% by weight) means that it osmotically ties up free water and "drains water" from wounds, helping to deprive pathogens of an ideal environment in which to grow. In addition, when honey encounters fluid from a wound, it slowly releases small quantities of hydrogen peroxide that are not damaging to skin tissues. It also speeds wound healing.

If that isn't enough, there also is evidence that honey protects against tooth decay. Professor Molan's group has shown that, in the lab, honey completely inhibits the growth of plaque-forming bacterial species, including *Streptococcus mitis, S. sobrinus,* and *Lactobacillus caseii.* Honey cut acid production to almost zero and stopped the bacteria from producing dextran, which is a component of dental plaque. Like its use for wound infections, hydrogen peroxide probably is, in part, responsible for the antimicrobial activity.

But beware! Not all honey is alike. The antibacterial properties of honey depend on the kind of nectar, or plant pollen, that bees use to make honey. At least manuka honey from New Zealand and honeydew from central Europe are thought to contain useful levels of antiseptic potency. Professor Molan is convinced that "honey belongs in the medicine cabinet as well as the pantry."

Wasabi

The green, pungent, Japanese horseradish called wasabi may be more than a spicy condiment for sushi. Professor Hedeki Masuda, director of the Material Research and Development Laboratories at Ogawa & Co. Ltd., in Tokyo, Japan, and his colleagues have found that natural chemicals in wasabi, called isothiocyanates, inhibit the growth of *Streptococcus mutans*—one of the bacterial species causing tooth decay. Researchers tested wasabi's tooth-decay fighting ability in test tubes and found the substance interferes with the way sugar affects teeth. At this point, these are only test-tube laboratory studies and the results will need to be proven in clinical trials.

So, are there products having genuine antimicrobial properties? It appears so—and there are many more than can be described here.

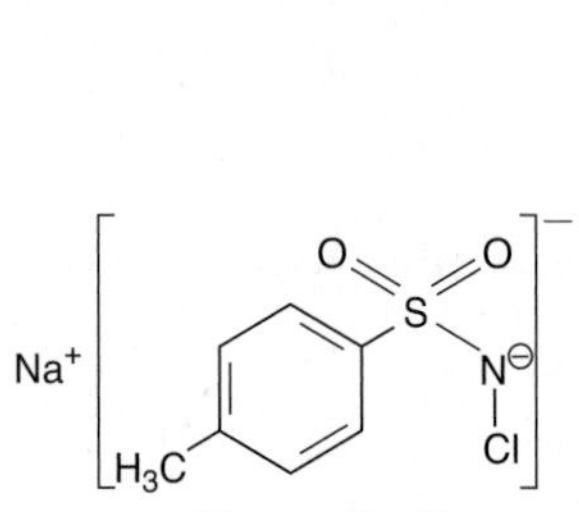

Hypochlorite compounds cause cellular proteins to clump together, destroying their function. To disinfect clear water for drinking, the Centers for Disease Control and Prevention (CDC) recommends a half-teaspoon of household chlorine bleach in two gallons of water, with 30 minutes of contact time before consumption. Hypochlorites also are useful in very dilute solutions for sanitizing commercial and factory equipment.

The chloramines, such as chloramine-T, are organic compounds used as bactericides and for the disinfection of drinking water.

The iodine atom (I) is slightly larger than the chlorine atom and is more reactive and more germicidal. A tincture of iodine, a commonly used antiseptic for wounds, consists of 2% iodine. For the disinfection of clear water, the CDC recommends five drops of an iodine tincture in one quart of water, with 30 minutes of contact time before consumption. Iodine compounds in different forms are also valuable sanitizers for restaurant equipment and eating utensils.

Iodophors are iodine linked to a solubilizing agent, such as a detergent or nondetergent carrier. These water-soluble complexes release iodine over a long period of time and have the added advantage of not staining tissues or fabrics. The solubilizing agent loosens the organisms from the surface and diatomic iodine (I_2) irreversibly damages the microbe by reacting with enzymes in the respiratory chain (see Chapter 6) and with proteins in the cell membrane and cell wall. Some examples of iodophors are Wescodyne, used in preoperative skin preparations; and Betadine, for local wounds. Iodophors also may be combined with nondetergent carrier molecules. The best known carrier is povidone, which stabilizes the iodine and releases it slowly. However, compounds like these are not self-sterilizing.

CONCEPT AND REASONING CHECKS

7.12 Compare the uses for chlorine and iodine chemical agents.

Phenol and Phenolic Compounds Denature Proteins

KEY CONCEPT

13. Many phenolic derivatives are used as disinfectants or antiseptics.

Phenol (carbolic acid) and phenolic compounds have played a key role in disinfection practices since Joseph Lister used them in the 1860s. Phenol

remains the standard against which other antiseptics and disinfectants are evaluated using the phenol coefficient test. It is active against gram-positive bacteria, but its activity is reduced in the presence of organic matter. Phenol and its derivatives act by denaturing proteins, especially in the cell membrane.

Phenol is expensive, has a pungent odor, and is caustic to the skin; therefore, the role of phenol as an antiseptic has diminished (**FIGURE 7.16**). However, phenol derivatives have greater germicidal activity and lower toxicity than the parent compound. Hexylresorcinol is used in some mouthwashes, topical antiseptics, and throat lozenges. It has the added advantage of reducing surface tension, thereby loosening bacterial cells from tissue and allowing greater penetration of the germicidal agent.

Combinations of two phenol molecules called **bisphenols** are prominent in modern disinfection

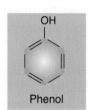

FIGURE 7.16 Phenol and Some Derivatives. The chemical structure of phenol and some important derivatives. »» Why are most phenolic compounds only used as disinfectants?

and antisepsis. Orthophenylphenol, for example, is used in Lysol and Amphyl. Another bisphenol, hexachlorophene, was used extensively during the 1950s and 1960s in toothpastes, underarm deodorants, and bath soaps. One product, pHisoHex, combined hexachlorophene with a pH-balanced detergent cream. Pediatricians recommended it to retard staphylococcal infections of the scalp and umbilical stump and for general cleansing of the newborn. However, studies indicate that excessive amounts can be absorbed through the skin and cause neurological damage, so hexachlorophene has been removed from over-the-counter products. The product pHisoHex is still available, but only by prescription.

An important bisphenol relative is **chlorhexidine**. This compound is used as a surgical scrub, hand wash, and superficial skin wound cleanser. A 4% chlorhexidine solution in isopropyl alcohol is commercially available as Hibiclens. Another bisphenol in widespread use is **triclosan**, a broad-spectrum antimicrobial agent that destroys bacterial cells by disrupting cell membranes (and possibly cell walls) by blocking the synthesis of lipids. Triclosan is fairly mild and nontoxic, and it is effective against pathogenic bacteria (but only partially against viruses and fungi). The chemical is included in antibacterial soaps, lotions, mouthwashes, toothpastes, toys, food trays, underwear, kitchen sponges, utensils, and cutting boards. A negative side to extensive triclosan use is the possibility of bacterial species developing resistance to the chemical, just as they have developed resistance to antibiotics.

CONCEPT AND REASONING CHECKS

7.13 Explain why bisphenols are preferred as disinfectants and antiseptics.

Heavy Metals Interfere with Microbial Metabolism

KEY CONCEPT

14. Mercury, copper, and silver compounds can be useful disinfectants.

Mercury, silver, and copper are called **heavy metals** because of their large atomic weights and complex electron configurations. They are very reactive with proteins, particularly at the protein's sulfhydryl groups (–SH), and they are believed to bind protein molecules together by forming bridges between the groups. Because many of the proteins involved are enzymes, cellular metabolism is disrupted, and the microorganism dies. However, heavy metals are not sporicidal.

Mercury (Hg) is very toxic to the host and the antimicrobial activity of mercury is reduced when other organic matter is present. In such products as merbromin (Mercurochrome) and thimerosal (Merthiolate), mercury is combined with carrier compounds and is less toxic when applied to the skin, especially after surgical incisions. Thimerosal was previously used as a preservative in vaccines (Chapter 22).

Copper (Cu) is active against chlorophyll-containing organisms and is a potent inhibitor of algae. As copper sulfate ($CuSO_4$), it is incorporated into algicides and is used in swimming pools and municipal water supplies.

Silver (Ag) in the form of silver nitrate ($AgNO_3$) is useful as an antiseptic and disinfectant. For example, one drop of a 1% silver nitrate solution used to be placed in the eyes of newborns to protect against infection by *Neisseria gonorrhoeae*. This gram-negative diplococcus can cause blindness if contracted by a newborn during passage through an infected mother's birth canal (Chapter 13).

CONCEPT AND REASONING CHECKS

7.14 Evaluate the use of heavy metals as antiseptics and disinfectants.

Alcohols Denature Proteins and Disrupt Membranes

KEY CONCEPT

15. Alcohols are widely used skin antiseptics.

For practical use, the preferred alcohol is ethyl alcohol (ethanol), which is active against vegetative bacterial cells, including the tubercle bacillus, but it has no effect on spores. It denatures proteins and dissolves lipids, an action leading to cell membrane disintegration. Ethyl alcohol also is a strong dehydrating agent.

Because ethyl alcohol reacts readily with organic matter, medical instruments and thermometers must be thoroughly cleaned before exposure. Usually, a 10-minute immersion in 50% to 80% alcohol solution is recommended to disinfect because water prevents rapid evaporation. Ethyl alcohol is used in many popular hand sanitizers.

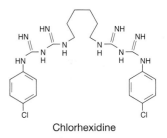

Merbromin

Thimerosal

Chlorhexidine

Ethanol

Venipuncture:
The piercing of a vein to take blood, to feed somebody intravenously, or to administer a drug.

Isopropyl alcohol

Alcohol is used to treat skin before a venipuncture or injection. It mechanically removes bacterial cells from the skin and dissolves lipids. Isopropyl alcohol, or rubbing alcohol, has high bactericidal activity in concentrations as high as 99%.

CONCEPT AND REASONING CHECKS

7.15 Why is 70% ethanol preferable to 95% ethanol as an antiseptic?

Soaps and Detergents Act as Surface-Active Agents

KEY CONCEPT

16. Cationic detergents are bacteriostatic.

Soaps are chemical compounds of fatty acids combined with potassium or sodium hydroxide. The pH of the compounds is usually about 8.0, and some microbial destruction is due to the alkaline conditions they establish on the skin. However, the major activity of soaps is as degerming agents for the mechanical removal of microorganisms from the skin surface.

Soaps, therefore, are surface-active agents called **surfactants**; that is, they emulsify and solubilize particles clinging to a surface and reduce the surface tension. Soaps also remove skin oils, further reducing the surface tension and increasing the cleaning action. MICROFOCUS 7.6 discusses the antibacterial soaps.

Detergents are synthetic chemicals acting as strong surfactants. Because they are actively attracted to the phosphate groups of cellular membranes, they also alter the membranes and encourage leakage from the cytoplasm. When used to clean cutting boards, for example, they can reduce the possibility of transmitting contaminants.

The most useful detergents to control microorganisms are cationic (positively charged) derivatives of ammonium chloride. In these detergents, four organic radicals replace the four hydrogens,

MICROFOCUS 7.6: Public Health/Being Skeptical
Are Antibacterial Soaps Worth the Money?

All of us want to be as clean as possible. In fact, hand washing is one of the best ways to protect oneself and prevent the spread of disease-causing microbes. To that end, numerous consumer product companies have provided us with many different types of antimicrobial cleaning and hygiene items. Perhaps the most pervasive are the antibacterial soaps, which usually contain about 0.2% triclosan.

It is estimated that 75% of liquid and 30% of bar soaps on the market today are of the antibacterial type. The question though is: Are these products any better than regular soaps? The short answer is—no.

Washing hands with soap and water is a key to preventing disease transmission.

Numerous studies have shown these antibacterial soaps do little against foodborne pathogens such as *Salmonella* and *Escherichia coli*. In addition, they do nothing to reduce the chances of picking up and harboring infectious microbes.

A 2005 study gathered together over 200 families with children. Each family was given cleaning and hygiene supplies—soaps, detergents, and household cleaners—to use for one year. Half of the families (controls) received regular products without added antibacterial chemicals, while the other half used products with the antibacterial chemicals.

When the families were surveyed after one year, those using the antibacterial products were just as likely to get sick, as identified by such symptoms as coughs, fevers, sore throats, vomiting, and diarrhea.

You may say that this is not surprising, as many of these symptoms are the result of a viral infection—and the antibacterial products are not effective on viruses. However, further analysis of the families indicated there were just as many bacterial infections in the antibacterial group as there were in the control group.

Antibacterial soaps may be useful in a hospital environment, but they certainly are not worth the extra cost for home use.

and at least one radical is a long hydrocarbon chain (**FIGURE 7.17**). Such compounds often are called **quaternary ammonium compounds** or, simply, **quats**. They react with cell membranes and can destroy some bacterial species and enveloped viruses.

Quats have rather long, complex names, such as benzalkonium chloride in Zephiran and cetylpyridinium chloride in Ceepryn. Quats are bacteriostatic, especially on gram-positive bacteria, and are relatively stable, with little odor. They are used as sanitizing agents for industrial equipment and food utensils; as skin antiseptics; as disinfectants in mouthwashes and contact lens cleaners; and as disinfectants for use on hospital walls and floors. Their use as disinfectants for food-preparation surfaces can help reduce contamination incidents. Mixing them with soap, however, reduces their activity, and certain gram-negative bacteria, such as *Burkholderia (Pseudomonas) cepacia*, can actually grow in them.

CONCEPT AND REASONING CHECKS

7.16 How do soaps differ from quats as chemical agents of control?

Peroxides Damage Cellular Components

KEY CONCEPT

17. Hydrogen peroxide can be used as an antiseptic rinse.

Peroxides are compounds containing oxygen-oxygen single bonds. **Hydrogen peroxide** (H_2O_2) has been used as a rinse in wounds, scrapes, and abrasions. However, H_2O_2 applied to such areas foams and effervesces, as catalase in the tissue breaks down hydrogen peroxide to oxygen and water. Therefore, it is not recommended as an antiseptic for open wounds. However, the furious bubbling loosens dirt, debris, and dead tissue, and the oxygen gas is effective against anaerobic bacterial species. Hydrogen peroxide decomposition also results in a reactive form of oxygen—the superoxide radical—which is highly toxic to microorganisms.

New forms of H_2O_2 are more stable than traditional forms, do not decompose spontaneously, and therefore can be used topically. Such inanimate materials as soft contact lenses, utensils, heat-sensitive plastics, and food-processing equipment can be disinfected within 30 minutes.

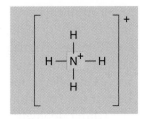

FIGURE 7.17 Cationic Detergents. The chemical structures of some important quaternary ammonium compounds (quats) used in disinfection and antisepsis. »» How has the basic ammonium ion been modified to generate these two quats?

Benzoyl peroxide is another peroxide chemical. At low concentrations (2.5%), it is used to treat acne and is an active ingredient in teeth whitening products.

CONCEPT AND REASONING CHECKS

7.17 Judge the advantages and disadvantages of using hydrogen peroxide as an antiseptic.

Benzoyl peroxide

Some Chemical Agents Combine with Nucleic Acids and/or Cell Proteins

KEY CONCEPT

18. Aldehydes and gases can be used for sterilization.

The chemical agents we discussed in previous sections are used as disinfectants and antiseptics. In addition, there are some chemicals that can be used for sterilization purposes, especially for modern high-technology equipment. Several such agents are considered.

Aldehydes. Aldehydes are agents that react with amino and hydroxyl groups of nucleic acids and proteins. The resulting cross linking inactivates the proteins and nucleic acids.

Formaldehyde is a gas at high temperatures and a solid at room temperature. As a 37% solution it is called **formalin**. For over a century, formalin was used in embalming fluid for anatomical specimens (though rarely used anymore) and by morticians for disinfecting purposes. In microbiology, formalin is used for inactivating viruses in certain vaccines and producing toxoids from toxins (Chapter 22).

Hydrogen peroxide

Formalin

O   O
‖   ‖
H—CH₂CH₂CH₂—H

Glutaraldehyde

Instruments can be sterilized by placing them in a 20% solution of formaldehyde in 70% alcohol for 18 hours. Formaldehyde, however, leaves a residue, and instruments must be rinsed before use. Many allergic individuals develop a contact dermatitis to this compound.

Glutaraldehyde is a small, organic molecule that destroys bacterial cells within 10 to 30 minutes and spores in 10 hours. As a 2% solution, glutaraldehyde can be used for sterilization purposes. Materials have to be precleaned, then immersed for 10 hours, rinsed thoroughly with sterile water, dried in a special cabinet with sterile air, and stored in a sterile container to ensure that the material remains sterile. If any of these parameters are altered, the materials may be disinfected but may not be considered sterile.

Glutaraldehyde does not damage delicate objects, so it can be used to disinfect or sterilize optical equipment, such as the fiber-optic endoscopes used for arthroscopic surgery. It gives off irritating fumes, however, and instruments must be rinsed thoroughly in sterile water.

Sterilizing Gases. The development of plastics for use in microbiology required a suitable method for sterilizing these heat-sensitive materials. In the 1950s, research scientists discovered the antimicrobial abilities of ethylene oxide, which essentially made the plastic Petri dish and plastic syringe possible.

Ethylene oxide is a small molecule with excellent penetration capacity, and is microbicidal as well as sporicidal by combining with cell proteins. However, it is carcinogenic and highly explosive. Its explosiveness is reduced by mixing it with Freon

O
/ \
H₂C — CH₂

Ethylene oxide

Cl
\
O   O
(Chlorine dioxide structure)

Chlorine dioxide

gas or carbon dioxide gas, but its toxicity remains a problem for those who work with it. The gas is released into a tightly sealed chamber where it circulates for up to four hours with carefully controlled humidity. The chamber then must be flushed with inert gas for 8 to 12 hours to ensure that all traces of ethylene oxide are removed; otherwise the chemical will cause "cold burns" on contact with the skin.

Ethylene oxide is used to sterilize paper, leather, wood, metal, and rubber products as well as plastics. In medicine, it is used to sterilize catheters, artificial heart valves, heart-lung machine components, and optical equipment. The National Aeronautics and Space Administration (NASA) uses the gas for sterilization of interplanetary space capsules. Ethylene oxide chambers, called "gas autoclaves," have become the chemical counterparts of heat- and pressure-based autoclaves for sterilization procedures.

Chlorine dioxide has properties very similar to chloride gas and sodium hypochlorite but, unlike ethylene oxide, it produces nontoxic by-products and is not a carcinogen. Chlorine dioxide can be used as a gas or liquid. In a gaseous form, with proper containment and humidity, a 15-hour fumigation can be used to sanitize air ducts, food and meat processing plants, and hospital areas. It was the gas used to decontaminate the 2001 anthrax-contaminated mail and office buildings (MICROFOCUS 7.7).

TABLE 7.3 and **FIGURE 7.18** summarize the chemical agents used in controlling microorganisms.

CONCEPT AND REASONING CHECKS

7.18 Summarize the uses for aldehydes, ethylene oxide, and chloride dioxide for sterilization.

MICROFOCUS 7.7: Tools/Bioterrorism

Decontamination of Anthrax-Contaminated Mail and Office Buildings

This chapter has examined the chemical procedures and methods used to control the numbers of microorganisms on inanimate and living objects. These control measures usually involve a level of sanitation, although some procedures may sterilize. Some examples were given for their use in the home and workplace. However, what about real instances where large-scale and extensive decontamination has to be carried out?

In October 2001, the United States experienced a bioterrorist attack. The perpetrator(s) used anthrax spores as the bioterror agent. Four anthrax-contaminated letters were sent through the mail on the same day, addressed to NBC newscaster Tom Brokaw, the New York Post, and to two United States Senators, Senator Patrick Leahy and Senate Majority Leader Tom Daschle (Figure A). The Centers for Disease Control and Prevention (CDC) confirmed that anthrax spores from at least the Daschle letter contaminated the Hart Senate Office Building and several post office sorting facilities in Trenton, New Jersey (from where the letters were mailed), and Washington, D.C., areas. This resulted in the closing of the Senate building and the postal sorting facilities. With the Senate building and postal sorting

(A)

(B) Mail sorting machines.

facilities closed, the CDC, the Environmental Protection Agency (EPA), other governmental agencies, and commercial companies had to devise and implement a strategy to decontaminate these buildings and the mail sorting machines (Figure B).

As mentioned in this chapter, most sanitation procedures do not require a high technology solution. In fact, all of the methods actually used for this situation are described in this chapter.

The Hart Senate Office Building and the post office sorting facilities were contaminated with *Bacillus anthracis* endospores. These are large, multi-room facilities with many pieces of furniture and instruments, including computers, copy machines, and mail sorting machines. To decontaminate these buildings, chemical disinfectants such as bleach or phenol solutions could have been used. However, spores may have gotten into the office machinery and sorting machines. Liquids would not work here.

Therefore, a gas was needed that could permeate the air ducts as well as all the office machinery and sorting machines. The gas chosen was chlorine dioxide. Essentially, the buildings were sealed as if they were going to be fumigated for termites. The gas was pumped in, and after a time that was believed to be sufficient to kill any anthrax spores, the gas was evacuated. Swabs were taken from the buildings and plated on nutrient media. If any spores were still alive, they would germinate on the plates and the vegetative cells would grow into visible colonies. Such results would require retreatment of the facility.

To protect the mail from similar attacks in the future, a system was devised using ultraviolet (UV) light to kill any spores that might be found in a piece of mail moving through the sorting machines.

It took months, and even years, for some of the postal facilities to be declared safe and free of anthrax spores. Still, simple physical and chemical methods worked to decontaminate the buildings and equipment.

TABLE

7.3 Summary of Chemical Agents Used to Control Microorganisms

Chemical Agent	Antiseptic or Disinfectant	Mechanism of Activity	Applications	Limitations	Antimicrobial Spectrum
Chlorine	Chlorine gas Sodium hypochlorite Chloramines	Protein oxidation Membrane leakage	Water treatment Skin antisepsis Equipment spraying Food processing	Inactivated by organic matter Objectionable taste, odor	Broad variety of bacteria, fungi, protozoa, viruses
Iodine	Tincture of iodine Iodophors	Reacts with proteins	Skin antisepsis Food processing Preoperative preparation	Inactivated by organic matter Objectionable taste, odor	Broad variety of bacteria, fungi, protozoa, viruses
Phenol and derivatives	Hexachlorophene Chlorhexidine Triclosan	Coagulates proteins Disrupts cell membranes	General preservatives Skin antisepsis with detergent	Toxic to tissues Disagreeable odor	Gram-positive bacteria Some fungi
Mercury	Mercuric chloride Merthiolate Merbromin	Combines with —SH groups in proteins	Skin antiseptics Disinfectants	Inactivated by organic matter Toxic to tissues Slow acting	Broad variety of bacteria, fungi, protozoa, viruses
Copper	Copper sulfate	Combines with proteins	Algicide in swimming pools Municipal water supplies	Inactivated by organic matter	Algae Some fungi
Silver	Silver nitrate	Binds proteins	Skin antiseptic Eyes of newborns	Skin irritation	Organisms in burned tissue Gonococci
Alcohol	70% ethyl alcohol	Denatures proteins Dissolves lipids Dehydrating agent	Instrument disinfectant Skin antiseptic	Precleaning necessary Skin irritation	Vegetative bacterial cells, fungi, protozoa, viruses
Cationic detergents	Quaternary ammonium compounds	Dissolve lipids in cell membranes	Industrial sanitization Skin antiseptic Disinfectant	Neutralized by soap	Broad variety of microorganisms
Peroxides	Hydrogen peroxide Benzoyl peroxide	Creates aerobic environment Oxidizes protein groups	Wound treatment Acne	Limited use	Anaerobic bacteria
Formaldehyde	Formalin	Reacts with functional groups in proteins and nucleic acids	Embalming Vaccine production Gaseous sterilant	Poor penetration Allergenic Toxic to tissues Neutralized by organic matter	Broad variety of bacteria, fungi, protozoa, viruses
Glutaraldehyde	Glutaraldehyde	Reacts with functional groups in proteins and nucleic acids	Sterilization of surgical supplies	Unstable Toxic to skin	All microorganisms, including spores
Ethylene oxide	Ethylene oxide gas	Reacts with functional groups in proteins and nucleic acids	Sterilization of instruments, equipment, heat-sensitive objects	Explosive Toxic to skin Requires constant humidity	All microorganisms, including spores
Chlorine dioxide	Chlorine dioxide gas	Reacts with functional groups in proteins and nucleic acids	Sanitizes equipment, rooms, buildings	Burns skin and eyes on contact	All microorganisms, including spores

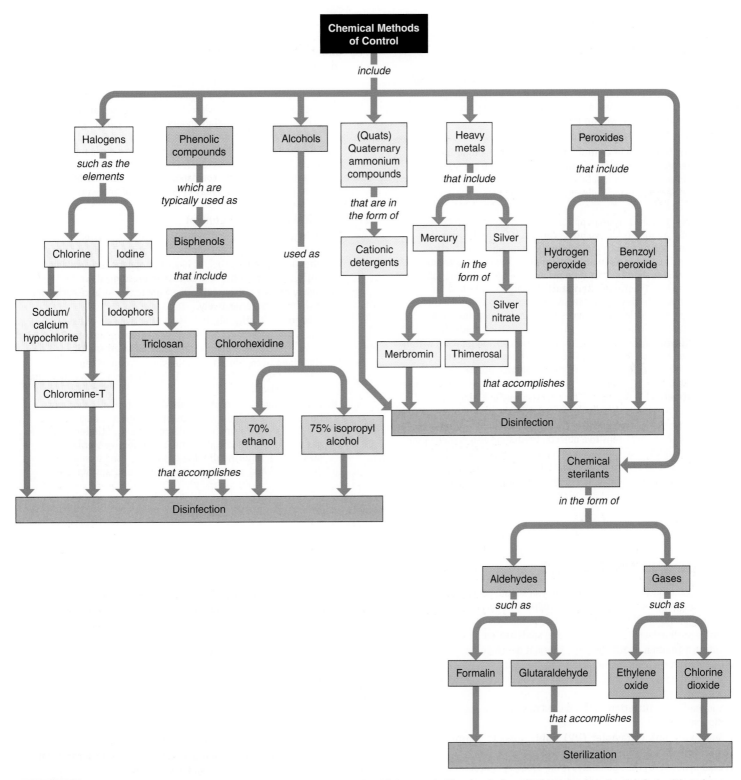

FIGURE 7.18 A Concept Map Summarizing the Chemical Methods of Microbial Control. The chemical methods predominantly disinfect, although a few can sterilize. »» If you wanted to sanitize a kitchen counter, which chemicals might be selected?

SUMMARY OF KEY CONCEPTS

7.1 General Principles of Microbial Control

1. The physical methods for controlling microorganisms are generally intended to achieve **sterilization**—the destruction or removal of all life-forms, including bacterial spores. **Sanitization** involves methods to reduce the numbers of, or inhibit the growth of, microbes.

7.2 Physical Methods of Control

2. Heat is a common control method. Used in the food industry, **thermal death point** and **thermal death time** are used to determine how long it takes to kill microbial cells.

3. **Incineration** using a direct flame achieves sterilization in a few seconds.

4. Moist heat has several applications: The exposure to boiling water at 100° C for two hours can result in sterilization, but spore destruction cannot always be assured; the **autoclave** sterilizes materials in about 15 minutes, while a **prevacuum sterilizer** shortens this time using higher temperatures and pressures. **Fractional sterilization** is another method for sterilization through successive exposures to steam on three days; **pasteurization** reduces the microbial population in a liquid and is not intended to be a sterilization method.

5. **Filtration** uses various materials to trap microorganisms within or on a filter. **Membrane filters** are the most common. Air can be filtered using a **high-efficiency particulate air (HEPA) filter**.

6. **Ultraviolet (UV) light** is an effective way of killing microorganisms on a dry surface and in the air.

7. **X rays** and **gamma rays** are two forms of **ionizing radiation** used to sterilize heat-sensitive objects. Irradiation also is used in the food industry to control microorganisms on perishable foods.

8. For food preservation, drying, salting, and low temperatures can be used to control microorganisms.

7.3 General Principles of Chemical Control

9. Chemical agents are effectively used to control the growth of microorganisms. A chemical agent used on a living object is an **antiseptic**; one used on a nonliving object is a **disinfectant**.

10. Both antiseptics and disinfectants are selected according to certain criteria, including an ability to kill microorganisms or interfere with their metabolism.

11. The **phenol coefficient test** can be used to evaluate antiseptics and disinfectants. Chemicals are contrasted based on their effectiveness compared to phenol. **In-use tests** are more practical for everyday applications of antiseptics and disinfectants.

7.4 Chemical Methods of Control

12. **Halogens** (chlorine and iodine) are useful for water disinfection, wound antisepsis, and various forms of sanitation.

13. **Phenol** derivatives, such as hexachlorophene, are valuable skin antiseptics and active ingredients in presurgical scrubs.

14. **Heavy metals** (silver and copper) are useful as antiseptics and disinfectants, respectively.

15. **Alcohol** (70% ethyl alcohol) is an effective skin antiseptic.

16. Soaps and detergents are effective degerming agents. **Quats** are more effective as a disinfectant than as an antiseptic.

17. **Hydrogen peroxide** acts by releasing oxygen to cause an effervescing cleansing action. It is better as a disinfectant than an antiseptic.

18. **Formaldehyde** and **glutaraldehyde** are sterilants that cross link amino and hydroxyl groups in proteins and nucleic acids to alter the biochemistry of microorganisms. **Ethylene oxide** gas under controlled conditions is an effective sterilant for plasticware. **Chlorine dioxide** gas can be used to sanitize air ducts, food and meat processing plants, and hospital areas.

LEARNING OBJECTIVES

After understanding the textbook reading, you should be capable of writing a paragraph that includes the appropriate terms and pertinent information to answer the objective.

1. Distinguish between **sterilization** and **sanitization**.
2. Contrast **thermal death time** and **thermal death point**.
3. Distinguish between **incineration** and dry heat.
4. Discuss the four ways that moist heat can be used to control microbial growth.
5. Summarize the **filtration** methods used to sterilize a liquid and decontaminate air.
6. Summarize how **ultraviolet (UV) light,** works to control microbial growth.
7. Explain how **X rays** and **gamma rays** are used as physical control agents.
8. Explain how dehydration and cold temperatures preserve foods.
9. Compare and contrast an **antiseptic** and a **disinfectant**.
10. List the desirable and chemical properties of antiseptics and disinfectants.
11. Describe how the effectiveness of a chemical agent can be measured.
12. Evaluate the usefulness of **halogens** as disinfectants.
13. Summarize the uses for phenolic derivatives as disinfectants and antiseptics.
14. Summarize the uses for **heavy metals** in the chemical control of microorganisms.
15. Justify why **alcohol** is not a method for skin sterilization.
16. Distinguish between a soap, a detergent, and **quats**.
17. Estimate the value of **hydrogen peroxide** as a bacteriostatic agent.
18. Identify the uses of aldehydes and gases as sterilants.

STEP A: SELF-TEST

Each of the following questions is designed to assess your ability to remember or recall factual or conceptual knowledge related to this chapter. Read each question carefully, then select the *one* answer that best fits the question or statement. Answers to even-numbered questions can be found in **Appendix C**.

1. All the following terms apply to microbial killing *except*:
   A. sterilization.
   B. microbicidal.
   C. bactericidal.
   D. fungistatic.

2. The thermal death time is
   A. the time to kill a microbial population at a given temperature.
   B. the time to kill a microbial population in boiling water.
   C. the temperature to kill all pathogens.
   D. the minimal temperature need to kill a microbial population.

3. At 160°C, it takes about _____ minutes to kill bacterial spores in a hot-air oven.
   A. 30
   B. 60
   C. 90
   D. 120

4. An autoclave normally sterilizes material by heating the material to _____ °C for _____ minutes at _____ psi.
   A. 100; 10; 30
   B. 121.5; 15; 15
   C. 100; 15; 0
   D. 110; 30; 5

5. Air filtration typically uses a _____ filter.
   A. HEPA
   B. membrane
   C. sand
   D. diatomaceous earth

6. For bactericidal activity, _____ has/have the ability to cause thymine dimer formation.
   A. X rays
   B. ultraviolet light
   C. gamma rays
   D. microwaves

7. The elimination of pathogens in foods by irradiation is called
   A. the D value.
   B. the pasteurizing dose.
   C. incineration.
   D. sterilization.

8. Preservation methods such as salting result in the _____ microbial cells.
   A. loss of salt from
   B. gain of water into
   C. loss of water from
   D. lysis of

9. Which one of the following statements does NOT apply to antiseptics?
   A. They are used on living objects.
   B. They usually are microbicidal.
   C. They should be useful as dilute solutions.
   D. They can sanitize objects.

10. All the following are chemical parameters considered when selecting an antiseptic or disinfectant *except*:
    A. dehydration.
    B. temperature.
    C. stability.
    D. pH.

11. If a chemical has a phenol coefficient (PC) of 63, it means the chemical
    A. is better than one with a PC of 22.
    B. will kill 63% of bacteria.
    C. kills microbes at 63°C.
    D. will kill all bacteria in 63 minutes.

12. Which one of the following is NOT a halogen?
    A. Iodine
    B. Mercury
    C. Clorox bleach
    D. Chlorine

13. Phenolics include chemical agents
    A. such as the iodophores.
    B. derived from carbolic acid.
    C. used as tinctures.
    D. such as formaldehyde.

14. Heavy metals, such as _____ work by _____.
    A. mercury; disrupting membranes
    B. copper; producing toxins
    C. iodine; denaturing proteins
    D. silver; binding protein molecules together

15. Alcohols are
    A. surfacants.
    B. heavy metals.
    C. denaturing agents.
    D. detergents.

16. All the following statements apply to quats *except*:
    A. they react with cell membranes.
    B. they are positively charged molecules.
    C. they are types of soaps.
    D. they can be used as disinfectants.

17. Hydrogen peroxide
    A. is an effective sterilant.
    B. cross-links proteins and nucleic acids.
    C. can emulsify and solubilize pathogens.
    D. is not recommended for use on open wounds.

18. Ethylene oxide can be used to
    A. kill bacterial spores.
    B. clean wounds.
    C. sanitize work surfaces.
    D. treat water supplies.

STEP B: REVIEW—PHYSICAL METHODS

Use the following syllables to form the term that answers the clue pertaining to physical methods of control. The number of letters in the term is indicated by the dashes, and the number of syllables in the term is shown by the number in parentheses. Each syllable is used only once. The answers to even-numbered terms are listed in **Appendix C**.

A AU BA BER BRANE CIL CLAVE CU DE DER DRY HOLD ING ING LET LO LUS MEM MO NA O OS PLAS POW SIS SIS SPORE TIC TION TO TRA TU TUR UL VI

19. Instrument for sterilization.   (3) __ __ __ __ __ __ __ __ __

20. Type of filter.   (2) __ __ __ __ __ __ __ __

21. Sterilized in an oven.   (2) __ __ __ __ __ __ __

22. Occurs with moist heat.   (5) __ __ __ __ __ __ __ __ __ __ __

23. Preserves meat, fish.   (2) __ __ __ __ __ __ __

24. Most resistant life-form.   (1) __ __ __ __ __

25. Method of pasteurization.   (2) __ __ __ __ __ __ __

26. Light for air sterilization.   (5) __ __ __ __ __ __ __ __ __ __ __

27. Melts in the autoclave.   (2) __ __ __ __ __ __ __

28. Water flow from salting.   (3) __ __ __ __ __ __ __

29. Genus of spore formers.   (3) __ __ __ __ __ __ __ __ __

30. Disease prevented by pasteurization.   (5) __ __ __ __ __ __ __ __ __ __ __

STEP B: REVIEW—CHEMICAL METHODS

Chemical agents are a broad and diverse group, as this chapter has demonstrated. To test your knowledge over the chemical methods of control, match the chemical agent on the right to the statement on the left by placing the correct letter in the available space. A letter may be used once, more than once, or not at all. The answers to even-numbered statements are listed in **Appendix C**.

Statement

____ 31. The halogen in bleach.

____ 32. Sterilizes heat-sensitive materials.

____ 33. Part of chloramine molecule.

____ 34. A 70% concentration is recommended.

____ 35. Active ingredient in Betadine.

____ 36. Quaternary compounds, or quats.

____ 37. Can induce a contact dermatitis.

____ 38. Often used as a tincture.

____ 39. Rinse for wounds and scrapes.

____ 40. Example of a heavy metal.

____ 41. Aids mechanical removal of organisms.

____ 42. Triclosan is a derivative.

____ 43. Used by Joseph Lister.

____ 44. Used for plastic Petri dishes.

____ 45. Broken down by catalase.

Chemical agent

A. Cationic detergent
B. Chlorine
C. Ethyl alcohol
D. Ethylene oxide
E. Formaldehyde
F. Glutaraldehyde
G. Hydrogen peroxide
H. Iodine
I. Phenol
J. Silver
K. Soap

STEP C: APPLICATIONS

Answers to even-numbered questions can be found in **Appendix C.**

46. When the local drinking water is believed to be contaminated, area residents are advised to boil their water before drinking. Often, however, they are not told how long to boil it. As a student of microbiology, what is your recommendation?

47. You need to sterilize a liquid. What methods could you devise using only the materials found in the average household?

48. Suppose you were in charge of a clinical laboratory where instruments are routinely disinfected and equipment is sanitized. A salesperson from a disinfectant company stops in to spur your interest in a new chemical agent. What questions might you ask the salesperson about the product?

49. A portable room humidifier can incubate and disseminate infectious microorganisms. If a friend asked for your recommendations on disinfecting the humidifier, what do you suggest?

50. A student has finished his work in the laboratory and is preparing to leave. He remembers the instructor's precautions to wash his hands and disinfect the lab bench before leaving. However, he cannot remember whether to wash first then disinfect, or to disinfect then wash. What advice would you give?

STEP D: QUESTIONS FOR THOUGHT AND DISCUSSION

Answers to even-numbered questions can be found in **Appendix C.**

51. Instead of saying that food has been irradiated, manufacturers indicate that it has been "cold pasteurized." Why do you believe they must use this deception?

52. The label on the container of a product in the dairy case proudly proclaims, "This dairy product is sterilized for your protection." However, a statement in small letters below reads: "Use within 30 days of purchase." Should this statement arouse your suspicion about the sterility of the product? Why?

53. In view of all the sterilization methods we have discussed in this chapter, why do you think none has been widely adapted to the sterilization of milk?

54. A liquid that has been sterilized may be considered pasteurized, but one that has been pasteurized may not be considered sterilized. Why not?

55. Before taking a blood sample from the finger, the blood bank technician commonly rubs the skin with a pad soaked in alcohol. Many people think that this procedure sterilizes the skin. Are they correct? Why?

56. Suppose you had just removed the thermometer from the mouth of your sick child and confirmed your suspicion of fever. Before checking the temperature of your other child, how would you treat the thermometer to disinfect it?

57. The water in your home aquarium always seems to resemble pea soup, but your friend's is crystal clear. Not wanting to appear stupid, you avoid asking him his secret. But one day, in a moment of desperation, you break down and ask, whereupon he knowledgeably points to a few pennies among the gravel. What is the secret of the pennies?

HTTP://MICROBIOLOGY.JBPUB.COM/9E

The site features eLearning, an online review area that provides quizzes and other tools to help you study for your class. You can also follow useful links for in-depth information, read more MicroFocus stories, or just find out the latest microbiology news.

PART 2

The Genetics of Microorganisms

CHAPTER 8 Microbial Genetics

CHAPTER 9 Gene Transfer, Genetic Engineering, and Genomics

False-color transmission electron microscope image of two *Escherichia coli* cells that are part of the human intestinal microbiota and are a model organism for genetic studies.

In Part 1, we learned that many of the microorganisms that make up most of the Earth's biomass have been evolving for billions of years. They exist in virtually every environment on planet Earth; many survive—and even thrive—in extremes of heat, cold, radiation, pressure, salt, and acidity. This diversity and its range of environmental conditions show that microbes long ago "solved" many of the problems of adaptation in these environments. In fact, whether we examine cell structure and function, growth and nutrition, microbial metabolism, or virtually any microbial characteristic we might wish to consider, including antibiotic resistance, all are the result of inherited information. This information is stored as deoxyribonucleic acid (DNA) and is passed on from generation to generation. The study is called microbial genetics.

The contributions of microbial genetics have been numerous, diverse, and far reaching. As described in Chapter 1, model microbial systems, such as *Escherichia coli*, have established many of the principles of molecular biology. In addition, many of the molecular techniques in the geneticist's toolbox (e.g., polymerases, restriction enzymes, cloning vectors) are derived from genetic studies of microbes. Despite this legacy and the huge clinical significance of microbial genetics, the potential of the field is only just beginning to be tapped. This is, in part, why we are in the third golden age of microbiology!

In Part 2, we will explore the microbial genome in detail. Chapter 8 is devoted to the basics of microbial genetics. We will examine how microbial DNA is replicated and how this information codes for and directs protein synthesis. We will also explore the mechanisms of genome regulation and how it is affected by mutation. Chapter 9 introduces us to the fields of genetic engineering, biotechnology, and microbial genomics. We will discover how DNA information can be transferred laterally from one microbe to another. Then, we will examine the techniques of genetic engineering, the applications of biotechnology, and finish with a discussion of microbial genomics; that is, the study of genes and their function. Genomic analyses of microorganisms have broad significance not only for microbiology, human health, industry, and the environment, but also in our daily lives.

Biotechnology

During the 1980s, the editors of *Time Magazine* referred to DNA technology as "the most awesome skill acquired by man since the splitting of the atom." Indeed, the work with DNA, begun in the 1950s and continuing today, has opened vistas previously unimagined. Scientists can now remove bits of DNA from organisms, snip and rearrange the genes, and insert them into different species, where the genes will express themselves. Practical results of these experiments have led to the mass production of hormones, blood-clotting factors, and other pharmaceutical products. They also have given us diagnostic methods based on DNA fingerprinting; advances in gene therapy; a revolution in agricultural research; barnyard animals producing human hemoglobin; and a colossal project that has mapped the entire human genome.

Industrial microbiology or microbial biotechnology (the terms "industrial microbiology" and "biotechnology" are often one and the same) applies scientific and engineering principles to the processing of materials by microorganisms and viruses, or plant and animal cells, to create useful products or processes. Because biotechnology essentially uses the basic ingredients of life to make new products, it is both a cutting-edge technology and an applied science.

If you would like to be part of what analysts predict will be one of the most important applied sciences of this century, then microbiology is the place to start. You would be well advised to take a course in biochemistry as well as one in genetics. Courses in physiology and cell biology are also helpful. Employers will be looking for individuals with good laboratory skills, so be sure to take as many lab courses as you can.

You may enter the biotechnology field with an associate's, bachelor's, master's, or doctoral degree. This is because there are so many levels at which individuals are hired. Most professional levels of employment require a college degree (BS) in biology, microbiology, or biotechnology with minors in one or more of the complementary sciences. Persons who have project responsibilities often have one or more advanced degrees (MS and/or PhD) in biology, microbiology, or some other allied field such as molecular biology, biochemistry, biotechnology, chemical engineering, or genetics.

An employer also will be looking for work experience, which you can obtain by assisting a senior scientist, doing an internship, or working summers in a biotech firm (usually for slave wages). The campus research lab is another good place to obtain work experience. It also might be a good idea to sharpen your writing skills, because you will be preparing numerous reports.

As Chapter 8 explains, the novel and imaginative research that established biotechnology was founded in microbiology, and it continues to call on microbiology for its continuing growth.

8

Microbial Genetics

We wish to suggest a structure for the salt of deoxyribose nucleic acid (DNA). This structure has novel features which are of considerable biological interest.
—In the first 1953 paper by Watson and Crick describing the structure of DNA

In our fast-paced world, we often measure time in minutes and seconds, so our minds find it difficult to imagine the colossal 4.5 billion years that the Earth has been in existence. It may help, however, to think of Earth's history as a single year.

In the months of January and February, Earth was a hot, volcanic, lifeless ball of rock bombarded by material left over from the formation of the solar system. As the earth cooled during March, water vapor condensed into oceans and seas, providing conditions more amenable for the origin of life. Around April, something akin to the *Bacteria* or *Archaea* first appeared (FIGURE 8.1). As they evolved, they thrived and diversified in environments without oxygen gas and, by mid-June (2.4 billion years ago), in environments with oxygen gas (see MicroFocus 2.1 and 5.3). Members of the *Bacteria* and *Archaea* were the only organisms on Earth until early August, when single-celled eukaryotes, such as the algae, emerged. These organisms flourished and represent ancestors of present-day species. About mid-September, multicellular eukaryotes arose, whose descendants would evolve into diverse plants, fungi, and animals. Not until mid-November did the first of the plants, fungi, and animals move out of the sea onto the land. The dinosaurs were in existence from December 19 to December 25, and by December 27, the Earth bore a resemblance to modern Earth. Finally, on December 31, close to midnight, humans appeared.

We take this trek through geologic time to help us appreciate why microorganisms have prospered genetically and in evolutionary terms. They have been successful primarily because they have been around the longest and have adapted well. In fact, the *Bacteria* and *Archaea* domains have been on Earth about 3.7 billion years (versus about 200,000 years for humans), as **FIGURE 8.2** shows. During this time, gene changes have been occurring regularly and nature has used the microbes to test its newest genetic traits. The detrimental traits have been eliminated (together with the organisms unlucky enough to have them),

FIGURE 8.1 **Fossil Microbes.** This photograph is looking down on an ancient sea floor in Western Australia's Pilbara region. The wavy markings and cone-shaped formations may be evidence for a microbial reef made of cyanobacteria that existed 3.4 billion years ago. »» What would it mean for the environment once cyanobacteria predominated?

while the beneficial traits have thrived and have been passed on to the next generation—and on to the present day.

Present day microorganisms enjoy the fruits of genetic change. Because of their diverse genes, they can thrive in the varied environments on Earth, whether it is the snows of the Arctic or the boiling hot volcanic vents of the ocean depths. No other organisms can compare in sheer numbers.

Finally, consider a bacterium's multiplication rate—a new generation every half hour in some cases—and it is easy to see how a useful genetic change (such as drug resistance) can be propagated quickly.

Any one of these factors—time on Earth, sheer numbers, multiplication rate—would be sufficient to explain how microorganisms have evolved to their current form. However, when taken together, the factors help us appreciate why they have done very well in the evolutionary lottery—very well, indeed.

In this chapter, we examine mutation, one of the two processes that have brought ancient microorganisms to the myriad forms we observe on Earth today. However, to understand the material in these topics, we first must look at DNA replication and how the information in DNA is processed and regulated in making proteins—something alluded to by the remarkable discovery made in 1953 by James Watson and Francis Crick (see chapter opening quotation).

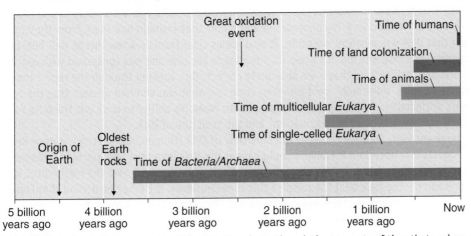

FIGURE 8.2 **The Appearance of Life on Earth.** This time line shows the relative amounts of time that various groups of organisms have existed on Earth. The *Bacteria* and *Archaea* have been in existence for a notably longer period than any other group, particularly humans. They have adapted well to Earth simply because they have had ample time and numbers. »» Why did it take so long for eukaryotes to appear on Earth?

8.1   DNA and Chromosomes

In 1953, James Watson and Francis Crick worked out DNA's double helix structure based, in part, on the X-ray studies of Rosalind Franklin (MicroFocus 8.1). This discovery has formed the foundation for all we know today about DNA rep-lication, protein synthesis, and gene control. In Chapter 2, we described the structure of the DNA molecule, so it might be helpful to review that material before proceeding too far in this chapter.

MICROFOCUS 8.1: History
The Tortoise and the Hare

We all remember the children's fable of the tortoise and the hare. The moral of the story was those who plod along slowly and methodically (the tortoise) will win the race over those who are speedy and impetuous (the hare). The race to discover the structure of DNA is a story of collaboration and competition—a science "tortoise and the hare."

Rosalind Franklin (the tortoise) was 31 when she arrived at King's College in London in 1951 to work in J. T. Randell's lab. Having received a Ph.D. in physical chemistry from Cambridge University, she moved to Paris where she learned the art of X-ray crystallography. At King's College, Franklin was part of Maurice Wilkins's group and she was assigned the job of using X-ray crystallography to work out the structure of DNA fibers. Her training and constant pursuit of excellence allowed her to produce excellent, high-resolution X-ray photographs of DNA.

Meanwhile, at the Cavendish Laboratory in Cambridge, James Watson (the hare) was working with Francis Crick on the struc-ture of DNA. Watson, who was in a rush for honor and greatness that could be gained by figuring out the structure of DNA, had a brash "bull in a china shop" attitude. This was in sharp contrast to Franklin's philosophy where you don't make conclusions until all of the experi-mental facts have been analyzed. Therefore, until she had all the facts, Franklin was reluctant to share her data with Wilkins—or anyone else.

Feeling left out, Wilkins was more than willing to help Watson and Crick. Because Watson thought Franklin was "incompetent in interpreting X-ray photographs" and he was better able to use the data, Wilkins shared with Watson an X-ray photograph and report that Franklin had filed. From these materials, it was clear that DNA was a helical molecule. It also seems clear Franklin knew this as well but, perhaps being a physical chemist, she did not grasp its importance because she was concerned with getting all the facts first and making sure they were absolutely correct. But, looking through the report that Wilkins shared, the proverbial "light bulb" went on when Crick saw what Franklin had missed; that the two DNA strands were antiparallel. This knowledge, together with Watson's ability to work out the base pairing, led Watson and Crick to their "leap of imagination" and the structure of DNA.

In her book entitled, *Rosalind Franklin: The Dark Lady of DNA* (HarperCollins, 2002), author Brenda Maddox suggests it is uncertain if Franklin could have made that leap as it was not in her character to jump beyond the data in hand. In this case, the leap of intuition won out over the methodical, data col-lecting in research—the hare beat the tortoise this time. However, it cannot be denied that Franklin's data provided an important key from which Watson and Crick made the historical discovery.

In 1962, Watson, Crick, and Wilkins received the Nobel Prize in Physiology or Medicine for their work on the structure of DNA. Should Franklin have been included? The Nobel Prize committee does not make awards posthumously and Franklin had died four years earlier from ovarian cancer. So, if she had lived, did Rosalind Franklin deserve to be included in the award?

Bacterial and Archaeal DNA Is Organized within the Nucleoid

KEY CONCEPT

1. The DNA exists as a single, circular chromosome.

Most of the genetic information in bacterial and archaeal cells is contained within the **chromosome**, the cell's intracellular source of genetic information. Usually, this is a single, circular molecule of DNA that is haploid, although a few species may have a single genome spread over multiple chromosomes.

The chromosome exists as thread-like fibers associated with some protein and is localized in the cytosol within a space called the **nucleoid** (see Chapter 4). Remember that one of the unique features defining the nucleoid area is the absence of a surrounding membrane envelope typical of the cell nucleus in eukaryotic cells.

The circular chromosome of *Escherichia coli* probably has been studied more thoroughly than that of any other microbe. The genome has about 4,400 genes coding for growth and metabolic activities. By contrast, many other bacterial and archaeal genomes are much smaller, especially those of obligate symbionts/parasites (**FIGURE 8.3**). At the other extreme are the genomes of the eukaryotes, which, along with other characteristics listed in **TABLE 8.1** , can be much larger is size. Note in this table some of the archaeal/eukaryote similarities.

Haploid:
Having a single set of genetic information.

CONCEPT AND REASONING CHECKS

8.1 Describe the basic structure of a bacterial chromosome.

TABLE

8.1 Characteristics of Bacteria, Archaeal, and Eukaryotic Chromosomes

Characteristic	*Bacteria*	*Archaea*	*Eukarya*
Organization	Nucleoid	Nucleoid	Nucleus
Chromosome morphology	Usually circular	Usually circular	Linear
Ploidy	Usually haploid	Usually haploid	Haploid, diploid, polyploid
Genome size (Mb)[1]	0.16–10	0.5–6	10–100,000
Average protein-coding genes	Few thousand	Few thousand	Tens of thousands
Presence of histone proteins	Histone-like	Yes	Yes
Presence of introns	No	Present in rRNA and tRNA genes	Yes
Replication	Just prior to binary fission	Just prior to binary fission	Few hours before mitosis
Replication origins	Single	Multiple	Multiple

[1]Mb = Millions of base pairs.

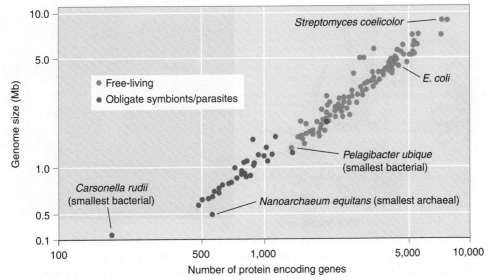

FIGURE 8.3 Genome Size among the *Bacteria* and *Archaea*. This graph illustrates the relationship between genome size (Mb = millions of base pairs) and number of protein-coding genes. Note the extremely small size of *Carsonella ruddii*.»» Produce a hypothesis to explain why the obligate symbionts/parasites have the smallest genomes.

DNA within a Chromosome Is Highly Compacted

KEY CONCEPT

2. Microbial DNA must be tightly packed to fit in the cell.

In most microbial cells, including *E. coli*, the DNA occupies about one-third of the total volume of the cell, and when extended its full length, it is about 1.5 millimeters (mm) long. This is approximately 500 times the length of the bacterial cell. So, how can a 1.5-mm-long circular chromosome fit into a 1.0 to 2.0 mm *E. coli* cell?

The answer is **supercoiling**, a twisting and tight packing caused by a number of abundant nucleoid-associated proteins. Thus, the DNA double helix twists on itself like a wound-up rubber band. The coils are folded further into loops of 10,000 bases, each forming a **supercoiled domain** (**FIGURE 8.4A**) and there are about 400 such domains in an *E. coli* chromosome, giving the molecule an overall "flower" structure called the **looped domain structure**. The high level of compaction is evident when the cell envelope is broken, releasing the DNA in a looped form (**FIGURE 8.4B**). How the loops are anchored in the nucleoid is not understood.

CONCEPT AND REASONING CHECKS

8.2 Justify the necessity for DNA supercoiling and looped domains.

Many Microbial Cells also Contain Plasmids

KEY CONCEPT

3. Plasmids carry nonessential, but often useful, information.

Many bacterial, archaeal, and fungal cells contain **plasmids**, which are stable extrachromosomal DNA elements that do not carry genetic information essential for normal structure, growth, and metabolism. This means a plasmid could be removed from a cell without affecting its viability, assuming the cell is in a nutrient-rich environment free of toxic materials.

Most plasmids are circular, and they are easily transferred between cells (Chapter 9). Plasmids exist and replicate as independent genetic elements in the cytosol where they typically contain about 2% of the total genetic information of the cell. Exceptions include some plasmids that can be quite large because they can integrate into a chromosome and excise from it many additional chromosomal genes, some of which may be essential for cell growth.

In most cases though, plasmids are not essential to the normal survival of the cell but they can confer selective advantages and provide genetic flexibility for those organisms possessing plasmids. For example, some bacterial plasmids, called **F plasmids**, allow for the trans-

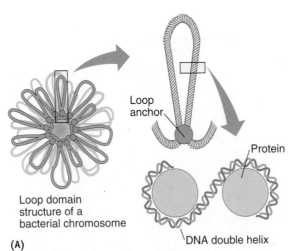

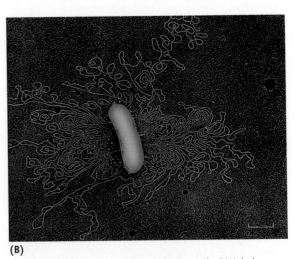

(A)   **(B)**

Loop anchor

Protein

Loop domain structure of a bacterial chromosome

DNA double helix

FIGURE 8.4 **DNA Packing. (A)** The loop domain structure of the chromosome as seen head-on. The loops in DNA help account for the compacting of a large amount of DNA in a relatively small cell. **(B)** An electron micrograph of an *E. coli* cell immediately after cell lysis. The uncoiled DNA fiber exists in loops attached to the disrupted cell envelope. (Bar = 1 μm.) »» How does plasmid structure compare to that of the chromosome in a bacterial cell?

fer of genetic material from donor to recipient through a recombination process (Chapter 9). Other plasmids may play protective roles. **R plasmids** ("resistance" factors), for example, carry genes for antibiotic resistance. Others contain genes for resistance to potentially toxic heavy metals (e.g., silver, mercury).

Some plasmids provide offensive abilities. For example, species of *Streptomyces* carry plasmids for the production of antibiotics while plasmids in other bacteria contain genes for the production of **bacteriocins**, a group of proteins that inhibit or kill other bacterial species.

Finally, there are plasmids containing genes coding for toxins affecting human cells and disease processes. The genes encoding the toxins responsible for anthrax are carried on a plasmid. We shall have much to say about these extrachromosomal units when we discuss recombination and genetic engineering in the next chapter.

CONCEPT AND REASONING CHECKS

8.3 What does it mean to say plasmids carry nonessential genetic information?

8.2 DNA Replication

Watson and Crick's 1953 paper on the structure of DNA provided a glimpse of how DNA might be copied. They concluded, *"It has not escaped our notice that the specific pairing we have postulated immediately suggests a possible copying mechanism for the genetic material."* In fact, the copying of the genetic material, called **DNA replication**, occurs with such precision that the two daughter cells from binary fission are genetically identical to the parent cell.

DNA Replication Occurs in "Replication Factories"

KEY CONCEPT

4. DNA replication is a three-phase event requiring an array of proteins working in sequence.

As described in Chapters 3 and 4, the archaeal organisms appear to be an interesting mosaic of bacterial, eukaryotic, and unique features. This applies to the chromosome (see Table 8.1) and to DNA replication as well. Most archaeal proteins involved in DNA replication are more similar in sequence to those found in eukaryotic cells than to analogous replication proteins in bacterial cells. The archaeal DNA replication apparatus also contains features not found in other organisms, which is probably a result of the broad range of environmental conditions in which members of this domain thrive. That being said, we will examine DNA replication in *E. coli*, which has been more thoroughly studied than most other microbes.

Chromosome replication requires the products of more than 20 genes and, although it occurs in a smooth process, we can separate it into three stages (**FIGURE 8.5**): **initiation** when the DNA unwinds and the strands separate; **elongation** when enzymes synthesize a new polynucleotide strand of DNA for each of the two old template (parental) strands; and **termination**, when each of the two DNA helices separate from one another. This combination of a new and old strand was first observed in *E. coli* in 1958 by Matthew J. Meselson and Franklin W. Stahl. It is called **semiconservative replication** because each old strand of the replicated DNA is conserved in each new chromosome and one strand is newly synthesized. Let's look at each stage in more detail using Figure 8.5 to guide us.

Initiation. DNA replication starts at a fixed region on the chromosome called the **replication origin** (*oriC*), which is a sequence of about 250 base pairs. A group of initiator proteins binds at the origin along with other enzymes, forming two "**replication factories**" in which DNA synthesis will occur. **Helicases** unwind and unzip the two polynucleotide strands, while **stabilizing proteins** keep the template strands separated for the replication of complementary strands. Because the replication factories are thought to be attached to the cell membrane, the yet to be replicated template strands move through a V-shaped **replication fork** in each factory.

Base pairs: The complementary pairing of A—T and G—C on the two opposite polynucleotide strands.

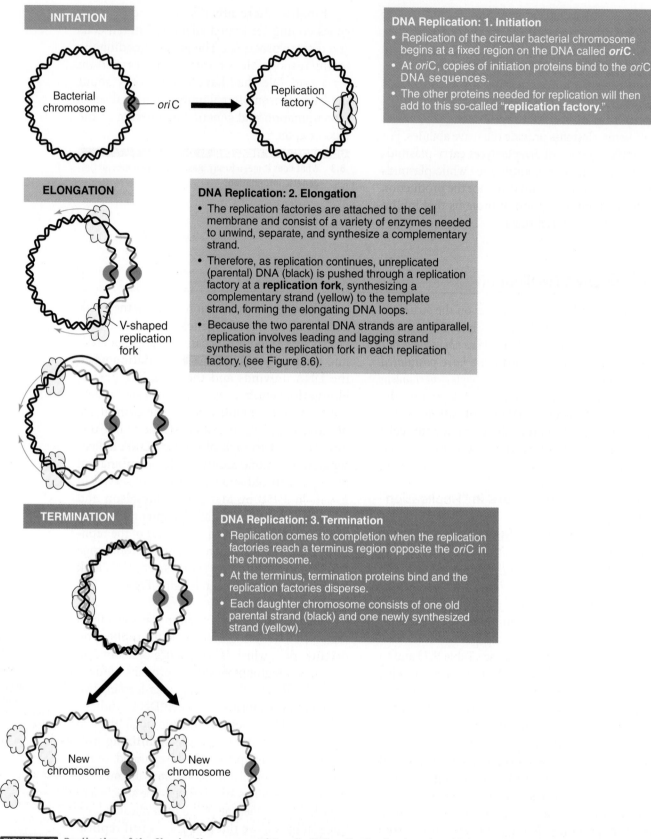

INITIATION

DNA Replication: 1. Initiation
- Replication of the circular bacterial chromosome begins at a fixed region on the DNA called *oriC*.
- At *oriC*, copies of initiation proteins bind to the *oriC* DNA sequences.
- The other proteins needed for replication will then add to this so-called "**replication factory.**"

Bacterial chromosome — *ori*C

Replication factory

ELONGATION

DNA Replication: 2. Elongation
- The replication factories are attached to the cell membrane and consist of a variety of enzymes needed to unwind, separate, and synthesize a complementary strand.
- Therefore, as replication continues, unreplicated (parental) DNA (black) is pushed through a replication factory at a **replication fork**, synthesizing a complementary strand (yellow) to the template strand, forming the elongating DNA loops.
- Because the two parental DNA strands are antiparallel, replication involves leading and lagging strand synthesis at the replication fork in each replication factory. (see Figure 8.6).

V-shaped replication fork

TERMINATION

DNA Replication: 3. Termination
- Replication comes to completion when the replication factories reach a terminus region opposite the *oriC* in the chromosome.
- At the terminus, termination proteins bind and the replication factories disperse.
- Each daughter chromosome consists of one old parental strand (black) and one newly synthesized strand (yellow).

New chromosome

New chromosome

FIGURE 8.5 **Replication of the Circular Chromosome of *E. coli*.** DNA replication involves the addition of complementary bases to the parental (template) strand within replication factories that are attached to the cell membrane. »» Why is DNA replication considered to be semiconservative?

Elongation. Synthesis of DNA in each factory then occurs on each old strand, which represents a template for the synthesis of a new complementary strand. Many proteins are involved in DNA synthesis. Besides the stabilizing protein, a **DNA polymerase III** moves along each strand, catalyzing the insertion of new complementary nucleotides to each template strand.

In *E. coli*, DNA synthesis takes about 40 minutes, which means at each replication fork DNA polymerase III is adding new complementary bases at the rate of about 1,000 per second! At this pace, errors occur where an incorrect base is added. Such potential mutations could be lethal, so there must be a mechanism to correct any errors. DNA polymerases III and I detect any mismatched nucleotides, remove the incorrect nucleotide in the pair, and add the correct nucleotide. Such proofreading reduces replication errors to about 1 in every 10 billion bases added. We will have more to say about mutations later in this chapter.

Termination. In about 40 minutes, the two replication forks meet 180° from *oriC*. At the terminus region, there are additional terminator proteins that block further replication, causing the replication factories to dissociate. Then, the two intertwined DNA molecules (chromosomes) are separated by other enzymes, guaranteeing that each daughter cell will inherit one complete chromosome after binary fission (see Chapter 5).

CONCEPT AND REASONING CHECKS
8.4 Describe the role for replication factories in DNA synthesis.

DNA Polymerase Only Reads in the 3′ to 5′ Direction

KEY CONCEPT

5. Continuous and discontinuous DNA synthesis occur in each replication fork.

Because DNA polymerase can "read" the template DNA only in the 3′ to 5′ direction and the two parental (template) strands are antiparallel, this means at each replication fork the complementary DNA strand is formed in two different ways (**FIGURE 8.6**).

Leading Strand Synthesis. One parental strand in each replication factory is the template for synthesizing a continuous complementary **leading strand**. Here the DNA polymerase reads the template in the 3′ to 5′ direction, bringing in triphosphate nucleotides (A, T, G, and C) that hydrogen bond with their complement in the template strand. The high-energy bonds in the triphosphate nucleotides provide the energy for the DNA polymerase to covalently join nucleotides into the continuous strand, forming an elongating chain of nucleotides from 5′ to 3′.

Lagging Strand Synthesis. The other template strand in each fork of a replication factory must be read "backwards"; that is, as the DNA polymerase moves away from the replication fork, a discontinuous process of starts and stops occurs, with the new strand always lagging behind the leading strand. This piecemeal strand, therefore, is called the **lagging strand**. These segments, which are about 1,000 nucleotides long, are called **Okazaki fragments**, after Reiji Okazaki, who discovered them in 1968. As these new polynucleotide segments are produced, the gaps between segments are eventually joined into a complete and elongating single strand with the help of an enzyme called **DNA ligase**.

Mutations: Permanent alterations in DNA base sequences.

CONCEPT AND REASONING CHECKS
8.5 Why are there leading and lagging strands in each replication fork?

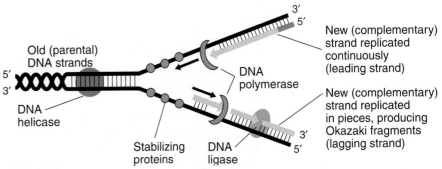

FIGURE 8.6 **A Replication Factory.** This diagram outlines the general events occurring at one replication fork, showing both the leading and lagging strand synthesis. The discontinuous synthesis on the lagging strand results from the DNA polymerase moving away from the replication fork, resulting in the formation of short DNA fragments, called Okazaki fragments, which are eventually joined by a DNA ligase. »» Why is the DNA polymerase on the lagging strand moving away from the replication fork?

8.3   Protein Synthesis

The discovery of the structure of DNA also provided a glimpse into understanding how a cell makes proteins. **Protein synthesis** is a process in which amino acids are precisely bound together in a three-dimensional structure determined by the hereditary information, the **genes**, in the cell. The process requires not only DNA, but also ribonucleic acid (RNA). Both DNA and RNA were described in Chapter 2 and are summarized in TABLE 8.2. A review of their structure is recommended so the discussion to follow can be fully comprehended.

One of the central truths in biology states that the genetic information in DNA first is expressed as RNA by a process called **transcription**. One type of RNA then functions as a messenger by carrying the genetic code to areas of the cytosol where the ribosomes are located. There, amino acids are fitted together in a precise sequence to form the protein. This sequencing process, called **translation**, reflects the genetic information in the DNA. This **central "dogma"** (*dogma* = "opinion") of biology is shown in FIGURE 8.7.

As we will see in Chapter 13, a few viruses modify this rule.

Genetic code:
The sequence of bases in the DNA or codons in the RNA that specify a specific polypeptide.

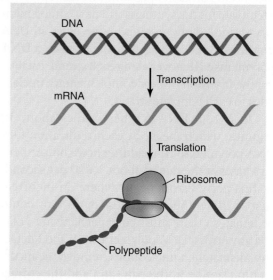

FIGURE 8.7 The Central Dogma. The flow of genetic information proceeds from DNA to RNA to protein (polypeptide). »» What is the name for the segment of DNA that is used to make a protein?

Transcription Copies Genetic Information into Complementary RNA

KEY CONCEPT

6. Different DNA segments are transcribed into one of three types of RNA.

TABLE

8.2   A Comparison of DNA and RNA

DNA (Deoxyribonucleic Acid)	RNA (Ribonucleic Acid)
In *Bacteria* and *Archaea*, found in the nucleoid and plasmids; in *Eukarya*, found in the nucleus, mitochondria, and chloroplasts	In all organisms, found in the cytosol and in ribosomes; in *Eukarya*, found in the nucleolus
Always associated with chromosome (genes); each chromosome has a fixed amount of DNA	Found mainly in combinations with proteins in ribosomes (ribosomal RNA) in the cytosol, as messenger RNA, and as transfer RNA
Contains a 5-carbon sugar called deoxyribose	Contains a 5-carbon sugar called ribose
Contains bases adenine, guanine, and cytosine, thymine	Contains bases adenine, guanine, cytosine, and uracil
Contains phosphorus (in phosphate groups) that connects deoxyribose sugars with one another	Contains phosphorus (in phosphate groups) that connects ribose sugars with one another
Functions as the molecule of inheritance	Functions in protein synthesis and gene regulation
Double stranded	Usually single stranded
Larger size	Smaller size

A microorganism's genome contains a significant number of genes (see Figure 8.3). **Transcription** is the process of expressing a number of these genes at a particular time. Thus, the gene DNA serves as a template for new RNA molecules (**FIGURE 8.8**).

RNA polymerase is the enzyme that carries out transcription. In the *Bacteria* and *Archaea*, there is but one RNA polymerase with the archaeal enzyme being more similar to the eukaryotic polymerases. Like DNA polymerases, the RNA polymerase "reads" the DNA template strand in the 3′ to 5′ direction. However, unlike DNA replication, only one of the two DNA strands within a gene is transcribed. The RNA polymerase recognizes this DNA template strand by a sequence of bases called the **promoter** located on the template strand. The polymerase binds to the promoter (initiation), unwinds the helix, and separates the two strands within the gene. As the enzyme moves along the DNA template strand (elongation), complementary pairing brings RNA triphosphate nucleotides to the template strand—guanine (G) and cytosine (C) pair with one another and thymine (T) in the DNA template pairs with adenine (A) in the RNA. However, an adenine base on the DNA template pairs with a uracil (U) base in the RNA because RNA nucleotides contain no thymine bases (see Chapter 2).

Like initiation, termination of transcription occurs at specific base sequences, called **terminators**, on the DNA template strand. The RNA transcription product released represents a complementary image to the sequence of bases in the DNA template strand.

Transcription produces three types of RNA, all of which are needed for translation.

Messenger RNA (mRNA). This RNA carries the genetic information or "blueprint" to

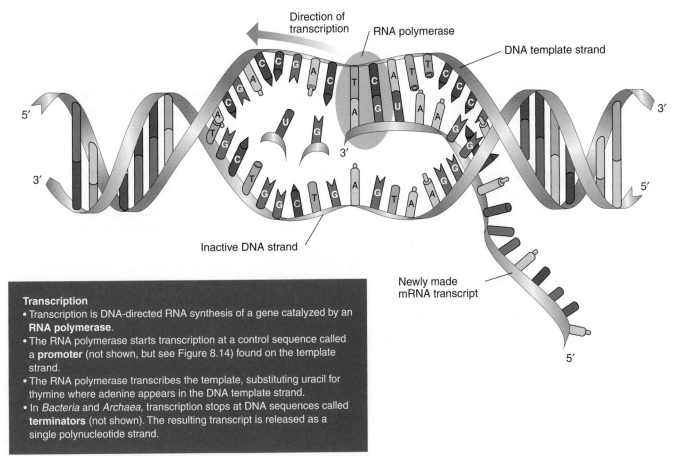

Transcription
- Transcription is DNA-directed RNA synthesis of a gene catalyzed by an **RNA polymerase.**
- The RNA polymerase starts transcription at a control sequence called a **promoter** (not shown, but see Figure 8.14) found on the template strand.
- The RNA polymerase transcribes the template, substituting uracil for thymine where adenine appears in the DNA template strand.
- In *Bacteria* and *Archaea*, transcription stops at DNA sequences called **terminators** (not shown). The resulting transcript is released as a single polynucleotide strand.

FIGURE 8.8 **The Transcription Process.** For transcription to occur, a gene must unwind and the base pairs separate. The enzyme RNA polymerase does this as it moves along the template strand of the DNA and adds complementary RNA nucleotides. Note that the other DNA strand of the gene is not transcribed. »» Justify the need for a promoter sequence in a gene.

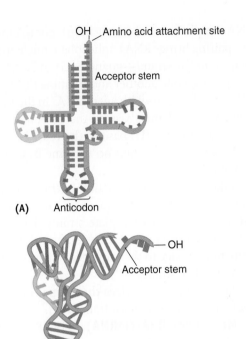

FIGURE 8.9 Structure of a Transfer RNA (tRNA).
(A) The traditional "cloverleaf" configuration for tRNA. The anticodon will pair up with a complementary codon in the mRNA. The appropriate amino acid attaches to the end of the acceptor arm. **(B)** A schematic diagram of the more correct three-dimensional structure of a tRNA. »» What part of the tRNA is critical for the complementary binding to a codon in a mRNA?

manufacture a polypeptide. Each mRNA transcribed from a different gene carries a different message; that is, a different sequence of nucleotides coding for a different polypeptide. The message is encoded in a series of three-base codes or **codons** found along the length of the mRNA. Each codon specifies an individual amino acid to be slotted into position during translation.

Ribosomal RNA (rRNA). Three rRNAs are transcribed from specific regions of the DNA. Together with protein, these RNAs serve a structural role as the framework of the ribosomes, which are the sites at which amino acids assemble into proteins (see Figure 4.19). They also serve a functional role in the translation process.

Transfer RNA (tRNA). The conventional drawing for a tRNA is in a shape roughly like a cloverleaf (**FIGURE 8.9**). One point presents a sequence of three nitrogenous bases, which func-

tions as an **anticodon**; that is, a sequence that complementary binds to an mRNA codon. The tRNAs have a structural role in delivering amino acids to the ribosome for assembly into proteins. Each tRNA has a specific amino acid attached through an enzymatic reaction involving ATP. For example, the amino acid alanine binds only to the tRNA specialized to transport alanine; glycine is transported by a different tRNA.

There is one important difference between microbial RNAs. In bacterial and some archaeal cells, all of the bases in a gene are transcribed and used to specify a particular protein (**FIGURE 8.10A**). However, in many archaeal and eukaryotic cells certain portions of a gene are not part of the final RNA and are removed from the RNA before the molecule can function (**FIGURE 8.10B**). These intervening DNA segments removed after transcription are called **introns**, while the remaining, amino acid-coding segments are called **exons**.

While the exons have the "standard" information coding for a polypeptide (or protein), the introns also may have roles in gene regulation, or in metabolic control through association with other RNAs or proteins. This may have helped in the evolution of multicellularity.

CONCEPT AND REASONING CHECKS

8.6 How is a bacterial gene processed differently from a eukaryotic gene?

The Genetic Code Consists of Three-Letter Words

KEY CONCEPT

7. More than one codon often specifies a specific amino acid.

By now you should have the idea that the information to specify the amino acid sequence for a polypeptide is encoded in the gene DNA. This specific sequence of nucleotide bases is called the **genetic code** and each sequence is made up of "three-letter words" that we called a codon. To synthesize a polypeptide then, the DNA codons must first be transcribed into RNA codons, as was depicted in Figure 8.8.

One of the startling discoveries of biochemistry is that the genetic code in most cases contains more than one codon for each amino acid. Because there are four nitrogenous bases, mathematics tells

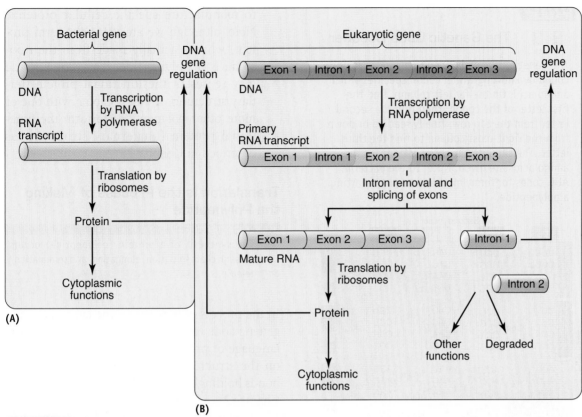

FIGURE 8.10 RNA Processing and Gene Activity. **(A)** In bacterial cells, gene DNA is almost entirely protein-coding information that is transcribed and translated into a protein having structural or functional roles in the cytoplasm, or gene regulation roles. **(B)** In eukaryotic cells, some of the intron RNA may be degraded, used to regulate gene function, or be associated with other RNAs or proteins in the cytoplasm. »» How do the roles of exonic RNA and intronic RNA differ?

us 64 possible combinations can be made of the four bases, using three at a time. But there are only 20 amino acids for which a code must be supplied. How do scientists account for the remaining 44 codes?

It is now known that 61 of the 64 codons are **sense codons** that specify an amino acid, and most of those amino acids have multiple codons (as shown in **TABLE 8.3**). For example, GCU, GCC, GCA, and GCG all code for the amino acid alanine (ala). This lack of a one-to-one relationship between codon and amino acid generates **redundancy**. In Table 8.3, notice one of the 64 codons, AUG, represents the **start codon** for making a protein. This codon usually specifies the amino acid methionine (met). Three additional codons, which do not code for an amino acid (UGA, UAG, UAA), are called **stop codons**

because they terminate the addition of amino acids to a growing polypeptide chain.

CONCEPT AND REASONING CHECKS

8.7 What is meant by the genetic code being redundant? Give two examples.

Before we proceed to the last stage, translation, let's summarize the protein synthesis process to this point (**FIGURE 8.11**).

1. Each gene of the DNA contains information to manufacture a specific form of RNA.

2. The information can be transcribed into:
   - mRNAs, which are produced from genes carrying the information as to what protein will be made during translation;
   - rRNAs, which form part of the structure of the ribosomes and help in the translation of the mRNA; and

TABLE 8.3 The Genetic Code Decoder

The genetic code embedded in an mRNA is decoded by knowing which codon specifies which amino acid. On the far left column, find the first letter of the codon; then find the second letter from the top row; finally read up or down from the right-most column to find the third letter. The three-letter abbreviations for the amino acids are given. Note: In the *Bacteria*, AUG codes for formylmethionine when starting a polypeptide.

Key

Ala = Alanine
Arg = Arginine
Asn = Asparagine
Asp = Aspartic acid
Cys = Cysteine
Gln = Glutamine
Glu = Glutamic acid
Gly = Glycine
His = Histidine
Ile = Isoleucine
Leu = Leucine
Lys = Lysine
Met = Methionine
Phe = Phenylalanine
Pro = Proline
Ser = Serine
Thr = Threonine
Trp = Tryptophan
Tyr = Tyrosine
Val = Valine

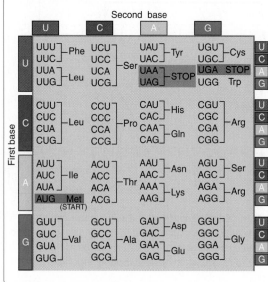

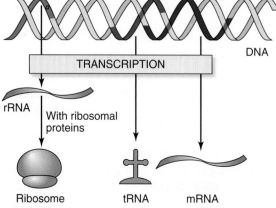

FIGURE 8.11 **The Transcription of the Three Types of RNA.** Genes in the DNA contain the information to produce three types of RNA: mRNA, rRNA, and tRNA. »» What is each type of RNA used for in a microbial cell?

Formylmethionine: The presence of a formyl group (H-CO—) attached to methionine.

Ribozyme: An RNA molecule capable of carrying out a chemical reaction.

- tRNAs, each of which carries a specific amino acid needed for the translation process.
3. With the tRNAs and mRNAs present in the cytosol, they can combine within ribosomes

to manufacture specific cellular proteins. Note: although we are using the term protein, what is actually made from the ribosome is a polypeptide. This polypeptide may represent the functional protein (tertiary structure) or first combine with one or more other polypeptides to form the functional protein (quaternary structure), as described in Chapter 2.

Translation Is the Process of Making the Polypeptide

KEY CONCEPT

8. The synthesis of a protein (polypeptide) occurs through chain initiation, elongation, and termination/release.

In the process of translation, the language of the genetic code (nucleotides) is translated into the language of proteins (amino acids). A refresher on the structure of proteins and the peptide bonds holding them together will be of value (see Chapter 2).

As with DNA replication and RNA transcription, translation occurs in three steps: chain initiation, elongation, and termination.

Chain Initiation. Translation begins with the association of a small ribosomal subunit with an initiator tRNA at the AUG start codon (**FIGURE 8.12A**). Then, the large ribosomal subunit was added to form the functional ribosome with three tRNA binding sites, called A, P, and E. In *Bacteria*, the first amino acid is formylmethionine (fmet) while in the *Archaea* and *Eukarya*, it is methionine (met). Once formed, a second tRNA can complementary bind at the A site and a ribozyme transfers the fmet to the amino acid on the second tRNA.

Chain Elongation. With the second tRNA attached, the first tRNA is released from the E site (**FIGURE 8.12B**). Moving right one codon, the ribosome exposes the next codon (GCC), and the appropriate tRNA with the amino acid alanine (ala) attached. Again, a ribozyme transfers the dipeptide fmet-Ser to alanine. The tRNA that carried serine exited the ribosome and the process of chain elongation continues as the ribosome moves to expose the next codon.

Chain Termination/Release. The process of adding tRNAs and transferring the elongating

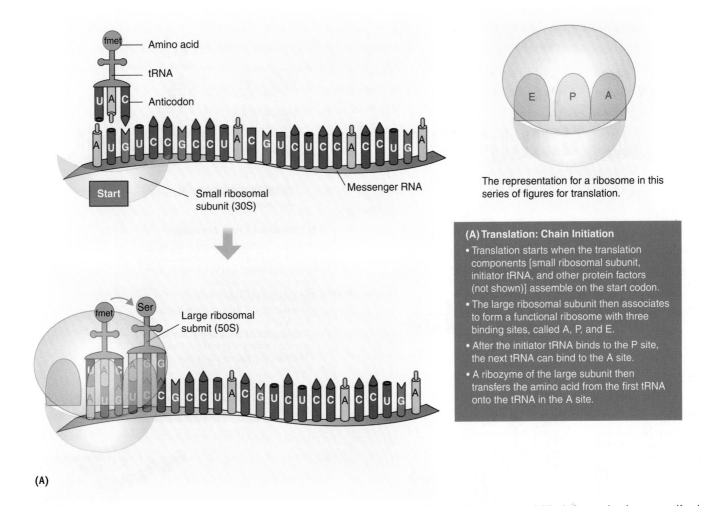

(A) Translation: Chain Initiation

- Translation starts when the translation components [small ribosomal subunit, initiator tRNA, and other protein factors (not shown)] assemble on the start codon.
- The large ribosomal subunit then associates to form a functional ribosome with three binding sites, called A, P, and E.
- After the initiator tRNA binds to the P site, the next tRNA can bind to the A site.
- A ribozyme of the large subunit then transfers the amino acid from the first tRNA onto the tRNA in the A site.

FIGURE 8.12 **Protein Synthesis in a Bacterial Cell.** The steps of (**A**) chain initiation, (**B**) chain elongation, and (**C**) chain termination are outlined. »» How do the P, A, and E sites in the ribosome differ?   *Continued*

polypeptide to the entering amino acid/tRNA at the A site continues until the ribosome reaches a stop codon (UGA in this example). There is no tRNA to recognize any of these stop codons (**FIGURE 8.12C**). Rather, proteins called **termination factors** bind where the tRNA would normally attach. This triggers the release of the polypeptide and a disassembly of the ribosome subunits, which can be reassembled for translation of another mRNA.

During synthesis, the polypeptide already may start to twist into its secondary and tertiary structure. For many polypeptides, groups of cytoplasmic proteins called **chaperones** ensure the folding process occurs correctly.

MicroFocus 8.2 describes how the understanding of protein synthesis has been used to block "harmful" proteins from being made.

Cells typically make hundreds, if not thousands, of copies of each protein. Producing such large amounts of a protein can be done efficiently and quite quickly. Remember, a cell contains thousands of identical ribosomes. Therefore, a single mRNA molecule can be translated simultaneously by several ribosomes (**FIGURE 8.13**). Once one ribosome has moved far enough along the mRNA, another small subunit can "jump on" and initiate translation. Such a string of ribosomes all translating the same mRNA at the same time is called a **polysome**.

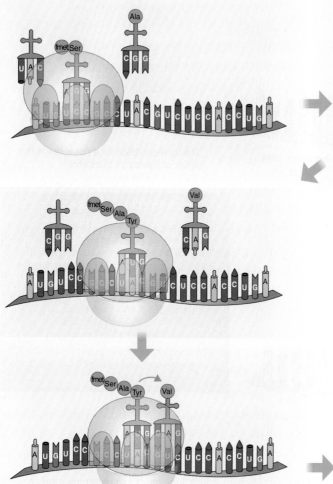

(B)

Ribosome movement

(B) Translation: Chain Elongation

- Once the functional ribosome has been formed, translation is a process of adding new amino acids, one at a time, to the forming polypeptide chain.
- The process involves:
  - The addition of the appropiate tRNA to the A site;
  - The transfer of the amino acid chain from the amino acid in the P site; and
  - The loss of a former tRNA from the E site
- The ribosome then moves down one codon and the process repeats; the result is an elongating amino acid sequence.

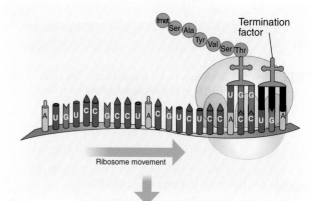

Ribosome movement

(C) Translation: Chain Termination

- Chain elongation continues until the A site covers a stop codon (UGA, UAG, or UAA) to which there is no corresponding tRNA.
- A termination factor binds to this site, causing:
  - The release of the polypeptide chain from the P site;
  - The disassociation of the ribosome into sub units; and
  - The release of the empty tRNA and termination factor.

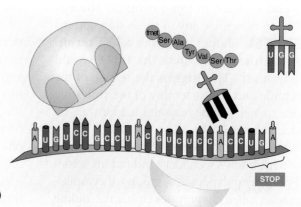

(C)

FIGURE 8.12 Continued.

MICROFOCUS 8.2: Biotechnology
Antisense and Interference Makes Sense

With the recent outbreaks of fatal encephalitis caused by the West Nile virus and atypical pneumonia caused by the severe acute respiratory syndrome (SARS) coronavirus, scientists have been trying to find ways to treat and cure these and other viral diseases, because they are not affected by antibiotics.

One of the potential approaches being considered is the use of antisense molecules as therapeutic agents. **Antisense molecules** are RNA fragments that are the complement of an mRNA that carries a specific genetic message for protein synthesis. By binding to the mRNA, antisense molecules should block the ability of ribosomes to translate the message and thus have the ability to shut off the production of unwanted or disease-causing proteins. To treat AIDS, for example, scientists could create an antisense strand that is complementary to specific mRNAs produced by the human immunodeficiency virus (HIV). In an infected individual, the antisense molecules should bind to these viral mRNAs and, as double-stranded RNA molecules, the mRNAs could not be translated by the cell's ribosomes. Without these essential viral proteins, no new HIV particles could be formed (Chapter 23). Although such strategies make sense on paper, they have yet to produce the successes that were hoped for in clinical trials.

More recently, another way has been discovered for turning off or silencing the expression of specific genes. This is called **RNA interference** (**RNAi**). This is a technique in which extracellular, double-stranded (ds) RNA that is complementary to a known target mRNA is introduced into a cell. The dsRNA in the cell is chopped into smaller pieces by cellular enzymes and these fragments then bind to the target mRNA. Again these new dsRNA pieces are degraded and the protein or polypeptide is not produced. Indirectly, the gene for that polypeptide has been silenced.

One potential use for RNAi is for antiviral therapy. Since many human diseases are caused by viruses that have an RNA genome (Chapter 14), RNAi may be valuable in inhibiting gene expression. For example, RNAi could silence viruses that induce human tumors, as well as the hepatitis A virus, influenza viruses, and other RNA viruses such as the measles virus. In all these examples, if the virus cannot replicate, new viruses cannot be produced—and disease development would be prevented.

The potential value of RNAi has recently been recognized. In 2006, Andrew Z. Fire (Stanford University School of Medicine) and Craig C. Mello (University of Massachusetts Medical School) were awarded the Nobel Prize in Physiology or Medicine "for their discovery of RNA interference—gene silencing by double-stranded RNA."

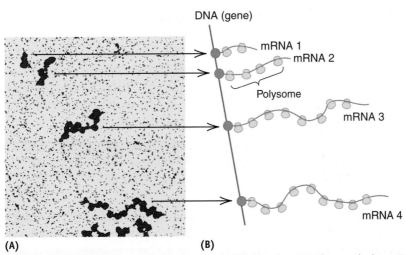

FIGURE 8.13 Coupled Transcription and Translation in *E. Coli.* (**A**) The electron micrograph shows transcription of a gene in *E. coli* and translation of the mRNA. The dark spots are ribosomes, which coat the mRNA. An interpretation of the electron micrograph is shown in (**B**). Each mRNA has ribosomes attached along its length. The large red dots are the RNA polymerase molecules; they are too small to be seen in the electron micrograph. The length of each mRNA is equal to the distance that each RNA polymerase has progressed from the transcription-initiation site. For clarity, the polypeptides elongating from the ribosomes are not shown. »» From the interpretation of the micrograph, (a) how many times has this gene been transcribed and (b) how many identical polypeptides are being translated?

CONCEPT AND REASONING CHECKS

8.8 Explain why the ribosome can be portrayed as a "cellular translator."

Antibiotics Interfere with Protein Synthesis

KEY CONCEPT

9. Many antibiotics can inhibit transcription or translation.

Many antibiotics affect protein synthesis in bacterial cells and therefore are clinically useful in treating human infections and disease. A few antibiotics interfere with transcription. Rifampin binds to the RNA polymerase so that transcription cannot initiate.

A very large number of antibiotics inhibit translation by binding to the bacterial 30S or 50S ribosomal subunit. For example, tetracycline prevents chain initiation by binding to the 30S subunit, while drugs like chloramphenicol and erythromycin inhibit chain elongation by binding to the 50S subunit. We will have much more to learn about antibiotics in Chapter 24.

CONCEPT AND REASONING CHECKS

8.9 Propose a hypothesis to explain why so many antibiotics specifically affect protein synthesis.

Protein Synthesis Can Be Controlled in Several Ways

KEY CONCEPT

10. Many genes are controlled by operons.

In Chapter 6, we described how negative feedback can control enzyme activity. Another control mechanism is to simply not make the enzyme (or any other protein in general) when it is not needed. Because transcription is the first step leading to protein manufacture in cells, another way to control what proteins and enzymes are present in bacterial and archaeal cells is to regulate the mechanisms that induce ("turn on") or repress ("turn off") transcription of a gene or set of genes.

In 1961, two Pasteur Institute scientists, Françoise Jacob and Jacques Monod, proposed such a mechanism for controlling protein synthesis. They suggested segments of bacterial DNA are organized into functional units called **operons** (FIGURE 8.14). Their pioneering research along with more recent studies indicates that each operon consists of a cluster of **structural genes** providing genetic codes for proteins often having metabolically related functions. In this way, bacterial and archaeal cells can co-regulate genes needed in the same functional or metabolic pathway (see Chapter 6). Adjacent to the structural genes is the **operator**, which is a sequence of bases controlling the expression (transcription) of the structural genes. Next to the operator is a **promoter**, which represents the sequence of bases to which the RNA polymerase binds to initiate transcription of the structural genes. Also important, but not part of the operon is a distant **regulatory gene** that codes for a **repressor protein**.

In the operon model, the repressor protein binds to the operator. Binding prevents the RNA polymerase from moving down the operon and thus cannot transcribe the structural genes. This is called **negative control** of protein synthesis because the repressor protein inhibits or "turns off" gene transcription within the operon. When the repressor in some way is prevented from binding to the operator, the RNA polymerase has clear sailing and transcribes the structural genes, which then are translated into the final polypeptides.

MicroInquiry 8 presents two contrasting examples of how an operon works to induce or repress gene transcription. Following the series of observations and explanations given in the MicroInquiry, you should have a firm understanding of how bacterial and archaeal cells can control protein synthesis through transcription.

CONCEPT AND REASONING CHECKS

8.10 Why is transcription described in this section referred to as negative control?

Transcription and Translation Are Compartmentalized

KEY CONCEPT

11. Transcription and translation occur in spatially separated compartments.

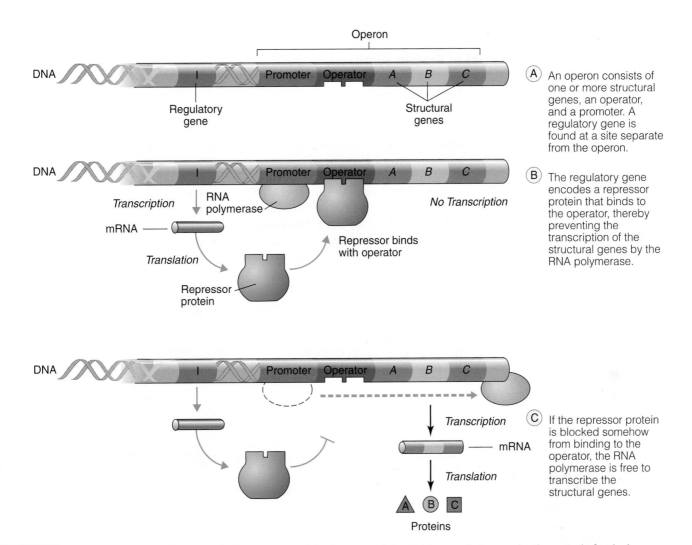

FIGURE 8.14 **The Operon and Negative Control.** An operon consists of a group of structural genes that are under the control of a single operator. Negative control exists if the operator prevents the RNA polymerase from transcribing the structural genes. »» How does the operator prevent structural gene transcription in negative control?

In Chapter 4, we described the nucleoid as an amorphous area containing the cell's chromosome. It also was noted that although the nucleoid lacked a nuclear envelope, nucleoid and cytoplasmic activities were segregated much as they are in eukaryotic cells.

Research studies have shown that, at least in *Bacillus subtilis*, RNA polymerases are concentrated within the nucleoid core in the central portion of the cell (**FIGURE 8.15**). If correct, then most of the cell's transcription presumably occurs in the same region. In fact, RNA polymerase was often localized to specific regions of the nucleoid, somewhat similar to its localization in eukaryotic nucleoli

carrying out rRNA synthesis. This is in contrast to the localization of ribosomes, which are absent in the nucleoid region and primarily concentrated at the cell poles. If the observation with *B. subtilis* holds true for other members of the *Bacteria* and *Archaea*, then the ability to segregate transcription and translation without the need for a nuclear envelope is analogous to that of the *Eukarya*.

Other research suggests that individual genes and DNA segments within the chromosome also have specific positions within the nucleoid. For example, the *oriC* region of the chromosome lies at one end of the nucleoid and is associated with the cell membrane. The precise mechanisms

MICROINQUIRY 8

The Operon Theory and the Control of Protein Synthesis

The best way to visualize and understand the operon model for control of protein synthesis is by working through a couple of examples.

The Lactose *(lac)* Operon

Here is a piece of experimental data. The disaccharide lactose represents a potential energy source for *E. coli* cells if it can be broken into its monomers of glucose and galactose. One of the enzymes involved in the metabolism of lactose is β-galactosidase. If *E. coli* cells are grown in the absence of lactose, β-galactosidase activity cannot be detected as shown in the graph (**Figure A**).

However, when lactose is added to the nutrient broth, very quickly enzyme activity is detected. How can this change from inhibition to expression be explained in the operon model?

Based on the operon theory, we would propose that when lactose is absent from the growth medium, the repressor protein for the *lac* operon binds to the operator and blocks passage of the RNA polymerase that is attached to the adjacent promoter (**Figure Bi**). Being unable to move past the operator, the polymerase cannot transcribe the structural genes, one of which (*lacZ*) codes for β-galactosidase.

When lactose is added to the growth medium, lactose will be transported into the bacterial cell, where the disaccharide binds to the repressor protein and inactivates it (**Figure Bii**). With the repressor protein inactive, it no longer can recognize and bind to the operator. The RNA polymerase now is not blocked and can translocate down the operon and transcribe the structural genes. Lactose is called an **inducer** because its presence has induced, or "turned on," structural gene transcription in the *lac* operon. It explains why β-galactosidase activity increases when lactose was present.

Now let's see if you can figure out this scenario.

Tryptophan *(trp)* Operon

E. coli cells have a cluster of structural genes that code for five enzymes in the metabolic pathway for the synthesis of the amino acid tryptophan (trp). Therefore, if *E. coli* cells are grown in a broth culture lacking trp, they continue to grow normally by synthesizing their own tryptophan, as shown in the graph (**Figure A**).

However, as the graph shows, when trp is added to the growth medium, new enzyme synthesis is repressed or "turned off" and cells use the trp supplied in the growth medium.

How can enzyme repression be explained by the operon model? The solution is provided in **Appendix D**.

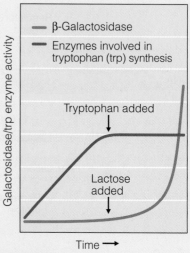

FIGURE A   Enzyme activity versus time.

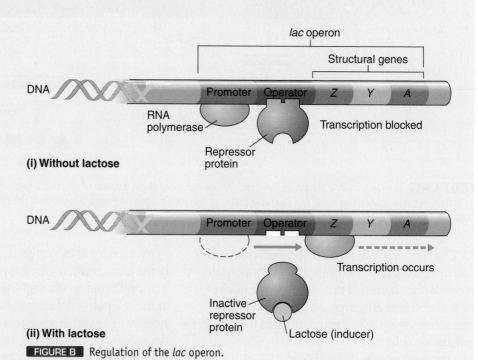

(i) Without lactose

(ii) With lactose

FIGURE B   Regulation of the *lac* operon.

responsible for nucleoid gene organization and the establishment of core and peripheral zones remain to be elucidated. **FIGURE 8.16** summarizes the protein synthesis process.

CONCEPT AND REASONING CHECKS

8.11 How is compartmentation in bacterial and eukaryotic cells similar in regard to transcription and translation?

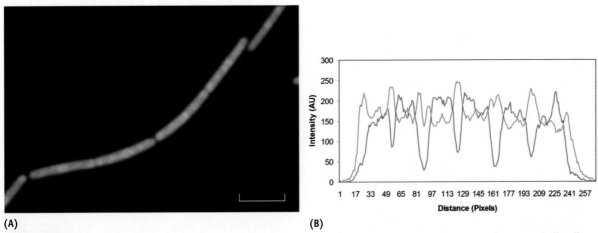

(A)   **(B)**

FIGURE 8.15 The Localization of Transcription and Translation in *Bacillus subtilis* Cells. (**A**) In these dividing *B. subtilis* cells, ribosomal subunits have been labeled with a green fluorescent protein (GFP) and RNA polymerase subunits with a label that fluoresces red. The RNA polymerase (transcription) is found mainly in the nucleoid core while the ribosomes (translation) are concentrated at the poles of the cell. (Bar = 3 μm.) (**B**) This linescan through the cells confirms the interpretation that where polymerase RNA polymerase fluorescence is high (red line), ribosomal fluorescence is low (green line) and vice-versa. »» How does fluorescence microscopy aid in the identification of spatially separated compartments?

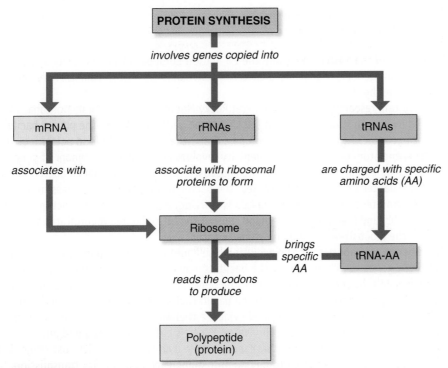

FIGURE 8.16 A Concept Map for Protein Synthesis. The relationships between transcription and translation, and the three types of RNA, are illustrated. »» In this concept map, circle those parts representing transcription and circle those parts representing translation.

8.4 Mutations

The information in a chromosome may be altered through a permanent change in the DNA, which is called a **mutation**. In most cases, a mutation involves a disruption of the nitrogenous base sequence in the DNA molecule. From this disruption, the production of a miscoded mRNA, and ultimately the insertion of one or more incorrect amino acids into the polypeptide during translation can occur. Because proteins govern numerous cellular activities, mutations may alter some aspect of these activities—for better or worse (MicroFocus 8.3).

Mutations Are the Result of Heritable Changes in a Genome

KEY CONCEPT

12. Mutations can be spontaneous or induced.

Spontaneous mutations are heritable changes to the base sequence in the DNA that result from natural phenomena. These changes could be from everyday radiation penetrating the atmosphere or errors made and not corrected by DNA polymerase III during replication. It has been estimated that one such mutation may occur for every 10^6 to 10^{10} divisions of a microbial cell.

A mutant cell arising from a spontaneous mutation usually is masked by the normal wild type cells in the population. However, should some agent be present from which the mutant survives, it may multiply and emerge as the predominant form. For many decades, for example, doctors used penicillin to treat gonorrhea. Then, in 1976, a penicillin-resistant strain of *Neisseria gonorrhoeae* emerged in human populations. Many investigators believe the resistant bacterial strain had been present for perhaps centuries, but only now with heavy use of penicillin could it arise and fill the niche once held by the penicillin-sensitive forms. MicroFocus 8.4 describes another example of an evolutionary shift "caught" in the research lab.

Wild type:
The common or native form of a gene or organism.

Niche:
The functioning of a species in relation to other species and its physical environment.

MICROFOCUS 8.3: Evolution
Evolution of An Infectious Disease

Could the Black Death of the 14th century and the 25 million Europeans that succumbed to plague have been the result of a few genetic changes to a bacterial cell? Could the entire course of Western civilization have been affected by these changes?

Possibly so, maintain researchers from the federal Rocky Mountain Laboratory in Montana. In 1996, a research group led by Joseph Hinnebusch reported that three genes missing in the plague bacillus *Yersinia pestis* are present in a related species *(Y. pseudotuberculosis)* that causes mild food poisoning. Thus, it is possible that the entire story of plague's pathogenicity revolves around a small number of gene changes.

Bubonic, septicemic, and pneumonic plague are caused by *Y. pestis,* a rod-shaped bacterium transmitted by the rat flea (Chapter 12). In an infected flea, the bacterial cells eventually amass in its foregut and obstruct its gastrointestinal tract. Soon the flea is starving, and it starts biting victims (humans and rodents) uncontrollably and feeding on their blood. During the bite, the flea regurgitates some 24,000 plague bacilli into the bloodstream of the unfortunate victim.

At least three genes are important in the evolution of plague. The nonpathogenic *Y. pseudotuberculosis* bacilli have these genes, which encourage the bacilli to remain harmlessly in the midgut of the flea. Pathogenic plague bacilli, by contrast, do not have the genes. Free of control, the bacteria migrate from the midgut to the foregut and form a plug of packed bacilli that are passed on to the victim in a flea bite.

In 2002, Hinnebusch and colleagues published evidence that another gene, carried on a plasmid, codes for an enzyme that is required for the initial survival of *Y. pestis* bacilli in the flea midgut. By acquiring this gene from another unrelated organism, *Y. pestis* made a crucial jump in its host range. It now could survive in fleas and became adapted to relying on its blood-feeding host for transmission. So, a few genetic changes may have been a key force leading to the evolution and emergence of plague. This is just another example of the flexibility that many microbes have to repackage themselves constantly into new and, sometimes, more dangerous agents of infectious disease.

Most of our understanding of mutations has come from experiments in which scientists purposely generate mutations. Such **induced mutations** are produced by chemical or physical agents called **mutagens**.

Physical Mutagens. Ultraviolet (UV) light is a physical mutagen whose energy induces adjacent thymine (or cytosine) bases in the DNA to covalently link together forming dimers (**FIGURE 8.17**). If these dimers occur in a protein-coding gene, the RNA polymerase cannot insert the correct bases (A—A) in mRNA molecules where the dimers are located.

Chemical Mutagens. Nitrous acid is an example of a chemical mutagen that converts DNA's adenine bases to hypoxanthine bases (**FIGURE 8.18A**). Adenine would normally base pair with thymine, but the presence of hypoxanthine causes a base pairing with cytosine after replication. Later, should replication occur from the gene with the cytosine mutation, the mRNA will contain a guanine rather than an adenine.

Mutations also are induced by **base analogs**, such as 5-bromouracil, which bears a close chemical resemblance to thymine (**FIGURE 8.18B**). During replication, the base analog could pair with adenine when thymine should be present.

Other base analogs resemble other DNA bases and are useful as antiviral agents in the treatment of diseases caused by DNA viruses, such as the herpesviruses (Chapter 15). Acyclovir, for example, is a base analog that can substitute for guanine during viral replication. The presence of acyclovir blocks viral replication, so new virus particles cannot be produced. As such, treatment with acyclovir can be effective in decreasing the frequency and severity of fever blisters (cold sores).

CONCEPT AND REASONING CHECKS

8.12 How do chemical mutagens interfere with DNA replication or protein synthesis?

Point Mutations Can Be Spontaneous or Induced

KEY CONCEPT

13. Point mutations affect one base pair in a DNA sequence.

Regardless of the cause of the mutation, one of the most common results is a **point mutation**, which affects just one point (base pair) in a gene. Such

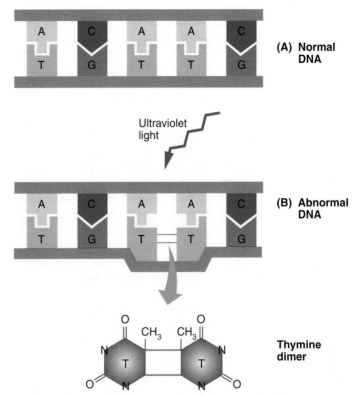

FIGURE 8.17 Ultraviolet Light and DNA. (**A**) When cells are irradiated with ultraviolet (UV) light either naturally or through experiment, the radiations may affect the cell's DNA. (**B**) UV light can cause adjacent thymine molecules to pair within the DNA strand to form a thymine dimer. »» How might a thymine dimer block the movement of RNA polymerase?

mutations may be a change to or substitution of a different base pair, or a deletion or addition of a base pair.

Base-Pair Substitutions. If a point mutation causes a base-pair substitution, then the transcription of that gene will have one incorrect base in the mRNA sequence of codons. Perhaps one way to see the effects of such changes is using an English sentence made up of three-letter words (representing codons) where one letter has been changed. As three-letter words, the letter substitution still reads correctly, but the sentence makes less sense.

Normal sequence: THE FAT CAT ATE THE RAT

Substitution: THE FAT CAR ATE THE RAT

As shown in **FIGURE 8.19A**, depending on the placement of the substituted base, when the mRNA is translated this may cause no change (**silent mutation**), lead to the insertion of the wrong amino acid (**missense mutation**), or generate a stop codon (**nonsense mutation**), prematurely terminating the polypeptide.

MICROFOCUS 8.4: Evolution
"Observing" Bacterial Evolution

Natural selection works to adapt populations to their prevailing environment. The evolutionary process generates heritable traits that make it more likely that over successive generations the organism will survive and reproduce in that particular environment. One such process for such heritable genetic variation is through random mutation. In the eukaryotic world of plants and animals, observing such heritable and evolutionary phenomena often is impossible because of the long time line between a heritable event and the subsequent evolutionary shift. However, in the bacterial world, where cells divide and populations grow at a much faster pace, modest evolutionary shifts can be "observed." One well studied example of such natural selection in bacteria is the development of antibiotic resistance.

Although generic variation does not necessarily affect survival, some heritable traits can improve the chances of survival of a particular individual. Such an example has been "observed" by Richard Lenski and his group at Michigan State University. In 1988, Lenski took one *Escherichia coli* cell and from its descendents established 12 separate laboratory populations (see diagram). These populations have continually been maintained and subcultured to where, in 2008, each population had gone through more than 44,000 generations. Over these 20 years, each population was "watched" to see what phenotypic, physiological, and biochemical changes occurred.

Through 44,000 generations, all 12 populations (grown aerobically on a glucose and citrate medium) went through millions of mutations and the patterns of changes were similar in all 12 populations. All produced larger cells, grew faster with glucose, had shorter lag phases when subcultured to a fresh medium, and remained unable to metabolize citrate (cit⁻). But then around the 31,500th generation, population 3 alone exhibited a dramatic change (green dot in figure)—the cells had gained the ability to metabolize citrate (cit⁺). How did this occur? Certainly, after 31,550 generations, most simple mutations had occurred making it likely that cit⁺ should have already occurred.

Every 500 generations, Lenski had saved samples from all 12 populations. Therefore, he went back and revived samples of each population to see if cit⁺ would again evolve and, if so, would it be from the same population 3 or from any of the other 11 populations (red dot in figure)? He discovered that only the original population 3 re-evolved cit⁺, it occurred only from generation 20,000 forward, and cit⁺ would not occur for another 11,550 generations. Lenski had replayed the original heritable evolutionary event.

The conclusion is that some genetic event occurred around generation 20,000 that formed the basis allowing for the eventual development of cit⁺ more than 10,000 generations later. Some heritable event around generation 20,000 would bring about an evolutionary shift in one population that remained unattainable in the 11 other populations (at least through more than 44,000 generations). Was it a very rare single mutational event, the last of many sequential mutations that started around generation 20,000, or some other genetic change? Lenski and his lab are now attempting to discover what that event was at 20,000 generations that set the stage for cit⁺ development years later.

Base-Pair Deletion or Insertion. Point mutations also can cause the loss or addition of a base in a gene, resulting in an inappropriate number of bases. Again, using our English sentence, we can see how a deletion or insertion of one letter affects the reading of the three-letter word sentence.

Normal sequence: THE FAT CAT ATE THE RAT

Deletion: THE F_TC ATA TET HER AT

Insertion: THE FAT ACA TAT ETH ERA T

As you can see, the "sentence mutations" are nonsense when reading the sentence as three-letter words. The same is true in a cell. Ribosomes always read three letters (one codon) at one time, generating potentially extensive mistakes in the amino acid sequence (FIGURE 8.19B). Thus, like our English sentence, the deletion or addition of a base will cause a "reading frameshift" because the ribosome always reads the genetic code in groups of three bases. Therefore, loss or addition of a base shifts the reading of the code by one base. The result is serious sequence errors in the amino acids, which will probably produce an abnormal protein (nonsense) unable to carry out its role in metabolism.

Transposable Genetic Elements Can Cause Mutations

KEY CONCEPT

15. Insertion sequences and transposons move from one DNA location to another.

Mutations of a different nature may be caused by fragments of DNA called **transposable genetic elements**. Two types are known to exist in all microbial cells: insertion sequences and transposons. **Insertion sequences (IS)** are small segments of DNA with about 1,000 base pairs. IS have no genetic information other than for the ability to insert into a chromosome; that is, they produce copies of themselves and the copies move into other areas of the chromosome. These events are rare, but when they occur they can interrupt the coding sequence in a gene, such that protein synthesis produces a nonfunctional protein, or more likely no protein at all. Thus, IS may be a prime force behind spontaneous mutations.

A second type of transposable genetic element is the **transposon**. These are the so-called "jumping genes" for which Barbara McClintock won the 1983 Nobel Prize in Physiology or Medicine (MicroFocus 8.6). Transposons are larger than IS and carry additional genes for various functions, such as antibiotic resistance, which can be conferred to the recipient cell. Like IS, they can interrupt the genetic code of a gene.

The movement of transposons appears to be nonreciprocal, meaning an element moves ("jumps") away from its location and nothing takes

MICROFOCUS 8.6: History
Jumping Genes

In the early 1950s, scientists assumed that genes were fixed elements, always found in the same position on the same chromosome. But in 1951, Barbara McClintock unveiled her research with corn plants at a symposium at Cold Spring Harbor Laboratory on Long Island, New York. McClintock described genes that apparently moved from one chromosome to another. The audience listened in respectful silence. There were no questions after her talk, and only three people requested copies of her paper.

McClintock grew Indian corn, or maize. In the 1940s, she noticed curious patterns of pigmentation on the kernels. Other scientists might have missed the patterns as random variations of nature, but McClintock's record keeping and careful analysis revealed a method to nature's madness. The pigment genes causing the splotches of color appeared to be switched on or off in particular generations. Still more remarkable, the "switches" seemed to occur at different places along the same chromosome. Some switches even showed up in different chromosomes. Such "controlling elements," as McClintock called them, were available whenever needed to turn the genes on or off.

In the modern lexicon of molecular genetics, McClintock's elements are recognized as a two-gene system. One is an activator gene, the other a dissociation gene. The activator gene, for reasons unknown, can direct a dissociation gene to "jump" along the arm of the ninth chromosome in maize plants where color is regulated. When the jumping gene reinserts itself, it turns off the neighboring pigmentation genes, thereby altering the color of the kernel. The jumping gene is identical to the transposon found in bacteria and certainly serves as a driving force in evolution.

For Barbara McClintock, recognition came 30 years after that symposium at Cold Spring Harbor. In 1981 (at the age of 79), she received eight awards, among them a $60,000-a-year lifetime grant from the MacArthur Foundation and the $15,000 Lasker prize. In 1983, she was awarded the Nobel Prize in Physiology or Medicine. When informed of the Nobel award, she humbly replied to an interviewer's question, "It seemed unfair to reward a person for having so much pleasure over the years, asking the maize plants to solve specific problems and then watching their response." Dr. McClintock died in 1992.

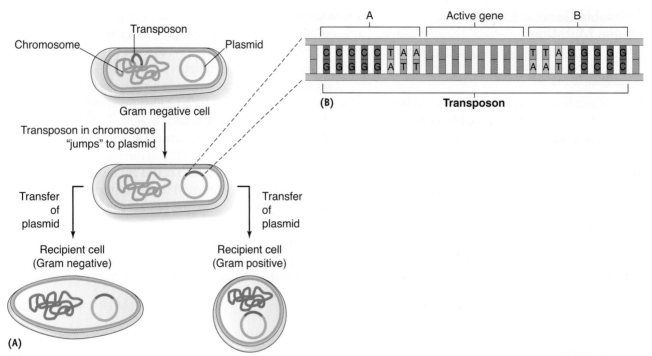

FIGURE 8.22 Transposon "Jumping" and Structure. (**A**) Transposons can "jump" to another DNA molecule, such as a plasmid, which then can be transferred to another cell conferring new genetic capabilities on the recipient. (**B**) A transposon contains one or more active genes bordered by inverted repetitive sequences. The base sequence in A is the reverse and complement of the base sequence in B. Inverted repetitive base sequences (C–G and G–C) are at the ends of the transposon. »» What is required for the transposon to "jump" to another DNA molecule?

its place. (This contrasts with insertion sequences, where copies move.) Transposons can move from plasmid to plasmid, from plasmid to chromosome, or from chromosome to plasmid. The presence of inverted repetitive base sequences at the ends of the element appears to be important in establishing the ability to move (**FIGURE 8.22**).

Of particular significance is the finding that many transposons contain genes for antibiotic resistance. For example, if a plasmid containing

a transposon is transferred from one bacterial cell to another, as plasmids are known to do, the transposon will move along with it, thus spreading the genes for antibiotic resistance. Moreover, the movement of transposons among plasmids helps explain how a single plasmid acquires numerous genes for resistance to different antibiotics.

CONCEPT AND REASONING CHECKS

8.15 What role do insertion sequences and transposons play?

8.5 Identifying Mutants

Any organism carrying a mutation is called a **mutant**, while the normal strain isolated from nature is the **wild type**. Some mutants are easy to identify because the **phenotype**, or physical appearance, of the organism or the colony has changed from the wild type. For example, some bacterial colonies appear red because they produce a red pigment. Treat the colonies with a mutagen and, after plating on nutrient agar, mutants form colorless colonies. However, not all mutants can be identified solely by their "looks."

Plating Techniques Select for Specific Mutants or Characteristics

KEY CONCEPT

16. Mutant identification can involve negative or positive selection techniques.

Selection is a very useful technique to identify and isolate a single mutant from among thousands of possible cells or colonies. Let's look at two selection techniques that make this search possible.

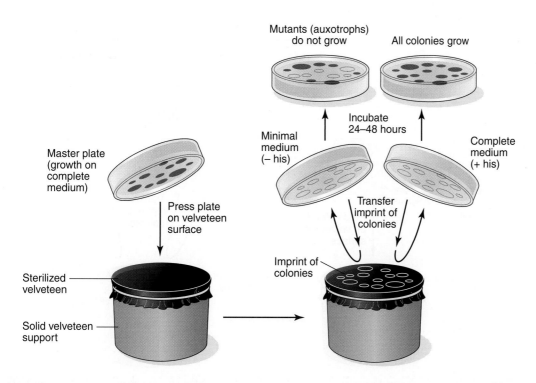

FIGURE 8.23 **Negative Selection Identifies Auxotrophs.** Negative selection plating techniques can be used to detect nutritional mutants (auxotrophs) that fail to grow when replica plated on minimal medium (in this example, a growth medium lacking histidine). Comparison to replica plating on complete medium visually identifies the auxotrophic mutants. »» In this example, which colonies transferred to the complete medium represent the auxotrophs (his⁻)?

First, in both techniques, the chemical composition of the transfer plate is key to visual identification of the colonies being hunted. The use of a replica plating device makes the identification possible. The device consists of a sterile velveteen cloth or filter paper mounted on a solid support. When an agar plate (master plate) with bacterial colonies is gently pressed against the surface of the velveteen, some cells from each colony stick to the velveteen. If another agar plate then is pressed against this velveteen cloth, some cells will be transferred (replicated) in the same pattern as on the master plate.

Now, suppose you want to find a nutritional mutant unable to grow without the amino acid histidine. This mutant (written his⁻) has lost the ability that the wild type strain (his⁺) has to make its own histidine. Such a mutant having a nutritional requirement for growth is called an **auxotroph** (*auxo* = "grow"; *troph* = "nourishment"), while the wild type is a **prototroph** (*proto* = "original"). Phenotypically, there is no difference between the two strains when they grow on a complete medium with histidine. However, you can visually identify

the auxotroph using a **negative selection** plating technique (**FIGURE 8.23**). Any colonies missing on the minimal medium plate (lacking histidine) must be his⁻.

As another example, suppose you want to see if there are any bacteria in a hospital ward that are resistant to the antibiotic tetracycline. Again, phenotypically there is no difference between those strains sensitive to tetracycline and those resistant to the antibiotic. However, a **positive selection** plating technique permits visual identification of such tetracycline resistant mutants (**FIGURE 8.24**).

CONCEPT AND REASONING CHECKS

8.16 How does negative selection differ from positive selection?

The Ames Test Can Identify Potential Mutagens

KEY CONCEPT

17. Ames test revertants suggest a chemical is a potential carcinogen in humans.

Some years ago, scientists observed that about 90% of human **carcinogens**—agents causing tumors in humans—also induce mutations in bac-

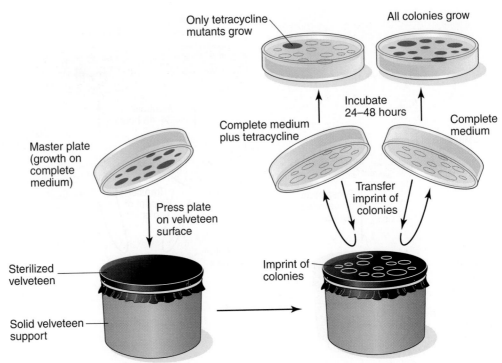

FIGURE 8.24 Positive Selection of Mutants. Positive selection plating techniques can be used to identify antibiotic resistant mutants. »» What do the "vacant spots" on the complete medium plus tetracycline represent?

Screening test:
A process for detecting mutants by examining numerous colonies.

terial cells. Working on this premise, Bruce Ames of the University of California developed a procedure to help identify potential human carcinogens by determining whether the agent can mutate bacterial auxotrophs. The procedure, called the **Ames test**, is a widely used, relatively inexpensive, rapid, and accurate screening test. For the Ames test, an auxotrophic, histidine-requiring strain (his⁻) of *Salmonella enterica* serotype Typhimurium is used. If inoculated onto a plate of nutrient medium lacking histidine, no colonies will appear because in this auxotrophic strain the gene inducing histidine synthesis is mutated and hence not active.

In preparation for the Ames test, the potential carcinogen is mixed with a liver enzyme preparation. The reason for doing this is because often chemicals only become tumor causing and mutagenic in humans after they have been modified by liver enzymes.

To perform the Ames test, the his⁻ strain is inoculated onto an agar plate lacking histidine

(**FIGURE 8.25**). A well is cut in the middle of the agar, and the potential liver-modified carcinogen is added to the well (or a filter paper disk with the chemical is placed on the agar surface). The chemical diffuses into the agar. The plate is incubated for 24 to 48 hours. If bacterial colonies appear, one may conclude the agent mutated the bacterial his⁻ gene back to the wild type (his⁺); that is, **revertants** were generated that could again encode the enzyme needed for histidine synthesis. Because the agent is a mutagen, it is therefore a possible carcinogen in humans. If bacterial colonies fail to appear, one assumes that no mutation took place. However, it is possible the mutation did occur, but was repaired by a DNA repair enzyme. This possibility has been overcome by using bacterial strains known to be inefficient at repairing errors.

CONCEPT AND REASONING CHECKS

8.17 Just because a potential carcinogen generates revertants in *Salmonella*, does that always mean it is cancer causing in humans? Explain.

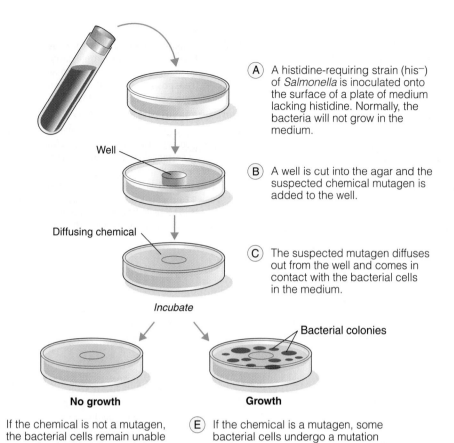

A histidine-requiring strain (his⁻) of *Salmonella* is inoculated onto the surface of a plate of medium lacking histidine. Normally, the bacteria will not grow in the medium.

Well

A well is cut into the agar and the suspected chemical mutagen is added to the well.

Diffusing chemical

The suspected mutagen diffuses out from the well and comes in contact with the bacterial cells in the medium.

Incubate

Bacterial colonies

No growth

Growth

If the chemical is not a mutagen, the bacterial cells remain unable to synthesize histidine and do not grow. Colonies fail to appear on the medium.

If the chemical is a mutagen, some bacterial cells undergo a mutation so they can again synthesize histidine. Visible colonies soon appear on the medium.

FIGURE 8.25 **Using the Ames Test.** The Ames test is a screening technique to identify mutants that reverted back to the wild type because of the presence of a mutagen. »» As a control, a plate with *Salmonella* (his⁻) but without mutagen is incubated. After 24–48 hours, a few colonies are seen on the plate. Explain this observation.

■ SUMMARY OF KEY CONCEPTS

8.1 DNA and Chromosomes

1. The bacterial and archaeal chromosomes are circular, haploid structures in the cell's nucleoid.

2. The DNA of a microorganism's **chromosome** is **supercoiled** and folded into a series of **looped domains**. Each loop consists of 10,000 bases.

3. Many microbial cells may have one or more **plasmids**. These small closed loops of DNA carry information that can confer selective advantages (e.g., antibiotic resistance, protein toxins) to the cells.

8.2 DNA Replication

4. DNA replicates by a **semiconservative** mechanism where each strand of the original DNA molecule acts as a template to synthesize a new strand. DNA replication starts with initiator proteins binding at the **replication origin**, forming two replication "factories." DNA polymerase III moves along the strands inserting the correct DNA nucleotide to complementary bind with the template strand.

5. At each **replication fork**, one of the two strands is synthesized in a continuous fashion while the other strand is formed in a discontinuous fashion, forming **Okazaki fragments**, which are joined by a **DNA ligase**.

8.3 Protein Synthesis

6. **Transcription** occurs when the **RNA polymerase** binds to a **promoter** sequence on the DNA template strand. Various forms of RNA, including **mRNA**, **tRNA**, and **rRNA**, are transcribed from the DNA and are important in the translation process.

7. The **genetic code**, a series of three-letter words to specify a polypeptide, is redundant because often more than one **codon** can specify an amino acid.

8. **Translation** occurs on ribosomes, which bring together mRNA and tRNAs. The ribosome "reads" the mRNA codons and inserts the correct tRNA to match codon and **anticodon**. Translation continues until the ribosome encounters a stop codon when the protein translation machinery disassembles and the protein is released. **Chaperone** proteins help the elongating polypeptide to fold properly during translation. A **polyribosome** is a string of ribosomes all translating the same mRNA.

9. Many antibiotics interfere with protein synthesis by binding to RNA polymerase, or to either the small or large ribosomal subunit.

10. Different control factors influence protein synthesis. The best understood is the bacterial **operon** model where binding of a repressor protein to the operon represses transcription.

11. Research studies suggest that at least in some bacterial species, transcription and translation are spatially separated.

8.4 Mutations

12. **Mutations** are permanent changes in the cellular DNA. This can occur spontaneously in nature resulting from a replication error or the effects of natural radiation. In the laboratory, physical and chemical mutagens can induce mutations.

13. Base pairs in the DNA can change in one of two ways. A base-pair substitution does not change the reading frame of an mRNA but can result in a **silent**, **missense**, or **nonsense mutation**. A **point mutation** also can occur from the loss or gain of a base pair. Such mutations change the reading frame and often lead to loss of protein function.

14. Replication errors or other damage done to the DNA often can be repaired. **Mismatch repair** replaces an incorrectly matched base pair with the correct pair. **Excision repair** removes a section of damaged DNA and replaces it with the correctly paired bases.

15. **Transposable genetic elements** exist in many microbial cells. **Insertion sequences** only carry information to copy the sequences and insert them into another location in the DNA. **Transposons** are genetic elements that "jump" from one location to another in the DNA.

8.5 Identifying Mutants

16. Auxotrophic mutants can be identified by **negative selection** plating techniques. **Positive selection** can be used to identify mutants having certain attributes, such as antibiotic resistance.

17. The **Ames test** is a method of using an auxotrophic bacterial species to identify mutagens that may be carcinogens in humans. The test is based on the ability of a potential mutagen to revert an auxotrophic mutant to its prototrophic form.

■ LEARNING OBJECTIVES

After understanding the textbook reading, you should be capable of writing a paragraph that includes the appropriate terms and pertinent information to answer the objective.

1. Describe the contents of the bacterial nucleoid.

2. Summarize the packing of the DNA and **chromosome** within the nucleoid.

3. Assess the role of **plasmids** in microbial cells.

4. Identify and explain the events of the three phases of **DNA replication**.

5. Explain how DNA synthesis occurs at a **replication factory**.

6. Describe the role of **RNA polymerase** in the **transcription** process.

7. Read the **genetic code** and identify **start** and **stop codons**.

8. Identify and describe the events occurring in the three stages of **translation**.

9. Evaluate the effect of antibiotics on the protein synthesis process.

10. Label the sequences composing a bacterial **operon** and compare the functions of each sequence to the transcription process.

11. Identify the spatial relationships between transcription and translation with the nucleoid and cytoplasm.

12. Compare and contrast **spontaneous** and **induced mutations**, and differentiate between physical and chemical **mutagens**.

13. Distinguish between a **base-pair substitution** and a **deletion** (or **insertion**) mutation.

14. Explain how **mismatch repair** works, and describe how cells use **excision repair** to correct UV-induced mutations.

15. Contrast how **insertion sequences** and **transposons** can cause **mutations**.

16. Compare and contrast **negative** and **positive selection** plating techniques for identifying mutants.

17. Evaluate the use of the **Ames test** to identify chemicals that are potential **carcinogens** in humans.

STEP A: SELF-TEST

Each of the following questions is designed to assess your ability to remember or recall factual or conceptual knowledge related to this chapter. Read each question carefully, then select the **one** answer that best fits the question or statement. Answers to even-numbered questions can be found in **Appendix C**.

1. Which one of the following statements is NOT true of the bacterial chromosome?
   A. It is located in the nucleoid.
   B. It usually is a single, circular molecule.
   C. Some genes are dominant to others.
   D. It usually is haploid.

2. DNA compaction involves
   A. a twisting and packing of the DNA.
   B. supercoiling.
   C. the formation of looped domains.
   D. All the above (**A–C**) are correct.

3. Plasmids are
   A. another name for transposons.
   B. accessory genetic information.
   C. domains within a chromosome.
   D. daughter chromosomes.

4. The enzyme _____ adds complementary bases to the DNA template strand during replication.
   A. ligase
   B. helicase
   C. DNA polymerase III
   D. RNA polymerase

5. At a chromosome replication fork, the lagging strand consists of _____ that are joined by _____.
   A. RNA sequences; DNA ligase
   B. Okazaki fragments; RNA polymerase
   C. RNA sequences; ribosomes
   D. Okazaki fragments; DNA ligase

6. In a eukaryotic microbe, those sections of a primary RNA transcript that will NOT be translated are called
   A. introns.
   B. anticodons.
   C. "jumping genes."
   D. exons.

7. Which one of the following codons would terminate translation?
   A. AUG
   B. UUU
   C. UAA
   D. UGG

8. The translation of a mRNA by multiple ribosomes is called _____ formation.
   A. Okazaki
   B. polysome
   C. plasmid
   D. transposon

9. If an antibiotic binds to a 50S subunit, what cellular process will be inhibited?
   A. DNA replication
   B. Intron excision
   C. Translation
   D. Transcription

10. Which one of the following is NOT part of an operon?
    A. Regulatory gene
    B. Operator
    C. Promoter
    D. Structural genes

11. Being compartmentalized, bacterial RNA polymerases are localized in the _____ and ribosomes are found _____.
    A. nucleoid; at the nucleoid periphery
    B. cytosol; in the cytosol
    C. cytosol; at the cell poles
    D. nucleoid; in the nucleoid

12. Spontaneous mutations could arise from
    A. DNA replication errors.
    B. atmospheric radiation.
    C. addition of insertion sequences.
    D. All the above (**A–C**) are correct.

13. Which one of the following could NOT cause a change in the mRNA "reading frame"?
    A. Insertion sequence
    B. Base-pair substitution
    C. Base addition
    D. Base deletion

14. Excision repair would correct DNA damage caused by
    A. antibiotics.
    B. UV light.
    C. a chemical mutagen.
    D. a DNA replication error.

15. Transposable genetic elements (transposons)
    A. were first discovered by Watson and Crick.
    B. are smaller than insertion sequences.
    C. are examples of plasmids.
    D. may have information for antibiotic resistance.

16. Nutritional mutants are referred to as
    A. prototrophs.
    B. wild type.
    C. revertants.
    D. auxotrophs.

17. The Ames test is used to
    A. identify potential human carcinogens.
    B. discover auxotrophic mutants.
    C. find pathogenic bacterial species.
    D. identify antibiotic resistant mutants.

STEP B: REVIEW

18. Construct a concept map for **translation** using the following terms.

amino acid	ribosome
chain elongation	sense codons
chain initiation	small subunit
chain termination	start codon
large subunit	stop codon
mRNA	termination factors
polypeptide	tRNAs

Answer the following questions that pertain to (1) transcription and translation, and (2) mutations. Use the genetic code (Table 8.3) on page 236 as needed. Answers to even-numbered questions can be found in **Appendix C**.

19. The following base sequence is a complete polynucleotide made in a bacterial cell.

AUGGCGAUAGUUAAACCCGGAGGGUGA

With this sequence, answer the following questions.

A. Provide the sequence of nucleotide bases found in the inactive DNA strand of the gene.

B. How many codons will be transcribed in the mRNA made from the template DNA strand?

C. How many amino acids are coded by the mRNA made and what are the specific amino acids?

D. Why isn't the number of codons in the template DNA the same as the number of amino acids in the polypeptide?

20. Use the base sequence to answer the following questions about mutations.

TACACGATGGTTTTGAAGTTACGTATT

A. Is the sequence above a single strand of DNA or RNA? Why?

B. Using the sequence above, show the translation result if a mutation results in a C replacing the T at base 12 from the left end of the sequence. Is this an example of a silent, missense, or nonsense mutation?

C. Using the sequence above, show the translation result if a mutation results in an A inserted between the T (base 12) and the T (base 13) from the left end of the sequence. Is this an example of a silent, missense, or nonsense mutation?

STEP C: APPLICATIONS

Answers to even-numbered questions can be found in **Appendix C**.

21. You are interested in identifying mutants of *E. coli*; specifically mutations occurring in the promoter and operator regions of the *lac* operon such that their respective molecules cannot bind to the region. You have agar plates containing lactose or glucose as the energy source. How can plating the potential mutants on these media help you identify (a) promoter region mutants and (b) operator region mutants?

22. A chemical is tested with the Ames test to see if the chemical is mutagenic and therefore possibly a tumor-causing chemical in humans. On the test plate containing the chemical, no his+ colonies are seen near the central well. However, many colonies are growing some distance from the well. If these colonies truly represent his+ colonies, why are there no colonies closer to the central well?

23. Bioremediation is a process that uses bacteria to degrade environmental pollutants. You want to use one of these organisms to clean up a toxic waste site (biochemical refinery) that contains benzene in the soil around the refinery. Benzene (molecular formula = C_6H_6) is a major contaminant found around many of these chemical refineries. You have a culture of bacterial cells growing on a culture plate that were derived from a soil sample from the refinery area. You also have a supply of benzene. Explain how you could visually identify chemoheterotrophic bacterial colonies on agar that use benzene as their sole carbon and energy source for metabolism.

24. Suppose you now have such benzene colonies growing on agar. However, it also has been discovered that material containing radioactive phosphorus (^{32}P) is in the soil around the refinery and this radioactive material can kill bacterial organisms. Because you want to identify colonies that might be sensitive to ^{32}P, you obtain a sample of the material containing ^{32}P. Explain how you could visually determine if any of your colonies are sensitive to ^{32}P.

STEP D: QUESTIONS FOR THOUGHT AND DISCUSSION

Answers to even-numbered questions can be found in **Appendix C**.

25. The author of a general biology textbook writes in reference to the development of antibiotic resistance, "The speed at which bacteria reproduce ensures that sooner or later a mutant bacterium will appear that is able to resist the poison." How might this mutant bacterial cell appear? Do you agree with the statement? Does this bode ill for the future use of antibiotics?

26. Many viruses have double-stranded DNA as their genetic information while many others have single-stranded RNA as the genetic material. Which group of viruses do you believe is more likely to efficiently repair its genetic material? Explain.

27. Some scientists suggest that mutation is the single most important event in evolution. Do you agree? Why or why not?

HTTP://MICROBIOLOGY.JBPUB.COM/9E

The site features eLearning, an online review area that provides quizzes and other tools to help you study for your class. You can also follow useful links for in-depth information, read more MicroFocus stories, or just find out the latest microbiology news.

9

Gene Transfer, Genetic Engineering, and Genomics

Genetic engineering is the most powerful and awesome skill acquired by man since the splitting of the atom.
—The editors of *Time* magazine describing the potential for genetic engineering

Medicinal microbe's genome sequenced. How often have we read in the newspaper or heard from the news media about an organism's genes (genome) being sequenced or its DNA being mapped? It must be significant, right? After all, it made the news! But what is the underlying significance of such sequencing? Here is a good example.

In late spring of 2003, a group of British scientists announced they had mapped the genome of a very important bacterial species, *Streptomyces coelicolor*. This is a gram-positive organism commonly found in the soil. The mapping project began in 1997 and took six years to complete partly because the organism's genome is one of the largest ever sequenced. It has 8.6 million base pairs and some 7,825 genes. On completion of the sequencing, one of the scientists on the project said, "*It is a fabulous resource for scientists.*" Why?

Well, here is where microbial genomics shows its power. *S. coelicolor* belongs to a phylum of *Bacteria*, the Actinobacteria, which are responsible for producing over 65% of the naturally-known antibiotics used today (see Chapter 4). This includes tetracycline and erythromycin. By analyzing the genome of *S. coelicolor* and other *Streptomyces* species, additional metabolic pathways may be discovered for the production of other, yet unidentified and perhaps novel antibiotics. In fact, the researchers have identified 18 gene clusters they suspect are involved with the production of antibiotics. If correct, knowing the genome and its organization might allow scientists to transform the organism into an "antibiotic factory" and add to the dwindling armada of usable antibiotics to which pathogens are not yet resistant (FIGURE 9.1). Using genetic engineering techniques, they could rearrange gene clusters and perhaps produce even more useful and potent antibiotics than naturally possible.

One example illustrating the need for newer antibiotics concerns "staph infections," most commonly caused by methicillin-resistant *Staphylococcus aureus* (MRSA). These pathogens are resistant to methicillin and other common antibiotics, such as oxacillin, penicillin, and amoxicillin. MRSA infections occur most frequently among persons in hospitals and healthcare facilities who have weakened immune systems (TABLE 9.1). According to MRSAInfection.org, "A patient with a hospital acquired [MRSA] infection is about 7 times more likely to die [in the hospital] than an uninfected patient." So, new and unique antibiotics might be able to inhibit or kill the drug resistant *S. aureus*.

But there is more. *S. coelicolor* is a close relative of the tuberculosis, leprosy, and diphtheria bacilli. By comparing genomes, scientists hope to learn why *S. coelicolor* are not pathogenic, while the other three are pathogens. What is different about their genomes might be important in understanding the pathogenicity of their relatives and perhaps even designing new antibiotics through genetic engineering to attack these pathogens.

Genetic engineering involves the manipulating of genes in organisms or between organisms in order to introduce new characteristics into the recipient to either produce a useful product or to actually generate **genetically modified organisms** (**GMOs**). **Genomics** is the study of an organism's genome; its study has the potential of offering new therapeutic methods for the treatment of several human diseases. So, this chapter addresses more than simply research techniques; rather, we discuss how their applications can have far-reaching consequences for all of us.

Before we can explore these topics, we need to understand the process of **genetic recombination**, a natural mechanism for DNA transfer from one microorganism to another. Its understanding provides a unique perspective on microbial evolution, ecology, and molecular biology while providing insights for the techniques of genetic engineering and the field of microbial genomics.

FIGURE 9.1 *Streptomyces coelicolor* **Colonies.** In this photograph of *S. coelicolor*, colonies growing on agar are secreting droplets of liquid containing antibiotics. »» What is the advantage to the bacterial cells to secrete a chemical with antibiotic properties?

TABLE 9.1	Hospital-Acquired Infection Statistics, Pennsylvania—2004*	
Number of hospital-acquired infections		11,668
Number of deaths associated with hospital-acquired infections		1,510
Extra number of hospital days associated with these infections		205,000
Additional hospital charges		$2 billion

*Data are from: MRSAInfection.org

9.1   Genetic Recombination in *Bacteria*

Traditionally, when one thinks about the inheritance of genetic information, one envisions genes passed from parent to offspring. However, imagine being able to transfer genes between members of your own family, or between you and one of your classmates. Although many bacterial and archaeal species are accomplished at doing both types of information transfer, we will focus on bacterial species in this section.

Genetic Information Can Be Transferred Vertically and Horizontally

KEY CONCEPT

1. Gene transfer can occur between generations and between individual organisms.

In Chapter 8, we discussed mutations, which were one of the ways by which the genetic material in a cell can be permanently altered. Because the permanent change occurred in the parent cell, all future generations derived by binary fission from the parent also will have the mutation. This form of genetic transfer is referred to as **vertical gene transfer** (FIGURE 9.2A).

The *Bacteria* lack sexual reproduction as a mechanism for genetic diversity. However, they still possess a process by which genetic recombination and diversity can arise. This is through the process of **horizontal gene transfer** (**HGT**), a type of genetic recombination that involves the lateral intercellular transfer of DNA from a donor cell to a recipient cell (FIGURE 9.2B). If, for example, the recipient cell receives from the donor a chromosomal DNA fragment, a plasmid, a transposon (see Chapter 8), or some combination of these elements containing a gene for antibiotic resistance, the new DNA pairs with a complementary region of recipient DNA and replaces it. In this case, there is no change in quantity of the recipient's chromosomal DNA,

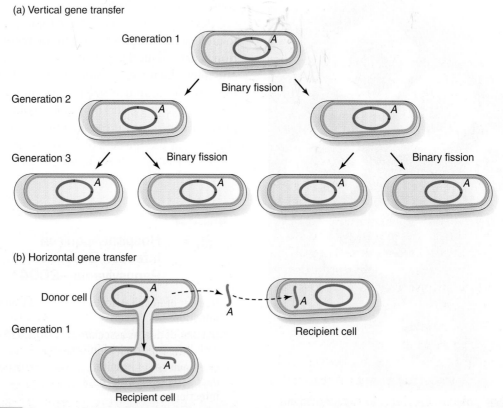

(a) Vertical gene transfer

Generation 1

Binary fission

Generation 2

Generation 3   Binary fission   Binary fission

(b) Horizontal gene transfer

Donor cell

Generation 1

Recipient cell

Recipient cell

FIGURE 9.2 Gene Transfer Mechanisms. Genes can be transferred between cells in two ways. (**A**) In vertical transfer, a cell undergoes binary fission and the daughter cells of the next generation contain the identical genes found in the parent. (**B**) In horizontal transfer, genes are transferred by various mechanisms to other individual cells. »» Determine which transfer mechanism (vertical or horizontal) provides the potential for the most genetic diversity.

MICROFOCUS 9.1: Environmental Microbiology
Gene Swapping in the World's Oceans

Many of us are familiar with the accounts of microorganisms in and around us, but we are less familiar with the massive numbers of microbes in the world's oceans. For example, microbial ecologists estimate there are an estimated 10^{29} *Bacteria* and *Archaea* in the world's oceans. At the Axial Seamount, a Pacific deep-sea volcano, researchers have discovered some 3,000 archaeal species and more than 37,000 different bacterial species. Also, there are some 10^{30} viruses called bacteriophages ("bacteria eaters") in the oceans that infect these oceanic microbes.

In the infection process, sometimes bacteriophages by mistake carry pieces of the bacterial chromosome (rather than viral DNA) from the infected cell to another recipient cell. In the recipient cell, the new DNA fragment can be swapped for an existing part of the recipient's chromosome. It is a fairly rare event, occurring only once in every 100 million (10^8) virus infections. That doesn't seem very significant until you now consider the number of bacteriophages and susceptible bacteria existing in the oceans. Working with these numbers and the potential number of virus infections, scientists suggest that if only one in every 100 million infections brings a fragment of DNA to a recipient cell, there are about 10 million billion (that's 10,000,000,000,000,000 or 10^{16}) such gene transfers per second in the world's oceans. That is about 10^{21} infections per day!

We do not understand what all this recombination means. What we can conclude is there's an awful lot of gene swapping going on!

---

but there is a substantial change in its quality and cell physiology—an antibiotic sensitive cell has become resistant. In fact, the increasing global resistance to antibiotics by human pathogens, such as *Streptococcus pneumoniae*, *Staphylococcus aureus*, and *Pseudomonas aeruginosa*, is one example of the prevalence of HGT. MICROFOCUS 9.1 provides another spectacular example for a very extensive rate for genetic recombination through HGT in the world's oceans.

Three distinctive mechanisms mediate the horizontal transfer of DNA between bacterial cells: transformation, conjugation, and transduction. All three processes involve a similar four steps. The donor DNA must be:

1. Readied for transfer.
2. Transferred to the recipient cell.
3. Taken up successfully by the recipient cell.
4. Incorporated in a stable state in the recipient.

Let's now look at each of the three recombination mechanisms involving HGT.

CONCEPT AND REASONING CHECKS

9.1 How does HGT differ from vertical gene transfer?

Transformation Is the Uptake and Expression of DNA in a Recipient Cell

KEY CONCEPT

2. Transformation involves the transfer of exposed DNA fragments from donor to recipient.

Transformation is the uptake of a free DNA fragment from the surrounding environment and the expression of the genetic information in the recipient cell; that is, by integration of the DNA fragment, the recipient (transformant) has gained some ability it previously lacked. Today, transformation represents a gene transfer process building genetic diversity.

This form of HGT was first described in 1928, when the English bacteriologist Frederick Griffith published the results of an interesting set of experiments with *S. pneumoniae*. This bacterial species, referred to as a pneumococcus (pl., pneumococci), is a major cause of pneumonia (Chapter 10). Pneumococci occur in two different strains: a wild-type encapsulated strain, designated S, because the organisms grow in smooth colonies and cause pneumonia; and an unencapsulated mutant strain,

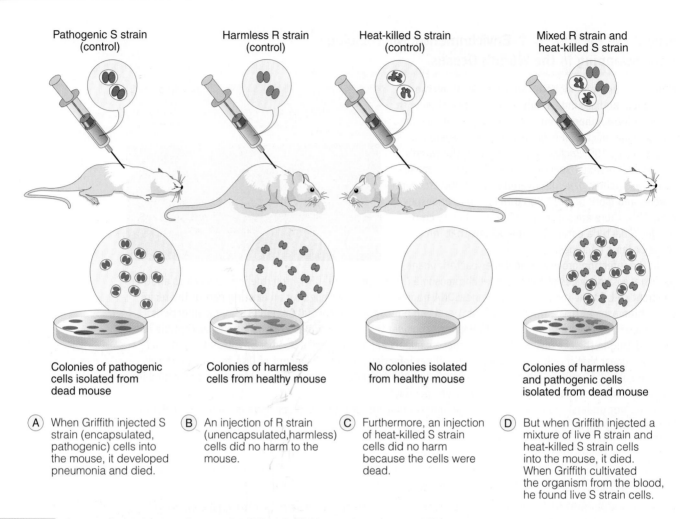

Pathogenic S strain (control)

Harmless R strain (control)

Heat-killed S strain (control)

Mixed R strain and heat-killed S strain

Colonies of pathogenic cells isolated from dead mouse

Colonies of harmless cells from healthy mouse

No colonies isolated from healthy mouse

Colonies of harmless and pathogenic cells isolated from dead mouse

(A) When Griffith injected S strain (encapsulated, pathogenic) cells into the mouse, it developed pneumonia and died.

(B) An injection of R strain (unencapsulated, harmless) cells did no harm to the mouse.

(C) Furthermore, an injection of heat-killed S strain cells did no harm because the cells were dead.

(D) But when Griffith injected a mixture of live R strain and heat-killed S strain cells into the mouse, it died. When Griffith cultivated the organism from the blood, he found live S strain cells.

FIGURE 9.3 **The Transformation Experiments of Griffith.** Griffith's experiments were the first to demonstrate transformation and the horizontal transfer of genetic information. »» What is unique about the S strain pneumococci that is responsible, in part, for making them pathogenic?

designated R, because the colonies appear rough and are harmless. Griffith showed that mice injected with living S strain die, while those mice injected with living R strain or heat-killed S strain survive (FIGURE 9.3).

What happened next surprised Griffith. He mixed heat-killed, S strain cells with live R strain cells (both of which were non-lethal) and let the mixture incubate; then he injected the mixture into mice. The mice died! Griffith wondered how a mixture of live, harmless R strain cells and debris from the dead pathogenic S strain cells could kill the mice. His answer came when he autopsied the animals: Microscopic examination of their blood showed the presence of live S strain pneumococci. Knowing that spontaneous generation does not occur, Griffith reasoned

that somehow the live R strain cells had been transformed into live S strain cells. In 1944 Oswald T. Avery and his associates Colin M. MacLeod and Maclyn N. McCarty of the Rockefeller Institute, purified and identified the transforming substance; it was DNA.

Modern scientists regard transformation as an important genetic recombination method even though it takes place in less than 1% of a cell population. Transformation occurs regularly when bacterial cells exist in crowded conditions, such as in rich soil or the human intestinal tract. Under natural conditions, about 40 known bacterial species are highly transformable when their DNA is very similar to the DNA being received.

The ability of a cell to be transformed depends on its **competence**, which refers to the ability of a recipient cell to take up extracellular DNA from the environment. Competence is an intriguing but variable property among bacterial species. For example, growing cells of *S. pneumoniae* secrete a competence factor to induce the competence state, while *Haemophilus influenzae* cells become competent when the culture is switched from a rich to a minimal growth medium. In both cases, several genes encode proteins for binding and uptake of DNA fragments.

The transformation process occurs as follows. In natural environments, when bacterial cells die and lyse, the chromosome typically breaks apart into fragments of DNA composed of about 10 genes to 20 genes (**FIGURE 9.4**). Uptake of DNA fragments (or a plasmid) by recipient cells appears to occur only at the cell poles, as this is where the competence factors are found and actual DNA uptake has been observed.

A competent cell usually incorporates only one or at most a few DNA fragments. Internalization of DNA is an ATP-dependent process and requires DNA-binding proteins, cell wall degradation proteins, and cell membrane transport proteins. Gram-positive bacterial species, such as *S. pneumoniae* and *Bacillus subtilis*, degrade one strand of the DNA fragment as it is being taken into the cell. Such single-stranded DNA, if stably associated with a similar region of the recipient's chromosome, will replace a similar chromosome sequence. In gram-negative *H. influenzae,* a double-stranded DNA fragment is transported into the periplasmic space, but then one strand is digested by a nuclease before transport into the bacterial cytoplasm.

One potential result of transformation is to increase an organism's ability to cause disease. In Griffith's pneumococci experiments, for example, the live R strain cells acquired the genes for capsule formation from the dead S strains, which allowed the organism to avoid body defenses and thus cause disease (see Chapter 4). Microbiologists also have demonstrated that when mildly pathogenic bacterial strains take up DNA from other mildly pathogenic strains, there is a cumulative effect, and the recipient becomes more dangerous. Observations such as these may help explain why highly pathogenic bacterial strains appear from time to time. Transformed bacterial cells also may display enhanced drug resistance from the

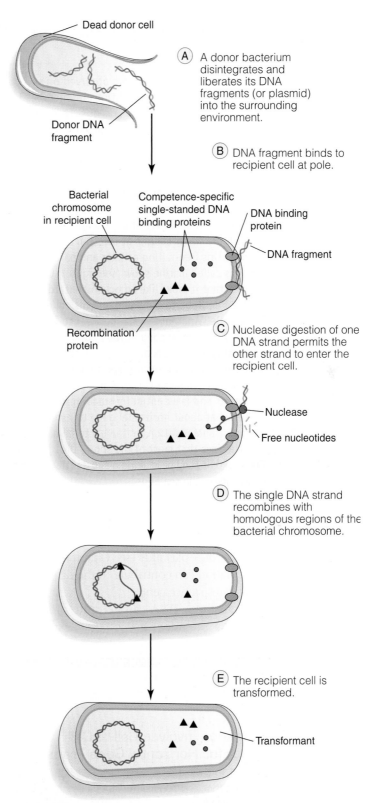

FIGURE 9.4 **Transformation in a Gram-Positive Cell.** Transformation is the process in which DNA fragments from the environment bind to a competent recipient cell, pass into the recipient, and incorporate into the recipient's chromosome. »» Using this diagram, illustrate how Griffith's experiment with mixed heat-killed S strain and live R stain pneumococci produced live S strain encapsulated cells.

MICROFOCUS 9.2: Environmental Microbiology

It's Snowing DNA!

Throughout this text, we have observed on several occasions the massive populations of microbes that live and thrive in the world's oceans. These marine environments are home to a huge variety of *Bacteria* and *Archaea*. So what happens to all the genetic material in these organisms when they die? Where does the DNA go?

A decade's worth of data suggest that as cyanobacteria and other eukaryotic phytoplankton die in the surface layers of the oceans, they sink through the water column at rates up to 100 meters per day. This free-falling cellular and particulate debris is called marine snow (see figure). Often it is so thick that it looks like an ocean blizzard!

In 2005, Roberto Danovaro and his colleague Antonio Dell'Anno of the Marine Science Institute of the University of Ancona in Italy published a paper

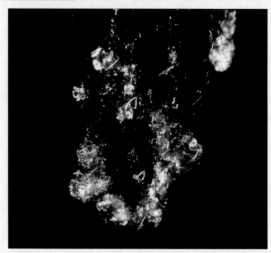

Marine snow represents microbial cellular debris.

suggesting about 65% of the DNA in the world's oceans is found in the sea-bed sediments; 90% of that DNA is extracellular, originating from organisms in the marine snow.

Danovaro suggests this DNA is a primary source for carbon, nitrogen, and phosphorus for sea-bed organisms and is essential for sustaining the microbial communities on and in the sediments.

But what about horizontal gene transfer? Could this DNA represent a source of genetic diversity through transformation for the bacterial and archaeal populations on the ocean bottom? Danovaro hypothesizes that most of the snowy DNA is of eukaryotic origin. Still, with all the marine cyanobacteria, why couldn't they be making a substantial contribution? And if so, is transformation a significant phenomenon on the sea beds?

---

acquisition of R plasmids (see Chapter 8). Thus, transformation can contribute to the dispersal of antibiotic-resistance genes.

Extracellular DNA can be quite prevalent. MICROFOCUS 9.2 identifies a massive source of such "dead" DNA.

CONCEPT AND REASONING CHECKS

9.2 Why is competence key to the transformation process?

Conjugation Involves Cell-to-Cell Contact for Horizontal Gene Transfer

KEY CONCEPT

3. Conjugation uses plasmids for DNA transfer from donor to recipient.

In the recombination process called **conjugation**, two live bacterial cells come together and the donor cell directly transfers DNA to the recipient cell. This process was first observed in 1946 by

Joshua Lederberg and Edward Tatum in a series of experiments with *Escherichia coli*. Lederberg and Tatum mixed two different strains of *E. coli* and found genetic traits could be transferred if contact occurred.

The process of conjugation requires a special conjugation apparatus called the **conjugation pilus** that was first described in Chapter 4 (**FIGURE 9.5**). For cell-to-cell contact, the donor cell, designated F+, produces the conjugation pilus that contacts the recipient cell, known as an F– cell. The donor cell is called F+ because it contains an **F factor**, which is a plasmid containing about 100 genes, most of which are associated with plasmid DNA replication and production of the conjugation pilus. The F– cell lacks the plasmid, and the pilus shortens to bring the two cells close together.

Following pair formation (**FIGURE 9.6A**), the F factor DNA replicates by a **rolling-circle mech-**

anism; one strand of plasmid DNA remains in a closed loop, while an enzyme nicks the other strand at a point called the **origin of transfer** (*ori*T). This single-stranded DNA then "rolls off" the loop and passes through the translocation channel to the recipient cell; the transfer takes about five minutes. As the horizontal transfer occurs, DNA synthesis in the donor cell produces a new complementary strand to replace the transferred strand. Once DNA transfer is complete, the two cells separate.

In the recipient cell, the new single-stranded DNA serves as a template for synthesis of a complementary polynucleotide strand, which then circularizes to reform an F factor. This completes the conversion of the recipient from F⁻ to F⁺ and this cell now represents a donor cell (F⁺) capable of conjugating with another F⁻ recipient. Transfer of the F factor does not involve the chromosome; therefore, the recipient does not acquire new genes other than those on the F factor.

The high efficiency of DNA transfer by conjugation shows that conjugative plasmids can spread rapidly, converting a whole population into plasmid-containing cells. Indeed, conjugation appears to be the major mechanism for antibiotic resistance transfer. In laboratory experiments, for example, bacterial strains carrying plasmid antibiotic-resistance genes were introduced into mice. HGT through conjugation of these R factors rapidly occurred. In nature, conjugation readily occurs in soil and in water.

CONCEPT AND REASONING CHECKS

9.3 What genes must be transferred to an F⁻ cell to convert it to F⁺?

Conjugation Also Can Transfer Chromosomal DNA

KEY CONCEPT

4. To transfer chromosomal DNA, conjugation requires chromosomally integrated plasmids.

Bacterial species also can undergo a type of conjugation that accounts for the horizontal transfer of chromosomal material from donor to recipient cell. Cells exhibiting the ability to donate chromosomal genes are called **high frequency of recombination (Hfr)** strains.

In Hfr strains, the F factor has attached to the chromosome (**FIGURE 9.6B**). This attachment

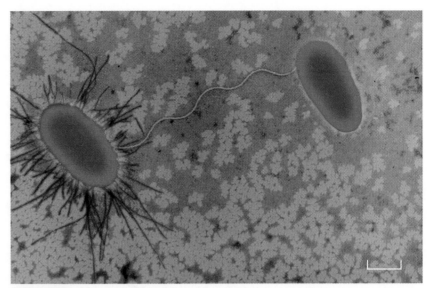

FIGURE 9.5 Bacterial Conjugation in *E. coli*. The direct transfer of DNA between live donor (F⁺; left cell) and recipient (F⁻) cells requires a conjugation bridge. In this false-color transmission electron micrograph, the F⁺ cell has produced a conjugation pilus that has contacted the F⁻ cell. Contraction of the pilus will bring donor and recipient close together. (Bar = 0.5 μm.) »» What are the other structures projecting from the F⁺ cell?

is a rare event requiring an insertion sequence to recognize the F factor. Once incorporated into the chromosomal DNA, the F factor no longer controls its own replication. However, the Hfr cell triggers conjugation just like an F⁺ cell. When a recipient cell is present, a conjugation pilus forms and attaches to the F⁻ cell. The two cells are brought together. One strand of the donor chromosome is nicked at *ori*T and a portion of the single-stranded chromosomal/plasmid DNA then passes into the recipient cell.

The *ori*T site in the chromosome is in the middle of the F plasmid genes. Therefore, in the recipient cell, the first genes to enter are only a part of the F factor and these genes are not the ones to make the cell an F⁺ donor. Rather, the last genes transferred to the recipient cell would control the donor state. However, these rarely enter the recipient because conjugation usually is interrupted by movements that break the bridge between cells before complete transfer is accomplished. An estimated 100 minutes is required for the transfer of a complete *E. coli* chromosome with plasmid genes—something rarely occurring in nature. Thus, the F⁻ cell usually remains a recipient, although it now has some recombined chromosomal genes, and is referred to as a **recombinant F⁻** cell.

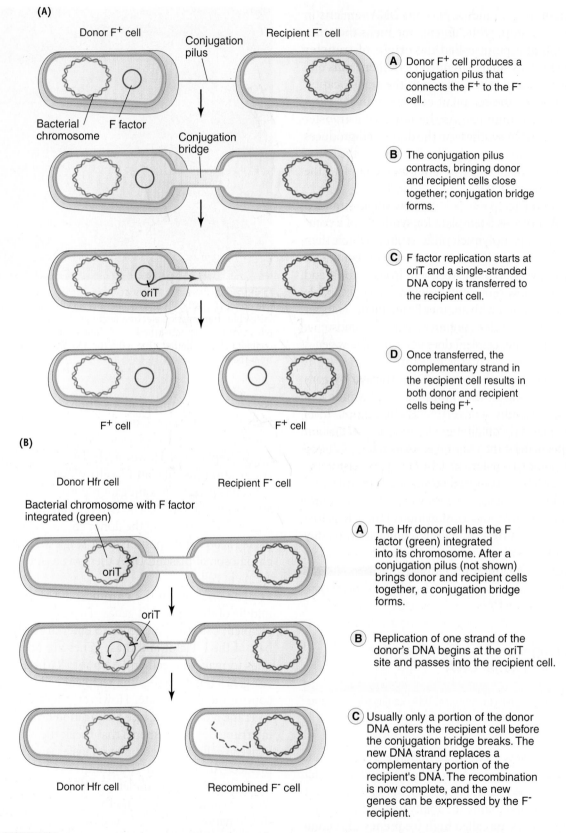

(A)

Donor F⁺ cell

Conjugation pilus

Recipient F⁻ cell

Bacterial chromosome   F factor

Conjugation bridge

oriT

F⁺ cell   F⁺ cell

(A) Donor F⁺ cell produces a conjugation pilus that connects the F⁺ to the F⁻ cell.

(B) The conjugation pilus contracts, bringing donor and recipient cells close together; conjugation bridge forms.

(C) F factor replication starts at oriT and a single-stranded DNA copy is transferred to the recipient cell.

(D) Once transferred, the complementary strand in the recipient cell results in both donor and recipient cells being F⁺.

(B)

Donor Hfr cell   Recipient F⁻ cell

Bacterial chromosome with F factor integrated (green)

oriT

oriT

Donor Hfr cell   Recombined F⁻ cell

(A) The Hfr donor cell has the F factor (green) integrated into its chromosome. After a conjugation pilus (not shown) brings donor and recipient cells together, a conjugation bridge forms.

(B) Replication of one strand of the donor's DNA begins at the oriT site and passes into the recipient cell.

(C) Usually only a portion of the donor DNA enters the recipient cell before the conjugation bridge breaks. The new DNA strand replaces a complementary portion of the recipient's DNA. The recombination is now complete, and the new genes can be expressed by the F⁻ recipient.

FIGURE 9.6 **Conjugation.** **(A)** Conjugation between an F⁺ cell and an F⁻ cell. When the F factor is transferred from a donor (F⁺) cell to a recipient (F⁻) cell, the F⁻ cell becomes an F⁺ cell as a result of the presence of an F factor. **(B)** Conjugation between an Hfr and an F⁻ cell allows for the transfer of some chromosomal DNA from donor to recipient cell. »» Propose a hypothesis to explain why only a single-stranded DNA molecule is transferred across the conjugation pilus.

Should the entire chromosome be transferred to the recipient, the F factor usually detaches from the chromosome, and enzymes synthesize a strand of complementary DNA. The F factor now forms a loop to assume an existence as a plasmid, and the recipient becomes a donor (F+) cell.

Occasionally, in an Hfr cell, the integrated F plasmid breaks free from the chromosome and, in the process, takes along a fragment of chromosomal DNA. The plasmid with its extra DNA is now called an **F′ plasmid** (pronounced "F-prime"). When the F′ plasmid is transferred during a subsequent conjugation, the recipient acquires those chromosomal genes excised from the donor. This process results in a recipient having its own genes for a particular process as well as additional genes from the plasmid DNA for the same process. In the genetic sense, the recipient is a partially diploid organism because there are two genes for a given function.

Conjugation has been demonstrated to occur between cells of various bacterial genera. For example, conjugation occurs between such gram-negative genera as *Escherichia* and *Shigella, Salmonella* and *Serratia,* and *Escherichia* and *Salmonella.* HGT has great significance because of the possible transfer of antibiotic-resistance genes carried on plasmids. Moreover, when the genes are attached to transposons, the transposons may "jump" from ordinary plasmids to F factors, after which transfer by conjugation may occur (see Chapter 8). TEXTBOOK CASE 9 describes one case with serious medical overtones.

Although conjugation pili are found only on some gram-negative bacteria, gram-positive bacteria also appear capable of conjugation. Microbiologists have experimented extensively with *Streptococcus mutans,* a common cause of dental caries. In this organism, conjugation appears to involve only plasmids, particularly those carrying genes for antibiotic resistance. Moreover, the conjugation does not involve pili. Rather, the recipient cell apparently secretes substances encouraging the donor cell to produce clumping factors composed of protein. The factors bring together (clump) the donor and recipient cell, and pores form between the cells to permit plasmid transfer. Chromosomal transfer has not been demonstrated.

CONCEPT AND REASONING CHECKS

9.4 Unlike conjugation between an F+ and F-, where the recipient cell becomes F+, why doesn't an Hfr and F- conjugation result in the recipient cell being an Hfr?

Transduction Involves Viruses as Agents for Horizontal Transfer of DNA

KEY CONCEPT

5. Bacterial viruses can transfer chromosomal DNA from donor to recipient.

Another form of genetic recombination was reported in 1952 by Joshua Lederberg and Norton Zinder. While working with mutant cells of *Salmonella*, Lederberg and Zinder observed recombination, but ruled out conjugation and transformation because the cells were separated by a thin membrane and DNA was absent in the extracellular fluid. Eventually, they discovered a virus in the fluid and uncovered the details of recombination.

Transduction is the third form of HGT and it requires a virus to carry a chromosomal DNA fragment from donor to recipient cell. The virus participating in transduction is called a **bacteriophage** (literally "bacteria eater") or simply **phage**. As with all viruses, the bacteriophages have a core of DNA or RNA surrounded by a coat of protein (Chapter 14).

In the replication cycle of a bacteriophage, different phages can interact with bacterial cells in one of two ways (FIGURE 9.7). In a **lytic cycle**, the phage DNA penetrates the cell, destroys the host chromosome, replicates itself within the cell, and then destroys (lyses) the cell as new phages are released. Because the phages killed the cell, they are called **virulent phages**.

Other phages interact with bacterial cells in a slightly different way, called a **lysogenic cycle**. These phages also invade the host but do not always directly cause cell lysis. Instead, the phage DNA integrates into the host chromosome as a **prophage** and the phages participating in this cycle are known as **temperate phages**. The host cell survives and, as it undergoes DNA replication and binary fission, the prophage is copied and vertically transferred to daughter cells. However,

Vancomycin-Resistant
Staphylococcus aureus

1 In June 2002, a 40-year-old Michigan resident with diabetes, peripheral vascular disease, and chronic renal failure developed a suspected catheter exit-site infection. A swab from the catheter exit-site was used to isolate vancomycin-resistant *Staphylococcus aureus* (VRSA).

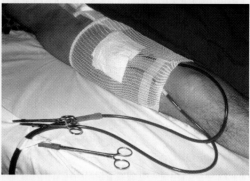

An arteriovenous graft for hemodialysis.

2 In April 2001, the patient had been treated for chronic foot ulcerations with multiple courses of antimicrobial therapy, some of which included the antibiotic vancomycin. In April 2002, the patient underwent amputation of a gangrenous toe and subsequently developed a methicillin-resistant *S. aureus* (MRSA) blood infection that resulted when hemodialysis was performed. In the form of hemodialysis used, an artificial vessel was used to join the artery and vein (called an arteriovenous graft; see figure), which often are subject to infection. The infection was treated with vancomycin, rifampin, and removal of the infected graft.

3 With the identification of VRSA from the swab, cultures from the exit site and catheter tip were made and subsequently grew *S. aureus* resistant to oxacillin and vancomycin.

4 A week after the patient's catheter was removed, the exit site appeared healed; however, the patient's chronic foot ulcer appeared infected. VRSA, vancomycin-resistant *Enterococcus faecalis* (VRE), and *Klebsiella oxytoca* also were recovered from a culture of the ulcer. Swab cultures of the patient's healed catheter exit site and anterior nares did not grow VRSA. The ulcer was cleaned of dead and contaminated tissue. The patient was urged to maintain aggressive wound care and, as an outpatient, put on systemic antimicrobial therapy with the sulfa drug trimethoprim/sulfamethoxazole (Bactrim).

5 The VRSA isolate recovered from the catheter exit site was identified initially at a local hospital laboratory and was confirmed by the Michigan Department of Community Health and the Centers for Disease Control and Prevention (CDC).

6 Further molecular analysis indicated the VRSA isolate contained the *vanA* vancomycin resistance gene typically found in enterococci.

7 Epidemiologic and laboratory investigations were undertaken to assess the risk for transmission of VRSA to other patients, health-care workers, and close family and other contacts. No VRSA transmission was identified.

8 Infection-control practices in the local dialysis center were assessed; all health-care workers followed standard precautions consistent with CDC guidelines.

Questions:

(Answers can be found in Appendix D.)

A. This report describes the first documented case of infection caused by vancomycin-resistant *S. aureus* (VRSA) in a patient in the United States. Why was this patient so susceptible to infection with *S. aureus*?

B. Because vancomycin resistance determinants had not previously been identified in clinical isolates of *S. aureus* in the United States, how did the *vanA* gene get "transferred" into *S. aureus* in this patient?

C. Why was a culture swab from the patient's anterior nares tested for VRSA?

D. Why was the patient put on a sulfa drug?

E. Besides standard precautions, what other procedures should be in place to prevent transmission of antimicrobial resistant microorganisms in health-care settings?

For additional information see http://www.cdc.gov/mmwr/preview/mmwrhtml/mm5126a1.htm

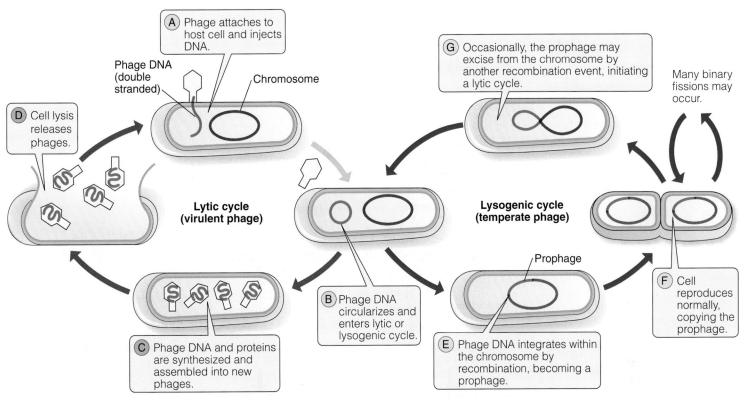

FIGURE 9.7 **Bacteriophage Replicative Cycles.** The consequences of infection by a virulent phage and a temperate phage are shown. »» How do the two replicative cycles differ between virulent and temperate phages?

eventually the prophage will excise itself and go through a lytic cycle (Chapter 14).

Let's examine how these two types of phage can transfer fragments of donor cell DNA to a recipient cell.

Generalized transduction is carried out by virulent phages, such as the P1 phage that infects *E. coli* (**FIGURE 9.8**). After injecting the DNA into the cell, the host cell's chromosome is digested into small fragments. When the new phage DNA is produced, the DNA normally is packaged into new phage particles. However, on rare occasions (1 in 100,000,000) a random (general) fragment of host cell DNA may accidentally be captured in the packaging process and end up in a phage head rather than phage DNA. These phages are fully formed though and they can infect another cell. However, they are called "defective particles" because they carry no phage genes and cannot replicate themselves after infection. This is the type of genetic recombination described in MicroFocus 9.1.

Following release from the lysed host cell, the defective phage attaches to a new (recipient) cell

and injects the donor chromosomal DNA into the host cell. Once in the recipient, new genes can pair with a section of the recipient's DNA and replace the section in a fashion similar to conjugation. The recipient has now been transduced (changed) using genes from the donor cell.

Specialized transduction occurs as a result of a lysogenic cycle and unlike generalized transduction, results in the transfer of specific genes. One of the most studied temperate viruses is phage lambda, which also infects *E. coli*. Being a temperate phage, the lambda phage DNA is integrated as a prophage into the chromosomal DNA (Figure 9.8).

At some time in the future, the prophage undergoes excision from the chromosome and enters the lytic cycle. Most of the time, the excision occurs precisely and the intact phage DNA is released. On rare occasions, an imprecise excision occurs, and the excised prophage takes along a few flanking *E. coli* genes while leaving behind a few phage genes. At the conclusion of phage replication, multiple copies of the phage, each with a donor gene, are produced. Again, these

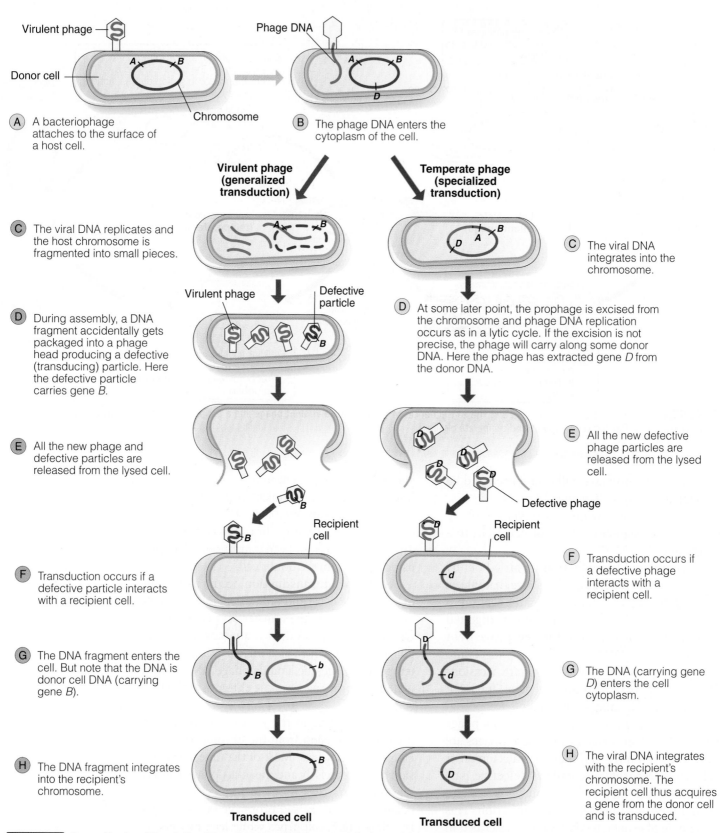

A A bacteriophage attaches to the surface of a host cell.

B The phage DNA enters the cytoplasm of the cell.

Virulent phage (generalized transduction)

Temperate phage (specialized transduction)

C The viral DNA replicates and the host chromosome is fragmented into small pieces.

C The viral DNA integrates into the chromosome.

D During assembly, a DNA fragment accidentally gets packaged into a phage head producing a defective (transducing) particle. Here the defective particle carries gene *B*.

D At some later point, the prophage is excised from the chromosome and phage DNA replication occurs as in a lytic cycle. If the excision is not precise, the phage will carry along some donor DNA. Here the phage has extracted gene *D* from the donor DNA.

E All the new phage and defective particles are released from the lysed cell.

E All the new defective phage particles are released from the lysed cell.

F Transduction occurs if a defective particle interacts with a recipient cell.

F Transduction occurs if a defective phage interacts with a recipient cell.

G The DNA fragment enters the cell. But note that the DNA is donor cell DNA (carrying gene *B*).

G The DNA (carrying gene *D*) enters the cell cytoplasm.

H The DNA fragment integrates into the recipient's chromosome.

H The viral DNA integrates with the recipient's chromosome. The recipient cell thus acquires a gene from the donor cell and is transduced.

FIGURE 9.8 **Generalized and Specialized Transduction.** Phage can transfer bacterial genes by generalized transduction (can involve any bacterial gene) or specialized transduction (can involve genes from a specialized region). »» How does the outcome of transduction differ between generalized and specialized?

are defective phages because they are missing a few phage genes needed for replication. Such a transducing phage can infect another cell and transfer its genes, which cannot encode a replicative cycle. Instead, the genes integrate into the recipient chromosome, carrying the donor's genes with them. As before, the recipient cell acquired genes from the original donor cell and the recipient is now considered transduced.

Specialized transduction is an extremely rare event in comparison to the generalized form. However, there are exceptions. For example,

the diphtheria bacillus, *Corynebacterium diphtheriae,* harbors proviral DNA providing the genetic code for a toxin causing diphtheria. Other toxins encoded by proviral DNA include staphylococcal enterotoxins in food poisoning, clostridial toxins in some forms of botulism, and streptococcal toxins in scarlet fever.

FIGURE 9.9 compares the three forms of genetic recombination through horizontal gene transfer.

CONCEPT AND REASONING CHECKS

9.5 What is the major difference between transformation and generalized transduction?

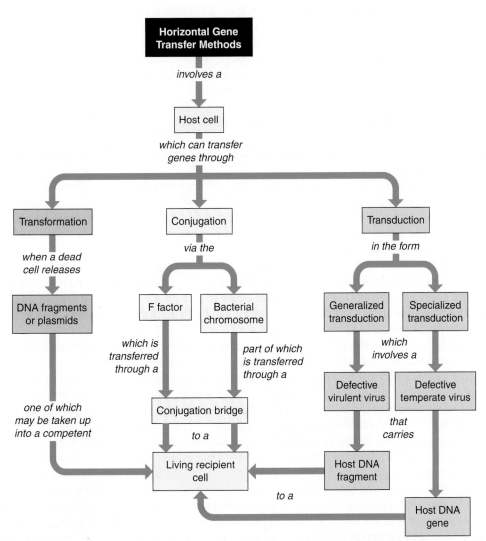

FIGURE 9.9 **A Summary of Genetic Recombination through Horizontal Gene Transfer.** This concept map summarizes the three mechanisms of transfer of genes from donor to recipient cells. »» In which of the HGT mechanisms is the donor cell dead?

9.2 Genetic Engineering and Biotechnology

Prior to the 1970s, bacterial species having special or unique metabolic properties were detected through mutant analysis or by screening for cells with certain metabolic talents (MicroFocus 9.3). Experiments in genetic recombination entered a new era in the late 1970s, when it became possible to insert genes into bacterial DNA and thereby establish a genetically identical population that would produce proteins from the inserted genes.

The use of microbial genetics, including the isolation, manipulation, and control of gene expression had far-reaching ramifications, leading to the entirely new field of **genetic engineering**.

Many of the products derived from genetic engineering have advanced the field of medicine and industrial production. **Biotechnology** is the name given to the commercial and industrial applications derived from genetic engineering.

MICROFOCUS 9.3: History/Biotechnology
Clostridium acetobutylicum and the Jewish State

In 1999, scientists completed sequencing the genome of *Clostridium acetobutylicum,* a nonpathogenic bacterial species. Because some other species of *Clostridium* are major pathogens (one produces the food toxin that causes botulism, and another is responsible for tetanus), the scientists hope their sequencing work will yield insights into what enables some species to become pathogens while others remain harmless. However, *C. acetobutylicum's* ability to convert starch into the organic solvents acetone and butanol is what has a prominent place in history.

In 1900, an outstanding chemist named Chaim Weizmann, a Russian-born Jew, completed his doctorate at the University of Geneva in Switzerland. He also was an active Zionist and advocated the creation of a Jewish homeland in Palestine. In 1904, Weizmann moved to Manchester, England, where he became a research fellow and senior lecturer at Manchester University. During this time, he was elected to the General Zionist Council.

Weizmann began working in the laboratory of Professor William Perkin, where he attempted to use microbial fermentation to produce industrially useful substances. He discovered that *C. acetobutylicum* converted starch to a mixture of ethanol, acetone, and butanol, the latter an important ingredient in rubber manufacture. The fermentation process seemed to have no other commercial value—until World War I broke out in 1914.

At that time, the favored propellant for rifle bullets and artillery projectiles was a material called cordite. To produce it, a mixture of cellulose nitrate and nitroglycerine was combined into a paste using acetone and petroleum jelly. Before 1914, acetone was obtained through the destructive distillation of wood. However, the supply was inadequate for wartime needs, and by 1915, there was a serious shell shortage, mainly due to the lack of acetone for making cordite.

After his inquiries to serve the British government were not returned, a friend of Weizmann's went to Lloyd George, who headed the Ministry of Munitions. Lloyd George was told about Weizmann's work and how he could synthesize acetone in a new way. The conversation resulted in a London meeting between Weizmann, Lloyd George, and Winston Churchill. After explaining the capabilities of *C. acetobutylicum,* Weizmann became director of the British admiralty laboratories where he instituted the full-scale production of acetone from corn. Additional distilleries soon were added in Canada and India. The shell shortage ended.

After the war ended, now British Prime Minister Lloyd George wished to honor Weizmann for his contributions to the war effort. Weizmann declined any honors but asked for support of a Jewish homeland in Palestine. Discussions with Foreign Minister Earl Balfour led to the Balfour Declaration of 1917, which committed Britain to help establish the Jewish homeland. Weizmann went on to make significant contributions to science—he suggested that other organisms be examined for their ability to produce industrial products and is considered the father of industrial fermentation. Weizmann also laid the foundations for what would become the Weizmann Institute of Science, one of Israel's leading scientific research centers. His political career also moved upward—he was elected the first President of Israel in 1949. Chaim Weizman died in 1952.

Genetic Engineering Was Born from Genetic Recombination

KEY CONCEPT

6. Plasmids can be spliced open and a gene of interest inserted to form a recombinant DNA molecule.

The science of **genetic engineering** involves an alteration to the genetic material in an organism to change its traits or to allow the organism to produce a biological product, usually a protein, that the organism was previously incapable of producing. The field surfaced in the early 1970s when the techniques became available to manipulate DNA. Among the first scientists to attempt genetic manipulation was Paul Berg of Stanford University. In 1971, Berg and his coworkers opened the circular DNA molecule from simian virus-40 (SV40) and spliced it into a bacterial chromosome. In doing so, they constructed the first **recombinant DNA molecule**—a DNA molecule containing DNA segments spliced together from two or more organisms. This human-manipulated genetic recombination process was extremely tedious though because the cut bacterial and viral DNAs had blunt ends, making sealing of the two DNAs difficult. Berg therefore had to use exhaustive enzyme chemistry to form staggered ends that would combine easily through complementary base pairing.

While Berg was performing his experiments, an important development came from Herbert Boyer and his group at the University of California.

Boyer isolated a **restriction endonuclease** enzyme that recognizes and cuts specific short stretches of nucleotides. Importantly, the enzyme leaves the DNA with mortise-like staggered ends. These dangling bits of single-stranded DNA extending out from the double-stranded DNA easily attached to complementary ends protruding from another fragment of DNA. Scientists quickly dubbed the single-stranded extensions "**sticky ends.**"

Today, there is a vast array of restriction enzymes from the *Bacteria* and *Archaea*, each recognizing a specific nucleotide sequence (TABLE 9.2). Enzyme designations are derived from the species from which they were isolated. For example, the restriction enzyme *Eco*RI stands for *Escherichia co*li **R**estriction enzyme **I**.

Each enzyme cuts both strands of the DNA because the sequences recognized are a palindrome. Each strand has the same complementary set of nucleotide bases. Thus, restriction enzymes are "molecular scissors" used by genetic engineers to open a bacterial chromosome or plasmid at specific locations and insert a DNA segment from another organism.

To seal the recombinant DNA segments, **DNA ligase** was used. This enzyme normally functions during the DNA replication and repair to seal together DNA fragments (see Chapter 8).

Meanwhile, Stanley Cohen, also at Stanford University, was accumulating data on the plasmids of *E. coli*. Cohen found he could isolate plasmids from bacterial cells and insert

Palindrome: A series of letters reading the same left to right and right to left.

TABLE

9.2 Examples of Restriction Endonuclease Recognition Sequences

Organism	Restriction Enzyme	Recognition Sequence*
Escherichia coli	*Eco*RI	G ↓ AATTC CTTAA ↑ G
Streptomyces albus	*Sal*I	G ↓ TCGAC CAGCT ↑ G
Haemophilus influenzae	*Hind*III	A ↓ AGCTT TTCGA ↑ A
Bacillus amyloliquefaciens	*Bam*HI	G ↓ GATCᵐC CCᵐTAG ↑ G
Providencia stuartii	*Pst*I	CTGCA ↓ G G ↑ ACGTC

*Arrows indicate where the restriction enzyme cuts the two strands of the recognition sequence; C^m = methylcytosine.

them into fresh bacterial cells by suspending the organisms in calcium chloride and rapidly heating them. This made the *E. coli* cells competent to take up the plasmid via the transformation process. Once inside the cells, the plasmids multiply independently and produce **clones;** that is, copies of themselves.

Working together, Boyer and Cohen isolated plasmids from *E. coli* and opened them with restriction enzymes (**FIGURE 9.10**). Next, they inserted a

segment of foreign DNA into the plasmids and sealed the segment using DNA ligase. Mimicking natural genetic recombination, they then inserted the plasmids (recombinant DNA molecules) into fresh *E. coli* cells. By 1973, their technique had successfully spliced genes across genera, from *S. aureus* into *E. coli*. These genetic engineering experiments intrigued the scientific community because for the first time they could manipulate genes from a wide variety of species and splice them together.

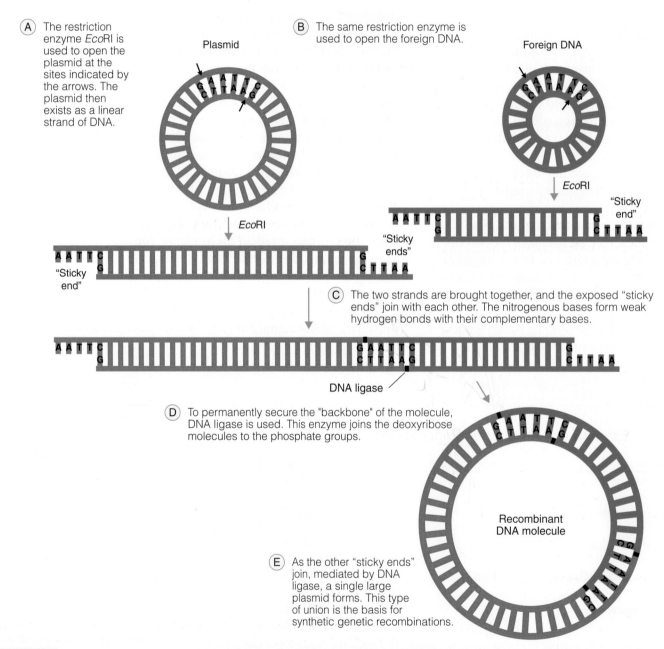

FIGURE 9.10 Construction of a Recombinant DNA Molecule. In this construction, two unrelated plasmids (loops of DNA) are united to form a single plasmid representing a recombinant DNA molecule. »» Why is the product of genetic engineering called a recombinant DNA molecule?

CONCEPT AND REASONING CHECKS

9.6 List the steps required to form a recombinant DNA molecule.

Genetic Engineering Has Many Commercial and Practical Applications

KEY CONCEPT

7. Bacterial cells can be transformed by the acquisition of a human gene.

Today, over 18 million people in the United States have **diabetes**, a group of diseases resulting from abnormally high blood glucose levels. One cause is from the inability of the body to produce sufficient levels of insulin (referred to as juvenile or insulin-dependent diabetes or type I diabetes) to control the blood glucose level. This means that diabetics must receive daily injections of insulin to survive. Before 1982, diabetics received purified insulin extracted from the pancreas of cattle and pigs, or even cadavers. This can pose a problem because animal insulin could trigger allergic reactions and possibly contain unknown viruses that had infected the animal. The solution was to produce insulin using genetic engineering techniques. Eli Lilly marketed the first such synthetic human insulin, called Humulin, in 1982. Since then, other genetically engineered insulin products have been developed. Such commercial successes of biotechnology were a sign of things to come.

The best way to understand how genetic engineering operates is to follow an actual procedure for the production of human insulin. MICROINQUIRY 9 describes one such method—by cloning the human gene for insulin into bacterial cells. This involves:

1. Isolating the piece of DNA containing the human insulin gene and precisely cutting the gene out;

2. Splicing the insulin gene into a bacterial plasmid;

3. Placing the recombinant plasmid in bacterial cells to form clones;

4. Screening for recombinant plasmids;

5. Identifying and isolating clones carrying the insulin gene.

Besides insulin, a number of other proteins of important pharmaceutical value to humans have been produced by genetically engineered micro-

organisms. Many of these proteins are produced in relatively low amounts in the body, making purification extremely costly. Therefore, the only economical solution to obtain significant amounts of the product is through genetic engineering.

The dairy industry was the first to feel the dramatic effect of the new DNA technology. In 1982, the U.S. Food and Drug Administration (FDA) licensed recombinant bovine somatotropin (rBST), a protein that can boost milk production in dairy cattle by 40%.

Another early application of genetic engineering to human disorders and diseases was the production of yet another growth hormone, human growth hormone (HGH). Genetically engineered HGH replaced the form that had been extracted from the pituitary of human cadavers. With its license by the FDA in 1985, HGH has been used to treat conditions that produce short stature, to improve muscle strength associated with some genetic disorders, and to maintain muscle mass in patients suffering from AIDS. Of course, it can be and has been used as an athletic enhancement to build muscle mass for bodybuilding.

In 2010, hundreds of biotechnology companies worldwide are working on the commercial and practical applications of genetic engineering (**FIGURE 9.11**). Many of the genetically

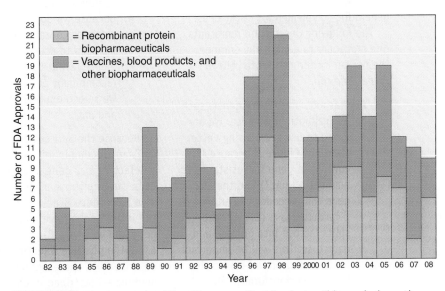

FIGURE 9.11 **FDA Approvals of New Pharmaceutical Products.** This graph shows the number of approvals of new biopharmaceutical products since 1982, when the first recombinant product was approved. Data redrawn from: Biotechnology Information Institute (http://www.biopharma.com/approvals_2008.html). »» Can you come up with a reason for the drop in recombinant protein biopharmaceuticals in 2007?

MICROINQUIRY 9

Molecular Cloning of a Human Gene into Bacterial Cells

Genetic engineering has been used to produce pharmaceuticals of human benefit. One example concerns the need for insulin injections in people suffering from diabetes (an inability to produce the protein insulin to regulate blood glucose level). Prior to the 1980s, the only source for insulin was through a complicated and expensive extraction procedure from cattle or pig pancreases. But, what if you could isolate the human insulin gene and, through transformation, place it in bacterial cells? These cells would act as factories churning out large amounts of the pure protein that diabetics could inject.

To do molecular cloning of a gene, besides the bacterial cells, we need three ingredients: a cloning vector, the human gene of interest, and restriction enzymes. Plasmids are the **cloning vector**, a genetic element used to introduce the gene of interest into the bacterial cells. Human DNA containing the insulin gene must be obtained. We will not go into detail as to how the insulin gene can be "found" from among 25,000 human genes. Suffice it to say that there are standard procedures to isolate known genes. Restriction enzymes will cut open the plasmids and cut the gene fragments that contain the insulin gene, generating complementary sticky ends.

The following description represents one procedure to genetically engineer the human insulin gene into cells of *Escherichia coli*.

1. Plasmids often carry genes, such as antibiotic resistance. We are going to use the cloning vector shown in **Figure A** because it contains a gene for resistance to ampicillin (amp^R) and the *lacZ* gene that encodes the enzyme β-galactosidase (B-gal) that splits lactose into glucose and galactose. This will be important for identification of clones that have been transformed. In addition, this vector has a single restriction sequence for the restriction enzyme *Sal*I (see Table 9.2). Importantly, this cut site is within the *lacZ* gene. Also, the plasmid will replicate independently in *E. coli* cells and can be placed in the cells by transformation.

2. The vector and human DNA are cut with *Sal*I to produce complementary sticky ends on both the opened vector and the insulin gene (**Figure B**). Vector and the insulin gene then are mixed together in a solution with DNA ligase, which will covalently link the sticky ends. Some vectors will be recombinant plasmids; that is, plasmids containing the insulin gene. Other plasmids will close back up without incorporating the gene.

3. The plasmids are placed in *E. coli* cells by transformation. The plasmids replicate independently in the bacterial cells, but as the bacterial cells multiply, so do the plasmids. By allowing the plasmids to replicate, we have cloned the plasmids, including any that contain the insulin gene.

4. Because we do not know which plasmids contain the insulin gene, we need to screen the clones to identify what colonies contain the recombinant plasmids. This is why we selected a plasmid with amp^R and *lacZ*.

Because the *Sal*I cut site is within *lacZ*, any recombinant plasmids will have a defective *lacZ* gene and the bacterial cells carrying those plasmids will not be able to produce B-gal. Plasmids without the insulin gene have an intact *lacZ* gene and will have B-gal activity. Thus, using the positive selection technique described in Chapter 8 and a selective medium as described in Chapter 5, we can identify which bacterial cells contain the recombinant plasmids.

Therefore, we plate all our bacterial cells onto an agar plate containing ampicillin and a substrate called X-gal that B-gal can hydrolyze.

$$\text{X-gal} \xrightarrow{\text{B-gal}} \text{X + galactose}$$

- The X product is blue in color, so any colonies having an intact *lacZ* gene will hydrolyze X-gal and appear blue on the agar plate;
- If the *lacZ* gene is inactive due to the presence of the insulin gene, then no product will be formed and those colonies on the plate will appear white.

Question 9a. Identify what colonies on the plate contain the insulin gene.

Question 9b. Explain why the bacteria without a plasmid did not grow on the plate. The answers can be found in **Appendix D**.

5. These colonies with the insulin gene can now be isolated and grown in larger batches of liquid medium. These batch cultures then are inoculated into large "production vats," called **bioreactors**, in which the cells grow to massive numbers while secreting large quantities of insulin into the liquid.

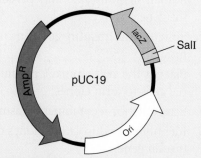

FIGURE A The Cloning Vector, pUC19.

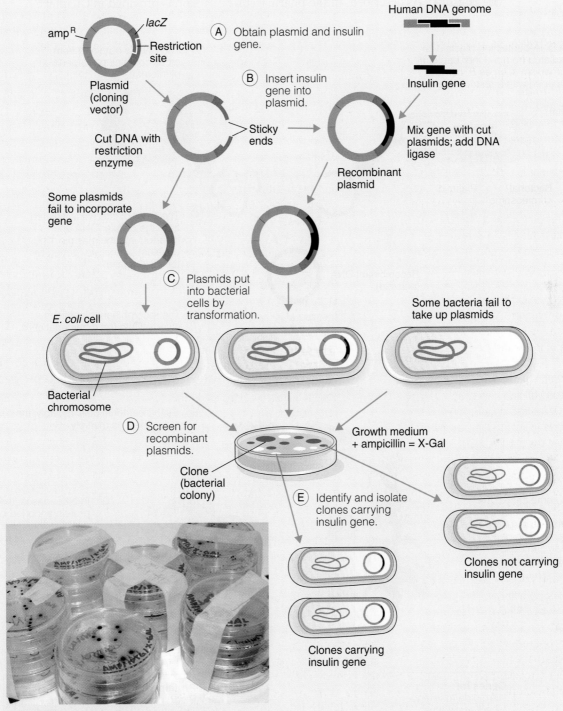

FIGURE B The Sequence of Steps to Engineer the Insulin Gene into *Escherichia coli* Cells.

engineered products are either proteins expressed by the recombinant DNA in the bacterial cells or the recombinant DNA from the cloned plasmid (FIGURE 9.12). As described, in the pharmaceuti-

cal industry, the protein products are numerous and diverse.

Let's look at a few examples as more discussion is provided in Chapter 27.

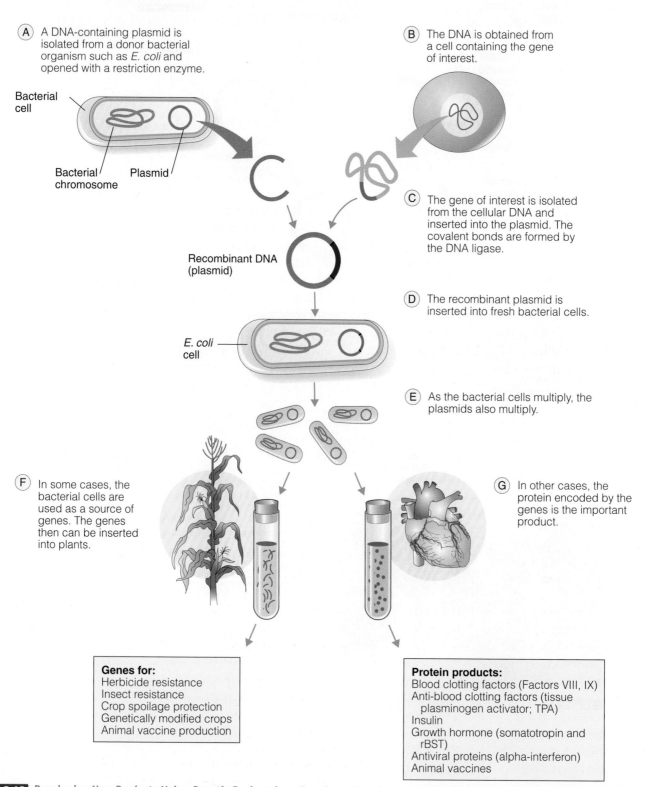

Ⓐ A DNA-containing plasmid is isolated from a donor bacterial organism such as *E. coli* and opened with a restriction enzyme.

Bacterial cell

Bacterial chromosome   Plasmid

Ⓑ The DNA is obtained from a cell containing the gene of interest.

Recombinant DNA (plasmid)

Ⓒ The gene of interest is isolated from the cellular DNA and inserted into the plasmid. The covalent bonds are formed by the DNA ligase.

Ⓓ The recombinant plasmid is inserted into fresh bacterial cells.

E. coli cell

Ⓔ As the bacterial cells multiply, the plasmids also multiply.

Ⓕ In some cases, the bacterial cells are used as a source of genes. The genes then can be inserted into plants.

Ⓖ In other cases, the protein encoded by the genes is the important product.

Genes for:
Herbicide resistance
Insect resistance
Crop spoilage protection
Genetically modified crops
Animal vaccine production

Protein products:
Blood clotting factors (Factors VIII, IX)
Anti-blood clotting factors (tissue plasminogen activator; TPA)
Insulin
Growth hormone (somatotropin and rBST)
Antiviral proteins (alpha-interferon)
Animal vaccines

FIGURE 9.12 Developing New Products Using Genetic Engineering. Genetic engineering is a method for inserting foreign genes into bacterial cells and obtaining chemically useful products. »» How do bacterial cells that have been genetically modified (F) differ from bacterial cells that encode genes for a product (G)?

Environmental Biology. We already have mentioned in this chapter and previous ones (see MicroFocus 8.4) the usefulness of microorganisms as a source of genes. The *Bacteria* and the *Archaea* represent a huge, mostly untapped gene pool representing metabolically diverse processes. Examples such as **bioremediation** have been discussed where genetically engineered or genetically recombined cells are provided with specific genes whose products will break down toxic pollutants, clean up waste materials, or degrade oil spills in an attempt to return the environment to its original condition. We have barely scratched the surface to take advantage of the metabolic diversity offered by these microbes.

Medicine. The presence or threat of infectious disease represents a high demand for antibiotics in the medical field. Although antibiotics are produced in nature, the bacterial or fungal organisms often do not produce these compounds in high yield. This means new antibiotic sources must be discovered (see chapter opener) and the microbes must be genetically engineered to produce larger quantities of antibiotics and/or to produce modified antibiotics to which infectious microbes have yet to show resistance.

Another product of genetic engineering is **interferon**, a set of three naturally produced antiviral agents produced by the human body, two of which block viral replication (Chapter 20). As with insulin and HGH, the body produces small amounts of these chemicals, so prior to the introduction of genetic engineering thousands of units of human blood were needed to obtain sufficient interferon to treat a patient. With genetic engineering, much larger amounts of pure protein can be produced to aid patients suffering from hepatitis B and C as well as some forms of cancer.

Vaccine production is now safer as a result of genetic engineering. By making a vaccine that only contains a part of the whole microbial agent, or isolating a gene that will stimulate the immune system to generate protective immunity, makes the vaccine much safer because the patient is not exposed to the active virus or bacterium that can cause the disease. For example, hepatitis B, a serious viral infection spread by contact with infected blood, kills 2 million people globally per year (Chapter 16). The first vaccine against hepatitis B was made by extracting the virus from infected blood of chronic carriers and then isolating a viral surface protein. Such a procedure was complex and posed the risk of possible contamination by other infectious agents. In 1986, the FDA approved the first recombinant hepatitis B vaccine (Recombivax HB®) that was made by inserting the gene for the viral surface protein into common baker's yeast, *Saccharomyces cerevisiae*. The protein produced by the yeast was identical to the natural viral protein and made for a much safer vaccine for humans. A second recombinant-type hepatitis B vaccine (Engerix-B®) was licensed in 1989. A more recent recombinant vaccine, Gardasil®, which generates protective immunity against many of the papilloma viruses responsible for cervical cancer and genital warts in women, was licensed by the FDA in 2006. In 2009, the vaccine was licensed for use in men to prevent penile cancer and genital warts.

Lastly, it should be mentioned that genetically engineered products are not always a "no brainer" in terms of their development. This is no clearer than in the attempts to develop a vaccine for AIDS. Since 1987, scientists and genetic engineers have tried to identify viral subunits that can be used to develop an AIDS vaccine. However, it is not so much that genetic engineering can't be done as it is the virus just seems to find ways to circumvent a vaccine and the immunity developing in the patient. Still, scientists hope a safe and effective genetically engineered vaccine can be developed. We will have much more to say about vaccines and AIDS in Chapters 22 and 23.

Agricultural Applications. Genetic engineering has extended into many realms of science. In agriculture, for example, genes for herbicide resistance have been transplanted from cloned bacterial cells into tobacco plants, demonstrating that these transgenic plants better tolerate the herbicides used for weed control. For tomato growers, a notable advance was made when researchers at Washington University spliced genes from a pathogenic virus into tomato plant cells and demonstrated the cells would produce viral proteins at their surface. The viral proteins blocked viral infection, providing resistance for the transgenic tomato plants.

Resistance to insect attack also has been introduced into plants using a plasmid carrying a bacterial gene that is toxic to beetle and fly larvae (Chapter 27).

Transgenic: Referring to an organism containing a stable gene from another organism.

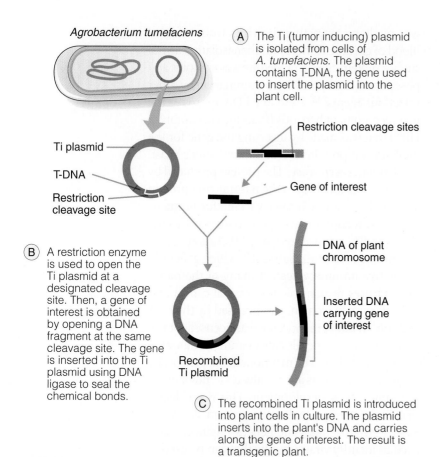

(A) The Ti (tumor inducing) plasmid is isolated from cells of *A. tumefaciens*. The plasmid contains T-DNA, the gene used to insert the plasmid into the plant cell.

Restriction cleavage sites

Gene of interest

(B) A restriction enzyme is used to open the Ti plasmid at a designated cleavage site. Then, a gene of interest is obtained by opening a DNA fragment at the same cleavage site. The gene is inserted into the Ti plasmid using DNA ligase to seal the chemical bonds.

DNA of plant chromosome

Inserted DNA carrying gene of interest

Recombined Ti plasmid

(C) The recombined Ti plasmid is introduced into plant cells in culture. The plasmid inserts into the plant's DNA and carries along the gene of interest. The result is a transgenic plant.

FIGURE 9.13 **The Ti Plasmid as a Vector in Plant Genetic Engineering.** *Agrobacterium tumefaciens* induces tumors in plants and causes a disease called crown gall. The catalyst for infection is a tumor-inducing (Ti) plasmid. This plasmid, without the tumor-causing gene, is used to carry a gene of interest into plant cells. »» Why must the tumor-inducing gene in the Ti plasmid be removed before the plasmid is used in genetic engineering procedures?

As much as 60% of the food we eat today has some connection to genetic engineering. By taking traits from one organism and introducing those traits into another organism, the food can be changed such that it tastes better, grows faster and larger, or has a longer shelf life. For gene transfer experiments in plants, the vector DNA often used is a plasmid from the bacterium *Agrobacterium tumefaciens*. This organism causes a plant tumor called crown gall, which develops when DNA from the bacterial cells inserts itself into the plant cell's chromosomes (**FIGURE 9.13**). Researchers remove the tumor-inducing (Ti) gene from the plasmid and then splice the desired gene into the plasmid and allow the bacterial cells to infect the plant. The Ti system works well with dicots (broadleaf plants) such as tomato, potato, soybeans, and cotton.

CONCEPT AND REASONING CHECKS

9.7 Give several examples of how genetic engineering has benefited the fields of environmental microbiology, medicine, and agriculture.

DNA Probes Can Identify a Cloned Gene or DNA Segment

KEY CONCEPT

8. DNA probes are single-stranded DNA segments.

The genes of an organism contain the essential information responsible for its behavior and characteristics. Bacterial and viral pathogens, for example, contain specific sequences of nucleotides that can confer on the pathogen the ability to infect and cause disease. Because these nucleotide sequences are distinctive and often unique, if detectable, they can be used as a definitive diagnostic determinant.

In the medical laboratory, diagnosticians are optimistic about the use of **DNA probes**, single-stranded DNA molecules that recognize and bind to a distinctive and unique nucleotide sequence of a pathogen. To use a DNA probe effectively, it is valuable to increase the amount of DNA to be searched. This can be done through the **polymerase chain reaction** (**PCR**); the technique is outlined in MicroFocus 9.4.

The DNA probe binds (hybridizes) to its complementary nucleotide sequence from the pathogen, much like strips of Velcro stick together. To make a probe, scientists first identify the segment (or gene) in the pathogen that will be the target of a probe. Using this segment, they construct the single-stranded DNA probe (**FIGURE 9.14**). More than 100 DNA probes have been developed for the detection of pathogens.

One example of where DNA probes and PCR have been useful is in the detection of the human immunodeficiency virus (HIV). T lymphocytes, in which HIV replicates, are obtained from the patient and disrupted to obtain the cellular DNA. The DNA then is amplified by PCR and the DNA probe is added. The probe is a segment of DNA that complements the DNA in the virus synthesized from the genome of HIV (Chapter 23). If the person is infected with HIV, the probe will locate the viral DNA, bind to it, and emit radioactivity. An accumulation of radioactivity constitutes a positive test.

A DNA probe also is available for detecting human papilloma virus (HPV). The test uses a DNA probe to detect viral DNA in a sample of tissue obtained from a woman's cervix. Because certain forms of HPV have been linked to cervical tumors, the test has won acceptance as an important preventive technique, and it has been licensed

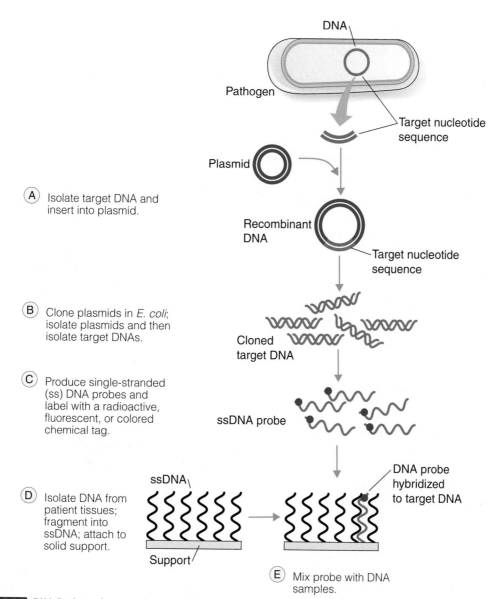

A Isolate target DNA and insert into plasmid.

B Clone plasmids in *E. coli*; isolate plasmids and then isolate target DNAs.

C Produce single-stranded (ss) DNA probes and label with a radioactive, fluorescent, or colored chemical tag.

D Isolate DNA from patient tissues; fragment into ssDNA; attach to solid support.

E Mix probe with DNA samples.

FIGURE 9.14 **DNA Probes.** Construction of a DNA probe and its use in disease detection and diagnosis. »» Why must the DNA probes be single stranded?

by the FDA. It is commercially available as the ViraPap test. Clearly, the use of DNA probes represents a reliable and rapid method for detecting and diagnosing many human infectious diseases (MICROFOCUS 9.5).

A similar technique can be used to conduct water-quality tests based on the detection of coliform bacteria such as *E. coli* (Chapter 26). Traditionally, *E. coli* had to be cultivated in the laboratory and identified biochemically. With DNA probe technology, a sample of water can be filtered, and the bacterial cells trapped on the filter can be broken open to release their DNA for PCR and DNA probe analysis. Not only is the process

time saving (many days by the older method, but only a few hours with the newer method), it is also extremely sensitive: a single *E. coli* cells can be detected in a 100-mL sample of water.

Another useful tool in biotechnology is the **DNA microarray**, a small slide surface on which genes or DNA segments are attached and arranged spatially in a known pattern that can be used to assess gene expression in microorganisms. The technique is described in MICROFOCUS 9.6.

CONCEPT AND REASONING CHECKS

9.8 Why are DNA probes such useful tools in the detection of disease-causing agents?

MICROFOCUS 9.4: Tools
The Polymerase Chain Reaction

The **polymerase chain reaction (PCR)** is a technique that takes a segment of DNA and replicates it millions of times in just a few hours. The technique was developed in 1984 by Kary Mullis working for the Cetus Corporation, a biotechnology company in Emeryville, California; in 1993, he shared the Nobel Prize in Chemistry for his discovery.

The PCR process is a repeating three-step process (see figure). Target DNA is mixed with DNA polymerase (the enzyme that synthesizes DNA), short strands of primer DNA, and a mixture of short chain nucleotides called oligonucleotides. The mixture is then alternately heated and cooled during which time the double-stranded DNA unravels, is duplicated, and then reforms the double helix.

The process is repeated over and over again in a highly automated PCR machine, which is the biochemist's equivalent of an office copier. Each cycle takes about five minutes, and each new DNA segment serves as a template for producing many additional identical copies, which in turn serve as templates for producing more identical copies.

PCR is now a common and often essential tool in medical and biological research. Applications for PCR are numerous and include its use for DNA cloning, organismal DNA phylogeny studies, and (as described in the text) diagnosis of infectious diseases.

9.3 Microbial Genomics

In April 2003, exactly 50 years to the month after Watson and Crick announced the structure of DNA (see Chapter 2), a publicly financed, $3 billion international consortium of biologists, industrial scientists, computer experts, engineers, and ethicists completed perhaps the most ambitious project in the history of biology. The Human Genome Project, as it was called, had succeeded in mapping the **human genome**—that is, the 3 billion nitrogenous bases (equivalent to 750 megabytes of data) in a human cell were identified and strung together in the correct order (sequenced). The completion of the project represents a scientific milestone with unimaginable health benefits.

Many Microbial Genomes Have Been Sequenced

KEY CONCEPT

9. Known genome sequences are rapidly expanding for many microorganisms.

If the human genome was represented by a rope two inches in diameter, it would be 32,000 miles long. The genome of a bacterial species like *E. coli* at this scale would be only 1,600 miles long or about 1/20 the length of human DNA. So, being substantially smaller, microbial genomes are easier and much faster to sequence.

In May 1995, the first complete genome of a free-living organism was sequenced: the 1.8 million base pairs (1.8 Mb) in the genome of the bacterial species *Haemophilus influenzae* (MicroFocus 9.7). In a few short months, the genome for a second organism, *Mycoplasma genitalium* was reported. This reproductive tract pathogen has one of the smallest known bacterial genomes, consisting of only 580,000 base pairs and 480 protein-coding genes. In 1996, the genome for the yeast *Saccharomyces cerevisiae* was sequenced. In a field already filled with milestones, the sequencing of *S. cerevisiae* marked the first glimpse into the eukaryotic genome. Sixteen chromosomes were analyzed, 12 million bases were sequenced, and 6,000 genes were identified. The sequencing revealed many genes wholly new to biology.

Since then, hundreds more genomes have been sequenced, including a variety of pathogens

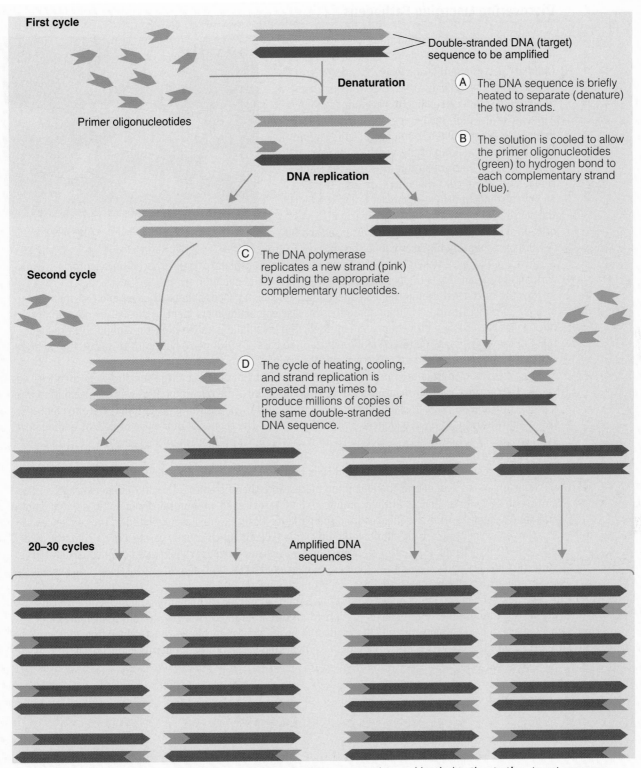

First cycle

Primer oligonucleotides

Denaturation

DNA replication

Double-stranded DNA (target) sequence to be amplified

(A) The DNA sequence is briefly heated to separate (denature) the two strands.

(B) The solution is cooled to allow the primer oligonucleotides (green) to hydrogen bond to each complementary strand (blue).

Second cycle

(C) The DNA polymerase replicates a new strand (pink) by adding the appropriate complementary nucleotides.

(D) The cycle of heating, cooling, and strand replication is repeated many times to produce millions of copies of the same double-stranded DNA sequence.

20–30 cycles

Amplified DNA sequences

The polymerase chain reaction produces billions of copies of a DNA sequence that are identical to the starting, target sequence.

MICROFOCUS 9.5: Tools
Discovering Emerging Pathogens

In April 2007, three Australian transplant patients received organs from a single 57-year-old donor, who had died of a stroke and was not thought to have had an infectious disease. Two women, aged 44 and 63, each received one of the donor's kidneys while a 64-year-old woman received the donor's liver. Each of the women soon developed fever and encephalitis, a swelling of the brain. Within six weeks of the operation, all three were dead. Traditional methods of culturing microbes from samples, and even standard DNA sequencing, failed to identify the cause of the patients' deaths. In fact, up to 40% of central nervous system infectious diseases and 30 to 60% of respiratory illnesses cannot be traced back to

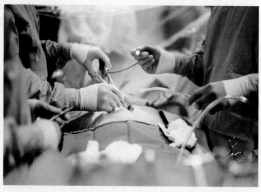

A surgical team performing a kidney transplant.

a specific pathogen. In the case of the three Australian transplant patients, based on symptoms, a virus with an RNA genome was the best guess the medical experts could offer.

Now the next generation in sequence technology is making it possible to discover and test for agents of infectious disease and is helping infectious disease epidemiologists identify the cause and origin of infections that had previously gone undiagnosed. The technique is called high-throughput DNA sequencing, and it can sequence up to 100 million nucleotide bases of DNA per 7-hour run. It was used to analyze genome sequences from the deceased transplant patients.

RNA was extracted from the infected tissues of two of the patients, and the samples were then treated with DNase, an enzyme that removes all traces of human DNA. The remaining RNA was then amplified into millions of copies of the corresponding single-stranded DNA. The resulting DNA strands were sequenced using high-throughput DNA sequencing, which determines the sequence of a piece of DNA by adding new complementary nucleotides in a reaction that gives off a burst of light when the complementary nucleotides bases are added.

Once the sequences were generated, additional techniques were used to eliminate any contaminating human genome sequences. The remaining pieces were then fitted together into longer strings. Of the more than 100,000 sequences initially produced, only 14 matched viral proteins in a database of all known microbial sequences. These sequences from the patients' tissues were closely related to the sequences of a pathogen called lymphocytic choriomeningitis virus (LCMV), which usually causes only a minor flu-like illness in healthy people. Once the LCMV-like virus was characterized, probes were designed to detect the virus in clinical samples. Using these probes, evidence of the LCMV-like virus was discovered in several tissue samples from all three Australian transplant recipients.

The new sequencing method is seen as a powerful new tool for pathogen surveillance and for diagnosing mysterious illnesses and emerging infectious diseases. In 2008, a 70-year-old woman died and a 57-year-old man became critically ill in a Boston hospital after each received a kidney from a donor infected with the LCMV-like virus. The virus was identified based on the probes developed from the Australian cases.

MICROFOCUS 9.6: Tools

Microarrays

Once a microbial genome had been sequenced, scientists needed a way to discover how the genes interact in that organism. So, in the early 1990s, biotechnologists discovered a way to put these DNA segments on a miniaturized surface, such as a glass slide, or a plastic or silicon surface. Thousands of spots could be contained on a single microarray, often called a **DNA chip**. Each spot is put in place by a mechanical robot and is unique because the spot contains many copies of a unique sequence of single-stranded nucleotides produced through the polymerase chain reaction (PCR) from the DNA of the sequenced organism. Such an array may contain DNA from one species or many species.

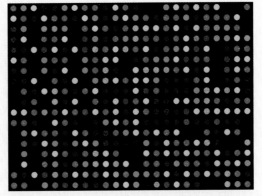

DNA microarray showing locations of differently labeled DNA molecules.

Scientists use single-stranded, fluorescently labeled DNA that is complementary to a DNA sequence as a probe. Often two different samples of probes are used, each labeled with a different colored fluorescent dye (often red and green). When exposed to (washed over) the microarray, a probe will bind to any complementary sequences (the target) and the spot will fluoresce the color (red or green) of the bound probe. If both probes bind, they will fluoresce the intermediate color, yellow; or if more of one probe binds than another, the spot may be more orange or light green. The results usually are scanned and analyzed by computer.

Microarrays can be used in many ways.

- **Gene expression:** Microarrays can be developed and used to study gene expression. For example, a microbiologist might want to study the effects of oxygen on gene expression in the facultative species *Escherichia coli*. Using a microarray containing the *E. coli* genome, the microbiologist could isolate *E. coli* cells grown under aerobic conditions and another sample under anaerobic conditions. From these samples, the mRNAs (from active genes under the two conditions) would be isolated and copied into single-stranded DNA. The DNA probes from the aerobic condition may be labeled with the red dye while the anaerobic condition may be labeled with the green dye. The two probes allow the scientist to study the same cell (genes) under two different conditions (see figure).

- **Organism detection:** Because of the constant threat posed by emerging infectious diseases and the limitations of existing approaches available to identify new pathogens, microarrays offer a rapid and accurate method for viral discovery. DNA microarray-based platforms have been developed to detect a wide range of known viruses as well as novel members of some viral families. For example, during the outbreak of severe acute respiratory syndrome (SARS) in March 2003 (Chapter 15), a viral isolate cultivated from a SARS patient was made into a DNA probe and, on the microarray, produced a spot representing the SARS virus.

  A microarray consisting of human DNA can be probed with a DNA sequence from an unknown pathogen to see if that person has been infected. For example, early infection with the malarial parasite *Plasmodium falciparum* can be identified by isolating DNA from the tissues of a suspected patient, fragmenting the DNA into single-stranded DNA segments, and attaching these to the solid support. This microarray is then washed with a fluorescently-labeled *P. falciparum* probe. Any fluorescent spots detected on the microarray indicate a match to *P. falciparum* DNA, confirming the patient is infected.

- **Phylogenetic relationships:** The extent of microbial diversity in an environment can be assessed by producing a microarray, called a **Phylochip**, containing oligonucleotides (short sequences of nucleotides) that are complementary to 16S rRNA sequence probes of different bacterial species. Any lit spot on the microarray is an indicator that that species is part of the microbial community in that environment.

MICROFOCUS 9.7: Biotechnology

Putting Humpty Back Together Again

Humpty Dumpty sat on a wall.
Humpty Dumpty had a great fall.
All the king's horses,
And all the king's men,
Couldn't put Humpty together again.

We all remember this nursery rhyme, but it can be an analogy for the efforts required in sequencing the genome of an organism. To sequence a whole genome, you have to take thousands of small DNA fragments and, after sequencing them, try to "put the fragments together again." The good news in genomics is that you can "put Humpty together again."

Small fragments must be used when sequencing a whole genome because current methods will not work with the tremendously long stretches of DNA, even those shorter ones found in bacterial genomes. Therefore, one strategy is to break the genome into small fragments. These fragments then are sequenced using sequencing machines and the fragments reassembled into the full genome. This technique is called the "whole-genome shotgun method." It can be extremely fast, but there are so many little pieces that it can be very difficult to put the whole genome together again.

The "shotgun" strategy first was used in 1995 by Craig Venter, Hamilton Smith, Claire Fraser, and their colleagues to sequence the genome of *Haemophilus influenzae* and *Mycoplasma genitalium*. To sequence these bacterial genomes, segments of the DNA were cut into 1,600 to 2,000 base pairs. The segments then were partly sequenced at both ends, using automated sequencing machines. These base-pair sequences—with their many overlaps—became the sequence information that was entered into the computer. Using innovative computer software, the thousands of DNA fragments generated were compared, clustered, and matched for assembling the genome of each organism.

Once assembled, the genes could be located, compared to known genes, and a detailed map developed (see figure). Sequencing of each genome took about a year but demonstrated that "the king's horses" (supercomputers and shotgun sequencing) and "the king's men" (the large group of collaborators) could "put Humpty together again"—and with speed and accuracy. Note: Since 1995, great strides have been made in sequencing technology. If *H. influenzae* were to be sequenced today, it would take about five days, rather than an entire year.

A linear map of the *Mycoplasma genitalium* genome. The horizontal arrows identify protein-coding genes. The direction of the arrow indicates the direction of transcription.

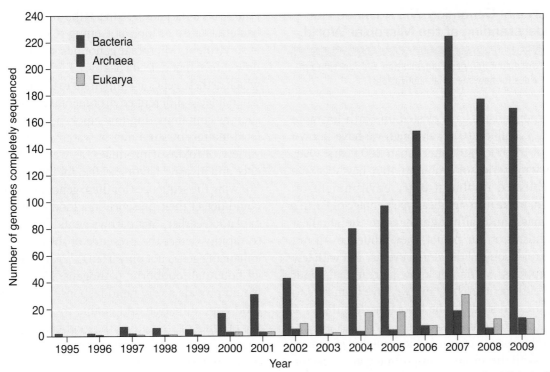

FIGURE 9.15 Microorganism Genomes Sequenced. The total number of genes completely sequenced and published will surpass 1,000 in 2009. »» Why have so few genomes been sequenced from the eukaryal organisms?
Source: Data from NCBI Database (Accessed Nov. 11, 2009). Available at http://www.ncbi.nlm.nih.gov/sites/entrez?db=genome.

(**FIGURE 9.15**). Sequences by themselves, although the result of very impressive work, do not tell us much. So, what do these sequences tell us and what practical use can be derived from this information?

CONCEPT AND REASONING CHECKS

9.9 Why have so many bacterial species been sequenced?

Segments of the Human Genome May Have "Microbial Ancestors"

KEY CONCEPT

10. Some human genes and human DNA sequences may have microbial origins.

With the sequencing of the human genome, one interesting development was to compare the human nucleotide gene sequences to known bacterial and viral sequences. Are there any similarities in the genes each contains?

Some comparisons indicate as many as 200 of our 25,000 genes are essentially identical to those found in members of the *Bacteria*; 25% or some 6,000 genes are found in yeast. However, these human genes were not acquired directly from these species, but rather were genes picked up by animals representing early ancestors of humans.

So important were these genes, they have been preserved and passed along from organism to organism throughout evolution; they are life's oldest genes. For example, several researchers suggest some genes coding for brain signaling chemicals and for communication between cells did not evolve gradually in human ancestors; rather, these ancestors acquired the genes directly from bacterial organisms. This provocative claim remains highly controversial though and more work is needed to better analyze this possibility.

Another discovery from the human genome project indicates that only about 5% to 10% of our DNA appears to code for proteins and regulatory RNAs. A number of scientists believe some of the non-gene DNA may be "genetic debris" from viruses that infected vertebrate ancestors hundreds of millions of years ago. Estimates suggest that 3% to 8% of the human genome is composed of self-replicating fragments of viral DNA. So, microbial genomes will have much to tell us about our past as comparisons continue.

CONCEPT AND REASONING CHECKS

9.10 How do microbial genomes compare in size and composition to the human genome?

Microbial Genomics Will Advance Our Understanding of the Microbial World

KEY CONCEPT

11. Understanding gene organization and function provides for new microbial applications.

Microorganisms have existed on Earth for more than 3.7 billion years, although we have known about them for little more than 300 years. Over this long period of evolution, they have become established in almost every environment on Earth, make up a significant percentage of Earth's biomass, and, although they are the smallest organisms on the planet, they influence—if not control—some of the largest events. Yet, with few exceptions, we do not know a great deal about any of these microbes, and we have been able to culture and study in the laboratory less than 1% of all microorganism species.

However, our limited knowledge is changing. With the advent of **microbial genomics**, the discipline of sequencing, analyzing, and comparing microbial genomes, we have begun the third Golden Age of microbiology, a time when remarkable scientific discoveries will be made toward understanding the workings and interactions of the microbial world. Some potential consequences from the understanding of microbial genomes are outlined below.

Safer Food Production. Since microorganisms play important roles in our foods both as contamination and spoilage agents, understanding how they get into the food product and how they produce dangerous foodborne toxins, will help produce safer foods. However, a major limitation with traditional food safety surveillance is that food-contaminating or spoiling microbes can only be identified after a long time-consuming approach that requires a number of days for colony counting of surviving microorganisms or bacterial toxin production using agar culture media. In addition, separate tests need to be run for each potential foodborne pathogen. This whole process can be greatly speeded up with microbial genomic technologies, enabling a quick and reliable prediction of the safety of our foods.

Microbial genomics can provide for a quick identification of microorganisms present in the (raw) food product. For example, until recently, there was little research into why *Campylobacter jejuni*, a bacterial pathogen that is the most common bacterial cause of food poisoning (Chapter 11), is so virulent. Microbial genomics studies have now shown that *C. jejuni* has over 1,700 genes. So, which of these genes are important to the organism when it faces different environmental challenges, such as contaminating raw chicken, residing in fecal matter, or surviving in water? There are a variety of toxins it produces, as well as adherence and invasive factors needed for infection. So knowing the sequences for these genes means one could detect their presence in a food sample. As explained earlier, microarrays would be one way to rapidly detect the presence of these disease-enhancing genes not only from *C. jejuni*, but from all potential foodborne pathogens. Microarrays would provide a very rapid response to potential food contamination and enable the removal of the product or strengthen the food-chain control strategies before the pathogen could cause illness to consumers.

Overall, genomics of food microbes generates valuable knowledge that can be used to protect our agriculture produce and meats, and also could be applied as a tool to trace potential food contamination between farm to table, a real problem today for fruits and vegetables coming from both within and outside the United States.

Identification of Unculturable Microorganisms. Because the vast majority of bacterial and archaeal species cannot be cultured (see Chapter 5), genomics offers a way to identify these organisms. Craig Venter is just one of many scientists studying the gene sequences of these **viable but not culturable (VBNC)** organisms. Venter and others are attempting to sequence and identify the collective genomes, called the **metagenome**, of all bacterial and archaeal species in a microbial community, such as the Sargasso Sea (see Chapter 1). This ability to identify unculturable organisms is opening up the new discipline of **metagenomics**. Such genomic information allows microbiologists to better understand how microbial communities function and how the organisms interact with one another.

Genomic information also is being used to discover if there are unculturable microbial representatives that could be used as alternative energy sources to solve critical environmental problems, including global warming and the development of renewable energy sources such as hydrogen and methane.

Microbial Forensics. The advent of the anthrax bioterrorism events of 2001 and the continued threat of bioterrorism has led many researchers to look for ways to more efficiently and more rapidly detect the presence of such bioweapons (see Chapter 1). Many of the diseases caused by these potential biological agents cause no symptoms for at least several days after infection, and when symptoms appear, initially they are flu-like. Therefore, it can be difficult to differentiate between a natural outbreak and the intentional release of a potentially deadly pathogen.

Such concerns have lead to a relatively new and emerging area in microbiology called **microbial forensics**, the discipline that attempts to recognize patterns in a disease outbreak, identify the responsible pathogen, control the pathogen's spread, and discover the source of the pathogenic agent. Investigative tools, like gene sequencing, DNA probes, microarrays, and PCR often are sufficient if such a disease outbreak is a natural one. However, if the "outbreak" is the result of a purposeful release—a bioterrorism attack—then tracking down the source of the microbe (and perpetrator) is critical. For example, the anthrax letter attacks of 2001 generated panic among the public and showed the need to establish "attribution" (who is responsible for the crime) for fear that another such attack might occur. The 2009 swine flu pandemic at one point was proposed by some to be the result of an accidental release of the virus from a research lab doing vaccine experimentation. Microbial forensics has not supported this claim.

The science behind microbial forensics includes classical microbiology, genetic engineering, microbial genomics, and phylogenetics. Forensic microbiological investigations are essentially the same as any other forensic investigation as they involve a crime scene(s) investigation, evidence collection, chain of custody for the collection, handling, and preservation of evidence, interpretation of results and—unique to the scientist—court presentation. Importantly, unlike research data that must hold up to scrutiny by peer reviewers and journal editors, microbial forensic data must be undeniable and must hold

up to the scrutiny of judges and juries in a court of law. Because the evidence presented must be beyond contention, the investigative tools of gene sequencing, DNA probes, microarrays, and PCR may not be enough to ensure the evidence will bring a perpetrator(s) to justice and to deter future attacks.

Microbial forensics also is involved in other types of medical and hygiene cases. Cases of medical negligence, where a hospital's inadequate or improper hygiene can lead to a patient contracting a post-surgical or hospital-acquired infection (and perhaps die), could be settled through forensics analysis. In addition, outbreaks of foodborne disease have brought lawsuits against companies alleging negligence in sanitary practices. And the potential for intentional contamination through foodborne terrorism is certainly possible. In all these cases, tracing the infecting microbe to the company or person(s) of origin will be critical. This places the fields of genetic engineering, biotechnology, and microbial genomics at the forefront of this emerging field.

CONCEPT AND REASONING CHECKS

9.11 How is microbial genomics contributing to a better world?

Comparative Genomics Brings a New Perspective to Defining Infectious Diseases and Studying Evolution

KEY CONCEPT

12. Comparative genomics compares the DNA sequences of related or unrelated species.

Sequencing the DNA bases of a microorganism (or any other organism) has and still does provide important information concerning the number of bases and genes comprising the organism. However, such sequences provide little understanding of how these genes work together to run the metabolism of an organism. One needs to understand how the organism uses its genome to form a functioning unit. Sequencing is only the first part of a deeper understanding.

Having sequenced a microbial genome, the next step is to discover the functions for the genes. Sequences need to be analyzed (called **genome annotation**) to identify the location of the genes and the function of their RNA or protein products.

For example, in most of the microbial genomes sequenced to date, nearly 50% of the identified genes encoding proteins have not yet been connected with a cellular function. About 30% of these proteins are unique to each species. The challenging discipline of **functional genomics** attempts to discover what these proteins do and how those genes interact with others and the environment to maintain and allow the microbe to grow and reproduce.

Pathogenicity: The ability of a pathogen to cause disease.

One of the most important areas beyond DNA sequencing is the field of **comparative genomics**, which compares the DNA sequence from one microbe with the DNA sequence of another similar or dissimilar organism. Comparing sequences of similar genes indicates how genomes have evolved over time and provides clues to the relationships between microbes on the phylogenetic tree of life (see Chapter 1).

Comparisons indicate, for example, that some strains of a bacterial species contain **genomic islands**, sequences of up to 25 genes that are absent from other strains of the same species. Many of these islands can be identified as having come from an altogether different species, suggesting some form of HGT, such as conjugation, occurred in the past. It is believed the nonpathogenic bacterial species *Thermotoga maritima* has acquired about 25% of its genome from HGT. In addition, sequence analysis indicates its genome is a mixture of bacterial and archaeal genes and suggests *T. maritima* evolved before the split of the *Bacteria* and *Archaea* domains.

One of the most interesting aspects of comparative genomics relates to infectious disease. By comparing the genomes of pathogenic and non-pathogenic bacterial species, or between pathogens with different host ranges, microbiologists are learning a lot about pathogen evolution. Here are a few examples.

There are three bacterial species of *Bordetella* (Chapter 10). *B. pertussis* causes whooping cough in humans, *B. parapertussis* causes whooping cough in infants, but also infects sheep, and *B. bronchiseptica* produces respiratory infections in other animals (**FIGURE 9.16A**). Comparative genomic analysis of these three species reveals that *B. pertussis* and *B. parapertussis* are missing large segments of DNA (1,719 genes), which are present in *B. bronchiseptica*. This analysis sug-

gests (1) *B. pertussis* and *B. parapertussis* evolved from a *B. bronchiseptica*-like ancestor; and (2) the adaptation of *B. pertussis* and *B. parapertussis* to their more restrictive hosts is due to the loss of the 1,719 genes. In fact, only *B. bronchiseptica* is capable of surviving outside the host. So, in this genome comparison between similar species, survival of *B. pertussis* and *B. parapertussis* requires they infect organisms supplying them with the materials they no longer can make; that is, pathogenicity has evolved from the loss of gene function.

At the opposite extreme is the evolution of pathogenicity through the acquisition of new genes (**FIGURE 9.16B**). *Corynebacterium diphtheriae* is the causative agent for diphtheria (Chapter 10). Genome analysis indicates this species in the not too distant past acquired through HGT 13 genetic regions, each representing a genomic island. These islands are called **pathogenicity islands** because they encode many of the pathogenic characteristics of the bacterial species (e.g., pili formation and iron uptake).

E. coli O157:H7 has recently become a dangerous threat to human health worldwide, causing severe gastrointestinal ailments (Chapter 11). One of the most recent outbreaks involved the contamination of bagged spinach. Some 200 Americans became ill and at least two died. When the genome of *E. coli* O157:H7 was compared to the nucleotide sequence of a non-pathogenic strain (K12), another example for the presence of pathogenicity islands was discovered (**FIGURE 9.16C**). Both strains have a large genome and have evolved from a common ancestor. Both have genomic islands acquired through HGT. However, the genomic islands in *E. coli* O157:H7 code for the known pathogenicity genes (e.g., pili and toxins) and therefore represent pathogenicity islands. The genomic islands in the non-pathogenic strain lack these pathogenicity genes. What is not clear is if the pathogenicity islands were acquired only by the O157:H7 strain or the non-pathogenic strain lost the pathogenicity islands.

These few examples represent examples of the power of comparative genomics to resolve differences between species and shed light on the evolution of bacterial pathogens.

CONCEPT AND REASONING CHECKS

9.12 Explain how comparative genomics helps explain some aspects of pathogenicity.

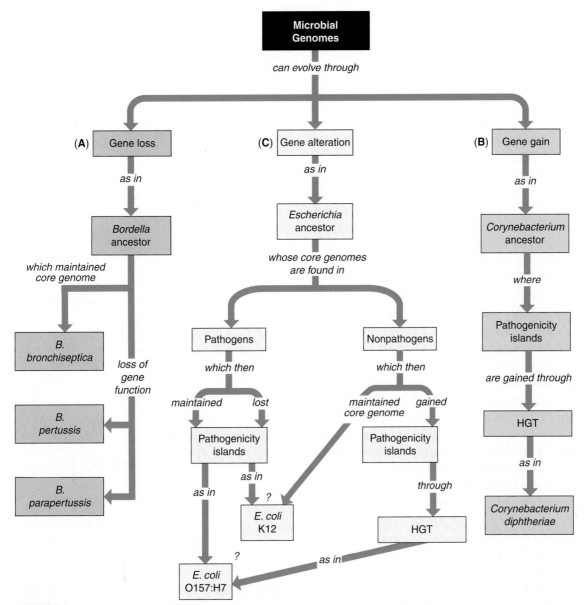

FIGURE 9.16 **Comparative Genomics Suggests How Microbial Genomes Can Evolve.** This concept map summarizes the evolution of microbial genomes through loss or gain of gene function. »» What advantage is there to losing or gaining genes, or whole sets of genes (pathogenicity islands)?

Metagenomics Is Identifying the Previously Unseen Microbial World

KEY CONCEPT

13. Techniques are now being developed to analyze and understand all the genomes within a microbial community.

Existing within, on, and around every living organism, and in most all environments on Earth, are microorganisms. Yet, as we mentioned in Chapter 5, some 99% of the bacterial and archaeal species found within us and in the environment will not grow on traditional growth media; we referred to these organisms as viable, but not culturable (VBNC). That means most organisms found in the soil, in oceans, and even in the human body have never been seen or named—and certainly not studied.

There is now a genetic process for analyzing this unculturable majority. As mentioned earlier, the process is called **metagenomics** (*meta* = "change"), and it refers to the change (1) in the way genes and genomes within mixtures of organisms within a community (the metagenome) are studied and

(2) in our understanding of the microbial world. Metagenomics has the potential to stimulate the development of advances in fields as diverse as medicine, agriculture, and energy production. Reread the introduction to Chapter 1, describing Craig Venter's ocean voyage of microbial sampling and finding new genes. Here is how the metagenomics process is carried out (FIGURE 9.17).

Samples of the desired community of microorganisms are collected. These samples could be from soil, water, or even the human digestive tract. The DNA is then extracted from the sample, which produces DNA fragments from all the microbes in the samples. The fragments can be amplified through PCR and cloned into plasmids, which are introduced into bacteria, such as *E. coli*. A metagenomic library results that represents the entire community DNA from the microbes sampled.

Analysis of the DNA fragments or plasmid clones can be used to do sequence analysis and functional analysis. In "sequence-based metagenomics," the random fragments are cloned to a level that produces vast amounts of DNA that can be linked back to the probable origin of the DNA. Genes and metabolic pathways of different organisms in the community and with other communities can be compared. With "functional-based metagenomics," the gene products from the cloned plasmids in the bacterial cells are searched for new enzymes, vitamins, antibiotics, or other potential chemicals of therapeutic or industrial use.

As Venter's explorations and those of others have shown, metagenomics already has opened our eyes to the diversity of microbes in ocean environments. The process also makes it possible to harness the power of microbial communities to help solve some of the most complex medical, environmental, agricultural, and economic challenges in today's world.

Medicine. Understanding how the microbial communities that inhabit our bodies affect human health could lead to new strategies for diagnosing, treating, and preventing diseases.

Ecology and the Environment. Exploring how microbial communities in soil and the oceans affect the atmosphere and environmental conditions could help us understand, predict, and address climate change.

Energy. Harnessing the power of microbial communities might result in sustainable and eco-

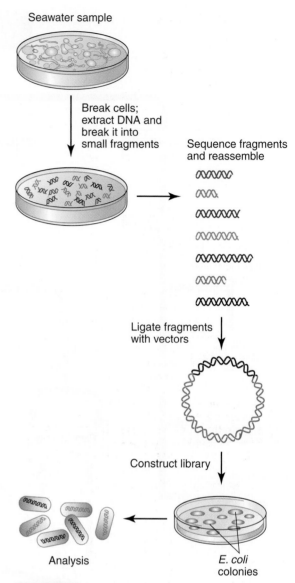

FIGURE 9.17 The Process of Metagenomics. Metagenomics allows the simultaneous sequencing of a whole community of microorganisms without growing each species in culture. Fragments can be either sequenced or subjected to functional analysis. »» What type of information is supplied by sequence-based metagenomics versus functional-based metagenomics?

friendly bioenergy sources, as exemplified by the explorations of Craig Venter and others.

Bioremediation. Adding to the arsenal of microorganism-based environmental tools can help in monitoring environmental damage and cleaning up oil spills, groundwater, sewage, nuclear waste, and other hazards.

Biotechnology. Taking advantage of the functions of microbial communities might lead to the development of newer and safer food and health products.

Agriculture. Understanding the roles of beneficial microorganisms living in, on, and around domesticated plants and animals could enable detection of diseases in crops and livestock, and aid in the development of improved farming practices.

Biodefense. The addition of metagenomics tools to microbial forensics will help to monitor

pathogens, create more effective vaccines and therapeutics against bioterror agents, and reconstruct attacks that involve microorganisms.

CONCEPTS AND REASONING CHECKS

9.13 What does metagenomics offer that previous sequencing techniques have not been able to provide?

SUMMARY OF KEY CONCEPTS

9.1 Genetic Recombination in Bacteria

1. Recombination implies a horizontal transfer of DNA fragments between bacterial cells and an acquisition of genes by the recipient cell. All three forms of recombination are characterized by the introduction of new genes to a recipient cell by **horizontal gene transfer**.

2. In **transformation**, a competent recipient cell takes up DNA fragments from the local environment. The new DNA fragment displaces a segment of equivalent DNA in the recipient cell, and new genetic characteristics may be expressed.

3. In one form of **conjugation**, a live donor (F⁺) cell transfers an **F factor** (plasmid) to a recipient cell (F⁻), which then becomes F⁺.

4. In another form of conjugation, **Hfr** strains contribute a portion of the donor's chromosomal genes to the recipient cell.

5. **Transduction** involves a virus entering a bacterial cell and later replicating within it. In **generalized transduction**, a bacterial DNA fragment is mistakenly incorporated into an assembling phage. In **specialized transduction**, the virus first incorporates itself into, then detaches from, the chromosome, taking a segment of chromosomal DNA with it. In both forms, the phage transports the DNA to a new recipient (transduced) cell.

9.2 Genetic Engineering and Biotechnology

6. **Genetic engineering** is an outgrowth of studies in bacterial genetic recombination. The ability to construct **recombinant DNA molecules** was based on the ability of **restriction endonucleases** to form **sticky ends** on DNA fragments.

7. **Plasmids** can be isolated from a bacterial cell, spliced with foreign genes, then inserted into fresh bacterial cells where the foreign genes are expressed as protein. The cells become

biochemical factories for the synthesis of such proteins as insulin, interferon, and human growth hormone.

8. **DNA probes** can be used to detect pathogens.

9.3 Microbial Genomics

9. Since 1995, increasingly more microbial genomes have been sequenced; that is, the linear sequence of bases has been identified.

10. A comparison of bacterial genomes with the human genome has shown there may be some 200 genes in common between these organisms. Comparisons between microbial genomes indicate almost 50% of the identified genes have yet to be associated with a protein or function in the cell.

11. With the understanding of the relationships between sequenced microbial DNA molecules comes the potential for safer food production, the identification of unculturable microorganisms, a cleaner environment, and improved monitoring of pathogens through **microbial forensics**.

12. Sequencing is only the first step in understanding the behaviors and capabilities of microorganisms. **Functional genomics** attempts to determine the functions of the sequenced genes and how those genes interact with one another and with the environment. **Comparative genomics** compares the similarities and differences between microbial genome sequences. Such information provides an understanding of the evolutionary past and how pathogens might have arose through the gain or loss of **pathogenicity islands**.

13. **Metagenomics** is providing new insights into the function of diverse genomes in microbial communities.

LEARNING OBJECTIVES

After understanding the textbook reading, you should be capable of writing a paragraph that includes the appropriate terms and pertinent information to answer the objective.

1. Contrast **vertical** and **horizontal gene transfer** mechanisms.

2. Describe and assess the role of **transformation** as a **genetic recombination** mechanism.

3. Explain how an **F factor** is transferred during **conjugation**.

4. Distinguish between an **Hfr strain**, and **F⁻ recombinant** cell, and an **F′ plasmid**.

5. Summarize the steps involved in (a) **generalized** and (b) **specialized transduction**.

6. Differentiate between **genetic engineering** and **biotechnology**.

7. Identify the role of **plasmids** and **restriction endonucleases** in the genetic engineering process.

8. Explain how **DNA probes** are used to (1) identify pathogens and (2) conduct water-quality tests.

9. Describe what it means to say that a bacterial genome has been "sequenced."

10. Assess the importance of **microbial genomics** in understanding the human genome.
11. Summarize how microbial genomics can contribute to food safety, microorganism identification, and **microbial forensics**.
12. Justify the need for **comparative genomics** as related to pathogen evolution.
13. Identify how information from **metagenomics** can contribute to medicine, energy production, agriculture, and biodefense.

STEP A: SELF-TEST

Each of the following questions is designed to assess your ability to remember or recall factual or conceptual knowledge related to this chapter. Read each question carefully, then select the **one** answer that best fits the question or statement. Answers to even-numbered questions can be found in **Appendix C**.

1. Which one of the following is NOT an example of genetic recombination?
   A. Conjugation
   B. Binary fission
   C. Transduction
   D. Transformation
2. Transformation refers to
   A. using a virus to transfer DNA fragments.
   B. DNA fragments transferred between live donor and recipient cells.
   C. the formation of an F⁻ recombinant cell.
   D. the transfer of naked fragments of DNA.
3. An F⁻ cell is unable to initiate conjugation because it lacks
   A. double-stranded DNA.
   B. a prophage.
   C. an F factor.
   D. DNA polymerase.
4. An Hfr cell
   A. has a free F factor in the cytoplasm.
   B. has a chromosomally integrated F factor.
   C. contains a prophage for conjugation.
   D. cannot conjugate with a F⁻ recombinant.
5. A _____ is NOT associated with specialized transduction.
   A. virulent phage
   B. lysogenic cycle
   C. prophage
   D. recipient cell
6. Which complementary sequence would NOT be recognized by a restriction endonuclease?
   A. GAATTC
      CTTAAG
   B. AAGCTT
      TTCGAA
   C. GTCGAC
      CAGCTG
   D. AATTCC
      TTAAGG

7. A _____ seals sticky ends of recombinant DNA segments.
   A. DNA ligase
   B. restriction endonuclease
   C. protease
   D. RNA polymerase
8. _____ are single-stranded DNA molecules that can recognize and bind to a distinctive nucleotide sequence of a pathogen.
   A. Prophages
   B. Plasmids
   C. Cloning vectors
   D. DNA probes
9. The first completely sequenced genome from a free-living organism was from
   A. humans.
   B. *E. coli*.
   C. *Haemophilus*.
   D. *Bordetella*.
10. What percentage of the human genome is identical to the yeast genome?
    A. 5%
    B. 10%
    C. 25%
    D. 50%
11. A metagenome refers to
    A. a large genome in an organism.
    B. the collective genomes of many organisms.
    C. the genome of a metazoan.
    D. two identical genomes in different species.
12. Genomic islands are
    A. gene sequences not part of the chromosomal genes.
    B. adjacent gene sequences unique to one or a few strains in a species.
    C. acquired by HGT.
    D. Both **B** and **C** are correct.
13. Craig Venter's sampling of ocean microorganisms is an example of
    A. microarrays.
    B. horizontal gene transfer.
    C. microbial forensics.
    D. metagenomics.

STEP B: REVIEW

Use the following syllables to compose the term that answers each of the clues below. The number of letters in each term is indicated by the blank lines, and the number of syllables is shown by the number in parentheses. Each syllable is used only once, and the answers to even-numbered statements are in **Appendix C.**

ASE BAC CLE COC COM CON CUS DO DROME EN FITH GA GASE GE GRIF HOR I I IN JU LENT LI MIDS MO NOME NU O PAL PE PHAGE PLAS PNEU TAL TENCE TER TION U VIR ZON

14.	Closed loops of DNA	(2)	__ __ __ __ __ __ __
15.	Restriction recognition sequence	(3)	__ __ __ __ __ __ __ __ __
16.	Transforming property	(3)	__ __ __ __ __ __ __ __ __ __
17.	Transduction virus	(5)	__ __ __ __ __ __ __ __ __ __ __ __
18.	Recombinant DNA enzyme	(5)	__ __ __ __ __ __ __ __ __ __ __
19.	Transformed bacterium	(4)	__ __ __ __ __ __ __ __ __ __
20.	DNA linking enzyme	(2)	__ __ __ __ __ __
21.	Discovered transformation	(2)	__ __ __ __ __ __
22.	Type of HGT	(4)	__ __ __ __ __ __ __ __ __ __ __
23.	Phage that causes lysis	(3)	__ __ __ __ __ __ __ __
24.	Complete set of genes	(2)	__ __ __ __ __ __
25.	Type of gene transfer	(4)	__ __ __ __ __ __ __ __ __ __

STEP C: APPLICATIONS

Answers to even-numbered questions can be found in **Appendix C.**

26. In 1976, an outbreak of pulmonary infections among participants at an American Legion convention in Philadelphia led to the identification of a new disease, Legionnaires' disease. The bacterial organism, *Legionella pneumophila,* responsible for the disease had never before been known to be pathogenic. From your knowledge of bacterial genetics, can you postulate how it might have acquired the ability to cause disease?

27. You are going to do a genetic engineering experiment, but the labels have fallen off the bottles containing the restriction endonucleases.

One loose label says *Eco*RI and the other says *Pvu*I. How could you use the plasmid shown in MicroInquiry 9 to determine which bottle contains the *Pvu*I restriction enzyme?

28. As a research member of a genomics company, you are asked to take the lead on sequencing the genome of *Legionella pneumophila* (see Question 26). (a) Why is your company interested in sequencing this bacterial species, and (b) what possible applications are possible from knowing its DNA sequence?

STEP D: QUESTIONS FOR THOUGHT AND DISCUSSION

Answers to even-numbered questions can be found in **Appendix C.**

29. Which of the recombination processes (transformation, conjugation, or transduction) would most likely occur in the natural environment? What factors would encourage or discourage your choice from taking place?

30. Some bacterial cells can take up DNA via the transformation process. From an evolutionary perspective, what might have been the original advantage for the cells taking up naked DNA fragments from the extracellular environment?

31. Since the 1950s, the world has been plagued by a broad series of influenza viruses that differ genetically from one another. For example, we have heard of swine flu, Hong Kong flu, Bangkok flu, and avian flu. How might the process of transduction help explain this variability?

32. It is not uncommon for students of microbiology to confuse the terms reproduction and recombination. How do the terms differ?

33. While studying for the microbiology exam covering the material in this chapter, a friend and biology major asks you why genomics, and especially microbial genomics, was emphasized. How would you answer this question?

⊙ HTTP://MICROBIOLOGY.JBPUB.COM/9E

The site features eLearning, an online review area that provides quizzes and other tools to help you study for your class. You can also follow useful links for in-depth information, read more MicroFocus stories, or just find out the latest microbiology news.

PART 3

Bacterial Diseases of Humans

False-color transmission electron microscope image of *Mycobacterium tuberculosis*, which has infected one-third of the human population.

Throughout history, bacterial diseases have posed a formidable challenge to humans and often swept through populations virtually unchecked. In the eighteenth century, the first European visitors to the South Pacific found the islanders robust, happy, and well adapted to their environment. But the explorers introduced syphilis, tuberculosis, and pertussis (whooping cough) to a susceptible population. Soon these diseases spread like wildfire. For example, the Hawaiian population was about 300,000 when Captain Cook landed in 1778; by 1860, disease had reduced the population to fewer than 37,000.

With equally devastating results, the Great Plague came to Europe from Asia, and cholera spread westward from India. Together with tuberculosis, diphtheria, and dysentery, these bacterial diseases ravaged European populations for centuries and insidiously wove themselves into the pattern of life. Infant mortality was particularly shocking: England's Queen Anne, who reigned in the early 1700s, lost 16 of her 17 babies to disease; and until the mid-1800s, only half the children born in the United States reached their fifth year.

Today, humans can cope better with bacterial diseases. Though credit often is given to antimicrobial drugs, the major health gains have resulted from understanding disease and the body's resistance mechanisms, coupled with modern sanitary methods to prevent microorganisms from reaching their targets. Immunization also has played a key role in preventing disease. Indeed, very few people in our society die of the bacterial diseases that once accounted for the majority of all deaths.

In Part 3 of this text, we study the bacterial diseases of humans. The diseases have been grouped according to their major mode of transmission. Airborne diseases are discussed in Chapter 10; foodborne and waterborne diseases in Chapter 11; soilborne and arthropodborne diseases in Chapter 12; and sexually transmitted and contact transmitted diseases in Chapter 13. Many of these diseases are of historical interest and are currently under control. However, the human body is continually confronted with newly emerging or resurgent infectious diseases. In this regard, disease has not changed; only the pattern of disease has changed.

Clinical Microbiology

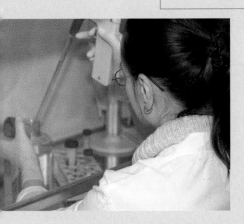

One of the most famous books of the twentieth century was *Microbe Hunters* by Paul de Kruif. First published in the 1920s, de Kruif's book describes the joys and frustrations of Pasteur, Koch, Ehrlich, von Behring, and many of the original microbe hunters. The exploits of these scientists make for fascinating reading and help us understand how the concepts of microbiology were formulated. I would urge you to leaf through the book at your leisure.

Microbe hunters did not come to an end with Pasteur, Koch, and their contemporaries, nor did the stories of microbe hunters end with the publication of de Kruif's book. Approximately 25% of all deaths worldwide and 60% of all deaths in children under four years of age are due to infectious agents. Today, clinical microbiology is concerned with the microbiology of infectious diseases, and the men and women working in hospital, public, and private laboratories are today's diseases detectives. These individuals search for the pathogens of disease. Many travel to far corners of the world studying organisms, and many more remain close to home, identifying the pathogens in samples sent by physicians, identifying their interactions with the immune system, and working out the diagnosis and epidemiology of these diseases.

In fact, a well-developed knowledge of clinical microbiology is critical for the physician and medical staff who are faced with the concepts of disease and antimicrobial therapy. Microbiologists even work in dental clinical labs, since many bacterial species are involved in tooth decay and periodontal disease. Microbiology is one of the few courses where much of the fundamentals of microbiology are used regularly. This includes the clinical aspects of infectious diseases: manifestations (signs and symptoms), diagnosis, treatment, and prevention.

A career in clinical microbiology usually requires a master of science in clinical microbiology. With such a degree, jobs include supervisory positions in medical centers or private reference laboratories, infection control positions in clinical settings, public health, marketing and sales in the pharmaceutical and biotechnology industries, teaching at community colleges or technical colleges, or research in academic, government or industry (pharmaceutical and biotechnology) settings.

Clinical microbiology also offers an outlet for the talents of those who prefer to tinker with machinery. New instruments and laboratory procedures are constantly being designed and developed in an effort to shorten the time between detection and identification of microorganisms. Many tests used in the clinical laboratory reflect human ingenuity. For example, there is a test that detects bacterial species by their interference with the passage of light and their ability to scatter light at peculiar angles. Such modern devices as laser beams are used in this kind of instrumentation.

The microbe hunters have not changed materially in the past 100 years. The objectives of the search may be different, but the fundamental principles of the detective work remain the same. The clinical microbiologist is today's version of the great masters of a bygone era.

Airborne Bacterial Diseases

Pertussis is the only vaccine-preventable childhood illness that has continued to rise since the 1980s with an increasing proportion of cases in adolescents and adults.
—Centers for Disease Control and Prevention

On August 14, 2002, a 39-year-old male oil refinery worker in Crawford County, Illinois, visited the refinery's health unit complaining of a two-week cough. Later that day, the worker's 50-year-old supervisor also visited the unit with a spastic cough, which had started three days earlier. Both patients were advised to see their own health care provider where blood samples indicated a recent infection with *Bordetella pertussis*. The Crawford County Health Department and Illinois Department of Public Health were contacted because a possible outbreak could be brewing.

In the early parts of the 20th century, one of the most common childhood diseases and causes of death in the United States was pertussis, commonly called whooping cough. Before the introduction of a pertussis vaccine in 1940, *B. pertussis* was responsible for infection and disease in 150 out of every 100,000 people. By 1980, the **incidence**, or frequency with which the disease occurs, had dropped to one in every 100,000 individuals. The vaccine had almost eliminated the pathogen.

At the oil refinery, active surveillance and case investigations were initiated by the health officials. Those workers with a persistent and spastic cough were sent to the local hospital for evaluation and interviews. Health department officials needed to know the time of illness onset, where workers worked in the refinery, work schedule, and individuals with whom they had close contacts. Local school officials and health care providers were alerted and given guidelines on ways to recognize pertussis and prevent its spread.

In the course of the epidemiological investigation, 17 cases of pertussis were identified at the refinery, 15 having had close contact with the supervisor originally diagnosed; 7 cases occurred among the community and had no apparent relation to the refinery. In all, 21 of the cases occurred in adults 20 years of age or older. Patients received an antibiotic effective against the pathogen and all recovered.

How the disease was passed from the supervisor remains unclear. *B. pertussis* is spread by airborne droplets (FIGURE 10.1). Other than an indoor, 5-minute morning meeting each day, work assignments were all outdoors, although workers often congregated in an indoor dining area at lunch.

Every 3 to 4 years, a pertussis outbreak occurs in the United States—and, as indicated above, many of these cases occur in adults. Although nearly all youngsters growing up receive the pertussis vaccine, vaccine-induced protection does not last a lifetime; therefore, adolescents and adults can become susceptible to disease when vaccine-induced immunity wanes, approximately 5 to 10 years after vaccination. As a result, college students and adults (like the refinery workers) may be vulnerable (FIGURE 10.2).

Pertussis is but one of a group of bacterial infectious diseases affecting the respiratory tract. We will divide these diseases into two general categories. The first category will include diseases of the upper respiratory tract, such as strep throat and diptheria. The second category will include diseases of the lower respiratory tract: pertussis, tuberculosis, and pneumonia. As we proceed, note that antibiotics are available for treating the bacterial diseases while immunizations are used for protecting the community at large.

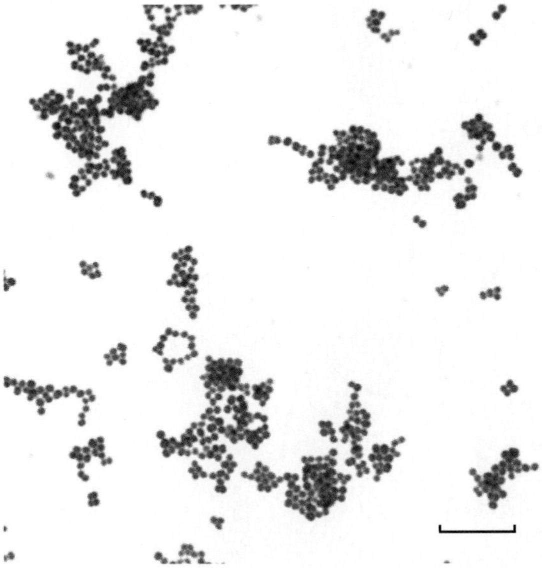

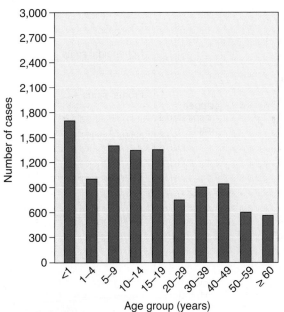

FIGURE 10.1 *Bordetella pertussis.* This Gram stain shows chains of small *B. pertussis* cells. (Bar = 10 μm.) »» What is the Gram reaction of these stained cells?

FIGURE 10.2 Number of Cases of Pertussis by Age Group—United States 2007. The actual number of pertussis cases (especially among adolescents [10–19 years] and adults) is substantially underreported because the illness resembles other conditions, so infected individuals might not seek medical care. »» Which age group accounts for the majority of cases: infants (<1); children (1–9 years), adolescents, or adults?
Source: http://www.cdc.gov/mmwr/preview/mmwrhtml/mm5653a1.htm; page 65.

10.1   Structure and Indigenous Microbiota of the Respiratory System

The **respiratory system** is composed of a conducting portion that brings oxygen to the lungs and a respiratory portion that exchanges oxygen and carbon-dioxide gases with the bloodstream. Because air typically contains microbes and viruses carried on dust and droplet nuclei, it should not be surprising that the respiratory system is the most common portal of entry for these infectious agents.

Upper Respiratory Tract Defenses Limit Microbe Colonization of the Lower Respiratory Tract

KEY CONCEPT

1. Microbial colonization usually is limited to the upper respiratory tract.

The respiratory system is divided into the upper respiratory tract and the lower respiratory tract (**FIGURE 10.3**). The **upper respiratory tract (URT)**

is composed of the nose, sinus cavities, and pharynx (throat), while the **lower respiratory tract (LRT)** is composed of the larynx, trachea, bronchi, and lungs. The lungs contain the alveoli where gas exchange occurs. The average adult inhales and exhales approximately 10,000 liters of air per day. Given that the inspired air contains microbes and microbe-laden particulate matter that could potentially bring microbes that cause infection, the respiratory system has evolved effective defense mechanisms to minimize such possibilities.

During breathing, the URT and bronchi play a critical role in filtering out foreign material, such as bacteria, viruses, and the dust particles that might carry these microbes. A process called **mucociliary clearance** involves the entrapment of microbes and particulate matter larger than 2 μm in a layer of mucus, which is then moved by ciliated epithelial cells toward the pharynx where it is

Mucus:
A slimy fluid containing antimicrobial compounds, proteins, and inorganic salts.

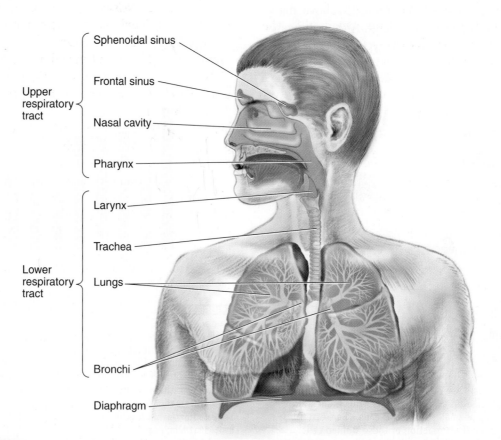

FIGURE 10.3 **Respiratory System Anatomy.** The major parts of the respiratory system are organized into the upper and lower respiratory tracts. »» Which part of the respiratory system would be the most susceptible to colonization and infection?

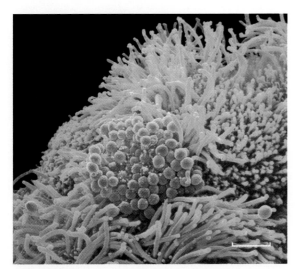

FIGURE 10.4 *Staphylococcus* **in the Ciliated Epithelium.** False-color scanning electron micrograph of a *Staphylococcus* colony (yellow) on the epithelial cells of the trachea. The cilia (blue hair-like projections) keep the trachea free of dust and other irritants. (Bar = 5 µm.) »» How does the ciliated epithelium work to eliminate these bacterial cells from the trachea?

either swallowed or expectorated (**FIGURE 10.4**). Mucociliary clearance is supplemented by the presence and activity of several antimicrobial substances, including lysozyme and lactoferrin. In addition, an anionic antimicrobial peptide, which is active against gram-positive and gram-negative bacteria, is present, along with IgA and IgG antibodies, and several human defensins. Thus, mucociliary clearance and the activity of antimicrobial substances are largely responsible for maintaining much of the LRT virtually free of microbes and particulate matter. In addition, recent studies suggest vitamin D plays an important role in immune defense, including the prevention of URT infections (Chapter 21).

In the nose, large particles present in inhaled air are removed by hairs in the nostrils, while smaller particles and suspended bacteria become trapped in the mucus covering the nasal mucosa. In the posterior two-thirds of the nasal mucosa, the mucociliary clearance also propels the mucus-entrapped particles into the pharynx. However, in the anterior region of the nasal mucosa the disposal of entrapped microbes is largely dependent on microbicidal substances present in nasal fluid. Several antimicrobial agents have been detected, including lysozyme and lactoferrin, and IgA anti-

bodies that block microbial adhesion to epithelial cells. The nasal fluids therefore are capable of killing or inhibiting potential pathogens such as *Staphylococcus aureus* and *Pseudomonas aeruginosa*. Sneezing and coughing are additional mechanical methods to eliminate microbes trapped in the mucus of the respiratory tract.

In the LRT, the epithelial cells lining the alveolar and respiratory bronchioles are not ciliated. However, the region is covered by alveolar fluid, which contains a number of antimicrobial components, including lysozyme and immunoglobulins. The predominant immunoglobulin is IgG, which facilitates the phagocytosis of any microbes lining this part of the respiratory tract. Should larger numbers of microbes enter the alveoli, the alveolar macrophages recruit neutrophils from the pulmonary capillaries to help clear the invaders.

The constant exposure to the environment means that many different microorganisms can form part of the normal **microbiota**, the microbial community found on and in the body. Among the microbes identified with the URT are species of *Streptococcus, Neisseria* (in the nasopharynx), *Haemophilus, Staphylococcus* (primarily in the anterior nares of the nose), *Corynebacterium,* and *Propionibacterium.* Several, including *Streptococcus, Neisseria, Haemophilus,* and *Staphylococcus,* are opportunistic species that have the potential to cause serious illnesses in immunocompromised individuals. These members of the microbiota can damage parts of the tract, such as the mucociliary clearance process, if given the opportunity.

Host microbiota have the upper hand for space and nutrients, thereby setting the limits on the ability of pathogens to compete for the space and nutrients, and thus to colonize and cause infections in the respiratory system.

The indigenous community of microbes will vary depending on the specific part of the tract because of the "environmental differences." For example, colonization of the external nares primarily involves corynebacteria and propionibacteria, bacteria normally found on the skin surface (Chapter 13). The anterior nares certainly are a source of *S. aureus.* Interestingly, about 20% of the population seldom if ever carries the bacterium, while another 20% are persistent carriers. That leaves about 60% of the population who are intermittent

Lactoferrin:
An iron-binding protein that inhibits bacterial growth.

Defensins:
Small, catronic proteins that form holes in bacterial membranes.

Immunocompromised:
Referring to a person with a weakened immune system.

carriers; the bacterium is a part of the transient microbiota.

In the nasopharynx (the upper portion of the pharynx), the mucosal surface is mainly colonized by streptococci, and species of *Neisseria* and *Haemophilus*. Ciliated cells are rarely colonized by members of the indigenous microbiota.

In the oropharynx (the middle portion of the pharynx), *Haemophilus* and *Neisseria* as well as streptococci are common inhabitants. The oropharynx also harbors potentially pathogenic bacterial species, such as *N. meningitidis*, *S. pneumoniae*, *S. pyogenes*, and *H. influenzae*.

In the LRT, there is little, if any, microbiota. Small numbers of bacterial cells may be found in the larynx and trachea, but they are very transient and are usually eliminated by the cellular and chemical immune defenses of the LRT.

CONCEPT AND REASONING CHECKS

10.1   Why are microorganisms primarily found in the upper respiratory tract?

10.2   Bacterial Diseases Affecting the Upper Respiratory Tract

The bacterial diseases of the upper respiratory tract (URT) can be severe, as several diseases in this section illustrate. One reason is because the respiratory tract is a portal of entry to the blood, and from there, bacterial pathogens can spread to other sensitive internal organs.

Pharyngitis Is an Inflammation of the Throat

KEY CONCEPT

2.   *Streptococcus pyogenes* causes strep throat and scarlet fever.

A sore throat, known medically as **pharyngitis**, is not a disease itself. Rather, it is a nonspecific inflammatory response to toxins or pathogens, and usually is a symptom of a viral infection, such as the common cold or the flu. However, sometimes bacteria, especially the streptococci, can be the cause. The **group A streptococci (GAS)** are bacteria often found on the skin. In particular, *Streptococcus pyogenes*, a facultative, gram-positive coccus, may be carried on the skin of individuals who either have no symptoms of illness or exhibit a relatively mild illness. *S. pyogenes* is also carried in the throat and is the most common cause of **tonsillitis**, an inflammation of the tonsils. The tonsils help the immune system defend against infection by harmful bacteria and viruses. When tonsillitis occurs, treatment typically involves simple pain relievers or, if indicated, antibiotics. Surgery is seldom necessary, although years ago, surgery was the standard treatment in young children for recurrent tonsillitis.

Another, and sometimes potentially more dangerous, example of pharyngitis is **strepto-coccal pharyngitis**, popularly known as **strep throat**.

The *S. pyogenes* cells are highly contagious and reach the URT within respiratory droplets expelled by infected persons during coughing and sneezing. If the cells grow and secrete toxins, these substances cause damage to surrounding human cells and lead to an inflammation of the oropharynx and tonsils. Besides a severe and sudden sore throat, patients may develop a fever, headache, swollen lymph nodes and tonsils, and a beefy red appearance to pharyngeal tissues owing to tissue damage. More than a million Americans, primarily children, suffer strep throat annually.

Interestingly, among children (4 to 16 years of age) who had a tonsillectomy, the risk of a subsequent strep throat infection is about 20% compared to a 58% risk for children who still have their tonsils. An oral antibiotic, such as penicillin, is often prescribed to lessen the duration and severity of the inflammation and to prevent possible complications. Hand hygiene is the best prevention.

Scarlet fever is a disease arising in about 10% of children with streptococcal pharyngitis. Some strains of *S. pyogenes* carry toxin-encoding prophages coding for **erythrogenic** (*erythro* = "red") exotoxins that cause a pink-red skin rash on the neck, chest, and soft-skin areas of the arms (**FIGURE 10.5A**). The rash, which usually occurs in children under 15 years of age, results from blood leaking through the walls of capillaries damaged by the toxins. Other symptoms include a sore throat, fever, and a strawberry-like inflamed tongue (**FIGURE 10.5B**). Normally, an individual

Respiratory droplets:
Relatively large mucus particles that travel less than one meter.

Streptococcus pyogenes

Exotoxins:
Poisonous proteins secreted by some gram-positive and gram-negative bacteria.

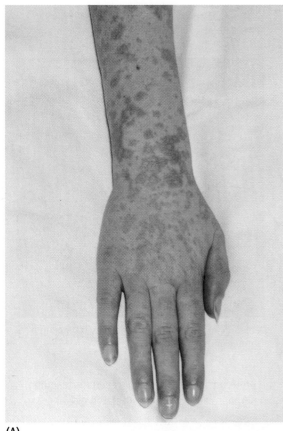

(A)

(B)

FIGURE 10.5 Scarlet Fever. Among the early symptoms of scarlet fever are (**A**) a pink-red skin rash and (**B**) a bright red tongue with a "strawberry" appearance.
»» What causes the skin rash seen with scarlet fever? What other symptoms are typical of scarlet fever?

experiences only one case of scarlet fever in a lifetime because recovery generates immunity.

Individuals with scarlet fever usually get better within two weeks without treatment

(MicroFocus 10.1). Treatment with antibiotics, such as penicillin or clarithromycin, can shorten the duration of symptoms and prevent serious complications.

A serious complication is **rheumatic fever**, which is most common in young school-age children. This condition, which is not an infection but rather an inflammation in response to the throat infection, primarily affects the joints and heart. It is characterized by fever and joint pain. The most significant long-range effect is permanent scarring and distortion of the heart valves, a condition called **rheumatic heart disease**. The damage arises from a response of the body's antibodies to streptococcal M proteins (pilus-like proteins) cross reacting with similar proteins on heart muscle. Rheumatic fever cases have been declining in the United States due to antibiotic treatment. In 1994, the last year the Centers for Disease Control and Prevention (CDC) required reporting of rheumatic fever cases, 112 cases were reported versus 10,000 reported cases in 1961. However, in developing nations, rheumatic fever remains a serious problem.

Another complication arising from streptococcal pharyngitis is **acute glomerulonephritis**, which is a rare inflammatory response to specific types of M proteins. The antigen-antibody complexes formed then accumulate in the glomerulus of the kidney. It is most common in young patients. Progressive, irreversible renal damage may occur in adults.

Infections of the LRT, causing streptococcal pneumonia, are described later in this chapter.

Glomerulus: The part of the kidney that controls filtering and excretion.

CONCEPT AND REASONING CHECKS

10.2 What makes *S. pyogenes* such a potentially dangerous pathogen in the upper respiratory tract?

Diphtheria Is a Life-Threatening Illness

KEY CONCEPT

3. *Corynebacterium diphtheriae* secretes a toxin that inhibits protein synthesis in epithelial cells.

Diphtheria is an infection of the URT that is caused by *Corynebacterium diphtheriae*, an aerobic, club-shaped, gram-positive rod (*coryne* = "club") that can spread rapidly through the URT.

When *C. diphtheriae* cells are stained with methylene blue, numerous blue cytoplasmic dots are seen in the cytoplasm that represent **metachromatic**

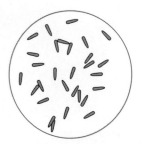

Corynebacterium diphtheriae

MICROFOCUS 10.1: Public Health
A Wakeup Call

In life, sometimes the best way to move past a road block is to simply face it head on—or attack the defensive team by running right at them. Often in the microbial world, bacterial species may have the same idea—but with a twist.

One of the most common host responses to a group A streptococci (GAS) infection in humans is to try to contain the infection by forming a blood clot around the infected area (see figure)—in other words, the host attempts to entomb the bacterial cells in the clot. Unfortunately, one of the abilities GAS possess is to simply break through the defensive wall set up to contain the infection. How do they do this?

Normally a blood clot stays as a clot because the protein plasmin that would dissolve the clot stays in an inactive form called plasminogen. What *Streptococcus pyogenes* does is to "wake up" or activate plasminogen.

Trapped within a clot, the *S. pyogenes* cells secrete an enzyme called streptokinase. Streptokinase then catalyzes the conversion of inactive plasminogen into active plasmin—in other words, it wakes the protein up. As plasmin, the protein triggers a series of reactions leading to the dissolving of the clot. Now the bacteria can escape and perhaps cause a serious infection elsewhere in the body.

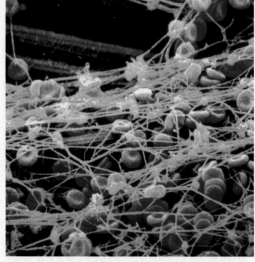

A blood clot, showing red blood cells trapped in a fibrin mesh.

■ **Incubation period:**
The time between exposure and the development of signs and symptoms.

■ **Epithelial cells:**
Cube-like cells lining the skin and body cavities such as the respiratory tract.

■ **Arrhythmia:**
An irregular heart beat.

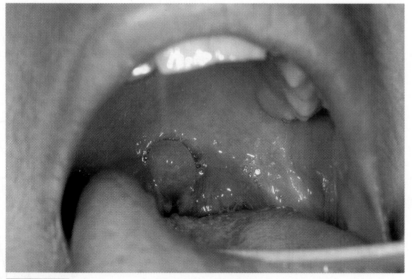

FIGURE 10.6 **Diphtheria Pseudomembrane.** Diphtheria is an upper respiratory tract infection of mucous membranes caused by a toxigenic strain of *Corynebacterium diphtheriae*. The infection is characterized by the formation of a pseudomembrane on the tonsils or pharynx. »» What is the pseudomembrane composed of?

granules (see Chapter 4). The bacterial cells remain in clumps after multiplying and form a picket fence-like arrangement called a "palisade arrangement."

Diphtheria is acquired by inhaling respiratory droplets from an infected person. Following a short incubation period (2-4 days), initial symptoms include a sore throat and low-grade fever. In epithelial cells, the bacterial cells secrete a potent exotoxin, which is encoded by a virus that infects *C. diphtheriae*. The exotoxin inhibits the translation process by ribosomes, resulting in the accumulation of dead tissue, mucus, white blood cells, and fibrous material, called a **pseudomembrane** ("pseudo" because it does not fit the definition of a true membrane) on the tonsils or pharynx (**FIGURE 10.6**). Mild cases fade after a week while more severe cases can persist for two to six weeks.

Complications can arise if the thickened pseudomembrane results in respiratory blockage, causing suffocation. If the exotoxin spreads to the bloodstream, heart and peripheral nerve destruction can lead to cardiac arrhythmia and coma. Left untreated, 5 to 10% of respiratory cases result in death.

Treatment requires antibiotics (penicillin or erythromycin) to eradicate the pathogen and antitoxins (antibodies) to neutralize the exotoxins. Immunization against diphtheria may be rendered by an injection of diphtheria **toxoid**, which, contained in the diphtheria-tetanus-acellular pertussis (DTaP) vaccine, consists of toxin molecules treated with formaldehyde or heat to destroy their toxic qualities. The toxoid induces the immune system to produce antibodies that circulate in the bloodstream throughout the person's life.

Due to immunization starting in early childhood, the number of cases of diphtheria in the United States is essentially zero (the last confirmed case was in 2003). However, the disease remains a health problem in many regions of the world and booster doses are required. The World Health Organization (WHO) reported 4,190 cases of diphtheria in 2007 with 80% of those cases in India.

CONCEPT AND REASONING CHECKS

10.3 In 17th-century Spain, diphtheria was called "el garatillo" = the strangler. Why was it given this name?

The Epiglottis Is Subject to Infection, Especially in Children

KEY CONCEPT

4. Swelling of the epiglottis can block the trachea.

A life-threatening but rare condition, called **epiglottitis**, involves infection of the epiglottis—the small flap of cartilage covering the trachea. If the area around the epiglottis becomes infected with *H. influenzae* or *S. pneumoniae,* the inflammation can spread to the epiglottis, which may swell such that it blocks the flow of air through the trachea and into the lungs.

Symptoms of epiglottitis include severe throat pain, fever, and a muffled voice. As the swelling of the epiglottis starts to narrow the airways, the affected person exhibits **stridor**, a high pitched wheezing sound when breathing in or out. The condition can progress rapidly making breathing even more difficult. In rare cases where medical intervention does not occur, complete airway blocking can result in death. The infection is most common and most dangerous in children because they have a smaller airway than adults. If, or once, the individual is breathing freely, intravenous antibiotics are given and the patient is closely observed in an intensive care unit until the swelling is reduced. Immunization with the Hib vaccine is the most effective way to prevent epiglottitis in children younger than age 5, which has made epiglottitis a rare infection in the United States.

CONCEPT AND REASONING CHECKS

10.4 How does an *H. influenzae* infection affect the epiglottis and potentially lead to a life-threatening condition?

The Nose Is the Most Commonly Infected Region of the Upper Respiratory Tract

KEY CONCEPT

5. Indigenous microbiota of the URT can cause sinus infections.

Because of its prominent position in the URT, the nose is a major portal of entry for infectious organisms and viruses. *S. aureus* is the most common agent of infection, which can result in boils (furuncles) that may develop into a more spreading infection under the skin (cellulitis) at the tip of the nose. Because this part of the face contains veins that lead to the brain, venous spread of bacteria can lead to a life-threatening condition called cavernous sinus thrombosis.

Sinusitis is an inflammation of the sinuses, the air-filled hollow cavities around the nose and nasal passages (**FIGURE 10.7**). The inflammation is most commonly caused by an allergy or infection, and represents one of the most common medical conditions. About 10 to 15 million people each year develop a so-called "sinus infection." It can occur in any of the four groups of sinuses: maxillary, ethmoid, frontal, or sphenoid. Sinusitis nearly always occurs in connection with inflammation of the nasal passages (**rhinitis**), and some doctors refer to the disorder as **rhinosinusitis**. It may be acute or chronic.

Acute sinusitis may be caused by a variety of indigenous microbiota of the URT. The condition often develops from a blockage at the openings to the sinuses, resulting from a common cold infection of the URT. Fluid trapped in the sinuses then becomes a nutrient growth medium for potential bacterial pathogens, including *S. pneumoniae* and *H. influenzae.* In addition, white blood cells and more fluid enter the sinuses to fight the infection and this only increases the pressure in the sinuses and causes more pain. Acute sinusitis

Thrombosis:
The formation of blood clots that partially or completely block a vein or artery.

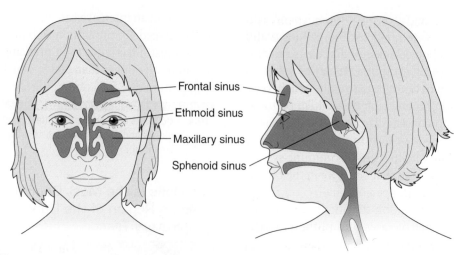

FIGURE 10.7 **The Sinuses.** The sinuses are hollow cavities within the facial bones. »» Which sinuses are most frequently "felt" by a person with a sinus infection?

usually results in pain, tenderness, and swelling over the affected sinuses. Yellow or green pus may be discharged from the nose. Fever and chills also can occur, but their presence is a likely indication that the infection has spread beyond the sinuses. Treatment of acute sinusitis is aimed at improving sinus drainage and curing the infection. Nasal sprays can be used for a short time, and antibiotics, such as amoxicillin or trimethoprim-sulfamethoxazole, can be prescribed for patients with a higher probability of a bacterial infection.

If untreated, acute sinusitis may develop into a chronic condition. **Chronic sinusitis** is defined as an infection that has been ongoing for 8 to 12 weeks. Doctors do not understand exactly what causes chronic sinusitis but it may follow a viral infection. The symptoms of chronic sinusitis are more subtle and pain occurs less often. The most common symptoms are nasal obstruction, nasal congestion, and post-nasal drip. The treatment is the same as with acute sinusitis, except antibiotic use, if bacterial, may be for a longer period of time. Preventing or reducing the risk of developing chronic sinusitis can be accomplished by good hygiene and frequent hand washing, and avoiding contracting viral, bacterial, or fungal infections. Treating cold symptoms immediately and using decongestants also may help prevent the development of a chronic condition.

Acute and chronic sinusitis affect more than 15% of Americans and together represent the fifth leading cause for antibiotic prescriptions.

CONCEPT AND REASONING CHECKS

10.5 How do acute and chronic sinusitis differ?

Ear Infections Are Common Illnesses in Early Childhood

KEY CONCEPT

6. Infections can occur in the outer and middle ear.

As part of the URT, the ears, nose, and throat are located near each other and, as such, allow infections to spread from one to the other. The ear, which is the organ of hearing and balance, consists of the outer, middle, and inner ear (**FIGURE 10.8**). The Eustachian tube vents the middle ear to the nasopharynx, which explains why URT infections often result in infections to the middle ear.

A few ear infections can occur in the outer ear. Such an inflammation, referred to as **otitis externa** (*oti* = "ear"), can affect the entire ear canal or just one small area, as in a boil (furuncle) or pimple. Normally, the ceruminous glands in the ear canal produce cerumen (earwax) that has antibacterial activity. However, outer ear infections commonly occur in children especially after extended swimming in fresh water pools. This can result in excessive moisture in the ear canal, which irritates and breaks down the skin in the canal, allowing bacterial cells to penetrate—such infections often are called **swimmer's ear**. A bacterial infection is most often caused by species of *Streptococcus*, *Staphylococcus*, or *Pseudomonas*.

The primary symptoms of otitis externa are itching followed by ear pain. Treatment involves the application of antibiotic ear drops to the ear several times a day for about one week. Swimmer's ear can be prevented by not swimming in polluted water and using ear plugs to keep the canal dry when swimming, showering, or bathing.

Short-term infections of the middle ear are called **acute otitis media** (*media* = "middle"). Such infections are among the most common illnesses of early childhood. The National Institute on Deafness and Other Communication Disorders reports that three out of four children contract at least one ear infection by age 3. Bacteria typically responsible for middle ear infections include *S. pneumoniae* and *H. influenzae*.

Middle ear infections usually start with a common cold infection of the URT. Inflammation of the Eustachian tube allows bacteria to infect the sterile environment of the middle ear where fluid buildup then provides an environment for bacterial growth. This is followed by ear pain with a red, bulging eardrum. Children with ear infections may develop a fever, produce a fluid that drains from the ears, or have headaches. They also may have trouble sleeping and be unusually irritable.

Most individuals get better without treatment. The American Academy of Pediatrics (AAP) and the American Academy of Family Physicians (AAFP) suggest that children older than 6 months who are otherwise healthy and have only mild signs and symptoms not be given antibiotics at least for the first 72 hours, as most cases of otitis media resolve on their own in a few days. In fact, the AAP and AAFP report that about 80% of children with otitis media recover without antibiotics. For chronic bacterial infections, a physician may decide to administer an antibiotic such as amoxicillin. Prevention can be difficult due to the behaviors of young children interacting with one another. Limiting the time a child spends in group childcare may help prevent common colds and ensuing middle ear infections.

Chronic otitis media (COM) is a condition involving long-term infection, inflammation, and damage to the middle ear. Children have persistent fluid in the ears that lasts for months in the absence of any other symptoms except hearing impairment. COM is a major global cause of hearing impairment and can have serious long-term effects on language, auditory and cognitive development, and educational progress; it is a major public health problem in many populations around the world, and a significant cause of morbidity and mortality.

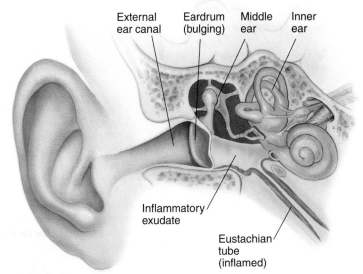

FIGURE 10.8 **Ear Anatomy.** The ear consists of external, middle, and inner structures. »» In a middle ear infection, how do the infecting microbes reach the middle ear?

It is now recognized that COM is not the result of re-infection, but rather stems from a persistent **biofilm** that has colonized the middle ear tissue. As described in Chapter 3, biofilms are antibiotic-resistant colonizations of bacteria that attach to surfaces and form a slime-like defensive barrier, protecting the bacteria from immune attack or antibiotic therapy. Treatment of COM depends upon the stage of the disease. Efforts are made to control the causes of Eustachian tube obstruction. If active infection is present in the form of ear drainage, antibiotic ear drops are prescribed, which may be supplemented with oral antibiotics. Once the active infection is controlled, surgery is usually recommended to clear the obstruction.

CONCEPT AND REASONING CHECKS

10.6 In most cases with otherwise healthy children, why is the use of antibiotics not recommended for acute otitis media?

We will end this section on URT diseases by discussing one of the most dangerous bacterial diseases spread through the air—meningitis. An infection can spread to the central nervous system where, without a resident microbiota, the bacterial cells may cause a life-threatening inflammation of the meninges.

Meninges: The membranes covering the brain and spinal cord.

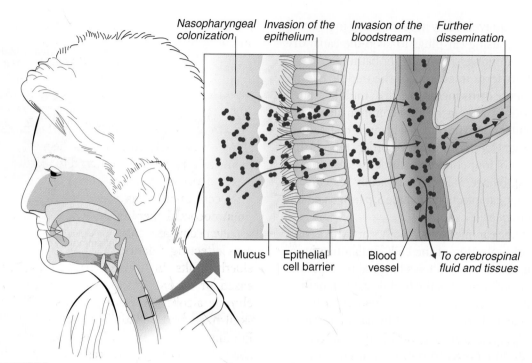

FIGURE 10.9 Pathogenic Steps Leading to Meningitis. The bacterial species capable of causing meningitis (*N. meningitidis*, *S. pneumoniae*, and *H. influenzae*) can colonize the nasopharynx, and then invade the epithelium causing respiratory distress. They then pass into the bloodstream. Finally, they disseminate to tissues near the spinal cord, causing inflammation and meningitis. »» What bacterial cell structures would facilitate (a) attachment to the nasopharynx and (b) survival in the bloodstream?

Acute Bacterial Meningitis Is a Rapidly Developing Inflammation

KEY CONCEPT

7. Acute bacterial meningitis is most common among children aged 1 month to 2 years.

Acute bacterial meningitis begins when a localized infection develops into a blood inflection that then invades the meninges (**FIGURE 10.9**). For example, it could start as: an upper respiratory tract infection caused by *N. meningitidis*, *S. pneumoniae*, or *H. influenzae*; lobar pneumonia due to *S. pneumoniae*; or otitis media caused by *H. influenzae*. Infection through the meninges, brain, and spinal cord can be very rapid, resulting in death within hours.

Neisseria meningitidis. A particularly dangerous form of meningitis is **meningococcal meningitis** caused by *N. meningitidis* (**FIGURE 10.10**). The pathogen is a small, encapsulated, aerobic, gram-negative diplococcus that attaches to the nasopharyngeal mucosa by pili. It also forms a capsule of which there are more than 14 serogroups. Most infections in the

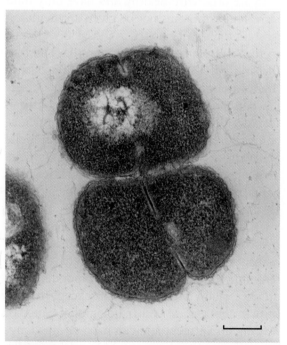

FIGURE 10.10 *Neisseria meningitidis*. A false-color transmission electron micrograph of *N. meningitidis* cells. The cytoplasm (blue) and cell membrane are surrounded by a capsule (green). (Bar = 0.5 μm.) »» What role does the capsule play in the disease process?

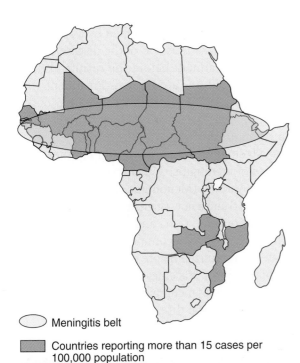

Meningitis belt

Countries reporting more than 15 cases per 100,000 population

FIGURE 10.11 **The African Meningitis Belt.** This map of Africa shows the so-called "meningitis belt" where deadly epidemics caused by *N. meningitidis* occur almost every year. »» How is *N. meningitidis* spread between individuals?

United States are due to serogroup B, C, Y, and W-135, with B the major cause of disease and mortality in infants. According to the World Health Organization (WHO), bacterial meningitis is responsible for more than 700,000 cases globally each year. Ninety percent of cases occurred in the 18 countries forming sub-Saharan Africa's so-called "meningitis belt" (**FIGURE 10.11**).

N. meningitidis is a fragile organism that does not survive easily in the environment and must be maintained in nature through person-to-person transfer of large droplet respiratory secretions. Meningococcal meningitis is therefore prevalent where people are in close proximity for long periods of time. Grade-school classrooms, military camps, college dorms, and prisons are examples. Though most people suffer nothing worse than a respiratory disease, the CDC reported 1,077 cases of meningococcal disease in 2007, a historic low. Globally, it is the agent responsible for about 50% of meningitis cases in 2- to 18-year-olds.

Cases of meningococcal meningitis in very young children are sometimes complicated by the formation of lesions in the adrenal glands and accompanying hormone imbalances. This condition, called the **Waterhouse-Friderichsen syndrome**, results from the release of a bacterial endotoxin into the blood. Death can occur within 10 to 12 hours.

Streptococcus pneumoniae. This bacterial species was described earlier as the major cause of bacterial pneumonia. Often referred to as the pneumococcus, it causes many cases of pneumococcal pneumonia that then develop into **pneumococcal meningitis**. The inflammation usually occurs as a community-acquired meningitis affecting infants, and middle-aged and elderly adults. In addition, patients with a diseased or absent spleen, sickle cell disease, chronic alcoholism, and patients with recent skull fractures or head injury are at high risk. Pneumoccocal meningitis is responsible for about 30% of meningitis cases and has a high mortality rate (20% to 30%).

Haemophilus influenzae **Type b.** In 1892, Richard Pfeiffer isolated a small, nonmotile, encapsulated, gram-negative rod he thought was the cause of influenza. However, during the great influenza epidemic of 1918, Pfeiffer's bacillus was identified as a secondary cause of the disease, and influenza was attributed to a virus rather than a bacterium.

Haemophilus influenzae type b (Hib) once was the most prevalent bacterial species causing meningitis (**Haemophilus meningitis**) in children between the ages of 2 months and 5 years. In 1986, about 18,000 such cases of *Haemophilus* meningitis occurred in the United States annually, killing about 1,000 children. At that time, however, a vaccine was licensed by the Food and Drug Administration (FDA) and the epidemic peaked. As of 1993, the vaccine was combined with the DTaP vaccine for distribution to children as Tetramune, and in 2007, the number of reported cases among children under 5 years of age in the United States was down to 22.

As mentioned, all three bacterial species discussed above can enter the body by respiratory droplets from prolonged contact, such as coughing, sneezing, or kissing. In the case of *N. meningitidis,* the disease consists of an influenza-like upper respiratory infection called **meningococcal pharyngitis**. However, should the organism invade into the nonciliated epithelium and spread to the blood,

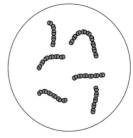

Streptococcus pneumoniae

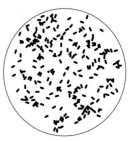

Haemophilus influenzae b

Serogroup: A group of microbes having similar surface proteins (antigens).

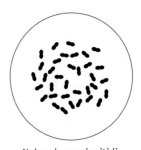

Neisseria meningitidis

a condition called **meningococcemia** occurs and the bacterial cells multiply rapidly.

Once in the blood, all three pathogens are capable of crossing the blood-brain barrier (Figure 10.9). The meninges then become inflamed, causing pressure on the spinal cord and brain. Patients normally experience a fever and stiff neck; symptoms rapidly evolve into a pounding headache, nausea and vomiting, and often sensitivity to bright light. With meningococcal meningitis, a rash also appears on the skin, beginning as bright-red patches, which progress to blue-black spots. Left untreated, within a few hours, 50% of cases of meningococcal meningitis result in coma and death. The disease also can produce lasting disabilities, such as deafness, blindness, and paralysis.

Acute bacterial meningitis, especially, meningococcal meningitis, represents a medical emergency. Early diagnosis and treatment are crucial to prevent disabilities or death. A principal criterion for diagnosis is the observation and/or cultivation of the *N. meningitidis* cells in samples of spinal fluid obtained by a spinal tap. However, the seriousness of the disease usually demands treatment before the results of diagnostic procedures are known. Treatment with antibiotics, such as penicillin, cefotaxime, or ceftriaxone, usually is recommended, often in large intravenous doses.

No single vaccine provides immunity to all causes of meningitis. A meningococcal polysaccharide vaccine (Menomune) has been available since 1978 and a new capsular vaccine (Menactra) to serogroups A, C, Y, and W–135 is available. The CDC strongly recommends all college freshman be immunized before taking up residence in a college dorm. Vaccines against pneumococcal and *Haemophilus* meningitis also are available. The airborne bacterial diseases of the URT are summarized in TABLE 10.1 .

CONCEPT AND REASONING CHECKS

10.7   Draw a concept map for bacterial meningitis using the bacterial species and bold-faced terms in this section.

10.3   Bacterial Diseases of the Lower Respiratory Tract

In the lower respiratory tract (LRT), a number of bacterial diseases affect the lung tissues. As injury occurs, fluid builds up in the lung cavity, and the space for obtaining oxygen and eliminating carbon dioxide is reduced. This is the basis for a possibly fatal pneumonia.

Pertussis (Whooping Cough) Is Highly Contagious

KEY CONCEPT

8.   *Bordetella pertussis* secretes toxins that destroy cells of the ciliated epithelium.

Pertussis (*per* = "through"; *tussi* = "cough"), also known as **whooping cough**, is caused by *Bordetella pertussis*, a small, aerobic, gram-negative rod. The bacilli are spread by respiratory droplets that adhere to and aggregate on the cilia of epithelial cells in the mouth and throat (FIGURE 10.12). Exotoxin production paralyzes the ciliated cells and impairs mucus movement, potentially causing pneumonia.

Pertussis is one of the more dangerous and highly contagious diseases of childhood years.

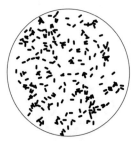

Bordetella pertussis

Typical cases of pertussis occur in two stages lasting 4 to 12 weeks. The initial (**catarrhal**) stage is marked by general malaise, low-grade fever, and increasingly severe cough. During the second (**paroxysmal**) stage, disintegrating cells and mucus accumulate in the airways and cause labored breathing. Patients experience multiple **paroxysms**, which consist of rapid-fire staccato coughs all in one exhalation, followed by a forced inhalation over a partially closed glottis. The rapid inhalation results in the characteristic "whoop" (hence, the name whooping cough). Ten to fifteen paroxysms may occur daily, and exhaustion usually follows each. Sporadic coughing continues during several weeks of convalescence, even after the pathogen has vanished. (Doctors call it the "100-day cough."). Convalescence depends on the speed at which the ciliated epithelium regenerates.

Eradication of the bacterial cells is generally successful when erythromycin is administered before the respiratory passageways become blocked. However, antibiotic treatment only reduces the duration and severity of the illness.

As with diphtheria, the low incidence of pertussis stems partly from use of a pertussis

TABLE 10.1	A Summary of the Major Bacterial URT Diseases				
Disease	Causative Agent	Signs and Symptoms	Transmission	Treatment	Prevention and Control
Streptococcal pharyngitis	*Streptococcus pyogenes*	Sore throat, fever, headache, swollen lymph nodes and tonsils	Respiratory droplets	Penicillin	Practicing good hand hygiene
Scarlet fever	*Streptococcus pyogenes*	Pink-red rash on neck, chest, arms Strawberry-like tongue	Respiratory droplets	Penicillin Clarithromycin	Practicing good hygiene
Diphtheria	*Corynebacterium diphtheriae*	Sore throat and low-grade fever	Respiratory droplets	Penicillin Erythromycin	Vaccinating with DTaP
Epiglottitis	*Haemophilus influenzae*	Severe throat pain, fever, muffled voice	Respiratory droplets	Intravenous antibiotics	Vaccinating with Hib
Sinusitis	Indigenous microbiota	Pain, tenderness, and swelling	—	Nasal sprays Antibiotics	Minimizing contact with individuals with colds
Otitis externa	*Streptococcus, Staphylococcus, Pseudomonas* species	Itching and ear pain	Contaminated water	Lifestyle modifications Topical and oral medications	Keeping ears dry
Acute otitis media	*Streptococcus pneumoniae Haemophilus influenzae*	Ear pain Red, bulging eardrum	Airborne Contact	Wait and see Antibiotics	Limiting time in childcare
Acute bacterial meningitis	*Neisseria meningitidis Streptococcus pneumoniae Haemophilus influenzae* type b	Fever, stiff neck, severe headache, vomiting and nausea, sensitivity to light	Respiratory droplets from prolonged contact	Antibiotics	Vaccination

vaccine. The older vaccine (diphtheria-pertussis-tetanus, or DPT) contained merthiolate (Thimerosal)-killed *B. pertussis* cells and was considered risky because about 1 in 300,000 vaccinees suffered high fevers and seizures. Now, public health officials recommend the newer acellular pertussis (aP) vaccine prepared from *B. pertussis* chemical extracts. Combined with diphtheria and tetanus toxoids, the triple vaccine has the acronym DTaP; commercially, it is known as Tripedia.

Although the incidence of pertussis has declined substantially since introduction of an effective vaccine in 1949, the number of cases in the United States has been rising since 1981. In fact, the CDC recorded 10,454 cases in 2007. Suggestions for this increase include a more virulent *B. pertussis* strain, a greater awareness of the disease, and the use of better and more accurate laboratory tests for the pathogen. Also the majority of reported cases occur in adolescents and adults who lose vaccine-induced immunity 5 to 10 years after vaccination (see chapter opener). In 2005, the U.S. Food and Drug Administration (FDA) licensed two new single-dose booster vaccines (Tdap) to provide adolescents and adults with protection against tetanus, diphtheria, and pertussis.

■ Merthiolate:
A mercury derivative compound formerly used in vaccines as a disinfectant and preservative.

CONCEPT AND REASONING CHECKS

10.8 Why have reported cases of pertussis been increasing in the United States?

*Mycobacterium
tuberculosis*

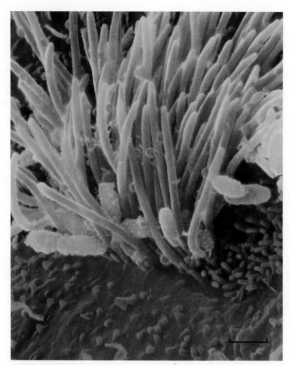

FIGURE 10.12 *Bordetella pertussis* **Associated with the Ciliated Epithelium.** False-color scanning electron micrograph of human tracheal epithelium. The pertussis cells (green) cause a dramatic loss of cilia function, which is to protect the respiratory tract from dust and particles. (Bar = 1 μm.) »» Propose a hypothesis to explain why cilia loss would lead to fits of coughing.

Tuberculosis Is One of the Greatest Challenges to Global Health

KEY CONCEPT

9. *Mycobacterium tuberculosis* causes a two-stage illness.

Tuberculosis (TB) is an ancient disease. Scientists analyzing spinal column fragments from Egyptian mummies more than 4,400 years old have found pathological signs of tuberculosis. Hippocrates, more than 2,000 years ago described a widespread illness that he called "phthisis," which was probably TB. Over the centuries, TB has continued to be a "slate wiper" in the human population. During the first half of the 20th century, TB was called "consumption" or "white plague" and it continued to be the world's leading cause of death from all causes, accounting for one fatality in every seven cases.

Today's statistics, though improved, are still threatening. Although the CDC reported that in 2007 there was an all-time low of 13,299 cases in the United States, in developing nations health officials report more deaths from TB than from any other infectious disease. In fact, today there are more people with TB than ever in history. The WHO estimates that some 2 billion individuals, one-third of the world's population, are infected with the TB bacillus; 2 million die of TB every year; and unless control measures are strengthened, another 1 billion people globally will become infected and 36 million will die of tuberculosis by 2020.

Tuberculosis is caused by *Mycobacterium tuberculosis,* the "tubercle" bacillus first isolated by Robert Koch in 1882. It is a small, aerobic, nonmotile rod whose cell wall forms a waxy cell surface that greatly enhances resistance to drying, chemical disinfectants, and many antibiotics. In the laboratory, Gram staining will not penetrate the waxy layer, so, when processing a sputum sample, the staining must be accompanied by heat to penetrate this barrier, or a lipid-dissolving material must be used. Once stained, however, the organisms resist decolorization, even when subjected to a 5% acid-alcohol solution. Thus, the bacilli are said to be acid-resistant or **acid-fast** (see Chapter 4).

Epidemiology. Tuberculosis is primarily an airborne disease and, as such, the bacilli are transmitted from person to person in small, aerosolized droplets when a person with active pulmonary disease sneezes, coughs, spits, or even sings. The infectious dose is quite small and the inhalation of even a single *M. tuberculosis* cell can lead to a new infection. However, individuals with prolonged, frequent, or intense contact with a diseased individual are at most risk of becoming infected, with an estimated 30% infection rate. Thus, crowded conditions and poor ventilation often contribute to disease spread and people who live in overcrowded, urban ghettoes often contract TB. Malnutrition and a generally poor quality of life also contribute to the establishment of disease.

Pathogenesis. Unlike many other infectious diseases where an individual becomes ill after several days or a week, the incubation period for TB is much longer. In addition, the illness has two separate stages: an infection stage and a disease stage (**FIGURE 10.13**). If a person has a pulmonary infection (85% of infections are respiratory), the bacterial cells

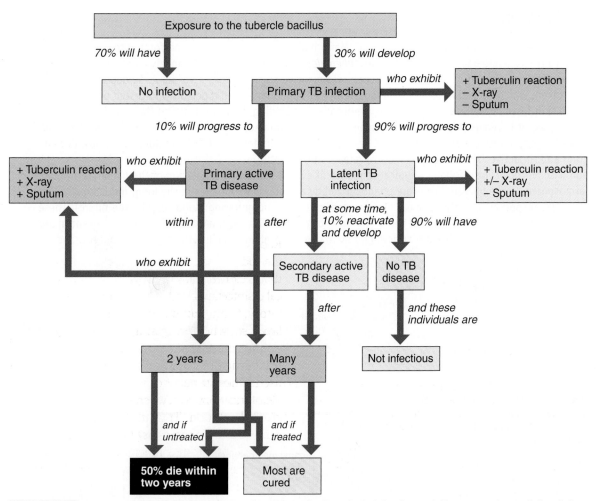

FIGURE 10.13 A Concept Map for Tuberculosis. The stages of tuberculosis infection and disease are shown. "+" or "−" equal positive or negative test results. »» How does TB infection differ from active TB disease?

enter the alveoli where pathogen interactions occur (**FIGURE 10.14**). This individual is now said to have a **primary TB infection.** If tested, the person would have a positive tuberculin reaction, but a chest X ray and sputum test would be negative (see Disease Detection, below).

In the alveoli, macrophages respond to the infection by ingesting the bacilli. Unfortunately, the bacilli are not killed in the macrophages and as more macrophages arrive, they too phagocytize bacilli but are incapable of destroying them. Eventually, lymphocytes and fibroblasts surround the mass in the lung, forming a hard nodule called a **tubercle** (hence the name tuberculosis), which may be visible in a chest X ray. MicroFocus 10.2 looks at the cell biology of tubercule formation and breakdown.

In 90% of primary TB infections, the infection becomes arrested; the tubercles heal and undergo

fibrosis and calcification. These individuals often are never aware they are infected, although they may now have a positive chest X ray. This dormant form of TB is referred to as a **latent TB infection** and is carried by 2 billion people worldwide. Of these, 90% will never develop active disease and will not be infectious.

Up to 10% of individuals who have a primary or latent TB infection will develop the second stage of the illness: a clinical disease. Primary TB infections can develop into **primary active TB disease** and latent TB infections can have a reactivation of the bacilli, which develop into **secondary active TB disease.** Individuals in both groups will become ill within three months, experience chronic cough, chest pain, and high fever, and they continue to expel sputum that accumulates in the LRT. (Often the sputum is rust colored, indicating that blood has entered

MICROFOCUS 10.2: Public Health
Anatomy of a Tubercle

Applying for a bachelor of science in nursing program, Maria had a tuberculin skin test performed to see if she had been exposed to the tubercle bacillus *Mycobacterium tuberculosis*, meaning she could conceivably have active tuberculosis (TB) or a latent infection. The result of the test was positive so she was sent to the X-ray department to have a chest X ray taken, which is used to rule out the possibility of pulmonary TB in a person who tested positive with the tuberculin skin test reaction but no symptoms of disease. The findings were normal and Maria had no signs of any TB granulomas, or tubercles. However, had she been infected, the bacterial cells would have caused an inflammation in the lungs, producing abnormal shadows on the chest X ray indicative of TB granulomas. So, how does a tubercle form and, in active TB disease, break up to release active bacilli?

Infection with *M. tuberculosis* follows a typical sequence of events. Following inhalation of the bacilli as droplets from the atmosphere, ① the bacterial cells are phagocytized by alveolar macrophages in the lung, inducing ② a localized inflammatory response that attracts more mononuclear cells from neighboring blood vessels (see figure). These white blood cells are the building blocks for the granuloma, or tubercle. As the granuloma develops, ③ a central area containing infected macrophages becomes surrounded by foamy macrophages, cells with high lipid content, and other immune cells. ④ As the granuloma matures, the surrounding layer of lymphocytes becomes associated with fibrous collagen and other extracellular components that mark the edge of the granuloma.

The surviving mycobacteria in the granuloma no longer replicate and switch to a dormant, non-replicative state as the immune cell aggregates restrict bacterial spreading. At this stage, then, the infection has been contained and there are no overt signs of disease. As a latent infection, the host cannot transmit the mycobacteria to others but the immune cells cannot kill all the bacilli, which can persist in the granuloma for years or a lifetime.

In secondary active TB disease, containment of the mycobacteria fails when the immune status of the host changes, usually as a result of old age, malnutrition, or co-infection with HIV—any condition that creates a weakened immune system. Following such a change, the immune defenses fail. ⑤ The granuloma decays into a formless mass of cellular debris that, to the naked eye, has a cheesy appearance, ruptures, and releases thousands of viable, infectious bacilli into the airways, ⑥ triggering a productive cough that facilitates the aerosol spread of infectious mycobacteria.

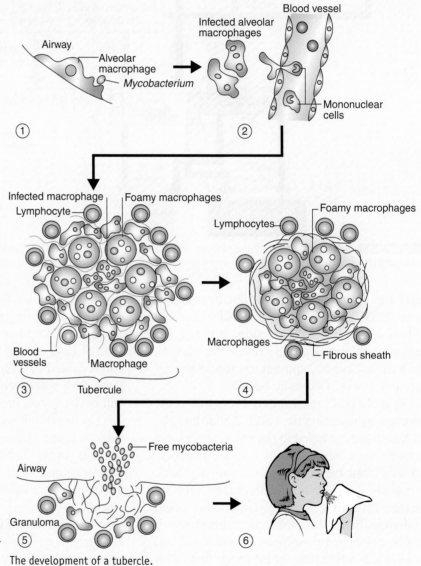

The development of a tubercle.

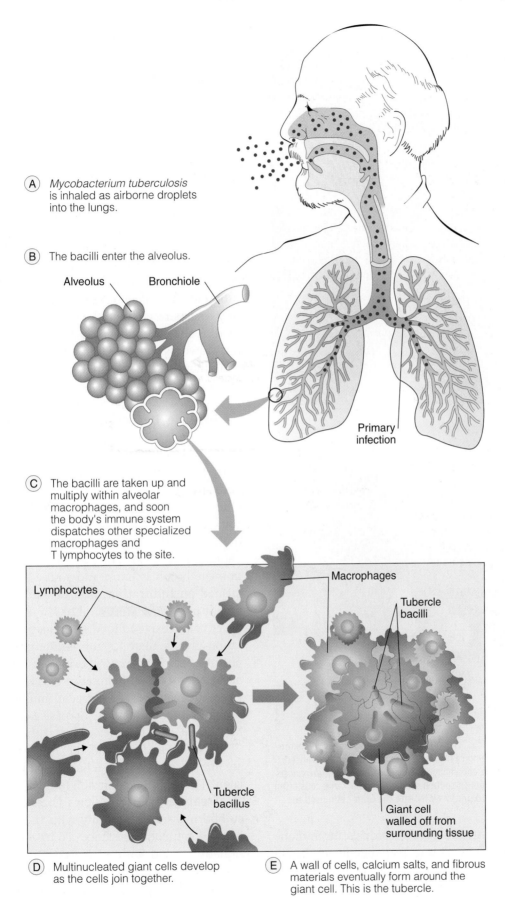

(A) *Mycobacterium tuberculosis* is inhaled as airborne droplets into the lungs.

(B) The bacilli enter the alveolus.

Alveolus   Bronchiole

Primary infection

(C) The bacilli are taken up and multiply within alveolar macrophages, and soon the body's immune system dispatches other specialized macrophages and T lymphocytes to the site.

Lymphocytes

Macrophages

Tubercle bacilli

Tubercle bacillus

Giant cell walled off from surrounding tissue

(D) Multinucleated giant cells develop as the cells join together.

(E) A wall of cells, calcium salts, and fibrous materials eventually form around the giant cell. This is the tubercle.

FIGURE 10.14 **The Progress of Tuberculosis.** Following invasion of the alveoli, the tubercle bacilli are taken up by macrophages and "walled off." »» What is the immune system attempting to do by forming tubercles?

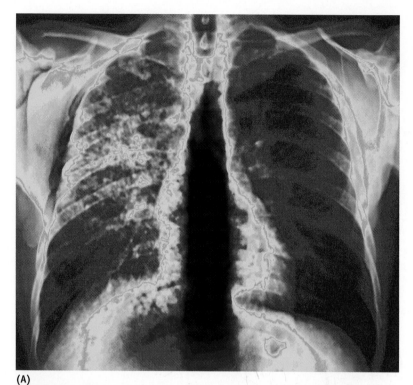

(A)

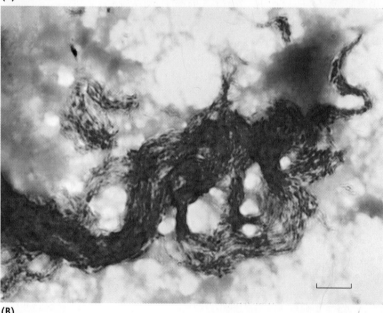

(B)

FIGURE 10.15 **Pulmonary Tuberculosis.** **(A)** This false color X ray shows the extensive fibrosis (fuzzy yellow color in the lung cavity) typical in patients with advanced, active tuberculosis. **(B)** This light microscope image shows *Mycobacterium tuberculosis* cells stained with the acid-fast procedure. In sputum samples, the bacterial cells often exhibit growth in thick strings. (Bar = 10 μm.) »» Why isn't the Gram stain used to identify *M. tuberculosis?*

the lung cavity.) In these individuals, the immune defenses were unable to keep the tubercle bacilli in check. Many of the infected macrophages die, releasing bacilli and producing a caseous (cheese-like) center in the tubercle. Live

bacterial cells rupture from the tubercles, and spread and multiply throughout the LRT as well as other body systems. These individuals will now have a positive tuberculin reaction, chest X ray, and sputum test (**FIGURE 10.15**). On average, the WHO estimates that a person with active TB disease will spread the infection to between 9 and 20 other susceptible individuals.

Because the bacilli in individuals with active TB disease are not killed, the bacterial cells can spread through the blood and lymph to other organs such as the liver, kidney, meninges, and bone. If active tubercles develop throughout the body, the disease is called **miliary (disseminated) tuberculosis** (*milium* = "seed;" in reference to the tiny lesions resembling the millet seeds in bird food). Tubercle bacilli produce no known toxins, but growth is so unrelenting that the respiratory and other body tissues are literally consumed, a factor that gave tuberculosis its alternate name of "consumption."

Disease Detection. Early detection of tuberculosis is aided by the tuberculin reaction, a delayed hypersensitivity test that begins with the application of a purified protein derivative (PPD) of *M. tuberculosis* to the skin. One method of application, called the **Mantoux test,** uses an injection of PPD intradermally into the forearm. Depending on the patient's risk of exposure, the skin at the injection site becomes thick, and a raised, red welt, termed an **induration,** of a defined diameter develops (**FIGURE 10.16**). For an individual never before exposed to *M. tuberculosis,* an induration greater than 15 mm is interpreted as a positive test. However, a positive test does not necessarily reflect the presence of active TB disease, but may indicate a recent immunization, previous tuberculin test, or past exposure to *M. tuberculosis.* Two newer tests, the QuantiFERON-TB Gold test and the T-SPOT.TB test, are more specific and reliable in diagnosing latent TB than tuberculin skin testing.

Treatment. Tuberculosis is an extremely stubborn disease especially with the development of antibiotic resistance. TB has been traditionally treated with such first-line drugs as isoniazid and rifampin. Ethambutol, pyrazinamide, and streptomycin also are used to help delay the emergence of resistant strains. Unfor-

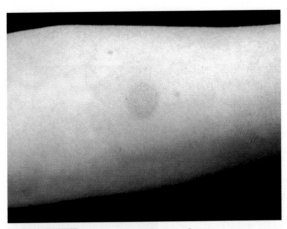

FIGURE 10.16 Tuberculin Skin Test for Tuberculosis. This is an example of a positive reaction to the Mantoux skin test. An induration of less than 5-mm is considered negative. »» What is an induration?

tunately, the appearance of **multidrug-resistant tuberculosis (MDR-TB)** has occured and now accounts for 5% of new TB cases. This has necessitated a switch to a group of second-line drugs, including fluoroquinolones and kanamycin. If drug therapy is effective for pulmonary TB, patients usually become noninfectious within three weeks as determined by bacteria-free sputum samples. Still, for such individuals, antimicrobial drug therapy is intensive and must be extended over a period of six to nine months or more, partly because the organism multiplies at a very slow rate (its generation time is about 18 hours). Early relief, boredom, and forgetfulness often cause the patient to stop taking the medication, and the disease flares anew. MicroFocus 10.3 describes the WHO treatment strategy to maintain patient compliance.

Very recently, cases of **extensively drug-resistant tuberculosis (XDR-TB)** have been reported in some 50 countries. This still rare form of TB is resistant to almost all drugs used to treat TB, including isoniazid, rifampin, fluoroquinolones, and kanamycin. Few treatment options remain for these individuals and a successful outcome depends upon the extent of *M. tuberculosis* drug resistance, the severity of the disease, and whether the patient's immune system is weakened.

A weakened immune system is especially worrisome because TB is a particularly insidious problem to those who have AIDS. In these co-infected patients, the T lymphocytes that normally mount a response to *M. tuberculosis* are being destroyed by HIV, and the patient cannot respond to the bacterial infection. Unlike most other TB patients, those with HIV usually develop miliary tuberculosis in the lymph nodes, bones, liver, and numerous other organs. Ironically, AIDS patients often test negative for the tuberculin skin test because without T lymphocytes, they cannot produce the telltale red welt signaling exposure. Today, TB is the leading cause of death in HIV-infected patients and is the causative agent in 13% of AIDS deaths worldwide.

Prevention. Vaccination against TB can sometimes be rendered by intradermal injections of an attenuated strain of *Mycobacterium bovis,* a species that causes tuberculosis in cows as well as humans. The weakened strain is called **Bacille Calmette-Guérin (BCG)**, after Albert Calmette and Camille Guérin, the two French investigators who developed it in the 1920s. Though the vaccine is used in parts of the world where the disease causes significant mortality and morbidity, health officials in the United States generally do not recommend the BCG vaccine because it has limited effectiveness for preventing TB in adults and is only moderately effective in children for 10 years. New vaccines consisting of subunits, molecules of DNA, and attenuated strains of mycobacteria are currently being developed.

Other *Mycobacterium* Species. Several other species of *Mycobacterium* deserve a brief mention. Although rare, *M. bovis* can be trasmitted person-to-person, especially to individuals with weakened immune systems. *M. chelonae,* another pathogenic species frequently found in soil and water, can cause lung diseases, wound infections, arthritis, and skin abscesses. *M. haemophilum* is a slow-growing pathogen often found in AIDS patients, causing cutaneous ulcerating lesions and respiratory illness. *M. kansasii* causes infections that are indistinguishable from tuberculosis and, in the United States, is most commonly found in the central states. The group known as *M. avium* complex (MAC) tends to cause a disseminated disease in individuals whose immune systems are weakened. Thus, in the United States, MAC represents an opportunistic infection that is

MICROFOCUS 10.3: Public Health
Tragic Endings but Hopeful Futures

Tuberculosis (TB) is a contagious disease caused by *Mycobacterium tuberculosis*. In fact, over 33% of the world's population currently is infected with the tubercle bacillus and TB is killing more people every year; in 2009 someone was dying of TB every 15 seconds. Among the reasons for the rise in TB cases is that infected individuals are not completing the full course of antibiotics once they start to feel better. Not only does this behavior fail to cure the disease, it also helps generate multidrug-resistant TB (MDR-TB).

To address these issues, the World Health Organization (WHO) has developed a TB treatment program called DOTS (Direct Observation Treatment System) to detect and cure TB. Once a patient with infectious TB is identified by a sputum smear, their treatment follows the DOTS strategy. A physician, community worker, or trained volunteer observes and records the patient swallowing four basic medications over a six- to eight-month period. A sputum smear is repeated after two months to check progress and again at the end of treatment.

DOTS appears to be very effective when it is used. Take for example the following case. In the early 1990s, a powerful strain of drug-resistant TB emerged in New York City. Affecting hundreds of people in hospitals and prisons, the outbreak killed 80% of the infected patients. The city responded with DOTS, which seemed to work because the outbreak subsided.

However, in 1997 it was back, this time in South Carolina. A New York patient (not on DOTS) moved to South Carolina. His TB had lingered and he infected three family members in his new community. Soon, another six members of the community were sick with TB. However, these six individuals had not had contact with the family; in fact, they did not even know the family. Investigators from the Centers for Disease Control and Prevention (CDC) were called in to investigate. They learned that one family member had been in the hospital, where he was examined with a bronchoscope (a lighted tube extended into the air passageways). Unfortunately, the bronchoscope was not disinfected properly after the examination and was used to examine these other individuals.

Many stories have happy endings, but this is not one of them. Of the six patients infected with the drug-resistant strain of *M. tuberculosis,* two died from TB and three died from other causes while battling TB. Only one recovered.

Thus, one sees the importance and effectiveness of DOTS. The WHO says that DOTS produces a 95% cure rate and thus prevents new infections. DOTS also prevents the development of MDR-TB by making sure TB patients take the full course of treatment.

Since DOTS was introduced in 1995, more than 10 million infectious patients have been treated successfully. In China, there has been a 96% cure rate and in Peru a 91% cure rate for new cases of TB. Overall, WHO has set a goal of detecting 70% of new infectious TB cases and curing 85% of the detected cases. Meeting the goal in all countries remains a formidable task, but one that offers a hopeful future.

responsible for most cases of miliary TB in AIDS patients. For all species mentioned here, there is no evidence for spread between individuals; rather, infection comes from contacting soil, or ingesting food or water contaminated with the organism.

CONCEPT AND REASONING CHECKS

10.9 Explain how a primary or latent tuberculosis infection is different from a primary or secondary tuberculosis disease.

Infectious Bronchitis Is an Inflammation of the Bronchi

KEY CONCEPT

10. Bronchitis produces excessive mucus and a narrowing of the bronchi.

Infectious bronchitis occurs most often during the winter and can be caused by bacterial species following a URT viral infection, such as the common cold. *Mycoplasma pneumoniae* and

Chlamydophila pneumoniae often cause bacterial bronchitis in young adults, while *S. pneumoniae* and *H. influenzae* are the primary agents among middle-aged and older individuals.

Infectious bronchitis generally begins with the symptoms of a common cold: runny nose, sore throat, chills, general malaise, and perhaps a slight fever. The onset of a dry cough usually signals the beginning of **acute bronchitis,** a condition that occurs when the inner walls lining the main airways of the lungs become infected and inflamed. Inflammation increases the production of mucus, which then narrows the air passages (**FIGURE 10.17**). Bacteria or viruses are present in the mucus. If the condition persists for more than three months, it is referred to as **chronic bronchitis.** The changing of the clear or white mucus to a yellow or green color usually indicates a bacterial infection.

Most cases of acute bronchitis disappear within a few days without any adverse effects, although a cough can linger for several weeks. Antibiotics may be prescribed if the bronchitis is caused by a bacterial infection. Because many cases of acute bronchitis result from influenza, getting a yearly flu vaccination may reduce the risk of

bronchitis. Other preventative measures include good hygiene, including hand washing, to reduce the chance of transmission.

CONCEPT AND REASONING CHECKS

10.10 Why does acute bronchitis often produce symptoms of breathlessness and wheezing?

Pneumonia Can Be Caused by Several Bacteria

KEY CONCEPT

11. Bacterial pneumonia can be community or hospital acquired.

The term **pneumonia** refers to microbial disease of the bronchial tubes and lungs. A wide spectrum of organisms, including viruses, fungi, and bacterial species, may cause pneumonia. Pneumonia is characterized by an inflammation or a build up of fluid in the alveoli. In the United States, some 3 million people develop pneumonia each year, 17% need hospital treatment, and 5% die. It is a very dangerous disease in older adults, and young children, as MICROFOCUS 10.4 describes.

"Typical" pneumonia refers to patients complaining of a cough, fever, and chest pain. Over 80% of bacterial cases of typical pneumonia are

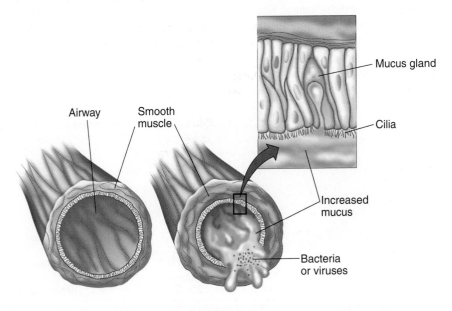

Airway   Smooth muscle

Mucus gland

Cilia

Increased mucus

Bacteria or viruses

Normal bronchus   Inflamed bronchus

FIGURE 10.17 Bronchus Inflammation. Infectious bronchitis develops as the bronchial wall narrows due to inflammation and swelling. Increased mucus production, which contains bacteria or viruses, also narrows the airway. »» How would the bacteria or viruses in the mucus be transmitted to another person?

Adapted from Merck, "Acute Bronchitis: Lung and Airway Disorders," *Merck Manual Home Edition,* December 09, 2008, http://www.merck.com/mmhe/sec04/ch041/ch041a.html.

MICROFOCUS 10.4: Public Health

The Killer of Children

Global Health Magazine recently reported the following: "Chitra Kumal knows the pain of losing a child. When her daughter, Sunita, was 15 months old, she developed a respiratory infection that quickly progressed into pneumonia. With no health facilities in her Nepalese village, Kumal depended on the advice and treatment of a traditional healer or shaman. After just three days of fever, fast breathing, and chest indrawing, her only daughter died."

Similar stories are reported everyday around the world. According to the World Health Organization (WHO), pneumonia kills 2 million children under five years of age each year—more than AIDS, malaria, and measles combined—accounting for nearly one in five child deaths globally (see figure). However, this number may be an underestimate as nearly half of all pneumonia cases occur in malarious parts of the world where pneumonia often is misdiagnosed as malaria.

The WHO estimates that more than 150 million episodes of pneumonia occur every year among children under five in developing countries, accounting for more than 95% of all new cases worldwide, and between 11 and 20 million of these episodes require hospitalization. The highest incidence of pneumonia cases among children under five occur in South Asia and Sub-Saharan Africa, accounting for more than half the total number of pneumonia episodes worldwide.

Preventing and treating childhood pneumonia obviously is critical to reducing childhood mortality. However, only about one in four caregivers knows the two key symptoms of pneumonia: fast breathing and difficult breathing (indrawing). Estimates suggest that if antibiotics were universally available and given to children with pneumonia, around 600,000 lives could be saved each year. But this represents only about 25% of the annual cases. Clearly, other control measures are needed.

At the beginning of the 20th century, pneumonia accounted for 19% of childhood deaths in the United States, a statistic remarkably similar to the rate in developing countries today. Control in the United States was achieved largely without antibiotics and vaccines. Therefore, other control measures and strategies are needed on a global scale.

Key prevention measures include promoting balanced nutrition (including breastfeeding, vitamin A supplementation, and zinc intake), reducing environmental air pollution, and increasing immunization rates with vaccines, such as those against *Haemophilus influenza* type b (Hib) and *Streptococcus pneumoniae* (pneumococcus). However, only about 50% of pneumonia cases in Africa and Asia are caused by these two organisms, so other vaccines need to be developed against other bacterial species (and viruses) that cause pneumonia. And of course—hand washing, like in all areas of infectious disease, can play a role in reducing the incidence of pneumonia.

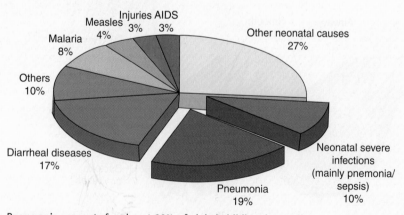

Pneumonia accounts for almost 20% of global childhood mortality.

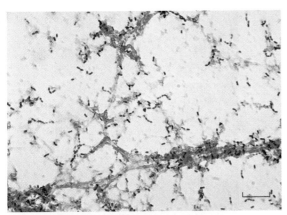

FIGURE 10.18 *Streptococcus pneumoniae.* A gram-stained preparation of *S. pneumoniae* cells (small, dark dots), the cause of pneumococcal pneumonia. (Bar = 10 μm.) »» What is the Gram reaction for *S. pneumoniae*?

due to *S. pneumoniae* (**FIGURE 10.18**); however, several other species also can cause the lung infection.

Streptococcus pneumoniae. Besides being the second leading cause of bacterial meningitis, *S. pneumoniae* also causes **pneumococcal pneumonia**. This form of pneumonia is community acquired and although it exists in all age groups, the mortality rate is highest among infants, the elderly, and those with underlying medical conditions. More than 500,000 cases are reported each year in the United States, resulting in approximately 40,000 deaths.

S. pneumoniae is usually acquired by aerosolized droplets or contact, and the pneumococci exist as a part of the normal microbiota in the URT of many individuals. However, the natural resistance of the body is high, and disease usually does not develop until the defenses are compromised. Malnutrition, smoking, viral infections, and treatment with immune-suppressing drugs most often predispose one to *S. pneumoniae* infections.

Patients with pneumococcal pneumonia experience high fever, sharp chest pains, difficulty breathing, and rust-colored sputum. The color results from blood seeping into the alveolar sacs of the lung as bacterial cells multiply and cause the tissues to fill with fluid and pus. The involvement of an entire lobe of the lung is called **lobar pneumonia**. If both left and right lungs are involved, the condition is called **double pneumonia**. Scattered patches of infection in the respiratory passageways are referred to as **bronchopneumonia**.

The antibiotic for pneumococcal pneumonia has been penicillin. However, increasing penicillin resistance has shifted antibiotic drug choice to cefotaxime or ceftriaxone.

Unfortunately, recovery from one serotype does not confer immunity to another serotype (over 90 capsular serotypes are known). An adult polyvalent antipneumococcal capsular polysaccharide vaccine immunizes against the 23 serotypes that are responsible for almost 90% of pneumococcal pneumonia cases.

Haemophilus influenzae. Some 10% of "typical" hospital-acquired pneumonia cases, especially in the elderly and compromised individuals, are caused by inhaling respiratory droplets containing unencapsulated *H. influenzae* strains. As described, the infection can become systemic and cause otitis media and sinusitis. These URT infections are treated with trimethoprim-sulfamethoxazole.

Staphylococcus aureus. One of the most common causes of hospital-acquired pneumonia results from an infection by *S. aureus,* a facultatively anaerobic, gram-positive coccus. If bacterial cells infect the lungs, a severe, necrotizing pneumonia may occur.

Pseudomonas aeruginosa. *Pseudomonas aeruginosa* is a bacterial species often found in soil and water, but also inhabiting plants and animals (including humans). The gram-negative rod is an opportunistic pathogen, causing infection in patents with a weakened immune system or chronic lung disease. Being a hospital-acquired infection, *P. aeruginosa* is commonly isolated from patients who have been hospitalized for more than a week. The pathogen is a frequent cause of pneumonia in patients whose breathing is being assisted with mechanical ventilation.

Klebsiella pneumoniae. *Klebsiella pneumoniae* is a nonmotile, gram-negative rod with a prominent capsule. The bacillus is acquired by droplets, and often it occurs naturally in the URT of humans. The *K. pneumoniae* may cause a primary disease or a secondary disease in alcoholics or people with chronic obstructive pulmonary disease. As a primary lobar pneumonia,

Serotype: A variation within a species that is detected by a specific set of antigens.

Necrotizing: Refering to cell or tissue death.

Predispose: Referring to an individual susceptible to a condition.

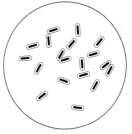

Klebsiella pneumoniae

it is characterized by sudden onset and gelatinous reddish-brown sputum. The organisms grow over the lung surface and rapidly destroy the tissue, often causing death. In its secondary form, *K. pneumoniae* occurs in already ill individuals and is a hospital-acquired disease spread by such routes as clothing, intravenous solutions, foods, and the hands of healthcare workers.

"Atypical" pneumonia is more insidious than "typical" pneumonia. Patient complaints include fever, cough, headache, and myalgia. Several bacterial species can cause this form of pneumonia.

Mycoplasma pneumoniae. *Mycoplasma pneumoniae* causes an illness known as **primary atypical pneumonia**—"primary" because it occurs in previously healthy individuals (pneumococcal pneumonia is usually a secondary disease); "atypical" because the organism differs from the typical pneumococcus and symptoms are unlike those in lobar pneumonia. Today, this community-acquired disease causes about 20% of pneumonias.

M. pneumoniae is recognized as one of the smallest bacterial species causing human disease. Mycoplasmas measure about 0.2 µm in size and are pleomorphic; that is, they assume a variety of shapes (FIGURE 10.19A .) Because they have no cell wall, they have no Gram reaction or sensitivity to penicillin. *M. pneumoniae* cells are very fragile and do not survive for long outside the human or animal host. Therefore, they must be maintained in nature by passage in droplets from host to host.

Most patients, who are usually between 6 and 20 years old, first experience a URT infection. Symptoms include headache, fever, fatigue, and a characteristic dry, hacking cough. Diagnosis is assisted by isolation of the organism on blood agar and observation of a distinctive "fried-egg" colony appearance (FIGURE 10.19B). The infection then progresses to pneumonia in 30% of the infections. Blood invasion does not occur, and the disease is rarely fatal. Often it is called **walking pneumonia** (even though the term has no clinical significance). Epidemics are common where crowded conditions exist, such as in college dormitories, military bases, and urban ghettoes. Erythromycin and tetracycline are commonly used as treatments.

MicroFocus 10.5 describes an interesting application for another bacterial species known to cause respiratory infections.

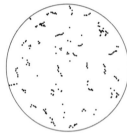

Mycoplasma pneumoniae

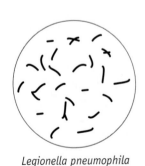

Legionella pneumophila

Insidious:
Referring to a disease with few or no specific symptoms.

Myalgia:
Pain in the muscles or muscle groups.

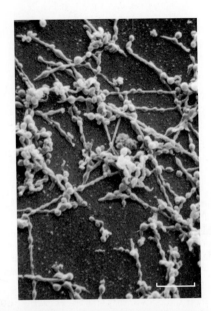

(A)

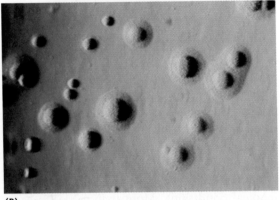

(B)

FIGURE 10.19 *Mycoplasma pneumoniae.* Two views of *M. pneumoniae*, the agent of primary atypical pneumonia. **(A)** A scanning electron micrograph, demonstrating the pleomorphic shape exhibited by mycoplasmas. (Bar = 2 µm.) **(B)** Colony morphology on agar shows the typical "fried egg" appearance.»» What structural feature is missing from the mycoplasma cells that allows for their pleomorphic shape?

Legionella pneumophila. From July 21 to July 24, 1976, the Bellevue-Stratford Hotel in Philadelphia was the site of the 58th annual convention of Pennsylvania's chapter of the American Legion. Toward the end of the convention, 140 conventioneers and 72 other people in or near the hotel became ill with headaches, fever, coughing, and pneumonia. Eventually, 34 individuals died of the disease and its complications. In January 1977, investigators from the CDC announced the isolation of the infecting bacterial species from the lung tis-

MICROFOCUS 10.5: History

"Keep It Short, Please!"

Defining, developing, and proving the germ theory of disease was one of the great triumphs of scientists in the late 1800s. Applying the theory to practical problems was another matter, however, because people were reluctant to change their ways. It would take some rather persuasive evidence to move them.

For many decades, microbiologists considered *Serratia marcescens* a nonpathogenic bacillus. Today, this motile, gram-negative rod is viewed as a cause of respiratory disease in immunocompromised patients.

In the summer of 1904, influenza struck with terrible force among members of Britain's House of Commons. Soon the members began wondering aloud whether they should ventilate their crowded chamber. They decided to hire British bacteriologist Mervyn Henry Gordon to determine whether "germs" were being transferred through the air and whether ventilation would help the situation.

Gordon devised an ingenious experiment. He selected as his test organism *Serratia marcescens* because the bacterium forms bright-red visible colonies in Petri dishes of nutrient agar. Gordon prepared a liquid suspension of the bacterial cells and gargled with it!

Gordon then stood in the chamber and delivered a two-hour oration consisting of selections from Shakespeare's *Julius Caesar* and *Henry V*. His audience was hundreds of open Petri dishes. The theory was simple: If bacteria were transferred during Gordon's long-winded speeches, then they would land on the agar plates and form red colonies.

And land they did. After several days, red colonies appeared on plates placed right in front of Gordon, as well as in distant reaches of the chamber. The members were impressed. They proposed a more constant flow of fresh air to the chamber, as well as shorter speeches. No one was about to object to either solution, especially the latter.

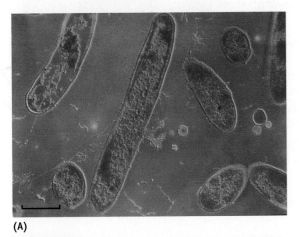

(A)

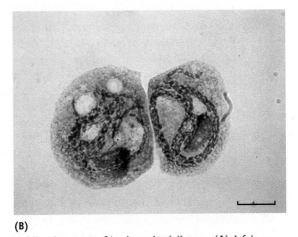

(B)

FIGURE 10.20 *Legionella pneumophila.* Two views of *L. pneumophila,* the agent of Legionnaires' disease. (**A**) A false-color transmission electron micrograph of *L. pneumophila* cells. (Bar = 1 μm.) (**B**) A false-color transmission electron micrograph of the protozoan *Tetrahymena pyriformis* infected with chains of *L. pneumophila* cells (dark red). (Bar = 10 μm.) »» How does infecting a protozoan benefit the bacterium when in its natural environment?

sue of one of the patients. The organism appeared responsible not only for the Legionnaires' disease (as the disease had come to be known), but also for a number of other unresolved pneumonia-like diseases.

Legionnaires' disease is caused by an aerobic, gram-negative rod named *Legionella pneumophila* (**FIGURE 10.20A** .) The bacillus exists where water collects, and apparently it becomes airborne in wind gusts and breezes. Cooling towers, industrial air-conditioning cooling water, lakes, stagnant pools, and puddles of water have been identified as sources of the pathogen (TEXTBOOK CASE 10). Older adults and people with a weakened immune system are most susceptible. After breathing the contaminated aerosolized droplets into the respiratory tract, the disease develops within a week. Human-to-human transmission does not occur.

Textbook CASE 10

Legionnaires' Disease Outbreak— Bogalusa, Louisiana

1 On October 31, 1989, two physicians in Bogalusa, Louisiana reported an outbreak of more than 50 cases of acute pneumonia to the state department of health. Most cases occurred within a 3-week interval in mid- to late October; six persons had died. All cases had occurred in older adults and 76% of the cases were female. Lab analysis confirmed 33 cases were caused by *Legionella pneumophila*, the agent of Legionnaires' disease.

2 When investigators from the Centers for Disease Control and Prevention (CDC) arrived, it was imperative to determine quickly the source and mode of transmission of *L. pneumophila*.

3 Most cases were among residents of Bogalusa. A total of 28 patients and 56 controls were interviewed. Patients and controls were asked about exposures to cooling towers and nearby buildings.

4 Of the 28 patients, three reported visiting Hospital B with a cooling tower; of the 56 controls, seven reported visiting Hospital B. Similarly, 7 of the patients and 12 of the controls reported having visited the nearby post office.

5 Microbiological analysis was unable to confirm any contamination in the hospital B cooling tower. In addition, visiting the post office was eliminated as a possible source.

6 Further interviews identified two other potential sources, butcher shop A and grocery store B. Butcher shop A was visited by 12 patients and 19 controls, while 25 patients and 28 controls had visited grocery store B.

7 A detailed microbiological investigation of grocery store B was undertaken. Several days later, the CDC investigators had the answer. *L. pneumophila* had been isolated from an ultrasonic misting machine close to where shoppers selected produce in the vegetable section (see figure). The machine's aerosol action had sprayed bacterial cells into the air to which shoppers were exposed. No cases were reported among employees.

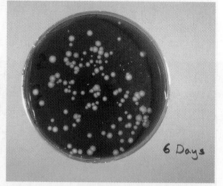

6 Days

Colonies of *L. pneumophila* growing on an enriched agar medium.

Questions:

(Answers can be found in Appendix D.)

A. Why was it important to know that most cases were in the city of Bogalusa? Who are the controls?

B. Why was a cooling tower first suspected as the source of the outbreak?

C. Why was the post office and butcher shop A eliminated as possible sources?

D. Who needs to know about these findings? How would you go about reporting the findings?

E. Explain the significance of (i) 76% of the cases being female and (ii) no cases occurring among grocery store employees.

For additional information see: http://www.cdc.gov/mmwr/preview/mmwrhtml/00001563.htm.

Legionnaires' disease usually causes a severe atypical pneumonia characterized by headache, fever, a dry cough with little sputum, and some diarrhea and vomiting. In addition, chest X rays show a characteristic pattern of lung involvement, and necrotizing pneumonia is the most dangerous effect of the disease. Erythromycin is effective for treatment. In 2007, the CDC reported 2,716 cases of legionellosis.

After *L. pneumophila* was isolated in early 1977, microbiologists realized the organism was responsible for another milder infection called **Pontiac fever**. This is an influenza-like illness that lasts 2 to 5 days but does not cause pneumonia. Symptoms disappear without treatment. The term "**legionellosis**" encompasses both Legionnaires' disease and Pontaic fever.

Since the discovery of *Legionella* in water, microbiologists have been perplexed how such fastidious bacilli survive in often hostile, aquatic environments. An answer was suggested by studies that showed the bacilli could live and grow within the protective confines of waterborne protozoa (FIGURE 10.20B). In a South Dakota outbreak, amoebas were found in abundance in the hospital's water supply. The water was treated by heating and adding chlorine to help quell the spread.

CONCEPT AND REASONING CHECKS

10.11 Hypothesize why some pneumonia-causing species are community acquired and others are hospital acquired?

Other Pneumonia-Causing Bacterial Species Are Obligate, Intracellular Parasites

KEY CONCEPT

12. Some pneumonia-causing bacteria are transmitted by dust particles or animal droppings.

Pneumonia also can be caused by a few bacterial organisms that are obligate, intracellular parasites, meaning they only grow inside host cells.

Coxiella burnetii. **Q fever** (the "Q" is derived from "query," originally referring to the unknown cause of the disease) is caused by *Coxiella burnetii*, a gram-negative rod phylogenetically related to *Legionella* (FIGURE 10.21).

Q fever is prevalent worldwide, especially in livestock, and therefore, outbreaks may occur wherever these animals are raised, housed, or transported. In 2007, the CDC reported 171

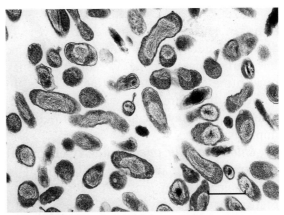

FIGURE 10.21 *Coxiella burnetii.* An electron micrograph of *C. burnetii,* the agent of Q fever. Note the oval-shaped rods of the organism. (Bar = 1 µm.) »» When Gram stained, what color would these cells become?

confirmed cases of Q fever. Transmission among diseased livestock and to humans occurs primarily by inhaling the organisms in contaminated barnyard dust or handling infected animals. In addition, humans may acquire the disease by consuming raw milk or cheese infected with *C. burnetii,* or in milk that has been improperly pasteurized. Although most individuals remain asymptomatic, others experience a bronchopneumonia characterized by severe headache, high fever, a dry cough, and occasionally, lesions on the lung surface. The mortality rate is low, and treatment with doxycycline is effective for chronic infections. A vaccine is available for workers in high-risk occupations.

Chlamydophila psittaci. **Psittacosis** is a zoonotic disease caused by *Chlamydophila* (formerly *Chlamydia*) *psittaci*. These obligate, intracellular pathogens are transmitted to humans by infected parrots, parakeets, canaries, and other members of the psittacine family of birds (*psittakos* = "parrot"). The disease also occurs in pigeons, chickens, turkeys, and seagulls, and some microbiologists prefer to call it **ornithosis** (*ornith* = "bird") to reflect the more widespread occurrence in bird species. Humans acquire *C. psittaci* by inhaling airborne dust or dried droppings from contaminated bird feces. Sometimes the disease is transmitted by a bite from a bird or via the respiratory droplets from another human. The symptoms of psittacosis resemble those of primary atypical pneumonia. Fever is accompanied by headaches, dry cough, and scattered patches of lung infection. Doxycycline is commonly used in therapy.

Zoonotic: Refers to a disease transmitted from animals to humans.

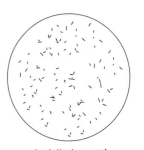

Coxiella burnetti

In 2007, the CDC reported 12 cases of psittacosis, the majority in women 25 to 49 years of age. The incidence in the United States remains low, partly because federal law requires a 30-day quarantine for imported psittacine birds. In addition, birds are given water treated with chlortetracycline hydrochloride (CTC) and CTC-impregnated feed.

Chlamydophila pneumoniae. Chlamydial **pneumonia** is caused by *Chlamydophila pneumoniae*. The organism is transmitted by respiratory droplets and causes a mild community-acquired pneumonia, principally in young adults and college students. It also causes bronchitis, sinusitis, and pharyngitis. The disease is clinically similar to psittacosis and primary atypical pneumonia, and is characterized by fever, headache, and nonproductive cough. Treatment with doxycycline or erythromycin hastens recovery from the infection. *C. pneumoniae's* relationship to cardiovascular disease is explored in MicroFocus 10.6.

The airborne bacterial diseases of the lower respiratory tract are summarized in TABLE 10.2. MicroInquiry 10 presents several case studies concerning some of the bacterial diseases discussed in this chapter.

CONCEPT AND REASONING CHECKS

10.12 Summarize (1) the unique characteristics and (2) the mode of transmission of the chlamydiae.

MICROFOCUS 10.6: Being Skeptical
Can Heart Disease Be Caused by an Infection?

Here's one to startle the senses: the "germ theory of cardiovascular diseases." Skeptical? Let's look deeper.

Coronary artery disease (CAD) remains a leading cause of death even with current medications and laser, angioplasty, or other innovative devices that are available. Perhaps antimicrobial drugs would help.

But, antibiotics for what? Why not start, scientists say, with *Chlamydophila pneumoniae?* In several studies, *C. pneumoniae* has been associated with heart attack patients and in numerous males with CAD. Researchers suggest the chlamydiae could injure the blood vessels, triggering an inflammatory response where immune system cells attack the vessel walls and induce the formation of large, fibrous lesions, or plaques. When pieces of plaque break free, they start blood clots that clog the arteries and cause heart attacks (the condition is known as atherosclerosis). Also, inoculation of mice and rabbits with *C. pneumoniae* accelerates atherosclerosis, even more so if the animals are fed cholesterol-enriched diets.

A second suspect is cytomegalovirus (CMV), an animal virus of the herpes family. In fact, more than 70% of the CAD population is CMV-positive. Scientists have known that people infected with CMV respond very poorly to arterial cleaning, the technique of angioplasty, and the arteries quickly close up. CMV also accelerates atherosclerosis in mice and rats.

A weaker association with CAD is seen with *Streptococcus sanguis,* an agent of periodontal disease. Some microbiologists suggest that poor oral hygiene and bleeding gums give the bacterial species access to the blood, where it produces blood-clotting proteins. It's no coincidence, they maintain, that people with unhealthy teeth and gums tend to have more heart trouble. Unfortunately, such groups also have other lifestyle factors that confound the association.

The final suspect is *Helicobacter pylori,* the cause of most peptic ulcers. Italian scientists have linked a virulent strain of *H. pylori* with increased incidence of heart disease. However, other trials do not find the organism as significant in predicting CAD as the other agents.

To account for these observations and research findings, many medical researchers and microbiologists are proposing the concept of a total pathogen burden. First proposed in 2000, the concept suggests that while a single infectious agent may only minimally increase the risk of atherosclerosis, the burden of several agents could greatly increase the risk. In fact, several recent studies suggest that exposure to several microbial suspects does correlate with increased risk for CAD and, in established CAD cases, incident death. The studies propose that *C. pneumoniae* and CMV probably play a direct role in atherosclerosis while other agents, like *S. sanguis* and *H. pylori,* contribute indirectly to inflammation. The concept does not prove causality, but suggests avenues for further study, including the possible use of antibiotics, antivirals, or vaccines to prevent or cure CAD.

Still skeptical? Many scientists are too but they keep an open mind to new evidence. So, for now, stick with the more likely anti-inflammatory factors of diet, exercise, and not smoking.

TABLE

10.2 A Summary of the Major Bacterial LRT Diseases

Disease	Causative Agent	Signs and Symptoms	Transmission	Treatment	Prevention and Control
Bacterial Pertussis	*Bordetella pertussis*	Malaise, low-grade fever, severe cough Multiple paroxysms	Respiratory droplets	Erythromycin	Vaccinating with DTaP, Tdap
Tuberculosis	*Mycobacterium tuberculosis*	Active TB: Cough, weight loss, fatigue, fever, night sweats, chills, breathing pain	Respiratory droplets	Antibiotics	Preventing exposure to active TB patients BCG vaccine
Infectious bronchitis	*Mycoplasma pneumoniae* *Chlamydophila pneumoniae* *Streptococcus pneumoniae* *Haemophilus influenzae*	Runny nose, sore throat, chills, general malaise, slight fever, and dry cough	Respiratory droplets	Antibiotics	Annual flu vaccination Good hygiene
Pneumococcal pneumonia	*Streptococcus pneumoniae*	High fever, sharp chest pains, difficulty breathing, rust-colored sputum	Respiratory droplets	Penicillin Cefotaxime	Vaccinating Hand hygiene
Other "typical" pneumonias	*Haemophilus influenzae* *Staphylococcus aureus* *Pseudomonas aeruginosa* *Klebsiella pneumoniae*	Chills, high fever, sweating, shortness of breath, chest pain, cough with thick, greenish or yellow sputum	Respiratory droplets	Antibiotics	Practicing good hand hygiene
"Atypical" pneumonia	*Mycoplasma pneumoniae* *Legionella pneumophila*	Headache, fever, fatigue, dry hacking cough	Respiratory droplets Via water systems, whirlpool spas, air conditioning systems	Antibiotics	Extreme cleaning and disinfecting of water systems, pools, and spas
Q fever	*Coxiella burnetii*	Headache, fever, dry cough	Dust particles Contact with infected animals	Doxycycline	Vaccine for high-risk occupations
Psittacosis	*Chlamydophila psittaci*	Headache, fever, dry cough	Contact with infected psittacine birds	Doxycycline	Keeping susceptible birds away from the infecting agent
Chlamydial pneumonia	*Chlamydophila pneumoniae*	Headache, fever, dry cough	Respiratory droplets	Doxycycline Erythromycin	Practicing good hygiene

MICROINQUIRY 10

Infectious Disease Identification

Below are several descriptions of infectious diseases based on material presented in this chapter. Read the case history and then answer the questions posed. Answers can be found in **Appendix D**.

Case 1

The patient, a 33-year-old male, arrives at a local health clinic complaining that he has felt "out of sorts," has a fever, and has lost over 10% of his body weight in the last month. He also has a cough that produced rust-colored sputum. The patient is referred to a local hospital where a chest X ray and sputum sample are taken. Upon further questioning, the patient admits to having tested HIV-positive about one year ago. A tuberculin test also is ordered. Additional questioning of the patient reveals he had been living with two roommates for two years. Before that, he had lived for eight years with another roommate who had tested positive for tuberculosis about 6 months before the onset of the patient's symptoms. The sputum samples are negative for the two roommates, but both have a positive tuberculin test result. Both test negative for HIV.

10.1a. Why was a chest X ray ordered?

10.1b. Why was a sputum sample taken?

10.1c. What should a positive tuberculin skin test look like?

10.1d. What does such a test result indicate?

10.1e. Based on the symptoms and laboratory results, from what infectious disease does the patient suffer? What is the agent?

10.1f. How did the patient contract the disease?

10.1g. Why is the infectious agent more virulent in HIV-infected patients?

Following diagnosis, the patient was placed on isoniazid (INH) for 12 months.

10.1h. Most treatment procedures call for a 6- to 8-month program. Why was the patient placed on INH for an extended period?

Case 2

The parents bring their 2-year-old daughter to the hospital emergency room. She appears to have an upper respiratory infection that her parents think started about one week previously. They say that their daughter had lost her appetite and appeared especially sleepy about four days ago. She complained of a sore throat. Examination indicates that she has a moderate fever but no chest congestion. Throat and blood cultures are taken and their daughter is put on penicillin.

On returning to the hospital three days later, her throat culture shows gram-positive rods. The blood culture is negative. Her parents remark that this morning their daughter started breathing harder. Examination of her pharynx indicates the presence of a leathery membrane. On questioning the parents, it is discovered that the child has had no immunizations. The child is admitted to the hospital and treatment immediately begun.

10.2a. What infectious agent does the child have?

10.2b. The medical staff is concerned about the seriousness of the disease. Why does the presence of gram-positive rods in the throat cause such concern?

10.2c. How could this disease have been prevented?

10.2d. What is the prescribed treatment protocol?

Case 3

A 63-year-old retired steel worker who is a heavy smoker and alcoholic comes to the emergency room complaining of having a fever and shortness of breath for the last two days. This morning he has developed a cough with rust-colored sputum. A chest X ray is taken and shows involvement in the left lower lobe of the lungs. A sputum sample and blood sample are taken for Gram staining, and the patient is checked into the hospital where he is placed on penicillin. Two days later the patient is feeling much improved. His physician tells him that both sputum and blood cultures indicate the presence of gram-positive diplococci. The patient is released from the hospital and recovers completely after finishing antibiotic therapy.

10.3a. What organism is responsible for the patient's infection?

10.3b. Why is this patient at high risk for becoming infected with the bacterial organism?

10.3c. How could the patient likely have prevented contracting the disease?

10.3d. What bacterial factors are responsible for virulence?

10.3e. If plated on blood agar, what type of hemolytic reaction should be seen?

■ SUMMARY OF KEY CONCEPTS

10.1 The Structure and Indigenous Microbiota of the Respiratory System

1. The respiratory system is divided into the **upper respiratory tract (URT)** and the **lower respiratory tract (LRT).**
   - The LRT is virtually free of microbes and particulate matter due to the mechanical **(mucociliary clearance)** and chemical defenses (lysozyme, lactoferrin, other antimicrobial peptides, IgA and IgG antibodies, and human defensins) of the URT;
   - In the LRT, lysozyme, IgG, and alveolar macrophages help eliminate any pathogens;
   - Among the **microbiota** of the URT are opportunist species that can cause serious illnesses in individuals with a weakened immune system. Microbes inhabiting the respiratory tract are easily disseminated in respiratory secretions during breathing, talking, coughing, sneezing, spitting, and kissing.

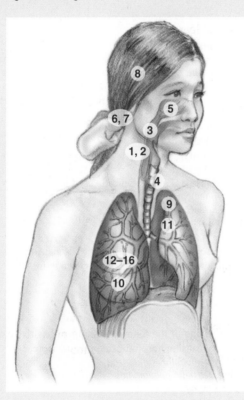

10.2 Bacterial Diseases Affecting the URT

- Streptococcal pharyngitis
  1. *Streptococcus pyogenes*
- Scarlet fever
  2. *Streptococcus pyogenes*
- Diphtheria
  3. *Corynebacterium diphtheriae*
- Epiglottitis
  4. *Haemophilus influenzae*
- Sinusitis
  5. Various bacteria species
- Otitis externa
  6. *Streptococcus, Staphylococcus, Pseudomonas* species
- Otitis media
  7. *Streptococcus pyogenes, Haemophilus influenzae*
- Acute meningitis
  8. *Neisseria meningitidis, Streptococcus pneumoniae, Haemophilus influenzae* type b

10.3 Bacterial Diseases of the LRT

- Pertussis
  9. *Bordetella pertussis*
- Tuberculosis
  10. *Mycobacterium tuberculosis*
- Infectious bronchitis
  11. *Mycoplasma pneumoniae, Chlamydophila pneumoniae, Streptococcus pneumoniae, Haemophilus influenzae*
- "Typical" pneumonia
  12. *Streptococcus pneumoniae, Haemophilus influenzae, Staphylococcus aureus, Pseudomonas aeruginosa, Klebsiella pneumoniae*
- "Atypical" pneumonia
  13. *Mycoplasma pneumoniae, Legionella pneumophila*
- Q fever
  14. *Coxiella burnetii*
- Psittacosis
  15. *Chlamydophila psittaci*
- Chlamydial pneumonia
  16. *Chlamydophila pneumoniae*

LEARNING OBJECTIVES

After understanding the textbook reading, you should be capable of writing a paragraph that includes the appropriate terms and pertinent information to answer the objective.

1. Explain (a) how the URT maintains sterility in the LRT and (b) what portions of the URT are normally colonized by **indigenous microbiota**.
2. Summarize the clinical aspects of **strep throat** and the complications arising from **streptococcal pharyngitis**.
3. Name the bacterial species responsible for, and describe the clinical aspects of treatment and prevention of, **diphtheria**.
4. Discuss why **epiglottitis** can be a life-threatening inflammation.
5. Distinguish between **acute** and **chronic sinusitis**.
6. Recognize the symptoms of outer and middle ear infections.

7. Distinguish between the bacterial species responsible for **acute meningitis**.
8. Justify why **pertussis** is viewed as one of the more dangerous contagious diseases.
9. Summarize (a) the clinical aspects of *Mycobacterium tuberculosis* as an infection and disease, and (b) the problems concerning antibiotic resistance.
10. Distinguish between **acute** and **chronic bronchitis**.
11. Distinguish between the bacterial species responsible for "**typical**" and "**atypical**" **pneumonia**.
12. Summarize the mode of transmission and types of pneumonia caused by obligate intracellular parasites.

STEP A: SELF-TEST

Answer each of the following questions by selecting the *one* answer that best fits the question or statement.

1. Which one of the following is NOT part of the lower respiratory system?
   A. Alveoli
   B. Pharynx
   C. Larynx
   D. Trachea
2. Which one of the following is a complication of streptococcal pharyngitis?
   A. Rheumatic fever
   B. Pseudomembrane blockage
   C. Strawberry tongue
   D. Chest, back, and leg pain
3. Methylene blue staining of metachromatic granules is diagnostic for which of the following bacteria?
   A. *Mycobacterium tuberculosis*
   B. *Corynebacterium diphtheriae*
   C. *Chlamydia pneumoniae*
   D. *Bordetella pertussis*
4. Severe throat pain, fever, muffled voice, and stridor are symptoms of
   A. sinusitis.
   B. epiglottitis.
   C. bronchitis.
   D. diphtheria.
5. Which one of the following illnesses is characterized by yellow or green pus discharged from the nose?
   A. Pertussis
   B. Diphtheria
   C. Bronchitis
   D. Acute sinusitis
6. Swimmer's ear is a common name for a _____ infection of the _____ ear.
   A. bacterial; outer
   B. viral; outer
   C. bacterial; middle
   D. viral; inner

7. Acute meningitis
   A. is an LRT infection.
   B. is a disease affecting the membranes of the heart.
   C. can be caused be *Corynebacterium diphtheriae*.
   D. often starts as a nasopharynx infection.
8. A catarrhal and paroxysmal stage is typical of which one of the following bacterial diseases?
   A. Tuberculosis
   B. Pneumonia
   C. Pertussis
   D. Q fever
9. Acid-fast staining is typically used to stain which bacterial genus?
   A. *Haemophilus*
   B. *Streptococcus*
   C. *Klebsiella*
   D. *Mycobacterium*
10. A person whose inner walls lining the main airways of the lungs become inflamed and who develops a dry cough for a few days, probably has
   A. acute bronchitis.
   B. epiglottitis.
   C. pneumonia.
   D. chronic bronchitis.
11. Which one of the following is a gram-positive bacterial species commonly causing hospital-acquired pneumonia?
   A. *Haemophilus influenzae*
   B. *Klebsiella pneumoniae*
   C. *Staphylococcus aureus*
   D. *Chlamydophila pneumoniae*
12. Humans can acquire which one of the following diseases from the droppings of infected birds?
   A. Q fever
   B. Legionellosis
   C. Tuberculosis
   D. Psittacosis

STEP B: REVIEW

Answer each of the following by filling in the blank with the correct word or phrase. Answers to even-numbered statements can be found in **Appendix C.**

13. Scarlet fever is caused by a species of _____.
14. _____ is caused by a species of *Mycobacterium*.
15. The bacterial species _____ causes psittacosis and pneumonia.
16. _____ is caused by transmission in droplets of airborne water.
17. _____ is a gram-negative rod that causes whooping cough.
18. The development of active tuberculosis throughout the body is called _____ tuberculosis.

19. _____ is a disease of parrots, parakeets, and canaries as well as humans.
20. Another name of *Streptococcus pneumoniae* is _____.
21. A bacterial form of _____ may be caused by *Neisseria* or *Haemophilus*.
22. *Mycoplasma* species have no _____, which often gives them a pleomorphic shape.
23. *Coxiella burnetti* is the causative agent of _____.
24. The genus _____ consists of acid-fast rods.
25. The agent of pneumococcal pneumonia is a gram-_____ organism.

STEP C: APPLICATIONS

Answers to even-numbered questions can be found in **Appendix C.**

26. A patient is admitted to the hospital with high fever and a respiratory infection. Pneumococci and streptococci were eliminated as causes. Penicillin was ineffective. The most unusual sign was a continually dropping count of red blood cells. Can you make the final diagnosis?

27. One of the major world health stories of 1995 was the outbreak of diphtheria in the New Independent and Baltic States of the former Soviet Union. If you were in charge of this international public health emergency, what would be your plan to help quell the spread of *Corynebacterium diphtheriae*?

28. The CDC reports that an estimated 40,000 people in the United States die annually from pneumococcal pneumonia. Despite this high figure, only 30% of older adults who could benefit from the pneumococcal vaccine are vaccinated (compared to over 50% who receive an influenza vaccine yearly). As an epidemiologist in charge of bringing the pneumonia vaccine to a greater percentage of older Americans, what would you do to convince older adults to be vaccinated?

STEP D: QUESTIONS FOR THOUGHT AND DISCUSSION

Answers to even-numbered questions can be found in **Appendix C.**

29. Between 1986 and 1996 *Haemophilus* meningitis was virtually eliminated in the United States. Indeed, at the beginning of the period there were 18,000 cases annually, but in 1996, only 254 cases were reported. What factors probably contributed to the decline of the disease?

30. A bacteriophage is responsible for the ability of the diphtheria bacillus to produce the toxin that leads to disease. Do you believe that having the virus is advantageous to the infecting bacillus? Why or why not?

31. A children's hospital in Salt Lake City reported a dramatic increase in the number of rheumatic fever cases. Doctors were alerted to start monitoring sore throats more carefully. Why do you suppose this prevention method was recommended?

HTTP://MICROBIOLOGY.JBPUB.COM/9E

The site features eLearning, an online review area that provides quizzes and other tools to help you study for your class. You can also follow useful links for in-depth information, read more MicroFocus stories, or just find out the latest microbiology news.

Foodborne and Waterborne Bacterial Diseases

In the aftermath of the Rwanda crisis in 1994, outbreaks of cholera caused at least 48,000 cases and 23,800 deaths within one month in the refugee camps in Goma, the Congo.
—Statement by the World Health Organization

On August 31, 1854, London was experiencing another epidemic of cholera. More than 500 people had died in just 10 days from a disease characterized by stomachache, vomiting, and diarrhea so profuse victims would die from dehydration. Previous epidemics had killed thousands of people on the European continent and in England. Through his pioneering epidemiological studies (see Chapter 1), John Snow, a London physician, pinpointed a water pump on Broad Street as the source of transmission for the current epidemic (**FIGURE 11.1**). Snow believed the only way to stop the spread was to remove the pump handle so no one could get water from the pump. On September 8, 1854, city officials took Snow's advice and had the pump handle removed. The action was successful, supporting Snow's belief that cholera was a waterborne, contagious disease. Few at the time believed Snow's theory for a waterborne disease, and it would not be until 1883 that the cholera bacterium, *Vibrio cholerae*, was isolated.

The year 2004 marked the 150th anniversary of John Snow's landmark epidemiological studies. Unfortunately, cholera still exists today. Before 2000, cholera epidemics had rampaged through Zambia on the African continent, causing 13,154 cases in 1991, 11,659 cases in 1992, and 11,327 cases in 1999. Instituting in-home chlorination of

FIGURE 11.1 **Cholera and the Broad Street Pump.**
John Snow was able to correlate the spread of cholera with a contaminated water pump on Broad Street in London in 1866. The outbreak was controlled by removing the pump handle. »» What is the significance of the "death" caricature in this piece of historical artwork from the time?

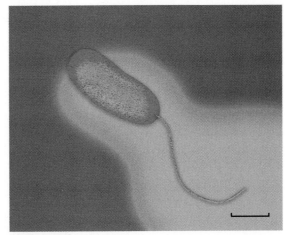

FIGURE 11.2 *Vibrio cholerae.* This false-color transmission electron micrograph shows a cell of *V. cholerae*. Its flagellum can be seen. (Bar = 1 μm.) »» What is the shape of this cell?

drinking water hopefully would bring an end to these epidemics—and no epidemics occurred between 2000 and 2002. Then, in November 2003, cases of cholera were confirmed in Lusaka, the capital city of Zambia with a population of 2 million people. Between November 28, 2003 and January 4, 2004, there were an estimated 2,529 cases of cholera and 128 deaths from the disease in the capital. The Centers for Disease Control and Prevention (CDC) was invited in by the Zambian government to help deal with the epidemic, which by March 1 had added another 2,101 cases and 25 deaths.

Had the water purification system broken down? Or was there another source for the outbreak? CDC officials undertook an extensive epidemiological field investigation by comparing many daily activities involving control (uninfected) individuals and case patients.

The CDC analysis indicated consumption of untreated drinking water and chlorine-treated water was essentially identical between the two groups, so a waterborne source was not the cause of this epidemic. Rather, analysis eventually pointed to raw vegetables as being associated with cholera. Somewhere during the transport, delivery, or use in the home, vegetables were contaminated with the *V. cholerae* bacterium (**FIGURE 11.2**). So, in this cholera epidemic, the mode of disease transmission was foodborne.

In addition, hand soap (indicative of hand washing) was found much more frequently in the homes of control-group individuals, suggesting hand washing reduced the risk for transmission of the disease. By instituting new food control methods and urging hand washing, the number of cholera cases in Lusaka declined dramatically.

This outbreak is remarkable because it reaffirms the 140-year old work of Snow and Semmelweis, who implicated contaminated water and poor hand hygiene in disease transmission. So, sometimes the more things change, the more they stay the same. Field epidemiology and simple prevention strategies, like hand washing, remain critical to meeting the public health challenges we face in today's modern world.

Cholera is just one of many illnesses affecting the digestive system. In this chapter, we will examine the intoxications and infections caused by the foodborne and waterborne bacteria. As we proceed, take note of the illnesses that have similar symptoms, sometimes making disease identification more difficult.

11.1 The Structure and Indigenous Microbiota of the Digestive System

When we eat a meal or drink fluids, our bodies take in and digest the nutrients necessary for survival.

During the ingestive and digestive processes, any pathogens that happen to be in the food or liquid we consume can upset the digestive system and lead to some form of intoxication or disease. Thus, the digestive system represents a major portal of entry for pathogens. In fact, acute infections of the digestive system are among the most frequent of all illnesses.

The Digestive System Is Composed of Two Separate Categories of Organs

KEY CONCEPT

1. Chemical, mechanical, and cellular defenses protect the GI tract from pathogen colonization.

The **digestive system** includes the organs that ingest the food, transport the food, digest the food

into smaller usable components, absorb the necessary nutrients into the bloodstream, and expel the waste products from the body. The **gastrointestinal (GI) tract,** also called the digestive tract or alimentary (*aliment* = "nourishment") canal, contains the digestive organs (**FIGURE 11.3**). These six organs, which consist of the oral cavity, pharynx, esophagus, stomach, small intestine, and large intestine, form a continual tube from mouth to anus.

The second set of digestive organs is the **accessory digestive organs** that are outgrowths from and are connected to the GI tract. These accessory digestive organs include the salivary glands, liver, gallbladder, and pancreas.

Because the GI tract is essentially an open tube running through the body from mouth to anus, digestive system defenses are essential for preventing pathogen colonization and potential infections and disease. In the oral cavity, the process of chewing food generates substantial

Portal of entry:
A site through which a pathogen can enter the body.

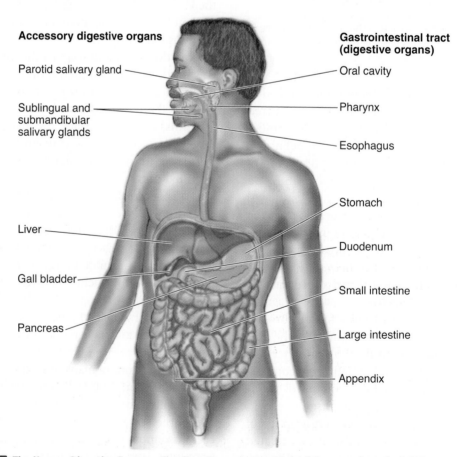

FIGURE 11.3 The Human Digestive System. The digestive system consists of the gastrointestinal (GI) tract and accessory digestive organs. »» Why are organs such as the teeth, liver, and pancreas considered accessory digestive organs?

mechanical and hydrodynamic forces within the oral cavity, which, along with salivary flow, dislodge many microbes from epidermal and tooth surfaces and so discourage colonization. The presence in saliva of defense proteins, including mucins, and the antimicrobial proteins lysozyme, lactoferrin, and defensins (see Chapter 10), bind to microbes, keeping them in suspension. These microbes along with those dislodged by chewing eventually are swallowed, where the low pH of the stomach fluid kills most microbes.

In the intestines, the surfaces are coated with a layer of mucus, which confers mechanical protection by forming a sticky gel that traps bacterial cells. By containing mucins, other defense proteins, and antibodies, these chemical defenses prevent colonization and invasion of the underlying epithelium. The peristaltic action of the intestinal walls of the tract is an important clearance mechanism that keeps food and microorganisms moving through the system.

Like all mucosal surfaces in the body, the intestinal epithelium is a shedding surface. If bacteria do overcome these defense mechanisms and colonize epithelial cells, the bacteria will be removed when the epithelial cells are shed. The small and large intestines also contain bile and proteolytic enzymes capable of killing a variety of microbes.

Finally, the mucosa of the small intestine, primarily in the adult ileum, contains large collections of lymphatic nodules called **Peyer patches**, which represent part of the **mucosa-associated lymphatic tissue** (**MALT**; Chapter 21) that can respond to foreign material through antibody secretion. Another MALT organ, the **appendix**, also may be important to maintaining an indigenous gut microbiota (MicroFocus 11.1)

The accessory digestive organs are sterile and normally free of microorganisms, similar to all other internal organs in the body.

Mucins:
Glycoproteins secreted on the surface of the gut (mucosa).

Bile:
A mixture of cholesterol, phospholipids, acids, and antibodies that are stored in the gallbladder.

CONCEPT AND REASONING CHECKS

11.1 Prepare a list of the chemical, mechanical, and cellular defense mechanisms found in the GI tract.

MICROFOCUS 11.1: Public Health

The Gut's Microbial Source

You probably have heard people say that if you have a case of diarrhea or even take antibiotics for a bacterial infection, you should eat products like yogurt to replenish the gut microbiota lost through diarrhea or killed by antibiotics. Although this cannot hurt, eating yogurt does not supply the microorganisms normally colonizing the gut. So, where do all the microbes come from to repopulate the gut? The answer—internally!

Published results from observations and experiments carried out in 2009 by investigators at Duke University Medical Center and Arizona State University suggest that the appendix serves as an internal "safe house" for the indigenous microbiota normally living in the gut. The appendix is a slender, hollow, blind-ended pouch that projects from the posterior-medial region of the cecum, near its junction with the small intestine (see figure). For a long time, the appendix was thought to be a vestigial or useless organ that may have had a purpose far back in our evolutionary past.

The university investigators suggest that the bacteria in the appendix form a very well developed biofilm that can survive a bout of diarrhea and emerge afterward to repopulate the gut. Further, the immune system cells found in the appendix are thought to protect the microbes from colonization by any pathogens. As one of the Duke researchers said, the appendix is "a place where the good bacteria can live safe and undisturbed until they are needed."

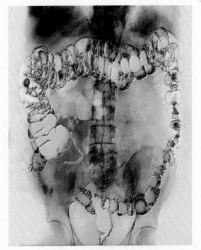

A computer-enhanced barium enema X ray showing the appendix (highlighted in red). The colon is blue-gray.

The Human Intestinal Microbiome Has Not Been Well Studied

KEY CONCEPT

2. The indigenous microbiota of mouth and large intestine represent extremely complex communities.

The indigenous microbiota inhabiting the human body is poorly understood. Surprisingly little is known about the role this astounding assortment of bacteria, fungi, and other microbes play in human health and disease. This is no clearer than in what we know (or don't know) about the microbial communities of the human intestines, which as a whole are referred to as the human intestinal **microbiome** (see Chapter 9). We do know that there is an enormous population of commensal microbiota in the GI tract, especially in the large intestine. And certainly these indigenous microorganisms provide protection through microbial antagonism (Chapter 19). However, they also have many positive roles to play in our ability to digest and process foods (see MicroFocus 19.1). Our understanding of the intestinal microbiome should increase substantially because the National Institutes of Health (NIH) announced in

December 2007 the official launch of the **Human Microbiome Project**, which will, over the next five years, sequence 600 human indigenous microbial genomes, completing a collection of more than 1,000 microbial genomes. Part of that project will emphasize the intestines and represent the **Human Gut Microbiome Initiative**.

The microbiota that first colonizes a newborn's sterile GI tract originates through contact with the parents' diverse microbiota and the infant's surroundings, including hospital and home. If breast fed, microbial colonization is stimulated by short saccharides in breast milk and through contact with the mother's skin microbiota. This succession of microbial species continues until weaning and forms the basis for a healthy microbiota (**FIGURE 11.4A**). By adulthood, the viable microbes in the adult intestine are ten times greater than the total number of intestinal cells.

Analysis of the adult human intestine microbiome suggests that there are trillions of bacterial cells, representing 500 to 1,000 different species. In the stomach, species number and diversity are relatively low due to the low pH **FIGURE 11.4B** . However, a few bacterial species, such as

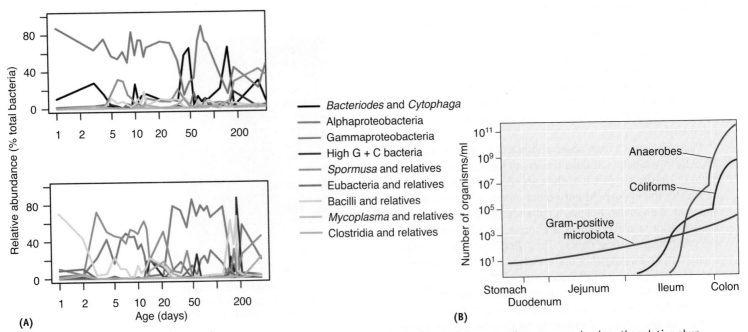

(A)   **(B)**

FIGURE 11.4 Microbiota of the GI Tract. **(A)** Temporal profiles of the most abundant bacterial groups. These two graphs show the relative abundance of phylogenetic groups populating the gut of two infants up to 200 days after birth. Modified from *Development of the Human Infant Intestinal Microbiota*, Palmer C., Bik E. M., DiGiulio D. B., Relman D. A., and Brown P. O., *PLoS Biology* Vol. 5(7): e177 doi:10.1371/journal.pbio.0050177. **(B)** The number of microbes found in the stomach, small intestine (duodenum, jejunum, and ileum) and large intestine (colon) increases from stomach toward colon. »» Why do the number of anaerobes and coliforms (facultative anaerobes) increase in the ileum and colon?

Helicobacter pylori, have mechanisms to survive the extreme acidity of the stomach fluid by embedding themselves in the mucus layer overlying the gastric epithelium. The stomach contains about 1,000 bacterial cells per gram of stomach contents. The small intestine also is relatively sparse in terms of indigenous microbiota, due to the large variety of antimicrobial substances found in the stomach and the short residence time of food in the small intestine. On the other hand, the large intestine has approximately 100 billion bacterial cells per gram of contents.

The predominant bacterial genera identified in the oral cavity as well as in the GI tract are shown in **FIGURE 11.5**. Many in the mouth are members of the phyla Bacteroidetes, Firmicutes, Actinobacteria, and Proteobacteria, while the adult GI tract, especially the large intestine, is just as diverse but is dominated by members of the Bacteroidetes and Firmicutes phyla.

Finally, as with the other body systems, some microbes within the indigenous microbiota may cause disease. As we will see in this chapter, the main organisms involved in dental caries (tooth decay) include a few members of the genus *Streptococcus,* which is a member of the indigenous oral microbiota. *H. pylori* is the causative agent of acute gastritis, gastric ulcers, and stomach cancer. Many of these indigenous organisms also are opportunistic pathogens that can cause a variety of infections in individuals with a weakened immune system.

In our survey of diseases affecting the digestive system, we will start with the oral cavity and proceed down the GI tract.

CONCEPT AND REASONING CHECKS

11.2 Why would the diversity of indigenous microbiota be highest in the oral cavity and in the large intestine?

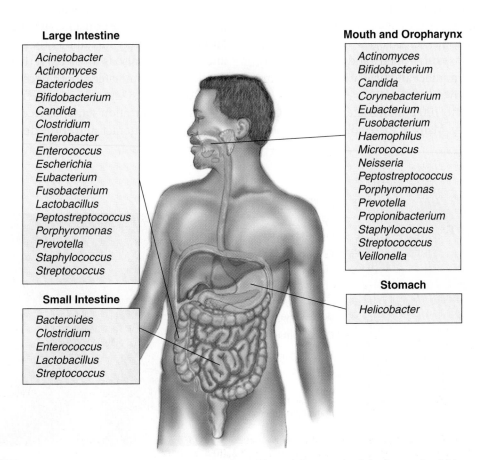

Large Intestine

Acinetobacter
Actinomyces
Bacteriodes
Bifidobacterium
Candida
Clostridium
Enterobacter
Enterococcus
Escherichia
Eubacterium
Fusobacterium
Lactobacillus
Peptostreptococcus
Porphyromonas
Prevotella
Staphylococcus
Streptococcus

Small Intestine

Bacteroides
Clostridium
Enterococcus
Lactobacillus
Streptococcus

Mouth and Oropharynx

Actinomyces
Bifidobacterium
Candida
Corynebacterium
Eubacterium
Fusobacterium
Haemophilus
Micrococcus
Neisseria
Peptostreptococcus
Porphyromonas
Prevotella
Propionibacterium
Staphylococcus
Streptococccus
Veillonella

Stomach

Helicobacter

FIGURE 11.5 Microbiota of the Digestive System. A variety of bacterial genera (and the fungus *Candida*) comprise the predominant indigenous microbiota of the digestive system. »» Why is species diversity and number of bacterial cells so much lower in the stomach and small intestine?

11.2 Bacterial Diseases of the Oral Cavity

Sequelae:
Pathological conditions resulting from a prior disease.

Dental caries, chronic marginal periodontal disease, and the sequelae of these two diseases constitute the majority of oral and dental infections. Because the source of the causative bacteria in these diseases is the microbial plaque that forms on the teeth, prevention and/or halting of the progression of these diseases rely upon the elimination of dental plaque from the tooth surfaces.

Dental Caries Causes Pain and Tooth Loss in Affected Individuals

KEY CONCEPT

3. Bacterial species in the human oral cavity can trigger tooth decay.

The **oral cavity** or mouth is composed of the cheeks, the hard and soft palates, and the tongue, and is bounded anteriorly by the teeth and lips and posteriorly by the oropharynx. It is a type of ecosystem, with complex interrelationships among the members of the resident population of microorganisms and the moist oral environment. The cavity has various ecological niches, each with a different physical property and nutrient supply dictating the number and type of microorganisms that can survive.

Some 50 to 100 billion bacterial cells inhabit the oral cavity. Many of these adhere to the tooth surface that is coated with a thin organic film, called the **dental pellicle** (FIGURE 11.6A). Oral bacteria can adhere to and colonize this pellicle above the gum line or free gingival margin. This is especially true after we eat foods and drinks containing sugars and starches. If these carbohydrates are not cleaned off the teeth through regular dental hygiene, the bacterial cells will adhere to the pellicle and start converting the carbohydrates into acids. It may only be time before tooth decay, or **dental caries** (*cario* = "rottenness"), begins (FIGURE 11.6B).

Dental caries affects more than 20% of American children aged 2 to 4, 50% of those aged 6 to 8, and nearly 60% of those aged 15. The disease develops if three factors are present: a caries-susceptible tooth with a buildup of plaque; dietary carbohydrate, usually in the form of sucrose (sugar); and acidogenic (acid-producing) bacterial species (FIGURE 11.7). The gram-positive streptococci, *Streptococcus mutans* and *S. sobrinus*, are the main acid producers, although a mixed species community including other streptococci and lactobacilli also is involved. If these bacterial species were eliminated, tooth decay could be a thing of the past (MicroFocus 11.2).

Once colonization starts, a succession of bacterial species interact and form **dental plaque**

Streptococcus mutans

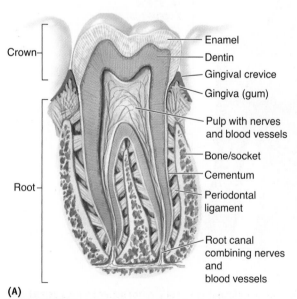

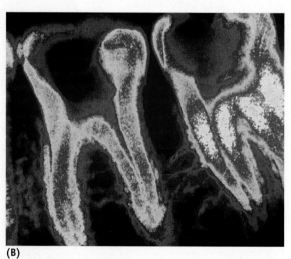

(A)   (B)

FIGURE 11.6 The Anatomy of a Tooth. (A) Different diseases can affect each part of the human molar shown in cross section. **(B)** In this dental X ray, the caries appear as black spots on the molars. »» What parts of the tooth surface would be most susceptible to colorization and infection?

Labels in Figure 11.6A:
Crown — Enamel, Dentin, Gingival crevice, Gingiva (gum)
Pulp with nerves and blood vessels
Bone/socket
Root — Cementum, Periodontal ligament
Root canal combining nerves and blood vessels

(**FIGURE 11.8A**). Plaque is essentially a **biofilm**, a deposit of dense gelatinous material consisting of salivary proteins, trapped food debris, and an enormous mass of bacterial cells and their products. The plaque is slightly rough and is more noticeable on the molars, especially along the gumline. As the plaque thickens, it becomes dominated by anaerobic species. It has been suggested that mature dental plaque contains more than 300 bacterial species (**FIGURE 11.8B**).

The acids in plaque attack minerals in the tooth's hard, outer surface, called the **enamel**

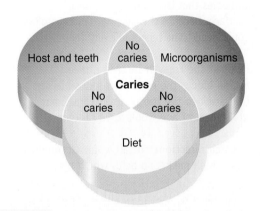

FIGURE 11.7 Dental Caries. Overlapping circles depicting the interrelationships of the three factors that lead to caries activity. »» What role does *S. mutans* play in the development of dental caries?

(Figure 11.6B). This demineralization eventually leads to a cavity, which is a tiny hole or lesion in the enamel. Once spots of enamel are worn away, the bacterial cells and acid can reach the next layer of the tooth, called **dentin**. This layer is softer and less resistant to acid than the enamel, so once tooth decay reaches this point, the decay process often speeds up. Untreated, the decay can progress into the pulp of the tooth. Because the pulp contains nerves and blood vessels, severe toothache pain and sensitivity are typical symptoms. The immune system may respond to the bacterial invaders by sending white blood cells to the blood vessels to fight the infection. This may result in a tooth abscess.

Treatment of dental caries depends on the progress and severity of the disease. Fluoride helps prevent cavities and helps teeth repair themselves by restoring the enamel. Most of us probably have one or more fillings, which have been the main treatment option to seal and restore tooth shape. Extensive decay may require a crown to replace the treated cavity, a root canal if the decay has reached the pulp, or tooth extraction.

Prevention means good oral and dental hygiene. Flossing each day and brushing with fluoride toothpaste twice a day are essential along with regular dental cleaning and examination. Frequent

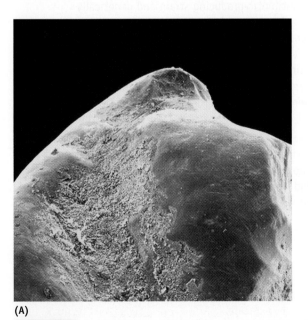

(A)

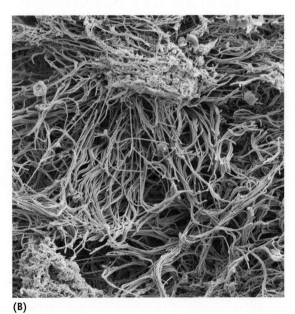

(B)

FIGURE 11.8 Dental Plaque. (**A**) This false-color scanning electron micrograph image of a tooth surface shows dental plaque (yellow) coating the gray enamel of the tooth. (**B**) A higher magnification image of the plaque biofilm (green), which consists of bacteria embedded in a glycoprotein matrix. »» Do these oral bacteria actually adhere to the tooth enamel? Explain.

MICROFOCUS 11.2: Biotechnology

Watch out *S. Mutans*—Your Days May Be Numbered!

In the war against tooth decay, people have armed themselves with floss, toothpaste, toothbrushes, and many innovative variations of these. Still, the bacterial cells seem to win, especially *Streptococcus mutans*. Modifying the diet helps and adding fluoride to the water tips the balance still further. And newer innovations may be just around the corner.

- British researcher Charles G. Kelly and his research team have produced an antibacterial coating that binds to the tooth surface where *S. mutans* would attach and prevents bacterial attachment. In their studies, they smeared the synthetic preparation on the teeth of volunteers whose mouths were cleansed of bacterial cells. Control volunteers were treated with a placebo. Those receiving the synthetic preparation remained free of *S. mutans* for over three months while the bacterial cells appeared in the mouths of control volunteers after only three weeks.

- Could a vaccine be the death knell for *S. mutans*? In London, Julian Ma and coworkers at Guy's Hospital are experimenting with a vaccine containing antibodies that latch onto *S. mutans*, preventing the bacterial cells from binding to the tooth surface. Because the vaccine, called CaroRx™, does not trigger an immune reaction, such treatments would need to be repeated every year or two. CaroRx™ is currently in phase two clinical trials.

- In Boston, Martin Taubman, Daniel Smith, and colleagues at the Forsyth Institute have developed another vaccine against *S. mutans*. This vaccine blocks the enzyme responsible for synthesizing the long-chain bacterial carbohydrates that stick to the tooth surface. Taubman and Smith believe "immunizing" children 18 months to 3 years of age would give lifelong protection.

   So, these three innovations come from different research directions but all would accomplish the same result: no attachment means no bacterial cells, which means no acid, which means no tooth decay.

- Jeffrey Hillman and his associates at the University of Florida and Oragenics, Inc. in Alachua, Florida, have come up with perhaps the most ingenious way to rid the mouth of *S. mutans*—chemically attack them. Hillman's goal is to replace the caries-causing *S. mutans* that secrete lactic acid with "good" *S. mutans* unable to produce the acid. By collecting mouth samples from hundreds of patients, his group isolated a strain of *S. mutans* that produced an antibiotic that kills the other strains. The group then took this antibiotic-producing strain and genetically engineered it so that it would not secrete lactic acid. Using this "replacement therapy," the strain was squirted onto the teeth of rats. By producing the antibiotic, the genetically engineered bacterial species killed and replaced the resident *S. mutans* population and the rats exhibited much reduced levels of cavities. Similarly, three human volunteers had their mouths rinsed for five minutes with the engineered strain. At this writing, they either have no *S. mutans* in their mouth or only the engineered strain. Again, replacement therapy on this very limited group of volunteers worked. Phase I clinical trials began in 2005 and a second human safety trial began in 2009.

For the present, the dental wars go on. It is comforting to know, however, that scientists have imaginative solutions that extend beyond a new flavor of toothpaste. Dental caries is the most widespread infectious disease in today's world. Putting it to an end would be a considerable feather in the scientific cap. So, watch out *S. mutans*, your days may be numbered!

snacking and drinking sugary beverages place the teeth under constant bacterial attack, so limiting such foods and drinks is important, and eating tooth-healthy foods can actually strengthen oral health.

CONCEPT AND REASONING CHECKS

11.3 From the bacterial perspective, how does good oral hygiene help prevent dental caries?

Periodontal Disease Can Arise from Bacteria in Dental Plaque

KEY CONCEPT

4. Bacterial species in plaque can trigger tooth and gum disease.

Swollen and tender gums (gingiva) that bleed when brushing can be a sign of **periodontal**

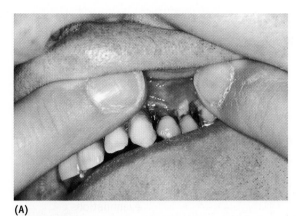
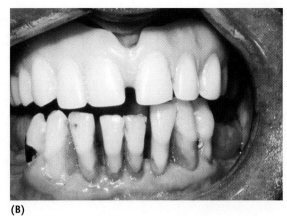

(A)   (B)

FIGURE 11.9 **Periodontal Disease.** (**A**) Gingivitis is an inflammation of the gums (gingiva) around the teeth, which is evident in this patient. (**B**) Periodontitis refers to conditions when gum disease affects the structures supporting the teeth. In this example, gum recession and bone loss have occurred, which can lead to loosening of the teeth and even tooth loss. »» What are the best ways to prevent periodontal diseases?

disease that in some form affects about 80% of American adults. One of the most common forms is **gingivitis**, which develops when plaque bacteria multiply and build up between the teeth and gums (**FIGURE 11.9A**). The bacteria include several gram-negative rods, such as *Fusobacterium, Porphyromonas,* and *Prevotella* as well as *Peptostreptococcus* species. These species secrete toxins and enzymes, including collagenase and hyaluronidase, which directly or indirectly injure the periodontal tissues. The resulting immune inflammatory response mobilizes neutrophils, lymphocytes, and antibody-secreting cells that damage connective tissue, resulting in irritation and bleeding. Because there is no direct bacterial invasion of the gingival tissue, gingivitis is both treatable and preventable. If an individual already has gingivitis, professional cleaning to remove the plaque can reverse the damage and the periodontal tissue will return to normal. As the most common cause of gingivitis is poor oral hygiene, daily brushing and flossing, along with regular professional cleanings, can greatly reduce the risk of developing more serious complications.

Gingivitis is an early-stage periodontal disease and, as such, is seldom painful and causes relatively minor signs. Therefore, it may not be easily recognized without professional examination. However, if left untreated, gingivitis can progress to **periodontitis**, a serious disease of the soft tissue and bone supporting the teeth; bone resorption occurs and the periodontal ligament may be lost (**FIGURE 11.9B**). This can result in loosening of the tooth and tooth loss may result. Left unchecked,

long-term periodontitis can lead to even more serious problems, including higher blood sugar levels, and an increased risk of heart attack and stroke, as bacteria now have access to the blood (see MicroFocus 10.6). Periodontitis may even affect a pregnant woman's fetus. Studies have shown that pregnant women with periodontitis are much more likely to give birth to premature babies than are women with healthy gums.

Treatment involves a thorough cleaning of the periodontal pockets to prevent further damage. Daily brushing and flossing, along with regular professional cleanings, greatly reduce the chances of developing periodontitis.

During World War I, soldiers often were confined to trenches for long periods of time without any means to properly care for their teeth. As a result, many soldiers developed **trench mouth**, a severe form of acute gingivitis where spirochete and fusiform bacteria directly invade the underlying tissues, causing painful, bleeding gums and ulcerations. The inflammatory condition can quickly lead to bone loss. Although trench mouth is rare today in developed nations, it is more common in developing nations with poor nutrition and poor living conditions. Trench mouth is also known as **Vincent's infection** or **acute necrotizing ulcerative gingivitis (ANUG)**.

Without treatment, trench mouth will worsen, and it may lead to other conditions that can cause serious infection that can spread to other areas of the body. Systemic antibiotics can be used for immediate relief of symptoms. As with all the periodontal diseases, practicing good oral hygiene,

TABLE

11.1   A Summary of Bacterial Diseases of the Oral Cavity

Disease	Causative Agent	Signs and Symptoms	Transmission	Treatment	Prevention and Control
Dental Caries	*Streptococcus mutans* *Streptococcus sobrinus* Other species	Toothache Sensitivity and pain when drinking and eating	Due to normal indigenous microbiota	Fluoride treatment Fillings Extraction	Practicing good oral hygiene Regular dental examinations
Gingivitis	*Bacteroides, Fusobacterium, Porphyromonas, Prevotella, Eikenella, Peptostreptococcus, Treponema*	Swollen, soft, and red gums Bleeding gums	Due to normal indigenous microbiota	Cleaning of teeth to remove plaque	Practicing good oral hygiene Regular dental examinations
Periodontitis	*Actinobacillus Porphyromonas gingivalis Bacteroides forsythus Treponema denticola*	Swollen, bright red gums that are tender when touched Gums pulled away from teeth New spaces between teeth	Due to normal indigenous microbiota	Cleaning pockets of bacteria	Practicing good oral hygiene Regular dental examinations
Trench Mouth	*Bacteroides Fusobacterium Treponema*	Painful, red swollen, and bleeding gums	Due to normal indigenous microbiota	Antibiotics	Practicing good oral hygiene Regular dental examinations

along with professional tooth cleaning, help prevent future problems.

TABLE 11.1 summarizes the bacterial diseases of the oral cavity.

11.3   Introduction to Bacterial Diseases of the GI Tract

Most GI tract diseases caused by bacterial pathogens or their toxins are spread through food or water. The CDC estimates 76 million people in the United States suffer foodborne illnesses each year, accounting for 325,000 hospitalizations and more than 5,000 deaths—primarily infants, the elderly, and immunocompromised individuals. According to the World Health Organization (WHO), waterborne diseases account for an estimated 1.7 million deaths worldwide each year. Most of these deaths are from diarrheal diseases, especially in children in developing nations.

Immunocompromised: Refers to the lack of an adequate immune response resulting from disease, exposure to radiation, or treatment with immunosuppressive drugs.

CONCEPT AND REASONING CHECKS

11.4   How do bacterial infections cause gingivitis and periodontitis?

GI Tract Diseases May Arise from Intoxications or Infections

KEY CONCEPT

5.   Several factors contribute to the occurrence of a GI tract disease.

Most illnesses of the GI tract represent some form of **gastroenteritis**, an inflammation of the stomach and the intestines, usually with vomiting and diarrhea. Such inflammations can arise from either intoxications or infections. **Intoxications** are illnesses in which bacterial toxins are ingested in food or water. Examples are the toxins causing botulism, staphylococcal food poisoning, and clostridial food poisoning. By contrast, **infections** refer to illnesses in which live bacterial pathogens in food and water are ingested and subsequently grow in the body. Salmonellosis, shigellosis, and cholera are examples. Toxins may be produced, but they are the result of infection.

Determining the etiology of a bacterial disease depends on several factors.

Incubation Period. If an individual ingests and swallows a contaminated food or beverage, there is a delay, called the **incubation period**, before the symptoms appear. This period can range from hours to days, depending on the bacterial species. During the incubation period, the toxins or microbes pass through the stomach into the intestine where they may directly affect gastrointestinal function or be absorbed into the bloodstream.

Clinical Symptoms. The symptoms produced by an intoxication or infection depend on the specific toxin or microbe, and the number of toxins or cells ingested (toxic or infectious dose). Although the intoxications and infections may have different symptoms, nausea, abdominal cramps, vomiting, and diarrhea often are common. Because these symptoms are so universal, it can be difficult to identify the microbe causing a disease unless the disease is part of a recognized outbreak, or laboratory tests are done to identify the causative agent.

Duration of Illness. Intoxications and infections can have very different lengths of time during which the symptoms persist. Some may appear and disappear quite quickly, while others may linger for a longer period of time.

Demographics. Certain individuals within a population may be more prone to infections or the effects of a toxin. Often infants, whose immune system is still maturing, or the elderly, whose immune system is waning, are more susceptible. Individuals who are immunocompromised due to another illness or chemotherapy also are vulnerable. In addition, populations living in unsanitary and overcrowded conditions where public health measures are lacking will be more likely to become ill from contaminated food or water.

CONCEPT AND REASONING CHECKS

11.5 What types of etiologic information may help in identifying the cause of an intoxication or infection?

There Are Several Ways Foods or Water Become Contaminated

KEY CONCEPT

6. Unsanitary processing procedures and fecal material often contaminate foods and water.

We live in a microbial world, and there are many opportunities for food to become contaminated as it is produced, prepared, packaged, or distributed. Many foodborne microbes are present in healthy animals (usually in their intestines) raised for food. The carcasses of cattle and poultry can become contaminated during slaughter if they are exposed to small amounts of intestinal contents. Fresh fruits and vegetables can be contaminated if they are washed or irrigated with water contaminated with animal manure or human sewage (Chapter 26).

Other foodborne microbes can be introduced through the fecal-oral route; from infected humans who handle the food, or by cross-contamination from some other raw agricultural product. For example, *Shigella* bacteria can contaminate foods from the unwashed hands of food handlers who are infected. In the home kitchen, microbes can be transferred from one food to another food by using the same knife, cutting board, or utensil to prepare both without washing the surface or utensil between uses. A food fully cooked can become recontaminated if it touches raw foods or drippings containing pathogens.

Water can become contaminated in several ways. A common example is when an ill child or adult swimmer with diarrhea has an "accident" in a public pool. If the pool is not sufficiently chlorinated, such recreational water diseases can be passed to other individuals if they swallow the feces-contaminated water. Disease pathogens also can be spread by surface or groundwater contaminated

Etiology:
The study of the causes (origins) of disease.

Toxic or infectious dose:
The number of toxins or organisms consumed to give rise to symptoms of an illness.

with untreated or poorly treated sewage. In either case, individuals can become sick.

Finally, some bacterial pathogens cause foodborne or waterborne bacterial diseases only when they are in large numbers. With warm, moist conditions and plenty of nutrients, lightly contaminated food left out overnight can be highly contaminated by the next day. *Escherichia coli,* for instance, dividing every 30 minutes can produce 17 million new cells in 12 hours. If the food had been refrigerated promptly, the bacterial cells would not multiply at all.

Several bacterial species are commonly found on many raw foods. Therefore, illness can be prevented by (1) controlling the initial number of bacteria present, (2) preventing the small number from growing, (3) destroying the bacteria by proper cooking, and (4) avoiding re-contamination. **FIGURE 11.10** identifies where many foodborne illnesses occur.

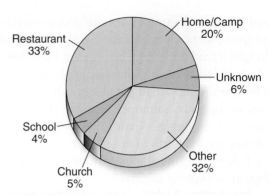

FIGURE 11.10 Sources of Foodborne Illnesses. The mishandling of food can lead to many cases of foodborne illness. Restaurants are the single most common source. (Data from CDC.) »» What might be some of the "other" sources that make up 32% of cases?

CONCEPT AND REASONING CHECKS

11.6   List the ways food and water can become contaminated with pathogens.

11.4   Foodborne Intoxications Caused by Bacteria

A few bacterial species, such as *Staphylococcus aureus, Clostridium perfringens,* and *Bacillus cereus* secrete preformed microbial toxins that when present in food result in **food poisoning**, which is a type of **noninflammatory gastroenteritis**. These foodborne intoxications involve a brief incubation period and quick resolution. Botulism, caused by *Clostridium botulinum,* also is an example of food poisoning. However, the toxins affect the nervous system.

Food Poisoning Illnesses Are the Result of Enterotoxins

KEY CONCEPT

7.   Bacterial intoxications are characterized by diarrhea and/or vomiting.

Staphylococcus aureus. Staphylococcal food **poisoning** is caused by *Staphylococcus aureus,* a facultatively anaerobic, gram-positive sphere that tends to grow in clusters (**FIGURE 11.11A**). Today, staphylococcal food poisoning ranks as the second most reported of all types of foodborne disease (*Campylobacter*-related illnesses are first). Because most staphylococcal outbreaks probably go unre-

ported, staphylococcal food poisoning could actually be the most common type.

Staphylococcal food poisoning is caused by exotoxin-contaminated foods. The incubation period is a brief 1 to 6 hours, so the individual usually can think back and pinpoint the source, which often is a protein-rich food, such as meats or fish. Contaminated dairy products, cream-filled pastries, or salads such as egg salad also can be a cause. Ham is particularly susceptible because staphylococci tolerate salt.

Because the symptoms are restricted to the intestinal tract, the toxin is called an **enterotoxin** (*entero* = "intestine"). Patients experience abdominal cramps, nausea, vomiting, prostration, and diarrhea as the preformed toxin encourages the release of water. (The word diarrhea is derived from the Greek stems *dia* = "through" and *rhein* = "to flow"; hence, water "flows through" the intestines). The symptoms last for several hours, and recovery is usually rapid and complete in 1 to 2 days.

A key reservoir of *S. aureus* in humans is the nose. Thus, an errant sneeze by a food handler may be the source of staphylococcal contamina-

Staphylococcus aureus

Reservoir:
The natural host or habitat of a pathogen.

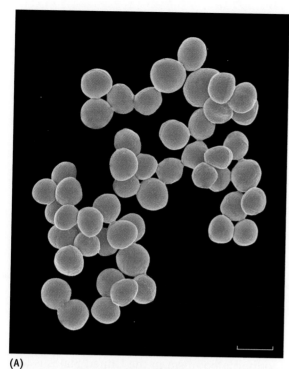

(A)

(B)

FIGURE 11.11 *Staphylococcus aureus.* **(A)** A false-color scanning electron micrograph of *S. aureus* illustrating the typical grape-like cluster of cocci. (Bar = 1 μm.) **(B)** Typing of *S. aureus* strains with bacteriophages. The plate of nutrient medium was seeded with the unknown strain of *Staphylococcus*, and numbered bacteriophages were then placed into different areas. The clear areas indicate which of the phages interacted specifically with the bacteria and killed them, leaving a clear circle. In this case, the strain of *S. aureus* is one that interacts with phages 3, 5, 7, and 14. »» Why is it important to type *S. aureus* strains?

tion. Studies indicate, however, the most common mode of transmission is from boils or abscesses (Chapter 13) on the skin that shed staphylococci into the food product.

Foods containing *S. aureus* lack any unusual taste, odor, or appearance. Thus, case reports often are based on symptoms, patterns of outbreak, and type of food eaten. When investigators locate staphylococci, they can identify the organisms by growth on mannitol salt agar, Gram staining, and testing with bacteriophages to learn the strain involved, as **FIGURE 11.11B** illustrates.

Staphylococci grow over a broad temperature range of 8°C to 45°C, and because refrigerator temperatures are generally set at about 5°C, refrigeration is not an absolute safeguard against contamination. The staphylococcal enterotoxin is among the most heat resistant of all exotoxins. Heating at 100°C for 30 minutes will not denature the protein toxin.

Clostridium perfringens. Since its recognition in the 1960s, **clostridial food poisoning** has risen to prominence—there are about 250,000 reported cases each year—as a common type of food poisoning in the United States. The causative organism, *Clostridium perfringens,* is an obligately anaerobic, spore-forming, gram-positive rod (**FIGURE 11.12**). Most commonly, it contaminates protein-rich foods such as beef, poultry, and fish. If the endospores survive the cooking process, they germinate to vegetative cells and produce an enterotoxin. Consumption of the toxin leads to illness.

The incubation period for clostridial food poisoning is 8 to 24 hours, a factor that distinguishes it from staphylococcal food poisoning. Clinical symptoms require a high infectious dose (10^8 or greater). Moderate to severe abdominal cramping and watery diarrhea are common symptoms. Recovery is rapid, often within 1 to 2 days, and antibiotic therapy is generally unnecessary.

Bacillus cereus. The spore-forming, gram-positive bacillus, *Bacillus cereus,* causes food poisoning after a 2- to 6-hour incubation period. If the enterotoxin is found in meats or cream sauces, it produces diarrhea but little vomiting. If present in starchy foods, substantial vomiting occurs; vomiting is frequently experienced after consuming contaminated cooked rice. Neither form involves

Clostridium perfringens

■ Neurotoxin:
A substance that damages, destroys, or impairs the functioning of nerve tissue.

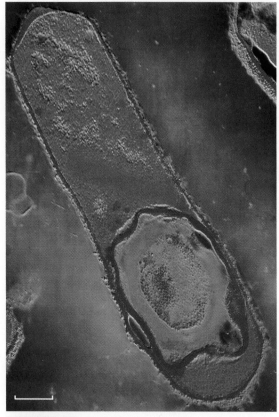

FIGURE 11.12 *Clostridium perfringens.* False-color transmission electron micrograph of a *C. perfringens* cell. It produces a heat-resistant enterotoxin that can contaminate protein-rich foods. (Bar = 1μm.) »» What is the "green" structure in the bottom portion of the cell?

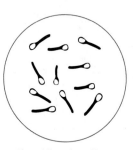

Clostridium botulinum

fever, and most patients recover within 2 days without treatment.

Clostridium botulinum. Of all the foodborne intoxications in humans, none is more dangerous than **botulism**. It is a rare but serious illness caused by *Clostridium botulinum*, a spore-forming, obligately anaerobic, gram-positive bacillus. The endospores exist in the intestines of humans as well as fish, birds, and barnyard animals. They reach the soil in manure, organic fertilizers, and sewage, and often, they cling to harvested products. When spores enter the anaerobic environment of cans or jars, they germinate to vegetative bacilli, and the bacilli produce the exotoxin, which is so powerful that one gram in an aerosolized form could kill one million people. Scientists have identified seven types of toxins (A–G) produced by different strains of *C. botulinum.* Types A, B, and E cause most human disease, with type E associated with most cases of foodborne transmission.

The symptoms of botulism usually develop within 12 to 48 hours after ingesting the toxin-contaminated food. Being a preformed neurotoxin, patients suffer neurologic manifestations, including blurred vision, slurred speech, difficulty swallowing and chewing, and labored breathing. Because the neurotoxin inhibits the release of the neurotransmitter acetylcholine, muscles do not contract and the limbs lose their tone, becoming flabby, a condition called **flaccid paralysis**. Failure of the diaphragm and rib muscles to function leads to respiratory paralysis and death within a day or two.

Because botulism is a type of foodborne intoxication, antibiotics are of no value as a treatment against the toxin. Instead, if treated early, large doses of specific antibodies called **antitoxins** can be administered to neutralize the unbound toxins. Life-support systems such as mechanical ventilators also are used.

Although the annual number of cases of foodborne botulism in the United States is low (34 cases in 2007), the number of deaths is about 8%. Complete recovery can take up to one year.

Botulism can be avoided by heating foods before eating them, because the toxin is destroyed on exposure to temperatures of 90°C for 10 minutes. However, experience shows that most outbreaks are related to home-canned foods having a low acid content, such as asparagus, green beans, beets, and corn, and from foods eaten cold. Other foods linked to botulism include mushrooms, olives, salami, and sausage. In fact, the word botulism is derived from the Latin *botulus,* for "sausage."

Although foodborne botulism is very dangerous, other forms of botulism exist. **Wound botulism** is caused by toxins produced in the anaerobic tissue of a wound infected with *C. botulinum.* Penicillin is an effective treatment. **Infant botulism** is the most common form of botulism in the United States, accounting for 85 reported cases in 2007. Unlike foodborne botulism where the botulinum toxin is ingested, infant botulism results from the ingestion of soil or food contaminated with *C. botulinum* endospores. Thus, parents should not give their infant honey, which is the most common food triggering infant botulism. Spores that are in about 10% of honey germinate and grow in the colon where the botulinum cells release toxin. This form of botulism typically affects infants 3 to 24 months old because they have not established the normal balance of bowel microbes. Because the toxin produces lethargy and poor muscle tone, infant botulism often is referred to as floppy baby syndrome. Hospitalization along with mechanically-assisted ventilation may be necessary, with antitoxin treatment reducing recovery time.

In recent years, one of the botulinum toxins has been put to practical use. In minute doses, **Botox®** or **Dysport®** (botulism toxin type A) can relieve temporarily a number of movement disorders (so-called dystonias) caused by involuntary sustained muscle contractions. For example, the toxin is used to treat strabismus, or misalignment of the eyes, commonly known as cross-eye; it also is used against blepharospasm, or involuntarily clenched eyelids. The toxin may be valuable in relieving stuttering, uncontrolled blinking, and musician's cramp (the bane of the violinist). The toxin also has been approved for use in the temporary relief of facial wrinkles and frown lines (FIGURE 11.13). In 2004, it was approved for temporary relief of hyperhidrosis (excessive body sweating).

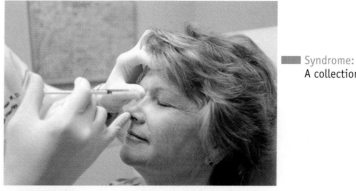

FIGURE 11.13 **Cosmetic Injection of Botulinum Toxin.** The botulinum toxin in controlled doses can be used to temporarily minimize wrinkles. »»How does the toxin work to "remove wrinkles"?

■ Syndrome:
A collection of symptoms.

Care must be used when using the drug. In 2008, four children suffering cerebral palsy died after injection of the toxin to treat spasticity in their legs.

CONCEPT AND REASONING CHECKS

11.7 What characteristics are shared by all foodborne intoxications?

11.5 Foodborne and Waterborne Infections

Bacterial GI infections have a longer incubation period than intoxications because bacterial cells must first establish themselves in the body after ingestion of the contaminated food or water.

Bacterial GI infections can be one of two types. **Inflammatory gastroenteritis** is characterized by diarrhea and/or vomiting, and usually a fever, but there is no blood in the stool. **Invasive gastroenteritis** involves bacterial invasion beyond the intestinal lumen. Signs and symptoms may include fever, diarrhea or vomiting, and dysentery (the passage of blood and mucus in the feces). We will examine several infections characteristic of each type of gastroenteritis.

Bacterial Gastroenteritis Often Produces an Inflammatory Condition

KEY CONCEPT

8. Dehydration is a common but serious complication of gastroenteritis.

The most familiar illnesses causing a bacterial inflammatory gastroenteritis are cholera, caused by *Vibrio cholerae*, and *Escherichia coli* gastrointestinal infections, such as traveler's diarrhea.

Vibrio cholerae. No diarrheal disease can compare with the extensive diarrhea associated with **cholera** (see chapter opener). The WHO estimates there are more than 100,000 cases and over 1,900 deaths annually. The organism, *Vibrio cholerae,* was first isolated by Robert Koch in 1883.

The *V. cholerae* cells are motile, aerobic, gram-negative, curved rods (see Figure 11.2). The bacilli enter the intestinal tract in contaminated food, such as raw oysters, or in water (FIGURE 11.14A). *V. cholerae* is extremely susceptible to stomach acid, so most cases are mild or go unnoticed. However, if millions are ingested, enough cells survive to colonize the intestines. There the bacilli move along the intestinal epithelium, secreting an enterotoxin (**cholera toxin**) that stimulates the unrelenting loss of fluid and electrolytes.

In the most severe cases, extreme dehydration may occur since a patient may lose up to one liter of fluid every hour for several hours. The fluid, referred to as **rice-water stools**, is

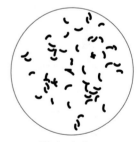

Vibrio cholerae

■ Electrolytes:
Any ions in cells, blood, or other organic material.

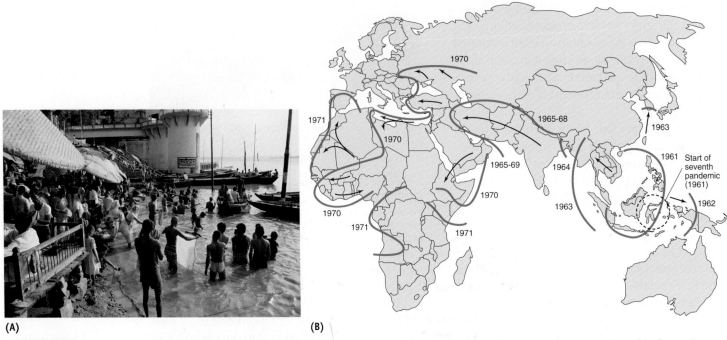

(A)   (B)

FIGURE 11.14 Cholera Disease Spread. (**A**) The spread of cholera is associated with poor sanitation and water supplies contaminated by feces. The Ganges River delta in India is a central point of endemic cholera. Cholera bacilli frequently inhabit the river and pass easily among unsuspecting bathers. (**B**) The spread of *V. cholerae* biotype El Tor started in 1961 and spread eastward and westward. »» How would *V. cholerae* be spread?

colorless and watery, reflecting the conversion of the intestinal contents to a thin liquid-like barley soup. The patient's eyes become gray and sink into their orbits. The skin is wrinkled, dry, and cold, and muscular cramps occur in the arms and legs. Despite continuous thirst, sufferers cannot hold fluids. The blood thickens, urine production ceases, and the sluggish blood flow to the brain leads to shock and coma. In untreated cases, the mortality rate may reach 70%. However, it is easily treated and prevented. Although prevalent in the United States in the 1800s, good sanitation and water treatment eliminated the pathogen as a health threat.

Antibiotics such as tetracycline may be used to kill the bacterial cells, but the key treatment is restoration of the body's water and electrolyte balance. Often, this entails intravenous injections of salt solutions. More commonly, patients can be treated with **oral rehydration solution (ORS)**, a solution of electrolytes and glucose designed to restore the normal balances in the body. At present, travelers to cholera regions of the world can be immunized with preparations of dead *V. cholerae* cells, thereby obtaining protection for about six months. The best prevention for cholera

is a careful and wise selection of foods and beverages when in countries experiencing cholera outbreaks.

Cholera has been observed for centuries in human populations, and seven pandemics caused by the O1 serogroup have been documented since 1817. The first six were caused by *V. cholerae* O1 Classic. The current seventh pandemic, caused by *V. cholerae* O1 El Tor, began in 1961 in Indonesia and now involves about 35 countries (**FIGURE 11.14B**). A major outbreak occurred in Peru and Ecuador early in 1991 and from there spread throughout the region, accounting for 731,000 cases by the end of 1992 and 6,300 deaths in 21 countries in the Western Hemisphere. Microfocus 11.3 describes how the responsible biotype may have emerged. An eighth pandemic emerged in India in 1992 and has spread across Asia. This serogroup, *V. cholerae* O139, likely is derived from El Tor because comparative genomics (see Chapter 9) indicates the two serogroups share 99% of their genes. The O1 vaccine does not protect against O139.

For generations, scientists believed the cholera bacillus existed only within a human host. That idea was refuted by University of Maryland

MICROFOCUS 11.3: Biotechnology
How the Sheep Got the Wolf's Clothes

The seventh pandemic of cholera began in Sulawesi, Indonesia, in 1961. Soon it spread to India, the Soviet Union, and the Middle East. By 1991, it had reached Latin America and affected all countries of South America, except Argentina. By 1993, over 6,000 people had died of cholera, and many hundreds of thousands had been terribly sick.

The pandemic was due to a toxin-producing serotype of *Vibrio cholerae* known as O1. The cholera toxin binds to intestinal cells, setting off a cascade of reactions, and water pours out of the cells—up to 5 gallons of water per day. Where did the toxin come from? Microbiologists from Harvard think they have the answer: a virus.

Matthew Waldor and John Mekalanos had been studying cholera for many years. They were impressed by the toxicity of the O1 serotype and wondered why other serotypes were far less lethal. Their interest centered on the cholera toxin gene in *V. cholerae,* and they speculated the gene might have been delivered by a bacteriophage through the process of transduction (see Chapter 9). To test their theory, they removed the entire toxin gene by sophisticated genetic engineering techniques, and then replaced it with an antibiotic-resistance gene. Now they cultured the new antibiotic-resistant cholera cells along with normal cholera cells susceptible to antibiotics. Bingo! The susceptible cells soon became antibiotic resistant. Something (a phage?) seemed to be leaving the antibiotic-resistant cells and ferrying the resistance gene to the susceptible cells.

But, maybe the bacteria were conjugating and exchanging their genes directly. To test this theory, Waldor and Mekalanos passed the genetically engineered (antibiotic-resistant) cells through a filter that would trap everything except phages. They took the clear, cell-free liquid and added it to a fresh batch of normal cells. Double bingo! The cells became resistant to antibiotics. The phage theory strengthened.

Still another test: They treated the clear, cell-free fluid with enzymes that destroy free-floating nucleic acids but have no effect on viruses. (This would eliminate any molecular DNA or RNA that might transform cells.) Then they combined the fluid with normal cells. Once again, the cells became antibiotic resistant. And the coup de grace: Electron micrographs revealed long, stringy phages in the cell-free fluid.

To be sure, the cholera bacterium is not the first to have its toxicity associated with a phage (the diphtheria and botulism organisms are others), but the finding helps explain how an organism can suddenly become lethal. The sheep had acquired the wolf's clothing.

---

researchers led by Rita Colwell. Colwell and her colleagues found *V. cholerae* in waters from the Chesapeake Bay, even though no cholera outbreaks were remotely close to the site. They postulated the organisms persist in a viable but non-culturable (VBNC) state (see Chapter 5). Apparently, in the cold, nutrient-poor water, the organism's metabolic rate diminishes, and it stops reproducing. This state, says Colwell, allows the cholera bacillus to survive in habitats and environments ranging from seawater to the human intestinal tract. In the sea, moreover, the bacillus appears to be a regular gut inhabitant of a tiny crustacean known as a copepod. The findings are novel, and they may signal a new outlook for cholera in the decades ahead.

CONCEPT AND REASONING CHECKS

11.8a How is the mechanism of action of cholera toxin similar to that of other enterotoxins mentioned in this chapter?

Escherichia coli. The human colon contains *E. coli* as part of its indigenous microbiota. However, other serotypes can be pathogenic. In fact, they are one of the major causes of infantile diarrhea.

The *E. coli* cells are facultatively anaerobic, gram-negative rods (**FIGURE 11.15**). Transmission follows the fecal-oral route where contaminated food or water represents the vehicle for transmission. The pathogenic serotypes may induce watery diarrhea in several ways.

Enterotoxigenic *E. coli* (ETEC) penetrates the intestinal epithelium and produces two enterotoxins that cause gastroenteritis. One toxin is heat labile and is similar to the cholera toxin. The other toxin is heat stable. The illness, which is the most common bacterial form of **traveler's diarrhea**, affects 20% to 50% of travelers within 6 to 48 hours. (A number of organisms including several bacterial species, viruses, and protozoa may cause

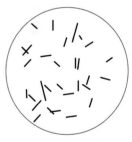

Escherichia coli

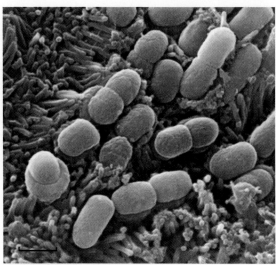

FIGURE 11.15 *Escherichia coli.* False-color scanning electron micrograph of *E. coli* cells inhabiting the surface of the intestines. When exogenous strains of *E. coli* enter the intestines through contaminated food or water, infections such as gastroenteritis may occur. (Bar = 2 μm.) »» Why doesn't the normal intestinal *E. coli* cause gastroenteritis?

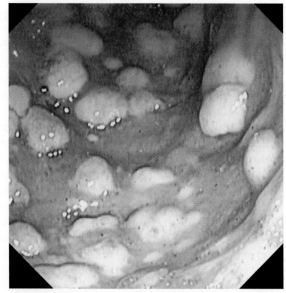

FIGURE 11.16 Pseudomembranous Colitis. An endoscope view of an inflamed colon caused by a *Clostridium difficile* infection. The white plaques represent mucus and dead cells that have built up on the colon walls. »» What defensive proteins might be present in the plaques?

traveler's diarrhea, but recent studies point to *E. coli* as the principal agent.) The volume of fluid lost is usually low, but vomiting, cramps, nausea, and a low-grade fever may occur. The diarrhea lasts 3 to 7 days. The possibility of traveler's diarrhea may be reduced by careful hygiene and attention to the food (uncooked vegetables) and water consumed during visits to other countries. Rifaximin, an antibiotic that affects only the intestinal tract, can be used to fight traveler's diarrhea caused by *E. coli*. However, it does not cure the infection or prevent complications.

Enteropathogenic *E. coli* (EPEC) is an important cause of diarrhea in infants, especially where sanitation conditions are poor. The infection occurs during birth and the *E. coli* cells attach to the intestinal mucosa, causing fever, nausea, watery diarrhea, and vomiting.

An enterohemorrhagic serotype of *E. coli* causes an invasive gastroenteritis and is described later in this chapter.

CONCEPT AND REASONING CHECKS

11.8b How is the cause of diarrhea different between ETEC and EPEC serotypes?

Clostridium difficile. Healthcare associated infections (HAIs) are illnesses acquired during a stay in a hospital or long-term care facility. Recently, such HAIs have reached epidemic

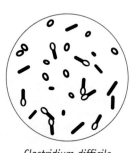

Clostridium difficile

proportions around the world. One of the most widespread and potentially serious is caused by *Clostridium difficile,* often simply called *C. diff.* The bacterium is an anaerobic, spore-forming, gram-positive rod that occurs in soil, air, water, human and animal feces, and on most surfaces. It causes diarrhea and is the major cause of **pseudomembranous colitis,** a severe infection of the colon that can lead to a grossly dilated bowel that could rupture or perforate (FIGURE 11.16).

Healthy individuals usually do not get infected with *C. difficile*. Those most at risk are the elderly and people who have been taking antibiotics for a prolonged time for another illness or condition. The bacterial cells are shed in feces, so any surface, instrument, or material contaminated with feces may serve as a reservoir for the *C. difficile* spores, which can survive for months on surfaces. In a healthcare facility, the endospores could be transferred to patients via the hands of healthcare personnel who have touched a contaminated surface or object.

The pathogen produces two toxins, an enterotoxin that causes fluid loss, and a **cytotoxin** that causes further mucosal injury. Symptoms of pseudomembranous colitis can begin within one to two days after beginning an antibiotic, or may not occur until several weeks after discontinuing the antibiotic. The most common symptoms include watery diarrhea, fever, bloody stools, and abdominal pain/tenderness. Because most patients develop only a mild diarrhea, stopping antibiotic therapy, if clinically possible, together with fluid replacement, usually results in rapid improvement. For more severe cases like pseudomembranous colitis, anticlostridial antibiotics, such as metronidazole or vancomycin, may be required.

In 2008, a new hypervirulent strain of *C. difficile* emerged both in the hospital and community that causes a more severe illness. Its hydervirulence may be due to its ability to produce greater quantities of the two toxins. Treatment for the new strain remains the same.

CONCEPT AND REASONING CHECKS

11.8c Why would *C. difficile* be considered an opportunistic pathogen?

Listeria monocytogenes. Listeriosis is caused by *Listeria monocytogenes*, a small, non-spore-forming, facultatively anaerobic, gram-positive rod that is motile at room temperature. The bacillus is commonly found in the soil, in water, and in the intestines of many animals, including birds, fish, barnyard animals, dairy cattle, and household pets. It is transmitted to humans by food contaminated with fecal matter, as well as by the consumption of contaminated meats, vegetables, and raw milk. Contaminated delicatessen cold cuts, as well as soft cheeses (e.g., Brie, Camembert, feta, and blue-veined cheeses) have been associated with a significant number of cases. The pathogen is considered **psychrotrophic**, meaning it can grow at refrigerator temperatures (see Chapter 5). Therefore, refrigerated foods contaminated with *L. monocytogenes* will not kill the bacterial cells that can remain in a suspended, dormant-like state.

The incubation period for the diarrheal diseases is 3 to 70 days. Healthy adults may experience diarrhea, fever, and abdominal cramps. However, in newborns, pregnant women, the elderly, and immunocompromised individuals, the illness can become invasive with symptoms of fever, malaise, arthritis, and jaundice.

The invasive disease occurs in many forms within 2 to 6 weeks. During severe infections, the pathogen crosses the blood-brain barrier,

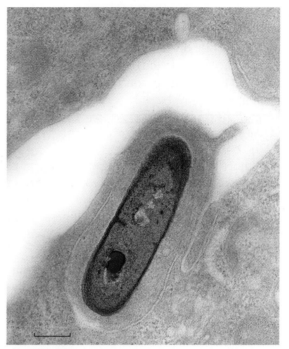

FIGURE 11.17 *Listeria monocytogenes.* A transmission electron micrograph of *L. monocytogenes* entering a monocyte. (Bar = 1 μm.) »» How does *L. monocytogenes* evade destruction by the monocyte's lysosomes?

causing **meningoencephalitis**, characterized by headaches, stiff neck, delirium, and coma. Another form is **septicemia**, a blood disease accompanied by high numbers of infected white blood cells called monocytes (hence the organism's name) (**FIGURE 11.17**). The intracellular bacterial cells secrete a toxin that damages the membrane enveloping the internalized bacterial cells so they cannot be destroyed by lysosomes (see Chapter 3).

A third form, occurring in females, is characterized by infection of the uterus, with vague flu-like symptoms. If contracted during pregnancy, the *L. monocytogenes* cells cross the placenta, causing miscarriage of the fetus or mental damage in the newborn (newborn meningitis). Other individuals suffer respiratory distress, diarrhea, back pain, skin itching, and other nonspecific symptoms, which makes diagnosing the disease difficult.

Although ampicillin is an effective treatment, relapses are common. The CDC estimates that there are 2,500 cases of listeriosis in the United States annually (808 reported cases in 2007) and 20% of those cases result in death. The illness does not appear to be transmissible from person to person. A notable outbreak of listeriosis occurred

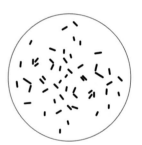

Listeria monocytogenes

▮ Jaundice:
A condition in which there is yellowing of the whites of the eyes, skin, and mucous membranes, caused by bile pigments in the blood.

in 2008 in Canada. Of 57 confirmed cases, 22 people died, all linked to a commercial brand of deli meats. Because the deli meats had been cooked, health officials suspected that contamination during packaging at the production plant may have been involved.

CONCEPT AND REASONING CHECKS
11.8d What are the three forms of listeriosis?

A number of other bacterial organisms transmitted by food or water merit brief consideration in this chapter.

Brucella species. A bacterial illness called **brucellosis** is an occupational hazard of farmers, veterinarians, dairy and meat plant workers, and others who work with large ruminant animals. Among the species responsible for the most common zoonotic infection worldwide are *Brucella abortus* (cattle), *B. suis* (swine), *B. melitensis* (goats and sheep), and *B. canis* (dogs). All species are small, nonmotile, gram-negative rods that can cause disease in humans.

Bacterial transmission can occur by splashing contaminated milk into the eye, by the accidental passage of contaminated fluids through skin abrasions, by contact with infected animals, and by the consumption of contaminated milk or other dairy products. In one example of food transmission, the CDC reported 29 cases in Mexican emigrants living in Houston, Texas. All had eaten goat cheese made from unpasteurized goat's milk. Human-to-human transmission is unknown.

On infection, the bacterial cells are transported to the blood-rich organs, such as the spleen and lymph glands. Patients experience flu-like weakness, as well as backache, joint pain, and a high fever (with drenching sweats) in the daytime and low fever (with chills) in the evening. This fever pattern gives the disease its alternate name, **undulant fever** (*undulat* = "wavy"). The disease seldom causes death in humans, and a combination of doxycycline and gentamicin speeds recovery.

B. abortus has been effectively eliminated from American livestock and most human cases of brucellosis are in travelers or immigrants from an endemic region. The consumption of unpasteurized milk from outside the United States remains a source of infection from *B. abortus* and *B. melitensis*. The CDC reported 131 cases in 2007.

Vibrio Species. *Vibrio*-caused food illnesses (**vibriosis**) are increasing with 549 cases reported to the CDC in 2007. A major cause of gastroenteritis in Japan and other areas of the world where seafood is the main staple of the diet is *Vibrio parahaemolyticus*, a curved, gram-negative rod. Within 4 to 30 hours of consuming contaminated seafood, the bacilli invade the mucosa, causing acute abdominal pain, vomiting, watery diarrhea, and fever and chills. In 2004, 54 people in Alaska became ill from eating raw oysters, and in 2006, 177 cases were reported in New York, Oregon, and Washington from eating raw shellfish contaminated with *V. parahaemolyticus*.

In 1996, the CDC reported an outbreak of intestinal illness due to another *Vibrio* species, *V. vulnificus*. This species, the most virulent of the vibrios, occurs naturally in brackish and seawaters, where oysters and clams live. People who consume these mollusks raw are at risk, especially immunocompromised individuals and those who suffer from liver disease or low stomach acid. (Indeed, taking an antacid after a meal of any contaminated food may neutralize stomach acid and facilitate the passage of the pathogen to the bloodstream.) The gastrointestinal infection involves fever, nausea, and severe abdominal cramps. The infection can become systemic and cause necrotic skin lesions and cellulitis. The mortality rate for the disease can be as high as 50%. There are some 100 reported cases in the United States each year, usually limited to the Gulf Coast. This has lead the US Food and Drug Administration (FDA) in 2009 to explore methods for post-harvest processing of raw Gulf Coast oysters to reduce infections.

CONCEPT AND REASONING CHECKS
11.8e How might shellfish become contaminated with a *Vibrio* species?

Several Bacterial Species Can Cause an Invasive Gastroenteritis

KEY CONCEPT
9. Some bacterial pathogens leave the intestinal lumen and infect submucosal layers.

Besides those bacterial species that cause inflammatory gastroenteritis, additional species possess virulence factors that allow them to invade layers of the GI tract beyond the epithelium.

Salmonella Species. **Typhoid fever** is among the classical diseases (the "slate wipers") that have ravaged human populations for generations. The disease was first studied by Karl Eberth

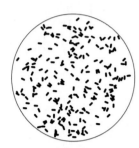

Brucella abortus

■ Zoonotic:
Refers to a disease transmitted from animals to humans.

■ Cellulitis:
A diffuse inflammation of the connective tissues of the skin.

■ Virulence factors:
Structures or molecules that increase a pathogen's ability to invade or cause disease.

and Georg Gaffky, both coworkers of Robert Koch. Eberth observed the causative organism in 1880 and Gaffky isolated it in pure culture four years later.

There are some 2500 unique serotypes of *Salmonella*. Serotypes are used for *Salmonella* instead of species because of the uncertain relationships existing among the organisms. Typhoid fever is caused by *S. enterica* serotype Typhi, a motile, nonspore-forming, facultatively anaerobic, gram-negative rod. For convention sake, we will refer to it as *S.* Typhi.

S. Typhi displays high resistance to environmental conditions outside the body, which enhances its ability to remain alive for long periods in water, sewage, and certain foods exposed to contaminated water. It is transmitted by the five Fs: flies, food, fingers, feces, and fomites (FIGURE 11.18). Humans are the only host for *S.* Typhi.

The organism is transmitted via the fecal-oral route and survives passage through the stomach. During the 5- to 21-day incubation period, it invades the small intestine, causing deep ulcers, bloody stools, and abdominal pain. Blood invasion leads to a systemic illness and the patient experiences mounting fever, lethargy, and delirium. (The word *typhoid* is derived from the Greek *typhos* = "smoke," a reference to the delirium.) In about 30% of cases, the chest and abdomen become covered with a faint rash (**rose spots**), indicating blood hemorrhage in the skin. Symptoms last for 3 to 4 weeks. If untreated, about 15% of individuals die.

There are more than 400 cases of typhoid fever reported annually to the CDC. However, globally there are 21 million cases and 200,000 deaths annually, so about 85% of infections in the United States are acquired during international travel to endemic areas. Treatment is generally successful with the antibiotic ceftriaxone. About 5% of recoverers are carriers and continue to harbor and shed the organisms for a year or more without symptoms. Because food handlers can be carriers of disease, the public health department usually monitors the activities of carriers. The experiences of one of history's most famous carriers, Typhoid Mary, are recounted in MicroFocus 11.4.

Traditional vaccines for typhoid fever have consisted of dead *S.* Typhi cells, but adverse reactions have introduced an element of risk. A newer oral Ty21a vaccine is composed of a weakened (attenuated) strain of *S.* Typhi. Another injectable

Serotypes:
Closely related groups of microorganisms or structures distinguished by their ability to bind to different antibodies.

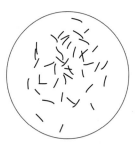

Salmonella Typhi

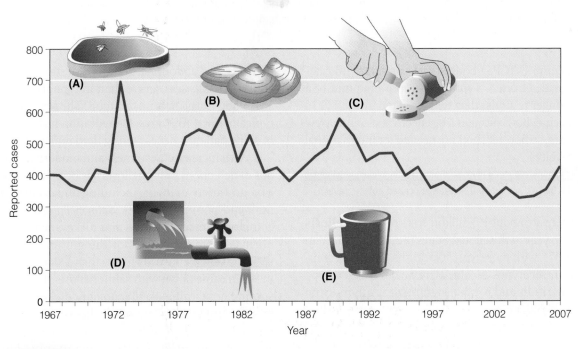

FIGURE 11.18 **The Incidence of Typhoid Fever.** Reported cases of typhoid fever in the United States by year, 1967 to 2007. For 2007, 434 cases were reported. »» What do the five images (A–E) portray? Name them. Data from CDC, *Summary of Notifiable Diseases,* 2007.

MICROFOCUS 11.4: History

Typhoid Mary

By 1906, typhoid fever was claiming about 25,000 lives annually in the United States. During the summer of that year, a puzzling outbreak occurred in the town of Oyster Bay on Long Island, New York. One girl died and five others contracted typhoid fever, but local officials ruled out contaminated food or water as sources. Eager to find the cause, they hired George Soper, a well-known sanitary engineer from the New York City Health Department.

Soper's suspicions centered on Mary Mallon, the seemingly healthy family cook. But she had disappeared three weeks after the disease surfaced. Soper was familiar with Robert Koch's theory that infections like typhoid fever could be spread by people who harbor the organisms. Quietly, he began to search for the woman who would become known as Typhoid Mary.

Soper's investigations led him back over the ten years' time during which Mary Mallon cooked for several households. Twenty-eight cases of typhoid fever occurred in those households, and each time, the cook left soon after the outbreak.

Soper tracked Mary Mallon through a series of leads from domestic agencies and finally came face-to-face with her in March 1907. She had assumed a false name and was now working for a family in which typhoid had broken out. Soper told her he believed she was a carrier and pleaded with her to be tested for typhoid bacilli. When she refused to cooperate, the police forcibly brought her to a city hospital on an island in the East River off the Bronx shore. Tests showed her stools teemed with typhoid organisms, but fearing her life was in danger, she adamantly refused the gall bladder operation that would eliminate them. As news of her imprisonment spread, Mary became a celebrity. Soon public sentiment led to a health department policy deploring the isolation of carriers. She was released in 1910.

But Mary's saga had not ended. In 1915, she turned up again at New York City's Sloane Hospital working as a cook under a new name. Eight people had recently died of typhoid fever, most of them doctors and nurses. Mary was taken back to the island, this time in handcuffs. Still, she refused the operation and vowed never to change her profession. Doctors placed her in isolation in a hospital room while trying to decide what to do. The weeks wore on.

Eventually, Mary became less incorrigible and assumed a permanent residence in a cottage on the island. She gradually accepted her fate and began to help out with routine hospital work. However, she was forced to eat in solitude and was allowed few visitors. Mary Mallon died in 1938 at the age of 70 from the effects of a stroke. She was buried without fanfare in a local cemetery.

vaccine (ViCPS) consists of capsular polysaccharides from *S.* Typhi. Its supporters point to a stronger response than that rendered by the Ty21a vaccine. Both vaccines lose effectiveness after a few years, so booster shots are needed for international travelers.

CONCEPT AND REASONING CHECKS

11.9a Why is typhoid fever often called enteric fever?

There are about 1.3 million cases of **salmonellosis** each year in the United States (48,000 reported in 2007), with more than 500 deaths. The most common serotypes involved are *S. enterica* serotype Enteritidis and *S. enterica* serotype Typhimurium (FIGURE 11.19A).

Transmitted via the fecal-oral route, a relatively large infectious dose is required to initiate an illness. After an incubation period of 6 to 48 hours, the individual experiences fever, nausea, diarrhea, and abdominal cramps. Intestinal ulceration is usually less severe than in typhoid fever, and blood invasion is uncommon. The symptoms typically last 4 to 7 days. Dehydration may occur in some patients, necessitating fluid replacement.

With increased awareness and modern methods of detection, salmonellosis has been linked to a broad variety of foods, including unpasteurized milk and poultry products. *Salmonella* serotypes commonly infect chickens and turkeys when the normal bacterial species of the gut are absent (FIGURE 11.19B). In 2005, more than 1,000 people in Spain came down with salmonellosis after consuming contaminated precooked chicken. This was the largest outbreak of the disease in Spain's history.

Contaminated produce eaten raw has increasingly become a vehicle for transmission of *Salmonella* and other pathogens. In April, 2008,

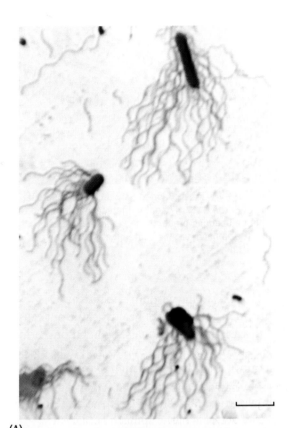

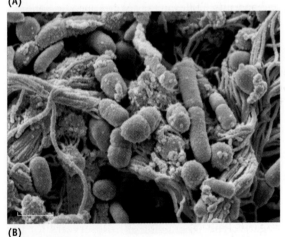

FIGURE 11.19 *Salmonella enterica.* Two views of *S. enterica.* (**A**) A light micrograph photo of *S. enterica* serotype Typhi cells. Note the long length of the flagella relative to the cell. (Bar = 2 μm.) (**B**) A false-color scanning electron micrograph of *S. enterica* serotype Typhimurium on muscle tissue from an infected chicken. (Bar = 3 μm.) »» How might *Salmonella* cells attach to the muscle tissue?

an outbreak of salmonellosis in the United States was traced to fresh, raw jalapeno or serrano peppers. More than 1,400 individuals were infected with the Saintpaul serotype of *S. enterica*. At least 286 persons were hospitalized, and the infection might have contributed to two deaths.

Further discussion of salmonellosis is presented in Chapter 25.

CONCEPT AND REASONING CHECKS

11.9b What is the best way to protect oneself from salmonellosis?

Shigella sonnei. Members of the genus *Shigella* were first described by the Japanese investigator Kiyoshi Shiga in 1898, and two years later by the European microbiologist Simon Flexner.

Shigellosis in the United States is caused primarily by *Shigella sonnei*, which continues to cause outbreaks in daycare centers and accounts for the majority of shigellosis cases. About 20,000 cases of this form of invasive gastroenteritis are reported annually in the United States, which represents about 3% of the estimated infections. Another species, *S. dysenteriae,* causes deadly epidemic dysentery in the developing world where the illness takes the lives of some 1 million people, more than half in children under 5 years of age.

Shigella species are small unencapsulated, gram-negative rods found mainly in humans and other primates. Humans ingest the organisms in contaminated water as well as in many foods, especially eggs, vegetables, shellfish, and dairy products contaminated by a food handler.

An infectious dose requires fewer than 200 organisms, which invade epithelial cells lining the intestine and, after 2 to 4 days, produce sufficient exotoxins (**Shiga toxin**) to trigger gastroenteritis. Infection of the large intestine results in fever, abdominal pain, and bloody mucoid stools that sometimes produce a fatal **bacterial dysentery**. Most cases of shigellosis subside within a week.

Usually there are few complications, but patients who lose excessive fluids must be given salt tablets, oral solutions, or intravenous injections of salt solutions for rehydration. Careful hygiene is most important although antibiotics are sometimes effective in reducing the duration of illness and the number of bacilli shed, but many strains of *Shigella* are becoming more resistant to antibiotics. Recoverers generally are carriers for a month or more and continue to shed the bacilli in their feces. Vaccines are not available.

CONCEPT AND REASONING CHECKS

11.9c What control measure (hygiene practice) would most likely decrease transmission rates of shigellosis?

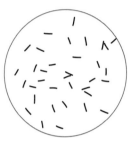

Shigella sonnei

Escherichia coli. Besides the enterotoxigenic (ETEC) and enteropathogenic (EPEC) strains of *E. coli* described earlier that were responsible for an inflammatory gastroenteritis, an **enterohemorrhagic *E. coli* (EHEC)** strain, referred to as *E. coli* O157:H7, has become the most recognizable to the general public—and the most dangerous.

Its major source of transmission is from undercooked ground beef, although other sources including consumption of unpasteurized milk and juice, sprouts, lettuce, and salami, and contact with cattle, have been documented. In fact, in 2004–2005, three outbreaks in North Carolina, Florida, and Arizona sickened 173 people attending agricultural fairs and petting zoos. Textbook Case 11 presents an example. Waterborne transmission occurs through swimming in contaminated lakes and pools or drinking water inadequately chlorinated.

The organism is particularly pathogenic because less than 1000 bacilli can establish infection, it is acid-tolerant, it produces toxins at an unusually high rate, and, because it colonizes in the intestines, it can deliver toxins to this area efficiently. The toxin blocks protein synthesis. After an incubation period of 1 to 4 days, O157:H7 produces sufficient cytotoxin to damage the intestinal lining, resulting in **hemorrhagic colitis**, a severe, bloody diarrhea. In uncomplicated cases, the symptoms resolve within 5 to 10 days.

Older adults, young children, or individuals with a weakened immune system may develop complications involving the kidneys and leading to kidney failure. It is referred to as **hemolytic uremic syndrome (HUS)**; seizures, coma, colonic perforation, liver disorder, and heart muscle infections have been associated with HUS.

Epidemiologists at the CDC estimated that in 2007 there were 4,847 reported cases of **Shiga-toxin-producing *E. coli* (STEC)**, of which most were due to O157:H7. All states require *E. coli* O157:H7 isolates be reported to the state health department and the CDC, and physicians are alerted to watch for cases of bloody diarrhea. Such a situation arose in 2006 when *E. coli* O157:H7 contaminated bagged spinach and sickened more than 200 Americans in 26 states (**FIGURE 11.20**). More than 100 were hospitalized and there were 31 reported cases of HUS that took the lives of three individuals.

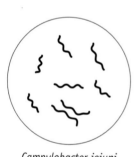

Campylobacter jejuni

FIGURE 11.20 Fresh Bagged Spinach. The 2006 outbreak of *E. coli* O157:H7 contaminated spinach forced the recall of all bagged products. »» Propose a mechanism by which the spinach became contaminated with the *E. coli* strain.

The prevailing wisdom is that *E. coli* O157:H7 exists in the intestines of healthy cattle but causes no disease in these animals. Slaughtering brings *E. coli* to beef products, and excretion to the soil accounts for transfer to plants and fruits. The source of the *E. coli* contamination of bagged spinach was traced to cattle feces from a cattle ranch within one mile of the spinach fields.

CONCEPT AND REASONING CHECKS

11.9d Why are children and the elderly more susceptible to EHEC?

Campylobacter jejuni. Since the early 1970s, **campylobacteriosis** has emerged from an obscure disease in animals to being the most commonly reported bacterial cause of invasive gastroenteritis in the United States. The illness affects over 2.5 million persons and causes about 100 deaths each year.

The pathogen, *Campylobacter jejuni,* is a microaerophilic, curved (*campylo* = "bending"), gram-negative rod (**FIGURE 11.21**). Reservoirs for the organism are the intestinal tracts of many warm-blooded animals, including dairy cattle, chickens, and turkeys. In fact, chickens raised commercially are colonized with *C. jejuni* by the fourth week of life. Unpasteurized dairy products also can be a source of infection.

C. jejuni primarily is transmitted via the fecal-oral route through contact or exposure to contaminated foods or water. During an incubation period of 2 to 7 days, the bacterial cells colonize the small or large intestine. Invasion of the mucosa leads to inflammation and occasional mild ulceration. However, the signs and symptoms of

Textbook CASE 11

Outbreak of *E. coli* O157:H7 Infection

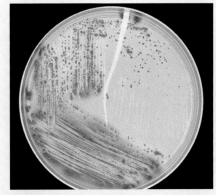

E. coli O157:H7 growing on a culture plate.

1 On October 6, 1996, family A from a large city in Connecticut decided to take a drive in the country. Along the way they stopped at a general store for a bite of lunch. The father and two daughters had apple cider; the mother had a soda instead.

2 Three days later, the father and children began to experience serious abdominal pains and vomiting. Moreover, there was blood in their stools. The mother had no symptoms.

3 One of the daughters of family A became worse and had to be admitted to the hospital. The presence of bloody diarrhea was noted by the doctor.

4 On October 11, the Connecticut Department of Public Health (DPH) was notified of the three illnesses plus five more with disease onset during the same period. A case definition was defined and a stool sample from the daughter was sent to the clinical lab for identification.

5 Meanwhile, the kidneys of family A's daughter were suffering and the doctor advised dialysis to assist the kidney function.

6 The laboratory results identified and confirmed *E. coli* O157:H7 as the cause of infection (see figure).

7 Health officials were notified, and they began a telephone survey to find out if anyone else was similarly infected. Over two dozen cases were found. All were asked about food consumption during the seven days preceding the illness.

8 Based on the interviews, increased risk of illness was associated with drinking fresh apple cider from cider plant A.

9 Inspectors visited the cider plant A. They were told the apples were taken from a pasture where cattle, sheep, and wild deer grazed. The apples were picked directly from the trees. What most interested investigators was hearing many apples also were picked up from the ground.

10 Appropriate control measures were instituted immediately to prevent further cases.

Questions:

(Answers can be found in Appendix D.)

A. What would be the case definition (illness criteria of a person directly affected by the outbreak) defined by the DPH?

B. Hearing that the cause of infection was *E. coli* O157:H7, what types of food might the DPH investigators be most interested in from the phone survey?

C. What is the infection complication exhibited by family A's daughter?

D. Why were the DPH investigators most interested in the "drop" apples collected from the soil surface of the pasture?

E. What control measures were instituted to prevent further outbreak cases?

For additional information, see: http://www.cdc.gov/mmwr/preview/mmwrhtml/00045558.htm.

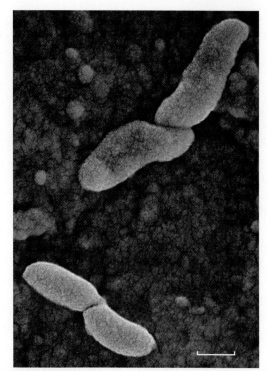

FIGURE 11.21 *Campylobacter jejuni.* A false-color scanning electron micrograph of *C. jejuni* cells. (Bar = 0.5 μm.) »» What shape are these cells?

campylobacteriosis are not unique as they range from mild diarrhea due to production of an enterotoxin to severe gastroenteritis with bloody diarrhea due to production of a Shiga-like toxin. Most patients recover in less than a week without treatment, but some have high fevers and bloody stools for prolonged periods. Erythromycin therapy hastens recovery.

Some people may develop a rare nervous system disease several weeks after the diarrheal illness. This disease, called **Guillain-Barré syndrome (GBS)**, results from the immune system attacking the body's own nerves. The resulting nerve damage can cause paralysis lasting several weeks and usually requires intensive care. Approximately 1 in every 1,000 reported cases of campylobacteriosis leads to GBS and up to 40% of GBS cases in the United States might be caused by campylobacteriosis.

Raw *Campylobacter*-contaminated poultry often is the source of infection. A common route of infection is to cut chicken on a cutting board and then use the unwashed cutting board to prepare other raw foods or vegetables. Other foods also can be contaminated. In 1998, 79 persons at a summer camp were infected with *C. jejuni* through ingestion of contaminated tuna salad. In 2003, an investigative study identified salad vegetables and bottled water as newly identified risk factors.

CONCEPT AND REASONING CHECKS

11.9e Explain how *C. jejuni* could be responsible for traveler's diarrhea.

Yersinia enterocolitica. Another emerging foodborne illness is **yersiniosis**, caused by *Yersinia enterocolitica*. This bacterium is motile at room temperature and is a member of a family of facultative, gram-negative rods that includes *Y. pseudotuberculosis*, which causes an illness similar to *Y. enterocolitica*, and *Y. pestis*, which causes plague (Chapter 12). *Y. enterocolitica* is widely distributed in animals, which often display no symptoms of the disease. Only a few strains of *Y. enterocolitica* cause illness in humans. Infections (primarily in children) occur by consuming food that came in contact with infected domestic animals, raw or undercooked pork products, or by ingesting contaminated water or milk.

Y. enterocolitica causes an invasive gastroenteritis through tissue destruction in the ileum, followed by multiplication in Peyer patches (**FIGURE 11.22**). After a 4- to 6-day incubation, affected individuals experience diarrhea and severe abdominal pain. The symptoms last 1 to 3 weeks and, unless the illness becomes systemic, antibiotic therapy is unnecessary.

CONCEPT AND REASONING CHECKS

11.9f Why is *Y. enterocolitica* considered an invasive pathogen?

Gastric Ulcer Disease Can Be Spread Person to Person

KEY CONCEPT

10. *Helicobacter pylori* causes gastritis and gastric ulcers.

Approximately 25 million Americans suffer from gastric ulcers during their lifetime. For

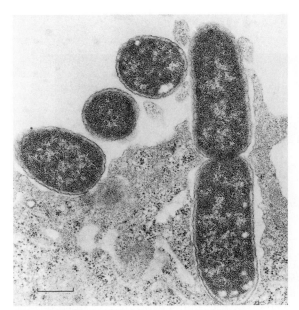

FIGURE 11.22 *Yersinia enterocolitica.* Transmission electron micrograph of invasive *Y. enterocolitica.* A number of bacterial cells are attached to the plasma membrane of a host cell. One bacterial cell appears to be undergoing cell division and entering the cell at the same time. (Bar = 1 μm.) »» What types of foods or beverages might be contaminated by *Y. enterocolitica*?

decades, scientists believed all ulcers resulted from "excess acid" due to factors such as nervous stress, smoking, alcohol consumption, diet, and physiological stress. However, the work of two Australian gastroenterologists, Barry Marshall and J. Robin Warren, made it clear that the bacterial species *Helicobacter pylori* is involved (MicroInquiry 11). The bacterium is closely related to *Campylobacter.*

This microaerophilic, gram-negative, curved rod infects half the world's population, yet only 2% are afflicted with **gastric ulcer disease**. One percent has stomach cancer, which also is associated with an *H. pylori* infection.

It is uncertain how *H. pylori* is transmitted. Most likely, it is spread person to person through contaminated food or water. How *H. pylori* manages to survive in the intense acidity of the stomach is interesting. When *H. pylori* penetrates the stomach mucous layer, it attaches to the stomach wall. There it secretes the enzyme urease that digests urea in the area and produces ammonia as an end product (**FIGURE 11.23**). The ammonia neutralizes acid in the vicinity of the infection. The ammonia and an *H. pylori* cytotoxin cause destruction of the mucous-secreting cells, exposing the underlying connective tissue to the stomach acid. In the stomach lining, a sore 0.6 to 12 cm in diameter appears (although some ulcers may be up to 30 cm in diameter). The pain is severe and is not relieved by food or an antacid.

In the past, physician biopsies of a patient's stomach tissue were used to detect *H. pylori*, but in 1996, a new and relatively simple, noninvasive urea breath test was approved by the FDA. The patient drinks a urea solution fortified with harmless carbon-13 isotopes (see Chapter 2). Because *H. pylori* breaks down urea rapidly, the carbon-13 is quickly expelled as CO_2 in the patient's breath, and it can be detected easily if the bacterial cells are present. If no isotope is detected, the organism is probably not present. The test also can be used to document eradication of the organism after therapy. Thus, doctors have revolutionized the treatment of ulcers by prescribing antibiotics such as amoxicillin, tetracycline, or clarithromycin (Biaxin), along with omeprazole (Prilosec) for acid suppression. They have achieved cure rates of up to 94%; relapses are uncommon.

The bacterial foodborne and waterborne infections discussed in this chapter are summarized in **TABLE 11.2**. MicroFocus 11.5 (p. 366) describes how some of these organisms have been used or could be used as biological weapons.

CONCEPT AND REASONING CHECKS

11.10 Summarize how an *H. pylori* infection causes stomach ulceration.

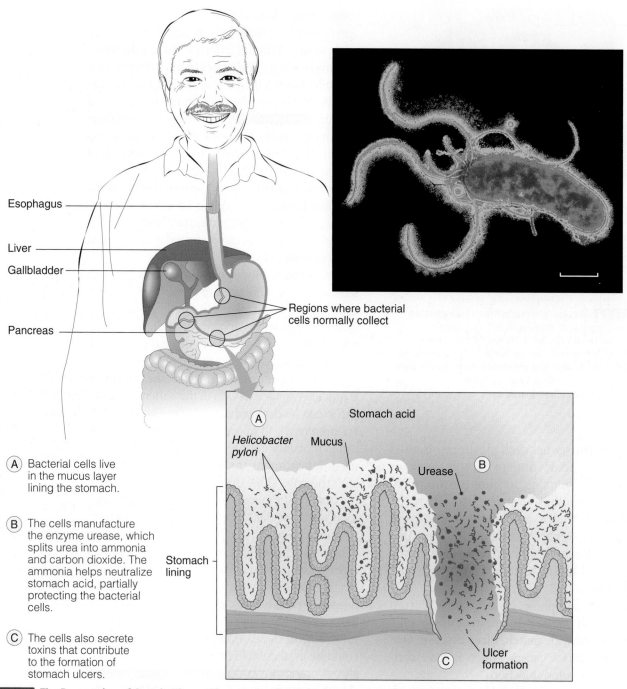

Esophagus

Liver

Gallbladder

Pancreas

Regions where bacterial cells normally collect

(A) Bacterial cells live in the mucus layer lining the stomach.

(B) The cells manufacture the enzyme urease, which splits urea into ammonia and carbon dioxide. The ammonia helps neutralize stomach acid, partially protecting the bacterial cells.

(C) The cells also secrete toxins that contribute to the formation of stomach ulcers.

Stomach acid

Helicobacter pylori

Mucus

Urease

(A)

(B)

Stomach lining

Ulcer formation

(C)

FIGURE 11.23 **The Progression of Gastric Ulcers.** The majority of gastric ulcers are caused by *Helicobacter pylori* [inset, false-color transmission electron micrograph (Bar = 1μm.)]. This figure illustrates how they cause an ulcer. »» How might gastric ulcer disease lead to stomach cancer?

MICROINQUIRY 11

Are Ulcers an Infectious Disease?

Could the old claim, "You're giving me an ulcer!" actually be true? Could ulcers be a transmissible disease?

In 1982, two Australian gastroenterologists, Barry J. Marshall and J. Robin Warren, identified bacterial cells living in the stomach lining of over 100 patients who had ulcers. The initial discovery was serendipitous because Marshall and Warren could not cultivate the organism under normal conditions. Only when they were swamped with work did they leave their culture plates in the incubator too long. And only then did bacterial colonies appear. Marshall (**Figure A**) and Warren identified the organism as the gram-negative curved rod *Campylobacter pyloridis*. Since then, the organism's name has been changed to *Helicobacter pylori* (**Figure B**).

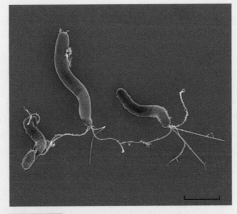

FIGURE B False-color scanning electron micrograph of *Helicobacter pylori cells*. (Bar = 1 µm.)

Marshall and Warren speculated that inflammation in the stomach (gastritis) and ulceration of the stomach or duodenum (peptic ulcer disease) was the result of an infection of the stomach caused by *H. pylori*. This speculation was met with widespread skepticism. Even additional research did not change many minds because the conventional wisdom said that stress, diet, or other factors trigger excess acid secretion and ulcer formation. Indeed, two of the top-selling drugs at the time in the United States were antacids cimetidine (Tagamet) and ranitidine (Zantac), both used to control acid secretion. So, most microbiologists and physicians doubted the duo's claims.

This so frustrated Marshall that he decided to do the ultimate experiment—become a guinea pig. So, in July 1984, he made a dilute solution of the *H. pylori*

bacterial cells and drank it. "I drank it down very quickly, like a tequila shot," he said. If his theory was correct, he should get an ulcer. And guess what? His big gulp paid off. Fourteen days later he had the telltale signs of stomach inflammation. Luckily, his immune system then knocked out the infection.

In 1993, a study published in the *New England Journal of Medicine* indicated that 48 of 52 peptic ulcer patients could be cured of their ulcers in six weeks if treated with two antibiotics over a 12-day period. Another 52 patients received a placebo and 39 seemed to be cured, but a year later, the ulcers had returned in all 39 patients. By comparison, only 4 of the patients receiving antibiotics experienced a recurrence.

In all, it took ten years for the medical community to agree with Marshall's assertion. But the ultimate reward came in 2005 when he and Warren won the 2005 Nobel Prize in Physiology or Medicine for their discovery. As the Nobel Assembly announced, "Thanks to the pioneering discovery by Marshall and Warren, peptic ulcer disease is no longer a chronic, frequently disabling condition, but a disease that can be cured by a short regimen of antibiotics and acid secretion inhibitors." Three cheers (not a glassful of *H. pylori* though) to Marshall and Warren!

FIGURE A Doctor Barry J. Marshall.

TABLE

11.2   A Summary of the Bacterial Foodborne and Waterborne Infections

Disease	Causative Agent	Signs and Symptoms	Toxin Involved	Transmission	Treatment	Prevention and Control
Noninflammatory Gastroenteritis						
Staphylococcal food poisoning	*Staphylococcus aureus*	Abdominal cramps, nausea, vomiting and diarrhea	Enterotoxin	Foodborne from foods improperly handled or stored	Illness usually resolves without treatment	Practicing good hand hygiene Avoiding suspect foods
Clostridial food poisoning	*Clostridium perfringens*	Abdominal cramping, watery diarrhea	Enterotoxin	Foodborne from protein-rich foods improperly handled or stored	Illness usually resolves without treatment	Practicing good hand hygiene Avoiding suspect foods
Food poisoning	*Bacillus cereus*	Diarrhea	Enterotoxin	Foodborne from meats, cream sauces	Recovery without treatment	Avoiding suspect foods
		Vomiting	Enterotoxin	Foodborne starchy foods		
Botulism	*Clostridium botulinum*	Foodborne and wound: difficulty swallowing, slurred speech, blurred vision, trouble breathing Flaccid paralysis	Neurotoxin	Contaminated canned food Contaminated wounds

Consuming spores | Antitoxin Breathing assistance | Practicing good home canning Preparing and storing food properly Avoiding honey in infants |
Inflammatory Gastroenteritis						
Cholera	*Vibrio cholerae*	Severe, watery diarrhea, nausea, vomiting, muscle cramps, and dehydration	Enterotoxin	Waterborne	Oral rehydration therapy Antibiotics	Practicing good hand hygiene Avoiding untreated water
ETEC, EPEC	*Escherichia coli*	Diarrhea, vomiting, cramps, nausea, and low-grade fever	Enterotoxins	Foodborne or waterborne	Illness usually resolves without treatment	Avoiding suspect foods and untreated water
Pseudo-membranous colitis	*Clostridium difficile*	Watery diarrhea, fever, loss of appetite, nausea, dehydration, and abdominal pain	Enterotoxin Cytotoxin	Indirect from contaminated hands or materials	Stopping antibiotic therapy Anticlostridial antibiotic therapy	Practicing good hand hygiene Keeping bathrooms and kitchens disinfected
Listeriosis	*Listeria monocytogenes*	Headache, stiff neck, confusion, loss of balance, convulsions	Not established	Food contaminated with fecal matter Contaminated animal foods	Ampicillin	Practicing good hand washing Washing and preserving food properly
Brucellosis with other foodborne infections	*Brucella abortus* Other *Brucella* species	Flu-like symptoms, backache, joint pain, chills	Not established	Consumption of raw dairy foods, passage through skin abrasions	Doxycycline with gentamicin	Avoiding raw dairy foods, cooking meat thoroughly, wearing gloves

TABLE

11.2   A Summary of the Bacterial Foodborne and Waterborne Infections—*Continued*

Disease	Causative Agent	Signs and Symptoms	Toxin Involved	Transmission	Treatment	Prevention and Control
Vibriosis	*Vibrio para-haemolyticus*	Acute abdominal pain, vomiting, watery diarrhea	None	Foodborne in contaminated seafood (oysters and raw shellfish)	None Antibiotic therapy for severe or prolonged illnesses	Cooking seafood thoroughly, especially oysters
	Vibrio vulnificus	Fever, nausea, severe abdominal cramps	None	See above	Immediate antibiotic therapy	Avoiding raw oysters and clams
Invasive Gastroenteritis						
Typhoid fever	*Salmonella* Typhi	Bloody stools, abdominal pain, fever, lethargy, delirium	Not established	Foodborne from person shedding S. Typhi Foodborne and waterborne from contaminated sewage	Antibiotics	Avoiding risky foods and drinks Getting vaccinated
Salmonellosis	*Salmonella* serotypes	Fever, diarrhea, vomiting, and abdominal cramps	Not established	Foodborne in a broad variety of foods	Fluid replacement Antibiotic therapy	Practicing good hand hygiene and food preparation
Shigellosis	*Shigella sonnei*	Diarrhea, dysentery	Exotoxin	Foodborne and waterborne	Antibiotics Fluid and salt replacement	Practicing good hand hygiene
EHEC	*Escherichia coli* 0157:H7	Severe, bloody diarrhea	Enterotoxin	Foodborne through ingestion of undercooked meat or contaminated fruits and vegetables	For HUS: red blood cell and platelet transfusions Kidney dialysis	Practicing good hand hygiene Cooking foods thoroughly Washing fruits and vegetables; avoiding unpasteurized milk
Campylo-bacteriosis	*Campylobacter jejuni*	Diarrhea, fever	Enterotoxin	Foodborne from contaminated foods or water	None Antibiotic therapy for severe or prolonged illnesses	Practicing good hand hygiene and food preparation
Yersiniosis	*Yersinia enterocolitica*	Fever, diarrhea, and abdominal pain	Not established	Foodborne from contaminated foods (pork and pork products)	None Antibiotic therapy for severe or prolonged illnesses	Avoiding raw or undercooked pork or pork products Practicing good hand hygiene
Gastric ulcer disease	*Helicobacter pylori*	Aching or burning pain in abdomen, nausea, vomiting, bloating, bloody vomit or stools	Cytotoxin	May be transmitted person to person through direct or indirect saliva contact	Antibiotics and acid suppression medications	Practicing good hand hygiene Not sharing utensils or glasses

MICROFOCUS 11.5

Foodborne and Waterborne Bacterial Diseases as Biological Weapons

There are several bacterial species and bacterial toxins the Centers for Disease Control and Prevention (CDC) and the United States government have identified as potential biological weapons that could be used through foodborne or waterborne routes. Category A agents are of highest risk to national security because they could (1) be easily disseminated or transmitted person to person, (2) cause high mortality with a major public health impact, and (3) cause public panic and social disruption. The toxins of botulism fall into this category.

Category B agents are moderately easy to disseminate and cause moderate morbidity and low mortality. Several foodborne and waterborne pathogens are in this category, including *Clostridium perfringens*, *Staphylococcus aureus*, *Salmonella* species, *Shigella dysenteriae*, *Escherichia coli* O157:H7, *Vibrio cholerae*, and *Brucella* species.

Some of these agents actually have been used to commit **biocrimes**, the intentional introduction of a biological agent into food or water to sicken small groups of people. The first known biocrime in the United States occurred in The Dalles, Oregon, in 1984 when the Rajneeshee religious cult, in an effort to influence and win seats in the local election, intentionally contaminated salad bars and a city water tank with *Salmonella enterica* serotype Typhimurium. The result was unsuccessful, although the crime did sicken over 750 citizens and hospitalized 40. Another example occurred in 1996 at a Texas Medical Center. A disgruntled employee deliberately contaminated doughnuts and muffins in a hospital workroom with *Shigella dysenteriae*. All 12 employees who ate the food became ill and 4 were hospitalized.

Bioterrorism has similar aims but on a much larger scale, because the use of such biological agents would cause fear in or actually inflict death or disease upon a large population. Botulinum toxin is one such biological agent that, disseminated in food or water, could be used as a bioterror weapon. The muscle-paralyzing botulinum toxins are among the most powerful toxins produced by living organisms.

If large amounts of the toxin could be produced and disseminated in food or water, the net result would be symptoms similar to typical foodborne botulism. A bioterror act should be relatively easy to identify as an act of bioterrorism rather than a natural outbreak because fewer than 150 cases of botulism are reported each year in the United States. Provided medical and health authorities act quickly, antitoxin treatment given before the onset of symptoms would be very effective against a botulism incident. Also, an investigational vaccine, available for high-risk exposures, is being tested.

The United States has more than 57,000 food processors and 1.2 million food retailers amounting to a $200 billion annual business. In May 2003, the Food and Drug Administration (FDA) proposed new regulations requiring food companies to keep better records for tracking foods involved in any future emergencies or terrorism-related contamination. The agency also plans to require advance information of food import shipments to intercept any contaminated products. In July 2003, the FDA began evaluating ways to prevent or reduce the risk of deliberate contamination of the nation's food supply. This might include chemical treatments, temperature controls, and technology intervention.

Still, for most of the agents identified at the top of this box, a large outbreak should signal bioterrorism because these agents normally do not produce large-scale outbreaks in the United States.

SUMMARY OF KEY CONCEPTS

11.1 The Structure and Indigenous Microbiota of the Digestive System

1. The **digestive system** includes the **gastrointestinal (GI) tract** and the **accessory digestive organs** (teeth, tongue, salivary glands, liver, gallbladder, and pancreas).
   - Digestive system defenses include: salivary mucins, lysozyme, lactoferrin, and defensins; antibodies; stomach acid; mucus on intestinal surfaces; peristalsis; exfoliation of epithelial cells; bile and proteolytic enzymes; **Peyer patches**.
2. The human intestinal **microbiome** contains an enormous population of normal microbiota that provides protection through microbial antagonism.

11.2 Bacterial Diseases of the Oral Cavity

- **Dental caries**
  1. *Streptococcus mutans. S. sobrinus*
- **Gingivitis** and **periodontitis**
  2. *Bacteroides* and several other genera
- **Trench mouth (acute necrotizing ulcerative gingivitis; ANUG)**
  3. *Bacteroides, Fusobacterium, Treponema*

11.3 Introduction to Bacterial Diseases of the GI Tract

5. **Intoxications** represent a form of **noninflammatory gastroenteritis** caused by bacterial toxins, while **inflammatory** and **invasive gastroenteritis** are infections and diseases arising from bacterial growth in the GI tract.
6. Contaminated food and water arising from unsanitary procedures often are the source of the intoxication or infection.

11.4 Foodborne Intoxications Caused by Bacteria

- **Staphylococcal food poisoning**
  4. *Staphylococcus aureus*
- **Clostridial food poisoning**
  5. *Clostridium perfringens*
- Food poisoning
  6. *Bacillus cereus*
- **Botulism**
  7. *Clostridium botulinum*

11.5 Inflammatory Foodborne and Waterborne Infections

- **Cholera**
  8. *Vibrio cholerae*
- **ETEC, EPEC**
  9. *Escherichia coli*
- **Pseudomembranous colitis**
  10. *Clostridium difficile*
- **Listeriosis**
  11. *Listeria monocytogenes*
- Other foodborne infections
  12. *Brucella abortus, Vibrio parachaemolyticus, V. vulnificus*

Invasive

- **Typhoid fever**
  13. *Salmonella* Typhi
- **Salmonellosis**
  14. *Salmonella* serotypes
- **Shigellosis**
  15. *Shigella sonnei*
- **EHEC**
  16. *Escherichia coli* 0157:H7
- **Campylobacteriosis**
  17. *Campylobacter jejuni*
- **Yersiniosis**
  18. *Yersinia enterocolitica*
- **Gastric ulcer disease**
  19. *Helicobacter pylori*

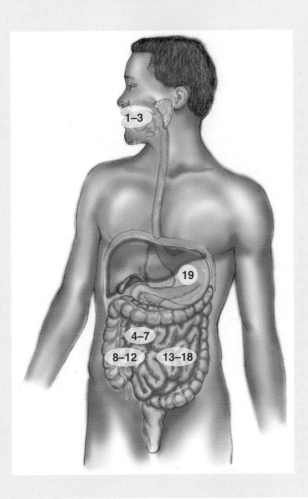

LEARNING OBJECTIVES

After understanding the textbook reading, you should be capable of writing a paragraph that includes the appropriate terms and pertinent information to answer the objective.

1. Identify the organs composing the (a) **gastrointestinal (GI) tract** and (b) the **accessory digestive organs**, and summarize the digestive system defenses against pathogen colonization and infection.
2. Estimate the population size and diversity of the human intestinal **microbiome**.
3. Assess the seriousness of **dental caries** occurring in the **oral cavity**.
4. Describe the role played by oral bacterial species in causing **periodontal disease**.
5. Differentiate between a bacterial **intoxication** and a bacterial **infection**.
6. Identify ways that foods and water become contaminated with bacterial pathogens.
7. Distinguish between the four bacterial species causing intoxications.
8a. Summarize the clinical significance of *Vibrio cholerae* and assess the impact of **cholera** pandemics.
8b. Compare and contrast the **ETEC** and **EPEC** forms of *E. coli* infections.
8c. Explain why *Clostridium difficile* is of concern today.
8d. Explain why a *Listeria* infection is most dangerous to newborns, the elderly, and pregnant women.
8e. Describe how the consumption of dairy products could lead to **brucellosis**.
9a. Summarize the clinical significance of *Salmonella* Typhi infections.
9b. Identify what foods are most likely to become contaminated with *Salmonella enterica* serotypes.
9c. Describe how *Shigella sonnei* is transmitted and how it infects humans.
9d. Describe the consequences of **EHEC** infection.
9e. Explain why **campylobacteriosis** has become the most commonly reported form of bacterial gastroenteritis.
9f. Discuss how *Yersinia entercolitica* infects the digestive system.
10. Diagram the steps involved in **gastric ulcer disease**.

STEP A: SELF-TEST

Each of the following questions is designed to assess your ability to remember or recall factual or conceptual knowledge related to this chapter. Read each question carefully, then select the *one* answer that best fits the question or statement. Answers to even-numbered questions can be found in **Appendix C**.

1. Which one of the following is NOT a digestive organ of the gastrointestinal (GI) tract?
   A. Large intestine
   B. Oral cavity
   C. Liver
   D. Small intestine
2. What part of the GI tract contains the largest population of microorganisms (microbiota)?
   A. Colon
   B. Jejunum
   C. Duodenum
   D. Stomach
3. Which one of the following statements does NOT apply to dental plaque?
   A. It is an example of a biofilm.
   B. It is most noticeable on the molars.
   C. Its buildup on teeth can lead to gingivitis and periodontal disease.
   D. It is dominated by aerobic bacterial species.
4. What type of periodontal disease occurs when plaque bacteria build up between teeth and gums?
   A. Trench mouth
   B. ANUG
   C. Gingivitis
   D. Periodontitis
5. Gastroenteritis can result in
   A. an intestinal inflammation.
   B. an infection.
   C. an intoxication
   D. All the above (**A–C**) are correct.

6. Foodborne microbes can be found in
   A. cattle carcasses.
   B. fresh fruits and vegetables.
   C. healthy animals.
   D. All the above (**A–C**) are correct.
7. Which one of the following bacterial species is NOT a cause of food poisoning (noninflammatory gastroenteritis)?
   A. *Escherichia coli*
   B. *Bacillus cereus*
   C. *Staphylococcus aureus*
   D. *Clostridium perfringens*
8a. One of the most excessive diarrheas of the GI tract is associated with which of the following poisonings or diseases?
   A. Staphylococcal food poisoning
   B. Typhoid fever
   C. Cholera
   D. Campylobacteriosis
8b. *Escherichia coli* is a common gram _____ that can be a cause of _____.
   A. positive rod; hemorrhagic colitis
   B. negative rod; traveler's diarrhea
   C. positive coccus; typhoid fever
   D. negative rod; cholera
8c. *Clostridium difficile* is
   A. the cause of pseudomembranous colitis.
   B. traveler's diarrhea.
   C. meningoencephalitis.
   D. undulant fever.
8d. This gram-positive rod can cause meningoencephalitis, septicemia, and newborn meningitis.
   A. *Bacillus cereus*
   B. *Listeria monocytogenes*
   C. *Clostridium perfringens*
   D. *Escherichia coli*

8e. This species is the most virulent of the vibrios.
  A. *V. vulnificus*
  B. *V. cholerae*
  C. *V. enterocolitica*
  D. *V. parahaemolyticus*

9a. Typhoid fever is characterized by
  A. the production of an exotoxin.
  B. hemolytic uremic syndrome.
  C. rose spots on chest and abdomen.
  D. All the above (**A–C**) are correct.

9b. The symptoms of salmonellosis usually last about
  A. 24 hours.
  B. 48 hours.
  C. 5 days.
  D. 14 days.

9c. What is the name of the syndrome of fever, abdominal cramps, and bloody mucoid stools caused by *Shigella* species?
  A. Bacterial dysentery
  B. Typhoid fever
  C. Pseudomembranous colitis
  D. HUS

9d. Enterohemorrhagic *E. coli* (EHEC) can cause
  A. undulant fever.
  B. hemolytic uremic syndrome.
  C. Guillain-Barré syndrome.
  D. stomach ulcers.

9e. The most commonly reported cause of invasive bacterial gastroenteritis is associated with which of the following bacterial genera?
  A. *Campylobacter*
  B. *Staphylococcus*
  C. *Shigella*
  D. *Clostridium*

9f. Yersiniosis is caused by
  A. *Yersinia pestis.*
  B. *Yersinia pseudotuberculosis.*
  C. *Yersinia entercolitica.*
  D. All of the above (**A–C**) cause the illness.

10. Gastric ulcer disease is caused by
  A. *Helicobacter pylori.*
  B. *Yersinia entercolitica.*
  C. *Escherichia coli.*
  D. *Salmonella* Typhi.

▌ STEP B: REVIEW

When you have completed your study of foodborne and waterborne bacterial diseases, test your knowledge of the important terms by circling the choices that best complete each of the following statements. The answers to even-numbered statements are listed in **Appendix C.**

11. To treat patients who have botulism, large doses of (antitoxin, antibiotic) must be administered.

12. Disease associated with *Shigella* species can produce (diarrhea, dysentery), which is identified by the presence of cramps and bloody stools.

13. (Neurotoxins, Cytotoxins), such as those found with botulism, can cause flaccid paralysis.

14. Many foodborne and waterborne bacterial diseases have ill-defined (symptoms, syndromes), making pathogen identification difficult.

15. Only a small percentage of those who recover from typhoid fever remain (carriers, free) of the bacterial cells.

16. Dental caries is caused by (*Streptococcus mutans, Staphylococcus aureus*), a gram-positive sphere.

17. The diarrheal disease (traveler's diarrhea, cholera) is caused by a motile, gram-negative, curved rod.

18. The genus *Salmonella* is made up of many (genotypes, serotypes); that is, there are many closely related groups that are identified by a specific antibody reaction.

19. An infection of the blood, which can occur with listeriosis, is referred to as (septicemia, gastroenteritis.)

20. If one has ingested botulism exotoxins, the individual is considered to be (infected, intoxicated).

▌ STEP C: APPLICATIONS

Answers to even-numbered questions can be found in **Appendix C.**

21. You are doing the supermarket shopping for the upcoming class barbecue. What are some precautions you can take to ensure that the event is remembered for all the right reasons?

22. You read in the newspaper that botulism was diagnosed in 11 patrons of a local restaurant. The disease was subsequently traced to mushrooms bottled and preserved in the restaurant. What special cultivation practice enhances the possibility that mushrooms will be infected with the spores *Clostridium botulinum?*

23. In preparation for a summer barbecue, your roommate cuts up chickens on a wooden carving board. After running the board under water for a few seconds, he uses it to cut up tomatoes, lettuce, peppers, and other salad ingredients. What sort of trouble may occur?

24. The state department of health received reports of illness in 18 workers at a local pork processing plant. All the affected employees worked on the plant's "kill floor." All had gram-negative rods in their blood. Their symptoms included fever, chills, fatigue, sweats, and weight loss. Which disease was pinpointed in the workers?

25. A classmate plans to travel to a tropical country for spring break. To prevent traveler's diarrhea, she was told to take 2 ounces or 2 tablets of Pepto-Bismol four times a day for 3 weeks before travel begins. Short of turning pink, what better measures can you suggest she use to prevent traveler's diarrhea?

STEP D: QUESTIONS FOR THOUGHT AND DISCUSSION

Answers to even-numbered questions can be found in **Appendix C**.

26. In 1997, researchers in Boston reported that *Helicobacter pylori* accumulates in the gut of houseflies after the flies feed on food containing the pathogen. What are the implications of this research?

27. Some years ago, the CDC noticed a puzzling trend: Reported cases of salmonellosis seemed to soar in the summer months and then drop radically in September. Can you venture a guess as to why this is so?

28. Most physicians agree that the illness called "stomach flu" is not influenza at all. They say the cramps, diarrhea, and vomiting can be due to a variety of bacterial species. Which organisms in this chapter might be good candidates?

29. A frozen-food manufacturer recalls thousands of packages of jumbo stuffed shells and cheese lasagna after a local outbreak of salmonellosis. Which parts of the pasta products would attract the attention of inspectors as possible sources of salmonellosis? Why?

HTTP://MICROBIOLOGY.JBPUB.COM/9E

The site features eLearning, an online review area that provides quizzes and other tools to help you study for your class. You can also follow useful links for in-depth information, read more MicroFocus stories, or just find out the latest microbiology news.

Soilborne and Arthropodborne Bacterial Diseases

Father abandoned child, wife husband, one brother another . . . And I, Agnolo di Tura . . . buried my five children with my own hands . . . . So many died that all believed that it was the end of the world.
—Agnolo di Tura, describing bubonic plague in his chronicle (*The Plague of Siena*) in 1348

As the above quote attests, bubonic plague, commonly known as "the Black Death," was probably the greatest catastrophe ever to strike Europe. It swept back and forth across the continent for almost a decade, each year increasing in ferocity. By 1348, two-thirds of the European population was stricken and half of the sick had died. Houses were empty, towns were abandoned, and a dreadful solitude hung over the land. The sick died too quickly for the living to bury them, so victims often were buried in "plague pits" (FIGURE 12.1). At one point, the Rhône River was consecrated as a graveyard for plague victims. Contemporary historians wrote that posterity would not believe such things could happen, because those who saw them were themselves appalled. The horror was almost impossible to imagine; to many people, including Agnolo di Tura, "it was the end of the world."

Before the century concluded, the Black Death visited Europe at least five more times in periodic reigns of terror. During one epidemic in Paris, an estimated 800 people died each day; in Siena, Italy, the population dropped from 42,000 to 15,000; and in Florence, almost 75% of the citizenry perished. Flight was the chief recourse for people who could afford it, but ironically, the escaping travelers spread the disease.

FIGURE 12.1 **The Plague Pit.** This painting shows the unloading of dead bodies during the plague of 1665. These pits were no more than mass graves used to bury the plague victims that had been gathered up from the streets on "dead carts." »» Why would these plague victims be buried in mass graves rather than by traditional funerals—and at night?

Those who remained in the cities were locked in their homes until they succumbed or recovered.

The immediate effect of the plague was a general paralysis in Europe. Trade ceased and wars stopped. Bewildered peasants who survived encountered unexpected prosperity because landowners had to pay higher wages to obtain help. Land values declined and class relationships were upset, as the system of feudalism gradually crumbled. However, medical practices became increasingly sophisticated, with new standards of sanitation and a 40-day period of detention (quarantine) imposed on vessels docking at ports.

The graveyard of plague left fertile ground for the renewal of Europe during the Renaissance. To many historians, the Black Death remains a major turning point in Western civilization.

Plague is a disease caused by *Yersinia pestis,* an organism found in rodents and their fleas in many areas around the world. Although there are sporadic outbreaks of plague today around the world, there is another reason to be concerned about the potential horrors of the disease. Many

Arthropods:
Animals having jointed appendages and segmented bodies (e.g., ticks, lice, fleas, mosquitoes).

microbiologists and government officials see *Y. pestis* as a possible bioterror agent. Used in an aerosol attack, the pathogen could cause cases of pneumonic plague, an infection of the lungs. One to six days after becoming infected with the bacilli, people would develop fever, weakness, and rapidly developing pneumonia with shortness of breath, chest pain, cough, and sometimes bloody or watery sputum. Nausea, vomiting, and abdominal pain may also occur. Without early treatment, pneumonic plague usually leads to respiratory failure, shock, and rapid death.

Because of the delay between being exposed to *Y. pestis* and becoming sick, people could travel over a large area before becoming contagious and possibly infecting others. Once people have the disease, the bacterial cells can spread to others who have close contact with them. Several types of antibiotics are effective for curing the disease provided they are given within 24 hours of the first symptoms.

A bioweapon carrying *Y. pestis* is possible because the pathogen occurs in nature and could be isolated and grown in an appropriately equipped laboratory, although weaponizing *Y. pestis*—that is, making it easily transmissible through the air—would require advanced knowledge and technology.

Besides plague, other diseases like typhus and relapsing fever are transmitted by arthropods, and both can be interrupted by arthropod control. Neither is a major problem in our society, but a substantial number of cases of other arthropod-borne diseases, such as tularemia, Rocky Mountain spotted fever, and Lyme disease are reported each year. We shall study each of these diseases in this chapter.

First we will examine a number of soilborne diseases where organisms enter the body through a cut, wound, or abrasion, or by inhalation. Among these are anthrax, another feared disease in bioterrorism, and tetanus, a concern to anyone who has stepped on a nail or piece of glass. We also will study other diseases receiving wider recognition as detection methods improve. The soilborne diseases, as well as the arthropodborne diseases, are primarily problems of the blood.

12.1 Soilborne Bacterial Diseases

Soilborne bacterial diseases are those whose agents are transferred from the soil to the unsuspecting individual. To remain alive in the soil, the bacterial cells must resist environmental extremes, and often the cells form endospores, as the first three diseases illustrate.

Anthrax Is an Enzootic Disease

KEY CONCEPT

1. *Bacillus anthracis* produces a capsule and three exotoxins.

Bacillus anthracis was the first bacterial species shown by Koch to be the causative agent of an infectious disease (see Chapter 1). **Anthrax** is primarily an enzootic disease of large, domestic herbivores, such as cattle, sheep, and goats. Animals ingest the spores from the soil during grazing, and soon they are overwhelmed with vegetative bacterial cells as their organs fill with bloody black fluid (*anthrac* = "coal"; the disease name is thus a reference to the blackening of the blood). About 80% of untreated animals die.

Anthrax is caused by *Bacillus anthracis*, a spore-forming, aerobic, gram-positive rod (**FIGURE 12.2**). Endospores germinate rapidly on contact with human tissues to produce vegetative cells (see Chapter 5). The thick capsule of the cells impedes phagocytosis and the organisms produce three exotoxins that work together to cause disease. Capsule and toxins are coded by genes carried on two plasmids.

Humans acquire anthrax from infected animal products or contaminated dust. This can happen in one of three ways.

Inhalation anthrax. Workers who tan hides, shear sheep, or process wool may inhale the spores and contract inhalational (pulmonary) anthrax as a form of pneumonia called **woolsorter's disease**. It initially resembles a common cold (fever, chills, cough, chest pain, headache, and malaise). After several days, the symptoms may progress to severe breathing problems and shock. Inhalation anthrax is usually fatal without early treatment.

Intestinal anthrax. Consumption of contaminated and undercooked meat may lead to gastrointestinal anthrax. It is characterized by an acute inflammation of the intestinal tract. Initial signs include nausea, loss of appetite, vomiting, and fever. This is followed by abdominal pain, vomiting of blood, and severe diarrhea. Intestinal anthrax results in death in 25% to 60% of untreated cases.

Enzootic:
Refers to a disease endemic to a population of animals.

Bacillus anthracis

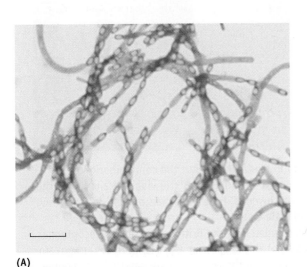

(A)

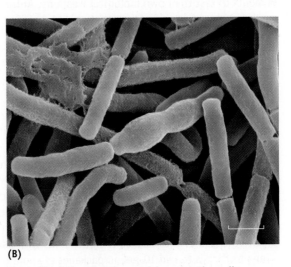

(B)

FIGURE 12.2 *Bacillus anthracis.* *B. anthracis* is the cause of anthrax. (**A**) Spores (white ovals) in vegetative cells can be seen in this photomicrograph. (Bar = 10 μm.) (**B**) A false-color scanning electron micrograph of vegetative cells. It is from such cells that the exotoxins are produced. (Bar = 2 μm.) »» What advantage is provided to the organism by producing endospores?

Cutaneous anthrax. Skin abrasions with spore-contaminated animal products, including violin bows, shaving bristles, goatskin drumheads, and leather jackets, can lead to infection. Skin infection begins as a papule, but within one to two days it develops into a pustule of black, necrotic (dying) tissue that eventually crusts over (**FIGURE 12.3**). Lymph glands in the adjacent area may be invaded and swell. Cutaneous anthrax accounts for more than 95% of all anthrax infections. About 20% of untreated cases will result in death.

B. anthracis infections can be treated with penicillin or ciprofloxacin. In 2000 to 2002, the total number of anthrax cases in the United States was two per year. At present, there is no vaccine for civilian use, although there is a cell-free vaccine for veterinarians and others who work with livestock.

Anthrax also is considered a threat in bioterrorism and in biological warfare (MicroFocus 12.1).

> **Papule:**
> A raised itchy bump that resembles an insect bite.
>
> **Pustule:**
> A papule containing pus.

The seriousness of using biological agents as a means for bioterrorism was underscored in October 2001 when *B. anthracis* spores were distributed intentionally through the United States

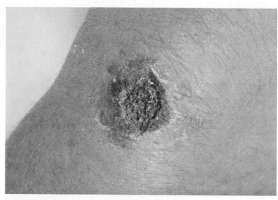

FIGURE 12.3 An Anthrax Lesion. This cutaneous lesion is a result of infection with anthrax bacilli. Lesions like this one develop when anthrax spores contact the skin, germinate to vegetative cells, and multiply. »» Why was a skin ulceration like this given the name "anthrax"?

MICROFOCUS 12.1: History
The Legacy of Gruinard Island

In 1941, the specter of airborne biological warfare hung over Europe. Fearing that the Germans might launch an attack against civilian populations, British authorities performed a series of experiments to test their own biological weapons. Anthrax spores were seen as an agent that could be aerosolized and released unobtrusively. Drifting over a large city, they would be undetectable and could infect thousands of individuals.

To test this possibility, investigators placed 60 sheep on Gruinard Island, a mile-long patch of land off the coast of Scotland. An anthrax spore-containing bomb was exploded over the island. Within days, all the sheep were dead.

Warfare with biological weapons never came to reality in World War II, but the contamination of Gruinard Island remained. A series of tests in 1971 discovered anthrax spores still viable at and below the upper crust of the soil. Fearing they could be spread by earthworms, British officials posted signs warning people not to set foot on the island (see figure) but did little else.

Then a strange protest occurred in 1981. Activists demanded that the British government decontaminate the island. They backed their demands with packages of soil taken from the island. Notes led government officials to two 10-pound packages of spore-laden soil, and the writers threatened that 280 pounds were hidden elsewhere.

Partly because of the protests, the British government instituted a decontamination of the island in 1986. Technicians used a powerful brushwood killer, combined with burning and treatment with formalin in seawater. Finally, they managed to rid the soil of anthrax spores. By April 1987, sheep were once again grazing on the island. However, people were somewhat reluctant to return. Gruinard Island remains a monument of sorts to the effects of biological warfare testing.

mail. In all, 22 cases of anthrax (11 inhalation and 11 cutaneous) were identified, making the case-fatality rate among patients with inhalation anthrax 45% (5/11). The six other individuals with inhalation anthrax and all the individuals with cutaneous anthrax recovered. Had it not been for antibiotic therapy, many more might have been stricken.

CONCEPT AND REASONING CHECKS

12.1 What cellular factors make *B. anthracis* a dangerous pathogen?

Tetanus Causes Hyperactive Muscle Contractions

KEY CONCEPT

2. *Clostridium tetani* produces a powerful neurotoxin.

Tetanus is one of the most dangerous human diseases. *Clostridium tetani,* the bacterial species causing tetanus, is an anaerobic, gram-positive bacillus first isolated in 1889 by the Japanese bacteriologist Shibasaburo Kitasato (see Chapter 1). *C. tetani* forms endospores typically found in barnyard and garden soils containing animal manure.

Spores in very small numbers enter the body through a wound. The spore-containing wounds may result from a fracture, gunshot, animal bite, or puncture by a piece of glass, a thorn, or a rusty nail. Even illicit drugs can contain spores (MICRO-FOCUS 12.2). In dead, oxygen-free tissue of the wound, spores germinate in about 10 days into vegetative bacilli that produce several toxins. The most important of these toxins is the neurotoxin **tetanospasmin**, the second most powerful toxin known to science (after the botulism toxin).

Once inside the tissue, the spores germinate to vegetative cells. At the neuromuscular junction, tetanospasmin prevents the release of neurotransmitters needed to inhibit muscle contraction. Without any inhibiting influence, volleys of spontaneous impulses arise in the motor neurons, causing muscle spasms and stiffness.

Symptoms of tetanus intoxication develop rapidly, often within hours of exposure. A patient first experiences generalized muscle stiffness, especially in the facial and swallowing muscles. Spasms of the jaw muscles cause the teeth to clench and bring on a condition called **trismus**, or **lockjaw**. Severe cases are characterized by a "fixed smile" (risus sardonicus) and muscle spasms cause an arching of the back (**opisthotonus**). Spasmodic inhalation and seizures in the diaphragm and rib muscles leads to reduced ventilation, and patients often experience violent deaths.

Patients are treated with sedatives and muscle relaxants and are placed in quiet, dark rooms as noise and bright light can trigger muscle spasms. Physicians prescribe penicillin to destroy the organisms and inject tetanus antitoxin into a vein to neutralize the toxin.

Immunization for children involves injections of tetanus toxoid in the **diphtheria-tetanus-acellular pertussis (DTaP) vaccine**. The toxoid is prepared by treating the toxin with formaldehyde to eliminate its toxic quality. Children usually receive injections at 2, 4, 6, and 18 months and 5 years of age. Booster injections of tetanus toxoid in the **Td vaccine** (a "tetanus shot") are recommended every 10 years to keep the level of immunity high.

The United States has had a steady decline in the incidence of tetanus, with 28 cases confirmed in 2007. Two individuals died. Older Americans are primarily affected because they either have not been immunized or kept up their booster immunizations. About 40% of cases occur in persons over 60 years of age.

In other parts of the world, tetanus remains a major health problem. Neonatal tetanus accounts for the majority of cases and deaths, often the result of the umbilical stump becoming infected from non-sterile instruments or dressings.

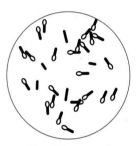

Clostridium tetani

CONCEPT AND REASONING CHECKS

12.2 What are the definitive symptoms of tetanus?

Gas Gangrene Causes Massive Tissue Damage

KEY CONCEPT

3. *Clostridium perfringens* produces a group of toxins and hydrolytic enzymes.

Gangrene (*gangren* = "a sore") develops when the blood flow ceases to a part of the body, usually as a result of blockage by dead tissue. The body part, generally an extremity, becomes dry and shrunken, and the skin color changes to purplish or black. The gangrene may spread as enzymes from broken cells destroy other cells, and the tissue may have to be debrided or the body part amputated.

Gas gangrene, or **myonecrosis** (*myo* = "muscle"; *necros* = "death") is caused by *Clostridium*

Debrided: Referring to the removal of dead, damaged, or infected tissue.

MICROFOCUS 12.2: Public Health

Tetanus Outbreak among Injecting Drug Users

When one thinks about tetanus, the typical image coming into mind is stepping on a rusty nail. Such nails do pose a threat because spores cling to the rough edges of the nail and the nail may cause extensive tissue damage as it penetrates. Actually, most cases in the developed nations of the world are found in older women who become infected by contaminated soil while gardening. However, there are other ways spores can be introduced into the body.

Between July 2003 and January 2004, 14 cases of tetanus in injecting drug users (IDUs) were reported to public health officials in the United Kingdom. The method of injection was subcutaneous injection of heroin. Nine women and five men between the ages of 20 and 53 years were identified as having clinical tetanus, which was defined as mild to moderate trismus and one or more of the following: muscle spasms, difficulty swallowing, and respiratory problems. The cases ranged from mild trismus to full blown tetanus and respiratory arrest, resulting in the death of one woman.

At last seven IDUs had not either been immunized against tetanus or kept up immunization boosters.

The source of the tetanus infection has remained unclear. Because all cases were clustered in a short period, the most likely source was the drug or the adulterant (a substance added to the drug to make it less pure).

Public health officials as well as intensive care facilities throughout England, Scotland, and Wales were notified of the outbreak as were drug action teams and hospitals.

Clostridium perfringens

perfringens, an obligately anaerobic, spore-forming, gram-positive rod. *C. perfringens* also was a common cause of food poisoning due to the production of an enterotoxin (see Chapter 11).

After endospores in contaminated dirt are introduced through a severe, open wound, the spores germinate and the vegetative cells multiply rapidly in the anaerobic environment. As they grow, they ferment muscle carbohydrates and decompose the muscle proteins (thus the term "myonecrosis"). Large amounts of gas may result from this metabolism, causing a crackling sound as the gas accumulates under the skin. The gas also presses against blood vessels, thereby blocking the flow and forcing cells away from their blood supply. In the infection process, the organisms secrete at least 12 exotoxins. The most important is α-toxin, which damages and lyses blood cells. Degradative enzymes, such as DNase and hyaluronidase, are produced that disrupt cell tissues, facilitating the passage of bacterial cells and spread of infection.

The symptoms of gas gangrene include a foul odor, and intense pain and swelling at the wound site. Initially the site turns dull red, then green, and finally blue black (**FIGURE 12.4**). Anemia is

FIGURE 12.4 Gas Gangrene of the Hand. A severely infected hand showing gangrene (blackened tissue necrosis). This gangrene developed from an infection from an accident while the patient was scaling fish. Antibiotic drugs may prevent the infection leading to gangrene, but in this advanced stage, amputation of the hand may be necessary. »» What properties of *Clostridium perfringens* would cause the tissue necrosis?

common, and bacterial toxins may damage the heart and nervous system. Treatment consists of antibiotic therapy as well as debridement, amputation, or exposure in a hyperbaric oxygen chamber. However, without treatment, disease spreads rapidly, and death frequently occurs within days of gangrene initiation.

CONCEPT AND REASONING CHECKS

12.3 What characteristic of *C. perfringens* is being "attacked" by placing a gas gangrene patient in a hyperbaric chamber?

Leptospirosis Is a Zoonotic Disease Found Worldwide

KEY CONCEPT

4. *Leptospira interrogans* infection produces a diverse set of symptoms.

Globally, **leptospirosis** (*lepto* = "thin"; *spir* = "spiral") is the most widespread **zoonosis**—a disease of animals that can spread to humans. The disease affects household pets such as dogs and cats as well as barnyard and wild animals. Humans acquire it by contact with these animals or from soil, food, or water contaminated with their urine.

The agent of leptospirosis is *Leptospira interrogans,* a thin, aerobic, coiled, gram-negative spirochete usually with a hook at one end resembling a question mark, hence the name interrogans (*roga* = "ask") (**FIGURE 12.5**). The undulating movements of these organisms result from contractions of submicroscopic fibers called **endoflagella** (see Chapter 4). In infected animals, the spirochetes colonize the kidney tubules and are excreted in the urine to the soil and water.

As a result of a person swimming or wading in contaminated water, *L. interrogans* enters the human body through the mucus membranes of the eyes, nose, and mouth, or through the skin, especially through abrasions and the soft parts of the feet. The bacterial cells multiply rapidly and most infected individuals experience vague flu-like symptoms. However, for 5% to 10% of these individuals, the illness progresses to a systemic form, which can be lethal. In the first (early) phase, there is an acute onset of headache, muscle aches, vomiting and nausea; the eyes become very red indicating conjunctivitis. Episodes of fever and chills occur for 4 to 9 days, but then disappear.

As the spirochetes directly invade and infect various organs, a second (late) phase occurs as

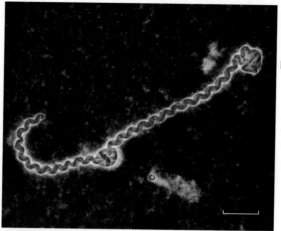

FIGURE 12.5 *Leptospira interrogans.* A false-color transmission electron micrograph of *L. interrogans,* the agent of leptospirosis. Note the tightly coiling spirals and the absence of flagella in this spirochete. (Bar = 2 μm.) »» Besides the coil shape, what are the other notable features of this bacterial species?

the immune system reacts to the infection. A fever returns and meningitis is common, which can lead to stupor and coma. Inflammation also may occur in the liver and lungs. In addition, kidney damage and jaundice (Weil disease) may be present. In other even more severe cases, the patient may vomit blood from gastric hemorrhages and have liver and kidney dysfunction. Despite the numerous tissues involved, the mortality rate from leptospirosis is low (about 10%), and doxycycline is used with success.

Leptospirosis incidence in the United States is low, with 40 to 100 cases identified annually (50% of cases occur in Hawaii). However, leptospirosis is emerging as a globally important infectious organism, especially in warm, subtropical regions. The disease is of particular concern to adventure travelers to distant, exotic locales (MicroFocus 12.3). Closer to home, in 1997 the largest ever outbreak of leptospirosis in the United States broke out at a triathlete competition in Wisconsin where 98 of 1,000 athletes became ill and 28 were hospitalized. The source of infection was from swimming in a lake used for the competition. However, health officials were unable to identify how the lake became a breeding ground for *L. interrogans.*

The soilborne bacterial diseases are summarized in **TABLE 12.1**.

CONCEPT AND REASONING CHECKS

12.4 What bacterial and host factors make *L. interrogans* a potent pathogen?

■ Hyperbaric oxygen: Refers to oxygen pressures (concentrations) higher than normally found in the body.

Leptospira interrogans

■ Conjunctivitis: An inflammation of the conjunctiva of the eye.

It started as a headache on the plane back from Borneo to the States. Within three days, Steve went to a hospital emergency room in Los Angeles complaining of fever and chills, muscle aches, vomiting, and nausea.

The Eco-Challenge 2000 in Sabah, Borneo was the site for the annual adventure race. Some 304 participants composing 76 teams from 26 countries competed in the 10-day endurance event, which was designed to push participants and their teams beyond their athletic limits. During Eco-Challenge 2000, teams would kayak on the open ocean, mountain bike into the rainforest, spelunk in hot caves, and swim in local rivers (see figure). Of the 76 teams starting the challenge, 44 teams finished.

A river in Borneo.

About the time Steve arrived at the emergency room, the Centers for Disease Control and Prevention (CDC) in Atlanta received calls from the Idaho Department of Health, the Los Angeles County Department of Health Services, and the GeoSentinel Network (a network of international travel clinics) reporting cases of a febrile illness similar to Steve's.

The CDC quickly carried out a phone questionnaire to 158 participants in the Eco-Challenge. Many reported symptoms similar to Steve's, including chills, fever, headache, diarrhea, and conjunctivitis. Twenty-five respondents had been hospitalized. Within a few days, antibiotic therapy had Steve recovering. In fact, all 135 affected participants completely recovered.

The similar symptoms suggested leptospirosis and laboratory tests either confirmed the presence of *Leptospira* antibodies or positive culture of the organism from serum samples collected from ill participants.

To identify the source and the exposure risk, information was gathered from participants about various portions of the race course. Analysis identified swimming in and kayaking on the Sagama River as the probable source.

Several participants who did not become ill had taken doxycycline as a preventative for malaria and leptospirosis, as race organizers had advised. Unfortunately, Steve had not heeded those words and suffered a real "eco-challenge." Asked if he would participate in the 2001 Eco-Challenge in New Zealand, he said, "Heck yes! It just adds to pushing the limits."

TABLE

12.1 A Summary of Soilborne Bacterial Diseases

Disease	Causative Agent	Signs and Symptoms	Transmission	Treatment	Prevention and Control
Anthrax	*Bacillus anthracis*	Fever, chills, cough, chest pain, headache, and malaise. Severe breathing and shock can develop	Airborne endospores	Penicillin Ciprofloxacin	Avoiding contact with infected livestock and animal products
Tetanus	*Clostridium tetani*	Muscle stiffness in jaw and neck, trismus	Wounds contaminated with soil, dust, and animal feces	Tetanus antitoxin Penicillin	DTaP vaccine
Gas gangrene	*Clostridium perfringens*	Foul odor and intense pain and swelling at the wound site	Soil Endogenous transfer	Surgery Cephalosporin Hyperbaric oxygen	Cleaning wounds Debridement
Leptospirosis	*Leptospira interrogans*	Acute headache Muscle aches Vomiting and nausea Fever and chills	From contaminated soil, food, water	Antibiotics	Avoiding contaminated water

12.2 Arthropodborne Bacterial Diseases

A living organism such as an arthropod that transmits disease agents is called a **vector** (*vect* = "carried"). Arthropods transmit diseases to humans usually by taking a blood meal from another animal and themselves becoming infected. Then they pass the organisms to another individual during the next blood meal. Arthropodborne diseases occur primarily in the bloodstream, and they often are characterized by a high fever and a body rash.

Plague Can Be a Highly Fatal Disease

KEY CONCEPT

5. *Yersinia pestis* is carried in rodents and their fleas.

Few diseases have had the rich and terrifying history of bubonic plague, nor can any match the array of social, economic, and religious changes wrought by this disease (see chapter introduction).

The first documented pandemic of plague probably began in Africa during the reign of the Roman emperor Justinian in AD 542. It lasted 60 years, killed millions, and contributed to the downfall of Rome. The second pandemic was known as the Black Death because of the purplish-black splotches on victims and the terror it evoked in the 1300s (MicroFocus 12.4). The Black Death killed an estimated 40 million people in Europe, almost one-third of the population on the continent. A deadly epidemic also occurred in London in 1665, where 70,000 people succumbed to the disease.

The third pandemic occurred in the late 1800s, when Asian warfare facilitated the spread of a Burmese focus of plague, and migrations brought infected individuals to China and Hong Kong. During an epidemic in 1894, the causative organism was isolated by Alexandre Yersin and, independently, by Shibasaburo Kitasato (see Chapter 1). Plague first appeared in the United States in San Francisco in 1900, carried by rats on ships from Asia. The disease spread to ground squirrels, prairie dogs, and other wild rodents, and it is now endemic in the southwestern states, where it is commonly called **sylvatic plague** (*sylva* = "forest").

Plague is caused by *Yersinia pestis* (*pestis* = "plague"). This nonmotile, gram-negative rod stains heavily at the poles of the cell, giving it a safety-pin appearance and a characteristic called **bipolar staining** when direct smears from infected specimens are observed (FIGURE 12.6A). The bacillus is transmitted by the oriental rat flea *Xenopsylla cheopis*. The bacterial cells in an infected flea often clot in the digestive system, starving the flea (FIGURE 12.6B). This causes the flea to become even more voracious in finding a blood meal. Normally, the fleas infest only rats, but as septicemic rats die, the fleas may jump to another animal, such as humans, in an attempt to feed. In this process, bacterial cells are regurgitated into the human bloodstream.

Bubonic plague is a blood disease. The bacterial cells multiply in the bloodstream and localize in the lymph nodes, especially those of the armpits, neck, and groin. Hemorrhaging in the lymph nodes causes painful and substantial swellings called **buboes** (*bubon* = "the groin"). Dark,

■ Endemic: Referring to a disease that is constantly present in a specific area or region.

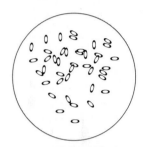

Yersinia pestis

■ Septicemic: Referring to growth of bacterial cells in the blood.

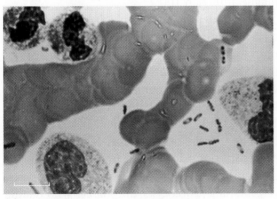

(A)

(B)

FIGURE 12.6 *Yersinia Pestis* and the Flea Vector. **(A)** This light micrograph shows the bipolar staining of the bacterial cells. (Bar = 1 μm.) **(B)** *Xenopsylla cheopis* (oriental rat flea) with clotted *Y. pestis* mass (red foregut). »» How does clotting in the foregut lead to human infections?

MICROFOCUS 12.4: History

Catapulting Terror

Bodies came flying across the walls. People in panic either fled in horror or tried to remove the dead, decaying bodies. Caffa was in chaos.

One of the most horrendous emerging infectious diseases was starting to spread to Europe, North Africa, and the Near East in the mid-fourteenth century. It was the Black Death—bubonic plague—which historians believe moved out of the lands north of the Caspian and Black Seas.

Caffa (today Feodosija, Ukraine) was a port city situated on the north shore of the Black Sea. Through an agreement with the local Tartars (Mongols) who controlled the area, Caffa was placed under control of Genoa, Italy, and Christian merchants were allowed to trade goods with the Far East. In 1343, a group of Italian merchants from Genoa found themselves trapped behind the walls of Caffa after a brawl between the Italians and Tartars. The dreaded Tartars laid siege to the city and over the next five years, Genoa lost and regained control of the city several times.

During the siege of 1346, the Tartars were unable to drive the Italians and other Christians from the city. Then, plague broke out. Large numbers of Tartars started dying. Losing interest in the siege, the Tartars had the bodies of their dead plague victims placed in catapults and lobbed over the walls of Caffa into the city. The Tartars hoped the stench would kill everyone in Caffa; and, in fact, soon plague was sweeping through Caffa. The townspeople were terrified: Either the plague would kill them inside the walls, or the Tartars would kill them outside the walls. But the Tartars were equally terrified of the plague, and they were withdrawing.

Sensing an opportunity to escape, the merchants ran for their ships and sailed off to Genoa, Venice, and other homeports in the Mediterranean. Unfortunately, their voyage home would be a voyage of death. Many died of the plague onboard, and the survivors spread the disease wherever they stopped to replenish their supply of food and water.

Could such a tale be true? Could the dead diseased bodies catapulted into Caffa transmit plague? Almost certainly they could. City defenders would have carried away the dead, mangled bodies, which would spread the disease by contact. Poor sanitation and health of citizens in Caffa would make transmission even easier and more widespread, especially if pneumonic plague broke out.

The attempted siege of Caffa in 1346 represents the most spectacular early episode of biological warfare, with the Black Death as its consequence. It demonstrates the very essence of terrorism as defined today—the intentional or threatened use of biological agents to cause fear in or actually inflict death or disease upon a large population for political, religious, or ideological reasons. The siege of Caffa shows us the horrifying consequences that can come from the use of infectious disease as a weapon.

purplish splotches from hemorrhages also can be seen through the skin (the "rosies" in "Ring-a-ring of rosies"; see Chapter 1). Blood pressure drops and, without treatment, mortality reaches 60%. From the buboes, the bacilli may spread to the bloodstream, where they cause **septicemic plague** or lead to **plague meningitis**. Nearly 100% of untreated cases are fatal.

Human-to-human transmission of plague during epidemics is spread by respiratory droplets because septicemic cases can progress to the lungs. In this form, the disease is called **pneumonic plague** and is highly contagious. Lung symptoms are similar to pneumonia, with extensive coughing and sneezing; hemorrhaging and fluid accumulation are common. Many suffer cardiovascular col-

lapse with death common within 48 hours of the onset of symptoms. Mortality rates for pneumonic plague approach 100%.

When detected early, plague can be treated with streptomycin or tetracycline, reducing mortality to about 18%. Diagnosis consists of the laboratory isolation of *Y. pestis*, together with tests for plague antibodies and typing with bacteriophages. A vaccine consisting of dead *Y. pestis* cells is available for high-risk groups.

The disease occurs sporadically in Native American populations and in travelers through the Southwest. Small-game hunters, taxidermists, veterinarians, zoologists, and others who handle small rodents must be aware of possibly contracting plague. About 12 cases are reported to the

MICROFOCUS 12.5: Public Health

"What a Tragedy!"

"Just an incredible day," he must have been thinking, as he trudged along. It was great to be out in the desert. Great to get away from the books for a while. Final exams were only a few weeks away, then summer, then college, then who knows? Maybe a career in physical education? Maybe a coaching job? Three miles completed, five miles to go. He had to keep moving to make it back home for dinner.

He didn't mind the solitude of the Arizona desert. It was cool up here in Flagstaff, just right for hiking. And there were plenty of animals to see and interesting plants to watch for. He would stop now and then for some water or to watch the horizon or feed a colony of prairie dogs as they bustled about the terrain. Then he continued on to home.

He enjoyed wrestling as well, and he was the team captain. Two days after the hike, he sustained a groin injury while wrestling. When the ache remained, he went to the doctor. The doctor noted the groin swelling and wondered whether it had anything to do with the fever the young man was experiencing. Was it just a coincidence—the ache, the swelling, the fever? This was no ordinary groin injury. "We'll watch it for a couple of days," the doctor said as he gave the young man a pain reliever.

Tragedy struck the next morning. All his mother remembered was a loud thud from the bathroom. She didn't remember her terrified shriek or dialing the emergency number. The emergency medical technicians were there in a flash, but it was too late. He was dead.

The investigation that followed took public health officials down many dead-ends. It wasn't the wrestling injury, they concluded. Still, the groin swelling made them suspicious. "That hiking trail," they asked his mother, "where is it?"

They set out in search of an elusive answer. About three and a half miles out, they came upon a colony of prairie dogs. The animals didn't look well. In fact, some were dead nearby. Carefully, they trapped a sick animal and carried it back to the lab. Two days later, the lab report was ready—the animal was sick with plague. Then the young man's tissues were tested—again, plague. The investigators shook their heads. "What a tragedy!"

Centers for Disease Control and Prevention (CDC) annually. In 2007, seven cases were reported, five from New Mexico; two cases were fatal. Overall, about 14% of all plague cases in the United States are fatal. MicroFocus 12.5 recounts a case associated with wildlife.

CONCEPT AND REASONING CHECKS

12.5 Explain how bubonic plague can develop into communicable pneumonic plague.

Tularemia Has More Than One Disease Presentation

KEY CONCEPT

6. *Francisella tularensis* infections can involve an extremely small infectious dose.

Tularemia is one of several microbial diseases first recorded in the United States (others include St. Louis encephalitis, Rocky Mountain spotted fever, Lyme disease, and Legionnaires' disease).

The causative organism, *Francisella tularensis* is a small, aerobic, encapsulated, gram-negative rod. It is extremely virulent because as few as 10 to 50 CFUs (see Chapter 5) can cause disease. It occurs in a broad variety of wild animals, especially rodents, and it is particularly prevalent in rabbits (in which case it is known as **rabbit fever**). Cats and dogs may acquire the bacillus during romps in the woods, and humans are infected by arthropods from the fur of animals. Ticks are important vectors in this regard, as evidenced by an outbreak of 20 tickborne cases in South Dakota in the 1980s. Other methods of transmission include consumption of infected rabbit meat, drinking contaminated water, or inhaling contaminated air.

Various forms of tularemia exist, depending on where the bacilli enter the body. An arthropod bite, for example, may lead to swollen lymph nodes and a skin ulcer at the bite site (**FIGURE 12.7**). Individuals typically experience flu-like symptoms. Inhalation of *F. tularensis* cells may lead to respiratory disease and produce swollen lymph nodes, a dry cough, and pain under the breast bone.

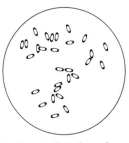

Francisella tularensis

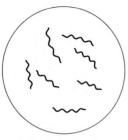

Borrelia burgdorferi

Tularemia usually resolves on treatment with streptomycin, and few people die of the disease. Epidemics are unknown, and evidence suggests that tularemia may not be communicable among humans despite the many modes of entry to the body. Physicians reported 137 cases to the CDC in 2007, half in Missouri, Oklahoma, and Arkansas.

CONCEPT AND REASONING CHECKS

12.6   Propose a hypothesis to explain why the *F. tularensis* infectious dose is so small.

Lyme Disease Can Be Divided into Three Stages

KEY CONCEPT

7.   *Borrelia burgdorferi* is transmitted by the bite of a small tick.

One of the major emerging infectious diseases is **Lyme disease**, currently the most commonly reported arthropodborne illness in the United States. Although 95% of cases occur in the northeastern, mid-Atlantic, and north-central states, cases also have been reported in the Pacific coast states and the Southeast. In 2007 it accounted for 27,444 U.S. cases of infectious disease, primarily in persons 5 to 14 years old and 35 to 64 years old.

Lyme disease is named for Old Lyme, Connecticut, the suburban community where a cluster of cases was observed in 1975. The disease is caused by the spirochete *Borrelia burgdorferi* (**FIGURE 12.8A**). It was named for Willy Burgdorfer, the microbiologist who studied the spirochete in the gut of infected ticks.

The tick that transmits most cases of Lyme disease in the Northeast and Midwest is the deer tick *Ixodes scapularis* (formerly *I. dammini*) (**FIGURE 12.8B**); in the West, the major vector is the western black-legged tick *I. pacificus*. It is smaller than the American dog tick, and it lives and mates in the fur of white-tailed deer.

FIGURE 12.7   **A Tularemia Lesion.** The lesions of tularemia occur where the bacilli enter the body. The disease can be acquired by handling infected rabbit meat or from the bite of an infected arthropod. »» What are the symptoms associated with lesion formation?

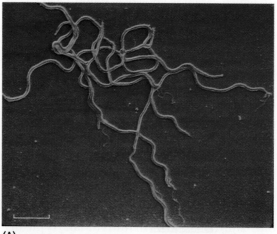

(A)

(B)

FIGURE 12.8   *Borrelia burgdorferi* **and Tick Vector.** **(A)** A false-color scanning electron micrograph showing an abundance of *B. burgdorferi* spirochetes, the agents of Lyme disease. (Bar = 1 μm.) **(B)** A photograph of *Ixodes scapularis*, a tick species that transmits Lyme disease. »» Although not evident in the electron micrograph, what structure typical of spirochetes contributes to their motility?

Eventually it falls into the tall grass, where it waits for an unsuspecting dog, rodent, or human to pass by. The tick then attaches to its new host and penetrates into the skin. During the next 24 to 36 hours, it takes a blood meal and swells to the size of a small pea. While sucking the blood, it also defecates into the wound, and, if the tick is infected, spirochetes are transmitted (FIGURE 12.9). If the tick is observed on the skin, it should be removed with forceps or tweezers, and the area should be thoroughly cleansed with soap and water before applying an antiseptic.

Lyme disease has a variable incubation period of 3 to 31 days and, untreated, typically has three stages. The **early localized stage** involves a slowly expanding red rash at the site of the tick bite. The rash is called **erythema** (red) **migrans** (expand-

ing), or **EM**. Beginning as a small flat or raised lesion, the rash increases in diameter in a circular pattern over a period of weeks, sometimes reaching a diameter of 10 to 15 inches (FIGURE 12.10). It has an intense red border and central clearing, termed the **bull's-eye rash**. It can vary in shape and is usually hot to the touch, but it need not be present in all cases of disease. Indeed, about 20% of patients do not develop EM. The tick bite can be distinguished from a mosquito bite because the latter itches, while a tick bite does not. Fever, aches and pains, and flu-like symptoms usually accompany the rash. Effective treatment can be rendered with amoxicillin or doxycycline.

Left untreated, an **early disseminated stage** begins weeks to months later with the spread of *B. burgdorferi* to the skin, heart, nervous system,

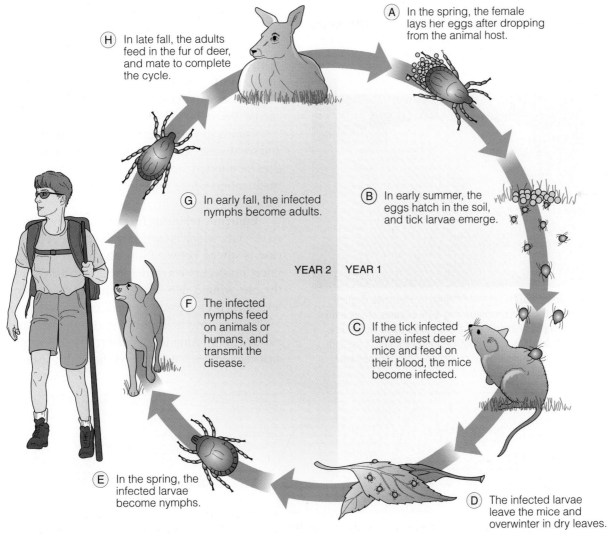

H In late fall, the adults feed in the fur of deer, and mate to complete the cycle.

A In the spring, the female lays her eggs after dropping from the animal host.

G In early fall, the infected nymphs become adults.

B In early summer, the eggs hatch in the soil, and tick larvae emerge.

YEAR 2   YEAR 1

F The infected nymphs feed on animals or humans, and transmit the disease.

C If the tick infected larvae infest deer mice and feed on their blood, the mice become infected.

E In the spring, the infected larvae become nymphs.

D The infected larvae leave the mice and overwinter in dry leaves.

FIGURE 12.9 **Life Cycle of Ixodid Ticks.** The life cycle of *Ixodes scapularis* and *I. pacificus,* the ticks that transmit Lyme disease. »» Why is Lyme disease considered a zoonotic disease?

and joints. On the skin, multiple smaller EMs develop while invasion of the nervous system can lead to meningitis, facial palsy, and peripheral nerve disorders. Cardiac abnormalities are the most common, such as brief, irregular heartbeats. Joint and muscle pain also occur.

If still left untreated, a **late stage** occurs months to years later. About 10% of patients develop chronic arthritis with swelling in the large joints, such as the knee. Although Lyme disease is not known to have a high mortality rate, the overall damage to the body can be substantial.

A vaccine for dogs (LymeVax) has been licensed and is in routine use. A vaccine for humans (LYMErix®) was removed from the market in February 2002 apparently because of poor sales. Because immunity lessens with time, if you received the vaccine before 2002, you probably are no longer immune to Lyme disease.

CONCEPT AND REASONING CHECKS

12.7 Describe the three clinical stages of untreated Lyme disease.

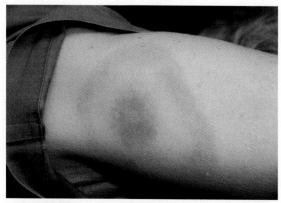

FIGURE 12.10 Erythema Migrans. Erythema migrans (EM) is the rash accompanying 80% of cases of Lyme disease. The rash consists of a large patch with an intense red border. It is usually hot to the touch, and it expands with time. »» What is the common name for the rash?

Borrelia recurrentis

Relapsing Fever Is Carried by Ticks and Lice

KEY CONCEPT

8. Other *Borrelia* species cause a relapsing, febrile illness.

Relapsing fever is caused by a long spirochete responsible for two forms of the disease. It is called relapsing fever because infected individuals go through periods of fever and chills when many spirochetes are present in the blood. As the spirochetes decline in number, the individual recovers for several days before a recurrence of the symptoms (**FIGURE 12.11**).

Endemic relapsing fever often is caused by *Borrelia hermsii* and *B. turicatae*. It is transmitted by the bite of the tick *Ornithodoros* from rodent hosts to humans. Ticks normally inhabit the rodent burrows and nests, where the natural infection cycle proceeds without apparent disease in the rodents. Humans are incidental hosts, often bitten briefly (5 to 20 minutes) at night and without notice by the infected ticks. With endemic relapsing fever, up to 13 relapses can occur and, untreated, mortality can be 2% to 5%.

Cabins in wilderness areas of the northwest and southwest United States are favorable nesting sites for infected rodents and their ticks, especially rustic cabins where rodents have access (TEXTBOOK CASE 12).

Epidemic relapsing fever is caused by *Borrelia recurrentis*, which is carried by body lice. The disease is spread between humans by the lice, and infection occurs when the louse is crushed into the bite wound. No cases of epidemic relapsing fever have been recorded in the United States since 1906. It mainly is found in Africa, China, and parts of South America in overcrowded, poverty-stricken regions where the public health systems

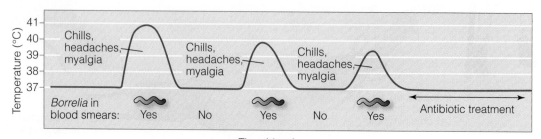

FIGURE 12.11 The Cycles of Relapsing Fever. In relapsing fever, chills, headache and fever peak with high numbers of *Borrelia* spirochetes in the blood. »» During which part of each cycle would the immune system be most active?

Textbook CASE 12

Endemic Relapsing Fever Outbreak

1 In late July 2002, a family reunion was planned at a remote, previously uninhabited cabin in the mountains of northern New Mexico. Three days before the reunion, three family members arrived to clean the cabin. After the reunion, about half of the 39 family members slept in the cabin overnight.

2 Four days after the reunion, one of the family members who cleaned the cabin arrived at a local hospital complaining of fever, chills, muscle aches, and a rash on the forearms. The symptoms had started two days previously.

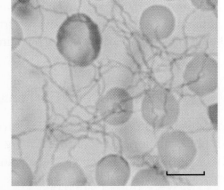

Endemic relapsing fever spirochetes. (Bar = 8 μm.)

3 A hospital lab technician took a blood sample from the patient and identified spirochetes on a blood smear (see figure). The immediate diagnosis was endemic relapsing fever (ERF). On August 2, 2002, the New Mexico Department of Health and the Indian Health Services were immediately notified.

4 By August 7, 2002, 39 family members had sought medical care or were visited by a public health nurse. A study was begun to determine risk factors for infection, better understand the outbreak, and provide prevention measures.

5 A case of ERF was defined. Based on the definition, 14 family members suffering the illness were identified.

6 Blood samples were taken from all 14 symptomatic patients. Spirochetes were identified in the blood smears of nine patients.

7 Samples from 13 patients were sent to the Centers for Disease Control and Prevention (CDC) in Atlanta. Eleven samples, including two samples negative for spirochetes by blood smear, were cultured and all grew *Borrelia hermsii*.

8 All 14 patients received antibiotic therapy and recovered.

9 Risk analysis revealed that all three family members who cleaned the cabin and eight of the other 36 were more likely to be ERF patients.

10 Inspection of the cabin by Indian Health Service workers identified abundant rodent nesting materials and droppings within the walls of the cabin.

Questions:

(Answers can be found in Appendix D.)

A. Why was such a prompt diagnosis and notification of health services necessary?

B. How would you define a case of ERF?

C. Why were spirochetes only found in 9 of the 14 patients?

D. Why did all three who cleaned the cabin become ill?

E. Of the other eight patients, what is most likely the reason for them contracting the disease?

For additional information see http://www.cdc.gov/mmwr/preview/mmwrhtml/mm5234a1.htm.

have failed. It also is a disease of war, such as in World War II when some 10 million people were infected.

Cases of louseborne relapsing fever are characterized by substantial fever, shaking, chills, headache, prostration, and drenching sweats. The symptoms last for a couple of days, disappear for about eight days, then reappear up to 5 times during the following weeks. Left untreated, mortality rates can be as high as 40%.

Both types of relapsing fever can be effectively treated with doxycycline or erythromycin to hasten recovery.

CONCEPT AND REASONING CHECKS

12.8 Explain the reason that infected individuals experience relapsing fevers.

12.3 Rickettsial and Ehrlichial Arthropodborne Diseases

In 1909, Howard Taylor Ricketts, a University of Chicago pathologist, described a new organism in the blood of patients with Rocky Mountain spotted fever and showed that ticks transmit the disease. A year later, he located a similar organism in the blood of animals infected with Mexican typhus, and discovered that fleas were the important vectors in this disease. Unfortunately, in the course of his work, Ricketts fell victim to the disease and died. When later research indicated that Ricketts had described a unique group of microorganisms, the name rickettsiae was coined to honor him.

Rickettsial Infections Are Transmitted by Arthropods

KEY CONCEPT

9. *Rickettsia* infections often involve a characteristic rash.

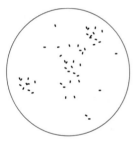

Rickettsia rickettsii

The rickettsiae are small, gram-negative, nonmotile, obligate, intracellular parasites. Most infections are transmitted by ticks, lice, or fleas (**FIGURE 12.12**). All illnesses can be treated effectively with doxycycline or chloramphenicol.

Rocky Mountain spotted fever (RMSF) is caused by *Rickettsia rickettsii* (**FIGURE 12.13A**). It is transmitted by hard ticks, especially those of the genera *Amblyomma* and *Dermacentor* (Figure 12.12A). Following a tick bite, the hallmarks of RMSF are a high fever lasting for many days, severe headaches, and a skin rash reflecting damage to the small blood vessels. The rash begins as pink spots (macules), and progresses to pink-red papules (**FIGURE 12.13B**). Where the spots fuse, they form a maculopapular rash, which becomes dark red and then fades without evidence of scarring. As the disease progresses, the rash appears on the palms of the hands and soles of the feet and progressively spreads

Maculopapular: Refers to a lesion with a broad base that slopes from a raised center.

to the body trunk. Mortality rates of untreated cases are about 30%, although an outbreak in Montana recorded a rate as high as 75%. Antibiotic treatment reduces this rate significantly.

About a thousand cases were reported annually in the United States in the early 1980s, but public education about the disease, along with improved methods of diagnosis and treatment, caused a drop until 1998. Reported cases have since increased with 2,221 cases reported in 2007. Contrary to its name, RMSF is not commonly reported in western states any longer, but it remains a problem in Oklahoma, Arkansas, Texas, and many southeastern and Atlantic Coast states. Children are its primary victims because of their contact with ticks.

Epidemic typhus (also called **typhus fever**) is one of the most notorious of all bacterial diseases. It is considered a prolific killer of humans, and on several occasions it has altered the course of history, as when it helped decimate the Aztec population in the 1500s. Historians report that Napoleon marched into Russia in 1812 with over 200,000 French soldiers but his forces were hit hard by typhus. Hans Zinsser's classic book *Rats, Lice, and History* describes events such as these and provides an engaging look at the effects of several infectious diseases on civilization (MicroFocus 12.6).

Epidemic typhus is caused by *Rickettsia prowazekii*. The rickettsiae are transmitted to humans by body lice of the genus *Pediculus* (Figure 12.12B). Lice are natural parasites of humans and flourish where sanitation measures are lacking and hygiene is poor. Often these conditions are associated with war, famine, poverty, and a generally poor quality of life.

The rickettsiae are excreted in feces from the lice, so scratching the bite will facilitate infection

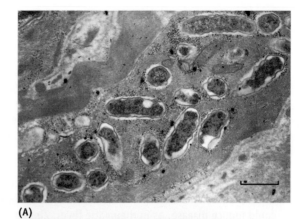

(A)

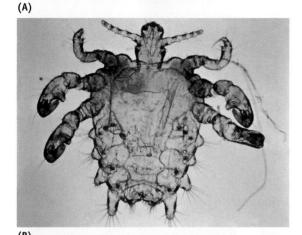

(A)

(B)

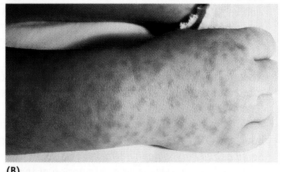

(B)

FIGURE 12.13 *Rickettsia rickettsii* and Symptoms of Rocky Mountain Spotted Fever. (**A**) *R. rickettsii* in the cytoplasm of a kidney cell. (Bar = 0.5 μm.) (**B**) Child's right hand and wrist displaying the characteristic spotted rash of Rocky Mountain spotted fever. »» What is happening in the host to produce a spotted rash?

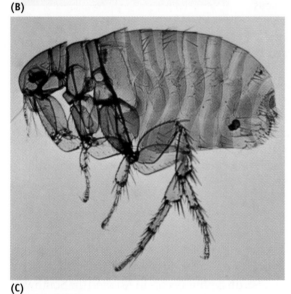

(C)

FIGURE 12.12 Three Arthropods that Transmit Rickettsial Diseases. (**A**) Rocky Mountain woodtick (*Dermacentor*). (**B**) The body louse (*Pediculus*). (**C**) The flea (*Xenopsylla*). »» Why are arthropods such excellent vectors for the transmission of rickettsial diseases?

into the wound. The characteristic fever and rash of rickettsial disease are particularly evident in epidemic typhus. There is a maculopapular rash, but unlike the rash of RMSF, it appears first on the body trunk and progresses to the extremities. Intense fever, sometimes reaching 40°C, remains for many days as the patient hallucinates and becomes delirious. Some patients suffer perma-

nent damage to the blood vessels and heart, and over 75% of sufferers die in epidemics. Again, antibiotic therapy reduces this percentage substantially, and the disease is rare in the United States.

Today microbiologists believe **Brill-Zinsser disease** is a relapse of an earlier case of epidemic typhus in which *R. prowazekii* remains dormant in the patient for many years. However, symptoms are usually milder.

The agent of **murine typhus** (*murin* = "of the mouse") is *Rickettsia typhi* and the disease is prevalent in rodent populations where fleas abound (such as rats and squirrels) (Figure 12.12C). Cats and their fleas are involved also, and lice may harbor the bacilli. When an infected flea feeds in the human skin, it deposits the organism into the wound. Murine typhus is usually characterized by a mild fever, persistent headache, and a maculopapular rash that spreads from trunk to extremities. Often the recovery is

MICROFOCUS 12.6: History

Students and Typhus Fever

Hans Zinsser once remarked, ". . .*[the students] force a teacher continually to renew the fundamental principles of the sciences from which his specialty takes off. So while we are, technically speaking, professors, we are actually older colleagues of our students, from whom we often learn as much as we teach them.*"

Hans Zinsser was born in New York City in 1878. He achieved fame for isolating the bacterial agent of epidemic typhus and was a pioneer in the study of autoimmunity and how microbes could induce disease, as in rheumatic fever. Zinsser was equally well known for his writing and wrote some of the most uniquely personal, wise, and witty prose of the twentieth century.

Zinsser also was a dedicated teacher. Writing at a time when most medical students were male, he remarked, "*. . . as we grow wiser we learn that the relatively small fractions of our time which we spend with well-trained, intelligent young men are more of a privilege than an obligation.*"

Perhaps the culmination of his writing was his most famous book entitled (with tongue in cheek) *Rats, Lice, and History: Being a Study in Biography, Which, After Twelve Preliminary Chapters Indispensable for the Preparation of the Lay Reader, Deals with the Life History of Typhus Fever.* It was an international best seller.

Dr. Zinsser.

Here is a sample of Zinsser's writing from his 1935 book: "*Soldiers have rarely won wars. They often mop up after the barrage of epidemics. And typhus, with its brothers and sisters—plague, cholera, typhoid, dysentery—has decided more campaigns than Caesar, Hannibal, Napoleon, and all the other generals of history. The epidemics get the blame for defeat, the generals the credit for victory. It ought to be the other way 'round. . . .*" Zinsser died in 1940 from lymphatic cancer at the age of 61.

If you think you'd enjoy learning more about disease and its effect on civilization, Zinsser's book is still in print. It would be a worthwhile investment of your time.

---

spontaneous, without the need of drug therapy. However, lice may transport the rickettsiae to other individuals and initiate an epidemic. Most recent cases have occurred in south Texas. In 2004, 27 people in Nueces County, Texas, contracted murine typhus. Then, in 2008 another 33 confirmed cases were reported in Austin and Travis counties with another 24 cases reported in late 2009. All recovered.

Scrub typhus occurs in Asia and the Southwest Pacific. It is so named because *Trombicula*, the mite transmitting the disease, lives in scrubland where the soil is sandy or marshy and vegetation is poor. Scrub typhus also is called **tsutsugamushi fever**, from the Japanese words tsutsuga for "disease" and mushi for "mite." The causative agent, *Orientia* (formerly *Rickettsia*) *tsutsugamushi*, enters the skin during mite infestations and soon causes fever and a rash, together with other typhus-like symptoms. Outbreaks may be significant, as evidenced by the 7,000 U.S. servicemen affected in the Pacific during World War II.

Rickettsialpox was first recognized in 1946 in an apartment complex in New York City. Investigators traced the disease to mites in the fur of local house mice, and named the disease rickettsialpox because the skin rash was similar to a chickenpox rash. Rickettsialpox is now considered a benign disease. It is caused by *Rickettsia akari* (*acari* = "mite"). Fever and rash are typical symptoms, and fatalities are rare.

CONCEPT AND REASONING CHECKS

12.9 What distinguishes the different rickettsial diseases from one another?

Ehrlichia and Anaplasma Infections Are Emerging Diseases in the United States

KEY CONCEPT

10. *Ehrlichia* and *Anaplasma* species infect different groups of leukocytes.

The final diseases we shall consider were first described in humans in 1986. Formerly believed to be confined to dogs, **ehrlichiosis** and **anaplasmosis** are now recognized as two similar rickettsial tickborne diseases. **Human monocytic ehrlichiosis** (**HME**) is caused by *Ehrlichia chaffeensis* (because the first case was observed at Fort Chaffee, Arkansas); and **human granulocytic anaplasmosis** (**HGA**) is caused by *Anaplasma phagocytophilum* (FIGURE 12.14).

HME is transmitted by the lone star tick (prevalent in the South), while HGA is transmitted by the blacklegged tick, which also transmits Lyme disease (prevalent in the Northeast). Patients with HME or HGA suffer from headache, malaise, and fever, with some liver disease and, infrequently, a maculopapular rash. Indeed, both HGA and HME are quite similar to Lyme disease, except

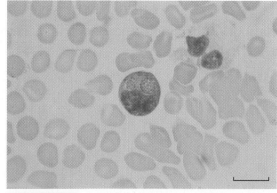

FIGURE 12.14 *Anaplasma phagocytophilum* in a Blood Smear. *A. phagocytophilum* causes human granulocytic anaplasmosis (HGA). This tickborne organism multiplies in human white blood cells (stained purple) called granulocytes. The granulocytes are neutrophils. (Bar = 10 μm.) »» What are the predominant pinkish-reddish cells in this blood smear?

the symptoms come on faster in HGA and HME, they clear more quickly, and the rash is infrequent. Because HME affects the body's monocytes (hence "monocytic") and HGA neutrophils (a type of granulocyte, hence "granulocytic") both bring about a lowering of the white blood cell count (**leukopenia**).

The CDC reported 834 cases of HGA and 828 cases of HME in 2007. However, this may be an underestimate because identification and reporting of human ehrlichioses are incomplete at the state level. Still, HGA cases predominate in the upper Midwest and Northeast while HME cases predominate in New York, North Carolina, and the central Midwest.

The arthropodborne bacterial diseases covered in this chapter are summarized in TABLE 12.2. MicroInquiry 12 presents four cases for study involving soilborne and arthropodborne diseases.

CONCEPT AND REASONING CHECKS

12.10 Draw a concept map for the rickettsial diseases indicating how the diseases differ from one another.

TABLE

12.2   A Summary of Arthropodborne Bacterial Diseases

Disease	Causative Agent	Signs and Symptoms	Transmission	Treatment	Prevention and Control
Plague	*Yersinia pestis*	Bubonic: Sudden onset of fever and chills, headache, fatigue, muscle aches, and buboes Septicemic: High fever, abdominal pain, diarrhea Pneumonic: Headache, malaise, extensive coughing	Infected flea bite Inhaled infectious droplets from person or animal	Streptomycin and gentamicin given intravenously or intramuscularly	Avoiding contact with sick or dead animals Flea control
Tularemia	*Francisella tularensis*	Flu-like symptoms Own set of symptoms depending on body site	Several modes of transmission: ticks, sick or dead animals, airborne, contaminated food or water	Streptomycin	Avoiding and preventing tick bites Avoiding sick or dead animals
Lyme disease	*Borrelia burgdorferi*	Rash, flu-like symptoms, joint pain	Bite of infected deer tick	Amoxicillin	Avoiding and preventing tick bites
Relapsing fever	Endemic: *Borrelia hermsii* *Borrelia turicatae* Epidemic: *Borrelia recurrentis*	Periods of fever and chills Fever, chills, headache, sweating	Bite of infected tick Entry of spirochete from crushed louse	Doxycycline	Avoiding and preventing tick bites Avoiding and preventing lice infestation
Rocky Mountain spotted fever	*Rickettsia rickettsii*	High fever, severe headache, skin rash of red spots	Bite of infected hard tick	Doxycycline or tetracycline	Avoiding and preventing tick bites
Epidemic typhus	*Rickettsia prowazekii*	Fever and rash	Scratching bites from body lice	Doxycycline	Avoiding and preventing lice infestation
Murine typhus	*Rickettsia typhi*	Mild fever, persistent headache, rash	Bite of infected flea	Doxycycline	Avoiding and preventing flea bites
Ehrlichiosis	(HME) *Ehrlichia chaffeensis* (HGA) *Anaplasma phagocytophilum*	Headache, malaise, fever	Bite of infected lone star tick Bite of blacklegged tick	Tetracycline or doxycycline	Avoiding and preventing tick bites

MICROINQUIRY 12

Soilborne and Arthropodborne Disease Identification

Below are several descriptions of soilborne and arthropodborne bacterial diseases based on material presented in this chapter. Read the case history and then answer the questions posed. Answers can be found in **Appendix D**.

Case 1

An 18-year-old girl who had been in good health goes to her physician in Seattle, Washington, complaining of a headache and flu-like symptoms. Further questioning indicates she had developed a red rash on her thigh that enlarged but then disappeared after two weeks. During this time, she also had a fever. The physician discovers the patient had been hiking in the hills east of Seattle ten days prior to developing the rash. The patient is placed on doxycycline and recovers.

12.1a. What important clues toward identifying the disease are indicated from the patient's case?

12.1b. What disease does she have and what organism is responsible for the disease?

12.1c. How was the organism responsible for the disease transmitted to the patient?

12.1d. What complications could occur if the patient had not visited her physician?

12.1e. How could this disease have been prevented?

Case 2

A 33-year-old indigent man comes to the emergency room of the county hospital. The emergency room nurse immediately notices that the man cannot open his mouth because of facial muscle spasms. The physician on duty detects right-sided face pain and trismus. She is able to ascertain from the patient that he has not been able to eat for two days because of jaw pain. Examination of his body shows necrotic, blackened areas on the bottom of his left foot. Based on the signs and physical exam, treatment is started.

12.2a. Based on these signs and the physical exam results, what organism has infected this man and what disease does he have?

12.2b. What treatment should be provided?

12.2c. Should treatment have waited until the diagnosis was confirmed by laboratory results? Explain.

12.2d. What is the significance of the necrotic areas of the patient's left foot?

12.2e. How might this indigent man have become infected?

Case 3

A 10-year-old boy is brought to the emergency room in Charleston, South Carolina, by his mother and father. He had been in good health when the family went for a July Fourth holiday camping trip nine days ago in the Appalachian Mountains. The day after returning from the trip, a tick was discovered in his scalp and was removed. In the emergency room, the boy complains of a headache. He has a fever and the physician observes a pink rash on the palms of his hands and soles of his feet. His white blood cell count is slightly elevated. The physician starts the patient on erythromycin therapy and the boy eventually recovers completely.

12.3a. Identify two organisms that could produce the finding described in this case study.

12.3b. What is the agent responsible for the patient's disease? What were the clues for the diagnosis?

12.3c. What infectious bacterial diseases are spread by ticks?

12.3d. What test(s) might be used to confirm the diagnosis?

12.3e. How would organisms like *Ehrlichia chaffeensis* be eliminated as the agent responsible for the patient's disease?

Case 4

A 34-year-old woman arrives in the emergency room complaining of extreme pain in the right shin with limited mobility in the leg. Her breathing is normal. Examination of the lower leg indicates trauma and the skin is discolored a greenish blue. An additional finding was a crackling sound in her lower leg. The patient tells the physician that one week ago she had a severe mountain biking accident and had several deep cuts to her leg. She is given antibiotics and taken to the operating room where necrotic muscle is discovered. A biopsy Gram stain from the tissue shows gram-positive rods.

12.4a. Based on the emergency room findings, what disease does the patient have and what bacterium is responsible for the infection?

12.4b. What physical conditions of growth should be used when incubating the blood culture?

12.4c. In the operating room, what should the surgeons do once they know the identity of the disease?

12.4d. To ensure a systemic infection does not develop, what physical treatment can be used to slow down or stop the growth of the infecting bacterial cells? Why would this treatment work?

SUMMARY OF KEY CONCEPTS

12.1 Soilborne Bacterial Diseases

1. **Anthrax** is an acute infectious disease caused by *Bacillus anthracis*. Human contact can be by inhalation, consumption, or skin contact with spores. Inhalation produces symptoms of respiratory distress and causes a blood infection. Consumption and skin contacts lead to boil-like lesions.

2. *Clostridium tetani* is the causative agent of **tetanus**. Symptoms of generalized muscle stiffness and **trismus** lead to convulsive contractions with an unnatural fixed smile. Antitoxin and antibiotics can be used to neutralize the toxin and kill the bacterial cells.

3. **Gas gangrene** is caused by the anaerobic endospore-forming *Clostridium perfringens*. Symptoms include intense pain, swelling, and a foul odor at the wound site.

4. **Leptospirosis** is a disease spread from animals to humans (**zoonosis**) by *Leptospira interrogans*. Infected individuals have flu-like symptoms. Up to 10% of patients experience a systemic form of the disease.

12.2 Arthropodborne Bacterial Diseases

5. *Yersinia pestis* is the causative agent of **plague**. This highly fatal infectious disease is transmitted to humans by the bites of infected fleas. **Bubonic plague** is characterized by the formation of **buboes**. Spreading of the bacilli to the blood leads to **septicemic plague**. Localization in the lungs is characteristic of **pneumonic plague**, which can be spread person to person. Without treatment, septicemic and pneumonic plague are nearly 100% fatal.

6. **Tularemia** is caused by *Francisella tularensis*, which is highly infectious at low doses. Various forms of the disease occur depending on where the bacilli enter the body. Skin ulcers, eye lesions, and pulmonary symptoms can result.

7. **Lyme disease** results from an infection by the spirochete *Borrelia burgdorferi*. It involves three stages: the **erythema migrans** (**EM**) rash; neurological and cardiac disorders of the central nervous system; and migrating arthritis.

8. **Relapsing fever** results in recurring attacks of high fever caused by ticks carrying *Borrelia hermsii* (**endemic relapsing fever**) or body lice carrying *Borrelia recurrentis* (**epidemic relapsing fever**). Symptoms include substantial fever, shaking chills, headache, and drenching sweats.

12.3 Rickettsial and Ehrlichial Arthropodborne Diseases

9. Rickettsial and ehrlichial infections usually involve a skin rash, fever, or both.

   - **Rocky Mountain spotted fever** is caused by *Rickettsia rickettsii*, an intracellular parasite. Carried by ticks, the symptoms include a high fever for several days, severe headache, and a maculopapular skin rash.

   - **Epidemic typhus** is a potentially fatal disease occurring in unsanitary conditions and in overcrowded living conditions. It is caused by *Rickettsia prowazekii* that is carried by body lice. A maculopapular rash progresses to the extremities and an intense fever, hallucinations, and delirium are characteristic symptoms of the disease. In epidemics, mortality can be as high as 75%.

   - **Murine typhus** is transmitted by fleas. The causative agent, *Rickettsia typhi*, causes a mild fever, headache, and maculopapular rash. Recovery often is spontaneous.

   - Other diseases caused by rickettsiae include **scrub typhus** (*Orientia tsutsugamushi*) and **rickettsialpox** (*Rickettsia akari*).

10. Ehrlichiosis has been recognized in two forms: **human monocytic ehrlichiosis** (*Ehrlichia chaffeensis*) and **human granulocytic anaplasmosis** (*Anaplasma phagocytophilum*).

LEARNING OBJECTIVES

After understanding the textbook reading, you should be capable of writing a paragraph that includes the appropriate terms and pertinent information to answer the objective.

1. Summarize the clinical significance of the three forms of **anthrax**.
2. Identify how *Clostridium tetani* causes **tetanus** and list the symptoms of the disease.
3. Explain why **gas gangrene** is referred to as **myonecrosis** and identify the toxins and enzymes involved with the disease.
4. Distinguish between the more mild and systemic forms of **leptospirosis**.
5. Contrast between **bubonic, septicemic,** and **pneumonic plague**.
6. Summarize the clinical significance of glandular and inhalation **tularemia**.
7. Distinguish between the three stages of untreated **Lyme disease**.
8. Compare and contrast the two forms of **relapsing fever**.
9. Identify the hallmarks of **Rocky Mountain spotted fever** and how it differs from **epidemic** and **murine typhus**.
10. Discuss the characteristics of the two forms of **ehrlichiosis**.

STEP A: SELF-TEST

Each of the following questions is designed to assess your ability to remember or recall factual or conceptual knowledge related to this chapter. Read each question carefully, then select the *one* answer that best fits the question or statement. Answers to even-numbered questions can be found in **Appendix C**.

1. Woolsorter disease applies to the _____ form of _____.
   A. inhalation; tularemia
   B. toxic; myonecrosis
   C. intestinal; anthrax
   D. inhalation; anthrax

2. Which one of the following describes the mode of action of tetanospasmin?
   A. It inhibits muscle contraction.
   B. It damages and lyses red blood cells.
   C. It disrupts cell tissues.
   D. It inhibits muscle relaxation.

3. A crackling sound associated with myonecrosis is due to
   A. respiratory distress due to plague.
   B. nerve contractions due to tetanus.
   C. lymph node swelling due to plague.
   D. gas produced by *C. perfringens*.

4. *Leptospira interrogans* has all the following characteristics *except*:
   A. endoflagella.
   B. aerobic metabolism.
   C. exotoxin production.
   D. a hook at one end of the cell.

5. A characteristic of cell staining of *Y. pestis* is a
   A. gram-positive staining.
   B. bipolar staining.
   C. gram-positive staining.
   D. gram-variable staining.

6. Skin ulcers are a common lesion resulting from being bitten by
   A. a tick infected with *B. burgdorferi*.
   B. fleas infected with *Y. pestis*.
   C. a tick infected with *F. tularensis*.
   D. lice infected with *C. tetani*.

7. Erythema migrans is typical of the _____ stage of Lyme disease.
   A. early localized
   B. early disseminated
   C. late
   D. recurrent

8. A brief tick bite and a small number of recurring periods of fever and chills is typical of
   A. louseborne relapsing fever.
   B. ehrlichiosis.
   C. Rocky Mountain spotted fever.
   D. epidemic typhus.

9. Rocky Mountain spotted fever is most common in the
   A. southeastern United States.
   B. Rocky Mountains.
   C. Pacific northwest.
   D. New England.

10. A lowering of the white blood cell count is characteristic of
    A. plague.
    B. anthrax.
    C. ehrlichlosis.
    D. RMSF.

STEP B: REVIEW

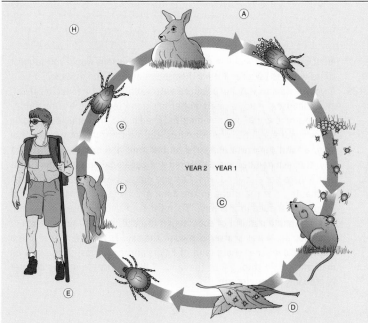

11. In the figure to the left, outline the life cycle of the Ixodid tick responsible for the transmission of Lyme disease.

▮ STEP B: REVIEW

The bacterial diseases transmitted by soil and arthropods are the main focus of this chapter. To test your understanding of the chapter contents, match the statement on the left with the disease on the right by placing the correct letter in the available space. A letter may be used once, more than once, or not at all. **Appendix C** contains the answers to even-numbered statements.

Statement

_____ 12. Accompanied by erythema migrans.

_____ 13. Transmitted by lice; caused by *Rickettsii prowazekii*.

_____ 14. Affects monocytes in the body; transmitted by ticks.

_____ 15. Caused by a spore-forming rod that produces hyaluronidase and α-toxin.

_____ 16. Primarily endemic in large herbivores, such as cattle, sheep, and goats.

_____ 17. Treated with antitoxins; caused by an anaerobic spore former.

_____ 18. Bubonic, septicemic, and pneumonic stages.

_____ 19. Maculopapular rash beginning on extremities and progressing to body trunk.

_____ 20. Caused by a spirochete that infects kidney tissues in pets and humans.

_____ 21. Occurs in small game animals, especially rabbits.

_____ 22. Caused by a gram-negative rod with bipolar staining; transmitted by the rat flea.

_____ 23. Up to 13 attacks of substantial fever, joint pains, and skin spots; *Borrelia* involved.

_____ 24. Pulmonary, intestinal, and skin forms possible; due to a *Bacillus* species.

_____ 25. Most commonly reported tickborne disease in the United States.

Disease

A. Anthrax

B. Ehrlichiosis

C. Epidemic typhus

D. Gas gangrene

E. Leptospirosis

F. Lyme disease

G. Murine typhus

H. Plague

I. Relapsing fever

J. Rickettsialpox

K. Rocky Mountain spotted fever

L. Tetanus

M. Tularemia

▮ STEP C: APPLICATIONS

Answers to even-numbered questions can be found in **Appendix C.**

26. In February 1980, a patient was admitted to a Texas hospital complaining of fever, headache, and chills. He also had greatly enlarged lymph nodes in the left armpit. A sample of blood was taken and Gram stained, whereupon gram-positive diplococci were observed. The patient was treated with cefoxitin, a drug for gram-positive organisms, but soon thereafter he died. On autopsy, *Yersinia pestis* was found in his blood and tissues. Why was this organism mistakenly thought to be diplococci, and what error was made in the laboratory? Why were the symptoms of plague missed?

27. Some estimates place epidemic typhus among the all-time killers of humans; one listing even has it in third place behind malaria and plague. In 1997, during a civil war, an outbreak of epidemic typhus occurred in the African country of Burundi. The outbreak was estimated to be the worst since World War II. What conditions may have led to this epidemic?

28. Leptospirosis has been contracted by individuals working in such diverse locales as subway tunnels, gold mines, rice paddies, and sewage-treatment plants. As an epidemiologist, what precautions would you suggest these workers take to protect themselves against the disease?

29. A young woman was hospitalized with excruciating headache, fever, chills, nausea, muscle pains in her back and legs, and a sore throat. Laboratory tests ruled out meningitis, pneumonia, mononucleosis, toxic shock syndrome, and other diseases. On the third day of her hospital stay, a faint pink rash appeared on her arms and ankles. By the next day, the rash had become darker red and began moving from her hands and feet to her arms and legs. Can you guess the eventual diagnosis?

30. In Chapter 9 of the Bible, in the Book of Exodus, the sixth plague of Egypt is described in this way: "Then the Lord said to Moses and Aaron, 'Take a double handful of soot from a furnace, and in the presence of Pharaoh, let Moses scatter it toward the sky. It will then turn into a fine dust over the whole land of Egypt and cause festering boils on man and cattle throughout the land.' " Which disease in this chapter is probably being described?

STEP D: QUESTIONS FOR THOUGHT AND DISCUSSION

Answers to even-numbered questions can be found in **Appendix C.**

31. Although the tetanus toxin is second in potency to the toxin of botulism, many physicians consider tetanus to be a more serious threat than botulism. Would you agree? Why?

32. Murine typhus was observed in five members of a Texas household. On investigation, epidemiologists learned that family members had heard rodents in the attic, and two weeks previously they had used rat poison on the premises. Investigators concluded that both the rodents and the rat poison were related to the outbreak. Why?

33. Centuries ago, the habit of shaving one's head and wearing a wig probably originated in part as an attempt to reduce lice infestations in the hair. Why would this practice also reduce the possibilities of certain diseases? Which diseases?

34. Even before October 2001, people from various government and civilian agencies were concerned about a terrorist attack using anthrax spores. In various scenarios, try to paint a picture of how such an attack might happen. Then, using your knowledge of microbiology, present your vision of how agencies might deal with such an attack.

35. At various times, local governments are inclined to curtail deer hunting. How might this lead to an increase in the incidence of Lyme disease?

36. In autumn, it is customary for homeowners in certain communities to pile leaves at the curbside for pickup. How might this practice increase the incidence of tularemia, Lyme disease, and Rocky Mountain spotted fever in the community?

HTTP://MICROBIOLOGY.JBPUB.COM/9E

The site features eLearning, an online review area that provides quizzes and other tools to help you study for your class. You can also follow useful links for in-depth information, read more MicroFocus stories, or just find out the latest microbiology news.

13

Sexually Transmitted and Contact Transmitted Bacterial Diseases

I just froze. Then I closed the door, and went in my room and cried.
—A soft-spoken, 38-year-old woman recalling her reaction when she was visited by a health official and told she had syphilis

What do Abraham Lincoln, Adolf Hitler, Friedrich Nietzsche, Oscar Wilde, Ludwig van Beethoven, and Vincent van Gogh have in common? Very likely all suffered from syphilis if Deborah Hayden's research is correct—and it more than likely is. In 2003, she wrote a book entitled *Pox: Genius, Madness, and the Mysteries of Syphilis* (New York: Basic Books) in which she looks at 14 eminent figures from the 15th to 20th centuries whose behavior, careers, or personalities were more than likely shaped by this sexually transmitted disease.

Syphilis originally was called the Great Pox to separate it from smallpox and until the introduction of penicillin in 1943, was untreatable. It caused a chronic and relapsing disease that could disseminate itself throughout the body, only to reappear later as so-called tertiary syphilis. In this most dangerous and terminal form, the disease produces excruciating headaches, gastrointestinal pains, and eventually deafness, blindness, paralysis, and insanity.

Yet sometimes ecstasy and fierce creativity were part of the "symptoms." As Deborah Hayden says, "one of the 'warning signs' of tertiary syphilis is the sensation of being serenaded by angels." In fact, writer Karen Blixen (Isak Dinesen) once said that "Syphilis sold her soul to the devil for the ability to tell stories." Deborah Hayden believes it is just such emotions that provided much of the creative spark for many of the notable historical figures she describes in her book. Perhaps the most intriguing is the debated proposal that syphilis may have driven Hitler mad and that he was dying of syphilis when he committed suicide in his Berlin bunker in the final days of World War II.

If Deborah Hayden's arguments are true, it is amazing how a bacterial organism has affected the body and mind in different ways, shaping the thoughts of writers and philosophers, the creative genius of artists, composers, and scientists—and yes, the madness of dictators.

***

Sexually transmitted diseases (STDs) remain a health problem in the United States and around the world, the seriousness of which is underscored by the number of infections. The Centers for Disease Control and Prevention (CDC) estimate that in 2007, more than 19 million Americans contracted an STD, almost half among the 15 to 24 year age group. Importantly, more than one STD can be acquired at the same time. In addition, individuals infected with an STD are two to five times more likely to acquire HIV than uninfected individuals if they are exposed to the virus through sexual contact.

The STDs discussed in this chapter make up four of the top 10 infectious diseases in the United States (FIGURE 13.1). However, the increase in STDs is but one example of how changing social patterns can affect the incidence of other bacterial diseases discussed in this chapter. For example, the incidence of leprosy in the United States has risen because in the last decades immigrant groups have brought the disease with them. Toxic shock syndrome was first recognized widely in 1980 when a new brand of high-absorbency tampon appeared on the commercial market. These and other contact and miscellaneous diseases will be addressed in this chapter. We will start with diseases associated with the reproductive system.

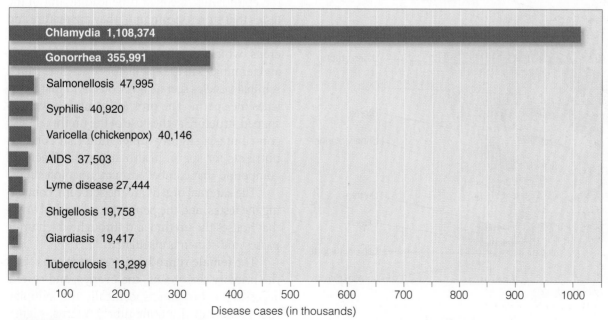

FIGURE 13.1   **Reported Cases of Notifiable Diseases in the United States, 2007.** Three of the top five most reported microbial diseases in the United States are sexually transmitted. »» How many of the top 10 diseases are of bacterial origin?

13.1 The Structure and Indigenous Microbiota of the Female and Male Reproductive Systems

Sexually transmitted diseases (STDs) are a group of infectious diseases usually transmitted through sexual contact and affect the male and female reproductive systems. They often are considered to be "hidden epidemics" because some individuals may not realize they are infected or are reluctant to disclose their condition because of the social stigma of these diseases. Look back again at Figure 13.1 to remind yourself of the significant numbers of reported infections causing chlamydia, gonorrhea, and syphilis.

To understand the diseases, we need to first briefly review the male and female reproductive systems.

The Male and Female Reproductive Systems Consist of Primary and Accessory Sex Organs

KEY CONCEPT

1. Male and female reproductive systems bring egg and sperm cells together.

The **male reproductive system** produces, maintains, and transports sperm cells and is the source of male sex hormones. The **primary sex organs** are the testes, which produce sperm cells (**FIGURE 13.2A**). The testes also produce sex hormones called androgens, such as testosterone, that are important for the maturation, development, and changes in activity of the reproductive system organs.

The **accessory reproductive organs** include the epididymis, a coiled tube leading out of the testes that stores sperm until they are mature and motile. The epididymis terminates in the vas deferens, which combines with the seminal vesicle and terminates in the ejaculatory duct. The paired seminal vesicles secrete a viscous fluid that nourishes the sperm. The prostate gland is a walnut-shaped structure at the base of the urethra. It also contributes a slightly milky fluid that contains nutrients for sperm health and an antibacterial compound that combats urinary tract infections.

The external organs are the scrotum, containing the testes, and the penis, the cylindrical organ that houses the urethra and through which urine passes and semen is ejaculated.

The **female reproductive system** consists of the primary sex organs, the ovaries, and several accessory sex organs, specifically the fallopian tubes (also called uterine tubes), uterus, vagina, and vulva (**FIGURE 13.2B**). During ovulation, an egg is released from one of the ovaries and enters

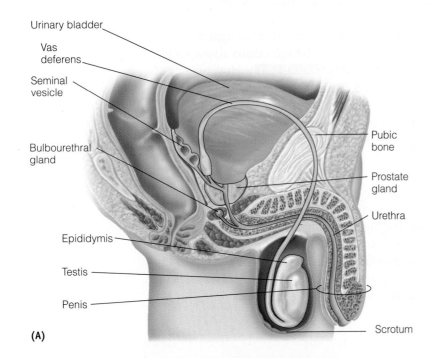

Urinary bladder
Vas deferens
Seminal vesicle
Bulbourethral gland
Epididymis
Testis
Penis

Pubic bone
Prostate gland
Urethra
Scrotum

(A)

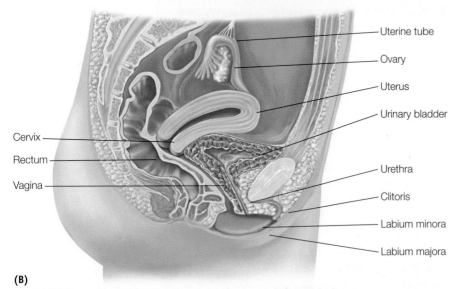

Uterine tube
Ovary
Uterus
Urinary bladder
Urethra
Clitoris
Labium minora
Labium majora

Cervix
Rectum
Vagina

(B)

FIGURE 13.2 Reproductive Systems of Males and Females. The **(A)** male reproductive system and **(B)** female reproductive system showing the primary and secondary sex organs. »» Identify the primary and secondary sex organs in males and females.

a fallopian tube. If sperm are present, fertilization may take place after which the fertilized egg moves through the fallopian tube to the uterus. It is in the uterine lining where the fertilized egg implants. If fertilization does not occur, the lining of the uterus degenerates and sloughs off; this is the process of menstruation. The terminal portion of the female reproductive tract is the vagina, which is a tube about 9 cm long. It represents the receptive chamber for the penis during sexual intercourse, the outlet for fluids from the uterus, and the passageway during childbirth. One very important tissue of the female reproductive tract is the cervix, the lower portion of the uterus and the part that opens into the vagina. The cervix is a common site of infection.

CONCEPT AND REASONING CHECKS

13.1 Trace the path of (a) a sperm cell and (b) an egg as they move through their respective reproductive systems.

Portions of the Male and Female Reproductive Systems Have an Indigenous Microbiota

KEY CONCEPT

2. Antimicrobial defenses depend on the mucosa.

In the male reproductive system, only the urethra is colonized by indigenous microbes. In addition, the reproductive system can mount local immune responses, providing antibodies along the entire length of the urethra and in the seminal fluid.

In the female reproductive tract, the vagina, vulva, and cervix harbor an indigenous population of microbes. The rest of the system usually remains sterile. The vagina produces an acidic pH in which *Lactobacillus* species thrive and contribute to the acid condition by converting sugars to lactic acid. The acidic environment discourages the growth of many potential pathogens.

The mucosa of the vagina and cervix contain immune defense mechanisms (Chapters 20 and 21). Because the cervix represents a potential portal of entry for microbes from the lower genital tract, the cervix contains several antimicrobial defense mechanisms, including: a mucociliary escalator (see Chapter 10); mucus that contains a variety of antimicrobial chemicals, including lysozyme and lactoferrin; and antibacterial peptides

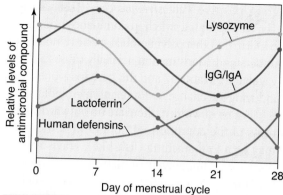

FIGURE 13.3 Fluctuations in Antimicrobial Chemicals and Proteins in the Vagina During the Menstrual Cycle. During the female menstrual cycle, as some antimicrobial compounds fall (IgG/IgA, lactoferrin), others rise (lysozyme, human defensin). »» Why are there fluctuations in the levels of these antimicrobial compounds? Adapted from Wilson, M. *Microbial Inhabitants of Humans: Their Ecology and Role in Health and Disease.* Cambridge University Press. 2005.

that can kill or inhibit the growth of many bacterial species. One interesting note with the female reproductive system is that these chemicals and peptides fluctuate in concentration in the cervix and vagina during the menstruation cycle. As the concentrations of lactoferrin and IgA antibodies drop, the concentrations of human defensins and lysozyme rise (**FIGURE 13.3**). Thus, there is a constant chemical defense against potential microbial invasion.

CONCEPT AND REASONING CHECKS

13.2 Why does there appear to be a much greater and diverse defense system in the female reproductive tract as compared to the male reproductive tract?

Common Vaginal Infections Come From Indigenous Microbiota

KEY CONCEPT

3. Changes in microbiota affect the vaginal environment.

Bacterial Vaginosis. Many women of childbearing age experience **bacterial vaginosis**, which occurs when there is a disruption in the normal balance of vaginal microbiota. Although most women are unaware of the condition, the lactobacilli that normally produce lactic acid and hydrogen peroxide in the vagina are replaced by a mixture of other indigenous bacteria. These include the anaerobic gram-negative rods *Gardnerella vaginalis* and *Prevotella*, and gram-positive *Peptostreptococcus* (FIGURE 13.4). The reason for the switch in dominance from lactobacilli to the other polymicrobial species is not clearly understood. Menstruation normally causes a rise in vaginal pH to about 7 and this perhaps is one factor inducing the change in dominant microbiota.

For most women, there are no outward symptoms of the condition, although for some there may be a foul-smelling (fish-like), grayish-white discharge. Clindamycin or metronidazole therapy generally provides relief. Good hygiene, such as avoiding douching and not using scented tampons or pads, and healthy eating including yogurt that contains active lactobacillus cultures, may prevent infections from recurring.

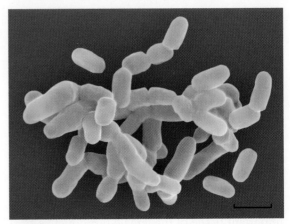

FIGURE 13.4 False-Color Scanning Electron Microscope Image of *Gardnerella vaginalis*. These *G. vaginalis* rods are part of the polymicrobial overgrowth during bacterial vaginosis. (Bar = 2 μm.) »» Could *G. vaginalis* be isolated from a healthy woman not suffering from bacterial vaginosis? Explain.

CONCEPT AND REASONING CHECKS

13.3 What factors appear to be responsible for the overgrowth of the vagina by other members of the indigenous microbiota?

13.2 Sexually Transmitted Diseases Caused by Bacteria

Until the 1990s, STDs were commonly known as **venereal diseases** (*venerea* = referring to Venus, the Roman goddess of love) because they are a group of infectious diseases usually transmitted through sexual contact. STDs also are known as **sexually transmitted infections (STIs)**, denoting the fact that a person may be infected—and can potentially infect others—without showing signs of disease.

STDs continue to be a major public health challenge in the United States. Although the incidence of syphilis and gonorrhea has remained relatively stable since 1996, the incidence of chlamydia over this period has continued to increase (FIGURE 13.5). The CDC estimates that almost 20 million new infections occur each year—and 50% of these infections occur in young people between the ages 15 to 24. This is in spite of the recent advances in preventing, diagnosing, and treating certain STDs. Despite the fact that STDs are extremely widespread, most people in the United States remain unaware of the risk and consequences of all but the most prominent STD—acquired immunodeficiency disease (AIDS). AIDS is discussed in Chapter 23.

In this section, we will consider the common STDs and the microbes that cause them. We will start with the big three: chlamydia, gonorrhea, and syphilis.

Chlamydial Urethritis Is the Most Frequently Reported STD

KEY CONCEPT

4. *Chlamydia trachomatis* can damage the female reproductive organs.

Within the phylum Chlamydiae, three species cause human illness. *Chlamydophila psittaci* and *C. pneumoniae* were discussed in Chapter 10. The

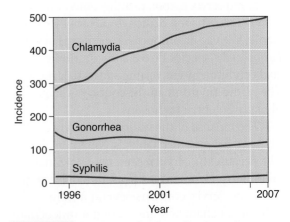

FIGURE 13.5 Incidence of Chlamydia, Gonorrhea, and Syphilis—United States and U.S. Territories, 1995–2007. The incidence refers to the number of cases per 100,000 population. »» If the population of the United States is 300 million, approximately how many actual cases of these STDs were there in 2006?

disease "**chlamydia**" or **chlamydial urethritis** is caused by *Chlamydia trachomatis,* an exceptionally small (0.35 μm), round to ovoid-shaped organism with a cell membrane and outer membrane, but without any peptidoglycan (see Chapter 4).

Chlamydia is the most common STD globally and the most frequently reported STD in the United States (see Figure 13.1). In 2007, more than one million new cases of chlamydia were reported to the CDC. This was the highest rate since reports began in the mid-1980s and the highest since mandatory reporting in 1995. Seventy-five percent of these cases were in individuals under 25 years of age. Part of the reason for the increase is due to the increased number of screening programs available and the development of better diagnostic tests.

C. trachomatis has a biphasic and unique reproductive cycle (FIGURE 13.6). There is a non-replicating, extracellular, infectious **elementary body** (**EB**) and a replicating, intracellular, non-infectious **reticulate body**. Humans appear to be the only host for the organism.

Chlamydial urethritis represents one of several diseases collectively known as **nongonococcal urethritis**, or **NGU**. NGU is a general term for a condition in which people without gonorrhea have a demonstrable infection of the urethra usually characterized by inflammation, and often accompanied by a discharge. Over 50% of cases of NGU are due to chlamydial urethritis, 25% to ureaplasmal urethritis (see below), and the remaining 25% to unknown causes.

C. trachomatis causes a gonorrhea-like disease transmitted by vaginal, anal, or oral sex. Any sexually active individual can be infected. The disease has an incubation period of about one to three weeks. Chlamydia often is referred to as the "silent disease" because the organism does not cause extensive tissue injury directly. Thus, some 85%

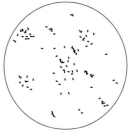

Chlamydia trachomatis

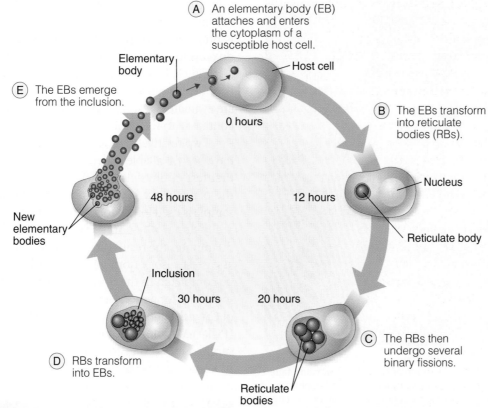

FIGURE 13.6 **The Chlamydial Life Cycle.** The reproductive cycle of the Chlamydiae involves two types of cells. After infection, nonreplicating elementary bodies (EBs) reorganize into reticulate bodies (RBs), which divide to form additional RBs. Within 30 hours, the RBs begin to reorganized into EBs within an inclusion. The EBs are released directly from the inclusion body or by the spontaneous lysis of the inclusion and host cell. »» Since *C. trachomatis* only reproduces inside cells, what cellular products might the bacterium require to survive and grow?

Ectopic pregnancy:
Development of a fertilized egg outside the womb, often in a fallopian tube.

to 90% of infected individuals have no symptoms, often do not seek treatment, and can unknowingly pass the disease on to others. If symptoms do occur, they are due to the inflammation caused by the immune system's response to limit the spread of the infection.

Females often note a slight vaginal discharge, as well as inflammation of the cervix. Burning pain also is experienced on urination, reflecting infection in the urethra. In complicated cases, the disease may spread to, block, and inflame the fallopian tubes, causing **salpingitis** (**FIGURE 13.7**). About 40% of untreated infections progress to **pelvic inflammatory disease (PID)**, which is an inflammation of the uterus, fallopian tubes, and/or ovaries. PID from chlamydia and gonorrhea is believed to affect about 50,000 women in the United States annually. Often, however, there are few symptoms of disease before the salpingitis manifests itself,

Proctitis:
An inflammation of the rectum.

thus adding to the danger of infertility and ectopic pregnancy (TEXTBOOK CASE 13). Chlamydial urethritis is the number one cause of first trimester pregnancy-related deaths in the United States.

In males, chlamydia is characterized by painful urination and a discharge that is more watery and less copious than in gonorrhea. The discharge often is observed after urinating for the first time in the morning. Tingling sensations in the penis are generally evident. Inflammation of the epididymis may result in sterility, but this complication is uncommon. Chlamydial pharyngitis or proctitis is possible as a result of oral or anal intercourse. Research suggests the urogenital infections in males may influence male fertility and reduce sperm quality. If this proves true in actual individuals, then *C. trachomatis* infections in males could lead to infertility.

Newborns may contract *C. trachomatis* during delivery from an infected mother and develop a disease of the eyes known as **neonatal conjunctivitis**.

Chlamydial infections may be treated successfully (95% cure rate) with one dose of azithromycin or 7 days with doxycycline. If a woman is pregnant, erythromycin is substituted because doxycycline affects bone formation in newborns.

Timely treatment of a patient's sexual partners is critical to prevent reinfection. Sexual partners are evaluated, tested, and treated if they have had contact with the patient in the previous 60 days.

Two relatively fast and simple laboratory tests have been available to detect *C. trachomatis*. In the first test, a physician takes a swab sample from the penis or the cervix (as in a Pap smear). The sample is used for a fluorescent antibody test using monoclonal antibodies (Chapter 21). Within 30 minutes the results are available. An immunoassay test, also performed with a swab sample, can be completed in the doctor's office and is similar to the test for gonorrhea. A test to detect the DNA of *C. trachomatis* using polymerase chain reaction (PCR) technology also is available. It uses a urine sample and is said to detect as few as five cells in a sample. MICROFOCUS 13.1 investigates why chlamydial genomes are so small.

Monoclonal antibodies:
Antibodies experimentally produced against a single type of cell or substance.

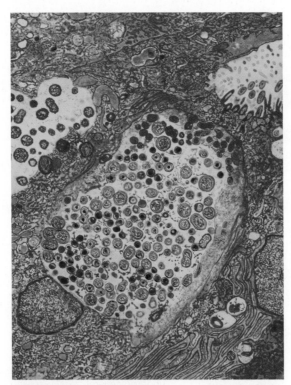

FIGURE 13.7 **A Chlamydial Inclusion.** False-color transmission electron micrograph of cells from a woman's fallopian tube infected with *Chlamydia trachomatis*. Spherical cells (green/brown) are elementary bodies and reticulate bodies seen inside an inclusion body (yellow). Cell cytoplasm is reddish-brown. »» What complications can arise from an infection of the fallopian tubes?

CONCEPT AND REASONING CHECKS

13.4 Why is chlamydia referred to as the "silent disease?"

Textbook CASE 13

A Case of Chlamydia Leading to Sterility

1. An educated, professional woman met a gentleman at a friend's house one evening. She was director of a New York law firm. The man was equally successful in his professional career.

2. The couple hit it off immediately. There were many evenings of quiet candlelit dinners, and soon, they became sexually intimate. Neither one used condoms or other means of protection.

3. Three days after having intercourse, the woman began experiencing fever, vomiting, and severe abdominal pains. She immediately made an appointment to see her doctor.

4. Assuming the illness was an intestinal upset, the physician prescribed appropriate medication. The woman was actually suffering from chlamydia, but the fact that she was sexually active did not come up during the examination.

5. Feeling better, the woman continued her normal routine. But six months later, with no apparent warning, she collapsed on a New York City sidewalk.

6. The woman was rushed to a local hospital, where doctors diagnosed a chlamydial infection by observing the dark inclusion bodies typical of chlamydia (see figure). They performed emergency surgery: her uterus and fallopian tubes were badly scarred. She recovered; however, the scarring left her unable to have children.

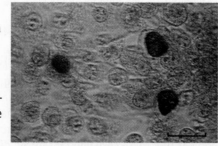

Dark inclusion bodies typical of a chlamydial infection. (Bar = 2 μm.)

Questions:

(Answers can be found in Appendix D.)

A. Would condom use block the transfer of *C. trachomatis* cells?

B. Does fever, vomiting, and abdominal pains indicate a chlamydial infection?

C. What "complications" were developing during the six months when the woman was in improved health?

D. What are inclusion bodies?

E. How would scarring lead to infertility?

Gonorrhea Can Be an Infection in Any Sexually Active Person

KEY CONCEPT

5. *Neisseria gonorrhoeae* infects the urogenital tract.

Gonorrhea is the second most frequently reported notifiable STD in the United States (see Figure 13.1). During the 1960s, the incidence of gonorrhea rose dramatically; since 1975, several hundred thousand cases have been reported annually, the highest percentage being in persons under 24 years of age. Epidemiologists suggest that 3 to 4 million cases go undetected or unreported each year. Despite effective antibiotic therapy and historically low rates, gonorrhea remains an epidemic.

The agent of gonorrhea is *Neisseria gonorrhoeae*, a small, unencapsulated, nonmotile, gram-negative diplococcus named for Albert Neisser, who isolated it in 1879 (**FIGURE 13.8**). The organism, commonly known as the **gonococcus**, has a characteristic double-bean shape. *N. gonorrhoeae* is a very fragile organism susceptible to most antiseptics and disinfectants. Being sensitive to

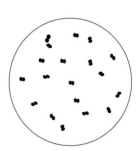

Neisseria gonorrhoeae

MICROFOCUS 13.1: Evolution
Why Do Intracellular Parasites Have Fewer Genes Than Free-Living Species?

The members of the phylum Chlamydiae are among the oldest described human pathogens, and have long fascinated researchers because of their obligate intracellular lifestyle, their small genome size, and the number of diseases they cause. *Chlamydia trachomatis* (see figure) is the causative agent of a number of diseases and has been divided into three groups: serotypes A–C can lead to trachoma, the leading cause of infectious blindness in the world; serotypes D–K cause sexually transmitted diseases of the urogenital tract, such as chlamydial urethritis; and serotype L causes another STD, lymphogranuloma venereum (LGV), which is a more invasive serotype than A–K.

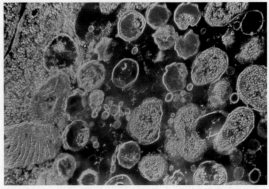

False-color transmission electron microscope image of *Chlamydia trachomatis* cells.

Genomics has provided important insights into the evolution of chlamydial types. Complete genome sequences have been produced for *C. trachomatis* biotypes A, D, and L. All three genomes consist of only about 1 million base pairs, while its free-living primitive cousin, *Protochlamydia amoebophila*, contains about 2.4 million base pairs. All three genomes have some 846 protein-coding genes—its primitive cousin has about 2,000 protein-coding genes. All three chlamydial serotypes also share similar gene locations and nucleotide sequence similarities, suggesting that horizontal gene transfer (HGT) has not had significant impact on the genomes or disease pathology within *C. trachomatis*—HGT has had a significant role in the evolution and pathogenesis of other free-living pathogens (see Chapter 9). So why the small genome?

Once an organism "settles in" as an obligate intracellular parasite, there are gene functions that can be eliminated if the organism can get by without those gene products or by accessing them ready made from the host. Indeed, analysis of the *C. trachomatis* genomes shows that gene loss has occurred relative to its ancestors. As one example, several important metabolic pathways, including the citric acid cycle, are incomplete. Yet the organism can still make ATP. These examples of metabolic streamlining are consistent with those in many other pathogens that depend on an intracellular lifestyle. So, all-in-all, sequence analysis supports the large-scale deletion of genes from the organism. But blanket statements can be dangerous to make. A puzzling discovery is that all *C. trachomatis* types have retained the set of genes to synthesize and assemble peptidoglycan—even though the organisms apparently have no peptidoglycan in their cell walls. If genome reduction is correlated with no longer needing a product, why keep these genes, especially because the organism remains susceptible to penicillin? One would have to conclude that peptidoglycan synthesis of some sort is still involved with chlamydial binary cell division.

So, sequence analysis suggests the evolutionary history of this obligate intracellular pathogen has undergone significant genome reduction accompanied by loss of metabolic functions. In the steady-state environment of a host cell, obligate pathogens may no longer have a need for the diverse genes that allow free-living organisms to rapidly adapt to environmental conditions or, if needed, can be obtained directly from the host.

To really stir up the pot—one of the most surprising observations from genome analyses has been the identification of genes with a sequence similarity to plant sequences! These "plant genes" appear to be derived not from plants directly but rather from a cyanobacterial endosymbiont, reflecting an ancient evolutionary relationship between the Chlamydiae, cyanobacteria, and the chloroplast. Perhaps further sequence analyses will uncover additional information regarding the evolution of the Chlamydiae from free-living bacteria to obligate parasites.

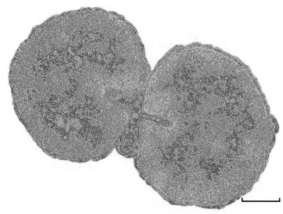

FIGURE 13.8 *Neisseria gonorrhoeae.* A false-color transmission electron micrograph of *N. gonorrhoeae* cells. (Bar = 5 µm.) »» What is the unique feature of these cells?

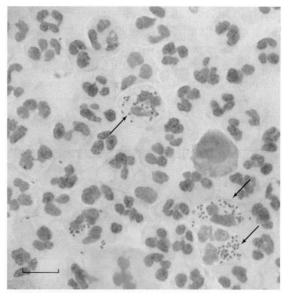

FIGURE 13.9 Gonococcal Urethral Smear. A gram-stained smear of discharge from the male urethra showing the gonococci (arrows) in the cytoplasm of white blood cells. (Bar = 10 µm.) »» What is the gram reaction for *N. gonorrhoeae*?

dehydration, it survives only a brief period outside the body and is rarely contracted from a dry surface such as a toilet seat. The great majority of cases of gonorrhea therefore are transmitted during sexual intercourse. (Gonorrhea is sometimes called "the clap," from the French "clappoir" for "brothel.")

The incubation period for gonorrhea ranges from 2 to 6 days. In females, the gonococci invade and attach by pili to the epithelial surfaces of the cervix and the urethra. The cervix may be reddened, and a discharge may be expressed by pressure against the pubic area. Patients often report abdominal pain and a burning sensation on urination, and the normal menstrual cycle may be interrupted.

In some females, gonorrhea also spreads to the fallopian tubes. As these thin passageways become riddled with pouches and adhesions, salpingitis and PID may occur. Sterility may result from scar tissue remaining after the disease has been treated, or a woman may experience an ectopic pregnancy. It should be noted that symptoms are not universally observed in females, and an estimated 50% of affected women exhibit no symptoms. Such asymptomatic women may spread the disease unknowingly.

Symptoms tend to be more acute in males than in females, and males thus tend to seek diagnosis and treatment more readily. When gonococci infect the mucus membranes of the urethra, symptoms include a tingling sensation in the penis, followed in a few days by pain when urinating. There is also a thin, watery discharge at first, and later a more obvious yellow, thick fluid resembling semen. Frequent urination and

an urge to urinate develop as the disease spreads further into the urethra. The lymph nodes of the groin may swell, and sharp pain may be felt in the testicles. Unchecked infection of the epididymis may lead to sterility.

Gonorrhea does not restrict itself to the urogenital organs. **Gonococcal pharyngitis**, for example, may develop in the pharynx if bacterial cells are transmitted by oral-genital contact; patients complain of sore throat or difficulty in swallowing. Infection of the rectum, or **gonococcal proctitis**, also is observed, especially in individuals performing anal intercourse. Transmission to the eyes may occur by fingertips or towels, developing into keratitis.

Gonorrhea is particularly dangerous to infants born to infected women. The infant may contract gonococci during passage through the birth canal and develop **neonatal conjunctivitis** (discussed later in chapter).

Gonorrhea can be treated effectively with ceftriaxone. However, antibiotic resistance has become an important challenge to controlling gonorrhea such that the cephalosporin antibiotics are the only class that can now be used.

Gonorrhea can be detected by observing gram-negative diplococci from cultivated swab samples as well as in white blood cells from the urogenital tract (**FIGURE 13.9**). For an immunological

Keratitis:
An inflammation and swelling of the cornea.

test, physicians take a swab sample and dip the swab into an antibody solution. An immunoassay reaction takes place (Chapter 22), and within a few hours, a color reaction indicates the presence or absence of gonococci. The test allows doctors to detect gonorrhea early so that treatment can start immediately. A test to detect the DNA of gonococci also is available.

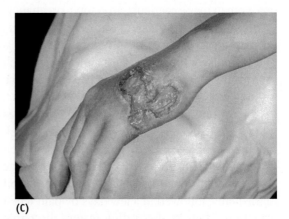

Treponema pallidum

CONCEPT AND REASONING CHECKS

13.5   Contrast the symptoms of gonorrhea in females and males.

Syphilis Is a Chronic, Infectious Disease

KEY CONCEPT

6.   *Treponema pallidum* transmission almost always is by sexual contact or from mother to fetus.

Over the centuries, Europeans have had to contend with four pox diseases: chickenpox, cowpox, smallpox, and the Great Pox, a disease now known as syphilis. The first European epidemic was recorded in the late 1400s, shortly after the conquest of Naples by the French army. For decades the disease had various names, but by the 1700s, it had come to be called **syphilis** (MicroFocus 13.2).

Syphilis is currently ranked among the top five most reported microbial diseases in the United States (see Figure 13.1). Statistics indicate more than 40,000 people are afflicted with the disease annually and the number has been rising since 2000. Twelve million cases are reported each year worldwide. Taken alone, these figures suggest the magnitude of the syphilis epidemic, but some pub-

lic health microbiologists believe for every case reported, as many as nine cases go unreported.

Syphilis is caused by *Treponema pallidum,* (*trepo* = "turn"; *nema* = "thread"; *pallid* = "pale"; thus literally "pale turning thread"). This spirochete moves by means of endoflagella (see Chapter 4). Humans are the only host for *T. pallidum,* so the organism must spread by direct human-to-human contact, usually during sexual intercourse. It penetrates the skin surface through the mucous membranes of the genitalia or via a wound, abrasion, or hair follicle. The variety of clinical symptoms accompanying the stages, and their similarity to other diseases, have led some physicians to call syphilis the "great imitator." Untreated, the disease can progress through a number of stages.

The incubation period for syphilis varies greatly, but it averages about three weeks. **Primary syphilis** is the first stage of the disease. It is characterized by a lesion, called a **chancre**, which is a painless circular, purplish ulcer with a small, raised margin with hard edges (FIGURE 13.10A). The chancre develops at the site of entry of the spirochetes, often the genital organs. However, any area of the skin may be affected, including the pharynx, rectum, or lips. The chancre teems with spirochetes and represents the stage that is most infectious. It persists for three to six weeks, and then heals spontaneously. However, the infection has not been eliminated, as the spirochetes have spread through the blood and lymph to other body organs.

Several weeks after the chancre has healed, the untreated patient experiences **secondary syphilis**.

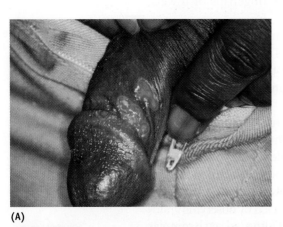

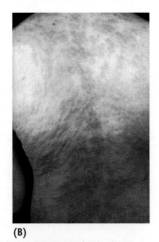

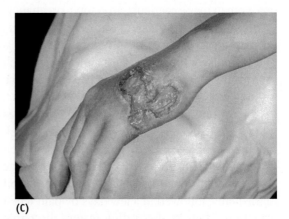

(A)   (B)   (C)

FIGURE 13.10 **The Stages of Syphilis. (A)** The chancre of primary syphilis as it occurs on the penis. The chancre has raised margins and is usually painless. **(B)** A skin rash is characteristic of secondary syphilis. **(C)** The gumma that forms in tertiary syphilis is a granular, diffuse lesion compared with the primary chancre. »» From which of these stages could the spirochete be contagious?

MICROFOCUS 13.2: History
The Origin of the Great Pox

Among the more intriguing questions in medical history are how and why syphilis suddenly emerged in Europe in the late 1400s. Writers of that period tell of an awesome new disease that swept over Europe and on to India, China, and Japan. But where did the actual disease originate?

Since the 16th century, historians have debated the geographic origins of syphilis. For most of the past 500 years, Christopher Columbus has been alternately blamed or exonerated for bringing syphilis to Europe. In the 20th century, critics of the "Columbian theory" proposed that syphilis had always been present in the Old World, perhaps coming from Spain and Portugal through the slave trade with Africa.

Now, a genetic analysis of the disease-causing *Treponema pallidum* carried out by Kristin Harper and colleagues at Emory University in Atlanta suggests that today's syphilis is a close cousin to another South American tropical disease called yaws, a tropical infection of the skin, bones, and joints.

To see if Columbus and his men introduced syphilis to Europe after "catching yaws" in the Americas, Harper and her colleagues in 2007 collected species of *T. pallidum* from across the globe (Africa, Europe, Asia, the Middle East, the Americas, and the Pacific Islands) to determine a phylogenetic family tree—which microbes gave rise to which—and perhaps determine where syphilis originated. They also sequenced and compared genes from *Treponema pertenue,* which causes yaws, from Africa and South America. By comparing genetic sequences between the *Treponema* species, the researchers found the species that cause syphilis originated relatively recently, with their closest relatives collected in South America and responsible for yaws.

However, yaws is spread through simple skin contact, not sexual intercourse. So how did a simple skin disease in South America become an STD in Europe? One suggestion is that syphilis became sexually transmitted only when it reached Europe, where the climate was not as hot and humid as it was in the tropics, and where people wore more clothing, limiting skin contact. Because skin contact would not be a viable transmission mechanism, the bacterium evolved to use sexual intercourse as the transmission mechanism.

This woodcut depicts a man infected with syphilis as evidenced by the lesions on the face and legs. The woodcut refers to syphilis as the result of an unfavorable planetary alignment in 1484.

To completely confirm the "New World" theory, more genetic analyses need to be done, including the sequencing of the entire DNA of the related *Treponema* species. Professor George Armelagos, one of co-workers on the project said, "Understanding [*T. pallidum's*] evolution is important not just for biology, but for understanding social and political history. It could be argued that syphilis is one of the important early examples of globalization and disease, and globalization remains an important factor in emerging diseases."

Symptoms include fever and a flu-like illness as well as swollen lymph nodes. The skin rash that develops, which may be mistaken for measles, rubella, or chickenpox, appears as reddish-brown spots on the palms, soles, face, and scalp (**FIGURE 13.10B**).

Transmission can occur if there are moist lesions. Loss of the eyebrows often occurs, and a patchy loss of hair results in "moth-eaten" areas commonly seen on the head. Involvement of the liver may lead to jaundice and suspicion of hepatitis.

In untreated patients, the symptoms resolve after several weeks. Most patients recover, but they bear pitted scars from the healed lesions and remain "pockmarked." In 2007, there were 11,466 cases of primary and secondary syphilis reported to the CDC—the highest level since 1997.

These individuals now enter a latent period that can last several years. Many patients will have relapses of secondary syphilis during which time they remain infectious. Within four years, the relapses cease and the disease is no longer infectious (except in pregnant females). Patients either remain asymptomatic or slowly progress to the third stage.

About 40% of untreated patients eventually develop **tertiary syphilis**. This stage occurs in many forms, but most commonly it involves the skin, skeletal, or cardiovascular and nervous systems. The hallmark of tertiary syphilis is the **gumma**, a soft, painless, gummy noninfectious granular lesion (FIGURE 13.10C). In the cardiovascular system, gummas weaken the major blood vessels, causing them to bulge and burst; in the spinal cord and meninges, gummas lead to degeneration of the tissues and paralysis; and in the brain, they alter the patient's personality and judgment, and cause insanity so intense that for many generations people with tertiary syphilis were confined to mental institutions (but read chapter introduction). Damage can be so serious as to cause death.

Syphilis is a serious problem in pregnant women because the spirochetes penetrate the placental barrier after the third or fourth month of pregnancy, causing **congenital syphilis** in the fetus. Infection can lead to death (stillbirth). Surviving infants develop skin lesions and open sores. Affected children often suffer poor bone formation, meningitis, or **Hutchinson's triad**, a combination of deafness, impaired vision, and notched, peg-shaped teeth. In 2007, there were 430 congenital syphilis cases reported to the CDC, a number that has been declining since 1991.

There is no vaccine for syphilis, so the cornerstone of syphilis control is safe sex practices, and the identification and treatment of the sexual contacts of patients. Penicillin is the drug of choice and a single dose usually is sufficient to cure primary and secondary syphilis. Because *T. pallidum* cannot be cultivated on laboratory media, diagnosis in the primary stage depends on the observation of spirochetes from the chancre using fluorescence or dark-field microscopy

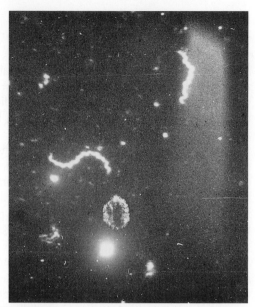

FIGURE 13.11 The *Treponema pallidum* Spirochete. A dark-field microscope view of the syphilis spirochete seen in a sample taken from the chancre of a patient. »» Why are all photos of *T. pallidum* taken from patient samples and not from samples grown in culture?

(FIGURE 13.11). As the disease progresses, a number of tests to detect syphilis antibodies becomes useful, including the rapid plasma reagin (RPR) test, the Venereal Disease Research Laboratory (VDRL) test, and others (Chapter 22).

CONCEPT AND REASONING CHECKS
13.6 Describe the three stages of syphilis.

Other Sexually Transmitted Diseases Also Exist

KEY CONCEPT
7. Other bacterial genera also are responsible for STDs that are not life-threatening.

A few other STDs merit brief mention.

Chancroid. An STD believed to be more prevalent worldwide (especially in tropical and semitropical climates) than gonorrhea or syphilis is **chancroid**. The disease is endemic in many developing nations, and it is common in tropical climates and where public health standards are low. The disease has been decreasing in the United States since 1987 and there were 23 cases reported in 2007. However, the organism is difficult to culture and therefore the disease could be substantially underdiagnosed.

The causative agent of chancroid is *Haemophilus ducreyi,* a small gram-negative rod named for Augusto Ducrey, who observed it in skin lesions in 1889. After a 3 to 5 day incubation period, a tender papule surrounded by erythema forms at the entry site. The papule quickly becomes pus filled and then breaks down, leaving a shallow, saucer-shaped ulcer that bleeds easily and is painful in men. The ulcer has ragged edges and soft borders, a characteristic distinguishing it from the primary lesion of syphilis. For this reason, the disease is often called **soft chancre**.

The lesions in chancroid most often occur on the penis in males and the labia or clitoris in females (see Figure 13.2). Substantial swelling in the lymph nodes of the groin may be observed. However, the disease generally goes no further; in fact, in women, it often goes unnoticed. The clinical picture in chancroid makes the disease recognizable, but definitive diagnosis depends on isolating *H. ducreyi* from the lesions.

The transmission of chancroid depends on contact with the lesion, although sexual contact is not required. Contact with open ulcers or their fluid can spread the disease. Azithromycin, erythromycin, or ceftriaxone drugs are useful for therapy, but the disease often disappears in 10 to 14 days without treatment. However, treatment is essential because the open lesion increases the risk of HIV transmission.

Ureaplasma urethritis. Another type of NGU, *Ureaplasma* **urethritis**, is caused by *Ureaplasma urealyticum,* a small, gram-negative bacterium so named because of its ability to digest urea in culture media. The organism often is referred to as a **T-mycoplasma** because "tiny" colonies of the organisms develop on laboratory media. Transmission is generally by sexual contact.

The symptoms of ureaplasmal urethritis are similar to those of gonorrhea and chlamydial urethritis. A distinction can be made between the diseases because in ureaplasmal urethritis, the discharge is variable in quantity, and the urethral pain is usually aggravated during urination. Symptoms are often very mild. Tetracycline is currently the drug of choice. Diagnosis often depends on eliminating gonorrhea or other types of NGU as possibilities, and for this reason, cases are not often recognized.

Infertility is one consequence of ureaplasmal urethritis because, like chlamydia, low sperm counts and poor movement of sperm cells have been observed in males. Salpingitis in females also has been described. Moreover, *U. urealyticum* is capable of colonizing the placenta during pregnancy, and reports have linked it to spontaneous abortions and premature births. As previously noted, 25% of NGU cases may represent ureaplasmal infections.

Lymphogranuloma venereum. A systemic STD caused by a different serotype of *C. trachomatis* is **lymphogranuloma venereum (LGV)**. LGV is more common in males than females, and is accompanied by fever, malaise, and swelling and tenderness in the lymph nodes of the groin. Females may experience infection of the rectum (proctitis), if the chlamydial cells pass from the genital opening to the nearby intestinal opening. LGV is prevalent in Southeast Asia and Central and South America. Sexually active individuals returning from these areas may show symptoms of the disease, but treatment with doxycycline leads to rapid resolution. It rarely occurs in the United States and other developed nations. However, in 2004 and 2005 outbreaks of LGV among men who had sex with men in Canada, the Netherlands, and other European countries were reported.

Granuloma inguinale. A rare STD in Europe and North America (typically in the Southeast) is **granuloma inguinale**. However, it remains an endemic problem in tropical and subtropical areas of the world, such as Caribbean countries and Africa. It is caused by *Klebsiella* (formerly called *Calymmatobacterium*) *granulomatis,* a large, gram-negative bacillus. The disease begins with a primary lesion starting as a nodule and progressing to a granular ulcer that bleeds easily. In most cases, this ulcer forms in the external genital organs but it may spread to other regions by contaminated fingers. The lymph nodes in the groin may swell, but fever and other body symptoms are usually absent, a factor distinguishing the disease from LGV. Tissue samples reveal masses of bacterial cells called **Donovan bodies** within white blood cells in the lesion. Anal intercourse, rather than vaginal intercourse, is the most frequent source of infection. About 50% of infected men and women have lesions in the anal area. The disease responds well to doxycycline. TABLE 13.1 summarizes the STDs caused by bacteria.

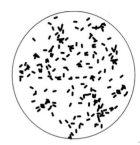

Haemophilus ducreyi

Erythema:
A redness of the skin resulting from inflammation.

CONCEPT AND REASONING CHECKS

13.7 Propose an explanation as to why all the STD-causing walled bacteria are gram-negative.

TABLE

13.1   A Summary of Sexually Transmitted Diseases Caused by Bacteria

Disease	Causative Agent	Signs and Symptoms	Transmission	Treatment	Prevention and Control
Chlamydia	*Chlamydia trachomatis*	Pain on urination Watery discharge Salpingitis	Sexual intercourse	Doxycycline Erythromycin	Practicing safe sex Limiting sex partners Screening for STDs
Gonorrhea	*Neisseria gonorrhoeae*	Pain on urination Discharge Salpingitis	Sexual intercourse	Ceftriaxone	Practicing safe sex Avoiding oral sex
Syphilis	*Treponema pallidum*	Chancre Skin lesions Gumma	Sexual intercourse	Penicillin	Practicing safe sex
Chancroid	*Haemophilus ducreyi*	Soft chancre Erythema Swollen inguinal lymph nodes	Sexual intercourse	Azithromycin Erythromycin Ceftriaxone	Practicing safe sex
Ureaplasma urethritis	*Ureaplasma urealyticum*	Pain on urination Variable discharge Salpingitis	Sexual intercourse	Tetracycline	Practicing safe sex
Lymphogranuloma venereum (LGV)	*Chlamydia trachomatis*	Swollen lymph nodes Proctitis	Sexual intercourse	Doxycycline	Practicing safe sex
Granuloma inguinale	*Klebsiella granulomatis*	Bleeding ulcer Swollen inguinal lymph nodes	Sexual intercourse	Doxycycline	Practicing safe sex

13.3   The Structure, Indigenous Microbiota, and Illnesses of the Female and Male Urinary System

If not for the urinary system, waste products from all the other body systems would accumulate in the blood producing such toxic levels that the waste products would kill us. However, the anatomy of the urinary system is such that contact with exogenous microbes, or through sexual intercourse, can provide an avenue for their dissemination.

Part of the Urinary Tract Harbors an Indigenous Microbiota

KEY CONCEPT

8. The distal region of the urethra is usually colonized by a variety of bacterial species.

The organs of the urinary system are two kidneys, two ureters, a urinary bladder, and a urethra (**FIGURE 13.12**). The kidneys filter waste products from the bloodstream and convert the filtrate into urine. The ureters, urinary bladder, and urethra are collectively known as the **urinary tract** because they transport the urine from the kidneys through the ureters to the urinary bladder. **Urination** involves expulsion of urine from the bladder through the urethra to the exterior of the body.

When considering infections and diseases of the urinary system, it is important to understand that: (1) in males, the urethra also constitutes part of the reproductive system and is involved in the ejaculation of semen; (2) the urethral opening in females is closer to the anus than in males and is also close to the vaginal opening, making these heavily colonized sites important sources of potential microbial colonizers. These factors combine to generate significant differences not only in the types of microbes colonizing the urinary system of males and females, but also their relative susceptibility to infection.

The kidneys, ureters, and bladder are normally sterile due to normal urine flow. The ure-

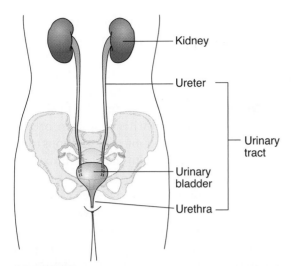

FIGURE 13.12 The Female Urinary Tract. After the kidneys remove excess liquid and waste from the blood as urine, the ureters carry the urine to the bladder in the lower abdomen. Stored urine in the bladder is emptied through the urethra. »» Explain how an infection of the kidney would arise.

thra, on the other hand, is colonized by microbes either along its whole length or certainly near the terminus. As already mentioned, the urethra of females is much shorter than that of males, making ascending infections into the bladder and kidneys more frequent in females.

One of the main antimicrobial defense mechanisms in the urethra is the shedding of the outermost cells of the mucosa with their adherent microbial populations. In addition, the secretion of mucus forms a layer on the epithelial surface that entraps microbes as well as prevents their adhesion to the urethral epithelium. Microbial colonization of the urethra also is hindered by the flushing action of urine, and its normally acidic pH and high urea content can be microbicidal and microbiostatic. The urine also contains a number of antimicrobial proteins, including defensins and secretory antibodies.

Many of the defense mechanisms operating in the male urethra are similar to those in the female urethra, including shedding of epithelial cells, urinary flow, and the presence in urine of antimicrobial and anti-adhesive factors. In addition, the prostate—the genital organ that surrounds and communicates with the first segment of the male urethra—secretes zinc, which in prostatic secretions can inhibit the growth of potential pathogens such as *Escherichia coli*.

The main indigenous bacterial genera colonizing both the female and male urethra are *Corynebacterium, Streptococcus, Staphylococcus, Bacteroides,* and a variety of lesser known genera including *Peptostreptococcus, Prevotella,* and *Porphyromonas.*

Elimination of bacteria from the urethra occurs mainly during urination. In females and males who do not have a urinary tract infection (UTI), the number of viable bacteria in the urine can be anywhere from 0 to approximately 10^5 cells per milliliter, with most individuals having between 10 and 1,000 cells per milliliter of urine. The diagnosis of a UTI depends on counting only the infecting bacteria in urine, not the indigenous microbiota. This is accomplished by allowing the initial urinary flow to flush out the urethral microbiota. Then a "mid-stream" urine sample is taken for analysis. Urethral microbes also can be transmitted to a partner during sexual intercourse, and to the hands and underclothing.

Although, like urine, semen is sterile when first produced, it will accumulate microbes during its passage through the urethra. In fact, the semen of approximately 80% of males contains between one and nine different bacterial species. Therefore, again sexual intercourse is likely to provide an avenue by which urethral microbes can be disseminated.

CONCEPT AND REASONING CHECKS

13.8 Why is the urethra the only part of the urinary system normally colonized by indigenous microbiota?

Because all parts of the urinary tract are joined by a fluid medium, infection at any site may spread to involve other areas of the system. *E. coli* is the most common cause of infections, although other bacterial genera such as *Chlamydia, Mycoplasma,* and *Proteus* can be involved (**FIGURE 13.13**). *Leptospira interrogans,* the causative agent of leptospirosis, represents an infection of the kidney tubules. Because the organism is often of zoonotic origin, it was described in Chapter 12.

Infections of the Urinary Tract Are the Second Most Common Type of Infection in the Body

KEY CONCEPT

9. Urinary tract infections occur primarily in the urethra and bladder.

It has been estimated that one-third to one-half of humans suffer a **urinary tract infection (UTI)** at

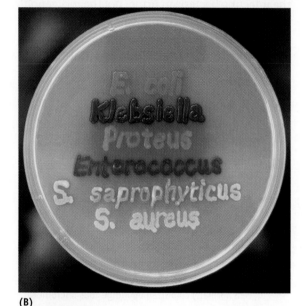

(A)

(B)

FIGURE 13.13 **Pathogens of the Urinary Tract and Their Detection.** **(A)** A false-color transmission electron micrograph of *Proteus mirabilis*. This large cell contains numerous flagella. (Bar = 5 μm.) **(B)** Various urinary tract pathogens are cultivated on a laboratory medium called CHROMagar. The different bacteria display different color reactions, which aid identification. »» What advantage is it to *P. mirabilis* to possess flagella?

Purulent:
Relating to, containing, or consisting of pus.

some time during their lives. In the United States, the CDC reports that UTIs account for about 4 million ambulatory-care visits each year, representing about 1% of all outpatient visits. UTIs account for about 10 million doctor visits each year.

Women are most at risk of developing a UTI. In fact, half of all women will develop a UTI during their lifetimes, and many will experience more than one (MicroFocus 13.3).

UTI infections in males remain rare through the fifth decade of life, when enlargement of the male prostate begins to interfere with emptying of the bladder. Infection rates can then rise 20%.

Most UTIs occur in the lower tract and involve the urethra, bladder, or prostate. Infections of the upper urinary tract involve the kidneys.

Urethritis. An inflammation and infection of the urethra is called **urethritis**. In women, bacteria can travel from the lower intestine and anus to the vagina. From there, due to the close proximity, the infecting organisms can travel to the urethra. Sexually transmitted organisms, such as *Neisseria gonorrhoeae* and *Chlamydia*, can spread to the urethra during sexual intercourse with an infected partner. Infections in men are far less frequent due to the greater distance between urethra and anus. When men develop urethritis, *Chlamydia* is often the cause resulting from sexual intercourse.

Infections confined to the urethra are characterized by painful urination in women and men, and discharge of mucoid or purulent material from the urethral orifice in men.

Treatment of urethritis depends on the cause of the infection. Antibiotics are given for bacterial infections. Sexually transmitted diseases that cause urethritis can be prevented by using a condom.

Cystitis. **Cystitis** is an infection of the bladder most often caused by *E. coli*. Community-acquired cystitis is particularly common and recurring in women between 30 and 50 years of age. Pregnancy can increase the chances of cystitis because the pregnancy can interfere with emptying of the bladder. Use of a diaphragm with a spermicide also can increase the chances of cystitis because the spermicide inhibits the vaginal microbiota and allows cystitis-causing organisms to grow.

Cystitis in men is less common. Community-acquired infections often begin in the urethra, move to the prostate, and then to the bladder. In fact, the most common cause of cystitis in men is a prolonged bacterial infection in the prostate.

Hospital-acquired (nosocomial) bladder infections can occur in patients in a medical care facility who have had a urinary catheter placed through the urethra and into the bladder to collect urine.

The symptoms of cystitis are similar to those of urethritis and result from an irritation of the mucosal surface of the urethra as well as the bladder. Unlike urethritis, cystitis is produced by the multiplication of *E. coli* in the bladder urine. It is clinically distinguished from urethritis by a more acute onset, more severe symptoms, the presence

MICROFOCUS 13.3: Research

What Goes Up Does Not Always Come Down

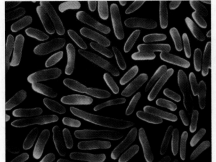

Typical rod-shaped *E. coli* cells.

Some 10 million Americans—mainly women—consult their doctors every year because of a urinary tract infection (UTI). These infections normally result from the transfer of *Escherichia coli* from the gut to the opening of the urethra, which often occur as a result of sexual intercourse. From the urethra, the bacterial cells can move up the urinary tract and into the bladder where they multiply, causing an inflammation of the bladder called cystitis. If a woman does not seek treatment, the *E. coli* cells may move further up the urinary tract and infect the kidneys.

Physicians have observed that 50% of the women who get a UTI will have a second infection within a year and the cause is usually the same bacterial strain as the first infection. Women who seek treatment usually are given antibiotics that cure the infection rapidly. Researchers therefore assumed the original *E. coli* strains were eliminated from the urinary tract but persisted in the gut such that the cells could reinfect the urinary tract at a later time. However, experiments with mice suggest another possibility.

As in humans, *E. coli* strains in mice also cause UTIs that invade the lining the bladder. When the bacterial cells replicate, they form so-called intracellular bacterial communities (IBCs). Within these biofilms, many of the infected cells on the bladder's surface slough off into the urine, but IBCs also release bacterial cells, many of which now have a long, slender filamentous appearance compared to the normal short, rod-shaped cells (see figure). The cells of the immune system normally kill and eliminate the short rods in the urine but they cannot destroy the bacterial cells if they are in a filamentous form. In mice, these filamentous bacterial cells can then reinfect the bladder lining and establish long-lasting intracellular pockets of bacterial cells that, like typical biofilms, are protected from antibiotics and the immune system. Therefore, do IBCs and the filamentous *E. coli* cells also occur in women?

To look for IBCs and filamentous cells, researchers at Washington University School of Medicine in St. Louis collected urine from 80 young women with cystitis and from 20 women with no symptoms but who had previously had cystitis. The researchers searched for IBCs and filamentous bacterial cells in each sample using light microscopy and electron microscopy, and they identified the bacterial species composing the IBCs and filamentous cells. The results: The researchers found no IBCs or filamentous bacterial cells in the urine of the cystitis-free women. However, IBCs were found in nearly 20%, and filamentous bacterial cells in nearly 50%, of urine samples from the women with cystitis. Analysis of the urine samples from all of the women with IBCs, and most of them with filamentous bacteria, contained *E. coli*.

The experimental results suggest the IBC/filamentous bacterial cells identified in mice also occur in at least some women with UTIs and probably are composed of *E. coli* cells. Because only one urine sample was collected from each woman, and the samples were taken at only one time point, the IBC phenomenon may be more common than these findings suggest. So, at least in these samples, the infection that moved up the urinary tract to the bladder does not need to move back down and out of the tract before reinfection at another time. Rather, the *E. coli* cells can establish a long-lasting, antibiotic resistant biofilm from which a future infection may propagate. Knowing this, researchers can try to develop some type of intervention that can disrupt the IBCs, making reinfection a less likely occurrence.

of bacteria in the urine, and, in approximately half of cases, hematuria.

Symptoms of cystitis often disappear without treatment. People with frequent bladder infections caused by bacteria often take antibiotics regularly at low doses. Symptoms usually disappear within a few days of treatment; however, antibiotics may need to be taken for up to a week, depending on the severity of the infection. The bacteria causing hospital-acquired bladder infections often are resistant to the common antibiotics used to treat community-acquired bladder infections. Therefore, different types of antibiotics and different treatment approaches may be needed.

Hematuria:
The presence of blood in the urine.

Prostatitis. An inflammation and infection of the prostate, called **prostatitis**, typically develops in young men and in older men following placement of an indwelling catheter. Prostatitis has been classified by the National Institutes of Health (NIH) into one of two categories if it involves a bacterial infection. In **acute bacterial prostatitis**, symptoms come on suddenly and may include fever and chills, and the signs and symptoms of cystitis, including increased urinary urgency and frequency. In **chronic bacterial prostatitis**, signs and symptoms develop more slowly and usually are not as severe as those of acute prostatitis. However, the signs and symptoms are similar to the acute condition and antibiotics are usually the choice of treatment for both categories of prostatitis.

Pyelonephritis. A UTI involving one or both kidneys is called **pyelonephritis**. If not treated properly, kidney infection can permanently damage the kidneys or spread to the bloodstream and cause a life-threatening infection. Thus, prompt medical attention is essential.

Again, pyelonephritis is more common in women than in men. About 90% of cases among individuals living in the community are caused by *E. coli*. Infections usually ascend from the genital area through the urethra to the bladder, then up through the ureters and into the kidneys. Importantly, it has been estimated that 20% to 50% of pregnant women with acute pyelonephritis give birth to premature infants, making pyelonephritis one of the most serious consequences of a UTI.

Symptoms of pyelonephritis often begin suddenly with chills, fever, pain in the lower part of the back, nausea, and vomiting. About one-third of people with pyelonephritis also have symptoms of cystitis. Antibiotics are the first line of treatment and are started as soon as the doctor suspects pyelonephritis. Usually, the signs and symptoms of kidney infection begin to clear up within a few days of treatment. However, patients need to continue antibiotics for a week or more.

The bacterial diseases of the urinary system are summarized in **TABLE 13.2**.

CONCEPT AND REASONING CHECKS
13.9   How can you diagnose the different UTIs based on signs and symptoms?

13.4   Contact Diseases Caused by Indigenous Bacterial Species

Numerous bacterial diseases are transmitted by contact other than sexual. Usually, some form of skin contact is required.

The skin is the largest organ of the body, accounting for more than 10% of body weight and having a surface area of almost 2 square meters. Being the interface between environment and internal body tissues and organs, the skin plays a critical role in protecting the body against trauma as well as physical and biological injury. Should anything go wrong with skin function, or should there be damage to—or wounding or puncturing of—the skin, inflammation, infection, and disease may occur.

The Skin Protects Underlying Tissues from Microbial Colonization

KEY CONCEPT
10. Each skin layer performs specific activities.

The skin, along with its accessory structures (nails, hair, sweat glands, and sebaceous glands) constitutes the **integumentary system** (**FIGURE 13.14**).

The tough top layer of skin that you see is called the **epidermis** (*epi* = "over"). It consists of layers of **keratinocytes**, the most common cell type in the epidermis. Keratinocytes are produced in the stratum basale and, as they push upward toward the surface, they produce a tough, fibrous protein called **keratin**. The keratinocytes eventually die in the stratum corneum, and are worn away and replaced by newer cells that are continually pushing up from below. This thickened, superficial layer is dry and waterproof, and, along with epidermal dendritic (Langerhans) cells that present an immunological barrier, prevents most bacterial cells, viruses, and other foreign substances from entering the underlying tissues.

Beneath the epidermis lies the **dermis**, a complex, thick layer of fibrous and elastic connective tissue. The dermis contains sweat glands, the sweat being composed of water, salt, and a few other chemical substances. Specialized sweat glands, called apocrine sweat glands, in the armpit and genital region secrete a thick, oily sweat that pro-

Dendritic cells: Immune cells that recognize and process foreign material and present it to other immune cells.

TABLE

13.2 Urinary Tract Infections Caused by Bacteria

Disease	Causative Agent	Signs and Symptoms	Transmission	Treatment	Prevention and Control
Urethritis	*Escherichia coli* *Chlamydia trachomatis* *Mycoplasma* Other bacterial species	Persistent urge to urinate Burning sensation when urinating Blood in the urine Cloudy, strong smelling urine	Bacterial entry from the large intestine to the urethra Sexual intercourse	Antibiotics	Drinking plenty of liquids (water) Urinating frequently Keeping the genital area clean
Cystitis	*Escherichia coli*	Strong, persistent urge to urinate Burning sensation when urinating Blood in the urine Passing cloudy or strong-smelling urine	Bacterial entry from the large intestine through the urethra to the bladder Sexual intercourse	Antibiotics	Drinking plenty of liquids (water) Urinating frequently Keeping the genital area clean
Prostatitis	*Escherichia coli* *Staphylococcus aureus* *Proteus* *Pseudomonas*	Fever and chills Flu-like symptoms Pain in the prostate gland, lower back, or groin Increased urinary urgency and frequency Pain when urinating Blood-tinged urine	Bacteria from the large intestine From bladder Sexual intercourse	Antibiotics	Practicing good hygiene Keeping the penis clean Drinking enough fluids to cause regular urination
Pyelonephritis	*Escherichia coli* *Enterococcus faecalis* *Klebsiella* *Pseudomonas aeruginosa*	Frequent urination Persistent urge to urinate Burning sensation or pain when urinating Abdominal pain or pressure Cloudy urine with a strong odor Pus or blood in urine	Bacteria from the large intestine enter urethra From bloodstream	Antibiotics	Practicing good hygiene Drinking enough fluids to cause regular urination

duces a characteristic body odor when the sweat is digested by the resident skin microbiota in those areas. Sweat glands usually are not colonized by microbes because antimicrobial proteins, including lysozyme, are continuously being produced.

Hair covers much of the skin surface and consists of dead keratinized cells emerging from hair follicles that originate deep in the dermis. Associated with these follicles are **sebaceous glands** that are primarily found on the face, scalp, and upper torso. These glands produce an oily substance, called **sebum**, which keeps the skin and hair soft and moist. Sebum normally produced by the sebaceous glands combines with dead cells being sloughed off within the hair follicle. As the follicle "fills up," the sebum spreads over the skin surface giving the skin an oily appearance. When this process works correctly, the skin is moisturized and remains healthy. By being open to the environment as infoldings of the epidermis, the

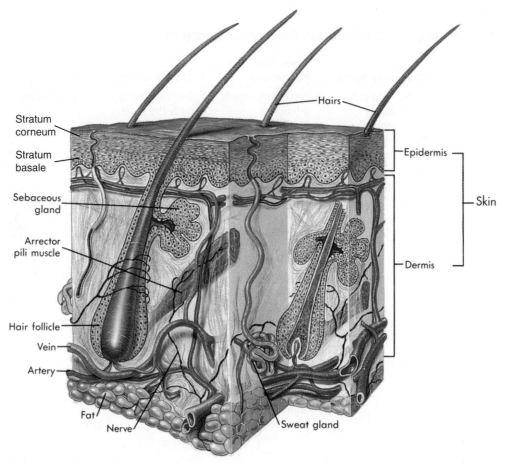

FIGURE 13.14 Layers of the Skin. A diagramatic sectional veiw through the epidermis and dermis layers. »» Identify the accessory structures most susceptible to infection.

most susceptible target in the dermis for microbial colonization is the hair follicle and its associated sebaceous gland.

CONCEPT AND REASONING CHECKS

13.10 How does the epidermis and dermis help protect the body from microbial colonization and infection?

The Skin Harbors Indigenous Microbes

KEY CONCEPT

11. The human resident microbiota of the skin helps protect deeper tissues from infection.

The skin is colonized by a resident **microbiota**, which consists of a community of indigenous microbes that can withstand the environment of the skin surface. This community is dominated by gram-positive bacterial species that, due to their cell wall structure, are better at enduring the skin environment than are gram-negative species.

The "**coryneforms**" are major colonizers on the skin. Also called "diphtheroids," the coryneforms refer to the cutaneous bacterial species that do not form spores, are non-acid-fast, and are gram-positive rods. This includes several species of *Corynebacterium* (*coryn* = "club") that usually do not cause disease. The propionibacteria also are found on the skin surface, although the most abundant species, *Propionibacterium acnes*, chiefly inhabits hair follicles where the microaerophilic environment is optimal for growth. In higher numbers, as we will discuss below, *P. acnes* can be associated with infections, especially acne.

The relatively high salt environment of the skin is home to *Staphylococcus* species. The predominant species, making up some 50% of the staphylococcal population on the skin, is *S. epidermidis*. Although *S. aureus* can be found in the anterior regions of the nose of many individuals, it is considered to be a part of the transient micro-

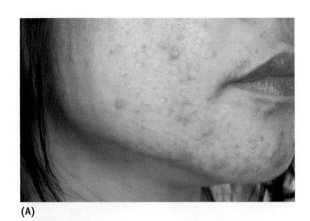

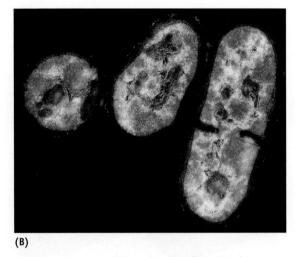

(A)

(B)

FIGURE 13.15 Acne Vulgaris. (**A**) Acne, a typical affliction of teenagers and young adults, primarily affects the face, upper torso, and back. (**B**) *Propionibacterium acnes* bacteria. This false-color transmission electron micrograph shows three of the rod-shaped bacterial cells. »» Why are the face, upper torso, and back the sites most likely to develop acne?

biota. Other halotolerant, gram-positive genera of the skin surface include *Micrococcus,* of which *M. luteus* is one of the most common.

Acinetobacter is a genus consisting of gram-negative coccobacillus-shaped cells often found on moist skin areas of some 25% of individuals especially those in the healthcare professions. Because it has become extremely resistant to antibiotics, nosocomial infections causing pneumonia or bloodstream diseases are of extreme concern.

The antagonistic activities of the indigenous microbiota toward exogenous or transient microbes also are the consequence of their community interactions. Take, for example, the microaerophilic propionibacteria. Their survival depends on other aerobic and facultative species, such as *Micrococcus, Acinetobacter,* and *Staphylococcus,* consuming oxygen and creating the optimal microaerophilic conditions for growth of the propionibacteria. Thus, most pathogens entering the cutaneous microbial environment can't compete effectively for space or nutrients; they find themselves in a very hostile environment and usually are out-competed by the resident microbiota.

Even with the physical, cellular, and microbial defenses provided by the skin, infections still occasionally occur. If a member of the indigenous microbiota has an opportunity to undergo excessive growth and spread, an infection can result.

CONCEPT AND REASONING CHECKS

13.11 What microbes most commonly compose the indigenous microbiota and why are they predominantly gram-positive species?

Acne Is the Most Common Skin Disease Worldwide

KEY CONCEPT

12. Acne is a chronic inflammatory condition involving *Propionibacterium acnes.*

During puberty, about 85% of adolescents and young adults develop **acne**, medically referred to as "acne vulgaris," but more commonly as zits, blackheads, whiteheads, or pimples (**FIGURE 13.15A**). It affects primarily the face and upper torso and, although the overall health of the affected individual is not impaired, acne can result in permanent cutaneous scars.

As already described, the skin is covered with numerous hair follicles in association with sebaceous glands. During puberty, sex hormone changes cause the sebaceous glands to enlarge and produce excessive amounts of sebum. The resident *P. acnes* bacteria metabolize the sebum and grow to higher population densities (**FIGURE 13.15B**). Therefore, acne is a chronic inflammatory condition and *P. acnes* is considered to be a contributing factor to, not the cause of, the inflammation because the bacterial cells also are abundant on the skin of acne-free individuals.

The first stage of the disease occurs when hair follicles become swollen with sebum and keratinocytes. *P. acnes* can exacerbate this process by causing the keratinocytes, which are normally sloughed off, to become sticky, plugging up the follicle. Such plugged sebaceous glands, called **comedones** (sing., comedo), can exist as one of

Nosocomial infections: An infection acquired during an individual's stay in the hospital or care facility.

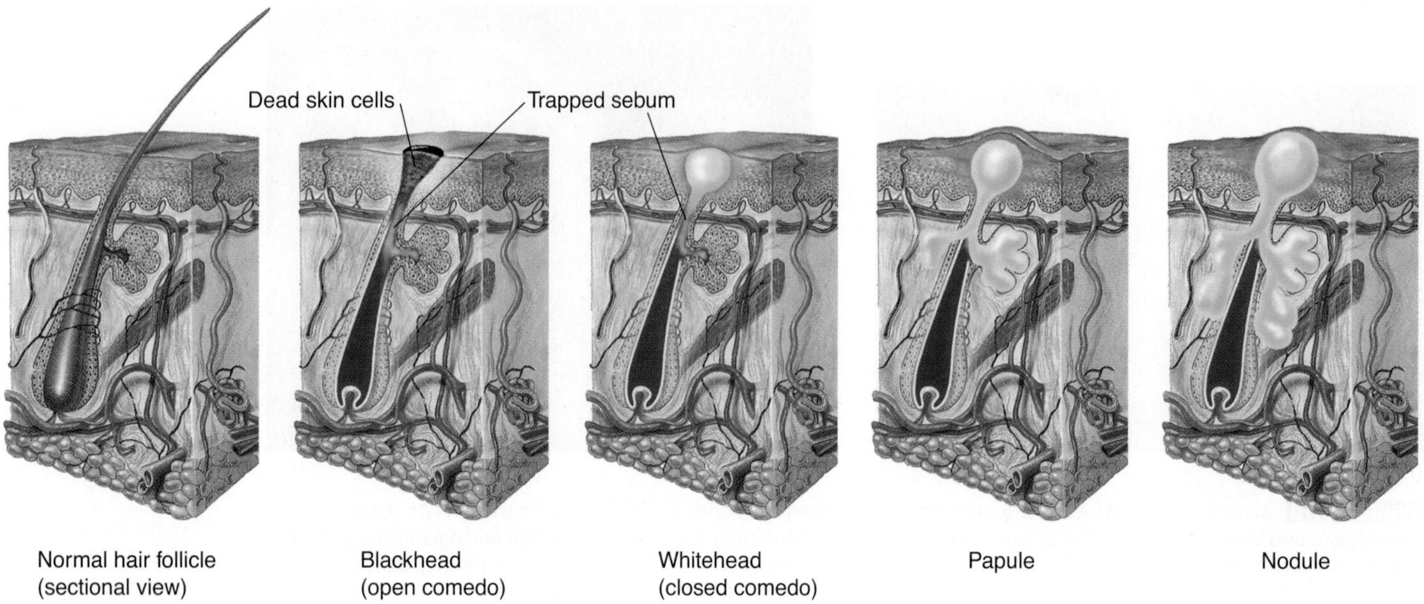

Dead skin cells   Trapped sebum

Normal hair follicle   Blackhead   Whitehead   Papule   Nodule
(sectional view)   (open comedo)   (closed comedo)

FIGURE 13.16 **Skin Lesions of Acne.** Plugged sebaceous glands can develop into open or closed comedones, papules, or nodules. »» Why does *P. acnes* prefer to grow in a hair follicle?
Adapted from McKinley, M. and V. O'Loughlin. *Human Anatomy, Second Edition*. McGraw-Hill, 2006.

two types (**FIGURE 13.16**). "Blackheads," known as open comedones, are follicles where the blockage is incomplete but are filled with plugs of sebum and sloughed-off cells. A chemical reaction oxidizes the melanin, giving the material in the follicle the typical black color. "Whiteheads," known as closed comedones, are completely blocked follicles that are filled with the same material as open comedones. Because air cannot reach the follicle, the melanin is not oxidized and the follicle remains white.

Within the comedones, the overgrowth of *P. acnes* bacterial cells attracts neutrophils and lymphocytes, which accumulate around the swollen follicle. The immune cells damage the follicle wall, allowing the fatty acids and microbial antigens to erupt into the dermis. This process results in an inflammatory response, forming papules and pustules. People with severe acne (acne pimples) develop numerous nodules or cysts deeper in the dermis. The cysts can be quite painful and join together into larger abscesses.

P. acnes is not transmissible and therefore acne is not a preventable disease. It is not caused by poor hygiene or eating such things as greasy foods or chocolate. As mentioned, acne usually wanes by about age 20 due to the stabilization of the sex hormones. However, flare-ups may occur in young

women during menstruation or pregnancy. The use of corticosteroids or anabolic steroids also can stimulate sebaceous glands and cause acne eruptions, while some cosmetics can clog pores and lead to an outbreak of acne.

Most mild cases of acne can be treated effectively at home, with good daily skin care with mild soap and, if necessary, over-the-counter treatments, such as **benzoyl peroxide**, that decrease the chemical reaction that changes the lining of the hair follicle. More severe cases may require antibiotics, such as clindamycin and erythromycin, which are effective against *P. acnes*.

In the most troublesome cases, topical retinoids, such as Retin-A, are effective because of the compound's ability to prevent the blockage of follicles and promote the extrusion of the plugged material in the follicle of whiteheads and blackheads. Systemic retinoids, such as isotretinoin (Accutane) is used only for severe cystic acne. It is very effective for comedones when used properly, but cannot be prescribed for pregnant women or women thinking of becoming pregnant as the compound can cause birth defects.

CONCEPT AND REASONING CHECKS

13.12 How does *Propionibacterium acnes* contribute to acne?

Papules:
Small, solid, round bumps on the skin.

Pustules:
Small, round, raised areas of inflamed skin filled with pus.

Indigenous Microbiota of the Skin Can Form Biofilms

KEY CONCEPT

13. Skin wounding through surgery can introduce indigenous microbes.

Should a wound occur, not only could exogenous microbes enter and cause an infection but also indigenous microbiota. **Chronic wounds** are the result of predisposing conditions, such as peripheral vascular or metabolic diseases where there is a lack of blood flow. Here, dead skin breaks and produces an open sore, resulting in a leg ulcer or pressure sores (bed sores) that become infected. Most such indigenous infections are **polymicrobial**; that is, they are the result of more than one infectious agent.

An **acute wound** arises from cuts, lacerations, bites, or surgical procedures on or to the skin. In such cases, indigenous microbes can enter the deeper layers of tissue and cause an infection. For example, each year in the United States intravenous catheters, used in about 50% of hospitalized patients, cause an estimated 80,000 bloodstream infections, resulting in almost 30,000 deaths among patients in intensive care units. Because the device must be inserted through the skin, there is a potential that skin bacteria, such as *S. epidermidis,* can be mechanically "injected" into the tissues when the catheter is inserted. *S. epidermidis* then can colonize the catheter and form a **biofilm** (see Chapter 3), which can be very hard to remove, resistant to immune defenses, and impervious to antibiotics. A local infection, septicemia, or an abscess may form in a distant part of the body.

Implanted medical devices, such as hip and knee prostheses, pacemakers, and heart valves, can be subject to the same biofilm scenario but as a result of the surgical invasion. Again, *S. epidermidis* along with corynebacteria and propionibacteria are common bacterial species associated with implant infections. To try to limit such infections, rigorous cleaning and disinfection of the skin surface must be done to minimize the resident microbiota.

Corynebacterium minutissimum can cause a superficial, scaly skin infection typically located in folds of the skin, such as between the toes. This infection, called **erythrasma**, is the result of an increased population density of the bacterium, and represents the most common bacterial infection of the foot. It is more common in tropical climates and can be diagnosed by shining ultraviolet light on the infected area, which will glow coral red. Erythrasma can be effectively treated with oral erythromycin.

CONCEPT AND REASONING CHECKS

13.13 How do indigenous bacteria colonize and survive on an invasive medical device?

Intravenous catheters: Tubes inserted into the patient's body cavity or a blood vessel where they deliver medications or fluids to the body or drain fluids from the body.

13.5 Contact Diseases Caused by Exogenous Bacterial Species

When exogenous skin pathogens infect the body, they either colonize the skin directly or reach the skin through the bloodstream. Such bloodborne involvement often is observed on the skin surface as a rash, often the result of a bacterial toxin. Therefore, a physician must consider many possible diseases when evaluating a skin infection and may need to identify the presence or absence of specific signs and symptoms before coming to a diagnosis.

Staphylococcal Contact Diseases Have Several Manifestations

KEY CONCEPT

14. *Staphylococcus aureus* infections can create abscesses and/or produce exotoxins.

Staphylococci are normal inhabitants of the human skin, mouth, nose, and throat. Although they generally live in these areas without causing harm, they can initiate disease when they penetrate the skin barrier or the mucous membranes. Skin penetration is assisted by open wounds, damaged hair follicles, ear piercing, or irritation of the skin by scratching. *Staphylococcus aureus,* the grape-like cluster of gram-positive cocci, is the species usually involved in these contact diseases.

Localized Skin Infections. The hallmark of staphylococcal skin disease is the production of pus-filled pockets in the skin. **Folliculitis** is a minor infection at the base of a hair follicle (**FIGURE 13.17A**). Because infection is often associated with sweat gland activity and areas of the

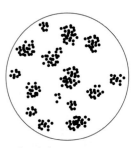

Staphylococcus aureus

body subjected to rubbing or abrasion, it typically occurs on the neck, face, armpits, and buttocks. *Pseudomonas aeruginosa,* an aerobic gram-negative rod, also can infect hair follicles and cause folliculitis. In fact, infection by *P. aeruginosa* has become more common with people who frequent hot tubs or whirlpool baths that are not kept clean and chlorinated.

Another infection caused by *S. aureus* involves the formation of an **abscess**, a circumscribed pus-filled lesion. A **furuncle** (boil) is a warm, painful abscess that develops in the region of a hair follicle. Rupturing of the abscess can lead to infection of the surrounding tissue. Treatment requires a physician to open and drain the pus, and then rinse the abscess with sterile saline solution to remove any remaining bacterial cells. The disease needs to be treated with caution because staphylococci in a trivial skin boil can invade the blood and be transported to other organs including the lungs, heart, brain, or kidneys.

Carbuncles are a group of connected, deeper abscesses (FIGURE 13.17B). Skin contact with other people spreads the disease. Infected individuals often are tired and may have a fever. Bacteremia can occur, requiring antibiotic treatment and debridement.

Food handlers should be aware that staphylococci from furuncles and carbuncles can be transmitted to food, where they can cause food poisoning (see Chapter 11).

A more widespread and highly contagious staphylococcal skin disease is **impetigo**. Here the infection, which is most often seen in children, is more superficial and appears as thin-walled blisters oozing a yellowish fluid that forms a yellowish-brown flaky crust. Usually the blisters occur on the exposed parts of the body, but they also may occur around the nose and upper lip. *S. aureus* produces two phage-coded exotoxins that cause the blistering skin.

Staphylococcal skin diseases are commonly treated with penicillin, but resistant strains of *S. aureus* are well known, and physicians may need to test a series of alternatives before an effective antibiotic is selected. In fact, today multidrug-resistant *Staphylococcus aureus* has appeared in many hospitals. These resistant strains are referred to as **methicillin-resistant** *S. aureus* **(MRSA)**. To help reduce the chances of antibiotic resistance,

Debridement:
The removal of dead, damaged, or infected tissue.

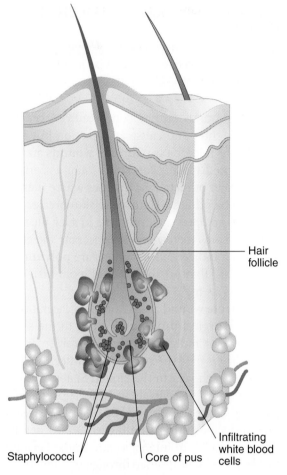

Hair follicle

Staphylococci

Core of pus

Infiltrating white blood cells

(A)

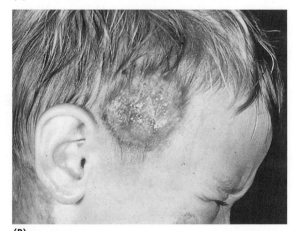

(B)

FIGURE 13.17 Staphylococci and Skin Abscesses.
(A) Staphylococci at the base of a hair follicle during the development of a skin lesion. White blood cells that engulf bacterial cells have begun to collect at the site as pus accumulates, and the skin has started to swell. **(B)** A severe carbuncle on the head of a young boy. Abscesses often begin as trivial skin pimples and boils, but they can become serious as staphylococci penetrate to the deeper tissues. »» How would a severe carbuncle be treated?

doctors may drain an abscess rather than treat the infection with drugs.

Toxin-Generated *S. aureus* Contact Diseases. Several *S. aureus* strains produce toxins. Some cause gastrointestinal illnesses (see Chapter 11) while other strains produce toxins that can initiate diseases coming from contact with the toxin.

Scalded skin syndrome is usually seen in infants and young children under the age of five. Exotoxins produced by staphylococci at the infection site travel through the bloodstream to the skin. The skin becomes red, wrinkled, and tender to the touch, with a sandpaper appearance. The epidermis may then peel off just above the stratum basale (Figure 13.14). Mortality rates in children are very low unless there is a bacterial invasion of the lungs or blood. Cefazolin is the drug of choice for treatment.

Toxic shock syndrome (TSS) is the term for a disease characterized by sudden fever and circulatory collapse. TSS remained in relative obscurity until the fall of 1980, when a major outbreak occurred in menstruating women who used a particular brand of highly absorbent tampons. News of TSS dominated the media for about six months and led to a recall of the tampons.

Toxic shock syndrome is a rare illness caused by another exotoxin-producing strain of *S. aureus*. The earliest symptoms of disease include a rapidly rising fever, accompanied by vomiting and watery diarrhea. Patients then experience a sore throat, severe muscle aches, and a sunburn-like rash. Between the third and seventh days peeling of the skin occurs, especially on the palms of the hands and soles of the feet (**FIGURE 13.18**). A sudden drop in blood pressure can possibly lead to shock and heart failure. Antibiotics, such as semisynthetic penicillins or clindamycin, may be used to inhibit bacterial growth, but measures such as blood transfusions must be taken to control the shock.

Although the staphylococci involved in TSS can be the result of a skin abscess, surgery, or even tattooing (MICROFOCUS 13.4), *S. aureus* cells exist in various places in the body, including the vagina. During the 1980 outbreak, scientists speculated that lacerations or abrasions of the tissue by tampon inserters gave the staphylococci access to deeper tissues. Others suggested

FIGURE 13.18 Toxic Shock Syndrome. The photograph shows the peeling skin often associated with TSS. »» What characteristic does *Staphylococcus aureus* possess to cause toxic shock?

staphylococci grew in the warm, stagnant fluid during the long period the tampon was in place. It appears certain that multiple factors play a role in TSS, because males, prepubertal girls, and postmenopausal women also have been stricken. In 2007, 92 cases of TSS were reported to the CDC. Most were due to ordinary *S. aureus* infections or resulted from a circulating endotoxin rather than from the exotoxigenic strain.

CONCEPT AND REASONING CHECKS

13.14 Summarize the characteristics of exotoxigenic *S. aureus* infections.

Streptococcal Diseases Can Be Mild to Severe

KEY CONCEPT

15. Group A streptococci attach to cells and secrete toxins.

Streptococci are a large and diverse group of encapsulated, nonmotile, facultatively anaerobic, gram-positive cocci. The bacterial cells divide in one plane and cling together to form pairs or chains of various lengths (**FIGURE 13.19A**).

Microbiologists classify the streptococci by two widely accepted systems. The first system divides streptococci into hemolytic (*hemo* = "blood") groups, depending on how they affect sheep red blood cells when plated on blood agar. The **α-hemolytic** streptococci turn blood agar an olive-green color as a secreted toxin partially destroys red blood cells in the medium; colonies of **β-hemolytic** streptococci produce clear, colorless zones around the colonies due to the complete

MICROFOCUS 13.4: Public Health

"It Seemed Like a Good Idea!"

The idea of a tattoo seemed okay. All her friends had them, and a tattoo would add a sense of uniqueness to her personality. After all, she was already 22. It took some pushing from her friends, but she finally made it into the tattoo parlor that day in Fort Worth, Texas.

Two weeks later the pains started—first in her stomach, then all over. Her fever was high, and now a rash was breaking out; it looked like her skin was burned and was peeling away. There was one visit to the doctor, then immediately to the emergency room of the local hospital. The gynecologist guessed it was an inflammation of the pelvic organs (pelvic inflammatory disease, they called it), so he gave her an antibiotic and sent her home.

But it got worse—the fever, the rash, the peeling, the pains. Back she went to the emergency room. This time they would admit her to the hospital, give her intravenous blood transfusions and antibiotics, keep her for 11 days, and discover a severe blood infection due to *Staphylococcus aureus*. And there was an unusual diagnosis: toxic shock syndrome. Don't women get that from tampons? Most do, she was told, but a few get it from staphylococci entering a skin wound—a wound that can be made by a contaminated tattooing needle.

Streptococcus pyogenes

Eczema:
A skin inflammation involving reddening and itching along with formation of scaly or crusty patches.

Serotypes:
Closely related groups of microorganisms or structures distinguished by their ability to bind to different antibodies.

destruction of red blood cells (FIGURE 13.19B); and **non-hemolytic** (gamma) streptococci have no effect on red blood cells and thus cause no change in blood agar.

The second classification system is based on variants of a carbohydrate located in the cell walls of some β-hemolytic streptococci. Groups A and B are the most important to human disease and *Streptococcus pyogenes* is the most common species. This β-hemolytic organism is generally implied when physicians refer to **group A streptococci (GAS)**.

The pathogenicity of *S. pyogenes* is enhanced by the presence of a capsule and pili consisting of **M protein**. This protein is anchored in the cell membrane, traverses the cell wall, and appears as a coat of dense fibrils protruding from the cell surface (FIGURE 13.19C). The protein helps the cells adhere to tissue and retards phagocytosis. Over 80 serotypes of M protein have been identified.

The GAS are among the most common human bacterial pathogens. They can be carried asymptomatically but also can be responsible for a variety of diseases; most are relatively mild, such as **streptococcal pharyngitis**, popularly known as **strep throat** (see Chapter 10).

Although *S. aureus* is the most common cause of impetigo, *S. pyogenes* also can cause the infection. The clinical signs are similar and treatment involves topical agents or systemic antibiotics, such as penicillin or cephalexin.

S. pyogenes is a major cause of **cellulitis**, an inflammation of the tissues beneath the skin. A superficial form of cellulitis involving a streptococcus infection of the dermis is **erysipelas** (*erysi* = "red"; *pela* = "skin"). The infection results in an inflammation extending into the underlying (subcutaneous) fat tissue and is most common among infants, children, and the elderly. Individuals who are immunocompromised or have impaired lymphatic drainage also are at risk. The streptococci enter the skin through minor trauma, insect bites, eczema, or surgical incisions; or from the individual's own nasal passages. Within 48 hours, symptoms appear and include a high fever, shaking and chills, and headache. The skin lesion enlarges rapidly, invading and spreading through the lymphatic vessels. It appears as a fiery red, swollen, warm, and painful rash usually on the lower limbs or face (FIGURE 13.20). Today, about 85% of cases occur on the legs and an increasing percentage of infections are caused by non–group A streptococci. Treatment usually involves oral or intravenous antibiotics, such as penicillin or clindamycin. The symptoms may resolve in a few days, although the skin may take weeks to return to normal.

GAS also cause **streptococcal toxic shock syndrome (STSS)**. Exotoxins structurally similar to those in staphylococcal TSS are involved and they also can trigger an abnormally strong immune response. In STSS, early signs and symptoms include fever, dizziness, confusion, and a large flat rash over several areas of the body. Infection can cause a sudden drop in blood pressure, causing organs such as the kidneys, liver, and lungs to fail. Because more than half of STSS patients

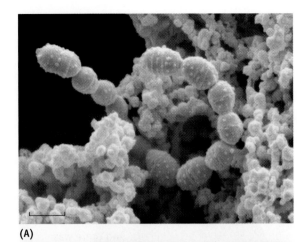

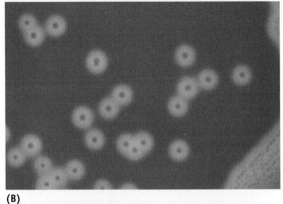

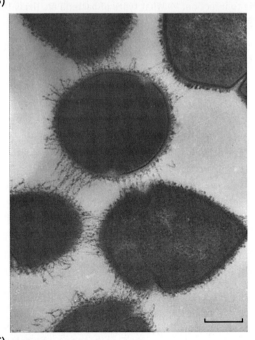

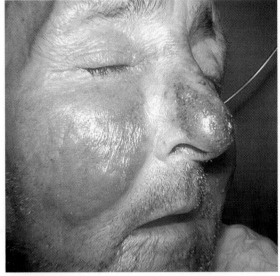

FIGURE 13.20 Facial Erysipelas. Infection of the skin and subcutaneous tissue with streptococcal bacteria produces a bright red rash of the affected areas, in this case the cheek and nose. »» What is most likely the source of this patient's infection?

FIGURE 13.19 Streptococci. (A) A false-color scanning electron micrograph of streptococci growing as long chains. (Bar = 2 μm.) (B) An example of β-hemolysis caused by a toxin released from *Streptococcus pyogenes* cells. (C) A close-up view of *S. pyogenes* cells showing strands of M protein protruding through the capsule. (Bar = 0.5 μm.) »» What function does M protein play in pathogenesis?

die without treatment, dealing with STSS requires prompt action with high doses of penicillin and clindamycin.

Necrotizing fasciitis (occasionally described in the media as "the flesh-eating disease") is a rare but dangerous infection that destroys muscles, fat, and skin tissue (**FIGURE 13.21**). GAS reach the subcutaneous tissue through a wound or trauma to the skin surface. The enzymes and toxins produced by the bacteria cause **necrosis** (cell death) of the subcutaneous tissue and adjacent fascia. Early signs and symptoms of necrotizing fasciitis include fever along with severe pain, swelling, and redness at the wound site. Widespread damage to the surrounding tissue along with blockage of small subcutaneous vessels produces additional dermal cell death. For persons with necrotizing fasciitis, debridement and surgery often are needed to remove damaged tissue; in severe cases, amputation may be the only recourse. Therefore, early diagnosis and treatment are critical to preventing devastating tissue destruction. Failing to recognize necrotizing fasciitis and its severity can lead to death; in fact, about 20% of patients with necrotizing fasciitis die.

Fascia:
The connective tissue covering or binding together parts of the body such as muscles or organs.

CONCEPT AND REASONING CHECKS

13.15 Assess the clinical significance of group A streptococci (GAS) in causing streptococcal toxic shock syndrome and necrotizing fasciitis.

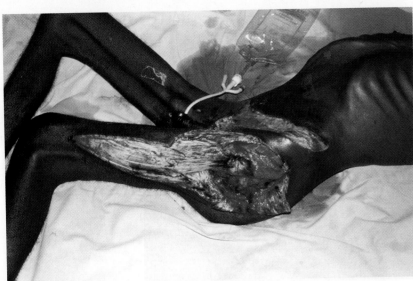

FIGURE 13.21 Necrotizing fasciitis. The extensive loss of connective tissue can be seen in the leg of a 15-year-old AIDS patient. Extensive wound debridement was required. »» What tissues are most affected by GAS?

Other Wounds Also Can Cause Skin Infections

KEY CONCEPT

16. Traumatic wounds to the skin surface can lead to localized infections.

The risk of infection due to a **traumatic wound**, such as a deep cut, compound fracture, or thermal burn, depends on several factors, including: the extent of potential contamination; the contaminating dose of bacterial cells; and their virulence. The physical and physiologic nature of the wound—that is, are there areas of necrosis, or poor blood and oxygen supply—also are important factors affecting bacterial colonization and infection.

Gas gangrene is often the result of a traumatic wound. Because it is a soilborne disease, it was discussed in Chapter 12.

Burns are one of the most common and devastating forms of trauma. In the United States, the National Center for Injury Prevention and Control reports approximately 2 million fires each year, resulting in 1.2 million burn injuries. In a burn injury, the skin has been mechanically damaged and the underlying tissues are open to potential infection (**FIGURE 13.22**). Thus, individuals with moderate to severe burn injuries require hospitalization, which accounts for 100,000 infectious cases each year.

Burn wounds can be classified as **wound cellulitis**, which involves the unburned skin at

Opportunistic:
Referring to a pathogen that invades tissues when body defenses are suppressed.

Eschar:
A dry scab formed on skin that has been burned.

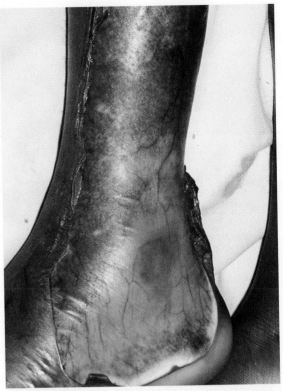

FIGURE 13.22 A Burn Trauma to the Leg. Burns to this patient's lower leg and ankle have reached underlying fat and muscle tissue, making the tissue extremely susceptible to infection. »» What types of bacteria are likely to infect this patient?

the margin of the burn, or as an **invasive wound infection**, which is characterized by microbial invasion of viable tissue beneath the burn. Gram-negative bacteria are the most common agents of an invasive infection due to the extensive range of virulence factors and antimicrobial resistance mechanisms they possess (**TABLE 13.3**). Although there is a variety of bacteria, fungi, and viruses that can cause infection, one of the most likely infective agents in many burn centers is *Pseudomonas aeruginosa*. This aerobic, gram-negative rod is widely distributed in soil, water, plants, and animals (including humans). As an opportunistic pathogen, it rarely causes disease in healthy individuals. However, *P. aeruginosa* has emerged as an important source of burn wound sepsis, especially when nosocomially acquired. About 5,000 hospitalized patients die each year from burn-related complications.

Invasive burn wound infections due to *P. aeruginosa* can form mature biofilms within about 10 hours after colonization. The area soon appears black or as a violet discoloration or eschar. Limiting burn wound infections and patient mor-

TABLE 13.3	Bacterial Genera Commonly Associated with Invasive Burn Wound Infections[a]
Gram-Positive Bacterial Genera	**Gram-Negative Bacterial Genera**
Staphylococcus	*Pseudomonas*
Enterococcus	*Escherichia*
	Klebsiella
	Serratia
	Enterobacter
	Proteus
	Acinetobacter
	Bacteroides

[a]Table modified from *Clin Microbiol Rev.*, 2006 **19**(2): p. 403–434, DOI and reproduced with permission from the American Society for Microbiology.

tality requires rapid burn debridement and wound closure. In addition, patients with serious invasive burns require immediate topical and systemic antibiotic therapy to further minimize morbidity and mortality. Double antibiotic therapy—a combination of a beta-lactam, such as penicillin or cephalosporin, and an aminoglycoside, such as gentamicin—usually is administered because hospital burn centers may harbor strains of *P. aeruginosa* that are resistant to multiple drugs.

CONCEPT AND REASONING CHECKS

13.16 Why is debridement necessary for most burn wounds?

Animal Bites Can Puncture the Skin

KEY CONCEPT

17. Bacterial disease can arise from animal bites or scratches.

Each year in the United States, about 3.5 million people are bitten by animals. Most of these wounds heal without complications, but in certain cases, bacterial skin disease may develop.

Pasteurellosis is caused by *Pasteurella multocida*, a small, aerobic, gram-negative rod common in the nasopharynx of cats and various other animals. The pathogen causes most wound infections resulting from a dog or cat bite. In humans, symptoms develop rapidly, with local redness, warmth, swelling, and tenderness at the wound site. Abscesses frequently form, especially if the wound has been sutured. The disease responds slowly to penicillin therapy.

Cat-scratch disease (CSD) affects an estimated 20,000 Americans each year, primarily children who have been scratched, bit, or licked by an infected cat. Most cases are associated with *Bartonella henselae*, a rickettsia (FIGURE 13.23A). Symptoms of CSD include red-crusted blisters at the site of entry, headache, malaise, and low-grade fever. Swollen lymph glands, generally on the side of the body near the bite, accompany the disease (FIGURE 13.23B).

Rat-bite fever can be caused by one of two different bacterial species. In the United States, *Actinobacillus muris* (formerly *Streptobacillus moniliformis*), a gram-negative rod that occurs in long chains, is found in the pharynx of wild rats and other rodents feeding on infected rodents. Patients experience a lesion at the site of the bite or scratch,

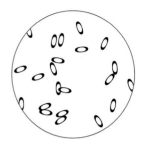

Pasteurella multocida

Actinobacillus muris and *Spirillum minus*

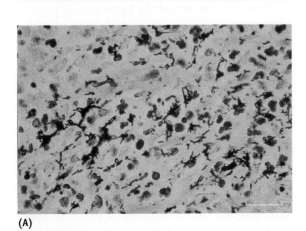

(A)

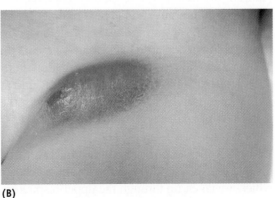

(B)

FIGURE 13.23 **Cat-Scratch Disease.** (**A**) A photomicrograph of *Bartonella henselae* (dark rods), the causative agent of cat-scratch disease. (Bar = 10 μm.) (**B**) A patient displaying in the groin the considerable lymph node swelling accompanying cat-scratch disease. »» Can *B. henselae* be cultured in microbiological growth media? Explain.

then a typical triad of prolonged fever, pain in the back and joints, and skin rash. In Asia, *Spirillum minus*, a rigid spiral cell with polar flagella, causes lesions at the wound site, and a maculopapular rash that spreads out from this point. In Japan and other parts of Asia, *Spirillum*-related rat-bite fever is known as **sodoku**. Antibiotic therapy with penicillin or erythromycin is recommended for either form of the disease.

Maculopapular rash:
A flat, red area on the skin with small, raised papules.

CONCEPT AND REASONING CHECKS

13.17 Describe the symptoms of the diseases caused by animal bites or scratches.

Leprosy (Hansen Disease) Is a Chronic, Systemic Infection

KEY CONCEPT

18. *Mycobacterium leprae* infects the skin and peripheral nerves.

For most of the past 2600 years, **leprosy** has been considered a curse of the damned. It did not kill, but neither did it seem to end. Instead, it lingered for years, causing the tissues to degenerate and deforming the body. In biblical times, the afflicted were ostracized from the community, though what was called leprosy in the Old Testament often was not that specific disease. Among the more heroic stories of medicine is the work of Father Damien de Veuster, the Belgian priest who in 1870 established a hospital for leprosy patients on Molokai, Hawaii. An equally heroic story was written more recently (MicroFocus 13.5).

The agent of leprosy is *Mycobacterium leprae,* an acid-fast rod related to *M. tuberculosis* (see Chapter 10). *M. leprae* was observed for the first time in 1874 by the Norwegian physician Gerhard Hansen. It is referred to as Hansen bacillus, and leprosy is commonly called **Hansen disease**. *M. leprae* is an obligate intracellular parasite, so it cannot be cultivated in artificial laboratory media. In 1960, researchers at the CDC succeeded in cultivating the bacillus in the footpads of mice, and in 1969, scientists discovered it would grow in the skin of nine-banded armadillos.

Leprosy is hard to transmit because about 95% of the world's population has a natural immunity to the disease. It is spread by contact with nasal secretions, which are taken up through the upper respiratory tract. The disease has an unusually long incubation period of three to six years, a factor making diagnosis very difficult. Because the organisms are heat sensitive, the symptoms occur in the skin and peripheral

Mycobacterium leprae

nervous system in the cooler parts of the body, such as the hands, feet, face, and earlobes. Severe cases also involve the eyes and the respiratory tract. Susceptibility is highest in childhood and decreases with age. More males appear to be infected than females.

Patients with leprosy experience disfiguring of the skin and bones, twisting of the limbs, and curling of the fingers to form the characteristic claw hand. The largest number of deformities develops from the loss of pain sensation due to nerve damage caused by lower numbers of bacilli. This form is called **paucibacillary** (*pauci* = "few") or **tuberculoid leprosy**. Inattentive patients, for example, might pick up a pot of boiling water without flinching.

Disease progression can lead to the loss of facial features accompanied by a thickening of the outer ear and collapse of the nose (**FIGURE 13.24A**). Many tumor-like growths called **lepromas** form on the skin and in the respiratory tract. This form is referred to as **multibacillary** or **lepromatous leprosy**. This is the most serious form of the disease because the immune system fails to react, meaning there can be millions of bacilli in the body.

For many years, the principal drug for the treatment and cure of leprosy was a sulfur compound known commercially as **Dapsone**. In many cases, such as the one shown in **FIGURE 13.24B**, the results were dramatic. Today, multidrug therapy with dapsone, rifampin, and clofazimine has cured more than 16 million people of leprosy worldwide.

In 1985, the WHO began a campaign to "eliminate" leprosy. In 2009, 119 countries achieved the WHO's elimination goal of fewer than one patient per 10,000 population.

The genomes of both *M. leprae* and *M. tuberculosis* have been sequenced and comparative genomics has revealed some interesting findings. The most interesting discovery so far is that more than half of the genes in *M. tuberculosis* are missing from *M. leprae*. This suggests at some time in the distant past *M. leprae* lost a substantial number of essential genes for metabolism, requiring the organism to depend on infection of host cells to provide its necessary metabolic and growth needs. More analysis may discover an "Achilles heel" by which leprosy can be eliminated.

TABLE 13.4 summarizes the bacterial contact diseases. MicroInquiry 13 presents four cases for study involving sexually transmitted and contact diseases.

CONCEPT AND REASONING CHECKS

13.18 Distinguish between tuberculoid and lepromatous leprosy.

MICROFOCUS 13.5: History
The "Star" of Carville

On December 1, 1894, seven leprosy patients arrived at an old plantation on a crook in the Mississippi River. Soon thereafter, four nuns of the Order of St. Vincent de Paul joined them. Together this small band formed the nucleus of what was to become the National Hansen's Disease Center at Carville, Louisiana.

Change came slowly. In 1921, the United States Public Health Service acquired the institution, but it remained essentially a prison, patrolled by guards and surrounded by a chain link fence with barbed wire. Then, in 1931, a leprosy patient named Stanley Stein arrived. Stanley Stein was not his real name—he had forsaken that for fear of bringing shame to his family. Soon, Stein instituted a weekly paper to bring a sense of community to the patients. Originally named *The Sixty-Six Star* (Carville was United States Marine Hospital Number 66), the name was eventually shortened to *The Star*.

As the circulation of *The Star* increased, Stein and others launched a campaign for change. In 1936, the patients acquired a telephone so they could hear the voices of their families. Three years later, the swamps were drained to reduce the incidence of malaria. Soon there came a better infirmary, a new recreation hall, and removal of the barbed wire. In 1946, the State of Louisiana allowed the patients to vote in local and national elections.

Stanley Stein stands next to the printing press as copies of *The Star* are printed.

Through all these years, Stein's leprosy worsened. Originally he had tuberculoid leprosy, the form in which the nerves are damaged. Afterward, however, he developed lepromatous leprosy, which causes lesions to form on the face, ears, and eyes. Soon he was totally blind. Without feeling in his fingers, he could not even learn Braille.

But Stein was not finished. He and his newspaper tirelessly fought for a new post office and weekend passes for patients. In 1961, President Kennedy paid tribute to *The Star* on its thirtieth anniversary and singled out its indomitable editor for praise. Stanley Stein died in 1968. By that time, *The Star* had a circulation of 80,500 in all 50 states and 118 foreign countries. *The Star* is still being published today by Carville patients.

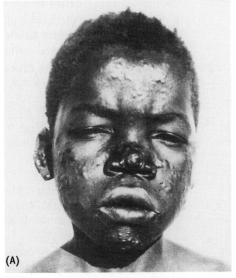
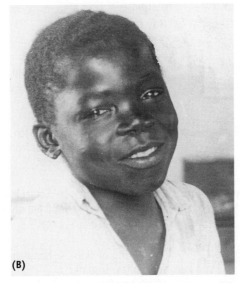

FIGURE 13.24 **Treating Leprosy.** The young boy with lepromatous leprosy is pictured (**A**) before treatment with dapsone and (**B**) some months later, after treatment. Note that the lesions of the ear and face and the swellings of the lips and nose have largely disappeared. »» Why is the disfiguring caused by *Mycobacterium leprae* limited to the body extremities?

TABLE

13.4 A Summary of the Major Contact Diseases Caused by Bacterial Species

Inflammation or Disease	Causative Agent	Signs and Symptoms	Transmission	Treatment	Prevention and Control
Acne	*Propionibacterium acnes*	Comedones (whiteheads and blackheads)	Not transmissible	Benzoyl peroxide Antibiotics Isotretinoin	Gentle washing of affected skin Benzoyl peroxide
Furuncle and Carbuncle	*Staphylococcus aureus*	Painful single or cluster of boils	Autoinfection Contact with infected person	Drainage Debridement	Practicing good hygiene
Impetigo	*Staphylococcus aureus* *Streptococcus pyogenes*	Thin-walled blisters forming a crust	Direct or indirect contact	Skin cleansing Topical antibiotic	Practicing good hygiene
Scalded skin syndrome	*Staphylococcus aureus*	Red, wrinkled and tender skin Epidermis may peel	Direct or indirect contact	Cefazolin	Practicing good hygiene
Toxic shock syndrome (TSS)	*Staphylococcus aureus*	Fever, vomiting, watery diarrhea, sore throat, muscle aches, and sunburn-like rash	Vaginal tampons Skin wounds Surgery	Supportive care Antibiotics	Avoiding highly absorbent vaginal tampons
Erysipelas	*Streptococcus pyogenes*	High fever, shaking and chills, headache leading to fiery rash on lower limbs and face	Minor skin trauma Eczema Surgical incisions	Oral or intravenous antibiotics	Avoiding dry skin Preventing cuts and scrapes
Streptococcal TSS (STSS)	*Streptococcus pyogenes*	Fever, dizziness, confusion, and flat body rash	Direct contact with patients or carriers	Clean any wounds Debridement	Cleaning wounds
Necrotizing fasciitis	*Streptococcus pyogenes*	Fever with pain and swelling at the wound site	Trauma to skin surface	Broad spectrum antibiotics Debridement Surgery	Cleaning skin after a cut, scrape, or other deep wound
Burn infections	*Pseudomonas aeruginosa* Other bacterial species	Difficult to diagnose May be absent, minimal or late developing	Nosocomial	Antibiotic therapy Debridement	Protecting burn patients Practicing high levels of disinfection and sterilization
Pasteurellosis	*Pasteurella multocida*	Redness, warmth, swelling, and tenderness at the wound site	Cat and dog scratch or bite	Penicillin Doxycycline	Cleaning animal scratches and bites
Cat-scratch disease	*Bartonella henselae*	Red-crusted blisters, headache, malaise, low-grade fever	Scratch, bite, or lick from healthy cat	Symptomatic management	Cleaning cat scratches and bites
Rat-bite fever	*Actinobacillus muris* *Spirillum minus*	Fever, pain in the back and joints, and a skin rash	Scratch or bite from an infected rat	Penicillin Tetracycline	Avoiding contact with rats and rat contaminated dwellings
Leprosy (Hansen disease)	*Mycobacterium leprae*	Disfiguring of skin and bones, loss of pain sensation, loss of facial features	Nasal secretions	Multidrug therapy with dapsone, rifampin, and clofazimine	Avoiding contact where endemic

MICROINQUIRY 13

Sexually Transmitted and Contact Disease Identification

Below are several descriptions of sexually transmitted, contact, and miscellaneous bacterial diseases based on material presented in this chapter. Read the case history and then answer the questions posed. Answers can be found in **Appendix D**.

Case 1

The patient is a 17-year-old woman who comes to the clinic indicating that several days ago she started feeling nauseous but had not experienced any vomiting. She tells the physician that the day before coming to the clinic she had a fever and chills; she also has been urinating more frequently and the urine has a foul smell. She is diagnosed as having a urinary tract infection.

13.1a. What types of bacterial species could be responsible for her illness?

13.1b. Why are these types of diseases more prevalent in women than they are in men?

13.1c. What types of urinary infections can occur?

13.1d. How could this patient attempt to avoid another UTI?

13.1e. What role do biofilms play in UTIs?

Case 2

A 19-year-old unwed mother arrives at the emergency room of the county hospital complaining of having cramps and abdominal pain for several days. She says she had never had a urinary tract infection and could not have gonorrhea, as she was treated and cured of that two years ago. She has not experienced nausea or vomiting. When questioned, she tells the emergency room nurse that she has a single male sexual partner and condoms always are used. Based on further examination, the patient is diagnosed with pelvic inflammatory disease (PID). An endocervical swab is used for preparing a tissue culture. Staining results indicate the presence of cell inclusions.

13.2a. What bacterial species can be associated with PID? What disease does she most likely have?

13.2b. Why was a tissue culture inoculum ordered? Describe the reproductive cycle of this organism.

13.2c. What other tests could be ordered for the patient's infection?

13.2d. Why was the emergency room concerned about her sexual activity?

13.2e. What misconception does the patient have about her past gonorrhea infection?

Case 3

A 71-year-old man visits the local hospital emergency room at 6:00 PM after noticing a red infection streak running up his left forearm. He tells the physician that he was playing with his cat this morning when it bit him on the left wrist. Thinking nothing of it and being an amateur photographer, he went about printing some photographs in his darkroom. At 3:00 PM, he finished and, in the daylight, noticed that his wrist was swollen and painful. The physician also notes that the patient experiences tenderness at the site. She then notices a small puncture wound on the wrist and a small abscess. A Gram stain smear from the abscess indicates the presence of gram-negative rods.

13.3a. What bacterial species is responsible for the patient's illness?

13.3b. What clues lead you to identify this specific organism?

13.3c. Where is this organism normally found in cats?

13.3d. What could the patient have done to make it less likely that an infection occurred?

Case 4

A 17-year-old man comes to a free neighborhood clinic. He says that he noticed some white pus-like discharge and a tingling sensation in his penis. Since yesterday, he has had pain on urinating. He tells the physician that he has been sexually active with several female partners over the past eight months, but no one has had any sexually transmitted disease. Examination determines that there is no swelling of the lymph nodes in the groin or pain in the testicles. A Gram stain indicates the presence of gram-negative diplococci. The patient is given antibiotics, instructed to tell his female partners they both should be medically examined, and then he is released.

13.4a. Based on the clinic findings, what disease does the patient have and what bacterial species is responsible for the infection?

13.4b. Why is it important for his sexual partners to be medically examined, even if they experience no symptoms? What complications could arise if they are infected?

13.4c. For which other organisms is this patient at increased risk? Why?

13.4d. What significance can be drawn from the fact that the patient does not have any swelling of the lymph nodes in the groin or pain in the testicles?

13.4e. What antibiotics most likely would be given to the patient?

13.6 Contact Diseases Affecting the Eye

Although the eye has a multilayered defense against infection, eye diseases do occur. As we will discover, trachoma is the most prevalent infectious eye disease in the world. However, in the United States and in other developed nations, eye infections typically involve inflammation of the eyelid, conjunctiva, or cornea (FIGURE 13.25). The most common diseases of the eye surface are described below.

Some Bacterial Eye Infections Can Cause Blindness

KEY CONCEPT

19. Bacterial eye infections can involve the eyelid, cornea, or conjunctiva.

Several eye infections are caused by a few bacterial species. *Staphylococcus aureus,* often a transient member of the skin, is a major cause of infections of the eyelid and the cornea. When the bacterium infects the eyelid margin, a painful red swelling develops. The inflammation, called **blepharitis**, makes the sufferer have a burning sensation and feeling that there is a foreign body in the eye. The inflammation sometimes leads to the formation of a **stye**. Treatment usually involves warm compresses and a topical antibiotic, such as one containing bacitracin.

One form of **bacterial conjunctivitis**, commonly called **pink eye** or **red eye**, is an eye redness due to dilation of the conjunctival blood vessels (FIGURE 13.26). It is common in childhood, but the inflammation can occur in people of any age. The organisms that most commonly cause bacterial conjunctivitis are staphylococci, pneumococci, or streptococci that are picked up by contact with the eye. Symptoms include eye pain, swelling, redness, and a moderate to large amount of discharge, usually yellow or greenish in color. After sleeping, some affected individuals have a situation where their "eyes are stuck shut," requiring a warm washcloth applied to the eyes to remove the dried discharge.

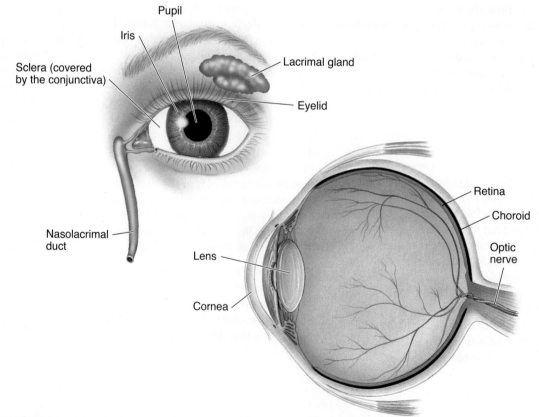

FIGURE 13.25 **Eye Anatomy.** External infections are most often associated with the eyelid, conjunctiva, or cornea. »» Why would the conjunctiva and cornea be most susceptible to infection?

Bacterial conjunctivitis is a fairly common condition and usually causes no long-term eye or vision damage. Still, it is important to see a doctor because some types require treatment with antibiotic eye drops or ointment. Prevention includes frequent hand washing with soap and warm water. Infected individuals should not share eye drops, tissues, eye makeup, washcloths, towels, or other objects that come in contact with the eye.

A form of **chronic bacterial conjunctivitis**, leading to keratitis, also is caused by *S. aureus*. The bacterium invades the cornea after some form of eye trauma that causes a break in the corneal epithelium. The resulting ulcers are painful and are treated with antibiotic drops.

The most common and severe form of **hyperacute bacterial conjunctivitis** is caused by *Neisseria gonorrhoeae*. Left untreated, it can progress to keratitis and corneal perforation. Ceftriaxone can be used to treat gonococcal conjunctivitis. Any delay in treatment can lead to corneal damage or eye loss. A perforation also can be a portal of entry leading to septicemia.

Neonatal conjunctivitis is an inflammation of the conjunctiva of the newborn. The inflammation, also called **ophthalmia of the newborn**, results from contact with the bacterium during passage through the birth canal of a mother infected with *N. gonorrhoeae* or *Chlamydia trachomatis*. Infection with *N. gonorrhoeae* is the most

serious and, if untreated, can lead to blindness (**FIGURE 13.27**). Infection with *C. trachomatis* is the most common form of ophthalmia. It usually heals without permanent eye damage. Treatment for both forms of conjunctivitis involves the antibiotics doxycycline or erythromycin. In the United States, prevention involves the use of antimicrobial drugs put into the eyes of all newborns after delivery. Silver nitrate drops are still used in some parts of the world.

Trachoma is the world's leading cause of preventable blindness. It occurs in hot, dry regions of the world and it is prevalent in Mediterranean countries, parts of Africa and Asia, and in the southwestern United States in Native American populations. There are 500 million infections, mostly children, worldwide and 7 to 9 million individuals have been blinded by trachoma.

Trachoma is caused by serotypes A, B, and C of *C. trachomatis* that are not sexually transmitted but rather by personal contact with contaminated fingers, towels, and optical instruments. Face-to-face contact and flies also are important modes of transmission.

The chlamydiae multiply in the conjunctiva of the eye. A series of tiny, pale nodules form on this membrane, giving it a rough appearance (*trach* = "rough"; *oma* = "tumor"). An initial infection typically heals without permanent damage. However, the initial infection sets up a hypersensitive state,

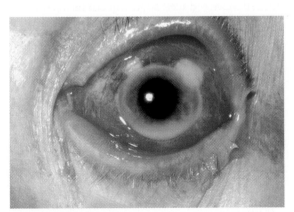

FIGURE 13.26 **Conjunctivitus.** The inflammation on the conjunctiva also called pink eye, causes swelling in the surface. »» What causes the redness in the sclera?

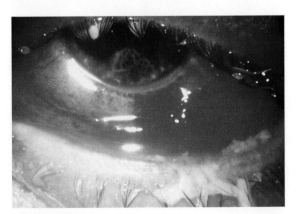

FIGURE 13.27 **Gonorrheal conjunctivitus.** This case of gonorrheal conjunctivitus resulted in partial blindness due to the spread of *N. gonorrhoaea* bacteria. »» How does one become infected with *N. gonorrhoeae*?

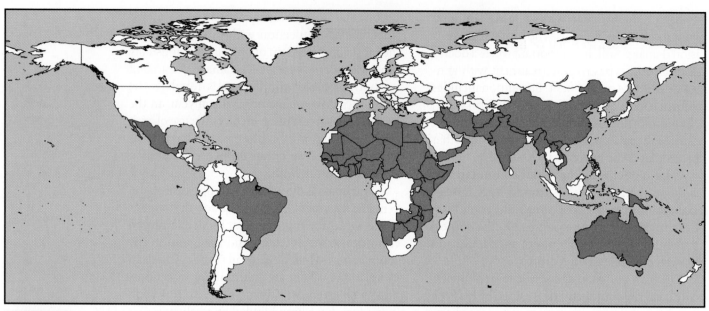

FIGURE 13.28 **Worldwide Trachoma.** Blinding or suspected blinding trachoma affects individuals worldwide, but especially in Africa, Southeast Asia, Mexico, and parts of South America. »» How can trachoma be treated? (Map by Silvio Mariotti/WHO.)

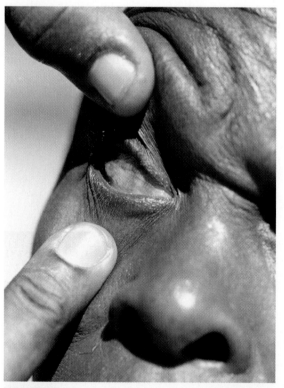

FIGURE 13.29 **Trachoma.** An opthalmic surgeon in rural Malawi, Africa, attends to a trachoma patient. »» How is trachoma spread?

such that repeated infections result in a chronic inflammation. Over many years, scarring of the conjunctiva occurs as the eyelashes turn inwards and abrade the cornea, eventually leading to blindness (**FIGURE 13.28**).

Azithromycin helps reduce the symptoms of trachoma, but in many patients, the relief is only temporary because chlamydiae reinfect the tissues (**FIGURE 13.29**). In 1997, the WHO established the Alliance for Global Elimination of Trachoma by 2020 (GET 2020). Since then, ten national programs, making up 50% of the global trachoma burden, have reduced acute infections in children by 50%.

This has involved using a "**SAFE strategy**"; that is, **S**urgery of the eyelids; **A**ntibiotics for acute infections; **F**acial hygiene improvements; and **E**nvironmental access to safe water.

TABLE 13.5 summarizes the bacterial eye diseases.

CONCEPT AND REASONING CHECKS

13.19 What are the challenges facing health officials trying to eliminate trachoma?

TABLE

13.5   A Summary of Infectious Bacterial Eye Diseases

Inflammation or Disease	Causative Agent	Signs and Symptoms	Transmission	Treatment	Prevention and Control
Blepharitis	*Staphylococcus aureus*	Burning sensation in eye Stye may form	Contaminated instruments Hands Shared towels Droplets	Warm compress Antibiotic medication	Practicing good hygiene
Bacterial conjunctivitis	*Staphylococcus aureus* *Streptococcus pyogenes* *Haemophilus influenzae* *Neisseria gonorrhoeae*	Eye pain, swelling, redness, and a yellow or greenish discharge	Direct or indirect contact Pink eye	Antibiotic eye drops	Practicing good hygiene
Neonatal conjunctivitis	*Neisseria gonorrhoeae* *Chlamydia trachomatis*	Eye swelling and pus discharge Watery discharge	Infected mother to child during childbirth	Topical and oral antibiotics	Using silver nitrate or antibiotics Screening mother
Trachoma	*Chlamydia trachomatis*	Tiny, pale nodules on the conjunctiva Upper eyelid abrasion can cause blindness	Direct or indirect contact	Topical or oral antibiotics	Washing face Controlling flies Freshwater source

■ SUMMARY OF KEY CONCEPTS

13.1 The Structure and Indigenous Microbiota of the Female and Male Reproductive Systems

1. The **primary sex organs** in the male are the testes, while the epididymis, vas deferens, seminal vesicles, prostate, and penis are **accessory reproductive organs**. In females, the primary sex organs are the ovaries, while the accessory organs consist of the fallopian tubes, uterus, vagina, and vulva.

2. Antimicrobial defenses in the reproductive tracts include the urethral mucosa, the vagina, vulva, and cervix in females. *Lactobacillus* species in the vagina as well as antibodies and other antimicrobial products of the systems produce an environment not favorable for pathogen colonization.

3. Nonsexually transmitted illnesses include:
   - **Bacterial vaginosis**
     – *Gardnerella vaginalis*
     – *Prevotella*
     – *Peptostreptococcus*
     – Other indigenous bacterial species

13.2 Sexually Transmitted Diseases Caused by Bacteria
   - **Chlamydial urethritis**
     1. *Chlamydia trachomatis*
   - **Gonorrhea**
     2. *Neisseria gonorrhoeae*
   - **Syphilis**
     3. *Treponema pallidum*
   - **Chancroid**
     4. *Haemophilus ducreyi*
   - ***Ureaplasma* urethritis**
     5. *Ureaplasma urealyticum*
   - **Lymphogranuloma venereum**
     6. *Chlamydia trachomatis*
   - **Granuloma inguinale**
     7. *Klebsiella granulomatis*

13.3 The Structure and Indigenous Microbiota of the Female and Male Urinary System

8. The organs of the **urinary system** that are susceptible to infection are the kidneys and the urinary tract (ureters, urinary bladder, and urethra). The kidneys, ureters, and bladder are normally sterile due to normal urine flow. However, the urethra is colonized by microbes either along its whole length or near the terminus.

9. Bacterial diseases include:
   - **Urethritis**
     8. *Escherichia coli*
     9. *Chlamydia trachomatis*
     10. *Mycoplasma*
     11. Other bacterial species
   - **Cystitis**
     12. *Escherichia coli*

- **Prostatitis** (males not shown)
  - *Escherichia coli*
  - *Staphylococcus aureus*
  - *Proteus*
  - *Pseudomonas aeruginosa*
- **Pyelonephritis**
  - 13. *Escherichia coli*
  - 14. *Staphylococcus aureus*
  - 15. *Klebsiella pneumoniae*
  - 16. *Pseudomonas aeruginosa*

13.4 Contact Diseases Caused by Indigenous Bacterial Species

10. Many bacterial diseases are caused by contact with the skin, which normally protects the underlying tissues from bacterial colonization and infection.

- **Acne**
  - 17. *Propionibacterium acnes*
- Burn infections
  - 18. *Pseudomonas aeruginosa*

13.5 Contact Diseases Caused by Exogenous Bacterial Species

- **Furuncles** (boils) and carbuncles
  - 19. *Staphylococcus aureus*
- **Impetigo**
  - 20. *Staphylococcus aureus*
  - 21. *Streptococcus pyogenes*
- **Scalded skin** and **toxic shock syndromes**
  - 22. *Staphylococcus aureus*
- **Erysipelas** and **necrotizing fasciitis**
  - 23. *Streptococcus pyogenes*
- Animal bite diseases
  - 24. *Pasteurella multocida*
  - 25. *Bartonella henselae*
  - 26. *Actinobacillus muris*
  - 27. *Spirillum minus*

13.6 Contact Diseases Affecting the Eye

- **Conjunctivitis**
  - 28. *Neisseria gonorrhoeae*
  - 29. *Chlamydia tachomatis*
- **Trachoma**
  - 30. *Chlamydia tachomatis*

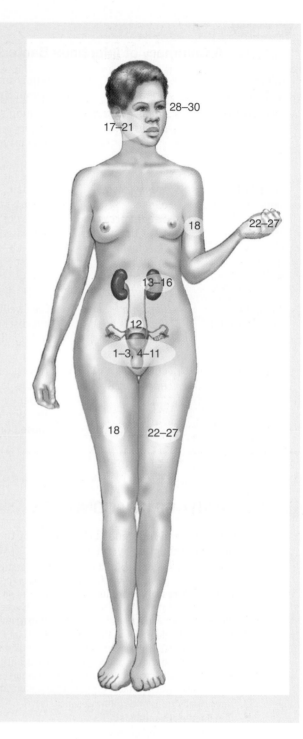

LEARNING OBJECTIVES

After understanding the textbook reading, you should be capable of writing a paragraph that includes the appropriate terms and pertinent information to answer the objective.

1. Trace the pathway of (a) a sperm cell through the **male reproductive system** and (b) an egg through the **female reproductive system**.
2. Identify the defenses in the male and female reproductive systems that normally prevent colonization and infection.
3. Describe the causes of **bacterial vaginosis**.
4. Distinguish between the signs and symptoms of **chlamydial urethritis** in males and females.
5. Describe (a) the possible complications resulting from **gonorrhea** in females and (b) explain the danger of gonorrhea in pregnant females.
6. Distinguish among the three possible stages of **syphilis**.
7. Describe the signs and symptoms of **chancroid** and **ureaplasmal urethritis**.
8. Assess the role of **urination** and indigenous microbiota to maintaining a healthy urinary system free from infection.
9. Differentiate among the various forms of UTIs: **urethritis**, **cystitis**, **prostatitis**, and **pyelonephritis**.
10. Identify the skin layers and other chemical skin defenses protecting underlying tissues from microbial colonization.
11. Name the significant members of the indigenous skin microbiota that normally out-compete most transient microbiota and pathogens.
12. Assess the role of *Propionibacterium acnes* in triggering **acne** and describe the follicle-associated lesions.
13. Estimate the significance of infections by indigenous microbiota resulting from surgical procedures.
14. Assess the role of *Staphylococcus aureus* as an agent for contact diseases.
15. Describe the contrast skin infections caused by *Streptococcus pyogenes*.
16. Summarize the types of diseases caused by **traumatic wounds**, including burns.
17. Identify and discuss the characteristics of the three animal bite diseases.
18. Summarize the clinical significance of **leprosy (Hansen disease)**.
19. Explain why **trachoma** is a major cause of blindness in many developing nations.

STEP A: SELF-TEST

Each of the following questions is designed to assess your ability to remember or recall factual or conceptual knowledge related to this chapter. Read each question carefully, then select the *one* answer that best fits the question or statement. Answers to even-numbered questions can be found in **Appendix C**.

1. The primary sex organs of the female reproductive system is/are the
   A. uterus.
   B. vagina.
   C. fallopian tubes.
   D. ovaries.
2. What part or parts of the male and female reproductive systems are typically colonized by indigenous microbiota?
   A. Male: ureters; female: vagina and ovaries
   B. Male: testes and epididymis
   C. Male: urethra; female: vagina, vulva, and cervix
   D. Male: bladder and ureters; female: fallopian tubes and cervix
3. Which one of the following microbes is NOT associated with vaginosis?
   A. *Peptostreptococcus*
   B. *Gardnerella*
   C. *Prevotella*
   D. *Staphylococcus*
4. Which one of the following statements is NOT correct concerning the reproductive cycle of *Chlamydia*?
   A. Reticulate bodies are infectious.
   B. Reticulate bodies reorganize into elementary bodies.
   C. Elementary bodies infect host cells.
   D. Elementary bodies transform into reticulate bodies.
5. Salpingitis is associated with _____ and can lead to _____.
   A. syphilis; gumma formation
   B. gonorrhea; sterility
   C. chlamydia; ophthalmia
   D. chancroid; soft chancre

6. A chancre is typical of which stage of syphilis?
   A. Primary
   B. Secondary
   C. Tertiary
   D. Chronic, latent
7. Besides chlamydia urethritis, what other STD is associated with another serotype of *Chlamydia trachomatis*?
   A. Lymphogranuloma venereum (LGV)
   B. Genital warts
   C. Granuloma inguinale
   D. Chancroid
8. Which one of the following is NOT part of the urinary tract?
   A. Urethra
   B. Bladder
   C. Kidneys
   D. Ureters
9. What bacterial species is most often associated with cystitis?
   A. *Treponema pallidum*
   B. *Escherichia coli*
   C. *Chlamydia trachomatis*
   D. *Pseudomonas aeruginosa*
10. What type of immune defensive cell is found in the sublayers of the epidermis?
    A. Keratinocyte
    B. Dendritic (Langerhans) cell
    C. Neutrophil
    D. Basophil
11. The skin is
    A. dominated by gram-negative bacterial cells.
    B. free of bacterial cells.
    C. without a microbiota.
    D. dominated by gram-positive bacterial cells.

12. Which one of the following statements is NOT true of acne?
  A. Acne is caused by *Propionibacterium acnes*.
  B. Plugged sebaceous glands are called erythemas.
  C. Whiteheads are completely blocked follicles.
  D. Acne is not a preventable disease.
13. An acute wound could be due to
  A. surgical procedures.
  B. cuts.
  C. lacerations.
  D. All the above (A–C) are correct.
14. In children, this skin disease is characterized by the production of thin-walled blisters oozing a yellowish fluid and forming yellowish-brown flakes.
  A. Toxic shock syndrome
  B. Scalded skin syndrome
  C. Erysipelas
  D. Impetigo
15. Which one of the following skin diseases is NOT caused by *Streptococcus pyogenes*?
  A. Necrotizing fasciitis
  B. Toxic shock syndrome
  C. Gas gangrene
  D. Erysipelas

16. The most common cause of an invasive wound infection, such as a burn, is
  A. gram-positive bacterial species.
  B. *Treponema pallidum*.
  C. *Pseudomonas aeruginosa*.
  D. *Escherichia coli*.
17. *Bartonella henselae* is the causative agent of this skin disease.
  A. Leprosy
  B. Cat-scratch disease
  C. Necrotizing fasciitis
  D. Rat-bite fever
18. Leprosy can be contracted by contact with
  A. contaminated water.
  B. insects.
  C. nasal secretions.
  D. contaminated food.
19. The SAFE strategy has greatly reduced the global burden of what disease?
  A. Trachoma
  B. Neonatal conjunctivitis
  C. Leprosy
  D. Blepharitis

STEP B: REVIEW

Answer each of the following by filling in the blank with the correct word or phrase. Answers to even-numbered questions can be found in **Appendix C**.

20. Ureaplasmal urethritis is a type of _____.
21. _____ is the disease caused by *Haemophilus ducreyi*.
22. Lymphogranuloma venereum (LGV) is prevalent in _____-Asia.
23. A _____ is absent in *Mycoplasma* species.
24. Granuloma inguinale is characterized by having bacterial masses called _____ bodies.
25. *Gardnerella* is the cause of _____.
26. *Mycoplasma* _____ is the agent responsible for mycoplasmal urethritis.
27. _____ are a group of connected, deep abscesses caused by *S. aureus*.
28. _____ is the most common agent of urinary tract infections.
29. Sufferers of UTIs have a _____ on urination.
30. Dapsone, rifampin, and clofazimine are drugs used to treat _____.

STEP C: APPLICATIONS

Answers to even-numbered questions can be found in **Appendix C**.

31. Suppose a high incidence of leprosy existed in a particular part of the world. Why is it conceivable that there might be a correspondingly low level of tuberculosis?
32. An African patient reports to a local hospital with an upper lip swollen to about three times its normal size. Probing with a safety pin at facial points where major nerve endings terminate showed that the area to the left of the nose and above the lip was without feeling. When a biopsy of the tissue was examined, it revealed round reservoirs of immune system cells called granulomas within the nerves. On bacteriological analysis, acid-fast rods were observed in the tissue. What disease do all these data suggest?
33. During a field trip, an undergraduate biology student is bitten on the left index finger by a wild rat. Within 12 hours, her finger is swollen and throbbing. Soon thereafter she is hospitalized with swollen lymph nodes, a skin rash, fever, and exquisite sensitivity of the finger. Gram-negative, branching rods were found in the tissue. What infectious disease had she contracted?
34. Certain microscopes have the added feature of a small hollow tube that fits over the eyepieces or oculars. Viewers are encouraged to rest their eyes against the tube and thereby block out light from the room. Why is this feature hazardous to health?
35. After a young man suffers an abrasion on the right arm, his affectionate cat licks the wound. Several days later, a pustular lesion appears at the site and a low-grade fever develops. He also experiences "swollen glands" on the right side of his neck. What disease has he acquired?
36. A woman suffers two miscarriages, each after the fourth month of pregnancy. She then gives birth to a child, but impaired hearing and vision become apparent as it develops. Also, the baby's teeth are shaped like pegs and have notches. What medical problem existed in the mother?

STEP D: QUESTIONS FOR THOUGHT AND DISCUSSION

Answers to even-numbered questions can be found in **Appendix C**.

37. One of the major problems of the current worldwide epidemic of AIDS is the possibility of transferring the human immunodeficiency virus (HIV) among those who have a sexually transmitted disease. Which diseases in this chapter would make a person particularly susceptible to penetration of HIV into the bloodstream?

38. Studies indicate that most cases of *Staphylococcus*-related impetigo occur during the summer months. Why do you think this is the case?

39. In some African villages, blindness from trachoma is so common that ropes are strung to help people locate the village well, and bamboo poles are laid to guide farmers planting in the fields. What measures can be taken to relieve such widespread epidemics as this?

40. At a specified hospital in New York City, hundreds of patients pay a regular visit to the "neurology ward." Some sign in with numbers; others invent fictitious names. All receive treatment for leprosy. Why do you think this disease still carries such a stigma?

41. Several years ago the Rockefeller Foundation offered a $1 million prize to anyone who could successfully develop a simple and rapid test to detect chlamydia and/or gonorrhea. The test had to use urine as a test sample, and be performed and interpreted by someone with a high school education. No one ever claimed the prize. Can you guess why?

HTTP://MICROBIOLOGY.JBPUB.COM/9E

The site features eLearning, an online review area that provides quizzes and other tools to help you study for your class. You can also follow useful links for in-depth information, read more MicroFocus stories, or just find out the latest microbiology news.

False-color scanning transmission microscope image of bacterial viruses (bacteriophages) attacking an *Escherichia coli* cell.

PART 4

Viruses and Eukaryotic Microorganisms

The bacterial species we have examined in the previous chapters are but one of several groups of microbial agents interwoven with the lives of humans. Other prominent groups are the viruses, fungi, and parasites. Knowledge of these groups developed slowly during the early 1900s, partly because they were generally more difficult to isolate and cultivate than bacterial organisms. Also, the established methods for research into bacterial growth were more advanced than for other microorganisms, and investigators often chose to build on established knowledge rather than pursue uncharted courses of study. Moreover, the urgency to learn about the other groups was not as great because they did not appear to cause such great epidemics and pandemics.

This perception changed in the second half of the 1900s. Many bacterial diseases came under control with the advent of vaccines and antibiotics, and the increased funding for biological research allowed attention to shift to other infectious agents. The viruses finally were identified and cultivated, and microbiologists laid the foundations for their study. Fungi gained prominence as tools in biological research, and scientists soon recognized their significance in ecology and industrial product manufacturing. As remote parts of the world opened to trade and travel, public health microbiologists realized the global impact of parasitic diseases. Moreover, as concern for the health of the world's people increased, observers expressed revulsion at the thought that hundreds of millions of human beings were infected with these parasites.

In Part 4, we examine the viruses and eukaryotic microorganisms over the course of five chapters. Chapter 14 is devoted to a study of the viruses and viral-like agents, while Chapters 15 and 16 outline the multiple diseases caused by these infectious particles. In Chapter 17, the discussion moves to fungi, while in Chapter 18, the area of interest is the protozoa and the multicellular parasites. Throughout these chapters, the emphasis is on human disease. You will note some familiar diseases, such as hepatitis, chickenpox, and malaria, as well as some less familiar ones, such as dengue fever, toxoplasmosis, and schistosomiasis. The spectrum of diseases continues to unfold as scientists develop new methods for the detection, isolation, and cultivation of viruses and eukaryotic microorganisms.

Virology

When Ed Alcamo was in college, he was part of a group of twelve biology majors. Each of them had a particular area of "expertise." John was going to be a surgeon, Jim was interested in marine biology, Walt was a budding dentist, and Ed was the local virologist. He was fascinated with viruses, the ultramicroscopic bits of matter, and at one time wrote a term paper summarizing arguments for the living or nonliving nature of viruses. (At the time, neither side was persuasive, and even his professor gracefully declined to take a stand.)

Ed never quite made it to being a virologist, but if your fascination with these infectious particles is as keen as his, you might like to consider a career in virology.

Virologists investigate dreaded diseases such as AIDS, polio, and rabies, while others (epidemiologists) investigate disease outbreaks. Virologists also concern themselves with many types of cancer, and others study the chemical interactions of viruses with various tissue culture systems and animal models.

Virologists also are working to replace agricultural pesticides with viruses able to destroy mosquitoes and other pests. Some virologists are inserting viral genes into plants and are hoping the plants will produce viral proteins to lend resistance to disease. One particularly innovative group is trying to insert genes from hepatitis B viruses into bananas. They hope that one day we can vaccinate ourselves against hepatitis B by having a banana for lunch.

If you wish to consider the study of viruses, an undergraduate major in biology would be a good choice. Because the biology of viruses is related to the biology of cells, courses in biochemistry and cell biology will be required.

Following completion of college, most virologists study for an M.D. or Ph.D. degree. M.D.s pursue virology research in the context of patients or disease and become investigators with an interest in infectious disease or epidemiology. Most Ph.D.s pursue more basic questions with academic institutions, or in industrial or governmental organizations.

A great way to find out if you have the "research bug" is to work in a college or university laboratory. Many colleges and universities with research programs employ undergraduate students (near minimal wage!) in the research laboratory. So, here is a good way to "get your feet wet." If you find it interesting, it also will enhance your chances to be accepted by a top-flight graduate school.

Not interested in the lab bench research? Virologists also can find careers in full-time teaching. In addition, your knowledge can be used to pursue a career in communications, serving as a science writer or reporter. They also pursue careers in business administration or law, especially involving the pharmaceutical industry or patent law.

But no matter what avenue chosen, you should be a curious and hardworking student with a passion for science, like Ed Alcamo and your author. Do you have these interests?

14

Chapter Preview and Key Concepts

The Viruses and Virus-Like Agents

It's just a piece of bad news wrapped up in protein.
—Nobel laureate Peter Medawar (1915–1987) describing a virus

Ed Alcamo remembers the spring of 1954.

"I was a lad of thirteen growing up in the Bronx and looking forward to a carefree summer. But I could feel the tension in my parents' voices as they anticipated the months ahead, for summer was the dreaded polio season.

And sure enough, by early July the tension had turned to outright fear. I was told to avoid the public pool and the lusciously cool air-conditioned movie house. I had to report any cough or stiff neck promptly. 'Stick with your old friends,' my father told me. 'You've already got their germs.' Most of the time I was indoors, and the only baseball I got to play was in my imagination, as I listened to the Yankees every afternoon on my portable radio.

Our family was one of the lucky few to have a television, and each night we watched row upon row of iron lungs, and we saw the faces of the kids whose bodies were captured forever in their iron prisons (**FIGURE 14.1**). Iron lungs, I was told, help you breathe when paralysis affects the respiratory muscles. We heard and read about the daily toll from polio, where the victims lived, and how many kids had died.

But there was hope. My mom and her friends were out collecting dimes to fight polio (they called it the Mothers' March Against Polio), and the National Foundation of Infantile Paralysis said it had 75 million dimes to help fund the tests of a new vaccine—Dr. Salk's vaccine. Two million children would be getting shots. Maybe next year would be different.

Boy, next year sure was different! On April 12, 1955, at a televised news conference, Dr. Salk declared, 'The vaccine works!' The celebration

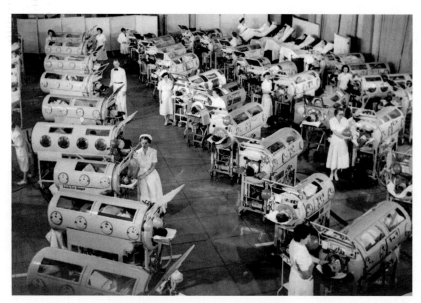

FIGURE 14.1   Iron Lung Ward—1953. This iron lung ward in Rancho, California is filled with rows of polio patients. The iron lung sealed the thoracic cavity in an air-tight chamber. The chamber created a negative pressure around the thoracic cavity, thereby causing air to rush into the lungs to equalize intrapulmonary pressure. »» What is the iron lung attempting to do for the patient?

was wild. Our school closed for the day. And the church bells rang, even though it was a Thursday. I could tell my mother was relieved—we had steak for dinner that night.

By summertime, things were back to normal. I was now fourteen and eager to show off my baseball skills to any girl who cared to watch. Down at the neighborhood pool I was learning how to dive (when no one was watching). And for a quarter, I got to cheer for the cavalry at the Saturday afternoon movie. Summer was back."

The polio virus is one of the smallest viruses, being about the same diameter as a cell ribosome (**FIGURE 14.2**). At the upper end of the spectrum is the smallpox virus, which approximates the size of the smallest bacterial cells, such as the chlamydiae and mycoplasmas (see Chapter 10). You will note a simplicity in viruses that has led many microbiologists to question whether they are living organisms or fragments of genetic material leading an independent existence.

Before we begin this chapter it is worth mentioning that viruses are, as Bernard Dixon said, "our most abundant co-terrestrials." Thus, the term **virosphere** has been coined to represent all the places where viruses are found or in which they interact with their hosts, be that in the *Bacteria*, *Archaea*, or *Eukarya*. Although we will emphasize their role in infections and disease, realize that viruses are more than a parasite and, in fact, are a key part of the living world. They are found in every environment on the planet and, as such, play a critical role in the recycling of elemental carbon; they contain more genetic material than the rest of life, consisting of more than 10,000 different viral genomes in 1 kg of ocean sediment and hundreds of thousands of genomes in the world's open oceans; and they probably represent a class of "biological entities" that may be older than the *Archaea* and *Bacteria*. A part of the human genome consists of viral genes (Chapter 9) and it is the ability of viruses to transfer genes by transduction and to spread those genes throughout the biological world that may have been an essential factor in the evolution of life on Earth. The virosphere is huge, it has incredible diversity, and it produces tremendous impact beyond infection and disease.

In this chapter, we study the properties of viruses, especially bacterial and animal viruses, focusing on their unique structure and mechanism for replication. We will see how they are classified, how they are cultured and identified, and how they may have evolved. The chapter then discusses even more bizarre virus-like agents that can cause disease in plants and animals.

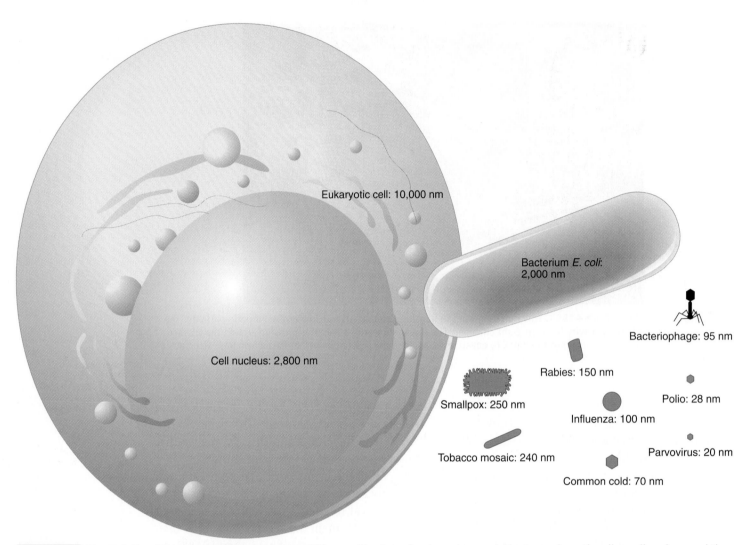

FIGURE 14.2 Size Relationships among Microorganisms and Viruses. The sizes of various viruses relative to a eukaryotic cell, a cell nucleus, and the bacterium *E. coli*. Viruses range from the very small poliovirus to the much larger smallpox virus. »» Propose a hypothesis to explain why viruses are so small.

14.1 Foundations of Virology

The development of the germ theory recognized disease patterns associated with a specific bacterial species (see Chapter 1). However, some diseases resisted identification and many of these would turn out to be viral diseases.

Many Scientists Contributed to the Early Understanding of Viruses

KEY CONCEPT

1. Viruses were first identified with diseases in plants.

In Chapter 1, we mentioned the work of the Russian pathologist Dimitri Ivanowsky who, in 1892, studied tobacco mosaic disease

(**FIGURE 14.3**). Ivanowsky filtered the crushed leaves of a diseased plant and found that the clear liquid passing through the filter (rather than the crushed leaves on the filter) contained the infectious agent. Unable to see any microorganisms, Ivanowsky suspected that a filterable virus (*virus* = "a poison") was the agent of disease. Six years later, Martinus Beijerinck repeated Ivanowsky's work and demonstrated the virus was inactivated by boiling. Beijerinck concluded the disease agent was a contagious, living fluid.

In 1898, foot-and-mouth disease was suspected as being a filterable virus, implying that a virus could be transmitted among animals as

Tobacco mosaic disease: A viral disease causing tobacco leaves to shrivel and assume a mosaic appearance.

Foot-and-mouth disease: A highly contagious viral disease of cloven-hoofed animals (i.e., cattle, sheep, deer).

(A)　　　　　　　　　　　　　　　　　　　　　　　　　　(B)

FIGURE 14.3 The Investigator, the Disease, and the Virus. In 1892, the Russian pathologist Dimitri Ivanowsky (**A**) used an ultramicroscopic filter to separate the clear juice from crushed leaves of tobacco plants suffering from tobacco mosaic disease. He placed the juice on healthy leaves and reproduced the disease as the leaves (**B**) became shriveled with a mosaic appearance. »» Why was it impossible for Ivanowsky and his contemporaries to see viruses with the light microscope?

well as plants. Three years later, Walter Reed and his group in Cuba provided evidence linking yellow fever with an unfilterable virus, and with this report, viruses were associated with human disease.

In 1915, English bacteriologist Frederick Twort discovered viruses that infected bacterial cells. Two years later, such viruses were identified by French-Canadian scientist Felix d'Herrelle. He called them **bacteriophages** (*phage* = "eat"), or simply **phages**, for their ability to destroy the bacterial cells they infected. When a drop of phages was placed in a broth culture of cells, the cells disintegrated within minutes.

By the early 1930s, it was generally assumed viruses were living microorganisms below the resolving power of available light microscopes. However, in 1935 the tobacco mosaic virus (TMV) was crystallized, suggesting viruses might be nonliving agents of disease (MicroFocus 14.1). Additional work with TMV revealed the virus was composed exclusively of nucleic acid and protein.

Because viruses will not grow on a nutrient agar plate the way bacterial cells do, some other form for virus cultivation was needed. In 1931, Alice M. Woodruff and Ernest W. Goodpasture described how fertilized chicken eggs could be

used to cultivate some viruses. The shell of the egg was a natural culture dish containing nutrient medium, and viruses multiplied within the chick embryo tissues.

By 1941, with the invention of the electron microscope, virologists were beginning to visualize viruses, including TMV (**FIGURE 14.4**).

Another key development occurred in the 1940s as a result of the national polio epidemic. Attempts at vaccine production were stymied by the inability to cultivate polioviruses outside

Yellow fever:
A mosquito-borne viral disease of the human liver and blood.

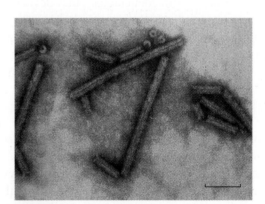

FIGURE 14.4 The Tobacco Mosaic Virus (TMV). This false-color transmission electron micrograph of TMV shows the rod-shaped structure of the virus particles. (Bar = 80 nm.) »» From this micrograph, why would virologists call viruses "crystallizable" particles?

MICROFOCUS 14.1: Being Skeptical
Are Viruses Living Organisms? Part 1

Viruses are on the edge of life. But on which side of the edge are they? Are they living or, being able to be crystallized, are they nonliving? If nonliving, how can they cause disease?

Most biology textbooks use certain emergent properties of life to define something as a living organism. These properties include an ability to:

- Grow and develop.
- Reproduce.
- Establish a complex organization.
- Regulate its internal environment.
- Transform energy.
- Respond to the environment.
- Evolve by adapting to a changing environment.

Viruses have a complex organization, respond to the environment, and evolve. As independent entities, they do not grow or develop, reproduce, regulate their internal environment, or transform energy.

So, are they alive? Hmm.

Many scientists consider them living because they can reproduce when they infect a host cell, and they do contain genetic information like all living organisms. Other scientists would say they are not living because they do not satisfy all seven emergent properties of life.

So, are they alive? Hmm.

Perhaps more important than debating whether they are living is to determine how they cause disease and how they relate to the phylogenetic tree of life. On these two points, they are in the mainstream of microbiological thought and investigation.

These agents of disease also are part of the genomic sequence of all life. Biologists and microbiologists estimate that viral gene sequences make up 8% of the human genome. Most of these genetic sequences are remnants of ancient viral infections, the sequences having been passed down from generation to generation. However, some make us human. One ancient viral protein, for example, is critical in placental formation.

Are they alive or not? Take your pick, but make sure you read MicroFocus 14.2 first. In any case, they certainly are intriguing agents for study.

---

the body, but John Enders, Thomas Weller, and Frederick Robbins of Children's Hospital in Boston solved that problem. Meticulously, they developed a test tube medium of nutrients, salts, and pH buffers in which living animal cells would remain alive. In these living cells, polioviruses replicated to huge numbers, and by the late 1950s, Jonas Salk and Albert Sabin had adapted the technique to produce massive quantities of virus for use in polio vaccines (Chapter 22).

CONCEPT AND REASONING CHECKS

14.1 Describe the major events leading to the recognition of viruses as pathogens.

14.2 What Are Viruses?

Today more than 5,000 viruses have been identified. Amazingly, this is only a small proportion of the estimated 400,000 different viruses virologists believe may exist—their total numbers making viruses the most abundant biological entities on Earth.

Viruses Are Tiny Infectious Agents

KEY CONCEPT

2. Viruses are submicroscopic and have either a DNA or RNA genome.

Viruses are small, obligate, intracellular particles; that is, most can be seen only with the electron microscope and they must infect and take over a host cell in order to replicate. This is because they lack the chemical machinery for generating energy and synthesizing large molecules. Viruses, therefore, must find an appropriate host cell in which they can replicate—and, as a result, often cause disease.

Viruses have some unique features not seen with the living microorganisms. They have no organelles, no cytoplasm, and no cell nucleus or nucleoid (see Chapter 4). Instead, they are comprised of two basic components: a nucleic acid core and a surrounding coat of protein; thus, as Peter Medawar remarked (chapter opening quote), a virus is "just a piece of bad news [meaning they cause disease] wrapped up in protein."

The viral **genome** of almost all viruses contains either DNA or RNA, but not both, and the nucleic acid occurs in either a double-stranded or a single-stranded form. Usually the nucleic acid is a linear or circular molecule, although in some instances (as in influenza viruses) it exists as separate, nonidentical segments. The viral genome is folded or coiled, which allows the viruses to maintain their extremely small size.

The protein coat of a virus particle, called a **capsid**, gives shape or symmetry to the virus (FIGURE 14.5). Generally, the capsid is subdivided into individual protein subunits called **capso-**

meres (the organization of capsomeres yields the viral symmetry) and the capsid with its enclosed genome is referred to as a **nucleocapsid**.

The capsid also provides a protective covering for the viral genome because the construction of its amino acids resists temperature, pH, and other environmental fluctuations. In some viruses, special capsid proteins called **spikes** help attach the virus to the host cell and facilitate penetration of the cell. Viruses composed solely of a nucleocapsid are sometimes referred to as "naked" viruses.

The nucleocapsids of many viruses are surrounded by a flexible membrane known as an **envelope**; the viruses are referred to as "enveloped" viruses (Figure 14.5B). The envelope is composed of lipids and protein, similar to the host cell membrane; in fact, it is acquired from the host cell during replication and is unique to each type of virus. These viruses may lose their infectivity if the envelope is destroyed. Also, when the envelope is present, the symmetry of the capsid may

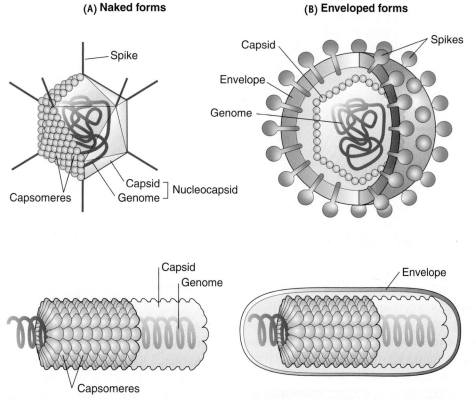

(A) Naked forms

Spike

Capsomeres

Capsid ⎤ Nucleocapsid
Genome ⎦

(B) Enveloped forms

Capsid

Envelope

Genome

Spikes

Capsid
Genome

Capsomeres

Envelope

FIGURE 14.5 The Components of Viruses. **(A)** Naked viruses consist of a nucleic acid genome (either DNA or RNA) and a protein capsid. Capsomere units are shown on one face of the capsids. Spikes may be present on the capsid. **(B)** Enveloped viruses have an envelope that surrounds the nucleocapsid. Again spikes usually are present. »» What important role do spikes play in the infective behavior of viruses?

MICROFOCUS 14.2: Environmental Microbiology
Are Viruses Living Organisms? Part 2

The news headline states, *"Virus Discovered That Is As Large As Some Bacteria!"* But how can that be? Dogma says that viruses are so small you cannot see them with the light microscope. So, could the news media have it wrong?

In 2002, researchers at the CNRS in France were looking for *Legionella* bacteria in water samples when they stumbled upon a virus in waterborne amoebae. The name mimivirus ('microbe-mimicking' virus) was given to the virus and a separate family, the Mimiviridae, established.

The mimivirus is a naked, double-stranded DNA virus about 400 nm in diameter. Its genome was sequenced in 2004 and found to contain 911 protein-coding genes, which is more than three times the size of other virus genomes. Like other viruses though, it cannot convert energy or replicate on its own. But that's where the similarities end.

The mimiviruses contain both DNA *and* RNA, something viruses by definition are not supposed to contain. They also have seven genes shared between all three domains of life—*Bacteria, Archaea,* and *Eukarya.* However, it is not yet known if the viruses use these genes.

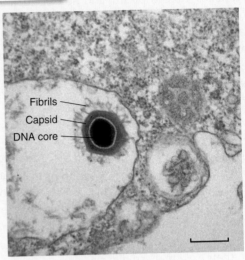

Transmission electron micrograph of the mimivirus. (Bar = 400 nm.)

If this isn't weird enough, in 2007 some of the same scientists isolated a previously unknown, 50 nm icosahedral dsDNA virus that is a parasite of mimivirus. Named Sputnik because it is like a satellite virus in that it only replicates in the presence of mimivirus replication, this new virus is an example of the first known "virophage"—a virus that causes an infection of another virus. In fact, when Sputnik replicates, it interferes with mimivirus replication, causing it to form abnormal virus particles. The Sputnik genome contains some genes from mimivirus and others from archaeal viruses, suggesting Sputnik can participate in horizontal gene transfer, yet another characteristic of living organisms.

So, are viruses living organisms? These discoveries with the mimivirus and Sputnik virus certainly blur the lines between viruses and single-celled organisms.

---

not be apparent because the envelope is generally a loose-fitting structure over the nucleocapsid. Many enveloped viruses also contain spikes projecting from the envelope. These proteins also function for attachment and host cell penetration.

A completely assembled and infectious virus outside its host cell is known as a **virion**. MicroFocus 14.2 revisits the question of whether viruses are living organisms.

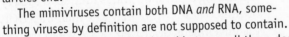

14.2 Identify the role of each structure found on (A) a naked and (B) an enveloped virus.

Viruses Are Grouped by Their Shape

KEY CONCEPT

3. Viruses can have helical, icosahedral, or complex symmetry.

Viruses can be separated into groups based on their nucleocapsid symmetry; that is, their three-dimensional shapes. Certain viruses, such as rabies and tobacco mosaic viruses, exist in the form of a helix and are said to have **helical** symmetry (FIGURE 14.6A). The helix is a tightly wound coil resembling a corkscrew or spring.

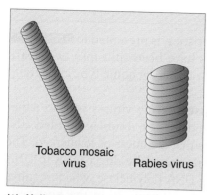
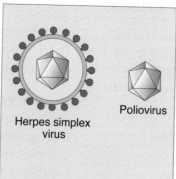
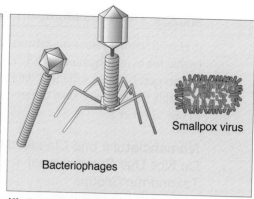

Tobacco mosaic virus   Rabies virus

Herpes simplex virus   Poliovirus

Bacteriophages   Smallpox virus

(A) Helical viruses   **(B) Icosahedral viruses**   **(C) Complex viruses**

FIGURE 14.6 **Various Viral Shapes.** Viruses exhibit numerous variations in symmetry. The nucleocapsid may have helical symmetry (**A**) in the tobacco mosaic and rabies viruses or (**B**) icosahedral symmetry typical of the herpesviruses and polioviruses. In other viruses (**C**) a complex symmetry exists. The smallpox virus has a series of rod-like filaments embedded within the membranous envelope at its surface. »» Although the bacteriophages are classified as complex, what two symmetries do these viruses exhibit?

Other viruses, such as herpes simplex and polioviruses, have the shape of an icosahedron and hence, **icosahedral** (*icos* = "twenty," *edros* = "side") symmetry (**FIGURE 14.6B**). The icosahedron then has 20 triangular faces and 12 corners.

A few viruses have a combination of helical and icosahedral symmetry, a construction described as **complex** symmetry (**FIGURE 14.6C**). Some bacteriophages, for example, have an icosahedral head with a collar and tail assembly in the shape of a helical sheath. Poxviruses, by contrast, are brick shaped, with submicroscopic filaments occurring in a swirling pattern at the periphery of the virus.

Before we leave viral shapes, it is worth mentioning that many archaeal viruses, all of which to date have a DNA genome, are morphologically unique, often having a spindle shape. Others have the typical icosahedral or helical shape, and most bind to host cells with their tail fibers.

CONCEPT AND REASONING CHECKS

14.3 What are the three viral shapes and what viral structure determines that shape?

Viruses Have a Host Range and Tissue Specificity

KEY CONCEPT

4. Viruses infect specific organisms and tissues within multicellular organisms.

As a group, viruses can infect almost any cellular organism. There are specific viruses able to infect bacterial cells, while others infect protozoa, fungi, plants, or animals. A virus' **host range** refers to what organisms (hosts) the virus can infect and it is based on a virus' capsid structure. Most viruses have a very narrow host range. A specific bacteriophage, for example, only infects specific bacterial species, the smallpox virus only infects humans, and the poliovirus infects only humans and primates. A few viruses may have a broader host range, as the rabies viruses infect humans and most warm-blooded animals.

Even within its host range, many viruses only infect certain cell types or tissues within a multicellular plant or animal. This limitation is called **tissue tropism** (tissue attraction). For example, the host range for the human immunodeficiency virus (HIV) is a human. In humans, HIV primarily infects a specific group of white blood cells called T helper cells because the envelope has protein spikes for binding to receptor molecules on these cells. The virus does not infect cells in other tissues or organs such as the heart or liver. Rabies virus is best at infecting cells of the nervous system and brain because its envelope contains proteins recognizing receptors only on these tissues. Therefore, a virus' host range and tissue tropism are linked to infectivity. If a potential host cell lacks the appropriate receptor or the virus lacks the complementary protein, the virus usually cannot bind to or infect that cell.

CONCEPT AND REASONING CHECKS

14.4 How does viral structure determine host range and tissue tropism?

14.3 The Classification of Viruses

A classification system for viruses that is similar to that for living organisms has been slow in coming, in part because sufficient data are not always available to determine how different viruses relate to one another.

Nomenclature and Classification Do Not Use Conventional Taxonomic Groups

KEY CONCEPT

5. Viruses can be organized by their nucleic acid type.

Viral nomenclature has used a variety of conventions. The measles virus and poxviruses, for example, are named after the disease they cause; the Ebola and Marburg viruses after the location from which they were originally isolated; and the Epstein-Barr virus after the researchers who studied it. Others are named after morphologic factors—the coronaviruses (*corona* = "crown") have a crown-like capsid and the picornaviruses (*pico* = "small"; *rna* = "ribonucleic acid") are very small viruses with an RNA genome.

A more encompassing classification system is being devised by the International Committee on Taxonomy of Viruses (ICTV). At this writing, higher order taxa (phyla and classes) have not been developed. In 2008, five orders were recognized that comprised 20 families, each ending with -viridae (e.g., Herpesviridae; Coronaviridae). Another 64 families have not yet been assigned to a family. Viruses have been categorized into hundreds of genera; each genus name ends with the suffix -virus (e.g., Herpesvirus; Coronavirus). Names (binomial nomenclature) have not yet been agreed upon for species. In this text, we mostly use the colloquial names (e.g., herpesviruses, coronaviruses).

Viruses from many different families cause disease in humans. A selection of viral families affecting humans, together with some of their characteristics, is presented in **TABLE 14.1**. These viruses have been split into two board classes based on their genome type and strand type.

DNA Viruses. Many viruses contain either single-stranded (ss) or double-stranded (ds) DNA genomes that are linear or segmented. The genomes are replicated by direct DNA-to-DNA copying using DNA polymerase, which requires most DNA viruses replicate in the host cell's nucleus. One exception is the poxviruses that replicate in the host cytoplasm, which means these viruses must carry the gene for their own DNA polymerase.

RNA Viruses. A large number of viruses contain either ssRNA or dsRNA genomes, which are replicated by direct RNA-to-RNA copying. Again, the genomes can be linear or segmented. Some of the single-stranded viruses, such as the picornaviruses and coronaviruses, have their RNA genome in the form of messenger RNA (mRNA). These RNA viruses are referred to as **positive-strand** (+ **strand**).

Other ssRNA viruses, such as the orthomyxoviruses and paramyxoviruses, have RNA genomes consisting of RNA strands that would be complementary to a mRNA; these genomes are referred to as **negative-strand** (− **strand**).

Although usually grouped with the RNA viruses, the retroviruses are replicated indirectly through a DNA intermediate (RNA-to-DNA-to-RNA). Each virion contains two copies of + strand RNA. During the infection process, a DNA intermediate will be formed using a reverse transcriptase enzyme carried within the virion.

As a rule, RNA virus genomes are smaller than DNA virus genomes and depend more heavily on host cell proteins and enzymes for replication.

CONCEPT AND REASONING CHECKS

14.5 How do DNA viruses differ from RNA viruses?

Reverse transcriptase:
An enzyme that copies single-stranded RNA into double-stranded DNA.

TABLE

14.1   Major Families of Human Viruses and Their Characteristics

Family	Strand Type*	Capsid Symmetry	Envelope or Naked Virion	Diameter (nm)	Disease Examples
DNA Viruses					
Poxviridae	Double	Complex	Envelope	170–300	Smallpox, monkeypox
Herpesviridae	Double	Icosahedral	Envelope	150–200	Cold sores, genital herpes, chickenpox, shingles, infectious mononucleosis
Adenoviridae	Double	Icosahedral	Naked	70–90	Common cold, viral meningitis
Papovaviridae	Double	Icosahedral	Naked	45–55	Warts, genital warts, cervical cancer
Hepadnaviridae	Double (w/RNA intermediate)	Icosahedral	Envelope	42	Hepatitis B, liver cancer
Parvoviridae	Single	Icosahedral	Naked	18–26	Fifth disease, gastroenteritis
RNA Viruses					
Reoviridae	Double	Icosahedral	Naked	60–80	Gastroenteritis
Picornaviridae	Single (+)	Icosahedral	Naked	28–30	Polio, some colds, hepatitis A
Caliciviridae	Single (+)	Icosahedral	Naked	35–40	Gastroenteritis
Togaviridae	Single (+)	Icosahedral	Envelope	60–70	Rubella, encephalitis
Flaviviridae	Single (+)	Icosahedral	Envelope	40–50	Yellow fever, dengue fever, hepatitis C, West Nile fever (encephalitis)
Coronaviridae	Single (+)	Helical	Envelope	80–160	SARS
Filoviridae	Single (−)	Helical	Envelope	80–10,000	Ebola and Marburg hemorrhagic fevers
Bunyaviridae	Single (−)	Helical	Envelope	90–120	Hantavirus pulmonary syndrome
Orthomyxoviridae	Single (−)	Helical	Envelope	90–120	Influenza
Paramyxoviridae	Single (−)	Helical	Envelope	150–300	Mumps, measles
Rhabdoviridae	Single (−)	Helical	Envelope	70–380	Rabies
Arenaviridae	Single (−)	Helical	Envelope	50–300	Lassa fever
Retroviridae	Single (+) (w/DNA intermediate)	Icosahedral	Envelope	80–130	AIDS, human adult T-cell leukemia

*(+) = positive-strand; (−) = negative-strand

14.4   Viral Replication and Its Control

The process of viral replication is one of the most remarkable events in nature. A virus invades a living host cell a thousand or more times its size, hijacks the metabolism of the cell to produce copies of itself, and often destroys the host cell when new virions are released.

Replication has been studied in a wide range of viruses and their host cells. We examine the bacteriophages first and then discuss the animal viruses.

The Replication of Bacteriophages Is a Five-Step Process

KEY CONCEPT

6. Bacteriophages undergo a lytic or lysogenic cycle of infection.

One of the best studied processes of replication is that carried out by bacteriophages of the T-even group (T for "type"). Bacteriophages T2,

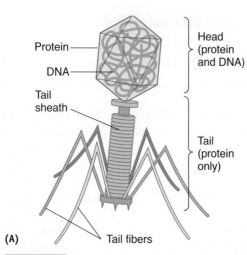

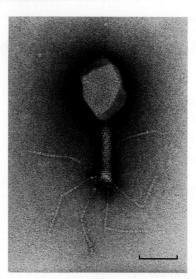

Protein

DNA

Tail sheath

Head (protein and DNA)

Tail (protein only)

(A)

Tail fibers

(B)

FIGURE 14.7 **Bacteriophage Structure. (A)** The structure of a bacteriophage consists of the head, inside of which is the nucleic acid, and the tail. The tail sheath is hollow to allow transfer of nucleic acid during infection. The tail fibers attach the phage to the host cell surface. **(B)** A false-color transmission electron micrograph of a T-even bacteriophage (Bar = 70 μm.) »» What is the purpose of the tail fibers?

T4, and T6 are in this group. They are large, complex, naked DNA virions with the characteristic head and tail of bacteriophages (**FIGURE 14.7**). They contain tail fibers, which function similar to spikes on animal viruses and identify what bacterial species the phage will be able to infect. The T-even phages are **virulent viruses**, meaning they lyse the host cell while carrying out a **lytic cycle** of infection.

It is important to note that the nucleic acid in a phage contains only a few of the many genes needed for viral synthesis and replication. It contains, for example, genes for synthesizing viral structural components, such as capsid proteins, and for a few enzymes used in the synthesis; but it lacks the genes for many other key enzymes, such as those used during nucleic acid synthesis. Therefore, its dependence on the host cell is substantial.

We shall use phage replication in *E. coli* as a model for the lytic cycle. An overview of the five-step process is presented in **FIGURE 14.8** .

1. Attachment. The first step in the replication cycle of a virulent phage occurs when phage and bacterial cells collide randomly. If sites on the phage's tail fibers match with a complementary receptor site on the cell wall of the bacterium, attachment will occur. The actual attachment consists of a weak chemical union between phage and receptor site. In some cases, the bacterial flagellum or pilus contains the receptor site.

2. Penetration. Following attachment, the tail of the phage releases lysozyme, an enzyme that dissolves a portion of the bacterial cell wall. The tail sheath then contracts and the tail core drives through the cell wall. As the tip of the core reaches the cell membrane below, the DNA is ejected through the hollow tail core and on through the cell membrane into the bacterial cytoplasm. The ejection process takes less than two seconds and the capsid remains outside.

3. Biosynthesis. Having entered the cytoplasm, production of new phage genomes and capsid parts begins. As phage genes code for the disruption of the host chromosome, the phage DNA uses bacterial nucleotides and enzymes to synthesize multiple copies of its genome. Messenger RNA molecules transcribed from phage DNA appear in the cytoplasm, and the biosynthesis of phage enzymes and capsid proteins begins. Bacterial ribosomes, amino acids, and enzymes are all enlisted for biosynthesis. Because viral capsids are repeating units of capsomeres, a relatively simple genetic code can be used over and over.

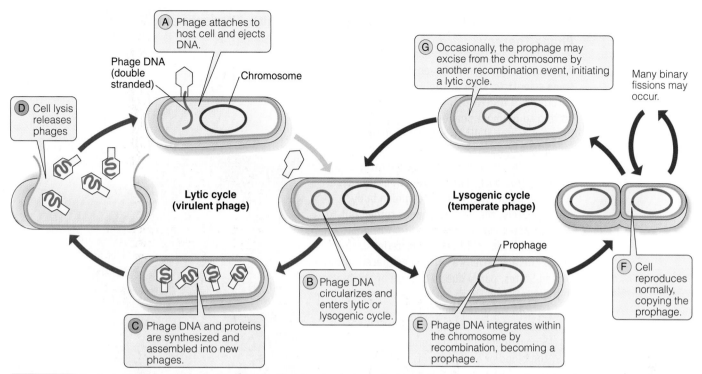

FIGURE 14.8 **Bacteriophage Replication.** The pattern of replication in bacteriophages (tail fibers not shown) can involve a lytic or lysogenic cycle. »» How can bacteriophages facilitate the horizontal transfer of genetic material?

4. Maturation. Once the phage parts are made, they are assembled into complete virus particles. The enzymes encoded by viral genes guide the assembly in step-by-step fashion. In one area of the host cytoplasm, phage heads and tails are assembled from protein subunits; in another area, the heads are packaged with DNA; and in a third area, the tails are attached to the heads.

5. Release. Mature phage particles now burst out from the ruptured bacterial shell. For some phages, lysozyme, encoded by the bacteriophage genes late in the replicative cycle, degrades the bacterial cell wall. The mature bacteriophages are set free to infect more bacterial cells.

MicroFocus 14.3 describes the use of phages as a way to combat bacterial diseases.

Other phages interact with bacterial cells in a slightly different way, called a **lysogenic cycle** (Figure 14.8). For example, lambda (λ) phage also infects E. coli but may not immediately cause cell lysis. Instead, the phage DNA integrates

into the bacterial chromosome as a **prophage** (Figure 14.8E). Bacteriophages participating in this cycle are known as **temperate phages**. The bacterial cell survives the infection and continues to grow and divide normally. As the bacterial cell undergoes DNA replication and binary fission, the prophage is copied and vertically transferred to daughter cells as part of the replicated bacterial chromosome (Figure 14.8F). Thus, as cells divide, each daughter cell is "infected"; that is, it contains the viral genome as a prophage.

Such binary fissions can continue for an undefined period of time. Usually at some point, the bacterial cells become stressed (e.g., lack of nutrients, presence of noxious chemicals). This triggers the prophage to excise itself from the bacterial chromosomes (Figure 14.8G) and switch to a lytic cycle, lysing the bacterial cells as new λ phage are released.

CONCEPT AND REASONING CHECKS

14.6 Why would it be advantageous for a phage to carry out a lysogenic cycle rather than a lytic cycle?

MICROFOCUS 14.3: History/Public Health/Biotechnology
Phage Therapy

When bacteriophages were identified in 1915, some scientists, such as their discoverer Felix d'Herrelle of the Pasteur Institute in Paris, started to promote phages as therapy for curing dreaded bacterial diseases such as cholera and bubonic plague.

D'Herrelle and others believed that if phages could destroy bacterial cells (see figure) in test tubes, why not try them in the human body? Unfortunately, the bacteriophages turned out to be highly specific viruses with a narrow host range and would attack only certain bacterial strains. Phage therapy never was deeply embraced in the United States, partly due to the rise of antibiotics in the 1940s and 1950s.

Fast-forward to the modern era.

- 1980s: The Polish microbiologist Stefan Slopek identified a bacterial pathogen in the blood of an ill patient, then searched out and isolated a bacteriophage specific for that pathogen. He injected a solution of the phages into the patient. Over a period of days, the infection gradually resolved. Slopek tried again—this time with 550 patients. Each patient benefited from the treatment, and several patients seemed completely cured.

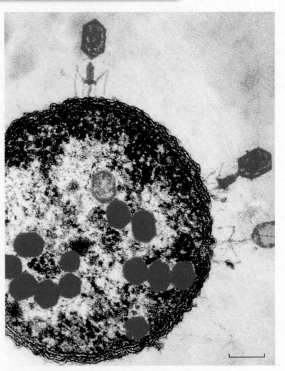

Phage infecting an *E. coli* cell. (Bar = 100 nm)

- 2004: Researchers in Vienna, Austria engineered bacteriophage that would kill bacteria but not lyse the cells. More surprising was the discovery that these phages were ten times more lethal than the natural phage.
- 2006: The U.S. Food and Drug Administration (FDA) approved a "cocktail" of six purified phages that can be safely sprayed on luncheon meats (cold cuts, hot dogs, sausages) to kill any of 170 strains of *Listeria monocytogenes,* which cause a serious foodborne disease in newborns, pregnant women, and immunocompromised individuals.
- 2007: James Collins and his team at Boston University engineered a phage to attack and destroy bacterial biofilms that are often the cause of human infections and disease.
- 2007: A group of Italian researchers identified a phage that cured 97% of mice infected with a deadly form of methicillin-resistant *Staphylococcus aureus* (MRSA).

So, the phage therapy envisioned by d'Herrelle is on the rebound and holds great promise for the treatment of acute and chronic infections, as well as for use as a preservative in the food industry.

Animal Virus Replication Often Results in a Productive Infection

KEY CONCEPT

7. Both naked and enveloped animal viruses share a similar series of infection and replication events.

Like bacteriophages, animal viruses also lead often brief but eventful "lives" as they produce more viruses as a result of infection. Such a **produc-** **tive infection** retains the five replication stages described for the bacteriophages. An overview of the dsDNA and ssRNA viruses is presented here.

1. **Attachment.** Animal viruses infect host cells by binding to receptors on the host cell's plasma membrane. This binding is facilitated by the spikes distributed over the surface of the capsid (e.g., adenovirus) or envelope (e.g., HIV). The spikes are what determine the host range for a virus.

2. Penetration. Some viruses, such as HIV and the adenoviruses, require a second receptor, called a **co-receptor**, for viral penetration into the cytoplasm. Viral entry also differs from that in phages, where only the DNA entered the bacterial cell cytoplasm. Animal viruses often are taken into the cytoplasm as intact nucleocapsids. For viruses like HIV, the viral envelope fuses with the plasma membrane and releases the nucleocapsid into the cytoplasm (**FIGURE 14.9A**).

For other animal viruses, such as the adenoviruses and influenzavirus, the virion is taken into the cell by endocytosis. At the attachment site, the cell enfolds the virion within a vacuole and brings it into the cytoplasm (**FIGURE 14.9B**). Once in the cell, the vacuole membrane breaks down, releasing the nucleocapsid or the genome into the cytoplasm. In both examples, the capsid disassembles from the genome in a process called

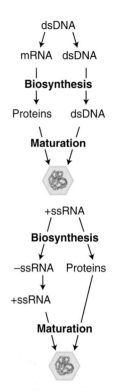

uncoating and the genome is transported to the site where transcription or replication will occur.

3. and 4. Biosynthesis and Maturation. One way we split the virus families was based on whether they have DNA or RNA as their genetic information. The DNA of a DNA virus supplies the genetic codes for enzymes that synthesize viral parts from available building blocks. Although the poxviruses replicate entirely in the host cell cytoplasm, most of the DNA viruses employ a division of labor: DNA genomes are synthesized in the host cell nucleus, and capsid proteins are produced in the cytoplasm (**FIGURE 14.10A**). The proteins are then transported to the nucleus and join with the nucleic acid molecules for maturation. Adenoviruses and herpesviruses follow this pattern.

RNA viruses follow a slightly different pattern. Because the +ssRNA viruses act as a messenger RNA, following uncoating, the RNA immediately

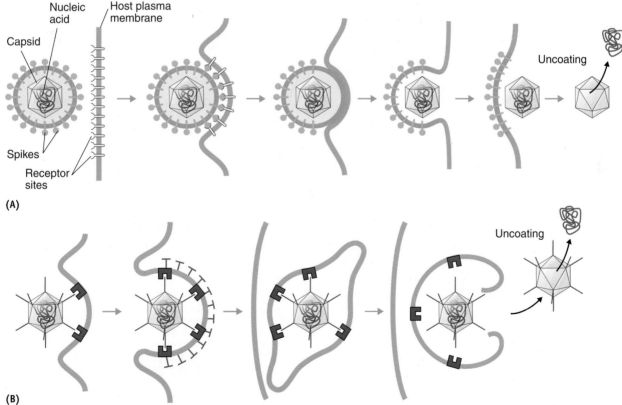

Nucleic acid **Host plasma membrane** **Capsid** **Spikes** **Receptor sites** Uncoating

(A)

Uncoating

(B)

FIGURE 14.9 **The Entry of Animal Viruses into Their Host Cells.** Animal viruses enter their host cells by one of two major routes. (**A**) An enveloped virus, such as HIV, contacts the plasma membrane and the spikes interact with receptor sites on the membrane surface. This action is highly specific for the viral spikes and receptor sites. The envelope fuses with the plasma membrane, and the nucleocapsid passes into the host cell's cytoplasm. Then, the nucleic acid is released (uncoated). (**B**) A second entry method also involves a specific interaction between spikes and receptor sites on the plasma membrane. However, for this naked adenovirus, it undergoes endocytosis into a vacuole forming around the virus. The vacuole pinches off into the cytoplasm. Loss of the vacuole membrane liberates the nucleocapsid into the cytoplasm and uncoating occurs. »» Propose a hypothesis to explain why two different entry mechanisms exist for virion entry into a host cell.

FIGURE 14.10 **Replication of a DNA and RNA Virus.**
The replicative cycle for a DNA virus, such as a herpesvirus (**A**), involves genome replication in the cell nucleus. The cycle for a +ssRNA virus, such as the poliovirus (**B**), occurs only in the cytoplasm. »» Why must the viral DNA of a herpesvirus (**A**) enter the cell nucleus to replicate?

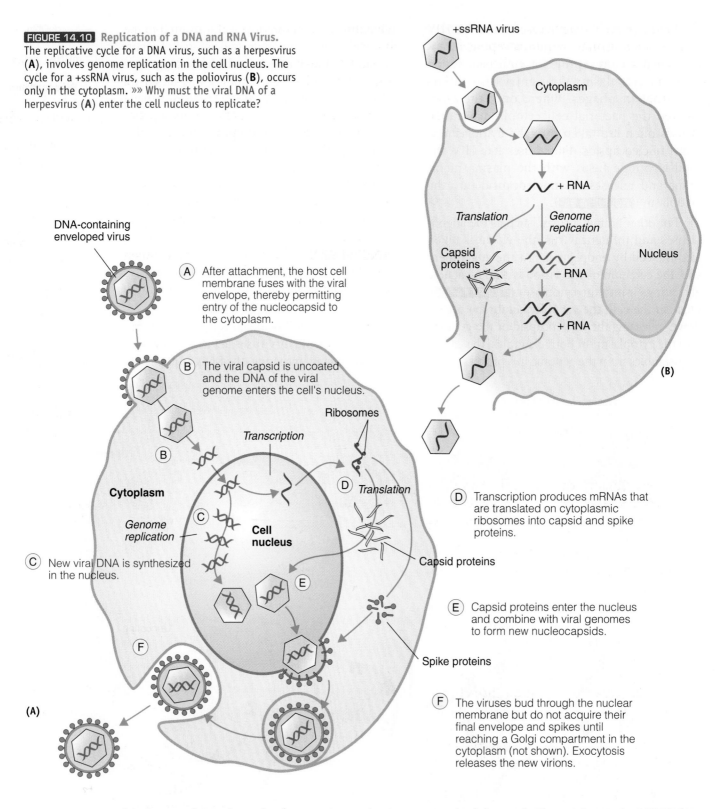

(A) After attachment, the host cell membrane fuses with the viral envelope, thereby permitting entry of the nucleocapsid to the cytoplasm.

(B) The viral capsid is uncoated and the DNA of the viral genome enters the cell's nucleus.

(C) New viral DNA is synthesized in the nucleus.

(D) Transcription produces mRNAs that are translated on cytoplasmic ribosomes into capsid and spike proteins.

(E) Capsid proteins enter the nucleus and combine with viral genomes to form new nucleocapsids.

(F) The viruses bud through the nuclear membrane but do not acquire their final envelope and spikes until reaching a Golgi compartment in the cytoplasm (not shown). Exocytosis releases the new virions.

begins supplying the codes for protein synthesis as genome replication occurs. (**FIGURE 14.10B**). Other −ssRNA viruses, such as the influenzavirus, use their RNA as a template to synthesize a complementary (+) strand of RNA. An RNA-dependent RNA polymerase is present in the virus to synthe-size the (+) strand. The synthesized +ssRNA then is used as a messenger RNA molecule for protein synthesis as well as the template to form the −ssRNA genome.

The final steps of maturation may include the acquisition of an envelope. In this step, envelope proteins (spikes) are synthesized and, depending on the virus, incorporated into a nuclear or cytoplasmic membrane, or the plasma membrane.

5. Release. In the final stage, enveloped viruses either: (1) push through the plasma membrane, forcing a portion of the membrane ahead of and around the virion, resulting in an envelope; or (2), as with the herpesvirus, a membrane-enclosed virus fuses with the plasma membrane, releasing the virion. This process, called **budding**, need not necessarily kill the cell during release. The same cannot be said for naked viruses. They leave the cell when the cell membrane ruptures, a process that generally results in cell death.

CONCEPT AND REASONING CHECKS

14.7 Distinguish between the replication events in DNA and RNA viruses.

Some Animal Viruses Produce a Latent Infection

KEY CONCEPT

8. Some animal viruses maintain their viral genome in the host cell nucleus.

Unlike most RNA viruses that go through a productive infection, many of the DNA viruses and the retroviruses can establish a **latent infection**, characterized by repression of most viral genes. Thus the virus lies "dormant." For example, some herpesviruses, such as herpes simplex virus-1 (HSV-1), can generate a productive or latent infection. In an infected sensory neuron, HSV-1 undergoes latency as the viral dsDNA enters the neuron's cell nucleus and circularizes. No viral particles are produced for months or years until some stress event reactivates the viral dsDNA and a new productive infection will be initiated (Chapter 15).

Retroviruses, such as HIV, also carry out a latent infection. However, in this group of viruses, the virus carries an enzyme, called **reverse transcriptase**, which is used to reverse transcribe its +ssRNA into dsDNA (**FIGURE 14.11**). The dsDNA then enters the host cell nucleus and, like the temperate phage DNA inserted into a bacterial chromosome, becomes integrated into the DNA of one chromosome. This integrated viral genome is referred to as a **provirus** and represents a unique and stable association between the viral DNA and the host genome.

The advantage for the virus is that every time the host cell divides, the provirus will be replicated along with the host genome and be present in all progeny cells. In addition, as a provirus, it is protected from attack by antiretroviral drugs (Chapter 24). However, at any time, the provirus can be reactivated and a productive infection involving biosynthesis and maturation of new virions will ensue.

MicroFocus 14.4 describes a hypothesis for the origin of cellular DNA arising from viruses. **FIGURE 14.12** summarizes the outcomes for animal virus infections.

CONCEPT AND REASONING CHECKS

14.8 Why don't RNA viruses, like the poliovirus or influenzavirus, undergo provirus formation?

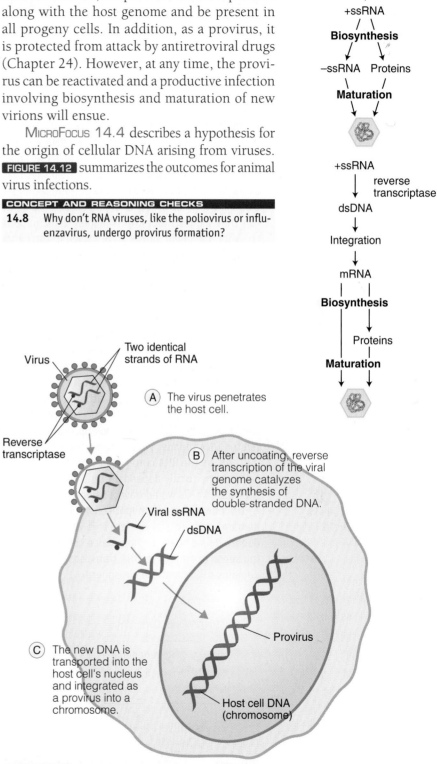

FIGURE 14.11 **The Formation of a Provirus by HIV.** Retroviruses can incorporate their genome into a host cell chromosome. Shown here, is the human immunodeficiency virus (HIV), a retrovirus causing HIV disease and AIDS. The single-stranded RNA genome can be reverse transcribed into double-stranded DNA by a reverse transcriptase enzyme. The DNA then enters the cell nucleus and integrates into a chromosome. »» Propose a reason why genome integration is an advantage for the virus.

MICROFOCUS 14.4: Evolution

A Hypothesis: DNA Arose from Viruses

As we have seen on many occasions in this text, there are three domains of life—the *Bacteria*, *Archaea*, and *Eukarya*. Their basic differences, including the gene sequences for 16S rRNA, have provided important information for assigning organisms to a domain. However, one characteristic they all have in common is the presence of a DNA genome. Because many researchers believe life started in an "RNA world" (see MicroFocus 2.6), where did DNA come from? Some evolutionary biologists hypothesize that we should look to viruses.

The hypotheses. Many evolutionary biologists believe that DNA-based organisms are descended from simpler RNA-based organisms. In fact, the **RNA world** hypothesis proposes that life was once RNA based. Realize, RNA can store information like DNA and act as an enzyme (see MicroFocus 6.2), catalyzing some chemical reactions. The **DNA world** hypothesis then says that at some point, DNA, with its greater chemical and mutational stability, took over the role of information storage while proteins, which are more flexible in catalysis through the great variety of amino acids, became the specialized catalytic molecules. The RNA world hypothesis thus suggests that RNA in modern cells, in particular rRNA, is the evolutionary link to the RNA world. If so, what role might viruses have played in the generation of DNA as the master information storage molecule?

Again evolutionary biologists propose that the "RNA-world organisms" were paralleled by the evolution of RNA viruses that parasitized the RNA-based organisms; that is, some of the genes in RNA-based organisms broke free and evolved into RNA viruses (see figure). Now, when these viruses infected an RNA-based cell, the RNA would be attacked and perhaps destroyed by defensive host cell enzymes. However, if a virus contained DNA, it could avoid host defenses because the host cell would not have yet evolved any defense against this novel molecule DNA. In this way, DNA viruses would have a selective advantage.

So, a hypothesis for the DNA world says that different DNA viruses became integrated into a host cell, where they could coexist (like a provirus or plasmid today) as a "founder virus" and, if they lost the ability to "escape" the host cell but replicate much like plasmids do today, they could become the depository for new genes as new RNA viruses attempted to infect the host cell (see figure). Since in many ways members of the *Archaea* and *Eukarya* are more similar to each other than to the *Bacteria*, perhaps two somewhat similar founder DNA viruses took up residence in the ancestors of these domains, while another, less related DNA founder virus became the resident in an ancestor of the *Bacteria* (see figure).

Then, as the RNA chromosome of the host cell diminished, the "DNA chromosome" enlarged until the cell became totally dependent on DNA for survival. Then, being more stable and less prone to mutation, DNA-based cells would have a selective advantage over their RNA predecessors.

Of course all this is hypothesis, but is based on logical arguments from what we know of the three domains and life and viruses. It is very controversial and subject to great interpretation, discussion, and debate by members of the science community. However, since we cannot go back to the RNA world, making sense of an RNA to DNA transition requires careful thinking based on what we know today about the domains of life and the virosphere.

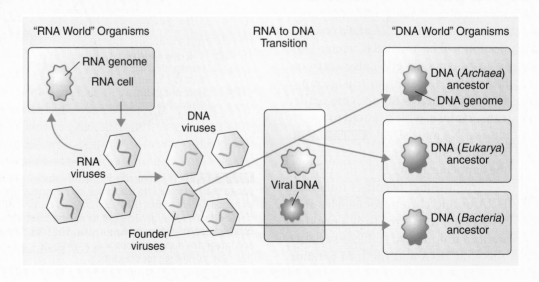

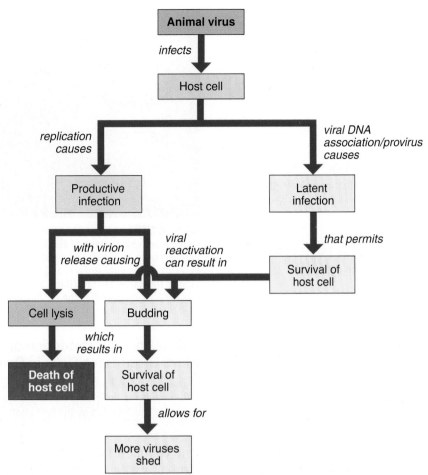

FIGURE 14.12 **Potential Outcomes for Animal Virus Infection of a Host Cell.** As depicted in this concept map, depending on the virus and host cell interaction, the host cell may be killed or survive, producing more viruses and/or being modified to form a tumor. »» Identify a specific virus that follows each of the three outcomes (productive infection; cell death or survival; latent infection).

14.5 The Cultivation and Detection of Viruses

If an individual contracts a viral disease, there are various ways the viral agent can be cultivated and detected for identification and eventual diagnosis. A prompt identification often is necessary for selecting possible antiviral therapy.

Detection of Viruses Often Is Critical to Disease Identification

KEY CONCEPT

9. Cytological analysis can provide a rapid initial diagnosis to identify an unknown viral infection.

The diagnosis of viral diseases like the flu and a cold are usually straightforward and do not require further laboratory confirmation. In some cases, viral infections leave their mark on the infected individuals. Measles is accompanied by **Koplik spots**, which are a series of bright red patches with white pimple-like centers on the lateral mouth surfaces. Swollen salivary glands and teardrop-like skin lesions are associated with mumps and chickenpox, respectively.

However, virus detection is not straightforward and the proper identification has a bearing on relating a particular virus to a particular disease. In the classical sense, Koch's postulates cannot be applied to a viral disease because, unlike bacterial cells, viruses cannot be cultivated in pure culture. To circumvent this problem, Thomas M. Rivers in 1937 expanded Koch's postulates to include viruses. He proposed filtrates of the infectious material isolated from the diseased host shown not to contain bacterial or other cultivatable organisms must produce the same disease as found in the original host; or, the filtrates must produce specific antibodies in appropriate animals. This concept has come to be known as **Rivers' postulates**.

TABLE

14.2 Examples of Virus Cytopathic Effects

Virus Family	Cytopathic Effect
	Changes in cell structure
Picornaviridae	Shrinking of cell nucleus
Papovaviridae	Cytoplasmic vacuoles
Paramyxoviridae, Coronaviridae	Cell fusion syncytia
Herpesviridae	Chromosome breakage
Herpesviridae, Adenoviridae, Picornaviridae, Rhabdoviridae	Rounding and detachment of cells from culture
	Cell inclusions
Adenoviridae	Virions in nucleus
Rhabdoviridae	Virions in cytoplasm (Negri bodies)
Poxviridae	"Viral factories" in cytoplasm
Herpesviridae	Nuclear granules

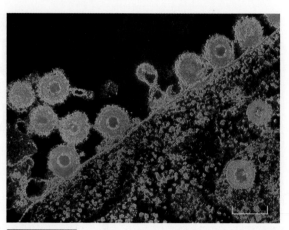

FIGURE 14.13 Cells and Viruses. A false-color transmission electron micrograph of tissue cells and associated herpesviruses (orange). The envelope of the virus is seen as a ring at the viral surface, and the nucleocapsid of the virus is the darker center. (Bar = 200 nm.) »» How might these herpesviruses enter the host cell?

The clinical laboratory diagnosis of a viral disease can be carried out using light microscopy to examine cells obtained from body tissues or fluids. When viruses replicate in host cells, often a noticeable deterioration or structural change occurs. This is called a **cytopathic effect** (CPE). Characteristic CPEs are listed in TABLE 14.2.

Viruses often cause a change in cell structure. For example, infectious mononucleosis is characterized by large numbers of lymphocytes with a "foamy-looking," highly vacuolated cytoplasm. Paramyxoviruses cause host cells to fuse together into multinucleate giant cells called **syncytia**.

Viruses sometimes produce cell inclusions. The brain tissue of a rabid animal can contain cytoplasmic nucleoprotein inclusions called **Negri bodies**, and cells from patients with herpes infections contain nuclear granules.

Not all viral infections cause CPEs, so other methods can be used to detect viral infections. One way is to search for viral antibodies in a patient's serum. This hemagglutination-inhibition test is described in Chapter 22.

Although it is not a common clinical laboratory technique, viruses can be observed directly by electron microscopy when virologists need to identify unknown viruses by comparison to known viruses (FIGURE 14.13).

CONCEPT AND REASONING CHECKS

14.9 How can some viruses be identified simply by observing a tissue sample from the infected patient?

Cultivation and Detection of Viruses Most Often Uses Cells in Culture

KEY CONCEPT

10. Viruses can be "grown" in various types of tissue culture and detected by the formation of plaques.

One common method once used to cultivate viruses was to inoculate them into fertilized (embryonated) chicken eggs. A hole is drilled in the egg shell and a suspension of viral material is introduced (FIGURE 14.14A). Today, only the influenza viruses are routinely cultivated by this method, usually to produce high concentrations of viruses for vaccine production.

The most common method of cultivating and detecting viruses is to infect cell cultures. To prepare the culture, animal cells are separated from a tissue with enzymes and suspended in a solution of nutrients, growth factors, pH buffers, and salts. In such a **primary cell culture**, the cells adhere to the bottom of a plastic dish or well, and reproduce to form a single layer, called a **monolayer**. The different cell types in a primary cell culture can be separated enzymatically and isolated as a single cell type, called a **cell line**. The type of cell culture used will depend on the virus species to be cultivated in the monolayer. Viruses are then introduced into the culture (FIGURE 14.14B).

Viruses can be detected by the formation of plaques in a monolayer. A **plaque** is a clear zone within the cloudy "lawn" of bacterial cells or monolayer of animal cells (FIGURE 14.14C).

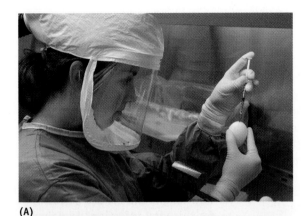

(A)

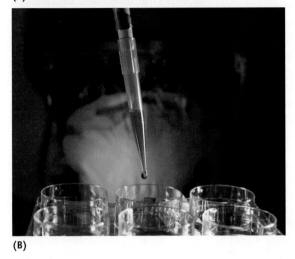

(B)

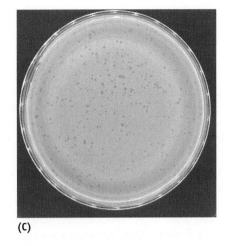

(C)

FIGURE 14.14 Infection of Cells in Embryonated Eggs and Cell Culture. (**A**) Inoculation of fertilized eggs. Techniques such as these are standard practice for growing the flu virus. (**B**) A masked researcher uses a pipette to infect a culture of human cells with a virus. The dishes contain a culture medium, which allows the cells to survive outside the body. The effect of the virus on the infected cells will then be studied. (**C**) Plaque formation in a cell culture. Plaques are evident as the clear areas in the dish. »» Why must a face guard or mask be worn when infecting embryonated eggs and cells in culture?

The viruses infect and replicate in the cells, thereby destroying them and forming plaques. Epidemiologic surveys of several bacterial diseases such as staphylococcal food poisoning are aided by plaque formation called **phage typing** (see Chapter 11). MICROINQUIRY 14 uses the plaque assay to monitor intracellular virus production and extracellular virus release during a productive infection of animal cells in culture.

CONCEPT AND REASONING CHECKS

14.10 Explain how plaques would be formed on a lawn of bacterial cells.

14.6 Tumors and Viruses

Cancer is indiscriminate. It affects humans and animals, young and old, male and female, rich and poor. In the United States, over 557,000 people die of cancer annually, making the disease the leading cause of death after heart disease and stroke. Worldwide, over 7 million people die of cancer each year.

Cancer Is an Uncontrolled Growth and Spread of Cells

KEY CONCEPT

11. Tumors are the result of uncontrolled cell divisions.

Cancer results, in part, from the uncontrolled reproduction (mitosis) of a single cell. The cell escapes the cell cycle's controlling factors and as it continues to multiply, a cluster of cells soon forms. Eventually, the cluster yields a clone of abnormal cells referred to as a **tumor**. Normally, the body will respond to a tumor by surrounding it with a capsule of connective tissue. Such a local tumor is designated **benign** because it usually is not life threatening.

Additional changes can occur to tumor cells that release them from their specific confines. Often they stick together less firmly than normal cells and fail to stop dividing when cells come in contact with one another. They may break out of the capsule and **metastasize**, a spreading of the

MICROINQUIRY 14

The One-Step Growth Cycle

In the research laboratory, we can follow the replication of viruses (productive infection) by generating a one-step growth curve. Realize that we are not really looking at growth, but the replication and increase in number of virus particles. There are several periods associated with a virus growth cycle that you should remember because you will need to identify the periods in the growth curve. The **eclipse period** is the time when no virions can be detected inside cells (intracellular) and the **latent phase** is the time during which no extracellular virions can be detected. Also, the **burst size** is the number of virions released per infected cell.

To generate our growth curve, we start by inoculating our viruses onto a suscep-

tible cell culture. There are 100,000 (10^5) cells in each of 10 cultures. We then add ten times the number of viruses (10^6) in a small volume of liquid to each culture to make sure all the cells will be infected rapidly. After 60 minutes of incubation, we wash each culture to remove any viruses that did not attach to the cells and add fresh cell growth medium. Then, at 0 hour and every four hours after infection, we remove one culture, pour off the growth medium, and lyse the cells. The number of viruses in the growth medium and in the lysed cells are determined. Because we cannot see viruses, we measure any viruses in the growth medium or in the lysed cells using a plaque assay. We assume that one plaque (plaque-forming

unit or PFU) resulted from one infected cell.

The curves drawn below plot the results from our growth experiment.

Answer the following questions based on the one-step growth curve. Answers can be found in **Appendix D**.

14.1a. How long is the eclipse period for this viral infection? Explain what is happening during this period.

14.1b. How long is the latent period for this infection? Explain what is happening during this period.

14.1c. What is the burst size?

14.1d. Explain why the growth curve shows a decline in phages between 0 and 4 hours.

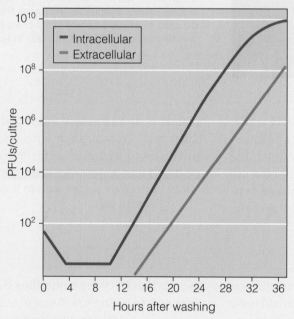

A One-Step Growth Cycle

cells to other tissues of the body. Such a tumor now is described as **malignant** and the individual now has **cancer** (*cancer* = "crab"; a reference to the radiating spread of cells, which resembles a crab). So, a series of changes must occur to a healthy cell before it can become a tumor or cancer cell.

How can such a mass of cells bring illness and misery to the body? By their sheer numbers, cancer cells invade and erode local tissues, interrupt normal functions, and choke organs to death by robbing them of vital nutrients. Thus, the cancer patient will commonly experience weight loss even while maintaining a normal diet.

CONCEPT AND REASONING CHECKS

14.11 How does a benign tumor differ from a malignant tumor?

Viruses Are Associated with About 20% of Human Tumors

KEY CONCEPT

12. A few animal viruses are agents of tumor development.

The World Health Organization (WHO) estimates that 60% to 90% of all human cancers are associated with **carcinogens**, chemicals and physical agents that produce cellular changes leading to cancer (**FIGURE 14.15**). Among the known chemical carcinogens are the hydrocarbons found in cigarette smoke as well as asbestos, nickel, certain pesticides and dyes, and environmental pollutants in high amounts. Physical agents include ultraviolet (UV) light and X rays (see Chapter 7).

A few viruses also act as carcinogens. In 1911, Peyton Rous demonstrated that a virus caused a sarcoma in chickens. Further experiments with animals also indicated other viruses can induce tumor formation.

A number of viruses now have been isolated from human cancers (**TABLE 14.3**). When these viruses are transferred to animals and cell cultures, an observable transformation of normal cells to tumor cells takes place. Examples of such **oncogenic** (tumor-causing) **viruses** are the herpesviruses associated with tumors of the human

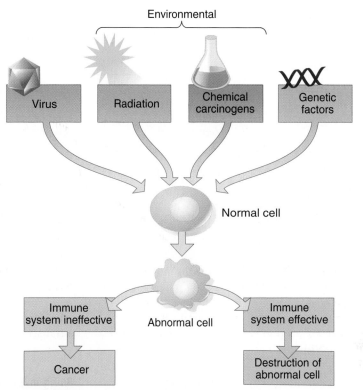

FIGURE 14.15 **The Onset of Cancer.** Viruses and other factors induce a normal cell to become abnormal. When the immune system is effective, it destroys the abnormal cell, and no cancer develops. However, when the abnormal cell evades the immune system, it develops first into a tumor and then may become malignant, spreading to other tissues in the body. »» What factors are required for a tumor to become malignant?

cervix and the Epstein-Barr virus, which is linked to **Burkitt lymphoma**, a tumor of the jaw. Of special note is **cervical cancer**, the second most common cancer in women under age 35. Some human papilloma virus (HPV) subtypes (family Papovaviridae) have a strong correlation with this cancer. In one study, 71% of the women developing cervical cancer had HPV present in their PAP smear. The good news is there is now a vaccine, called Gardasil, that provides almost 100% protection against the two most common HPV strains causing 70% of cervical cancers and two other strains responsible for 90% of genital warts (Chapter 24).

CONCEPT AND REASONING CHECKS

14.12 Name several viruses involved in the development of human cancers.

Sarcoma:
A malignant tumor that begins growing in connective tissue such as muscle, bone, fat, or cartilage.

PAP smear:
A test to detect cancerous or precancerous cells of the cervix, allowing for early diagnosis of cancer.

TABLE

14.3   Human Oncoviruses and Their Effects on Cell Growth

Oncogenic Virus	Benign Disease	Effect	Associated Cancer
DNA Tumor Viruses			
Papovaviridae			
• Human papilloma virus (HPV)	Benign warts Genital warts	Encodes genes that inactivate cell growth regulatory proteins	Cervical, penile, and oropharyngeal cancer
Polyomaviridae			
• Merkel cell polyoma virus (MCPV)	Unknown	Being investigated	Merkel cell carcinoma
Herpesviridae			
• Epstein-Barr virus (EBV)	Infectious mononucleosis	Stimulates cell growth and activates a host cell gene that prevents cell death	Burkitt lymphoma Hodgkin lymphoma
• Herpesvirus 8 (HV8)	Growths in lymph nodes	Forms lesions in connective tissue	Kaposi's sarcoma
Hepadnaviridae			
• Hepatitis B virus (HBV)	Hepatitis B	Stimulates overproduction of a transcriptional regulator	Liver cancer
RNA Tumor Viruses			
Flaviviridae			
• Hepatitis C virus (HCV)	Hepatitis C	Being investigated	Liver cancer
Retroviridae			
• Human T-cell leukemia virus (HTLV-1)	Weakness of the legs	Encodes a protein that activates growth-stimulating gene expression	Adult T-cell leukemia/lymphoma
• Xenotropic murine leukemia virus-related virus (XMRV)	Chronic fatigue syndrome (?) Unknown	Effect unclear	Unknown Prostate cancer (?)

Oncogenic Viruses Transform Infected Cells

KEY CONCEPT

13. Oncogenes represent tumor-causing genes.

The road to cancer usually involves many cell changes, and oncogenic viruses may play a role in some of these steps that disable the controls of normal cell growth. For example, some viral effects are indirect. In the case of hepatitis B and hepatitis C, it is the long-term liver inflammation caused by the viruses that can eventually trigger tumor formation. In other cases, oncogenic viruses play a more insidious role. Some of these viruses carry genes whose protein products throw a monkey wrench into cellular growth control. When HPV infects a cell and abnormal growth begins, HPV genes produce protein products that prevent the infected cells from committing suicide. Growth continues and a tumor may develop.

Other oncogenic viruses can carry what are called **oncogenes**, genes capable, when activated, of transforming a cell. First postulated by Robert Huebner and George Todaro in 1969, the researchers suggested that oncogenes normally reside in the chromosomal DNA of a cell. In the late 1970s, researchers J. Michael Bishop and Harold Varmus, of the University of California at San Francisco, located oncogenes in a wide variety of organisms from fruit flies to humans. Bishop and Varmus also made the astonishing discovery that practically the same genes exist in certain viruses, and they hypothesized these genes could have been captured by the viruses from previous infections. It appeared that the oncogenes were not viral in origin but part of the genetic endowment of every living cell.

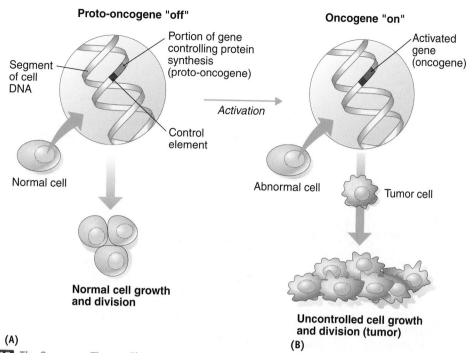

Proto-oncogene "off"

Oncogene "on"

Portion of gene
controlling protein
synthesis
(proto-oncogene)

Segment
of cell
DNA

Control
element

Activation

Activated
gene
(oncogene)

Normal cell

Abnormal cell

Tumor cell

**Normal cell growth
and division**

**Uncontrolled cell growth
and division (tumor)**

(A)

(B)

FIGURE 14.16 **The Oncogene Theory.** The oncogene theory helps explain the process of cancer development. (**A**) The normal cell grows and divides without complications. Within its DNA it contains proto-oncogenes that are "turned off." When the genes are activated by viruses or other factors, they revert to oncogenes, which "turn on." An abnormal cancer cell (**B**) results. The oncogenes encode proteins that regulate the development of a tumor and may contribute to transformation from a normal cell to a cancer cell. »» What properties are common to tumor cells?

The discovery of oncogenes demonstrated some forms of cancer have a genetic basis. As research continued, oncology researchers extracted DNA from tumors and used it to turn healthy cells into tumorous ones in the culture dish. Moreover, they surmised the transforming substance was in a small segment of the tumor cell DNA—probably a single gene. Finally, in 1981, three separate research groups isolated an oncogene residing in a human bladder cancer. At this writing, over 60 different oncogenes have been identified.

In recent years, the theory of oncogene activity has been revised slightly. Researchers now propose that normal genes, called **proto-oncogenes**, are the forerunners of oncogenes (**FIGURE 14.16**). Proto-oncogenes may have important functions as regulators of growth and mitosis. Because proto-oncogenes exist in diverse forms of life, they must be important in cell metabolism, perhaps as growth regulators. In fact, proto-oncogenes can be converted to oncogenes by radiation, by chemi-cal carcinogens, by chromosomal breakage and rearrangement, or by viruses after which tumor formation begins.

But how do viruses trigger the transformation? Virologists have observed that when some viruses, such as the retroviruses, enter a cell, they may become a provirus. Sometimes the integration of the provirus will be adjacent to a proto-oncogene, altering its expression such that it discrupts normal growth control and a tumor may result (**FIGURE 14.17**).

In addition, when virus replication is triggered, the provirus may not only replicate its own viral DNA, but also a few neighboring host genes. Thus, when new viruses are produced, the viral DNA contains the proto-oncogene as part of its genome. Such proto-oncogenes "captured" in the viral genome are called **v-oncogenes**.

When these oncogenic viruses infect another cell, the v-oncogene is under control of the virus—not the host cell. So, expression of the proto-oncogene or v-oncogene can influence

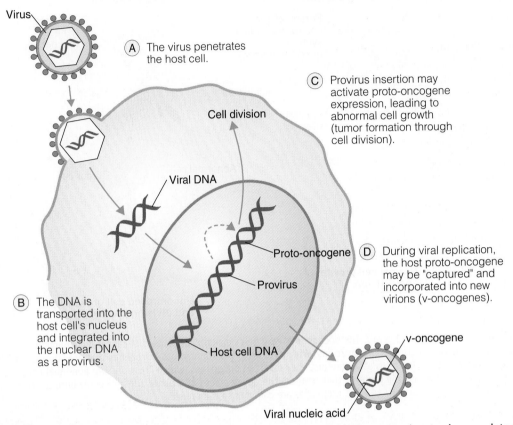

FIGURE 14.17 **Retrovirus Stimulation of a Tumor and v-Oncogene Formation.** An oncogenic retrovirus may integrate its genome into a host chromosome adjacent to a proto-oncogene. This may activate the expression of the proto-oncogene. During replication, the provirus along with the proto-oncogene will be replicated and incorporated into new virions. »» How would the oncogenic virus with the v-oncogene lead to tumor formation?

cellular growth and mitosis. For example, v-oncogenes may provide the genetic sequences for growth factors stimulating uncontrolled cell development and reproduction.

Another mechanism of transformation has been studied with the virus of **Burkitt lymphoma**. In this cancer of the lymphoid connective tissues of the jaw, the viral genome causes transformation of **B lymphocytes**, a type of white blood cell important in immunity. The event triggers the proto-oncogene *c-myc*, which is involved in cell growth, to move from chromosome 8 to chromosome 14, far from the influence of its control genes. The proto-oncogene, now an oncogene, appears to produce elevated amounts of its protein product, leading to tumor formation.

Infection by an oncogenic virus is more common than the cancer that may develop. So, infection may cause a tumor, but additional genetic and/or environmental events are needed before the cell is truly a cancer cell. The reason the viruses are called oncogenic viruses is because the v-oncogene or activated oncogene in many cases is telling the infected cells, "Divide, divide, divide!" In a short period, a tumor forms. The excessive divisions provide an opportunity for other mutations to occur that can lead the infected cells down the path to malignancy and cancer formation.

This ability of viruses to get inside cells and deliver their viral information has been manipulated by medical research teams using genetic engineering techniques to deliver genes to cure genetic illnesses. MicroFocus 14.5 looks at the pros and cons of using viruses for this purpose.

CONCEPT AND REASONING CHECKS

14.13 Explain how the oncogene theory is related to virus infection.

MICROFOCUS 14.5: Biotechnology
The Power of the Virus

How would you like to be injected with an adenovirus (see figure) or herpesvirus? How about the virus used to make the smallpox vaccine or a retrovirus? This doesn't sound like a great idea, but in fact it might be a way to treat genetic diseases and many forms of cancer. **Virotherapy**, as it is called, may be useful in curing disease rather than causing it.

Ever since the power of genetic engineering made it possible to transfer genes between organisms, scientists and physicians have wondered how they could use viruses to cure disease.

One potential way is to use viruses against cancer. In 1997, a mutant herpesvirus was produced that could only replicate in tumor cells. Because viruses make ideal cellular killers, perhaps they would kill the infected cancer cells. Used on a terminal patient with a form of brain cancer, the herpesvirus seems to have worked, as the patient was still alive in 2009.

Additional cancer-killing viruses are being developed; some cause the cancer cells to commit cell suicide, while others alert the immune system to the cancer danger.

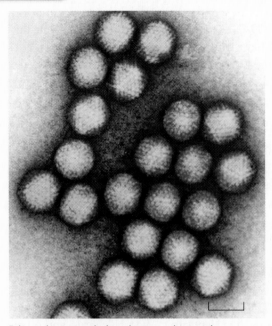

False-color transmission electron micrograph of adenoviruses. (Bar = 50 nm.)

Viral gene therapy has shown promise, but it has had its misfires as well. In 1999, an 18-year-old patient, Jesse Gelsinger, died as a result of an immune response that developed in response to the adenovirus used to treat a rare metabolic disorder. Importantly, 17 patients had been successfully treated using the same adenovirus prior to Gelsinger. In 2002, French scientists used virus-inserted genes to cure four young boys suffering from a nonfunctional immune system. Again, the therapy worked, but the virus-inserted gene disrupted other cellular functions, leading to leukemia (cancer of the white blood cells) in two of the boys.

So, the idea of using virotherapy certainly works, but delivering the virus without complications and targeting the genes to the correct place can be a difficult assignment. With regard to cancer, the only adverse effects have been patients reporting flu-like symptoms, again due to the immune system's response to the detection of virus in the body.

In this regard, one problem of virotherapy is the immune system response. Like any infection, the immune system mounts an attack on the injected viruses, which, therefore, may be destroyed before the viruses reach their target. To overcome this problem, researchers have developed a version of the vaccinia virus and adenovirus coated with an "extracellular envelope" that is invisible to the immune system yet able to search out and find its target—the cancer cells.

Another approach is to load up cancer-targeted viruses with radioactive materials, a toxic drug, or a gene that codes for an anti-cancer protein. These "stealth viruses" would deliver their cargo only to cancer cells and their "payload" would destroy the cells.

In 2009, virotherapy was in high gear with many early clinical trials in progress to examine the power of the virus to directly or indirectly kill cancer cells.

14.7   Emerging Viruses and Virus Evolution

Almost every year a newly emerging influenza virus descends upon the human population, 2009–2010 not withstanding with the "swine flu." Other viruses not even heard of a few decades ago, such as HIV, Ebolavirus, Hantavirus, West Nile virus, and SARS virus, are often in the news. Where are these viruses coming from?

Emerging Viruses Usually Arise Through Natural Phenomena

KEY CONCEPT

14. Viral recombination and mutation can give rise to new viruses.

Many **emerging infectious diseases** are the result of viruses appearing for the first time in a population or rapidly expanding their host range with a corresponding increase in detectable disease (**TABLE 14.4**). Most of these viruses are not new. Some, like Hantavirus, West Nile virus, and others are just new to certain populations or geographical areas. Others, like HIV, have crossed host ranges and resulted in the development of new human diseases.

One way "new" viruses arise is through **genetic recombination** (see Chapter 9). Take for example influenza. Genetic recombination allows different influenza viruses to reassort genome segments (Chapter 16). The "swine flu" that broke out in Mexico in 2009 was the result of the reassortment of genome segments from a strain of avian flu virus, a human flu virus, and a swine flu virus. This produced a virus with the capability of infecting humans.

Viruses also arise as a result of the second force driving evolution—**mutation** (see Chapter 8). For example, when a single nucleotide is altered (point mutation) in an RNA virus genome in the host cytoplasm, there is no way to "proofread" or correct the mistake during replication. Occasionally, one of these mutations may be advantageous. In the case of HIV, a beneficial mutation could generate a new virus strain resistant to an antiviral drug. With a rapid replication rate and burst size, it does not take long for a beneficial mutation to establish itself within a population.

Even if a new virus has emerged, it must encounter an appropriate host to replicate and spread. It is believed that smallpox and measles both evolved from cattle viruses, while flu probably originated in ducks and pigs. HIV almost certainly has evolved from a monkey (simian) immunodeficiency virus (Chapter 23). Consequently, at some time such viruses had to make a species jump. What could facilitate such a jump?

Our proximity to animals and pathogens makes such a jump possible. Today, population pressure is pushing the human population into new areas where potentially virulent viruses may be lurking. Evidence shows the Machupo and Junin viruses (Arenaviridae viruses that cause hemorrhagic fevers in South America) jumped from rodents to humans as a result of increased agricultural practices that, for the first time, brought infected rodents into contact with humans.

TABLE

14.4   Examples of Emerging Viruses

Virus	Family	Emergence Factor
Influenza	Orthomyxoviridae	Mixed pig and duck agriculture, mobile population
Dengue fever	Flaviviridae	Increased population density, environments that favor breeding mosquitoes
Sin Nombre (Hantavirus)	Bunyaviridae	Large deer mice population and contact with humans
Ebola/Marburg	Filoviridae	Human contact with fruit bats
HIV	Retroviridae	Increased host range, blood and needle contamination, sexual transmission, social factors
West Nile	Flaviviridae	Vector transported unknowingly to New York City
Machupo/Junin	Arenaviridae	Rodent contact via agricultural practices
Nipah/Hendra	Paramyxoviridae	Human contact with flying foxes (bats)
SARS-associated	Coronaviridae	Contact with horsehoe bat

MICROFOCUS 14.6: Public Health/Enivornmental Microbiology
Jungle Viruses and Exotic Pets

In June 2003, a couple from Wisconsin along with their 3-year-old daughter went to a swap meet. However, this was not just any old swap meet; this one was selling exotic pets. The family purchased two prairie dogs (see figure), one of which bit the daughter two days later. Within 10 days the girl had a fever and developed a skin pustule, a small, round, raised area of inflamed skin filled with pus. The hospital could not diagnose the girl's disease but she recovered.

A prairie dog in New Mexico.

Then, 10 days later, the mother developed the same type of skin lesion. When clinicians examined her lesion and the lymph nodes from the now-sick prairie dogs, they discovered the presence of a poxvirus. Samples sent to the Centers for Disease Control and Prevention (CDC) identified the virus as a monkeypox virus, typically found in Central and West Africa—not Wisconsin! CDC epidemiologists then traced the illness back to the swap meet and to the man, also now suffering from monkeypox, who sold the prairie dogs. The investigation discovered that the man had caged the prairie dogs with Gambian rats he had received from an African animal importer. The Gambian rats, normally found in the jungles of Ghana, were infected and transmitted the virus to the prairie dogs.

Significantly, the monkeypox case is not an isolated incident.

- The 2003 SARS (severe acute respiratory syndrome; see Chapter 16) epidemic in China and then Canada was traced back to a suspected infected Chinese horseshoe bat that had been caged with other exotic animals in a market. Often these animal markets are very unsanitary with animal parts, blood, and fecal matter strewn about. Before the epidemic was quelled, there were some 8,100 known human cases of SARS and 774 deaths.
- West Nile fever/encephalitis, first discovered in North America in 1999 and which has now spread across the continental United States, probably arose from birds imported into New York from the Middle East. By the end of 2008, there were almost 30,000 reported human cases and more than 1,100 deaths in the United States.

The World Wildlife Fund (WWF) estimates that the global trade in exotic animals is a $20 million a year business and involves some 350 million exotic animals. Although most of the trade is legal, about 25% is illegal according to the WWF. From an infectious disease standpoint, this is very scary as most of the nearly 1.5 billion live animals imported from 190 countries, 92% are for sale as pets and are not tested for disease. So, how likely is disease contact and spread? Of the more than 1,400 known human pathogens, 61% are known to be zoonotic—transmitted from animals to humans—and about 50% of those are viruses.

An increase in the size of the animal host population carrying a viral disease also can "explode" as an emerging viral disease. Spring 1993 in the American Southwest was a wet season, providing ample food for deer mice. The expanding deer mice population brought them into closer contact with humans. Leaving behind mouse feces and dried urine containing the hantavirus made infection in humans likely. The deaths of 14 people with a mysterious respiratory illness in the Four Corners area that spring eventually were attributed to this newly recognized virus.

So, emerging viruses are not really new. They are simply evolving from existing viruses and, through human changes to the environment, are given the opportunity to spread or to increase their host range.

MICROFOCUS 14.6 describes the transmission of viruses via the pet trade and the emerging disease that can result.

CONCEPT AND REASONING CHECKS

14.14 Distinguish the ways that emerging viral diseases arise.

There Are Three Hypotheses for the Origin of Viruses

KEY CONCEPT

15. Viruses may have preceded cellular life.

Although no one can know for certain how viruses originated, scientists and virologists have put forward three hypotheses.

Regressive Evolution Hypothesis. Viruses are degenerate life-forms; that is, they are derived from intracellular parasites that have lost many functions other organisms possess and have retained only those genes essential for their parasitic way of life.

Cellular Origins Hypothesis. Viruses are derived from fragments of cellular genetic material and functional assemblies of macromolecules that have escaped their origins inside cells by being able to replicate autonomously in host cells (see MicroFocus 14.4).

Independent Entities Hypothesis. Viruses coevolved with cellular organisms from the self-replicating molecules believed to have existed in the primitive prebiotic earth.

While each of these theories has its supporters, the topic generates strong disagreements among experts. In the end, it is not so much a matter of how viruses arose, but rather how we become infected with them.

CONCEPT AND REASONING CHECKS

14.15 Which came first—viruses or cells? Support your answer.

14.8 Virus-Like Agents

When viruses were discovered, scientists believed they were the ultimate infectious particles. It was difficult to conceive of anything smaller than viruses as agents of disease in plants, animals, and humans. However, the perception was revised as scientists discovered new disease agents—the subviral particles referred to as **virus-like agents**.

Viroids Are Infectious RNA Particles

KEY CONCEPT

16. Viroids lack a protein coat.

Viroids are tiny fragments of nucleic acid known to cause diseases in crop plants. In the 1960s, Theodore O. Diener and colleagues at the U.S. Department of Agriculture in Beltsville, Maryland were investigating a suspected viral disease, potato spindle tuber (PST), which results in long, pointed potatoes shaped like spindles. Nothing would destroy the disease agent except an RNA-degrading enzyme, and in 1971, the group postulated that a fragment of single-stranded RNA was involved. Diener called the agent a viroid, meaning "virus-like."

Today, more than two dozen crop diseases have been related to viroids. The largest of these particles is about one-fifteenth of the size of the smallest virus (**FIGURE 14.18**). The RNA chain of the PST viroid has a known molecular sequence

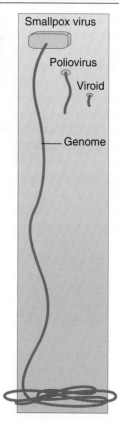

FIGURE 14.18 Genome Relationships. The size relationships of a smallpox virus, poliovirus, and viroid. The genome of the potato spindle tuber viroid has 359 nucleotides, while that of the smallpox virus has almost 200,000. »» Estimate whether viroid RNA is large enough to code for a variety of proteins.

(359 nucleotides), but it contains so few genetic sequences that the replication cycle is not understood. Diener speculates the viroids originated as introns, the sections of RNA spliced out of messenger RNA molecules before the messengers are able to function (see Chapter 8). Because the viroid RNA encodes no proteins, another hypothesis suggests the viroid RNA interacts with host cell RNA, inactivating proteins that bring about disease through loss of cell function. A similar proposal suggests the viroid RNA "silences" host cell "target RNA," again bringing about disease through loss of cell regulation. Disease causation remains to be determined.

CONCEPT AND REASONING CHECKS

14.16 How do viroids differ from viruses?

Prions Are Infectious Proteins

KEY CONCEPT

17. Prions lack nucleic acid.

In 1986, cattle in Great Britain began dying from a mysterious illness. The cattle experienced weight loss, became aggressive, lacked coordination, and were unsteady on their hooves. These detrimental effects became known as "**mad cow disease**" and were responsible for the eventual death of these animals. A connection between mad cow disease and a similar human disease surfaced in Great Britain in the early 1990s when several young people died of a human brain disorder resembling mad cow disease. Symptoms included dementia, weakened muscles, and loss of balance. Health officials suggested the human disease was caused by eating beef that had been processed from cattle with mad cow disease. It appeared the disease agent was transmitted from cattle to humans.

Besides mad cow disease, similar neurologic degenerative diseases have been discovered and studied in other animals and humans. These include scrapie in sheep and goats, wasting disease in elk and deer, and Creutzfeldt-Jakob disease in humans. All are examples of a group of rare diseases called **transmissible spongiform encephalopathies (TSEs)** because, like mad cow disease, they can be transmitted to other animals of the same species and possibly to other animal species, including humans, and the disease causes the formation of "sponge-like" holes in brain tissue (**FIGURE 14.19A**).

Many scientists originally believed these agents were a new type of virus. However, in the early 1980s, Stanley Prusiner and colleagues isolated an unusual protein from scrapie-infected tissue, which they thought represented the infectious agent. Prusiner called the proteinaceous infectious particle a **prion** (pronounced pree-on). The sequencing of the protein led to the identification of the coding gene, called *PrP*. The *PrP* gene is primarily expressed in the brain.

This led Prusiner and colleagues to propose the **protein-only hypothesis**, which predicts that prions are composed solely of protein and contain no nucleic acid. The protein-only hypothesis further proposes there are two types of prion proteins (**FIGURE 14.19B, C**). Normal cellular

■ Dementia:
A loss of memory.

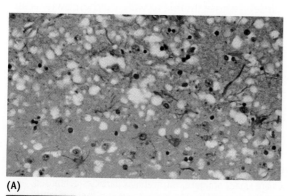

(A)

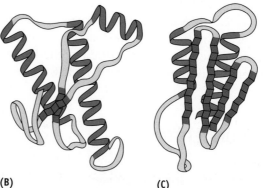

(B)   (C)

FIGURE 14.19 **Prion "Infection" and Structure.** (**A**) A photomicrograph showing the vacuolar ("spongy") degeneration of gray matter characteristic of human and animal prion diseases. (**B**) This drawing shows the tertiary structure of the normal prion protein (PrP^C). The helical regions of secondary structure are green. (**C**) A misfolded prion (PrP^{SC}). This infectious form of the prion protein results from the helical regions in the normal protein unfolding and an extensive pleated sheet secondary structure (blue ribbons) forming within the protein. The misfolding allows the proteins to clump together and react in ways not completely understood that contribute to disease. »» How does a change in tertiary structure affect the functioning of a protein?

prions (PrP^C) are found on the surfaces of brain cells while abnormal, misfolded prions, as found in scrapie (PrP^{SC}), have a different shape. These latter forms are the suspected infectious agents of diseases like mad cow and scrapie. More definitive proof came in 2004 when Prusiner's group demonstrated that purified prions can cause disease when injected into brains of genetically engineered mice.

Many researchers believe prion diseases—that is, TSEs—are spread by the infectious PrP^{SC} binding to normal PrP^C, causing the latter to change shape (**FIGURE 14.20**). In a domino-like scenario, the newly converted PrP^{SC} proteins, in turn, would cause more PrP^C proteins to become abnormal. The PrP^{SC} proteins then form insoluble protein fibers that aggregate, forming sponge-like holes left where groups of nerve cells have died. Importantly, PrP^{SC} does not trigger an immune response. Death of the animal occurs from the numerous nerve cell deaths that lead to loss of speech and brain function.

The human form of TSE appears similar to the classical and spontaneous form of Creutzfeldt-Jakob disease (CJD). The new form of CJD, called **variant CJD (vCJD)** is characterized clinically by neurologic abnormalities such as dementia. Neuropathology shows a marked spongiform appearance throughout the brain and death occurs within 3 to 12 months after symptoms appear.

As of January 2010, the number of vCJD definite or probable deaths in Great Britain was 170, although there have been more than 185,000 cases of BSE reported in farmed cattle in the United Kingdom.

In North America, the first mad cow was discovered in Canada in 2003, and in December the first such cow was reported in the United States. In all, there have been 17 BSE cattle in

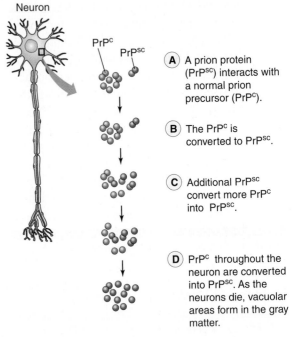

FIGURE 14.20 **Prion Formation and Propagation.** Prion proteins (PrP^{SC}) bind to normal prion precursors (PrP^C) in neurons, causing these proteins to the misfolded PrP^{SC} conformation. This process is repeated over and over, leading to the accumulation of PrP^{SC} protein fibers that cause progressive neuron cell death. »» How could these areas of PrP^{SC} protein fiber accumulation be observed in a brain autopsy from a person who has died from variant CJD?

Canada and three in the United States (through May 2009). No human cases of vCJD have been reported. Protection measures include condemning those animals having signs of neurologic illness and holding any cows suspected of having BSE until test results are known. In addition, "downer cattle"—those unable to walk on their own—cannot be used for human food or animal feed.

CONCEPT AND REASONING CHECKS

14.17 How are prions (a) similar to and (b) different from viruses?

SUMMARY OF KEY CONCEPTS

14.1 Foundations of Virology

1. The concept of a **virus** as a filterable agent of disease was first suggested by Ivanowsky, while Twort and d'Herrelle identified viruses as infective agents in bacterial organisms. In the 1940s, the electron microscope enabled scientists to see viruses and innovative methods allowed researchers to cultivate them.

14.2 What Are Viruses?

2. Viruses are small, obligate, intracellular particles composed of nucleic acid as either DNA or RNA and in either a **single-stranded** or **double-stranded** form. The genome is surrounded by a protein **capsid**, and many viruses have an **envelope** surrounding the **nucleocapsid**. **Spikes** protruding from the capsid or envelope are used for attachment to host cells.

3. Viruses have **helical**, **icosahedral**, or **complex symmetry**.

4. The **host range** of a virus refers to what organisms (hosts) the virus can infect. Many viruses only infect certain cell types or tissues within the host, referred to as a **tissue tropism** (tissue attraction).

14.3 The Classification of Viruses

5. Two broad classes of viruses can be organized based on their genome and strand type: the single-stranded or double-stranded DNA viruses and the single-stranded or double-stranded RNA viruses.

14.4 Viral Replication and Its Control

6. **Bacteriophages** undergo either a **lytic cycle**, involving attachment, penetration, biosynthesis, maturation, and release stages, or a **lysogenic cycle** where the phage genome integrates into the bacterial chromosome as a **prophage**.

7. Animal viruses also progress through the same stages of replication. However, penetration can occur by membrane fusion or endocytosis, and the biochemistry of nucleic acid synthesis varies among DNA and RNA viruses.

8. Many of the DNA viruses and the retroviruses can incorporate their viral DNA independently within the host's cell nucleus, or as a **provirus**.

14.5 The Cultivation and Detection of Viruses

9. Various detection methods for viruses are based on characteristic cytologic changes referred to as the **cytopathic effect**.

10. The cultivation of animal viruses is maintained by using cell cultures. The detection of viruses can be determined by **plaque** formation.

14.6 Tumors and Viruses

11. **Tumor** formation is a complex condition in which cells multiply without control and possibly develop into **cancers**.

12. There are at least seven **carcinogenic viruses** known to cause human tumors. Many of these tumors may develop into **cancer**.

13. Viruses may bring about tumors by converting **proto-oncogenes** into tumor **oncogenes**. If such genes are carried in a virus, they are called **viral oncogenes**.

14.7 Emerging Viruses and Virus Evolution

14. Viruses use **genetic recombination** and **mutations** as mechanisms to evolve. Human population expansion into new areas and increased agricultural practices in previously forested areas have exposed humans to existing animal viruses.

15. Three hypotheses have been proposed for the origin of viruses. These include the **regressive evolution hypothesis**, the **cellular origins hypothesis**, and the **independent entities hypothesis**.

14.8 Virus-Like Agents

16. **Viroids** are infectious particles made of RNA. They infect a few plant species and cause disease. Viroids lack protein and a capsid, but can replicate themselves inside the host.

17. **Prions** are infectious particles made of protein. They are capable of causing a number of animal diseases, including **mad cow disease** in cattle and **variant Creutzfeldt-Jakob disease (vCJD)** in humans. Prions cause disease by folding improperly and in the misfolded shape, cause other prions to misfold. These and other similar diseases in animals are examples of a group of rare diseases called **transmissible spongiform encephalopathies (TSEs)**.

▮ LEARNING OBJECTIVES

After understanding the textbook reading, you should be capable of writing a paragraph that includes the appropriate terms and pertinent information to answer the objective.

1. Identify the important historical developments in the identification and role of viruses.

2. Distinguish between the structures common to all viruses and those that separate **naked** from **enveloped** viruses.

3. Identify the three shapes viruses may exhibit.

4. Contrast viral **host range** with **tissue tropism**.

5. Explain the difference between **DNA** and **RNA viruses**.

6. Identify the five stages of and explain the events occurring in a phage lytic replicative cycle. State how a **lysogenic cycle** differs from a **lytic cycle** in **bacteriophages**.

7. Describe how animal viruses cause a **productive infection**.

8. Explain how some DNA viruses and retroviruses establish a **latent infection**.

9. Assess the role of the **cytopathic effect** for virus detection.

10. Explain how animal cell cultures are produced and how viruses can be visually detected in these infected cultures.

11. Define how **benign** and **malignant tumors** arise.

12. Identify three viruses that are involved in causing human tumors.

13. Explain how viruses can cause tumors.

14. Identify the two mechanisms by which emerging viruses arise and how an animal virus could jump to humans.

15. Contrast the three hypotheses for the origin of viruses.

16. Summarize the properties of **viroids**.

17. Summarize the properties of and diseases resulting from **prions**.

STEP A: SELF-TEST

Each of the following questions is designed to assess your ability to remember or recall factual or conceptual knowledge related to this chapter. Read each question carefully, then select the *one* answer that best fits the question or statement. Answers to even-numbered questions can be found in **Appendix C**.

1. Which one of the following scientists was NOT involved with discovering viruses?
   A. Felix d'Herrelle
   B. Dimitri Ivanowsky
   C. Robert Fleming
   D. Martinus Beijerinck
2. Viral genomes consist of
   A. DNA only.
   B. RNA only.
   C. DNA or RNA.
   D. DNA and RNA.
3. A nucleocapsid can have _____ symmetry.
   A. radial
   B. icosahedral
   C. vertical
   D. bilateral
4. Tissue tropism refers to
   A. what tissues grow due to a viral infection.
   B. what tissues are resistant to viral infection.
   C. what organisms a virus infects.
   D. what cells or tissues a virus infects.
5. An RNA virus genome in the form of messenger RNA is referred to as a
   A. + strand RNA.
   B. double-stranded RNA.
   C. – strand RNA.
   D. reverse strand RNA.
6. A virulent bacteriophage will
   A. carry out a lytic cycle.
   B. integrate its genome in the host cell.
   C. remain dormant in a bacterial cell.
   D. exist as a prophage.
7. The release of the viral genome from the capsid is called
   A. uncoating.
   B. endocytosis.
   C. penetration.
   D. maturation.
8. Provirus formation is possible in members of the
   A. single-stranded DNA viruses.
   B. retroviruses.
   C. double-stranded RNA viruses.
   D. single-stranded (– strand) RNA viruses.
9. Cytopathic effects would include all the following *except*
   A. changes in cell structure.
   B. provirus formation.
   C. vacuolated cytoplasm.
   D. syncytia formation.
10. A plaque is a
    A. change in cell structure due to viral infection.
    B. viral cell inclusion.
    C. clear zone within a lawn of bacteria.
    D. cellular aggregation of phage heads.
11. A benign tumor
    A. will metastasize.
    B. represents cancer.
    C. is malignant.
    D. is a clone of dividing cells.
12. Which of the following is NOT a carcinogen?
    A. Genetic factors
    B. UV light
    C. Certain chemicals
    D. X rays
13. The oncogene theory states that transforming genes
    A. normally occur in the host genome.
    B. can exist in viruses.
    C. are not of viral origin.
    D. All of the above (**A**–**C**) are correct.
14. Newly emerging viruses causing human disease can arise from
    A. species jumping.
    B. mutations.
    C. genetic recombination.
    D. All of the above (**A**–**C**) are correct.
15. Viruses derived from fragments of cellular genetic material and macro-molecules forms the basis of the
    A. cellular origins hypothesis.
    B. independent entities hypothesis.
    C. RNA world hypothesis.
    D. regressive evolution hypothesis.
16. Viroids contain
    A. RNA and DNA.
    B. only RNA.
    C. DNA and a capsid.
    D. RNA and an envelope.
17. Which one of the following statements about prions is FALSE?
    A. Prions are infectious proteins.
    B. Prions have caused disease in Americans.
    C. Human disease is called variant CJD.
    D. Prions can be transmitted to humans from infected beef.

HTTP://MICROBIOLOGY.JBPUB.COM/9E

The site features elearning, an online review area that provides quizzes and other tools to help you study for your class. You can also follow useful links for in-depth information, read more MicroFocus stories, or just find out the latest microbiology news.

STEP B: REVIEW

18. Construct a concept map for virus types using the virus family names and the following terms.

complex	icosahedral
dsDNA	naked
dsRNA	ssDNA
enveloped	ssRNA
helical	

Use the following syllables to compose the term that answers each clue from virology. The number of letters in the term is indicated by the dashes, and the number of syllables in the term is shown by the number in parentheses. Each syllable is used only once. The answers are listed in Appendix D.

A BAC CAP CAP C CO COS DRON EN GE GENE HE I I I LENT MERES MOR NOME O ON ON ONS OPE PHAGE PRI SID SO TER TU U VEL VIR VIR

19. Viral protein coat   (2) __ __ __ __ __ __
20. Bacterial virus   (5) __ __ __ __ __ __ __ __ __ __ __ __ __
21. Clone of abnormal cells   (2) __ __ __ __ __ __
22. Cancer gene   (3) __ __ __ __ __ __ __ __
23. Phages that lyse bacterial cells   (3) __ __ __ __ __ __ __ __
24. Shape of poliovirus   (5) __ __ __ __ __ __ __ __ __ __ __ __
25. Viral genetic information   (2) __ __ __ __ __
26. Completely assembled virus   (3) __ __ __ __ __ __ __
27. Surrounds the capsid   (3) __ __ __ __ __ __ __ __
28. Infectious protein particles   (2) __ __ __ __ __
29. Capsid subunits   (3) __ __ __ __ __ __ __

STEP C: APPLICATIONS

Answers to even-numbered questions can be found in **Appendix C**.

30. As part of a lab exercise, you and your lab partner are to estimate the number of bacteriophages in a sample of sewer water using a plaque assay with *Escherichia coli*. You add the phages to a lawn of bacteria and let the plates incubate for 24 hours before counting the number of plaques. This then got your lab partner thinking. She says, "If all these bacteriophages are in the sewer, they must have come from human waste flushed into the sewer. If they indeed did come from humans, why haven't the phage wiped out all the microbiota in the human digestive tract?" How would you respond to her question?

31. A person appears to have died of rabies. As a coroner, you need to verify that the cause of death was rabies. What procedures could you use to confirm the presence of rabies virus in the brain tissue of the deceased?

32. As a young genetic engineer with a biotech company, you have been given a virotherapy project. The hepatitis B virus is a dsDNA virus that during its replicative cycle often produces different forms of virus particles. The ones that are filamentous or spherical completely lack a viral genome and are thus defective; they are just assembled capsids but with the same spikes as the infective virions. The biotech company wants you to come up with a plan to use these defective particles for biotechnology applications. What might you include on your list of applications?

STEP D: QUESTIONS FOR THOUGHT AND DISCUSSION

Answers to even-numbered questions can be found in **Appendix C**.

33. If you were to stop 1,000 people on the street and ask if they recognize the term "virus," all would probably respond in the affirmative. If you were then to ask the people to describe a virus, you might hear answers like, "It's very small" or "It's a germ," or a host of other colorful but not very descriptive terms. As a student of microbiology, how would you describe a virus?

34. Oncogenes have been described in the recent literature as "Jekyll and Hyde genes." What factors may have led to this label, and what does it imply? In your view, is the name justified?

35. Researchers studying bacterial species that live in the oceans have long been troubled by the question of why these microorganisms have not saturated the oceanic environments. What might be a reason?

36. When discussing the multiplication of viruses, virologists prefer to call the process replication, rather than reproduction. Why do you think this is so? Would you agree with virologists that replication is the better term?

37. Bacterial species can cause disease by using their toxins to interfere with important body processes; by overcoming body defenses, such as phagocytosis; by using their enzymes to digest tissue cells; or other similar mechanisms. Viruses, by contrast, have no toxins and produce no digestive enzymes. How, then, do viruses cause disease?

38. How have revelations from studies on viruses, viroids, and prions complicated some of the traditional views about the principles of biology?

15

Chapter Preview and Key Concepts

Viral Infections of the Respiratory Tract and Skin

Sometime soon we will face a biological attack that has nothing to do with terrorists and everything to do with viruses mutating in the chicken farms of southeast Asia.
—*NewScientist magazine*, January 10, 2004

Every year there are seasonal flu outbreaks that may become **epidemic**; that is, they spread quickly through a large population. However, once in awhile there is a **pandemic**, a worldwide epidemic. The biggest influenza pandemic, called the "Spanish flu," occurred in 1918 to 1919 (**FIGURE 15.1**). A fifth of the world's population was infected by a virus thought to have originated in birds. Estimates suggest between 20 and 100 million people died, young adults being the worst hit. As we see in this chapter, such pandemics can occur when a new infectious virus appears for which the human population has no immunity. Thus, many of us get a flu shot every year to protect us from the current dominant strains of circulating flu viruses.

Influenza continues to be an ongoing problem in the twenty-first century. Health experts and microbiologists have been saying that we are overdue for another serious and perhaps very deadly flu pandemic. In fact, one

FIGURE 15.1 U.S. Soldiers—1918. This photo of American soldiers marching near the end of World War I shows them wearing gauze masks to protect against the flu. The masks were totally ineffective. »» Why would these masks be useless against spreading or contacting the flu virus?

may be "brewing" and another is occurring. An avian flu virus has been circulating throughout Asia and into Europe and Africa. By December 2009, this influenza virus had killed at least 282 people of the 467 infected in Asia and Southeast Asia since 2003. At this time, it is still difficult for people to contract the disease because it is spread mainly from infected fowl to humans. Human-to-human transmission has not yet been documented. However, it might only take a mutation or two to transform the avian flu virus into a form that can become highly contagious, as the chapter opening quote suggests.

The second potential flu strain, which caught the world's attention in 2009, is the novel H1N1 flu (the so-called "swine flu"). This is a new flu virus of swine origin that was first detected in Mexico and the United States in March and April 2009. Unlike the avian flu, the novel H1N1 virus can spread from person-to-person. However, it appears to spread just like the normal, seasonal flu viruses spread; that is, through coughs and sneezes from people who are sick with the virus. By June 2009, all 50 states in the United States, the District of Columbia, and Puerto Rico had reported cases of novel H1N1 infection to the Centers for Disease Control and Prevention (CDC). At this writing (December 2009), most of the approximately 50 million people stricken with H1N1 in the United States had recovered without

needing medical treatment and experienced only typical flu symptoms. Somewhere between 3% to 5% have required hospitalization and about 0.2% (10,000), primarily young adults and individuals with other medical complications, had died. Globally, the World Health Organization (WHO) estimates there have been more than 12,000 deaths, which considering the U.S. figures, is certainly underestimated.

The H1N1 virus quickly spread to pandemic levels. By June 2009, 69 countries had officially reported almost 22,000 cases of novel H1N1 virus infections, and by December 2009, virtually every country had reported H1N1, amounting to more than 6 million cases. Because this is a new virus, most people have little or no immunity against it. The flu season is not yet over (at this writing), so by the time you read this, we may have a better understanding of the pandemic.

There are several other viruses with tropisms to the respiratory tract or skin. Chickenpox is still among the most commonly reported diseases of childhood years. Genital herpes has become so rampant that, by some estimates, 10 to 20 million Americans are currently infected.

The news on viruses is not all bad. We now have vaccines to many viral diseases that once wreaked havoc. For example, two skin diseases, mumps and rubella, were part of the fabric of life only a generation ago, but the annual case reports have dropped from hundreds of thousands to mere hundreds (and some officials are bold enough even to whisper the word "eradication").

In this chapter we focus on several viral diseases, besides influenza, which affect the respiratory tract—the so-called pneumotropic diseases. We also examine viral diseases affecting the skin—the dermotropic viral diseases. Note: the division of the pneumotropic and dermotropic viral diseases is an artificial classification simply for grouping convenience (see Chapter 14). Therefore, the symptoms described may go beyond the respiratory tract or skin, respectively. In this chapter, each disease will be presented as an independent essay, so you can establish an order that best suits your study needs.

Contagious: Capable of being easily transmitted from one person to the next.

15.1   Viral Infections of the Upper Respiratory Tract

As described and illustrated in Chapter 10, the **upper respiratory tract** (**URT**) consists of the nose, pharynx (throat), and the middle ear and auditory tubes. The various versions of the common cold are the most common viral infections of the URT, accounting for more than 1 billion infections each year in the United States alone. In fact, it is the world's most common infectious disease.

Rhinovirus Infections Produce Inflammation in the Upper Respiratory Tract

KEY CONCEPT

1. Rhinoviruses are often responsible for the common-cold syndrome.

Rhinovirus

Croup:
An inflammation of the larynx and trachea, causing a cough, hoarseness, and breathing difficulties.

FIGURE 15.2   **A Rhinovirus.** A computer-generated image of the surface (capsid) of human rhinovirus 16. (Bar = 3 nm.) »» What is unique about the rhinoviruses that makes vaccine development impractical?

Rhinoviruses (*rhino* = "nose") are a broad group of over 100 different naked, icosahedral, +ssRNA viruses, (**FIGURE 15.2**). They belong to the family Picornaviridae (*pico* = "small"; hence, small-RNA-viruses).

Rhinoviruses thrive in the human nose, where the temperature is a few degrees cooler (33°C) than in the rest of the body. They are transmitted through airborne droplets or by contact with an infected person or contaminated objects. Rhinoviruses account for 30% to 50% of **common colds**, also called "head colds." Adults typically suffer two or three colds and children six to eight colds per year. The rhinoviruses are most common in the fall and spring (**FIGURE 15.3**).

A common cold is a viral infection of the lining of the nose, sinuses, throat, and upper airways. One to three days after infection, a sequence of symptoms (common-cold syndrome) begins. These include sneezing, a sore throat, runny or stuffy nose, mild aches and pains, and a mild-to-moderate hacking cough. Some children suffer from croup. The illness usually is of short duration of 7 to 10 days. One of the old wives tales was that becoming chilled could cause colds (MicroFocus 15.1).

So, why cannot scientists find a cure for the common cold? The answer is quite simple. There are more than 200 viruses and strains—some 100 rhinoviruses alone—that cause common colds. Therefore, it would be impractical to develop a vaccine to immunize people against all different types of cold viruses.

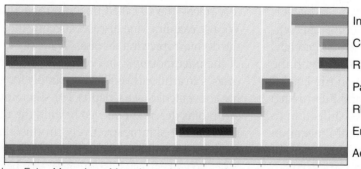

Influenza (5–15%)

Coronavirus infections (10–15%)

RS disease (5%)

Parainfluenza (5%)

Rhinovirus infections (30–50%)

Enterovirus diseases (<5%)

Adenovirus infections (<5%)

Jan Feb Mar Apr May June July Aug Sept Oct Nov Dec

FIGURE 15.3   **The Seasonal Variation of Viral Respiratory Diseases.** This chart shows the seasons associated with various viral diseases of the respiratory tract (and their annual percentage). Enteroviruses (Chapter 16) cause diseases of the gastrointestinal tract as well as respiratory disorders. »» Hypothesize as to why different cold viruses cause diseases at different times of the year.

MICROFOCUS 15.1: Being Skeptical
Catching a Chill: Can It Cause a Cold?

How many times can you remember your mom or a family member saying to you, "Bundle up or you will catch a cold!" Can you actually "catch" a cold from a body chill?

During the last 50 years of research, scientists have claimed that one cannot get a cold from a chill. Rather, colds result from more people being cooped up indoors during the winter months, making virus transmission from person-to-person very likely. At least, that was the scientific dogma until 2005.

In the November 2005 issue of *Family Practice*, British researchers Claire Johnson and Professor Ron Eccles at Cardiff University's Common Cold Center announced that a drop in body temperature can allow a cold to develop.

The researchers signed up 180 volunteers between October and March to participate in the study. Split into two groups, one group put their bare feet into bowls of ice cold water for 20 minutes. The other group put their bare feet in similar but empty bowls.

Over the next five days, 29% of individuals who had their feet chilled developed cold symptoms, while just 9% of the control group developed symptoms.

Professor Eccles suggests that colds may develop not because the volunteers actually "caught" a cold virus, but rather that they harbored the virus all along. Chilling lowers the person's immunity, providing the opportunity for the virus to produce more severe symptoms.

Chilling causes a constriction of blood vessels in the nose, limiting the warm blood flow supplying white blood cells to eliminate or control the cold viruses present. Professor Eccles states that, "A cold nose may be one of the major factors that causes common colds to be seasonal."

The verdict: Well, the results of one study do not make for a general consensus. More definitive studies may help verify or refute the "cold nose" claim.

But not all is lost. In 2009, researchers reported for the first time the genome sequences for more than 100 rhinoviruses. Having this information will allow scientists to do **comparative genomics** (see Chapter 9) to identify any common features to all rhinoviruses that might aid in vaccine development. In addition, they may discover why some rhinovirus strains cause a more severe illness than others. In fact, the researchers uncovered a new class of rhinoviruses, called the human rhinovirus class C (HRV-C), that causes very serious, flu-like symptoms in affected individuals. Their preliminary work also indicates that at least a few rhinovirus strains can simultaneously infect the same cell and in so doing exchange genome segments during replication. Such an exchange, which can occur quite readily in flu viruses (see below), had never been described in rhinoviruses.

Antibiotics will not prevent or cure a cold. Antihistamines can sometimes be used to treat the symptoms of a cold; however, they do not shorten the length of the illness. For other remedies, such as vitamin C, zinc, and herbs like echinacea, there is no verified scientific evidence these substances limit or prevent colds. Because rhinoviruses spread by respiratory droplets, washing hands remains an important hygiene measure to decrease transmission of the viruses.

CONCEPT AND REASONING CHECKS

15.1 Why do we get colds over and over again?

Adenovirus Infections Also Produce Symptoms Typical of a Common Cold

KEY CONCEPT

2. Adenovirus respiratory infections typically cause a severe sore throat (pharyngitis).

Adenovirus

Adenoid tissue:
Refers to the pharyngeal tonsil located in the upper rear of the pharynx.

Febrile:
Relates to fever.

Conjunctivitis:
An inflammation of the conjunctiva of the eye.

Adenoviruses (family Adenoviridae) are a group of over 50 types of nonenveloped, icosahedral virions having double-stranded DNA (FIGURE 15.4). They multiply in the nuclei of several human host tissues and induce the formation of **inclusion bodies**, which are a series of bodies composed of numerous virions arranged in a crystalline pattern (see Chapter 14). The viruses take their name from the adenoid tissue from which they were first isolated.

Adenoviruses can be highly infectious and are frequent causes of URT diseases often symptomatic of a common cold in infants and young children. Transmitted through respiratory droplets, the viruses most often cause distinctive symptoms because the fever is substantial, the throat is very sore (**acute** febrile **pharyngitis**), and the cough is usually severe. In addition, the lymph nodes of the neck swell and a whitish-gray material appears over the throat surface.

Some adenoviruses may produce a form of conjunctivitis called **pharyngoconjunctival fever**, which is most commonly contracted by swimming in virus-contaminated water. There were more than 42,000 cases reported in Japan in 2006. New military recruits may suffer **acute respiratory disease (ARD)** as a result of adenovirus transmission in crowded locations. Any of these conditions can progress to **viral pneumonia**.

No antiviral agents currently available treat adenoviral infections. An adenovirus vaccine is available for ARD in military recruits. In recent

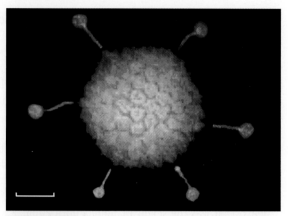

FIGURE 15.4 **An Adenovirus.** This false-color transmission electron micrograph of an adenovirus shows the icosahedral symmetry. The rough surface of the virus is due to the presence of capsomeres. (Bar = 25 nm.) »» What are the projections from the capsid surface?

years, adenoviruses have developed a more positive image as vectors (carriers) for genes in viral gene therapy experiments (see MicroFocus 14.5).

Although bacterial infections of the larynx are extremely rare, the most common cause of **laryngitis** is a viral infection of the URT caused by a common cold or flu virus. Symptoms are an unnatural change of voice, such as hoarseness, or even loss of voice that develops within hours to a day or so after infection. The throat may tickle or feel raw, and a person may have a constant urge to clear the throat. Fever, malaise, difficulty swallowing, and a sore throat may accompany severe infections. Treatment of viral laryngitis depends on the symptoms. Resting the voice (by not speaking), drinking extra fluids, and inhaling steam relieve symptoms and help healing. The viral diseases of the URT are summarized in TABLE 15.1 .

CONCEPT AND REASONING CHECKS

15.2 Why isn't an adenovirus vaccine available to the general public?

TABLE

15.1 A Summary of the Major Viral URT Diseases

Disease	Causative Agent	Signs and Symptoms	Transmission	Treatment	Prevention and Control
Common colds (rhinitis)	Rhinoviruses Adenoviruses Other viruses	Sneezing, sore throat, runny and stuffy nose, hacking cough	Respiratory droplets	Pain relievers Decongestants	Practicing good hygiene
Laryngitis	Cold viruses	Hoarseness Loss of voice	Respiratory droplets	Rest voice	Drinking plenty of water

15.2 Viral Infections of the Lower Respiratory Tract

The **lower respiratory tract** (**LRT**) in humans consists of the larynx (voice box), trachea (windpipe), bronchial tubes, and the alveoli (see Chapter 10).

Influenza Is a Highly Communicable Acute Respiratory Infection

KEY CONCEPT

3. Influenza viruses are involved with seasonal (winter) epidemics and occasional pandemics.

Influenza (the "flu") is a highly contagious acute disease that is transmitted by airborne respiratory droplets. Beginning in the URT, most cases, including H1N1, involve the LRT as well. The disease is believed to take its name from the Italian word for "influence," a reference either to the influence of heavenly bodies, or to the *influenza di freddo*, "influence of the cold." Since the first recorded epidemic in 1510, scientists have described 31 pandemics. The most notable pandemic of the twentieth century was the "Spanish" flu in 1918–1919; others took place in 1957 (the "Asian" flu) and in 1968 (the "Hong Kong" flu). Today, there are about 35,000 deaths in the United States and 250,000 to 500,000 deaths worldwide annually related to seasonal influenza infections.

The enveloped influenza virion belongs to the Orthomyxoviridae family. It is composed of eight −ssRNA segments each wound helically and surrounded by a nucleocapsid (**FIGURE 15.5**). An additional structural protein, the **matrix protein**, surrounds the core of RNA segments and an envelope covers the matrix protein.

Projecting through the envelope are two types of spikes. One type contains the enzyme **hemagglutinin** (**H**), a substance facilitating the attachment and penetration of influenza viruses into host cells. Its shape determines the virus' host range and tropism. The second type contains another enzyme, **neuraminidase** (**N**), a protein assisting in the release of the virions from the host cell when replication is complete.

Three types of influenza viruses are recognized.

Influenza A strikes every year and causes most "flu" epidemics. It circulates in many animals, including birds, pigs, and humans. This was made most evident from the H1N1 virus ("swine flu") that caused a global pandemic in 2009–2010.

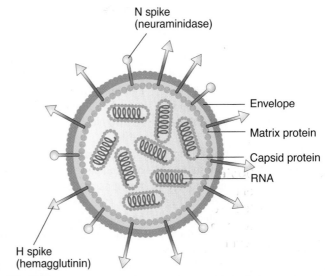

FIGURE 15.5 The Influenzavirus. This diagram of the influenzavirus shows its eight segments of RNA, matrix protein, and envelope with hemagglutinin and neuraminidase spikes protruding. »» What is the role of the hemagglutinin and neuraminidase spikes?

Type A is divided into subtypes based on the H and N surface glycoproteins. There are 15 known H subtypes and 9 different N subtypes. The current seasonal flu subtypes in humans are A(H1N1) and A(H3N2).

Influenza B also strikes every year but is less widespread than type A. It only circulates between humans and is not divided into subtypes. Each year's flu vaccine is a mixture of the most prevalent A and B subtypes.

Influenza C causes a mild respiratory illness but not epidemics.

Each year a slightly different seasonal flu strain evolves, based, in part, on changes to H and/or N spike proteins; thus, there is a need for a new flu vaccine each year. Sometimes the new strain is quite mild, while in other years, a predominant strain will be more dangerous, such as the "Spanish" flu. How do new flu strains arise? MicroInquiry 15 examines their evolution.

The onset of influenza A is abrupt after an incubation period of one to four days. The individual with an uncomplicated illness develops sudden chills, fatigue, headache, and pain most pronounced in the chest, back, and legs. Over a 24-hour period, body temperature can rise to 40°C, and a severe cough develops. Individuals may experience sore throat, nasal congestion, sneezing, and tight chest, the latter a probable

Influenza virus

MICROINQUIRY 15

Drifting and Shifting—How Influenza Viruses Evolve

Influenza viruses evolve in two different ways. Both involve **antigenic variation**, a process in which chemical and structural changes occur periodically in hemagglutinin (H) and neuraminidase (N) spike proteins (antigens), thereby yielding new virus strains.

Antigenic drift involves small changes to the virus (**Panel A**, below). These changes involve minor point mutations resulting from RNA replication errors. The mutations will be expressed in the new virions produced. Although such mutations may be detrimental, on occasion one might confer an advantage to the virus, such as being resistant to a host's immune system. For example, a spike protein might have a subtle change in shape (that is, the structural shape has "drifted") so they are not recognized by the host's immune system. This is what happens prior to most flu seasons. The virus spikes are different enough from the previous season that the host's antibodies fail to recognize the new strain. Both influenza A and B viruses can undergo antigenic drift.

Antigenic shift is an abrupt, major change in structure to influenza A viruses. Antigenic shift may give rise to new strains that can now jump to another spe-

cies, including humans (**Panel B**, below). (that is, the spike structure has "shifted") to which everyone is totally defenseless, and from which a pandemic may ensue (see figure, next page).

Two mechanisms account for antigenic shift.

The "Spanish" flu was the introduction of a completely new flu strain (H1N1) into the human population from birds. The H1N1 strain jumped directly to humans and adapted quickly to replicate efficiently in humans (**Panels B, C**).

The second mechanism involves "gene swapping" or reassortment between different flu viruses (**Panels B, D, E**). The 1957 influenza virus ("Asian" influenza; H2N2), for example, was a reassortment, where the human H1N1 virus acquired new spike genetic segments (H2 and N2) from an avian species. The 1968 influenza virus ("Hong Kong" influenza; H3N2) was the result of another reassortment; the human H2N2 strain acquired a new hemagglutinin genetic segment (H3) from another avian species.

In this second transmission mechanism, pigs usually are the reassortment "vessels" (or intermediate host) because

they can be infected by both avian and human flu viruses. The 2009 swine flu is another example (**Panel F**).

Future pandemic strains could arise through either antigenic drift or shift. The 2003 H5N1 avian flu strain could mutate further or recombine with H3N2 to produce a potentially deadly new strain in humans.

Discussion Point

The current avian (or bird) flu (H5N1 strain) is lethal to domestic fowl and can be transmitted to humans. Through 2009, at least 282 of the 467 people infected 60% in Southeast Asia have died.

If a human pandemic should occur, health care providers would play a crucial role in minimizing the pandemic. Therefore, planning for pandemic influenza is crucial as it was for the 2009 H1N1.

Discuss the specific steps that should be taken by individuals, colleges and universities, and communities in planning for a pandemic outbreak.

Once you have come up with a plan, check out what the CDC suggests (http://www.pandemicflu.gov/) and what your state is doing.

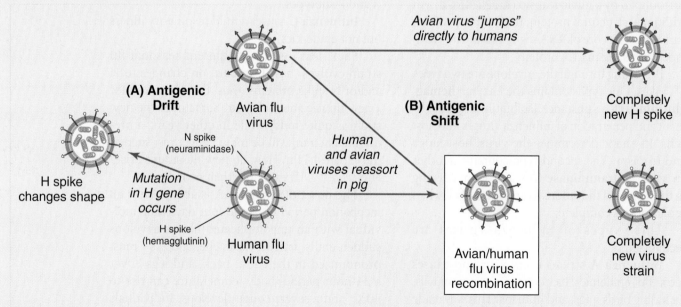

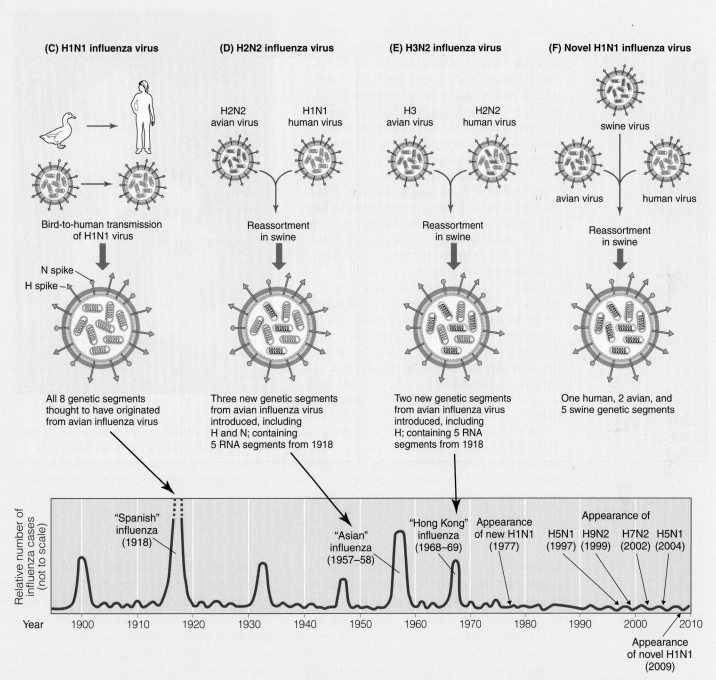

The Major Influenza Pandemics and Novel H1N1 Appearance in Humans. The four pandemics were the result of antigenic shifts.

MICROFOCUS 15.2: Public Health
Preschoolers Drive Flu Outbreaks

Every October to November in the Northern Hemisphere, we brace for another outbreak or epidemic spread of the seasonal flu. Many people go to their physician, clinic, or even grocery store for a flu shot. Others of us will "catch" the flu and suffer through several days of agony. For some 35,000, especially the elderly and immunocompromised, contracting the flu will lead to pneumonia and death. Wouldn't it be great if there was some way to predict pneumonia and flu deaths in a population? That now may be possible.

Researchers at the Children's Hospital of Boston and Harvard Medical School reported biosurveillance data suggesting that otherwise healthy preschoolers (3- to 4-year-olds) drive flu epidemics. The researchers found that by late September, kids in this age group were the first to develop flu symptoms.

Current immunization policies suggest that infants 6 to 23 months old be vaccinated against the flu because their immune systems are not yet fully developed. Policy also suggests that older children, including preschoolers, only be vaccinated if they have high medical-risk conditions. That means most will be vulnerable to the flu. And with many being in daycare and preschool, close contact makes spreading the flu effortless as those exposed bring the infection home.

The data also demonstrate that when preschoolers start sneezing, flu viruses are transmitted to the unvaccinated elderly who then become ill. Thus, the research suggests vaccinating those individuals who are driving and transmitting flu to others—the preschoolers. They are the sentinels by which an ensuing flu outbreak or epidemic can be identified. Immunizing preschoolers will decrease flu transmission and limit adult mortality in the unvaccinated.

reflection of viral invasion of tissues of the trachea and bronchi. Despite these severe symptoms, influenza is normally short-lived and has a favorable prognosis. The disease is self-limiting and usually resolves in 7 to 10 days.

Most of the annual deaths from seasonal influenza A are due to pneumonia caused by the virus spreading into the lungs. Secondary complications in unvaccinated individuals, especially those over 65 years, infants, or with underlying medical conditions (immunocompromised), may lead to bacterial pneumonia if *S. aureus* or *H. influenzae* invade the damaged respiratory tissue. MicroFocus 15.2 describes the role of preschoolers in driving flu outbreaks that can lead to pneumonia deaths in the community.

Influenza infection in rare cases is associated with two serious complications. **Guillain-Barré syndrome** (**GBS**) occurs when the body mistargets the infection and instead damages its own peripheral nerve cells, causing muscle weakness and sometimes paralysis. **Reye syndrome** usually makes its appearance in young people after they are given aspirin to treat fever or pain associated with influenza. It begins with nausea and vomiting, but the progressive mental changes (such as confusion or delirium) may occur. Thankfully, very few children develop Reye syndrome (less than 0.03–1 case per 100,000 persons younger than 18 years).

Because influenza-like symptoms can be caused by bacterial and other viral infections, a

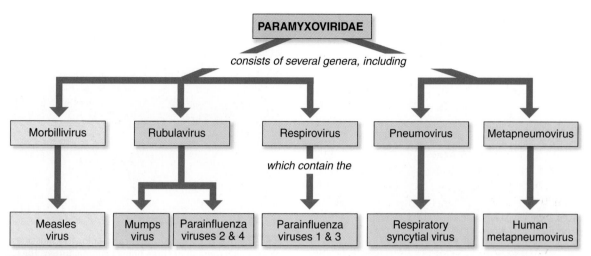

FIGURE 15.6 The Relationships between the Human Paramyxoviruses. This concept map illustrates the relationships between the paramyxoviruses that cause human respiratory and skin diseases. Note the relationship between measles and mumps (skin diseases). »» Which virus is most closely related to the respiratory syncytial virus?

diagnosis of influenza, if necessary, is based on several factors. This includes the pattern of spread in the community, observation of disease symptoms, laboratory isolation of viruses, or hemagglutination of red blood cells (Chapter 22).

The best prevention for the flu is an annual flu vaccination. Each year's batch of vaccine (see MicroFocus 22.1) is based on the previous year's predominant influenza A and B viruses and is about 75% effective. Because the viruses are grown in chicken embryos (see Chapter 14), people who are allergic to chickens or eggs should not be given the vaccine.

Treatment of uncomplicated cases of influenza requires bed rest, adequate fluid intake, and aspirin (or acetaminophen in children) for fever and muscle pain. Two antiviral drugs, zanamivir (Relenza) and, in the United States, oseltamivir (Tamiflu), are available by prescription. These drugs target the neuraminidase spikes projecting from the influenza virus envelope and block the release of new virions. If given to otherwise healthy adults or children early in disease onset, these drugs can reduce the duration of illness by one day and make complications less likely to occur. However, these drugs should not be taken in place of vaccination, which remains the best prevention strategy.

CONCEPT AND REASONING CHECKS

15.3 From MicroInquiry 15, why is the influenza A virus the cause of most flu epidemics and pandemics?

Paramyxovirus Infections Affect the Lower Respiratory Tract

KEY CONCEPT

4. The paramyxoviruses are a group of viruses causing similar symptoms.

A number of viruses, primarily in the Paramyxoviridae, are associated with the LRT (**FIGURE 15.6**). All these viruses are enveloped, −ssRNA viruses. The measles and mumps viruses will be discussed in the next section.

Respiratory syncytial (RS) disease is caused by the respiratory syncytial virus (RSV). Since 1985, RS disease has been the most common lower respiratory tract disease affecting infants and young children. RSV is transmitted by respiratory droplets or virus-contaminated hands.

Infection takes place in the bronchioles and air sacs of the lungs, and the disease is often described as **viral pneumonia**. When the virus infects tissue cells, the latter tend to fuse together, forming giant multinucleate cells called **syncytia** (see Chapter 14).

RS disease can occur in adults as an influenza-like syndrome and as severe bronchitis with pneumonia in the elderly. Outbreaks occur yearly throughout the United States, but most cases are misdiagnosed or unreported. Some virologists believe that up to 95% of all children have been exposed to the disease by the age of five, and CDC epidemiologists estimate there are 51,000 to 82,000 hospitalizations and 4,500 deaths in

Respiratory syncytial virus

■■■ Bronchioles:
The narrow tubes inside of the lungs that branch off of the main air passages (bronchi).

■■■ Bronchitis:
An acute inflammation of the bronchi (main air passages) in the lungs.

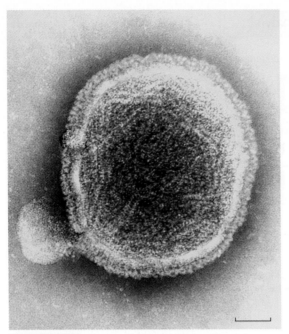

FIGURE 15.7 The Parainfluenza Virus. False-color transmission electron micrograph of a parainfluenza virus. The envelope of individual virions often gives the virus a pleomorphic shape, although they do have a helical capsid. (Bar = 50 nm.) »» Why is the virus referred to as parainfluenza?

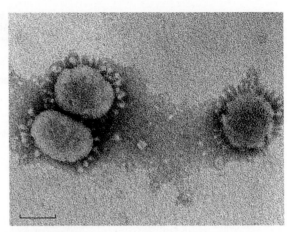

FIGURE 15.8 Coronaviruses. False-color transmission electron micrograph of three human coronaviruses. The spikes can be seen clearly extending from the viral envelope. Viruses similar to these are responsible for severe acute respiratory syndrome (SARS). (Bar = 100 nm.) »» By looking at the micrograph, explain why the virus is referred to as a "corona" virus.

infants and children each year in the United States as a result of RS disease.

Maternal antibodies passed from mother to child probably provide protection during the first few months of life, but the risk of infection increases as these antibodies disappear. Indeed, researchers have successfully used preparations of antibodies to lessen the severity of established cases of RS disease. Aerosolized ribavirin, an antiviral drug, also has been used with success.

Parainfluenza (*para* = "near") infections are caused primarily by human parainfluenza viruses 1 and 3 (FIGURE 15.7). They account for 40% of acute respiratory infections in children. Although as widespread as influenza, parainfluenza is a much milder disease and is transmitted by direct contact or aerosolized droplets. It is characterized by minor upper respiratory illness, often referred to as a cold. Bronchiolitis and croup may accompany the disease, which is most often seen in children under the age of six. The disease predominates in the late fall and early spring, and is seasonal, as Figure 15.3 indicated. No specific therapy exists.

RSV-like illnesses may be caused by the **human metapneumovirus (hMPV)**. Just about

every child in the world has been infected by the virus by age 5. Human MPV is responsible for 12% of LRT infections and 15% of common colds in children. It appears to be milder than RS disease and, like RS disease, can be treated with ribavirin.

CONCEPT AND REASONING CHECKS

15.4 Identify the common relationships between the paramyxoviruses.

Other Viruses Also Produce Pneumonia

KEY CONCEPT

5. The SARS coronavirus and hantaviruses cause unique forms of pneumonia.

Several other viruses can cause LRT infections in adults. These include the SARS coronavirus and the hantaviruses.

SARS Coronavirus. Severe acute respiratory syndrome (SARS), an emerging infectious disease of the LRT, was first reported in southeastern China in spring 2003, and quickly spread through Southeast Asia and to 29 countries. It is an example of how fast an emerging disease can spread (Textbook Case 15).

Scientists at the CDC and other laboratories identified in SARS patients a previously unrecognized coronavirus, which they named the SARS coronavirus (SARS-CoV). Being a member of the Coronaviridae, it is a +ssRNA virus with helical symmetry and a spiked envelope (FIGURE 15.8). The mortality rate from SARS appears to be about

Textbook CASE 15

The Outbreak of SARS: 2002–2003

1 On February 11, 2003, the Chinese Ministry of Health notified the World Health Organization (WHO) of a mystery respiratory illness that had been occurring since November 2002 in Guangdong province in southern China. However, officials of China refused to allow WHO officials to investigate.

2 February 21, a 64-year-old doctor from Guangdong came to Hong Kong to attend a wedding. He stayed at a Hong Kong hotel, infecting 16 people who spread the disease to Hanoi, Vietnam; Singapore; and Toronto, Canada.

3 February 23, a Canadian tourist checked out of the same Hong Kong hotel and returned to Toronto, where her family greeted her. She died 10 days later as five family members were hospitalized.

4 February 28, a WHO doctor in Hanoi, Carlo Urbani, treated one of the people infected in Hong Kong and realized this was a new disease, which he named "severe acute respiratory syndrome" (SARS). He died of the disease 29 days later.

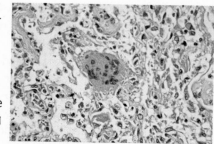

Light micrograph of a section through tissue from the lung of a patient with SARS. A giant cell is seen at center, which has numerous nuclei (dark purple). The white spaces are alveoli, tiny sacs in the lungs through whose walls gaseous exchange takes place. Several red blood cells (red) are seen among the tissue. Alveolar damage causes a cough and breathing difficulties.

5 March 15, the WHO declared SARS a worldwide health threat. To block the chain of transmission, isolation and quarantine were instituted. Over half of those infected were health care providers.

6 April 16, the identification was made by 13 laboratories around the world that a new, previously unknown, coronavirus caused SARS (see figure).

7 June 30, the WHO announced there had been no new cases of SARS for two weeks. During the 114-day epidemic, more than 8,000 people from 29 countries were infected and 774 died.

Questions:

(Answers can be found in Appendix D.)

A. Why might Chinese officials be reluctant to allow a WHO investigation into the mystery illness?

B. Justify the use of quarantine and isolation to break the chain of SARS transmission.

C. Propose an explanation as to why there was such a disproportionately high number of infections in health care providers.

D. How could such a large number of infections in health care workers have been prevented?

E. Why is SARS a textbook case for an emerging infectious disease?

For additional information see http://www.cdc.gov/mmwr/preview/mmwrhtml/mm5226a4.htm.

10%. During the 2002 to 2003 epidemic, the WHO identified some 8,100 cases from 29 countries. There were 774 deaths, the majority being in China, Taiwan, Vietnam, Singapore, and Canada.

SARS-CoV can be spread through close person-to-person contact by touching one's eyes, nose, or mouth after contact with the skin of someone with SARS. Spreading also comes from contact with objects contaminated through coughing or sneezing with infectious droplets by a SARS-infected individual. Whether SARS can spread through the air or in other ways remains to be discovered. Bats are the reservoir of SARS-CoV.

Many people remain asymptomatic after contacting SARS-CoV. However, in affected individuals, moderate URT illness may occur and include

Reservoir: The location or organism where pathogens exist and maintain their ability for infection.

fever (greater than 38°C), headache, an overall feeling of discomfort, and body aches. After two to seven days, SARS patients may develop a dry cough and have trouble breathing. In those patients progressing to a severe LRT illness, pneumonia develops with insufficient oxygen reaching the blood. In 10% to 20% of cases, patients require mechanical ventilation.

Most of the cases of SARS in the United States in 2003 occurred among travelers returning from other parts of the world where SARS was present. Transmission of SARS to healthcare workers appears to have occurred after close contact with sick people before recommended infection control precautions were put into place.

Because this is a newly emerging disease, treatment options remain unclear.

Hantaviruses. In the autumn of 1992, the El Niño oscillation of the ocean-atmosphere system caused heavy precipitation in the Four Corners region of the United States (New Mexico, Arizona, Colorado, and Utah), resulting in the increased growth of berries, seeds, and nuts in the spring of 1993. The increased food supply brought an explosion in the rodent population in this area. Then, a cluster of sudden and unexplained deaths in previously healthy young adults occurred in rural New Mexico and the Four Corners region. The CDC identified a hantavirus, now called Sin Nombre virus, as the infectious agent and termed the pulmonary disease **hantavirus pulmonary syndrome (HPS)**. Between 1993 and 2009, there have been some 500 cases of HPS and about 35% of reported cases have resulted in death.

The hantaviruses are members of the Bunyaviridae. Their name is derived from the Hantaan River in South Korea where the virus was first isolated in 1978. These are enveloped, −ssRNA viruses with helical symmetry

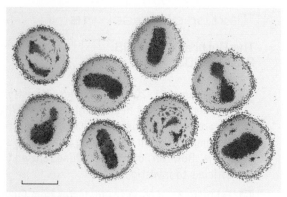

FIGURE 15.9 **The Hantavirus.** False-color transmission electron micrograph of hantaviruses. (Bar = 90 nm.) »» What do the red geometric shapes represent in the hantaviruses?

(**FIGURE 15.9**). The genome consists of three segments.

The deer mouse is the reservoir for the Sin Nombre virus, and it sheds the virus in saliva, urine, and feces. Humans usually are infected by breathing the infectious aerosolized dried urine or feces.

In one to five weeks after exposure, early symptoms of infection include fatigue, fever, and muscle aches. About half of all HPS patients experience headaches, dizziness, difficulty breathing, and low blood pressure that can lead to respiratory failure as the lungs fill with fluid.

Prevention consists of eliminating rodent nests and minimizing contact with them. There is no vaccine for hantavirus infection.

HPS has now become an established disease in the lexicon of medicine not only in the United States but throughout much of the Americas. **TABLE 15.2** summarizes the viral diseases affecting the LRT.

CONCEPT AND REASONING CHECKS

15.5 How do the SARS coronavirus and the hantavirus differ in their spread between individuals?

15.3 Diseases of the Skin Caused by Herpesviruses

Several viral diseases of the skin are caused by members of the Herpesviridae. Some, such as herpes simplex, remain epidemic in contemporary times; others such as chickenpox are being brought under control through effective vaccination programs. However, there are relatively few antiviral drugs for, and prevention programs remain a major course of action in, dealing with these diseases.

Presently, there are eight known viral species in the family Herpesviridae that infect humans (**FIGURE 15.10**). However, the small number of viruses should not be an indication of their significance. For example, some virologists believe over 90% of Americans have been exposed to the herpes simplex virus (HSV) by age 18 and more than 90% of adults worldwide have been infected with the Epstein-Barr virus (EBV). In addition,

TABLE

15.2   A Summary of the Major Viral LRT Diseases

Disease	Causative Agent	Signs and Symptoms	Transmission	Treatment	Prevention and Control
Influenza	Influenza A, B, and C viruses	Chills, fatigue headache Chest, back, and leg pain	Respiratory droplets	Bed rest and fluids Antiviral medications (oseltamivir or zanamivir)	Getting annual flu vaccination
Respiratory syncytial (RS) disease	Respiratory syncytial virus (RSV)	Influenza-like	Respiratory droplets Hand contact	Fever-reducing medications Ribavirin for severe cases	Practicing good hygiene
Parainfluenza	Human parainfluenza viruses 1 and 3	Cold-like	Respiratory droplets Direct contact	No specific therapy	Practicing good hygiene
RSV-like illness	Human metapneumovirus	Cold-like	Respiratory droplets Direct contact	Fever-reducing medications Ribavirin for severe cases	Practicing good hygiene
SARS	SARS coronavirus	Fever, headache, body aches, dry cough, and breathing difficulty	Respiratory droplets and airborne particles Direct contact	No effective treatment	Practicing good hygiene
Hantavirus pulmonary syndrome	Hantavirus (Sin Nombre virus)	Fatigue, fever, muscle aches, headache, dizziness, breathing difficulty	Aerosolized droplets of rodent saliva, urine, feces	Supportive care	Eliminating rodent nests Minimizing contact

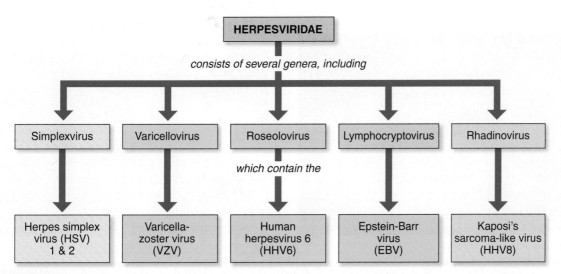

FIGURE 15.10 **The Relationships between the Human Herpesviruses.** This concept map illustrates the relationships between six of the eight viral species in the family Herpesviridae that cause human disease. Only EBV and HHV8 are known to be oncogenic. »» What does it mean for a virus to be oncogenic?

genital herpes is one of the most common sexually transmitted diseases (STDs) in the world. The word *herpes* is Greek for "creeping," referring to the spreading of herpes infections through the body after contact has been made.

All human herpesviruses are large virions with a double-stranded DNA genome. They have icosahedral symmetry and an envelope with spikes. Another shared characteristic is their ability to establish a latent infection (see Chapter 14) and then reactivate at some later period. The herpesviruses can infect several types of cells in epithelial and neural tissue, causing a variety of skin and related diseases.

Human Herpes Simplex Infections Are Widespread and Often Recurrent

KEY CONCEPT

6. Human herpes simplex viruses cause an array of viral diseases.

Two of the most prevalent herpesviruses are herpes simplex viruses 1 and 2 (HSV-1, HSV-2; **FIGURE 15.11A**).

Cold sores are caused primarily by HSV-1 and they are contagious. Such sores and blisters of herpes infections have been known for centuries. In ancient Rome, an epidemic was so bad that the Emperor Tiberius banned kissing and Shakespeare, in *Romeo and Juliet*, writes of "blisters o'er ladies lips."

Cold sores, also called fever blisters, start as a tingling sensation and the presence of a small, hard spot on the lip. Within a couple of days red, fluid-filled blisters appear (**FIGURE 15.11B**). The blisters eventually break, releasing the fluid containing infectious virions. A yellow crust forms and then peels off without leaving a scar. Cold sores generally clear up without treatment in 7 to 10 days. A person is most likely to transmit the infection from the time the blisters appear until they have completely dried and crusted over.

After the primary infection, the viruses enter the sensory neurons and become latent in the nearby sensory ganglia. Viral reactivation and movement to the epithelia can trigger another round of cold sores. Reactivation of the dormant viruses often occurs after some form of trauma, often in response to stressful triggers, such as fever, menstruation, or emotional disturbance. Even environmental factors like sunburn (exposure to ultraviolet light) can trigger reactivation (**FIGURE 15.12**).

Preventing the transmission of cold sores means not kissing others while the blisters are present. Washing one's hands often and not touching other areas of the body also help limit virus spread. For example, the eyes can become infected through touching with a contaminated finger. Infection of the eye, called **herpes keratitis**, causes scarring of the cornea and is a leading cause of blindness in the United States. Some 400,000 Americans suffer from a form of ocular herpes.

Genital herpes is one of the most common STDs in the United States, accounting for

Ganglia:
Structures containing a dense cluster of nerve cells.

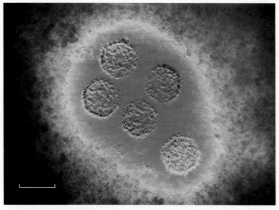

Herpes simplex virus

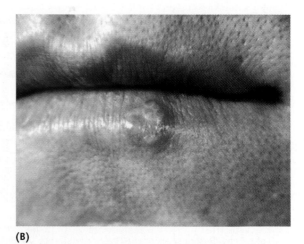

(A)

(B)

FIGURE 15.11 Herpes Simplex Virus and Cold Sores. (**A**) A false-color transmission electron micrograph of a cell infected with herpes simplex viruses. (Bar = 150 nm.) (**B**) The cold sores (fever blisters) of herpes simplex erupting as tender, itchy papules and progressing to vesicles that burst, drain, and form scabs. Contact with the sores accounts for spread of the virus. »» How would the release phase of the HSV replicative cycle lead to blisters and vesicles on the lips?

more than 600,000 new cases diagnosed each year. Globally, more than 500 million people are infected with HSV-2, although the incidence of genital herpes caused by HSV-1 has been increasing. Genital herpes is spread primarily by sexual contact. It is highly unlikely a person could contract an infection through contact with toilets or other objects used by an infected person because the virus "dies" quickly outside the body.

Signs generally appear within a few days of sexual contact, often as itching or throbbing in the genital area. This is followed by reddening and swelling of a small area where painful, thin blisters erupt. The blisters crust over and the sores disappear, usually within about three weeks. The signs often are very mild and may go unnoticed. Fifty percent of infected individuals experience only one outbreak in their lifetime.

However, for many infected individuals, a latent period occurs during which time the virus remains in nerve cells near the blisters. In the majority of cases, the virus becomes reactivated again by stressful situations and symptoms reappear. The cycle of latency and recurrent infections can occur three to eight times a year. People with active herpes lesions are highly infectious and can pass the viruses to others during sexual contact.

In healthy individuals, genital herpes usually causes no serious complications. Importantly though, a person with the infection has an increased risk of transmitting or contracting other sexually transmitted diseases, including AIDS.

Prevention is similar to that for any STD—abstain from sexual activity or limit sexual contact to only one person who is infection-free. Although there is no cure for genital herpes, antiviral drugs such as acyclovir can help heal the sores sooner and reduce the frequency of recurrent infections.

Neonatal herpes is a devastating and life-threatening disease transmitted by infected

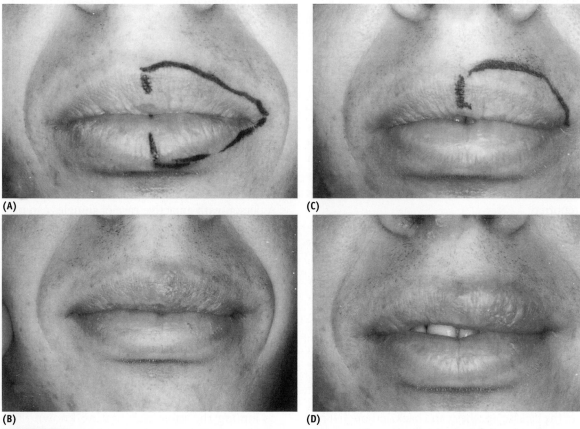

(A)   (B)   (C)   (D)

FIGURE 15.12 Ultraviolet Light and Herpes. An experiment showing the effects of ultraviolet (UV) light on the formation of herpes simplex sores of the lips. This patient usually experienced sores on the left upper lip. (**A**) The patient was exposed to UV light from a retail cosmetic sunlamp on the left upper and lower lips in the area designated by the line. The remainder of the face was protected by a sunscreen. (**B**) Sores formed on the left upper lip. (**C**) The patient was later exposed a second time to the sunlamp, but only on the left upper lip. (**D**) Sores formed and were larger than the previous ones. »» What would you conclude from these experiments regarding the reactivation of herpesviruses?

mothers to newborns during childbirth. Although most pregnant women with genital herpes have healthy babies, a small number spread the infection to their newborns during labor and delivery. Even with adequate treatment for the newborn, mental development can be delayed, blindness can occur, and persistent seizures can result. If symptoms and a diagnosis of an active genital herpes infection are made prior to delivery, the obstetrician may recommend birth by cesarean section.

In recent years, the acronym **TORCH** has been coined to focus attention on diseases with congenital significance and induced by microbial teratogens: **T** for toxoplasmosis, **O** is for other diseases, such as syphilis; **R** for rubella, **C** for cytomegalovirus, and **H** for herpes simplex virus.

■ Teratogens:
Agents that interfere with the normal development of the fetus.

CONCEPT AND REASONING CHECKS

15.6 Explain how a primary infection and a latent infection differ for HSV.

Chickenpox Is No Longer a Prevalent Disease in the United States

KEY CONCEPT

7. The varicella-zoster virus causes chickenpox and shingles.

In the centuries when pox diseases regularly swept across Europe, people had to contend with the Great Pox (syphilis), the smallpox, the cowpox, and the chickenpox. Prior to the availability of a vaccine for chickenpox in 1995, about 4 million children contracted chickenpox each year in the Unites States. In 2007, there were 40,146 cases reported.

Chickenpox (varicella) is a highly communicable disease caused by the varicella-zoster virus (VZV), another virus species in the family Herpesviridae. It is transmitted by respiratory droplets and skin contact, and it has an incubation period of two weeks.

The disease begins in the respiratory tract, with fever, headache, and malaise. Viruses then pass into the bloodstream and localize in the peripheral nerves and skin. As it multiplies in the cutaneous tissues, VZV produces a red, itchy rash on the face, scalp, chest, and back, although it can spread across the entire body. The rash quickly turns into small, teardrop-shaped, fluid-filled vesicles (**FIGURE 15.13A**). The vesicles in chickenpox develop over three or four days in a succession of "crops." They itch intensely and eventually break open to yield highly infectious virus-laden fluid. Although many refer to the vesicles as pox, the latter term is more correctly reserved for the pitted scars of smallpox. In chickenpox, the vesicles form crusts that fall off. A person who has chickenpox can transmit the virus for up to 48 hours before the telltale rash appears and remains contagious until all spots crust over. Chickenpox usually lasts two weeks or less and rarely causes complications.

The most common complication of chickenpox is a bacterial infection of the skin. Chickenpox may also lead to pneumonia or an inflammation of the brain (encephalitis), both of which can be very serious if not treated. Reye syndrome may occur during the recovery period, so aspirin should not be used to reduce fever in both children and adults.

Acyclovir has been used in high risk groups to lessen the symptoms of chickenpox and hasten recovery. A varicella-zoster immune globulin (VZIG), which contains antibodies to the chickenpox virus, may also be used. The chickenpox vaccine (Varivax) is the safest, most effective way to prevent chickenpox and its possible complications. The varicella vaccine is 85% effective in disease prevention.

Shingles (zoster) is usually an adult disease produced by the same virus causing chickenpox. Anyone who has had chickenpox as a child is at risk for the latent illness, although only about 10% of adults actually develop shingles. Still, this amounts to some one million new cases each year. A person with an active case of shingles can pass VZV to anyone who has not had chickenpox; that person will develop chickenpox, not shingles.

After having chickenpox, some VZV may remain in nerve cells. Many years later, the virus can reactivate, travel down the nerves to the skin of the body trunk, and resurface as shingles (**FIGURE 15.13B**). Here they cause blisters with blotchy patches of red that appear to encircle the trunk (*zoster* = "girdle"). Many sufferers also experience a series of headaches as well as facial paralysis and sharp "ice-pick" pains described as among the most debilitating known. The condition can occur repeatedly and is linked to emotional and physical stress (such as radiation therapy) as well as to a suppressed immune system or aging.

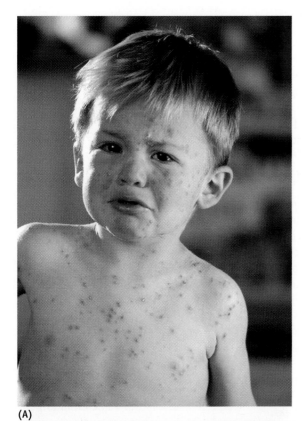

(A)

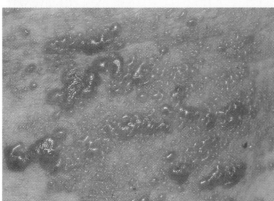

(B)

FIGURE 15.13 The Lesions of Chickenpox and Shingles. (**A**) A typical case of chickenpox. The lesions may be seen in various stages, with some in the early stage of development and others in the crust stage. (**B**) Dermal distribution of shingles lesions on the skin of the body trunk. The lesions contain less fluid than in chickenpox and occur in patches as red, raised blotches. »» Summarize how varicella latency could lead to shingles.

Shingles can lead to its own complication—a condition in which the pain of shingles persists for years after the blisters disappear. This complication, called **postherpetic** neuralgia, can be severe. Most cases occur in people over age 60.

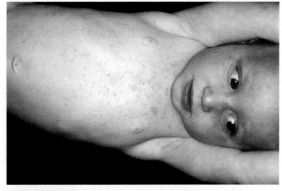

FIGURE 15.14 Roseola. Following a high fever, a red body rash may appear on the neck and trunk of infants or young children. »» Why should aspirin not be given to infants or young children suffering a viral infection such as roseola?

In 2006, a live, attenuated vaccine, called zostavax, was approved to help prevent or reduce the length and severity of the disease in people over 60 years of age. For herpes zoster, acyclovir therapy lessens the symptoms and some unconventional methods also may help (MICROFOCUS 15.3).

CONCEPT AND REASONING CHECKS

15.7 Assess the consequences of someone having shingles being able to transmit the virus to a susceptible person.

Human Herpesvirus 6 Infections Primarily Occur in Infancy

KEY CONCEPT

8. Human herpesvirus 6 infection primarily causes a short-term fever and rash in infants.

Human herpesvirus 6 (HHV-6) belongs to a different genus of herpesviruses from HSV-1 and HSV-2. HHV-6 primarily affects infants and young children six months to three years old. The virus causes **roseola**, an acute, self-limiting condition marked by high fever. This often is followed by a red body rash (**FIGURE 15.14**). Roseola can spread from person to person through contact with an infected person's respiratory secretions or saliva. Interestingly, it appears that about 50% of children suffering roseola inherited the virus congenitally; that is, it was passed as a chromosomal provirus (see Chapter 14) from either the mother or father. The infection usually lasts about one week and treatment requires bed rest, fluids, and fever-reducing (nonaspirin) medications.

Neuralgia:
Severe pain in a part of the body through which a particular nerve runs.

MICROFOCUS 15.3: Being Skeptical

Can Chinese Tai Chi Prevent Shingles?

We all know that exercise is a good thing. It builds strength, endurance, and cardiovascular health. One Chinese form of physical and aerobic exercise for attaining good health is through Tai Chi. Tai Chi involves a series of 20 slow and deliberate body movements that purport to produce a calm and tranquil mind. In 2003, researchers announced that Tai Chi Chih, a low-impact form of Tai Chi, boosts shingles immunity in the elderly. Can Tai Chi actually do this?

The theory goes that when a person's immunity is compromised or wanes as one ages, the individual is susceptible to more infections. For example, it is true that individuals over 50 are more prone to recurrent shingles attacks as the varicella-zoster virus (VZV) reactivates because it may no longer be kept in check by the immune system. Stress and other trauma also can intensify the reactivation. Therefore, Michael R. Irwin and colleagues at UCLA's Neuropsychiatric Institute decided to examine if Tai Chi could reduce the levels of stress and thereby boost a person's immunity—and reduce the chances for shingles.

Thirty-six men and women over 60 were enrolled in the study. Half were started on a 15-week program of Tai Chi Chih. The other half were asked to postpone the program for 15 weeks.

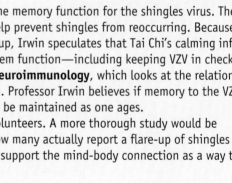

After the 15-week period, analysis indicated the Tai Chi group had a 50% increase over the control group in immune memory function for the shingles virus. The researchers state the increase was sufficient to actually help prevent shingles from reoccurring. Because Tai Chi did not improve physical movement within the group, Irwin speculates that Tai Chi's calming influence reduces stress levels, which could boost immune system function—including keeping VZV in check.

Such studies are part of a growing field called **psychoneuroimmunology**, which looks at the relationships between the nervous system and the immune system. Professor Irwin believes if memory to the VZV can be maintained, memory to other infections might also be maintained as one ages.

The verdict: This was a small study involving only 36 volunteers. A more thorough study would be to follow groups of Tai Chi Chih volunteers and monitor how many actually report a flare-up of shingles compared to the general population. Still, this study does support the mind-body connection as a way to maintain good health.

Monocytes:
A type of white blood cell formed in the bone marrow.

Viremia:
The presence of viruses in the blood.

The prevalence of the virus is seen by the fact that most children have been infected by the time they enter kindergarten and up to 80% of adults have antibodies against HHV-6. Like all herpesviruses, it can remain latent in saliva and monocytes. As such, 30% to 60% of bone marrow transplant recipients suffer a HHV-6 viremia during the first few weeks after transplantation. The recurrence can potentially lead to pneumonia or encephalitis. No drugs have been approved for HHV-6 infections.

CONCEPT AND REASONING CHECKS

15.8 Determine why a bone-marrow transplant recipient is at risk of contracting an HHV-6 infection.

A Few Herpesvirus Infections Are Oncogenic

KEY CONCEPT

9. The Epstein-Barr virus and human herpesvirus 8 have oncogenic potential.

Only two human herpesviruses, the Epstein-Barr virus (EBV) and human herpesvirus 8 (HHV-8), have been identified as oncogenic agents. EBV more commonly is associated with benign infections, such as infectious mononucleosis, and will be discussed in Chapter 16.

Kaposi sarcoma (KS) is a highly angiogenic tumor of the blood vessel walls. It is most commonly seen in individuals with a weakened immune system, such as those with HIV/AIDS. The malignancy is caused by human herpesvirus 8 (HHV-8) or Kaposi sarcoma–associated herpesvirus. KS, which is marked by dark or purple skin lesions, has become one of the most common tumors in AIDS patients (**FIGURE 15.15**). The DNA of HHV-8 is present in most biopsies of tissue from KS patients and antibodies against the virus are invariably detected in those with the disease or at risk of developing it.

Treatment of smaller skin lesions involves the use of liquid nitrogen, low-dose radiation, or chemotherapy applied directly to the lesion.

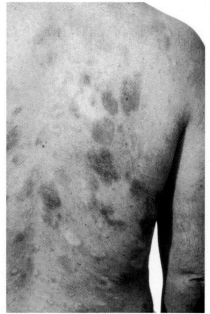

Angiogenic: Referring to the generation of many blood vessels.

FIGURE 15.15 **Kaposi's Sarcoma.** This man is suffering from AIDS. Kaposi's sarcoma is an opportunistic disease, appearing as dark lesions on the skin. »» What is meant by an opportunistic disease?

TABLE 15.3 summarizes the characteristics of the herpesviruses that affect the skin.

CONCEPT AND REASONING CHECKS

15.9 Although all herpesviruses appear to have a latent state, why are only EBV and HHV-8 potentially carcinogenic?

15.4 Other Viral Diseases of the Skin

There are a number of other viral infections and diseases associated with the skin. Some are common childhood diseases, such as mumps and measles, while another, smallpox, has been eradicated but was once a human scourge worldwide.

Paramyxovirus Infections Can Cause Typical Childhood Diseases

KEY CONCEPT

10. Two genera of paramyxoviruses cause measles and mumps.

Measles is a highly contagious disease caused by a viral species of the Paramyxoviridae family (see

Figure 15.6). Through global immunization efforts primarily in developing nations, since 2000 the number of reported measles cases worldwide has decreased 67% (279,000 cases in 2007) and global mortality from measles has been reduced by 74% (197,000 deaths in 2007).

Transmission of the measles virus usually occurs by respiratory droplets during the early stages of the disease. Symptoms commonly include a hacking cough, sneezing, nasal discharge, eye redness, sensitivity to light, and a high fever. Red patches with white grain-like centers appear along the gum line in the mouth two to four days after the onset of symptoms. These diagnostic patches

TABLE

15.3 A Summary of Viral Diseases Caused by the Herpesviruses

Disease	Causative Agent	Signs and Symptoms	Transmission	Treatment	Prevention and Control
Cold sores	Herpes simplex virus 1	Small, red hard spot on lip	Contact	Lidocaine Benzyl alcohol	Avoiding skin contact Not sharing personal items
Genital herpes	Herpes simplex virus 2	Itching or throbbing in the genital area Reddening and swelling of a small area where painful, thin blisters erupt	Sexual intercourse	Antiviral drugs (acyclovir)	Abstaining from sexual activity Limiting sexual contact to only one person who is infection free
Neonatal herpes	Herpes simplex virus 2	Mental development can be delayed Blindness can occur Persistent seizures	From infected mother to newborn during childbirth	Acyclovir Supportive therapy	Cesarean section if the maternal infection is recognized Screening women considered to be at high risk
Chickenpox (varicella)	Varicella-zoster virus	Fever, headache, malaise with red, itchy rash on face, scalp, chest, and back	Droplets Contact	Supportive care Acyclovir	Chickenpox vaccine
Shingles (zoster)	Varicella-zoster virus	Blisters on body trunk with intense pain		Acyclovir	Shingles vaccine Acyclovir
Roseola	Human herpesvirus 6	Red rash on neck and trunk	Contact	Supportive care	Avoiding exposure to infected child
Kaposi sarcoma	Human herpesvirus 8	Dark lesion on skin	Contact (sexual and nonsexual)	Anti-retroviral therapy	Using anti-HIV medications

Measles virus

are called **Koplik spots** after Henry Koplik, the New York pediatrician who described them in 1896 (FIGURE 15.16A).

The characteristic red rash of measles appears about two days after the first evidence of Koplik spots. Beginning as pink-red pimple-like spots (maculopapules), the rash breaks out at the hairline, then covers the face and spreads to the trunk and extremities (FIGURE 15.16B). **Rubeola**, the alternative name for measles, is derived from the Latin *rube* for "red." Rashes resemble those in scarlet fever, but the severe sore throat of scarlet fever generally does not develop. Within a week, the rash turns brown and fades.

Measles usually is characterized by complete recovery. In rare cases, **subacute sclerosing panencephalitis (SSPE)**, a brain disease characterized by a decrease in cognitive skills and loss of nervous function, may occur, primarily in men, some 1 to 10 years after recovery from measles, due to a persistent measles infection.

Prevention is accomplished with the measles vaccine, which usually is given as a measles-mumps-rubella (MMR) inoculation. Immunization of children entering grade school is now mandatory in all 50 states. In 1978, the U.S. Public Health Service launched a campaign to eliminate measles in the United States. By 1983, the total number of reported cases was

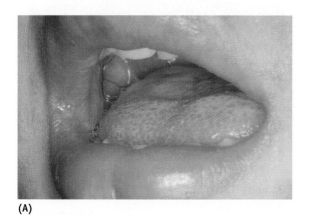

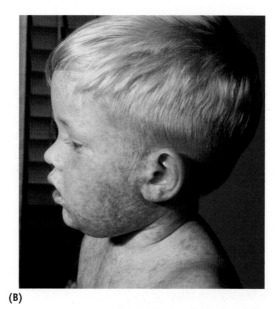

(A)   **(B)**

FIGURE 15.16 Koplik Spots and Measles. (**A**) Koplik's spots in the mouth of a child suffering from measles. The red spots with white centers are a frequent symptom in this highly infectious viral disease. (**B**) A child with measles, showing the typical rash on face and torso that mainly affects children. »» Based on the rash, how would you distinguish measles from scarlet fever?

1,463, a reduction of 99.7% from the pre-vaccine era. In 2007, there were 43 reported cases, most imported. Still, measles can be cause for alarm, as MicroFocus 15.4 points out.

Mumps is caused by another member of the Paramyxoviridae. Its name comes from the English "to mump," meaning to be sullen or to sulk. The characteristic sign of the disease is enlarged jaw tissues arising from swollen salivary glands, especially the parotid glands (**FIGURE 15.17**). **Infectious parotitis** is an alternate name for the disease.

Mumps is spread by respiratory droplets or contact with contaminated objects; it is considered less contagious than measles or chickenpox. The virus is found in human blood, urine, and cerebrospinal fluid, even though its effects are observed primarily in the parotid glands.

About 20% of people infected with the mumps virus have no signs or symptoms. When signs and symptoms do appear, swollen and painful salivary glands are typical, which causes the puffy cheek appearance. Obstruction of the ducts leading from the parotid glands retards the flow of saliva, which causes the characteristic swelling. The skin overlying the glands is usually taut and shiny, and patients experience pain when the glands are touched as well as when chewing or swallowing.

In male patients, the mumps virus may pose a threat to the reproductive organs causing swelling and damage to the testes, a condition called **orchitis** (*orchi* = "the testicles"). The sperm count may be reduced, but sterility is not common. An estimated 25% of mumps cases in post-adolescent males develop into orchitis.

Prevention of mumps is achieved in developed nations through vaccination with the MMR (measles-mumps-rubella) vaccine. The average number of mumps cases reported nationwide by the CDC from 2001 to 2005 was 265. However, in 2006, the largest mumps outbreak in more than 20 years occurred, with more than

■ Parotid glands:
One of three pairs of salivary glands, located below and in front of the ears.

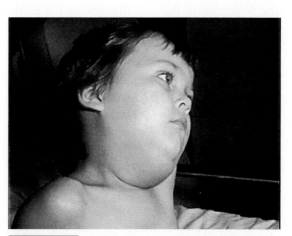

FIGURE 15.17 Mumps. Close-up of a young child with mumps (infectious parotitis). »» What causes the swelling of the jaw region?

MICROFOCUS 15.4: Public Health
The Wanderings of a Measles Infection

Very few measles cases are reported in the United States anymore. However, imported cases could spread the disease to at-risk individuals.

The following example, which occurred in Arizona in January 2005, demonstrates how one infectious individual could spread the disease to a community including thousands of students at Arizona State University (ASU). Times and locations are estimates.

January 9
- 3:30 PM: McDonald's (Camp Verde)
- 4:00 PM: Target (Flagstaff)
- 5:00 PM: Burger King (Flagstaff)
- Evening: Beaver Street Brewery (Flagstaff)
- Quality Inn (Flagstaff)

January 10
- AM: Chapel of the Holy Cross (Sedona)
- ??: Subway/gas station (Camp Verde)
- Evening: Safeway (Tempe)

January 11
- ASU campus buildings (Tempe)

January 12
- ASU campus buildings (Tempe)
- 4:00 PM: Phoenix City Hall (Phoenix)
- 5:00 PM: Hotel San Carlos (Phoenix)

January 13
- ASU campus buildings (Tempe)

January 14–17
- ??

January 18
- 9:00 AM: Advanced Urgent Care (Phoenix)
- 1:00 PM: Emergency room, Tempe St. Luke's Hospital (Tempe)

No measles cases were reported, but the potential to infect unvaccinated individuals was of great concern.

6,500 cases reported in six Midwestern states (Illinois, Iowa, Kansas, Nebraska, South Dakota, and Wisconsin). Because the disease primarily affected young adults aged 18 to 24 years of age—a large percentage of them in high school and college environments—the CDC has stated that the immunity conferred by the vaccine may have waned. Therefore, anyone 18 years of age or older who was born after 1956 should get at least one dose of MMR vaccine. In 2007, there were 1,800 reported cases.

Rubella virus

CONCEPT AND REASONING CHECKS

15.10 How do measles and mumps differ as diseases?

Rubella (German Measles) Is an Acute, Mildly Infectious Disease

KEY CONCEPT

11. The rubella virus produces a pale-pink rash on the trunk and extremities.

For generations, **rubella** (*rube* = "reddish"; *ella* = "small") or **German measles** (*germanus* = "similar") was thought to be a mild form of measles. In 1829, the German physician Rudolph Wagner noted the differences between the diseases. The disease occurs in several countries, although more than half now use a rubella vaccine.

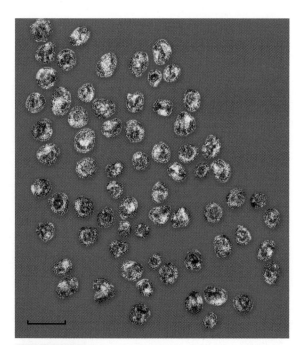

FIGURE 15.18 The Rubella Virus. This false-color transmission electron micrograph shows a group of rubella viruses. (Bar = 160 nm.) »» What visual characteristic would assign the virus to the togaviruses?

Rubella is caused by a +ssRNA virus of the Togaviridae (*toga* = "cloak") family (**FIGURE 15.18**). The virion is icosahedral with a coat-like envelope and spikes. Viral transmission generally occurs by contact or respiratory droplets, and the disease is usually mild. Rubella—the **R** in the **TORCH** group of diseases—is accompanied by occasional fever with a variable, pale-pink maculopapular rash beginning on the face and spreading to the body trunk and extremities. The rash develops rapidly, often within a day, and fades after another two days. Recovery is usually prompt, but relapses appear to be more common than with other diseases, possibly because the viruses remain active within body cells.

Rubella is dangerous to the developing fetus in a pregnant woman. Called **congenital rubella**, transplacental infection of the fetus can lead to destruction of the fetal capillaries, and blood insufficiency follows. The organs most often affected are the eyes, ears, and cardiovascular organs, and children may be born with cataracts, glaucoma, deafness, or heart defects.

Since its introduction in 1969, the rubella vaccine, which is combined with the measles and mumps vaccines (MMR), has had a dramatic effect on the rate of incidence of the disease in the United States. In 1969, physicians reported 58,000 cases of rubella, but by 2007, the number was down to 12 reported cases and rubella is no longer considered an endemic disease in the United States. Rubella symptoms are so mild that no treatment is required.

CONCEPT AND REASONING CHECKS

15.11 Hypothesize why rubella has been called three-day measles.

Fifth Disease (Erythema Infectiosum) Produces a Mild Rash

KEY CONCEPT

12. Human parvovirus B19 is most common among elementary school-age children.

In the late 1800s, Roman numerals were assigned to diseases accompanied by skin rashes. Disease I was measles, II was scarlet fever, III was rubella, IV was Duke's disease (also known as roseola and now recognized as any rose-colored rash), and V was erythema infectiosum. This so-called fifth disease remained a mystery until the modern era.

The agent of **fifth disease** is human parvovirus B19, which is a small, single-stranded DNA virion of the Parvoviridae (*parv* = "small") family, having icosahedral symmetry. Like all parvoviruses, B19 is dependent on the host cell or other viruses (adenoviruses, herpesviruses) to replicate.

Community outbreaks of fifth disease occur worldwide and transmission appears to be by respiratory droplets. Fifth disease primarily affects children although most infections are asymptomatic. If symptoms occur, the outstanding characteristic is a fiery red rash on the cheeks and ears (**FIGURE 15.19**). The rash may spread to the trunk and extremities, but it fades within several days, leaving a "lacy" rash on the skin.

Parvovirus B19 only infects humans (dog and cat parvoviruses do not infect humans). However, fifth disease is not limited to children. Adults suffer from painful joints similar to the symptoms

Endemic: Referring to a disease occurring within a specific area, region, or locale.

Fifth disease virus

Human papilloma virus

Transplacental: Refers to movement from mother to fetus.

FIGURE 15.19 Fifth Disease. The fiery red rash of a child with fifth disease (erythema infectiosum). The confluent red rash is seen. »» Why is fifth disease sometimes called "slapped-cheek disease"?

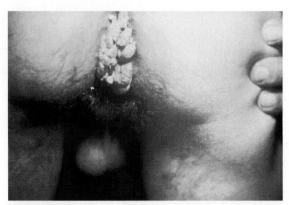

FIGURE 15.20 Genital Warts. Genital warts are caused by the human herpes simplex virus types 1 or 2. The warts typically appear as bumps on the genitalia, or in this case on the anal region of this male patient. »» What must the herpesvirus infection do to the infected skin and mucosal surfaces to produce the wart appearance?

of rheumatoid arthritis, especially in the fingers, wrists, knees, and ankles. Infection of the bone marrow also may lead to anemia, and pregnant women may suffer miscarriage (but birth defects generally do not occur). Although antibody preparations (immune globulin) are available for treatment, the symptoms usually resolve spontaneously.

CONCEPT AND REASONING CHECKS

15.12 What is the unique feature of parvovirus B19?

Some Human Papillomavirus Infections Cause Warts

KEY CONCEPT

13. Human papillomaviruses can cause common or genital warts.

Common Warts. Common warts are small, usually benign skin growths resulting from infection by a specific strain of the human papillomavirus (HPV). These viruses represent a collection of over 100 different types of icosahedral, naked, double-stranded DNA virions of the Papovaviridae family.

Common warts typically appear on the hands or fingers, and **plantar warts** occur on the soles of the feet. In most cases, these skin warts are white or pink and cause no pain. Although rare, common warts can be acquired through direct contact with HPV from another person or by direct contact with a towel or object used by someone who has the virus. Therefore, prevention requires maintaining proper cleanliness and not picking at them, which can lead to their spread.

Common warts can be difficult to treat and eradicate. A physician may try freezing or minor surgery, or prescribe a chemical treatment. Recently, a new and innovative treatment has shown great success in eliminating common warts (MicroFocus 15.5).

Genital Warts. According to the CDC, there are some 1 million new cases of **genital warts** each year in the United States. Although a third of the known HPV types is sexually transmitted, about 90% of genital warts are caused by only two types, HPV 6 and 11.

Being an STD, the virus is most commonly transmitted through oral, vaginal, or anal sex with someone who has an HPV infection. Most people who acquire these types never develop warts or any other symptoms. When the warts do occur, they appear as small, flat, flesh-colored bumps or tiny, cauliflower-like bumps anywhere in the genital region or areas around the anus (**FIGURE 15.20**). The warts are sometimes called **condylomata** (*condylo* = "knob"), a reference to the bumpy appearance of the warts.

Although genital warts are not life threatening, there is no cure for the HPV infection because it can remain dormant in a latent form (see Chapter 14). Visible warts can be removed by freezing or electrosurgical excision, or by using specific chemicals available only to doctors. No one method is better than the others. Abstinence is the only sure way to prevent infection (although see below), as all other methods carry some risk.

MICROFOCUS 15.5: Being Skeptical
Skin Test Antigens Eliminate Warts

Many of us suffer from allergies and have had skin tests done to discover what substances (allergens) we are sensitive to. Researchers now suggest that injections of these same allergens can cure common or genital warts. So, what's the evidence?

Researchers at the University of Arkansas School for Medical Sciences reported in 2005 that injecting substances representing skin test antigens could eliminate warts—all of them! All that was required was to inject a single wart.

The researchers stated that 50% of volunteers injected with the antigens had complete eradication of all body warts, including genital warts. The volunteers were wart free!

How could this be? The research team suggests that the antigen injection into a single wart (see figure) stimulates an attack by the immune system. This attack is not only on the antigens injected, but also the human papillomaviruses in the mix of antigens.

The verdict: It is hard to argue against results that completely abolish the infection. And besides being effective, it is safe and relatively painless.

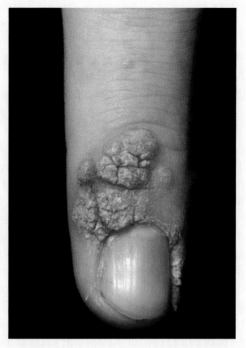

Skin warts on finger.

But, even after treatment, genital warts can come back. In fact, 25% of cases recur within 3 months.

Cervical/Penile Cancer. Cervical cancer develops in tissues of the cervix. Although HPV 6 and 11 are considered a "low risk" potential as cancer-causing agents, other HPV types are strongly associated with precancerous changes and cervical cancers (**FIGURE 15.21**). HPV 16 is responsible for about 50% of cervical cancers, and together with types 18, 31, and 45, account for 80% of the cancers. Once cervical cells begin to change or become precancerous, it typically takes 10-15 years before invasive cervical cancer develops. All HPV viruses can result in abnormal Pap smears, so Pap tests are a critical screening procedure for all women, especially women who may be infected with HPV.

A recombinant vaccine, called Gardasil, has now been approved for use in females 9 to 26 years of age. The vaccine is aimed at the most prominent types of HPV causing genital warts and cancer; in fact, the U.S. Food and Drug Administration (FDA) reports the vaccine to be essentially 100% effective against HPV types 16 and 18, which cause

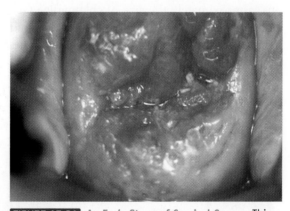

FIGURE 15.21 An Early Stage of Cervical Cancer. This photograph shows the precancerous cervix of a patient typified by erosion to her cervix. Such abnormal changes can lead to invasive cervical cancer. »» Why is it most likely that this patient is in her 20s rather than her 50s?

approximately 70% of cervical cancers and against HPV types 6 and 11, which cause approximately 90% of genital warts. In 2009, the vaccine was approved for boys and men, as HPV infections can lead to penile cancer. Importantly, the vaccine is not a treatment as it does not eliminate the virus in already infected individuals.

CONCEPT AND REASONING CHECKS

15.13 Why would a woman be considered to be at risk of developing cervical cancer if she has genital warts?

Poxvirus Infections Have Had Great Medical Impacts on Populations

KEY CONCEPT

14. The poxviruses produce contagious and dangerous diseases.

Though most dermotropic viral diseases tend not to be life-threatening, a few, such as smallpox, have exacted heavy tolls of human misery.

Smallpox is a contagious and sometimes fatal disease that until recently had ravaged people around the world since pre-biblical times. Few people escaped the pitted scars accompanying the disease, and children were not considered part of the family until they had survived smallpox. Thanks to Edward Jenner, the first attempts at a vaccine were begun in the late 1700s (see Chapter 1).

Smallpox is caused by a brick-shaped double-stranded DNA virus of the Poxviridae family (**FIGURE 15.22**). It is one of the largest virions, approximately the size of chlamydiae (see Chapter 13). The nucleocapsid is surrounded by a series of fiber-like rods with an envelope. Transmission is by contact.

The earliest signs of smallpox are high fever and general body weakness. Pink-red spots, called **macules**, soon follow, first on the face and then on the body trunk. The spots become pink pimples, called **papules** followed by fluid-filled vesicles so large and obvious that the disease is also called variola (*varus* = "vessel"; **FIGURE 15.23**). The vesicles become deep **pustules**, which break open and emit pus. If the person survives, the pustules leave pitted scars, or **pocks**. These are generally smaller than the lesions of syphilis (the Great Pox) or chickenpox. **TABLE 15.4** summarizes the stages of smallpox.

Vaccination has been hailed as one of the greatest medical and social advances because it was the first attempt to control disease on a national scale. It was also the first effort to protect the community rather than the individual. In 1966, the WHO received funding to attempt the global eradication of smallpox. With help from international service organizations, such as Rotary International, surveillance containment methods were used to isolate every known smallpox victim, and all contacts were vaccinated. The eradication was aided by the fact that smallpox viruses do not exist anywhere in nature except in humans. On October 26, 1977, health care workers reported isolation of the last case. In 1979, the WHO announced worldwide smallpox eradication, the first such claim made for any disease.

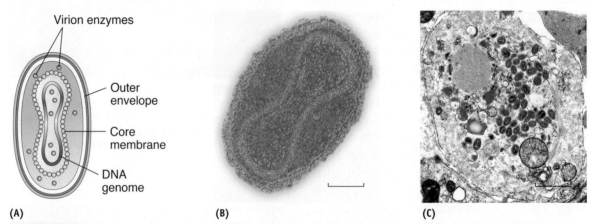

(A)   (B)   (C)

FIGURE 15.22 **The Smallpox Virus. (A)** A drawing of the smallpox virus, showing its complex features. **(B)** A false-color transmission electron micrograph of the smallpox virus cultivated in cell culture. (Bar = 50 nm.) **(C)** A false-color transmission electron micrograph of a cell infected with smallpox viruses. Rectangular mature virions (red) can be observed. (Bar = 800 nm.) »» Although variola virus has a DNA genome, it replicates totally in the cell cytoplasm. What enzyme must it carry to ensure proper DNA replication in the cytoplasm?

Labels in figure A: Virion enzymes; Outer envelope; Core membrane; DNA genome

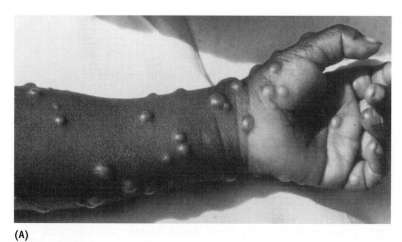
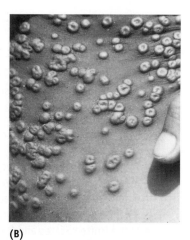

(A)   **(B)**

FIGURE 15.23 **The Lesions of Smallpox. (A)** The smallpox lesions are raised, fluid-filled vesicles similar to those in chickenpox. For this reason, cases of chickenpox sometimes are misdiagnosed as smallpox. Later, the lesions will become pustules (**B**), and then form pitted scars, the pocks. »» What is the difference between a vesicle and a pustule?

TABLE

15.4 Stages of Smallpox

Stage	Explanation	Duration	Contagious?
Incubation period	Following exposure to the virus, people do not have any symptoms and usually feel fine during the incubation period.	7–17 days	No
Initial symptoms	First symptoms include high fever (38°–40°C), malaise, head and body aches, and sometimes vomiting. Affected individuals are too sick to carry on their normal activities.	2–4 days	Sometimes
Early rash	**Days 1 & 2:** A rash emerges as small red spots on the tongue and in the mouth, and develop into sores that break open and spread large amounts of the virus into the mouth and throat. About the time the sores in the mouth break down, a rash appears on the skin, starting on the face, and spreads to the arms and legs and then to the hands and feet. Usually the rash spreads to all parts of the body within 24 hours. As the rash appears, the fever usually falls and the person may start to feel better. **Day 3:** The rash becomes raised bumps. **Day 4:** The bumps fill with thick, opaque fluid and often have a depression in the center that looks like a belly button. (This is a major distinguishing characteristic of smallpox.) Fever often occurs again until scabs form over the bumps.	4 days	Very contagious
Pustular rash	The bumps become pustules—sharply raised, usually round and firm to the touch, as if there is a small round object under the skin. People often said the bumps feel like BB pellets embedded in the skin.	5 days	Yes
Pustules and scabs	By the end of the second week after the rash appears, the pustules begin to form a crust and then scab.	5 days	Yes
Resolving scabs	Within three weeks after the rash appears, the scabs begin to fall off, leaving marks on the skin that become pitted scars.	6 days	Yes
Scabs resolved	Scabs have fallen off.		No

Source: Centers for Disease Control. Available at: http://www.cdc.gov.

MICROFOCUS 15.6: Public Health
"Should We or Shouldn't We?"

One of the liveliest debates in microbiology is whether the last remaining stocks of smallpox viruses should be destroyed. Here are some of the arguments.

For Destruction
- People are no longer vaccinated, so if the virus should escape the laboratory, a deadly epidemic could ensue.
- The DNA of the virus has been sequenced and many cloned fragments are available for performing research experiments; therefore, the whole virus is no longer necessary.
- Eradicating the disease means eradicating the remaining stocks of laboratory virus, and the stocks must be destroyed to complete the project.
- If the United States and Russia destroy their smallpox stocks, it will send a message that biological warfare will not be tolerated.

Against Destruction
- Future studies of the virus are impossible without the whole virus. Indeed, certain sequences of the viral genome defy deciphering by current laboratory means.
- Studying the genome of the virus without the whole virus will not provide insights into how the virus causes disease.
- Mutated viruses could cause smallpox-like diseases, so continued research on smallpox is necessary in order to be prepared.
- Smallpox viruses may be secretly retained in other labs in the world for bioterrorism purposes, so destroying the stocks may create a vulnerability. Smallpox viruses also may remain active in buried corpses.
- Destroying the virus impairs the scientists' right to perform research, and the motivation for destruction is political, not scientific.

Now it's your turn. Can you add any insights to either list? Which argument do you prefer? P.S. In April 2002, the World Health Assembly of the World Health Organization recommended postponing the destruction of all remaining smallpox stocks until all research and drug development is concluded. This will allow time to prepare for a natural outbreak or potentially deliberate (bioterrorist) release of variola virus.

There are two known stocks of smallpox virus, one at the CDC in Atlanta and the other at a similar facility in Russia. However, the former Soviet Union produced massive amounts of smallpox virus during the Cold War years, so there may be stocks that "walked away" to rogue nations or terrorist organizations after the fall of the Soviet Union in 1991. The destruction of the remaining smallpox stocks at the CDC and in Russia has been planned by the WHO. However, it has been postponed several times because of the controversy over the value of keeping smallpox stocks (MicroFocus 15.6).

Since vaccinations against smallpox stopped in the United States in 1972 and elsewhere soon after, a majority of people lack immunity to the disease. This makes smallpox one of the most dangerous weapons of bioterrorism, even though those who were vaccinated prior to 1972 still have some level of immunity. In addition, the United States has stated that it now has adequate stockpiles of smallpox vaccine to vaccinate every American, if necessary.

Other diseases caused by poxviruses also are potentially dangerous (MicroFocus 15.7).

Molluscum contagiosum is another viral disease that forms mildly contagious, wart-like skin lesions. The virus of molluscum contagiosum is an enveloped, double-stranded DNA virion of the Poxviridae family. Transmission is generally by sexual contact.

Smallpox virus

MICROFOCUS 15.7: Public Health
The First American Case of Monkeypox

Monkeypox is a rare viral disease occurring mostly in central and western Africa. It is called "monkey-pox" because it was first found in 1958 in laboratory primates. Blood tests of animals in Africa later found other types of animals also had monkeypox. In 1970, the first case of monkeypox was reported in humans.

The monkeypox virus is a double-stranded DNA virus in the same group of Poxviridae as the smallpox and cowpox viruses. People get monkeypox from an infected animal through bites or contact with the animal's blood, body fluids, or its rash. The disease also can spread from person to person through large respiratory droplets during long periods of face-to-face contact, or by touching body fluids of a sick person or objects such as bedding or clothing contaminated with the virus.

The first outbreak of monkeypox in the Western Hemisphere occurred in the Midwest in June 2003. On April 9, 2003, a Texas animal distributor received a shipment of some 800 small mammals from Accra, Ghana. This included rope squirrels, Gambian giant rats, and dormice.

Twelve days later, an Illinois distributor received Gambian rats and dormice from the Texas distributor. An infected Gambian rat was housed with prairie dogs that the distributor had before the prairie dogs were sold to distributors in six states. Additional prairie dogs were sold at animal swap meets.

Starting on May 15, doctor reports from the Midwest came in that people were exhibiting symptoms similar to but much milder and less contagious than smallpox. Individuals became ill with a fever, respiratory symptoms, and swollen lymph nodes. A rash developed, which progressed into raised bumps filled with fluid. The rash started on the face and spread across the body. Eventually the rash crusted over and the scabs fell off. The illness lasted for two to four weeks.

Over the next few weeks, outbreaks of monkeypox were identified in Wisconsin, Illinois, and Indiana where the Texas distributor had sent prairie dogs to pet dealers. On June 20, the last case was reported.

On June 30, 2003, a final report to the CDC identified 71 cases of monkeypox from Wisconsin, Illinois, Indiana, Missouri, Kansas, and Ohio. There were no deaths, although 18 were hospitalized. Of the 71 cases, 35 were laboratory confirmed and 36 were suspect and probable cases.

This outbreak of monkeypox underlined the need to closely screen and protect the public from exotic animals exported to the United States. This time the infection was fairly mild as it was caused by a weak monkeypox strain; the next time, it or another disease might be contagious and lethal.

The lesions are firm, waxy, and elevated with a depressed center. When pressed, they yield a milky, curd-like substance. Although usually flesh toned, the lesions may appear white or pink. Possible areas involved include the facial skin and eyelids in children, and the external genitals in adults. The lesions may be removed by excising them (cutting them out) and pose no public health threat. A characteristic feature of the disease is the presence of large cytoplasmic bodies called **molluscum bodies** in infected cells from the base of the lesion.

TABLE 15.5 presents a summary of these other dermotropic viral diseases.

CONCEPT AND REASONING CHECKS

15.14 Describe how one could identify smallpox from chickenpox based on the rash formed.

TABLE

15.5 A Summary of Other Viral Diseases of the Skin

Disease	Classification of Virus	Signs and Symptoms	Transmission	Treatment	Prevention and Control
Measles (rubeola)	Paramyxoviridae	Cough, nasal discharge, eye redness, and high fever; Koplik spots	Droplets Contact	Supportive care	MMR vaccine
Mumps	Paramyxoviridae	Swollen and painful parotid glands Pain on chewing and swallowing	Person-to-person in infected saliva	Supportive	MMR vaccine
Rubella (German measles)	Togaviridae	Fever with pale-pink maculopapular rash spreading to body trunk	Droplets Contact	Supportive care	MMR vaccine
Fifth disease (erythema infectiosum)	Parvoviridae	Maculopapular rash on cheeks and ears	Droplets (?)	Supportive hygiene	Practicing good care
Warts	Papovaviridae	White or pink skin growth on hands and feet	Contact	Freezing Minor surgery	Practicing proper care of affected areas
Genital warts	Human papillomavirus types 6 and 11	Small, flat, flesh-colored bumps or tiny, cauliflower-like bumps anywhere in the genital region or areas around the anus	Oral, vaginal, or anal sex with someone who has HPV	Removal by freezing, electro-surgical excision, or by using specific chemicals	Abstaining from sexual activity HPV vaccine
Cervical/Penile cancer	Human papillomavirus types 16 and 18	Vaginal bleeding and pelvic pain Precancerous growth on the cervix or penis	Oral, vaginal, or anal sex with someone who has HPV	Surgery Radiation therapy Chemotherapy	HPV vaccine Pap test
Smallpox (variola)	Poxviridae	Fever, macules that became papules, then vesicles and pustules	Contact Droplets Fomites	None	Vaccine if available
Molluscum contagiosum	Poxviridae	Flesh toned, wart-like lesions	Contact	Removal of papules	Avoiding touching papules Avoiding sexual contact

SUMMARY OF KEY CONCEPTS

15.1 Viral Infections of the URT
- **Common colds**
  1. Rhinoviruses, adenoviruses, and others
- **Laryngitis**
  2. Rhinoviruses

15.2 Viral Infections of the LRT
- **Influenza**
  3. Influenza A and B viruses
- **Respiratory syncytial (RS) disease**
  4. Respiratory syncytial virus
- **Parainfluenza**
  5. Human parainfluenza viruses 1 and 3
- RSV-like illness
  6. Human metapneumovirus
- SARS
  7. SARS coronavirus
- **Hantavirus pulmonary syndrome (HPS)**
  8. Hantavirus

15.3 Diseases of the Skin Caused by Herpesviruses
- **Cold sores**
  9. Herpes simplex 1

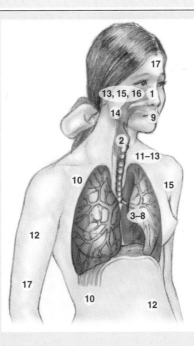

- **Genital herpes** (not shown)
  – Herpes simplex 2
- **Chickenpox** and **shingles**
  10. Varicella zoster
- **Roseola** (children)
  11. Human herpes virus 6
- **Kaposi sarcoma**
  12. Human herpes virus 8

15.4 Other Viral Diseases of the Skin
- **Measles**
  13. Measles virus
- **Mumps**
  14. Paramyxoviruses
- **Rubella**
  15. Rubella virus
- **Fifth disease**
  16. Parvovirus B19
- **Warts** (extremities; not shown)
  Papillomaviruses
- **Smallpox**
  17. Variola

LEARNING OBJECTIVES

After understanding the textbook reading, you should be capable of writing a paragraph that includes the appropriate terms and pertinent information to answer the objective.

1. Explain why a vaccine against **rhinoviruses** is not feasible.
2. List the types of diseases associated with **adenovirus** infections.
3. Identify the major **influenza viruses** and the structures involved in generating subtypes.
4. Organize the **paramyxoviruses** into related species.
5. Distinguish how **SARS** differs from other respiratory tract infections including **hantavirus pulmonary syndrome (HPS)**.
6. Describe the infections caused by herpes simplex virus-1 and herpes simplex virus-2.

7. Explain why the incidence of **chickenpox** has declined in the United States and how the varicella-zoster virus causes **shingles**.
8. Summarize the diseases caused by human herpesvirus-6.
9. Describe the relationship between human herpesvirus-8 and HIV/AIDS.
10. Summarize the characteristics of the paramyxovirus infections causing (1) **measles** and (2) **mumps**.
11. State the potential complication to a pregnant mother who has contracted **rubella**.
12. Identify the characteristics associated with **fifth disease**.
13. Distinguish between **common** and **genital warts**, including possible complications.
14. Summarize the clinical and social significance of **smallpox**.

STEP A: SELF-TEST

Each of the following questions is designed to assess your ability to remember or recall factual or conceptual knowledge related to this chapter. Read each question carefully, then select the **one** answer that best fits the question or statement. Answers to even-numbered questions can be found in **Appendix C**.

1. There are more than _____ different rhinoviruses, which belong to the _____ family.
   A. 50; Orthomyxoviridae
   B. 100; Adenoviridae
   C. 30; Paramyxoviridae
   D. 100; Picornaviridae

2. All of the following are diseases caused by the adenoviruses *except*
   A. viral pneumonia.
   B. acute respiratory disease.
   C. parainfluenza.
   D. common cold.

3. Which one of the following statements is NOT true of the influenza viruses?
   A. They have a segmented genome.
   B. The genome is double-stranded DNA.
   C. The viruses have an envelope.
   D. There are three types of flu viruses.

4. Which one of the following is NOT a member of the Paramyxoviridae.
   A. RSV
   B. Human metapneumovirus
   C. SARS-CoV
   D. Parainfluenza virus
5. SARS is
   A. a skin infection.
   B. spread through close person-to-person contact.
   C. a mild, respiratory infection.
   D. most often seen in young children.
6. Cold sores and genital herpes can be caused by
   A. HSV-1.
   B. HHV-6.
   C. VZV.
   D. HHV-8.
7. A red, itchy rash that forms small, teardrop-shaped, fluid-filled vesicles is typical of
   A. measles.
   B. rubella.
   C. chickenpox.
   D. mumps.
8. HHV-6 that causes roseola primarily affects
   A. the elderly.
   B. pregnant mothers.
   C. infants.
   D. teenagers.
9. Kaposi sarcoma is a tumor of the
   A. blood vessels.
   B. liver.
   C. lymph nodes.
   D. kidneys.

10. _____ are diagnostic for measles.
    A. Koplik spots
    B. Wart-like lesions
    C. Swollen lymph nodes
    D. Blisters on the body trunk
11. The characteristic sign of rubella is
    A. orchitis.
    B. pale-pink maculopapular rash.
    C. fiery red rash on cheeks and ears.
    D. high fever and sensitivity to light.
12. Fifth disease
    A. is hospital acquired.
    B. causes white skin warts.
    C. causes benign skin growths.
    D. is transmitted by respiratory droplets.
13. Some papillomaviruses are capable of causing
    A. a "lacy" rash on the skin.
    B. lung cancer.
    C. pneumonia.
    D. cervical cancer.
14. Which one of the following statements applies to smallpox?
    A. The disease is associated with animal contact.
    B. The disease has been eradicated worldwide.
    C. It can be sexually-transmitted.
    D. The virus can lie dormant in host cells.

STEP B: REVIEW

On completing your study of viral diseases of the respiratory tract and skin, test your comprehension of the chapter contents by circling the choices that best complete each of the following statements. The answers to even-numbered statements are listed in **Appendix C**.

15. Rhinoviruses are a collection of (RNA, DNA) viruses having (helical, icosahedral) symmetry and the ability to infect the (air sacs, nose), causing (mild, serious) respiratory symptoms.
16. Herpes simplex is a viral disease transmitted by (breathing contaminated air, contact) and is characterized by thin-walled (blisters, ulcers) that often appear during periods of (emotional stress, exercising).
17. In children, the skin lesions of chickenpox occur (all at once, in crops) and resemble (teardrops, pitted scars), but in adults the lesions are known as (shingles, erythemas).
18. The complications of influenza include (Reye, Koplik) syndrome; for mumps, the complication is a disease of the (testes, pancreas) called (colitis, orchitis).
19. One of the early signs of (smallpox, measles) is a series of (Koplik spots, inclusion bodies) occurring in the (lungs, mouth) and signaling a (red, purple) rash is forthcoming.
20. Antigenic variation among (mumps, influenza) viruses seriously hampers the development of a highly effective (vaccine, treatment), and a life-threatening situation can occur if secondary infection due to (fungi, bacteria) complicates the primary infection.
21. Respiratory syncytial disease is caused by a (DNA, RNA) virus infecting the (lungs, intestines) of (adults, children) and inducing cells to (fuse together, cluster) and form giant cells called (syncytia, tumors).
22. SARS is caused by a (coronavirus, orthomyxovirus), a/an (naked, enveloped) virus spread by (sexual, person-to-person) contact.
23. Adenoviruses include a collection of (DNA, RNA) viruses and are responsible for (yellow fever, common colds), as well as infections of the (eye, ear).
24. Genital herpes is caused by a (helical, icosahedral) virus most often (HSV-1, HSV-2) and causes blisters with (thick, thin) walls that disappear in about three (days, weeks), only to reappear when (stress, physical injury) occurs.
25. The TORCH diseases are a set of (infectious, physiological) diseases transmitted by (airborne droplets, transplacental passage), occurring in (the elderly, newborns), and including (rubeola, rubella) and (herpes simplex, humoral disease).

STEP C: APPLICATIONS

Answers to even-numbered questions can be found in **Appendix C.**

26. The CDC reports an outbreak of measles at an international gymnastics competition. A total of 700 athletes and numerous coaches and managers from 51 countries are involved. What steps would you take to avert a disastrous international epidemic?

27. A child experiences "red bumps" on her face, scalp, and back. Within 24 hours, they have turned to tiny blisters and become cloudy, some developing into sores. Finally, all become brown scabs. New "bumps" keep appearing for several days, and her fever reaches 39°C by the fourth day. Then the blisters stop coming and the fever drops. What disease has she had?

28. A man experiences an attack of shingles and you warn him to stay away from children as much as possible. Why did you give him this advice?

STEP D: QUESTIONS FOR THOUGHT AND DISCUSSION

Answers to even-numbered questions can be found in **Appendix C.**

29. Thomas Sydenham was an English physician who, in 1661, differentiated measles from scarlet fever, smallpox, and other fevers, and set down the foundations for studying these diseases. How would you go about distinguishing the variety of look-alike skin diseases discussed in this chapter?

30. A Boeing 737 bound for Kodiak, Alaska, developed engine trouble and was forced to land. While the airline rounded up another aircraft, the passengers sat for 4 hours in the unventilated cabin. One passenger, it seemed, was in the early stages of influenza and was coughing heavily. By the week's end, 38 of the 54 passengers on the plane had developed influenza. What lessons about infectious disease does this incident teach?

31. In the United Kingdom, the approach to rubella control is to concentrate vaccination programs on young girls just before they enter the childbearing years. In the United States, the approach is to immunize all children at the age of 15 months. Which approach do you believe is preferable? Why?

HTTP://MICROBIOLOGY.JBPUB.COM/9E

The site features eLearning, an online review area that provides quizzes and other tools to help you study for your class. You can also follow useful links for in-depth information, read more MicroFocus stories, or just find out the latest microbiology news.

16

Viral Infections of the Blood, Lymphatic, Gastrointestinal, and Nervous Systems

Somewhere in the world, a person dies almost every 10 minutes from rabies.
—The Alliance for Rabies Control

The following is a digest of a story posted on the Centers for Disease Control and Prevention's (CDC) Rabies Web Page "That's Just for Kids!"

Sean was 11 years old when he went on a class trip to the Okeefenokee National Wildlife Refuge in Georgia with his teachers and fifth grade class. One the first day, they set up tents in which they would sleep and helped the teachers unpack the vans with all the food and other camping gear. The class went on many expeditions that included canoeing up the river and just "hanging out around the camp."

On the fourth night, Sean woke up in the middle of the night with a sharp pain in his arm. When he looked at his arm, there were two bite marks that drew blood. Hearing a rustling sound, Sean looked up and saw a raccoon running out of his tent. He immediately called for his teacher and they cleaned out the bite with soap, hot water, and a disinfectant.

First thing in the morning Sean was taken to the hospital in Homerville, Georgia. The attending physician took Sean into a room where he would get the immune globulin shots for rabies. While waiting, various hospital staff

High-risk area

Moderate-risk area

FIGURE 16.1 Rabies Distribution Worldwide. Rabies causes more than 55,000 deaths worldwide, half of which are in India. Countries completely free of rabies include: Australia, New Zealand, Hong Kong, Singapore, Great Britain, and some Scandinavian countries. »» How do some countries stay rabies-free?

stopped in to see Sean and tell him, "Well, you know that you are a lucky boy because 10 years ago you had to get 26 shots in the stomach and boy did they hurt."

Sean was given two shots of immune globulin, one shot for tetanus, and the rabies vaccine. Sean said he had to wait one hour "to make sure that I did not have an allergic reaction to any of it." With no adverse effects, he got to go home. After the initial shots, Sean received four more shots over a period of one month and he fully recovered.

Fortunately for Sean, today there is a vaccine that, given in time, is 100% effective in preventing rabies. Unfortunately, these vaccines are not available worldwide and rabies takes the lives of more than 55,000 people each year (**FIGURE 16.1**). Notably, 30% to 50% of the deaths occur in children under the age of 15 years.

The most common global source of rabies in humans is from uncontrolled rabies in dogs, which is why children are often at greatest risk. To get the message out concerning rabies, in 2006 a group of researchers and medical professionals,

along with the CDC, formed the global Alliance for Rabies Control (ARC). The ARC was created to alleviate the burden of rabies across the world by promoting and implementing rabies control and human rabies prevention. This culminated in the creation of World Rabies Day, which now occurs every September 28. (This is the date in 1895 when Louis Pasteur, who produced the first rabies vaccine, died.) In the 2007 inaugural World Rabies Day, nearly 400,000 people from 74 countries participated. The ARC says, "Through the World Rabies Day initiative, partners will be . . . *Working Together to Make Rabies History!*"

Rabies is by no means the only human viral infection, so we also examine several other diseases of concern. Overall, the diseases addressed in this chapter fall into three general categories. Some illnesses, such as hepatitis B, yellow fever, and mononucleosis are diseases of the blood, while others, including the noroviruses, affect the digestive system. Still others, such as rabies, polio, and West Nile encephalitis affect the nervous system. As in Chapter 15, each disease is presented as a separate essay, and you may select the order of study most suitable to your course needs.

16.1 Viral Diseases of the Blood and the Lymphatic Systems

Several viral diseases are of the blood and the lymphatic systems. To reach these areas, the viruses generally are introduced into the body tissues by sexual contact, mechanical means, or arthropods.

Two Herpesviruses Cause Blood Diseases

KEY CONCEPT

1. Infectious mononucleosis and cytomegalovirus disease affect adolescents and produce congenital defects, respectively.

Although most members of the human herpesviruses primarily affect the skin (see Chapter 15), two species are associated with blood infections and disease.

Infectious mononucleosis is a blood disease, especially of antibody-producing B lymphocytes (a type of mononuclear white blood cell) in the lymph nodes and spleen. The name infectious mononucleosis (or "mono") is familiar to young adults because the disease is common in this age group. It is sometimes called the "kissing disease" because it is spread by contact with saliva. The disease strikes an estimated 100,000 people annually in the United States.

Infectious mononucleosis is caused by the Epstein-Barr virus (EBV) of the Herpesviridae. This is one of the most common human viruses; up to 90% of the world population between 35 and 40 years of age has been infected with EBV. Infants usually are susceptible to EBV as soon as their maternal antibodies (present at birth) disappear. Many children who become infected with EBV show no symptoms or the symptoms are indistinguishable from other typical childhood illnesses.

If a person is not infected as an infant or young child, an infection with EBV during adolescence or young adulthood runs a 35% to 50% chance of causing infectious mononucleosis. **EBV disease**, involving infection of B lymphocytes, is assumed to be a precursor to mononucleosis.

EBV is spread person-to-person via saliva or by saliva-contaminated objects, such as table utensils and drinking glasses. Major symptoms are sore throat, enlargement of the lymph nodes ("swollen glands"), and fever, giving the disease another name—**glandular fever**. Mononucleosis usually runs its course in 3 to 4 weeks and is not highly contagious.

Among the most dangerous complications are defects of the heart, paralysis of the face, and rupture of the spleen. The liver may be involved and jaundice may occur, a condition some physicians refer to as hepatitis. Those who recover usually become carriers for several months and shed the viruses into their saliva.

The diagnostic procedures for mononucleosis include detection of an elevated lymphocyte count and the observation of **Downey cells**, the damaged B cells with vacuolated and granulated cytoplasm. The **Monospot test** can be used to detect heterophile antibodies to EBV (**FIGURE 16.2**). No vaccine is available for EBV infections and no drugs have proven effective in treating mononucleosis.

EBV in immunocompromised patients may lead to **Burkitt lymphoma**, a tumor of the connective tissues of the jaw that is prevalent in areas of Africa. In fact, EBV was the first human virus associated with a malignancy (see Chapter 14). Contemporary virologists continue to search for reasons why EBV is associated with tumors on one continent and infectious mononucleosis on another. Some cancer specialists theorize the malaria parasite, common in central Africa, acts as an irritant of the lymph gland tissue, thereby stimulating tumor development. All Burkitt lymphoma cells have a chromosomal translocation that activates an oncogene (see Chapter 14). Therefore, EBV reactivation may stimulate or increase the frequency of the translocation.

EBV also has been associated with T-cell malignancies, including **nasopharyngeal carcinoma**. In immune-suppressed transplant patients, EBV can cause B cell lymphomas. In addition, there is growing evidence suggesting a possible link between EBV and **Hodgkin disease**, a lymphoma of the lymph nodes and spleen. Also, evidence is mounting for an association between EBV and an increased risk of developing **multiple sclerosis**, a muscle-weakening disease of the central nervous system.

Cytomegalovirus disease can produce serious birth defects. The cytomegalovirus (CMV) is the largest member of the Herpesviridae. The virus takes its name from the enlarged cells (*cyto* = "cell";

Heterophile antibodies: Antibodies nonspecifically reacting with proteins or cells from unrelated animal species.

Epstein-Barr virus

Lymphomas: Malignant tumors originating in a lymph node.

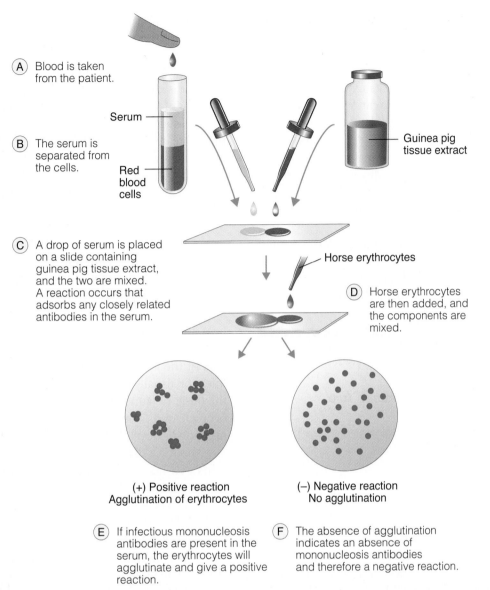

(A) Blood is taken from the patient.

Serum

(B) The serum is separated from the cells.

Red blood cells

Guinea pig tissue extract

(C) A drop of serum is placed on a slide containing guinea pig tissue extract, and the two are mixed. A reaction occurs that adsorbs any closely related antibodies in the serum.

Horse erythrocytes

(D) Horse erythrocytes are then added, and the components are mixed.

(+) Positive reaction
Agglutination of erythrocytes

(−) Negative reaction
No agglutination

(E) If infectious mononucleosis antibodies are present in the serum, the erythrocytes will agglutinate and give a positive reaction.

(F) The absence of agglutination indicates an absence of mononucleosis antibodies and therefore a negative reaction.

FIGURE 16.2 **The Monospot Slide Test for Infectious Mononucleosis.** About one week after the onset of infection by EBV, many patients develop heterophile antibodies. The antibodies peak at weeks two to five and may persist for several months to one year. The Monospot test is performed by first mixing samples of the patient's serum with a guinea pig tissue extract, which binds to and removes any closely related antibodies. Horse erythrocytes then are added and the red blood cells are observed for agglutination. Suitable controls (not shown) must also be included. »» What types of controls should be included?

megalo = "large") found in infected tissues. Usually, these are cells of the salivary glands, epithelium, or liver. After the primary infection, CMV undergoes lifelong latency.

CMV disease in a healthy individual may be among the most common diseases in American communities. Infection may be asymptomatic or produce a mononucleosis-like syndrome, involving fever and malaise. Most patients recover uneventfully.

However, if a CMV-infected woman is pregnant, a serious congenital disease may ensue in 5% to 10% of infants if the virus passes into the fetal bloodstream and damages the fetal tissues. Mental impairment is sometimes observed among the more than 17,000 congenital CMV-associated infected infants in the United States and Europe every year. The **C** in the **TORCH** group of diseases refers to congenital CMV-associated disease (see Chapter 14).

Endoscopes:
Instruments consisting of a fiber-like strand that are inserted through an incision for diagnostic or surgical procedures.

Retinitis:
A serious infection of the retina that can lead to blindness.

Cytolytic:
Refers to the destruction (lysis) of cells.

Jaundice:
A condition in which bile pigments seep into the circulatory system, causing the skin and whites of the eyes to have a dull yellow color.

Hepatitis B virus

CMV infection and latency generates an immune response that keeps the virus in check. However, in immunocompromised individuals, CMV can reactivate. Prior to antiretroviral therapy, up to 25% of AIDS patients experienced CMV-induced retinitis. CMV also can accelerate the progression to AIDS and infect the lungs, liver, brain, and kidneys and cause death. Patients undergoing cancer therapy or receiving organ transplants may be susceptible to CMV disease because immunosuppressive drugs often are administered to "knock out" the immune system's ability to reject the transplant.

A vaccine for congenital CMV-associated infections is being researched. Several drugs, including acyclovir and valacyclovir provide effective treatment for transplant patients and as a preemptive therapy to CMV disease.

CONCEPT AND REASONING CHECKS

16.1 Identify the serious complication that may result from (a) EBV and (b) CMV infections.

Several Hepatitis Viruses Are Bloodborne

KEY CONCEPT

2. Hepatitis B and C can become chronic and cause liver disease.

Hepatitis B (formerly called "serum hepatitis") is a global health problem, accounting for 1 million deaths every year. Two billion people, one-third of the world's population, have been exposed to the hepatitis B virus (HBV) and some 350 million have chronic (lifelong) HBV infections. About 20% of these individuals are at risk of dying from HBV-related liver disease.

HBV is in the family Hepadnaviridae (*hepa* = "liver") and is the smallest known DNA virus. It has a partially double-stranded, circular DNA genome. Normal virions consist of a nucleocapsid surrounded by a **hepatitis B core antigen (HBcAg)** and envelope containing **hepatitis B surface antigen (HBsAg)**.

Transmission of hepatitis B usually involves direct or indirect contact with an infected body fluid such as blood or semen. For example, transmission may occur by contact with blood-contaminated needles, such as hypodermic syringes or those used for tattooing, acupuncture, or ear piercing (**FIGURE 16.3**). Blood-contaminated

objects such as endoscopes, saliva-contaminated (non-sterile) dental instruments, and renal dialysis tubing also are implicated.

Hepatitis B is also an important sexually transmitted disease through vaginal, anal, or oral sex with an infected partner. HBV can enter through small tissue tears, allowing the virus to enter the bloodstream of the receptive partner from blood, saliva, semen, or vaginal secretions from the infected individual.

Hepatitis B has a long incubation period of four weeks to six months during which time HBV infects the liver but is not cytolytic. Among children and adults, the primary (acute) infection may be asymptomatic. Symptoms are more likely in adults and include fatigue, loss of appetite, nausea and vomiting, and dark urine. Patients experience jaundice weeks later. Abdominal pain and tenderness are felt in the upper-right quadrant of the abdomen (the liver is located on the upper right side of the abdomen). Recovery from an acute infection usually occurs about 3 to 4 months after the onset of jaundice. By this time, the virus has been cleared from the blood and liver, and the individuals develop immunity to reinfection.

About 5% of patients develop persistent infections and may or may not have symptoms. In rare cases, **cirrhosis**, an extensive hepatocellular injury, can occur due to immune system reactions to the infection. In addition, chronically-infected carriers run a 100-times higher chance of developing liver cancer or **hepatocellular carcinoma (HCC)** than non-carriers. The cellular and molecular reasons for carcinoma development are not completely understood.

The risk of developing a persistent, chronic state decreases with age. Newborns have a 90% risk, and 25% of infants and young children with persistent infections will eventually die from cirrhosis or HCC. The mortality rate drops to 15% if a persistent infection develops as adolescents or young adults.

Hepatitis B can be prevented by immunization with any of several hepatitis B vaccines. These vaccines consist of HBsAg produced by genetically-engineered yeast cells (see Chapter 9). Recommended for all age groups (including infants), they are particularly valuable for health care workers who might be exposed to blood from patients. For infant use, it is combined

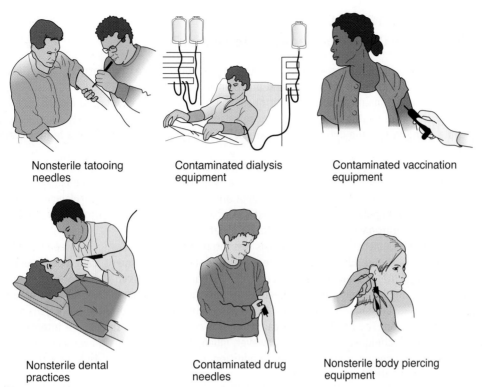

Nonsterile tatooing needles

Contaminated dialysis equipment

Contaminated vaccination equipment

Nonsterile dental practices

Contaminated drug needles

Nonsterile body piercing equipment

FIGURE 16.3 Some Methods for the Transmission of Hepatitis B. »» What is the common denominator in all these methods of transmission?

with the Hib vaccine as Comvax. As a result of child and adolescent vaccinations, there has been a 68% decrease in reported cases of hepatitis B in the United States (4,519 cases in 2007).

Injections of interferon alfa-2b (Intron A) can influence the course of hepatitis B. Moreover, injections of hepatitis B immune globulin, which consists of antibodies concentrated from the serum of blood donors, can be used for persons without known immunity who have come in contact with HBV.

Hepatitis C causes a major, chronic liver disease in 123 million people worldwide, with the highest proportion in Asia and Africa. Almost 4 million Americans are infected with the hepatitis C virus and 2.7 million of those have chronic infections. Still, new infections have declined almost 90% since 1992 to 845 in 2007.

The hepatitis C virus (HCV) is an enveloped, +ssRNA virus of the Flaviviridae family and is primarily transmitted by blood, injection drug use, or blood transfusions (in countries without a blood screening program). However, it also has been spread inadvertently through medical intervention, as MicroFocus 16.1 clearly demonstrates.

There are few symptoms associated with the acute infection and 20% to 50% of patients fully recover, although no permanent immunity is generated. Most cases (50% to 80%) develop a symptomless, insidious chronic infection, which takes the lives of almost 10,000 Americans who die from cirrhosis, HCC, or other complications every year. Indeed, damage from hepatitis C is the primary reason for liver transplants in the United States. Alcoholism and intravenous drug use are among the major cofactors accelerating the incidence of chronic hepatitis C.

Interferon alfa and ribavirin, standard therapies for the disease, cure less than 50% of those infected. No vaccine is available.

Hepatitis D is caused by two viruses: HBV and the hepatitis D virus (HDV). The latter virus consists of a protein fragment called the delta antigen and a segment of RNA, and can only cause liver damage when HBV is present. HDV requires the outside coat of HBV to infect cells. Therefore, one cannot become infected with hepatitis D unless he or she already is infected with HBV. Chronic liver disease with cirrhosis is two to six times more likely in a co-infection.

MICROFOCUS 16.1: History
What's Worse—the Disease or Medical Intervention?

Egypt has a population of 62 million and contains the highest prevalence of hepatitis C in the world. The Egyptian Ministry of Health estimates a national prevalence rate of at least 12% (or 7.2 million people). Chronic hepatitis C is the main cause of liver cirrhosis and liver cancer in Egypt, and indeed, one of the top five leading causes of death. The highest concentration of the hepatitis C virus (HCV) appears in farming people living in the Nile delta and rural areas. So, why does Egypt have such a high prevalence rate of HCV?

Several recent studies suggest that HCV was transmitted through the contamination of reusable needles and syringes used in the treatment of schistosomiasis, a disease caused by a blood parasite (Chapter 18). Schistosomiasis is a common parasitic disease in Egypt and can cause urinary or liver damage over many years. Farmers and rural populations are at greatest risk of acquiring the disease through swimming or wading in contaminated irrigation channels or standing water.

Prior to 1984, the treatment for schistosomiasis was intravenous tartar emetic. Between the 1950s and the 1980s, hundreds of thousands of Egyptians received this standard treatment, called parenteral anti-schistosomal therapy (PAT). Today, drugs for schistosomiasis are administered in pill form.

Evidence suggests inadequately sterilized needles used in the PAT campaign contributed to the transmission of HCV and set the stage for the world's largest transmission of blood-borne pathogens resulting from medical intervention. The PAT treatment campaign was conducted with the best of intentions, using accepted sterilization techniques of the time. However, much of the PAT campaign was carried out before disposable syringes and needles were available. In addition, no one was aware of HCV prior to the 1980s or the dangers associated with blood exposure.

Further evidence for the correlation between the PAT campaign and hepatitis C was the drop in the hepatitis C rate when PAT injections were replaced with oral medications. Sadly, in part because of the high number of people who were infected, the risk of transmission remains high in the Egyptian population today.

Emetic:
A chemical substance that induces vomiting.

Hepatitis G also is a chronic liver illness. The hepatitis G virus (HGV) is an enveloped, +ssRNA virus of the Flaviviridae. Like HBV and HCV, HGV is transmitted by blood, blood products, or sexual intercourse. It appears to cause persistent infections in 15% to 30% of infected adults.

TABLE 16.1 summarizes the viral diseases associated with the blood and the lymphatic system.

CONCEPT AND REASONING CHECKS

16.2   Summarize the similarities in symptoms between hepatitis B and C.

16.2 Viral Diseases Causing Hemorrhagic Fevers

In March 2005, the Angola Ministry of Health and the World Health Organization (WHO) reported 63 hemorrhagic deaths (mostly children along with three health care workers) at the Uige Provincial Hospital. By late July, there were a total of 374 cases of which 329 died from an outbreak of Marburg hemorrhagic fever. This is just one of several illnesses called **viral hemorrhagic fevers (VHFs)** caused by four families of RNA viruses. The illnesses, characterized by vascular system damage (rash, bleeding gums and mucous membranes, internal bleeding), occur sporadically and are rare in the United States.

Flaviviruses Can Cause a Terrifying and Severe Illness

KEY CONCEPT

3. Yellow fever and dengue fever can produce fever, bleeding, and/or circulatory failure.

The Flaviviridae are enveloped, icosahedral, +ssRNA virions. They also are referred to as **arboviruses** because they are *arthropod-borne* viruses. The two most virulent are the yellow fever and dengue fever viruses.

Yellow fever was the first human disease associated with a virus. As a result of the slave trade

Yellow fever virus

TABLE

16.1   Viral Diseases Associated with the Blood and Lymphatic Systems

Disease	Causative Agent	Signs and Symptoms	Transmission	Treatment	Prevention and Control
Infectious mononucleosis	Epstein-Barr virus	Sore throat, fever, swollen lymph nodes in neck and armpits	Person-to-person via saliva or saliva-contaminated objects	Bed rest and adequate fluid intake	Avoiding kissing Not sharing food or personal objects
Cytomegalovirus disease	Cytomegalovirus	Asymptomatic Fever, malaise, swollen glands	Contact with the body fluids of an infected person	None	Practicing good hand hygiene Avoiding kissing Not sharing food or personal objects
Hepatitis B	Hepatitis B virus	Fatigue, loss of appetite, nausea, vomiting Jaundice and abdominal tenderness	Direct or indirect contact with body fluids	Interferon and other drugs	Receiving the hepatitis B vaccine
Hepatitis C	Hepatitis C virus	Early: no signs or symptoms May experience fatigue, loss of appetite, nausea, vomiting, low-grade fever, jaundice	Direct or indirect contact with body fluids Blood transfusions	Interferon and other drug therapies	Avoiding illegal drugs, tattooing

from Africa, the disease spread rapidly in large regions of the Caribbean and tropical Americas, and was common in the southern and eastern United States for many generations in the 18th and 19th centuries. In 1901, a group led by the American Walter Reed identified mosquitoes as the agents of transmission. With widespread mosquito control, the incidence of the disease gradually declined. Today, yellow fever is endemic in 33 countries in Africa and South America, and causes over 200,000 cases and 30,000 deaths annually.

"Jungle yellow fever" occurs in monkeys and other jungle animals, where the virus is taken up by blood-feeding mosquitoes. Epidemics occur when a person is bitten by an infected mosquito in the forest and then travels to urban areas where the virus is passed via a blood meal to another mosquito species, *Stegomyia* (formerly *Aedes*) *aegypti*. This species then transmits the virus, now causing an "urban yellow fever" among humans.

The *S. aegypti* mosquito injects the virus into the human bloodstream, causing abrupt symp-

toms of headache, fever, and muscle pain lasting 3 to 5 days. Then, the symptoms abate as the virus is cleared by antibodies and the immune system. Many patients recover at this stage.

In 15% to 25% of patients, the illness reappears in a much more terrifying form, causing severe nausea, uncontrollable hiccups, and a violent, black vomit. Liver damage produces jaundice (the disease often is called "yellow jack") and major hemorrhaging from the gums, mouth, and nose occur as the patient becomes delirious. Up to 50% of patients go into a coma and die from internal bleeding.

As a zoonosis, yellow fever cannot be eradicated. It can be prevented by immunization with either of two vaccines. To this end, the WHO and health charities plan to immunize almost 50 million people in West Africa by 2015 to reestablish immunity to yellow fever. Except for supportive therapy, no treatment exists for yellow fever.

Dengue fever takes its name from the Swahili word *dinga*, meaning "cramp-like attack," a reference

Endemic:
Referring to a disease occurring within a specific area, region, or locale.

Zoonosis:
A disease transmitted from animals to humans.

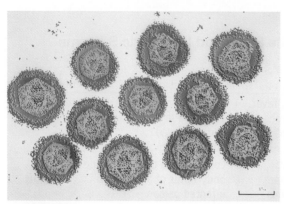

FIGURE 16.4 The Dengue Fever Virus. False-color transmission electron micrograph of dengue fever viruses. (Bar = 50 nm.) »» What do the yellow geometric shapes represent in the dengue fever viruses?

FIGURE 16.5 The Ebolavirus. False-color transmission electron micrograph of an ebolavirus. Here the viral filament is seen looping back on itself. (Bar = 140 nm.) »» Why are these viruses placed in the Filoviridae family?

to the symptoms. It is the most common arboviral disease of humans, as there are 50 million new infections and over 12,000 deaths annually. Dengue fever traditionally was confined to Southeast Asia. However, in 1963, the disease broke out in Central America and continues to occur sporadically in the Americas. Outbreaks have occurred in Puerto Rico in 2007 and Brazil in 2008. Dengue fever is now endemic in the United States. A 2005 outbreak in Brownsville, Texas resulted in 25 hospital cases.

There are four types of dengue fever virus (**FIGURE 16.4**). Transmission is by the *S. aegypti* mosquito and the tiger mosquito *S. albopicta*. While taking a blood meal, infected mosquitoes inject viruses into the blood where they multiply in white blood cells and platelets.

Sudden high fever and prostration are early signs of infection. These are followed by sharp pain in the muscles and joints. Patients often report intense joint and muscle pain, sensations feeling like their bones are breaking; thus, the disease has been called **breakbone fever**. After about a week, the symptoms fade.

Serious complications or death are uncommon unless one of the other three types of dengue virus later enters the body of a recovered patient. Then, a condition called **dengue hemorrhagic fever** may occur. In this condition, the immune system reacts to the memory of the first dengue infection, allowing the new one to replicate unchallenged. A rash from skin hemorrhages appears on the face and extremities, and severe vomiting and shock ensue (**dengue shock syndrome**) as blood pressure decreases dramatically.

Prostration:
Drained of strength.

Several vaccines are in clinical trials. No successful antiviral therapy has been identified. Vector control, using insecticides and draining standing water, are the main methods of prevention. A newly emerging dengue-like hemorrhagic fever is described in MicroFocus 16.2.

CONCEPT AND REASONING CHECKS

16.3 Compare the similarities and differences between yellow fever and dengue fever.

Members of the Filoviridae Produce Severe Hemorrhagic Lesions of the Tissues

KEY CONCEPT

4. Ebola and Marburg hemorrhagic fevers are among the most deadly.

Marburg and Ebola hemorrhagic fevers are severe illnesses caused by viruses in the Filoviridae (*filo* = "thread") family. They consist of long thread-like, −ssRNA viruses (**FIGURE 16.5**).

Ebola hemorrhagic fever (EHF) captured headlines in 1976 and 1979 when the first outbreaks occurred in Sudan and Zaire. When it was over, 88% (280) of infected people died. Through 2007, there have been 19 confirmed outbreaks of EHF in Africa. Over 2,200 cases have been reported and 67% have died.

EHF is caused by infection with one of the two known strains of ebolavirus (Ebola is the river in

MICROFOCUS 16.2: Public Health

A Newly Emerging Hemorrhagic Fever

In 2006, the Centers for Disease Control and Prevention (CDC) reported 37 cases of a unique hemorrhagic fever in U.S. travelers returning from destinations in the Indian Ocean and India—that is 34 more cases than had occurred in the previous 15 years combined. These travelers experienced fever, headache, fatigue, nausea, vomiting, muscle pain, and a skin rash—typical symptoms of dengue fever. However, unlike dengue fever, these patients also had incapacitating joint pain. The symptoms typically lasted a few days to a few weeks, although the joint pain sometimes lasted for many months. In the CDC cases, all recovered.

The disease experienced by these travelers was chikungunya (CHIK) fever (chikungunya means "to walk bent over," referring to the severe joint pain). CHIK fever is caused by the chikungunya virus (CHIKV), the only arbovirus in the family Togaviridae and is endemic to tropical East Africa and regions rimming the Indian Ocean. This single-stranded (+ strand), enveloped RNA virus is transmitted by mosquitoes and, since its first isolation in 1953 in Tanzania, has caused numerous CHIK fever epidemics in humans. The 2006 outbreak on Reunion Island in the Indian Ocean affected more than 266,000 of the 775,000 inhabitants. It then spread to India where more than 1 million cases were reported. Of most concern was that, for the first time, CHIK fever had claimed a substantial number of lives; 254 fatalities were attributed directly or indirectly to CHIKV. There is no specific antiviral treatment for CHIK fever and care is based on symptoms. Prevention consists of protecting individuals from mosquito bites and controlling the vector through insecticide spraying.

What makes this emerging disease especially worrisome is the report in July 2007 of a CHIK fever outbreak in northern Italy, a temperate country far from the tropical confines of the Indian Ocean. More than 200 cases were reported; one died from complications. It is believed an infected person from a state in India where the disease was epidemic brought it to Italy. There, he became ill and was bitten by a *Stegomyia* (tiger) mosquito, which, once infected, transmitted CHIKV to other humans. Many believe CHIK fever may now be endemic in southern Europe.

Such movements only underscore the ease with which an emerging disease can be transported in the era of jet travel. The possibility of CHIKV reaching further into Europe and even to North America, where competent tiger mosquitoes also exist, represents yet another health threat from a mosquito-borne disease.

---

the Democratic Republic of the Congo, formerly Zaire, where the illness was first recognized). Recent evidence suggests the virus is zoonotic and is normally maintained in an animal host native to Africa. Fruit bats have been identified as a possible reservoir.

Uninfected individuals, including health care workers, can be exposed to the virus by direct contact with the blood and/or secretions of an infected person or animal, or through contact with contaminated objects, such as needles. Following an incubation period of a few days to two weeks, symptoms include fever, headache, joint and muscle aches, sore throat, and weakness. This is followed by diarrhea, vomiting, and stomach pain. A rash, red eyes, hiccups and internal and external bleeding may be seen in some patients.

The ebolavirus damages endothelial cells, causing massive internal bleeding and hemorrhaging. Infection also suppresses the immune response.

Thus, the patient bleeds to death internally before a reasonable immunologic defense can be mounted.

Marburg hemorrhagic fever (MHF), named for Marburg, West Germany, where an outbreak occurred in 1967, was first identified in tissues of green monkeys imported from Africa. How transmission of the marburgvirus occurs is unclear, although fruit bats may be the reservoir.

After an incubation period of five to ten days, MHF develops symptoms similar to EHF but usually has a lower fatality rate. In 2005, the biggest outbreak of MHF occurred in Angola. The virus infected 374 people and killed 329.

For both EHF and MHF, prevention measures have not been established. A subunit vaccine is being tested that completely protects monkeys from infection by both viruses.

Reservoir:
The natural host or habitat of a pathogen.

Endothelial:
Refers to the layer of cells lining body cavities, such as the veins and arteries.

CONCEPT AND REASONING CHECKS

16.4 Why is EHF such a deadly disease?

Members of the Arenaviridae Are Associated with Chronic Infections in Rodents

KEY CONCEPT

5. Lassa fever can lead to hemorrhage and shock.

Lassa fever is so named because it was first reported in the town of Lassa, Nigeria, in 1969. The disease is caused by a zoonotic, –ssRNA virus of the Arenaviridae family that has sandy-looking granules in the virion (*arena* = "sand"). Lassa fever is responsible for about 5,000 deaths per year in West Africa.

Transmission of the Lassa fever virus is shown in **FIGURE 16.6**. Infection leads to severe fever, exhaustion, and patchy blood-filled hemorrhagic lesions of the throat. The fever persists for weeks and profuse internal hemorrhaging is common. Overall, only about 1% of infections result in death.

Prevention involves avoiding contact with rodents and keeping homes clean. Ribavirin is effective in treating the disease.

Other viral hemorrhagic fevers, all caused by Bunyaviridae or Arenaviridae are Congo-Crimea hemorrhagic fever, which occurs worldwide; Oropouche fever, which affects regions of Brazil; and Junin and Machupo, the hemorrhagic fevers of Argentina and Bolivia, respectively. An arenavirus called the Sabia virus has caused hemorrhagic illnesses in Brazil, while the Guanarito virus is associated with Venezuelan hemorrhagic fever.

TABLE 16.2 summarizes the diseases caused by the hemorrhagic fever viruses.

CONCEPT AND REASONING CHECKS

16.5 How are humans most likely infected by the Lassa fever virus?

A The bush rat is a staple in the diet of certain African natives. The traditional rat hunt begins with a fire in the grasslands that drives the rats into the open, where they are clubbed.

B Occasionally the rats will run into local houses for shelter.

C Lassa fever viruses are transmitted by arthropods in rat fur or rat excretions to susceptible people visiting the area.

D When infected rats are prepared as food, infection can occur.

FIGURE 16.6 **Transmission of Lassa Fever Virus.** Transmission of Lassa fever is primarily through aerosol or direct contact with excreta from infected rodents, or from contaminated food. »» Why might Lassa fever-infected rodents be a health hazard around homes or villages?

TABLE

16.2   Viral Diseases Causing Hemorrhagic Fevers

Disease	Causative Agent	Signs and Symptoms	Transmission	Treatment	Prevention and Control
Yellow fever	Yellow fever virus	Acute phase: Headache, fever muscle pain Toxic phase: Severe nausea, black vomit, jaundice, hemorrhaging	Bite from a *Stegomyia aegypti* mosquito	No antiviral medications Supportive care	Vaccination Avoiding mosquito bites in endemic areas
Dengue fever	Dengue fever virus	Dengue fever: Sudden high fever, headache, nausea, vomiting Dengue Hemorrhagic fever: decrease in platelets, skin hemorrhage Dengue shock syndrome: heavy bleeding, drop in blood pressure	Bite from an infected *Stegomyia aegypti* mosquito Bite from an infected *Stegomyia* mosquito with another dengue virus	No specific treatment available	Avoiding mosquito bites in endemic areas
Ebola/Marburg hemorrhagic fevers	Ebola and Marburg viruses	Fever, headache, joint and muscle aches, sore throat, weakness Internal bleeding and hemorrhaging	Bite of infected fruit bat Blood transfer through cut, abrasion, or infected animal bite	No specific treatment available	Avoiding dead animals and bats in outbreak areas
Lassa fever	Lassa fever virus	Severe fever, exhaustion, hemorrhagic lesions on throat	Aerosol and direct contact with excreta from infected rodents	Ribavirin	Avoiding dead or infected rodents Maintaining good home sanitary conditions

16.3   Viral Infections of the Gastrointestinal Tract

Several human viruses are responsible for illnesses and diseases of the digestive (gastrointestinal) system (see Chapter 11). These illnesses include hepatitis and viral gastroenteritis.

Hepatitis Viruses A and E Are Transmitted by the Gastrointestinal Tract

KEY CONCEPT

6.   Hepatitis A and E are spread by the fecal-oral route.

Hepatitis A is an acute inflammatory disease of the liver most commonly transmitted by food or water contaminated by the feces of an infected individual. Approximately 1.5 million cases of hepatitis A occur each year worldwide and it remains one of the most frequently reported vaccine-preventable diseases in the United States. In 2007, there were 2,979 cases reported to the CDC.

Hepatitis A is caused by a small, naked, icosahedral, +ssRNA virion belonging to the Picornaviridae family (**FIGURE 16.7**). The hepatitis A virus (HAV) is very resistant to chemical and physical agents, and several minutes of exposure to boiling water may be necessary to inactivate the virion.

Hepatitis A virus

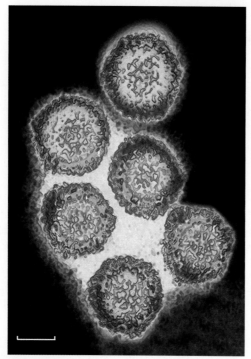

Fulminate:
Refers to sudden and severe symptoms of short duration.

FIGURE 16.7 **Hepatitis A Viruses.** A false-color transmission electron micrograph of hepatitis A viruses. (Bar = 15 nm.) »» In what organ would one most likely find these viruses replicating?

Transmission of hepatitis A, sometimes referred to as "infectious hepatitis," often involves an infected food handler (fecal-oral route) and outbreaks have been traced also to day-care centers where workers contacted contaminated feces. Interestingly, children often serve as the principal reservoir because their infections usually are asymptomatic. In addition, the disease may be transmitted by raw shellfish such as clams and oysters, because these animals filter and concentrate the viruses from contaminated seawater.

The incubation period is usually between 2 to 4 weeks. Because the primary site of replication is the gastrointestinal tract, initial symptoms in older children and adults include anorexia, nausea, vomiting, and low-grade fever. The virus then is transported to the liver, its major site of replication. Discomfort in the upper-right quadrant of the abdomen follows as the liver enlarges. Considerable jaundice usually follows the onset of symptoms by 1 or 2 weeks (the urine darkens, as well), but many cases are without jaundice. Passage into the bile occurs and then new virions are shed through the intestines and into the feces.

The symptoms may last for several weeks, and relapses commonly occur in 3% to 20% of cases. A long period of convalescence generally is required, during which alcohol and other liver irritants are excluded from the diet. Recovery brings life-long immunity. However, about 100 Americans die each year from fulminate hepatitis A infections.

Diagnostic procedures for hepatitis A are based on liver enzyme tests, observation of characteristic symptoms, and the demonstration of hepatitis A antibodies in the serum. The largest ever outbreak of hepatitis A in the United States occurred in late 2003 (TEXTBOOK CASE 16).

Maintaining high standards of personal hygiene (hand washing), and removing the source of contamination are essential to preventing the spread of hepatitis A. Two safe and highly effective vaccines, known commercially as Havrix and Vaqta, are available in the United States for people over one year of age. Also available for individuals over 18 years of age is Twinex, a combination vaccine for hepatitis A and B.

There is no treatment for hepatitis A except for prolonged rest and relieving symptoms. In those exposed to HAV, it is possible to prevent development of the disease by administering hepatitis A immune globulin within two weeks of infection. This preparation consists of antiviral antibodies obtained from blood donors (Chapter 22). Blood is routinely screened for hepatitis antibodies.

Hepatitis E is an opportunistic, emergent disease caused by a naked, +ssRNA virus of the Caliciviridae family. It shares many clinical characteristics and symptoms with hepatitis A. The hepatitis E virus is primarily transmitted via the fecal-oral route. Evidence also suggests other strains may be a porcine zoonosis.

The disease affects immunocompromised individuals, with pregnant women in developing nations being particularly susceptible to illness, with mortality being as high as 30%. No evidence of chronic infection has been noted.

The impact of hepatitis E can be appreciated from two 2004 outbreaks coming from contaminated water. One outbreak was in the Greater Darfur region of Sudan, where almost 7,000 cases and 87 deaths were reported; the other was in neighboring Chad, where there were some 1,500 cases and 46 deaths in refugee camps.

CONCEPT AND REASONING CHECKS

16.6 Compare and contrast hepatitis A and E.

Textbook CASE   16

Hepatitis A Outbreak— Monaca, PA, 2003

1 On November 5, 2003, an outbreak of hepatitis A was confirmed through lab analysis reported by the Pennsylvania Health Department. At that time, there were 34 cases confirmed at a mall restaurant (restaurant A) with 10 customers and 12 restaurant employees reporting symptoms of hepatitis A infection.

2 By Friday, November 7, 130 people had contracted hepatitis A (see figure) and the health department provided injections of immunoglobulin as a precaution for anyone who had eaten at the restaurant between September 22 and November 2.

3 At this time, state officials and arriving CDC investigators suspected the virus was spread by an infected worker who failed to wash his or her hands before handling food.

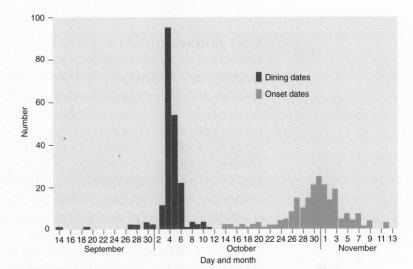

Number of hepatitis A cases by date of eating at restaurant A and illness onset.

4 On November 8, the first fatality from the hepatitis A outbreak was reported. The person died from liver failure. The outbreak had risen to 240 confirmed cases. Officials still believed the problem centered on an infected food worker.

5 On November 12, health officials announced the confirmed case count had risen to 340. Mention also was made of a recent multi-state (Tennessee, Georgia, and North Carolina) outbreak of hepatitis A in late September and early October. These 250 cases resulted from eating contaminated green onions at a few local restaurants.

6 By November 13, 410 illnesses were reported. Transmission from an infected food worker was ruled out as all employees became ill after the outbreak started. Interviews of restaurant A patrons began in order to identify what and how much they ate at the mall restaurant.

7 By November 15, two more people had died from the hepatitis A outbreak and more than 500 illnesses had been reported.

8 Patron interviews and further menu item investigations pointed to Mexican salsa containing green and white onions as the prime source of illness.

9 Ninety-eight percent of restaurant A patrons who became ill reported eating salsa containing raw green onions. Illness was not associated with eating salsa containing raw white onions.

10 In all, more than 550 people had been stricken during the hepatitis A outbreak. Genetic analysis of the virus implicated raw green onions imported from three firms in Mexico.

Questions:

(Answers can be found in Appendix D.)

A. Why would the infections of restaurant employees be significant?

B. Explain why immunoglobulin injections were recommended.

C. How common are deaths from hepatitis A?

D. Why was it important to discover what and how much food the affected restaurant customers ate?

E. Provide some ways that the green onions may have initially become contaminated with the hepatitis A virus.

For additional information see http://www.cdc.gov/mmwr/preview/mmwrhtml/mm5247a5.htm.

Viral Gastroenteritis Is Caused by Several Unrelated Viruses

KEY CONCEPT

7. Infectious gastroenteritis can cause substantial morbidity in children and adults.

Viral gastroenteritis is a general name for a common illness occurring in both epidemic and endemic forms. It affects all age groups worldwide and may include some of the frequently encountered traveler's diarrheas. Public health officials believe gastroenteritis is second only to the common cold in frequency among infectious illnesses affecting people in the United States. In developing nations, gastroenteritis is estimated to be the second leading killer of children under the age of five, accounting for 23% of all deaths in this age group.

Clinically the disease varies, but usually it has an explosive onset with varying combinations of watery diarrhea, nausea, vomiting, low-grade fever, cramping, headache, and malaise. It can be severe in infants, the elderly, and patients whose immune systems are compromised by other illnesses. Some people mistakenly call it "stomach flu."

Rotavirus infections represent one of the world's deadliest forms of gastroenteritis in children. The diarrhea-related illness is associated with 25 million clinic visits, 2 million hospitalizations, and more than 500,000 deaths worldwide among children younger than five years of age.

The rotavirus (*rota* = "wheel") is a naked, circular-shaped virion whose genome contains eleven segments of dsRNA (**FIGURE 16.8**). It is a member of the Reoviridae family. Rotavirus infections tend to occur in the cooler months (October–April) in the United States and are thus called "winter diarrhea."

Transmission occurs by ingestion of contaminated food or water (fecal-oral route), or from contaminated surfaces. The rotaviruses make their way to the small intestine where they infect enterocytes, inducing diarrhea, vomiting, and chills. The disease normally runs its course in 3 to 8 days. Treatment requires oral rehydration salt solutions.

Recovery from the infection does not guarantee immunity, as many children have multiple rounds of reinfection. The CDC considers rotaviruses the single most important cause of diarrhea in infants and young children admitted to

hospitals. Some 75,000 hospitalizations and 20 to 40 deaths occur each year in the United States.

These numbers, but especially the 500,000 childhood deaths worldwide make development of a safe and effective rotavirus vaccine a high priority objective. In early 2006, a new oral vaccine, RotaTeq, was licensed in the United States while another, Rotarix, was licensed in 2008. With the use of these vaccines, hospitalizations for severe diarrhea have been reduced 70% to 80%.

Norovirus infections are the most likely cause of nonbacterial gastroenteritis in adults. Noroviruses (formerly called the Norwalk-like viruses) are transmitted primarily through the fecal-oral route, either by consumption of contaminated food or water, or by direct person-to-person spread especially through aerosols produced from a vomiting episode. Contamination of surfaces also may act as a source of infection, where the virus can survive for days if not a week or longer. The CDC estimates at least 50% of all foodborne outbreaks (23,000,000) of viral gastroenteritis are caused by norovirus infections. In 2006, more than 45 outbreaks of norovirus infections occurred on cruise ships in European waters, similar to those on U.S. ships between 2001 and 2004. The infections occur most often during the summer months, giving the illness the name "summer diarrhea."

The noroviruses belong to the family Caliciviridae. These viruses are naked, icosahedral virions with a +ssRNA genome. Currently, there are at least seven norovirus groups. Noroviruses

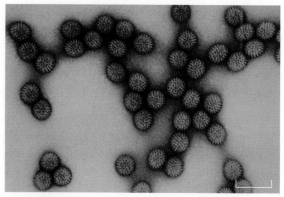

FIGURE 16.8 Rotaviruses. False-color transmission electron micrograph of rotaviruses. (Bar = 140 nm.)
»» From this micrograph, why were these viruses called rotaviruses?

Rotavirus

Norovirus

Enterocytes: Cells of the intestinal epithelium.

TABLE

16.3   Viral Infections of the Gastrointestinal Tract

Disease	Causative Agent	Signs and Symptoms	Transmission	Treatment	Prevention and Control
Hepatitis A	Hepatitis A virus	Nausea, vomiting, low-grade fever	Indirect through food and raw shellfish	No specific treatment	Receiving the hepatitis A vaccine Practicing good hygiene
Hepatitis E	Hepatitis E virus	Nausea, muscle pain, low-grade fever	Indirectly through water Zoonosis	No effective treatment	Avoiding untreated water
Viral gastroenteritis	Rotavirus	Diarrhea, nausea, vomiting, low-grade fever, stomach cramping, headache, malaise	Indirect through food or water Contaminated surfaces	Self care	Practicing good hand hygiene Avoiding shared items
	Norovirus	Nausea, vomiting, diarrhea, stomach cramping	Indirect through food or water Person-to-person	Oral rehydration fluids	Practicing good hand hygiene Washing fruits and vegetables
	Enterovirus	Fever, mild rash, mild upper respiratory tract illness	Indirect through food or water Person-to-person		Practicing good hand hygiene Cleaning contaminated surfaces

are highly contagious, and as few as 10 virions can cause illness in an individual.

The incubation period for norovirus gastroenteritis is 24 to 48 hours. Typical gastrointestinal symptoms of fever, diarrhea, abdominal pain, and extensive vomiting last 24 hours and recovery is complete. Dehydration is the most common complication. Washing hands and having safe food and water are the only preventions and the only treatment for norovirus gastroenteritis is fluid and electrolytes.

Enterovirus infections also can cause viral gastroenteritis. Enteroviruses are small, icosahedral +ssRNA virions of the Picornaviridae family.

One enterovirus is the Coxsackie virus, first isolated from a patient residing in Coxsackie, New York. The virus occurs in many strains, with B4 and B5 associated most commonly with a form of gastroenteritis called **hand, foot, and mouth disease**. It typically affects infants and young children during the summer and fall. Group B viruses also are implicated in triggering type-1 diabetes in genetically predisposed children. In addition, the viruses may be responsible for the so-called "24-hour diarrhea."

The second enterovirus is the echovirus, which gets its name from *e*nteric (intestinal), *c*ytopathogenic (pathogenic to cells), *h*uman (human host), and *o*rphan (a virus without a famous disease). Echoviruses occur in many strains and cause infantile diarrhea.

TABLE 16.3 summarizes the characteristics of the virally caused gastrointestinal illnesses.

CONCEPT AND REASONING CHECKS

16.7   Explain why oral rehydration salt solutions are the treatment for most forms of gastroenteritis.

Coxsackie virus

16.4 Viral Diseases of the Nervous System

Several viral diseases affect the human nervous system, which can suffer substantial damage when viruses replicate in the tissue. Rabies, polio, and West Nile fever are perhaps the most recognized diseases.

The Rabies Virus Is of Great Medical Importance Worldwide

KEY CONCEPT

8. Rabies is a highly fatal disease once symptoms arise.

Rabies virus

Rabies (*rabies* = "madness") is notable for having the highest mortality rate of any human disease, once the symptoms have fully materialized; there are an estimated 55,000 deaths annually, mostly in rural areas of Africa and Asia. Few people in history have survived rabies, and in those who did, it is uncertain whether the symptoms were due to the disease or the therapy (MicroFocus 16.3).

The rabies virus is a −ssRNA virion of the Rhabdoviridae family with a meager five genes in its genome. It is rounded on one end, flattened on the other, and looks like a bullet (**FIGURE 16.9**).

Animal rabies can occur in most warm-blooded animals, including dogs, cats, prairie dogs, and bats. In 2007, more than 5,862 wildlife cases were reported throughout the United States,

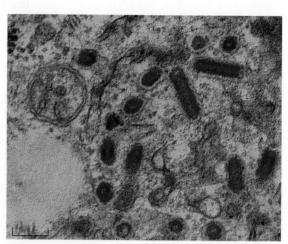

FIGURE 16.9 The Rabies Virus. An accumulation of rabies viruses in the salivary gland tissue of a canine. In this false-color transmission electron micrograph, the viruses are red. (Bar = 100 nm.) »» In this micrograph, identify the viruses that most closely have the shape characteristic of rhabdoviruses.

MICROFOCUS 16.3:
The "Milwaukee Protocol"

On September 4, 2004, 15-year-old Jeanne Giese of Fond du Lac, Wisconsin, was bitten by a rabid bat outside her church. Jeanne did not seek immediate treatment and became gravely ill a month later.

Normally, once the symptoms of rabies set in, recovery is not an option and death occurs within a week. Jeanne was rushed to Children's Hospital in Milwaukee where doctors administered an unproven treatment. She was placed in a chemically-induced coma and given a combination of antiviral drugs (ketamine, midazolam, ribavirin, and amantadine), sedatives, and anesthetics in their effort to save her life.

About a week later, Jeanne was brought out of the coma. She was paralyzed, unable to speak or walk, and without sensation, as she suffered from the effects of rabies on her nervous system. Physicians detected brain-wave activity, but were unsure what was ahead after the drugs would wear off. However, when they did wear off, Jeanne demonstrated some eye movements and reflexes—a real positive sign. Progress continued and after 76 days, Jeanne was released from the hospital in a wheelchair. She had a long road to recovery, if indeed recovery would be complete. She would need to regain her faculties, including her ability to speak and walk.

On Christmas of 2005, Jeanne celebrated her improbable survival at home with her family. Since then she has regained her ability to walk and talk, and nerves damaged by rabies have recommunicated with muscles. Jeanne has returned to school, graduated, and is now a junior in college where she is majoring in biology.

Jeanne's lead physician, Dr. Rodney E. Willoughby, Jr. at Children's Hospital and the Medical College of Wisconsin, has had his so-called "Milwaukee protocol" used a dozen times but Jeanne was the only survivor—until 2008. In October, doctors in Cali, Colombia announced that the Milwaukee protocol had saved the life of an 8-year-old girl who had been bitten by a rabid cat. Then, in November, Brazilian doctors used the protocol to save a 16-year-old boy who had been bitten by a rabid bat.

Although this is good news, medical experts agree that the focus needs to remain in rabies prevention.

with raccoons and bats accounting for the majority of the cases (**FIGURE 16.10**).

The virus enters the tissue and peripheral nervous system through a skin wound contaminated with the saliva from a rabid animal. The incubation period for rabies varies according to the amount of virus entering the tissue and the wound's proximity to the central nervous system. As few as six days or as long as a year may elapse before symptoms appear. A bite from a rabid animal does not ensure transmission, however, because experience shows that only 5% to 15% of bitten individuals develop the disease.

Early signs of rabies encephalitis are abnormal sensations such as tingling, burning, or coldness at the site of the bite. Fever, headache, and increased muscle tension develop, and the patient becomes alert and aggressive. Soon there is paralysis, especially in the swallowing muscles of the pharynx, and saliva drips from the mouth. Brain degeneration, together with an inability to swallow, increases the violent reaction to the sight, sound, or thought of water. The disease therefore has been called **hydrophobia**—literally, the fear of water. Death usually comes within days from respiratory paralysis.

An estimated 10 million people worldwide receive **post-exposure immunization** each year after being exposed to rabies-suspect animals. As described in the chapter introduction, this entails five injections on days 0, 3, 7, 14, and 28 in the shoulder muscle of the arm. These injections are preceded by thorough cleansing and one dose of rabies immune globulin to provide immediate

Encephalitis: An inflammation of the brain.

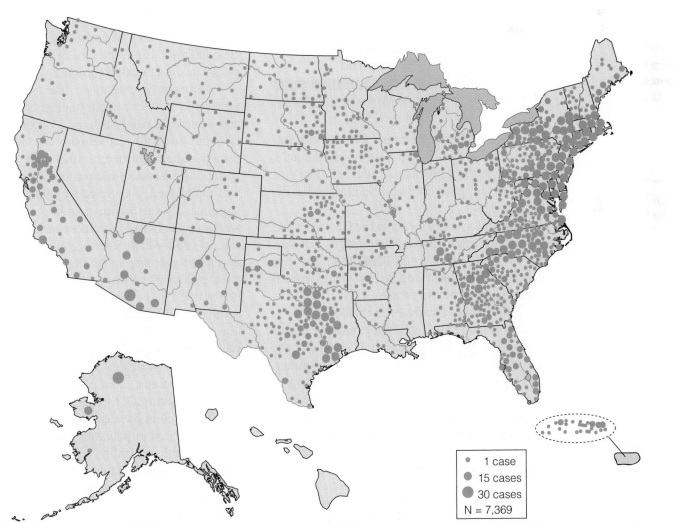

•	1 case
●	15 cases
⬤	30 cases
	N = 7,369

FIGURE 16.10 Reported Cases of Animal Rabies. This geographic map identifies the location of animal rabies reported in 49 states, the District of Columbia, and Puerto Rico. »» Why are so many of the animal rabies cases found in the Northeast?
Source: CDC. Available at http://www.cdc.gov/ncidod/dvrd/rabies/Epidemiology/Epidemiology.htm.

antibodies at the site of the bite. This usually is accompanied by a tetanus booster, but the latter is omitted if exposure is not certain. For high-risk individuals such as veterinarians, trappers, and zoo workers, a preventive immunization of three injections may be given.

Rabies historically has been a major threat to animals. One form, called **furious rabies**, is accompanied by violent symptoms as the animal becomes wide eyed, drools, and attacks anything in sight. In the second form, **dumb rabies**, the animal is docile and lethargic, with few other symptoms. Health departments now are conducting a novel campaign to immunize wild animals using vaccine air-dropped within biscuit-sized baits of dog food and fish meal.

CONCEPT AND REASONING CHECKS

16.8   Why is the incubation period so variable for the rabies virus?

The Polio Virus May Be the Next Infectious Disease Eradicated

KEY CONCEPT

9.   Polio historically has been a severe paralytic disease.

The name **polio** is a shortened form of **poliomyelitis** (*polio* = "gray"; *myelo* = for "spinal cord"), referring to the "gray matter," which is the nerve tissue of the spinal cord and brain in which the virus infects.

The polioviruses, being in the Picornaviridae family, are among the smallest virions, measuring 27 nm in diameter (**FIGURE 16.11**). They are composed of a naked icosahedron-shaped capsid containing a +ssRNA genome.

Polioviruses usually enter the body by contaminated water and food. They multiply first in the tonsils and then in lymphoid tissues of the gastrointestinal tract, causing nausea, vomiting, and cramps. In many cases, this is the extent of the problem. Sometimes, however, the viruses pass through the bloodstream and localize on the meninges, where they cause **meningitis**. Rarely, the virus causes paralysis of the arms, legs, and body trunk which could lead to paralytic poliomyelitis.

In the most severe cases, the viruses infect the medulla of the brain, causing **bulbar polio** (the medulla is bulb-like). Nerves serving the upper

Poliovirus

Meninges:
The membranes surrounding and protecting the brain and spinal cord.

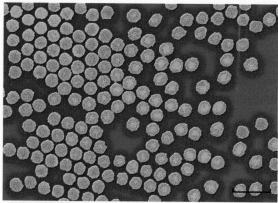

FIGURE 16.11 **Polioviruses.** A false-color transmission electron micrograph of polioviruses. Although the particles appear circular, their symmetry actually is icosahedral. With a diameter of about 27 nm, these are among the smallest viruses that cause human disease. (Bar = 60 nm.) »» How can such a small virus cause such a paralytic disease?

body torso are affected. Swallowing is difficult, and paralysis develops in the tongue, facial muscles, and neck. Paralysis of the diaphragm muscle causes labored breathing and may lead to death.

Virologists have identified three types of poliovirus: type 1 causes a major number of epidemics and sometimes paralysis; type 2 occurs sporadically but invariably causes paralysis; and type 3 usually remains in the intestinal tract. Once the method of laboratory cultivation of polioviruses was established, a team led by Jonas Salk grew large quantities of the viruses and inactivated them with formaldehyde to produce the first polio vaccine in 1955. Albert Sabin's group subsequently developed a vaccine containing attenuated (weakened) polioviruses. This vaccine was in widespread use by 1961 and could be taken orally (OPV) as compared with Salk's vaccine, which had to be injected (IPV). Both are referred to as **trivalent** vaccines because they contain all three strains of virus. One drawback of the Sabin vaccine is that being a live vaccine, a few cases of vaccine-caused polio have occurred (1 in 2.4 million).

The vaccines have contributed substantially to the reduction of polio. In 1988, the forty-first World Health Assembly, with funds raised by Rotary International, launched a global initiative to eradicate polio by the end of the year 2000. To meet this goal, progress was made in the 1980s to eliminate the poliovirus from the Americas, and to protect all children from the disease. In the

MICROFOCUS 16.4: History/Public Health

Blitz

This was the National Immunization Day for India, one in a series of such events across the world. The campaigns were coordinated by the WHO, with help from UNICEF, Rotary International, and the CDC. On this day in December 1997, two million people set out to eradicate polio in India. Coming from every conceivable corner of a vast country, they arrived at 650,000 Indian villages, where they set up immunization posts. Children came to them by the thousands, then the hundreds of thousands, and then the millions (see figure). By the end of the day, 127 million children had received polio immunizations—just one of the success stories from around the world to eradicate polio.

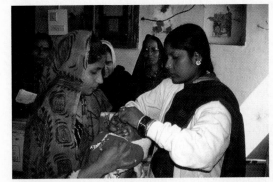
A child being immunized against polio in India.

There are also setbacks in the vaccination blitz. In 2004, polio cases broke out in Nigeria, which had been polio-free. Back in 2003, rumors of "contaminated" polio vaccine caused an immediate cessation of immunization. Polio returned, paralyzing 491 Nigerian children. It was not until mid-2004 that the vaccine was deemed safe and the Nigerian government relaunched the immunization effort.

Wars and civil strife also make eradication difficult. There are the Days of Tranquility, where pauses in wars and civil strife allow children to be immunized. In Sri Lanka, for example, polio vaccine was passed across front lines during Days of Tranquility in 1995 and 1996.

In other countries it has not been so easy. In Somalia, a country plagued by ongoing conflicts, 131 polio cases were reported in January 2006. In 2008, 98% of the reported cases occurred in India, Nigeria, and Pakistan.

But optimism remains high. With sufficient funding and additional immunization activities—and perseverance—the day may be near when polio may be nothing more than a memory.

15 years since the Global Polio Eradication Initiative was launched, the number of cases of polio has fallen by over 99%, from more than 350,000 cases in 1988 to less than 1,700 reported cases in 2008. In the same period, the number of polio-infected countries was reduced from 125 to 20 (MICROFOCUS 16.4).

Widely endemic on five continents in 1988, polio now is limited to parts of Africa and south Asia. Obviously, the goal to eradicate polio by year 2000 has not been met. In fact, there was a major outbreak in India in 2006 and in 2009 a runaway outbreak occurred in Nigeria due to the type 2 vaccine virus mutating back to its virulent form.

A discouraging legacy of the polio epidemics is **postpolio syndrome** (**PPS**). Apparently, many people who had polio decades ago now are experiencing the initial ailments they had with polio, including muscle weakness and atrophy, general fatigue and exhaustion, muscle and joint pain, and breathing or swallowing problems. The National Center for Health Statistics estimates there are 440,000 polio survivors in the United States at risk of developing PPS, of whom 25% to 50% may be affected.

Several theories have been put forward to explain the cause of PPS, including a reactivated virus or an autoimmune reaction (Chapter 23), which over the years has caused the body's immune system to attack motor neurons as if they were foreign substances.

CONCEPT AND REASONING CHECKS

16.9 Explain why the American polio vaccination schedule has dropped the OPV as part of the vaccination.

Arboviral Encephalitis Is a Result of a Primary Viral Infection

KEY CONCEPT

10. Cases of arboviral encephalitis are zoonoses.

Viruses that are transmitted by mosquitoes and ticks (arboviruses; arbo = *ar*thropod-*bo*rne) have been the cause of a fairly rare primary encephalitis called **arboviral encephalitis**. Arboviral encephalitis is an example of a zoonosis, a disease transmitted by a vector from another vertebrate host to humans. Mosquitoes are the most common vector and most cases occur during warmer weather, when the insects are more active.

Arboviral encephalitis may be sporadic or epidemic, and the diseases have a global distribution (**FIGURE 16.12**). In the United States, the most reported types are St. Louis encephalitis (SLE), Eastern equine encephalitis (EEE), Powassan virus disease, La Crosse (LAC) encephalitis, and most notably, West Nile virus disease.

In humans, arboviral encephalitis is characterized by sudden, very high fever and a severe headache. Normally, the patient experiences pain and stiffness in the neck, with general disorientation. Patients become drowsy and stuporous, and may experience a number of convulsions before lapsing into a coma. Paralysis and mental disorders may afflict those who recover. Mortality rates are generally high.

Arboviral encephalitis is also a serious problem in horses, causing erratic behavior, loss of coordination, and fever. The disease can be transmissible from horses to humans by ticks, mosquitoes, and other arthropods. Important forms are EEE, WEE, and Venezuelan equine encephalitis (VEE).

West Nile virus disease is an emerging disease in the Western Hemisphere. It is caused by the West Nile virus (WNV), another member of the Flaviviridae (**FIGURE 16.13**).

Before the outbreak in the United States in 1999, WNV was established in Africa, western Asia, and the Middle East. It is closely related to St. Louis encephalitis virus. The virus has a somewhat broad host range and can infect humans, birds, mosquitoes, horses, and some other mammals. Since the first outbreak in 1999 in New York City, each year WNV moved farther west

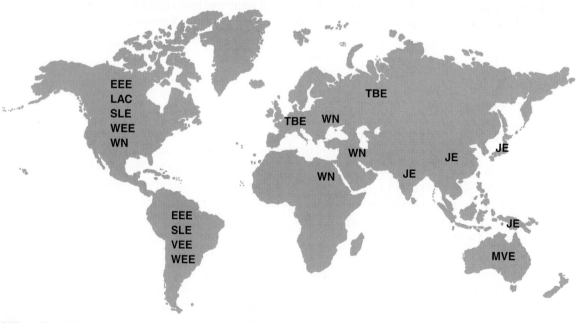

EEE	Eastern equine encephalitis	**SLE**	St. Louis encephalitis
JE	Japanese encephalitis	**TBE**	Tick-borne (Powassan) virus disease
LAC	La Crosse encephalitis	**WEE**	Western equine encephalitis
MVE	Murray Valley encephalitis	**WN**	West Nile virus disease
		VEE	Venezuelan equine encephalitis

FIGURE 16.12 **Global Distribution of Major Arboviral Encephalitides.** The various disease caused by arboviruses are shown. »» Propose a reason why there is such a large variety of arboviral encephalitides in the Western Hemisphere.
Source: Centers for Disease Control and Prevention, http://www.cdc.gov/ncidod/dvbid/arbor/worldist.pdf.

across the United States. By 2006, cases had been reported in every state except Alaska and Hawaii. Now, experts believe the virus is here to stay and will cause seasonal epidemics that flare up in the summer and continue into the fall. In 2009, there were 663 cases reported and 30 deaths.

WNV is spread by the bite of an infected mosquito, which itself becomes infected when it feeds on infected birds (FIGURE 16.14) Infected mosquitoes then spread WNV to humans and other animals when they bite. In 2002, health officials reported a very small number of cases where WNV was transmitted through blood transfusions, organ transplants, breast-feeding, or even during pregnancy from mother to baby. The virus is not spread through casual contact such as touching or kissing a person who is infected.

WNV causes a potentially serious illness. People typically develop symptoms between 3 and 14 days after they are bitten by the infected mosquito. Approximately 80% of people who are infected remain asymptomatic. Most of the remaining infected individuals display **West Nile fever**, consisting of a mild fever, headache, body aches, and extreme fatigue. They also may develop a skin rash on the chest, stomach, and back. Symptoms typically last a few days.

About one in 150 people infected with WNV will develop **West Nile encephalitis** or **meningitis** that affects the central nervous system. Symptoms can include high fever, headache, stiff neck, stupor, disorientation, coma, convulsions, muscle weak-

ness, vision loss, numbness, and paralysis. These symptoms may last several weeks to over a year, and the neurologic effects may be permanent. Death can result.

There is no vaccine yet available nor is there any specific treatment for WNV infections. In cases of encephalitis and meningitis, people may need to be hospitalized so they can receive supportive treatment including intravenous fluids, help with breathing, and nursing care. People who spend a lot of time outdoors are more likely to be bitten by an infected mosquito. These individuals should take special care to avoid mosquito bites (MicroFocus 16.5). Also, people over the age of 50 are more likely to develop serious symptoms from a WNV infection if they do get sick, so they too should take special care to avoid mosquito bites.

TABLE 16.4 summarizes the neurotropic viral diseases. MicroInquiry 16 presents several case studies concerning some viral diseases discussed in this chapter.

CONCEPT AND REASONING CHECKS

16.10 Assess the consequences of being infected with the West Nile virus.

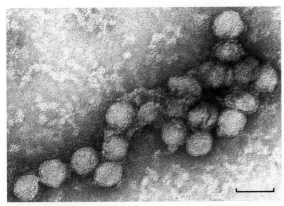

FIGURE 16.13 **West Nile Virus.** False-color transmission electron micrograph of West Nile viruses (WNV). WNV is transmitted by mosquitoes and is known to infect both humans and animals (such as birds). (Bar = 100 nm.) »» How would you describe the shape of the West Nile virus?

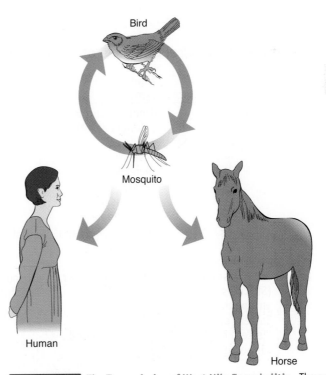

FIGURE 16.14 The Transmission of West Nile Encephalitis. The generalized pattern of viral encephalitis transmission involves various animals, including humans. »» Why would mosquito elimination be the best bet for interruption of the transmission cycle?

MICROFOCUS 16.5: Public Health

Avoiding Mosquitoes

Have you ever felt that you were "picked on" by mosquitoes while others were essentially left alone? Well, your feelings are correct and avoiding the diseases carried by them can be difficult because mosquitoes "love" some of us more than others (see figure).

There are more than 170 species of mosquitoes in the United States. Luckily, not all of them like humans and, of those that do, only the females take a blood meal from their victim. So, what makes for a good victim?

Compounds, such as lactic acid and carbon dioxide gas, as well as some perfumes attract mosquitoes. If you are hot and sweaty, you are a more likely target than someone whose body temperature is cooler and dry. Mosquitoes tend to be drawn to the face, ankles, and hands. Men are more attractive than women and adults more often are bitten than children.

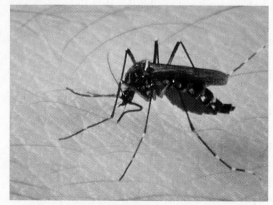

Mosquito feeding on human.

One's daily physiology may influence a mosquito bite. Women, for example, at certain points in their menstrual period are more likely to be attacked. Walking outside in the early morning or late evening makes one more likely to be bitten as that is when many mosquito species are most active.

So, it is best to wear insect repellent, such as one containing at least 20% to 35% DEET, and cover the extremities as best as possible. Also, eliminate any standing water around your home where mosquitoes may breed. As the CDC says, "Tell mosquitoes to buzz off!"

TABLE

16.4   A Summary of the Viral Diseases of the Nervous System

Disease	Causative Agent	Signs and Symptoms	Transmission	Treatment	Prevention and Control
Rabies	Rabies virus	Tingling, burning, coldness at bite site Fever, headache, increased muscle tension Paralysis and hydrophobia	Bite from rabid animal	Rabies immune globulin Rabies vaccine	Avoiding rabid animals Thoroughly washing the bitten area Pre-exposure vaccination when needed
Polio	Poliovirus	Often no signs and symptoms	Fecal-oral route	Supportive treatments	Polio vaccine Practicing good personal hygiene
Arboviral encephalitis	Arboviruses (St. Louis encephalitis, Eastern equine, Western equine, La Crosse, West Nile)	Sudden high fever, severe headache, stiff neck, disorientation, paralysis, mental disorders Convulsions and coma	Mosquitoes, ticks, sandflies	Supportive treatment	Avoiding mosquitoes in endemic areas Insect repellant Covering exposed skin

MICRO**INQUIRY** 16
Viral Disease Identification

Below are descriptions of several viral diseases based on material presented in this chapter. Read the case history and then answer the questions posed. Answers can be found in **Appendix D**.

Case 1

A 27-year-old male with a history of intravenous and over-the-counter drug abuse comes to the emergency room complaining of nausea, vomiting, headache, and abdominal pain. He indicates he has had these symptoms for two days. His vital signs are normal and he presents with slight jaundice. Physical examination indicates discomfort in the upper right quadrant. Serologic tests for HBsAg and serum antibodies to HAV are negative.

16.1a. Based on these symptoms and clinical findings, what disease does the patient have and what virus is the cause?

16.1b. What factors lead you to this conclusion?

16.1c. What is the significance of discomfort in the upper right quadrant?

16.1d. What is the probable source of the infection?

16.1e. To what other diseases is the patient at risk?

Case 2

A 19-year-old female comes to a neighborhood clinic complaining of difficulty swallowing. She tells the physician she has had a sore throat and fever for about a week and has been extremely tired. Examination indicates she has enlarged tonsils and a high count of B lympho-cytes. Her airways are clear and she has a clean chest X ray. Throat cultures for a bacterial infection are negative but a Monospot test is positive.

16.2a. What viral disease would be consistent with the patient's symptoms and clinical tests? What virus is the causative agent?

16.2b. What is the significance of the elevated B lymphocyte count?

16.2c. What dangerous complications can occur from this disease?

16.2d. In developing nations, what tumor often is associated with an infection by the same virus? How might the virus influence tumor formation?

Case 3

A 59-year-old man visits the local hospital emergency room complaining of nausea, abdominal discomfort with vomiting, and fever. On physical examination, he has a temperature of 38°C and appears jaundiced. He indicates the symptoms appeared abruptly two weeks ago after he had returned from a tour to Egypt. While there, the patient indicated he lived like the locals, ate in local restaurants, and often drank the local water. He did not have any vaccinations before leaving on his trip.

16.3a. Name two viral diseases discussed in this chapter that the patient might have.

16.3b. What is this patient's most likely disease? Why would you make that conclusion?

16.3c. Identify a possible transmission route for this virus.

16.3d. What serologic tests could be done to pinpoint the illness?

16.3e. If the patient had sought medical intervention earlier in the illness, what treatment could have been prescribed?

Case 4

A 27-year-old woman comes to a neighborhood clinic with a fever, backache, headache, and bone and joint pain. She indicates these symptoms appeared about two days ago. She also complains of eye pain. Questioning the woman reveals that she has just returned from a trip to Bangladesh where she was doing ecological research in the tropical forests. Skin examination shows remnants of several mosquito bites, which the woman corroborates. She indicates this was her first trip to Southeast Asia. She also reports that she had been taking some antibiotics that she had been given.

16.4a. Based on the woman's symptoms, what viral disease does she most likely have?

16.4b. What clues lead you to this diagnosis?

16.4c. Explain why the antibiotics did not help the woman's condition.

16.4d. Why might the woman want to seriously consider not returning to Southeast Asia to continue her ecological work?

■ SUMMARY OF KEY CONCEPTS

16.1 Viral Diseases of the Blood and the Lymphatic Systems
- **Infectious mononucleosis**
  1. Epstein-Barr virus
- **Cytomegalovirus disease**
  2. Cytomegalovirus
- **Hepatitis B**
  3. Hepatitis B Virus
- **Hepatitis C**
  4. Hepatitis C Virus

16.2 Viral Diseases Causing Hemorrhagic Fevers
- **Hemorrhagic fevers**
  5. Yellow fever virus
  6. Dengue fever virus
  7. Ebola virus/Marburg virus
  8. Lassa fever virus

16.3 Viral Infections of the Gastrointestinal Tract (not shown)
- **Hepatitis A**
  Hepatitis A virus
- **Hepatitis E**
  Hepatitis E virus
- **Viral gastroenteritis**
  Rotavirus, norovirus, enterovirus

16.4 Viral Diseases of the Nervous System
- **Rabies**
  9. Rabies virus
- **Polio**
  10. Poliovirus
- **Arboviral encephalitis**
  11. St. Louis encephalitis virus
  12. Eastern equine encephalitis virus
  13. Western equine encephalitis virus
  14. La Crosse virus
  15. West Nile virus

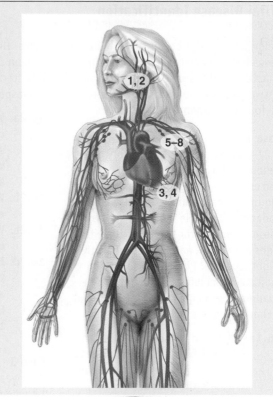

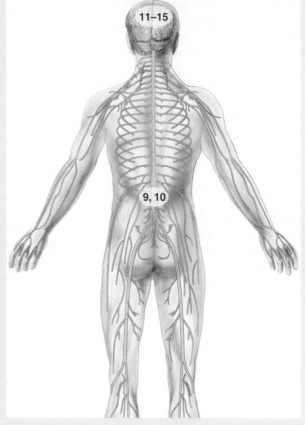

LEARNING OBJECTIVES

After understanding the textbook reading, you should be capable of writing a paragraph that includes the appropriate terms and pertinent information to answer the objective.

1. Describe the symptoms of **mononucleosis** and the potential complications arising from the illness.
2. Compare the similarities and differences between the nature of the hepatitis B and C viruses and the illnesses they cause.
3. Discuss the potential complications of a **yellow fever** illness and explain why **dengue fever** is considered the most important **arboviral disease** of humans.

4. Summarize the symptoms of **Ebola** and **Marburg hemorrhagic fevers**.
5. Describe how **Lassa fever** is transmitted.
6. Explain how the hepatitis A virus is spread and prevented.
7. Explain why rotavirus infections are so deadly in children and describe how noroviruses are transmitted.
8. Assess the outcome to someone bit by a rabid animal and recommend treatment if **rabies** symptoms have not yet appeared.
9. Explain how the polio virus causes disease and identify the two types of polio vaccines.
10. List the possible outcomes from an infection with the West Nile virus.

STEP A: SELF-TEST

Each of the following questions is designed to assess your ability to remember or recall factual or conceptual knowledge related to this chapter. Read each question carefully, then select the *one* answer that best fits the question or statement. Answers to even-numbered questions can be found in **Appendix C**.

1. Mononucleosis is an infection of _____ cells by the _____.
   A. T; cytomegalovirus
   B. B; Epstein-Barr virus
   C. lung; cytomegalovirus
   D. red blood; Epstein-Barr virus
2. Which of the following is NOT a transmission mechanism for hepatitis B?
   A. Sexual contact
   B. Non-sterile body piercing equipment
   C. Fecal-oral route
   D. Blood-contaminated needles
3. Symptoms of headache, fever, and muscle pain lasting 3 to 5 days, followed by a 2 to 24 hour abating of symptoms is characteristic of
   A. Yellow fever.
   B. Hepatitis C.
   C. Dengue fever.
   D. Ebola hemorrhagic fever.
4. A long thread-like RNA virus is typical of the _____ viruses.
   A. hepatitis C
   B. Ebola
   C. polio
   D. West Nile
5. The reservoir for Lassa fever is
   A. rats.
   B. mosquitoes.
   C. ticks.
   D. sandflies.

6. Which one of the following characteristics pertains to hepatitis A?
   A. Transmission is by the fecal-oral route.
   B. The incubation period is 2 to 4 weeks.
   C. It is an acute, inflammatory liver disease.
   D. All of the above (**A–C**) are correct.
7. _____ are the single most important cause of diarrhea in infants and young children admitted to American hospitals.
   A. Noroviruses
   B. Echoviruses
   C. Hepatitis A viruses
   D. Rotaviruses
8. Hydrophobia is a term applied to
   A. rotavirus infections.
   B. West Nile fever.
   C. arboviral encephalitis.
   D. rabies.
9. These viruses multiply first in the tonsils and then the lymphoid tissues of the gastrointestinal tract.
   A. Rabies virus
   B. Rotavirus
   C. Polio virus
   D. Hepatitis A virus
10. Arboviral encephalitis is an example of a
    A. disease causing gastroenteritis.
    B. disease spread by the fecal-oral route.
    C. zoonosis.
    D. type of hepatitis.

HTTP://MICROBIOLOGY.JBPUB.COM/9E

The site features eLearning, an online review area that provides quizzes and other tools to help you study for your class. You can also follow useful links for in-depth information, read more MicroFocus stories, or just find out the latest microbiology news.

STEP B: REVIEW

On completing your study of these pages, test your understanding of their contents by deciding whether the following statements are true or false. If the statement is true, write "T" in the space. If false, write "F" and substitute a word for the *underlined* word or phrase to make the statement true. The answers to even-numbered statements are listed in **Appendix C.**

11. _____ Both yellow fever and dengue fever are caused by a <u>DNA</u> virus transmitted by the mosquito.
12. _____ Eighty percent of people infected by West Nile virus experience <u>flu-like symptoms.</u>
13. _____ Downey cells are a characteristic sign of the viral disease <u>infectious mononucleosis.</u>
14. _____ The term "hydrophobia" means "fear of water," and it is commonly associated with patients who have <u>encephalitis.</u>
15. _____ Norovirus and rotavirus are both considered to be agents of viral <u>encephalitis.</u>
16. _____ The <u>Epstein-Barr</u> virus is the cause of infectious mononucleosis.
17. _____ Hepatitis is primarily a disease of the <u>liver</u>.
18. _____ A vaccine is available to prevent <u>yellow fever</u>.
19. _____ The Salk and Sabin vaccines are used for immunizations against <u>hepatitis</u>.
20. _____ <u>Hepatitis B</u> is most commonly transmitted by contact with infected semen or infected blood.
21. _____ One of the most important causes of diarrhea in infants and young children admitted to hospitals is the <u>Epstein-Barr virus</u>.
22. _____ <u>Filoviruses</u> are long, thread-like viruses that cause hemorrhagic fevers and include the marburgvirus.
23. _____ Enlargement of the lymph nodes, sore throat, mild fever, and a high count of B-lymphocytes are characteristic symptoms in people who have <u>polio</u>.
24. _____ The C in the TORCH group of diseases stands for the <u>cytomegalovirus</u>, which can be transmitted from a pregnant woman to her unborn child.

STEP C: APPLICATIONS

Answers to even-numbered questions can be found in **Appendix C.**

25. Written on some blood donor cards is the notation "CMV." What do you think the letters mean, and why are they placed there?
26. Sicilian barbers are renowned for their skill and dexterity with razors (and sometimes their singing voices). French researchers studied a group of 37 Sicilian barbers and found that 14 had antibodies against hepatitis C, despite never having been sick with the disease. By comparison, when a random group of 50 blood donors was studied, none had the antibodies. As an epidemiologist, what might account for the high incidence of exposure to hepatitis C among the barbers?

27. As a state health inspector, you are suggesting all restaurant workers should be immunized with the hepatitis A vaccine. Why would restaurant owners agree or disagree with your idea?
28. An epidemiologist notes that India has a high rate of dengue fever but a very low rate of yellow fever. What might be the cause of this anomaly?

STEP D: QUESTIONS FOR THOUGHT AND DISCUSSION

Answers to even-numbered questions can be found in **Appendix C.**

29. Health authorities panicked when an outbreak of Ebola hemorrhagic disease occurred among imported macaques in a quarantine facility in Reston, Virginia, in 1989. What sparks such a dramatic response when a disease like Ebola fever breaks out?
30. A diagnostic test has been developed to detect hepatitis C in blood intended for transfusion purposes. Obviously, if the test is positive, the blood is not used. However, there is a lively controversy as to whether the blood donor should be informed of the positive result. What is your opinion? Why?

31. In the southwestern United States, abundant rain and a mild winter often bring conditions that encourage a burgeoning rodent population. Under these circumstances, what viral disease would health officials anticipate and what precautions should they give residents?
32. Disney World and 20 swampy counties in Florida use "sentinel chickens" strategically placed on the grounds to detect any signs of viral encephalitis. Why do you suppose they use chickens? Why are Disney World and many Florida counties particularly susceptible to outbreaks of viral encephalitis? What recommendations might be offered to tourists if the disease broke out?

Eukaryotic Microorganisms: The Fungi

That mold. It smells like death.
—Veronica Randazzo after returning to her house following hurricane Katrina

In August 2005, hurricane Katrina left unimaginable devastation everywhere along the south-central coast of the United States. In New Orleans, thousands of homes were flooded and left sitting in feet of water for weeks. Many health experts were concerned about outbreaks of infectious diseases like cholera, West Nile fever, and gastrointestinal illnesses. Homes sitting in stagnant water could become breeding grounds for microorganisms.

Thankfully, most of these infectious disease scenarios did not occur. However, what did break out in many of the parishes in New Orleans was mold. There was mold on walls, ceilings, cabinets, clothes, and just about anything that provided a source of moisture and nutrients (**FIGURE 17.1**). It formed carpets of spore-forming colonies everywhere.

If you see small spots of mold in your home, a bleach solution will do a great job to kill and eliminate the problem. But what if an entire home and its contents are one giant "culture dish"? More than likely, most of these homes have been demolished (or stripped to the framing) and furniture and other home contents destroyed. Many home items, like beds, couches, or cabinets that were above the water level contained mold due to the prolonged humidity, and had to be replaced. Most health officials told residents to

(A)

(B)

FIGURE 17.1 A Wall of Mold. (A) Every circular spot seen on this wall and (B) on the ceiling and other objects represents a mold colony. Each colony started from a single spore; now each mature colony contains millions of spores. »» How do you think molds reproduce?

Biosphere:
The whole area of Earth's surface, atmosphere, and sea that is inhabited by living things.

Immunocompromised:
Refers to the lack of an adequate immune response resulting from disease, exposure to radiation, or treatment with immunosuppressive drugs.

well as offices and hospitals, molds were discovered growing in ventilation systems and the ventilation ducts. If the ventilation fans were turned on, literally hundreds of billions of spores would be blown and spread to new areas to germinate and grow. In fact, six weeks after Katrina, the mold spore count in the air at various sites in New Orleans was as high as 102,000 spores per cubic meter—twice the number considered normal for New Orleans.

What about illnesses or disease from breathing the mold spores? By November 2005, many New Orleans residents who had returned were suffering from upper respiratory problems—the residents "affectionately" called it the "Katrina cough." In fact, residents with asthma, bronchitis, and allergies who had left New Orleans were asked not to return just yet. Although no respiratory outbreaks have occurred, in 2008 a five-year study began to investigate if workers in New Orleans are at risk of inhaling mold spores or bacteria stemming from the Hurricane Katrina floods.

This discussion indicates that molds grow in many natural and constructed environments. They often break down dead or decaying matter, such as the cases described in New Orleans. Many molds though are of great importance to the natural recycling in the biosphere. Still others can act as pathogens and cause some dangerous and debilitating diseases.

From the description of the situation in New Orleans, you can appreciate fungi as producers of massive numbers of **spores**, representing microscopic cells for disseminating the organisms to new territories and environments where they germinate, grow, and again reproduce.

Molds and yeasts are fungi, and they contain many species some of which we survey and study in this chapter. We will encounter many beneficial fungi such as those used to make antibiotics or in commercial and industrial processes. We also will identify and discuss several widespread human diseases caused by fungi. Many are of concern to immunocompromised individuals.

Our study begins with a focus on the structures, growth patterns, and life cycles of fungi—something quite unique from the other groups of microorganisms.

follow the same slogan used for potentially spoiled food: "When in doubt, throw it out."

At Tulane Hospital, the first floor was covered with mold, which made cleanup and reopening of many wards a very difficult chore. In homes as

17.1   Characteristics of Fungi

The **fungi** (sing., fungus) are a diverse group of eukaryotic microorganisms. Some 75,000 species have been described, although as many as 1.5 million may exist. For many decades, fungi were classified as plants, but laboratory studies have revealed at least four properties that distinguish fungi from plants:

- Fungi lack chlorophyll, while plants have this pigment.
- Fungal cell walls contain a carbohydrate called chitin; plant cell walls have cellulose.
- Most fungi are not truly multicellular like plants.
- Fungi are heterotrophic, while plants are autotrophic.

Mainly for these reasons, fungi are placed in their own kingdom Fungi, within the domain *Eukarya* of the "tree of life" (see Chapter 3). The study of fungi is called **mycology** (*myco* = "fungus") and a person who studies fungi is a mycologist.

Fungi Share a Combination of Characteristics

KEY CONCEPT

1. Most fungi grow as a branching filamentous network.

Fungi generally have life cycles involving two phases: a growth (vegetative) phase and a reproductive phase. A major group of fungi, the **molds**, which you have come to appreciate already from the chapter opener, grow as long, tangled filaments of cells that give rise to visible colonies (**FIGURE 17.2A**). Another group, the **yeasts**, are unicellular organisms whose colonies on agar visually resemble bacterial colonies (**FIGURE 17.2B**). Yet other forms are **dimorphic**; usually at ambient temperature (25°C) they grow as filamentous molds, but at body temperature (37°C) they convert to unicellular, pathogenic, yeast-like forms.

The molds consist of masses of intertwined filaments called **hyphae** (sing., hypha). The hyphae are the morphological unit of a filamentous fungus and individual hyphae usually are visible only with the aid of a microscope (**FIGURE 17.3A**). Hyphae have a broad diversity of forms and can be highly branched. A thick mass of hyphae is called a **mycelium** (pl., mycelia). This mass is

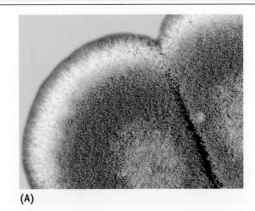

(A)

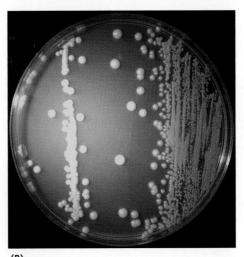

(B)

FIGURE 17.2 Fungal Colonies. **(A)** On growth media, molds, such as this *Penicillium* species, grow as colonies visible to the naked eye. The spores are the gray-green regions of the colonies. **(B)** Petri dish culture showing colonies of the yeastlike fungus *Torulopsis glabrata*.
»» What does the white fuzzy growth in (A) represent?

usually large enough to be seen with the unaided eye, and generally it has a rough, cottony texture (Figure 17.2A). The mycelium along with any reproductive structures would represent the fungal organism.

Being eukaryotic organisms, fungi have one or more nuclei as well as a range of organelles including mitochondria, an endomembrane system, ribosomes, and a cytoskeleton. The cell wall is composed of large amounts of chitin. **Chitin** is a carbohydrate polymer of acetylglucosamine units; that is, glucose molecules containing amino and acetyl groups (see Chapter 2). The cell wall provides rigidity and strength, which, like the cell wall of all microorganisms, allows the cells to resist bursting due to high internal water pressure.

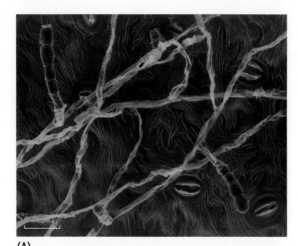

(A)

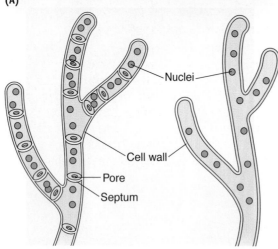

Septate hyphae   Nonseptate hyphae

(B)

FIGURE 17.3 Hypha Structure. (**A**) A false-color scanning electron micrograph of fungal hyphae growing on a leaf surface. (Bar = 40 μm.) (**B**) Molds have hyphae that are either septate or nonseptate. Septa compartmentalize hyphae into separate cells, although the septa have a pore through which cytoplasm and nuclei can move. »» What unique structural features are presented in this figure?

In many species of fungi, hyphal cross walls, called **septa** (sing., septum), divide the cytoplasm into separate cells (**FIGURE 17.3B**). Such fungi are described as **septate**. In the blue-green mold *Penicillium chrysogenum,* the septa are incomplete, however, and pores in the cross walls allow adjacent cytoplasms to mix. In other fungal species, such as the common bread mold *Rhizopus,* the filaments are without septa. In both examples, such hyphae are considered **coenocytic,** meaning they contain many nuclei in a common cytoplasm.

Because fungi absorb preformed organic matter, they are described as **heterotrophs**. Most are

saprobes, feeding on dead or decaying organic matter. Together with many bacterial species, these fungi make up the **decomposers** (see Chapter 3), recycling vast quantities of organic matter (MICROFOCUS 17.1). In industrial settings, these decompositions can be profitable. The fungus *Trichoderma,* for example, produces enzymes to degrade cellulose and give jeans a "stone-washed" appearance.

Other fungi are pathogens, living on plants or animals, and often causing disease. Most such fungi are opportunistic and only cause serious disease in immunocompromised individuals.

CONCEPT AND REASONING CHECKS

17.1 Assess the role of hyphae to fungal growth.

Fungal Growth Is Influenced by Several Factors

KEY CONCEPT

2. Fungal growth is dependent on moisture, adequate chemical nutrients, and physical factors.

Fungi acquire their nutrients through absorption, either as saprobes or pathogens. Being mostly terrestrial organisms, the molds and yeasts secrete enzymes into the surrounding environment that break down (hydrolyze) complex organic compounds into simpler ones. As a result of this extracellular digestion, simpler compounds, like glucose and amino acids, can be absorbed by the hyphae.

The mycelium formed by a mold represents the "feeding network" for these fungi. In some cases the mycelium can form a tremendously large surface area for nutrient absorption (MICROFOCUS 17.2). For the yeasts, nutrients are absorbed across the cell surface, similar to the way bacterial cells obtain their nutrients.

Mycelial extension occurs by the elongation or growth of the hyphae. The nuclei divide by mitosis and growth of these mycelial fungi involves lengthening and branching of the hyphae at the hyphal tips. Yeast growth in number occurs by cell division as the cells undergo many rounds of mitosis and cytokinesis to form a large population of individual cells.

Fungal growth is influenced by many factors in the environment. Besides the availability of chemical nutrients, oxygen, temperature, and pH influence growth.

MICROFOCUS 17.1: Evolution

When Fungi Ruled the Earth

About 250 million years ago, at the close of the Permian period, a catastrophe of epic proportions visited the Earth. Apparently, over 90% of animal species in the seas vanished. The great Permian extinction, as it is called, also wreaked havoc on land animals and cleared the way for dinosaurs to inherit the planet.

But land plants managed to survive, and before the dinosaurs came, they spread and enveloped the world. At least, that is what paleobiologists traditionally believed.

Now, however, they are revising their outlook and finding a significant place for the fungi. In 1996, Dutch scientists from Utrecht University presented evidence that land plants were decimated by the

A mushroom species growing on a rotting log.

extinction and that for a brief geologic span, dead wood covered the planet. During this period, they suggest the fungi emerged and wood-rotting species experienced a powerful spike in their populations (see figure). Support for their theory is offered by numerous findings of fossil fungi from the post-Permian period. The fossils are bountiful and they come from all corners of the globe. Significantly, they contain fungal hyphae, the active feeding forms rather than the dormant spores.

And so it was that fungi proliferated wildly and entered a period of feeding frenzy where they were the dominant form of life on Earth. It's something worth considering next time you kick over a mushroom growing in a lawn.

MICROFOCUS 17.2: Environmental Microbiolgy

A Humongous and Ancient Individual

What's bigger than a blue whale and older than a California redwood? A fungus, of course! *Armillaria*, a plant pathogen, infects evergreen trees. The visible part of the fungus is the golden mushrooms it produces (see figure). However, underground in the soil lies an invisible mycelium invading the roots of the evergreen tree and absorbing its water and carbohydrates, which often spells death for the infected tree.

In 2001, scientists studying in Oregon's Malheur National Forest discovered a single *Armillaria* mycelium spreading through 2,200 acres of forest soil. It measures over nine square kilometers—about the area of 1,600 football fields! To be of this size, scientists believe the fungus

Armillaria mushrooms growing on a tree trunk.

germinated from a single spore somewhere between 2,000 and 8,500 years ago. Now that's old and humongous, considering the mycelium of most *Armillaria* species covers only 20 or 30 acres.

This discovery is not unique. Another species of *Armillaria* was discovered in a Michigan hardwood forest that measured about 37 acres in size and, in eastern Washington, an *Armillaria* mycelium measuring about 1,500 acres in size has been discovered.

The world's biggest fungus is challenging the concept of what constitutes an individual. Is the Oregon *Armillaria* a single individual organism? "If you could take away the soil and look at it, it's just one big heap of fungus with all of these filaments that go out under the surface," said Dr. Catherine Parks, who was one of its discoverers.

Oxygen. The majority of fungi are aerobic organisms, with the notable exception of the facultative yeasts, which can grow in either the presence of oxygen or under fermentation conditions (see Chapter 6).

Temperature. Most fungi grow best at about 23°C, a temperature close to normal room temperature. Notable exceptions are the pathogenic fungi, which grow optimally at 37°C, which is body temperature. As mentioned, dimorphic fungi grow as yeastlike cells at 37°C and a mycelium at 23°C. Psychrophilic fungi grow at still lower temperatures, such as the 5°C found in a normal refrigerator.

> **Psychrophilic:**
> Refers to organisms that prefer to grow at cold temperatures.

pH. Many fungi thrive under mildly acidic conditions at a pH between 5 and 6. Acidic soil therefore may favor fungal turf diseases, in which case lime (calcium carbonate) is added to neutralize the soil. Mold contamination also is common in acidic foods such as sour cream, citrus fruits, yogurt, and most vegetables. Moreover, the acidity in breads and cheese encourages fungal growth. Blue (Roquefort) cheese, for example, consists of milk curds in which the mold *Penicillium roqueforti* has been added for flavor and texture (**FIGURE 17.4**).

Normally, high concentrations of sugar are conducive to growth, and laboratory media for fungi usually contain extra glucose; examples include Sabouraud dextrose (glucose) agar and potato dextrose agar.

Fungal growth in nature forms important links in ecological cycles because fungi, along with bacterial species, rapidly decompose animal and plant matter (Chapter 26). Working in immense numbers, fungi release carbon and minerals back to the environment, making them available for recycling.

Many fungi also live in a mutually beneficial relationship with other species in nature through a symbiotic association called mutualism (see Chapter 3). Fungi called **mycorrhizae** (*rhiza* = "root") live harmoniously with plants where the hyphae of these fungi invade or envelop the roots of plants (**FIGURE 17.5A**). These mycorrhizae consume some of the carbohydrates produced by the plants, but in return act as a second root system, contributing essential minerals and water to promote plant growth (**FIGURE 17.5B**). Mycorrhizae have been found in plants from salt marshes, deserts, and pine forests. In fact, these beneficial fungi

> **Mutualism:**
> A symbiotic relationship between two organisms of different species that benefits both.

FIGURE 17.4 **Roquefort Cheese.** Roquefort cheese is made from cow's milk and contains the mold *Penicillium roquefortii*. »» Why would a mold be added to the ripening process of cheeses such as this?

live in and around the roots of 95% of examined plant species.

Besides the mycorrhizae, most plants examined also contain fungal **endophytes**, which are fungi living and growing entirely within plants, especially leaf tissue. They do not cause disease but, rather like mycorrhizae, they provide better or new growth opportunities for the plant. In the southwestern Rocky Mountains, for instance, a fungus thrives on the blades of a species of grass, producing a powerful poison that can put horses and other animals to sleep for about a week (the grass is called "sleepy grass" by the locals). Thus, the grass survives while other species are nibbled to the ground. MicroFocus 17.3 describes a few more remarkable examples.

CONCEPT AND REASONING CHECKS

17.2 Estimate the value of mycorrhizae and fungal endophytes to plant survival.

Reproduction in Fungi Involves Spore Formation

KEY CONCEPT

3. Fungal reproduction may take place by asexual as well as by sexual processes.

Sporulation is the process of spore formation. It usually occurs in structures called **fruiting bodies**, which represent the part of a fungus in

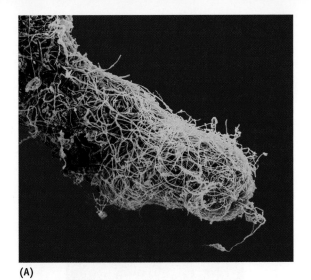

(A)   (B)

FIGURE 17.5 **Mycorrhizae and Their Effect on Plant Growth.** (**A**) Mycorrhizae surround the root of a *Eucalyptus* tree in this false-color scanning electron micrograph. These fungi are involved symbiotically with their plant host, such as aiding in mineral metabolism. (**B**) An experiment analyzing the mycorrhizal effects on plant growth. »» Which plant or plants (CK, GM, GE) represent(s) the control and experimental set ups?

MICROFOCUS 17.3: Environmental Microbiology
Fungi as Protectors

What would you think if you saw wheat growing in the deserts of California or Arizona? You probably would think it was a mirage or you were hallucinating from the lack of water. Well, in the near future, such an apparition might indeed be real.

It turns out that mycorrhizae are not the only fungi providing protection and growth benefits for hundreds of species of plants. The hyphae of endophytes grow within plant tissue and between cells of many healthy leaves where nutrients are exchanged between host and fungus. No damage is done to the leaves and only the reproductive structures make it to the surface to release spores into the wind.

One example of endophyte benefit was discovered by Regina Redman and Russell Rodriguez of the U.S. Geological Survey in Seattle, Washington. They took

In cacao leaves, the more mature (green) leaves gain more benefit from endophyte associations than young (red) leaves.

perennial grasses normally growing in hot soils around geysers and grew them in the lab. Those plants exposed to hot soil but lacking endophytes died while those living with endophytes survived. Most interesting, the fungi by themselves also died in the hot soils. Apparently, there needs to be a give and take between host and fungus—representing a true mutualistic relationship.

Redman and Rodriguez then teamed up with Joan Henson at Montana State University to show that these same endophytes protect other plants. At least in the lab, when endophyte spores were placed on tomato, watermelon, or wheat seedlings, the tomato and watermelon seedlings with endophytes survived the stresses of high temperatures (50°C) or drought conditions. Although the wheat seedlings died, the plants survived about a week longer than those without endophytes.

Meanwhile in Panama, another group of researchers, led by A. Elizabeth Arnold, now at the University of Arizona, discovered that chocolate-tree (cacao) leaves harboring endophytes are more resistant to pathogen attack, while fungus-free leaves become diseased. In their studies, when leaves were purposely inoculated with one of the major pathogens of cacao, leaves associated with several different endophytes were less likely to be invaded and would most likely survive. Although young leaves benefited from the symbiosis, the older leaves appeared to benefit the most from the endophyte association (see figure).

So, at least in the lab, endophytic fungi play various but important roles in plant survival. As more is discovered, perhaps the day will come when wheat will grow in the desert.

which spores are formed and from which they are released. These structures may be asexual and invisible to the naked eye, or sexual structures, such as the macroscopic mushrooms.

Asexual reproduction. Asexual reproductive structures develop at the ends of specialized hyphae. As a result of mitotic divisions, thousands of spores are produced, all genetically identical.

Many asexual spores develop within sacs or vessels called **sporangia** (sing., sporangium; *angio* = "vessel") (FIGURE 17.6A). Appropriately, the spores are called **sporangiospores**. Other fungi produce spores on supportive structures called **conidiophores** (FIGURE 17.6B). These unprotected, dust-like spores are known as **conidia** (sing., conidium; *conidio* = "dust"). Fungal spores are extremely light and are blown about in huge numbers by wind currents. In yet other fungi, spores may form simply by fragmentation of the hyphae yielding **arthrospores** (*arthro* = "joint"). The fungi that cause athlete's foot multiply in this manner.

Many yeasts reproduce asexually by **budding**. In this process, the cell becomes swollen at one edge, and a new cell called a **blastospore**, (*blasto* = "bud") develops (buds) from the parent cell (FIGURE 17.7). Eventually, the spore breaks free to live independently. The parent cell can continue to produce additional blastospores.

Once free of the fruiting body, spores landing in an appropriate environment have the capability of germinating to reproduce new unicellular yeast cells or a new hypha (FIGURE 17.8). Continued growth will eventually form a mycelium.

Sexual reproduction. Many fungi also produce spores by sexual reproduction. In this process, opposite mating types come together and fuse (FIGURE 17.9). Because the nuclei are genetically different in each mating type, the fusion cell represents a **heterokaryon** (*hetero* = "different"; *karyo* = "nucleus"); that is, a cell with genetically dissimilar nuclei existing for some length of time in a common **dikaryotic** cytoplasm. Eventually the nuclei fuse and a diploid cell is formed. The chromosome number soon is halved by meiosis, returning the cell or organism to a haploid condition.

A visible fruiting body often results during sexual reproduction and it is the location of the

Mating types:
Separate mycelia of the same fungus or separate hyphae of the same mycelium.

(A)

(B)

FIGURE 17.6 **Fungal Fruiting Bodies.** False-color scanning electron micrographs of sporangia and conidia. **(A)** Sporangia of the common bread mold *Rhizopus*. Each round sporangium contains thousands of sporangiospores. (Bar = 20 μm.) **(B)** The conidiophores and conidia in the mold-like phase of *Penicillium roquefortii*. Many conidiophores are present within the mycelium. Conidiophores (orange) containing conidia (blue) are formed at the end of specialized hyphae (green). (Bar = 20 μm.) »» Why must fungal spores be elevated on the tips of hyphae?

haploid spores. Perhaps the most recognized fruiting body from which spores are produced is the mushroom.

CONCEPT AND REASONING CHECKS

17.3 Assess the role of asexual and sexual reproduction in fungi.

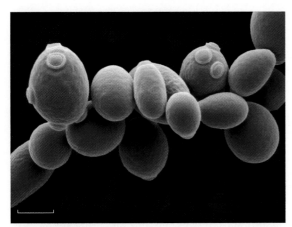

FIGURE 17.7 Yeast Budding. False-color scanning electron micrograph of the unicellular yeast *Saccharomyces cerevisiae*. (Bar = 3 μm.) »» Propose what the circular areas represent on the parent cell at the left and right center of the micrograph.

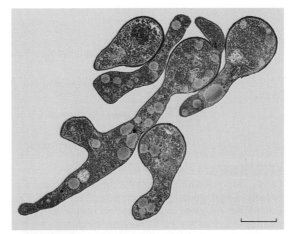

FIGURE 17.8 Germinating Fungal Spores. A false-color transmission electron micrograph of germinating fungal spores. (Bar = 3 μm.) »» What do the elongated structures protruding from the round spore represent?

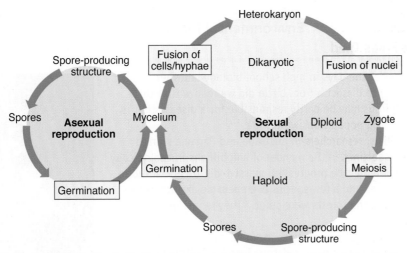

FIGURE 17.9 A Typical Life Cycle of a Mold. Many fungi have both an asexual and sexual reproduction characterized by spore formation. The unique phase in the life cycle is the presence of a heterokaryon where nuclei from two different mating types remain separate (dikaryotic) in a common cytoplasm. »» How could an organism like a mold survive without a sexual cycle?

17.2 The Classification of Fungi

Originally, fungi were considered part of the plant kingdom. Then in 1968, they received their own kingdom status with Whittaker's classification scheme (see Chapter 3). Currently, they are still considered a kingdom, but under the domain *Eukarya* in Woese's three-domain system.

Fungi Can Be Classified into Five Different Phyla

KEY CONCEPT

4. Fungal classification usually is based on the form of sexual reproduction.

Historically, fungal distinctions were made on the basis of either structural differences, or physiological and biochemical patterns. However, DNA analyses and genome sequencing are becoming important tools for drawing evolutionary relationships among various fungi (**FIGURE 17.10**).

A fungus can be cataloged into one of five phyla, depending on its mode of sexual reproduction. These phyla are the Chytridiomycota, Glomeromycota, Zygomycota, Ascomycota, and the Basidiomycota. If the fungus lacks a recognized sexual cycle, it is placed into an informal group called the mitosporic fungi.

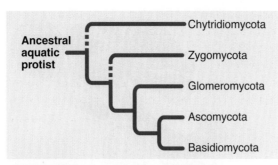

FIGURE 17.10 A Phylogeny for the Kingdom **Fungi.** This generalized scheme for the relationships between fungal groups is based on molecular findings. The dashed lines represent an as yet unclear lineage for the chytrids and zygomycetes. »» Because the chytrids are the only fungal group with flagella, indicate on the phylogenetic tree where you would place the phrase *"Loss of flagella."*

The Chytridiomycota and Glomeromycota.

The oldest known fungi are related to certain members of the **Chytridiomycota**, commonly called **chytrids**. Members of the phylum give us clues about the possible origin of fungi. First, chytrids are predominantly aquatic, and not terrestrial, organisms. This means the fungi originated in water probably from a flagellated, protistan ancestor. Secondly, being aquatic, chytrids have flagellated reproductive cells. No other fungi have motile flagellate cells, suggesting the other fungi lost this trait at some point in their evolutionary history. Finally, like other fungi, chytrids have chitin strengthening their cell walls. Until recently, few chytrids had any noticeable impact—for good or bad (MicroFocus 17.4).

MICROFOCUS 17.4: Environmental Microbiology
The Day the Frogs Died

Many of us remember the days in high school biology lab when we had an opportunity to dissect a frog. Little did we realize that these imported African frogs may be partly responsible for a disease that is decimating other frog populations worldwide.

In the early 1990s, researchers in Australia and Panama started reporting massive declines in the number of amphibians in ecologically pristine areas. As the decade progressed, massive die-offs occurred in dozens of frog species and a few species even became extinct. Once filled with frog song, the forests were quiet. "They're just gone," said one researcher.

Hypotheses to explain these declines included, among others, habitat modification, introduction of new predators, increased ultraviolet radiation, pollution, adverse weather changes, and infectious disease.

By 1998, infectious disease was identified as one of the reasons for the decline. More than 100 amphibian species on four continents, including North and South America, were infected with a chytrid called *Batrachochytrium dendrobatidis*. This chytrid, the only one known to infect vertebrates, uses the frog's keratinized skin as a nutrient source;

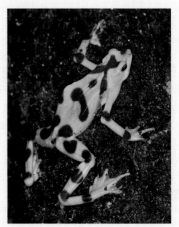

A species of harlequin frog, one of many species being wiped out by a chytrid infection.

epidermal sloughing of the frog's skin is one sign of the disease called chytridiomycosis. Roughly one-third of the world's 6,000 amphibian species are now considered under threat of extinction (see figure). But, why the sudden die off in recent years?

Alan Pounds, a researcher at the Golden Toad Laboratory for Conservation in Costa Rica, suggests that global warming is creating optimal conditions for the fungus. His work indicates climate change has led to cooler days and warmer nights on tropical mountainsides, creating the ideal growth—and infection—conditions for *B. dendrobatidis*.

Pounds also suggests the chytrid may have come from some of the African frogs exported around the world. His hypothesis is that some escaped frogs passed the fungus to hardier ones, like bullfrogs, which in turn infected more susceptible frog species.

Are frog deaths from chytridiomycosis a sign of a yet unseen shift in the ecosystem, much like a canary in a coal mine? Some believe this type of "pathogen pollution" may be as serious as chemical pollution.

"Disease is killing the frogs, but climate change is pulling the trigger," Dr. Pounds said.

The **Glomeromycota** form what some consider the most extensive symbiosis on Earth. These fungi represent a group of **endomycorrhizae** that exist within the roots of more than 80% of the world's land plants. They do not kill the plants but rather interact mutualistically by providing essential phosphate and other nutrients to the plant. In return, the fungi receive needed organic compounds from the plant. In fact, some mycologists believe that plant evolution onto land more than 400 million years ago depended on the symbiosis with the ancestral Glomeromycota, which were able to provide plants with needed nutrients from the soil.

CONCEPT AND REASONING CHECKS

17.4A What are the unique features of the Chytridiomycota and Glomeromycota?

The Zygomycota. The phylum **Zygomycota** consists of a group of fungi (zygomycetes) inhabiting terrestrial environments. Familiar representatives include fast-growing bread molds and other molds typically growing on spoiled fruits with high sugar content or on acidic vegetables (**FIGURE 17.11**). On these and similar materials, the heterotrophic fungi typically grow inside their food, dissolving the substrate with extracellular enzymes, and taking up nutrients by absorption.

Members of the phylum make up about 1% of the described species of fungi. The zygomycetes have chitinous cell walls and grow as a coenocytic mycelium. During sexual reproduction, sexually opposite mating types fuse, forming a unique, heterokaryotic, diploid **zygospore** (**FIGURE 17.12**). After a period of dormancy, the zygospore germinates and releases haploid sporangiospores from a sporangium. Elsewhere in the mycelium, thousands of asexually produced sporangiospores are produced within sporangia. Both sexually produced and asexually produced spores are dispersed on wind currents. Several members can cause infections and disease in humans.

CONCEPT AND REASONING CHECKS

17.4B Describe the unique properties of the Zygomycota.

The Ascomycota. Members of phylum **Ascomycota** (*asco* = "sac") or sac fungi, commonly are called the ascomycetes. They are very diverse and account for about 75% of all known fungi. The phyla contains many common and useful fungi, including *Saccharomyces cerevisiae* (Baker's yeast), *Morchella esculentum* (the edible morel), and *Penicillium chrysogenum* (the mold that produces penicillin) (**FIGURE 17.13**). The phylum also has several members associated with illness and disease. *Aspergillus flavus*, produces **aflatoxin**, a fungal contaminant of nuts and stored grain that is both a toxin and the most potent known natural carcinogen; *Cryphonectria parasitica,* responsible for the death of 4 billion chestnut trees in the eastern United States; and *Candida albicans,* cause of thrush, diaper rash, and vaginitis.

Penicillium and *Candida* species lack a sexual reproductive cycle. However, comparative

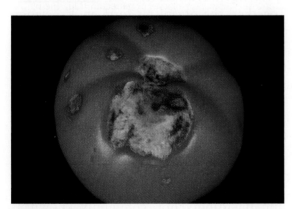

FIGURE 17.11 **Mold Growing on a Tomato.**
Zygomycete molds typically grow and reproduce on over-ripe fruits or vegetables, such as this tomato. »» Identify the black structures and the gray fuzzy growth on the tomato.

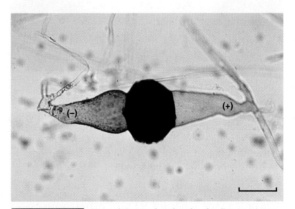

FIGURE 17.12 Sexual Reproduction in the Zygomycota.
Sexual reproduction between hyphae of different mating types (+ and –) produces a darkly pigmented zygospore. (Bar = 30 μm.) »» When the zygospore germinates, what will be produced from the structure?

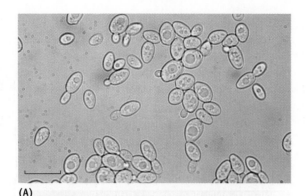

(A)

(B)

(C)

FIGURE 17.13 **Representative Ascomycetes. (A)** Light micrograph of yeast cells. Several of the cells are undergoing budding. (Bar = 10 μm.) **(B)** The edible morel, an ascomycete prized for its delicate taste. **(C)** A *Penicillium* species growing on an orange. »» Where is the growing edge of the mycelium on the orange?

genomics (see Chapter 9) and other nonsexual phenotypic characters have shown that these fungi actually are members of the Ascomycota.

The Ascomycota is a sister group to the Basidiomycota because the hyphae of both phyla

are septate; both have cross-walls dividing the hyphae into segments, but with large pores allowing a continuous flow of cytoplasm. The hyphae, like other filamentous fungi, form a mycelium to obtain nutrients from dead or living organisms. In fact, their biggest ecological role is in decomposing and recycling plant material.

As a group, the ascomycetes have the ability to form conidia through asexual reproduction or **ascospores** through sexual reproduction. Ascospores are formed within a reproductive structure called an **ascus** (pl., asci), within which eight haploid ascospores form (**FIGURE 17.14**). Many asci and other hyphae form the fruiting body.

Some ascomycetes form symbioses with plant roots (mycorrhizae) or the leaves and stems of plants (endophytes). Ascomycetes also are the most frequent fungal partner in lichens. A **lichen** is a mutualistic association between a fungus and a photosynthetic organism (**photobiont**) such as a cyanobacterium or green alga (**FIGURE 17.15A, B**). Most of the visible body of a lichen is the fungus with the hyphae penetrating the cells of the photosynthetic partner to receive carbohydrate nutrients. The photosynthetic organism receives fluid from the water-husbanding fungus.

Lichen asexual reproduction occurs through the formation of **soredia** that consist of a group of photobionts surrounded by hyphae. Soredia are carried by wind currents and deposited on a new surface, where a new lichen then forms.

Lichens often are grouped by appearance into leafy lichens (foliose), shrubbery lichens (fruticose), and crusty lichens (crustose) (**FIGURE 17.15C**). Together, the organisms form an association that readily grows in environments where neither organism could survive by itself (e.g., rock surfaces). Indeed, in some harsh environments, lichens support entire food chains. In the Arctic tundra, for example, reindeer graze on carpets of reindeer moss, which actually is a type of lichen.

CONCEPT AND REASONING CHECKS

17.4C Summarize the properties of the Ascomycota.

The Basidiomycota. Members of the phylum **Basidiomycota**, commonly known as basidiomycetes, are club fungi. The Basidiomycota contains about 30,000 identified species, representing 37% of the known species of true fungi. Members of the Basidiomycota can be

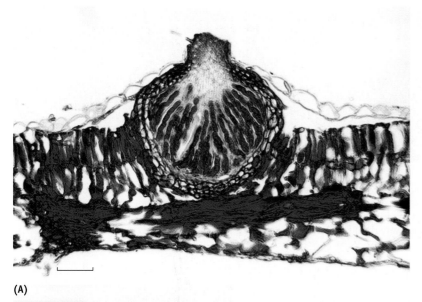

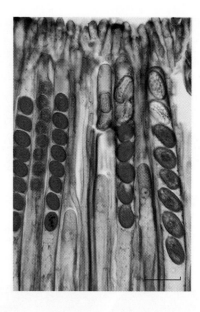

(A)   (B)

FIGURE 17.14 **Ascomycetes and Spores.** (**A**) Cross section of a fruiting body on an apple leaf. (Bar = 20 μm.) (**B**) A higher magnification of several asci, each containing eight ascospores. (Bar = 5 μm.) »» Ascospores are representative of what stage of the ascomycete life cycle?

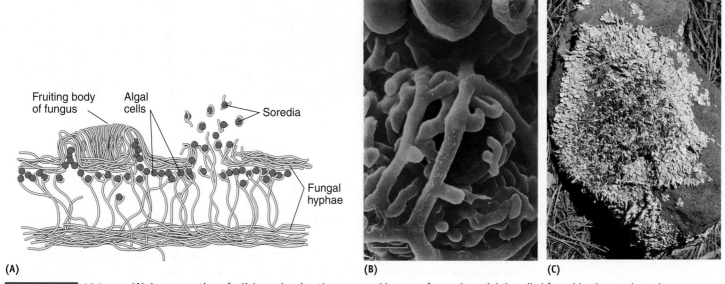

FIGURE 17.15 **Lichens.** (**A**) A cross section of a lichen, showing the upper and lower surfaces where tightly coiled fungal hyphae enclose photosynthetic algal cells. On the upper surface, a fruiting body has formed. Airborne clumps of algae and fungus called soredia are dispersed from the mycelium to propagate the lichen. Loosely woven fungi at the center of the lichen permit the passage of nutrients, fluids, and gases. (**B**) A false-color scanning electron micrograph of the close, intimate contact between fungal hyphae (orange) and an alga cell (green). (Bar = 2 μm.) (**C**) A typical lichen growing on the surface of a rock. Lichens are rugged organisms that can tolerate environments where there are few nutrients and extreme conditions. The brown fruiting bodies can be seen. »» What attributes of the fungus and alga permit the lichen to withstand extreme environmental conditions?

unicellular or multicellular, sexual or asexual, and terrestrial or aquatic. The most recognized members are the mushrooms and puffballs (**FIGURE 17.16A, B**). Some Basidiomycota are important saprobes, decomposing wood and other plant products. Other members form mycorrhizae while still others are important plant pathogens, such as the so-called "rusts" and "smuts" that infect cereals and other grasses. Rust fungi are so named because of the rusty, orange-red color of the infected plant, while smut fungi are characterized by sooty

FIGURE 17.16 **The Basidiomycota.** The basidiomycetes are characterized by sexual reproductive structures that usually are macroscopic. These include (**A**) the mushrooms, such as this species of *Amanita muscaria,* and (**B**) the puffballs, such as this species of *Calvatia,* which is over 12 inches in diameter. (**C**) The outward spreading of a fungal mycelium can be visually detected by the formation of a ring of mushrooms, often called a "fairy ring." »» Knowing how a mycelium grows, where on the mycelium (fairy ring) in (C) do the mushrooms form?

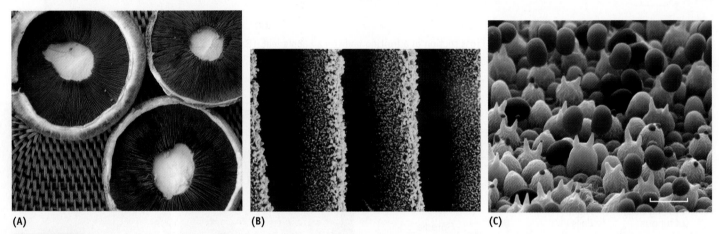

FIGURE 17.17 **Mushroom Gills and Basidiospores.** (**A**) A group of *Agaricus bisporis* mushroom (Portobello) caps, showing the gills on their underside. (**B**) A false-color scanning electron micrograph of the mushroom gills. The rough appearance represents forming basidia on which basidiospores will form. (**C**) Another scanning EM showing the basidiospores (dark brown) attached to basidia. (Bar = 20 μm.) »» Why is the specific epithet for this species of *Agaricus* called *bisporis*?

black masses of spores forming on leaves and other plant parts. Some Basidiomycota cause serious diseases in animals and humans.

The name basidiomycete refers to the reproductive structure on which sexual spores are produced. In many mushrooms, the underside of the cap is lined with "gills" on which club-shaped **basidia** (sing., basidium; *basidium* = "small pedestal") are formed (**FIGURE 17.17**). Within these basidia, the haploid, sexual spores, called **basidiospores**, are produced.

In the soil, the basidiospores germinate and grow as a mycelium. When mycelia of different mating types come in contact, they fuse into a heterokaryon containing genetically different haploid nuclei (**FIGURE 17.18**). Under appropriate environmental conditions, some of the hyphae become tightly compacted and force their way to the surface and grow into a fruiting body (basidiocarp) typically called a **mushroom**. Often a ring of mushrooms forms, which has been called a "fairy ring" because centuries ago people thought the mushrooms appeared where "fairies" had danced the night before (**FIGURE 17.16C**).

MICROINQUIRY 17 looks at the relationship of the fungi to other eukaryotic kingdoms.

CONCEPT AND REASONING CHECKS

17.4D Identify the properties of the Basidiomycota.

The Mitosporic Fungi. Certain fungi lack a known sexual cycle of reproduction; consequently, they are labeled with the term **mitosporic fungi** because the asexual spores are the product of mitosis.

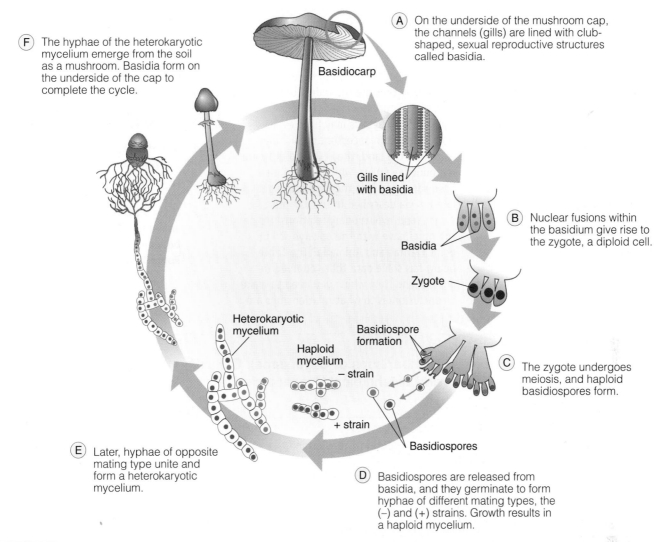

F The hyphae of the heterokaryotic mycelium emerge from the soil as a mushroom. Basidia form on the underside of the cap to complete the cycle.

Basidiocarp

A On the underside of the mushroom cap, the channels (gills) are lined with club-shaped, sexual reproductive structures called basidia.

Gills lined with basidia

B Nuclear fusions within the basidium give rise to the zygote, a diploid cell.

Basidia

Zygote

Heterokaryotic mycelium

Haploid mycelium

Basidiospore formation

– strain

C The zygote undergoes meiosis, and haploid basidiospores form.

+ strain

Basidiospores

E Later, hyphae of opposite mating type unite and form a heterokaryotic mycelium.

D Basidiospores are released from basidia, and they germinate to form hyphae of different mating types, the (–) and (+) strains. Growth results in a haploid mycelium.

FIGURE 17.18 **The Sexual Reproduction Cycle of a Typical Basidiomycete.** The basidiomycetes include the mushrooms from which basidiospores are produced. »» Identify the portion of the cycle that exists as a heterokaryon.

Many mitosporic fungi are reclassified when a sexual cycle is observed or comparative genomics identifies a close relationship to a known phylum. We already mentioned this situation for *Penicillium* and *Candida.*

Another case in point is the fungus *Histoplasma capsulatum,* which causes a human respiratory disease we will discuss below. When the organism was found to produce ascospores, it was reclassified with the ascomycetes and given the new name *Emmonsiella capsulata.* However, some traditions die slowly, and certain mycologists insisted on retaining the old name because it was familiar in clinical medicine. Thus, mycologists decided to use two names for the fungus: the new name, *Emmonsiella capsulata,* for the sexual stage; and the old name, *Histoplasma capsulatum,* for the asexual stage.

Among the mitosporic fungi, a few are pathogenic for humans. These fungi usually reproduce by asexual spores, budding, or fragmentation, where the hyphal segments are commonly blown about with dust or deposited on environmental surfaces. For example, fragments of the athlete's foot fungus are sometimes left on towels and shower room floors.

MICROINQUIRY 17
Evolution of the Fungi

The fossil record provides evidence that fungi and plants "hit the land" about the same time. There are fossilized fungi that are over 450 million years old, which is about the time that plants started to colonize the land. In fact, as described in this chapter, some fossilized plants, perhaps representing some of the first to colonize land, appear to have mycorrhizal associations.

Taxonomists believe that a single ancestral species gave rise to the five fungal phyla (Chytridiomycota, Glomeromycota, Zygomycota, Ascomycota, and Basidiomycota). Only the

Chytridiomycota (chytrids) have flagella, suggesting they are the oldest line and that fungal ancestors were flagellated and aquatic. The chytrids may have evolved from a protistan ancestor and, as fungi colonized the land, they evolved different reproductive styles, which taxonomists have separated into the four nonflagellated phyla described in this chapter.

So, what evolutionary relationships do the fungi have with the eukaryotic kingdoms Plantae and Animalia? The **Table** below has some data that you need to analyze and from which you need to make a conclusion as to what relationships are

best supported by the evidence. You do not need to understand the function or role for every characteristic listed. The answers can be found in **Appendix D**.

17.1. Based on the data presented, are fungi more closely related to animals or plants? Justify your answer.

17.2. Draw a taxonomic scheme showing the relationships between Plantae, Animalia, and Fungi kingdoms.

TABLE

Comparisons of Organismal, Cellular, and Biochemical Characteristics

Characteristic	Fungi	Plants	Animals
Protein sequence similarities	✓		✓
Elongation factor 3 protein (translation) similarities	✓		✓
Types of polyunsaturated fatty acids			
Alpha linolenic		✓	
Gamma linolenic	✓		✓
Cytochrome system similarities	✓		✓
Mitochondrial UGA codes for tryptophan	✓		✓
Plate-like mitochondrial cristae	✓		✓
Presence of chitin	✓		some
Type of glycoprotein bonding			
O-linked	✓		✓
N-linked	✓		✓
Absorptive nutrition	✓		some
Glycogen reserves	✓		✓
Pathway for lysine biosynthesis			
Aminoadipic acid pathway	✓		
Diaminopimelic pathway		✓	
Type of sterol intermediate			
Lanosterol	✓		✓
Cycloartenol		✓	
Mitochondrial ribosome 5S RNA present		✓	
Presence of lysosomes	✓		✓

TABLE

17.1 Comparisons of the Fungal Phyla Important to Human Disease

Phylum	Common Name	Cross Walls (Septa)	Sexual Structure	Sexual Spore	Asexual Spore	Representative Genera
Zygomycota	Zygomycetes	No	Zygospore	Zygospore	Sporangiospore	*Rhizopus*
Ascomycota	Ascomycetes, sac fungi	Yes	Ascus	Ascospore	Conidia	*Saccharomyces* *Aspergillus* *Penicillium*
Basidiomycota	Basidiomycetes, club fungi	Yes	Basidium	Basidiospore	Conidia Hyphal fragments	*Agaricus* *Amanita*

The three phyla of fungi important to human health are compared in **TABLE 17.1** .

CONCEPT AND REASONING CHECKS

17.4E Distinguish the characteristics of the mitosporic fungi.

Yeasts Represent a Term for Any Single-Celled Stage of a Fungus

KEY CONCEPT

5. Baker's yeast and other yeasts are important eukaryotic organisms in industry and research.

The word "yeast" refers to a large variety of unicellular fungi (as well as the single-cell stage of any fungus). Included in the group are nonspore–forming yeasts of the mitosporic fungi, as well as certain yeasts belonging to the Basidiomycota and Ascomycota. The yeasts we consider here are the species of *Saccharomyces* used extensively in industry and research. Pathogenic yeasts will be discussed with human diseases.

Saccharomyces (saccharo = "sugar") is a fungus with the ability to ferment sugars. The most commonly used species of *Saccharomyces* are *S. cerevisiae* and *S. ellipsoideus,* the former used for bread baking (Baker's yeast) and alcohol production, the latter for alcohol production. Yeast cells reproduce chiefly by budding, as described previously (see Figure 17.13A), but a sexual cycle also exists in which cells fuse and form an enlarged structure (an ascus) containing spores (ascospores). *S. cerevisiae* therefore is an ascomycete.

The cytoplasm of *Saccharomyces* is rich in B vitamins, a factor making yeast tablets valuable nutritional supplements (ironized yeast) for people with iron-poor blood. The baking industry relies heavily upon *S. cerevisiae* to supply the texture in breads.

During the dough's rising period, yeast cells break down glucose and other carbohydrates, producing carbon dioxide gas (see Chapter 6). The carbon dioxide expands the dough, causing it to rise. Protein-digesting enzymes in yeast partially digest the gluten protein of the flour to give bread its spongy texture.

Yeasts are plentiful where there are orchards or fruits (the haze on an apple is a layer of yeasts). In natural alcohol fermentations, wild yeasts of various *Saccharomyces* species are crushed with the fruit; in controlled fermentations, *S. ellipsoideus* is added to the prepared fruit juice. The fruit juice bubbles profusely as carbon dioxide evolves. When the oxygen is depleted, the yeast metabolism shifts to fermentation and the pyruvate from glycolysis changes to consumable ethyl alcohol (see Chapter 6). The huge share of the American economy taken up by the wine and spirits industries is testament to the significance of the fermentation yeasts. A fuller discussion of fermentation processes is presented in Chapter 27.

S. cerevisiae probably is the most understood eukaryotic organism at the molecular and cellular levels. Its complete genome was sequenced in 1997, the first eukaryotic model organism to be completely sequenced. *S. cerevisiae* contains about 6,000 genes. It might appear initially that *S. cerevisiae* would have little in common with human beings. However, both are eukaryotic organisms with a cell nucleus, chromosomes, and a similar mechanism for cell division (mitosis).

S. cerevisiae, therefore, has been used to better understand not only cell function in animals, but also as a model for human disease. In fact, about 20% of human disease genes have

Model organism: A relatively simple organism used to study general principles of biology.

counterparts in yeast. For example, the chemical and signaling process by which a cell prepares itself for mitosis was either first discovered using yeast cells or major contributions to the understanding came from research with yeast cells. Research with yeast has identified the presence of prions, which have helped in the understanding of human prion diseases (see Chapter 14).

Potential drugs useful in disease treatment also can be screened using yeast cells. A yeast mutant, for example, with the equivalent of a human disease-causing gene, can be treated with potential therapeutic drugs to identify a compound able to restore normal function to the yeast cell gene. Such drugs, or modifications of them, might also be useful in humans.

CONCEPT AND REASONING CHECKS

17.5   Summarize the importance of yeasts to commercial interests and research.

17.3 Fungal Intoxications

Although there are no major infectious diseases of the digestive system caused by fungi, some mushrooms, molds, and yeasts produce toxins, called **mycotoxins**. Although these toxins are ingested, they may have poisonous effects distant from the digestive system. We will describe several mycotoxin-associated intoxications here.

Some Fungi Can Be Poisonous or Even Deadly When Consumed

KEY CONCEPT

6.   Fungal toxins can cause a variety of conditions from vomiting to lethal liver or kidney failure.

Two closely related fungi, *Aspergillus flavus* and *A. parasiticus,* produce mycotoxins called **aflatoxins**. The molds are found primarily in warm, humid climates, where they contaminate agricultural products such as peanuts, grains, cereals, sweet potatoes, corn, rice, and animal feed. Aflatoxins are deposited in these foods and ingested by humans where they are thought to be carcinogenic, especially in the liver. Contaminated meat and dairy products are also sources of the toxins. Half of the cancers in sub-Saharan Africa are liver cancers and 40% of the analyzed foods contain aflatoxins, highlighting the threat.

Ergotism is caused by *Claviceps purpurea,* an ascomycete fungus producing a powerful toxin. *C. purpurea* grows as hyphae on kernels of rye, wheat, and barley. As hyphae penetrate the plant, the fungal cells gradually consume the substance of the grain, and the dense tissue hardens into a purple body called a **sclerotium**. A group of peptide derivatives called alkaloids are produced by the sclerotium and deposited in the grain as a substance called **ergot**. Products such as bread made from rye grain may cause ergot rye disease, or ergotism.

Symptoms may include numbness, hot and cold sensations, convulsions with epileptic-type seizures, and paralysis of the nerve endings. Lysergic acid diethylamide (LSD) is a derivative of an alkaloid in ergot. Commercial derivatives of these alkaloids are used to cause contractions of the smooth muscles, such as to induce labor or relieve migraine headaches. MicroFocus 17.5 relates another possible effect of ergotism.

CONCEPT AND REASONING CHECKS

17.6   Summarize the unique features of aflatoxins and ergot.

Some Mushrooms Produce Mycotoxins

KEY CONCEPT

7.   Some mushroom species produce deadly toxins and/or hallucinogenic effects.

Amateur mushroom hunting is a popular pastime in many parts of the world. However, every year in the United States such hunting results in some 9,000 cases of mushroom poisoning being reported to the American Association of Poison Control Centers; children under 6 years of age account for the majority of cases. A poisonous mushroom refers to any mushroom that produces various toxins that can cause adverse or harmful reactions when eaten. By this definition, of the more than 5,000 species of identified mushrooms in the United States, about 100 are responsible for mushroom poisoning but less than a dozen species are considered deadly.

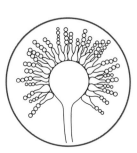

Aspergillus flavus

MICROFOCUS 17.5: History

"The Work of the Devil"

As an undergraduate, Linda Caporael was missing a critical history course for graduation. Little did she know that through this class she was about to provide a possible answer for one of the biggest mysteries of early American history: the cause of the Salem Witch Trials. These trials in 1692 led to the execution of 20 people who had been accused of being witches in Salem, Massachusetts (see figure).

Linda Caporael, now a behavioral psychologist at New York's Rensselaer Polytechnic Institute, in preparation of a paper for her history course had read a book where the author could not explain the hallucinations among the people in Salem during the witchcraft trials. Caporael made a connection

A witch trial in Massachusetts, 1692.

between the "Salem witches" and a French story of ergot poisoning in 1951. In Pont-Saint-Esprit, a small village in the south of France, more than 50 villagers went insane for a month after eating ergotized rye flour. Some people had fits of insomnia, others had hallucinogenic visions, and still others were reported to have fits of hysterical laughing or crying. In the end, three people died.

Caporael noticed a link between these bizarre symptoms, those of Salem witches, and the hallucinogenic effects of drugs like LSD, which is a derivative of ergot. Could ergot possibly have been the perpetrator?

During the Dark Ages, Europe's poor lived almost entirely on rye bread. Between 1250 and 1750, ergotism, then called "St. Anthony's fire," led to miscarriages, chronic illnesses in people who survived, and mental illness. Importantly, hallucinations were considered "work of the devil."

Toxicologists know that eating ergotized food can cause violent muscle spasms, delusions, hallucinations, crawling sensations on the skin, and a host of other symptoms—all of which, Linda Caporael found in the records of the Salem witchcraft trials. Ergot thrives in warm, damp, rainy springs and summers, which were the exact conditions Caporael says existed in Salem in 1691. Add to this that parts of Salem village consisted of swampy meadows that would be the perfect environment for fungal growth and that rye was the staple grain of Salem—and it is not a stretch to suggest that the rye crop consumed in the winter of 1691–1692 could have been contaminated by large quantities of ergot.

Caporael concedes that ergot poisoning can't explain all of the events at Salem. Some of the behaviors exhibited by the villagers probably represent instances of mass hysteria. Still, as people reexamine events of history, it seems that just maybe ergot poisoning did play some role—and, hey, not bad for an undergraduate history paper!

Mushroom poisoning, or **mycetism**, can occur from mushrooms that produce mycotoxins that affect the human body. Some are neurotoxins that: (1) mimic the action of acetylcholine, affecting the peripheral parasympathetic nervous system; (2) generate psychotropic effects (hallucinations); or (3) produce a condition resembling alcohol intoxication. Deaths are rare, but in severe cases the intoxication may cause cardiac or respiratory failure. Further, in children, accidental consumption of large quantities of some mushrooms may cause convulsions, coma, and other neurologic problems for up to 12 hours.

Amanita phalloides is referred to as the "death cap" because it accounts for about 90% of deaths from eating poisonous mushrooms (FIGURE 17.19A). About 10 to 14 hours after ingesting just one to three mushrooms, enough of the protein toxins, called **amatoxins** and **phallotoxins**, are present in the body to produce gastrointestinal symptoms of severe abdominal cramping, vomiting, and watery diarrhea. The symptoms last for about 24 hours and can cause dehydration and low blood pressure. A period of remission of symptoms lasts for 1 to 2 days during which the patient feels better, but blood

(A)

(B)

FIGURE 17.19 Poisonous Mushrooms. **(A)** *Amanita phalloides*. Referred to as the "death cap" mushroom, these fungi can produce a deadly mycotoxin. **(B)** *Boletus santanas*. Referred to as the "Satan's mushroom," if eaten raw, this mushroom can cause gastrointestinal symptoms of nausea and vomiting. »» To what phylum do these mushrooms belong?

tests begin to show evidence of liver and kidney deterioration. Liver and kidney failure then develops and either results in death within about a week or recovery within 2 to 3 weeks.

Numerous species of mushrooms, including *Agaricus, Amanita, Boletus, Lactarius, Lepiota, Lycoperdon, Polyporus,* and *Russula,* contain mycotoxins that can cause a noninflammatory gastroenteritis (**FIGURE 17.19B**). Within 30 to 90 minutes of ingestion, sudden severe vomiting and mild to severe diarrhea with abdominal cramps occur. The symptoms typically last about six hours. The main diagnostic identification of these intoxications is a rapid onset of symptoms. Fatalities are rare but if they do occur they are associated with dehydration and electrolyte imbalances in debilitated, very young, or very old patients. In children, dehydration may be severe enough to require hospital treatment.

If people suspect they have eaten a poisonous mushroom, it is not wise to wait for symptoms to appear, as symptoms may not develop until several days later. It should be treated as a medical emergency.

As far as prevention goes, all mushrooms not bought at the grocery store should be considered potentially dangerous—and no antidote exists. Early replacement of lost body fluids (water and electrolytes) is important in improving recovery rates. Still, according to some medical reports death rates are 20% to 30%, with a higher mortality rate of 50% in children less than 10 years old. According to the Minnesota Mycological Society, *"There are old mushroom hunters and bold mushroom hunters. But there are no old, bold mushroom hunters."*

CONCEPT AND REASONING CHECKS

17.7 In the case of *A. phalloides* mushroom poisoning, what is probably happening during the two days when symptoms are in remission?

17.4 Fungal Diseases of the Skin

In humans, fungal diseases, called **mycoses**, often affect many body regions. Some of these affect the skin or body surfaces. For example, several diseases, including ringworm and athlete's foot, involve the skin areas. One disease, candidiasis, may take place in the oral cavity, intestinal tract, skin, vaginal tract, or other body locations depending on the conditions stimulating its development.

Dermatophytosis Is an Infection of the Skin, Hair, and Nails

KEY CONCEPT

8. Dermatophytes produce cutaneous lesions.

Dermatophytosis (*dermato* = "skin"; *phyto* = "plant," referring to the days when fungi were grouped with plants) is a general name for a fungal disease of the hair, skin, or nails. The diseases are commonly known as **tinea infections** (*tinea* = "worm") because in ancient times, worms were thought to be the cause. Tinea infections compose several forms of **ringworm**, including: athlete's foot (tinea pedis); head ringworm (tinea capitis); body ringworm (tinea corporis); groin ringworm or "jock itch" (tinea cruris); and nail ringworm (tinea unguium).

The causes of dermatophytosis are a group of fungi called **dermatophytes**. *Epidermophyton* currently is considered a mitosporic fungus, while species of *Trichophyton* (sexual stage *Arthroderma*) and *Microsporum* (sexual stage *Nannizzia*) are ascomycetes (FIGURE 17.20A).

Trichophyton

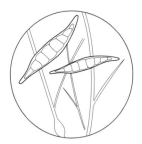

Microsporum

If protected from dryness, the dermatophytes live for weeks on wooden floors of shower rooms or on mats. People transmit the fungi by contact and on towels, combs, hats, and numerous other types of **fomites** (inanimate objects). Because tinea diseases affect cats and dogs, they can transmit the fungi to humans (FIGURE 17.20B).

Dermatophytosis is commonly accompanied by blister-like lesions appearing along the nail plate, in the webs of the toes (*Epidermophyton*), on the scalp or skin (*Microsporum*), or in the nail plate (*Trichophyton*). Often a thin, fluid discharge exudes when the blisters are scratched or irritated. As the blisters dry, they leave a scaly ring. There also can be loss of hair, change of hair color, and local inflammatory reactions.

Treatment of dermatophytosis often is directed at changing the conditions of the skin environment. Commercial powders dry the diseased area, while ointments change the pH to make the area inhospitable for the fungus. Certain acids such as undecylenic acid (Desenex) and mixtures of acetic acid and benzoic acid (Whitfield's ointment) are active against the fungi. Also, tolnaftate (Tinactin) and miconazole (Micatin) are useful as topical agents for infections not involving the nails and hair. Griseofulvin, administered orally, is a highly effective chemotherapeutic agent for severe dermatophytosis.

CONCEPT AND REASONING CHECKS

17.8 Describe the characteristics of dermatophytosis.

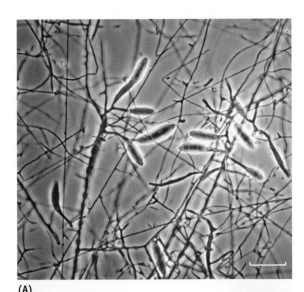

(A)

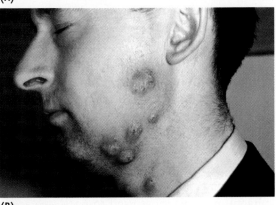

(B)

FIGURE 17.20 Ringworm. Ringworm of the skin or body (tinea corporis). **(A)** Light microscope photograph of *Microsporum,* one of the fungi causing ringworm on the scalp and body. (Bar = 15 μm.) **(B)** A case of body ringworm on the face and neck that was "caught" from a cat. »» Why is the shape of each ringworm skin lesion in **(B)** circular?

Candidiasis Often Is a Mild, Superficial Infection

KEY CONCEPT

9. *Candida albicans* most commonly causes vaginitis or thrush.

Candida albicans often is present in the skin, mouth, vagina, and intestinal tract of healthy humans, where it lives without causing disease. The organism is a small mitosporic yeast that forms filaments called pseudohyphae when cultivated in laboratory media. When immune system defenses are compromised, or when changes occur in the normal microbial population in the body, *C. albicans* flourishes and causes numerous forms of **candidiasis**.

One form of candidiasis occurs in the vagina and is often referred to as **vulvovaginitis**, or a "yeast infection." There are some

Pseudohyphae:
Cells formed by budding that are more elongated than typical oval cells.

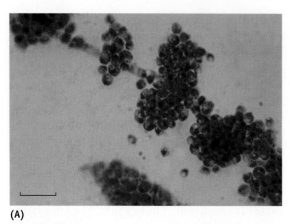

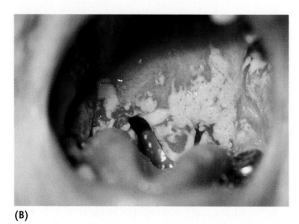

(A)   **(B)**

FIGURE 17.21 *Candida albicans.* **(A)** A light micrograph of stained *C. albicans* cells from a vaginal swab. (Bar = 60 μm.) **(B)** A severe case of oral candidiasis, showing a thick, creamy coating over the tongue. »» From photo (A), how do you know this specimen of *C. albicans* was from a human infection?

Candida albicans

20 million cases reported every year in the United States. Symptoms include itching sensations (pruritis), burning internal pain, and a white "cheesy" discharge. Reddening (erythema) and swelling of the vaginal tissues also occur. Diagnosis is performed by observing *C. albicans* in a sample of vaginal discharge or vaginal smear (**FIGURE 17.21A**), and by cultivating the organisms on laboratory media. Treatment is usually successful with nystatin (Mycostatin) applied as a topical ointment or suppository. Miconazole, clotrimazole, and ketoconazole are useful alternatives.

Vulvovaginitis is considered a sexually transmitted disease. In addition, studies have shown that excessive antibiotic use may encourage loss of the rod-shaped lactobacilli normally present in the vaginal environment. Without lactobacilli as competitors, *C. albicans* flourishes. Other predisposing factors are the contraceptive intrauterine device (IUD), corticosteroid treatment, pregnancy, diabetes, and tight-fitting garments, which increase the local temperature and humidity.

Oral candidiasis is known as **thrush**. This disease is accompanied by small, white flecks that appear on the mucous membranes of the oral cavity and then grow together to form soft, crumbly, milk-like curds (**FIGURE 17.21B**). When scraped off, a red, inflamed base is revealed. Oral suspensions of gentian violet and nystatin ("swish and spit") are effective for therapy. The disease is common in newborns, who acquire it during passage through the vagina (birth canal) of infected mothers. Children also may contract thrush from nursery utensils, toys, or the handles of shopping

carts. Candidiasis may be related to a suppressed immune system. Indeed, thrush may be an early sign of AIDS in an adult patient.

Candidiasis in the intestinal tract is closely tied to the use of antibiotics. Certain drugs destroy the bacterial cells normally found there and allow *C. albicans* to flourish. In the 1950s, yogurt became popular as a way of replacing the bacterial cells. Today when intestinal surgery is anticipated, the physician often uses antifungal agents to curb *Candida* overgrowth. Moreover, people whose hands are in constant contact with water may develop a hardening, browning, and distortion of the fingernails called **onychia**, also caused by *C. albicans*.

CONCEPT AND REASONING CHECKS

17.9 Describe the different forms of candidiasis based on body location.

Sporotrichosis Is an Occupational Hazard

KEY CONCEPT

10. *Sporothrix schenkii* forms lesions under the skin.

People who work with wood, wood products, or the soil may contract **sporotrichosis**. Handling sphagnum (peat) moss used to pack tree seedlings or skin punctures with rose thorns (rose thorn disease) can lead to the disease as the result of infection by conidia from *Sporothrix schenkii* (**FIGURE 17.22A**). Pus-filled purplish lesions form at the site of entry, and "knots" may be felt under the skin (**FIGURE 17.22B**). Dissemination, though

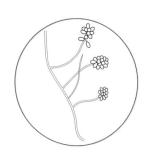

Sporothrix schenkii

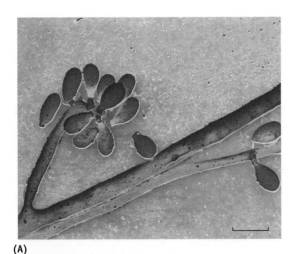

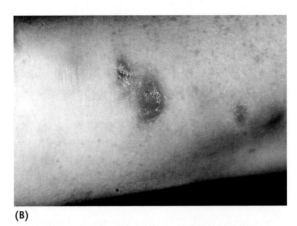

(A)

(B)

FIGURE 17.22 *Sporothrix schenkii.* (**A**) A false-color scanning electron micrograph of hyphae (orange) and conidia (purple) formed on conidiophores. (Bar = 8 μm.) (**B**) A patient showing the lesions of sporotrichosis on an infected arm. Characteristic "knots" can be felt under the skin. »» Explain how a rose thorn harboring *S. schenkii* conidia can cause a skin disease.

TABLE

17.2 A Summary of the Fungal Skin Diseases

Inflammation or Disease	Causative Agent	Signs and Symptoms	Transmission	Treatment	Prevention and Control
Dermatophytosis (Tineas)	*Epidermophyton Microsporum Trichophyton*	Blister-like lesions	Fragments of skin on floors or surfaces	Antifungal medications	Keeping skin dry Not sharing personal items
Candidiasis	*Candida albicans*	Itching, burning pain, "cheesy" discharge	Sexually	Nystatin	Avoiding baths and hot tubs Avoiding douches
Thrush	*Candida albicans*	White flecks on mucous membranes	Passage through the birth canal	No treatment in children Unsweetened yogurt	Practicing good oral hygiene Limiting sugar intake
Sporotrichosis	*Sporothrix schenkii*	Pus-filled, purplish lesions	From plant material harboring the fungus	Itraconazole Amphotericin B	Wearing long sleeves and gloves while working with suspect materials or vegetation

rare, may occur to the bloodstream, where blockages may cause swelling of the tissues (edema). In one outbreak, 84 cases of cutaneous sporotrichosis occurred in people who handled conifer seedlings packed with sphagnum moss from Pennsylvania. Cutaneous infections are controlled with itraconazole, but systemic infections require amphotericin B therapy.

TABLE 17.2 summarizes the fungal diseases of the skin.

CONCEPT AND REASONING CHECKS

17.10 Describe the characteristics of sporotrichosis.

17.5   Fungal Diseases of the Lower Respiratory Tract

Additional mycoses affect other body parts in humans, with a primary infection in the lungs that can spread to other body areas. In many of these fungal diseases, a weakened immune system contributes substantially to the occurrence of the infection.

Cryptococcosis Usually Occurs in Immunocompromised Individuals

KEY CONCEPT

11. *Cryptococcus neoformans* causes a rare, but dangerous respiratory disease.

Cryptococcosis is among the most dangerous fungal diseases in humans. It affects the lungs and the meninges and is estimated to account for over 25% of all deaths from fungal disease.

Cryptococcosis is caused by an oval-shaped yeast known as *Cryptococcus neoformans* (sexual stage *Filobasidiella neoformans*) and is a member of the Basidiomycota. The organism is found in the soil of urban environments and grows actively in the droppings of pigeons, but not within the pigeon tissues. Cryptococci may become airborne with gusts of wind, and the organisms subsequently enter the respiratory passageways of humans.

The *C. neoformans* cells, which have a diameter of about 5 to 6 μm, are embedded in a thick, gelatinous capsule that provides resistance to phagocytosis (**FIGURE 17.23A**). The cells penetrate to the air sacs of the lungs, but symptoms of infection are generally rare. However, if the cryptococci pass into the bloodstream and localize in the meninges and brain, the patient experiences piercing headaches, stiffness in the neck, and paralysis. Diagnosis is aided by the observation of encapsulated yeasts in respiratory secretions or cerebrospinal fluid (CSF) obtained by a spinal tap (**FIGURE 17.23B**).

Untreated cryptococcal meningitis may be fatal. However, intravenous treatment with the antifungal drug amphotericin B is usually successful, even in severe cases. Because this drug has toxic side effects, such as kidney damage and anemia, the patient should be monitored continually.

Resistance to cryptococcal meningitis appears to depend upon the proper functioning of the branch of the immune system governed by T lymphocytes (T cells). When these cells are absent in sufficient quantities, the immune system becomes

severely compromised, and cryptococci can invade the tissues as opportunists. In AIDS patients cryptococcosis can be life-threatening.

CONCEPT AND REASONING CHECKS

17.11 Explain why cryptococcosis is such a dangerous fungal disease.

Histoplasmosis Can Produce a Systemic Disease

KEY CONCEPT

12. *Histoplasma capsulatum* most often causes a mild flu-like illness.

In 1988, a group of 17 American university students crawled into a cave in Costa Rica to observe the numerous bats whose droppings covered the

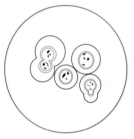

Meninges:
The membranes surrounding and protecting the brain and spinal cord.

Cryptococcus neoformans

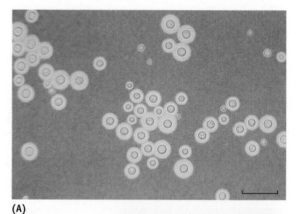

(A)

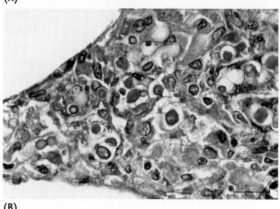

(B)

FIGURE 17.23 *Cryptococcus neoformans.* **(A)** A light microscope photomicrograph of *C. neoformans* cells. Internalization of the fungal cells is retarded by the capsules (white halo) they possess. (Bar = 20 μm.) **(B)** A stained photomicrograph of *C. neoformans* cells (red) from lung tissue of an AIDS patient. The capsule surrounding the cells provides resistance to phagocytosis and enhances the pathogenic tendency of the fungus. (Bar = 20 μm.) »» Why would *C. neoformans* be a serious health threat to AIDS patients?

floor. Within three weeks, 15 students developed fever, headache, cough, and severe chest pains. Twelve patients tested positive for *Histoplasma capsulatum*, and all were treated for histoplasmosis.

Histoplasmosis is a lung disease most prevalent in the Ohio and the Mississippi River valleys where it is often is called "summer flu." The causative agent is *Histoplasma capsulatum*. The lungs are the primary portal of entry; thus infection usually occurs from the inhalation of spores present in dry, dusty soil or found in the air of chicken coops and bat caves (FIGURE 17.24A). Prolonged exposure to the air therefore may be hazardous, as the Costa Rica outbreak illustrates. Being a dimorphic fungus, it grows as a yeast form at 37°C.

Most people remain asymptomatic or experience mild influenza-like illness and recover without treatment. However, in immunocompromised people a disseminated form with tuberculosis-like lesions of the lungs and other internal organs may occur, making AIDS patients especially vulnerable. Amphotericin B or ketoconazole may be used in treatment.

CONCEPT AND REASONING CHECKS

17.12 How does histoplasmosis differ in "healthy" individuals versus in an immunocompromised individual?

Blastomycosis Usually Is Acquired Via the Respiratory Route

KEY CONCEPT

13. *Blastomyces dermatitidis* can cause lung and skin infections.

Blastomycosis occurs principally in Canada, the Great Lakes region, and areas of the United States from the Mississippi River to the Carolinas. The pathogen is *Blastomyces dermatitidis* (sexual stage *Ajellomyces dermatitidis*). The fungus is dimorphic and produces conidia that are inhaled. Within the lungs, the conidia germinate as the yeast form.

Acute blastomycosis is associated with dusty soil and bird droppings, particularly in moist soils near barns and sheds. Inhalation leads to lung lesions with persistent cough and chest pains. Entry to the body also may occur through cuts and abrasions, and raised, wart-like lesions often are observed on the face, legs, or hands (FIGURE 17.24B). Healing is generally spontaneous (in 2 to 3 weeks).

Although blastomycosis is rare, it can affect immunocompromised patients, such as those with

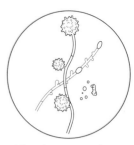

Histoplasma capsulatum

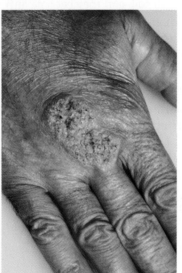

(A)

(B)

FIGURE 17.24 *Histoplasma* Mycelium and *Blastomyces* Skin Infection. (**A**) *H. capsulatum* grows as a mycelium in soil enriched by animal excrement. The spores are produced from the hyphal tips. (Bar = 20 μm.) (**B**) A skin infection of blastomycosis. The fungus usually affects the lungs after inhalation of fungal spores, but may become disseminated to the skin. »» Which dimorphic form of these fungi is associated with human infections?

Blastomyces dermatitidis

AIDS. Chronic pneumonia is the most common clinical manifestation. The disseminated form of blastomycosis may involve many internal organs (bones, liver, spleen, or central nervous system) and may prove fatal. Amphotericin B is used in therapy.

CONCEPT AND REASONING CHECKS

17.13 Why is blastomycosis a dangerous disease in immunocompromised individuals?

Coccidioidomycosis Can Become a Potentially Lethal Infection

KEY CONCEPT

14. *Coccidioides immitis* infections can disseminate and be life-threatening.

Travelers to the San Joaquin Valley of California and dry regions of the southwestern United States may contract a fungal disease known as **coccidioidomycosis**, known more commonly as valley or desert fever. It is caused by *Coccidioides immitis* (in California) or *C. posadasii* (in Arizona). The ascomycete fungus produces highly infectious arthrospores by a unique process of endospore and spherule formation (**FIGURE 17.25**). Coccidioidomycosis is usually transmitted by dust particles laden with fungal spores. Cattle, sheep, and other animals deposit the spores in soil, and they become airborne with gusts of wind. TEXTBOOK CASE 17 describes an outbreak in northeastern Utah.

When inhaled into the human lungs, *C. immitis* induces within 7 to 28 days an influenza-like disease, with a dry, hacking cough, chest pains, and high fever. During most of the 1980s, about 450 annual cases of coccidioidomycosis were reported to the CDC. In 1991, that number jumped to over 1,200 cases, and in 2009, the number of reports exceeded 12,000, the majority occurring in California and Arizona.

Although most cases are self-limiting, a small number of patients (1 out of 1,000) develop a chronic pulmonary infection or a disseminated infection involving internal organs and structures, including the meninges of the spinal cord. Recovery brings lifelong immunity. Amphotericin B is prescribed for severe cases and a vaccine is in the early stages of development.

CONCEPT AND REASONING CHECKS

17.14   What is the unique feature of a coccidioidomycosis infection?

Pneumocystis Pneumonia Can Cause a Lethal Pneumonia

KEY CONCEPT

15. *Pneumocystis jiroveci* is a fungal pathogen in immunocompromised hosts.

***Pneumocystis* pneumonia** (**PCP**) currently is the most common cause of nonbacterial pneumonia in Americans with suppressed immune systems. The causative organism, *Pneumocystis jiroveci* (previously called *Pneumocystis carinii*) produces an atypical pneumonia that remained in relative obscurity until the 1980s, when it was recognized as the cause of death in over 50% of patients dying from the effects of AIDS.

P. jiroveci has a complex life cycle taking place entirely in the alveoli of the lung. A feeding stage, called the **trophozoite**, swells to become a precyst stage, in which up to eight sporozoites develop in forming a mature cyst. The cyst then ruptures and liberates the sporozoites, which enlarge and undergo further reproduction and maturation to trophozoites.

P. jiroveci is transmitted person-to-person by droplets from the respiratory tract, although transmission from the environment also can occur. A wide cross section of individuals harbors the organism without symptoms, mainly because of the control imposed by the immune system.

When the immune system is suppressed, as in AIDS patients, *Pneumocystis* trophozoites and cysts fill the alveoli and occupy all the air spaces. A nonproductive cough develops, with fever and difficult breathing. Progressive deterioration leads

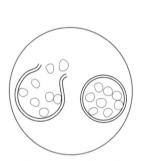

Pneumocystis jiroveci

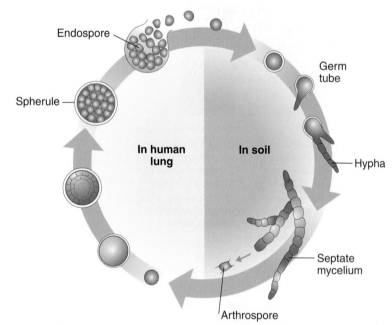

Endospore

Germ tube

Spherule

In human lung

In soil

Hypha

Septate mycelium

Arthrospore

FIGURE 17.25 **The Life Cycle of *Coccidioides immitis*.** Outside the body, the organism exists as a septate mycelium. It fragments into airborne arthrospores, which are inhaled. In the respiratory tract, the arthrospores swell to yield a large body, the spherule, that segments and breaks down to release endospores. When released to the environment, the endospores form germ tubes and then the mycelium. »» Explain why *C. immitis* is considered a dimorphic fungus.

Textbook CASE 17

Coccidioidomycosis Outbreak at Dinosaur National Monument, Utah

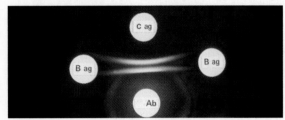

1 On June 18, 2001, two archeologists with the National Park Service (NPS) directed a team of six student volunteers and two leaders at an archeological site in Dinosaur National Monument (DNM), a 320 square mile area in northeastern Utah and northwestern Colorado.

2 The volunteers and leaders laid stone steps, built a retaining wall, and sifted dirt for artifacts. Peak dust exposure occurred on June 19, the day most sifting occurred. Workers did not wear protective facemasks.

3 Between June 29 and July 3, all eight team members and the two NPS archeologists sought medical care at a local hospital emergency department for a flu-like illness with fever, cough, headache, rash, and muscle aches. Chest X rays showed that all 10 individuals had fluid filling the pulmonary airspaces; eight individuals were hospitalized with pneumonia.

4 NPS closed the work site to all visitors and staff, and the TriCounty Health Department alerted the public. On July 2, the TriCounty Health Department, the Utah Department of Health, and the Centers for Disease Control and Prevention (CDC) initiated an investigation to identify the risk factors, cause, and extent of the outbreak.

5 During July 2–4, a total of 18 persons (the 8 team members and 10 archeologists) with potential exposure to dust at the work site in June were interviewed to determine symptoms and previous activities. Hospital records were reviewed to ascertain clinical information. A case definition was defined for a person working at DNM.

6 Results of blood cultures from the hospitalized persons were negative for bacterial pathogens. Initial serologic tests were negative for antibodies to *Francisella tularensis*, *Yersinia pestis*, *Mycoplasma* species, *Histoplasma capsulatum*, and *Blastomyces dermatitidis*. Further serological tests to detect IgM antibodies (typically found in an early immunological response) to *Coccidioides immitis* showed positive precipitin lines indicative of IgM antibodies in 9 of the 10 acute serum specimens from patients, confirming the diagnosis of acute coccidioidomycosis (see figure). All hospitalized patients were treated with fluconazole.

7 A serosurvey of 40 other park employees was conducted to identify other infected persons and to guide prevention and control measures. None of the 40 had detectable antibodies to *C. immitis*.

8 The DNM site, which reopened on September 28, is located approximately 200 miles north of areas where *C. immitis* is normally found.

Questions:

(Answers can be found in Appendix D.)

A. What would be present in the dust from the sifted dirt that would lead to the described illness?

B. A case definition is the method public health professionals use to define who is included as a "case" in an outbreak investigation. Provide a case definition for the illness described in this textbook case.

C. Why was a serosurvey seen as an important part of the outbreak analysis? What would health officials be looking for in a serosurvey; that is, in serum collected from the 40 other individuals?

D. What conclusions can you draw from the fact that this coccidioidomycosis outbreak was much farther north than where the disease is usually found?

E. How could these cases of coccidioidomycosis have been prevented?

For additional information see http://www.cdc.gov/mmwr/preview/mmwrhtml/mm5045a1.htm

Consolidation: Formation of a firm dense mass in the alveoli.

to consolidation of the lungs and, respiratory failure. The current treatment for severe PCP is trimethoprim-sulfamethoxazole (co-trimoxazole) and corticosteroid therapy.

CONCEPT AND REASONING CHECKS

17.15 What are the unique features of *P. jiroveci*?

Other Fungi Also Cause Mycoses

KEY CONCEPT

16. *Aspergillus* species can cause human illnesses.

A few other ascomycete fungal diseases deserve brief mention because they are important in certain parts of the United States or they affect individuals in certain professions. Generally the diseases are mild, although complications may lead to serious tissue damage in severely immunocompromised patients.

Invasive **aspergillosis** is a unique disease because the fungus enters the body as conidia and then grows as a mycelium. Disease usually occurs in an immunosuppressed host or where an overwhelming number of conidia has entered the tissue. The most common cause is *Aspergillus fumigatus*, which is found in decaying leaves, rotting vegetables, and stored grain. An oppor-

tunistic infection of the lung may yield a round ball of mycelium called a **pulmonary aspergilloma**, requiring surgery for removal. In addition, conidia in the earwax can lead to a painful ear disease known as **otomycosis** while disseminated *Aspergillus* causes blockage of blood vessels, inflammation of the inner lining of the heart, or clots in the heart vessels.

The most deadly form of aspergillosis— **invasive aspergillosis**—occurs when the fungal infection spreads beyond the lungs to the other organs, such as the skin, heart, kidneys, or brain. Signs and symptoms depend on which organs are affected, but, in general, include headache, fever with chills, bloody cough, shortness of breath, and chest or joint pain.

Treatment usually involves antifungal drugs such as voriconazole. It can be difficult to avoid *Aspergillus* spores in the environment. Staying away from obvious sources of mold, such as compost piles and damp places, can help prevent infection in susceptible individuals.

TABLE 17.3 summarizes the fungal diseases of the lower respiratory tract.

CONCEPT AND REASONING CHECKS

17.16 Summarize the unique features of aspergillosis.

TABLE

17.3 A Summary of the Major Fungal LRT Diseases

Disease	Causative Agent	Signs and Symptoms	Transmission	Treatment	Prevention and Control
Cryptococcosis	*Cryptococcus neoformans*	Asymptomatic Opportunistic infection leads to severe headache, stiff neck, paralysis	Airborne cells	Amphotericin B	Maintaining strong immune system
Histoplasmosis	*Histoplasma capsulatum*	Mild influenza-like illness Can disseminate to other organs	Airborne spores	Amphotericin B or ketoconazole for systemic disease	Wearing face mask in contaminated areas
Blastomycosis	*Blastomyces dermatitidis*	Persistent cough Chest pains Chronic pneumonia	Airborne spores	Amphotericin B	Wearing face mask in contaminated areas
Coccidioidomycosis	*Coccidioides immitis* *Coccidioides posadasii*	Dry, hacking cough Chest pains High fever	Airborne arthrospores	Amphotericin B	Limiting exposure where infection is highest
Pneumocystis pneumonia	*Pneumocystis jiroveci*	Nonproductive cough Fever Breathing difficulty	Airborne droplets	Trimethoprim-sulfamethoxazole	Maintaining strong immune system
Aspergillosis	*Aspergillus fumigatus*	Bloody cough Chest pain Wheezing Shortness of breath	Airborne spores	Voriconazole	Staying away from sources of mold

SUMMARY OF KEY CONCEPTS

17.1 Characteristics of Fungi

1. Fungi are eukaryotic microorganisms with **heterotrophic** metabolism. Most fungi consist of masses of **hyphae** that form a **mycelium**. **Cross-walls** separate the cells of hyphae in many fungal species. Most fungi are **coenocytic**.

2. Fungi secrete enzymes into the surrounding environment and absorb the breakdown products. Tremendous absorption can occur when there is a large mycelium surface. Most fungi are aerobic, grow best around 25°C, and prefer slightly acidic conditions.

3. Reproductive structures generally occur at the tips of hyphae. Masses of asexually- or sexually-produced **spores** are formed within or at the tips of **fruiting bodies**.

17.2 The Classification of Fungi

4A. The phylum **Chytridiomycota** is characterized by motile cells, while the **Glomeromycota** represent fungi living symbiotically with land plants.

4B. In the phylum **Zygomycota**, the sexual phase is characterized by the formation of a **zygospore**, which releases haploid spores that germinate into a new mycelium.

4C. The phylum **Ascomycota** includes the unicellular yeasts and filamentous molds. **Ascospores** are produced that germinate to form a new haploid mycelium, while asexual reproduction is through the dissemination of **conidia**. **Lichens** are a mutualistic association between an ascomycete and either a cyanobacterium or a green alga.

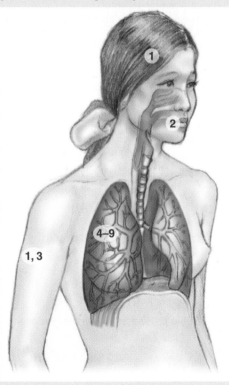

4D. The phylum **Basidiomycota** includes the **mushrooms**. Within these fruiting bodies, **basidiospores** are produced. On germination, they produce a new haploid mycelium. "Rusts" and "smuts" that cause many plant diseases are additional members of the phylum.

4E. The **mitosporic fungi** lack a sexual phase. Many human fungal diseases involve fungi in this informal group.

5. *Saccharomyces* is a notable unicellular ascomycete **yeast** involved in baking and brewing, and scientific research.

17.3 Fungal Intoxications

6. *Aspergillus flavus* produces a **mycotoxin**, called **aflatoxin**, which is carcinogenic. *Claviceps purpurea* produces alkaloid derivatives in grain called **ergot**. Eating ergotized breads can cause **ergotism**, which produces a variety of symptoms including convulsions with epileptic-type seizures and paralysis of nerve endings.

7. **Mushroom poisoning** (**mycetism**) can affect the gastrointestinal tract. *Amanita phalloides* produces **mycotoxins** that cause severe abdominal cramping, vomiting, and watery diarrhea when the raw mushrooms are consumed. Numerous other species cause a noninflammatory gastroenteritis that produces a sudden severe vomiting and mild to severe diarrhea with abdominal cramps.

17.4 Fungal Diseases of the Skin

- **Dermatophytosis**
  1. *Microsporum*
     Trichophyton (nail plate; not shown)
     Epidermophyton (webs of toes; not shown)
- **Candidiasis** (**vulvovaginitis**)
     Candida albicans (not shown)
- **Thrush** (children)
  2. *Candida albicans*
- **Sporotrichosis**
  3. *Sporothrix schenkii*

17.5 Fungal Diseases of the LRT

- **Cryptococcosis**
  4. *Cryptococcus neoformans*
- **Histoplasmosis**
  5. *Histoplasma capsulatum*
- **Blastomycosis**
  6. *Blastomyces dermatitidis*
- **Coccidioidomycosis**
  7. *Coccidioides immitis, Coccidioides posadasii*
- ***Pneumocystis* pneumonia**
  8. *Pneumocystis jirovecii*
- Aspergillosis
  9. *Aspergillus fumigatus*

LEARNING OBJECTIVES

After understanding the textbook reading, you should be capable of writing a paragraph that includes the appropriate terms and pertinent information to answer the objective.

1. Differentiate between **molds**, **yeasts**, and **dimorphic** fungi and summarize the structure and function of fungal **hyphae**.
2. Identify and describe the physical factors governing fungal growth and discuss examples of fungal symbioses, including the **mycorrhizae** and the fungal **endophytes**.
3. Distinguish between the forms of asexual and sexual spores produced by fungi and describe the generalized sexual life cycle of a mold.
4. Summarize the characteristics of the **Chytridiomycota** and their relevance to fungal taxonomy. Identify the key characteristics of the **Zygomycota**, **Ascomycota**, and **Basidiomycota**.
5. Assess the usefulness of yeasts, such as *Saccharomyces*, to industry and scientific research.
6. Summarize the effects of (a) *Aspergillus flavus* intoxication and (b) *Claviceps purpurea* ingestion.
7. Justify why it can be dangerous to eat non-store-bought mushrooms.
8. Summarize the symptoms and treatment of **dermatophytosis**.
9. Describe the major types of *Candida* infections.
10. Identify common mechanisms for transmission of *Sporothrix schenkii*.
11. Summarize the symptoms and complications of **cryptococcosis**.
12. Discuss the consequences of a **histoplasmosis** infection in an immunocompromised individual.
13. Explain the dimorphic nature of *Blastomyces dermatitidis*.
14. Review the symptoms of and the complications arising from **coccidioidomycosis**.
15. Evaluate the potential seriousness of a *Pneumocystis* pneumonia infection.
16. Summarize the possible affects of an *Aspergillus* species infection.

STEP A: SELF-TEST

Each of the following questions is designed to assess your ability to remember or recall factual or conceptual knowledge related to this chapter. Read each question carefully, then select the *one* answer that best fits the question or statement. Answers to even-numbered questions can be found in **Appendix C.**

1. Which one of the following statements about fungi is NOT true?
   A. Some fungi are dimorphic.
   B. Fungi have cell walls made of chitin.
   C. Fungi are photosynthetic organisms.
   D. Fungi consist of the yeasts and molds.
2. Which one of the following best describes the growth conditions for a typical fungus?
   A. pH 3; 23°C; no oxygen gas present
   B. pH 8; 30°C; no oxygen gas present
   C. pH 6; 30°C; oxygen gas present
   D. pH 3; 23°C; oxygen gas present
3. All the following are examples of asexual spore formation *except:*
   A. arthrospores.
   B. conidia.
   C. sporangiospores.
   D. basidiospores.
4. An organism without a known sexual stage would be classified in the
   A. Mitosporic fungi.
   B. Zygomycota.
   C. Basidiomycota.
   D. Ascomycota.
5. Yeasts of the species *Saccharomyces*
   A. are used in bread making.
   B. reproduce by budding.
   C. are members of the ascomycetes.
   D. All of the above (**A–C**) are correct.
6. Aflatoxin is produced by _____ and is _____.
   A. *Sporothrix schenkii;* a hallucinogen
   B. *Aspergillus flavus;* carcinogenic
   C. *Claviceps purpurea;* a hallucinogen
   D. *Aspergillus niger;* a mycotoxin
7. Mushroom poisoning is
   A. always fatal.
   B. characterized by immediate symptoms.
   C. most common in children under 6 years of age.
   D. treatable with antifungal drugs.
8. This fungal disease can cause a blister-like lesion on the scalp.
   A. Candidiasis
   B. Dermatophytosis
   C. Cryptococcosis
   D. Histoplasmosis
9. _____ causes more than 20 million cases each year and symptoms include an itching sensation and burning internal pain.
   A. Thrush
   B. "Jock itch"
   C. Vulvovaginitis
   D. Sporotrichosis
10. From the following, sporotrichosis would most likely be transmitted from
    A. peat moss.
    B. bat caves.
    C. mushrooms.
    D. dusty soil.
11. Which one of the following fungi would most likely be found in pigeon droppings?
    A. *Pneumocystis*
    B. *Cryptococcus*
    C. *Coccidioides*
    D. *Sporothrix*
12. Moving to the Ohio or Mississippi River valleys might make one susceptible to
    A. PCP.
    B. valley fever.
    C. dermatophytosis.
    D. histoplasmosis.

13. This dimorphic fungus produces conidia that are inhaled from dusty soil or bird droppings.
   A. *Aspergillus fumigatus*
   B. *Pneumocystis jiroveci*
   C. *Amanita phalloides*
   D. *Blastomyces dermatitidis*

14. The formation of arthrospores and spherules is characteristic of
   A. coccidioidomycosis.
   B. histoplasmosis.
   C. candidiasis.
   D. aspergillosis.

15. This fungal disease is the most common cause of nonbacterial pneumonia in immunocompromised individuals.
   A. Blastomycosis
   B. *Pneumocystis* pneumonia
   C. Aspergillosis
   D. Coccidioidomycosis

16. The most deadly form of aspergillosis is
   A. a pulmonary form.
   B. a toxigenic form.
   C. a blood form.
   D. an invasive form.

▓ STEP B: REVIEW

17. Using the diagram below, explain why these fungal diseases are considered to be systemic mycoses.

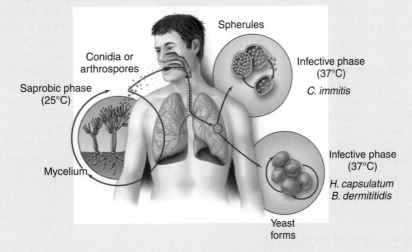

To test your knowledge of the important fungi, match the statement on the left to the organism on the right by placing the correct letter in the available space. A letter may be used once, more than once, or not at all. Answers to even-numbered statements are listed in **Appendix C**.

_____ **18.** Produces a widely used antibiotic.

_____ **19.** Used for bread baking.

_____ **20.** Causes "valley fever" in the southwestern U.S.

_____ **21.** Common white or gray bread mold.

_____ **22.** Agent of rose thorn disease.

_____ **23.** Associated with the droppings of pigeons.

_____ **24.** Agent of ergot disease in rye plants.

_____ **25.** Cause of vaginal yeast infections.

_____ **26.** One of the causes of dermatophytosis.

_____ **27.** Often found in chicken coops and bat caves.

_____ **28.** Produces a toxic aflatoxin.

_____ **29.** Reproduction includes a spherule.

_____ **30.** Involves a trophozoite stage.

A. *Agaricus* species
B. *Aspergillus flavus*
C. *Aspergillus* species
D. *Blastomyces dermatitidis*
E. *Candida albicans*
F. *Claviceps purpurea*
G. *Coccidioides immitis*
H. *Cryptococcus neoformans*
I. *Epidermophyton* species
J. *Histoplasma capsulatum*
K. *Penicillium notatum*
L. *Pneumocystis jiroveci*
M. *Rhizopus stolonifer*
N. *Saccharomyces cerevisiae*
O. *Saccharomyces ellipsoideus*
P. *Sporothrix schenkii*

STEP C: APPLICATIONS

Answers to even-numbered questions can be found in **Appendix C**.

31. You decide to make bread. You let the dough rise overnight in a warm corner of the room. The next morning you notice a distinct beer-like aroma in the air. What did you smell, and where did the aroma come from?

32. In a Kentucky community, a crew of five workers demolished an abandoned building. Three weeks later, all five required treatment for acute respiratory illness, and three were hospitalized. Cells obtained from the patients by lung biopsy revealed oval bodies and epidemiologists found an accumulation of bat droppings at the demolition site. As the head epidemiologist, what disease did the workers contract?

33. A woman comes to you complaining of a continuing problem of ringworm, especially of the lower legs in the area around the shins. Questioning her reveals she has five very affectionate cats at home. What disease does she have and what would be your suggestion to her?

STEP D: QUESTIONS FOR THOUGHT AND DISCUSSION

Answers to even-numbered questions can be found in **Appendix C**.

34. In a suburban community, a group of residents obtained a court order preventing another resident from feeding the flocks of pigeons that regularly visited the area. Microbiologically, was this action justified? Why?

35. Mr. A and Mr. B live in an area of town where the soil is acidic. Oak trees are common, and azaleas and rhododendrons thrive in the soil. In the spring, Mr. A spreads lime on his lawn, but Mr. B prefers to save the money. Both use fertilizer, and both have magnificent lawns. Come June, however, Mr. B notices that mushrooms are popping up in his lawn and that brown spots are beginning to appear. By July, his lawn has virtually disappeared. What is happening in Mr. B's lawn and what can Mr. B learn from Mr. A?

36. Residents of a New York community, unhappy about the smells from a nearby composting facility and concerned about the health hazard posed by such a facility, had the air at a local school tested for the presence of fungal spores. Investigators from the testing laboratory found abnormally high levels of *Aspergillus* spores on many inside building surfaces. Is there any connection between the high spore count and the composting facility? Is there any health hazard involved?

37. On January 17, 1994, a serious earthquake struck the Northridge section of Los Angeles County in California. From that date through March 15, 170 cases of coccidioidomycosis were identified in adjacent Ventura County. This number was almost four times the previous year's number of cases. What is the connection between the two events?

HTTP://MICROBIOLOGY.JBPUB.COM/9E

The site features eLearning, an online review area that provides quizzes and other tools to help you study for your class. You can also follow useful links for in-depth information, read more MicroFocus stories, or just find out the latest microbiology news.

Eukaryotic Microorganisms: The Parasites

It races through the bloodstream, hunkers down in the liver, then rampages through red blood cells before being sucked up by its flying, buzzing host to mate, mature, and ready itself for another wild ride through a two-legged motel.
—The editor of *Discover* magazine describing, in flowery terms, the life cycle of the malarial parasite

Approximately 2 million people die every year from malaria, an infection caused by a protozoan parasite and carried from person to person by mosquitoes. The disease is one of the most severe public health problems worldwide and is a leading cause of death and disease in many developing countries. Yet in 1957, a global program to eradicate the parasite commenced only to end in failure 21 years later. What happened?

Believe it or not, malaria once was an infectious killer in the United States. In the late 1880s, malaria was quite common in the American plains and southeast with epidemics reaching as far north as Montreal, Canada. Malaria was a major source of casualties in the American Civil War and until the 1930s, malaria remained endemic in the southern states.

To eliminate malaria, American officials established the National Malaria Eradication Program on July 1, 1947. This was a cooperative undertaking between the newly established Communicable Disease Center (the original CDC and a new component of the U.S. Public Health Service) and state and local health agencies of the 13 malaria-affected southeastern states. In 1947, 15,000 malaria cases were reported. However, by 1950, after more than 4,650,000 homes had been sprayed with the DDT (dichlorodiphenyltrichloroethane) pesticide to kill the mosquitoes, only 2,000 malaria cases were reported. By 1952, the United States was malaria free and the program ended.

Encouraged by the success of the American eradication effort, in 1957 the World Health Organization (WHO) began a similar effort to eradicate malaria worldwide. These efforts involved house spraying with insecticides, antimalarial drug treatment, and surveillance. Successes were made in some countries, but the emergence of drug resistance, widespread mosquito resistance to DDT, wars, and massive population movements made the eradication efforts unsustainable. The eradication campaign was eventually abandoned in 1978.

Now, more than 30 years later, the WHO estimates 40% of the world's population is at risk of malaria (**FIGURE 18.1**). The good news is another global campaign, initiated by the WHO, several United Nations agencies, and the World Bank, is underway. Named "Roll Back Malaria," the program calls for a 50% reduction in the burden of malaria by 2010. Let's hope there is great success this time around.

Malaria is just one of a number of human diseases caused by **parasites**, organisms that must live in or on a different species to get their nourishment. There are two different groups of eukaryotic parasites of concern to microbiology because of their ability to cause infectious disease.

One group contains single-celled **protists** (see Chapter 3). Some of the diseases they cause are familiar to us, such as malaria. Others affect the intestine, blood, or other organs of the body.

The second group are multicellular parasites, referred to as **helminths** (*helminth* = "worm"). These include the flatworms and roundworms, which together probably infect more people worldwide than any other group of organisms. In the strict sense, flatworms and roundworms are animals, but they are studied in microbiology because of their ability to cause disease. Together with the parasitic protists, they are the subject of study of the biological discipline known as **parasitology**.

Our study begins with a focus on the characteristics and classification of the parasitic protists, followed by a survey of human diseases they cause. Our review of the helminths is brief but we include descriptions of several parasites, their life cycles, and the types of organisms and tissues they infect.

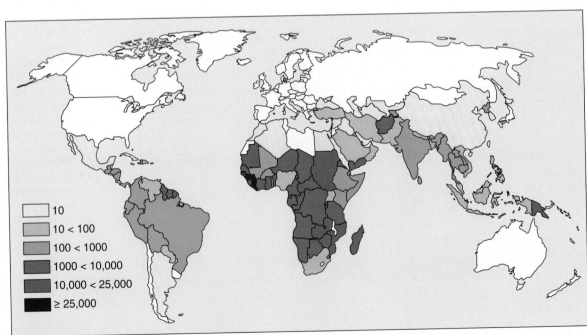

FIGURE 18.1 Malaria Cases (per 100,000) by Country—2008. The WHO has stated that 40% of the world's population is at risk of malaria. *Source*: Data from the WHO/Malaria Department. »» How can 40% of the world's population be at risk?

Legend:
- 10
- 10 < 100
- 100 < 1000
- 1000 < 10,000
- 10,000 < 25,000
- ≥ 25,000

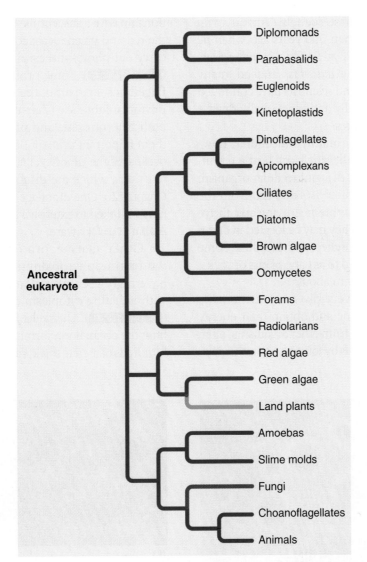

FIGURE 18.2 **A Tentative Phylogeny for the Kingdom *Eukarya*.** The phylogeny includes the four eukaryotic kingdoms. Everything besides the land plants and animals can be considered microbial and everything besides the land plants, animals, and fungi would be considered in the kingdom Protista. »» What kingdom of organisms is most closely related to the animals?

18.1 The Classification and Characteristics of the Protista

There are about 200,000 named species of protists, some seen by Leeuwenhoek over 300 years ago. The protists are an extremely diverse group of eukaryotes that often are very difficult to classify. For example, some algal species are multicellular, and the slime molds have unique life cycles with unicellular, colonial, and multicellular stages. In addition, some species are more like animal or plant cells than they are like other members of the kingdom Protista. As such, their taxonomic relationships are diverse and not always well understood.

For us, the phylogenetic framework presented in **FIGURE 18.2** simply provides a scheme for cataloging some of the protistan groups that we will survey in the first part of this chapter.

The Protista Are a Perplexing Group of Microorganisms

KEY CONCEPT

1. Protista include plant-, fungal-, and animal-like microorganisms.

The protists were first seen by Antony van Leeuwenhoek more than 300 years ago when he wrote in a letter to the Royal Society, *"No more pleasant sight has met my eye than this."* Indeed, many natural philosophers and scientists have continued to study this structurally and functionally diverse assortment of eukaryotes.

The protists are primarily unicellular; however, the functions of the single cell bear a resemblance to the functions of a multicellular organism rather than to those of an isolated cell from the organism. Most protists are free-living and thrive where there is water. They may be located in damp soil, mud, drainage ditches, and puddles. Some species remain attached to aquatic plants or rocks, while other species swim about.

Protists also are very diverse nutritionally. Many are heterotrophic and obtain their energy and organic molecules from other organisms, often obtaining these materials by forming a parasitic rela-

Primary producers:
Organisms that produce organic compounds from carbon dioxide gas.

Red tides:
Brownish-red discoloration in seawater that is caused by an increased presence of dinoflagellates.

tionship with a susceptible host. Others, including the red and green algae, contain chloroplasts and carry out photosynthesis similar to green plants (**FIGURE 18.3A**). Some protists, such as the **dinoflagellates**, are part of the freshwater and marine phytoplankton (see Chapter 3) and are able to be both heterotrophic and photosynthetic, making them important primary producers in the world's oceans. Some dinoflagellates cause the infamous red tides, which are discussed in more detail in Chapter 26. One dinoflagellate, *Pfiesteria piscicida*, may be linked to extensive fish kills in waters from Alabama to Delaware.

Other marine protists also are part of the marine phytoplankton. The **radiolarians** have highly sculptured and glassy silica plates with radiating cytoplasmic arms to capture prey (**FIGURE 18.3B**). The skeletons of dead radiolarians litter the ocean floor, forming deposits, sometimes hundreds of meters thick, called "radiolarian ooze."

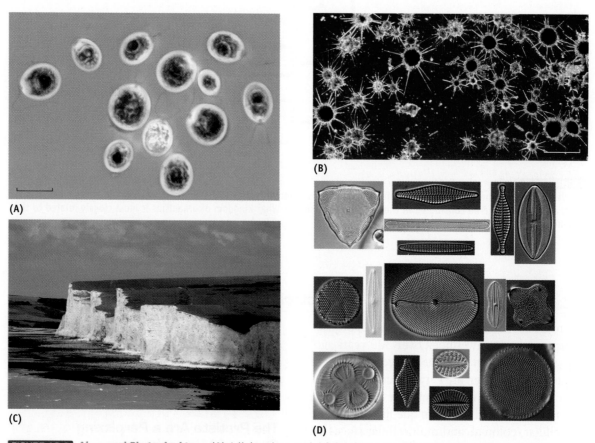

(A)

(B)

(C)

(D)

FIGURE 18.3 **Algae and Phytoplankton.** (**A**) A light micrograph of the green alga *Chlamydomonas*. (Bar = 10 μm.) (**B**) Light micrograph of an assortment of radiolarians. Radiolarians build hard skeletons made of silica around themselves as they float in seas with other plankton. (Bar = 120 μm.) (**C**) The white chalk cliffs at Beachy Head in Sussex, England, consist of the ancient shells of foraminiferans. (**D**) The diversity in shape of the diatoms. »» What characteristics are similar in the algae and phytoplankton?

The **foraminiferans** (**forams**) are marine protists that have chalky skeletons, often in the shape of snail shells with openings between sections (*foram* = "little hole"). The shells of dead forams form sediments hundreds of meters thick. When brought to the surface by geologic upthrusting, massive white cliffs have formed (**FIGURE 18.3C**).

The **diatoms** are another group of single-celled protists. By being encased in a two-part, hard-shelled, silica wall (**FIGURE 18.3D**), the cells can withstand great pressures and are not easily crushed or destroyed by predators. Diatoms carry out photosynthesis and compose an important part of the phytoplankton found in marine and freshwater environments. The massive accumulations of fossilized diatom walls are mined and ground up into diatomaceous earth, which has many useful applications, such as a filtering agent in swimming pools and a mild abrasive in household products, including toothpastes and metal polishes. Diatomaceous earth also can be used as a pesticide, because it grinds holes in the exoskeleton of crawling insects, causing the animals to desiccate.

Yet other protists are fungal-like. The Oomycetes are completely heterotrophic and absorb extracellularly digested food materials. However, these protists resemble fungi because they produce a filamentous growth characteristic of the molds. Some are plant pathogens. *Phytophthora ramorum*, for example, infects the California coastal live oak and causes sudden oak death. *P. infestans* is the agent of late blight in potatoes. It was responsible for the Irish famine of the 19th century and its infection had great ecological impact on humans and society (MICROFOCUS 18.1).

MICROFOCUS 18.1: History
The Great Irish Potato Famine

Ireland of the 1840s was an impoverished country of 8 million people. Most were tenant farmers paying rent to landlords who were responsible, in turn, to the English property owners. The sole crop of Irish farmers was potatoes, grown season after season on small tracts of land. What little corn was available was usually fed to the cows and pigs.

Early in the 1840s, heavy rains and dampness portended calamity. Then, on August 23, 1845, *The Gardener's Chronicle and Agricultural Gazette* reported: "*A fatal malady has broken out amongst the potato crop. On all sides we hear of the destruction. In Belgium, the fields are said to have been completely desolated.*"

Potatoes infected with *Phytophthora infestans* (dark brown regions).

The potatoes had suffered before. There had been scab, drought, "curl," and too much rain, but nothing was quite like this new disease. It struck down the plants like frost in the summer. Beginning as black spots, it decayed the leaves and stems, and left the potatoes a rotten, mushy mass with a peculiar and offensive odor (see figure). Even the harvested potatoes rotted.

The winter of 1845 to 1846 was a disaster for Ireland. Farmers left the rotten potatoes in the fields, and the disease spread. The farmers first ate the animal feed and then the animals. They also devoured the seed potatoes, leaving nothing for spring planting. As starvation spread, the English government attempted to help by importing corn and establishing relief centers. In England, the potato disease had few repercussions because English agriculture included various grains. In Ireland, however, famine spread quickly.

After two years, the potato rot seemed to slacken, but in 1847 ("Black '47") it returned with a vengeance. Despite relief efforts by the English, over 2 million Irish people died from starvation. At least 1 million Irish people left the land and moved to cities or foreign countries. During the 1860s, great waves of Irish immigrants came to the United States—and in the next century, their Irish American descendants would influence American culture and politics. And to think—it all resulted from *Phytophthora infestans* that caused late blight of potatoes.

Bilateral symmetry:
A form of symmetry where an imaginary plane divides an object into right and left halves, each side being a mirror of the other.

The protists also include many motile, predatory, or parasitic species that absorb or ingest food. These protists traditionally have been called the **protozoa** (*proto* = "first; *zoo* = "animal"), referring to their animal-like properties that incorrectly suggested to biologists that protozoa were close evolutionary ancestors of the first animals. Though often studied by zoologists, protozoa also interest microbiologists because they are unicellular, most have a microscopic size, and several are responsible for infectious disease. Therefore, these are the microbial parasites we will emphasize next.

CONCEPT AND REASONING CHECKS

18.1 Summarize the characteristics of the plant-like and fungal-like protists.

The Protozoa Encompass a Variety of Lifestyles

KEY CONCEPT

2. Protozoa represent several different branches on the "tree of life."

When Robert Whittaker assigned protozoa to the kingdom Protista in 1969, he did so as a matter of convenience, rather than on the basis of evolutionary relationships (see Chapter 3). Today, the protozoa are a group of about 65,000 species of single-celled microorganisms. A tentative taxonomy, based on comparative studies involving genetic analysis and genomics, places the protists in one of six informal super groups. Here, we briefly consider the biological features of the three groups containing protozoa that are human parasites.

Super Group Excavata. This group contains species that are single-celled and possess flagella for motility. Some members in the group may represent organisms whose ancestors were the earliest forms of eukaryotes.

Members of the **parabasalids** lack mitochondria and, as such, live in low oxygen or anaerobic environments. Several species, including *Trichonympha,* are found in the guts of termites where the symbionts participate in a mutualistic relationship (FIGURE 18.4A). The cells contain hundreds of flagella with the typical 9+2 arrangement of microtubules found in all eukaryotic flagella (see Chapter 3). Undulations sweep down the flagella to the tip, and the lashing motion forces water outward to provide locomotion. Another species,

Trichomonas vaginalis, is parasitic in humans and is transmitted through sexual intercourse.

The **diplomonads** have two haploid nuclei and three pairs of flagella at the anterior end and one pair at the posterior end, giving the cell bilateral symmetry. Reproduction is only asexual by binary fission.

The most notable species is *Giardia intestinalis* (FIGURE 18.4B). It is spread through contaminated water and, thus, affects the gastrointestinal tract. The diplomonad can survive outside the anaerobic environment of the intestine by forming a **cyst,** which is a dormant, highly resistant stage. So many lakes and rivers in the United States are contaminated with the cysts that hikers and campers must first boil or filter the water before drinking.

Another set of protists in the Excavata is the **Euglenozoa.** Among these are the **kinetoplastids,** another ancient lineage of heterotrophic species. A unique characteristic of these species is a single, posterior flagellum that is attached to the cell's wavy, undulating membrane (FIGURE 18.4C). The kinetoplastids have the typical array of eukaryotic organelles and the single mitochondrion contains a mass of DNA called the **kinetoplast.**

Some 60% of the kinetoplastid species are trypanosomes (*trypano* = "hole"; *soma* = "body"), referring to the hole the organism bores to enter and infect the host. Two *Trypanosoma* species are transmitted by insects and cause forms of human sleeping sickness in Africa and South America, affecting millions of people. Species of *Leishmania,* which is also transmitted by insects, can produce a skin disease or an often fatal visceral infection.

CONCEPT AND REASONING CHECKS

18.2A What features can be used to separate the parabasilids, diplomonads, and kinetoplastids?

Super Group Amoebozoa. Another group of protozoa, the **amoebas,** are mostly free-living, single-celled organisms that can be as large as 1 mm in diameter. They usually live in freshwater or marine environments and reproduce by binary fission. The amoebas are soft bodied organisms that have the ability to change shape (*amoeba* = "change") as their cytoplasm flows into temporary formless cytoplasmic projections called **pseudopods** (*pseudo* = "false"; *pod* = "a foot"); thus, the motion is called **amoeboid motion.** Pseudopods also capture bacteria,

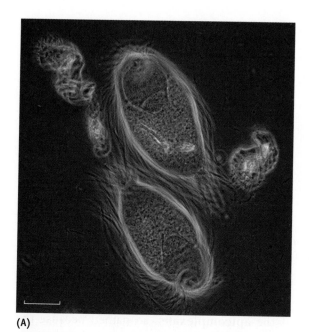

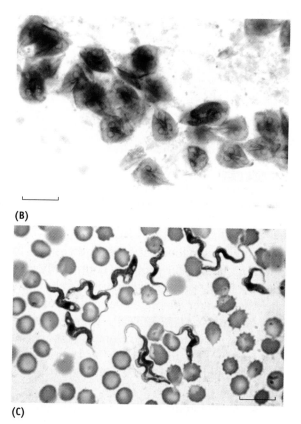

FIGURE 18.4 Parabasalid, Diplomonad, and Kinetoplastid Parasites. **(A)** A light micrograph of *Trichonympha*, a parabasalid parasite found in the gut of termites. Each thin, wispy line represents a flagellum used for motility. (Bar = 25 µm.) **(B)** Another light micrograph of stained *Giardia intestinalis* cells. The pear-shaped cell body of this diplomonad and flagella are typical features. (Bar = 5 µm.) **(C)** Light micrograph of *Trypanosoma* in a blood smear. The wavy cell appearance is due to the wavy membrane and flagellum. (Bar = 10 µm.) »» If the parabasalids and diplomonads lack mitochondria, what metabolic pathway remains for synthesizing ATP?

small algae, and other protozoa through the ingestive process of **phagocytosis** (**FIGURE 18.5A**). The pseudopods enclose the particles to form an organelle called a **food vacuole**, which then joins with lysosomes. The lysosomes contain digestive enzymes to digest the material in the captured prey. Nutrients are absorbed from the vacuole, and any undigested residue is eliminated from the cell.

The genus *Entamoeba* can be far more serious, as all species are parasitic. In humans, amoebic dysentery or encephalitis may result from drinking water or consuming food contaminated with amoebal cysts. Amoebic dysentery is the third leading cause of death due to a parasitic infection.

CONCEPT AND REASONING CHECKS

18.2B What unique cellular structures and behaviors are associated with the Amoebozoa?

Super Group Chromalveolata. This group is very diverse, and includes the dinoflagellates and diatoms described earlier. The ciliated protozoans, or **ciliates**, are among the most complex cells on Earth and have been the subject of biological investigations for many decades. They are found in almost any pond water sample. They have a variety of shapes and can exhibit elaborate and controlled behavior patterns. The cytoplasm contains the typical eukaryotic organelles.

Ciliates range in size from a microscopic 10 µm to a huge 3 mm. All ciliates are covered with **cilia** (sing., cilium) in longitudinal or spiral rows. Cilia beat in a synchronized and coordinated pattern, the organized "rowing" action moving the ciliate along in one direction.

Ciliates, such as *Paramecium*, are heterotrophic by ingestion through a primitive gullet,

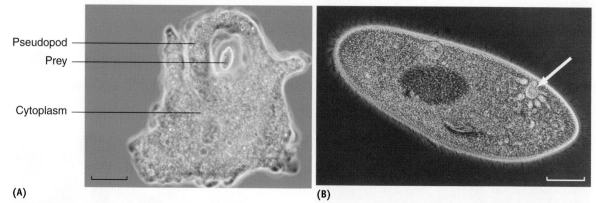

Pseudopod
Prey

Cytoplasm

(A)   (B)

FIGURE 18.5 **An Amoeba and a Ciliate.** **(A)** A light micrograph of *Amoeba proteus*. The pseudopodia are extending around the prey, beginning the process of phagocytosis. (Bar = 100 μm.) **(B)** This light micrograph of the ciliate *Paramecium* shows the contractile vacuole (arrow). (Bar = 50 μm.) »» What is the function of the contractile vacuole?

which sweeps in food particles for digestion. In addition, freshwater protozoa continually take in water by the process of osmosis and eliminate the excess water via organelles called **contractile vacuoles** (**FIGURE 18.5B**). These vacuoles expand with water drawn from the cytoplasm and then appear to "contract" as they release water through a temporary opening in the cell membrane.

Asexual reproduction in the ciliates occurs by binary fission. The complexity of ciliates however is illustrated by the nature of sexual reproduction. Ciliates have two types of nuclei; there is a single large **macronucleus** that only has the genes for cell metabolism, and one **micronucleus** that contains a complete set of genes. During sexual recombination, called **conjugation**, two cells make contact, and a cytoplasmic bridge forms between them (**FIGURE 18.6**). A micronucleus from each cell undergoes two divisions to form four micronuclei, of which three disintegrate and only one remains to undergo mitosis. Now a "swapping" of micronuclei takes place, followed by a union to re-form the normal micronucleus.

This genetic recombination is outwardly analogous to what occurs in bacterial cells. It is observed during periods of environmental stress, a factor that suggests the formation of a genetically different and, perhaps, better-adapted organism.

The last group of protozoa we will mention is the **apicomplexans**, so named because the *api*cal tip of the cell contains a *complex* of organelles used for penetrating host cells. Adult apicomplexan cells have no cilia or flagella, although a few species have flagellated gametes.

These animal parasites of the Chromalveolata have a complex life cycle including alternating sexual and asexual reproductive phases. These phases often occur in different hosts. Two parasitic species, *Plasmodium* and *Toxoplasma*, are of special significance because the first causes one of the most prolific killers of humans—malaria—and the second is associated with AIDS.

Although most protozoa require only one host for the completion of their life cycle, most apicomplexans require two or more different hosts to complete their life cycle. The host organism in which the sexual cycle occurs is called the **definitive host** while the host in which the asexual cycle occurs is the **intermediate host** (**FIGURE 18.7**).

For example, the protozoan *Plasmodium* produces infective **sporozoites** in mosquitoes (definitive host). These cells enter the human body. Later, within the human host (intermediate host), plasmodial gametes, called **gametocytes**, are produced. These are taken up by mosquitoes in a blood meal where sexual reproduction forms more sporozoites. **TABLE 18.1** summarizes some of the attributes of the protozoal groups of protists.

CONCEPT AND REASONING CHECKS

18.2C How does conjugation in ciliates differ from that in bacterial cells?

CONCEPT AND REASONING CHECKS

18.2D What are some unique features of the apicomplexans?

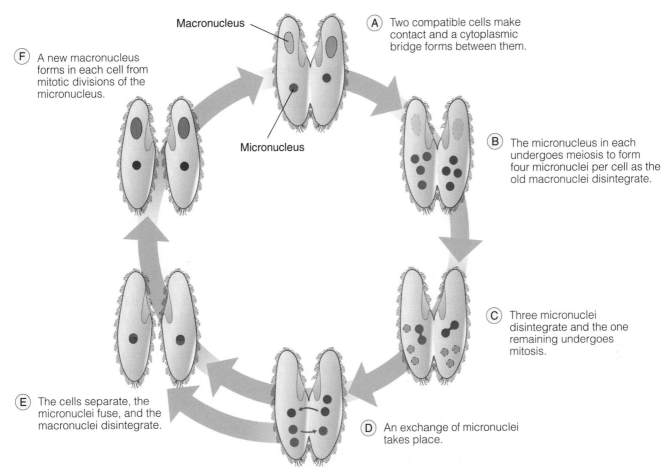

Macronucleus

Micronucleus

(A) Two compatible cells make contact and a cytoplasmic bridge forms between them.

(B) The micronucleus in each undergoes meiosis to form four micronuclei per cell as the old macronuclei disintegrate.

(C) Three micronuclei disintegrate and the one remaining undergoes mitosis.

(D) An exchange of micronuclei takes place.

(E) The cells separate, the micronuclei fuse, and the macronuclei disintegrate.

(F) A new macronucleus forms in each cell from mitotic divisions of the micronucleus.

FIGURE 18.6 Conjugation and Reproduction in *Paramecium*. Ciliates, such as *Paramecium,* reproduce sexually by the process of conjugation. In this process, an exchange of micronuclei gives rise to new macronuclei. »» Provide a hypothesis to explain why ciliates, unlike most other organisms, must have two types of nuclei.

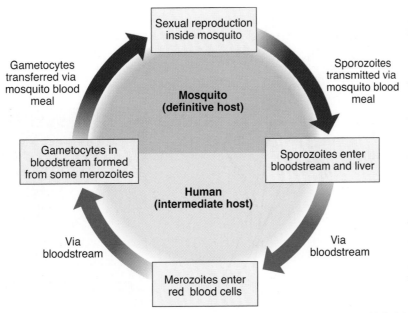

Sexual reproduction inside mosquito

Sporozoites transmitted via mosquito blood meal

Gametocytes transferred via mosquito blood meal

Mosquito (definitive host)

Sporozoites enter bloodstream and liver

Gametocytes in bloodstream formed from some merozoites

Human (intermediate host)

Via bloodstream

Via bloodstream

Merozoites enter red blood cells

FIGURE 18.7 The Life Cycle of *Plasmodium*. The malaria parasite requires two different hosts (definitive and intermediate) in which specific stages of the life cycle occur. »» In trying to prevent malaria, which stage of the *Plasmodium* life cycle would appear to be the most amenable to blocking transmission of the parasite?

TABLE 18.1 Comparison of the Protozoan Groups of the Protists

Super Group	Motility	Other Characteristics	Representative Genera
Excavata			
Parabasalids	Flagella	Thousands of flagella; no mitochondria	*Trichomonas*
Diplomonads	Flagella	Two nuclei; no mitochondria	*Giardia*
Kinetoplastids	Flagella	Kinetoplast DNA	*Trypanosoma, Leishmania*
Amoebozoa			
Amoebas	Amoeboid movement	Pseudopodia	*Entamoeba*
Chromalveolata			
Ciliates	Cilia	Macro- and micronuclei	*Paramecium*
Apicomplexans	Flagella (gametes only)	Apical complex; multiple hosts	*Plasmodium, Toxoplasma*

18.2 Protozoal Diseases of the Skin, and the Digestive and Urinary Tracts

Protozoal diseases occur in a variety of systems of the human body. For example, some diseases, such as amoebiasis and giardiasis, take place in the digestive system, while others, such as trichomoniasis, develop in the urogenital tract. We start with a cutaneous protozoal disease.

Leishmania Can Cause a Cutaneous or Visceral Infection

KEY CONCEPT

3. Leishmaniasis is a vector-born disease.

Leishmaniasis is a rare disease in the United States, but it occurs in 88 countries on four continents with an at-risk population of 350 million people. The responsible protozoa are in the kinetoplastid group and include several species of *Leishmania*, including *L. major* and *L. donovani* (**FIGURE 18.8A**). Transmission is by the bite of an infected female sandfly of the genus *Phlebotomus* (**FIGURE 18.8B**). The sandfly **vector**, which is only about one-third the size of a mosquito, becomes infected by biting an infected animal, such as a rodent, dog, or another human.

There are two main forms of leishmaniasis. *L. major* causes a disfiguring **cutaneous** (skin) **disease**. Within a few weeks after being bitten, a sore appears on the skin. The sore then expands

Leishmania donovani

Vector:
An insect that transmits pathogens or parasites from an infected animal to humans.

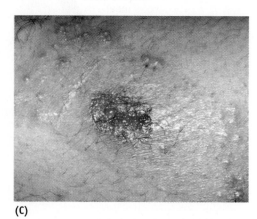

(A)   (B)   (C)

FIGURE 18.8 **Leishmaniasis.** (A) A light micrograph showing a cluster of *Leishmania* cells, which are long, thin, and flagellated. (Bar = 20 μm.) (B) The sandfly, *Phlebotomus*, which is the vector for transmission of leishmaniasis. (C) Skin lesion due to cutaneous leishmaniasis. »» Explain how the parasite can bring about the physical skin lesion.

and ulcerates to resemble a volcano with a raised edge and central crater (**FIGURE 18.8C**). Some sores may be painless and become covered by a scab. There are about 500,000 new cases of cutaneous leishmaniasis each year worldwide.

The other form of leishmaniasis is a **visceral** (body organ) **disease** called **kala azar**, meaning "black fever." It is caused by *L. donovani*. Symptoms do not appear until several months after being bitten by a sandfly. Infection of the white blood cells leads to irregular bouts of fever, swollen spleen and liver, progressive anemia, and emaciation. About 90% of cases are fatal, if not treated. There are about 1.5 million new cases globally each year, 90% occurring in India, Bangladesh, Nepal, Sudan, and Brazil. American soldiers have been infected during the Iraq conflicts with the cutaneous or visceral form (MicroFocus 18.2).

Control of the sandfly remains the most important method for preventing outbreaks of leishmaniasis. The antimony compound, stibogluconate, is used to treat established cases.

CONCEPT AND REASONING CHECKS

18.3 Summarize the two types of leishmaniasis and their health consequences.

MICROFOCUS 18.2: Public Health
The "Baghdad Boil"

Between August 2002 and February 2004, the United States Department of Defense (DoD) announced they had identified 522 cases of cutaneous leishmaniasis (CL) among military personnel serving in Afghanistan and Iraq. The disease was "affectionately" called the "Baghdad boil" (see figure).

Leishmania major, which is endemic in Southwest/Central Asia, was the parasitic species identified in the 176 cases analyzed. Patients were treated with sodium stibogluconate.

The DoD has implemented prevention measures to decrease the risk of CL. These procedures included improving hygiene conditions, instituting a CL awareness program among military personnel, using permethrin-treated clothing and bed nets to kill or repel sandflies, and applying insect repellent containing 30% DEET to exposed skin. These measures, according to the Department of Defense's Medical Surveillance Monthly Report, have reduced the number of reported cases from 52 per month in the fall of 2003 to less than 3 per month in the fall of 2009.

Cutaneous leishmaniasis.

Over the period from 2002–2008, only four cases of visceral leishmaniasis (VL) were reported in military personnel—two in Afghanistan and two in Iraq. All patients exhibited a persistent fever, enlargement of the spleen, and progressive anemia. The seriousness of the disease required aggressive treatment. On hospitalization and identification of VL, three patients received antileishmanial therapy, which consisted of a liposome [artificial lipid vesicle] formulation of amphotericin B. After one week, the fever dissipated and the patients resumed their duties after one additional week.

The fourth patient had an unusual 14-month incubation period, during which time his deployment ended. His self-reported onset required the administration of lipid amphotericin B and symptoms briefly improved before relapsing. The patient was rehospitalized and received a 28-day treatment with sodium stibogluconate.

This latter case highlights the need for civilian health care workers in the United States to be aware of potential VL in persons who had been deployed to Southwest/Central Asia.

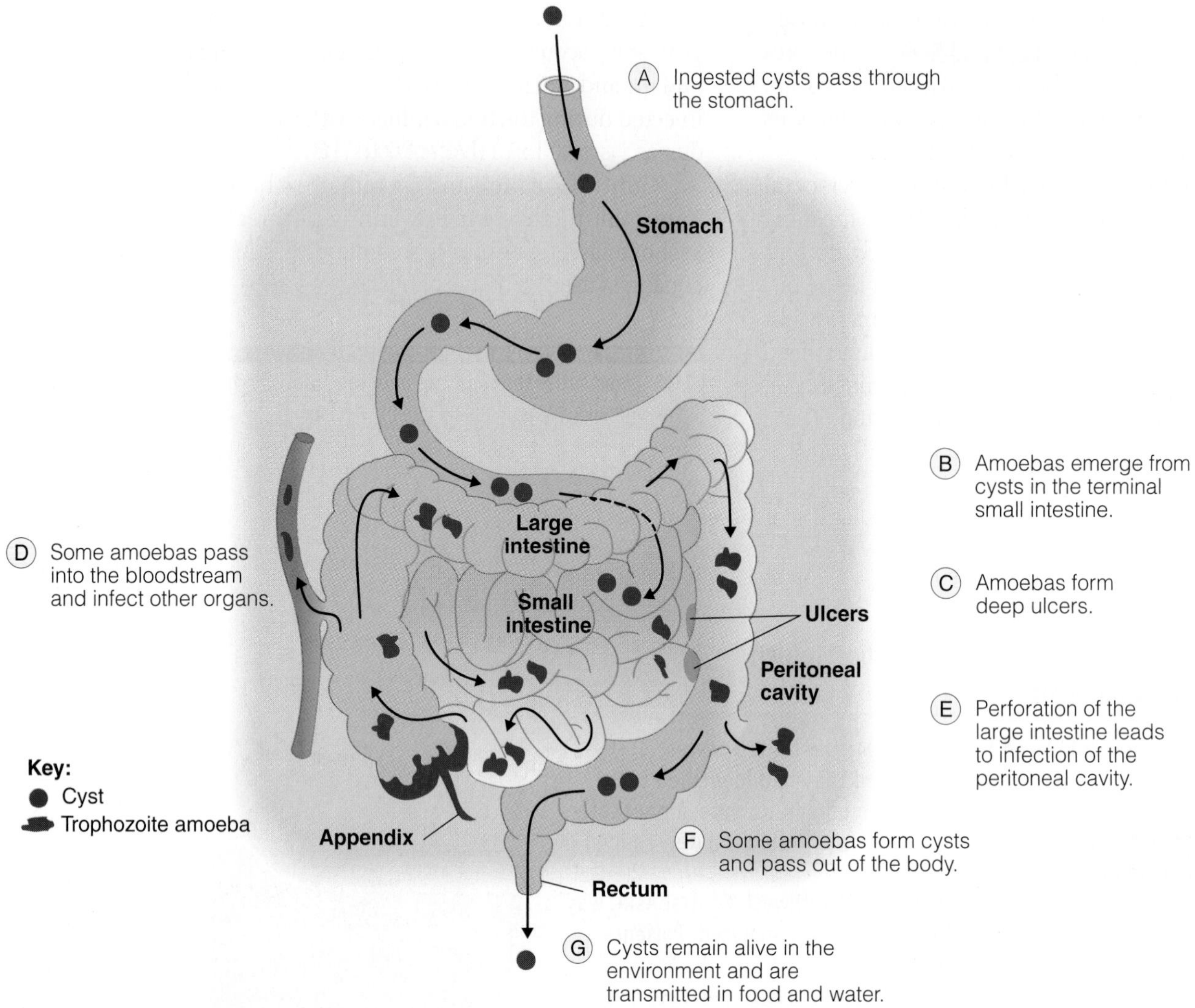
A Ingested cysts pass through the stomach.
Stomach
B Amoebas emerge from cysts in the terminal small intestine.
Large intestine
Small intestine
D Some amoebas pass into the bloodstream and infect other organs.
Ulcers
C Amoebas form deep ulcers.
Peritoneal cavity
E Perforation of the large intestine leads to infection of the peritoneal cavity.
Key:
Cyst
Trophozoite amoeba
Appendix
F Some amoebas form cysts and pass out of the body.
Rectum
G Cysts remain alive in the environment and are transmitted in food and water.

FIGURE 18.9 The Course of Amoebiasis Due to *Entamoeba histolytica*. The parasite enters the body as a cyst, which develops into an amoeboid form that causes deep ulcers. »» What is the advantage for the parasite to enter the body as a cyst?

Several Protozoal Parasites Cause Diseases of the Digestive System

KEY CONCEPT

4. Amoebiasis, giardiasis, cryptosporidiosis, and cyclosporiasis result from ingestion of contaminated food or water.

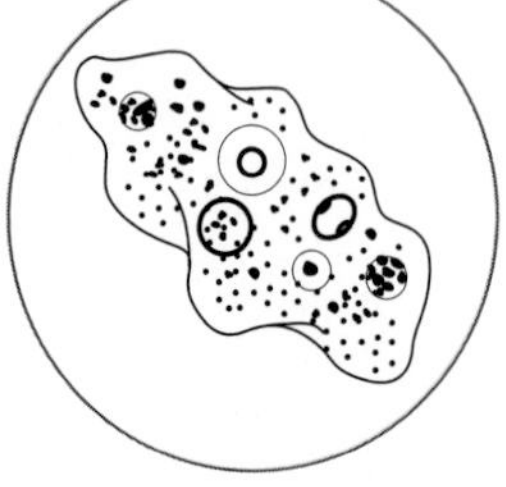

Entamoeba histolytica

There are three groups of parasite diseases caused by different super groups of protists, including the amoebozoans, the apicomplexans, and the kinetoplastids.

Amoebiasis. The second leading cause of death from parasite diseases, only surpassed by malaria, is **amoebiasis**. This parasitic form of **gastroenteritis** occurs worldwide and primarily affects children and adults who are undernourished and living in unsanitary conditions. Although an intestinal illness at first, it can spread to various organ systems. Some 40,000 to 100,000 people die each year from amoebiasis.

The causative agent of amoebiasis is *Entamoeba histolytica*. In nature, the protozoan exists in the cyst form, which enters the body by food or water contaminated with human or animal feces, or by direct contact with feces. The cysts pass through the stomach and emerge as amoebal trophozoites in the distant portion of the small intestine and in the large intestine (FIGURE 18.9).

On average, about 1 in 10 people who are infected with *E. histolytica* becomes sick from the infection. The symptoms often are quite mild and can include diarrhea, stomach pain, and stomach

MICROFOCUS 18.3: Public Health
What's Growing in Your Plumbing?

Since 1986, the Centers for Disease Control and Prevention (CDC) has been receiving reports of intermittant cases of eye infections specifically in individuals who wore contact lenses. Testing identified the free-living amoebozoan *Acanthamoeba* as the infecting agent and the disease was *Acanthamoeba* keratitis (AK).

AK is a rare but very painful and potentially blinding infection of the cornea, the transparent covering at the front of the eye (see figure). *Acanthamoeba* has been found in virtually every environment, including soil, dust, fresh water, and seawater. The parasite sometimes resides in untreated swimming pools, hot tubs, and even in bottled water.

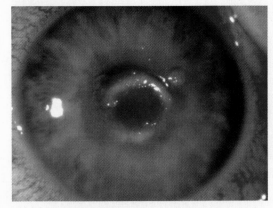

Acanthamoeba keratitis.

The cases reported to the CDC resulted from contact lenses becoming contaminated after improper cleaning and handling. In fact, people who make their own lens cleaning solution are more at risk because proper sterile conditions often are not followed.

In the United Kingdom (UK), an infection of AK occurs about once in every 30,000 contact lens wearers, which is a rate 15-times higher than in the United States. Why?

That is exactly what John Dart, an ophthalmologist at Moorefields Eye Hospital in London, England, wanted to know. Dart knew that until recently all homes in the UK had to have cold-water storage tanks. Could this be the breeding ground for *Acanthamoeba*?

Dart and his colleagues compared the *Acanthamoeba* mitochondrial DNA from eight patients with that from their home water supply. In six cases, the DNAs were identical, indicating the organisms must have come from the home water supply. However, not all patients stored their contact lenses using the home tap water, so Dart believes the source of the eye infections comes either from the patients having washed their faces with water while wearing their contact lenses or handling the lenses with wet hands. A program for better hygiene practices has been introduced.

---

cramping. The trophozoites multiply by binary fission and produce cysts, which are passed in the feces.

The *E. histolytica* amoebae have the ability to destroy tissue (*histo* = "tissue"; *lyt* = "loosened"). Using their protein-digesting enzymes, the trophozoites can penetrate the wall of the large intestine, causing lesions and deep ulcers. Patients who experience stomach pain, bloody stools, and fever have a more severe form of the disease called **amoebic dysentery**. In rare cases, the parasites invade the blood and spread to the liver, lung, or brain, where fatal abscesses may develop.

Prevention is a matter of not eating potentially contaminated food or drinking unpasteurized milk or other diary products. Bottled water or boiled water should be consumed in countries where amoebiasis occurs. Metronidazole and paromomycin commonly are used to treat amoebiasis,

but the drugs do not affect the cysts, and repeated attacks of amoebiasis may occur for months or years. The patient often continues to shed cysts in the feces to infect other people.

MicroFocus 18.3 describes a parasitic amoebal infection of the eye.

CONCEPT AND REASONING CHECKS

18.4A Why is amoebiasis considered an example of "invasive" gastroenteritis?

Giardiasis. In 2007, the Centers for Disease Control and Prevention (CDC) received reports of 19,417 cases of **giardiasis**, making it the most commonly detected protozoal disease of the intestinal tract in the United States. Because not all cases are reported to the CDC, the disease is estimated to cause 100,000 to 2.5 million infections, primarily in the summer and early fall. The disease is sometimes

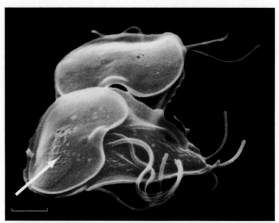

FIGURE 18.10 *Giardia intestinalis.* False-color scanning electron micrograph of *G. intestinalis* in the human small intestine. The pear-shaped cell body and sucker device (arrow) are evident, as are the flagella of this diplomonad. (Bar = 1 μm.) »» What is the function of the sucker device?

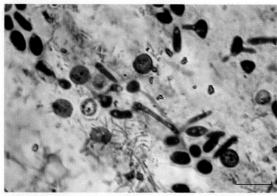

FIGURE 18.11 *Cryptosporidium* Oocysts. A light micrograph of a stained fecal smear. *Cryptosporidium* oocysts are red. (Bar = 2 μm.) »» What is an oocyst?

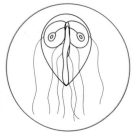

Giardia intestinalis

Duodenum:
The first short section of the small intestine immediately beyond the stomach.

Oocyst:
The thick-walled fertilized gamete of apicomplexans.

mistaken for viral gastroenteritis and is considered a type of traveler's diarrhea.

The causative agent is the diplomonad *Giardia intestinalis* (also called *G. lamblia*). This organism is distinguished by four pairs of anterior flagella and two nuclei that stain darkly to give the appearance of eyes on a face. The protozoan can be divided equally along its longitudinal axis and is therefore said to display bilateral symmetry.

Giardiasis is commonly transmitted by food or water containing *Giardia* cysts stemming from cross-contamination of drinking water with sewage as well as by the fecal-oral route. The cysts pass through the stomach and the trophozoites emerge as flagellated cells in the duodenum. They multiply rapidly by binary fission and adhere to the intestinal lining using a sucking disk located on the lower cell surface (FIGURE 18.10).

Acute giardiasis develops after an incubation period of about seven days. The patient feels nauseous, experiences gastric cramps and flatulence, and emits a foul-smelling watery diarrhea sometimes lasting for one or three weeks. Infectious cysts are excreted in the feces.

Treatment of giardiasis may be administered with drugs such as metronidazole or tinidazole. However, these drugs have side effects and the physician may wish to let the disease run its course

without treatment. Those who recover often become carriers and excrete the cysts for years.

CONCEPT AND REASONING CHECKS

18.4B Identify why most cases of giardiasis are reported in the summer to early fall.

Cryptosporidiosis. Since 1976, outbreaks of **cryptosporidiosis** have been reported in several countries. The most remarkable outbreak occurred in 1993 in Milwaukee, Wisconsin, where more than 400,000 people were affected (and 54 died), making it the largest waterborne infection ever recorded in the United States. In 2007, there were 11,170 reported cases, with the highest number of reported cases in children 1 to 9 years of age.

Human cryptosporidiosis is caused by the apicomplexans *Cryptosporidium parvum* and *C. hominis.* The organisms have a complex life cycle involving trophozoite, sexual, and oocyst stages (FIGURE 18.11). Transmission occurs mainly through contact with contaminated water (such as drinking or recreational water containing oocysts). Physical contact also can transmit *Cryptosporidium* oocysts, making children in daycare centers at risk.

Cryptosporidiosis has an incubation period of about one week. Patients with competent immune systems appear to suffer limited diarrhea lasting 1 or 2 weeks during which time newly-formed infectious oocysts are excreted in the feces.

In immunocompromised individuals, such as AIDS patients, *Cryptosporidium* is an opportunistic infection. Patients experience cholera-like profuse diarrhea that can be severe and irreversible. These patients undergo dehydration and emaciation, and often die of the disease.

Cyclosporiasis. In the late 1990s, public health officials in the United States identified a series of clusters of intestinal disease related to the consumption of raspberries imported from Guatemala. In 2004 and 2005, outbreaks in Texas, Illinois, and Florida sickened over 400 people after eating raw basil. In all cases, the outbreaks were related to the apicomplexan *Cyclospora cayetanensis*.

Fresh produce and water can serve as vehicles for transmission of oocysts, which are ingested in contaminated food or water. The oocysts are similar to, but larger than, those of *Cryptosporidium parvum*. Differential diagnosis is important because *C. cayetanensis* responds to the drug combination of trimethoprim-sulfamethoxazole, whereas *Cryptosporidium* does not.

Cyclosporiasis has an incubation period of one week. Symptoms of the disease include watery diarrhea, nausea, abdominal cramping, bloating, and vomiting. Treatment is successful with the drugs noted above, but the symptoms often return. Moreover, the symptoms often remain for over one month during the first illness.

How fresh produce becomes contaminated during outbreaks is not clear. Tainted water used for washing the produce may be the source, the produce may be handled by someone whose hands are contaminated, or the produce may be contaminated during shipping. Regardless of the source, better control measures focusing on improved water quality and better sanitation methods on local farms have resulted in only 93 reported cases by the CDC in 2007.

CONCEPT AND REASONING CHECKS

18.4C Compare and contrast cryptosporidiosis and cyclosporiasis.

A Protozoan Parasite Also Infects the Urinary Tract

KEY CONCEPT

5. Trichomoniasis affects over 10% of sexually active individuals.

Trichomoniasis is among the most common pathogenic protozoan diseases in men and women

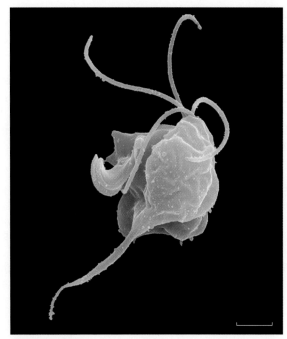

FIGURE 18.12 *Trichomonas vaginalis.* False-color scanning electron micrograph of a *T. vaginalis* trophozoite. (Bar = 5 μm.) »» What are the whip-like structures evident in the micrograph?

in industrialized countries, including the United States, where an estimated 7.4 million new cases occur annually. The disease is transmitted primarily by sexual contact and is considered a sexually transmitted disease (STD).

Trichomonas vaginalis, the causative agent, is a pear-shaped, flagellated protozoan of the parabasalids (**FIGURE 18.12**). The organism has only a trophozoite stage and its only host is humans, where it thrives and replicates by binary fission in the slightly acidic environment of the female vagina and the male urethra. Establishment may be encouraged by physical or chemical trauma, including poor hygiene, drug therapy, diabetes, or mechanical contraceptive devices such as the intrauterine device (IUD).

The incubation period is 5 to 28 days. In females, trichomoniasis is accompanied by intense itching (pruritis) and discomfort during urination and sexual intercourse. Usually, a yellow-green, frothy discharge also is present. The symptoms are frequently worse during menstruation, and erosion of the cervix may occur. In males, the disease may be asymptomatic. Symptoms occur primarily

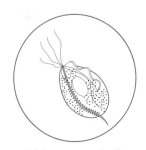

Trichomonas vaginalis

TABLE

18.2 A Summary of Protozoal Diseases of the Skin, and the Digestive and Urinary Tracts

Disease	Causative Agent	Signs and Symptoms	Transmission	Treatment	Prevention and Control
Cutaneous leishmaniasis	*Leishmania major*	Skin sore that ulcerates	Sandfly bites	Stibogluconate	Protecting skin from sandfly bites
Visceral leishmaniasis (kala azar)	*Leishmania donovani*	Fever, swollen spleen and liver, anemia, emaciation	Sandfly bites	Stibogluconate	Protecting skin from sandfly bites
Amoebiasis	*Entamoeba histolytica*	Loose stools, stomach pain and cramping Amoebic dysentery	Indirect through food or water Direct contact with feces	Metronidazole Paromomycin	Avoiding contaminated food, unpasteurized milk
Giardiasis	*Giardia intestinalis*	Nausea, gastric cramps, flatulence, foul-smelling watery diarrhea	Indirect through food or water	Metronidazole Tinidazole	Practicing good hand hygiene Avoiding untreated water
Cryptosporidiosis	*Cryptosporidium parvum Cryptosporidium hominis*	Watery diarrhea, dehydration, vomiting, nausea, stomach cramps, fever, malaise	Contact with contaminated water	Nitazoxanide, paromomycin Fluid replacement	Practicing good hand hygiene Avoiding untreated water
Cyclosporiasis	*Cyclospora cayetanensis*	Watery diarrhea, nausea, abdominal cramping, bloating, vomiting	Contaminated fresh produce or water	Trimethoprim-sulfamethoxazole Fluid replacement	Avoiding untreated water and contaminated fresh produce
Trichomoniasis	*Trichomonas vaginalis*	Intense itching Discomfort on urination and sexual intercourse Yellow-green frothy discharge with strong odor	Oral, vaginal, or anal sex with someone who is infected	Metronidazole or tinidazole	Abstaining from sexual activity Limiting sexual partners Patient and partner treatment

in the urethra, with slight pain on urination and a thin, mucoid discharge. The disease can occur concurrently with gonorrhea.

The drug of choice for treatment is orally administered metronidazole or tinidazole, and both patient and sexual partners should be treated concurrently to prevent transmission

or reinfection. Drug resistance has become of increasing concern.

TABLE 18.2 summarizes the protozoal diseases of the skin, and the digestive and urinary tracts.

CONCEPT AND REASONING CHECKS

18.5 Why is trichomoniasis such a common protozoan disease?

18.3 Protozoal Diseases of the Blood and Nervous System

We finish our discussion of the protozoal parasites by examining those protozoa that cause infections in the blood or in the nervous system. This includes two of the most prevalent diseases, malaria and sleeping sickness. Visceral leishmaniasis was discussed in the previous section with its cutaneous form.

The *Plasmodium* Parasite Infects the Blood

KEY CONCEPT

6. Malaria is caused by four different apicomplexan species.

Malaria is a disease that has been known since at least 1000 BC. During the 1700s, Europeans suffered wave after wave of malaria and few regions were left untouched. Even American pioneers settling in the Mississippi and Ohio valleys suffered great losses from the disease.

Today, between 300 and 500 million of the world's population suffer from malaria, which exacts its greatest toll in Africa. The WHO estimates over 1 million children under the age of 5 die from malaria annually; that is equivalent to one child dying every 30 seconds! No infectious disease of contemporary times can claim such a dubious distinction. And it is worldwide. In 2006, more than 18,000 cases were reported in China's Anhui Province. This is a 90% increase from 2005. Even the United States is involved in the malaria pandemic—over 1,400 imported cases were reported in 2007.

Malaria is caused by four species of the apicomplexan genus *Plasmodium: P. vivax, P. ovale, P. malariae,* and *P. falciparum.* All are transmitted by the female *Anopheles* mosquito, which consumes human blood to provide chemical components for her eggs. The life cycle of the parasites has three important stages: the sporozoite, the merozoite, and the gametocyte. Each is a factor in malaria. The most serious infections that can be life threatening are caused by *P. falciparum.*

The mosquito (definitive host) sucks blood from a person with malaria and acquires gametocytes, the form of the protozoan found in the blood (**FIGURE 18.13**). Within the insect sexual reproduction occurs and a transition to sporozoites takes place. The sporozoites then migrate to the salivary gland.

Sporozoite infection causes the female *Anopheles* mosquito to increase its biting frequency. When the mosquito bites another human (intermediate host), several hundred sporozoites enter the person's bloodstream and quickly migrate to the liver. After several hours, the transformation of sporozoites to **merozoites** is completed, and the merozoites emerge from the liver to invade the red blood cells (RBCs). This triggers the blood infection stage and the pathology characteristic of malaria.

While in the RBCs, the merozoites can synthesize any of about 150 proteins that attach to RBC membranes and cause the RBCs to cluster in the blood vessels. By constantly switching among 150 genes (for 150 proteins), the malarial parasite avoids detection by the body's immune system.

Within RBCs, the merozoites undergo another series of transformations resulting in several gametocytes and thousands of new merozoites being formed. In response to a biochemical signal, thousands of RBCs rupture simultaneously releasing the parasites and their toxins in a 48 to 72 hour cycle.

Now the excruciating malaria attack begins. First, there is intense cold, with shivers and chattering teeth. The temperature then rises rapidly to 40°C, and the sufferer develops intense fever, headache, and delirium. After two or three hours, massive perspiration ends the hot stage, and the patient often falls asleep, exhausted. During this quiet period, the merozoites enter a new set of RBCs and repeat the cycle of transformations.

Death from malaria is due to a number of factors related to the loss of red blood cells. Substantial anemia develops, and the hemoglobin from ruptured blood cells darkens the urine; malaria is, therefore, sometimes called **black-water fever**. Cell fragments and RBC clustering accumulate in the small vessels of the brain, kidneys, heart, liver,

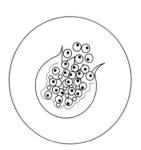

Plasmodium falciparum

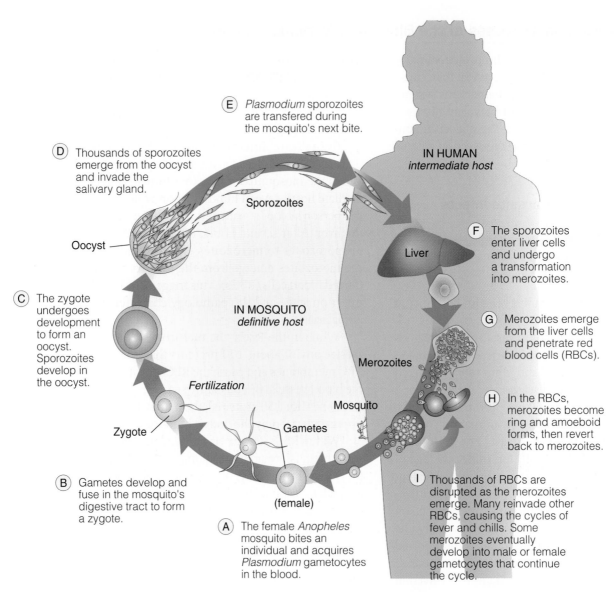

E *Plasmodium* sporozoites are transfered during the mosquito's next bite.

D Thousands of sporozoites emerge from the oocyst and invade the salivary gland.

IN HUMAN
intermediate host

Sporozoites

Oocyst

Liver

F The sporozoites enter liver cells and undergo a transformation into merozoites.

C The zygote undergoes development to form an oocyst. Sporozoites develop in the oocyst.

IN MOSQUITO
definitive host

G Merozoites emerge from the liver cells and penetrate red blood cells (RBCs).

Merozoites

Fertilization

Zygote

Mosquito

Gametes

H In the RBCs, merozoites become ring and amoeboid forms, then revert back to merozoites.

B Gametes develop and fuse in the mosquito's digestive tract to form a zygote.

(female)

I Thousands of RBCs are disrupted as the merozoites emerge. Many reinvade other RBCs, causing the cycles of fever and chills. Some merozoites eventually develop into male or female gametocytes that continue the cycle.

A The female *Anopheles* mosquito bites an individual and acquires *Plasmodium* gametocytes in the blood.

FIGURE 18.13 **Malaria and the *Plasmodium* Life Cycle.** The propagation of the *Plasmodium* parasite requires two hosts, mosquitoes and humans. »» Why are two hosts required to complete the *Plasmodium* life cycle?

and other vital organs and cause clots to form. Heart attacks, cerebral hemorrhages, and kidney failure are common.

Since its discovery around 1640, quinine has been the mainstay for treating malaria. During World War II, American researchers developed the drug chloroquine, which remained an important mode of therapy until recent years, when drug resistance began emerging in *Plasmodium* species. Since 1989, an alternative drug called mefloquine has been recommended for individuals entering malaria regions of the world. However, serious medical side effects, including cognitive functioning problems, have been associ-

ated with some people taking the drug. Another drug, artemisinin, is effective in curing malaria especially when combined with other drugs to limit the development of drug resistance. Many other drugs are in clinical trials. Experimental vaccines directed against the sporozoite or merozoite stage are being tested.

Though progress has been made in the control of malaria, the mortality and morbidity figures remain appallingly high. MicroFocus 18.4 looks at a program to reduce this "malaria burden."

CONCEPT AND REASONING CHECKS

18.6 Summarize the life cycle of the *Plasmodium* parasite.

MICROFOCUS 18.4: Public Health
Roll Back Malaria

The World Health Organization (WHO) estimates there were 247 million cases of malaria in 2006, resulting in almost 1 million deaths.

The United Nations General Assembly has designated 2001–2010 the decade to "Roll Back Malaria" in developing nations. Roll Back Malaria is a global partnership designed to cut by 50% the world's malaria burden by 2010. The program will enable countries to take effective and sustained action against malaria by providing their citizens with rapid and effective treatment, and attempting to prevent and control malaria in pregnant women.

Because over 70% of all malarial deaths occur in children under five years of age, a child's most vulnerable period for contracting malaria starts at six

Children in Guayoquil, Ecuador, receiving malaria tablets.

months, when the mother's protective immunity wears off and before the infant has established its own fully functional immune system. During this window of vulnerability, a child's condition can deteriorate quickly and the child can die within 48 hours after the first symptoms appear. Therefore, a key part of Roll Back Malaria is aimed at children (see figure) and a series of interim goals have been put forward. They propose that:

- At least 60% of those suffering from malaria should have access to inexpensive and correct treatment within 24 hours of the onset of symptoms.
- At least 60% of those at risk of malaria, particularly pregnant women and children less than five years old, should have access to suitable personal and community protective measures, such as insecticide-treated mosquito nets.
- At least 60% of all pregnant women who are at risk of malaria, especially those in their first pregnancies, should receive intermittent preventive treatment.

Through worldwide partnerships, the Roll Back Malaria program will allow countries that experience a high malaria burden to take effective and sustainable action to halve the malaria burden by 2010.

The *Trypanosoma* Parasites Can Cause Life-Threatening Systemic Diseases

KEY CONCEPT

7. Trypanosomiasis embodies African sleeping sickness and Chagas disease.

Trypanosomiasis is a general name for two diseases caused by parasitic species of the kinetoplastid *Trypanosoma* (FIGURE 18.14A). The two diseases caused by trypanosomes are traditionally known as human African sleeping sickness and Chagas disease.

Human African sleeping sickness is endemic in 36 African countries and exerts a level of mortality greater than that of HIV disease/AIDS. The WHO estimates there are more than 500,000 new cases every year with some 65,000 deaths.

Trypanosoma cycles between humans and the tsetse fly (FIGURE 18.14B). The insect bites an infected patient or animal, and the trypanosomes localize in the insect's salivary gland. After a two-week development, transmission occurs during a bite. The point of entry becomes painful and swollen in several days and a chancre similar to that in syphilis is observed. Invasion of, and multiplication in, the bloodstream follows (stage 1). It then spreads to the central nervous system (stage 2).

Two types of African sleeping sickness exist. A chronic form, common in central and western Africa, is caused by *Trypanosoma brucei* variety *gambiense*. It is accompanied by chronic bouts of fever, as well as severe headaches, changes in sleep patterns and behavior, and a general wasting away. As the trypanosomes invade the brain, the patient slips into a coma (hence the name, "sleeping

Trypanosoma brucei

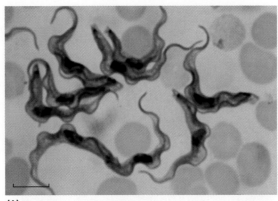

Babesia microti

FIGURE 18.14 *Trypanosoma*. **(A)** A light micrograph of stained *Trypanosoma* among red blood cells. (Bar = 5 μm.) **(B)** The tsetse fly, a vector for *Trypanosoma*. »» What is unique about the *Trypanosoma* cells?

sickness"). The second form, common in eastern and southern Africa, is due to *Trypanosoma brucei* variety *rhodesiense*. The disease is more acute with high fever and rapid coma preceding death.

Prevention involves clearing brushlands and treating areas where the tsetse flies breed. Patients are treated with either the drug pentamidine, melarsoprol, or eflornithine, depending on the form and stage of the disease.

American trypanosomiasis or **Chagas disease** is found in Mexico and 17 countries in Central and South America. A recent estimate put the number of cases in South and Central America at 18 million with approximately 200,000 new cases every year and some 50,000 deaths.

Chagas disease is caused by *Trypanosoma cruzi* and is transmitted by the reduviid bug. The insect feeds at night and bites where the skin is thin, such as on the forearms, face, and lips. For this reason, it is called the "kissing bug."

Once in the blood, the trypanosomes invade many cell types and undergo multiple rounds of binary fission. During this acute phase, the individual experiences high parasite numbers even though most infections are asymptomatic. Following the acute phase, the parasite number declines. In 20% to 30% of infected individuals, a chronic irreversible disease occurs that in 10 to 30 years can develop clinical symptoms, which vary by geographical region. Individuals may experience widespread tissue damage including intestinal tract abnormalities and extensive cardiac nerve destruction that is so thorough the victim experiences sudden heart failure. Benznidazole and nifurtimox have proved useful for acute disease; no effective drug is available for chronic infections.

In 2006, two American heart transplant patients developed acute Chagas disease after receiving donor hearts. So in January 2007, the American Red Cross and Blood Systems started screening the U.S. blood supply for Chagas disease and *T. cruzi*. About one in 30,000 donors test positive.

CONCEPT AND REASONING CHECKS

18.7 What is the vector difference between African and American trypanosomiasis?

Babesia Is an Apicomplexan Parasite

KEY CONCEPT

8. Babesiosis is a malaria-like disease.

Babesiosis, found in the northeastern United States, is a malaria-like disease caused by *Babesia microti*. The protozoa live in ticks of the genus *Ixodes*, and are transmitted when these arthropods feed in human skin. Areas of coastal Massachusetts, Connecticut, and Long Island, New York, have experienced outbreaks in recent years.

Babesia microti penetrates human red blood cells. As the cells disintegrate, a mild anemia develops. Piercing headaches accompany the disease and, occasionally, meningitis occurs. A suppressed immune system appears to favor establishment of the disease. However, babesiosis is rarely fatal and drug therapy is not recommended. Carrier conditions may develop in recoverers, and spread by blood transfusion is possible. Travelers returning from areas of high incidence therefore are advised to wait several weeks before donating blood to blood banks. Tick control is considered the best method of prevention.

Babesia has a significant place in the history of American microbiology. In the late 1800s, Theobald Smith located *B. bigemina* in the blood of cattle suffering from Texas fever. His report was one of the first linking protozoa to disease, and, in part, it necessitated that the then-prevalent "bacterial" theory of disease be modified to include eukaryotic microorganisms.

CONCEPT AND REASONING CHECKS

18.8 Why is babesiosis considered to be a malaria-like disease?

Toxoplasma Causes a Relatively Common Blood Infection

KEY CONCEPT

9. Toxoplasmosis is extremely contagious.

Toxoplasmosis affects up to 50% of the world's population, including 50 million Americans. Thus, the causative agent, *Toxoplasma gondii,* is regarded as a universal parasite. Some researchers believe it is the most common parasite of humans and other vertebrates.

T. gondii exists in three forms: the trophozoite, the cyst, and the oocyst. Trophozoites are crescent-shaped or oval organisms without flagella (**FIGURE 18.15**). Located in tissue during the acute stage of disease, they force their way into all mammalian cells, with the notable exception of erythrocytes. To enter cells, the parasites form a ring-shaped structure on the host cell membrane and then pull the membrane over themselves, much like pulling a sock over the foot. Cysts develop from the trophozoites within host cells and may be the source of repeated infections. Muscle and nerve tissue are common sites of cysts. Oocysts are oval bodies that develop from the cysts by a complex series of asexual and sexual reproductive processes.

This apicomplexan can exist in nature in the cyst and oocyst forms. Grazing animals acquire these forms from the soil and pass them to humans via contaminated beef, pork, or lamb (**FIGURE 18.16**). Rare hamburger meat is a possible source. Domestic cats acquire the cysts from the soil or from infected birds or rodents. Oocysts then form in the cat. Humans are exposed to the oocysts when they forget to wash their hands after

FIGURE 18.15 *Toxoplasma gondii.* A false-color scanning electron micrograph of numerous crescent-shaped *T. gondii* trophozoites (orange). (Bar = 5 μm.) »» What infective stage is represented by these cells?

Toxoplasma gondii

contacting cat feces while changing the cat litter or working in the garden. Touching the cat also can bring oocysts to the hands, and contaminated utensils, towels, or clothing can contact the mouth and transfer oocysts.

Toxoplasmosis develops after trophozoites are released from the cysts or oocysts in the host's gastrointestinal tract. *T. gondii* rapidly invades the intestinal lining and spreads throughout the body via the blood. However, for most healthy individuals, the parasite causes no serious illness even though one remains infected for life. The highly contagious nature of the parasite stems from its need to get back into its definitive host, the cat, where oocysts are produced.

Pregnant women, however, are at risk of developing a dangerous toxoplasmosis infection because the protozoa may cross the placenta and infect the fetal tissues (Figure 18.16). Neurologic damage, lesions of the fetal visceral organs, or spontaneous abortion may result. Congenital infection is least likely during the first trimester, but damage may be substantial when it occurs. By contrast, congenital infection is more common if

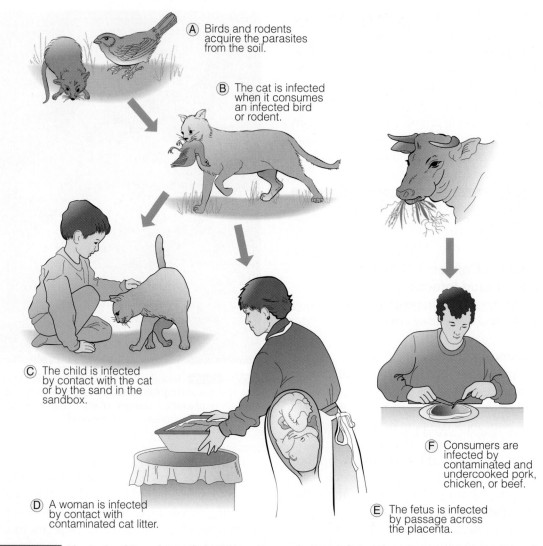

FIGURE 18.16 **The Cycle of Toxoplasmosis in Nature.** Humans become infected through contact with the feces of an infected cat or consumption of undercooked, contaminated beef. »» What form of the *Toxoplasma* parasite is passed to humans?

A Birds and rodents acquire the parasites from the soil.

B The cat is infected when it consumes an infected bird or rodent.

C The child is infected by contact with the cat or by the sand in the sandbox.

D A woman is infected by contact with contaminated cat litter.

E The fetus is infected by passage across the placenta.

F Consumers are infected by contaminated and undercooked pork, chicken, or beef.

the woman is infected in the third trimester, but fetal damage is less severe. Lesions of the retina are the most widely documented complication in congenital infections. The **T** in the **TORCH** group of diseases refers to toxoplasmosis (the others are rubella, cytomegalovirus, and herpes simplex; O is for other diseases, such as syphilis).

T. gondii also is known to cause severe disease in immunosuppressed individuals. For example, in patients with AIDS the parasite attacks the brain tissue, causing an inflammation and swelling that often results in cerebral lesions, seizures, and death.

CONCEPT AND REASONING CHECKS

18.9 Assess the impact of *Toxoplasma* to pregnant women and immunocompromised people.

Naegleria Can Infect the Central Nervous System

KEY CONCEPT

10. Primary amoebic meningoencephalitis (PAM) is a rare but deadly infection of the brain.

Primary amoebic meningoencephalitis (PAM) is a rare disease with less than 200 cases reported worldwide since 1965. However, it is the most deadly disease of the central nervous system after rabies. Ninety-five percent of patients die within 4 to 5 days of infection (TEXTBOOK CASE 18).

PAM is caused by several species of thermophilic parasites in the genus *Naegleria*, especially *N. fowleri* (FIGURE 18.17). It also can be caused by *Acanthamoeba* and *Hartmannella* species.

Naegleria fowleri

Textbook CASE 18

Primary Amoebic Meningoencephalitis—Georgia 2002

1 In late August, a previously healthy boy was evaluated in a local emergency department for a 2-day history of headache and vomiting; he had a fever and was lethargic. There were no signs of meningitis, and a lumbar puncture identified no infectious organism. Four days before onset of illness, the patient had attended a social event and had swum in a freshwater river with a group of friends in southern Georgia.

2 The patient was started on intravenous antibiotics for suspected bacterial meningitis. Within several hours of admission, he had spontaneous nonpurposeful movements, was unable to follow verbal commands, and was transferred to a children's hospital intensive care unit (ICU). En route to the ICU, he had a 30-minute right-sided seizure. A computed tomography (CT) scan of the head on admission to the ICU showed edema of the midbrain, and cranial magnetic resonance imaging (MRI) demonstrated areas of meningeal enhancement in the brainstem suggestive of meningitis.

3 No organisms were observed on a Gram-stained smear of cerebrospinal fluid (CSF); CSF antigen-detection tests were negative for bacterial pathogens. Fresh preparation of CSF revealed: no amebae, a red blood cell count of 1,550/mm^3 (normal: 0/mm^3); and a white blood cell count of 13,650/mm^3 (normal: 0-5/mm^3).

4 Approximately 12 hours after admission to the ICU, the patient had episodes when his breathing stopped and was tracheally intubated. Treatment included hyperventilation, hypertonic sodium chloride infusion, mannitol infusion, and surgically placing a tube through the skull to monitor the intracranial pressure as well as to drain CSF.

5 The patient's condition worsened, with progressive neurologic deterioration. On the fourth hospital day, the patient died. A postmortem lumbar puncture demonstrated a few motile amoebae.

6 Autopsy findings revealed acute primary amoebic meningoencephalis (PAM) caused by *N. fowleri* (see figure).

7 An epidemiologic investigation discovered that the boy was one of five children who had participated in water activities that included swimming, swimming under water, wrestling in the water, and diving into the water.

8 The environmental investigation revealed a high ambient temperature (>90°F [>32°C]) and water temperature (91°F [33°C]) in the river at the time of the exposure. In addition, the river level was low and the river was flowing slowly. Bacteriologic testing of the river water demonstrated that fecal coliform levels were within acceptable limits.

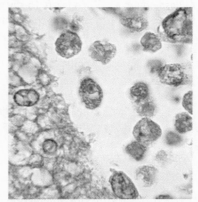

A light photomicrograph of brain tissue showing the infection with *Naegleria fowleri*.

Questions:

(Answers can be found in Appendix D.)

A. Would the absence of microbes in the initial lumbar puncture mean the young boy did not have meningitis? Explain.

B. Why was a follow-up lumbar puncture ordered?

C. How might the patient have become infected with the protozoal parasite?

D. Why were no other children infected?

E. What precautions should be taken to prevent another incident?

For additional information see http://www.cdc.gov/mmwr/preview/mmwrhtml/mm5240a4.htm.

Meningoencephalitis:
An inflammation of the
brain and meninges.

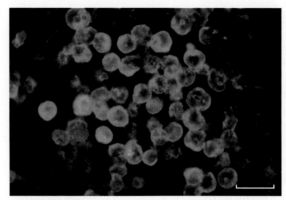

FIGURE 18.17 *Naegleria fowleri*. A fluorescent antibody stain of *N. fowleri* amoebae in tissue from a brain autopsy. (Bar = 40 μm.) »» What is meant by fluorescent antibody staining?

Naegleria is an opportunistic pathogen of humans, causing meningoencephalitis when inhaled, often by swimming in contaminated warm surface water. The free-living trophozoites appear to enter the body through the mucous membranes of the nose and then follow the olfactory tracts to the brain.

The symptoms resemble those in other forms of encephalitis and meningitis. Nasal congestion precedes piercing headaches, fever, delirium, neck rigidity, and occasional seizures if the victim is not treated with amphotericin B combined with miconazole and rifampin.

TABLE 18.3 summarizes the protozoal diseases of the blood and nervous system.

CONCEPT AND REASONING CHECKS

18.10   What similarities exist between *Naegleria* and *Entamoeba*?

TABLE 18.3   A Summary of Protozoal Diseases of the Blood and Nervous System

Disease	Causative Agent	Signs and Symptoms	Transmission	Treatment	Prevention and Control
Malaria	4 *Plasmodium* species	Moderate to severe shaking chills, high fever, profuse sweating with chills, malaise	Bite of infected *Anopheles* mosquito	Chloroquine, quinine, mefloquine, malarone	Mefloquine, malarone
African trypanosomiasis (African sleeping sickness)	*Trypanosoma brucei*	Chronic bouts of fever, severe headache, change in sleep patterns and behavior	Bite of infected tsetse fly	Pentamidine, melarsoprol, eflornithine	Clearing areas where tsetse flies breed
Chagas disease (American trypanosomiasis)	*Trypanosoma cruzi*	Acute phase: Redness and swelling at the site of infection, asymptomatic Chronic phase: widespread tissue damage	Bite of an infected reduviid bug	Acute phase: benznidazole and nifurtimox Chronic phase: depends on signs and symptoms	Avoiding endemic residences Insecticides
Babesiosis	*Babesia microti*	Mild anemia, piercing headache	Bite of infected tick	Azithromycin and atovaquone	Avoiding ticks in endemic areas
Toxoplasmosis	*Toxoplasma gondii*	Usually asymptomatic	Contact with contaminated cat feces, ingestion of contaminated food, water, or fomites	Pyrimethamine and sulfadiazine for acute disease	Practicing good hygiene Avoiding raw or undercooked meat and unpasteurized goat's milk
Primary amoebic meningoencephalitis (PAM)	*Naegleria fowleri*	Piercing headache, fever, delirium, neck rigidity	Inhalation of contaminated water	Early with amphotericin B, miconazole, and rifampin	Avoiding swimming in contaminated water

18.4   The Multicellular Helminths and Helminthic Infections

The **helminths** are among the world's most common animal parasites. For example, 2 billion people—approximately 33% of the human population—are infected with soil-transmitted helminths! Therefore, in concluding this chapter on the parasites, we are concerned with medical helminthology and the diseases caused by the parasitic worms.

As mentioned in the introduction to this chapter, such parasites are of interest to microbiologists because of the parasites' ability to cause an enormous level of morbidity worldwide. However, different from many of the viral and bacterial pathogens we have discussed, most parasitic helminths are dependent on the host or hosts for sustenance, so it is to the helminth's benefit that the host stays alive. Therefore, the helminths tend to cause diseases of debilitation and chronic morbidity resulting from physical factors related to the **helminthic load** (number of worms present) or location in the body.

There Are Two Groups of Parasitic Helminths

KEY CONCEPT

11. Parasitic helminths include the flatworms and the roundworms.

The helminths of medical significance are the flatworms and the roundworms.

Flatworms. Animals in the phylum Platyhelminthes (*platy* = "flat"; "*helmin*" = "worm") are the **flatworms**. As multicellular animals, they have tissues functioning as organs in organ systems. However, they have no specialized respiratory or circulatory structures, and they lack a digestive tract. The gut (gastrovascular cavity) simply consists of a sac with a single opening, thus placing the worm in close contact with its surroundings. Complex reproductive systems are found in many species within the phylum, and a large number of species are hermaphroditic. Two groups of flatworms are of concern regarding human disease.

The **trematodes**, includes the **flukes**, which have flattened, broad bodies (**FIGURE 18.18A**). The animals exhibit bilateral symmetry. Trematodes have a complex life cycle that may include encysted egg stages and temporary larval forms. Sucker devices are commonly present to enable the parasite to attach to its host. In many cases, two hosts exist: an intermediate host, which harbors the larval form, and a definitive host, which harbors the mature adult form. In this chapter, we shall be mostly concerned with parasites whose definitive host is a human.

The life cycle of a fluke often contains several phases. In the human host, the parasite

Hermaphroditic: Refers to an organism having both male and female reproductive organs.

(A)

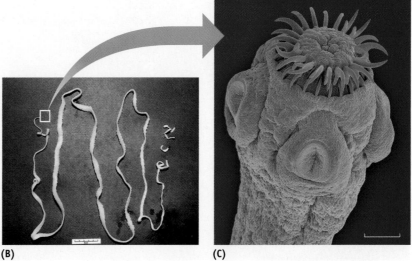
(B)   (C)

FIGURE 18.18 **Flukes and Tapeworms.** **(A)** Light micrograph of a liver fluke, *Fasciola hepatica,* a parasite infecting sheep, cattle, and humans. (Bar = 50 μm.) **(B)** Photograph of a coiled beef tapeworm, *Taenia saginata,* which has grown to several meters in length. **(C)** A false-color scanning electron micrograph of the head, scolex, of *T. saginata* showing the suckers and hooks. (Bar = 1 mm.) »» What does the dark branched structures form in the liver fluke (A) and what is the purpose of the suckers and hooks on the scolex head of the tapeworm (C)?

produces fertilized eggs generally released in the feces. When the eggs reach water, they hatch and develop into tiny ciliated larvae called **miracidia** (sing., miracidium). The miracidia penetrate snails (the intermediate host) and go through a series of asexual reproductive stages. The cyst makes its way back to humans.

The trematode life style requires the parasite to evade the host's immune system. It accomplishes this by having its surface resemble the surface of the host cells, so the immune system "sees" the worm as a "normal" cell, not an invader. This mimicry is quite effective as some flukes can remain in a human host for 40 years or more.

The other group of flatworms is the **cestodes**, which includes the **tapeworms**. These parasitic worms have a head region, called the **scolex**, and a ribbon-like body consisting of segments called **proglottids** (FIGURE 18.18B, C). The proglottids most distant from the scolex are filled with fertilized eggs. As the proglottids break free, they spread the eggs.

Tapeworms generally live in the intestines of a host organism. In this environment, they are constantly bathed by nutrient-rich fluid, from which they absorb food already digested by the host. Tapeworms have adapted to a parasitic existence and have lost their intestines, but they still retain well-developed muscular, excretory, and nervous systems.

Tapeworms are widespread parasites infecting practically all mammals, as well as many other vertebrates. Because they are more dependent on their hosts than flukes, tapeworms have precarious life cycles. Tapeworms have a limited range of hosts, and the chances for completing the cycle are often slim. With rare exceptions, tapeworms require at least two hosts. Humans often become infected by eating undercooked meat containing tapeworm cysts, which then develop into mature adult worms.

Roundworms. Among the most prevalent animals are the **roundworms** in the phylum Nematoda (*nema* = "thread"). These parasites have a thread-like body and occupy every imaginable habitat on Earth. They live in the sea, in freshwater, and in soil from polar regions to the tropics. Good topsoil, for example, may contain billions of nematodes per acre. They parasitize every conceivable type of plant and animal, causing both economic crop damage and serious disease in animals. Yet, they may even have a beneficial effect for humans (MicroFocus 18.5).

Roundworms have separate sexes. Following fertilization of the female by the male, the eggs hatch to larvae that resemble miniature adults. Growth then occurs by cellular enlargement and mitosis. Damage in hosts is generally caused by large worm burdens in the blood vessels, lymphatic vessels, or intestines (FIGURE 18.19). Also, the infestation may result in nutritional deficiency or damage to the muscles.

CONCEPT AND REASONING CHECKS

18.11 Compare the body plan of the flatworms with that of the roundworms.

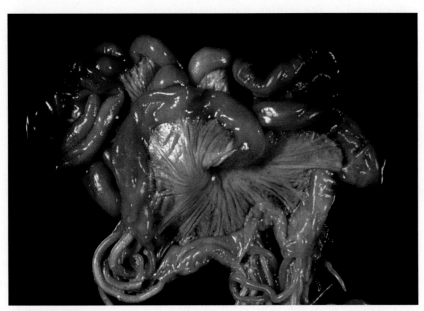

FIGURE 18.19 Roundworms. Photograph of threadworms seen on the surface of, and burrowing into, the gut wall of a pig. »» How do the nematodes differ from the flatworms?

Several Trematodes Can Cause Human Illness

KEY CONCEPT

12. Schistosomiasis, and human lung and liver fluke diseases, are due to trematode infections.

Schistosomiasis. The WHO estimates that 200 million people in 74 countries suffer from schistosomiasis, which kills approximately 200,000 every year. There are even about 400,000 individuals in the United States who suffer from a mild form of the disease.

MICROFOCUS 18.5: Being Skeptical
Eat Worms and Cure Your Ills—and Allergies!

Doctors and researchers have shown that eating worms (actually drinking worm eggs) can fight disease. What! Are you nuts? Drink worm eggs!

Joel Weinstock, a gastroenterologist at the University of Iowa, discovered that as allergies and other diseases have increased in Western countries, infections by roundworm parasites have declined. This is not the case in other countries where allergies are rare and worm infections are quite common. Weinstock wondered if there was a correlation between allergy increase and parasite decline.

To test his hypothesis, Weinstock "brewed" a liquid concoction consisting of thousands of pig whipworm eggs (ova). The whipworms are called *Trichuris*

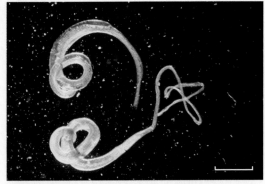

Adult whipworms (male top; female bottom). (Bar = 1 cm.)

suis, so his product with ova is called TSO. This roundworm (see figure) was chosen because once the ova hatch, they will not survive long in the human digestive system and will be passed out in the feces.

One woman in Iowa was suffering from incurable ulcerative colitis, which is caused by the immune system overreacting. Immune cells start attacking the person's own gut lining, making it bleed. The symptoms are severe cramps and acute, intense, diarrhea. In a trial run, Weinstock gave the woman a small glass full of his TSO. Every three weeks she downed another glass. Guess what? Her ulcerative colitis is in remission and she no longer suffers any disease symptoms.

Further trials in 2004 involved 100 people suffering the same disease and a further 100 suffering Crohn disease, which is another type of inflammatory bowel disease related to immune function. In this study, 50% of the volunteers suffering ulcerative colitis and 70% of those suffering Crohn disease went into remission, as identified by no symptoms of abdominal pain, ulcerative bleeding, and diarrhea.

Weinstock believes some parasites are so intimately adapted with the human gut that if they are eradicated, bad things may happen, such as the bowel disorders mentioned. He says the immune system has become so involved with defending against parasites that if you take them away, the immune system overreacts to other events.

Is there something to this? Alan Brown, an academic researcher in the United Kingdom (UK) picked up a hookworm infection while on a field trip outside the UK. Being that he was a well-nourished Westerner, the 300 hookworms in his gut caused no major problem. However—since being infected, his hayfever allergy has disappeared!

The result: I need more evidence before I would drink worm eggs for a gastrointestinal disease.

Schistosomiasis is caused by several species of blood flukes, including *Schistosoma mansoni* (Africa and South America), *S. japonicum* (Asia), and *S. haematobium* (Africa and India). In some regions, the term **bilharziasis** is still used for the disease; it comes from the older name for the genus, *Bilharzia.*

Species of *Schistosoma* measure about 10 mm in length. The eggs hatch in freshwater to produce miracidia, which then make their way to snails. In the snails, miracidia convert to a second larval form called **cercariae** (FIGURE 18.20).

The cercariae escape from the snails and attach themselves to the bare skin of humans wading in contaminated water. Cercariae infect the blood and mature into adult flukes, which cause fever and chills. However, the major effects of disease are due to eggs: carried by the bloodstream to the liver, they cause substantial liver damage; in the intestinal wall, ulceration, diarrhea, and abdominal pain occur; and in the bladder, egg infection causes bloody urine and pain on urination. Male and female species mate in the human liver and produce eggs that are released in the feces.

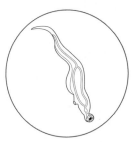

Schistosoma mansoni

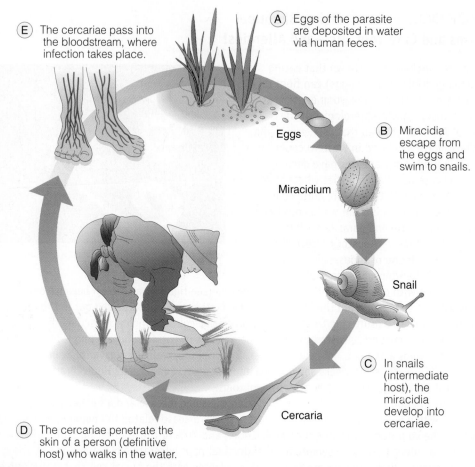

E The cercariae pass into the bloodstream, where infection takes place.

A Eggs of the parasite are deposited in water via human feces.

Eggs

B Miracidia escape from the eggs and swim to snails.

Miracidium

Snail

C In snails (intermediate host), the miracidia develop into cercariae.

Cercaria

D The cercariae penetrate the skin of a person (definitive host) who walks in the water.

FIGURE 18.20 **The Life Cycle of the Blood Fluke *Schistosoma mansoni*.** The life cycle of the trematode *S. mansoni* involves an intermediate host, the snail, and a definitive host, a human. »» What defines a definitive and intermediate host?

The antihelminthic drug praziquantel is used for treatment. MicroFocus 16.1 discussed how treatment for schistosomiasis spread hepatitis C through much of Egypt.

Certain species of *Schistosoma* penetrate no farther than the skin because the definitive hosts are birds instead of humans. Often after swimming in schistosome-contaminated water, the cercariae penetrate the skin but are attacked and destroyed by the body's immune system. However, they release allergenic substances that cause an itching and body rash, commonly known as **swimmer's itch**. The condition, not a serious threat to health, is common in northern lakes in the United States.

MicroInquiry 18 examines some amazing phenomena of how parasitic flukes can manipulate their host for their own benefit—and survival.

CONCEPT AND REASONING CHECKS

18.12   Summarize how schistosomiasis causes such great morbidity worldwide.

Tapeworms Survive in the Human Intestines

KEY CONCEPT

13. Tapeworm diseases result from eating undercooked meat contaminated with cysts.

Beef and Pork Tapeworm Diseases. Approximately 50,000 people die each year from beef and pork tapeworm diseases. Humans are the definitive hosts for both the beef tapeworm *Taenia saginata* and the pork tapeworm *Taenia solium*. Humans acquire the tapeworm cysts by eating poorly cooked beef or pork. The beef

MICROFOCUS 18.6: Public Health

Man Leaps Off Speeding Car, Report Says

It was truly a bizarre death. The chief financial officer (CFO) for the city of Phoenix, Arizona, was a very professional, quiet man who had a way with numbers. He cared about others and worked with the underprivileged. In fact, he and his wife had formed a foundation to help working women and children in need find health insurance. Life was good.

So, why in 2004, wearing tattered jeans and no shoes, would he climb out the window of his car going 55 mph on a crowded street, stand on top of the car with his arms outstretched—and then jump to his death?

A few days before his death, the finance officer had taken a sick day to see his doctor complaining of feeling "tired and worn out." But other than that, there was no indication of any abnormal behavior.

However, it happens that two years earlier, he had fallen ill after returning from a trip to Mexico. At that time, nobody outside of the family knew of his illness.

According to his wife, he had been suffering from cysticercosis, an infection caused by the larval stage of the pork tapeworm *Taenia solium*. He had been taking medication for the parasitic infection and had been considering changing the dosage. In fact, after his death it was discovered he had been having flare-ups of the parasite.

Apparently, the CFO had been infected in Mexico probably from eating vegetables or fruits contaminated with pig feces containing the parasite. On very rare occasions, the larvae can move to the brain and cause frontal lobe disinhibition, which means that they can make an individual do bizarre things, including impairing one's decision-making abilities.

So, it appears, like MicroInquiry 18 describes, parasites can cause strange behaviors. In this case, the result was a very sad and tragic consequence.

tapeworm may reach 10 meters in length, while the pork tapeworm length is 2 to 8 meters.

Attachment via the scolex occurs in the small intestine and obstruction of this organ may result. In most cases, however, there are few symptoms other than mild diarrhea, and a mutual tolerance may develop between parasite and host. Each tapeworm may have up to 2,000 proglottids and infected individuals will expel numerous gravid proglottids daily. The proglottids accumulate in the soil and are consumed by cattle or pigs. Embryos from the eggs travel to the animal's muscle, where they form cysts.

In rare instances, infection can lead to very bizarre behaviors (MICROFOCUS 18.6).

Echinococcosis. Dogs and other canines such as wolves, foxes, and coyotes are the definitive hosts for dog tapeworms belonging to the genus *Echinococcus* (FIGURE 18.21A). Eggs reach the soil in feces and spread to numerous intermediary hosts, one of which is humans. Contact with a dog also may account for transmission. In humans, the parasites travel by the blood to the liver, where they form thick-walled **hydatid cysts** (FIGURE 18.21B). Common symptoms include abdominal and chest pain, and coughing up blood.

CONCEPT AND REASONING CHECKS

18.13 Draw simple life cycles for *Taenia* and *Echinococcus* parasites.

Humans Are Hosts to at Least 50 Roundworm Species

KEY CONCEPT

14. Roundworm diseases affect the digestive system and muscles.

Here, we discuss a few significant parasitic infections caused by roundworms.

Pinworm Disease. The most prevalent helminthic infection in the United States is **pinworm disease**, where an estimated 30% of children and 16% of adults serve as hosts.

Pinworm disease is caused by *Enterobius vermicularis*. The male and female worms live in the distant part of the small intestine and in the large

Taenia saginata

Echinococcus granulosus

■ Gravid:
Refers to carrying eggs.

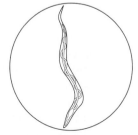

Enterobius vermicularis

MICROINQUIRY 18

Parasites as Manipulators

Over the last several chapters, we have been discussing viral, bacterial, and fungal pathogens, and protozoan and multicellular parasites. For the latter parasites, you might conclude they make their living "sponging off" their hosts with the ultimate aim to reproduce. In fact, for a long time scientists thought these organisms simply evolved so they could take advantage of what their hosts had to offer. In reality, the parasites may be quite a bit more sophisticated than previously thought. Two cases involving multicellular parasites are submitted for your inquiry.

Case 1

The lancet fluke, *Dicrocoelium dendriticum*, infects cows. Because the infected cow is the definitive host, the fluke needs to get out of the cow to form cysts in another animal. If the cysts can then be located on grass blades, another cow may eat those cyst-containing grass blades and another round of infection by *D. dendriticum* can occur.

18.1a. As a parasite, how do you ensure cysts will be on grass blades that cows eat?

Infected cows excrete dung containing fluke eggs. Snails (intermediate host #1) forage on the dung and in the process ingest the fluke eggs. The eggs hatch and bore their way from the snail gut to the digestive gland where fluke larvae are produced. In an attempt to fight off the infection, the snail smothers the parasites in balls of slime coughed up into the grass.

Ants (intermediate host #2) come along and swallow these nutrient-rich slime balls. Now, here is the really interesting part, because the lancet fluke has yet to get cysts onto grass blades so they will be eaten by more cows. In infected ants, the larvae travel to the ant's head and specifically the nerves that control the ant's mandibles. While most of the larvae then return to the abdomen where they form cysts, a few flukes remain in the head and control the ant's behavior. As evening comes and the temperature cools, the head-infected ants with

a belly full of cysts leave their colony, climb to the tip of grass blades, and hold on tight with their mandibles (see figure). The lancet fluke has taken control of the ants and placed them in a position most favorable to be eaten by a grazing cow.

18.1b. What happens if no cow comes grazing that evening?

If a grazing cow does not come by, the next morning the flukes release their influence on the ants, which return to their "normal duties." But, the next evening, the flukes again take control and the ants march back up to the tips of grass blades. Should a cow eat the grass with the ants, the fluke cysts in the ant abdomen quickly hatch and another cycle of reproduction begins in the cow. Now, that is quite a behavioral driving force to control your destiny; in this case, to ensure fluke cysts are positioned so they will be eaten by the cow.

Case 2

(This case is based on research carried out in a coastal salt marsh by Kevin Lafferty's group at the University of California at Santa Barbara.) Another fluke, *Euhaplorchis californiensis*, uses shorebirds as its definitive host. Infected birds drop feces loaded with fluke eggs into the marsh. Horn snails (intermediate host) eat the droppings containing the eggs. The eggs hatch and castrate the snails. Larvae are produced in the water and latch onto the gills of their second intermediate host, the California killifish. In the killifish, the larvae move from the blood vessels to a nerve that carries the larvae to the brain where the larvae form a thin layer on top of the brain. There they stay, waiting for the fish to be eaten by a shorebird. Once in the bird's gut, the adults develop and produce another round of fertilized eggs. Here, we have two questions.

18.2a. What is the purpose for the larvae castrating the snails?

An experiment was set up to answer this question. Throughout the salt marsh,

An ant clutching a grass blade by its feet and mandibles.

Lafferty set up cages with uninfected snails and other cages with infected ones. The results were as expected—the uninfected snails produced many more offspring than did the infected snails. Lafferty believes a marsh full of uninfected snails would reproduce so many offspring as to deplete the algae and increase the population of crabs that feed on the snails. By castrating the snails, the parasite actually is controlling the snail population and keeping the salt marsh ecosystem balanced—again for its benefit.

18.2b. Why do the larvae in the killifish take up residence on top of the fish's brain?

Through a series of experiments, Lafferty's group discovered that infected fish underwent a swimming behavior not observed in the uninfected fish. The infected killifish darted near the water surface, a very risky behavior that makes it more likely the fish would be caught by shorebirds. In fact, experiments showed infected fish were 30 times more likely to be caught by shorebirds than uninfected fish. Now, from the parasite's perspective, they want to be caught so another round of reproduction can start in the shorebird. So again, the fluke has manipulated the situation to its benefit.

Parasites may cause disease, but they also maximize their chances for ensuring infection. Pretty amazing!

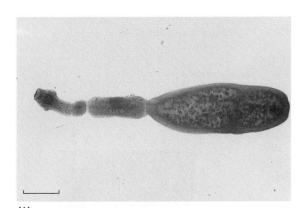
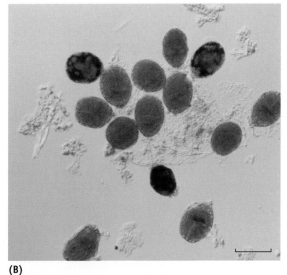

(A)   **(B)**

FIGURE 18.21 The Dog Tapeworm. (**A**) A light micrograph of the dog tapeworm, *Echinococcus*. A large section of the worm (right) contains numerous eggs. (Bar = 2 mm.) (**B**) In humans, the tapeworm migrates to the liver or lungs, where they form slow-growing hydatid cysts (red). (Bar = 200 μm.) »» Identify the scolex in the light micrograph of *Echinococcus*. What was the identifying feature?

intestine, where the symptoms of infection include diarrhea and itching in the anal region. The female worm is about 10 mm long, and the male is about half that size.

The life cycle of the pinworm is relatively simple. Females migrate to the anal region at night and lay a considerable number of eggs. The area itches intensely and scratching contaminates the hands and bed linens with eggs. Reinfection can take place if the hands are brought to the mouth or if eggs are deposited in foods by the hands. The eggs are swallowed, whereupon they hatch in the duodenum and mature in the regions beyond.

Diagnosis of pinworm disease may be made accurately by applying the sticky side of cellophane tape to the area about the anus and examining the tape microscopically for pinworm eggs (**FIGURE 18.22**). Mebendazole is effective for controlling the disease, and all members of an infected person's family should be treated because transfer of the parasite probably has taken place. Even without medication, however, the worms will die in a few weeks, and the infection will disappear as long as reinfection is prevented.

Trichinellosis. Most of us are familiar with the term **trichinellosis** because packages of pork usually contain warnings to cook the meat thoroughly to avoid this disease. The disease is rare in the United States.

Trichinellosis often is caused by the small roundworm *Trichinella spiralis*. The worm lives in the intestines of pigs and several other mammals. Larvae of the worm migrate through the blood and penetrate the pig's skeletal muscles, where they remain in cysts. When raw or poorly cooked pork is consumed, the cysts pass into the human intestines and the worms emerge. Intestinal pain, vomiting, nausea, and constipation are common symptoms.

Complications of trichinellosis occur when *T. spiralis* adults mate and the female produces larvae in the intestinal wall. The tiny larvae migrate to the muscles primarily in the tongue, eyes, and ribs where they form cysts (**FIGURE 18.23**). The patient commonly experiences pain in the breathing muscles of the ribs and loss of eye movement.

The cycle of trichinellosis is completed as cysts are transmitted back to nature in the human feces. Consumption of human waste and garbage then brings the cysts to the pig. Drugs have little effect on cysts, although mebendazole can be used to kill larvae.

CONCEPT AND REASONING CHECKS

18.14A Why do you suppose pinworm disease is so common in the United States?

Soil-Transmitted Diseases. These diseases are caused by the most significant parasites in

Trichinella spiralis

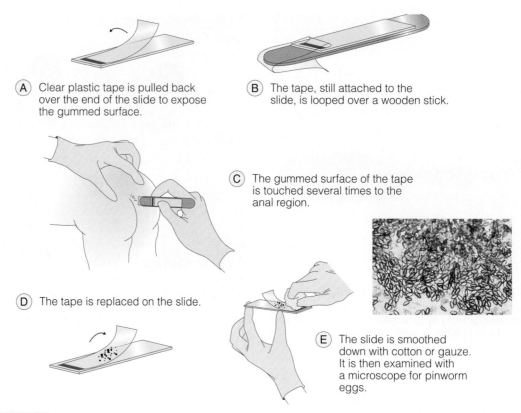

(A) Clear plastic tape is pulled back over the end of the slide to expose the gummed surface.

(B) The tape, still attached to the slide, is looped over a wooden stick.

(C) The gummed surface of the tape is touched several times to the anal region.

(D) The tape is replaced on the slide.

(E) The slide is smoothed down with cotton or gauze. It is then examined with a microscope for pinworm eggs.

FIGURE 18.22 **Diagnosing Pinworm Disease.** The transparent tape technique used in the diagnosis of pinworm disease. »» What time of day might be best to use the tape technique?

Ascaris lumbricoides

humans. They are associated with poverty, lack of adequate sanitation and hygiene, and over-population. One parasite, caused by whip-worms, was mentioned in MicroFocus 18.5.

Ascariasis is an infection with *Ascaris lumbricoides*, a parasite that is the second most prevalent multicellular parasite in the United States. Globally, the WHO estimates there are 1.4 billion infections and 380 million cases worldwide, leading to about 60,000 deaths every year, especially in tropical and subtropical regions.

The parasite resembles an earthworm and one of the largest intestinal nematodes; females may be up to 30 cm long, and males 20 cm long. A female *Ascaris* is a prolific producer of eggs, sometimes generating over 200,000 per day. The eggs are fertilized and passed to the soil in the feces, where they can remain viable for several months. Unfortunately, in many parts of the world, human feces, called "nightsoil," are used as fertilizer for crops, which adds to the spread of the parasite.

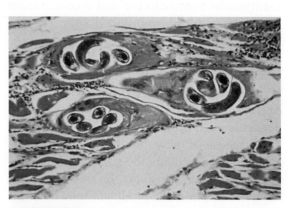

FIGURE 18.23 *Trichinella spiralis.* A false-color light micrograph of a *T. spiralis* larva (dark red) coiled inside a cyst in muscle tissue (red). »» What is the purpose of forming a cyst in muscle, which represents a good nutrient source for a parasite?

Contact with contaminated fingers and consumption of water containing soil runoff are other possible modes of transmission.

When ingested in contaminated food, each egg releases a larva that grows but does not multiply in

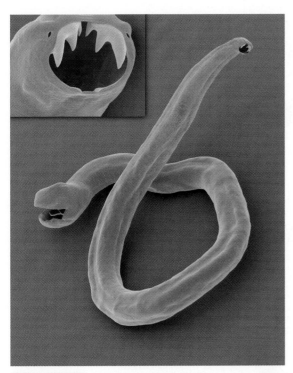

FIGURE 18.24 **A Hookworm.** False-color scanning electron micrograph of the parasitic hookworm *A. duodenale*. Inset: the head of the hookworm. »» Why are these parasites called hookworms?

the small intestine. Abdominal symptoms develop as the worms reach maturity in about two months. Intestinal blockage may be a consequence when tightly compacted masses of worms accumulate, and perforation of the small intestine is possible. In addition, roundworm larvae may pass to the blood and infect the lungs, causing pneumonia. If the larvae are coughed up and then swallowed, intestinal reinfection occurs. Mebendazole is the drug for treatment.

Hookworms are roundworms with a set of hooks or sucker devices for firm attachment to tissues of the host's upper intestine (**FIGURE 18.24**). Approximately 1.3 billion people around the globe are infected by hookworms. There are 150 million cases and 65,000 deaths each year from this disease.

Two hookworms, both about 10 mm in length, may be involved in human disease. The first is the Old World hookworm, *Ancylostoma duodenale,* which is found in Europe, Asia, and the United States; the second is the New World

hookworm, *Necator americanus,* which is prevalent in the Caribbean islands.

These parasites live in the human intestine, where they suck blood from the ruptured capillaries. **Hookworm disease** therefore is accompanied by blood loss and is generally manifested by anemia. Cysts also may become lodged in the intestinal wall, and ulcer-like symptoms may develop.

The life cycle of a hookworm involves only a single host, the human (**FIGURE 18.25**). Female hookworms can release 5,000 to 20,000 eggs, which are excreted to the soil and remain viable for months. Eventually, larvae emerge as long, rod-like **rhabditiform** (*rhabdo* = "rod"; *form* = "shape") larvae. These later become thread-like **filariform** (*filum* = "thread") larvae that attach themselves to vegetation in the soil. When contact with bare feet is made, the filariform larvae penetrate the skin layers and enter the bloodstream. Soon, they localize in the lungs and are carried up to the pharynx in secretions, and then swallowed into the intestines.

Hookworms are common where the soil is warm, wet, and contaminated with human feces and the disease is prevalent where people go barefoot. Mebendazole may be used to reduce the worm burden and the diet may be supplemented with iron to replace that in the blood loss. It should be noted that dogs and cats also harbor hookworm eggs and pass them in the feces.

CONCEPT AND REASONING CHECKS

18.14B Why are ascariasis and hookworm disease so prevalent worldwide?

Roundworms Also Infect the Lymphatic System

KEY CONCEPT

15. Filariasis is a swelling of the lymphatic tissues.

Filariasis is a parasitic disease affecting over 130 million people in 80 countries throughout the tropics and subtropics. The disease in Africa and South America is caused by *Wuchereria bancrofti*, and in Asia by *Brugia malayi*. Both are transmitted by mosquitoes.

The female roundworm is about 100 mm long and carries larvae called **microfilariae**. Injected in a blood meal, the larvae grow to

Necator americanus

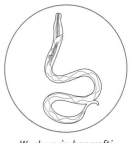

Wuchereria bancrofti

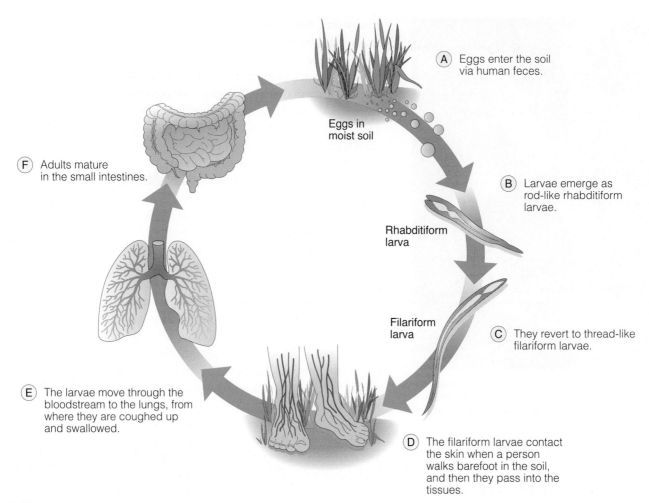

(A) Eggs enter the soil via human feces.

Eggs in moist soil

(B) Larvae emerge as rod-like rhabditiform larvae.

Rhabditiform larva

(F) Adults mature in the small intestines.

Filariform larva

(C) They revert to thread-like filariform larvae.

(E) The larvae move through the bloodstream to the lungs, from where they are coughed up and swallowed.

(D) The filariform larvae contact the skin when a person walks barefoot in the soil, and then they pass into the tissues.

FIGURE 18.25 The Life Cycle of the Hookworms *Ancylostoma duodenale* and *Necator americanus*. The filariform larvae are the infective form. »» What other helminth can be coughed up and swallowed?

adults and infect the lymphatic system where they can survive for up to seven years, causing an extensive inflammation and damage to the lymphatic vessels and lymph glands. After years of infestation, the arms, legs, and scrotum swell enormously and become distorted with fluid (**FIGURE 18.26**). This condition is known as **elephantiasis** because of the gross swelling of lymphatic tissues, called **lymphedema**, and the resemblance of the skin to elephant hide.

The adult worms mate and release millions of microfilariae into the blood, which are ingested in a mosquito's blood meal. There they develop into infective larvae, ready to be passed along to another human during the next blood meal.

TABLE 18.4 summarizes the human diseases caused by the helminths.

CONCEPT AND REASONING CHECKS

18.15 What is the relationship between the nematode infection and limb swelling?

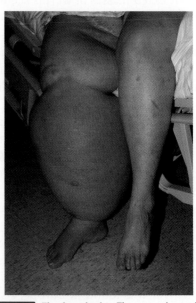

FIGURE 18.26 Elephantiasis. The severely swollen leg in a patient with lymphatic filariasis (elephantiasis) caused by *Wuchereria bancrofti*. »» Why is the disease called elephantiasis?

TABLE

18.4   A Summary of Human Helminthic Diseases

Disease	Causative Agent	Signs and Symptoms	Transmission	Treatment	Prevention and Control
Schistosomiasis	*Schistosoma* species	Ulceration, diarrhea, abdominal pain Bloody urine and pain on urination	Larval spread through bloodstream	Praziquantel	Avoiding wading in snail-contaminated water
Beef tapeworm disease	*Taenia saginata*	Mild diarrhea	Eating poorly cooked beef or pork	Praziquantel	Eliminating livestock
Pork tapeworm disease	*Taenia solium*				Contact with tapeworm eggs Thoroughly cook or freeze all meat and pork
Dog tapeworm disease	*Echinococcus granulosus*	Abdominal chest pain, coughing up blood	Transmitted from domestic dogs and livestock	Praziquantel	Avoiding infected animals
Pinworm disease	*Enterobius vermicularis*	Anal or vaginal itching, irritability and restlessness, intermittent abdominal pain and nausea	Ingestion of pinworm eggs from contaminated food, drink, or hands	Albendazole	Practicing good hygiene and household cleaning
Ascariasis	*Ascaris lumbricoides*	Vague abdominal pain, nausea, vomiting, diarrhea or bloody stools	Ingestion of contaminated food	Mebendazole Albendazole	Practicing good hygiene
Trichinellosis	*Trichinella spiralis*	Intestinal pain, vomiting, nausea, constipation	Consumption of raw or undercooked pork	Mebendazole	Avoiding undercooked pork or wild animal meat
Hookworm disease	*Ancylostoma duodenale Necator americanus*	Raised rash at skin entry site Abdominal pain, loss of appetite, diarrhea, weight loss	Larvae in soil	Mebendazole	Improving sanitation Avoiding contact with contaminated soil
Lymphatic filariasis	*Wuchereria bancrofti*	Swelling of arms, legs, scrotum	Lymphatic vessels from bite of infected mosquito	Diethyl-carbamazine Albendazole	Avoiding mosquitoes in endemic areas

SUMMARY OF KEY CONCEPTS

18.1 Classification and Characteristics of Protists

1. Protists are a diverse group. The majority are unicellular and free-living organisms inhabiting moist areas or water. Some protists, such as the **green algae** are photosynthetic and other protists, like the **dinoflagellates**, **radiolarians**, and **foraminiferans**, are part of the marine **phytoplankton**. Other protists are heterotrophic and have a fungus-like structure.

2. The protozoa have many life styles but can be cataloged into one of three super groups that contain human parasites. The **parabasalids** and **diplomonads** are motile but lack mitochondria; the **kinetoplastids** are motile, have an undulating membrane, and a mass of DNA called the **kinetoplast**; the amoebas move by means of pseudopods and reproduce by binary fission; the **ciliates** have their cell surfaces covered by cilia and contain two different types of nuclei—**macronuclei**, which control metabolic events and **micronuclei**, which play a critical role in genetic recombination through **conjugation**; and **apicomplexans**, which are nonmotile as adults and are obligate parasites. Some protozoa (and most helminths) require at least two hosts to complete their life cycle, the **definitive host**, where sexual reproduction occurs, and the **intermediate host**, where asexual reproduction takes place.

18.2 Protozoal Diseases of the Skin, and Digestive and Urinary Tracts

- **Cutaneous leishmaniasis**
  1. *Leishmania major*
- **Visceral leishmaniasis**
  2. *Leishmania donovani*
- **Amoebiasis**
  3. *Entamoeba histolytica*
- **Giardiasis**
  4. *Giardia intestinalis*
- **Cryptosporidiosis**
  5. *Cryptosporidium parvum, C. hominis*
- **Cyclosporiasis**
  6. *Cyclospora cayetanensis*
- **Trichomoniasis**
  Trichomonas vaginalis

18.3 Protozoal Diseases of the Blood and Nervous System

- **Malaria**
  7. *Plasmodium* species
- **Trypanosomiasis (Human African sleeping sickness)**
  8. *Trypanosoma brucei*
- **Trypanosomiasis (American trypanosomiasis; Chagas disease)** (not shown)
  – *Trypanosomiasis cruzi*
- **Babesiosis** (not shown)
  – *Babesia microti*
- **Toxoplasmosis**
  9. *Toxoplasma gondii*
- **Primary amoebic meningoencephalitis**
  10. *Naegleria fowleri*

18.4 The Multicellular Helminths and Helminthic Infections

11. Among the **helminths**, there are two groups responsible for parasitic diseases in humans—**flatworms** and **roundworms**. Flatworms are multicellular animals lacking respiratory and circulatory structures, and having a gastrovascular cavity. **Tapeworms** can infect all mammals and generally are transmitted to humans in foods. Roundworms (nematodes) are among the most prevalent animals worldwide. They are multicellular with separate sexes. Disease usually is the result of large worm burdens within the individual.

- **Schistosomiasis**
  11. *Schistosoma mansoni*
- **Beef/Pork Tapeworm disease**
  12. *Taenia saginata, T. solium*
- **Echinococcosis**
  13. *Echinococcus granulosus*
- **Pinworm disease**
  14. *Enterobius vermicularis*
- **Trichinellosis**
  15. *Trichinella spiralis*
- **Ascariasis**
  16. *Ascaris lumbricoides*
- **Hookworm disease**
  17. *Ancylostoma duodenale, Necator americanus*
- **Filariasis** (not shown)
  – *Wuchereria bancrofti, Brugia malayi*

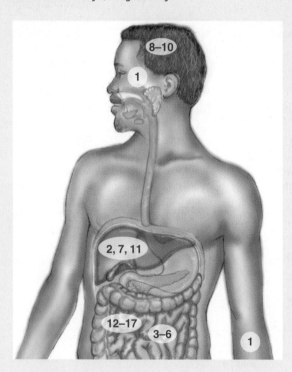

◼ LEARNING OBJECTIVES

After understanding the textbook reading, you should be capable of writing a paragraph that includes the appropriate terms and pertinent information to answer the objective.

1. List the characteristics of the **protista**, including those of the **green algae** and members of the **phytoplankton**.
2A. Identify the characteristics of the **parabasalids**, **diplomonads**, and **kinetoplastids**.
2B. Describe the mechanism of locomotion and food capture by the **Amoebozoa**.
2C. Explain how **ciliates** move in a watery environment.
2D. Differentiate between a **definitive host** and an **intermediate host**.
3. Describe the mode of transmission and forms of **leishmaniasis**.
4. Summarize the characteristics of **amoebiasis, giardiasis, cryptosporidiosis,** and **cyclosporiasis** as human diseases.
5. Explain how **trichomoniasis** is sexually transmitted.
6. Construct a simple life cycle for *Plasmodium*.
7. Compare and contrast **human African sleeping sickness** and **Chagas disease**.
8. Infer why **babesiosis** is a malaria-like disease.
9. Identify the role of **trophozoites, cysts,** and **oocysts** to the infection cycle of *Toxoplasma*.
10. Name the agent and summarize the characteristics of **primary amoebic meningoencephalitis (PAM)**.
11. Identify the two groups of **helminths** of medical importance and summarize their characteristics.
12. Explain the life cycle and infection process of *Schistosoma*.
13. Compare and contrast **beef** and **pork tapeworm diseases**.
14A. Assess the significance of **pinworm disease** and **trichinellosis** in the United States, and describe their infectious cycles.
14B. Describe the soil-transmitted helminthic diseases, **ascariasis** and **hookworm disease**.
15. Discuss how *Wuchereria bancrofti* causes **filariasis (elephantiasis)**.

◼ STEP A: SELF-TEST

Each of the following questions is designed to assess your ability to remember or recall factual or conceptual knowledge related to this chapter. Read each question carefully, then select the *one* answer that best fits the question or statement. Answers to even-numbered questions can be found in **Appendix C**.

1. The _____ are members of the _____.
   A. green algae; protozoa
   B. radiolarians; apicomplexans
   C. dinoflagellates; phytoplankton
   D. radiolarians; fungus-like protists
2A. This group of protozoa has a single mitochondrion with a mass of DNA.
   A. Kinetoplastids
   B. Ciliates
   C. Apicomplexans
   D. Diplomonads
2B. The _____ group of protozoa have food vacuoles and pseudopods.
   A. Kinetoplastid
   B. Excavata
   C. Chromalveolata
   D. Amoebozoa
2C. Which one of the following is NOT found in the ciliates?
   A. A contractile vacuole
   B. Macronuclei and micronuclei
   D. A complex of organelles in the tip
   D. Mitochondria
2D. An intermediate host is
   A. where parasite asexual cycle occurs.
   B. always a nonhuman host.
   C. where parasite sexual cycle occurs.
   D. the human host between two other animal hosts.
3. The vector transmitting leishmaniasis is the
   A. mosquito.
   B. sandfly.
   C. tsetse fly.
   D. sand flea.

4A. _____ enters the human body as a cyst and develops into a trophozoite in the small intestine; a severe form of dysentery may occur.
   A. *Cryptosporidium parvum*
   B. *Entamoeba histolytica*
   C. *Giardia intestinalis*
   D. *Cyclospora cayetanensis*
4B. Sucker-like devices allow this protozoan to adhere to the intestinal lining.
   A. *Cyclospora cayetanensis*
   B. *Entamoeba histolytica*
   C. *Giardia intestinalis*
   D. *Cryptosporidium parvum*
4C. This disease sickened more than 400,000 residents of Milwaukee in 1993.
   A. Giardiasis
   B. Cryptosporidiosis
   C. Trypanosomiasis
   D. Cyclosporiasis
5. This genus of parabasalid affects over 7.4 million Americans annually and is considered a sexually transmitted disease.
   A. *Toxoplasma*
   B. *Cryptosporidium*
   C. *Cyclospora*
   D. *Trichomonas*
6. The _____ form of the malarial parasite enters the human blood while the _____ enters the mosquito.
   A. sporozoites; merozoites
   B. merozoites; gametocytes
   C. merozoites; sporozoites
   D. sporozoites; gametocytes
7. Chagas disease is caused by
   A. *Trypanosome cruzi*.
   B. *Toxoplasma gondii*.
   C. *Babesia microti*.
   D. *Trypanosoma brucei*

8. Babesiosis is carried by _____ and infects _____.
   A. mosquitoes; RBCs
   B. fleas; the kidneys
   C. ticks; RBCs
   D. mosquitoes; the liver
9. The T in TORCH stands for the disease caused by
   A. *Trypanosome brucei.*
   B. *Toxoplasma gondii.*
   C. *Taenia spiralis.*
   D. *Trypanosoma cruzi.*
   E. *Trichomonas vaginalis.*

10. This opportunistic protozoan causes primary meningoencephalitis (PAM).
    A. *Plasmodium vivax*
    B. *Toxoplasma gondii*
    C. *Naegleria fowleri*
    D. *Paragonimus westermani*
    E. *Entamoeba histolytica*

STEP B: REVIEW

Read the statement concerning the helminths, and then select the answer or answers that best apply to the statement. Place the letter(s) next to the statement. The answers to even-numbered statements are listed in **Appendix C.**

_____ 11. Transmitted by an arthropod.
   A. Filariasis
   B. Trichinosis
   C. Hookworm disease

_____ 12. Beef tapeworm species.
   A. *Echinococcus granulosus*
   B. *Schistosoma mansoni*
   C. *Taenia saginata*

_____ 13. Type of fluke.
   A. *Schistosoma*
   B. *Necator*
   C. *Echinococcus*

_____ 14. Cause of pinworm disease.
   A. *Trichinella*
   B. *Enterobius*
   C. *Ascaris*

_____ 15. Infects the human intestines.
   A. *Trichinella spiralis*
   B. *Ascaris lumbricoides*
   C. *Taenia saginata*

_____ 16. Type of tapeworm.
   A. *Taenia*
   B. *Echinococcus*
   C. *Necator*

_____ 17. Snail is the intermediate host.
   A. Blood fluke
   B. *Echinococcus*
   C. Intestinal fluke

_____ 18. Attaches to host tissue by hooks.
   A. *Necator*
   B. *Trichinella*
   C. *Enterobius*

_____ 19. Affects pigs as well as humans.
   A. *Wuchereria*
   B. *Trichinella*
   C. *Schistosoma*

_____ 20. Life cycle includes miracidia and cercaria.
   A. *Schistosoma*
   B. *Echinococcus*
   C. *Ascaris*

_____ 21. Acquired by consuming contaminated pork.
   A. *Taenia solium*
   B. *Echinococcus*
   C. *Necator americanus*

_____ 22. Classified in the phylum Nematoda.
   A. *Ascaris*
   B. *Wuchereria*
   C. *Schistosoma*

_____ 23. Causes inflammation and damage to the lymphatic vessels.
   A. *Echinococcus*
   B. *Ascaris*
   C. *Wuchereria*

_____ 24. Male and female forms exist.
   A. *Taenia*
   B. *Ascaris*
   C. *Schistosoma*

_____ 25. Forms hydatid cysts.
   A. Blood fluke
   B. *Enterobius*
   C. *Echinococcus*

STEP B: REVIEW—PARASITE LIFE CYCLES

26. Label the two parasite life cycles, (1) malaria and (2) schistosomiasis, using the term lists provided.

(1) *Plasmodium* Life Cycle
Term List
Definitive host
Gametes
Gameyocyte
Intermediate host
Merozoites
Oocyst
Sporozoites
Zygote

IN HUMAN

Liver

IN MOSQUITO

Mosquito

(2) *Schistosoma* Life Cycle
Term List
Cercaria
Definitive host
Eggs
Intermediate host
Miracidium

Snail

STEP C: APPLICATIONS

Answers to even-numbered questions can be found in **Appendix C**.

27. You and a friend who is three months pregnant stop at a hamburger stand for lunch. Based on your knowledge of toxoplasmosis, what helpful advice can you give your friend? On returning to her home, you notice she has two cats. What additional information might you share with her?

28. Cardiologists at a local hospital hypothesized that a few patients with a certain protozoal disease easily could be lost among the far larger population of heart disease sufferers patronizing county clinics. They proved their theory by finding 25 patients with this protozoal disease among patients previously diagnosed as having coronary heart disease. Which protozoal disease was involved?

29. Federal law stipulates that food scraps fed to pigs must be cooked to kill any parasites present. It also is known that feedlots for swine are generally more sanitary than they have been in the past. As a result of these and other measures, the incidence of trichinellosis in the United States has declined, and the acceptance of "pink pork" has increased. Do you think this is a dangerous situation? Why?

STEP D: QUESTIONS FOR THOUGHT AND DISCUSSION

Answers to even-numbered questions can be found in **Appendix C**.

30. A newspaper article written in the 1980s asserted that parasitology is a "subject of low priority in medical schools because the diseases are exotic infections only occurring in remote parts of the world." Would you agree with that statement today? Explain.

31. Many historians believe malaria contributed to the downfall of ancient Rome. Over the decades, malaria incidence increased with expansion of the Roman Empire, which stretched from the Sahara desert to the borders of Scotland, and from the Persian Gulf to the western shores of Portugal. How do you suspect the disease and the expansion are connected?

32. It has been said that until recent times, many victims of a particular protozoal disease were buried alive because their life processes had slowed to the point where they could not be detected with the primitive technology available. Which disease was probably present?

33. The World Health Organization has reported that, after malaria, schistosomiasis and filariasis (200 million and 90 million annual cases, respectively) are the most prevalent tropical diseases. How do you believe the incidence of these diseases can be reduced on a global scale?

34. Some restaurants offer a menu item called steak tartare, which is a dish served with raw ground beef. What hazard might this meal present to the restaurant patron?

35. Justify this statement: "Perhaps the most important reason for discussing parasitical diseases is that they highlight just how enmeshed we are in the web of life."

36. Many of the parasitic diseases described in this chapter, including schistosomiasis, ascariasis, filariasis, trypanosomiasis, leishmaniasis, and some bacterial diseases mentioned in previous chapters, such as trachoma and leprosy, are often considered "neglected diseases." Why do you think these diseases have been somewhat "neglected" by the medical community, especially in the developed world?

HTTP://MICROBIOLOGY.JBPUB.COM/9E

The site features eLearning, an online review area that provides quizzes and other tools to help you study for your class. You can also follow useful links for in-depth information, read more MicroFocus stories, or just find out the latest microbiology news.

PART 5

Disease and Resistance

False-color transmission electron micrograph image of a T lymphocyte (T cell), which plays a critical role in the human immune response against viruses, bacteria, and tumors.

In past centuries, the spread of disease appeared to be willfully erratic. Illnesses would attack some members of a population while leaving others untouched. A disease that for many generations had taken small, steady tolls would suddenly flare up in epidemic proportions. And strange, horrifying plagues descended unexpectedly on whole nations.

Scientists now know humans live in a precarious equilibrium with the microorganisms surrounding them. Generally, the relationship is harmonious, because humans can come in contact with most microorganisms and develop resistance to them. However, when the natural resistance is unable to overcome the aggressiveness of microorganisms, disease sets in. In other instances, the resistance is diminished by a pattern of human life that gives microorganisms the edge. For example, during the Industrial Revolution of the 1800s, many thousands of Europeans moved from rural areas to the cities. They sought new jobs, adventure, and prosperity. Instead, they found endless labor, unventilated factories, and wretched living conditions—and they found disease.

In Part 5 of this text, we shall explore the infectious disease process and the mechanisms by which the body responds to disease. Chapter 19 opens with an overview of the host-microbe relationship and the factors contributing to the establishment of disease. Epidemiology and diseases within populations also will be explored. In Chapters 20 and 21, the discussion turns to nonspecific and specific methods by which body resistance develops, with emphasis on the immune system. Various types of immunity are explored in Chapter 22, together with a survey of laboratory methods using the immune reaction in the diagnosis of disease. In Chapter 23 the discussion centers on immune disorders leading to serious problems in humans. This includes an extensive discussion of AIDS. Finally, in Chapter 24, we move to treating the patient by discussing antimicrobial drugs, including antibiotics. In these chapters, we uncover the roots of infectious disease and resistance and come to understand the interactions and impact that microorganisms and viruses have on humans at the fundamental level.

Epidemiology

Flying to an impoverished African country on your second day of work to battle Ebola, one of the most deadly viruses, isn't most people's idea of a dream assignment. But it was for Marta Guerra. In fact, the trip to Uganda in 2000 was the assignment she had been coveting. "I wasn't that worried," Guerra says. "This particular strain has only a 65% death rate instead of the Congo strain which is 85%."

Guerra is a disease epidemiologist, popularly known as a "disease detective," with the Epidemic Intelligence Service (EIS) of the Centers for Disease Control and Prevention (CDC) in Atlanta.

Growing up in the multicultural environments of Havana, Cuba, and Washington, D.C., Marta Guerra developed a keen interest in other cultures and teaching people about health risks. She was fascinated by stories her professors told about working overseas. After seeing the movie *Outbreak,* "I thought, that's what I want to do—help contain a deadly epidemic," recalls Guerra. So she obtained a Master's in public health and a Ph.D. in tropical medicine.

Like all EIS officers, Guerra's job is to isolate the cause of an outbreak, prevent its spread, and get out public health messages to people who could have been exposed. When Guerra flew to Uganda in November 2000, the Ebola outbreak had already been identified, so her task was to go from village to village, trying to locate family members and friends, and educate them about symptoms and treatment. "In every corner of Africa people know the word Ebola, and they are terrified of it," Guerra explains. "Sometimes they hide sick family members; sometimes they're frightened of survivors."

The job of a disease detective can be difficult—and dangerous. Although Guerra seldom considered she might acquire a disease she was investigating, she was concerned about political violence. In Uganda, Guerra's team needed military escorts on their travels through villages. In Ethiopia, while on a polio-eradication mission in the summer of 2001, she recalls, "There was rebel activity in all the areas we traveled through—plus land mines. It was pretty scary."

Perhaps you might be interested in a similar career. "You have to be highly motivated, with the ability to think fast on your feet and make quick decisions. You have to be able to walk into a chaotic situation and deal with whatever is thrown at you," Guerra says. "Sometimes I barely drop my bags at home before I'm called out again. Being adaptable is really essential."

To get started, you need a bachelor's degree in a biological science. In addition to required courses in chemistry and biology, undergraduates should study microbiology, mathematics, and computer science. A master of science in epidemiology or public health also is required; many have a Ph.D. or a medical or veterinary degree. Most American disease epidemiologists then apply to EIS's two-year, post-graduate program of service and on-the-job training, where they work with mentors like Marta. She says, "I like the fact that I am contributing to science in the sense that what I do will affect people far into the future."

(Essay modified from *Disease Detective* by Carol Sonenklar in MedHunters.com)

Infection and Disease

Chapter Preview and Key Concepts

"Health care matters to all of us some of the time, public health matters to all of us all of the time"
—C. Everett Koop, former Surgeon General of the United States

By late July 1999, crows were literally dropping out of the sky in New York City and dead crows were found in surrounding areas as well. By early September, officials at the Bronx Zoo discovered other birds, including a cormorant, two red Chilean flamingoes, and an Asian pheasant at the zoo had died of the same brain and heart inflammations as found in the crows.

On August 23, 1999, an infectious disease physician at a hospital in northern Queens reported to the New York City Department of Health that two patients had been admitted with encephalitis. In fact, on further investigation, the health department identified a cluster of six patients with encephalitis. Testing by the Centers for Disease Control and Prevention (CDC) of these initial cases for antibodies to common North American **arboviruses**—viruses transmitted by arthropods, such as insects—was positive for St. Louis encephalitis (SLE)—a virus carried by mosquitoes. These findings prompted the health department to begin aerial and ground application of insecticides.

News of SLE and spraying caught the attention of the Bronx Zoo officials. If the birds were dying from the same encephalitis disease as in humans, it could not be caused by the SLE virus because birds do not contract SLE.

So, a reinvestigation by the CDC of virus samples taken from humans, birds, and mosquitoes was undertaken. Results indicated all viral isolates were closely related to West Nile virus (WNV), which had never been isolated in the western hemisphere. It soon became evident that, indeed, the disease in birds and humans was caused by WNV (see Chapter 16).

By early fall, mosquito activity waned and the number of human cases declined (**FIGURE 19.1**). In all, 61 people would be infected and 7 would die. Although New York City and the surroundings could

Epidemiology:
The scientific study
of the causes and transmission of disease within a
population.

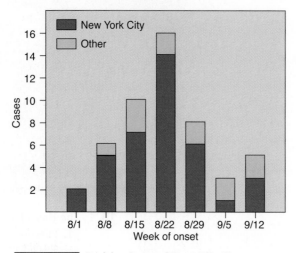

FIGURE 19.1 Positive Cases of West Nile Virus Infection, New York—1999.
Source: www.cdc.gov/mmwr/preview/mmwrthml/ mm4839a5.htm. »» What does this graph tell you about the frequency with which WNV occurred?

breathe a sigh of relief, the WNV outbreak was only the beginning of a virus march across the United States. By 2008, human cases of West Nile encephalitis would be reported across America (**FIGURE 19.2**).

The outbreak of West Nile encephalitis is one example describing the epidemiology of infection and disease; that is, the scientific (and medical) study of the causes, transmission, and prevention of disease within a population. As you remember from Chapter 1, modern epidemiology can trace its origins back to John Snow, the English surgeon who studied a cholera outbreak in the Soho district of London in 1854. His paper, *On the Mode of Communication of Cholera*, is a model of epidemiological detective work that included a map of cholera cases in Soho. By marking with rectangles

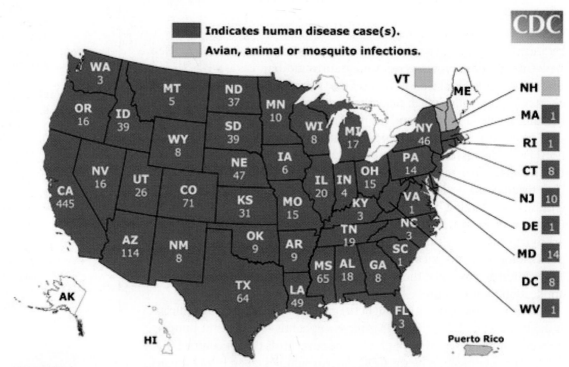

FIGURE 19.2 West Nile Virus Activity, United States—2008. The map indicates the distribution of avian, animal, or mosquito infections in 2008. It also identifies the number of reported human cases in each state (a state is shaded no matter in what area of the state the disease case(s) occurred). *Source*: www.cdc.gov/ncidod/dvbid/westnile/ mapsactivity/surv&contol08Maps.htm »» Why is this epidemiological map referred to as a geographical distribution?

where each cholera death occurred, Snow showed they all pointed to the Broad Street pump, which supplied drinking water for the victims and, Snow realized, was the source of the disease. In the end, the outbreak was quelled by simply removing the handle from the pump. Snow's investigation was the classic of epidemiology: what's the source of the outbreak; how is it being transmitted; and how does one prevent further spread?

Today, the CDC, the World Health Organization (WHO), and other agencies throughout the world have built on the historical work of John Snow, Ignaz Semmelweis, Joseph Lister, Louis Pasteur, Robert Koch, and many others. Yet, the epidemiologists employed by these organizations seek the same goals as John Snow: to identify and investigate disease outbreaks; conduct research to enhance prevention; and then devise prevention strategies.

In this chapter, we discuss the mechanisms underlying the spread and development of infectious disease. Our purpose is to bring together many concepts of disease and synthesize an overview of the host–microbe relationship. We summarize much of the important terminology used in medical microbiology and outline some of the factors used by microorganisms to establish themselves in tissues. An understanding of the topics concerning the host–microbe relationship will be essential preparation for the detailed discussion of host resistance (immune) mechanisms in Chapters 20 and 21.

19.1 The Host–Microbe Relationship

By the early 1970s, the Surgeon General of the United States claimed we could "close the books on infectious diseases" (see Chapter 1). The development and use of antibiotics, and vaccines, had made the threat of infectious disease of little consequence. However, antibiotic resistance and new emerging diseases, including Legionnaires' disease, AIDS, Lyme disease, hantavirus pulmonary syndrome, and SARS, have thwarted such optimism. In 2008, of the approximately 57 million humans who died worldwide, more than 25% (15 million) died from infectious diseases, making them the second leading cause of death (behind cardiovascular disease) (FIGURE 19.3). In fact, infectious diseases are the leading cause of death in children under 5 years of age.

The Human Body Maintains a Symbiosis with Microbes

KEY CONCEPT

1. Infection and disease occur when a host–microbe relationship tilts in favor of the microbe.

Infection refers to the multiplication of a microbe in a host and the competition for supremacy taking place between them. (Note: in this chapter, for simplicity of discussion, "microbe" includes the viruses.) A host whose resistance is strong remains healthy, and the microbe is either driven from the host or assumes a benign relationship with the

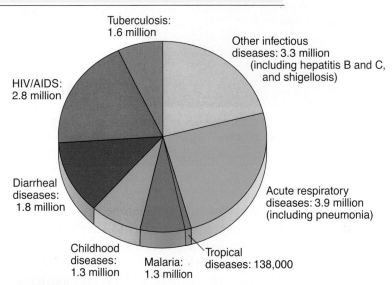

FIGURE 19.3 Infectious Disease Deaths Worldwide. This pie chart depicts the leading causes of infectious diseases and the number of worldwide deaths as reported by the World Health Organization. Tropical diseases: trypanosomiasis, Chagas disease, schistosomiasis, leishmaniasis, filariasis, and onchocerciasis. Childhood diseases: diphtheria, measles, pertussis, polio, and tetanus.
»» Only one of the "pie slices" has grown explosively in mortality numbers since 1993. Identify the disease and explain why that is so.

host. By contrast, if the host loses the competition, disease develops. The term **disease** refers to any change from the general state of good health. It is important to note that disease and infection are not synonymous; a person may be infected without suffering a disease.

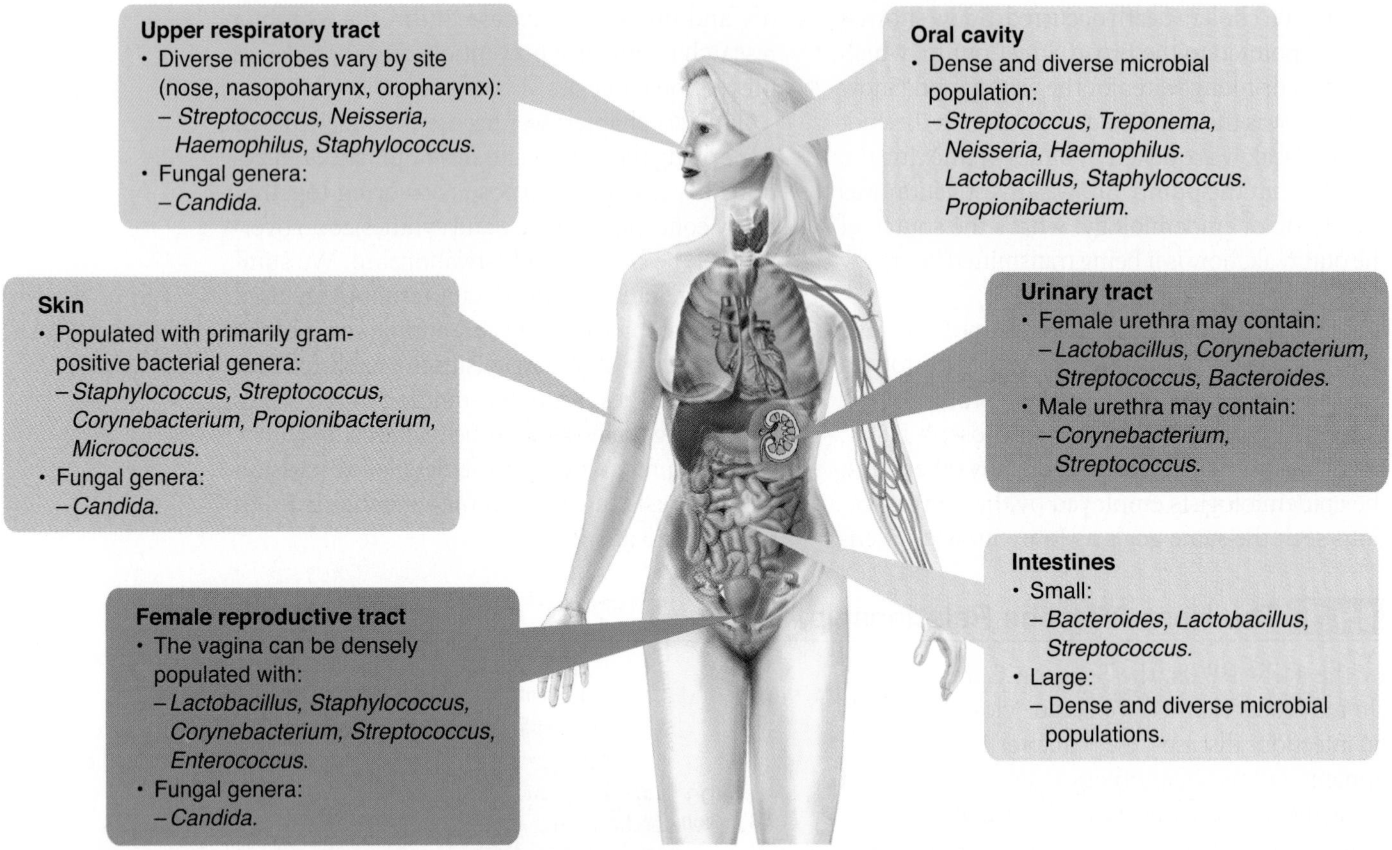

Upper respiratory tract
- Diverse microbes vary by site (nose, nasopoharynx, oropharynx):
  – *Streptococcus, Neisseria, Haemophilus, Staphylococcus.*
- Fungal genera:
  – *Candida.*

Oral cavity
- Dense and diverse microbial population:
  – *Streptococcus, Treponema, Neisseria, Haemophilus. Lactobacillus, Staphylococcus. Propionibacterium.*

Skin
- Populated with primarily gram-positive bacterial genera:
  – *Staphylococcus, Streptococcus, Corynebacterium, Propionibacterium, Micrococcus.*
- Fungal genera:
  – *Candida.*

Urinary tract
- Female urethra may contain:
  – *Lactobacillus, Corynebacterium, Streptococcus, Bacteroides.*
- Male urethra may contain:
  – *Corynebacterium, Streptococcus.*

Female reproductive tract
- The vagina can be densely populated with:
  – *Lactobacillus, Staphylococcus, Corynebacterium, Streptococcus, Enterococcus.*
- Fungal genera:
  – *Candida.*

Intestines
- Small:
  – *Bacteroides, Lactobacillus, Streptococcus.*
- Large:
  – Dense and diverse microbial populations.

FIGURE 19.4 A Sampling of the Indigenous Microbiota of the Human Body. In reality there are thousands of microbial species and viruses on and within some human body systems. »» Explain why some body systems (e.g., circulatory system, nervous system) normally remain sterile?

Whether host or microbe gets the upper hand often is due in part to the 100 trillion microbes found on and in the human body. This remarkable number amounts to almost 3 pounds of human weight! These microbes represent the **microbiota** (*biota* = "life"), a population of microorganisms and viruses residing in the body without causing disease. Some, called the **indigenous microbiota** establish a permanent relationship with various parts of the body, while others, the **transient microbiota**, are more temporary and found only for limited periods of time. In the large intestine of humans, for example, *Escherichia coli* and *Candida albicans* are almost always found, but species of *Streptococcus* are transient.

The relationship between the body and its microbiota is an example of a **symbiosis**, or living together. If the symbiosis is beneficial to both the host and the microbe, the relationship is called **mutualism**. For example, species of *Lactobacillus* live in the female vagina and derive nutrients from the environment while producing acid to prevent the overgrowth of other organisms.

A symbiosis also can be beneficial only to the microbe and the host is unaffected, in which case the symbiosis is called **commensalism**. *E. coli* is generally presumed to be a commensal in the human intestine, although some evidence exists for mutualism because the bacterial cells produce nonessential amounts of vitamins B and K. In addition, the microbiota usually will outcompete invading microbial pathogens, thereby protecting the body from dangerous infections.

Microbiota may be found in several body tissues (**FIGURE 19.4**). These include the skin, the external ears and eyes, and upper respiratory tract. Most of the digestive tract, from oral cavity to rectum, is heavily populated with indigenous microbiota. In fact, many of these are thought to play an important functional role for humans (MicroFocus 19.1). Microbiota also make up a population at the urogenital orifices in both males and females.

MICROFOCUS 19.1: Being Skeptical
Can Gut Bacteria Control Human Metabolism?

There is a great diversity of indigenous microbiota present on and within the human body. Adults contain somewhere between 500 and 1,000 different bacterial species in their gut alone. What do all these microbes do? We know *Escherichia coli* can help with water reabsorption and produce some of the vitamin K we use in our diet (see figure). Can these species actually be essential to our metabolism?

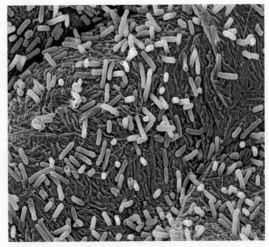

False-color scanning electron micrograph of *E. coli* cells on the rat intestinal lining.

Since the 1950s, investigators have manipulated and engineered special strains of mice that have germ-free guts; that is, their guts are sterile. So, one way to determine if various members of the normal microbiota help us in some way is to introduce each species separately into germ-free mice and see what happens. One assumes what happens in a mouse may mirror what happens in a human.

Jeffrey Gordon and colleagues at Washington University School of Medicine in St. Louis studied *Bacteroides thetaiotaomicron,* a gram-negative anaerobe that exists in the human gut at concentrations 1,000 times greater than *E. coli.* At these concentrations, it must be doing something. Gordon's team introduced *B. thetaiotaomicron* into the guts of germ-free mice and monitored what happened. The team quickly discovered the mice synthesized fucose, a monosaccharide sugar, onto the surface of the intestinal cells. Germ-free mice did not. Apparently, *B. thetaiotaomicron* provided a stimulus to the intestinal cells "telling" them to turn on the genes for fucose synthesis. Why fucose? This is the sugar that *B. thetaiotaomicron* uses for energy and metabolism.

Using DNA techniques that allow large numbers of genes to be monitored all at once, Gordon's group realized *B. thetaiotaomicron* actually triggered the intestinal cells to turn on or turn off some 100 of the 25,000 genes in the cell's genome. Some of these genes helped the mice absorb and metabolize sugars and fats. Other genes activated by the bacterial cells produced products helping protect the intestinal wall from being penetrated by other normal microbiota or pathogens. So, the indigenous microbiota may do more than simple commensals. Yet other genes stimulated the growth of new blood vessels, explaining why germ-free mice had to eat 30% more calories to maintain body weight than ordinary mice—germ-free mice have a less well developed blood vessel system and are inefficient at absorbing nutrients. *B. thetaiotaomicron* cells made the human digestive metabolic processes more efficient.

Conclusion: Just from this one bacterial species, the physical development of the normal gut in mice (and extrapolating to the human gut, too) appears to depend on the normal microbiota. Microbes might not only rule the world, they may control our gut physiology and development as well.

Most other tissues of the body remain sterile; the blood, cerebrospinal fluid, joint fluid, and internal organs, such as the kidneys, liver, muscles, bone, and brain, are sterile unless disease is in progress.

The first nine months of human development within a mother's womb is the only time when the human body is truly a sterile organism. Indigenous microbiota are introduced when the newborn passes through the birth canal or from a cesarean birth (**FIGURE 19.5**). Additional organisms enter upon first feeding where nursing or formula can influence what microbes colonize the newborn's gut. During the next weeks, additional contact with the mother and other individuals will expose the infant to additional intestinal microbes. Besides the gut, the skin will be colonized by many different bacterial and fungal species. The upper respiratory tract will be covered with a diverse group of bacterial species while the lower urethra

Sterile:
Devoid of living microorganisms, viruses, and spores.

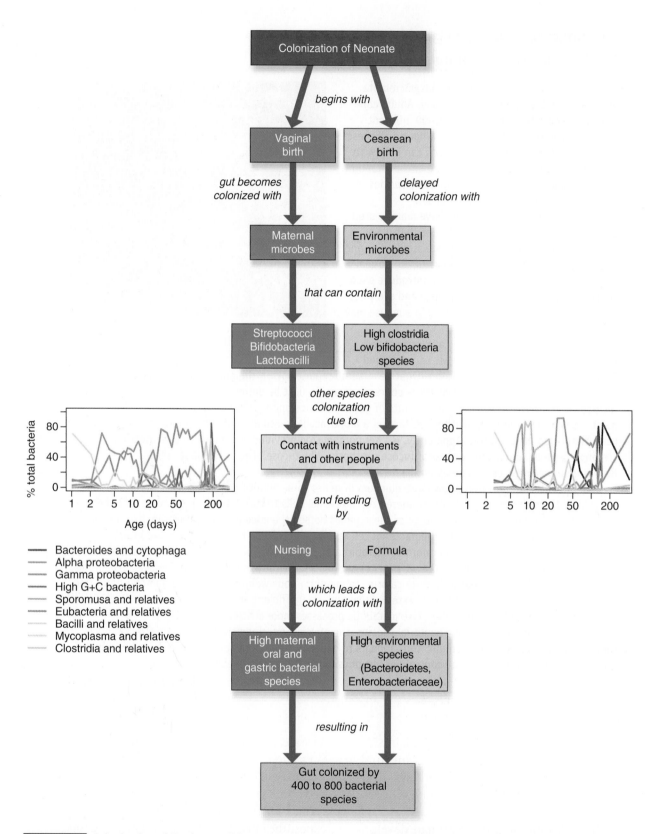

FIGURE 19.5 Colonization of Newborns. This concept map diagrams possible colonization of two newborns based on form of birth (vaginal or cesarean) and feeding (nursing or formula). The graphs (inserts) plot fluctuations in bacterial microbiota (left: vaginal birth; right: cesarean birth) over the first 200 days. »» How would each form of birth and feeding introduce microbiota into the newborn?
Graphs modified from *Development of the Human Infant Intestinal Microbiota*, Palmer C., Bik E.M., DiGiulio D.B., Relman D.A., and Brown P.O., *PLoS Biology* Vol. 5, No. 7, el77 doi:10.1371/journal.pbio.0050177

will be populated by bacterial and fungal organisms, as well as a few potential pathogens.

By one year, the infant's indigenous microbiota is adult-like and remains throughout life, undergoing small changes in response to the internal and external environment of the individual.

CONCEPT AND REASONING CHECKS

19.1 Distinguish between a mutualistic relationship and a commensalistic one.

Pathogens Differ in Their Ability to Cause Disease

KEY CONCEPT

2. Microbes vary greatly in their pathogenicity.

There also are symbiotic relationships, called **parasitism**, where the pathogen causes damage to the host and disease can result. Microbiologists once believed microbes were either pathogenic or nonpathogenic; they either caused disease or they did not. From the previous section, we know that is not true.

Pathogenicity refers to the ability of a microorganism to gain entry to the host's tissues and bring about a physiological or anatomical change, resulting in altered health and leading to disease. Certain pathogens, such as the cholera, plague, and typhoid bacilli, are well known for their ability to cause serious human disease. Others, such as common cold viruses, are considered less pathogenic because they induce milder illnesses. Still other microorganisms are opportunistic.

Whether a disease is mild or severe depends on the pathogen's ability to do harm to the body. Thus, the degree of pathogenicity is called **virulence** (*virul* = "poisonous"). For example, an organism invariably causing disease, such as the typhoid bacillus, is said to be "highly virulent." By comparison, an organism sometimes causing disease, such as *Candida albicans,* is labeled "moderately virulent." Certain organisms, described as **avirulent**, are not regarded as disease agents. The lactobacilli and streptococci found in yogurt are examples. However, it should be noted that any microorganism has the ability to change genetically and become virulent.

In recent years, a new term, **pathogenicity islands**, has been used to refer to clusters of genes responsible for virulence (see Chapter 9). The genes, present on the bacterial cell's chromosome or plasmids, encode many of the virulence factors making a microbe more virulent. These unstable islands are fairly large segments of a pathogen's genome and are absent in nonpathogenic strains. Pathogenicity islands have many of the properties of intervening sequences (see Chapter 8), suggesting they move by horizontal gene transfer. A copy of these blocks of genetic information can move from a pathogenic strain into an avirulent (harmless) organism, converting it to a pathogen. Such horizontal transfer processes show how the evolution of pathogenicity can make quick jumps.

Before examining disease progression, realize that an infection and a resulting disease can be caused by a single microbe. The diseases identified by Pasteur, Koch, and their contemporaries are examples. However, it is now clear that some diseases are caused by two or more microbes acting together or in succession. Such **polymicrobial diseases** include tooth decay, gastroenteritis, urinary tract infections, otitis media, and HIV/AIDS.

CONCEPT AND REASONING CHECKS

19.2 Distinguish between pathogenicity and virulence.

Several Events Must Occur for Disease to Develop in the Host

KEY CONCEPT

3. Contact with a potential pathogen can have several outcomes.

For disease to occur, a potential pathogen must first come in contact with exposed parts of the body (**FIGURE 19.6**). Several outcomes are possible: the pathogen may be lost to the environment or it may colonize the normal microbiota and remain as a transient member. Depending on the nature of the pathogen, it could also become a commensal.

An **exogenous infection** is established if a pathogen from the environment breaches the host's external defenses and enters the host. Likewise, if a microbial member of the normal microbiota should gain access to sterile tissue, an **endogenous infection** ensues. In both cases, the infection may trigger additional host defenses capable of eliminating the invader.

Should the pathogen cause injury or dysfunction to host tissues, then a disease is established. Again, additional host defenses may eliminate the pathogen, in which case the disease declines and the host recovers. In other cases, the pathogen

Virulence factors: Pathogen-produced molecules or structures that allow the cell to invade the host (or evade the immune system) and possibly cause disease.

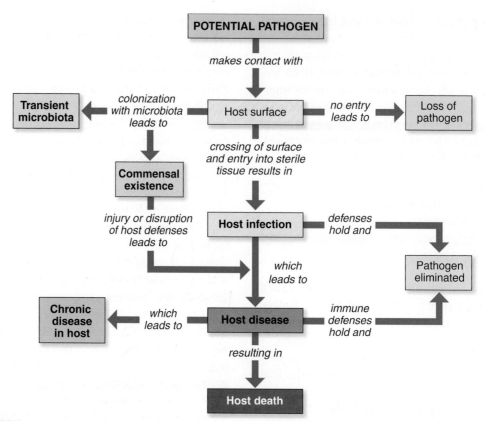

FIGURE 19.6 **The Progression and Outcomes of Infection and Disease.** A concept map illustrating possible outcomes resulting from the contact between host and pathogen. »» Propose some ways that a pathogen would gain entry into sterile tissue.

and host reach a stalemate where neither has the advantage. Tuberculosis is an example of such a chronic state. Lastly, the inability to eliminate the pathogen may lead to death.

Opportunistic infections often are caused by commensals taking advantage of a shift in the body's delicate balance to one favoring the microbe. If the indigenous microbiota is reduced or the host's immune system is weakened, some commensals seize the "opportunity" to invade the tissues and cause disease. AIDS is an example where crippling of the immune system makes the patient highly susceptible to opportunistic organisms. Thus, an upset in resistance mechanisms or microbiota control may enhance the ability of organisms to establish disease.

Infections may develop in one of two ways. A **primary infection** occurs in an otherwise healthy body, while a **secondary infection** develops in an individual weakened by the primary infection. In the influenza pandemic of 1918 and 1919, hundreds of millions of individuals contracted influenza as a primary infection and many developed

pneumonia as a secondary infection. Numerous deaths in the pandemic were due to pneumonia's complications.

As the names imply, **local diseases** are restricted to a single area of the body, while **systemic diseases** are those disseminating to the deeper organs and systems. Thus, a staphylococcal skin boil beginning as a localized skin lesion may become more serious when staphylococci spread and cause systemic disease of the bones, meninges, or heart tissue.

The transient appearance of living bacterial cells in the blood is referred to as **bacteremia**. **Septicemia** refers to an infection of bacterial cells in the blood which can be a life-threatening condition (MICROFOCUS 19.2). Other microbes also are disseminated. **Fungemia** refers to the spread of fungi, **viremia** to the spread of viruses, and **parasitemia** to the spread of protozoa and multicellular worms through the blood.

CONCEPT AND REASONING CHECKS

19.3 Contrast exogenous, endogenous, and opportunistic infections.

19.2 Establishment of Infection and Diseases

Disease is the result of a dynamic series of events expressing the competition between host and pathogen. To overcome host defenses and bring about the anatomic or physiologic changes leading to disease, the pathogen must possess unusual abilities. Disease therefore is a complex series of interactions between pathogen and host.

In this section, we outline the stages of disease from which we can explore the processes and factors determining whether disease can occur.

Diseases Progress through a Series of Stages

KEY CONCEPT

4. Disease progression involves incubation, prodromal, acute, decline, and convalescent stages.

In most instances, there is a recognizable pattern in the progress of the disease following the entry of the pathogen into the host. Often these periods are identified by **signs**, which represent evidence of disease detected by an observer (e.g., physician). Fever or bacterial cells in the blood would be examples. Disease also can be noted by **symptoms**, which represent changes in body function sensed by the patient. Sore throat and headache are examples. Diseases also may be characterized by a specific collection of signs and symptoms called a **syndrome**. AIDS is an example.

Disease progression is distinguished by five stages (**FIGURE 19.7**). The episode of disease begins with an **incubation period**, reflecting the time elapsing between the entry of the microbe into the host and the appearance of the first symptoms. For example, an incubation period may be as short as 2 to 4 days for the flu; one to two weeks for measles; or three to six years for leprosy. Such factors as the number of organisms, their generation time and virulence, and the level of host resistance determine the incubation period's length. The location of entry also may be a determining factor. For instance, the incubation period for rabies may be as short as several days or as long as a year, depending on how close to the central nervous system the viruses enter the body.

The next phase in disease is a time of mild signs or symptoms, called the **prodromal phase**. For many diseases, this period is characterized by indistinct and general symptoms such as nausea, headache, and muscle aches, which indicate the

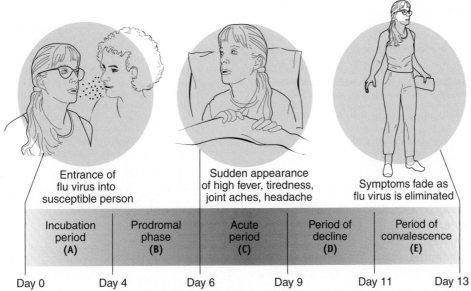

Entrance of flu virus into susceptible person		Sudden appearance of high fever, tiredness, joint aches, headache		Symptoms fade as flu virus is eliminated
Incubation period (A)	Prodromal phase (B)	Acute period (C)	Period of decline (D)	Period of convalescence (E)
Day 0	Day 4	Day 6	Day 9	Day 11 Day 13

FIGURE 19.7 **The Course of an Infectious Disease, as Typified by the Flu.** (A) A susceptible person could be exposed to flu viruses in respiratory droplets at which time the incubation period begins. (B) The prodromal phase is characterized by mild signs and symptoms, such as a headache and fever. (C) The acute period is characterized by sudden symptoms of high fever with chills, cough, tired muscles and joint pain, and loss of appetite. (D) As the virus is eliminated from the body, the fever breaks and appetite returns as recovery begins. (E) With the period of convalescence, the body returns to normal. »» How would the course of the flu differ from that for a common cold?

MICROFOCUS 19.2: Evolution
Sepsis and Septic Shock

The presence of living bacteria in the bloodstream is called **bacteremia**. Because the blood is sterile, any bacterial cells detected in the blood are cause for alarm. In most situations, however, only a small number of bacterial cells gain entry and no symptoms develop because the invaders are rapidly removed by white blood cells (Chapter 20). Such temporary bacteremia may occur in healthy individuals during dental procedures or tooth brushing, when bacterial species living on the gums around the teeth enter the bloodstream through trauma to the gums. However, in a vulnerable host, such as a person with heart valve disease, prevention of bacteremia may include taking antibiotics prior to surgery or dental procedures to prevent any bacterial cells from colonizing the heart.

If more cells enter the bloodstream than can be effectively removed, an infection will develop. A more serious, but rare condition is **septicemia** (or **sepsis**) that occurs when the bacterial cells multiply and spread throughout the bloodstream. Often sepsis results from another infection in the body or from surgery on an infected area. In the United States, sepsis is the leading cause of death in non-coronary intensive care unit (ICU) patients, and the tenth most common cause of death overall according to data from the CDC. Sepsis affects 750,000 Americans each year of which about 30% die.

Sepsis can be caused by several gram-positive and gram-negative bacterial species (see figure A). Common gram-positive bacteria include *Staphylococcus aureus*, *Streptococcus pneumoniae*, *Enterococcus* species that are microbiota of the intestines, and *Streptococcus* pyogenes. Common gram-negative bacteria causing septic shock include opportunistic microbiota of the intestines such as *Escherichia coli*, *Klebsiella* species, and *Pseudomonas aeruginosa*. The most common obligate anaerobe is *Bacteroides fragilis*.

Sepsis is also called **systemic inflammatory response syndrome (SIRS)** because it involves an exaggerated immune response to released bacterial toxins (see figure B). Such a systemic response is very serious and can be life threatening. The manifestations of SIRS include two or more of the following conditions: high or low body temperature, increased heart rate, rapid breathing, and an elevated leukocyte count. Besides the inflammatory cascade that develops, the bacterial toxins attack the walls of the small blood vessels, causing them to become leaky so that fluid is lost from the blood into the surrounding tissues. Leakage and swelling also can develop in the lungs, causing difficulty breathing (respiratory distress).

Another dangerous effect of bacterial toxins is widespread clotting of the blood within the small blood vessels. This is called **disseminated intravascular coagulation (DIC)**, and can be fatal if not treated quickly. As clotting factors are used up, blood vessel leakage can lead to hemorrhaging. Often the person now has a condition called **severe sepsis**, which is associated with at least one acute organ dysfunction, decreased blood flow, or low blood pressure.

The risk of death now becomes high and requires immediate, aggressive treatment with antibiotics. A delay starting antibiotic treatment greatly decreases the person's chances of survival. Often two or three antibiotics will be given together to increase the chances of killing the bacterial cells.

Without treatment, septicemia often develops into **septic shock**, which is characterized by a dangerously low drop in blood pressure and multiple organ system failure (see figure B). In the United States, septic shock is the number one cause of death in ICUs and the 13th most common cause of death. It occurs most often in newborns and people with a weakened immune system.

The loss of fluid from the blood may be so great that the normal circulation (the rate the heart pumps at) cannot be maintained and blood pressure drops. Persistent hypotension reduces the blood flow and supply of oxygen to major organs such as the heart, kidneys, and brain. Signs of septic shock include a rapid and very weak pulse, reduced urine flow, confusion, and collapse—that is, multiple organ failure.

Septic shock is a medical emergency and is normally treated in an ICU. Large doses of antibiotics, along with infusions of fluids, are given to fight off the infection and maintain blood pressure. Drugs are given to increase blood flow to the brain, heart, and other organs. If the lungs fail, the person may need a mechanical ventilator to help breathing. Despite all efforts, more than 25% of people with septic shock die.

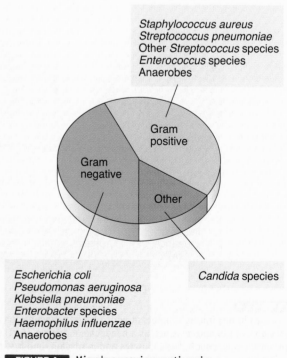

Staphylococcus aureus
Streptococcus pneumoniae
Other *Streptococcus* species
Enterococcus species
Anaerobes

Gram positive

Gram negative

Other

Escherichia coli
Pseudomonas aeruginosa
Klebsiella pneumoniae
Enterobacter species
Haemophilus influenzae
Anaerobes

Candida species

FIGURE A   Microbes causing septicemia.

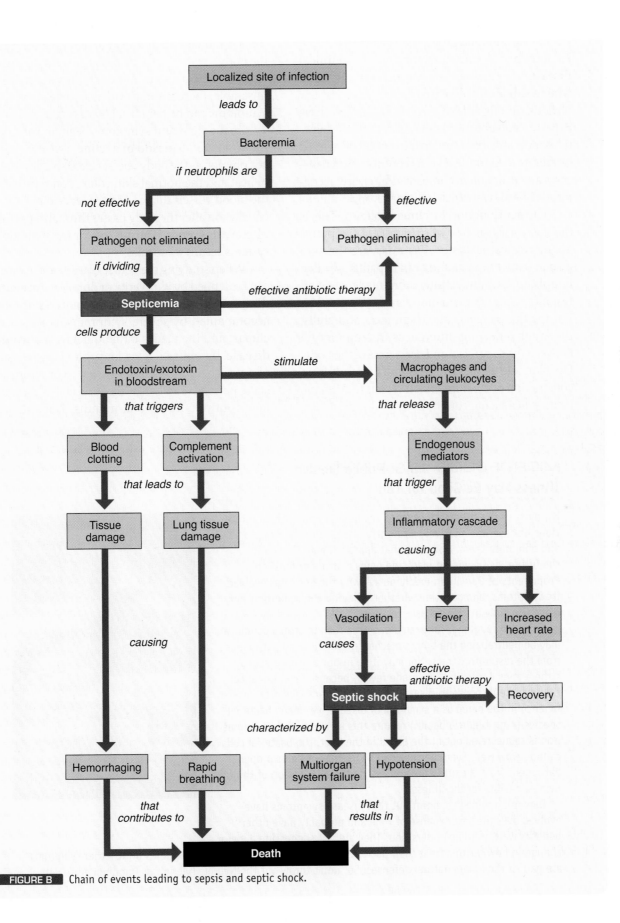

FIGURE B Chain of events leading to sepsis and septic shock.

competition between host and microbe has begun. During the onset and progression of a disease, it can be described as clinical or subclinical. A **clinical disease** is one in which the symptoms are apparent, while a **subclinical disease** is accompanied by few obvious symptoms. Many people, for example, have experienced subclinical cases of mumps or infectious mononucleosis, and in the process developed immunity to future attacks. By contrast, certain diseases are invariably accompanied by clearly recognized clinical symptoms. Influenza is one example.

The **acute period** or **climax** follows. This is the stage of the disease when signs and symptoms are of greatest intensity. Examples are the skin rash in scarlet fever, jaundice in hepatitis, swollen lymph nodes in infectious mononucleosis, and Koplik spots and rash in measles (see Chapter 15). For the flu, patients suffer high fever and chills, the latter reflecting differences in temperature between the superficial and deep areas of the body. Dry skin and a pale complexion may result from constriction of the skin's blood vessels to conserve heat. A headache, cough, body and joint aches,

and loss of appetite are common. The length of this period can be quite variable, depending on the body's response to the pathogen and the virulence of the pathogen. Although the patient feels miserable, there is some evidence some signs and symptoms can be beneficial (MicroFocus 19.3).

As the signs and symptoms begin to subside, the host enters a **period of decline**. Sweating may be common as the body releases excessive amounts of heat and the normal skin color soon returns as the blood vessels dilate. The sequence comes to a conclusion after the body passes through a **period of convalescence**. During this time, the body's systems return to normal.

When studying the course of a disease, it often can be defined by its severity or duration. An **acute disease**, like the flu, develops rapidly, is usually accompanied by severe symptoms, comes to a climax, and then fades rather quickly. A **chronic disease**, by contrast, often lingers for long periods of time. The symptoms are slower to develop, an acute period is rarely reached, and convalescence may continue for several months. Hepatitis A,

MICROFOCUS 19.3: Public Health
Illness May Be Good for You

For most of the twentieth century, medicine's approach to infectious disease was relatively straightforward: Note the symptoms and eliminate them. However, that approach may change in the future, as Darwinian medicine gains a stronger foothold. Proponents of Darwinian medicine ask why the body has evolved its symptoms, and question whether relieving the symptoms may leave the body at greater risk.

Consider coughing, for example. In the rush to stop a cough, we may be neutralizing the body's mechanism for clearing pathogens from the respiratory tract. Nor may it be in our best interest to stifle a fever (at least a low grade fever), because fever enhances the immune response to disease. Many physicians view iron insufficiency in the blood as a symptom of disease, yet many bacterial species (e.g., tubercle bacilli) require this element, and as long as iron is sequestered out of the blood in the liver, the bacterial cells cannot grow well. Even diarrhea can be useful—it helps propel pathogens from the intestine and assists the elimination of the toxins responsible for the illness.

Darwinian biologists point out that disease symptoms have evolved over the vast expanse of time and probably have other benefits waiting to be understood. They are not suggesting a major change in how doctors treat their patients, but they are pushing for more studies on whether symptoms are part of the body's natural defenses. So, don't throw out the Nyquil, Tylenol, or Imodium quite yet.

trichomoniasis, and infectious mononucleosis are examples of chronic diseases. Sometimes an acute disease may become chronic when the body is unable to rid itself completely of the microbe. For example, one who has contracted a parasitic disease, such as giardiasis or amoebiasis, may experience sporadic symptoms for many years.

With this understanding of the disease stages, we now can examine several factors required for the establishment of disease, as outlined in **FIGURE 19.8**.

CONCEPT AND REASONING CHECKS

19.4 Assign the following signs and symptoms of the flu to the appropriate disease stage (fever, headache, chills, and muscle pain).

Pathogen Entry into the Host Depends on Cell Adhesion and the Infectious Dose

KEY CONCEPT

5. Pathogens gain access to the host through portals of entry.

A portal of entry refers to the characteristic route by which an exogenous pathogen enters the host. It varies considerably for different organisms and is a key factor leading to the establishment of disease. Abrasions or mechanical injury to the skin can be a portal of entry. For example, tetanus may occur if *Clostridium tetani* spores on a sharp object in the soil puncture the skin and enter the anaerobic tissue of a wound. Tetanus will not develop if spores are consumed with food because the spores do not germinate in the human intestinal tract.

The ability of a pathogen to establish an infection and possible disease usually depends on the **infectious dose**, the numbers of microbes taken into the body. The consumption of a few hundred thousand typhoid bacilli will lead to disease. By contrast, many millions of cholera bacilli must be ingested if cholera is to be established. One explanation for the difference is the high resistance of typhoid bacilli to the acidic conditions in the stomach, in contrast to the low resistance of cholera bacilli. Also, it may be safe to eat fish when the water contains hepatitis A viruses, but eating raw clams from the same water can be dangerous because clams are filter-feeders, concentrating hepatitis A viruses in their bodies.

Often the host is exposed to low doses of a pathogen and, as a result, develops immunity. For instance, many people can tolerate low numbers of

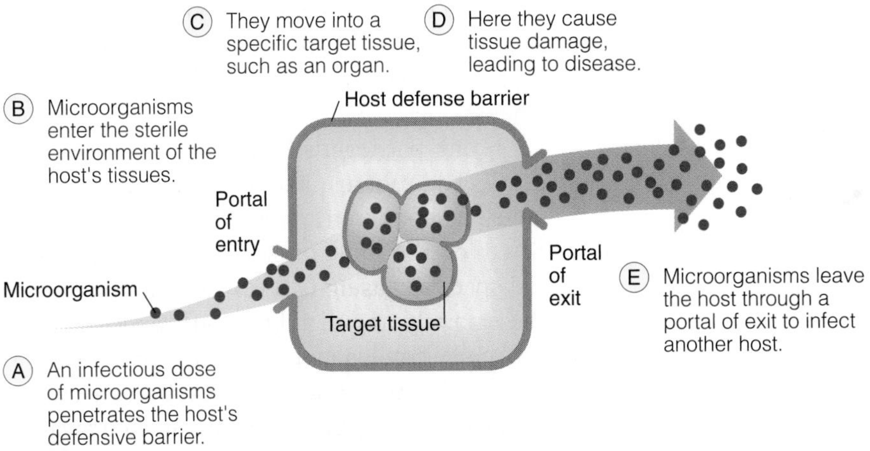

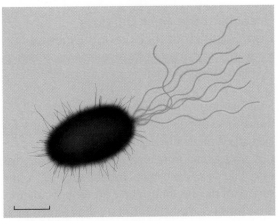

Ⓒ They move into a specific target tissue, such as an organ.

Ⓓ Here they cause tissue damage, leading to disease.

Host defense barrier

Ⓑ Microorganisms enter the sterile environment of the host's tissues.

Portal of entry

Portal of exit

Microorganism

Target tissue

Ⓔ Microorganisms leave the host through a portal of exit to infect another host.

Ⓐ An infectious dose of microorganisms penetrates the host's defensive barrier.

FIGURE 19.8 **The Generalized Events on the Establishment of a Local Disease.** The infectious dose and adhesion to cells or tissues are required to initiate infection and disease. »» Identify which events would not occur if an infection but not a disease occurred.

FIGURE 19.9 **Pili on *Escherichia coli*.** A false-color transmission electron micrograph of *E. coli* displaying the pili used for adhesion to the tissue. Adhesins in pili such as these increase the pathogenicity of the organism by allowing it to localize at its appropriate tissue site. (Bar = 1 μm.) »» What are the long structures projecting from the right side of the cell?

mumps viruses without exhibiting disease. They may be surprised to find they are immune to mumps when it breaks out in their family at some later date.

Many pathogens enter at specific, natural portals of entry because these microbes contain on their surface "sticky" factors, called **adhesins**, that allow pathogens to adhere to appropriate tissue sites. A variety of adhesins often are associated with bacterial capsules, flagella, or pili (see Chapter 4) (**FIGURE 19.9**). For example, the gonococci and some other agents of sexually transmitted diseases often attach by means of pili to specific receptor sites only found on tissues of the urogenital system. The host cell is often an active partner in the adhesion because the pathogen triggers it to express target receptor sites for adhesin binding. Also, many viruses have spikes on the capsid or envelope, allowing for attachment (see Chapter 14).

Some pathogens have multiple portals of entry. The tubercle bacillus, for instance, may enter the body in respiratory droplets, contaminated food or milk, or skin wounds. The bacterial species causing Q fever can enter by any of these portals, as well as by an arthropod bite. The tularemia bacillus may enter the eye by contact, the skin by an abrasion, the respiratory tract by droplets, the intestines by contaminated meat, or the blood by an arthropod bite.

CONCEPT AND REASONING CHECKS

19.5 Assess the role of the infectious dose and adhesion in establishing an infection.

Breaching the Host Barriers Can Establish Infection and Disease

KEY CONCEPT

6. Invasiveness is critical for many pathogens.

Some pathogens do not need to penetrate cells or tissues to cause disease. The pertussis bacillus, for example, adheres to the surface layers of the respiratory tract while producing the toxins causing disease. Likewise, the cholera bacillus attaches to the surface of the intestine, where it produces toxins.

However, the ability of a pathogen to penetrate tissues and cause structural damage is an important component for the virulence of many pathogens. The ability to penetrate and spread is called **invasiveness** and the bacilli of typhoid fever and the protozoan causing amoebiasis are examples of pathogens that depend on their invasiveness. By penetrating the tissue of the gastrointestinal tract, these microorganisms cause ulcers and sharp, appendicitis-like pain characteristic of the respective diseases.

Invasiveness often is facilitated through the pathogen's internalization by immune cells (**FIGURE 19.10**). These cells, including macrophages, undergo **phagocytosis** by engulfing pathogens, taking them into the cell cytoplasm in vacuoles, and then attempting to destroy them in lysosomes. In addition, some bacterial pathogens are internalized by inducing nonphagocytic cells to undergo phagocytosis. If the pathogen can evade destruction by lysosomes, the cell interior provides a protective niche or a vehicle to pass

Actin:
A cytoskeletal protein essential for cell movement and the maintenance of cell shape in most eukaryotic cells.

Macrophages:
Large white blood cells that remove waste products, microorganisms, and foreign material from the bloodstream.

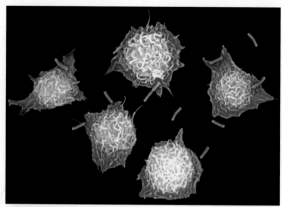

FIGURE 19.10 Macrophages Undergoing Phagocytosis. In this false-color scanning electron micrograph, macrophages, a type of immune cell, are capturing bacterial cells for phagocytosis. »» Why might phagocytosis be an important activity of the human immune system when infection occurs?

through otherwise impenetrable defenses, such as the blood-brain barrier.

Some cytoplasm-invading pathogens, such as *Shigella flexneri* (shigellosis), *Rickettsia prowazekii* (epidemic typhus), and *Listeria monocytogenes* (listeriosis) have cell membrane adhesive proteins that form a zipper-like binding of pathogen to the host cell. As a result of this molecular adhesion and cross-talk, the pathogen triggers the host cell to undergo phagocytosis. Once in the cell, the pathogens escape from the vacuole, eliminating any chance of their destruction by host lysosomes (**FIGURE 19.11**). In the cytoplasm, the bacterial cells trigger the host cell to synthesize an actin tail on the bacterial cells, which propels the cells through the cell's cytoplasm. When a cell bumps against the host's plasma membrane, it distorts and indents the adjacent cell, bridges the junction between the two cells, and enters the next cell (somewhat like moving from train car to train car through connecting doors). The system allows bacterial invasion to occur without the bacterial cells leaving the cellular environment.

CONCEPT AND REASONING CHECKS

19.6 Assess phagocytosis as an invasiveness mechanism used by pathogens.

Successful Invasiveness Requires Pathogens to Have Virulence Factors

KEY CONCEPT

7. Virulence factors include enzymes and toxins.

It should be noted that upon entry into a host, a pathogen is confronted with a profoundly dif-

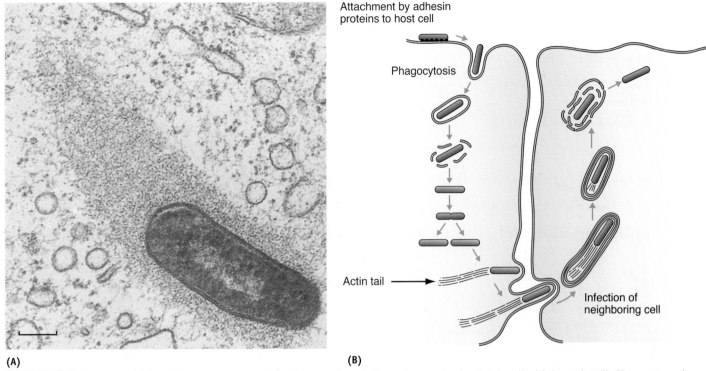

(A)

(B)

FIGURE 19.11 Invasion by *Listeria monocytogenes*. (**A**) A transmission electron micrograph of a *Listeria* cell with its actin tail. (Bar = 0.3 μm.) (**B**) After entering a host cell by phagocytosis, the bacterial cell loses the surrounding membrane and divides to form a larger population. Invasion of neighboring cells uses the actin tail to "drive" the cells into the adjacent cell. In that cell, the cell again loses the surrounding cell membranes and initiates another infection. »» What invasion process is negated by the *Listeria* cell's ability to generate an actin tail?

ferent environment. Adaptation to this environment requires the genetic machinery enabling the pathogen to adapt and withstand the resistance put forward by the host. Several factors may be present to overcome host defenses.

Enzymes. The virulence of a microbe depends to some degree on its ability to produce a series of extracellular enzymes to help the pathogen resist body defenses. A few examples illustrate how bacterial enzymes act on host cells and interfere with certain functions or barriers meant to retard invasion.

One bacterial enzyme is the **coagulase** produced by virulent staphylococci (**FIGURE 19.12A**). Coagulase catalyzes the formation of a blood clot from fibrinogen proteins in human blood. The clot sticks to staphylococci, protecting them from phagocytosis. Part of the walling-off process observed in a staphylococcal skin boil is due to the clot formation. Coagulase-positive *Staphylococcus aureus* may be identified in the laboratory by combining the cells with human or rabbit plasma. The formation of a clot in the plasma indicates coagulase activity.

Many streptococci have the ability to produce the enzyme **streptokinase**. This substance dissolves fibrin clots used as a defense by the body to restrict and isolate an infected area. Streptokinase thus overcomes an important host defense and allows further tissue invasion by the bacterial cells.

Hyaluronidase is sometimes called the "spreading factor" because it enhances penetration of a pathogen through the tissues. The enzyme digests hyaluronic acid, a polysaccharide that binds cells together in a tissue (**FIGURE 19.12B**). The term tissue cement is occasionally applied to this polysaccharide. Hyaluronidase is an important virulence factor in pneumococci and certain species of streptococci and staphylococci. In addition, gas gangrene bacilli use it to facilitate spread through the muscle tissues.

Some pathogens also produce enzymes that destroy blood cells. **Leukocidins** are products of staphylococci, streptococci, and pneumococci. The enzymes destroy circulating neutrophils and tissue macrophages, both of which are immune system cells designed to phagocytize and destroy the pathogens. The enzymes attach to the immune

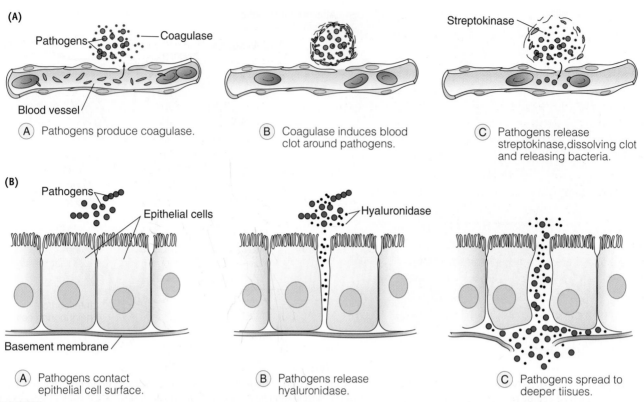

(A)

Pathogens — Coagulase

Blood vessel

Ⓐ Pathogens produce coagulase.

Ⓑ Coagulase induces blood clot around pathogens.

Streptokinase

Ⓒ Pathogens release streptokinase, dissolving clot and releasing bacteria.

(B)

Pathogens

Epithelial cells

Hyaluronidase

Basement membrane

Ⓐ Pathogens contact epithelial cell surface.

Ⓑ Pathogens release hyaluronidase.

Ⓒ Pathogens spread to deeper tiisues.

FIGURE 19.12 Enzyme Virulence Factors. **(A)** Some bacterial cells produce the enzyme coagulase, which triggers clotting of blood plasma. Bacterial cells within the clot can break free by producing streptokinase. **(B)** Some invasive bacterial species produce the enzyme hyaluronidase, which degrades the cementing polymer holding cells of the intestinal lining together. »» From these examples, how do virulence factors protect bacterial cells and increase their virulence?

cells' membrane and trigger changes leading to the release of lysosomal enzymes in the cytoplasm. The phagocytes quickly lyse.

Hemolysins combine with the membranes of erythrocytes, causing lysis to take place. Staphylococci and streptococci produce these virulence factors, which, by lysing red blood cells, gives the pathogens the iron in hemoglobin that the bacterial cells need for metabolism. In the laboratory, hemolysin producers can be detected by **hemolysis**, a destruction of blood cells in a blood agar medium (see Chapter 10).

Furthermore, if a pathogen exists in a **biofilm**, its virulence can be enhanced because here it can resist body defenses and drugs. A biofilm is a sticky layer of extracellular polysaccharides and proteins enclosing a colony of bacterial cells at the tissue surface (see Chapter 3). Phagocytes and antibodies have difficulty reaching the microor-

ganisms in this slimy conglomeration of armor-like material. Moreover, microorganisms often survive without dividing in a biofilm. This makes them impervious to the antibiotics that attack dividing cells. (Indeed, the antibiotics do not penetrate the biofilm easily.) CDC officials have estimated that 65% of human infections involve biofilms. **TABLE 19.1** summarizes the activities of these enzymes.

Toxins. Microbial poisons, called **toxins**, can profoundly affect the establishment and course of disease. The ability of pathogens to produce toxins is referred to as **toxigenicity**, while toxins present in the blood is called **toxemia** and the person is considered "intoxicated." Two types of toxins are recognized: exotoxins and endotoxins.

Exotoxins are heat-sensitive protein molecules, manufactured during bacterial metabolism.

They are produced by gram-positive and gram-negative bacterial cells and released into the host environment. Alternatively, some gram-negative bacteria inject toxins directly into the host cell. The toxins act locally or diffuse to their site of activity where symptoms of disease soon develop.

The exotoxin produced by the botulism bacillus *Clostridium botulinum* is among the most lethal toxins known (see Chapter 11). Botulism toxin is a neurotoxin that inhibits the release of acetylcholine at the synaptic junction, a process leading to a type of "flaccid" paralysis. Another neurotoxin is produced by the tetanus bacillus, *C. tetani* (see Chapter 12). In this case, the exotoxin blocks the relaxation pathway that follows muscle contraction, thereby permitting volleys of spontaneous nerve impulses and uncontrolled muscular contractions causing a "rigid" paralysis. Other types of exotoxins are identified in TABLE 19.2 .

The body responds to exotoxins by producing antibodies called **antitoxins**. When toxin and antitoxin molecules combine with each other, the toxin is neutralized (Chapter 21). This process represents an important defensive measure in the body. Therapy for people who have botulism, tetanus, or diphtheria often includes injections of antitoxins (immune globulin) to neutralize the toxins.

Because exotoxins are proteins, they are susceptible to the heat and chemicals that normally denature proteins. A chemical such as formaldehyde may be used in the laboratory to alter the toxin and destroy its toxicity without hindering its ability to elicit an immune response. The result is a **toxoid**. When the toxoid is injected into the body, the immune system responds with antitoxins, which circulate and provide a measure of defense against disease. Toxoids are used for diphtheria and tetanus immunizations in the diphtheria-tetanus-acellular pertussis (DTaP) vaccine.

Endotoxins all have similar effects and usually are released only upon disintegration of gram-negative cells. They are present in the outer membrane in many gram-negative bacilli and are part of the lipid-polysaccharide (LPS) complex (see Chapter 4). The lipid portion of the LPS is the toxic agent. Endotoxins do not stimulate an immune response in the body, nor can they

TABLE 19.1	A Summary of Some Bacterial Enzymes that Contribute to Virulence		
Enzyme	Source	Action	Effect
Coagulase	*Staphylococcus aureus*	Forms a fibrin clot	Provides resistance to phagocytosis
Streptokinase	Streptococci Staphylococci	Dissolves a fibrin clot	Prevents isolation of infection
Hyaluronidase	Streptococci Staphylococci	Digests hyaluronic acid	Allows tissue penetration
Leukocidin	Staphylococci Streptococci Pneumococci	Destroys phagocytes	Limits phagocytosis
Hemolysins	Clostridia Staphylococci Streptococci	Lyses red blood cells	Provides pathogens with source of iron for growth

be altered to prepare toxoids. They function by activating a blood-clotting factor to initiate blood coagulation and by influencing the complement system (Chapter 20). The toxins of plague bacilli are especially powerful.

Endotoxins all have similar toxic effects. At high concentrations, they manifest their presence by certain signs and symptoms. Usually an individual experiences an increase in body temperature, substantial body weakness and aches, and general malaise. Damage to the circulatory system and shock may occur. In this case, the permeability of the blood vessels changes and blood leaks into the intercellular spaces. The tissues swell, the blood pressure drops, and the patient may lapse into a coma. This condition, commonly called **endotoxin shock**, may accompany antibiotic treatment of diseases due to gram-negative bacilli because endotoxins are released as the bacilli are killed by the antibiotic.

Endotoxins usually play a contributing rather than a primary role in the disease process. They may reduce platelet counts in the host and thereby increase hemorrhaging elsewhere in the body. Like exotoxins, endotoxins add to the virulence of a microbe and enhance its ability to establish disease. TABLE 19.3 summarizes the characteristics of the bacterial toxins.

TABLE

19.2   Characteristics and Effects of Some Bacterial Exotoxins

Exotoxin	Organism	Gene Location	Disease	Effect
Anthrax toxin	*Bacillus anthracis*	Plasmid	Anthrax	Altered host cell communication; cell death
Botulism toxin	*Clostridium botulinum*	Prophage	Botulism	Flaccid paralysis
Cholera toxin	*Vibrio cholerae*	Prophage	Cholera	Water and electrolyte loss
Diphtheria toxin	*Corynebacterium diphtheriae*	Prophage	Diphtheria	Inhibits protein synthesis; cell death
Enterotoxin	*Clostridium perfringens*	Chromosomal	Food poisoning	Permeability of intestinal epithelia
Enterotoxin	*Escherichia coli*	Plasmid	Diarrhea	Water and electrolyte loss
Enterotoxin A	*Staphylococcus aureus*	Prophage	Food poisoning	Diarrhea and nausea
Erythrogenic toxin	*Streptococcus pyogenes*	Prophage	Scarlet fever	Capillary destruction
Exfoliative toxin	*Staphylococcus aureus*	Prophage	Scalded skin syndrome	Massive skin blistering
Exotoxin A	*Pseudomonas aeruginosa*	Chromosomal	Pneumonia (?)	Inhibits protein synthesis; cell death
Perfringens toxin	*Clostridium perfringens*	Chromosomal	Gas gangrene	Hemolysis; membrane lysis
Pertussis toxin	*Bordetella pertussis*	Chromosomal	Whooping cough (pertussis)	Interferes with host cell communication
Pyrogenic toxin	*Staphylococcus aureus*	Prophage	Toxic shock syndrome	Fever, shock
Tetanus toxin	*Clostridium tetani*	Plasmid	Tetanus	Rigid paralysis

TABLE

19.3   A Comparison of Exotoxins and Endotoxins

Characteristic	Exotoxins	Endotoxins
Source	Living gram-positive and gram-negative bacteria	Lysed gram-negative bacteria
Location	Released from cell	Part of cell wall
Chemical composition	Protein	Lipopolysaccharide
Heat sensitivity	Labile (60–80°C)	Stable (250°C)
Immune reaction	Strong	Weak
Conversion to toxoid	Possible	No
Fever	No	Yes
Toxigenicity	High	Low
Representative effects	Interfere with synaptic activity (botulism) Interrupt protein synthesis (diphtheria) Increase capillary permeability Increase water elimination (cholera)	Increase body temperature Increase hemorrhaging Increase swelling in tissues Induce vomiting, diarrhea

CONCEPT AND REASONING CHECKS

19.7 Evaluate the role of enzymes and toxins as important virulence factors in the establishment of disease.

Pathogens Must Be Able to Leave the Host to Spread Disease

KEY CONCEPT

8. Pathogens leave the host through portals of exit.

At the conclusion of its pathogenicity cycle, pathogens or their toxins exit the host through some suitable **portal of exit** (**FIGURE 19.13**). This is of more than passing importance because easy transmission permits the pathogen to continue its pathogenic existence in the world.

CONCEPT AND REASONING CHECKS

19.8 Are portal of entry and exit always the same? Explain.

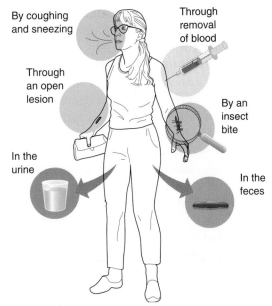

FIGURE 19.13 Six Different Portals of Exit from the Body. »» Why do pathogens need specific portals of exit?

19.3 Infectious Disease Epidemiology

Infectious disease epidemiology is concerned with how infectious diseases are distributed in a population and the factors influencing or determining that distribution. In this final section, we examine the factors putting population groups at risk of contracting infectious disease.

We also look at special environments, such as health care settings, and the public agencies saddled with the job of disease identification, control, and prevention.

Epidemiologists Often Have to Identify the Reservoir of an Infectious Disease

KEY CONCEPT

9. Reservoirs are places in the environment where a pathogen can be found.

To cause an infection, pathogens have to be transferred from a source to a susceptible host (**FIGURE 19.14**). For many diseases to perpetuate themselves, the disease-causing microbes must exist somewhere in the environment. These ecological niches or sources where a microbe lives and multiplies are called **reservoirs** of infection. Animals are one type of reservoir. A domestic house cat that is infected with *Toxoplasma gondii,* for example, usually shows no symptoms of

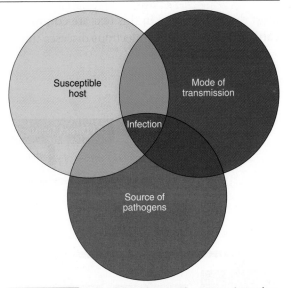

FIGURE 19.14 Infectious Disease Elements. Assuming there is a susceptible host, to cause an infection the pathogenic organism must have a way to be transmitted to that susceptible host. »» What is meant by a susceptible host?

toxoplasmosis but it can transmit the protozoan to humans, where the disease manifests itself. Water and soil also can be reservoirs because they often are contaminated with disease agents, such as the cholera bacterium or *Giardia* protozoan.

Niches: Environmental areas that ensure an organism's survival.

Not all diseases have a nonhuman reservoir though. The smallpox virus only exists in humans. This is why the World Health Organization (WHO) was able to limit the spread of the virus through vaccination and eradicate smallpox from the world by locating all human reservoirs.

A special type of reservoir is a **carrier**, which is a person who has recovered from the disease but continues to shed the disease agents. For instance, a person who has recovered from typhoid fever or amoebiasis becomes a carrier for many weeks after the symptoms of disease have left. The feces of this individual may spread the disease to others via contaminated food or water.

CONCEPT AND REASONING CHECKS

19.9 Identify the different reservoirs of disease transmission.

Epidemiologists Have Several Terms that Apply to the Infectious Disease Process

KEY CONCEPT

10. Diseases have certain behaviors in populations.

Most diseases studied in this text are **communicable diseases**; that is, infectious diseases trans-missible among hosts in a population. Certain communicable diseases are described as being **contagious** because they pass with particular ease among hosts and are highly infectious. Chickenpox and measles fall into this category.

Noncommunicable diseases are singular events in which the agent is acquired directly from the environment and are not easily transmitted to the next host. In tetanus, for example, penetration of soil containing *Clostridium tetani* spores to the anaerobic tissue of a wound must occur before this disease develops. It cannot be spread person-to-person.

CONCEPT AND REASONING CHECKS

19.10 Explain the difference between a communicable and contagious disease. Provide examples beyond the ones mentioned above.

Infectious Diseases Can Be Transmitted in Several Ways

KEY CONCEPT

11. Disease transmission can involve direct or indirect contact.

Diseases can be transmitted by a broad variety of methods involving **direct contact**

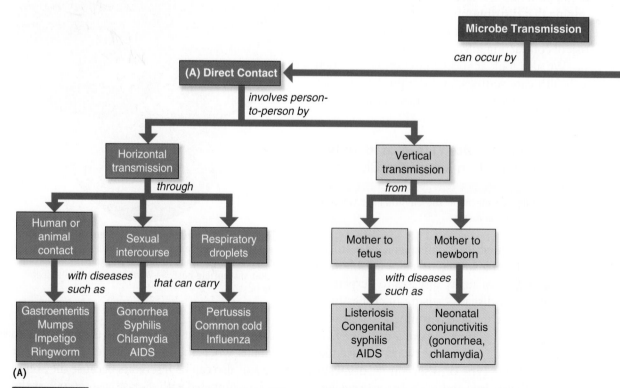

FIGURE 19.15 **Transmission of Microorganisms and Viruses.** **(A)** Pathogens can be transmitted by direct contact involving horizontal or vertical transmission. **(B)** Pathogens also can be transmitted by fomites, contaminated food and water, and vectors. »» Would indirect contact transmission represent horizontal or vertical transmissions? Explain.

(FIGURE 19.15A); that is, close contact that results in exposure to skin or body secretions.

Direct Contact. Person-to-person or **horizontal transmission** implies close or personal contact with someone who is infected or who has the disease. Hand-shaking or kissing an infected person can spread bacterial cells, viruses, or other pathogens to an uninfected person. For some diseases, such as rabies, leptospirosis, and toxoplasmosis, direct contact with an animal is necessary. An animal bite or scratch by an infected animal can spread the disease to an uninfected person. The exchange of body fluids, such as through sexual contact, is another example of direct contact transmission for diseases like gonorrhea and AIDS.

Direct transmission also can involve the violent expulsion of **respiratory droplets** through sneezing, coughing, or simply talking (FIGURE 19.16). **Droplet transmission** through the air requires the "recipient" be close to an infected individual. In a sneeze, the droplets can travel 150 feet per second. However, the droplets are fairly large and fall out of the air within about 1 meter of their source. If an uninfected person is within that distance, the eyes, mouth, or nose may be portals of entry for the airborne pathogens.

Direct contact called **vertical transmission** includes the spread of pathogens, such as HIV or *Toxoplasma gondii,* from a pregnant mother to her unborn child. Transmission of gonorrhea from mother to newborn can occur during labor or delivery.

CONCEPT AND REASONING CHECKS
19.11A Identify four ways by which infectious disease can be transmitted directly.

Indirect Contact. Indirect transmission can be the result of contact with a non-living object or medium, or a vector (FIGURE 19.15B). **Fomites** are inanimate objects on which or in which disease organisms linger for some period of time. For instance, bed linens may be contaminated with pinworm eggs, and contaminated syringes and needles may passively transport the viruses of hepatitis B or AIDS.

Vehicle transmission involves the indirect spread of disease through contaminated food and water, or air. Foods can be contaminated during processing or handling, or they may be dangerous when made from diseased animals. Poultry products, for example, are often a source of salmonellosis because *Salmonella* species frequently infect chickens, while pork may spread trichinellosis because *Trichinella* parasites may

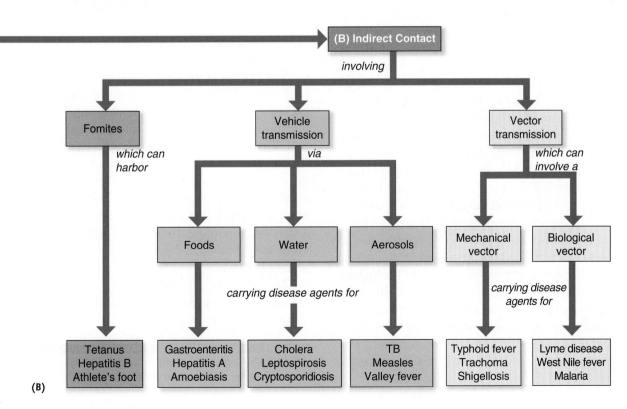

(B)

FIGURE 19.16 Droplet Transmission. Sneezing or coughing represents a method for airborne transmission of pathogens. »» Identify some portals of entry for pathogen-containing droplets.

live in muscles of the pig. Other examples include the cholera bacterium, *Vibrio cholerae,* which can contaminate many water supplies in developing nations lacking proper water sanitation and the parasite causing giardiasis, *Giardia intestinalis,* which is found in some recreational waters. MICROFOCUS 19.4 describes two transmission modes: airborne and fomites.

Pathogens also are transmitted through the air on smaller particles called **aerosols**. With **particle transmission**, the particles can remain suspended in the air for longer periods of time and can be moved some distance by air currents. The virus of SARS and the bacterial cells of tuberculosis are two pathogens that can be carried in the air by droplets or aerosols. MICROFOCUS 19.5 describes long-range effects of airborne transmission.

Arthropods represent another indirect method of transmission. Many pathogens hitch a ride on arthropods, such as mosquitoes, ticks, fleas, and lice, which act as **vectors**, living organisms carrying disease agents from one host to another. **Mechanical vectors** represent arthropods passively transporting microbes on their legs and other body parts. For example, house flies can carry diseases picked up on their feet. In other cases, arthropods represent **biological vectors**, where the pathogen must multiply in the insect before it can infect another host. The malarial protozoan and the West Nile virus infect and reproduce in mosquitoes and accumulate in their salivary glands, from which the pathogens are injected during the next blood meal.

CONCEPT AND REASONING CHECKS

19.11B Identify the methods by which infectious disease can be transmitted indirectly.

Diseases Also Are Described by How They Occur Within a Population

KEY CONCEPT

12. Diseases are identified as being endemic, epidemic, or pandemic.

When epidemiologists investigate an infectious disease, they need to determine if it is localized or spread through a community or region. **Endemic** refers to a disease habitually present at a low level in a certain geographic area. Plague in the American Southwest is an example.

By comparison, an **epidemic** refers to a disease that occurs in a community or region in excess of what is normally found within that population. Influenza often causes widespread epidemics. This should be contrasted with an **outbreak**, which is a more contained epidemic. An abnormally high number of measles cases in one American city would be classified as an outbreak. Not only do epidemiologic investigations look at current outbreaks, they also consider future outbreaks, including bioterrorism (MICROFOCUS 19.6).

A **pandemic** is a worldwide epidemic, affecting populations around the globe. The most obvious example here would be AIDS and the H1N1 flu.

As in the opening quote, "*Health care matters to all of us some of the time, public health matters to all of us all of the time,*" maintaining vigilance against infectious disease is extremely important. For this reason, national and international public health organizations, such as the CDC and the WHO, learn a lot about diseases by analyzing disease data reported to them. The CDC, for example, has a list of infectious diseases that must be reported to state health departments, which then report them to the CDC (TABLE 19.4). These are published in the *Morbidity and Mortality Weekly Report.*

MicroInquiry 19 explores the use of epidemiological data as a tool for understanding disease occurrence.

CONCEPT AND REASONING CHECKS

19.12 Why do you think the term "outbreak" is typically used in news releases rather than epidemic?

MICROFOCUS 19.4: Environmental Microbiology

Riders on the Storm

Dust storms can be a relatively common meteorological phenomenon in arid and semi-arid regions of the world. Sometimes called sandstorms, they arise when a gust front passes or when the wind force is strong enough to remove loose sand and dust from the dry soil surface. The result can be an awesome wall of sand that can obscure visibility within seconds. On a more global scale, the major dust storms arise in the Sahara Desert and arid lands around the Arabian Peninsula. In Asia, the Gobi Desert is a major source. Annually, such storms carry more than 3 billion metric tons of dust aloft into the atmosphere.

The long-range movement of dust and suspended particles can certainly have an impact on air quality that often can be observed. Between May and October, strong winds blow off the Sahara Desert and the west coast of North Africa, carrying soil and dust westward across the Atlantic Ocean (see figure). Although much of the dust settles in the ocean, large amounts stay suspended and make the journey to the Caribbean islands and Florida. It is not unusual to wake up in the morning and see a fiery orange sunrise—what the locals call a "tequila sunrise." The orange is the African dust. In fact, traces of such dust have been detected as far west as New Mexico!

Today, these storms are becoming of more concern as worldwide deforestation, overgrazing, and climate change combine to generate massive dust clouds that can carry particles aloft. Besides causing respiratory distress, such as asthma, these storms may have another impact on human health—microbes can be "riders on the storm."

Scientists had believed most of these microbes would be killed by the intense ultraviolet light in the atmosphere and the dry, desiccating conditions of a dust storm. However, recent studies have shown that hundreds of bacterial and fungal species can be cultured from samples of dust clouds moving across the mid-Atlantic Ocean. Exactly what microbes are carried aloft and which might be pathogens is not yet clear. *Pseudomonas aeruginosa* has been detected and many species of *Aspergillus* have been identified. In fact, the researchers suggest that 20% to 30% of the microbes in the dust clouds are animal or plant pathogens.

More research is needed to define the role of dust storms as an indirect transmission mechanism spreading transatlantic pathogens across the globe. Still, these "riders on the storm" might not only renew reservoirs for some plant and animal pathogens in the United States, but also, on occasion, may bring "riders" capable of new diseases.

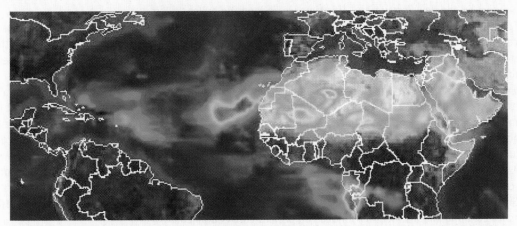

Satellite image of African dust storm spreading westward across the Atlantic Ocean towards Central and North America.

MICROFOCUS 19.5: Public Health

Planes, Trains, and — Ambulances

Vehicles that can take us to great destinations at supersonic speeds, or save our lives, can also harbor and transmit infectious disease.

In August, 2004, a New Jersey man returned home from a trip to West Africa. Within hours after taking the train home from the airport, the man was stricken with fever, chills, a severe sore throat, diarrhea, and back pain. The family rushed him to a local hospital. However, despite intensive care, the gentleman continued to decline and died a few days later. Clinical and postmortem specimens were sent to the Centers for Disease Control and Prevention (CDC) for a specific cause of death. The finding: Lassa fever.

The alarms went off! Lassa fever is an acute viral disease, rarely seen outside West Africa where the disease is endemic (see Chapter 15). In West Africa, the virus, which is carried in rodents, infects 100,000 to 300,000 people every year, and kills 5,000. The case of the New Jersey man was a lethal health risk as the virus can be spread person to person if a susceptible person comes into contact with virus in the blood, tissue, or excretions from infected individuals. Although the virus cannot be spread through casual contact (including skin-to-skin contact without exchange of body fluids), CDC epidemiologists contacted all passengers seated nearby the man (see figure) and others who could have been exposed. These individuals were asked to monitor their body temperatures for fever. Luckily, there were no reported cases.

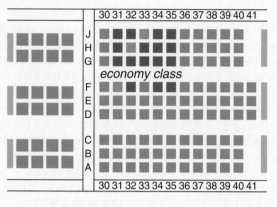

■ Patient
■ Passenger reported healthy
■ Passenger could not be contacted

Passenger seating and contact status.

The scary part is that the man was traveling while ill and potentially exposed some 200 people to the virus: 19 family members, 139 healthcare workers, 16 lab workers, 19 airplane passengers, and numerous commuters on the New Jersey train. This time we dodged the bullet.

From November 2004 to April 2005, a decontamination firm in the United Kingdom examined the ambulances from 12 firms for microbial contamination. They swabbed several fomites, including stretcher rails, the stretcher tracks below the stretcher, the paramedic's utility bag, and five other sites within the vehicle. The swabs were streaked on nutrient agar plates to see what bacterial species would grow.

Examination of the plates indicated the ambulances were heavily contaminated with a diverse group of bacterial species. In fact, in many cases, there were so many bacterial colonies present, they could not be counted. The bacterial species included antibiotic-resistant *Staphylococcus aureus* and a variety of species typically found in the human colon. More surprising, after the ambulances were cleaned by standard procedures, there was little reduction in the numbers of bacterial cells present. In fact, another study showed that cleaning actually spread the bacterial cells onto previously "clean" surfaces. Such contaminated fomites could be dangerous to a person with open wounds in the ambulance.

Since that initial 2005 report, emergency medical services have evaluated and improved their best practices for cleaning and disinfecting surfaces on patient care equipment.

TABLE

19.4 CDC's Summary of Notifiable Diseases in the United States in 2007

Acquired immunodeficiency syndrome (AIDS)

Anthrax

Botulism

Brucellosis

Chancroid

Chlamydia trachomatis, genital infections

Cholera

Coccidioidomycosis

Cryptosporidiosis

Cyclosporiasis

Diphtheria

Domestic arboviral diseases, neuroinvasive and
   nonneuroinvasive
   California serogroup virus disease
   Eastern equine encephalitis virus disease
   Powassan virus disease
   St. Louis encephalitis virus disease
   West Nile virus disease
   Western equine encephalitis virus disease

Ehrlichiosis
   Human granulocytic
   Human monocytic
   Human, other or unspecified agent

Giardiasis

Gonorrhea

Haemophilus influenzae, invasive disease

Hansen disease (leprosy)

Hantavirus pulmonary syndrome

Hemolytic uremic syndrome, postdiarrheal

Hepatitis A, acute

Hepatitis B, acute

Hepatitis B, chronic

Hepatitis B, perinatal infection

Hepatitis C, acute

Hepatitis C, infection (past or present)

Human immunodeficiency virus (HIV) infection
   Adult (age ≥ 13 yrs)
   Pediatric (age < 13 yrs)

Influenza-associated pediatric mortality

Legionellosis

Listeriosis

Lyme disease

Malaria

Measles

Menigococcal disease

Mumps

Pertussis

Plague

Poliomyelitis, paralytic

Psittacosis

Q fever

Rabies
   Animal
   Human

Rocky Mountain spotted fever

Rubella

Rubella, congenital syndrome

Salmonellosis

Severe acute respiratory syndrome-associated
   coronavirus (SARS-CoV) disease

Shiga toxin-producing *Escherichia coli* (STEC)

Shigellosis

Smallpox

Streptococcal disease, invasive, group A

Streptococcal toxic-shock syndrome

Streptococcus pneumoniae, invasive disease
   Age <5 yrs
   Drug-resistant, all ages

Syphilis

Syphilis, congenital

Tetanus

Toxic-shock syndrome (other than streptococcal)

Trichinellosis

Tuberculosis

Tularemia

Typhoid fever

Vancomycin-intermediate *Staphylococcus
aureus* infection (VISA)

Vancomycin-resistant *Staphylococcus aureus*
infection (VRSA)

Varicella infection (morbidity)

Varicella mortality

Yellow fever

Bioterrorism: The Weaponization and Purposeful Dissemination of Human Pathogens

The anthrax attacks that occurred on the East Coast in October 2001 confirmed what many health and governmental experts had been saying for over 10 years—it is not if bioterrorism would occur but when and where. Bioterrorism represents the intentional or threatened use of primarily microorganisms or their toxins to cause fear in or actually inflict death or disease upon a large population for political, religious, or ideological reasons.

Is Bioterrorism Something New?

Bioterrorism is not new, and two other MicroFocus boxes in this text (MicroFocus 12.1 and 12.4) mention historical examples. Such bioterrorism agents also have been used as biowarfare agents. In the United States, during the aftermath of the French and Indian Wars (1754–1763), British forces, under the guise of goodwill, gave smallpox-laden blankets to rebellious tribes sympathetic to the French. The disease decimated the Native Americans, who had never been exposed to the disease before and had no immunity. Between 1937 and 1945, the Japanese established Unit 731 to carry out experiments designed to test the lethality of several microbiological weapons as biowarfare agents on Chinese soldiers and civilians. In all, some 10,000 "subjects" died of bubonic plague, cholera, anthrax, and other diseases. After years of their own research on biological weapons, the United States, the Soviet Union, and more than 100 other nations in 1973 signed the Biological and Toxin Weapons Convention, which prohibited nations from developing, deploying, or stockpiling biological weapons. Unfortunately, the treaty provided no way to monitor compliance. As a result, in the 1980s the Soviet Union developed and stockpiled many microbiological agents, including the smallpox virus, and anthrax and plague bacteria. After the 1991 Gulf War, the United Nations Special Commission (UNSCOM) analysts reported that Iraq had produced 8,000 liters of concentrated anthrax solution and more than 20,000 liters of botulinum toxin solution. In addition, anthrax and botulinum toxin had been loaded into SCUD missiles.

In the United States, several biocrimes have been committed. **Biocrimes** are the intentional introduction of biological agents into food or water, or by injection, to harm or kill groups of individuals. The most well known biocrime occurred in Oregon in 1984 when the Rajneeshee religious cult, in an effort to influence local elections, intentionally contaminated salad bars of several restaurants with the bacterium *Salmonella*. The unsuccessful plan sickened over 750 citizens and hospitalized 40. Whether biocrime or bioterrorism, the 2001 events concerning the anthrax spores mailed to news offices and to two U.S. congressmen only increases our concern over the use of microorganisms or their toxins as bioterror agents.

What Microorganisms Are Considered Bioterror Agents?

A considerable number of human pathogens and toxins have potential as microbiological weapons. These "select agents" include bacterial organisms, bacterial toxins, fungi, and viruses. The seriousness of the agent depends on the severity of the disease it causes (virulence) and the ease with which it can be disseminated. The pathogens of most concern, called the Category A Select Agents, are those that can be spread by aerosol contact, such as anthrax and smallpox, and toxins that can be added to food or water supplies, such as the botulinum toxin (see the table below).

Why Use Microorganisms?

At least 15 nations are believed to have the capability of producing bioweapons from microorganisms. Such microbiological weapons offer clear advantages to these nations and terrorist organizations in general. Perhaps most important, biological weapons represent "The Poor Nation's Equalizer." Microbiological weapons are cheap to produce compared to chemical and nuclear weapons and provide those nations with a deterrent every bit as dangerous and deadly as the nuclear weapons possessed by other nations.

TABLE

Category A Select Agents and Perceived Risk of Use

Type of Microbe	Disease (Microbe Species or Virus Name)	Perceived Risk
Bacteria	Anthrax (*Bacillus anthracis*)	High
	Plague (*Yersinia pestis*)	Moderate
	Tularemia (*Francisella tularensis*)	Moderate
Viruses	Smallpox (Variola)	Moderate
	Hemorrhagic fevers (Ebola, Marburg, Lassa, Machupo)	Low
Toxins	Botulinum toxin (*Clostridium botulinum*)	Moderate

With biological weapons, you get high impact and the most "bang for the buck." In addition, microorganisms can be deadly in minute amounts to a defenseless (nonimmune) population. They are odorless, colorless, and tasteless, and unlike conventional and nuclear weapons, microbiological weapons do not damage infrastructure, yet they can contaminate such areas for extended periods. Without rapid medical treatment, most of the select agents can produce high numbers of casualties that would overwhelm medical facilities. Lastly, the threatened use of microbiological agents creates panic/anxiety, which often is at the heart of terrorism.

How Would Microbiological Weapons Be Used?

All known microbiological agents (except smallpox) represent organisms naturally found in the environment. For example, the bacterium causing anthrax is found in soils around the world (see left figure). Assuming one has the agent, the microorganisms can be grown (cultured) easily in large amounts. However, most of the select agents must be "weaponized"; that is, they must be modified into a form that is deliverable, stable, and has increased infectivity and/or lethality. Nearly all of the microbiological agents in category A are infective as an inhaled aerosol. Weaponization, therefore, requires the agents be small enough in size so inhalation would bring the organism deep into the respiratory system and prepared so that the particles do not stick together or form clumps. Several of the anthrax letters of October 2001 involved such weaponized spores.

Dissemination of biological agents by conventional means would be a difficult task. Aerosol transmission, the most likely form for dissemination, exposes microbiological weapons to environmental conditions to which they are usually very sensitive. Excessive heat, ultraviolet light, and oxidation would limit the potency and persistence of the agent in the environment. Although anthrax spores are relatively resistant to typical environmental conditions, the bacterial cells causing tularemia become ineffective after just a few minutes in sunlight. The possibility also exists that some nations have developed or are developing more lethal bioweapons through genetic engineering and biotechnology. The former Soviet Union may have done so. Commonly used techniques in biotechnology could create new, never before seen bioweapons, making the resulting "designer diseases" true doomsday weapons.

Conclusions

Ken Alibek, a scientist and defector from the Soviet bioweapons program, has suggested the best biodefense is to concentrate on developing appropriate medical defenses that will minimize the impact of bioterrorism agents. If these agents are ineffective, they will cease to be a threat; therefore, the threat of using human pathogens or toxins for bioterrorism, like that for emerging diseases such as SARS and West Nile fever, is being addressed by careful monitoring of sudden and unusual disease outbreaks. Extensive research studies are being carried out to determine the effectiveness of various antibiotic treatments (see right figure) and how best to develop effective vaccines or administer antitoxins. To that end, vaccination perhaps offers the best defense. The United States has stated it has stockpiled sufficient smallpox vaccine to vaccinate the entire population if a smallpox outbreak occurred. Other vaccines for other agents are in development.

This primer is not intended to scare or frighten; rather, it is intended to provide an understanding of why microbiological agents have been developed as weapons for bioterrorism. We cannot control the events that occur in the world, but by understanding bioterrorism, we can control how we should react to those events—should they occur in the future.

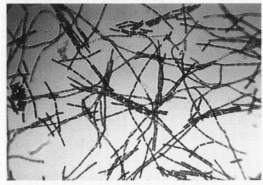

Light micrograph of gram-stained *Bacillus anthracis,* the causative agent of anthrax. There is concern that terrorists could release large quantities of anthrax spores in a populated area, which potentially could cause many deaths.

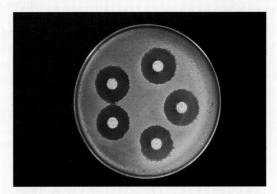

Antibiotic drugs in paper discs are used to test the sensitivity of anthrax bacteria (*Bacillus anthracis*) cultured on an agar growth medium. The clear zone surrounding each disc indicates the bacterial cells are sensitive to the antibiotic.

MICROINQUIRY 19
Epidemiological Investigations

Infectious disease epidemiology is a scientific study from which health problems are identified. In this inquiry, we are going to look at just a few of the applications and investigative strategies for analyzing the patterns of illness. Answers can be found in **Appendix D.**

One of the important measures is to assess disease occurrence. The **incidence** of a disease is the number of reported cases in a given time frame. **Figure A** is a line graph showing the number of new cases of AIDS per year in the United States.

The **prevalence** of a disease refers to the percentage of the population that is affected at a given time.

19.1a. What was the incidence of AIDS in 1993 and 2003?

19.1b. Assuming that 264,000 were living with AIDS in 1993 and 380,000 in 2003, how has the prevalence of AIDS changed between 1993 and 2003? (Assume the population of the United States has remained at 290 million).

Descriptive epidemiology describes activities (time, place, people) regarding the distribution of diseases within a population. Once some data have been collected on a disease, epidemiologists can analyze these data to characterize disease occurrence. Often a comprehensive description can be provided by showing the disease trend over time, its geographic extent (place), and the populations (people) affected by the disease.

Characterizing by Time

Traditionally, drawing a graph of the number of cases by the date of onset shows the time course of an epidemic. An epidemic curve, or "epi curve," is a histogram providing a visual display of the magnitude and time trend of a disease.

Look at the epi curve in **Figure B** for an Ebola outbreak in Africa. One important aspect of a bar graph is to consider its overall shape. An epi curve with a single peak indicates a single source (or "point source") epidemic in which people are exposed to the same source over a relatively short time. If the duration of exposure is prolonged, the epidemic is called a continuous common source epidemic, and the epi curve will have a plateau instead of a peak. Person-to-person transmission is likely and its spread may have a series of plateaus one incubation period apart.

19.2a. Identify the type of epi curve drawn in Figure B and explain what the onset says about the nature of disease spread.

19.2b. Is there more than one plateau? Explain the significance that multiple plateaus might have in interpreting the spread of the Ebola hemorrhagic fever outbreak.

Characterizing by Place

Analysis of a disease or outbreak by place provides information on the geographic extent of a problem and may show clusters or patterns that provide clues to the identity and origins of the problem. It is a simple and useful technique to look for geographic patterns where the affected people live, work, or may have been exposed. A geographic distribution for Lyme disease is shown in **Figure C**. This is a spot map, where each reported case of a disease in a county or state may be shown to reflect clusters or patterns of disease. Figure C identifies cases of Lyme disease by county, where each dot represents one reported case in that county in 2007.

19.3a. From this spot map, what inferences can you draw with regard to the reported cases of Lyme disease?

Characterizing by Person

Populations at risk for a disease can be determined by characterizing a disease or outbreak by person. Persons also refer to populations identified by personal characteristics (e.g., age, race, gender) or by exposures (e.g., occupation, leisure activities, drug intake). These factors are important because they may be related to disease susceptibility and to opportunities for exposure.

Age and gender often are the characteristics most strongly related to exposure and to the risk of disease. For example, **Figure D** is a histogram showing the incidence of pertussis (whooping cough) in the United States in 2003.

19.4a. Look at the histogram and describe what important information is conveyed in terms of the majority of cases and relative incidence in 2003.

19.4b. As a health care provider, what role do you see for vaccinations and booster shots with regard to this disease?

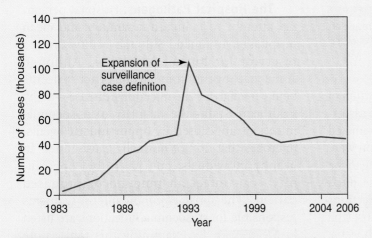

FIGURE A Acquired Immunodeficiency Syndrome (AIDS). Number of cases reported by year in the United States and U.S. territories for the years 1983 to 2006. *Source:* CDC.

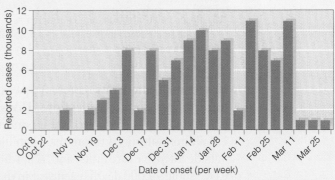

FIGURE B Ebola Hemorrhagic Fever (Congo and Gabon). Number of cases of Ebola hemorrhagic fever by week from October 2001 to March 2002. *Data from: Weekly Epidemiological Record,* No. 26, June 27, 2003.

FIGURE C Lyme Disease. Each dot represents one reported case of Lyme disease in 2007. *Source:* CDC.

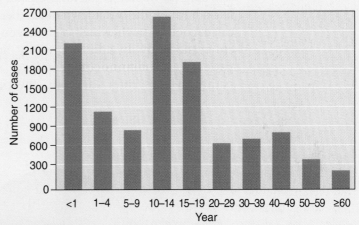

FIGURE D Pertussis. The reported number of cases of pertussis by age group in the United States in 2003. *Source:* CDC.

Nosocomial Infections Are Serious Health Threats within the Health Care System

KEY CONCEPT

13. Nosocomial infections are contracted as a result of being treated for another illness in a hospital or other health care setting.

Nosocomial infections represent that portion of **healthcare-associated infections** (**HAIs**) associated with hospitals and account for an estimated 1.7 million infections (more than 1 million were outside intensive care units) and, according to the CDC, some 99,000 deaths each year in the United States. Like all infections, nosocomial infections involve three elements: a compromised host (the hospital patient), a source of hospital pathogens, and a chain of transmission (**FIGURE 19.17**).

The Compromised Host. Most hospital patients have some form of physical injury, such as a surgical wound, some form of skin trauma like a burn, or the breakdown of another physical barrier through which a pathogen could enter the body. In addition, many hospital patients are **immunocompromised**, so if a pathogen does enter the body, the patient's immune system may be unable to mount an attack and eliminate it. The most common sites of infection are listed in **FIGURE 19.18**.

The Hospital Pathogens. Hospital personnel attempt to maintain a sanitary and clean hospital environment. Still, the facility can be a reservoir for human pathogens. Although some pathogens come from other patients being treated for an infectious disease—or by healthcare staff—the majority of nosocomial infections are caused by **opportunistic** agents, microbes that do not normally cause illness in healthy individuals, but, given the "opportunity," can infect an immunocompromised patient. The most common microorganisms responsible for nosocomial infections are listed in **TABLE 19.5**. By examining this table, note that there is now another dimension to the virulence of these potential pathogens—antibiotic resistance. With the patient's immune system compromised, the use of antibiotics often is necessary to fight an infection. Unfortunately, many of these hospital pathogens are becoming resistant to several generations of antibiotics, meaning more toxic and expensive drugs must be used (see Chapter 24).

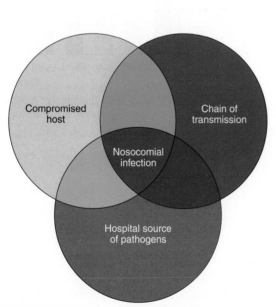

FIGURE 19.17 Nosocomial Infection Elements. For a nosocomial infection to occur, there needs to be a susceptible (compromised) host, pathogenic organisms within the hospital setting, and a chain of transmission. »» How does this nosocomial infection figure compare to that for an infectious disease (see Figure 19.14)?

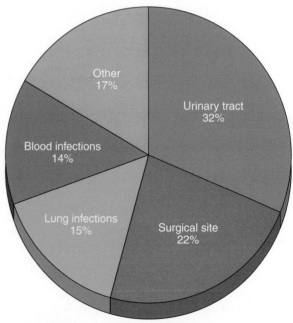

FIGURE 19.18 Sites of Nosocomial Infections. This pie chart shows the sites for the estimated 1.7 million healthcare-associated infections in American hospitals. »» What might be the primary source for the infections for the top four sites? *Source:* CDC. Division of Healthcare Quality Promotion (DHQP).

TABLE 19.5 Top 10 Infectious Agents Involved in Nosocomial Infections			
Microorganism	% of Total Infections	% Antibiotic Resistant[1]	Nosocomial Infections
Coagulase-negative staphylococci	15.3%	Not reported (historically greater than 80%)	Blood infections
Staphylococcus aureus	14.4%	56%	Urinary, blood, lung, and surgical site infections
Enterococcus species	12.1%	3–90%	Urinary tract, blood, and surgical site infections
Candida species	10.7%	Not reported	Urinary tract infections
Escherichia coli	9.6%	1–30%	Urinary and surgical site infections
Pseudomonas aeruginosa	7.9%	6–33%	Urinary, blood, and surgical site infections
Klebsiella pneumoniae	5.8%	3–27%	Lung, blood, and surgical site infections
Enterobacter species	4.8%	Not reported	Urinary and surgical site infections
Acinetobacter baumannii	2.7%	26–37%	Urinary, blood, lung, and surgical site infections
Klebsiella oxytoca	1.1%	3–17%	Urinary, blood, lung, and surgical site infections

[1]Means for different species and different nosocomial infections.

Source: CDC. National Healthcare Safety Network Annual Update (2008).

The Chain of Transmission. The key to nosocomial disease and its prevention stems from the way the agents are transmitted to the patient. These chains of transmission may involve direct contact between patients or between healthcare staff and patient. Indirect contact can also be part of the chain of transmission (**FIGURE 19.19**). Perhaps one of the most common chains of transmission is through the use of indwelling instruments that are not sterile or have not been cleaned thoroughly. Intravenous catheters, respirators, and other medical instruments can be the source.

Therefore, the key to reducing nosocomial infections is to break the chain of transmission. Besides the use of **standard precautions** when working with blood or other body fluids (Figure 19.19), the CDC has published preferred methods for cleaning, disinfecting, and sterilizing patient-care medical devices and general methods for cleaning and disinfecting the healthcare environment. The proper use of chemical disinfectants is essential. This includes alcohols, glutaraldehyde, formaldehyde, hydrogen peroxide, iodophors, phenolics, quaternary ammonium compounds (quats), and chlorine—all chemical methods described in Chapter 7. The sterilization methods recommended include steam sterilization, ethylene oxide, hydrogen peroxide gas, and liquid peracetic acid. The CDC stresses that these chemical and physical methods must be used properly to reduce the risk for infection associated with both invasive and noninvasive medical and surgical devices. And, importantly, it all starts with good hand hygiene on the part of healthcare providers and visitors while in the hospital.

CONCEPT AND REASONING CHECKS

19.13 How do standard precautions limit the chains of transmission?

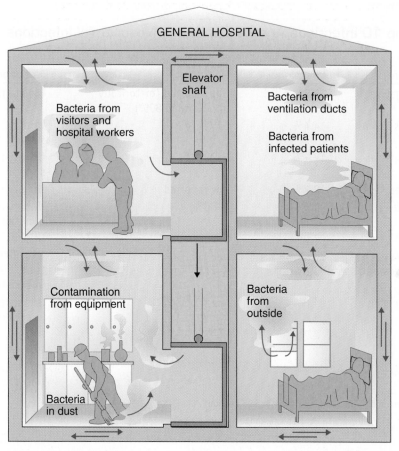

FIGURE 19.19 Microbe Transmission in a Hospital and Standard Precautions. Potential pathogens can be spread through several means within the hospital or health care environment. Protecting patients and other health care workers also means using standard precautions if working with a patient who may be infected. The precautions when handling blood or other body fluids that may harbor pathogens (HIV, hepatitis B or C) include:

- Washing hands
- Wearing personal protective equipment (gloves, mask, and eye protection).
- Handling and disposing of sharps (hypodermic needles) properly.
- Disposing of all hazardous and contaminated materials in approved and labeled biohazard containers.
- Cleaning up all spills with disinfectant or diluted bleach solution to kill any pathogens present.

»» Why is hand washing always at the top of the list for preventing disease transmission?

Infectious Diseases Continue to Challenge Public Health Organizations

KEY CONCEPT

14. Diseases emerging or reemerging anywhere in the world can become a global health menace.

In an era of supersonic jet travel and international commerce, it is not possible to adequately protect the health of any nation without focusing on diseases and epidemics elsewhere in the world. In 2009, we only have to look at H1N1 (swine) flu and the potential bird flu to realize the seriousness of infectious diseases and their threats to global health.

FIGURE 19.20 identifies recent emergent and resurgent (re-emergent) infectious diseases. Globally, there have been more than 40 new diseases and at least 20 resurgent diseases identified since 1980.

There are more than 1,400 known human pathogens (**FIGURE 19.21**). However, less than 100 are specialized within humans. Over half of the 1,400 represent **zoonoses** (*zoo* = "animal"; *noso* = "disease"), diseases transmitted from animal reservoirs to humans. Some 177 (13%) represent emerging or resurgent diseases, with the largest single number (65%) being RNA viruses (see Chapter 14).

Emerging and re-emerging infectious diseases in North, Central, and South America

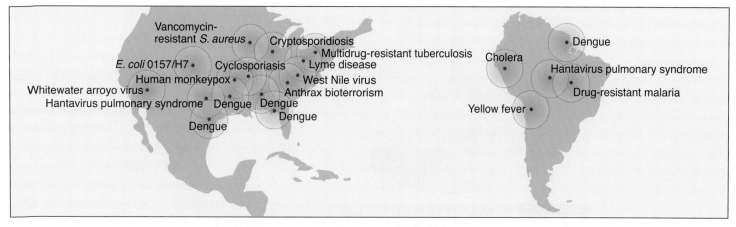

Emerging and re-emerging infectious diseases in Europe, Africa, Asia, and Australia

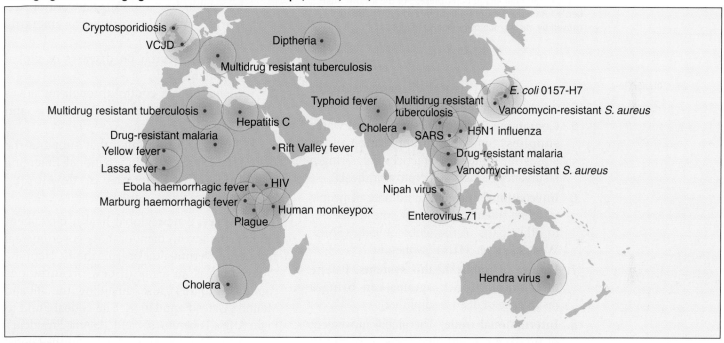

FIGURE 19.20 The Global Occurrence of Emerging and Re-emerging Disease. »» Can you suggest a reason why many of the diseases have appeared in North America?

Modified from American College of Microbiology. *Clinical Microbiology in the 21st Century: Keeping the Pace*. ASM Press, 2008, Washington, D.C.

Numerous reasons help explain how disease emergence and resurgence are driven (in decreasing rank of disease involvement).

1. **Changes in land use or agriculture practices.** Urbanization, deforestation, and water projects can bring new or re-emergent diseases, such as Dengue fever and schistosomiasis.

2. **Changes in human demographics.** The migration of many peoples or whole societies from agrarian to urban lifestyles has brought new diseases to a susceptible population (malaria).

3. **Poor population health.** In many developing nations, large numbers of people suffer from malnutrition or poor public health infrastructure, making disease eruption much more likely (cholera).

4. **Pathogen evolution.** Pathogens have developed resistance to antibiotics and antimi-

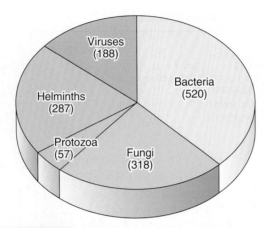

FIGURE 19.21 Known Species of Human Pathogens. The pie chart shows the relative proportions of diseases caused by bacteria, fungi, protozoa, helminths, and viruses. The numbers in parentheses are the known number. »» Name several prominent infections or diseases caused by each group of human pathogens.

crobial drugs, typical of many resurgent diseases (tuberculosis, yellow fever).

5. **Contamination of food sources and water supplies.** Substandard application of, or lapses in, sanitation practices can bring a resurgence of disease (cryptosporidiosis).

6. **International travel.** The number of people traveling internationally can spread diseases to other parts of the globe quickly (SARS, West Nile fever, H1N1 [swine] flu).

7. **Failure of public health systems.** Failure of immunization programs can bring a resurgence of disease (diphtheria).

8. **International trade.** The global movement of produce can introduce new or resurgent diseases (hepatitis A, cyclosporiasis). Likewise, wildlife trade (legal and illegal) provides new mechanisms for disease transmission (monkeypox).

9. **Climate change.** Global changes in weather patterns bring new diseases to new latitudes and elevations (hantavirus pulmonary syndrome).

By understanding these drivers, health organizations such as the CDC and WHO can develop plans to limit or stop emerging disease threats. The CDC has had a history of serving American public health. Regarding emerging and resurgent diseases, the CDC's priority areas include:

- International outbreak assistance to host-countries to maintain control of new pathogens when an outbreak is over.
- A global approach to disease surveillance by establishing a global "network of networks" for early warning of emerging health threats.
- Applied research on diseases of global importance.
- Global disease control through initiatives to reduce HIV disease/AIDS, malaria, and tuberculosis.
- Public health training that supports the establishment of International Emerging Infections Programs in developing nations.

However, a new global health movement requires the involvement of more governmental and non-governmental organizations. Happily, today there is an array of new organizations committed to this cause, including: the Bill and Melinda Gates Foundation; The Global Fund to Fight AIDS, Tuberculosis, and Malaria; President's Emergency Plan for AIDS Relief; and the Global Alliance for Vaccines and Immunization.

CONCEPT AND REASONING CHECKS

19.14 How might climate change alter the occurrence of emerging and re-emerging diseases?

SUMMARY OF KEY CONCEPTS

19.1 The Host–Microbe Relationship

1. **Infection** refers to competition between host and microbe for supremacy, while **disease** results when the microbe wins the competition. However, the human body contains large resident populations of **indigenous microbiota**, which usually out-compete invading pathogens. **Transient microbiota** are temporary residents of the body.

2. **Parasitism** is the symbiotic relationship occurring when the microbe does harm to the host. **Pathogenicity**, the ability of the pathogen to cause disease, and resistance go hand-in-hand. **Virulence** refers to the degree of pathogenicity a microbe displays.

3. Infections may come from an exogenous or endogenous source. If the immune system is compromised, microbes normally acting as commensals may cause **opportunistic infections**. A **primary infection** is an illness caused by a pathogen in an otherwise healthy host, while a **secondary infection** involves the development of other diseases as a result of the primary infection lowering host resistance. **Local diseases** are restricted to a specific part of the body while **systemic diseases** spread to several parts of the body and deeper tissues.

19.2 Establishment of Infection and Disease

4. Most diseases have certain **signs** and **symptoms**, making it possible to follow the course of a disease.

5. To cause disease, most pathogens must enter the body through an appropriate **portal of entry**. The **infectious dose** represents the number of pathogens taken into the body that can cause a disease.

6. The possibility of disease is enhanced if a microbe can penetrate host tissues. The pathogen's ability to penetrate tissues and cause damage is referred to as **invasiveness**. Virulence and invasiveness are strongly dependent on the spectrum of **virulence factors** a pathogen possesses. Bacterial **adhesins** or viral spikes allow bacterial cells or viruses to adhere to specific cells. Adhesion or attachment usually leads to **phagocytosis** of the pathogens.

7. Many bacterial species produce enzymes to overcome the body's defenses. These include **coagulase**, **streptokinase**, and **hyaluronidase** enzymes. In addition, some species produce lytic enzymes, such as **leukocidins** and **hemolysins**. **Exotoxins** are proteins released by gram-positive and gram-negative cells. Their effects depend on the enzyme produced and host system affected. **Endotoxins** are the lipopolysaccharides released from dead gram-negative cells. Their effects on the host are more universal.

8. To efficiently spread the disease to other hosts, the pathogen also must leave the body through an appropriate **portal of exit**.

19.3 Infectious Disease Epidemiology

9. **Reservoirs** include humans, who represent carriers of a disease, arthropods, and any food and water in which some parasites survive.

10. A disease may be **communicable**, such as measles, or **noncommunicable**, such as tetanus.

11. Diseases may be transmitted by **direct** or **indirect** methods. Indirect methods include consumption of contaminated food or water, contaminated inanimate objects (**fomites**), and arthropods. Arthropods can be **mechanical** or **biological vectors** for the transmission of disease.

12. The occurrence of diseases falls into three categories: **endemic**, **epidemic**, and **pandemic**. An "outbreak" is essentially the same as an epidemic, although usually an outbreak is more confined in terms of disease spread.

13. **Nosocomial infections**, or **healthcare-associated infections**, are infections acquired as a result of being treated for some other injury or medical problem. **Standard precautions** limit the **chain of transmission**.

14. Public health organizations, such as the CDC, are charged with the duty of limiting or stopping disease threats, including **emerging** and **resurgent diseases**.

LEARNING OBJECTIVES

After understanding the textbook reading, you should be capable of writing a paragraph that includes the appropriate terms and pertinent information to answer the objective.

1. Distinguish between **infection** and **disease**, and between **indigenous** and **transient microbiota**.

2. Contrast **pathogenicity** and **virulence**, explaining how each affects the establishment of disease.

3A. Discuss the consequences of **exogenous** and **endogenous** (including opportunistic) infections on the progression and outcomes of infection and disease.

3B. Distinguish between (a) **primary** and **secondary infections**, (b) **local** and **systemic diseases**, and (c) **bacteremia** and **septicemia**.

4A. Explain the differences between **signs**, **symptoms**, and **syndromes**.

4B. Identify the characteristics that compose the five stages in the course of disease development.

5. Assess the role of the **infectious dose** and pathogen **adhesion** to establishing an infection and disease.

6. Discuss the importance of **invasiveness** to the establishment of an infection.

7A. Name five enzymes and describe their roles as **virulence factors**.

7B. Summarize the differences between **exotoxins** and **endotoxins** as virulence factors associated with disease.

8. Identify six **portals of exit** from the human body.

9. Summarize the characteristics of **reservoirs** as applied to infectious disease.

10. Distinguish between a **communicable**, **contagious**, and **noncommunicable disease** and give an example of each.

11A. Identify the **direct contact** methods of disease transmission.

11B. Evaluate the **indirect contact** methods of disease transmission.

12. Discuss the three types of disease occurrence within populations.

13. Explain how **nosocomial infections** can be controlled or eliminated through using the **standard precautions** to break the **chain of transmission**.

14. Identify the drivers responsible for **emerging** and **resurgent infectious diseases**.

STEP A: SELF-TEST

Each of the following questions is designed to assess your ability to remember or recall factual or conceptual knowledge related to this chapter. Read each question carefully, then select the *one* answer that best fits the question or statement. Answers to even-numbered questions can be found in **Appendix C**.

1. A newborn
   A. contains indigenous microbiota before birth.
   B. remains sterile for many weeks after birth.
   C. becomes colonized soon after conception.
   D. is colonized with many common microbiota within a few days after birth.

2. Factors affecting virulence may include
   A. the presence of pathogenicity islands.
   B. their ability to penetrate the host.
   C. the infectious dose.
   D. All the above (**A–C**) are correct.

3. A healthy person can be diagnosed as having a _____ infection with _____, the multiplication of bacterial cells in the blood.
   A. primary; bacteremia
   B. primary; viremia
   C. primary; septicemia
   D. secondary; parasitemia

4. Changes in body function sensed by the patient are called
   A. symptoms.
   B. syndromes.
   C. prodromes.
   D. signs.

5. Adhesins can be found on
   A. host cells.
   B. viruses.
   C. bacterial pili and capsules.
   D. cells at the portal of entry.

6. In the body, bacterial invasiveness can be limited by
   A. fever.
   B. phagocytosis.
   C. enzyme production.
   D. toxic production.

7. Which one of the following is NOT true of exotoxins?
   A. They are proteins.
   B. They are part of cell wall structure.
   C. They are released from live bacterial cells.
   D. They trigger antibody production.

8. A portal of exit would be
   A. the feces.
   B. an insect bite.
   C. blood removal.
   D. All of the above (**A–C**) are correct.

9. If a person has recovered from a disease but continues to shed disease agents, that person is a
   A. vector.
   B. fomite.
   C. vehicle.
   D. carrier.

10. All of the following are examples of communicable diseases *except*:
    A. chickenpox.
    B. measles.
    C. the common cold.
    D. tetanus.

11. Which one of the following is an example of an indirect method of disease transmission?
    A. Coughing
    B. Droplet transmission
    C. A mosquito bite
    D. An animal bite

12. Fifty cases of hepatitis A during one week in a community would most likely be described as a/an
    A. outbreak.
    B. pandemic.
    C. endemic disease.
    D. epidemic.

13. The most common nosocomial infection involves
    A. blood.
    B. lungs.
    C. urinary tract.
    D. a surgical site.

14. A zoonosis is a disease
    A. transmitted from humans to animals.
    B. spread from animals to humans.
    C. transmitted between wild and domestic animals.
    D. spread between wild animals.

HTTP://MICROBIOLOGY.JBPUB.COM/9E

The site features eLearning, an online review area that provides quizzes and other tools to help you study for your class. You can also follow useful links for in-depth information, read more MicroFocus stories, or just find out the latest microbiology news.

STEP B: REVIEW

Test your knowledge of this chapter's contents by determining whether the following statements are true or false. If the statement is true, write "True" in the space. If false, substitute a word for the underlined word to make the statement true. The answers to even-numbered statements are listed in **Appendix C**.

_____ 15. An <u>epidemic</u> disease occurs at a low level in a certain geographic area.

_____ 16. Among the microbial enzymes able to destroy blood cells are <u>hemolysins</u> and leukocidins.

_____ 17. The term <u>disease</u> refers to a symbiotic relationship between two organisms and the competition taking place between them for supremacy.

_____ 18. Organs of the human body lacking a normal microbiota include the blood and the <u>small intestine</u>.

_____ 19. <u>Commensalism</u> is a form of symbiosis where the microbe benefits and causes no damage to the host.

_____ 20. A <u>biological</u> vector is an arthropod that carries pathogenic microorganisms on its feet and body parts.

_____ 21. Organisms causing disease when the immune system is depressed are known as <u>opportunistic</u> organisms.

_____ 22. The human body responds to the presence of exotoxins by producing <u>endotoxins</u>.

_____ 23. The term <u>bacteremia</u> refers to the spread of bacteria through the bloodstream.

_____ 24. A <u>toxoid</u> is an immunizing agent prepared from an exotoxin.

_____ 25. Few symptoms are exhibited by a person who has a <u>subclinical disease</u>.

_____ 26. <u>Indirect</u> methods of disease transmission include kissing and handshaking.

_____ 27. A <u>chronic</u> disease develops rapidly, is usually accompanied by severe symptoms, and comes to a climax.

_____ 28. The <u>acme period</u> is the time between the entry of the pathogen into the host and the appearance of symptoms.

_____ 29. An organism with <u>high</u> virulence generally is unable to cause disease.

_____ 30. <u>Symptoms</u> are changes in body function detected by a physician.

STEP C: APPLICATIONS

Answers to even-numbered questions can be found in **Appendix C**.

31. The transparent covering over salad bars is commonly called a "sneeze guard" because it helps prevent nasal droplets from reaching the salad items. As a community health inspector, what other suggestions might you make to prevent disease transmission via the salad bar?

32. While slicing a piece of garden hose, your friend cut himself with a sharp knife. The wound was deep, but it closed quickly. Shortly thereafter, he reported to the emergency room of the community hospital, where he received a tetanus shot. What did the tetanus shot contain, and why was it necessary?

33. After reading this chapter, you decide to make a list of the ten worst "hot zones" in your home. The title of your top-ten list will be "Germs, Germs Everywhere." What places will make your list, and why?

34. As a state epidemiologist responsible for identifying any disease occurrences, would an epidemic disease or an endemic disease pose a greater threat to public health in the community? Explain.

STEP D: QUESTIONS FOR THOUGHT AND DISCUSSION

Answers to even-numbered questions can be found in **Appendix C**.

35. In 1840, Great Britain introduced penny postage and issued the first adhesive stamps. However, politicians did not like the idea because it deprived them of the free postage they were used to. Soon, a rumor campaign was started, saying that these gummed labels could spread disease among the population. Can you see any wisdom in their contention? Would their concern "apply" today?

36. In 1892, a critic of the germ theory of disease named Max von Pettenkofer sought to discredit Robert Koch's work by drinking a culture of cholera bacilli diluted in water. Von Pettenkofer suffered nothing more than mild diarrhea. What factors may have contributed to the failure of the bacilli to cause cholera in von Pettenkofer's body?

37. A man takes a roll of dollar bills out of his pocket and "peels" off a few to pay the restaurant tab. Each time he peels, he wets his thumb with saliva. What is the hazard involved?

38. When Ebola fever broke out in Africa in 1995, disease epidemiologists noted how quickly the responsible virus killed its victims and suggested the epidemic would end shortly. Sure enough, within three weeks it was over. What was the basis for their prediction? What other conditions had to apply for them to be accurate in their guesswork?

39. A woman takes an antibiotic to relieve a urinary tract infection caused by _Escherichia coli_. The infection resolves, but in two weeks, she develops a _Candida albicans_ ("yeast") infection of the vaginal tract. What conditions may have caused this to happen?

20

Resistance and the Immune System: Innate Immunity

". . . medical science today has set itself the task of attempting to prevent disease. In order to achieve this aim one must attempt, on the one hand to find the disease germ and destroy it, and on the other hand to give the body the strength to resist attack."
—K. A. H. Mörner, Rector of the Royal Caroline Institute on presenting the 1908 Nobel Prize in Physiology or Medicine to Elie Metchnikoff and Paul Ehrlich for their work on the theory of immunity

Edward Jenner's epoch-making moment in history came in 1798 when he demonstrated the protective action of vaccination against smallpox (see Chapter 1). As great as Jenner's discovery was, it did not advance the development or understanding of **immunity**, which is how the human body can generate resistance to a particular disease, whether by recovering from the disease or as a result of vaccination, such as devised by Jenner.

It would be another 90 years before Russian zoologist and immunologist Elie Metchnikoff devised the first experiments to study immunity by investigating how an organism could destroy a disease-causing microbe in the body. Through landmark studies with invertebrates, Metchnikoff proposed that certain types of cells could attack, engulf, and destroy foreign material, including infectious microbes. These were among many experiments from a chain of investigations that studied immunity and then applied that knowledge to mammals and humans. Metchnikoff's work culminated in the theory of phagocytosis, which suggested certain types of human white blood cells could capture and destroy disease-causing microbes that had penetrated the body.

Around the same time, Louis Pasteur developed vaccines for chicken cholera and rabies, further stimulating an interest in the mechanism for protective immunity. In Pasteur's lab, Émile Roux and Alexandre Yersin discovered that the bacterial species causing diphtheria actually produced a toxin causing the disease. Then, Koch's coworker, Emil von Behring modified the diphtheria toxin to produce a substance with "antitoxic activity" that successfully immunized animals for a short period against diphtheria. The substance gave the body the power to resist infection.

With these early discoveries, the science of immunology was off and running. Today, Metchnikoff's work forms part of the basis for "innate immu-

nity," which consists of several nonspecific defenses present in all humans from the time of birth. The discoveries of Roux, Yersin, and von Behring are the basis for part of what is referred to today as "acquired immunity." This form of protective immunity or resistance is a response to a specific microbe and is directed only against that microbe. Together, innate and acquired immunity are equally important arms that form a "microbiological umbrella" protecting us against the torrent of potential microbial pathogens to which we are exposed daily (**FIGURE 20.1**).

This chapter examines the immune resistance to microbes (again, for brevity, we will include viruses as microbes), toxins, and other foreign substances by studying innate immunity. As we will see, the opening quote to this chapter made in 1908 is as true today as it was then.

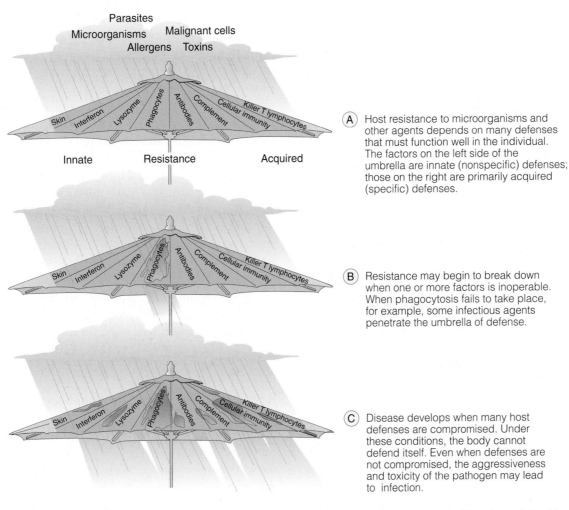

A Host resistance to microorganisms and other agents depends on many defenses that must function well in the individual. The factors on the left side of the umbrella are innate (nonspecific) defenses; those on the right are primarily acquired (specific) defenses.

B Resistance may begin to break down when one or more factors is inoperable. When phagocytosis fails to take place, for example, some infectious agents penetrate the umbrella of defense.

C Disease develops when many host defenses are compromised. Under these conditions, the body cannot defend itself. Even when defenses are not compromised, the aggressiveness and toxicity of the pathogen may lead to infection.

FIGURE 20.1 The Relationship between Host Resistance and Disease. Host resistance can be likened to a "microbiological umbrella" that forms a barrier or defense against infectious disease. »» Why are the defenses on the left side of the umbrella considered innate and those on the right acquired?

20.1   An Overview to Host Immune Defenses

The work of Metchnikoff, Roux, Yersin, and von Behring as well as that of their contemporaries and followers, opened up the whole field of **immunology**, the scientific study of how the immune system functions in the body to prevent or destroy foreign material, including pathogens.

Blood Cells Form an Important Defense for Innate and Acquired Immunity

KEY CONCEPT

1. Leukocytes perform a variety of defensive roles in the host.

In the circulatory system, blood consists of three major components: the fluid, the clotting agents, and the formed elements or cells. The fluid portion, called **serum**, is an aqueous solution of minerals, salts, proteins, and other organic substances. When clotting agents, such as fibrinogen and prothrombin, are present, the fluid is referred to as **plasma**. Blood platelets are small, disk-shaped cells that originate in the bone marrow. Platelets have no cell nucleus and function chiefly in the blood-clotting mechanism.

The other formed elements in the blood are the erythrocytes (red blood cells) and leukocytes, the latter being significant to innate and acquired immunity.

Leukocytes (White Blood Cells). As their name suggests, **leukocytes** (*leuko* = "white") have no pigment in their cytoplasm and therefore appear gray when unstained. These cells also are produced in the bone marrow. They number about 4,000 to 12,000 per microliter of blood and have different lifespans, depending on the type of cell. Most can be identified in a blood smear or tissue sample by the size and shape of the cell nucleus, and by the staining of visible cytoplasmic granules, if present, when observed with the light microscope. **TABLE 20.1** summarizes the six major types of white blood cells.

Neutrophils have a multilobed cell nucleus and therefore are referred to as "polymorphonuclear" cells (PMNs). Their cytoplasm contains many granules that contain digestive enzymes and other antimicrobial agents capable of killing pathogens. Neutrophils function chiefly as **phagocytes**, cells that carry out phagocytosis. Although usually found in the blood, neutrophils can move out of the circulation to engulf foreign particles

or pathogens causing infection. Their lifespan is short, only about one to two days, so they are continually replenished from the bone marrow.

Eosinophils exhibit red or pink cytoplasmic granules when a dye such as eosin is applied. Substances in the granules contain cytotoxic proteins to defend against multicellular parasites, such as flukes and tapeworms. They also are involved in the development of allergies and asthma.

Basophils are leukocytes whose cytoplasmic granules stain blue with methylene blue. Basophils, and their counterparts in tissues, the **mast cells**, also function in allergic reactions and defenses against parasites.

Eosinophils, basophils, and mast cells are concentrated in the skin, lungs, and the gastrointestinal tract, locations where breaks in the skin, inhalation, or ingestion might be portals of entry for pathogens. Because neutrophils, eosinophils, and basophils contain cytoplasmic granules, they often are called **granulocytes**.

A fourth major group of leukocytes is the **monocytes**. These phagocytic cells have a single, bean-shaped cell nucleus and lack visible cytoplasmic granules. In the tissues, monocytes mature into **macrophages**. In contrast to the two-week lifespan of monocytes, macrophages may live for several months. Macrophages (*macro* = "big"; *phage* = "eat") are more effective as phagocytes than neutrophils and are one of the key cells in both innate and acquired immunity. As such, they are tissue residents in the spleen, lymph nodes, and thymus, and are common immune cells in the respiratory and digestive tracts.

The **lymphocytes** migrate from the bone marrow to the lymph nodes after maturation. They have a single, large nucleus and no granules. One cell type, the **natural killer (NK) cells**, plays a key role in innate immunity by destroying virus-infected cells and tumor cells. The two other cell types, **B lymphocytes** and **T lymphocytes**, are key cells of acquired immunity. They increase in number dramatically during the course of an infectious disease but have very different functions, as we will describe in the next chapter. Lacking visible granules, monocytes and lymphocytes often are called **agranulocytes**.

The last group of leukocytes is the **dendritic cells**. The name comes from their resemblance to

Eosin:
A red anionic (acidic) dye.

Methylene blue:
A blue cationic (basic) dye.

TABLE

20.1 Major Leukocytes of the Human Immune System

Types of Leukocytes		Morphology	Approximate Percentage	Effector Function
Granulocytes	Neutrophil		50–70	Phagocytosis
	Eosinophil	Red-stained granules	2–4	Defense against parasites; part of allergic reactions
	Basophil and Mast cell	Blue-stained granules	<1	Release of chemical mediators of inflammation; part of allergic reactions
Agranulocytes	Monocyte and Macrophage		2–8	Phagocytosis and intracellular killing of pathogens; initiate acquired immunity
	Lymphocyte		20–30	Antibody production; cytotoxic properties
	Dendritic Cell		Tissue specific	Activate lymphocytes; initiate acquired immunity

the long, thin extensions (dendrites) seen on neurons. Dendritic cells are found in the skin, where they are called Langerhans cells, and in tissues where pathogens may enter. As active phagocytic cells, they are crucial to innate immunity and the activation of acquired immunity.

CONCEPT AND REASONING CHECKS

20.1 Draw a simple concept map for leukocytes; besides the different leukocytes, include the terms granulocytes, agranulocytes, and phagocytes in your map.

The Lymphatic System Is Composed of Cells and Tissues Essential to Immune Function

KEY CONCEPT

2. The lymphatic system maintains and distributes lymphocytes necessary for defense against pathogens.

In the human body, the clear fluid surrounding the tissue cells and filling the intercellular spaces is called tissue fluid, or **lymph** (*lymph* = "water").

Lymph bathes the body cells, supplying oxygen and nutrients while collecting wastes. To carry these materials, the lymph must be pumped through tiny vessels by the contractions of skeletal muscle cells. Eventually the lymph vessels unite to form larger lymphatic vessels. Along the way, lymph nodes filter the lymph before the lymph returns to the bloodstream. Should lymphatic vessels be blocked due to filariasis, swelling of the extremities can result in a disease called elephantiasis (see Figure 18.26).

It is the lymphatic tissues in which lymphocytes mature, differentiate, and divide (**FIGURE 20.2A**). The **primary lymphoid tissues** consist of the **thymus**, which lies behind the breast bone, and the **bone marrow**. Both are sites where T and B lymphocytes form or mature.

The **secondary lymphoid tissues** are sites where mature immune cells interact with pathogens and carry out the acquired immune response. The **spleen**, a flattened organ at the upper left of the abdomen, contains immune cells to monitor and fight infectious microbes entering the body. Likewise, the **lymph nodes**, prevalent in the neck, armpits, and groin, are bean-shaped structures containing macrophages and dendritic cells that engulf pathogens in the lymph, and lymphocytes, which respond specifically to foreign substances in the circulation (**FIGURE 20.2B**). Because resistance mechanisms are closely associated with the lymph nodes, it is not surprising they become enlarged (often called "swollen glands") during infections.

Other organ systems in the body contain additional secondary lymphoid tissues. In the intestine, the Peyer patches and appendix form part of the **mucosa-associated lymphoid tissue** (**MALT**), while secondary lymphoid tissues associated with the respiratory tract are the **tonsils**. Specialized lymphocytes and dendritic cells of the MALT help defend against pathogen infection.

CONCEPT AND REASONING CHECKS

20.2   Assess the role of the lymphatic system to defense against pathogens.

Innate and Acquired Immunity Are Essential Components of a Fully Functional Human Immune System

KEY CONCEPT

3.   The interactions between innate and acquired host defenses make infection and disease establishment difficult.

When one thinks about the body fighting an infection, one usually envisions antibodies and white blood cells as the "knights in shining armor." Although these molecules and cells are critical, they represent only the second half of the total immune system defense. First, and foremost, immune defense depends on nonspecific resistance.

Nonspecific resistance is referred to as **innate immunity** because genetically encoded molecules, present in the body from birth, are capable of recognizing microbial features common to many pathogens and foreign substances. Metchnikoff's studies were the first to suggest innate immunity is evolution's defense mechanism against infectious disease and represents an ancient one common to most animals. Therefore, innate immunity, with its set of preformed effector molecules, represents the host's early-warning system against potentially harmful pathogens, foreign cells, or even our own cells should they become damaged or cancerous. This defense system responds within minutes or a few hours after infection.

Upon recognition of microbes by innate immune defenses, these defenses try to eliminate or hold an infection in check while sending chemical signals to tissues involved with initiating **acquired immunity**. These chemical signals, called **cytokines**, are small proteins released by various defensive cells in response to an activating substance, such as an invading microbe. Cytokines are produced by many cells, including macrophages, lymphocytes, mast cells, and dendritic cells. We will see specific examples as we discuss innate and acquired immunity.

The knights in shining armor are part of acquired immunity because the response produces lymphocytes and antibodies specific only to the pathogen or foreign substance causing the infection. This form of immunity, which evolved more recently and only exists in vertebrates, is relatively slow compared to the innate response; in fact, acquired immunity develops over the course of the infection, taking several days to more than a week to mount an effective response and a protective defense. In the rest of this chapter, we will examine innate immunity in more detail. Chapter 21 will describe the acquired immune response.

CONCEPT AND REASONING CHECKS

20.3   Summarize the roles for innate and acquired immunity in the human host.

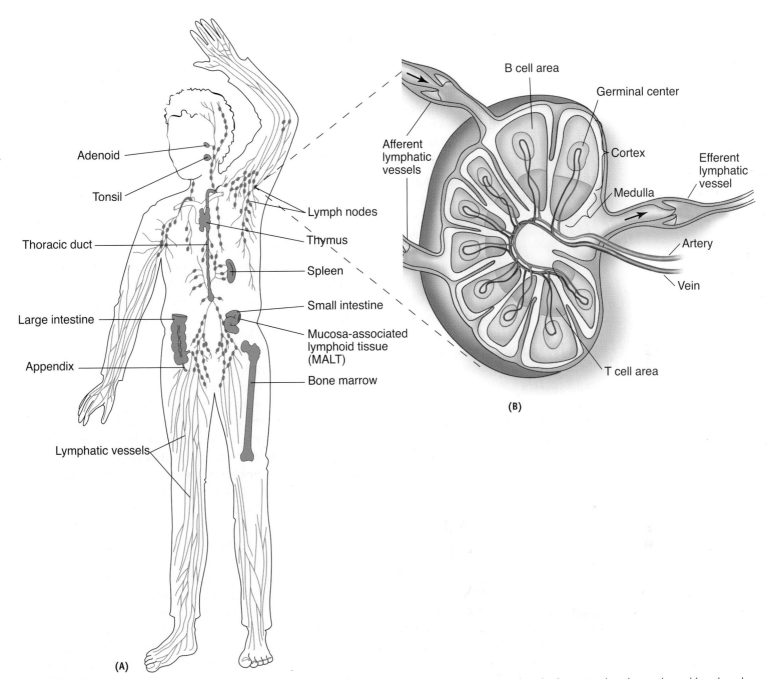

FIGURE 20.2 The Human Lymphatic System. (**A**) The human lymphatic system consists of lymphocytes, lymphatic organs, lymph vessels, and lymph nodes located along the vessels. The lymphatic organs are illustrated, and the preponderance of lymph nodes in the neck, armpits, and groin is apparent. (**B**) A lymph node is a filtering site for circulation of lymph. Any microbes present are removed by macrophages and dendritic cells before the "cleaned" lymph exists at an efferent lymphatic vessel. The outer cortex and germinal centers are sites rich in B cells while the deep cortex and medulla contain large numbers of T cells. »» Why are there so many lymph nodes distributed throughout the body?

20.2 The Innate Immune Response

The innate immune response to infection involves a broad group of components. It depends upon the general well-being of the individual and proper functioning of the body's systems. Under such conditions, innate defenses can contain pathogens, if not eliminate them, while an acquired immune response is generated or reestablished.

Physical, Chemical, and Microbiological Barriers Limit Entry of Pathogens

KEY CONCEPT

4. The skin, mucous membranes, chemical inhibitors, and microbes represent barriers to infection.

There are several portals of entry by which pathogens can enter the body (see Chapter 19). The intact skin and the mucous membranes, which extend over the body surface or into the body cavities, and their indigenous microbiota, are defenses providing barriers to infection at these exposed sites.

Physical Barriers. The keratin in the skin's outer layer, the epidermis, forms a tough and impenetrable barrier to infection. In addition, the thin outer epidermis is constantly being sloughed off, and with it goes any attached microorganisms. The skin also is a poor source of nutrition for invading microbes and, from the viewpoint of a pathogen, the low water content of the skin presents a veritable desert. Unless microbes can penetrate the skin barrier, disease is rare, although some dermatophytes can lead to diseases like athlete's foot (see Chapter 17).

Cuts or abrasions are a fact of everyday life. These breaches to the skin may allow microbes to enter the blood; bites from infected arthropods, acting as hypodermic needles, may inject the pathogens responsible for such diseases as West Nile fever, malaria, or plague. Other means of penetrating this mechanical barrier include puncture wounds, tooth extractions, burns, shaving nicks, war wounds, and injections.

Possible penetration of the skin barrier is patrolled by dendritic cells (called Langerhans cells in the skin tissues). If a pathogen is detected by these cells, they phagocytize the pathogen and migrate to a regional lymph node where they induce an acquired immune response. The dendritic cells also secrete cytokines, which influence both innate and acquired immune responses.

The **mucous membranes** (**mucosa**) are the moist epithelial lining of the digestive, urogenital, and respiratory tracts. Because these exposed sites represent potential portals of entry, the mucous membranes provide resistance to microbial invasion. For example, cells of the mucous membranes that line the respiratory passageways secrete **mucus**, a sticky secretion of glycoproteins that traps heavy particles and microorganisms (see Chapter 10). The epithelial cilia then move the microbe-laden mucus up to the throat, where the material is swallowed or coughed out (MicroFocus 20.1).

Chemical Barriers. Several host chemicals also provide an antimicrobial defense to infection by inhibiting microbial growth or viral replication. Resistance in the vaginal tract is enhanced by the low pH. This develops when *Lactobacillus* species in the microbiota break down glycogen to various acids. Many researchers believe the disappearance of lactobacilli during antibiotic treatment encourages opportunistic diseases such as candidiasis and trichomoniasis to develop (see Chapters 17 and 18). In the urinary tract, the slightly acidic pH of the urine discourages pathogen growth, and the flow of urine flushes the microbes away.

A natural barrier to the gastrointestinal tract is provided by stomach acid, which, with a pH of approximately 2.0, destroys most pathogens. However, there can be notable survivors, including polio and hepatitis A viruses, the typhoid and tubercle bacilli, and *Helicobacter pylori,* a major cause of peptic ulcers (see Chapter 11). Bile from the gallbladder enters the system at the duodenum and serves as an inhibitory substance. In addition, duodenal enzymes hydrolyze the proteins, carbohydrates, lipids, and other large molecules of microorganisms.

The chemical barriers also consist of numerous small, antimicrobial peptides, called **defensins**, which are found in various secretions throughout the body. They are produced by phagocytic cells as well as epithelial cells of the respiratory, gastrointestinal, and urogenital tracts. The sebum on the skin surface, for example, produces at least 12 defensins. Some defensins are continually produced, while others are induced by microbial products. They are active against many gram-negative and gram-positive bacterial

Bile:
A yellowish-green fluid, which in the small intestine plays an essential role in emulsifying fats.

Sebum:
An oily substance secreted by the sebaceous glands to lubricate the hair and skin.

MICROFOCUS 20.1: Miscellaneous
Who Turned On the Spigot?

Every time we get a cold, the flu, or a seasonal allergy, we often end up with the sniffles or a truly raging runny nose. When this happens, it is simply your body's response to what it believes is an infection—or a hypersensitivity to cold temperatures or spicy food. As a first line of defense against infection, innate immunity uses mucus flow as the best way to wash respiratory pathogens out of the airways.

You always are producing—and swallowing—mucus. Most people are not aware of their mucus production until their body revs up mucus production in response to a cold or flu virus, or allergens. So how much mucus or phlegm is produced?

In a healthy individual, glands in the nose and sinuses are continually producing clear and thin mucus—often more than 200 milliliters each day! An individual is not aware of this production because the mucus flows down the throat and is swallowed. Now if that individual comes down with a cold or the flu, the nasal passages often become congested, forcing the mucus to flow out through the nostrils of the nose. This requires clearing by blowing the nose or (to put it nicely) expectorating from the throat. With a serious cold or flu, the mucus may become thicker and gooier, and have a yellow or green color. The revved up mucus flow in such cases can amount to about 200 milliliters (a little less than a cup) every hour; if you blow your nose 20 times an hour, each blow could amount to anywhere from 2 to 10 milliliters of mucus. If you have watery eyes as well, then that tear liquid can enter the nasal passages and combine with the mucus, producing an even larger "flow per blow."

So, although a runny nose is usually just an annoyance, make sure you drink plenty of water to make up for the lost phlegm from the runny spigot.

cells, fungi, and viruses such as HIV. Many of the defensins damage membranes, killing the pathogens through cell lysis.

In a somewhat similar manner, **lysozyme** is a chemical inhibitor found in human tears, mucus, and saliva. The enzyme disrupts the cell walls of gram-positive bacterial cells by weakening the peptidoglycan. Osmotic lysis results in cell death (see Chapter 4).

Other soluble proteins, such as the complement components and interferon, are important physiological barriers in innate immunity and will be described later in this chapter.

Cellular Barriers. The blood monocytes, tissue macrophages, and neutrophils are key cellular defenses of innate immunity. These cells can carry out phagocytosis and trigger an inflammatory response to destroy invading pathogens. These more complex processes will be described below.

Besides these leukocytes, defensive barriers also include the normal microbiota of the body surfaces (see Chapter 19). These indigenous, non-pathogenic microbes form a cellular barrier by outcompeting pathogens for nutrients and attachment sites on the skin and mucosa. **FIGURE 20.3** summarizes the innate barrier mechanisms.

CONCEPT AND REASONING CHECKS

20.4 Explain how the skin and mucous membranes provide a physical and chemical defense against infection.

Phagocytosis Is a Nonspecific Defense Mechanism to Clear Microbes from Infected Tissues

KEY CONCEPT

5. Cellular defenses can phagocytize pathogens.

One of Metchnikoff's key observations involved the larvae of starfish. When jabbed with a splinter of wood, motile cells in the starfish gathered

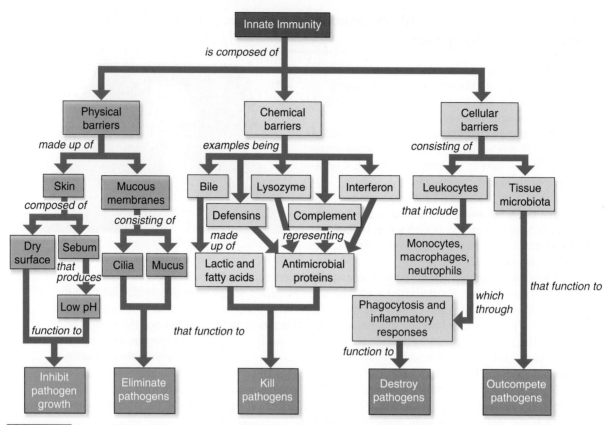

FIGURE 20.3 **Innate Defense Barriers to Pathogens.** This concept map summarizes the ways innate immunity is poised to inhibit, eliminate, destroy, and compete against an invading pathogen. »» Why does there need to be such a diversity of defense mechanisms?

around the splinter. From observations like these, Metchnikoff suggested cells could actively seek out and engulf (ingest) foreign particles in the body. Metchnikoff's theory of phagocytosis laid the foundation for one of the most important functions of innate immunity.

Phagocytosis (*phago* = "eat") is the capturing and digesting of foreign particles, including pathogens, by phagocytes. This may occur at the site of the infection as well as in the lymphoid tissues (**FIGURE 20.4**).

If a bacterial or viral pathogen breaches the exterior defenses, the intruder usually is recognized, ingested, and killed by neutrophils and macrophages residing in the tissue. On binding the pathogen, the phagocytes release cytokines that will trigger an inflammatory response (to be described next) as well as activate an acquired immune response.

Some cytokines, called **chemokines**, stimulate the migration of neutrophils to the infection site.

The process begins with the attachment of the microbe to the cell surface of the phagocyte. Macrophages do not simply "bump" into their victim. Rather, they extend thin cytoplasmic protrusions, called **filopodia**, that wave like a fishing lure. If a filopodium "catches" its victim, the filopodium rapidly contracts. The plasma membrane then surrounds the microbe and an invagination, or folding in, of the membrane produces an internalized vacuole, called a **phagosome**. The phagosome now becomes acidified and the lowered pH aids in killing or inactivating the pathogen. In addition, the phagosome fuses with several lysosomes. Within this **phagolysosome**, lysosome enzymes, such as lysozyme and acid hydrolases,

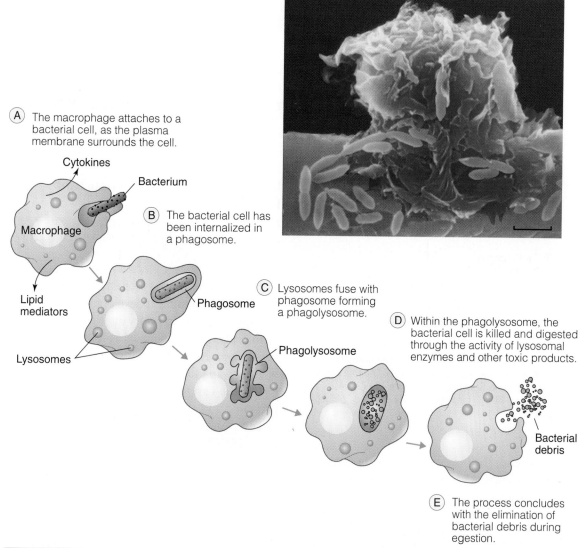

(A) The macrophage attaches to a bacterial cell, as the plasma membrane surrounds the cell.

Cytokines

Bacterium

Macrophage

(B) The bacterial cell has been internalized in a phagosome.

Lipid mediators

Lysosomes

Phagosome

(C) Lysosomes fuse with phagosome forming a phagolysosome.

(D) Within the phagolysosome, the bacterial cell is killed and digested through the activity of lysosomal enzymes and other toxic products.

Phagolysosome

Bacterial debris

(E) The process concludes with the elimination of bacterial debris during egestion.

FIGURE 20.4 **The Mechanism of Phagocytosis.** The stages of phagocytosis are shown. Inset: a false-color scanning electron micrograph of a macrophage (blue) engulfing *E. coli* cells (green) on the surface of a blood vessel (red). (Bar = 5 μm.) »» What might the bacterial debris or waste products be in (E)?

digest the pathogen. In addition, phagocytosis triggers a so-called **respiratory burst** that produces toxic metabolites, including hydrogen peroxide (H_2O_2), nitric oxide (NO), superoxide anions (O_2^-), and hypochlorous acid (bleach).

Neutrophils are usually the first cell type arriving at the infection site. Because neutrophils are short-lived cells, the longer-lived macrophages arriving at the site are more likely to kill the majority of pathogens. For the neutrophils, death is not the end of their role in innate defense against pathogens. When living neutrophils con-

tact pathogens, some of the cells may transform themselves into extracellular fibers called **neutrophil extracellular traps (NETs)**. In these situations, the neutrophils become deformed, and the nuclear envelope and granule membranes disintegrate (**FIGURE 20.5**). When the plasma membrane of the dying neutrophil ruptures, the NET components—nuclear DNA and granule antimicrobial proteins—mix. The invading pathogens get caught in these NET fibers and the antimicrobial agents degrade and kill the pathogens without the need for phagocytosis.

Antibodies:
Proteins produced by plasma cells in response to the presence of a foreign material or cell, such as a bacterium or virus.

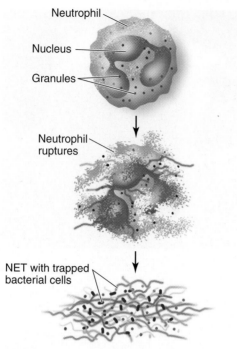

Neutrophil

Nucleus

Granules

Neutrophil ruptures

NET with trapped bacterial cells

FIGURE 20.5 Formation of NETs. A neutrophil becomes activated, leading to the release of DNA and granule proteins that, on rupture, assemble into NETs. (Adapted from Lee, Warren, L., and Grinstein, S., *Science* 303 (2004): 1477–1478.) »» How is innate immunity improved by generating NETs?

In the end, the host clears the pathogen and the infection ends, as waste materials are eliminated from the phagocytes. However, the host-microbe relationship is always evolving to give either the host or the microbe the advantage. Sometimes the pathogen gets the upper hand by "using" phagocytosis to its advantage (MICROFOCUS 20.2).

Although phagocytes are effective at killing pathogens and clearing many infections, collateral damage often occurs. During the fierce phagocytic activity occurring at the infection site, lysosome enzymes and other products may be released inadvertently into the surrounding tissue. In addition, neutrophils release lysosome enzymes when dying. The released materials may have damaging effects on surrounding tissue and often are the major reason for tissue damage from an infection. Indeed, **pus**, which is composed of dead and dying neutrophils and damaged tissue arising from lysosome damage, often is a sign of infection.

The internalization of microbes into phagocytes often is enhanced when the microbes are coated with certain serum proteins, such as antibodies or complement (discussed below). These protein molecules, called **opsonins**, attach to microbes and increase the ability of phagocytes to adhere to the pathogen. The enhanced phagocytic process is called **opsonization**.

CONCEPT AND REASONING CHECKS

20.5 Discuss the importance of phagocytosis to innate immunity.

Inflammation Plays an Important Role in Fighting Infection

KEY CONCEPT

6. The inflammatory response brings effector molecules and cells to the infection site.

Inflammation is a nonspecific defensive response by the body to trauma. It develops after a mechanical injury, such as an injury or blow to the skin, or from exposure to a chemical agent, such as acid or bee venom. An inflammatory response also may be due to an infection by a pathogen.

In the case of an infection, the pathogen's presence and tissue injury sets into motion an innate process to limit the spread of the infection (**FIGURE 20.6**). At the site of tissue damage (infection), resident tissue macrophages secrete several cytokines triggering a local dilation of the blood vessels (**vasodilation**) and increasing capillary permeability. This allows the flow of plasma into the injured tissue and fluid accumulation (**edema**) at the site of infection. Chemokines attract additional phagocytes (neutrophils and monocytes) toward the infected tissue. These cells adhere to the blood vessels near the infection and then migrate between capillary cells (**diapedesis**) into the infected tissue. At the infection site, the neutrophils augment phagocytosis as new monocytes differentiate into more macrophages. Chemokines also attract mast cells, which secrete histamine, bolstering the inflammatory response.

The inflamed area thus exhibits four cardinal signs: redness (from blood accumulation); heat (from the warmth of the blood); swelling (from the accumulation of fluid); and pain (from local tissue damage). Sometimes itching may occur, as MICROFOCUS 20.3 describes.

Histamine:
A chemical mediator causing contraction of smooth muscle and dilation of blood vessels.

MICROFOCUS 20.2: Evolution
Avoiding the "Black Hole" of the Phagocyte

Lysosomes contain quite a battery of enzymes, proteins, and other substances capable of efficiently digesting, or being toxic to, almost anything with which they come in contact. However, not all bacterial species are automatically engulfed by phagocytes.

Some species, such as *Streptococcus pneumoniae,* have a thick capsule (see Chapter 4), which is not easily internalized by phagocytes. Other pathogens simply invade in such large numbers they overwhelm the host defenses and the ability of phagocytes to "sweep" them all up. More interesting are pathogens that have evolved strategies to evade the lysosome—the "black hole" of the phagocyte (see figure).

Listeria monocytogenes is internalized by phagocytes. However, before lysosomes can fuse with the phagosome membrane, the bacterial cells release a pore-forming toxin that lyses the phagosome membrane. This allows the *L. monocytogenes* cells to enter and reproduce in the phagocyte cytoplasm. The bacterial cells then spread to adjacent host cells by the formation of actin tails (see Chapter 19).

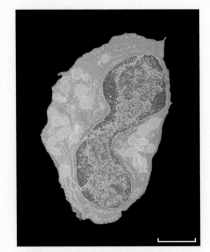

False-color transmission electron micrograph of lysosomes (yellow) in a macrophage. (Bar = 4 μm.)

Other pathogens prevent lysosome fusion with the phagosome in which they are contained. The causative agent of tuberculosis, *Mycobacterium tuberculosis,* is internalized by macrophages. However, once in a phagosome, the bacterial cells prevent fusion of lysosomes with the phagosome. In fact, the tubercle bacteria actually reproduce within the phagosome. A similar strategy is used by *Legionella pneumophila.*

Toxoplasma gondii, the protozoan responsible for toxoplasmosis (see Chapter 18), evades the lysosomal "black hole" by enclosing itself in its own membrane vesicle that does not fuse with lysosomes. Avoiding digestion by lysosomes is key to survival for these pathogens.

Eventually the formation of a fibrin clot occurs, which among other things prevents the spread of pathogens to the blood. Again, pus may be a product during inflammation. When pus becomes enclosed in a wall of fibrin through the clotting mechanism, an **abscess**, or boil, may form.

To limit tissue damage, the body tries to regulate the number of phagocytes and the level of cytokine production in the defense against infection. Therefore, the inflammatory response must be managed so it is only short-lived. Should the inflammatory response become long-lived, chronic inflammation may occur. The persisting cytokines can then produce excessive damage, leading to a number of dissimilar diseases, including rheumatoid arthritis, coronary heart disease, atherosclerosis, Alzheimer disease, or some forms of cancer.

Another consequence of an overwhelming "cytokine storm" is **septic shock**, characterized by a collapse of the circulatory and respiratory systems (see Chapter 19). In some gram-negative bacterial infections, large amounts of cell wall lipopolysaccharide (LPS) are released, which is referred to as an endotoxin. The systemic spread of LPS through the body triggers macrophages to secrete the "overdoses" of cytokines. Shock can lead to death.

Atherosclerosis: An arterial disease in which degeneration and cholesterol deposits (plaques) form on the inner surfaces of the arteries.

CONCEPT AND REASONING CHECKS

20.6 Explain how inflammation serves as an innate immune defense.

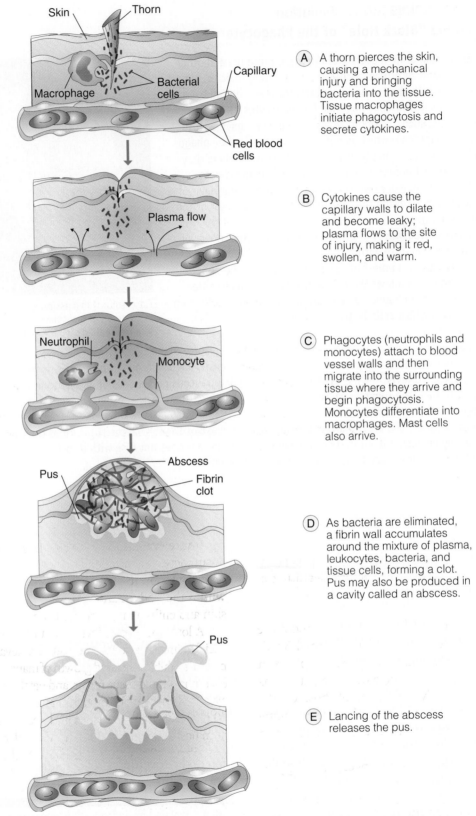

A A thorn pierces the skin, causing a mechanical injury and bringing bacteria into the tissue. Tissue macrophages initiate phagocytosis and secrete cytokines.

B Cytokines cause the capillary walls to dilate and become leaky; plasma flows to the site of injury, making it red, swollen, and warm.

C Phagocytes (neutrophils and monocytes) attach to blood vessel walls and then migrate into the surrounding tissue where they arrive and begin phagocytosis. Monocytes differentiate into macrophages. Mast cells also arrive.

D As bacteria are eliminated, a fibrin wall accumulates around the mixture of plasma, leukocytes, bacteria, and tissue cells, forming a clot. Pus may also be produced in a cavity called an abscess.

E Lancing of the abscess releases the pus.

FIGURE 20.6 **The Process of Inflammation.** A series of nonspecific steps generates an inflammatory response to some form of trauma or infection, such as being pierced by a plant thorn harboring bacterial cells. »» Why does the inflammatory response generate the same set of events no matter what the nature of the foreign substance?

MICROFOCUS 20.3: Evolution
But It Doesn't Itch!

How many of us at some time have had a small red spot or welt on our skin that looks like an insect bite—but it doesn't itch? When most of us are bitten by an insect, such as an ant, bee, or mosquito, the site of the bite starts itching almost immediately. Yet, there are cases where bites do not itch, such as a tick bite. So, what's the difference?

Most insect bites trigger a typical inflammatory reaction. Take a mosquito as an example. When it bites and takes a blood meal, saliva injected into the skin contains substances so the blood does not coagulate. The bite causes tissue damage and activation of chemical mediators that trigger the typical characteristics of an inflammatory response: redness, warmth, swelling, and pain at the injured site.

One of the major chemical mediators is histamine, released by the damaged tissue and immune cells within the tissue. An itch actually is a form of pain and arises when histamine affects nearby nerves. Itching is the result.

In a tick bite, the scenario is slightly different. The bite triggers the same set of inflammatory reactions and histamine is produced. However, in the tick saliva is another molecule called histamine-binding protein (HBP). HBP binds to histamine, preventing the mediator from affecting the nearby nerves—and no itching occurs.

So why is that an advantage for the tick? Probably because ticks take long blood meals that can last for three to six days. By the bite being painless and not itching, there is less chance of the ticks being noticed and removed.

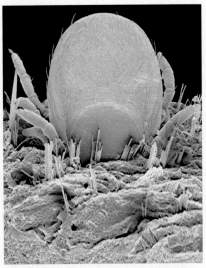

False-color scanning electron micrograph of a tick (orange) feeding head-down in human skin.

Moderate Fever Benefits Host Defenses

KEY CONCEPT

7. Fever, usually one sign of a possible infection, can help immune defenses.

Fever is an abnormally high body temperature that remains elevated above the normal 37°C. Fever supports the immune system response in trying to gain an advantage over the pathogens by making the body a less favorable host. On the other hand, fever is metabolically costly; keeping the body temperature elevated is an energy drain on the body.

Fever-producing substances, called **pyrogens**, represent endogenous cytokines produced by activated macrophages and other leukocytes, and exogenous microbial fragments from bacterial cells, viruses, and other microorganisms. Pyrogens move through the blood and affect the anterior hypothalamus such that the body temperature becomes elevated. As this takes place, cell metabolism increases and blood vessels constrict, thus denying blood to the skin and keeping its heat within the body. Patients thus experience cold skin and chills along with the fever.

A low to moderate fever in adults may be beneficial to immune defense because the elevated temperature inhibits the rapid growth of many microbes, encourages rapid tissue repair, and heightens phagocytosis. However, if the temperature rises above 40.6°C (105°F), convulsions and death may result from host metabolic inhibition. Also, infants with a fever above 38°C (100°F), or older children with a fever of 39°C (102°F), may need medical attention. MICROFOCUS 20.4 examines the saying that it is good to "feed a cold and starve a fever."

Hypothalamus: The part of the brain controlling involuntary functions.

CONCEPT AND REASONING CHECKS

20.7 Assess the advantage of a low grade fever during an infection.

MICROFOCUS 20.4: Being Skeptical
Feed a Cold, Starve a Fever?

One of the typical symptoms of the flu is a high fever (39 to 40°C) while a cold seldom produces a fever. So, is there any truth to the old adage, "You should feed a cold and starve a fever?" The definite answer is—yes!

There are three reasons you should reduce food intake with a fever. First, food absorbed by the intestines could be misidentified by the body as an allergen, worsening the illness and fever. Also, with the body already under stress, heavy eating can contribute on rare occasions to body seizures, collapse, and delirium.

Perhaps most common is that eating stimulates digestion of the foods. During times of increased physiologic stress, digestion can overstimulate the parasympathetic nervous system, when the sympathetic nervous system (involved in constricting blood vessels and conserving heat = fever) already is active. In other words, eating tends to drop body temperature, possibly eliminating fever as a protective mechanism.

So, with the flu or other illness-induced fever, stick to bed rest and just drink plenty of liquids.

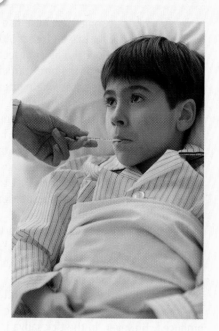

Natural Killer Cells Recognize and Kill Abnormal Cells

KEY CONCEPT

8. Natural killer cells are nonspecific defensive lymphocytes.

Besides the phagocytes, **natural killer (NK) cells** are another type of defensive leukocyte. These lymphocytes are formed in the bone marrow and then migrate to the tonsils, lymph nodes, and spleen, where they await activation. On stimulation, they secrete several cytokines, triggering acquired immune responses by macrophages and other immune cells. While those events develop, the NK cells move into the blood and lymph where they act as potent nonspecific killers of tumor cells and virus-infected cells.

NK cells are not phagocytic; rather, they contain on their surfaces a set of special receptor sites capable of forming cell-to-cell interactions with a target cell (**FIGURE 20.7**). If the receptor sites match up with a group of class I MHC proteins on the target cell, the NK cell recognizes the target cell as one of the body's own and leaves it alone. The NK cell does not release the cytolytic mediators.

Apoptosis:
A type of cell death activated by an internal death program.

MHC proteins:
A class of cell-surface proteins found on mammalian cells.

However, when these MHC proteins are absent or in reduced amounts (as on a cancer cell or virus-infected cell), then the matchup is incomplete and the NK cell secretes the cytotoxic mediators that damage the tumor or virus-infected cell. The cytolylic mediators include **perforins**, which drill holes in the target cell. They also secrete **granzymes**, which are cytotoxic enzymes inducing the target cell to undergo apoptosis. The DNA condenses, the cell shrinks, and breaks into tiny pieces that can be mopped up by wandering phagocytes.

CONCEPT AND REASONING CHECKS

20.8 How do natural killer cells assist the other innate defense mechanisms during a viral infection?

Complement Marks Pathogens for Destruction

KEY CONCEPT

9. Complement proteins assist innate immunity in identifying and eliminating pathogens.

Complement (also called the complement system) is a group of nearly 30 inactive proteins circulating in the bloodstream that represents another innate defense to disease. If a microbe penetrates the

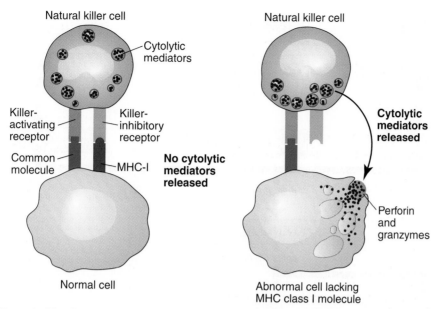

FIGURE 20.7 **Natural Killer (NK) Cell Recognition.** NK cells have the ability to destroy any cell they "see" as abnormal (for example, tumor cells or virus-infected cells). (Adapted from Delves, P. J., and Roitt, I. M. (2000). *N Engl J Med* 343: 37–49.) »» How does the presence or absence of class I MHC receptors affect NK cell activity?

body, complement proteins become sequentially activated through a cascade of steps that assist in the inflammatory response and phagocytosis. Additional complement proteins in the cascade bring about the destruction of the invading bacterial cells or viruses through cell or virion lysis. The three complementary pathways for complement activation are illustrated in **FIGURE 20.8** .

All these pathways—the classical, alternative, and lectin pathway—culminate in the activation of an enzyme called C3 convertase, which splits C3 into a C3a and C3b fragment. C3b activates another complement protein C5, which also is hydrolyzed into C5a and C5b fragments. Complement fragments C3a and C5a trigger an inflammatory response through vasodilation and increased capillary permeability. C3b also acts as an opsonin by binding to the pathogen surface (opsonization) and enhancing phagocytosis.

C5b plays one more important role. The complement fragment triggers another complement cascade that assembles into what are called **membrane attack complexes** (**MACs**). These complexes of complement proteins (C5b-9) form large holes in the membranes of many microbes, especially gram-negative bacteria and enveloped viruses. The MACs are so prevalent that lysis

occurs, killing the pathogen. This teamwork within the complement system and inflammation provides a formidable obstacle to pathogen invasion.

CONCEPT AND REASONING CHECKS

20.9 Explain the outcome to someone who has an infectious disease but has a genetic defect making the person unable to produce C5 complement protein.

Innate Immunity Depends on Receptor Recognition of Common Pathogen-Associated Molecules

KEY CONCEPT

10. Pattern-recognition receptors trigger the innate immune response.

In the description of the phagocytic process, how does a phagocyte "know" a particular cell or virus is a pathogen and needs to be engulfed? The answer is that macrophages, dendritic cells, and epithelial cells recognize pathogens by identifying unique, highly conserved microbial molecular sequences not found on host cells. This system of recognition is called **pathogen-associated molecular patterns** (**PAMPs**) and includes small molecular sequences in such structures as the lipopolysaccharide (LPS) layer of gram-negative cell

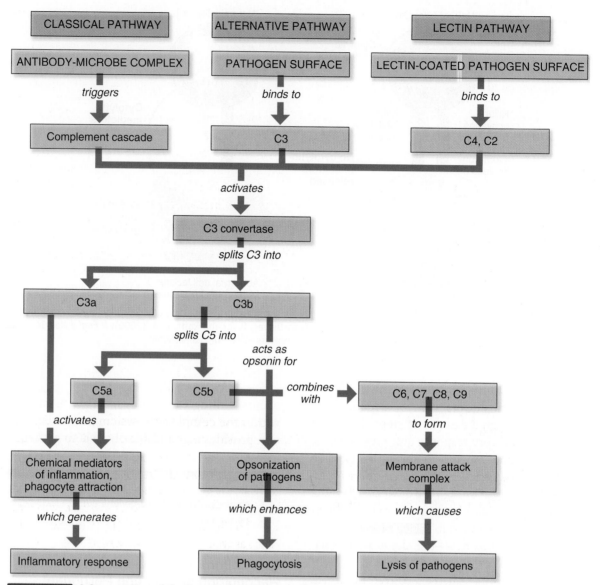

FIGURE 20.8 **A Concept Map of the Complement Pathways.** The early events of all three pathways of complement activation culminate in the activation of C3 convertase. Complement fragment C3b binds to the membrane and opsonizes bacterial cells, allowing phagocytes to internalize them (phagocytosis). The small fragments, C5a and C3a, are mediators of local inflammation. C5b also activates another complement cascade, whereby the terminal components of complement assemble into a membrane-attack complex (MAC) that can damage the membrane of gram-negative cells and enveloped viruses. »» What is the advantage of having all three pathways leading to activation of C3 convertase?

walls, peptidoglycans of gram-positive cell walls, flagella, fungal cell walls, and bacterial and viral nucleic acids. Currently, microbiologists believe innate immunity recognizes at least 1,000 different PAMPs.

PAMPs are recognized by protein receptors called **toll-like receptors** (**TLRs**) that are present on macrophages, dendritic cells, and endothelial cells. In fact, TLRs also are present in many other animals and plants, suggesting they, like the defen-

sins, have an ancient evolutionary origin. (*Toll* is German for "amazing," referring to the toll receptors first discovered in fruit flies where they are a fly's "amazing" defense against fungal infections).

The TLR family in humans consists of at least ten members, each working to mediate a specific response to distinct PAMPs. Specific TLRs are located on the immune defense cell's plasma membrane and the phagosome membrane (**FIGURE 20.9**). For example, plasma membrane

TLRs can bind to unique bacterial lipopeptides, zymosan, LPS, or bacterial flagellar proteins. TLRs found in the phagosome membrane accelerate the formation of phagolysosomes and encourage faster killing of engulfed microbes. Other PAMPs bind to double-stranded viral RNA and single-stranded viral RNA. TABLE 20.2 lists the known TLRs and their PAMP targets.

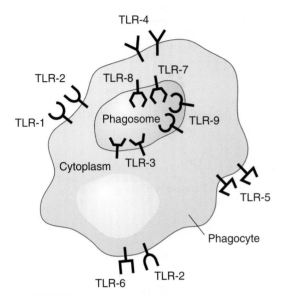

FIGURE 20.9 Pattern-Recognition Receptors on Phagocytes. Proteins called toll-like receptors (TLRs) are located on the plasma and phagosome membranes of phagocytes. The TLRs initiate important innate defense mechanisms by recognizing unique molecular patterns on pathogens. »» Why do some TLRs have to exist in the phagosome?

In humans, when TLRs bind to a particular PAMP, chemical signals are sent to the cell nucleus turning on the synthesis of cytokine genes. Following transcription and translation, the cytokines are secreted and sensed by immune cells and other nonimmune cells essential to the innate immune response. Some cytokines stimulate liver cells to synthesize and secrete defensive blood proteins called **acute phase proteins**, which elevate inflammation and activate the lectin pathway of the complement cascade. TLR activation to PAMPs in macrophages and dendritic cells also induce these cells to differentiate into cells active in the acquired immune response.

CONCEPT AND REASONING CHECKS

20.10 Justify the need for the spectrum of toll-like receptors found on and in human immune cells.

Interferon Puts Cells in an Antiviral State

KEY CONCEPT

11. Interferons are natural proteins capable of triggering proteins to block viral replication.

Infection of cells with viruses triggers yet another nonspecific, innate immune response. Virus-infected cells produce a set of cytokine protein called **interferons (IFNs)** that "alerts" surrounding cells to the viral threat. Two IFNs, IFN-alpha and IFN-beta, trigger a nonspecific reaction via TLR-3 designed to protect against the stimulating virus, as well as many other viruses.

Zymosan: A carbohydrate–protein complex in fungal cell walls.

TABLE		
20.2	**The Recognized Human Toll-Like Receptors**	
Toll-Like Receptor	Location	PAMP (Molecular Sequences)
1 and 2	Plasma membrane	Bacterial cells (lipoproteins) Parasites (membrane proteins)
2 and 6	Plasma membrane	Gram-positive bacterial cell walls (peptidoglycan) Gram-positive bacterial cell walls (lipoteichoic acids) Fungal cell walls (zymosan)
3	Phagosome membrane	Viruses (double-stranded RNA)
4	Plasma membrane	Gram-negative bacterial cell walls (LPS)
5	Plasma membrane	Bacterial flagella (flagellar proteins)
7 and 8	Phagosome membrane	Viruses (single-stranded RNA)
9	Phagosome membrane	Bacterial and viral DNA

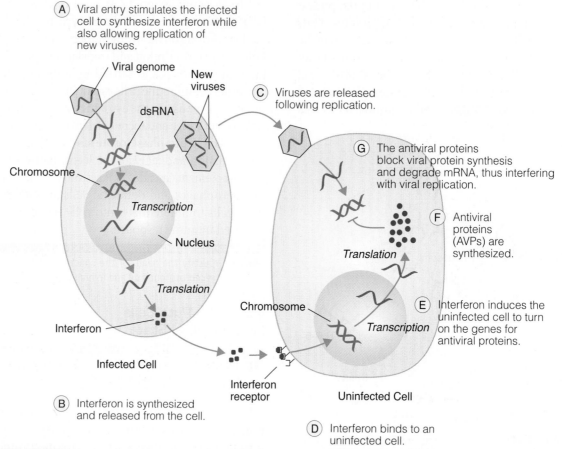

(A) Viral entry stimulates the infected cell to synthesize interferon while also allowing replication of new viruses.

Viral genome

New viruses

(C) Viruses are released following replication.

dsRNA

Chromosome

Transcription

Nucleus

Translation

Interferon

Infected Cell

(B) Interferon is synthesized and released from the cell.

Interferon receptor

(G) The antiviral proteins block viral protein synthesis and degrade mRNA, thus interfering with viral replication.

(F) Antiviral proteins (AVPs) are synthesized.

Translation

Chromosome

Transcription

(E) Interferon induces the uninfected cell to turn on the genes for antiviral proteins.

Uninfected Cell

(D) Interferon binds to an uninfected cell.

FIGURE 20.10 **The Production and Activity of Interferon.** Left: A host cell produces interferon following viral infection. The virus replicates in the same cell. Right: The interferon reacts with receptors at the surface of a neighboring uninfected cell, inducing the cell to produce antiviral proteins. The uninfected cell enters an activated, antiviral state capable of inhibiting viral replication. »» Why would interferon itself not be considered an antiviral compound?

IFN-alpha and IFN-beta do not interact directly with viruses, but rather with the cells they alert. The IFNs are produced when a virion releases its genome into the cell. High concentrations of double-strand RNA formed during the replicative process in the infected cells induce the cell to synthesize and secrete IFNs (**FIGURE 20.10**). These bind to specific IFN receptor sites on the surfaces of adjacent cells. Binding triggers a signaling pathway that turns on new gene transcription involved in the synthesis of **antiviral proteins (AVPs)** within those cells. Such cells are now in an **antiviral state** and capable of inhibiting viral replication. The AVPs inhibit translation and degrade mRNA—both viral and host.

Obviously, the interferons are not a magic bullet toward preventing viral infections; we all still get colds, the flu, and other virally caused illnesses. Today, the use of recombinant DNA technology (see Chapter 9) has made it possible for pharmaceutical companies to mass produce synthetic IFN-alpha for disease treatments. The FDA has approved INF-alpha (alfa) 2a for use, along with other drugs, in treating hepatitis B and C, and genital warts—all caused by viruses.

In summary, physical barriers, immune cells, antimicrobial chemicals, and signaling chemicals are needed to protect the human host, generate an effective innate immune response, and induce the acquired immune response. MICROINQUIRY 20 presents an overview linking innate and acquired immunity.

CONCEPT AND REASONING CHECKS

20.11 Distinguish between the role of interferon and antiviral proteins.

Resistance and the Immune System: Acquired Immunity

Identifying HIV was the critical first step in defining the cause of AIDS, but, as Robert Koch so elegantly pointed out more than a century ago, showing that a particular infectious agent causes a specific disease can be an arduous process.
—Stanley Prusiner in *Historical Essay: Discovering the Cause of AIDS*

The great breakthroughs in any field of science are infrequent. Although most experiments advance our understanding of a phenomenon, more often than not they are the result of a "yes" or "no" answer. But sometimes an experiment or observation not only gives a yes or no answer, it also is unexpected and makes a giant leap forward in scientific understanding. The observations and work carried out by Pasteur and Koch on the germ theory are just a few historical breakthroughs that come to mind.

In 1981, an immunodeficiency disease among gay men was reported. Once there was an alarming rise in the number of new cases of what became known as acquired immunodeficiency syndrome (AIDS), the race was on to discover its cause and how the disease propagated itself.

The eventual appearance of AIDS in distinctly different populations including homosexual and heterosexual individuals, children, intravenous drug users, hemophiliacs, and blood transfusion recipients suggested it must be an infectious agent. And as the number of AIDS cases continued to rise, so did the hypotheses about its possible causation.

Luc Montagnier, Françoise Barré-Sinoussi, and their colleagues at the Pasteur Institute in Paris were the first to report the discovery of the virus now called the human immunodeficiency virus (HIV) associated with AIDS. Coupled with this breakthrough was the work of Robert Gallo and his research team at the National Cancer Institute in Maryland. Gallo's group showed that the virus identified by Montagnier was the cause of AIDS.

So, what was the virus doing to cause the disease? As the clinical descriptions of AIDS solidified, it became apparent the virus caused a dramatic decrease in the number of T lymphocytes in infected individuals. This supported an earlier discovery by Gallo's group showing that the cytokine interleukin-2 was necessary to support HIV replication. In fact, interleukin-2 stimulation made possible the growth and proliferation of HIV in cultured T lymphocytes (**FIGURE 21.1**).

Scientific breakthroughs not only advance a field but also spin offs, in this case, many medical applications. Finding the cause of AIDS led to the development of a blood test to identify HIV, preventing millions of people from being infected through the transfusion of tainted blood. Knowing how the virus interacts and replicates in T cells stimulated the identification and design of many antiretroviral drugs to combat the disease, allowing infected individuals to live longer and more productive lives.

Most scientists today would agree that the discoveries made by Montagnier and Gallo rank as one of the major scientific breakthroughs of the 20th century and set the stage for influential discoveries made by many others. In 1987, both Montagnier and Gallo were given equal credit for the discovery. However, in 2008, the Nobel Prize in Physiology or Medicine went to Barré-Sinoussi and Montagnier.

If one set out to design a virus to cripple the immune system and the very cells designed to defend against the virus, one could not do better than the naturally evolved HIV. By destroying what are called CD4 T cells, the entire immune

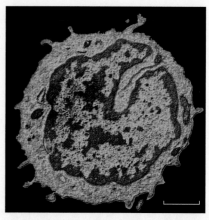

(A)

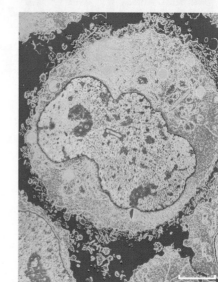

(B)

FIGURE 21.1 Uninfected and HIV-Infected T Lymphocytes. False-color transmission electron micrographs of T lymphocytes. (**A**) An uninfected T cell showing the typical large cell nucleus. (**B**) An infected T cell showing HIV virions (red) budding off the cell surface. (A & B: Bar = 2 μm.) »» Why is it significant that HIV primarily infects CD4 T cells?

system is crippled and unable to respond to any infectious disease exposure.

How is this possible? How can a simple virus inactivate a complex and powerful immune defense?

This chapter provides the answer as we examine how the actions of innate immunity, discussed in the previous chapter, supply the acquired immune response with the needed boost to fight pathogens.

21.1 An Overview of the Acquired Immune Response

With regard to infectious disease, innate immunity, the topic of the previous chapter, is designed to eliminate pathogens—or at least limit their spread until the other half of the immune response, **acquired** (specific) **immunity**, can mobilize the body's cells and molecules to eliminate the pathogen. Before we examine the components and events of acquired immunity, it is best to consider its key features.

The Ability to Eliminate Pathogens Requires a Multifaceted Approach

KEY CONCEPT

1. Acquired immunity responds to, distinguishes between, and remembers specific pathogens it has encountered.

Acquired immunity defends against a tremendous variety of potential pathogens. This defense is based on four important attributes.

1. Specificity. As we saw in the previous chapter, innate immunity has the ability to recognize a diversity of foreign substances, including pathogens. Such microbes or their molecular parts that are capable of mobilizing the immune system and provoking an immune response by B and T cells are called **antigens.** For example, a typical bacterial cell would contain several antigens (FIGURE 21.2).

Antigens do not stimulate the immune system directly. Rather, acquired immunity usually involves protein receptors on B and T cells that recognize discrete regions of the antigen called **antigenic determinants**, or **epitopes** (FIGURE 21.3). An antigen may have numerous epitopes. For example, a structure such as a bacterial flagellum may have hundreds of epitopes, each having a characteristic and distinct three-dimensional shape. It is the ability of acquired immunity to recognize these specific microbial "fingerprints" and generate the specificity needed.

In addition, the molecular makeup and assembly of epitopes represents tremendous diversity. Proteins are the most potent antigens because their amino acids have the greatest array of building blocks. Carbohydrates are less potent antigens than proteins because they lack chemical diversity and rapidly break down in the body. Nucleic acids and lipids also can be antigenic. It is estimated that the human immune system can respond to about 10^{14} diverse epitopes. If the body loses its ability to respond to antigens and antigenic determinants, an **immune deficiency** will occur (Chapter 23).

2. Tolerance of "Self." Under normal circumstances, one's own cells and molecules with their own molecular determinants do not stimulate an acquired immune response. This tolerance occurs because the epitopes are interpreted

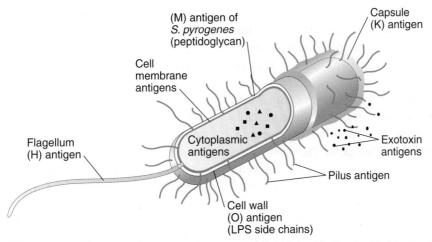

FIGURE 21.2 **The Various Antigens Possible on a Bacterial Cell.** Each different antigen in this idealized bacterial cell is capable of stimulating an immune response. »» Using this figure, what do you know about the *Escherichia coli* strain called 0157:H7?

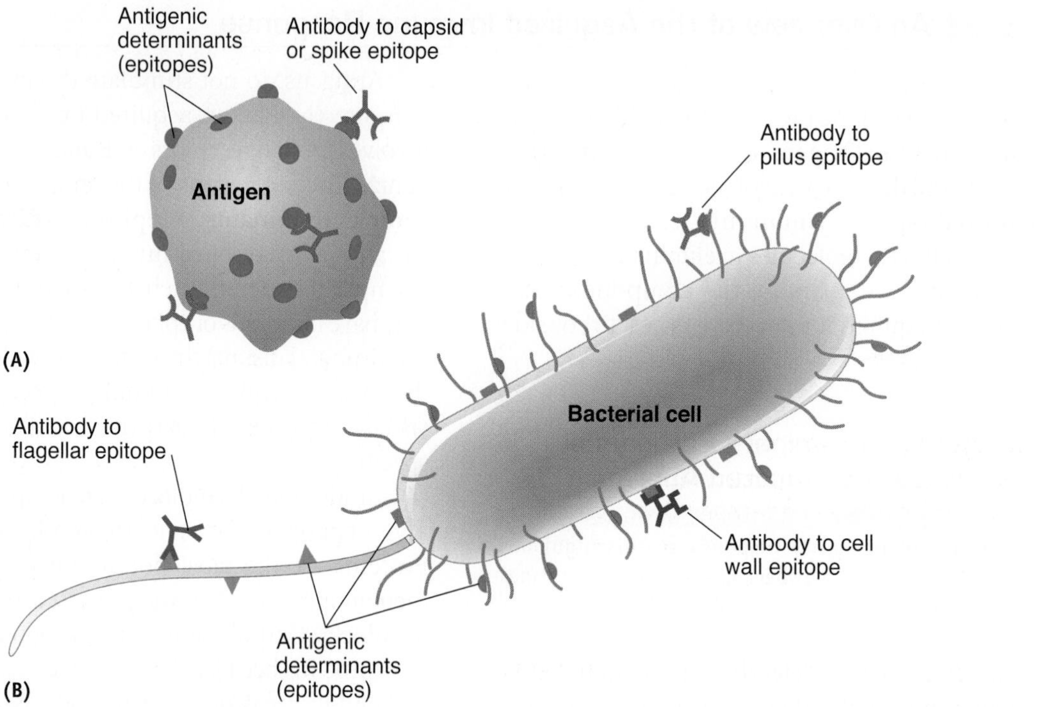

FIGURE 21.3 **Antigens and Antigenic Determinants.** An antigen, such as (**A**) a capsid spike or (**B**) a bacterial flagellum or pilus, contains a number of antigenic determinants or epitopes to which the immune system responds by producing antibodies. The antibody structures drawn in this chapter are based on the structure shown in Figure 21.7. »» Why are there different shaped antibodies seen on these antigens?

Apoptosis:
A process of programmed cell death.

Anaphylactic shock:
A sudden drop in blood pressure, itching, swelling, and difficulty in breathing typical of some allergic reactions.

as "self." A population of T cells called **regulatory T cells** prevents autoreactive T cells from attacking self. In fact, those T cells recognizing self are normally destroyed through apoptosis. The immune system thereby develops a tolerance of "self" and remains extremely sensitive to "nonself" antigens present on pathogens, tumor cells, and cells of other individuals of the same species. Should self-tolerance break down, an **autoimmune disorder**, such as lupus erythematosus, may occur (see Chapter 23).

Many nucleotides, hormones, peptides, and other molecules are recognized by immune cells but do not stimulate an immune system response; that is, they are not **immunogenic**. However, when these nonimmunogenic molecules, called **haptens** (*hapt* = "fastened") are linked (fastened) to proteins in the body, the larger combination may be recognized as "nonself" and trigger an immune system response to both hapten and the antigen. Examples of haptens include penicillin molecules, molecules in poison ivy plants, and molecules in certain cosmetics and dyes. These often are the cause of allergies. Some vaccine antigens also are haptens but

when chemically linked to a protein carrier, they become immunogenic (see Chapter 22).

3. Minimal "Self" Damage. In an acquired immune response, it is important that the response be strong enough to eliminate the pathogen, yet controlled so as not to cause extensive damage to the tissues and organs of the body. That is why most of the encounters between microbe and host occur at local sites, such as the actual infection site and in lymph nodes. In fact, many of the symptoms of an infectious disease are due to the innate and acquired immune responses and not directly to the invading pathogen. One reason the 1918 flu pandemic killed so many "healthy" adults was because the virus overstimulated the immune system (MICROFOCUS 21.1). In other cases, if the immune system extremely overreacts, a potentially fatal anaphylactic shock can occur (see Chapter 23).

4. Immunological Memory. Most of us during our lives have become immune to certain diseases, such as chickenpox and measles, from which we have recovered or to which we have been immunized. This precise long-term

MICROFOCUS 21.1: Infectious Disease
The "Over-Perfect" Storm

You get a cold or the flu, or a bacterial infection, and you suffer for several days. We think of pathogens as our worst enemy during an outbreak of influenza or pneumonia and we rely on our immune system to get us over the illness. In the end, hopefully you recover no less for wear; the controlled acquired immune response did its job to eliminate the infectious agent. However, sometimes the immune system over reacts to an infection and can result in a potentially fatal immune reaction.

When our body detects foreign microorganisms indicating an infection, cells of innate and acquired immunity produce a variety of cells and cytokines, protein signaling molecules that are used extensively in cellular communication to modulate the immune response to the invasion; that is, on encountering a pathogen, immune cells secrete cytokines that activate and recruit additional immune cells to battle the pathogen—without causing extensive collateral damage to the host. This efficient and controlled reaction represents the "perfect storm" in terms of disease elimination.

However, on occasion our body might respond by over-reacting to an infection. For reasons scientists do not completely understand, too many immune cells can be sent to the infection site and the cells produce more cytokines, which brings even more immune cells to the site of infection. This uncontrolled immune response propagates what is referred to as a **cytokine storm** where excessive numbers of immune cells are caught in an endless positive feedback loop that brings more and more immune cells to fight the infection. This "over-perfect" storm causes inflammation in the tissue surrounding the infection and this is what makes the immune reaction potentially so deadly.

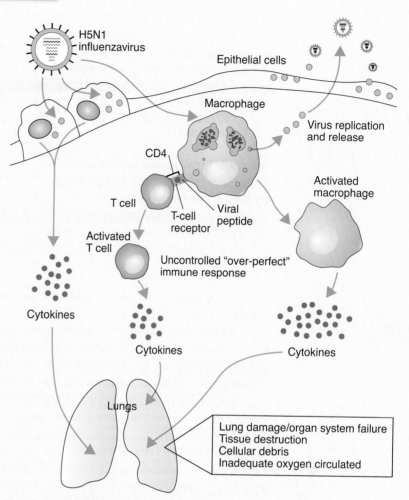

How the H5N1 influenzavirus triggers a cytokine storm.
Adapted from Osterholm, M.T., *N Engl J Med*. 352 (2005): 1839–1842.

For example, a cytokine storm in response to a lung infection can potentially cause permanent lung damage, resulting in multiple organ system failure due, in part, to an inadequacy of oxygen reaching the body's tissues. It explains why the 1918 Spanish flu pandemic (see Chapter 15), which killed an estimated 50 to 100 million people worldwide, took the lives of a disproportionate number of young adults—their healthy immune systems triggered a cytokine storm. A large number of the 774 individuals killed during the 2003 SARS epidemic also were the result of a cytokine storm.

Today, we are seeing two more flu illnesses: the 2009 pandemic of H1N1 (swine) flu, which does not, at this time, appear to be triggering a cytokine storm; and the H5N1 avian flu that does trigger a flood of cytokines and has been responsible for many of the deaths of the more than 280 people (60%) that have been infected with H5N1 (see figure). For some reason, H5N1 viruses are "more potent inducers" of cytokines in infected epithelial cells, macrophages, and T cells than are the H1N1 viruses responsible for seasonal flu. In fact, autopsies of H5N1 flu victims in Southeast Asia have revealed that these victims had lungs filled with debris from the excessive inflammation triggered by the virus and cytokines. In lab studies, the H5N1 virus produced 10-times higher levels of cytokines from human cells in culture than did seasonal flu viruses.

Research is continuing to understand the nature of this "over-perfect" storm and how best to treat affected individuals.

ability to "remember" past pathogen exposures is called **immunological memory** and is a hallmark of acquired immunity. Thanks to immunological memory, a second or ensuing exposure to the same pathogen produces such a rapid and vigorous immune response that the person does not even know they were exposed.

CONCEPT AND REASONING CHECKS

21.1 Justify the statement, "The four characteristics of acquired immunity are really variations of specificity."

Acquired Immunity Generates Two Complementary Responses to Most Pathogens

KEY CONCEPT

2. Acquired immunity involves humoral and cell-mediated responses through receptor recognition of "nonself."

Immunocompetent:
Referring to lymphocytes capable of reacting with a specific epitope.

The cornerstones of acquired immunity are the **lymphocytes**, small 10 to 20 μm in diameter cells, each with a large cell nucleus taking up almost the entire space of the cytoplasm. Viewed with the microscope, all lymphocytes look similar. However, two types of lymphocytes can be distinguished on the basis of developmental history, cellular function, and unique surface markers. The two types are B lymphocytes and T lymphocytes. **B lymphocytes (B cells)** are involved in the process by which antibodies are produced against epitopes. **T lymphocytes (T cells)** provide resistance through direct cell-to-cell contact with, and lysis of, infected or otherwise abnormal cells (for example, tumor cells). MicroFocus 21.2 suggests how people might help boost their T cell population and stay healthier.

Pathogens within the host trigger the innate response, which, in turn, launches two complementary acquired immune responses: the humoral and cell-mediated immune responses.

Humoral Immune Response. One of the immune behaviors toward antigens is called the **humoral immune response** (*humor* = "a fluid"). The result of this response is the activation of B cells and the ensuing production of soluble antibodies that recognize epitopes on the identified antigen in the blood or lymphatic fluids (see Figure 21.3). The response is so specific that the body can generate antibodies to just about any antigen or epitope it encounters.

Cell-Mediated Immune Response. Should microbes or antigens leave the body fluids and enter cells, antibodies are useless. Therefore, another immune response, called the **cell-mediated immune response**, becomes activated to eliminate "nonself" cells, such as virus-infected cells or cancer cells. T cells control and regulate these activities.

Both B cells and T cells have surface receptors anchored in the plasma membrane (**FIGURE 21.4**). After these receptors recognize and bind an epitope, the lymphocyte is said to be "committed." In B cells, the receptor protein is an antibody molecule. In T cells, the receptor protein is composed of two antibody-like chains of glycoproteins. About 100,000 identical surface receptors are found on a particular B or T cell.

The presence of highly specific receptor proteins on the lymphocyte surface implies that even before an antigen enters the body, immunocompetent B and T cells are already waiting "in the wings." Moreover, the genetic code for synthesizing the surface receptor is present in an individual even if the individual has not been exposed to that antigen. We may never experience malaria, for example, yet we have the genetic capability of producing B and T cells with surface receptor proteins for recognizing and binding to the antigens of malaria parasites.

CONCEPT AND REASONING CHECKS

21.2 How do the two arms of acquired immunity differ from one another?

Clonal Selection Activates the Appropriate B and T Cells

KEY CONCEPT

3. Immune recognition only activates the appropriate B and T cell clones that recognize "nonself" epitopes.

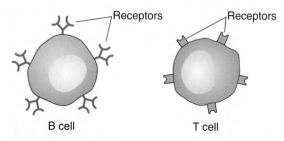

FIGURE 21.4 Receptors of B and T Lymphocytes.
(A) B cells have antibodies on their surface that perform receptor functions by binding to antigen (epitope).
(B) T cells have an antibody-like receptor to bind antigenic fragments. »» How does the characteristic of specificity apply to B and T cells?

MICROFOCUS 21.2: Being Skeptical
Tea Protects Us from Infectious Disease

The healthy aspect of tea has been appreciated for centuries, but within the last 15 years scientific evidence has accumulated to back up the age-old claim. It is known that tea contains flavonoids and antioxidants that researchers believe can inhibit the formation of cancer cells and have a positive protective effect on the cardiovascular system by preventing atherosclerosis.

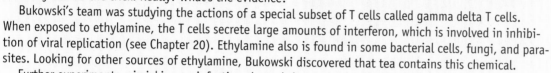

In 2003, Jack Bukowski at Harvard Medical School's Brigham and Woman's Hospital in Boston published a paper suggesting tea also can help defend against infectious disease. It seems just three to five cups of ordinary tea made from tea leaves each day will do the trick. Really? What's the evidence?

Bukowski's team was studying the actions of a special subset of T cells called gamma delta T cells. When exposed to ethylamine, the T cells secrete large amounts of interferon, which is involved in inhibition of viral replication (see Chapter 20). Ethylamine also is found in some bacterial cells, fungi, and parasites. Looking for other sources of ethylamine, Bukowski discovered that tea contains this chemical.

Further experiments mimicking an infection showed that exposing blood to these chemicals in the test tube could increase the numbers of gamma delta T cells by up to five times. In contrast, human blood cells not exposed to the chemical showed a much less significant cell response.

Using human volunteers, the investigators had 21 non-tea-drinkers drink either five to six small cups of black tea or five to six small cups of instant black coffee daily for four weeks. Coffee was used as a control because it doesn't contain ethylamine. The results indicated the blood of the tea-drinking volunteers had a fivefold increase in interferon over the course of the study; drinking coffee had no such effect.

So, it appears drinking tea can bring gamma delta T cells to a state of readiness. Should infection then occur, these T cells are available and waiting to defend the body.

Whether such stimulation would actually be useful in fighting a real infection remains to be discovered. But these early findings do present intriguing evidence that tea can affect the immune system.

The immune system of each individual contains a tremendously large array of uncommitted (naive) B and T cells capable of recognizing diverse antigens and epitopes. How do only the appropriate B and T cells become committed?

Acquired immunity is called a "specific response" because only the specific B and T cells having receptors that recognize epitopes or antigenic fragments are committed. The vast majority of the B and T cells remain unreactive. The explanation for commitment was first proposed in the 1950s by Nobel laureates Frank MacFarlane Burnet and Peter Medawar. The theory, called **clonal selection**, suggested that exposure to an antigen only activates those naive B and T cells with receptors recognizing specific epitopes on the antigen. A detailed discussion of the contemporary theory for B cells follows.

Activation of the appropriate B cells begins when antigens enter a lymphoid organ, bringing the antigenic determinants close to the appropriate B cells (**FIGURE 21.5**). Antigenic determinants must bind with B cell receptor proteins. This, along with chemical stimulation from T cells (see below), triggers B cells to divide and form a **clone**; that is, a population of genetically identical B cells. A similar set of events occurs to activate T cells, which will be described in Section 21.3.

Both B- and T-cell clones contain activated lymphocytes that will develop into effector cells and memory cells. **Effector cells** target the pathogen. For example, the B cell clones develop into **plasma cells**, which then synthesize and secrete antibodies against the invading pathogen.

The clonal selection process also is important because the events are critical to the memory

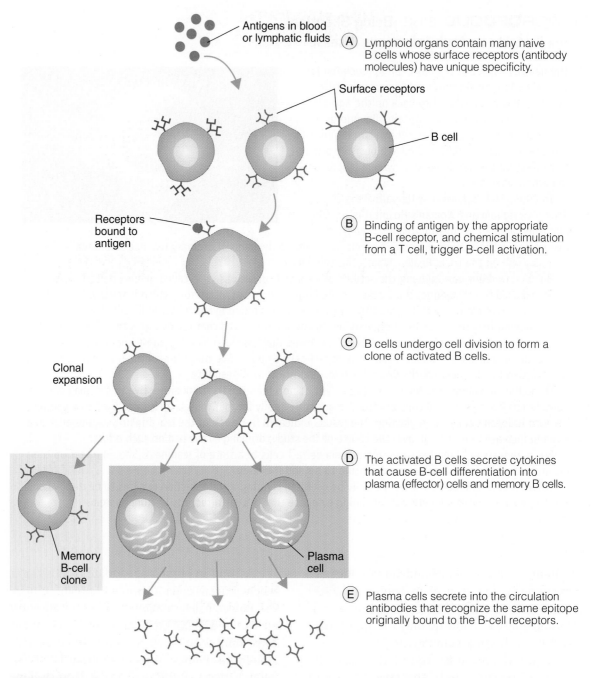

FIGURE 21.5 Clonal Selection of B Cells. In clonal selection, only those B cells that have a surface receptor that matches the shape of an antigenic determinant will be stimulated to divide, producing a clone of identical B cells. »» Why are the plasma cells called effector cells?

characteristic of immunity. In the immune response, **memory cells** are produced, which are long-lived B and T cells capable of dividing on short notice to produce more effector cells and additional memory cells. Memory cells can survive for decades in the body.

Should a second or ensuing contact be made with the same antigen, the acquired immune response will kick in much faster because of the presence of memory cells. This explains why most people are immune to many diseases once they have recovered from diseases like chickenpox or the measles. Yet having had measles does not prevent a child from contracting chickenpox. Likewise, immunization against a specific pathogen like the diphtheria bacterium or hepa-

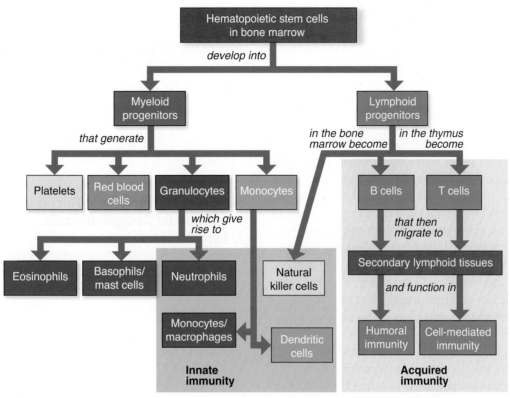

FIGURE 21.6 **The Origin of B and T Lymphocytes.** Lymophocytes arise in the bone marrow and mature in the bone marrow (B cells) or thymus (T cells). »» Why are two different progenitor lineages used to produce all blood cells?

titis B virus only stimulates an acquired immune response to the agent in the vaccine. Again, the protection acquired by experiencing one of these agents is specific for that agent alone.

CONCEPT AND REASONING CHECKS

21.3 Compare and contrast effector cells and memory cells in the clonal selection of B cells.

The Immune System Originates from Groups of Stem Cells

KEY CONCEPT

4. Hematopoietic stem cells give rise to all blood cells.

All immune cells, including the phagocytes, dendritic cells, and lymphocytes, arise in the fetus about two months after conception. At this time, lymphocytes originate from hematopoietic **stem cells** (*hema* = "blood"; *poiet* = "make") in the yolk sac but end up in the bone marrow (**FIGURE 21.6**). These undifferentiated cells develop into two types of cells: **myeloid progenitors**, which become red blood cells and most of the white blood cells (that is, neutrophils, basophils, mast cells, eosinophils, and

monocytes); and **lymphoid progenitors**, which become lymphocytes of the immune system.

The lymphoid progenitors take either of two courses. From the bone marrow, some of the cells proceed to the thymus. The thymus is large in size at birth and increases in size until the age of puberty, when it begins to shrink. Within the thymus, the progenitor cells mature over a two- or three-day period and are modified by the addition of surface receptor proteins. Those that survive the selection process emerge from the organ as naive T lymphocytes (T for thymus) that are now immunocompetent and ready to engage in acquired immunity. The T lymphocytes also colonize the lymph nodes, spleen, tonsils, and other secondary lymphoid tissues where potential interactions with pathogens and other immune cells may occur.

The B lymphocytes (B for bone) mature in the bone marrow. Like the T lymphocytes, B lymphocytes mature with surface receptor proteins on their membranes and become immunocompetent. They then move through the circulation to colonize organs of the lymphoid system, where they join the T cells.

Hematopoietic:
Referring to cellular components formed in the blood.

Undifferentiated:
Referring to cells that have yet to acquire specific characteristics and functions.

A large percentage of developing T and B cells, which normally would react with self antigens, are destroyed, respectively in the thymus and bone marrow, through apoptosis. Therefore, the T and B lymphocytes emerging are the cells capable of interacting only with nonself antigens.

CONCEPT AND REASONING CHECKS

21.4   Trace the origins of B and T lymphocytes.

21.2 Humoral Immunity

In the late 1800s, the mechanisms of specific resistance to infectious disease were largely obscure because no one was really sure how the body responded when infected. However, Paul Ehrlich (who received the 1908 Nobel Prize with Elie Metchnikoff) and others knew that certain proteins of the blood unite specifically with chemical compounds of microorganisms. By the 1950s, specific resistance to disease was virtually synonymous with immunity. Vaccines were available for numerous diseases, and immunologists saw themselves as specialists in disease prevention. In addition, the groundwork was laid for deciphering the nature and function of the antitoxins first used by von Behring to treat diphtheria (see Chapter 1). This work led to the elucidation of antibody structure in the 1960s, and the maturing of immunology to one of the key scientific disciplines of our modern times.

In this section, we study the immune system as it relates to humoral immunity and to the structure and function of antibodies. How the interaction of antibody and antigen leads to elimination of the antigen is also discussed.

Humoral Immunity Is a Response Mediated by Antigen-Specific B Lymphocytes

KEY CONCEPT

5. Antibodies are a group of protein molecules circulating in the body's fluids that recognize specific epitopes.

Antibodies are a class of proteins called **immunoglobulins**. They react with epitopes on toxin molecules in the bloodstream, on antigen surfaces or microbial structures (such as flagella, pili, capsules), and with epitopes on viruses in the extracellular fluid.

Antibodies recognize specific shapes associated with an antigen's epitopes (**FIGURE 21.7**). The basic antibody molecule consists of four polypeptide chains: two identical **heavy (H) chains** (each

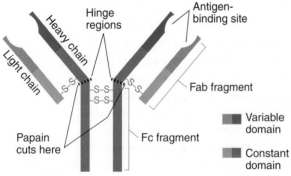

(A) The antibody molecule is a protein consisting of four polypeptide chains: two light chains and two heavy chains connected by disulfide (S-S) bridges. The heavy chains bend at a hinge point.

(C) On treatment with papain enzyme, cleavage occurs at the hinge point, and three fragments result: two identical Fab fragments and one Fc fragment.

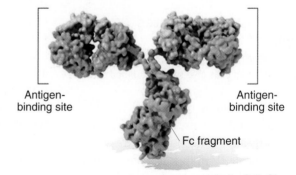

(B) The variable region is where the amino acid compositions of various antibodies differ. In the constant region, the amino acid compositions are similar in different antibodies.

FIGURE 21.7 Structure of an Antibody. (**A**) An immunoglobulin monomer consists of two light and two heavy polypeptide chains. The variable domains in a light and heavy chain form a pocket called the antigen-binding site. (**B**) Molecular model of the antibody immunoglobulin G (IgG). »» Using the illustration in (A), identify the light and heavy chains in (B).

heavy chain consists of about 400 amino acids) and two identical **light (L) chains** (each light chain has about 200 amino acids). These chains are joined together by disulfide bridges to form a Y-shaped structure, which represents a monomeric unit.

Each light and heavy chain has both a constant and variable domain. The **constant domain**, which contains virtually identical amino acids in both light and heavy chains, determines the destination (body location) and functional class to which the antibody belongs. However, the amino acids of the **variable domain** are different across the hundreds of thousands of antibodies produced in response to antigens (epitopes). The variable domains in a light and heavy chain form a highly specific, three-dimensional structure, called the **antigen-binding site**. It is to this region on the antibody that the antigen with its epitope binds.

The antigen-binding site is uniquely shaped to "fit" a specific epitope. Moreover, the two "arms" of the antibody are identical, so a single antibody molecule can combine with two identical epitopes on the same or separate pathogens. As we will see, these combinations may lead to a complex of antibody and antigen molecules, which become targets for elimination by phagocytes.

When an antibody molecule is treated with papain, an enzyme from the papaya fruit, the enzyme splits the antibody at the hinge joint, and two functionally different segments are isolated (Figure 21.7). One segment (actually two identical segments), called the **Fab fragment** (for *fragment-antigen-binding*) is the portion that will combine with the epitope. The second segment, called the **Fc fragment** (for *fragment* able to be *crystallized*) performs various functions, depending on the antibody class. It can combine with phagocytes, activate the complement system, or attach to certain cells in allergic reactions.

CONCEPT AND REASONING CHECKS

21.5   How is an antibody's antigen binding site similar to the active site on an enzyme?

There Are Five Immunoglobulin Classes

KEY CONCEPT

6.   Immunoglobulin classes differ in their heavy chains of the constant region.

Classes of immunoglobulins are based on whether the antibody will be secreted into the bloodstream, attached onto a cell as a receptor, or deposited in body secretions. Using the abbreviation Ig (immunoglobulin), the five classes are designated IgG, IgM, IgA, IgE, and IgD. The five classes are outlined in TABLE 21.1.

TABLE

21.1   The Five Immunoglobulin Classes and Functional Properties

Property	Immunoglobulin Class				
	IgG	IgM	IgA	IgE	IgD
Number of monomers	1	5	2	1	1
Antibody in serum (%)	80	5–8	10–15	<1	<1
Activates complement	Yes	Yes	No	No	No
Crosses placenta	Yes	No	No	No	No
Neutralizes bacterial toxins	Yes	Yes	Yes	No	No
Binds to phagocytes by Fc	Yes	No	No	No	No
Binds to basophils and mast cells	No	No	No	Yes	No
Additional properties	Principal antibody of secondary antibody response	First antibody formed in a primary antibody response	Monomer in serum; dimer secreted onto epithelial surfaces	Role in allergic reactions; effective against parasitic worms	Receptor on B cell surface

IgG. The antibody commonly referred to as gamma globulin represents the **IgG** class. This antibody is the major circulating antibody, comprising about 80% of the total antibody content in normal serum. IgG appears about 24 to 48 hours after antigenic stimulation and continues the antigen-antibody interaction begun by IgM. Booster injections of a vaccine raise the level of this antibody considerably in the serum. IgG also is the maternal antibody that crosses the placenta and confers additional immunity to the fetus.

IgM. The **IgM** (M stands for macroglobulin) class is the largest antibody molecule and, in serum, consists of a pentamer (five monomers) whose tail segments are connected by a glycoprotein. It is the first, albeit short-lived, immunoglobulin to appear in the circulation after antigenic stimulation. Thus, the presence of IgM in the serum of a patient indicates a very recent infection. Because of its size, most IgM remains in circulation.

IgA. About 10% of the total antibody in normal serum is **IgA**. However, it makes up more than 60% of the total immunoglobulin produced in a healthy person each day. This dimeric form of IgA is secreted through specialized epithelial cells into the external environment at mucosal surfaces such as the gastrointestinal tract. Thus, IgA is an important part of **mucosal immunity** where it provides a defense against potential pathogens (MICROINQUIRY 21). IgA also is located in tears and saliva, and in the colostrum, the first milk secreted by a nursing mother. When consumed by an infant, the antibodies provide added resistance to potential gastrointestinal pathogens.

IgE. The **IgE** class is another monomeric immunoglobulin. It plays a major role in allergic reactions by sensitizing cells to certain antigens. This process is discussed in Chapter 23.

IgD. The last class of immunoglobulins is the **IgD** antibody, which exists as a cell surface receptor on B cells. It is important to the activation of B cells.

CONCEPT AND REASONING CHECKS

21.6 Why are there five different classes of antibodies in the body?

Antibody Responses to Pathogens Are of Two Types

KEY CONCEPT

7. Memory cells produce a rapid antibody response.

The first time the body encounters a pathogen or antigen, a **primary antibody response** occurs (FIGURE 21.8). B cells are activated and effector cells, the plasma cells, start producing and secreting antibody. There is a lag of several days before

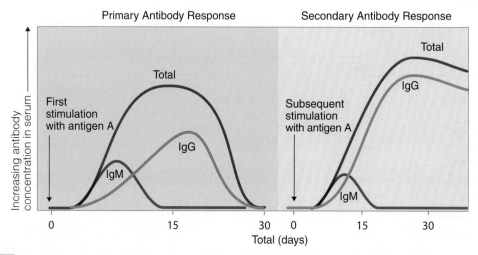

FIGURE 21.8 The Primary and Secondary Antibody Responses. After the initial antigenic stimulation, IgM is the first antibody to appear in the circulation. Later, IgM is supplemented by IgG. On a subsequent exposure to the same antigen (antigen A), the production of IgG is more rapid, and the concentration in the serum reaches a higher level than in the primary response. »» Why is the total antibody concentration greater than the sum of the IgM and IgG concentrations?

any measurable antibodies appear in the serum. IgM are the first on the scene, but they are soon replaced by a longer-lasting IgG response. During the primary antibody response, memory cells also are produced, which provide the immunological memory needed for subsequent encounters with the same antigen.

A second or subsequent infection by the same pathogen or antigen produces a more powerful and longer lasting **secondary antibody response**. Due to the presence of memory cells, a rapid response to antigen leads to the production of IgG, the principal antibody. The secondary antibody response occurs faster and is more vigorous than the primary response. It also provides a much longer period of resistance against disease.

CONCEPT AND REASONING CHECKS

21.7   Justify the need for the two antibody responses to pathogens.

Antibody Diversity Is a Result of Gene Rearrangements

KEY CONCEPT

8.   The mixing and matching of immunoglobulin gene segments produces unique antibodies.

For decades, immunologists were puzzled as to how an enormous variety of antigen-binding sites (capable of recognizing millions of different epitopes) could be encoded by the limited number of genes associated with the immune system—the human genome has only about 25,000 genes for all their functions (see Chapter 9).

The antibody diversity problem has been solved by showing that embryonic cells contain about 300 genetic segments, which are shuffled (like transposons) and combined in each B cell as it matures. This process, known as **somatic recombination**, is a random mixing and matching of gene segments to form unique antibody genes. Information encoded by these genes then is expressed in the surface receptor proteins of B cells, and in the antibodies later expressed by the stimulated clone of plasma cells.

The process of somatic recombination was first postulated in the late 1970s by Susumu Tonegawa, winner of the 1987 Nobel Prize in Physiology or Medicine. According to the process, the gene segments coding for the light and heavy chains of an antibody are located on different chromosomes. The light and heavy chains are synthesized separately, then joined to form the antibody. One of 8 constant gene, one of 4 joiner gene, one of 50 diversity gene, and one of up to 300 variable gene clusters can be used to form a heavy-chain gene. A 300 variable gene cluster is selected and combined with one of a 5 joiner gene cluster and a constant gene to form the active light-chain gene. After deletion of intervening sequences (introns), the new gene can function in protein synthesis, as shown in **FIGURE 21.9**.

Current evidence indicates more than 600 different antibody gene segments exist per cell. Additional diversity is generated through imprecise recombination and somatic mutation. Therefore, the total antibody diversity produced by the B cells is in the range of 10^{14} immunoglobulin possibilities for the variable region of an antibody.

CONCEPT AND REASONING CHECKS

21.8   Why don't the constant genes for an antibody show the same diversity as the variable genes?

Antibody Interactions Mediate the Disposal of Antigens (Pathogens)

KEY CONCEPT

9.   Antigen-antibody complexes are cleared by phagocytes.

In order for specific resistance to develop, antibodies must interact with antigens so that the antigen is changed in some way. Thus, the formation of **antigen-antibody complexes** may result in death to the microorganism possessing the antigen, inactivation of the antigen, or increased susceptibility of the antigen to other body defenses. **FIGURE 21.10** summarizes these mechanisms.

Certain antibodies can inhibit viral attachment to host cells through **viral inhibition**. By reacting with and covering capsid proteins or spikes, antibodies prevent viruses from attaching to and entering their host cells.

Neutralization represents a mechanism of defense against toxins and microbes. Since binding to a host cell surface is a necessary step toward infection, so-called neutralizing IgG and IgA antibodies can prevent microbes or

MICROINQUIRY 21

Mucosal Immunity

Most of this chapter discusses the systemic immune system; that is, the body's immune cells and chemical signals located in the blood and lymphatic systems. Until recently, however, not much was known about immunity at the body surfaces, where specialized cells and antibodies protect the body at its vulnerable portals of entry, such as the mucous membranes lining the respiratory, urinary, and gastrointestinal tracts. This defense is called **mucosal immunity** and is constructed to guard *against infection* rather than the systemic immune system's response to *resolve an infection* that is already occurring. Let's briefly look at the structure and function of mucosal immunity, using the gastrointestinal (GI) tract as our example.

The Mucosal Surface

Mucosal immunity represents an integrated network of tissues, cells, and effector molecules to protect the host from infection of the mucous membrane surfaces (see **Figure** and numbers as we proceed). If the epithelium is damaged by gastrointestinal pathogens such as *Vibrio cholerae*, enterotoxigenic *Escherichia coli* (ETEC), *Shigella* sp., or noroviruses (see Chapters 11 and 16), the pathogen may gain access to the mucosal tissues ①. Therefore, the mucosal surface must be protected against such possibilities.

In the GI tract, mucosal immunity involves the **mucosa-associated lymphoid tissue** (**MALT**). It forms the largest immune organ of the body and contains specialized regions. The intestinal epithelium contains specialized **M cells** that are important to uptake, transport, processing, and possibly presentation of microbial antigens within the intestinal lumen. The epithelium also contains **intraepithelial lymphocytes** (**IELs**). The lamina propria includes secondary lymphoid structures, such as the **Peyer's patches**, and a large concentration of macrophages, dendritic cells (DCs), plasma cells, and B and T lymphocytes. In a healthy human adult, MALT houses almost 80% of all immune cells.

Although not discussed here, the mucosa must exhibit "oral tolerance." In an environment where immune cells are continuously exposed to large numbers of foreign but nonpathogenic antigens from foods and endogenous microbiota, inflammatory responses would damage and destroy the intestinal lining. The phenomenon of oral tolerance is a unique feature of the mucosal immune system.

MALT Function: Epithelial Layer

One function of the mucosal immune system is to protect the mucous membranes and intestinal epithelium against invasion and colonization by potentially dangerous infectious agents that may be encountered in the intestinal lumen.

If a pathogen makes it through the acidic environment of the stomach, most often it will end up in the intestinal lumen or trapped in the mucus layer ②. The MALT, therefore, forms a defensive line against the pathogenic microbes, which tend to invade intestinal spaces devoid of microbiota, such as epithelial crypts and epithelial cells. The epithelial cells make up most of the cells of the intestinal lining, so they have the most contact with pathogens in the lumen. The epithelial cells can ingest and present antigen fragments to IELs scattered through the epithelium and secrete cytokines to stimulate development of the IELs. The IELs contribute to the removal of injured or infected epithelial cells.

The M cells are located over Peyer's patches and contain channels or passageways through which lymphocytes and DCs get closer to the luminal surface ③. Antigens taken up by M cells can be shuttled to, or directly captured by, DCs, B cells, and macrophages. These antigen fragments can be presented to typical helper T cells and cytotoxic T cells located nearby in the Peyer's patches.

MALT Function: Lamina Propria

The lamina contains large numbers of plasma cells, phagocytes, DCs, and B and T cells. The DCs have the ability to sample the contents of the intestinal lumen by extending finger-like projections between epithelial cells ④.

Peyer's patches are somewhat like lymph nodes in that they contain several follicles populated with B and T cells ⑤. The B cells are primarily committed to producing **secretory IgA** (**S-IgA**). The IgA-expressing plasma cells move to the crypts where they secrete the dimeric IgA that is then transported across the epithelial cell and deposited into the intestinal lumen ⑥. More S-IgA is produced at the mucosal surface than all other antibody classes (i.e., IgG, IgM, IgD, and IgE) in the body combined.

S-IgA protects the mucosa by binding to potential antigens. For example, S-IgA reduces influenza virus attachment and prevents virus penetration at epithelial surfaces. S-IgA also neutralizes bacterial toxins.

⑦ B cells, T cells, and DCs can leave the site of initial encounter with antigen, transit through the lymph, enter local lymph nodes, and then re-establish themselves in the mucosa of origin, or at other mucosal sites, where they differentiate into memory or effector cells. Thus, there appears to be a common mucosal immune system where immune cells activated at one site can disseminate that immunity to other mucosal tissues.

Discussion Point

Now that scientists are beginning to better understand mucosal immunity, they are considering ways to produce newer and better vaccines to boost mucosal immunity against enteric infections caused by bacteria and viruses. Traditional vaccines targeting the systemic immune system do not always confer strong mucosal immunity; that is, the IgA antibody response is not as strong as one would hope. Discuss what would be the best way to administer a mucosal vaccine and identify the "human physiology" hurdles in getting a vaccine to the mucosal surface.

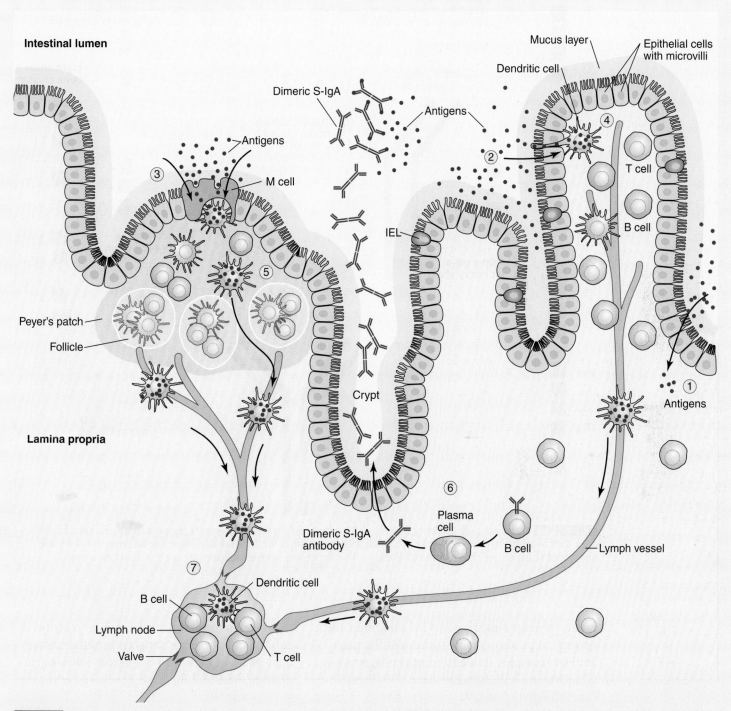

Intestinal lumen

Dimeric S-IgA

Antigens

Mucus layer

Dendritic cell

Epithelial cells with microvilli

Antigens

M cell

IEL

T cell

B cell

Peyer's patch

Follicle

Crypt

Lamina propria

Antigens

Dimeric S-IgA antibody

Plasma cell

B cell

Lymph vessel

B cell

Dendritic cell

Lymph node

Valve

T cell

FIGURE **The Mucosa of the Digestive Tract.** Mucosal immunity helps protect the body from colonization and infection by pathogenic and commensal microorganisms (Adapted from Wells, Jerry M. and Mercenier, Annick, *Nat Rev Microbiol* 6 (2008): 349–362.)

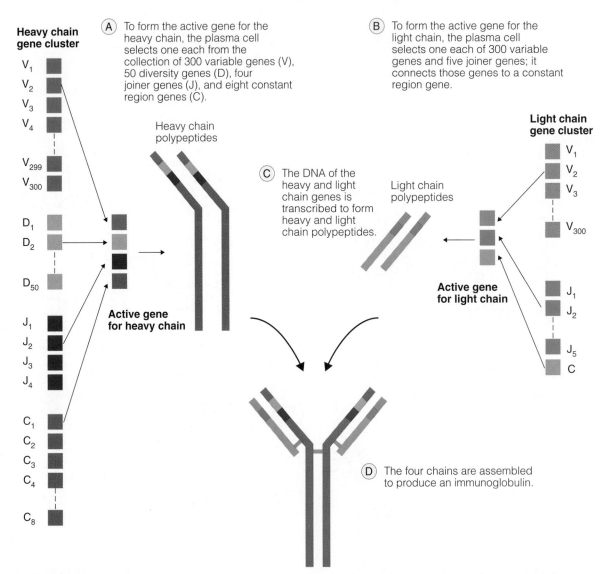

Heavy chain gene cluster

(A) To form the active gene for the heavy chain, the plasma cell selects one each from the collection of 300 variable genes (V), 50 diversity genes (D), four joiner genes (J), and eight constant region genes (C).

(B) To form the active gene for the light chain, the plasma cell selects one each of 300 variable genes and five joiner genes; it connects those genes to a constant region gene.

Heavy chain polypeptides

(C) The DNA of the heavy and light chain genes is transcribed to form heavy and light chain polypeptides.

Light chain polypeptides

Light chain gene cluster

Active gene for heavy chain

Active gene for light chain

(D) The four chains are assembled to produce an immunoglobulin.

FIGURE 21.9 **The Source of Antibody Diversity.** Almost any substance can stimulate the production of a specific antibody. Immunoglobulin gene rearrangements provide the mechanism for this diversity. Note: Not all potential heavy and light chain segments are shown in the diagram. »» How can gene arrangements generate such antibody diversity?

toxins from binding to cells. Viral inhibition and neutralization also increase the size of the antigen-antibody complex, thus encouraging phagocytosis while lessening the infective agent's ability to diffuse through the tissues.

As discussed earlier in the chapter, antibodies can coat bacterial cells (**opsonization**), preventing bacterial attachment to host cells. Because antibodies are bivalent, they can cross-link two separate antigens with the same epitope. This action causes clumping, or **agglutination**, of the bacterial cells. Movement of motile bacteria is inhibited if antibodies react with antigens on

the flagella of microorganisms. The reaction of antibodies with pilus antigens prohibits attachment of an organism to the tissues while agglutinating them and increasing their susceptibility to phagocytosis.

Antibodies also can react with dissolved antigens and convert them to solid precipitates (**precipitation**). In all these interactions, the antigen-antibody complexes formed enhance phagocytosis and thus the removal of the infecting pathogen or antigen from the body.

A final example of antigen-antibody interaction involves the three pathways of the comple-

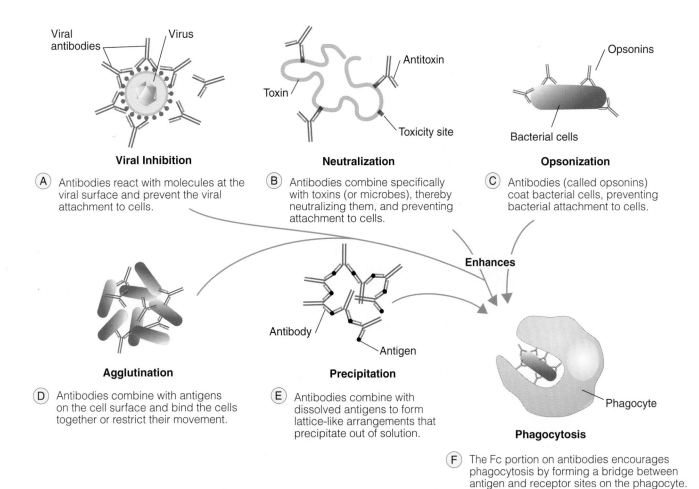

FIGURE 21.10 **Mechanisms of Antigen Clearance.** In all cases, the interaction facilitates phagocytosis by macrophages. »» Reorganize these mechanisms for antigen clearance into two basic strategies.

ment system, introduced in Chapter 20. The complement system is a group of nearly 30 proteins that functions in a cascading set of reactions. One set of reactions leads to lysis of bacterial cells through a complement cascade at the cell surface, resulting in the formation of a **membrane attack complex** (**MAC**) (**FIGURE 21.11**). This complex forms pores in the membrane, increasing cell membrane permeability and inducing the cell to undergo lysis through the unregulated flow of salts and water.

CONCEPT AND REASONING CHECKS

21.9   To this point, summarize the roles that phagocytes have had in the acquired immune response.

21.3 Cell-Mediated Immunity

The humoral immune response is one arm of acquired immunity. The production of antibodies and their association in antigen-antibody complexes can effectively clear an infection from the fluids of the body. However, many pathogens such as viruses and certain bacterial species (including *Mycobacterium tuberculosis*), have the ability to enter inside cells. Once in the host cell cytoplasm, they are hidden from the onslaught that would otherwise be leveled by antibodies, which are too large to enter cells.

The body's defense against microorganisms infecting cells and other "nonself" cells is centered in the cell-mediated immune response. In this final section, we describe the roles for the T cells in recognizing and eliminating virus-infected cells and "nonself" cells, as well as their part in "priming the pump" for the humoral immune response.

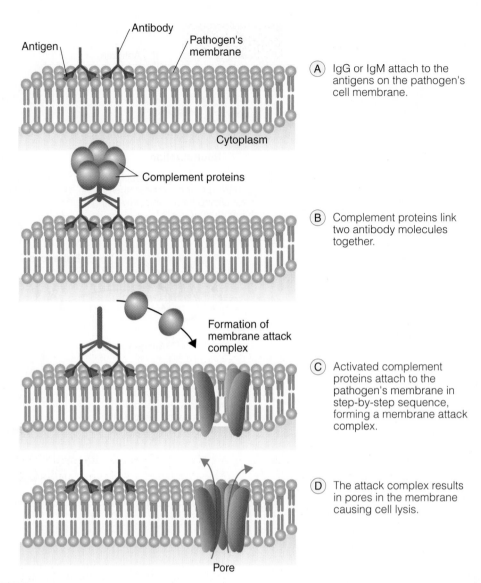

A IgG or IgM attach to the antigens on the pathogen's cell membrane.

B Complement proteins link two antibody molecules together.

C Activated complement proteins attach to the pathogen's membrane in step-by-step sequence, forming a membrane attack complex.

D The attack complex results in pores in the membrane causing cell lysis.

FIGURE 21.11 **The Formation of a Membrane Attack Complex (MAC).** Through the activation of a complement cascade, MACs are formed in the cell membrane of a pathogen. »» Which of the complement steps in Figure 20.8 must occur before complement proteins can be assembled into MACs?

Cellular Immunity Relies on T Lymphocyte Receptors and Recognition

KEY CONCEPT

10. Protein receptors found on different populations of T cells define their role in cellular immunity.

As we have seen, B-cell receptors directly recognize epitopes on an antigen. However, the interaction between antigenic determinants and T cells depends on sets of surface receptors capable of recognizing a multitude of antigenic (peptide) fragments. T-cell receptors are capable of such recognition because, like B cells and antibodies, they too are the result of somatic recombination.

There are two major subpopulations of effector T cells, and their receptors determine how the cells will function in cellular immunity. **Cytotoxic T cells** (**CTLs**) have T-cell receptors (TCRs) and co-receptor proteins, called **CD8**, on their cell surface (**FIGURE 21.12**). This combination allows the CTLs to recognize and eliminate non-self antigens, such as virus-infected cells and tumor cells.

The other group of T cells, the so-called **helper T cells**, has TCRs and co-receptors called **CD4**, which is the molecule to which HIV attaches to enable infection. These T helper (T_H) cells are divided into two effector classes, one that "helps" eliminate bacteria from infected antigen-presenting

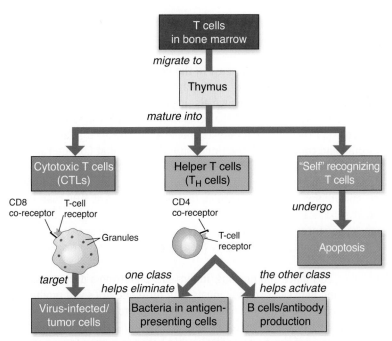

FIGURE 21.12 T-Cell Receptors. In the thymus, cytotoxic T cells gain T-cell receptors and CD8 co-receptors that recognize antigen peptides on virus-infected cells and tumor cells. Helper T cells gain T-cell receptors and CD4 co-receptors, one class helping eliminate pathogens from infected antigen-presenting cells and the other class helping activate B cells and antibody production. T cells that recognize "self" are destroyed through apoptosis. »» What is common to the receptors of all T cells?

cells, and the other that "helps" B cells produce antibody in the humoral acquired response.

The combination of TCRs and co-receptors allows T cells to recognize and bind to another set of glycoprotein molecules called the **major histocompatibility complex (MHC)**. MHC proteins are embedded in the membranes of all cells of the body. At least 20 different genes encode MHC proteins, and at least 50 different variants of those genes exist. Thus, the variety of MHC proteins existing in the human population is enormous (MHC is also called **human leukocyte antigen; HLA**), and the chance of two people having the same MHC proteins is incredibly small. (The notable exception is in identical twins.) The MHC proteins define the uniqueness of the individual and play a role in the immune response. Recent evidence also suggests MHC proteins may play more "amorous" roles (MICROFOCUS 21.3).

There are two important classes of MHC proteins, and it is to these that the CD8 and CD4 T cells bind. **MHC Class II (MHC-II)** proteins are found primarily on the surface of B cells, macrophages, and dendritic cells. These cells are called **antigen-presenting cells (APCs)** because the MHC-II molecules fold to form a pocket in which small antigen (peptide) fragments can bind. Thus, APCs "present" antigen fragments to CD4 T cells, which, as we will see, will bind to them with their TCR/CD4 receptor complex.

MHC Class I (MHC-I) molecules are found on all nucleated cells of the body. Again, the MHC-I molecules fold such that they can bind a small antigen fragment. If these proteins contain an antigen fragment from an infecting virus, CTLs will bind to it with their TCR/CD8 receptor complex.

Confused? The next section explains how these responses occur and what they accomplish. As a break, read MICROFOCUS 21.4, which talks about the possibility of "thinking healthy."

CONCEPT AND REASONING CHECKS

21.10 How can T cells be separated based on receptors and function?

Naive T Cells Mature into Effector T Cells

KEY CONCEPT

11. Naive T-cell activation stimulates memory cell and helper T-cell production.

The actual acquired immune response originates with the innate immune response and the entry

MICROFOCUS 21.3: Being Skeptical
Of Mice and Men—Smelling a Mate's MHC

Would you pick a mate by sight—or smell? At least in the nonhuman world, it might be by smell.

The major histocompatibility complex (MHC) is a group of immune system molecules essentially unique to each individual. Recent research suggests mammals, including humans, might consciously or unconsciously select a mate who has a different MHC than themselves. What's the evidence?

In mice, there is a region of the nose that detects chemicals in the air that are important to mouse reproduction. Frank Zufall and colleagues at the University of Maryland at College Park have discovered that mice also use the nasal region to recognize MHC molecules found in mouse urine. The researchers believe mice might use such "markers" to identify an individual through smell—and perhaps a suitable mate as well. In fact, mice prefer to smell other mice that have a different MHC than themselves and prefer mating with mice of another MHC.

Now to the human "experience." In a human study, people were given sweaty T-shirts to smell. In recorded responses, the participants indicated they preferred smells that the researchers then linked back to different MHCs.

Zufall's theory is that in both mice and humans, individuals preferring a different MHC are preventing inbreeding. Therefore, mating between two individuals with different MHCs will produce offspring having a unique set of MHCs that might make that individual more resistant to factors like infectious disease.

However, most humans don't pick their mates by smelling sweaty T-shirts. So, Craig Roberts at the University of Liverpool wondered if there was a relationship between human faces and MHC. After determining the MHC for a large group, he selected 92 women and showed them photographs of six men, three with somewhat similar MHCs and three with very different MHCs as compared to the women. The women were asked if they would prefer a long- or short-term relationship with each.

Roberts was surprised to find that the women preferred the faces of men who had similar MHCs for a long-term relationship.

The verdict? There certainly is much, much more research to be done. However, Roberts proposes two mechanisms may apply in the selection of a mate. First, select a mate facially with a similar MHC. Then, smell will make sure the mate is not too similar, so more "successful" offspring will be produced. Read MicroFocus 23.3 to learn how similar MHCs might work against successful pregnancies.

---

of antigens into the body. At their site of entry or within the lymphatic system, the antigens are phagocytized by the APCs (dendritic cells and macrophages). As the cells migrate to the lymph nodes or other secondary lymphoid tissues (see Chapter 20), the antigens are broken down and peptide fragments are displayed on the surface as **MHC-II/peptide complexes**.

When the APCs enter the lymphoid tissue, they mingle among the myriad groups of naive T cells, searching for the cluster having the surface receptors that recognize the MHC-II/peptide. This process requires considerable time and energy because only one cluster of T lymphocytes may have matching T cell/CD4 receptors (FIGURE 21.13A).

It is important to remember that recognition is between the MHC-II proteins on the APC surface and receptors on the T-cell surface. It is as if the T lymphocyte must first ensure it and the APC are from the same body (same MHC proteins) before it will respond. The CD4 coreceptor enhances binding of the naive T cell to the APC because the CD4 protein recognizes the peptide associated with the class II MHC protein.

Bound to the T cell, the APC secretes specific cytokines, such as **interleukin-1** (**IL-1**), which binds with the naive T cell and stimulates T-cell activation. Interleukin-1 causes the naive T cell to secrete other cytokines, including **interleukin-2** (**IL-2**), which stimulates cell division of that T cell and others activated by the APCs. The result is the production of a clone of antigen-specific activated T cells (FIGURE 21.13B). Some of these mature into **memory T cells** and await a future encounter with the same epitope (FIGURE 21.13C).

MICROFOCUS 21.4: Being Skeptical
Can Thinking "Well" Keep You Healthy?

The idea that mental states can influence the body's susceptibility to, and recovery from, disease has a long history. The Greek physician Galen thought cancer struck more frequently in melancholy women than in cheerful women. During the twentieth century, the concept of mental state and disease was researched more thoroughly, and a firm foundation was established linking the nervous system and the immune system. As a result of these studies, a new field called **psychoneuroimmunology** has emerged.

One such link exists between the hypothalamus and the T lymphocytes. The hypothalamus is a portion of the brain located beneath the cerebrum. It produces a chemical-releasing factor inducing the pituitary gland, positioned just below the hypothalamus, to secrete the hormone ACTH, which targets the adrenal glands. The adrenal glands, in turn, secrete steroid hormones (glucocorticoids) that influence the activity of T cells in the thymus gland.

Another link between the nervous and immune systems involves thymosins, a family of substances originating in the thymus gland. When experimentally injected into brain tissue, thymosins stimulate the pituitary gland, via the hypothalamus, to release hormones, including the one stimulating the adrenal gland. Although the precise functions of thymosins are yet to be determined, they may serve as specific molecular signals between the thymus and the pituitary gland. A circuit apparently stimulates the brain to adjust immune responses and the immune system to alter nerve cell activity.

The outcome of these discoveries is the emergence of a strong correlation between a patient's mental attitude and the progress of disease. Rigorously controlled studies conducted in recent years have suggested that the aggressive determination to conquer a disease can increase the lifespan of those afflicted. Therapies consist of relaxation techniques, as well as using mental imagery suggesting that disease organisms are being crushed by the body's immune defenses. Behavioral therapies of this nature can amplify the body's response to disease and accelerate the mobilization of its defenses.

Few reputable practitioners of behavioral therapies believe such therapies should replace drug therapy. However, the psychological devastation associated with many diseases, such as AIDS, cannot be denied, and it is this intense stress that the "thinking well" movement attempts to address. Very often, for instance, a person learning of a positive HIV test goes into severe depression, and because depression can adversely affect the immune system, a double dose of immune suppression ensues. Perhaps by relieving the psychological trauma, the remaining body defenses can adequately handle the virus.

As with any emerging treatment method, there are numerous opponents of behavioral therapies. Some opponents argue that naive patients might abandon conventional therapy; another argument suggests therapists might cause enormous guilt to develop in patients whose will to live cannot overcome failing health. Proponents counter with the growing body of evidence showing that patients with strong commitments and a willingness to face challenges—signs of psychological hardiness—have relatively greater numbers of T cells than passive, nonexpressive patients. To date, no study has proven that mood or personality has a life-prolonging effect on immunity. Still, doctors and patients are generally inspired by the possibility of using one's mind to help stave off the effects of infectious disease.

The majority of the activated T cells mature into one of two types of effector T cells, depending on the influence of specific cytokines. Some mature into **helper T2 (T_H2) cells** (FIGURE 21.13D). T_H2 cells "help" in the activation of humoral immunity, which we will discuss just ahead.

Other immature T cells mature into **helper T1 (T_H1) cells**. These cells recognize and bind to infected APCs displaying the appropriate MCH-I/ peptide. For example, in macrophages infected with *Mycobacterium tuberculosis* (see Chapter 10), the tubercle bacillus resides and survives in the phagosomes. T_H1 cells stimulate ("help") lysosome fusion in these cells, resulting in the destruction of the bacterial invaders (FIGURE 21.13E).

CONCEPT AND REASONING CHECKS

21.11 Justify the need for two populations of "helper T cells."

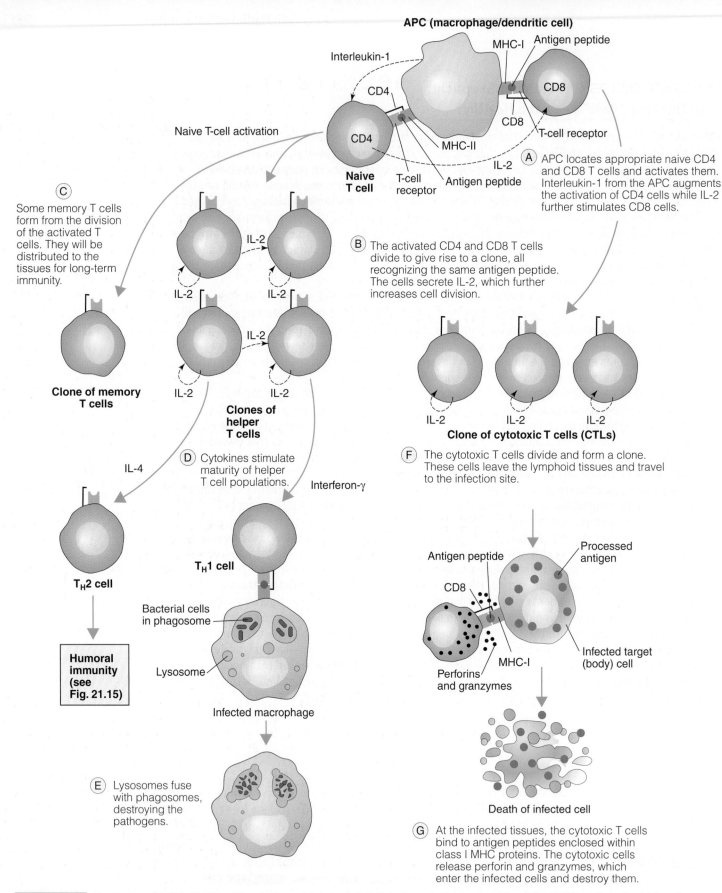

FIGURE 21.13 **The Process of Cell-Mediated Immunity and Cytotoxic T-Cell Action.** Antigen-presenting cells (APCs), such as dendritic cells or macrophages, activate naive T cells. The activated immature T cells stimulate humoral immunity and destruction of intracellular pathogens, and help activate the cytotoxic T cells. These cells secrete proteins (perforins and granzymes) to kill virus-infected cells or cancer cells through apoptosis.
»» What is the key recognition difference between helper T cells and cytotoxic T cells?

Cytotoxic T Cells Recognize MHC-I Peptide Complexes

KEY CONCEPT

12. Activated cytotoxic T cells seek out and kill virus-infected and abnormal cells.

Because body cells also can become infected with viruses, the immune system has evolved a way to attack virus-infected cells. During the infection, the host cells manage to degrade some of the viral antigens into small peptides. These are attached onto MHC-I proteins and transported to the cell surface, where they are displayed like "red flags" to denote an infected cell.

Activation of naive CD8 T cells occurs through interaction with dendritic cells or macrophages displaying a MHC-I/peptide (Figure 21.13A). Activation also requires IL-2 stimulation. Both events activate CD8 T cells to divide into mature cytotoxic T cells (CTLs) able to recognize the antigen being "red flagged" on infected body cells (**FIGURE 21.13F**).

The CTLs leave the lymphoid tissue and enter the lymph and blood vessels. They circulate until they come upon their target cells—the infected cells displaying the tell-tale MHC-I/peptide on their surface (**FIGURE 21.13G**). The CTLs bind to the MHC-I/ peptide on the virus-infected cell surface and release a number of active substances.

Toxic proteins, called **perforins**, insert into the membrane of the infected cell, forming cylindrical pores in the membrane. This "lethal hit" releases ions, fluids, and cell structures. In addition, the CTLs release **granzymes** that enter the target cell and trigger apoptosis. Cell death not only deprives the viral pathogen of a place to survive and replicate, but through cell lysis, also exposes the pathogen to antibodies in the extracellular fluid.

CTLs also are active against tumor cells because these cells often display distinctive molecules on their surfaces. The molecules are not present in other body cells, so they are viewed as foreign antigen peptides. Harbored within MHC-I proteins at the cell surface, the antigen peptides react with receptors on CTLs and the tumor cells are subsequently killed through apoptosis (**FIGURE 21.14**). However, some tumors reduce the level of MHC-I proteins at the cell surface, which impedes the ability of CTLs to "find" the abnormal

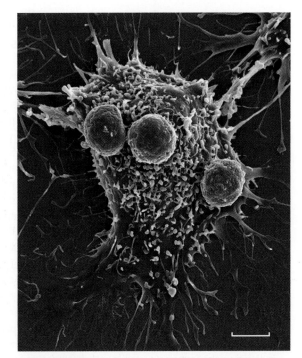

FIGURE 21.14 A "Lethal Hit." A false-color scanning electron micrograph of three cytotoxic T cells (pink) attacking a tumor cell. (Bar = 10 μm.) »» How do cytotoxic T cells "find" a tumor cell?

cells. Thus, tumor cells can escape immunologic surveillance and survive.

MicroFocus 21.5 describes how the "brakes" can be applied to the cell-mediated response.

CONCEPT AND REASONING CHECKS

21.12 Summarize the events leading to apoptosis of targeted cells.

T_H2 Cells Initiate the Cellular Response to Humoral Immunity

KEY CONCEPT

13. The active T$_H$2 cells co-stimulate B-cell activation.

As described earlier, the humoral immune response begins with an encounter between pathogen or toxin and B cells. Antigens stimulating the process are usually derived from the bloodstream, such as those associated with bacterial cells, viruses, and certain organic substances, as we have noted. The antigen epitopes interact with B cells (**FIGURE 21.15**).

Once the antigenic determinant has bound with its corresponding surface receptor protein, the combination of receptor protein and antigenic determinant is taken into the cytoplasm of the B cell. Then, the B cell displays the antigen peptide on its surface within a MHC-II molecule.

MICROFOCUS 21.5: Miscellaneous
Putting on the "Immune System" Brakes

Why does the misery of a common cold or the flu only last so long? The obvious answer would be that the body has eliminated the cold or flu virus. Yet the symptoms of a cold or flu are on the wane long before the virus has been eliminated. As mentioned in the beginning of this chapter, immune system responses must be strong enough to hopefully eliminate the pathogen yet not be so extreme as to harm the body (see MicroFocus 21.1). So there must be more to an immune response than simply an "on and off switch."

In 2007, scientists at Johns Hopkins University School of Medicine reported that they had identified a chemical necessary to control the immune machinery used to fight infections and foreign invaders. The researchers discovered a protein molecule called Carabin that is produced by T cells during a viral infection. When someone gets a cold or the flu, the virus infects the cells of the respiratory tract and uses the cells as viral factories to make more viruses. The immune system mounts a response to the infection by activating cytotoxic T cells that attack and destroy the infected cells. Viral infection also triggers Carabin synthesis in the same T cells to moderate the immune response to infection. In fact, in lab experiments, the researchers found that the more Carabin in a cell, the more severely was its action to "damp down" T-cell activity. So, Carabin acts like a "built-in timer" for the immune system. The immune response only lasts so long and then Carabin acts like an internal brake to slow down the speed and intensity of the immune response so that it does not go to an extreme or get out of control and attack healthy cells.

Carabin appears to act on the same protein receptor target as immune suppressing drugs used in transplant patients to block transplant rejection. If further research proves promising, Carabin may be useful in limiting the immune system's attack following transplant surgery. In addition, it could help control "autoimmune" disorders, such as multiple sclerosis and rheumatoid arthritis, by again dampening down the immune system's attack on self.

The activated T_H2 cells recognize the same MHC-II/peptide molecule complex on the B-cell surface and bind to the B cells. This immunologic cooperation between the B lymphocyte and the helper T lymphocyte continues the immune response. Interleukin-4 and other cytokines assist B-cell activation.

Antigens evoking this sort of response are called **T-dependent antigens** because they require the services of T_H2 cells. However, a few antigens are T-independent; these substances (such as in bacterial capsules and flagella) do not require the intervention of T_H2 cells. Instead, they bind directly with the receptor proteins on the B-lymphocyte surface and stimulate the cells. However, the immune response is generally weaker, primarily generates IgM antibodies, and no memory cells are produced.

Another group of antigens worth mentioning are the **superantigens**. "Regular" antigens must be broken down and processed to antigen peptides before they are presented on the APC's cell surface. Superantigens bind directly to the MHC proteins and to the T-cell receptor without any internal processing. Thus, massive numbers of activated T cells form, with an unusually high secretion of cytokines. The result is an extremely vigorous and excessive immune response called a **cytokine storm**, that can lead to shock and death (also see MicroFocus 21.1). Superantigens and massive cytokine release are associated with the staphylococcal toxins of toxic shock syndrome and scalded skin syndrome (see Chapter 13).

FIGURE 21.16 summarizes the humoral and cell-mediated immune responses of acquired immunity. Note that the naive/immature CD4 T cells are at the heart of almost all responses. Should these cells be infected and killed by HIV, humoral immunity is adversely affected, stimulation of bacterially infected macrophages does not occur, and the ability to fully activate cytotoxic T cells is limited. That is why the chapter opener referred to HIV as the perfectly designed virus to cripple the immune response. If you look forward at Figure 23.20 in Chapter 23, notice how the CD4 T-cell population declines with time in an HIV infection. As the T-cell population drops, cell-mediated immunity becomes less able to respond to pathogens. Eventually, there are so few T cells that the entire immune response collapses.

CONCEPT AND REASONING CHECKS

21.13 Describe the essential role played by T_H2 cells during acquired immunity.

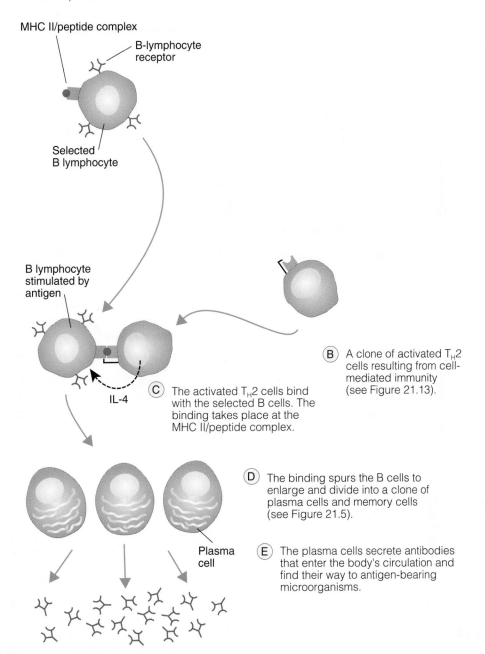

(A) The antigen-receptor complex has been taken into the selected B lymphocyte, and the antigen peptide is now displayed on the B cell's surface within the class II MHC protein.

MHC II/peptide complex

B-lymphocyte receptor

Selected B lymphocyte

B lymphocyte stimulated by antigen

IL-4

(B) A clone of activated T_H2 cells resulting from cell-mediated immunity (see Figure 21.13).

(C) The activated T_H2 cells bind with the selected B cells. The binding takes place at the MHC II/peptide complex.

(D) The binding spurs the B cells to enlarge and divide into a clone of plasma cells and memory cells (see Figure 21.5).

Plasma cell

(E) The plasma cells secrete antibodies that enter the body's circulation and find their way to antigen-bearing microorganisms.

FIGURE 21.15 **T Cell Activation of Humoral Immunity.** The secretion of specific antibodies from plasma cells depends on B-cell presentation of MHC-II/peptides and binding of T_H2 activated cells to the MHC-II/peptide complex. »» What is the similarity between the MHC-II/peptides complex on a B cell and the antigen presenting cell that activated the T_H2 cell?

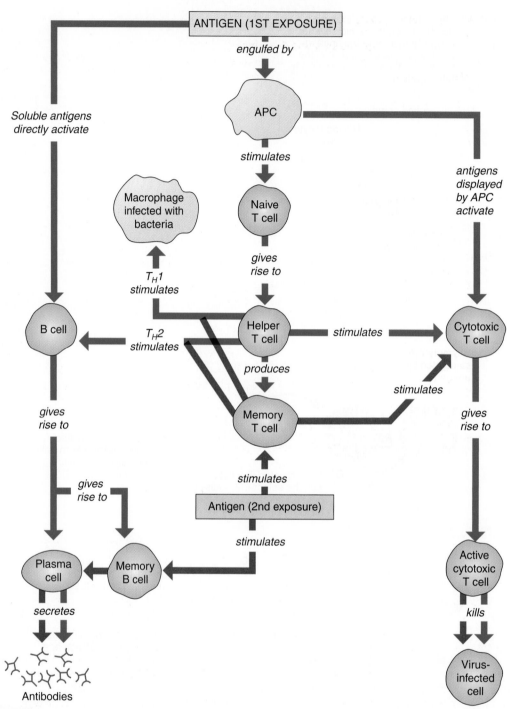

FIGURE 21.16 **A Concept Map Summarizing Acquired Immunity.** Humoral immunity produces antibodies that respond and bind to antigens. Cell-mediated immunity stimulates helper T cells that activate B cells and cytotoxic T cells (blue lines). Cytotoxic T cells bind to and kill infected cells or other abnormal cells. Memory B and T cells are important in a second or ensuing exposure to the same antigen (red lines). »» How would an HIV infection affect the immune response illustrated in this figure?

SUMMARY OF KEY CONCEPTS

21.1 An Overview of the Acquired Immune Response

1. **Acquired immunity** requires specificity for **epitopes**, tolerance of and minimal damage to "self," and memory of past infections.

2. Acquired immunity consists of **humoral immunity** maintained by **B cells** and **antibodies**, and **cell-mediated immunity** controlled by **T cells**.

3. Appropriate B and T cell populations are activated through **clonal selection** involving recognition of specific epitopes on antigens or antigen fragments.

4. Immune system cells arise from stem cells in the bone marrow. **Lymphoid progenitor cells** give rise to T cells that mature in the thymus and B cells that mature in the bone marrow. Immunocompetent B and T cells colonize the lymphoid organs.

21.2 Humoral Immunity

5. Monomeric antibodies consist of two identical **light chains** and two identical **heavy chains**. The **variable regions** of the light and heavy chains form two identical antigen-binding sites where the epitope (antigen) binds.

6. Of the five classes of antibodies, **IgM** and **IgG** are primary disease fighters, while secretory **IgA** is found on body (mucosal) surfaces.

7. A **primary antibody response** produces IgM and IgG antibodies and memory cells, the latter providing immunity in a **secondary antibody response**.

8. Antibodies can be produced that recognize almost any antigen. To do this, **somatic recombination** between a large number of gene clusters (variable, diversity, joiner, constant) for the heavy and light chains occurs.

9. The actual clearance of antigen is facilitated by antibody-antigen interactions (**inhibition, neutralization, opsonization, agglutination, precipitation**) that result in phagocytosis by phagocytes. In addition, the formation of **membrane attack complexes** by complement directly lyses and kills pathogens.

21.3 Cell-Mediated Immunity

10. There are two subpopulations of T cells that carry T-cell receptors. The **cytotoxic T cells** have a **CD8** co-receptor while the **helper T cells** have a **CD4** co-receptor. The receptors are responsible for recognition of antigen fragments presented in **major histocompatibility complex (MHC)** proteins.

11. Cellular reactions and cytokines of cell-mediated immunity drive the attack of virus-infected cells by activated cytotoxic T cells and bacterially-infected antigen-presenting cells by **helper T1 (T_H1) cells**. **Helper T2 (T_H2) cells** assist in the stimulation of B cells and antibodies.

12. **Cytotoxic T cells** are activated by APC/CD4 T cell combination. Active cells recognize abnormal cells (virus-infected cells or tumors) presented as MHC-I proteins with bound antigen peptide. Binding of cytotoxic T cells triggers the release of **perforins** and **granzymes**, which lyse and kill the abnormal cells.

13. T_H2 cells bind to MHC-II/peptides presented on the surface of B cells. Binding along with cytokines co-stimulate B-cell activation and the humoral response.

LEARNING OBJECTIVES

After understanding the textbook reading, you should be capable of writing a paragraph that includes the appropriate terms and pertinent information to answer the objective.

1. Identify and describe the four characteristics of an **acquired immune response**.

2. Distinguish between the cells responsible for a **humoral immune response** and a **cell-mediated immune response**.

3. List the steps in **clonal selection** of B and T cells.

4. Trace the origins of lymphocytes versus other leukocytes.

5. Draw the structure of a monomeric antibody and label the parts.

6. Summarize the characteristics for each of the five **immunoglobulin** classes.

7. Differentiate between a **primary** and **secondary antibody response** in terms of antibodies and responding cells.

8. Explain how antibody diversity is generated through gene arrangements.

9. Distinguish between the antibody mechanisms used to clear antigens (pathogens) from the body.

10. Compare receptors, receptor binding, and function for **cytotoxic T cells** and **helper T cells**.

11. Summarize how naive T cells are activated and identify how these effects contribute to cell-mediated immunity.

12. Discuss how cytotoxic T cells are activated and how they eliminate virus-infected cells.

13. Explain how T_H2 cells co-stimulate B-cell activation and the humoral response.

STEP A: SELF-TEST

Each of the following questions is designed to assess your ability to remember or recall factual or conceptual knowledge related to this chapter. Read each question carefully, then select the *one* answer that best fits the question or statement. Answers to even-numbered questions can be found in **Appendix C.**

1. All the following are immunogenic *except:*
   A. bacterial flagella.
   B. haptens.
   C. bacterial pili.
   D. viral spikes.

2. _____ cells are associated with _____ immunity while _____ cells are part of _____ immunity.
   A. B; cell-mediated; T; innate
   B. T; humoral; B; cellular
   C. T; cell-mediated; B; humoral
   D. T; humoral; B; nonspecific

3. Clonal selection includes
   A. antigen-receptor binding on B cells.
   B. antibody secretion recognizing same epitope as on B cell receptors.
   C. differentiation of B cells into plasma cells and memory cells.
   D. All the above (**A–C**) are correct.

4. Which one of the following cell types is NOT derived from myeloid progenitors?
   A. Lymphocytes
   B. Monocytes
   C. Eosinophils
   D. Neutrophils

5. An antigen binding site on the IgG antibody is a combination of
   A. one variable region from a light chain and one from a heavy chain.
   B. two variable regions from two light chains.
   C. two variable regions from two heavy chains.
   D. one variable region from a constant region and one from a variable region.

6. This dimeric antibody class often occurs in secretions of the respiratory and gastrointestinal tracts.
   A. IgE
   B. IgM
   C. IgA
   D. IgG

7. The presence of IgM antibodies in the blood indicates
   A. an early stage of an infection.
   B. a chronic infection.
   C. an allergic reaction is occurring.
   D. humoral immunity has yet to start.

8. Antibody diversity results from
   A. apoptosis.
   B. antigenic shift.
   C. somatic recombination.
   D. complement binding.

9. A/an _____ mechanism facilitates the clearance of toxins from the body.
   A. opsonization
   B. precipitation
   C. agglutination
   D. neutralization

10. MHC class I proteins would be found on _____ whereas MHC class II proteins would be found on _____.
    A. nucleated cells; plasma cells
    B. nucleated cells; macrophages
    C. dendritic cells; neutrophils
    D. only white blood cells; red blood cells

11. T_H1 cells activate
    A. B cells.
    B. killing of pathogens in macrophages.
    C. cytotoxic T cells.
    D. humoral immunity.

12. Perforins and granzymes are found in
    A. helper T cells.
    B. antigen-presenting cells.
    C. cytotoxic T cells.
    D. B cells.

13. T_H2 cells bind to
    A. MHC-II/peptide complex on APC cells.
    B. B cell receptors.
    C. MHC-I/peptide complex on infected cells.
    D. MHC-II/peptide complex on B cells.

STEP B: REVIEW

Test your knowledge of this chapter's contents by determining whether the following statements are true or false. If the statement is true, write "True" in the space. If false, substitute a word for the underlined word to make the statement true. The answers to even-numbered statements are listed in **Appendix C.**

_____ 14. <u>Cell-mediated</u> immunity involves T lymphocyte activity.

_____ 15. Small molecules called <u>antigens</u> are not immunogenic.

_____ 16. <u>Cytokine</u> is an alternate name for an antibody.

_____ 17. The end of an antibody molecule where an antigen binds is called the <u>Fc</u> fragment.

_____ 18. Antigenic materials are classified as "<u>nonself.</u>"

_____ 19. Cells that secrete antibody molecules are <u>B cells.</u>

_____ 20. <u>IgM</u> consists of five monomers.

_____ 21. A secondary antibody response primarily involves <u>IgG.</u>

_____ 22. IgD has <u>four</u> heavy chains in the antibody molecule.

_____ 23. <u>Epitope</u> is an alternate name for an antigenic determinant.

_____ 24. <u>Basophils</u> phagocytize microorganisms and begin an immune response.

_____ 25. <u>Dendritic cells</u> are involved in an immune response.

_____ 26. Antibodies are transported in the <u>blood.</u>

_____ 27. There are <u>four</u> polypeptide chains in a monomeric antibody molecule.

_____ 28. Secretory IgA has <u>two</u> antigen binding sites.

_____ **29.** The <u>humoral</u> immune response depends on antibody activity.

_____ **30.** The Fc region of an antibody consists of <u>light</u> chains.

_____ **31.** <u>Lysozyme</u> is secreted by cytotoxic T cells.

_____ **32.** <u>IgA</u> crosses the placenta.

_____ **33.** <u>Monocytes</u> secrete antibody.

_____ **34.** The <u>IgE</u> antibody is found on the surface of B lymphocytes.

_____ **35.** T_H2 cells activate lysosome killing of intracellular bacterial cells.

STEP C: APPLICATIONS

Answers to even-numbered questions can be found in **Appendix C**.

36. A microbiology professor suggests that an antigenic determinant arriving in the lymphoid tissue is like a parent searching for the face of a lost child in a crowd of a million children. Do you agree with this analogy? Why or why not?

37. In the book and classic movie, *Fantastic Voyage,* a group of scientists is miniaturized in a submarine (the Proteus) and sent into the human body to dissolve a blood clot. The odyssey begins when the miniature submarine carrying the scientists is injected into the bloodstream. Today, microscopic robots called nanorobots are being designed that would be injected into the body to fight diseases, including infectious ones. What do you think about this future microscopic robot technology?

38. As a consultant for the company Acme Nanobots, what immunological hurdles need to be considered before such microscopic robots could be fully developed?

STEP D: QUESTIONS FOR THOUGHT AND DISCUSSION

Answers to even-numbered questions can be found in **Appendix C**.

39. The ancestors of modern humans lived in a sparsely settled world where communicable diseases were probably very rare. Suppose that by using some magical scientific invention, one of those individuals was thrust into the contemporary world. How do you suppose he or she would fare in relation to infectious disease? What is the immunological basis for your answer?

40. Your brother's high school biology text contains the following statement: "T cells do not produce circulating antibodies. Rather, they carry cellular antibodies on their surface." What is fundamentally incorrect about this statement?

41. Some time ago, an immunologist reported that cockroaches injected with small doses of honeybee venom develop resistance to future injections of venom that would ordinarily be lethal. Does this finding imply that cockroaches have an immune system? What might be the next steps for the research to take? What does this research tell you about the cockroach's ability to survive for three or four years, far longer than most other insects?

🌐 HTTP://MICROBIOLOGY.JBPUB.COM/9E

The site features eLearning, an online review area that provides quizzes and other tools to help you study for your class. You can also follow useful links for in-depth information, read more MicroFocus stories, or just find out the latest microbiology news.

22

Immunity and Serology

"Vaccination is one of the most powerful means of protecting the public health."
—Marta A. Balinska, Institut National de Prévention et d'Education pour de Santé

Prior to the age of modern vaccines, the only way one could become immune to a disease was to contract the disease and hope for recovery. Unfortunately, the symptoms often were severe and perhaps disfiguring. There also was the risk of complications, where if the disease did not kill, a complication might. In addition, ill individuals could be contagious and spread the disease to other unsuspecting and susceptible individuals. Epidemics, and even pandemics, could result.

As early as the eleventh century, Chinese doctors ground up smallpox scabs and blew the powder into the noses of healthy people to protect them against the ravages of smallpox (see Chapter 1). Jenner improved on this technique by using a preparation from cowpox. Other diseases of past or recent history, including diphtheria and polio, were equally deadly to large numbers of people.

However, with the development of modern vaccines, these and many other infectious diseases have been either eradicated (smallpox), almost eradicated (polio), or brought under control (diphtheria) in much of the world. Equally important, vaccination is a technique to prevent disease from occurring—and preventing a disease from occurring is much safer and cheaper than trying to cure a disease after it has already struck.

Vaccination is the most cost-effective medical intervention. Cases of diseases like diphtheria and rubella have been controlled or nearly eliminated globally by vaccinating infants and children and maintaining high vaccine coverage from infancy through childhood (**FIGURE 22.1A, B, C**).

Unfortunately, many people have not been vaccinated against vaccine-preventable diseases. Many live in regions of the world where vaccines are not available, while others have elected not to use the opportunity of vaccination to protect themselves and their family. Yet vaccines are a key to public health and to the goal of preventing infectious disease.

However, some vaccine-preventable diseases once under control in the United States are resurfacing. Thanks to vaccination against pertussis, cases in the United States reached an all-time low in the mid-1970s (**FIGURE 22.1D**). Then between 1981 and 2004, for a set of reasons (see Chapter 10), there was a slow but steady increase in the incidence of pertussis, which in 2006 was finally beginning once again to fall.

In this chapter, we will study the role of vaccines in the immune response and examine the four major mechanisms by which immunity comes about. We also will examine how antibodies may be detected in a patient by a variety of laboratory tests. These diagnostic procedures help the physician understand the disease and prescribe a course of treatment. Immune mechanisms generally protect against disease, but when they fail, these laboratory tests provide a clue about what is taking place in the patient.

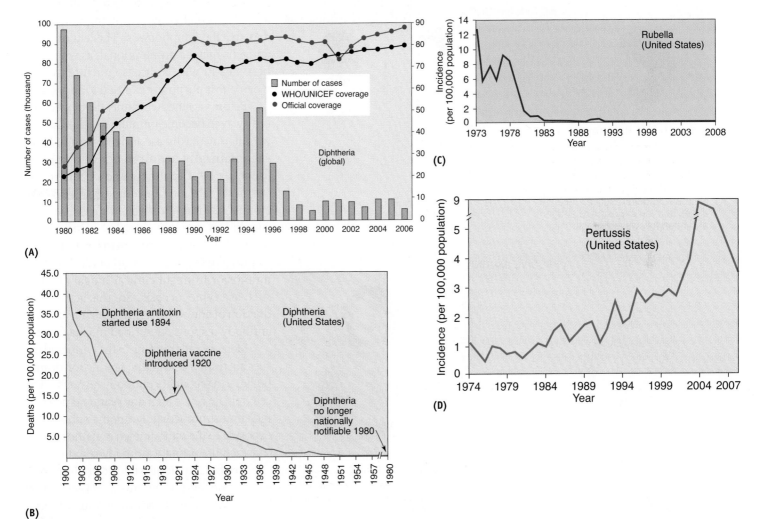

FIGURE 22.1 Vaccine Effects on Incidence and Mortality. (**A**) Global cases of diphtheria (bars) and vaccine coverage. With the establishment of vaccine coverage, the number of cases has dropped to all-time lows. *Source*: Data from the World Health Organization. Deaths from diphtheria (**B**) and rubella (**C**) in the United States. After use of the antitoxin vaccine and then the diphtheria and rubella vaccines, mortality rates per 100,000 population have dropped to zero. *B Source*: Data from www.alternativehealth.co.nz. *C Source*: Data from the CDC, *Summary of Notifiable Diseases*, US, 2008. (**D**) However, even with a pertussis vaccine, pertussis incidence until recently has been at its highest level since 1964. *Source*: Data from the CDC, *Summary of Notifiable Diseases*, US, 2007. »» From these graphs, how do vaccines and vaccinations contribute to community health in preventing infectious diseases?

22.1 Immunity to Disease

In biology, **immunity** (*immuno* = "safe") refers to a condition under which an individual is protected from disease. However, it does not mean one is immune to all diseases, but rather to a specific disease or perhaps a group of very similar diseases. It is the result of **acquired immunity** and depends on the presence of antibodies, T lymphocytes, and other factors originating in the immune system in response to a specific "nonself" antigen (see Chapter 21). Although the types of acquired immunity discussed here focus on humoral immunity, remember that cell-mediated immunity also is an important and essential arm of resistance to infectious disease. Four types of acquired immunity are recognized.

Acquired Immunity Can Result by Actively Producing Antibodies to an Antigen

KEY CONCEPT

1. The form of active immunity depends on whether the host experiences a naturally occurring or artificially exposed antigen.

Active immunity occurs when antigens enter the body and the individual's immune system actively responds by producing antibodies and specific lymphocytes. This exposure to antigens may be unintentional, as when one becomes ill with a disease, or intentional, when one is purposely exposed to an antigen.

Before vaccines were fully developed, the only way to become immune to a disease was by suffering the disease and recovering. Thus, such **naturally acquired** active immunity follows a bout of illness and occurs in the "natural" scheme of events (**FIGURE 22.2A**). However, this is not always the case, because subclinical diseases also may bring on the immunity. For example, many people have acquired immunity from subclinical cases of mumps or from subclinical fungal diseases such as cryptococcosis. So, this active production of antibodies represents the primary antibody response described in Chapter 21.

Memory cells residing in the lymphoid tissues are responsible for the production of antibodies in a secondary antibody response. The cells remain active for many years and produce IgG almost immediately upon a subsequent exposure to the same antigen or pathogen that triggered the primary antibody response.

Artificially acquired active immunity is less risky and represents an easier way to become immune to an infectious disease. This form of active immunity develops after the immune system produces antibodies following an intentional exposure to antigens;

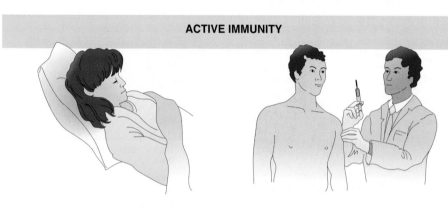

ACTIVE IMMUNITY

(A) Naturally acquired active immunity arises from an exposure to antigens and often follows a disease.

(B) Artificially acquired active immunity results from a vaccination.

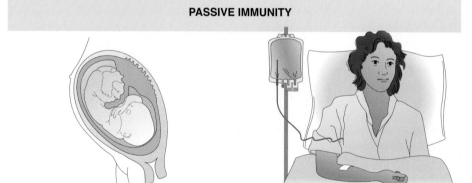

PASSIVE IMMUNITY

(C) Naturally acquired passive immunity stems from the passage of IgG across the placenta from the maternal to the fetal circulation.

(D) Artificially acquired passive immunity is induced by an injection of antibodies (antitoxins) taken from the circulation of an animal or another person.

FIGURE 22.2 The Four Types of Acquired Immunity. Immunity can be generated by natural or artificial means and acquired in an active or passive form. »» Why do these forms of immunity represent an acquired response rather than an innate response to an antigen?

that is, through **vaccination** (FIGURE 22.2B). Because the antigens usually are contained in an immunizing agent, such as an inactivated or toxoid vaccine, the exposure is called "artificial."

Vaccines are composed of treated microorganisms or viruses, chemically altered toxins, or chemical parts of microorganisms. Such vaccines work by mimicking a "natural" infection. By exploiting the immune system's ability to recognize antigens and respond with antibodies and lymphocytes, a vaccine triggers a primary antibody response. However, because the vaccine has been altered in some way (see below), the pathogen or toxin usually does not trigger the disease and the person vaccinated does not become ill. Importantly, memory cells are formed, which now establish active immunity. If the vaccinated person is exposed to the same antigen at some later time, the immune system acts swiftly and produces a secondary antibody response, stopping the infection before it can make the individual sick. Therefore, the person stays healthy.

Vaccines may be administered by injection, oral consumption, or, as is used for some influenza vaccines, nasal spray.

Let's now examine the different types of vaccines and find out how they work.

CONCEPT AND REASONING CHECKS

22.1 How do the naturally acquired and artificially acquired forms of active immunity differ?

There Are Several "Generations" of Vaccines

KEY CONCEPT

2. All vaccines are designed to ultimately generate memory cells.

The viral and bacterial vaccines currently in use in the United States are summarized in TABLE 22.1. The first generation of vaccines were developed by using the whole bacterium, virus, or toxin as the antigen.

Attenuated Vaccines. Some microbes can be weakened in the lab such that they should not cause disease. They are still able to grow or replicate, but **attenuated** means they will multiply only at low rates in the body and fail to cause symptoms of disease. Because the attenuated microbes mul-

tiply or replicate for a period of time within the body, they increase the dose of antigen to which the immune system will respond. Such vaccines are the closest to the natural pathogens and, therefore, they generate the strongest immune response. Often, the person vaccinated will have lifelong immunity. Also, attenuated organisms can be spread to other people and reimmunize them, or immunize them for the first time.

The downside of attenuated vaccines results from their continued multiplication. Because the vaccines contain dividing or replicating bacteria or viruses, there is a remote chance one of them could mutate (see Chapter 8) and revert back to a virulent form capable of causing disease. Usually a healthy person with a fully functioning immune system (**immunocompetent**) clears the infection without serious consequence. However, individuals with a compromised immune system, such as patients with AIDS, should not be given attenuated vaccines, if possible.

Today, there are many viral vaccines that consist of attenuated viruses. The Sabin oral polio vaccine, as well as the measles, mumps, and chickenpox vaccines, contain attenuated viruses. To avoid multiple injections of immunizing agents, it is sometimes advantageous to combine vaccines into a **single-dose vaccine**. The measles-mumps-rubella (MMR) vaccine is one example. In 2005, the U.S. Food and Drug Administration (FDA) approved a combination vaccine (Proquad) for children 12 months to 12 years old. This single-dose vaccine protects against chickenpox, measles, mumps, and rubella.

Making a vaccine with attenuated bacterial cells is more difficult. In fact, there are no such vaccines routinely used in the United States. The BCG tuberculosis vaccine (see Chapter 10), which is used in some countries with a high burden of tuberculosis, is composed of attenuated *Mycobacterium bovis* bacterial cells.

On a global scale, attenuated vaccines may not be the vaccine strategy of choice. These vaccines require refrigeration to retain their effectiveness, which could present a problem in many developing nations lacking widespread refrigeration facilities.

Inactivated Vaccines. Another strategy for preparing vaccines is to "kill" the pathogen. These

TABLE

22.1   The Principal Bacterial and Viral Vaccines Currently in Use

Disease	Route of Administration	Recommended Usage/Comments
Contain Killed Whole Bacteria		
Cholera	Subcutaneous (SQ) injection	For travelers; short-term protection
Typhoid	SQ and intramuscular (IM)	For travelers only; variable protection
Plague	SQ	For exposed individuals and animal workers; variable protection
Contain Live, Attenuated Bacteria		
Tuberculosis (BCG)	Intradermal (ID) injection	For high-risk occupations only; protection variable
Subunit Bacterial Vaccines (Capsular Polysaccharides)		
Meningitis (meningococcal)	SQ	For protection in high-risk individuals, such as military recruits; short-term protection
Meningitis (*H. influenzae*)	IM	For infants and children; may be administered with DTaP
Pneumococcal pneumonia	IM or SQ	Important for people at high risk: the young, elderly, and immunocompromised; moderate protection
Pertussis	IM	For newborns and children
Subunit Bacterial Vaccine (Protective Antigen)		
Anthrax	SQ	For lab workers and military personnel
Toxoids (Formaldehyde-Inactivated Bacterial Exotoxins)		
Diphtheria	IM	A routine childhood vaccination; highly effective in systemic protection
Tetanus	IM	A routine childhood vaccination; highly effective
Botulism	IM	For high-risk individuals, such as laboratory workers
Contain Inactivated Whole Viruses		
Polio (Salk)	IM	Routine childhood vaccine; highly effective; safer than Sabin vaccine
Rabies	IM	For individuals sustaining animal bites or otherwise exposed; highly effective
Influenza	IM	Persons at high risk and those living with or caring for persons at high risk
Hepatitis A	IM	Protection for travelers and anyone at risk; effectiveness not established
Contain Attenuated Viruses		
Adenovirus infection	Oral	For immunizing military recruits
Measles (rubeola)	SQ	Routine childhood vaccine; highly effective
Mumps (parotitis)	SQ	Routine childhood vaccine; highly effective
Polio (Sabin)	Oral	Routine childhood vaccine; highly effective; possible vaccine-induced polio
Smallpox (vaccinia)	Pierce outer layers of skin	For lab workers, military personnel, health care workers
Rubella	SQ	Routine childhood vaccine; highly effective
Chickenpox (varicella)	SQ	Routine childhood vaccine; immunity can diminish over time
Shingles (zoster)	SQ	Prevention in individuals 60 and older
Yellow fever	SQ	For travelers, military personnel in endemic areas
Influenza	IM	Persons at high risk and those living with or caring for persons at high risk
Rotavirus infection	Oral	Childhood protection against gastroenteritis
Recombinant Viral Vaccine		
Hepatitis B	IM	Medical, dental, laboratory personnel; newborns, others at risk; highly effective
Genital warts/ cervical cancer	IM	Preventative vaccine against certain types of human papilloma viruses

MICROFOCUS 22.1: Public Health/Tools
Preparing for Battle

Each flu season approximately 10% to 20% of Americans get the flu. Of this number, more than 114,000 are hospitalized and some 36,000 die from the complications of flu. Many of these hospitalizations and deaths could be prevented with a yearly flu shot, especially for those people at increased risk (people over 50, immunocompromised individuals, and healthcare workers in close contact with flu patients). Because influenza viruses change often (see Chapter 15), the influenza vaccine is updated each year to make sure it is as effective as possible. How is each year's vaccine designed?

Each flu season, information on circulating influenza strains and epidemiological trends is gathered by the World Health Organization (WHO) Global Influenza Surveillance Network. The network consists of 128 national influenza centers in 99 countries and four WHO Collaborating Centres for Reference and Research on Influenza located in Atlanta, United

Chicken eggs being "inoculated" with a flu virus.

States; London, United Kingdom; Melbourne, Australia; and Tokyo, Japan. The national influenza centers sample patients with influenza-like illness and submit representative isolates to WHO Collaborating Centres for immediate strain identification.

Twice a year (February: northern hemisphere; September: southern hemisphere), WHO meets with the Directors of the Collaborating Centres and representatives of key national laboratories to review the results of their strain identifications and to recommend the composition of the influenza vaccine for the next flu season.

In the United States, the Food and Drug Administration (FDA) or the Centers for Disease Control and Prevention (CDC) provide the identified viral strains to vaccine manufacturers in February. Each virus strain is grown separately in chicken eggs (see figure). After the virus has replicated many times, the fluid containing the viruses is removed, and the viruses purified and attenuated or inactivated. Then, the appropriate strains are mixed together with a carrier fluid. Production usually is completed in August and ready for shipment in October.

---

vaccines are relatively easy to produce because the pathogen is killed by simply using certain chemicals, heat, or radiation. However, the inactivation process alters the antigen so it produces a weaker immune response.

The Salk polio vaccine and the hepatitis A vaccine typify such preparations of inactivated whole viruses. For protection from diseases like hepatitis A, **booster shots** are required to maintain immunity (memory cells) for long periods of time. In the case of influenza, the virus changes genetically from year to year (see Chapter 15), so a different vaccine must be provided annually. MicroFocus 22.1 describes how the components of the flu vaccine are decided each year.

Some whole organism (bacterial) vaccines are used for short-term protection. For instance, bubonic plague and cholera vaccines are available to limit an epidemic. In these cases, the immunity lasts only for several months because the material in the vaccine is weakly antigenic.

Compared to attenuated vaccines, inactivated vaccines are safer as they cannot mutate and therefore cannot cause the disease in a vaccinated individual. The vaccines can be stored in a freeze-dried form at room temperature, making them a vaccine of choice in developing nations. However, the need for booster shots can be a drawback when people do not keep up their booster schedule.

Toxoid Vaccines. For some bacterial diseases, such as diphtheria and tetanus, a bacterial toxin is the main cause of illness. So, a third immunization strategy is to inactivate these toxins and use them as a vaccine. Such toxins can be inactivated with formalin, and the resulting inactivated toxin is called a **toxoid**. Immunity induced by a toxoid vaccine allows the body to generate antibodies and memory cells to recognize the natural toxin, should the individual again come in contact with it. Because toxoid vaccines are inactivated products, booster shots are necessary.

Single-dose vaccines include diphtheria-pertussis-tetanus vaccine (DPT) and the newer diphtheria-tetanus-acellular pertussis (DTaP) vaccine. For other vaccines, however, a combination single-dose vaccine may not be useful because the antibody response may be lower for the combination than for each vaccine taken separately.

To minimize the risks of vaccination, second-generation vaccines have been developed that contain only a fragment of the bacterium or virus. These subunit and conjugate vaccines generate acquired immunity that lacks a cytotoxic T-cell response.

Subunit Vaccines. Unlike the whole agent attenuated or inactivated vaccines, the strategy for a subunit vaccine is to have the vaccine contain only those parts or subunits of the antigen that stimulate a strong immune response. These subunits may be epitopes (see Chapter 21). For example, the subunit vaccine for pneumococcal pneumonia contains 23 different polysaccharides from the capsules of 23 strains of *Streptococcus pneumoniae*.

One way of producing a subunit vaccine is to use recombinant DNA technology (see Chapter 9), where the resulting vaccine is called a **recombinant subunit vaccine**. The hepatitis B vaccine (Recombivax HB or Engerix-B) is an example. Several hepatitis B virus genes are isolated and inserted into yeast cells, which then synthesize the antigens. These antigens are collected and purified to make the vaccine.

Adverse reactions to such subunit vaccines are very rare because only the important subunits of the antigen are included in the vaccine. These subunits cannot produce disease in the person vaccinated. Also, the vaccine is not made from blood fragments (as was a previous hepatitis B vaccine), so it allays the fear of contracting human immunodeficiency virus (HIV) from contaminated blood.

Conjugate Vaccines. *Haemophilus influenzae* b (Hib), which is responsible for a form of childhood meningitis (see Chapter 10), produces an external glycocalyx coat called a **capsule** (see Chapter 4). Since the capsular polysaccharides by themselves are not strongly immunogenic, the strategy here is to conjugate (attach) capsular polysaccharides to tetanus or diphtheria toxoid, which will stimulate a strong immune response. The result is the Hib vaccine, which has been a critical factor in reducing the incidence of *Haemophilus* meningitis from 18,000 cases annu-

> **Formalin:**
> A solution of formaldehyde in water.

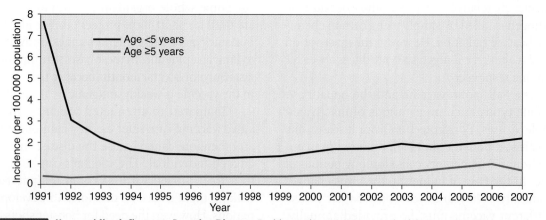

FIGURE 22.3 *Haemophilus influenzae*, Invasive Disease. Incidence, by age group—United States, 1991–2007.
»» Although the Hib vaccine was introduced in 1987, why didn't the incidence rate start to dramatically drop until 1992?
Source: Data from the CDC, *Summary of Notifiable Diseases—United States*, 2007.

ally in 1986 to 22 cases in 2007 (**FIGURE 22.3**). In 2005, a conjugate vaccine (MCV4) was licensed for a meningococcal vaccine against meningitis caused by *Neisseria meningitidis*.

By 6 years of age, children should be vaccinated against more than a dozen infectious diseases. **FIGURE 22.4** outlines the licensed vaccines recommended by the Centers for Disease Control and Prevention (CDC).

A third-generation of vaccine development is in the investigational stages.

DNA Vaccines. Part of the renaissance in vaccine development is a strategy to use DNA as a vaccine. It appears some cells in the body will take up injected foreign DNA, commence to make proteins encoded by the DNA, and display these antigens on its surface, much like an infected cell presents antigen fragments to T cells (see Chapter 21). Such a display should stimulate a strong antibody (humoral) and cell-mediated immune response.

These investigational **DNA vaccines** consist of plasmids engineered to contain one or more protein-encoding genes from a viral or bacterial pathogen. They are simply injected in a saline solution. Unlike replicating viruses or live bacteria, plasmids are not infectious or replicative, nor do they encode any proteins other than those specified by the plasmid genes, so they have an unparalleled measure of safety; someone vaccinated with a DNA vaccine could not contract the disease. They are relatively easy to construct and produce, and the vaccines are more stable than conventional vaccines at low and high temperatures, making shipping of these vaccines easier.

Thus far, few experimental trials have produced an immune response equivalent to first- or second-generation vaccines. However, in July 2005, a veterinary DNA vaccine to protect horses from West Nile virus became the world's first licensed DNA vaccine. In June 2006, scientists reported they had developed a DNA vaccine for bird flu and in August 2007, a preliminary study with a DNA vaccine against multiple sclerosis was reported as being effective.

Another version of the DNA vaccine strategy is to inject a vector (attenuated virus or bacterium) carrying the DNA segment into the person being vaccinated. Similar to virotherapy (see MicroFocus 14.5) an experimental recombinant vector vaccine uses a virus, such as an adenovirus, to ferry the DNA into the body cells whereas a bacterial vector in the body will incorporate and display the antigen on its surface. Both vectors will be seen as foreign and hopefully trigger a strong immune response.

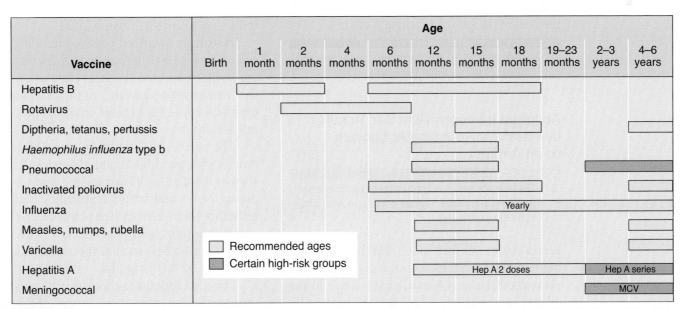

FIGURE 22.4   Childhood (0–6 years) Immunization Schedule—2010. Each year the Advisory Committee on Immunization Practices of the CDC reviews the childhood immunization schedule. This schedule indicates the recommended ages for routine administration of licensed vaccines. Additional recommendations concerning the schedule are available from the CDC at www.cdc.gov/vaccine/recs/schedules/downloads/child/2010/10_0_6yrs-schedule-pr.pdf.

TABLE 22.2 Vaccinations for Adults	
Bacterial (commercial name)	**Vaccination**
• Pneumococcal pneumonia (Pneumovax 23)	• Those over 65
• Tetanus and diphtheria booster (Decavac)	• Every 10 years for those over 65
• Tetanus, diphtheria, pertussis booster (Boostrix; Adacel)	• Every 10 years for those under 65
Viral (commercial name)	**Vaccination**
• Chickenpox (Varivax)	• Anyone who has never had chickenpox
• Shingles (Zostavax)	• Those over 60
• Hepatitis A (Havrix, Vaqta)	• Travelers to areas where hepatitis A is common
• Hepatitis B (Engerix-B; Recombivax HB) (Twinrix = combination Hep A and Hep B vaccine)	• Gay men, heterosexuals with multiple partners, health care workers, travelers to high-risk areas, drug users sharing needles
• Human papilloma virus (Gardasil)	• Women 19–26
• Influenza (Fluarix, Fluvirin, Fluzone, Flulaval, Flumist)	• Every fall for anyone over 50, health care workers, anyone who can benefit from the shot
• Rubella (Meruvax)	• Women of childbearing years

MicroFocus 22.2 discusses substances often added to vaccines to increase their effectiveness.

TABLE 22.2 lists the adult vaccines recommended by the CDC in 2008. A missing vaccine on this list is one for HIV disease/AIDS. It has been 25 years since the virus was discovered, yet there still is no vaccine. MicroFocus 22.3 outlines some reasons for the lack of a vaccine.

CONCEPT AND REASONING CHECKS

22.2   How do attenuated, inactivated, subunit, and conjugate vaccines differ from one another?

Acquired Immunity Also Can Result by Passively Receiving Antibodies to an Antigen

KEY CONCEPT

3. The form of passive immunity depends on the patient receiving antibodies from an outside source, either naturally or artificially.

Recall that in this section we have been discussing how one develops acquired immunity. The previous discussion looked at the active forms of acquired immunity. We finish this section by examining the ways one acquires antibodies in a passive and temporary manner.

Passive immunity develops when antibodies enter the body from an outside source (in contrast to active immunity, in which individuals synthesize their own antibodies). Again, the source of antibodies may be unintentional, such as a fetus receiving antibodies from the mother, or intentional, such as the transfer of antibodies from one individual to another (Figure 22.2).

Naturally acquired passive immunity, also called "congenital immunity," develops when antibodies pass into the fetal circulation from the mother's bloodstream via the placenta and umbilical cord. The process occurs in the "natural" scheme of events.

The maternal IgG antibodies (see Chapter 21) remain with the child for approximately three to six months after birth and play an important role during these months of life in providing additional resistance to diseases such as pertussis, staphylococcal infections, and viral respiratory diseases. Certain antibodies, such as measles antibodies, remain for 12 to 15 months.

Maternal antibodies also pass to the newborn through the first milk, or **colostrum**, of a nursing mother as well as during future breast-feedings. In this instance, IgA is the predominant antibody,

MICROFOCUS 22.2: Tools
Enhancing Vaccine Effectiveness

Often it may be necessary to add a substance to a vaccine to improve its ability to stimulate the immune system. One of those substances is an **adjuvant**, which helps stimulate the immune system and makes the vaccine more effective.

The adjuvants licensed for general use in humans in the United States are aluminum salts, specifically aluminum hydroxide or phosphate formulations simply called alum.

The particles of adjuvant linked to antigen are taken up by macrophages and presented to lymphocytes more efficiently than are dissolved antigens by themselves. Experiments also suggest that adjuvants may stimulate the macrophage to produce interleukin-1 (see Chapter 21), a lymphocyte-activating factor, and thereby reduce the necessity for helper T-cell activity. Adjuvants also provide slow release of the antigen from the site of entry and provoke a more sustained immune response than the antigen would by itself.

Besides adjuvants, some vaccines may contain antibiotics. These may be added during the manufacturing process to prevent bacterial contamination. The flu vaccine shortage of 2004, for example, was due to contamination by *Serratia marcescens* during the manufacturing process in Liverpool, England. About half of the American flu vaccine was to come from that source, so the supply had to be thrown out.

Sometimes stabilizers are added to vaccines to maintain their effectiveness in less-than-optimal conditions. One of these substances has been a point of concern. **Thimerosal**, a substance that had been put in some vaccines (not MMR) as a preservative contains high concentrations of ethylmercury, which is a neurotoxin. Although considered a low-level exposure, the American Academy of Pediatrics, the U.S. Public Health Service, and vaccine manufacturers stated in 1999 that, as a precautionary measure, thimerosal should be eliminated in all vaccines except some flu vaccines and the tetanus-diphtheria (Td) vaccine, which is given to children 7 years and older. Thimerosal is no longer found in any of the pediatric vaccines given to preschool children.

Although thimerosal was removed as a precautionary measure, it should be noted that almost all major studies on thimerosal, involving hundreds of thousands of children around the world, have not found any association between thimerosal exposure and harm to children. Only minor reactions, including redness and swelling at the injection site, typical of many vaccines, have been documented.

---

although IgG and IgM also have been found in the milk. The antibodies accumulate in the respiratory and gastrointestinal tracts of the child and lend increased disease resistance.

Artificially acquired passive immunity arises from the intentional injection of antibody-rich serum into the patient's circulation. The exposure to antibodies is thus "artificial." In the decades before the development of antibiotics, such an injection was an important therapeutic tool for the treatment of disease. The practice still is used for viral diseases such as Lassa fever and arthropodborne encephalitis, and for bacterial diseases in which a toxin is involved. For example, established cases of botulism, diphthe-

ria, and tetanus are treated with serum containing their respective antitoxins.

Various terms are used for the serum that renders artificially acquired passive immunity. **Antiserum** is one such term. Another is **hyperimmune serum**, which indicates the serum has a higher-than-normal level of a particular antibody. If the serum is used to protect against a disease such as hepatitis A, it is called prophylactic serum. When the serum is used in the therapy of an established disease, it is called **therapeutic serum**, and when taken from the blood of a convalescing patient, physicians refer to it as **convalescent serum**. Another common term, **gamma globulin**, takes its name from the fraction of blood serum in

■ **Prophylactic:**
Refers to a drug or agent preventing the development of a disease.

MICROFOCUS 22.3: Public Health

An AIDS Vaccine—Why Isn't There One after All These Years?

Producing an AIDS vaccine might appear rather straightforward: Cultivate a huge batch of human immunodeficiency virus (HIV), inactivate it with chemicals, purify it, and prepare it for marketing. In fact, this was the mentality and approach in the mid-1980s. Unfortunately, things are not quite so simple when HIV is involved. Here are the major reasons why a vaccine remains elusive after all these years.

The effects of a bad batch of weakened or inactivated vaccine would be catastrophic, and people generally are reluctant to be immunized with whole HIV particles, no matter how reassuring the scientists may be. In addition, how the body actually protects itself from pathogens such as HIV is not yet completely understood. Without that understanding, an effective vaccine cannot be developed.

Vaccine development also has been slow because of the high mutation rate of the virus. Some HIV strains around the world vary by as much as 35% in the capsid and envelope proteins they possess. Thus, HIV is more mutable than the influenza viruses, and no vaccine has been developed yet to make one completely immune to influenza. It is hard to make a vaccine targeted at one strain when the virus keeps changing its coat.

Another reason why a vaccine has not been successfully developed is because a short-lived vaccine would protect an individual only for a short time, necessitating an endless series of scheduled booster shots to which few would adhere. In addition, a vaccine also might act like dengue fever, where vaccination could actually make someone more at risk if they actually were to contract the disease. In 2002, an HIV patient who was holding the virus in check became infected with another strain of HIV through unprotected sex. This patient subsequently became "superinfected," meaning his immune system could keep the original strain in check, but was powerless to control the new strain. So, could a vaccine produce the same result if the individual was infected with another strain of HIV? Almost every AIDS vaccine researcher around the world is concerned about the unknown factors concerning an AIDS vaccine.

As of 2010, no vaccine trials have been shown to stimulate cell-mediated immunity to a level necessary to destroy HIV. The number of cytotoxic T cells and memory T cells produced is not up to the job of combating HIV. In March 2002, Anthony Fauci, Director of the National Institute for Allergy and Infectious Diseases, reported to the Presidential Advisory Council on HIV and AIDS that a "broadly effective AIDS vaccine could be a decade or more away." By 2010, minimal progress has been reported.

which most antibodies are found. Gamma globulin usually consists of a pool of sera from different human donors, and thus contains a mixture of antibodies (usually IgG), including those for the disease to be treated.

Passive immunity must be used with caution because in many individuals, the immune system recognizes foreign serum proteins as non-self antigens and synthesizes antibodies against them in an allergic reaction. When antibodies interact with the proteins, a series of chemical molecules called **immune complexes** may form and, with the activation of complement, the person develops a condition called **serum sickness** (Chapter 23). This often is characterized by a hive-like rash at the injection site, accompanied by labored breathing and swollen joints.

Although artificially acquired passive immunity provides substantial and immediate protection against disease, it is only a temporary measure. The immunity developing from antibody-rich serum usually wears off within weeks or months. For example, a person traveling to a country where hepatitis A is prevalent can obtain a serum preparation of hepatitis A antibodies several weeks prior to departing. The antiserum usually comes from outdated blood that is assumed to have high levels of antibodies.

CONCEPT AND REASONING CHECKS

22.3 How do the naturally acquired and artificially acquired forms of passive immunity differ?

Herd Immunity Results from Effective Vaccination Programs

KEY CONCEPT

4. When most of a population is vaccinated, disease spread is effectively stopped.

A population without a vaccination program is vulnerable to disease epidemics. Many people will suffer from the disease; some may die, while others could be left with a permanent disability. Even with a vaccination program, if insufficient numbers of citizens get the vaccination, the pathogen still can infect those who are not protected (FIGURE 22.5).

Vaccinations are never meant to reach 100% of the population. This shortfall is partly due to some people simply not being vaccinated or because some individuals simply respond poorly to a vaccine; all such individuals remain susceptible and unprotected. Still, the lack of 100% vaccination is not seen as a problem as long as most of the population is immune to an infectious disease. This makes it unlikely a susceptible person will come in contact with an infected individual. This phenomenon, called **herd (community) immunity**, implies that if enough people in a population are immunized against certain diseases, then it is very difficult for those diseases to spread.

Microbiologists and epidemiologists believe that when greater than 85% of the population is vaccinated (**herd immunity threshold**), the spread of the disease is effectively stopped. The rest of the "herd" or population remains susceptible, which is allowable because it then becomes very hard for pathogens "to find someone" who isn't vaccinated. Susceptible individuals are protected from catching the disease, and if one of these individuals should catch the disease, there are so many vaccinated people in the "herd" that it is unlikely the person could easily spread it.

Herd immunity can be affected by several factors. One is the environment. People living in crowded cities are more likely to catch a disease if they are not vaccinated than non-vaccinated people living in rural areas because of the constant close contact with other people in the city.

Another factor is the strength of an individual's immune system. People whose immune systems are compromised, either because they currently have a disease or because of medical treatment (anticancer or antirejection drugs), may not be able to be immunized. They are at a greater risk of catching the disease to which others are immunized.

MicroFocus 22.4 looks at the consequences of declining vaccinations and herd immunity.

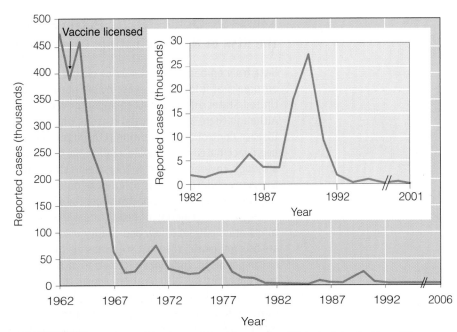

FIGURE 22.5 Measles Epidemics and Herd Immunity. With the introduction of the measles vaccine, vaccinations quickly built up a high herd immunity and few measles epidemics occurred. However, between 1989 and 1991 (inset), a measles epidemic occurred among college students and some groups of preschool children. »» Propose a hypothesis to explain the measles epidemic in 1989–1991.
Source: Data from the CDC, *Summary of Notifiable Diseases*, US, 2006.

CONCEPT AND REASONING CHECKS

22.4 What are the major factors required for a population to maintain high herd immunity?

Do Vaccines Have Dangerous Side Effects?

KEY CONCEPT

5. Some individuals' immune system can react differently to a vaccine.

If you look back at Figure 22.4, notice the current immunization schedule in the United States for young children consists of up to 12 shots of five vaccines by six months of age and an additional four shots of six more vaccines by 18 months. Records show that vaccinations have been very successful at eliminating or greatly reducing the incidence of childhood diseases in the United States. Still, many parents wonder if this number of vaccines can have adverse effects on a developing young child, and if so, should they avoid taking their child to the doctor for routine vaccinations?

MICROFOCUS 22.4: Public Health

The Risk of Not Vaccinating: Community and Health Care Settings

Data from epidemiological studies clearly show that community vaccinations essentially eliminate outbreaks of infectious disease by maintaining high herd immunity. Yet, about half the states in the United States allow parents to apply for vaccination waivers prior to enrolling their children in school. Although there are some valid reasons why a few children should not be vaccinated, the Centers for Disease Control and Prevention (CDC) has reported in 2010 that about 75% of American children have been vaccinated on schedule against the nine diseases mandated by federal law. The CDC said that although coverage varied widely among states and major cities, there were areas in the country where too many children are not up-to-date on their shots. In addition, only a small percentage of adults in the community receive booster shots for diseases like tetanus and diphtheria.

Likewise, vaccination of healthcare workers (HCWs) against diseases like influenza is just as important to maintain patient safety and quality-of-care. HCWs are frequently exposed to influenza and even if they are asymptomatic, they can still transmit the virus to patients, especially the vulnerable elderly and immunocompromised. Influenza vaccination of HCWs not only reduces the disease burden by maintaining high herd immunity but also has been shown to reduce the rate of influenza disease and overall mortality in the patients cared for by HCWs. Therefore, since 1981, the CDC has recommended priority vaccinations for all HCWs, unless they have an allergic reaction to eggs or have had a previous adverse reaction to the inactivated influenza vaccine.

Yet, despite local and national efforts to encourage influenza vaccination, the overall vaccination rate among HCWs in the United States remains unacceptably low at approximately 40%. Unvaccinated HCWs surveyed said they did not get the influenza vaccine because they feared the side effects and needles, they were skeptical of vaccine efficacy, they believed in their own innate ability to resist infection, or they reported an inability to access the vaccine. As with community vaccinations, unless there are extenuating circumstances, HCWs, as part of their "duty of care," should be vaccinated against influenza to prevent illness to themselves and transmission to others.

Intussusception: A rare, potentially life-threatening condition where bowel sliding creates swelling and intestinal obstruction.

The FDA requires vaccine manufacturers to follow extensive safety procedures when producing vaccines in the United States to ensure they are safe and effective. After lab and animal tests have been completed, a promising vaccine must undergo thorough human clinical trials before being licensed by the FDA for use. Still, a reaction may occur that can cause mild fever, soreness at the injection site, or malaise from a vaccination.

It is estimated that one individual in 100,000 vaccinated may develop a serious reaction to a vaccination. To "catch" these rare occurrences, the FDA and the CDC have established the **Vaccine Adverse Events Reporting System (VAERS)** to which anyone, including doctors, patients, and parents, can report adverse vaccine reactions. The FDA monitors the system weekly.

One vaccine that was permanently removed from the U.S. market was RotaShield®, which was administered to infants to provide immunity against rotavirus infections (see Chapter 16). In 1999, the Advisory Committee on Immunization Practices (ACIP) suggested that the vaccine licensed in the United States should no longer be recommended for infants based on a review of scientific studies indicating the virus might be associated with intussusception among some infants during the first one to two weeks following vaccination. Although these scientific studies may have been incorrect, a new and better vaccine, Rotateq, is now available and very effective against rotavirus disease.

People who are allergic to eggs should not be vaccinated against the flu virus because the virus is replicated in chicken eggs. In 2002–2003, there was much discussion weighing the risks of the smallpox vaccine's potential side effects (mostly mild, but potentially deadly for a few) versus the risk of a smallpox bioterrorist attack. The smallpox vaccine certainly is not recommended for young children.

Perhaps the vaccine of most concern regarding standard childhood vaccines is MMR. In 1998, a paper was published suggesting there might be a link between the MMR vaccine and autism in children. The idea was put forward that perhaps the sheer number of vaccinations could overwhelm a child's immune system and cause neurological damage. To make a long story short, as of 2008, numerous studies carried out by other scientists and researchers around the world have found no evidence to support the autism claim.

Through vaccination and booster shots, vaccines have been essential to maintaining public health and controlling infectious diseases. Although a few of the millions of people vaccinated each year suffer a serious consequence from vaccination, the risks of contracting a disease (especially in infants and children) from *not* being vaccinated are thousands of times greater than the risks associated with any vaccine. In addition, the licensed vaccines are always being re-examined for ways to improve their safety and effectiveness.

MicroFocus 22.5 looks at reasons for a lack of new vaccines being developed.

CONCEPT AND REASONING CHECKS

22.5 Assess the need for and effectiveness of maintaining vaccinations.

■ Autism:
A neurodevelopmental disturbance in which the use of communication, as well as normal behavior and social relationships, are not fully established and follow unusual patterns.

22.2 Serological Reactions

Antigen-antibody interactions studied under laboratory conditions are known as **serological reactions** because they commonly involve serum from a patient. In the late 1800s, serological reactions first were adapted to laboratory tests used in the diagnosis of disease. The principle was simple and straightforward: If the patient had an abnormal level of a specific antibody in the serum, a suspected disease agent probably was present. Today, **serology**, the study of blood serum and its constituents, and especially its role in protecting the human body against disease, have diagnostic significance as well as more broad-ranging applications. For example, they are used to confirm identifications made by other procedures and to detect organisms in body tissues. In addition, they help the physician to follow the course of disease and determine the immune status of the patient, and they aid in determining taxonomic groupings (serotypes) of microorganisms below the species level.

Serological Reactions Have Certain Characteristics

KEY CONCEPT

6. Serological reactions generally consist of an antigen and a serum sample.

Diagnostic microbiology makes great use of the serological reactions between antigen and antibody. Antisera can be used to determine the nature of the antigens, or the antigens can be used to detect antibodies present in a blood sample as evidence of a current or previous infection by that antigen.

In some cases, the unknown can be determined merely by placing the reactants on a slide and observing the presence or absence of a reaction. However, serological tests are not always quite so direct. For example, an antigen may have only one antigenic determinant, and the combination with an antibody molecule on a one-to-one basis may be invisible. To solve this dilemma, a second-stage reaction using an indicator system may be required, or a labeled molecule may be necessary. We shall see how this works to amplify detection in the tests that follow.

A successful serological reaction may require the antigen or antibody solution to be greatly diluted in order to reach a concentration at which a reaction will be most favorable. The process, called **titration**, may be used to the physician's advantage because the dilution series is a valuable way of determining the titer of antibodies. The **titer** is the most dilute concentration of serum antibody yielding a detectable reaction with its specific antigen (**FIGURE 22.6**). This number is expressed as a ratio of antibody to total fluid (for example, 1:50) and is used to indicate the amount of antibodies in a patient's serum. For instance, the titer of influenza antibodies may rise from 1:20 to 1:320 as an episode of influenza progresses, then continue upward and stabilize at 1:1,280 as the disease reaches its peak. A rise in the titer

MICROFOCUS 22.5: Public Health
Vaccine Development: An Upward Climb

Even though most agree that vaccination is the most cost-effective way to prevent infectious disease, there has recently been a lack of enthusiasm for vaccine development among the pharmaceutical companies. In fact, the number of companies in the United States supplying vaccines has dropped from 30 to 5 since 1980. Such a drop has been partially responsible in recent years for shortages of influenza, tetanus-diphtheria, MMR, and other vaccines.

Like any drug, vaccine development is a multimillion-dollar (often billion-dollar) investment. Therefore, a pharmaceutical company that invests its time and money in development expects a large return on its investment. Unfortunately, vaccines (and antibiotics too; Chapter 24) often do not provide that payback as compared to other phar-

A child receives vaccination against tuberculosis, Benin, Africa.

maceutical drugs. A vaccine may be administered only once or twice in a lifetime to an individual, whereas many pharmaceutical drugs, such as Lipitor (lowers "bad" cholesterol), Singulair (treats or helps prevent asthma), and Fosamax (prevents and treats osteoporosis), are long-term, often lifetime, medications for individuals.

In addition, an adverse reaction to a vaccination may bring legal action against the company. The potential exposure to legal liability is not an attractive business opportunity, so pharmaceutical companies have been reluctant to invest the time and money in vaccine development.

Many developing nations suffer from the burden of malaria, AIDS, cholera, meningitis, tuberculosis, and many other diseases less well known in the developed world (see figure). These developing nations cannot afford the costs of vaccines, and because the developed world already has vaccines available for most of their vaccine-preventable diseases, the incentive is not there for drug companies to spend the huge sums of money in vaccine development that are needed for treating and preventing infectious disease in the developing world.

Today, much of the vaccine research and development occurs through public and private partnerships and organizations. For example, the Bill and Melinda Gates Foundation supports the Medicines for Malaria Venture with its work to develop more affordable and effective malaria treatments and vaccines; the Global Alliance for Vaccines and Immunizations and their efforts to provide new vaccines to the millions of children who die every year from potentially vaccine-preventable diseases; the Aeras Global TB Vaccine Foundation to aid in the development of new vaccines to prevent tuberculosis; and the Advance Market Commitment (AMC) for Pneumococcal Vaccines, which is a new initiative launched by the Gates Foundation and five nations to promote the development of life-saving vaccines.

Recently, increased funding and higher profits are bringing the pharmaceutical companies back into the game. In 2006, three new vaccines were approved: for the human papilloma virus, rotavirus, and herpes zoster; so "big pharma" is a player once again. The elimination of disease can be accomplished with new vaccines, so getting vaccine design and development back on track at the top of the global public health agenda is key.

---

is evidence that an individual has a disease, an important factor in diagnosis.

Haptens may pose a problem in a serological reaction because of their small size (see Chapter 21). This problem has been solved by conjugating the haptens to carrier particles, such as polystyrene beads. When the hapten unites with an antibody, the entire bead is involved in the complex, and a visible reaction occurs.

Serology has become a highly sophisticated, and often automated, branch of immunology. As the following tests illustrate, the serological reac-

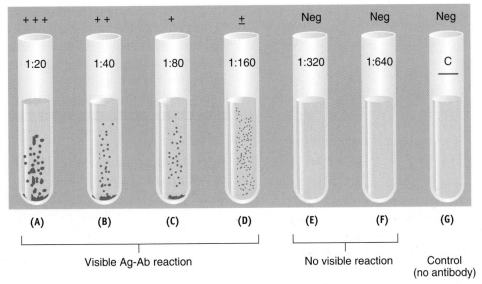

FIGURE 22.6 **The Determination of Titer.** A sample of antibody (Ab)-containing serum was diluted in saline solution to yield the dilutions shown. An equal amount of antigen (Ag) was then added to each tube, and the tubes were incubated. An antigen-antibody interaction may be seen in tubes (**A**) through (**D**), but not in tubes (**E**) or (**F**), or the control tube (**G**). The titer of antibody is the highest dilution of serum antibody in which a reaction is visible. »» What is the titer of antibody in this example?

tions have direct application to the clinical laboratory as well as other fields.

CONCEPT AND REASONING CHECKS

22.6 What does titration accomplish?

Neutralization Involves Antigen-Antibody Reactions

KEY CONCEPT

7. Neutralization is a serological reaction in which antigens and antibodies neutralize each other.

Neutralization is a serological reaction used to identify toxins and antitoxins, as well as viruses and viral antibodies (see Chapter 21). Normally, little or no visible evidence of a neutralization reaction is present, so the test mixture must therefore be injected into a laboratory animal to determine whether neutralization has taken place.

The detection of botulism toxin in food is an example of a neutralization test. Normally, the toxin is lethal to a laboratory animal, and if a sample of the food contains the toxin, the animal will succumb after an injection. However, if the food is first mixed with botulism antitoxins, the antitoxin neutralizes the toxin, and the mixture has no effect on the animal.

Conversely, if the toxin was produced by some other organism, no neutralization will occur, and the mixture will still be lethal to the animal. A similar test for diphtheria is the **Schick test**, in which a person's immunity to diphtheria can be determined by injecting diphtheria toxin intradermally. No skin reaction will occur if the person has neutralizing antibodies. A local edema occurs if no neutralizing antibodies are present, indicating the person is susceptible to diphtheria.

CONCEPT AND REASONING CHECKS

22.7 Explain why the Schick test is an example of neutralization.

Precipitation Requires the Formation of a Lattice between Soluble Antigen and Antibody

KEY CONCEPT

8. Precipitation is a serological reaction in which antigens and antibodies form a visible precipitate.

Precipitation reactions are serological reactions involving thousands of antigen and antibody molecules cross-linked at multiple determinant sites, forming a lattice. The lattices are so large that they precipitate in a form that can be observed visually.

Precipitation tests are performed in either fluid media or gels. In fluids, the antibody and antigen solutions are layered over each other in a

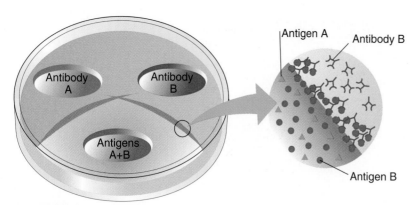

FIGURE 22.7 A Precipitation Test. Wells are cut into a plate of purified agar. Different known antibodies then are placed into the two upper wells, and a mixture of unknown antigens is placed into the lower well. During incubation, the reactants diffuse outward from the wells, and cloudy lines of precipitate form. The lines cross each other because each antigen has reacted only with its complementary antibody. »» Why does the precipitin line form where it does?

Agarose:
A polysaccharide derived from seaweed and part of the structure of agar.

thin tube. The molecules then diffuse through the fluid until they reach a **zone of equivalence**, the ideal concentration for precipitation to occur. A visible mass of particles now forms either at the interface or at the bottom of the tube. Fluid precipitation is used frequently in forensic medicine to learn the origin of albumin proteins in bloodstains.

In **immunodiffusion**, the diffusion of antigens and antibodies takes place through a semisolid gel, such as agar. As the molecules diffuse through the gel, they eventually reach the zone of equivalence, where they interact and form a visible precipitate. Variations of this technique are called the Ouchterlony plate technique, named for Orjan Ouchterlony, who devised it in 1953, and the **double diffusion assay**, so named because both reactants diffuse.

In immunodiffusion, antigen and antibody solutions are placed in wells cut into agar in Petri dishes. The plates are incubated and precipitation lines form at the zone of equivalence (**FIGURE 22.7**). The test has been used to detect fungal antigens of *Histoplasma, Blastomyces,* and *Coccidioides* (see Chapter 17).

In the antigen-detecting procedure known as **immunoelectrophoresis**, gel electrophoresis, and diffusion are combined. A mixture of antigens is placed in a reservoir on an agarose slide, and an electrical field is applied to the ends of the slide. The different antigens then move through the agarose at different rates of speed, depending on

their electrical charges (**FIGURE 22.8**). A trough is then cut into the agarose along the same axis, and a known antibody solution is added. During incubation, antigens and antibodies diffuse toward each other and precipitation lines form, as in the immunodiffusion technique.

CONCEPT AND REASONING CHECKS

22.8   What do lines of precipitation indicate?

Agglutination Involves the Clumping of Antigens

KEY CONCEPT

9.   Agglutination is a serological reaction in which antibodies interact with antigens on the surface of particular objects.

The amount of antibody or antigen needed to form a visible reaction can be reduced if either is attached to the surface of an object. The result is a clumping together, or **agglutination**, of the linked product. Agglutination procedures are performed on slides or in tubes. For bacterial agglutination, emulsions of unknown bacterial species are added to drops of known antibodies on a slide, and the mixture is observed for clumping.

Passive agglutination is a modern approach to traditional agglutination methods. Most often, antigens are adsorbed onto the surface of latex spheres or polystyrene particles (**FIGURE 22.9A**). Serum antibodies can be detected rapidly by observing agglutination of the carrier particle. Bacterial infections, such as those caused by *Streptococcus,* can be detected rapidly by mixing the bacterial sample with latex spheres attached to streptococcus antibody (**FIGURE 22.9B, C**).

Hemagglutination refers to the agglutination of red blood cells. This process is particularly important in the determination of blood types prior to blood transfusion (**FIGURE 22.9D**). In addition, certain viruses, such as measles and mumps viruses, agglutinate red blood cells. Antibodies for these viruses may be detected by a procedure in which the serum is first combined with laboratory-cultivated viruses and then added to the red blood cells. If serum antibodies neutralize the viruses, agglutination fails to occur. This test is called the **hemagglutination inhibition (HAI) test** (MICROFOCUS 22.6). A hemagglutination test called the **Coombs test** is used to detect Rh antibodies involved in hemolytic disease of the newborn (Chapter 23). A **slide**

(A) Gel electrophoresis:
On a gel-coated slide, an antigen sample is placed in a central well. An electrical current is run through the gel to separate antigens by their electrical charge (electrophoresis). Unlike the depiction in the diagram, the separate antigens cannot be detected visually at this point.

(B) Addition of antibodies:
A trough is made on the slide and a known antibody solution is added.

(C) Diffusion of antigens and antibodies:
As antigens and antibodies diffuse toward one another through the gel, precipitin lines are seen where optimal concentrations of antigen and antibodies meet.

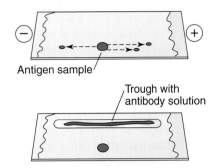

Antigen sample

Trough with antibody solution

Precipitin lines

FIGURE 22.8 **Immunoelectrophoresis.** By applying an electrical field, immunoelectrophoresis separates antigens by their electrical charge before antibodies are added. »» In (A), what is the electrical charge (+ or −) on the three antigens shown?

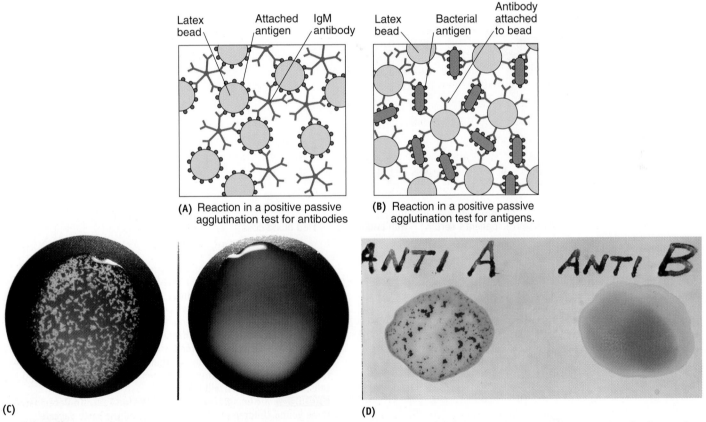

(A) Reaction in a positive passive agglutination test for antibodies

(B) Reaction in a positive passive agglutination test for antigens.

(C)

(D)

FIGURE 22.9 **Passive Agglutination Tests.** (A) When particles are bound with antigens, agglutination indicates the presence of antibodies, such as the IgM shown here. (B) When particles are bound with antibodies, agglutination indicates the presence of antigens. (C) Latex agglutination test for *Staphylococcus*. If group A antigen is present, it will bind to antibody-coated latex beads and agglutinate them for a positive test (left). A negative test is shown for comparison on the right. (D) Blood type can be determined using the agglutination test. Red blood cells are diluted with saline and mixed with anti-A agglutinin (at left), and anti-B agglutinin (at right). For some minutes they are allowed to react. The clumping (agglutination) is due to antibody-antigen reactions. These tests use antigens or antibodies adsorbed onto the surface of latex spheres. »» What is the blood type in the example shown in (D)?

The Hemagglutination-Inhibition Test

An indirect method for detecting viruses is to search for viral antibodies in a patient's serum. This can be done by combining serum (the blood's fluid portion) with known viruses.

Certain viruses—such as those of influenza, measles, and mumps—have the ability to agglutinate (clump) red blood cells. This phenomenon, called **hemagglutination** (**HA**), will produce a thin layer of cells over the bottom of a plastic dish (**Figure A**). If the cells do not agglutinate (no hemagglutination), they fall to the bottom of the well accumulating as a small "button."

HA can be used for detection and identification purposes. It is possible to detect antibodies against certain viruses because antibodies react with viruses and tie up the reaction sites that otherwise would bind to red blood cells, thereby inhibiting hemagglutination. This property allows a laboratory test called the **hemagglutination-inhibition** (**HAI**) **test** to be performed. In the test, the patient's serum is combined with known viruses, to which red blood cells are then added (**Figure B**). Hemagglutination inhibition indicates antibodies are present in the serum. Such a finding implies the patient has been exposed to the virus.

Figure C presents a scenario for detecting exposure to the measles virus.

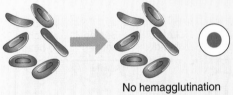

Viruses   Red blood cells   Hemagglutination

No hemagglutination

(A)

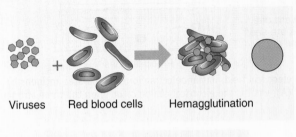

Viruses | Antibodies from patient's serum | Viruses coated with antibodies | Red blood cells | Hemagglutination inhibition

(B)

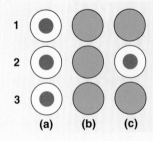

1   2   3
(a)   (b)   (c)

(C)

FIGURES A AND B Hemagglutination and the Hemagglutination-Inhibition Test. (**A**) Viruses of certain diseases, such as measles and mumps, can agglutinate (clump) red blood cells, forming a film of cells over the bottom of a plastic well. No agglutination produced a small pellet or "button" at the bottom of the cell. (**B**) In the hemagglutination-inhibition (HAI) test, known viruses are combined with serum from a patient. If the serum contains antibodies for that virus, the antibodies coat the virus, and when the viruses are then combined with red blood cells, no agglutination will take place.

FIGURE C A Hemagglutination-Inhibition Test. In December, three young children (1, 2, 3) who attended the same daycare center were admitted to the hospital with non-bacterial pneumonia. All three were tested for the presence of influenza A antibodies. Which child or children likely had the flu? (a) = positive control; (b) = negative control; (c) = patient's serum.

agglutination test called the **Venereal Disease Research Laboratory (VDRL) test** is used for the rapid screening of patients to detect syphilis.

CONCEPT AND REASONING CHECKS

22.9 Summarize the uses for neutralization, precipitation, and agglutination tests.

Complement Fixation Can Detect Antibodies to a Variety of Pathogens

KEY CONCEPT

10. Complement fixation depends on complement inactivation and hemolysis of sheep red blood cells.

The **complement fixation test** is performed in two parts. The first part—the test system—uses the patient's serum (± antibodies), a preparation of antigen from the suspected pathogen, and complement derived from guinea pigs. The second part—the indicator system—requires sheep red blood cells and a preparation of antibodies that recognizes the sheep red blood cells. The first step in the test is to heat the patient's serum to destroy any complement present in the serum. Next, carefully measured amounts of antigen and guinea pig complement are added to the serum (**FIGURE 22.10**). This test system then is incubated at 37°C for 90 minutes. If antibodies specific for the antigen are present in the patient's serum, an antibody-antigen interaction takes place, and the complement is used up, or "fixed." However, there is no visible sign of whether or not a reaction has occurred.

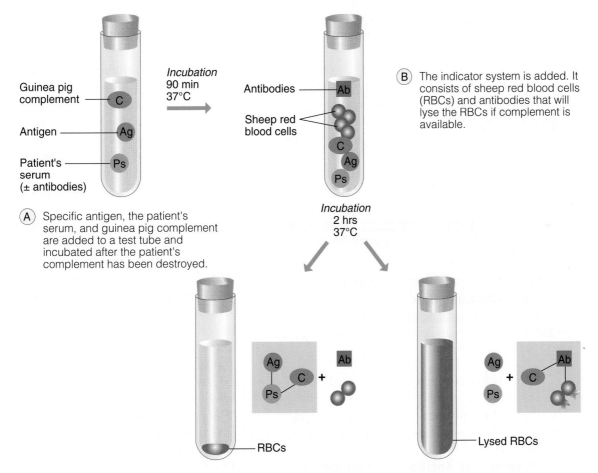

(A) Specific antigen, the patient's serum, and guinea pig complement are added to a test tube and incubated after the patient's complement has been destroyed.

(B) The indicator system is added. It consists of sheep red blood cells (RBCs) and antibodies that will lyse the RBCs if complement is available.

(C) If the patient's serum contains antibodies to the antigen antibodies, a reaction will take place between the antibodies, antigen, and complement. Because the complement has been used up, no lysis of the sheep RBCs will occur when the indicator system is added.

(D) If specific antibodies from the patient are not present in the serum, no reaction occurs during the first incubation, and the complement is left free to unite with antibodies attached to the sheep RBCs. Lysis of the red blood cells results.

FIGURE 22.10 **The Complement Fixation Test.** Complement fixation tests are carried out in two stages. »» What would happen if the patient's complement proteins were not inactivated?

Now the indicator system (sheep red blood cells and antisheep antibodies) is added to the tube, and the tube is reincubated. If the complement was previously fixed, lysis of the sheep red blood cells cannot take place. The blood cells therefore would remain intact, and when the tube is centrifuged, the technician observes clear fluid with a "button" of blood cells at the bottom. Conclusion: The patient's serum contained antibodies that reacted with the antigen and fixed the complement.

If the complement was not fixed in the test system, it will still be available to react with the antibodies bound to sheep red blood cells and, as a result, the sheep red blood cells will lyse. When the tube is centrifuged, the technician sees red fluid, colored by the hemoglobin of the broken blood cells, and no evidence of blood cells at the bottom of the tube. Conclusion: The serum lacked antibodies for the antigen tested.

The complement fixation test is valuable because it may be adapted by varying the antigen. In this way, tests may be conducted for such diverse diseases as encephalitis, Rocky Mountain spotted fever, meningococcal meningitis, and histoplasmosis. The versatility of the test, together with its sensitivity and relative accuracy, has secured its continuing role in diagnostic medicine.

CONCEPT AND REASONING CHECKS

22.10 Describe the uses for the complement fixation test.

Labeling Methods Are Used to Detect Antigen-Antibody Binding

KEY CONCEPT

11. Fluorescent or radioactive antibodies can identify antigens (pathogens).

The detection of antigen-antibody binding can be enhanced (amplified) visually by attaching a label to the antigen or antibody. This tag may be a fluorescent dye, a radioisotope, or an enzyme.

Fluorescent Antibody Technique. The detection of antigen-antibody binding can be done on a slide using a **fluorescent antibody technique**. Two commonly used dyes are fluorescein, which emits an apple-green glow, and rhodamine, which gives off orange-red light.

Fluorescent antibody techniques may be direct or indirect. In the direct method, the fluorescent dye is linked to a known antibody. After combining with particles having complementary antigens, the three components react, causing the complex to glow on illumination with ultraviolet (UV) light and viewed with a fluorescence microscope (see Chapter 3).

For example, suppose you want to know if spirochetes were present in a serum sample. The sample would be combined with antispirochete-labeled antibodies. If spirochetes are present, the tagged antibodies accumulate on the particle surface and the particle glows when viewed with fluorescence microscopy (FIGURE 22.11).

With the indirect method, the fluorescent dye is linked to an antibody recognizing human antibodies in a patient's serum. An example of this indirect method is the diagnostic procedure used for detecting syphilis antibodies in the blood of a patient (FIGURE 22.12). A sample of commercially available syphilis spirochetes is placed on a slide, and the slide is then flooded with the patient's serum. Next, a sample of fluorescein-labeled antiglobulin (antihuman) antibodies is added. These are the antibodies recognizing human antibodies. The slide then is observed using the fluorescence microscope.

The test is interpreted as follows. If the patient's serum contains antisyphilis antibodies, the antibodies bind to the surfaces of spirochetes and the labeled antiglobulin antibodies are attracted to them. The spirochetes then glow from the dye. However, if no antibodies are present in the serum, nothing accumulates on the spirochete's surface, and labeled antiglobulin antibodies also fail to gather on the surface. The labeled antibodies remain in the fluid and the spirochetes do not glow.

Fluorescent antibody techniques are adaptable to a broad variety of antigens and antibodies, and are widely used in serology. Antigens may be detected in bacterial smears, cell smears, and viruses fixed to carrier particles. The value of the technique is enhanced because the materials are sold in kits and are readily available to small laboratories.

Radioimmunoassay. An extremely sensitive serological procedure used to measure the

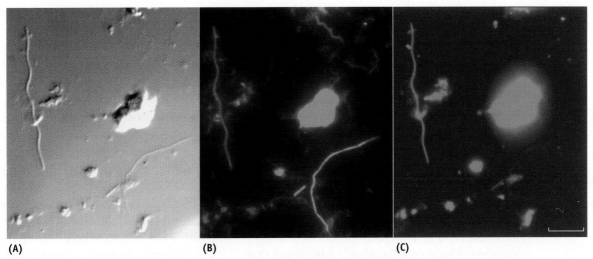

FIGURE 22.11 **Direct Fluorescent Antibody Staining.** Three views of spirochetes in the hindgut of a termite identified by a direct fluorescent antibody technique. (**A**) An unstained area displayed by differential interference contrast microscopy. (**B**) The same viewing area seen by fluorescence microscopy using rhodamine B as a stain. (**C**) The same area viewed after staining with fluorescein. (Bar = 10 μm.) »» Why is this method referred to as direct fluorescent antibody technique?

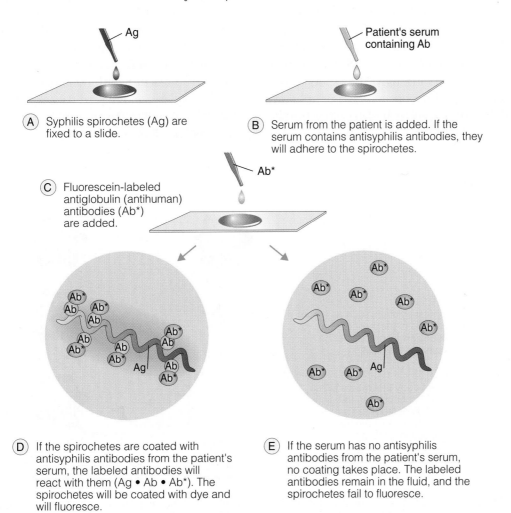

Ag

Patient's serum containing Ab

(A) Syphilis spirochetes (Ag) are fixed to a slide.

(B) Serum from the patient is added. If the serum contains antisyphilis antibodies, they will adhere to the spirochetes.

Ab*

(C) Fluorescein-labeled antiglobulin (antihuman) antibodies (Ab*) are added.

(D) If the spirochetes are coated with antisyphilis antibodies from the patient's serum, the labeled antibodies will react with them (Ag • Ab • Ab*). The spirochetes will be coated with dye and will fluoresce.

(E) If the serum has no antisyphilis antibodies from the patient's serum, no coating takes place. The labeled antibodies remain in the fluid, and the spirochetes fail to fluoresce.

FIGURE 22.12 **The Indirect Fluorescent Antibody Technique for Diagnosing Syphilis.** With the indirect technique, a larger visible signal (fluorescence) is observed. »» Why is this method referred to as the indirect fluorescent antibody technique?

concentration of very small antigens, such as haptens, is the **radioimmunoassay** (**RIA**). Since its development in the 1960s, the technique has been adapted for quantitating hepatitis antigens as well as reproductive hormones, insulin, and certain drugs. One of its major advantages is that it can detect trillionths of a gram of a substance.

The RIA procedure is based on the competition between radioactive-labeled antigens and unlabeled antigens for the reactive sites on antibody molecules. A known amount of the radioactive (labeled) antigens is mixed with a known amount of specific antibodies and an unknown amount of unlabeled antigens. The antigen-antibody complexes that form during incubation then are separated out, and their radioactivity is determined. By measuring the radioactivity of free antigens remaining in the leftover fluid, one can calculate the percentage of labeled antigen bound to the antibody. This percentage is equivalent to the percentage of unlabeled antigen bound to the antibody because the same proportion of both antigens will find spots on antibody molecules. The concentration of unknown unlabeled antigen then can be determined by reference to a standard curve.

Radioimmunoassay procedures require substantial investment in sophisticated equipment and carry a certain amount of risk because radioactive isotopes are used (see Chapter 2). For these reasons, the procedure is not widely used in routine serological laboratories.

The **radioallergosorbent test** (**RAST**) is an extension of the radioimmunoassay. The test may be used, for example, to detect IgE antibodies in the serum of a person possibly allergic to compounds like penicillin.

To detect IgE against penicillin, penicillin antigens are attached to a suitable plastic device, such as a plastic well (**FIGURE 22.13**). Serum is then added. If the serum contains antipenicillin IgE, it will combine with the penicillin antigens on the surface of the plastic well. Now another antibody—one that will react with human antibodies—is added. This antiglobulin antibody carries a radioactive label. The entire complex, therefore, will become radioactive if the antiglobulin antibody combines with the IgE. By contrast, if no IgE was present in the serum, no reaction with the antigen on the well will take place, and the radioactive antibody will not be attracted to the antigen. When tested, the well will not show radioactivity.

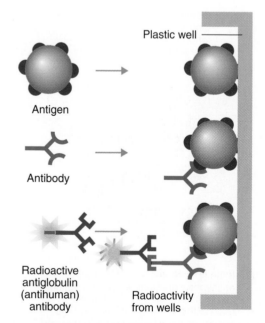

(A) The antigen to the suspected antibody is bound to a plastic device, such as a plastic plate with wells.

(B) The test solution, such as serum, is added to the device. The antibody may or may not be present in the solution.

(C) An antibody that reacts with a human antibody (an antiglobulin antibody) is added. The antiglobulin antibody carries a radioactive label. If radioactivity is detected in the wells, the technician concludes that the solution contained the suspected antibody.

Plastic well

Antigen

Antibody

Radioactive antiglobulin (antihuman) antibody

Radioactivity from wells

FIGURE 22.13 The Radioallergosorbent Test (RAST). This technique provides an effective way of detecting very tiny amounts of antibody in a preparation. A known antigen is used. The objective is to determine whether the complementary antibody is present. »» Would this test represent direct or indirect radioactive antibody labeling? Explain.

The RAST is commonly known as a "sand-wich" technique. There is no competition for an active site as in RIA, and the type of unknown antibody, as well as its amount, may be learned by determining the amount of radioactivity deposited.

Enzyme-Linked Immunosorbent Assay. Another test to detect antigens or antibodies is the **enzyme-linked immunosorbent assay (ELISA).** It has virtually the same sensitivity as radioim-munoassay and the RAST, but does not require expensive equipment or involve radioactivity. The procedure involves attaching antibodies or anti-gens to a solid surface and combining (immuno-sorbing) the coated surfaces with the test material. An enzyme system then is linked to the complex, the remaining enzyme is washed away, and the extent of enzyme activity is measured. This gives an indication that antigens or antibodies are pres-ent in the test material.

An application of the ELISA is found in the highly efficient laboratory test used to detect antibodies against the human immunodeficiency virus (HIV) (**FIGURE 22.14**). A serum sample is obtained from the patient and mixed with a solu-tion of plastic or polystyrene beads coated with antigens from an HIV antigen. If antibodies to HIV are present in the serum, these "primary antibodies" will adhere to the antigens on the surface of the beads. The beads then are washed and incubated with antiglobulin (antihuman) antibodies chemically tagged with molecules of horseradish peroxidase or a similar enzyme. These antibodies are referred to as the "second-ary antibody." The preparation is washed, and a solution of substrate molecules for the peroxidase enzyme is added. Initially the solution is clear, but if enzyme molecules react with the substrate, the solution will become yellow-orange in color. The enzyme molecules will be present only if HIV antibodies are present in the serum. If no HIV antibodies are in the serum, no enzyme molecules could concentrate on the bead surface, no change in the substrate molecules could occur, and no color change would be observed.

ELISA procedures may be varied depend-ing on whether one wishes to detect antigens or antibodies. The solid phase may consist of beads, paper disks, or other suitable supporting mecha-nisms. Additionally, alternate enzyme systems, such as the alkaline phosphatase system may be used, which produce a different color solution in a positive reaction. In addition, the results of the test may be quantified by noting the degree of enzyme-substrate reactions as a measure of the amount of antigen or antibody in the test sample. The availability of inexpensive ELISA kits has brought the procedure into the doctor's office and serological laboratory. Importantly, a positive ELISA for HIV or for any pathogen being tested only indicates that antibodies to the pathogen are present. A positive ELISA does not necessarily mean the person has the disease. The presence of antibodies only says the person has been exposed to the pathogen. However, in the case of HIV, a positive ELISA probably does indi-cate an infection, as no one has ever been known to be cured of an HIV infection. MicroInquiry 22 looks at the ELISA test.

CONCEPT AND REASONING CHECKS

22.11 Assess the need to label antigen or antibody in the RAST and ELISA procedures.

22.3 Monoclonal Antibodies

The variety of laboratory tests used to diag-nose infectious disease continues to expand as new biochemical and molecular techniques are adapted to special clinical situations. Some, like monoclonal antibodies, still depend on antibody-antigen binding, while others involve the analy-sis of microorganism DNA. Both methods are touched upon here.

Monoclonal Antibodies Are Becoming a "Magic Bullet" in Biomedicine

KEY CONCEPT

12. Monoclonal antibodies recognize specific cells or substances.

Usually pathogens contain several different epitopes (see Chapter 21). Therefore, when an

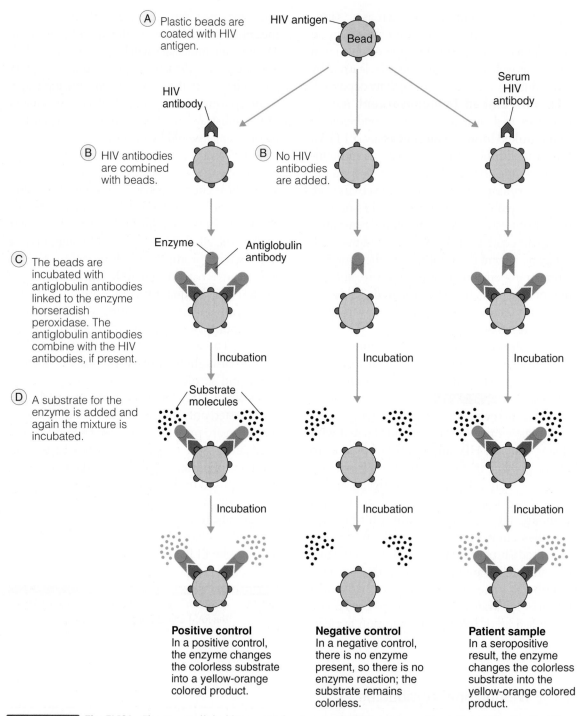

FIGURE 22.14 The ELISA. The enzyme-linked immunosorbent assay (ELISA) as it is used to detect antibodies (in this case HIV) in a patient's serum. »» What are positive and negative controls, and why are they needed in the test?

infection occurs, a different B-cell population is activated for each epitope and different antibodies specific to each of the epitopes are produced. Serum from such a patient would contain **polyclonal antibodies** because the antibodies are derived from several different clones of B cells. However, in the laboratory it is possible to pro-

duce populations of identical antibodies that bind to only one epitope.

In 1975, César Milstein of Argentina and Georges J. F. Köhler of then–West Germany developed a method for the laboratory production of **monoclonal antibodies** (**mAbs**); that is, antibodies recognizing only one epitope. The key

MICROINQUIRY 22

Applications of Immunology: Disease Diagnosis

Ancient Egyptian medical papyri from 1500 BC refer to many different disease symptoms and treatments. Some of these symptoms can still be used to identify diseases today. Thus, one of the most traditional "tools" of diagnosis over the centuries has been a patient's signs and symptoms. However, there are problems when relying solely on these signs and symptoms. Initially, many diseases display common symptoms. For example, the initial symptoms of a hantavirus or anthrax infection are very similar to those of the flu. Some diseases do not display symptoms for perhaps weeks or months—or years in the case of AIDS. Yet, it is important to identify these diseases rapidly so appropriate treatment, if possible, can be started.

Serological (blood) tests have been used in the United States since about 1910 to both diagnose and control infectious disease. Today's understanding of immunology has brought newer tests that rely on identifying antibody-mediated (humoral) immune responses; that is, antibody reactions with antigens. Serology laboratories work with serum or blood from patients suspected of having an infectious disease. The lab tests look for the presence of antibodies to known microbial antigens. Serological tests that are seropositive indicate antibodies to the microbe were detected, while a sero-negative result means no antibodies were detected in a patient's serum.

Let's use a hypothetical person, named Pat, who "believes" that 12 days ago he may have been exposed to the hepatitis C virus through unprotected sex. Afraid to go to a neighborhood clinic to be tested, Pat goes to a local drugstore and purchases an over-the-counter, FDA-approved, hepatitis C home testing kit (see figure). Using a small spring-loaded device that comes with the kit, Pat pricks his little finger and puts a couple of drops of blood onto a paper strip included with the kit. He fills out the paperwork and mails the paper strip in a prepaid mailer (supplied in the kit) to a specific blood testing facility where the ELISA test is performed. In ten business days, Pat can call a toll-free number anonymously, identify himself by a unique 14-digit code that came with the testing kit, and ask for his test results. If necessary, he can receive professional post-test counseling and medical referrals.

Two days later, Pat runs into a good friend of his who is a nurse. Pat explains his "predicament" and that he is anxious about having to wait ten days for the test results. He asks his friend some questions. If you were the nurse, how would you respond to his questions? Answers can be found in **Appendix D.**

22.1a. What is ELISA and how is it performed?
22.1b. You mention positive and negative controls in your

explanation of the ELISA process. Pat asks, "What are positive and negative controls and why are they necessary?"

After your explanations, Pat still has more questions for you to answer.

22.1c. Pat says, "So suppose I am seropositive. Does it mean I have hepatitis C?"
22.1d. Pat says, "Okay, so if I am seronegative, then I must be free of the virus. Right?"
22.1e. If, indeed, Pat was seronegative, what advice would you give him?

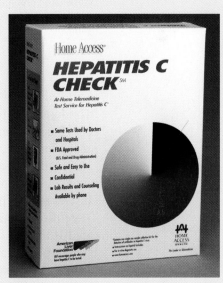

An FDA-approved home testing kit for hepatitis C.

---

was using **myelomas**, cancerous tumors that arise from the uncontrolled division of plasma cells. Although these myeloma cells had lost the ability to produce antibodies, they could be grown indefinitely in culture. Milstein and Köhler were able to fuse myeloma cells with B cells from which mAbs were produced. In 1984, Milstein and Köhler (along with Niels Jerne) shared the Nobel Prize in Physiology or Medicine for the development of the monoclonal antibody technique.

There are two basic steps to generating mAbs. First, a myeloma cell is fused to an activated (antibody secreting) B cell (plasma cell) to form a hybrid cell called a **hybridoma** (FIGURE 22.15). Then, each hybridoma is screened for the desired mAb and cultured to produce hybridoma clones.

To produce hybridomas, a mouse is injected with the antigen of interest, against which a mAb is needed (FIGURE 22.16). Injection triggers humoral immunity in the mouse. Plasma cells from the mouse's spleen are removed and mixed with myeloma cells from another mouse. By forcing many cells to fuse together, hybridomas are produced. Hybridomas are placed in a special

Screened:
Referring to testing or examining something for a particular characteristic or disease.

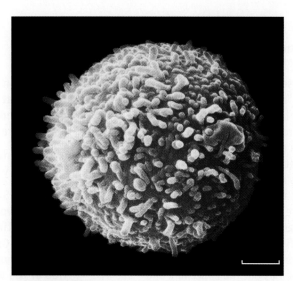

FIGURE 22.15 **A Hybridoma Cell.** False-color scanning electron micrograph of a hybridoma cell screened for its ability to produce a monoclonal antibody to a human cytoskeleton protein. (Bar = 3 μm.) »» What antibody must have been produced by the B cell used to form this hybridoma cell?

tissue culture medium and in seven to ten days, hybridomas form small clusters while any unfused cells (B cells or myelomas) cannot survive in the medium and die.

Because the antigen originally injected may have several epitopes, different hybridoma clusters may be producing different antibodies. Therefore, individual hybridoma cells are separated from one another and allowed to grow. Each hybridoma cell can multiply indefinitely (a myeloma characteristic), producing antibodies recognizing but one epitope (a B-cell characteristic).

Each hybridoma clone is screened for the synthesis of the desired antibody. The hybridoma producing the desired mAb can be propagated (cloned) in tissue culture flasks.

Initially, mAbs were of mouse origin, meaning the mAbs produced would be seen as foreign when injected into humans. Therefore, the antibodies would be destroyed and, in the process, kidney damage might occur. More recently, "humanized" mAbs were developed, whose structure is about 90% to 95% human immunoglobulin. Then, in 2002, the first totally human mAb was produced, finally solving the problem of immune system recognition.

Currently, more than 100 mAbs are in drug clinical trials and some 20 have been approved for use in the United States. So far, the most prominent mAb targeted at infectious disease is the humanized palivizumab (Synagis), which is designed to prevent respiratory syncytial virus infections (see Chapter 15) in high risk groups. Monoclonal antibodies also are in the pipeline for treating hepatitis B and C infections.

Monoclonal antibodies also are being used for **immunomodulation**, which is controlling overactive inflammatory responses (see Chapter 20). In fact, the first truly human mAb product, called adalimumab (Humira), is directed against inflammatory responses resulting in rheumatoid arthritis.

Perhaps most prominent are those mAb drugs targeted at some forms of cancer. Monoclonal antibody products have been approved for treating malignant lymphoid tumors, leukemia, and both colorectal and pancreatic cancers. Many additional products are in clinical trials.

More recent developments have devised ways to add a "payload" to a mAb. For example, scientists have developed a technique in which tumor cells are removed from a patient and injected into a mouse, whereupon the mouse's spleen begins producing tumor antibodies. Spleen cells then are fused with myeloma cells to create a hybridoma, producing mAbs for that specific tumor. Toxins (the payload) are attached to the tail (Fc fragment) of the antibodies (see Chapter 21). When the antibodies are injected back into the patient, they act like "stealth missiles." They react specifically with the tumor cells, and the toxin kills the cells without destroying other tissue cells.

The use of mAbs represents one of the most elegant expressions of modern biotechnology—and their potential uses in the clinic and medicine remain to be fully appreciated.

CONCEPT AND REASONING CHECKS

22.12 Explain how monoclonal antibodies are produced.

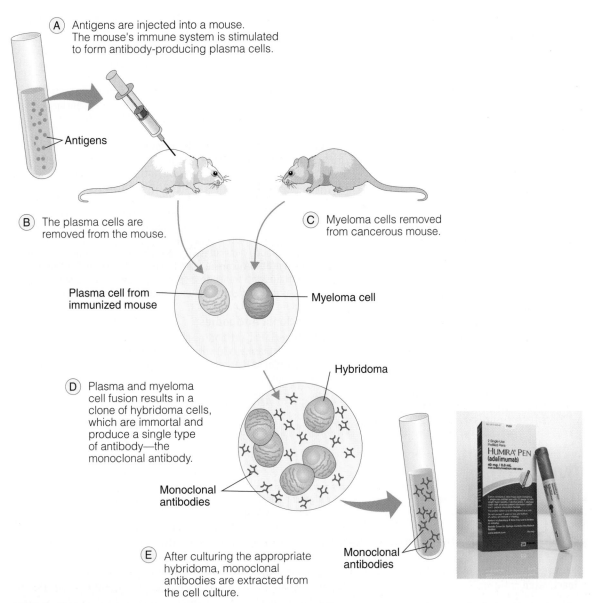

(A) Antigens are injected into a mouse. The mouse's immune system is stimulated to form antibody-producing plasma cells.

Antigens

(B) The plasma cells are removed from the mouse.

(C) Myeloma cells removed from cancerous mouse.

Plasma cell from immunized mouse

Myeloma cell

Hybridoma

(D) Plasma and myeloma cell fusion results in a clone of hybridoma cells, which are immortal and produce a single type of antibody—the monoclonal antibody.

Monoclonal antibodies

Monoclonal antibodies

(E) After culturing the appropriate hybridoma, monoclonal antibodies are extracted from the cell culture.

FIGURE 22.16 **The Production of Monoclonal Antibodies.** The production of a monoclonal antibody (mAb) requires fusing a myeloma cell with plasma cells. After screening for the mAb needed, those clones can be cultivated to produce more of the same mAb. (inset) Humira is the first fully human mAb approved for use in the United States. »» What are the unique features of the plasma cell from the normal mouse and the myeloma cell from the cancerous mouse?

■ SUMMARY OF KEY CONCEPTS

22.1 Immunity to Disease

1. In **naturally acquired active immunity** antibodies are produced, and lymphocytes activated, in response to an infectious agent. Immunity arises from either contracting the disease and recovering, or from a subclinical infection. **Artificially acquired active immunity** is when antibodies

and lymphocytes are produced as a result of a vaccination and immunity is established for some length of time.

2. There are several strategies for producing vaccines, including **attenuated**, **inactivated**, **toxoid**, **subunit**, and **conjugate vaccines.** Often booster shots are required to maintain immunological memory.

3. **Naturally acquired passive immunity** comes from the passage of antibodies from the mother to fetus, or mother to newborn through **colostrum**, which confers immunity for a short period of time. **Artificially acquired passive immunity** is when one receives an **antiserum** (antibodies) produced in another human or animal. This form confers immunity for a short period of time. Serum sickness can develop if the recipient produces antibodies against the antiserum.

4. **Herd immunity** results from a vaccination program that lowers the number of susceptible members within a population capable of contracting a disease. If there are few susceptible individuals, the probability of disease spread is minimal.

5. Established vaccines are quite safe, although some people do experience side effects that may include mild fever, soreness at the injection site, or malaise after the vaccination. Few people suffer serious consequences.

22.2 Serological Reactions

6. Serological reactions consist of antigens and antibodies (serum). Successful reactions require the correct dilution (**titer**) of antigens and antibodies.

7. **Neutralization** is a serological reaction in which antigens and antibodies neutralize each other. Often there is no visible reaction, so injection into an animal is required to see if the reaction has occurred.

8. In a **precipitation** reaction, antigens and antibodies react to form a matrix that is visible to the naked eye. Different microbial antigens can be detected using this technique.

9. **Agglutination** involves the clumping of antigens. The cross-linking of antigens and antibodies causes the complex to clump. Some infectious agents can be detected readily by this method.

10. **Complement fixation** can detect antibodies to a variety of pathogens. This serological method involves antigen-antibody complexes that are detected by the fixation (binding) of complement.

11. The **fluorescent antibody technique** involves the addition of fluorescently tagged antibodies to a known pathogen to a slide containing unknown antigen (pathogen). If the antibody binds to the antigen, the pathogen will glow (fluoresce) when observed with a fluorescence microscope, and the pathogen is identified. The **radioimmunoassay** (**RIA**) uses radioactivity to quantitate radioactive antigens through competition with nonradioactive antigens for reactive sites on antibody molecules. The **enzyme-linked immunosorbent assay** (**ELISA**) can be used to detect if a patient's serum contains antibodies to a specific pathogen, or to detect or measure antigens in serum. A positive ELISA is seen as a colored reaction product.

22.3 Monoclonal Antibodies

12. **Monoclonal antibodies** result from the fusion of a specific B cell (plasma cell) with a **myeloma cell** to produce a **hybridoma** that secretes an antibody recognizing a single epitope.

◼ LEARNING OBJECTIVES

After understanding the textbook reading, you should be capable of writing a paragraph that includes the appropriate terms and pertinent information to answer the objective.

1. Identify how **artificially acquired active immunity** produces immunity.
2. Differentiate between vaccines consisting of live, **attenuated** antigens and those containing **inactivated** antigens. Provide examples for each.
3. Distinguish between **naturally acquired passive immunity** and **artificially acquired passive immunity**.
4. Assess the importance of **herd immunity** to community and national health.
5. Judge the safety of vaccines and the reporting system to identify possible adverse affects.
6. Identify the role of **titration** in identifying disease.
7.–9. Compare and contrast (7) **neutralization**, (8) **precipitation**, and (9) **agglutination** as examples of serological reactions.
10. Explain how the **complement fixation test** works.
11. Differentiate between **fluorescent antibody**, **RAST**, and **ELISA** as labeling methods to detect antigen-antibody binding.
12. Summarize the characteristics of **monoclonal antibodies** and assess their role in identifying infectious disease.

◼ STEP A: SELF-TEST

Each of the following questions is designed to assess your ability to remember or recall factual or conceptual knowledge related to this chapter. Read each question carefully, then select the *one* answer that best fits the question or statement. Answers to even-numbered questions can be found in **Appendix C**.

1. Exposure to the flu virus, contracting the flu, and recovering from the disease would be an example of
   A. artificially acquired passive immunity.
   B. naturally acquired active immunity.
   C. artificially acquired active immunity.
   D. naturally acquired passive immunity.

2. An attenuated vaccine contains
   A. inactive toxins.
   B. living, but slow growing (replicating) antigens.
   C. killed bacteria.
   D. noninfective antigen subunits.

3. Immune complex formation and serum sickness are dangers of
   A. artificially acquired passive immunity.
   B. naturally acquired active immunity.
   C. artificially acquired active immunity.
   D. naturally acquired passive immunity.

4. Herd immunity is affected by
   A. the percentage of a population that is vaccinated.
   B. the strength of an individual's immune system.
   C. the number of susceptible individuals.
   D. All the above (A–C) are correct.

5. Approximately _____ of 100,000 vaccinated individuals are likely to suffer a serious reaction to the vaccination.
   A. 1
   B. 50
   C. 100
   D. 500

6. Titer refers to
   A. the most concentrated antigen-antibody concentration showing a reaction.
   B. the first diluted antigen-antibody concentration showing a reaction.
   C. the precipitation line formed between an antigen-antibody reaction.
   D. the most dilute antigen-antibody concentration showing a reaction.

7. _____ is a serological reaction that produces little or no visible evidence of a reaction.
   A. Precipitation
   B. ELISA
   C. Neutralization
   D. Agglutination

8. The serological reaction where antigens and antibodies form an extensive lattice of large particles is called
   A. fixation.
   B. precipitation.
   C. neutralization.
   D. agglutination.

9. When antigens are attached to the surface of latex beads and then reacted with an appropriate antibody, a/an _____ reaction occurs.
   A. inhibition
   B. agglutination
   C. neutralization
   D. precipitation

10. What serological test requires sheep red blood cells and a preparation of antibodies that recognizes the sheep red blood cells?
    A. ELISA
    B. Radioimmunoassay
    C. Immunodiffusion
    D. Complement fixation test

11. In an ELISA, the primary antibody represents
    A. the patient's serum.
    B. the antibody recognizing the secondary antibody.
    C. the enzyme-linked (labeled) antibody.
    D. the antibodies having been washed away.

12. A hybridoma cell
    A. secretes monoclonal antibodies.
    B. presents antigens on its surface.
    C. secretes polyclonal antibodies.
    D. is an antigen-presenting cell.

STEP B: REVIEW

On completing Section 22.1 (Immunity to Disease), test your comprehension of the section's contents by filling in the following blanks with two terms that answer the description best. **Appendix C** contains the answers to the even-numbered statements.

13. Two general forms of immunity: _____ and _____.

14. Two types of natural immunity: _____ and _____.

15. Two diseases that MMR is used against: _____ and _____.

16. Two diseases that DPT is used against: _____ and _____.

17. Two types of passive immunity: _____ and _____.

18. Two names for antibody-containing serum: _____ and _____.

19. Two antibody types formed on antigen stimulation: _____ and _____.

20. Two ways newborns have acquired maternal antibodies: _____ and _____.

21. Two types of viruses in viral vaccines: _____ and _____.

22. Two bacterial diseases for which toxoids are used: _____ and _____.

23. Two methods for administering vaccines: _____ and _____.

24. Two functions of antibodies in antiserum: _____ and _____.

25. Two bacterial diseases where passive immunity is used: _____ and _____.

▮ STEP C: APPLICATIONS

Answers to even-numbered questions can be found in **Appendix C**.

26. A friend is heading on a trip to a foreign country where hepatitis A outbreaks are quite common. For passive immunity, he gets a shot of an antiserum containing IgG antibodies. Why do you suppose IgM is not used, especially since the immunoglobulins are the important components of the primary antibody response?

27. A complement fixation test is performed with serum from a patient with an active case of syphilis. In the process, however, the technician neglects to add the syphilis antigen to the tube. Would lysis of the sheep red blood cells occur at the test's conclusion? Why?

28. Suppose the titer of mumps antibodies from your blood was higher than that for your fellow student. What are some of the possible reasons that could have contributed to this? Try to be imaginative on this one.

29. Given a choice, which of the four general types of acquired immunity would it be safest to obtain? Why?

30. An immunocompromised child is taken by his mother to the local clinic to receive a belated measles-mumps-rubella (MMR) vaccination. As the nurse practitioner, you decline to give the child the vaccine. Explain to the mother why you refused to vaccinate the child.

▮ STEP D: QUESTIONS FOR THOUGHT AND DISCUSSION

Answers to even-numbered questions can be found in **Appendix C**.

31. It is estimated that when at least 90% of the individuals in a given population have been immunized against a disease, the chances of an epidemic occurring are very slight. The population is said to exhibit "herd immunity," because members of the population (or herd) unknowingly transfer the immunizing agent to other members and eventually immunize the entire population. What are some ways by which the immunizing agent can be transferred?

32. When children are born in Great Britain, they are assigned a doctor. Two weeks later, a social services worker visits the home, enrolls the child on a national computer registry for immunization, and explains immunization to the parents. When a child is due for an immunization, a notice is automatically sent to the home, and if the child is not brought to the doctor, the nurse goes to the home to learn why. Do you believe a method similar to this can work (or should be used) in the United States to achieve uniform national immunization?

33. Since 1981, the incidence of pertussis has been increasing in the United States, with the greatest increase found in adolescents and adults. Why are adolescents and adults targets of the bacterial pathogen, considering these individuals were usually considered immune to the disease?

34. Children between the ages of 5 and 15 are said to pass through the "golden age of resistance" because their resistance to disease is much higher than that of infants and adults. What factors may contribute to this resistance?

⚛ HTTP://MICROBIOLOGY.JBPUB.COM/9E

The site features eLearning, an online review area that provides quizzes and other tools to help you study for your class. You can also follow useful links for in-depth information, read more MicroFocus stories, or just find out the latest microbiology news.

23

Immune Disorders and AIDS

I don't think we are losing the war, but we're certainly not finished with the war.
—Ronald Valdiserri, Deputy Director of the National Center for HIV, STD, and TB Prevention

2010 marked 29 years into the epidemic of acquired immunodeficiency syndrome (AIDS), and there still are only somber numbers reported. The Joint United Nations Programme on HIV/AIDS (UNAIDS) reports that globally there are over 33 million people living with AIDS (FIGURE 23.1) and 20 million have died of AIDS since 1981. Currently, there are some 3 million new infections and 2 million deaths each year. UNAIDS has projected that up to 70 million people will die from AIDS in the next 20 years—if a cure is not found. Indeed, there still is no cure or vaccine, and antiretroviral therapies can be toxic and quickly become less useful as HIV develops resistance.

The epidemic is worst in sub-Saharan Africa where more than 22 million people have an HIV infection. In some African cities, a staggering 30% to 50%—almost half the population—is HIV-positive. Without a cure, by 2020 more than 25% of the workforce in some African cities will be lost because of AIDS. In July 2003, then President Bush pledged $15 billion to fight AIDS in Africa and the Caribbean where almost half of the world's HIV infections are located. The plan to combat AIDS calls for antiviral treatment for 2 million HIV-infected people who cannot afford the costly cocktail of drugs that can prolong and improve their lives.

Chapter Preview and Key Concepts

23.1 Type I IgE-Mediated Hypersensitivity
1. A type I hypersensitivity can vary from simply a runny nose and watery eyes to a life-threatening disorder.
2. Anaphylactic reactions can be sudden, systemic allergic reactions that are life-threatening.
3. Common allergies are examples of a localized anaphylaxis.
4. Asthma is characterized by a constriction of the airways.
5. Genetics and the environment are responsible for most allergies.
6. Physical avoidance and drug treatment can limit allergic attacks.
   MicroInquiry 23: Allergies, Dirt, and the Hygiene Hypothesis

23.2 Other Types of Hypersensitivity
7. Antibodies can be the cause of cytotoxic (type II) hypersensitivities.
8. Immune complexes are responsible for type III hypersensitivities.
9. Type IV hypersensitivity is an exaggeration of a delayed cell-mediated immune response.

23.3 Autoimmune Disorders and Transplantation
10. Autoimmune disorders generate an immune attack on self cells and tissues.
11. Immune-mediated rejection must be overcome for successful tissue transplants or grafts.
12. Immunosuppressive drugs have several drawbacks.

23.4 Immunodeficiency Disorders
13. Immunodeficiency disorders lead to recurrent and often severe, longer lasting infections.
14. Acquired immunodeficiency syndrome (AIDS) results from immune system dysfunction.

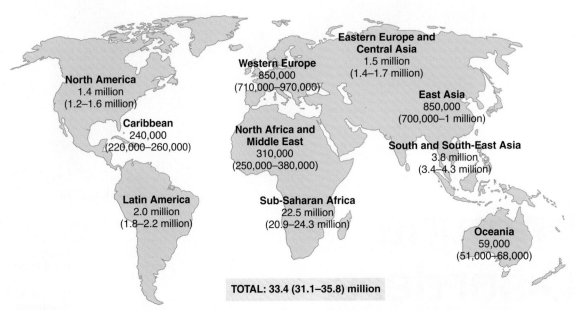

FIGURE 23.1 **Adults and Children Living with HIV Infection—2009.** The numbers of HIV infections is an estimate made by the Joint United Nations Program on HIV/AIDS. »» Provide some reasons why these numbers represent only estimates. *Source*: UNAIDS.org.

Africa is not the only worry. In South and Southeast Asia, there are an estimated 4 million people infected, 2.5 million in India alone. In some areas, UNAIDS reports the HIV infection rate among drug users is greater than 70%. In Eastern Europe, sexually transmitted diseases have been doubling every year since 1998, which means HIV infections also are increasing at alarming rates.

Compared to Africa, Asia, and Eastern Europe, the AIDS prevalence in most developed nations has remained flat, in part, because of the number of people on antiretroviral drugs. In 2009, there were almost 60,000 new AIDS infections in adults and children in the United States, bringing the total number to more than 1 million. Most alarming, 25% of those infected with HIV do not know it. There were 14,000 deaths in 2009, bringing the total to more than 590,000 since 1981.

In 2002, federal health authorities announced there was a 2.2% increase in HIV infections from the previous year, which represented the first rise since 1993. American health officials attributed this disturbing upswing to a growing complacency about the dangers of AIDS. In addition, younger people do not remember the devastation of the AIDS epidemic and perhaps feel safer with the advent of life-extending antiretroviral drugs now available. Since 2003, the number of cases and deaths has remained constant.

Although AIDS is a major topic in this chapter, it is by no means the only immune disorder. Hypersensitivities and allergies represent immune disorders that many believe can be at least partly the result of children not fighting the types of infectious diseases experienced by past generations. Also included in the broad category of immune disorders are the autoimmune diseases that often have a microbial component associated with them. We survey each type of immune disorder in the following sections.

23.1 Type I IgE-Mediated Hypersensitivity

Hypersensitivity is a multistep phenomenon triggered by exposure to an antigen. It consists of a dormant (latent) stage, during which an individual becomes sensitized, and an active stage (hyper-sensitivity) following a subsequent exposure to the antigen. The process may involve elements of humoral or cell-mediated immunity, or sometimes both.

Hypersensitivities can be classified into one of four types. **Immediate hypersensitivity** refers to an antibody response to the second or subsequent dose of the same antigens (**FIGURE 23.2**). Immediate hypersensitivities include:

- Type I IgE-mediated hypersensitivity, which is a process involving IgE, mast cells, basophils, and cell mediators inducing smooth muscle contraction;
- Type II cytotoxic hypersensitivity, which involves IgG, IgM, complement, and the destruction of host cells; and
- Type III immune complex hypersensitivity, which involves IgG, IgM, complement, and the formation of antigen-antibody aggregates in the tissues.

The fourth type is a **delayed hypersensitivity**, where a cell-mediated immune response develops over two to three days. This type IV cellular hypersensitivity involves cytokines and T lymphocytes.

Realize that a hypersensitivity is an exaggerated or inappropriate immune defense that is causing the problems in an affected individual.

Type I Hypersensitivity Is Induced by Allergens

KEY CONCEPT

1. A type I hypersensitivity can vary from simply a runny nose and watery eyes to a life-threatening disorder.

A **type I hypersensitivity** has all the characteristics of a humoral immune response. It begins with the entry of a substance, called an **allergen**, which triggers an allergic reaction in the body. The allergen may be any of a wide variety of materials such as plant pollen, certain foods, bee venom, serum proteins, or a drug, such as penicillin. In the case of penicillin, the drug molecule itself is the allergen, but the molecule does not stimulate the immune system until it has combined with tissue proteins to form an allergenic complex (see Chapter 21). Doses of antigen as low as 0.001 mg have been known to sensitize a person. Allergists refer to this first dose of antigen as the **sensitizing dose**.

The immune system responds to the allergen as if it was a dangerous antigen, such as a pathogen. The allergen is taken up by antigen-presenting

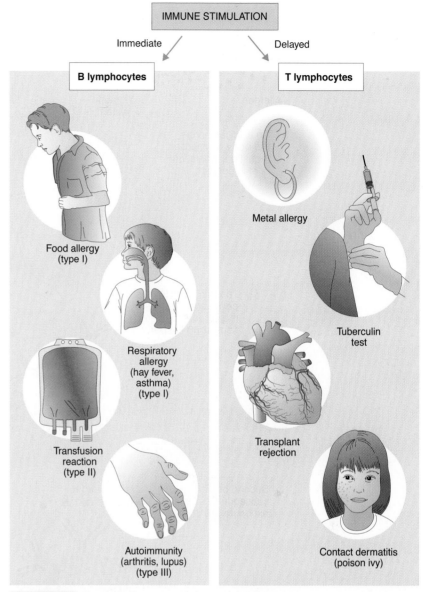

FIGURE 23.2 **The Various Forms of Hypersensitivities.** Immune system hypersensitivities may be related to B lymphocytes or T lymphocytes. Stimulation of the immune system is the starting point for all these disorders. »» Why are all these examples of immune stimulation referred to as hypersensitivities?

cells (APCs; see Chapter 21) and fragments presented to **helper T2 (T$_H$2) cells** (**FIGURE 23.3**). As in humoral immunity, T$_H$2 cells stimulate B cells to mature into plasma cells, which produce IgE antibodies. This antibody, formerly known as reagin, enters the circulation and attaches by its Fc tail to the surface of mast cells and basophils (Figure 23.3). **Mast cells** are connective tissue cells found in the respiratory and gastrointestinal tracts, and near the blood vessels (see Chapter 20). They measure about 10 μm to 15 μm in diameter and are filled with 500 to 1,500 granules containing histamine and other physiologically active substances.

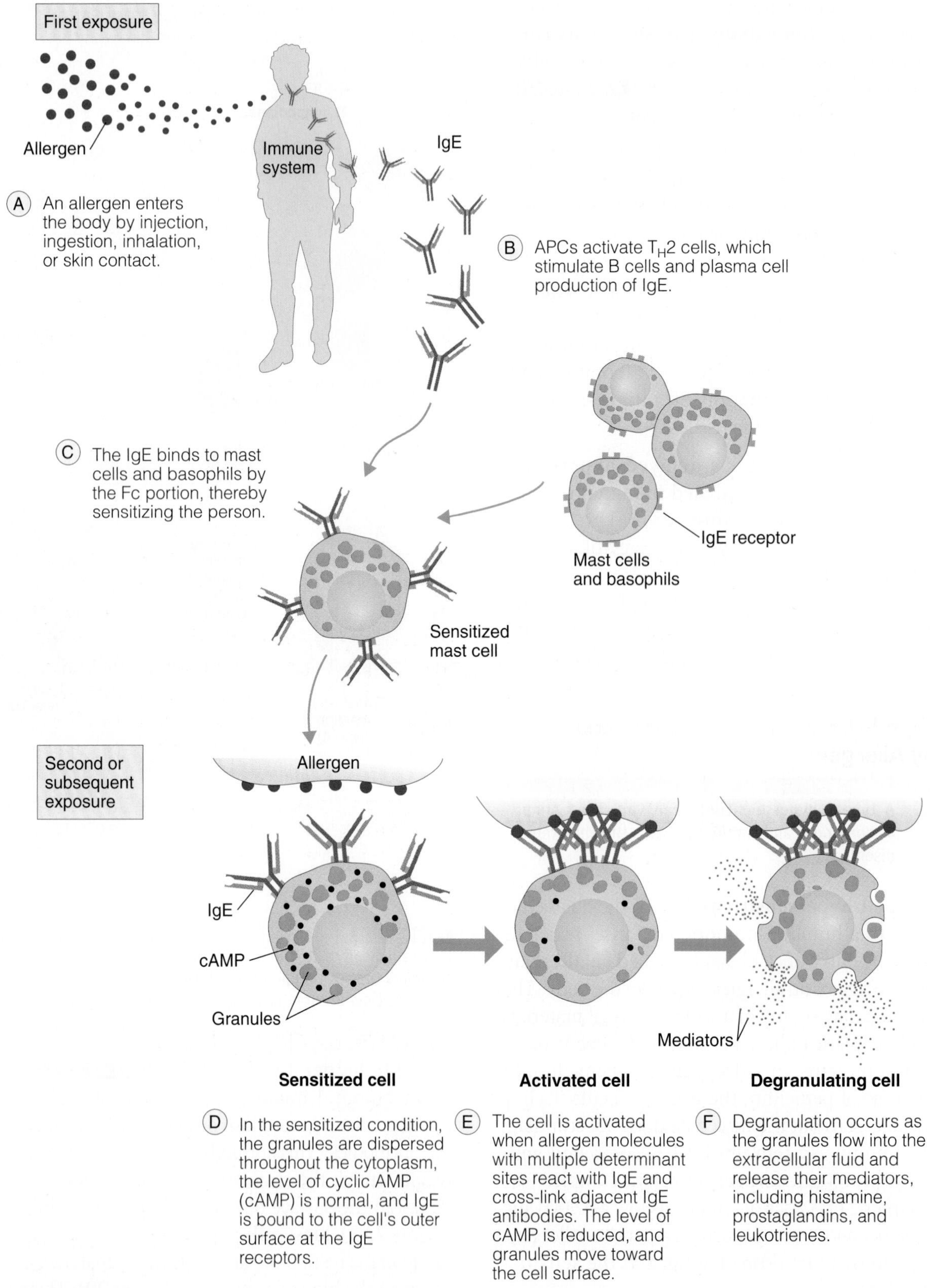

First exposure

Allergen

Immune system

IgE

(A) An allergen enters the body by injection, ingestion, inhalation, or skin contact.

(B) APCs activate T$_H$2 cells, which stimulate B cells and plasma cell production of IgE.

(C) The IgE binds to mast cells and basophils by the Fc portion, thereby sensitizing the person.

IgE receptor

Mast cells and basophils

Sensitized mast cell

Second or subsequent exposure

Allergen

IgE

cAMP

Granules

Mediators

Sensitized cell

Activated cell

Degranulating cell

(D) In the sensitized condition, the granules are dispersed throughout the cytoplasm, the level of cyclic AMP (cAMP) is normal, and IgE is bound to the cell's outer surface at the IgE receptors.

(E) The cell is activated when allergen molecules with multiple determinant sites react with IgE and cross-link adjacent IgE antibodies. The level of cAMP is reduced, and granules move toward the cell surface.

(F) Degranulation occurs as the granules flow into the extracellular fluid and release their mediators, including histamine, prostaglandins, and leukotrienes.

FIGURE 23.3 Type I IgE-Mediated Hypersensitivity. Basophils, mast cells, and IgE antibodies are involved in typical allergies like hay fever. »» Does a person experience the symptoms of an allergy on the first or ensuing exposure to an allergen? Explain.

Basophils are circulating leukocytes in the blood that also are rich in granules. They represent about 1% of the total leukocyte count in the circulation and measure about 15 μm in diameter. Mast cells and basophils each have over 100,000 receptor sites where IgE antibodies can attach by the Fc tail portion (see Chapter 21). As IgE antibodies attach to mast cells and basophils, the individual becomes **sensitized**. Multiple stimuli by allergen molecules may be required to sensitize a person fully. This is why penicillin often must be taken several times before a penicillin allergy manifests itself.

Sensitization usually requires a minimum of one week, during which time millions of molecules of IgE attach to thousands of mast cells and basophils. Because attachment occurs at the Fc end of the antibody, the Fab ends point outward from the cell as shown in Figure 23.3.

On subsequent exposure to the same allergen, the allergen molecules bind to the Fab ends and cross-link IgE antibodies. This cross-linking triggers **degranulation**, a release of granule contents at the cell surface.

Degranulation first requires an inflow of calcium ions (Ca^{2+}) and a brief stimulation of the enzyme **adenylyl cyclase**, which catalyzes the conversion of ATP to cyclic AMP (cAMP). This leads to the swelling of the granules. Adenylyl cyclase activity then is repressed and there is a sudden decrease in cAMP, which is necessary for degranulation to occur. An understanding of this biochemistry is important because drugs such as epinephrine are given to individuals suffering systemic anaphylaxis (see below). Epinephrine stimulates adenylyl cyclase, increasing the level of cAMP and inhibiting degranulation.

As granules fuse with the plasma membrane of the basophils and mast cells, they release a number of mediator substances having substantial pharmacologic activity (**FIGURE 23.4**). The most important preformed mediator of allergic reactions is **histamine**, a derivative of the amino acid histidine. Once in the bloodstream, histamine circulates to the body cells and attaches to the histamine receptors present on most body cells.

Along with serotonin and bradykinin, the principal effect of these mediators is to constrict smooth muscle cells.

Still other mediators must be synthesized after the antigen-IgE reaction (Figure 23.4). One example is a series of substances called **leukotrienes** (so named because they are derived from leukocytes and have a triene [triple] chemical bond). Leukotrienes result from a complex set of interactions. The end result is leukotriene D4, an extremely potent smooth muscle constrictor. It also causes leakage in blood vessels and attracts eosinophils to continue the inflammatory reaction.

The second family of synthesized mediators is the group of human hormones called **prostaglandins**. One prostaglandin, prostaglandin D2, is a powerful constrictor of the bronchial tubes.

Cytokines also are thought to be involved in a late allergic response and have actions that both stimulate and inhibit inflammation. One cytokine, interleukin-4 (IL-4), promotes IgE production by B cells; another, interleukin-5 (IL-5), encourages the maturation and activity of eosinophils; and a third, tumor necrosis factor alpha (TNF-α), is released from mast cells and may be responsible for shock and tissue damage in systemic reactions.

CONCEPT AND REASONING CHECKS

23.1 Summarize the events occurring during a first and second exposure to an allergen.

Systemic Anaphylaxis Is the Most Dangerous Form of a Type I Hypersensitivity

KEY CONCEPT

2. Anaphylactic reactions can be sudden, systemic allergic reactions that are life-threatening.

Systemic anaphylaxis (*ana* = "throughout"; *phylaxi* = "watch") involves allergens, such as bee venom, penicillin, nuts, and seafood, in the bloodstream triggering degranulation of mast cells throughout the body. In affected individuals, the principal activity of the released mediators is to contract smooth muscles in the small veins and increase vascular permeability,

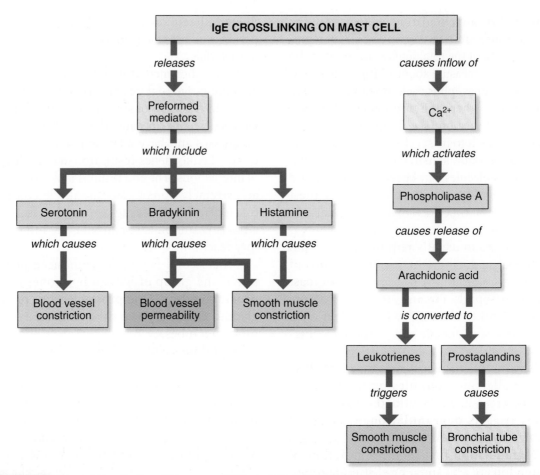

FIGURE 23.4 Concept Map for Type I Hypersensitivity and Mediator Substance Production. Several chemical mediators are stimulated upon IgE cross-linking on mast cells and basophils. »» Because IgE is an antibody normally targeting parasitic helminths, what might these mediator substances have done if it was a parasite infection rather than an allergen?

forcing fluid into the tissues. The resulting drop in blood pressure can lead to unconsciousness. In addition, the skin may become swollen around the eyes, wrists, and ankles, a condition called **edema**. The edema is accompanied by a hive-like rash, along with burning and itching in the skin, as the sensory nerves are excited. Contractions also occur in the gastrointestinal tract and bronchial muscles, leading to sharp cramps and shortness of breath, respectively. The individual inhales rapidly without exhaling and traps carbon dioxide in the lungs, an ironic situation in which the lungs are fully inflated but lack oxygen. Death may occur in 10 to 15 minutes as a result of asphyxiation if prompt action is not forthcoming (hence the name "immediate" hypersensitivity). Many doctors suggest that people subject to systemic anaphylaxis should carry a self-injecting syringe containing

epinephrine (adrenalin). Epinephrine stabilizes mast cells and basophils, dilates the airways, and constricts the capillary pores, keeping fluids in the circulation.

CONCEPT AND REASONING CHECKS

23.2 Explain why systemic anaphylaxis is life-threatening.

Atopic Disorders Are the Most Common Form of a Type I Hypersensitivity

KEY CONCEPT

3. Common allergies are examples of a localized anaphylaxis.

Type I hypersensitivity reactions need not result in the whole-body (systemic) involvement. In fact, the vast majority of hypersensitivity reactions are accompanied by limited production of IgE and the sensitization of mast cells only in localized areas of the body.

Common Allergies. According to public health estimates, more than 50 million Americans (almost 20% of the population) have some type of common (seasonal) allergy or **atopic disease** (*atopo* = "out of place") (FIGURE 23.5). An example of an atopic disease is hay fever, technically referred to as **allergic rhinitis.** Contrary to its common name, hay fever is neither caused by hay nor does it cause a fever! The condition develops from seasonal inhalations of tree, grass, or weed pollens. Immune stimulation by pollen antigens leads to IgE production, and a sensitization of mast cells then follows in the eyes, nose, and upper respiratory tract. Subsequent exposures bring on sneezing, tearing, swollen mucous membranes, and other well-known symptoms.

There also are year-round allergies. These perennial hypersensitivities usually result from chronic exposure to such substances as house dust, mold spores, dust mites, detergent enzymes, and the particles of animal skin and hair ("dander"). Actually, dander itself is not the allergen; the actual allergens are proteins deposited in the dander from the animal's saliva when it grooms itself. Included here are cockroaches and the substances carried on their bodies.

Some people experience a **late-phase anaphylaxis.** In this case, it takes several hours for the tissue to become hot, tender, red, and swollen. The mast cells induce this reaction by releasing chemokines that attract other cells to the site to bring about the changes. For example, eosinophils exist in unusually high numbers in allergic individuals; they arrive at the site and release leukotrienes as well as toxic substances contributing to tissue damage. Neutrophils are normally the phagocytes of the bloodstream, but in allergic reactions, they too liberate a number of enzymes causing local tissue damage. T lymphocytes produce IL-4, which augments the allergic response.

Food Allergies. A variety of foods can cause allergies, which usually are accompanied by symptoms in the gastrointestinal tract, including swollen lips, abdominal cramps,

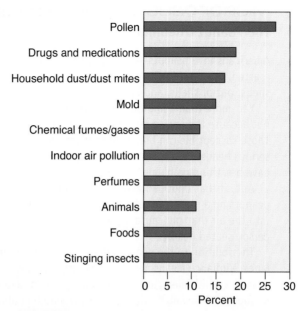

FIGURE 23.5 **The Top 10 Causes of Allergies.** These allergens were identified by allergy sufferers as causing the most allergic reactions. »» Of these allergens, which ones, if any, cause your allergies?
Source: Marist Institute for Public Opinion.

nausea, and diarrhea. The skin may break out in a rash consisting of a central puffiness, called a **wheal,** surrounded by a zone of redness known as a **flare.** Such a hive-like rash is called **urticaria** (*urtica* = "stinging needle"). Allergenic foods include fish, shellfish, eggs, wheat, cow's milk, soy, tree nuts, and even peanuts (MICROFOCUS 23.1). Some allergic reactions to foods can cause a potentially lethal anaphylactic reaction.

Allergenic: A substance causing an allergic reaction.

Physical and Exercise-Induced Allergies. Physical factors such as cold, heat, sunlight, or sweating also can cause allergies. Exactly what causes the reaction is unknown. Some immunologists believe a physical stimulus causes a change in a protein in the skin, which the immune system then "sees" as foreign and mounts an immune response. Exercise sometimes also causes allergies, usually in the form of an asthma attack.

CONCEPT AND REASONING CHECKS

23.3 Distinguish between the different forms of atopic disorders.

MICROFOCUS 23.1: Miscellaneous
The Peanut Dilemma

Americans love peanuts—salted, unsalted, oil roasted, dry roasted, Spanish, honey crusted, in shells, out of shells, and on and on.

Yet, more than 1 million Americans fear the peanut because they risk a rather nasty allergic reaction: Those susceptible can break out in hives, develop a serious headache, experience a racing heartbeat, or have breathing problems.

And, if that's not bad enough, someone eating peanuts on a plane can release small peanut particles into the air that can cause a reaction in a sensitive person seated nearby.

The solution to the peanut allergy dilemma can best be summed up as "V plus V." The first V is for vigilance. Vigilance means avoiding peanuts or peanut butter; but it also means being cautious about egg roll wrappers, chili fillers, and protein extenders in cake mixes, all of which may contain peanuts in one form or another. Drinking liquid charcoal (available in pharmacies) will absorb peanut particles and prevent the triggering of increased symptoms. A 2003 report in the *Journal of Allergy and Clinical Immunology* suggested that parents should keep liquid charcoal at home in case of a peanut allergy in young children. It means vigilance that manufacturers clearly label their products and insist that peanut-detection tests be performed routinely.

The second V is for vaccine. In 1999, investigators from Johns Hopkins University tested a peanut vaccine and showed that it protects sensitized mice against peanut proteins. The vaccine consists of DNA segments that encode the peanut proteins. Encased in protective molecules and delivered orally, the vaccine decreased the mice's capacity for producing peanut-related IgE. The developers postulated that the vaccine elicits the so-called "blocking antibodies" that bind the peanut antigens before they reach the animal's immune system.

So, does that mean we can expect health officials to distribute vaccine injections where we buy peanut butter, peanut brittle, or peanut oil–based skin care products? Not likely, say the researchers—at least not in the immediate future.

Allergic Reactions Also Are Responsible for Triggering Many Cases of Asthma

KEY CONCEPT

4. Asthma is characterized by a constriction of the airways.

Besides allergies and anaphylaxis, **asthma** is part of the same syndrome—the tendency of the immune system to overreact to common allergens. Some 300 million people live with asthma worldwide (20 million in the United States), and the disorder is responsible for one in 250 deaths. The Centers for Disease Control and Prevention (CDC) has reported that since 1980, the prevalence of asthma in American children has risen nearly 60%.

Asthma, which is a chronic condition, is characterized by wheezing, coughing, and stressed breathing. It has been blamed on airborne allergens such as pollen, mold spores, or products from insects like dust mites. In fact, about 50% of asthma cases have an environmental origin. Attacks also can be induced by physical exercise or cold temperatures, situations that irritate and inflame the airways (**FIGURE 23.6**).

Asthma primarily is an inflammatory disorder that occurs in two parts. First, there is an early response to allergen exposure. T_H2 cells (see Chapter 21) and natural killer cells release a variety of chemical mediators. However, unlike a common allergy the mediators are not released in the nose or eyes, but rather in the lower respiratory tract, resulting in bronchoconstriction, vasodilation, and mucus buildup (**FIGURE 23.7**).

FIGURE 23.7 **Airway Narrowing During an Asthma Attack.** During an asthma attack, chemical mediators (cytokines) cause the smooth muscle layers to spasm, constrict, and form a thick mucus that can block the airway. »» What types of cells are responsible for releasing the asthma-causing chemical mediators?

Why Do Only Some People Have IgE-Mediated Hypersensitivities?

KEY CONCEPT

5. Genetics and the environment are responsible for most allergies.

Not everyone suffers from allergies. An interesting avenue of research was opened when it was discovered that the B cells responsible for IgE and IgA production lie close to one another in the lymphoid tissue, and that the IgA level and its corresponding lymphocytes are greatly reduced in atopic individuals. Immunologists have suggested that in nonallergic individuals, the B lymphocytes and plasma cells that produce IgA shield lymphocytes producing IgE from antigenic stimulation, but atopic people may lack sufficient IgA-secreting lymphocytes to block the antigens.

Another theory of atopic disease maintains an allergy results from a breakdown of feedback mechanisms in the immune system. Research findings indicate that B cells and the plasma cells that synthesize IgE may be controlled by **suppressor T lymphocytes**, and regulated in turn by IgE. Under normal conditions, IgE may limit its own production by stimulating suppressor T-cell activity. However, in atopic individuals the mechanism malfunctions, possibly because the T cells are defective, and IgE is produced in massive quantities. Allergic people are known to possess almost 100 times the IgE level of people who

FIGURE 23.6 Using an Asthma Inhaler. For some asthma sufferers, simple exercise might bring on an asthma attack. Using an inhaler that delivers medication directly to the lungs allows sufferers to live active lives without fear of an attack. »» How could physical exercise trigger an asthma attack?

The synthesis and release of other cytokines result in the recruitment of eosinophils and neutrophils into the lower respiratory tract. These events represent a second or late response because they occur hours after the initial exposure to an allergen. The presence of the eosinophils and neutrophils can cause tissue injury and potentially cause blockage of the airways (bronchioles). Bronchodilators can be used to open the airways by widening the bronchioles. More recently, the use of anti-inflammatory agents such as inhaled steroids or nonsteroidal cromolyn sodium have been prescribed. Cromolyn sodium blocks Ca^{2+} inflow into mast cells and helps prevent degranulation.

CONCEPT AND REASONING CHECKS

23.4 How can asthma be a life-threatening allergic reaction?

do not have allergies. Radioimmunoassay (RIA) techniques and radioallergosorbent tests (RAST) are used to detect the nature and quantity of IgE to specific substances known to cause allergies (see Chapter 22).

Others suggest allergy was once a survival mechanism, and atopic individuals are the modern generation of people who developed this ability to resist pathogens and have passed the trait along. They suggest, for example, that sneezing expels respiratory pathogens and contractions of the gastrointestinal tract force parasites out of the body. Hygiene also might affect allergic responses and would thus demonstrate a connection between allergies and microorganisms (MICROINQUIRY 23).

CONCEPT AND REASONING CHECKS

23.5 Summarize the factors that can contribute to the development of allergies.

Therapies Sometimes Can Control Type I Hypersensitivities

KEY CONCEPT

6. Physical avoidance and drug treatment can limit allergic attacks.

First, the best way to avoid hypersensitivities is to identify and avoid contact with the allergen. Immunotherapy may then help control allergies.

An allergy skin test can be performed to determine to what allergens a person is sensitive. The skin test can be applied to an individual's forearm or back. In the example shown in **FIGURE 23.8**, the nurse first prepares the individual's back by wiping it with alcohol and then makes a series of number marks on the back with a pen that corresponds with a series of numbered allergens that are to be tested. The test allergens include grasses; molds; common tree, plant, and weed pollens; cat and animal dander; foods; and other common substances that people encounter. Each allergen is applied by pricking the skin with a separate sterile lancet containing a drop of the allergen.

It takes about 15 to 20 minutes for possible reactions to develop. Positive reactions appear as small circles of inflammation (wheal and flare reactions), which itch and look like mosquito bites. The nurse then measures the diameter of

any wheals with a millimeter ruler. To be sensitive, a wheal must be at least 3 mm larger than the reaction to the negative control. Based on wheal size, allergen sensitivity can be assessed as a mild, moderate, or severe reaction. Once all the test allergens are tabulated, the specific allergens to which the person is sensitive can be deduced and possible treatments discussed with the allergist.

Allergy blood tests are often used for people who are not able to have skin tests or for those who have had severe allergic reactions (anaphylaxis) in the past. The most common type of blood test used is the **enzyme-linked immunosorbent assay** (**ELISA**; see Chapter 22). It measures the level of IgE in a person's blood in response to certain allergens. IgE levels four times the normal level usually indicate sensitivity to the allergen or set of allergens.

Obviously, avoiding the allergen or allergens is the most effective way of preventing an allergic reaction. However, often that might not be possible, so various medications are available that "cover up" allergic symptoms or prevent them temporarily. There are numerous over-the-counter (OTC) and prescription drugs used to relieve the symptoms of allergies or prevent the release of the chemical mediators that cause the allergies (Figure 23.4).

Antihistamines are OTC and prescription drugs, such as Allegra and Claritin, that are used systemically to relieve or prevent the symptoms of hay fever and other types of common allergies by blocking the effects of histamine. Steroids, known medically as **corticosteroids**, help reduce inflammation associated with allergies. These medicines, such as Flonase and Advair (for asthma) are sprayed or inhaled into the nose or mouth. Mast cell stabilizers, such as Cromolyn, help control allergic symptoms by inhibiting the release of chemical mediators from mast cells. These products, therefore, are only effective where they are applied (nasal passages or eyes). Leukotriene modifiers are used to treat asthma and nasal allergy symptoms. The drugs, such as Singulair, block the effects of leukotrienes that cause smooth muscle contractions.

When medications fail to control adequately allergy symptoms, an allergist may

MICROINQUIRY 23

Allergies, Dirt, and the Hygiene Hypothesis

For reasons that are not completely understood, the incidence and severity of asthma—and allergies in general—are increasing in developed nations. In fact, between 1980 and 1994, the prevalence of asthma rose 71% in the United States. Today, about 20 million Americans suffer from asthma, including more than 5 million children and adolescents. Many scientists have suggested the increase is in large part due to our overly clean lifestyle. We use disinfectants for almost everything in the home, and antibacterial products have flooded the commercial markets (Chapter 18). In other words, maintaining overly good hygiene is making us sick. We need to eat dirt! Well, not literally.

The hygiene hypothesis, first proposed in 1989, suggests that a lack of early childhood exposure to dirt, microbes, and other infectious agents can lead to immune system weakness and an increased risk of developing asthma and allergies. In the early 1900s, infants and their immune systems had to battle all sorts of infectious diseases—from typhoid fever and polio to diphtheria and tuberculosis—as well as ones that were more mundane. Such interactions and recovery "pumped up" the immune system and prepared it to act in a controlled manner. Today, most children in developed nations are exposed to far fewer pathogens, and their immune systems remain "wimpy," often unable to respond properly to non-pathogenic substances like pollen and cat dander. Their immune systems have not had the proper "basic training."

The hygiene hypothesis has been the subject of debate since it was proposed in 1989. But new research studies are providing evidence that may make the hygiene hypothesis a theory. In 2003, researchers at the National Jewish Medical and Research Center in Denver reported that mice infected with the bacterium *Mycoplasma pneumoniae* had less severe immunological responses when challenged with an allergen. However, if mice were exposed to allergens first, they developed more severe allergic responses. Also, the allergy-producing mediators were at lower levels in mice first exposed to *M. pneumoniae*. So, early "basic training" of the immune system seems to temper allergic responses.

Two other studies also bolster the hygiene hypothesis. One study used data from the Third National Health and Nutrition Examination Survey conducted from 1988 through 1994 by the Centers for Disease Control and Prevention (CDC). The survey included 33,994 American residents ranging from 1 year to older than 90. The CDC analysis concluded that humans who were seropositive for hepatitis A virus, *Toxoplasma gondii*, and herpes simplex virus type 1—that is, markers for previous microbial exposures—were at a decreased risk of developing hay fever, asthma, and other atopic diseases.

A second study used data collected from 812 European children ages 6 to 13 who either lived on farms or did not live on farms. Using another marker—an endotoxin found in dust samples from bedding—the investigators reported that children who did not live on farms were more than twice as likely to have asthma or allergies as children growing up on farms. Presumably, on farms, children are exposed to more "immune-strengthening" microbes.

So, all in all, microbial challenges to the immune system as it develops in young children can drive the system to a balanced response to allergens. If the hypothesis proves correct, eating dirt or moving to a farm is not a practical solution, nor is a return to pre-hygiene days. However, a number of environmental factors can help lower incidence of allergic disease early in life. These include the presence of a dog or other pet in the home before birth, attending day care during the first year of life, and simply allowing children to do what comes naturally—play together and get dirty.

Dirt may be good for you.

Discussion Point

Do you suffer from common allergies? Discuss whether or not your allergies fit the description in this MicroInquiry.

recommend **allergen immunotherapy** (commonly called **allergy shots**) to reduce sensitivity to allergens. It is especially helpful for individuals sensitive to seasonal allergens (allergic rhinitis), indoor allergens, and insect stings; the shots do not work for food allergies. The therapy involves a series of shots containing the allergen extract given regularly for several years. The shots over the first three to seven months contain very tiny, but increasing amounts of the allergen(s) that are just sufficient to stimulate the immune system but not to cause an allergic reaction. Once the effective level is reached, for the next three to five years or longer, the person receives a maintenance dose every month. Allergy shots

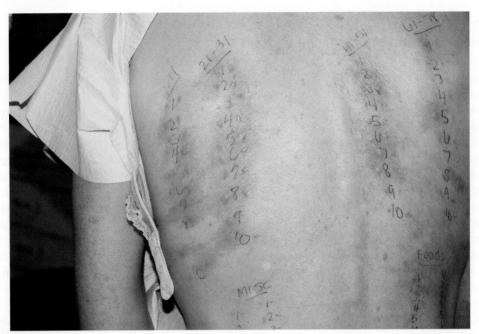

FIGURE 23.8 **An Allergy Skin Test.** The back of the individual is marked and then pricked with a large variety of test allergens. In about 20 minutes, the nurse notes if a skin wheal has developed in response to each allergen. The diameter of the wheal is then measured. »» Would you consider this patient to have an overall skin reaction that is mild, moderate, or severe?

hopefully allow the individual's body to stop producing as much IgE antibody, and, therefore, the individual will not have as severe an allergic response when exposed to the natural allergen in the environment. This process is called **desensitization**. After a course of

allergy shots, 80% to 90% of patients have less severe allergic reactions and may even have their allergies completely resolved.

CONCEPT AND REASONING CHECKS

23.6 Evaluate the medications and treatments available for allergies.

23.2 Other Types of Hypersensitivity

Immunological responses involving IgG antibodies or T cells also can lead to adverse hypersensitivity reactions.

Type II Hypersensitivity Involves Antibody-Mediated Cell Destruction

KEY CONCEPT

7. Antibodies can be the cause of cytotoxic (type II) hypersensitivities.

A **cytotoxic hypersensitivity** is a cell-damaging, humoral immune response occurring when IgG reacts with antigens on the surfaces of cells (**FIGURE 23.9**). Complement often is activated and IgM may be involved, but IgE does not participate, nor is there any degranulation of mast cells. Rather,

antibody-bound cells are subject to the formation of complement-stimulated membrane attack complexes (MACs; see Chapter 20). Alternatively, natural killer (NK) cells recognize antibody bound to cells, causing perforin and granzyme release that causes cell lysis (see Chapter 20).

A well-known example of cytotoxic hypersensitivity is the transfusion reaction arising from the mixing of incompatible blood types (**TABLE 23.1**).

A person receiving blood of the incorrect type can have a series of reactions that destroy the red blood cells and produce an inflammatory response that can lead to kidney failure or even death. For example, if a person with type A blood donates to a recipient with type O blood, the A antigens

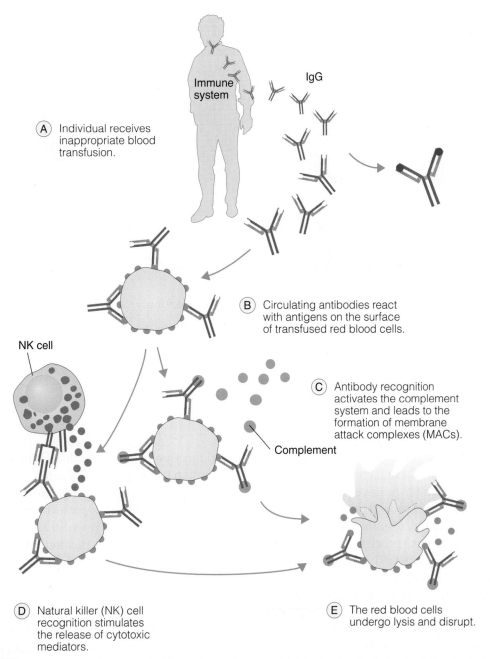

Immune system

IgG

(A) Individual receives inappropriate blood transfusion.

(B) Circulating antibodies react with antigens on the surface of transfused red blood cells.

NK cell

(C) Antibody recognition activates the complement system and leads to the formation of membrane attack complexes (MACs).

Complement

(D) Natural killer (NK) cell recognition stimulates the release of cytotoxic mediators.

(E) The red blood cells undergo lysis and disrupt.

FIGURE 23.9 **Type II Cytotoxic Hypersensitivity.** Type II hypersensitivities involve the lysis of red blood cells due to incorrect blood cell typing. »» How does complement cause cell lysis?

on the donor's erythrocytes will react with anti-A antibodies in the recipient's plasma, and the cytotoxic effect will be expressed as agglutination of donor erythrocytes and activation of complement in the recipient's circulatory system. Most blood banks cross-match the donor's erythrocytes with the recipient's serum to ensure compatibility.

MICROFOCUS 23.2 examines a microbial reason why there are different human blood types.

Another expression of cytotoxic hypersensitivity is **hemolytic disease of the newborn**, or **Rh disease**. This problem arises from the fact that erythrocytes of approximately 85% of Caucasian Americans contain a surface antigen, first described in rhesus monkeys and therefore known as the Rh antigen. Such individuals are said to be Rh-positive. The 15% who lack the antigen are considered Rh-negative.

TABLE 23.1   Some Characteristics of the Major Blood Groups

	Type A	Type B	Type AB	Type O
	Antigen A	Antigen B	Antigens A and B	Neither A nor B antigens
Red blood cells				
Serum	B antibody	A antibody	Neither A nor B antibody	A and B antibodies
Approximate incidence, U.S. caucasian population	40%	10%	5%	45%

The ability to produce the Rh antigen is a genetically inherited trait. When an Rh-negative woman marries an Rh-positive man, there is a 3 to 1 chance (or 75% probability) that the trait will be passed to the child, resulting in an Rh-positive child. During the birth process, a woman's circulatory system is exposed to her child's blood, and if the child is Rh-positive, the Rh antigens enter the woman's blood and stimulate her immune system to produce Rh antibodies (FIGURE 23.10). If a succeeding pregnancy results in another Rh-positive child, IgG antibodies (from memory cell activation) will cross the placenta (along with other resident IgG antibodies) and enter the fetal circulation. These antibodies will react with Rh antigens on the fetal erythrocytes and cause complement-mediated lysis of the cells. The fetal circulatory system rapidly releases immature erythroblasts to replace the lysed blood cells, but these cells are also destroyed. The result may be stillbirth or, in a less extreme form, a baby with jaundice.

Modern treatment for hemolytic disease of the newborn consists of the mother receiving an injection of Rh antibodies (**RhoGAM**). The injection is given within 72 hours of delivery of an Rh-positive child (no injection is necessary if the child is Rh-negative). Antibodies in the preparation interact with Rh antigens on any fetal red blood cells in the mother and remove them from the circulation, thereby preventing the cells from stimulating the woman's immune system.

Although cytotoxic hypersensitivity is generally cast in a negative role with deleterious effects on the body, the cytotoxic activity may contribute to the body's resistance to disease. For example, the antigen-antibody interaction occurring on the surface of a parasite leads to destruction of the parasite. The interaction encourages histamine release through the activity of C3a and C5a components of the complement system (see Chapter 20) and increases phagocytosis and membrane damage from the complement membrane attack complexes.

CONCEPT AND REASONING CHECKS

23.7 Summarize the factors responsible for type II hypersensitivities.

MICROFOCUS 23.2: Evolution

Why Are There Different Blood Types in the Human Population?

The human population contains four different blood types: A, B, AB, and O. Is this just a matter of divergent evolution producing different populations with different blood types? Or is there another reason for the difference?

Robert Seymour and his group at University College London believe that blood types are an evolutionary response to balancing defenses against viruses and bacteria. Remember: As shown in Table 23.1, type A people have anti-B antibodies and type B people have anti-A antibodies; Type AB have neither while type O have both antibodies in the circulation.

A photograph of blood types being stored.

Seymour's group associates blood types with virus transmission. For example, when measles viruses break out of infected cells, they carry on their envelope the chemical group (antigen) identifying the blood type of that individual. Therefore, measles viruses from people with blood type A or B would be neutralized by type O blood because O blood has antibodies to antigens A and B. In reverse, measles viruses emerging from someone with blood type O carries neither the A nor B antigen, so the viruses would not be neutralized by the blood of people with blood types A, B, or AB. Therefore, people with blood type O are better prepared to defend against viruses coming from people with other blood types—and they are better at transmitting viruses to those blood types.

To make matters more complex, Alexandra Rowe and her colleagues at the University of Edinburgh have reported that, at least in Africa, children suffering from mild forms of malaria are three times more likely to be blood type O than those with severe malaria. The malaria parasite (see Chapter 23) needs to attach to specific sugars on the surface of red blood cells, and type O blood, compared to A and B, has far fewer of these sugars.

If this is all true, then why isn't type O blood universal? Apparently, because of other pathogens. Seymour's group suggests that because there would be more type O individuals, probability says they would more likely be attacked by other pathogens, especially bacterial. That is supported by Rowe's group who says that type O individuals are more susceptible to other diseases, such as stomach ulcers and cholera (Chapter 21). In fact, in Latin America where the majority of people are type O, people suffering from severe cholera are eight times more likely to be type O than those suffering a milder form of the disease.

If all this is validated with much more work, here is yet another way that microbes and viruses have affected the human species. Once again, they do rule the world!

Type III Hypersensitivity Is Caused by Antigen-Antibody Aggregates

KEY CONCEPT

8. Immune complexes are responsible for type III hypersensitivities.

Immune complex hypersensitivity develops when IgG antibodies combine with antigens, forming soluble immunocomplexes that accumulate in blood vessels or on tissue surfaces (**FIGURE 23.11**). These immunocomplexes are deposited on the cell surface of several organs and activate complement. The C3a and C5a components increase vascular permeability and exert a chemotactic effect on phagocytic neutrophils, drawing them to the target site. The neutrophils release lysosomal enzymes, which cause tissue damage. Local inflammation is common, and fibrin clots may complicate the problem. As with type II hypersensitivities, tissue damage results from the formation of MACs or cytotoxic mediators released by NK cells.

Serum sickness is a common manifestation of immune complex hypersensitivity. It develops

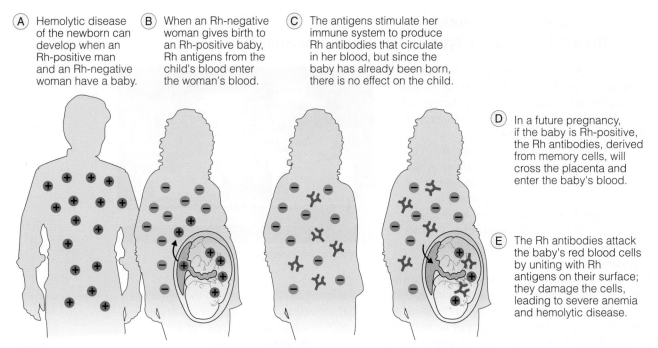

(A) Hemolytic disease of the newborn can develop when an Rh-positive man and an Rh-negative woman have a baby.

(B) When an Rh-negative woman gives birth to an Rh-positive baby, Rh antigens from the child's blood enter the woman's blood.

(C) The antigens stimulate her immune system to produce Rh antibodies that circulate in her blood, but since the baby has already been born, there is no effect on the child.

(D) In a future pregnancy, if the baby is Rh-positive, the Rh antibodies, derived from memory cells, will cross the placenta and enter the baby's blood.

(E) The Rh antibodies attack the baby's red blood cells by uniting with Rh antigens on their surface; they damage the cells, leading to severe anemia and hemolytic disease.

FIGURE 23.10 **Hemolytic Disease of the Newborn.** If a mother and a second child are Rh incompatible, maternal antibodies would cause the lysis and removal of the fetal red blood cells. »» Why is this type II hypersensitivity also called erythroblastosis fetalis?

when the immune system produces IgG against residual proteins in serum preparations. The IgG then reacts with the proteins, and immune complexes gather in the kidney over a period of days. The problem is compounded when IgE, also from the immune system, attaches to mast cells and basophils, thereby inducing a type I anaphylactic hypersensitivity. The sum total of these events is kidney damage, along with hives and swelling in the face, neck, and joints.

Another form of immune complex hypersensitivity is the **Arthus phenomenon**, named for Nicolas Maurice Arthus, the French physiologist who described it in 1903. In this process, excessively large amounts of IgG form complexes with antigens, either in the blood vessels or near the site of antigen entry into the body. Antigens in dust from moldy hay and in dried pigeon feces are known to cause this phenomenon. The names "farmer's lung" and "pigeon fancier's disease" are applied to the conditions, respectively. Thromboses in the blood vessels may lead to oxygen starvation and cell death.

Several microbial diseases also are complicated by immune complex formation. For exam-

Thromboses:
The formation or presence of blood clots that partially or completely block one or more arteries.

ple, rheumatic fever and glomerulonephritis that follow streptococcal diseases (see Chapters 10 and 13) appear to be consequences of immune complex formation in the heart and kidneys, respectively. In these cases, the deposit of complexes relates to common antigens in streptococci and the tissues. Other immune complex complications and inflammatory injury are associated with autoimmune disorders such as rheumatoid arthritis and systemic lupus erythematosus (see below).

CONCEPT AND REASONING CHECKS

23.8 Summarize the factors responsible for type III hypersensitivities.

Type IV Hypersensitivity Is Mediated by Antigen-Specific T Cells

KEY CONCEPT

9. Type IV hypersensitivity is an exaggeration of a delayed cell-mediated immune response.

Type IV hypersensitivity or "delayed-type hypersensitivity" is an exaggeration of the process of cell-mediated immunity, discussed in Chapter 21. The adjective "cell mediated" is used

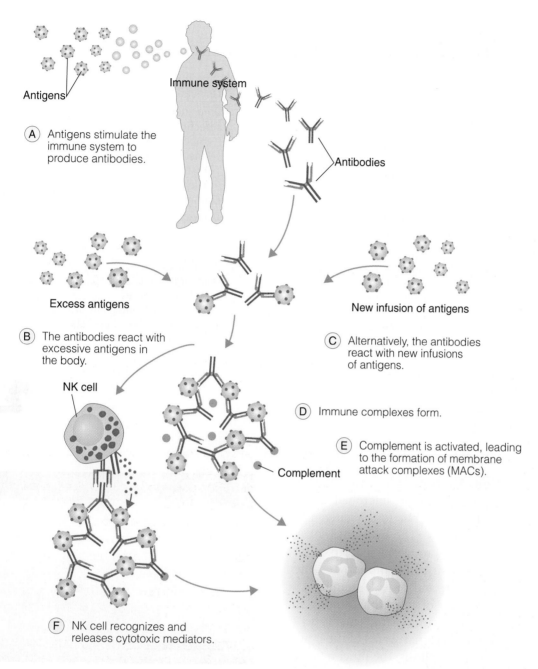

A Antigens stimulate the immune system to produce antibodies.

Antigens

Immune system

Antibodies

Excess antigens

New infusion of antigens

B The antibodies react with excessive antigens in the body.

C Alternatively, the antibodies react with new infusions of antigens.

NK cell

D Immune complexes form.

E Complement is activated, leading to the formation of membrane attack complexes (MACs).

Complement

F NK cell recognizes and releases cytotoxic mediators.

G Cell lysis causes tissue damage.

FIGURE 23.11 **Type III Immune Complex Hypersensitivity.** Type III hypersensitivities occur when antibodies combine with antigens to form aggregates (immune complexes) that accumulate in blood vessels or on tissue surfaces. »» What types of physical problems can develop from immune complex formation?

because T_H1 cells interact with antigen leading to cytokine secretion. The hypersensitivity is a delayed reaction because maximal effect is not seen until 24 to 72 hours after exposure to soluble antigen. It is characterized by a thickening and drying of the skin tissue, a process called **induration**,

and a surrounding zone of **erythema** (redness). **TABLE 23.2** compares this delayed hypersensitivity with type I immediate hypersensitivity.

Two major forms of type IV hypersensitivity are recognized: infection allergy and contact dermatitis.

TABLE 23.2   Immediate and Delayed Hypersensitivities Compared

	Type I Immediate Hypersensitivity	Type IV Delayed Hypersensitivity

Clinical state:	Hay fever
	Asthma
	Urticaria
	Allergic skin conditions
	Anaphylactic shock
		Drug allergies
		Infectious allergies
		    Tuberculosis
		    Rheumatic fever
		    Histoplasmosis
		    Trichinosis
		Contact dermatitis

Onset:	Immediate	Delayed
Duration:	Short: hours	Prolonged: days or longer
Allergens:	Pollen Molds House dust Danders Drugs Antibiotics Soluble proteins and carbohydrates Foods	Drugs Antibiotics Microorganisms: bacteria, viruses, fungi, animal parasites Poison ivy/oak and plant oils Plastics and other chemicals Fabrics, furs Cosmetics
Passive transfer of sensitivity:	With serum	With cells or cell fractions of lymphoid series

Infection Allergy. When the immune system responds to certain microbial agents, sensitized T_H1 cells migrate to the antigen site and release cytokines. The cytokines attract phagocytes and encourage phagocytosis (see Chapter 20). Sensitized lymphocytes then remain in the tissue and provide immunity to successive episodes of infection. Among the microbial agents stimulating this type of immunity are the bacterial agents of tuberculosis, leprosy, and brucellosis; the fungi involved in blastomycosis, histoplasmosis, and candidiasis; the viruses of smallpox and mumps; and the chlamydiae of lymphogranuloma.

The classic delayed hypersensitivity reaction is demonstrated by injecting an extract of the microbial agent into the skin of a sensitized individual. As the immune response takes place, the area develops induration and erythema, and fibrin is deposited by activation of the clotting system. An important application of infection allergy is the **tuberculin test** for tuberculosis

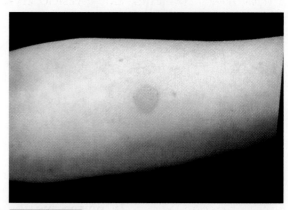

FIGURE 23.12 **A Positive Tuberculin Test.** The raised induration and zone of inflammation indicate that antigens have reacted with T cells, probably sensitized by a previous exposure to tubercle bacilli. »» Does this test mean the person has tuberculosis? Explain.

(see Chapter 10; **FIGURE 23.12**). A purified protein derivative (PPD) of *Mycobacterium tuberculosis* is applied to the skin by intradermal injection (the Mantoux test) or multiple punctures (tines). Individuals sensitized by a previous exposure to

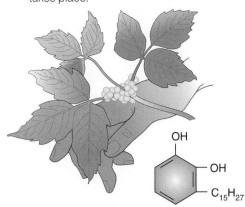

(A) When a sensitized person touches poison ivy, a substance called urushiol stimulates T lymphocytes in the skin; within 48 hours, a type IV reaction takes place.

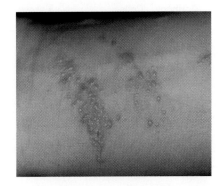

(B) The reaction is characterized by pinhead-sized blisters that usually occur in a straight row.

OH

OH

C₁₅H₂₇

The chemical structure of urushiol

FIGURE 23.13 **The Poison Ivy Rash.** When sensitized skin makes contact with substances like urushiol in poison ivy, the chemical combines with tissue proteins to form allergenic compounds. The result is a rash on the skin surface. »» Why is this form of hypersensitivity called delayed hypersensitivity?

Mycobacterium species develop a vesicle, erythema, and induration.

Skin tests based on infection allergy are available for many diseases. Immunologists caution, however, that a positive result does not constitute a final diagnosis. Rather, sensitivity may have developed from a subclinical exposure to the organisms, from clinical disease years before, or from a former screening test in which the test antigens elicited a T-cell response.

Contact Dermatitis. After exposure to a variety of antigens, a delayed-type hypersensitivity reaction occurs. In poison ivy, the allergen has been identified as urushiol, a low-molecular-weight chemical on the surface of the leaf. In the body, urushiol attaches to tissue proteins to form allergenic compounds. Within 48 hours, a rash appears, which consists of very itchy pinhead-sized blisters usually occurring in a straight row (**FIGURE 23.13**).

The course of contact dermatitis is typical of the type IV reaction. Repeated exposures cause a drying of the skin, with erythema and scaling.

Examples of other type IV responses are seen on the scalp when allergenic shampoo is used, on the hands when contact is made with detergent enzymes, and on the wrists when an allergy to costume jewelry exists. Contact dermatitis also may occur on the face where contact is made with cosmetics, on areas of the skin where chemicals in permanent-press fabrics have accumulated, and on the feet when there is sensitivity to dyes in leather shoes. Factory workers exposed to photographic materials, hair dyes, or sewing materials may experience allergies. The list of possibilities is endless.

Applying a sample of the suspected substance to the skin (a skin patch test) and leaving it in place for 24 to 48 hours will help pinpoint the source of the allergy. Relief generally consists of avoiding the inciting agents. **TABLE 23.3** summarizes the four types of hypersensitivities.

CONCEPT AND REASONING CHECKS

23.9 Summarize the factors responsible for type IV hypersensitivities.

TABLE
23.3   Overview of the Hypersensitivity Reactions

Hypersensitivity Type	Origin of Hypersensitivity	Antibody Involved	Cells Involved	Mediators Involved	Transfer of Sensitivity	Evidence of Hypersensitivity	Skin Reaction	Examples
Type I IgE-mediated	B lymphocytes	IgE	Mast cells Basophils	Histamine Serotonin Leukotrienes Prostaglandins	By serum	30 minutes or less	Urticaria	Systemic anaphylaxis Hay fever Asthma
Type II Cytotoxic	B lymphocytes	IgG	RBC WBC	Complement	By serum	5–8 hours	Usually none	Transfusion reactions Hemolytic disease of newborns
Type III Immune complex	B lymphocytes	IgG	Host tissue cells	Complement	By serum	2–8 hours	Usually none	Serum sickness Arthus phenomenon SLE Rheumatic fever Rheumatoid arthritis
Type IV Cellular	T lymphocytes	None	Host tissue cells	Cytokines	By lymphoid cells	1–3 days	Induration Tissue death	Contact dermatitis Infection allergy

23.3   Autoimmune Disorders and Transplantation

In Chapter 21, we mentioned that one of the properties of the immune system was tolerance of "self;" that is, one's own cells and molecules with their antigenic determinants do not stimulate an immune response. Should self-tolerance break down, an **autoimmune disorder** may occur.

An Autoimmune Disorder Is a Failure to Distinguish Self from Non-self

KEY CONCEPT

10. Autoimmune disorders generate an immune attack on self cells and tissues.

The idea that an individual's immune system is incapable of recognizing "self" antigens is not new. At the beginning of the 20th century, the German immunologist Paul Ehrlich suggested that the human body has an innate aversion to immunological self-destruction, what he called *horror autotoxicus* (terror of self-toxicity). Any autoimmune response therefore must be abnormal and due to human disease.

Today, we know autoimmunity is critical to the normal development and functioning of the human immune system—and essential to the development of immunological tolerance. Each individual's immune system must have a mechanism to differentiate "self" from "non-self," so it will only react to foreign antigens, such as pathogens. Such tolerance is thought to develop in the following ways (**FIGURE 23.14**):

- **Clonal Deletion Theory**. This theory says that self-reactive lymphoid cells are destroyed during the development of the immune system in an individual.
- **Clonal Anergy Theory**. This theory proposes that self-reactive T cells become inactivated in the normal individual and cannot differentiate into effector cells when presented with antigen.
- **Regulatory T Cell Theory**. According to this theory, specific regulatory T cells function to suppress exaggerated immune responses.

In fact, all of these theories may be correct and, as such, several mechanisms could actively contribute to the development of immunological tolerance.

Up to 8% of the American population (23.5 million individuals) suffers from an autoimmune disorder, and far more women than men are affected. So, what causes the loss of tolerance? Part of the answer lies in human heredity. Various gene mutations have been identified that affect cell division and apoptosis—and therefore clonal deletion, anergy, and regulatory T cell activity. There also are several other ways autoimmune disorders can be triggered.

- **Access to privileged sites.** Certain parts of the body (e.g., brain, eye) are "hidden" from the immune system (Figure 23.14).

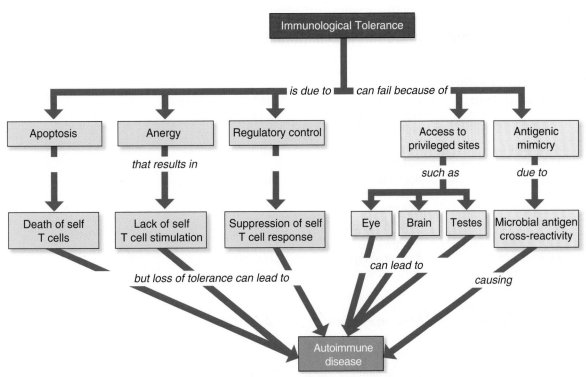

FIGURE 23.14 Mechanisms of Tolerance and Failure. Immunological tolerance involves killing self T cells, or blocking stimulation or suppressing a self T cell response. Autoimmune diseases result from a loss of tolerance, T cell access to privileged sites, or antigenic mimicry through cross reaction between and exogenous and self-antigen. »» Why might T cells have gained access to privileged sites?

If such sites become "accessible" to the immune system through injury, an immune response will be mounted.

- **Antigenic mimicry.** A foreign substance or microbe may enter the body that closely resembles (mimics) a similar body substance. The immune system then sees the "self" substance as foreign and attacks it. Rheumatoid arthritis and type I diabetes may be examples.

As a result of this loss of tolerance, the immune system attempts to mount an immune response against its own cells and tissues. Although many such disorders cause fever, specific symptoms depend on the disorder and the part of the body affected. The resulting inflammation and tissue damage can cause pain, deformed joints, jaundice, itching, breathing difficulties, and even death. To date, some 100 clinically distinct autoimmune disorders have been identified (**TABLE 23.4**).

Among the more notable organ-specific disorders is **myasthenia gravis**. In this disorder, **autoantibodies** react with acetylcholine recep-

tors on membranes covering the muscle fibers. This interaction reduces nerve impulse transfer to the fibers and results in a loss of muscle activity, manifested as weakness and fatigue.

In **type I diabetes**, which usually arises in children and teenagers, the islet cells of the pancreas are destroyed, resulting in a lack of insulin production. Without insulin, the blood glucose level skyrockets and cells are starved for energy; blindness, kidney failure, or other complications can occur. Because type I diabetes typically appears after an infection, the immune response to insulin-producing cells may result from the immune system's response to an earlier viral infection. More than 300 million adults and 430,000 children worldwide have type 1 diabetes.

A systemic autoimmune disorder primarily in women is **systemic lupus erythematosus** (**SLE**), also known as "lupus." With SLE, plasma cells produce IgG upon stimulation by nuclear components of disintegrating white blood cells. Immune complexes accumulate in the skin and body organs, and complement is activated. The patient experiences a butterfly rash, a facial skin

TABLE

23.4   A Summary of Some Autoimmune Disorders

Disorder	Target	Effect	Female:Male Ratio
Autoantibody Mechanism			
Thrombocytopenia	Blood platelet extracellular matrix protein	Red blood cell lysis and anemia	3:1
Goodpasture syndrome	Kidney extracellular matrix collagen	Glomerulonephritis	1:6
Myasthenia gravis	Neuromuscular junction acetylcholine receptor	Muscle weakness	3:1
Graves disease	Thyroid stimulating hormone	Hyperthyroidism	5–10:1
Systemic lupus erythematosus	Nuclear antigens in skin, kidneys, joints	Skin rash, glomerulonephritis, arthritis	9:1
Self T Cell Mechanism			
Multiple sclerosis	Nerve myelin basic protein	Partial or complete paralysis	Children—3:1 Adults over 50—1:1
Self T Cell/Autoantibody Mechanism			
Hashimoto disease	Thyroglobulin	Destruction of thyroid gland	10–20:1
Type 1 (insulin dependent) diabetes	Pancreatic islet cells	Failure to produce insulin	1:1
Rheumatoid arthritis	Synovium of joints	Joint inflammation Loss of movement	3–5:1

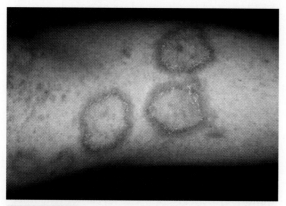

FIGURE 23.15 Skin Lesions of Systemic Lupus Erythematosus. This autoimmune disorder can affect the skin and other body organs. »» What causes the inflammation seen with lupus?

condition across the nose and cheeks, and body rashes (**FIGURE 23.15**). Lesions also form in the heart, kidneys, and blood vessels. Another systemic disorder is **rheumatoid arthritis** (**RA**). Unlike osteoarthritis, which results from wear and tear on joints, RA is an inflammatory condition resulting in the accumulation of immune complexes in the joints. Some researchers suspect rheumatoid arthritis is triggered through infection by a virus or bacterial pathogen.

Treatment of most autoimmune disorders requires suppressing the immune system, which means interrupting the system's ability to fight infectious disease. Immunosuppressants include corticosteroids, such as prednisone. Some disorders resolve as spontaneously as they appear while others become chronic, life-long disorders. Thus, the prognosis depends on the particular disorder.

CONCEPT AND REASONING CHECKS

23.10   Identify the common attributes to all autoimmune disorders.

Transplantation of Tissues or Organs Is an Important Medical Therapy

KEY CONCEPT

11.   Immune-mediated rejection must be overcome for successful tissue transplants or grafts.

Modern techniques for the transplantation of tissues and organs trace their origins to Jacques

MICROFOCUS 23.3
Acceptance

Ordinarily, a woman's body will reject a foreign organ, such as a kidney or heart, but it will accept the fetus growing within her womb. This acceptance exists even though half of the fetus' genetic information has come from a "foreigner"—namely, the father. Has her immune system failed?

Apparently not. It seems that the sperm carries an antigenic signal that induces the woman's immune system to produce a series of so-called blocking antibodies. The blocking antibodies form a type of protective screen that protects the fetus and prevents its antigens from stimulating the production of rejection antibodies by the mother. In addition, regulatory T cells (Treg), which congregate in the womb, dampen the activity of the lymphocytes, helping protect the fetus from rejection.

Sometimes, however, a rejection in the form of a miscarriage occurs. Research indicates the level of blocking antibodies and Treg cells in some pregnant women is too low to protect the fetus. Ironically, the low immune resistance may be because the father's tissue is very similar to the mother's. In such a case, the sperm's antigens elicit a weak antibody and Treg response, too low to protect the fetus. Perhaps the selection of a mate with a similar MHC should be avoided (see MicroFocus 21.2).

Physicians are now attempting to boost the level of blocking antibodies as a way of preventing miscarriage. They inject white blood cells from the father into the mother, thereby stimulating her immune system to produce antibodies to the cells. These antibodies exhibit the blocking effect.

In other experiments, injections of blocking antibodies are administered to augment the woman's normal supply. Both approaches have been successful in trial experiments, and continuing research has given cause for optimism that cases of fetal rejection can give way to acceptance and a full-term delivery.

Reverdin, who in 1870 successfully grafted bits of skin to wounded tissues. Enthusiasm for the technique rose after his reports were published, but it waned when doctors found that most transplants were rapidly rejected by the body. Then, in 1954, a kidney was transplanted between identical twins, and again, interest grew. The graft survived for several years, until ultimately it was destroyed by a recurrence of the recipient's original kidney disease. During that time, attempts to transplant kidneys between unrelated individuals were less successful.

Transplantation technology improved considerably during the next few decades, and today, four types of transplantations, or grafts, are recognized, depending upon the genetic relationship between donor and recipient. A graft taken from one part of the body and transplanted to another part of the same body is called an **autograft**. This graft is never rejected because it is the person's own tissue. A tissue taken from an identical twin and grafted to the other twin is an **isograft**. This, too, is not rejected because the genetic constitutions of identical twins are the same.

Rejection mechanisms become more vigorous as the genetic constitutions of donor and recipient become more varied (MicroFocus 23.3). For instance, grafts between brothers and sisters, or between fraternal twins, may lead to only mild rejection because of their genetic similarity. Grafts between cousins may be rejected more rapidly, and as the relationship becomes more distant, the vigor of rejection increases proportionally. **Allografts**, or grafts between genetically different members of the same species, such as two humans, have variable degrees of success. Most transplants are allografts. **Xenografts**, or grafts between members of different species, such as a

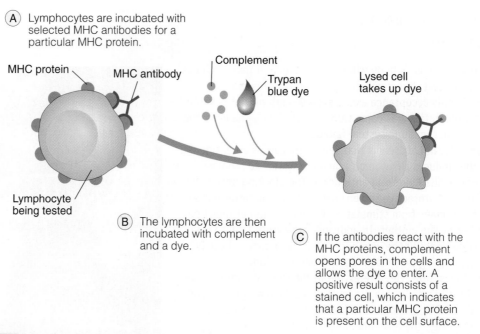

(A) Lymphocytes are incubated with selected MHC antibodies for a particular MHC protein.

MHC protein

MHC antibody

Complement

Trypan blue dye

Lysed cell takes up dye

Lymphocyte being tested

(B) The lymphocytes are then incubated with complement and a dye.

(C) If the antibodies react with the MHC proteins, complement opens pores in the cells and allows the dye to enter. A positive result consists of a stained cell, which indicates that a particular MHC protein is present on the cell surface.

FIGURE 23.16 Tissue Typing for MHC Proteins. Dye staining and antibodies can be used to identify potentially successful tissue transplants. »» What color will lymphocytes appear if an antibody triggers complement activity?

pig and a human, are rarely successful without a high degree of immunosuppression.

Mechanism of Allograft Rejection. With an allograft, the transplanted donor tissue or organ is rejected by the recipient if the immune system of the recipient interprets the transplant as "non-self." This recognition of non-self is stimulated by the recipient's immune system recognizing foreign **major histocompatibility complex** (**MHC**) proteins on the surface of the transplant (see Chapter 21). These proteins, typically referred to as **human leukocyte antigens** (**HLAs**) in the transplantation field, confer uniqueness to each individual. In the case of an allograft, the HLAs of donor and recipient can be quite diverse.

The mechanism of rejection is defined by the speed of the rejection mechanism. **Hyperacute rejection** can occur within minutes after transplantation as preformed antibodies attack the foreign HLA transplant. The antibodies activate complement and immediately block and destroy blood vessel in the donor transplant. In **acute rejection**, over a period of 10 to 30 days, T_H1 cells respond to the foreign HLA transplant by activating cytotoxic T cells (CTLs) that search out and destroy transplant cell tissue. T_H2 cells activate B cells and antibody production, which, as in hyperacute rejection, activate complement and lead to transplant destruction.

The closer the match between donor and recipient MHC proteins, the greater the chance of a successful transplant. The matching of donor and recipient is performed by tissue typing (**FIGURE 23.16**). In this procedure, the laboratory uses standardized MHC antibodies for particular MHC proteins. Lymphocytes from the donor are incubated with a selected type of MHC antibodies. Complement and a dye, such as trypan blue, are then added. If the selected MHC antibodies react with the MHC proteins of the lymphocytes, the cell becomes permeable and dye enters the cells (living cells do not normally take up the dye). Similar tests then are performed with recipient lymphocytes to determine which MHC proteins are present and how closely the tissues match one another. The blood types of donor and recipient must also be identical.

Even with tissue matching, **chronic rejection** can occur over a period of months to years as T cells and antibodies respond to very minor HLA differences. The result can be tissue lysis and eventual transplant rejection.

A rejection mechanism of a completely different sort is sometimes observed in bone marrow transplants. In this case, the transplanted marrow may contain immune system cells that form immune products against the host after the host's immune system has been suppressed during trans-

TABLE		
23.5 The Types of Drugs Used to Limit Transplant Rejection		
Type	Drug	Comments
Corticosteroids		
Anti-inflammatory drugs suppress T cell activity	Prednisone	Given by intravenous (IV) at time of transplantation at high dosage; then monthly at reduced dosage indefinitely
Immunoglobulins		
Antibodies lower circulating lymphocyte populations	Antilymphocyte Antithymocyte	Given by IV or used with other immunosuppressants so that these latter drugs can be started later or at reduced dosage
Mitotic inhibitors		
Drugs suppress cell division of lymphocytes	Azathioprine Methotrexate	Given by IV or orally at time of transplantation at high dosage; then monthly at reduced dosage indefinitely
Fungal metabolites		
Drugs inhibit T cell activity	Cyclosporine	Given first by IV and later orally; used with prednisone or azathioprine
Monoclonal antibodies		
Target and suppress specific immune cells	Basiliximab Infliximab Muromonab	Given by IV at time of transplantation or if a rejection event is initiated

plant therapy. Essentially, the graft is rejecting the host; it is a "reverse rejection." This phenomenon is called a **graft-versus-host reaction (GVHR)**. It sometimes can lead to fatal consequences in the host body.

CONCEPT AND REASONING CHECKS

23.11 Assess the need for tissue typing before a transplant is performed.

Immunosuppressive Agents Prevent Allograft Rejection

KEY CONCEPT

12. Immunosuppressive drugs have several drawbacks.

To inhibit rejection, it is necessary to suppress activity of the immune system. Because there is no clinical procedure that can be used to eliminate the rejection of allografts, all transplant patients require daily treatment with **immunosuppressive agents**, which are drugs that in some way suppress or inhibit T cells and antibody production. Because these drugs are injected and spread systemically through the body, they all have drawbacks that are related either to their toxicity and side effects in the body or to a failure to produce sufficient immune system suppression. In addition, immunosuppression can leave the body open to tumor (cancer) formation and opportunistic infections, so often transplant patients also need to be on antibiotics to minimize the latter risk. **TABLE 23.5** lists some immunosuppressive agents used to prevent transplant rejection.

CONCEPT AND REASONING CHECKS

23.12 Explain why immunosuppression leaves a transplant patient susceptible to opportunistic infections.

23.4 Immunodeficiency Disorders

The spectrum of immunodeficiency disorders ranges from relatively minor deficiency states to major abnormalities that are life-threatening. The latter may be serious in populations where malnutrition and frequent contact with pathogenic organisms are common. Diagnostic techniques for determining immune deficiency diseases include measurements of antibody types, detection tests for B cell function, and enumeration of T cells. Assays of complement activity and phagocytosis are also useful in diagnosis.

Immunodeficiencies Can Involve Any Aspect of the Immune System

KEY CONCEPT

13. Immunodeficiency disorders lead to recurrent and often severe, longer lasting infections.

If there is an immune system malfunction or developmental abnormality in immune system function, an **immunodeficiency disorder** may result. Such disorders hamper the immune system's ability to provide a strong response to any viral or microbial infection. Immunodeficiency disorders are identified by the part of the innate or acquired immune system affected (see Chapters 20 and 21). Thus, the disorders may involve B cells and antibodies, T cells, both B and T cells, phagocytes, or complement proteins. The affected immune component may be absent, present in reduced numbers or amounts, or functioning abnormally.

The immunodeficiency may be congenital (**primary immunodeficiency**) as a result of a genetic abnormality. Such disorders present from birth are rare, although more than 70 different congenital disorders have been documented.

An immunodeficiency may be acquired later in life (**secondary immunodeficiency**). The most common acquired immunodeficiency is AIDS.

Primary Immunodeficiencies. People with primary immunodeficiencies have multiple infections, especially recurring respiratory infections. Bacterial infections are common, often severe, and lead to complications. Several such congenital disorders are described next.

An example of a humoral immunodeficiency is **X-linked (Bruton) agammaglobulinemia**, first described by Ogden C. Bruton in 1952. In this disease, B cells fail to develop from pre-B cells in the bone marrow. The patient's lymphoid tissues lack mature B cells and plasma cells, all five classes of antibodies are either low in level or absent, and antibody responses to infectious disease are undetectable. Infections from staphylococci, pneumococci, and streptococci are common between the ages of six months and two years. As the name suggests, the disorder is a sex-linked inherited trait, much more frequently observed in males than in females. Artificially acquired passive immunity is used to treat infectious disease in these patients.

DiGeorge syndrome is a cellular immunodeficiency in which T cells fail to develop. The deficiency is linked to failure of the thymus gland to mature in the embryo. Cellular immunity is defective in such individuals, and susceptibility is high to many fungal and protozoal diseases and certain viral diseases. Grafts of thymus tissue correct the disorder.

Perhaps the most severe and life-threatening primary immunodeficiency is one combining both humoral and cell-mediated immunity disorders. **Severe combined immunodeficiency disease (SCID)** involves lymph nodes depleted in both B and T cells. Without a population of B and T cells, all acquired immune functions are suppressed. One form of the disorder is caused by an enzyme deficiency, and individuals with the disorder have to be kept in strict isolation to prevent infections and diseases from occurring. If such measures are not taken, most children die before the age of two. Gene therapy is now being used to correct the gene defect (**FIGURE 23.17**).

Other immune deficiency diseases are linked to leukocyte malfunction. In the rare disease known as **Chédiak-Higashi syndrome**, the phagocytes fail to kill microorganisms because of the inability of lysosomes to release their contents into a phagosome (see Chapter 20). In **chronic granulomatous disease**, the phagocytes fail to produce hydrogen peroxide and other substances needed to kill ingested bacterial cells and fungi. Antibiotics and interferon are given to reduce the number and severity of infections.

Deficiencies in the complement system may be life-threatening. As noted in Chapter 20, complement is a series of proteins, any of which the body may fail to produce. For reasons not currently understood, many patients with complement component deficiencies suffer systemic lupus erythematosus (SLE) or an SLE-like syndrome. Meningococcal and pneumococcal diseases are often observed in patients who lack C3, probably as a result of poor opsonization.

Secondary Immunodeficiencies. Autoimmune disorders acquired later in life are more common than primary immunodeficiency disorders. Cases often result from immunosuppressive treatments such as chemotherapy and radiation therapy that reduce or eliminate populations of white blood cells. Often such disorders are the result of a viral infection. Certainly acquired immunodeficiency syndrome (AIDS),

Gene therapy:
A genetic treatment to insert a normal or genetically altered gene into cells in order to replace or make up for the nonfunctional or missing gene.

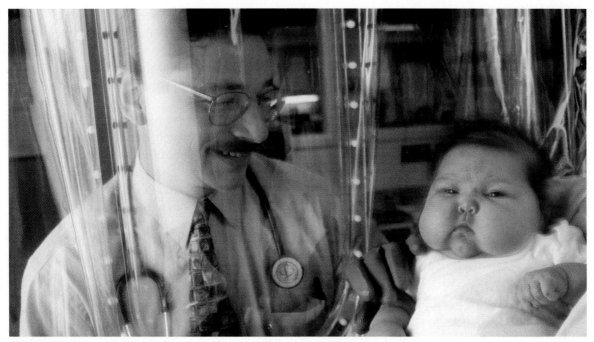

FIGURE 23.17 SCID and Gene Therapy. A five-month-old baby suffering from severe combined immunodeficiency disease (SCID) is protected in a sterile tent to prevent infection. He is receiving gene therapy to insert a gene for this enzyme into stem cells from his bone marrow. The stem cells are then transplanted back into the baby. With this enzyme restored, these stem cells may produce normal immune system blood cells. »» Why are stem cells in the bone marrow being used?

caused by the human immunodeficiency virus (HIV), is the most recognizable and is the subject of the last section of this chapter.

CONCEPT AND REASONING CHECKS

23.13 Compare and contrast the symptoms of the primary discussed immunodeficiency disorders.

The Human Immunodeficiency Virus (HIV) Is Responsible for HIV Disease and AIDS

KEY CONCEPT

14. Acquired immunodeficiency syndrome (AIDS) results from immune system dysfunction.

In 1981, physicians in the United States first reported a syndrome involving the development of certain opportunistic infections, including fungal pneumonia and an unusual type of skin cancer called Kaposi sarcoma. These illnesses, along with sudden weight loss, swollen lymph nodes, and immune system deficiencies represented the signs and symptoms for a disease that became known as **acquired immunodeficiency syndrome** (**AIDS**).

By 1983, the most plausible factor responsible for AIDS was a virus. In that year, Luc Montagnier and his French group isolated the infectious agent from a patient and Robert Gallo's group in the United States discovered how to grow the virus in culture and published convincing evidence that HIV causes AIDS. By 1986, the virus was given its current name of **human immunodeficiency virus type 1** (**HIV-1**).

Structure of HIV. HIV is a member of the Retroviridae. The virion contains two copies of a single-stranded (+ strand) RNA (**FIGURE 23.18**). Unique to these RNA viruses, the genome is packaged with several enzymes, one of which is called **reverse transcriptase** that is needed to copy single-stranded RNA into double-stranded DNA (see Chapter 14). This reversal of the usual mode of genetic information transfer (transcription) gives the virus its name, retrovirus (*retro* = "backward") and the enzyme its name, reverse transcriptase.

The HIV genome is surrounded by a cone-shaped icosahedral capsid. Between the capsid and envelope is a **matrix protein** that facilitates viral penetration. Like other enveloped viruses, the HIV envelope contains protein **spikes** for attachment and entry into the host cell. One spike

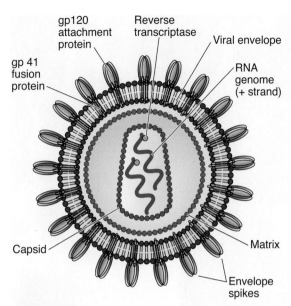

FIGURE 23.18 A Diagram of the Human Immunodeficiency Virus (HIV). The virus consists of two molecules of RNA and molecules of reverse transcriptase. A protein capsid surrounds the genome and an envelope with spikes of protein lies outside the capsid and matrix. »» What are the unique features about the structure of HIV?

protein, gp120, is used to attach to an appropriate host cell (see below) while the other, gp41, promotes fusion of the viral envelope with the host plasma membrane.

In 1986, a second type of HIV, called HIV-2, was isolated from AIDS patients in West Africa. Comparative studies indicate both types have the same mode of transmission and are associated with similar infections. However, HIV-2 appears to develop more slowly so people infected with HIV-2 are less infectious early in the course of infection. There are few reported cases of HIV-2 in the United States. It is important to note that for both types, AIDS is the end result of an HIV infection. MicroFocus 23.4 discusses the origins of HIV.

Replication of HIV. The viral replicative cycle begins when gp120 proteins on the HIV spikes contact CD4 receptors present on the plasma membrane of a host cell (**FIGURE 23.19**). If the infection is transmitted through sexual intercourse, the initial cells encountered are the dendritic cells (DCs) and CD4 T cells associated with the mucosa of the genital tract (see Micro-Inquiry 21). At least 50% of the body's CD4 T cells are located in the mucosal lining and

within two to three weeks after infection, HIV has destroyed half of the immune cell population. If transmission occurs via the blood, helper T cells, macrophages, and DCs are among the first cells infected, primarily in the lymph nodes.

Following entry of the capsid into the host cell cytoplasm, which is facilitated by the gp41 spike proteins, uncoating occurs and the reverse transcriptase synthesizes a molecule of DNA from the viral single-stranded RNA. The reverse transcriptase makes many replication errors, which occurs to such a degree that late into an infection, the virus is genetically different from the one that initiated the infection (see treatment below).

Within 72 hours of infection of a host cell, the DNA molecule has integrated into one of the host chromosomes as a **provirus.** If the infected cell is actively dividing, the provirus initiates DNA transcription and translation, resulting in the biosynthesis and maturation of new virions. The new virions then "bud" from the host cell to infect other cells. If the infected cell is not dividing, the prophage remains dormant. In either situation, the infected individual now has a "primary HIV infection."

Although the rate of clinical disease progression varies among individuals, realize the following stages are a continuum.

Stage I: Primary HIV Infection. Many people are asymptomatic when they first become infected with HIV. However, about 70% experience an acute HIV infection, which produces a flu-like illness within a month or two of HIV exposure. The symptoms of fever, chills, rashes, and night sweats usually last no longer than a few days.

During this period, the immune system is at war with HIV as the virus infects lymph nodes ①, replicates, and releases virions Ⓐ into the bloodstream (**FIGURE 23.20**).

The immune system starts responding to the infection by producing antibodies and T cells directed against HIV. Such a process in the body is referred to as **seroconversion,** meaning HIV antibodies can be detected in the blood. This usually has occurred 1–3 months post-infection Ⓑ.

Stage II: Asymptomatic Stage. Without treatment, the infected individual often remains free of major diseases and is quite healthy for

MICROFOCUS 23.5: Tools
Checking for HIV Infection—At Home

If you thought you might be infected with HIV, would you be more likely to: (a) be too afraid to be tested; (b) go to a clinic or family physician to be tested; (c) go to the drugstore and purchase an HIV testing kit that is sent to a lab for processing; or (d) go to your medicine cabinet and do the test right at home? The latter may soon be possible.

A quick HIV test, called the OraQuick Advance Rapid HIV-1/2 Antibody Test, currently is being used in clinics and hospitals with great success. It is greater than 99% accurate in detecting antibodies to HIV, meaning the person tested has been exposed to HIV.

The test works as follows: The person uses a specimen collection loop supplied in the kit to collect a drop of blood from a finger-prick. The loop with blood drop is stirred in a vial of a special solution for 20 minutes. An indicator device (see figure) is then inserted into the vial. The indicator device contains a strip of synthetic proteins that can detect HIV antibodies.

If a reddish band appears at the control (C) location, the test result is negative for HIV antibodies. If reddish bands appear at both the control (C) location and the test (T) location, the test is "reactive," meaning the result is a preliminary positive for HIV antibodies. Importantly, this is a "preliminary" result and does not definitely mean the person is HIV-positive. Rather, the individual would need to go to a clinic to have a more definitive test done to confirm the result. *The person is considered HIV-positive only if the confirmatory test result is positive.*

As of late 2009, the OraQuick test was not yet available for purchase by the general public. The necessary clinical trials required by the Food and Drug Administration for an over-the-counter test are being completed.

The test may identify many of the 300,000 individuals in the United States who do not know they are infected. In fact, these individuals are responsible for 65% of all new HIV infections each year. In addition, individuals who know they are infected are 50% less likely to transmit HIV to another uninfected individual.

The OraQuick test stick.

early to detect HIV antibodies. A repeat test at a later time is recommended. The Western blot analysis is used to confirm a positive test result.

In 2006, the U.S. Food and Drug Administration (FDA) approved the **APTIMA assay**, which amplifies and detects HIV RNA, as a diagnosis of primary HIV-1 infection. The **viral load test** detects the RNA of HIV and is available to monitor HIV-1 virus circulating in the blood of patients with established infections.

Treatment. The first drug used for treatment was azidothymidine, commonly known as AZT. AZT interferes with reverse transcriptase activity and acts as a chain terminator as it inhibits DNA synthesis. Other antiretroviral drugs are discussed in Chapter 24. They include reverse transcriptase inhibitors and protease inhibitors. These drugs interfere with viral genome replication and the processing step of capsid production, respectively. Other antiretroviral drugs include the fusion inhibitors, which work by blocking viral entry into the CD4 cells and integrase that block provirus formation.

HIV can become resistant to any of these drugs when the drug is used singly; that is, as **monotherapy** (FIGURE 23.22). So a more effective treatment requires a combination (or "cocktail") of drugs. When three or more drugs are used together, the

Western blot:
A technique to identify protein constituents, such as from a virus.

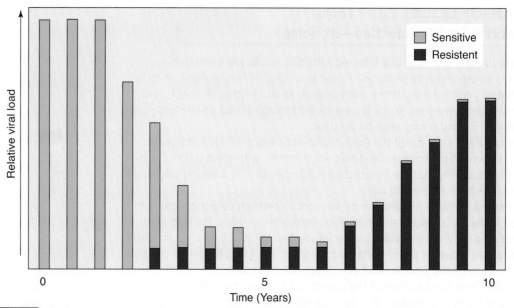

FIGURE 23.22 Viral Load and Drug Resistance. Because HIV replicates at a high rate and the reverse transcriptase enzyme is error prone, mutations will occur. After several years of drug monotherapy, the antiretroviral drug may have eliminated most of the original strain. However, some mutations will have conferred drug resistance and these viruses can now replicate free of any inhibition to that drug. »» What is the most logical way to combat drug resistance to monotherapy treatment?

combination is referred to as **highly active anti-retroviral therapy** (**HAART**). Although HAART is not a cure, it has been significant in reducing the risk of HIV transmission as well as the number of deaths. Overall, antiretroviral therapy has extended the life of HIV/AIDS patients. In fact, many patients are being told that if they stay on their medications, they can expect a near normal life expectancy. Unfortunately, the drugs are scarce globally and available to too few individuals. In addition, HIV can still hide out from drug attacks (MicroFocus 23.6).

Prevention. The only way to prevent HIV infection is to avoid behaviors putting a person at risk of infection, such as sharing needles or having unprotected sex.

A successful vaccine for HIV disease and AIDS has yet to be developed. Two types of vaccines are being considered. **Preventive vaccines** for HIV-negative individuals (health care providers) could be given to prevent infection. **Therapeutic vaccines,** for HIV-positive individuals, would help the individual's immune system control HIV either

by blocking virus entry, provirus formation, or replication such that the disease could not progress or be transmitted to others (see Chapter 14).

There are two major reasons why a successful vaccine has been lacking. First, HIV continually mutates and recombines. Such behavior by the virus means that a vaccine has to protect individuals against a moving target. In addition, because HIV infects CD4+ T cells, the vaccine needs to activate the very cells infected by the virus. Although at times it appears that vaccine research is an uphill battle, scientists are optimistic that a safe, effective, affordable, and stable HIV vaccine can be produced. Importantly, progress in basic and clinical research is moving forward and scientists are inching closer to identifying products suitable for a successful HIV vaccine. More information about a possible AIDS vaccine was presented in MicroFocus 22.3.

CONCEPT AND REASONING CHECKS

23.14 Distinguish between the three stages of HIV disease and AIDS.

MICROFOCUS 23.6: Public Health

HIV: The Escape Artist

Even though the levels of HIV in an individual on HAART are extremely low, the person is not cured because antiretroviral therapy cannot completely eradicate the virus from the body. The virus remains at low levels in locations where antiretroviral drugs appear unable to reach. These hiding places include: the T cells; the brain; semen; and, being a retrovirus, integrated into the DNA molecule itself.

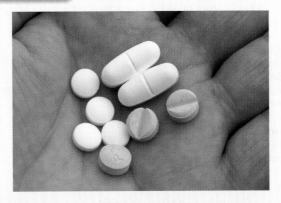

Because of this ability of HIV to escape attack, some researchers, such as the HIV co-discoverer Robert Gallo, believe there never will be a cure for the disease. With HIV's ability to hide out in the body, patients would have to keep taking drugs for their entire life. If they stop, the virus comes back with a vengeance. Others believe it might be possible to seek out and destroy the viruses in their hideouts.

One place HIV hides is in memory T cells, a subgroup of T cells that spring into action in a subsequent infection. (This is why you only get diseases like the mumps once.) For some reason, HIV does not destroy these cells—and no drug currently available attacks them. Being integrated into the DNA as a provirus, these memory T cells represent a potent hideout. Researchers are working on ways to activate the cells and make them divide, so antiretroviral drugs will then destroy these cells.

Resting T cells also will not be destroyed by antiretroviral drugs because the drugs only are effective in actively dividing cells. In these cells, HIV replicates slowly but surely.

In addition, many immune cells have the ability to pump out toxins that have been imported. To these cells, antiretroviral drugs look like toxins and they are effectively pumped out of the cell.

Other sites also need to be attacked. One is the brain where HIV also takes refuge. Even individuals on HAART can develop HIV-associated dementia, a neurological problem due to HIV infection. Here, the blood-brain barrier prevents the entry of potentially effective drugs. HIV also hides out in the male semen, where immune cells appear to be inefficient at activating antiviral drugs.

If new drugs can be developed to seek out and destroy the HIV hideouts, future HAART medications may become much more effective. Still others, like Gallo, believe there are just too many hideouts and they cannot all be eliminated.

Time will tell.

■ SUMMARY OF KEY CONCEPTS

23.1 Type I IgE-Mediated Hypersensitivity

1. **Type I hypersensitivity** is caused by **IgE antibodies** produced in response to certain antigens called **allergens**. The antibodies can attach to the surface of **mast cells** and **basophils**, and trigger the release of the mediators on subsequent exposure to the same allergen.

2. Type I hypersensitivities can involve **systemic anaphylaxis**, a whole-body reaction in which a series of mediators induces vigorous and life-threatening contractions of the smooth muscles of the body.

3. **Atopic diseases** involve **localized anaphylaxis** (common allergies), such as hay fever or a food allergy.

4. **Asthma** is an inflammatory disease involving an early response to an allergen and a late response by eosinophils and neutrophils that cause tissue injury and airway blockage.

5. Several ideas have been put forth concerning the nature of allergies. These include the shielding of IgE-secreting B cells (plasma cells) by IgA-secreting plasma cells and feedback mechanisms where IgE limits its own production in nonatopic individuals.

6. **Desensitization therapy** attempts to limit the possibility of an anaphylactic reaction through the injection of tiny amounts of allergen over time. Therapy also can involve the presence of IgG antibodies as **blocking antibodies**.

23.2 Other Types of Hypersensitivity

7. In **type II cytotoxic hypersensitivity**, the immune system produces IgG and IgM antibodies, both of which react with the body's cells and often destroy the latter.

8. No cells are involved in **type III hypersensitivity**. Rather, the body's IgG and IgM antibodies interact with dissolved antigen molecules to form visible **immune complexes**. The accumulation of immune complexes in various organs leads to local tissue destruction in such illnesses as **serum sickness**, **Arthus phenomenon**, and **systemic lupus erythematosus**.

9. **Type IV cellular hypersensitivity** involves no antibodies, but is an exaggeration of cell-mediated immunity based in T lymphocytes. **Contact dermatitis** is a manifestation of this hypersensitivity.

23.3 Autoimmune Disorders and Transplantation

10. **Autoimmune disorders** can occur through defects in **clonal deletion**, **clonal anergy**, or **regulatory T-cell activity**. Human heredity, as well as access to privileged sites and **antigenic mimicry**, also can trigger an autoimmune response.

11. Four types of grafts or transplants can be performed: **autografts**, **isografts**, **allografts**, and **xenografts**. Allografts are the most common. Rejection of grafts or transplants involves cytotoxic T cells and antibodies. The graft also can be rejected by immune cells in the graft that reject the recipient (**graft-versus-host reaction**).

12. Prevention of rejection is strengthened by using antirejection strategies, including anti-inflammatory drugs, antilymphocyte antibodies, antimitotic drugs, cyclosporine, monoclonal antibodies, or radiation.

23.4 Immunodeficiency Disorders

13. **Immunodeficiency disorders** may be congenital (**primary immunodeficiencies**) or acquired later in life (**secondary immunodeficiencies**).

14. **AIDS** (**acquired immunodeficiency syndrome**) is the final stage of the HIV (human immunodeficiency virus) infection. HIV infects and destroys CD4 T cells, eventually leading to an inability of the immune system to fend off opportunistic diseases. Transmission is through blood, blood products, contaminated needles, or unprotected sexual intercourse. Many **antiretroviral drugs** are available to slow the progression of disease.

LEARNING OBJECTIVES

After understanding the textbook reading, you should be capable of writing a paragraph that includes the appropriate terms and pertinent information to answer the objective.

1. Summarize the events occurring in and the role of humoral immunity in **type I hypersensitivities**.
2. Explain the initiation and outcomes of **systemic anaphylaxis**.
3. Distinguish between the different forms of **atopic disease**.
4. Compare **asthma** to other forms of type I hypersensitivities.
5. Discuss the reasons why allergies develop.
6. Identify the therapies and treatments available for type I hypersensitivities.
7. Summarize the characteristics of **type II hypersensitivity** and its relationship to blood transfusions and **hemolytic disease of the newborn**.
8. Summarize the characteristics of **type III hypersensitivity** and its relationship to **serum sickness** and the **Arthus reaction**.
9. Summarize the characteristics of **type IV hypersensitivity** and its relationship to infection allergies and **contact dermatitis**.
10. Identify the ways in which an **autoimmune disease** can arise.
11. Describe the immunological reasons why **organ transplants** are rejected and list the four types of grafts (transplants).
12. Assess the usefulness of **immunosuppressive drugs** by transplant patients.
13. Contrast **primary** and **secondary immunodeficiencies** and list several primary disorders and their accompanying immunological deficiencies.
14. Diagram how the **human immunodeficiency virus** (**HIV**) infects a cell and identify the prevention and treatment methods used for HIV disease and **AIDS**.

STEP A: SELF-TEST

Each of the following questions is designed to assess your ability to remember or recall factual or conceptual knowledge related to this chapter. Read each question carefully, then select the *one* answer that best fits the question or statement. Answers to even-numbered questions can be found in **Appendix C**.

1. All the following are types of immediate hypersensitivities *except*:
   A. asthma.
   B. contact dermatitis.
   C. food allergies.
   D. hay fever.

2. Systemic anaphylaxis is characterized by
   A. contraction of smooth muscles.
   B. a red rash.
   C. blood poisoning.
   D. hives.

3. Which of the following is NOT a type I hypersensitivity?
   A. Food allergies
   B. Contact dermatitis
   C. Allergic rhinitis
   D. Exercise-induced allergies

4. The early response in asthma is due to _____ activity.
   A. cytotoxic T cell
   B. basophil
   C. T$_H$2 cell and NK cell
   D. dendritic cells

5. What type of immune cell may control IgE-mediated hypersensitivities?
   A. Suppressor T cells
   B. Plasma cells
   C. Cytotoxic T cells
   D. Neutrophils

6. Desensitization therapy can involve
   A. the use of blocking antibodies.
   B. injections of small amounts of allergen.
   C. allergen injections of several months.
   D. All the above (A–C) are correct.

7. A cytotoxic hypersensitivity would occur if blood type _____ is transfused into a person with blood type _____.
   A. A; AB
   B. O; AB
   C. A; O
   D. O; B

8. Serum sickness is a common symptom of
   A. contact dermatitis.
   B. hemolytic disease of the newborn.
   C. immune complex hypersensitivity.
   D. food allergies.

9. Which one of the following allergens is NOT associated with contact dermatitis?
   A. Foods
   B. Cosmetics
   C. Poison ivy
   D. Jewelry

10. Immunological tolerance to "self" is established by
    A. destruction of self-reactive lymphoid cells.
    B. clonal anergy.
    C. clonal deletion.
    D. All the above (A–C) are correct.

11. A _____ is a graft between genetically different members of the same species.
    A. xenograft
    B. autograft
    C. allograft
    D. isograft

12. Immunosuppressive agents used in preventing transplant rejection primarily affect
    A. macrophages.
    B. neutrophils.
    C. dendritic cells.
    D. T cells.

13. What immunodeficiency disorder is associated with a lack of T and B cells and complete immune dysfunction?
    A. DiGeorge syndrome
    B. Severe combined immunodeficiency disease
    C. Chronic granulomatous disease
    D. Chédiak-Higashi syndrome

14. An HIV patient with swollen lymph nodes and CD4 T cell count of _____ would be in stage _____ of HIV disease/AIDS.
    A. 1,000; stage II
    B. 700; stage I
    C. 400; stage III
    D. 400; stage IV

▌ STEP B: REVIEW

This chapter has summarized some of the disorders associated with the immune system. To gauge your understanding, rearrange the scrambled letters to form the correct word for each of the spaces in the statements. The answers to the even-numbered statements are in **Appendix C**.

15. The simple compound _____ is one of the major mediators released during allergy reactions.
    ITMEHIASN

16. An immune deficiency called _____ syndrome is characterized by the failure of T lymphocytes to develop.
    IDOGGERE

17. Cases of rheumatoid arthritis are accompanied by immune complex formation in the body's _____.
    NISTJO

18. In a _____ hypersensitivity, antibodies unite with cells and trigger a reaction that results in cell destruction.
    XYOTCCITO

19. Hay fever is an example of an _____ disease, one in which a local allergy takes place.
    OCTIAP

20. Immune complex hypersensitivities develop when antibody molecules interact with _____ molecules and form aggregates in the tissues.
    ENIGNAT

21. The skin test for _____ relies on a response by T lymphocytes to PPD placed in the skin tissues.
    UUICTRLSBEOS

22. Mast cells and _____ are the two principal cells that function in anaphylactic responses.
    SSBIOHPAL

23. A key element in transplant acceptance or rejection is a set of molecules abbreviated as _____ proteins.
    LAH

24. Urticaria is a form of skin _____ occurring in a person having an allergic reaction.
    AHSR

25. The HIV DNA integrated into a chromosome is called a _____.
    SRVRUPOI

STEP C: APPLICATIONS

Answers to even-numbered questions can be found in **Appendix C**.

26. During war and under emergency conditions, a soldier whose blood type is O donates blood to save the life of a fellow soldier with type B blood. The soldier lives, and after the war becomes a police officer. One day he is called to donate blood to a brother officer who has been wounded and finds that it is his old friend from the war. He gladly rolls up his sleeve and prepares for the transfusion. Should it be allowed to proceed? Why?

27. Coming from the anatomy lab, you notice that your hands are red and raw and have begun peeling in several spots. This was your third period of dissection. What is happening to your hands, and what could be causing the condition? How will you solve the problem?

28. "He had a history of nasal congestion, swelling of his eyes, and difficulty breathing through his nose. He gave a history of blowing his nose frequently, and the congestion was so severe during the spring he had difficulty running." The person in this description is former President Bill Clinton, and the writer is an allergist from Little Rock, Arkansas. What condition (technically known as allergic rhinitis) is probably being described?

STEP D: QUESTIONS FOR THOUGHT AND DISCUSSION

Answers to even-numbered questions can be found in **Appendix C**.

29. As part of an experiment, one animal is fed a raw egg while a second animal is injected intravenously with a raw egg. Which animal is in greater danger? Why?

30. A woman is having the fifth injection in a weekly series of hay fever shots. Shortly after leaving the allergist's office, she develops a flush on her face, itching sensations of the skin, and shortness of breath. She becomes dizzy, then faints. What is taking place in her body, and why has it not happened after the first four injections?

31. You may have noted that brothers and sisters are allowed to be organ donors for one another, but that a person cannot always donate to his or her spouse. Many people feel bad about being unable to help a loved one in time of need. How might you explain to someone in such a situation the basis for becoming an organ donor and why it may be impossible to serve as one?

32. The immune system is commonly regarded as one that provides protection against disease. This chapter, however, seems to indicate that the immune system is responsible for numerous afflictions. Even the title is "Immune Disorders." Does this mean that the immune system should be given a new name? On the other hand, is it possible that all these afflictions are actually the result of the body's attempts to protect itself? Finally, why can the phrase "immune disorder" be considered an oxymoron?

33. In many diseases, the immune system overcomes the infectious agent, and the person recovers. In other diseases, the infectious agent overcomes the immune system, and death follows. Compare this broad overview of disease and resistance to what is taking place with AIDS, and explain why AIDS is probably unlike any other disease encountered in medicine.

⊛ HTTP://MICROBIOLOGY.JBPUB.COM/9E

The site features eLearning, an online review area that provides quizzes and other tools to help you study for your class. You can also follow useful links for in-depth information, read more MicroFocus stories, or just find out the latest microbiology news.

Antimicrobial Drugs

We are facing a crisis because doctors are pressured to prescribe antibiotics for the common cold and inner ear infection, yet we know that it is not prudent to do so. We must collectively inform our patients about the reasons why overprescribing antibiotics will not help patients return to work sooner, and that in the long run, could make them more susceptible to drug-resistant diseases.

—Richard E. Besser, M.D., Centers for Disease Control and Prevention

For centuries, physicians believed heroic measures were necessary to save patients from the ravages of infectious disease. They prescribed frightening courses of purges (bowel emptying), enormous doses of strange chemical concoctions, blood-curdling ice water baths, deadly starvations, and bloodlettings. These treatments probably complicated an already bad situation by reducing the natural body defenses to the point of exhaustion. In fact, the death of George Washington in 1799 is believed to have been due to a streptococcal infection of the throat, perhaps exacerbated by the bloodletting treatment that removed almost 2 liters of his blood within a 24-hour period.

When the germ theory of disease emerged in the late 1800s, insights about microorganisms added considerably to the understanding of disease and increased the storehouse of knowledge available to the doctor. However, it did not change the fact that little, if anything, could be done for the infected patient. Then, in the 1940s, antimicrobial agents, including antibiotics, burst on the scene, and another revolution in medicine began.

Doctors were astonished to learn they could kill microorganisms in the body without doing substantial harm to the body itself. Medicine had a period of powerful, decisive growth, as doctors found they could successfully alter the course of infectious disease. Antibiotics effected a radical change in medicine and charted a new course for treating infectious disease. Since the 1940s, millions of lives have been saved. In 1969, then U.S. Surgeon General William Stewart and many pharmaceutical companies believed it was time to "close the books on infectious diseases."

Chapter Preview and Key Concepts

24.1 The History and Properties of Antimicrobial Agents
1. Ehrlich believed in a selective toxicity for specific chemical agents against microbes.
2. Penicillin was the first clinically effective antibiotic.
3. Antimicrobial agents vary in their form and activity.
4. Antibiotics can be toxic to sensitive bacterial competitors.

24.2 The Synthetic Antibacterial Agents
5. Sulfonamides affect DNA synthesis.
6. Cell walls or DNA are the targets for other synthetic drugs.

24.3 The Beta-Lactam Family of Antibiotics
7. Penicillin is a bactericidal agent effective against gram-positive cells.
8. Cephalosporins and carbapenems also contain a beta-lactam ring and inhibit peptidoglycan cross-linking.

24.4 Other Bacterially Produced Antibiotics
9. Vancomycin is a glycopeptide antibiotic and cell wall assembly inhibitor.
10. Antibiotics targeting the cell membrane either interfere with cell wall synthesis or disrupt membrane structure.
11. Many antibiotics target the ribosome's ability to translate mRNA.
12. Rifampin inhibits transcription by inhibiting RNA synthesis.

24.5 Antiviral, Antifungal, and Antiparasitic Drugs
13. Antiviral drugs interfere with specific stages of the viral replication cycle.
14. Selective toxicity is important when treating fungal infections and diseases.
15. Antiprotozoal drugs target unique aspects of parasites.
16. Antihelminthic drugs affect helminth neuromuscular coordination, carbohydrate metabolism, or microtubule structure.

24.6 Antibiotic Assays and Resistance
17. Antibiotic susceptibility assays are used to study the inhibition of a test organism by one or more antimicrobial agents.
18. Bacterial species have several ways to generate resistance to antibacterial agents.
19. Antibiotic misuse and abuse encourage the emergence of resistant forms.
20. The discovery of new targets and the development of new antibiotics may help fight antibiotic resistance.

MicroInquiry 24: Testing Drugs—Clinical Trials

Unfortunately, it soon became clear that bacterial species and other microorganisms could quickly develop resistance to antimicrobials. Today, antibiotic resistance is a major concern throughout the world as an increasing number of antibiotic resistant organisms arise (FIGURE 24.1A).

There have been several reasons for this increase in antibiotic resistance. The general public often misuses or abuses antibiotics. Individuals, for example, might stop taking an antibiotic as soon as they start feeling better; they do not finish the course of antibiotic treatment; they take an antibiotic for a viral infection like a cold or the flu; they take an antibiotic that was prescribed for some other illness; or they take an antibiotic that was prescribed for someone else.

In addition, many doctors still prescribe antibiotics for diseases that are untreatable with such drugs. Again, this includes viral infections such as colds and the flu. Often this is at the insistence of the patient, as the opening quote for this chapter states.

Smart use of antibiotics is the key to controlling the spread of resistance. The Centers for Disease Control and Prevention (CDC) has started a campaign to prevent antimicrobial resistance. The campaign focuses on four main strategies: preventing infections, diagnosing and correctly treating infections, preventing transmission, and using antimicrobials wisely (FIGURE 24.1B).

In this chapter, we discuss the antimicrobial drugs as the mainstays of our health-care delivery system to treat bacterial, viral, fungal, and parasitic infections and diseases. Some drugs have been known for generations, but most are of recent development. We explore their discovery and examine their uses—and abuses.

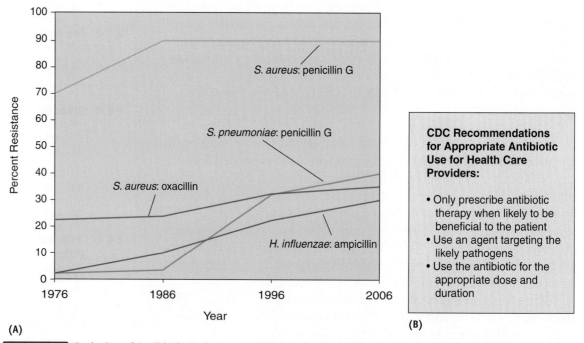

(A)

(B)

CDC Recommendations for Appropriate Antibiotic Use for Health Care Providers:

- Only prescribe antibiotic therapy when likely to be beneficial to the patient
- Use an agent targeting the likely pathogens
- Use the antibiotic for the appropriate dose and duration

FIGURE 24.1 Evolution of Antibiotic Resistance. **(A)** This graph shows examples of the evolution of antibiotic resistance by *Staphylococcus aureus*, *Streptococcus pneumoniae*, and *Haemophilus influenzae* to penicillin G and some of its derivatives (oxacillin and ampicillin) between 1976 and 2006, arising in part from inappropriate antibiotic use. **(B)** The CDC recommendations to curb the increase in antibiotic resistant strains of microbes. »» Why is there an increase in the percent of resistance by these and other bacteria?
Source: Seattle Times

24.1 The History and Properties of Antimicrobial Agents

In the previous chapters, we discussed how our immune system defends and protects the body from pathogen invasion. As described in this chapter opener, beginning in the late 1940s, a group of chemicals became available to assist the immune system. These **chemotherapeutic agents** were used to treat infections, diseases, and other disorders, such as cancer. In microbiology, the **antimicrobial agents**, those chemotherapeutic agents used to treat infectious disease were key.

The History of Chemotherapy Originated with Paul Ehrlich

KEY CONCEPT

1. Ehrlich believed in a selective toxicity for specific chemical agents against microbes.

In the drive to control and cure infectious disease, the efforts of microbiologists in the early 1900s were primarily directed toward enhancing the body's natural defenses. Sera containing antibodies lessened the impact of diphtheria, typhoid fever, and tetanus; and effective antibody-inducing vaccines for smallpox and rabies (and later, diphtheria and tetanus) reduced the incidence of these diseases (see Chapter 22).

Among the leaders in the effort to control disease was an imaginative German investigator named Paul Ehrlich. Ehrlich knew that specific dyes would stain specific bacterial species. Therefore, he believed there must be specific chemicals that would be toxic to these species. This selective toxicity concept was developed in the early 1900s when Ehrlich thought he could discover molecules that would be "magic bullets"—specific chemicals that would seek out and destroy specific disease organisms in infected tissues without harming those tissues.

Ehrlich and his staff had synthesized hundreds of arsenic-phenol compounds. One of Ehrlich's collaborators, the Japanese investigator Sahachiro Hata, set out to test the chemicals for their ability to destroy the syphilis spirochete *Treponema pallidum*. After months of painstaking study, Hata's attention focused on arsphenamine, compound #606 in the series. Hata and Ehrlich successfully tested arsphenamine against *T. pallidum* in animals and human subjects, and in 1910, they gave a derivative of the drug to doctors for use against syphilis. Arsphenamine, the first modern synthetic antimicrobial agent, was given the brand name **Salvarsan** because it offered salvation from syphilis and contained arsenic.

Salvarsan met with mixed success during the ensuing years. Its value against syphilis was without question, but local reactions at the injection site, and indiscriminate use by some physicians, brought adverse publicity. Ehrlich's death in 1915, together with the general ignorance of organic chemistry and the impending first World War, further eroded enthusiasm for chemotherapy. However, the team approach to drug discovery used by Ehrlich would become the model for modern pharmaceutical research.

Over the next 20 years, German chemists continued to synthesize and manufacture dyes for fabrics and other industries, and they routinely tested their new products for antimicrobial qualities. In 1932, one of these products was a red dye, trademarked as **Prontosil**.

Prontosil had no apparent effect on bacterial cells in culture. However, things were different in animals, where the drug is converted into an active antimicrobial form. When Gerhard Domagk, a German pathologist and bacteriologist, tested Prontosil in animals, he found a pronounced inhibitory effect on staphylococci, streptococci, and other gram-positive bacterial species. In February 1935, Domagk injected the dye into his daughter Hildegard, who had become gravely ill with septicemia after pricking her finger with a needle. Hildegard's condition gradually improved and her arm did not have to be amputated. Many historians see her recovery as setting into motion the age of modern chemotherapy. For his discovery, Domagk was awarded the 1939 Nobel Prize in Physiology or Medicine.

Arsphenamine forms

Chemotherapy: The use of chemical agents to treat diseases or disorders.

Prontosil

CONCEPT AND REASONING CHECKS

24.1 Evaluate the early discoveries of and uses for chemotherapeutic agents for selective toxicity.

Fleming's Observation of the Penicillin Effect Ushered in the Era of Antibiotics

KEY CONCEPT

2. Penicillin was the first clinically effective antibiotic.

One of the first to postulate the existence and value of antibiotics was the British microbiologist Alexander Fleming (**FIGURE 24.2A**). During his early years, Fleming experienced the excitement of the classical Golden Age of microbiology and spoke up for the therapeutic value of Salvarsan. In 1928, Fleming was performing research on staphylococci at St. Mary's Hospital in London. On one of these nutrient agar plates, he noted a green mold contaminating the plate. Surrounding the mold no bacteria were growing (**FIGURE 24.2B**). Intrigued by the failure of staphylococci to grow near the mold, Fleming isolated the mold, identified it as a species of *Penicillium,* and found it produced a substance that kills gram-positive organisms. Though he failed to isolate the elusive substance, he named it "penicillin."

Fleming was not the first to note the antibacterial qualities of *Penicillium* species. Joseph Lister had observed a similar phenomenon in 1871; John Tyndall did likewise in 1876; and a French medical student, Ernest Duchesne, wrote a research paper on the subject in 1896. However, it was Fleming who first proposed that penicillin could be used to eliminate gram-positive bacteria from mixed cultures. Further, he unsuccessfully tried to use a filtered broth from the fungus on infected wound tissue. Unfortunately, biochemistry was not sufficiently advanced to make complex separations possible, and Fleming's discovery soon was forgotten.

In 1935, Gerhard Domagk's dramatic announcement of the antimicrobial effects of Prontosil fueled speculation that chemicals could be used to fight disease in the body. Then, in 1939, a group at England's Oxford University, led by pathologist Howard Florey and biochemist Ernst Chain, reisolated Fleming's penicillin and conducted trials with highly purified samples. They discovered that penicillin was effective against a large variety of diseases, including gonorrhea, meningitis, tetanus, and diphtheria. However, England was already involved in World War II, so a group of American companies devel-

(A)

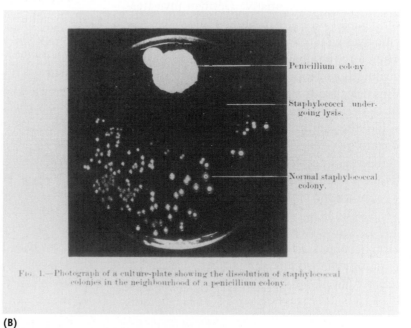
(B)

FIGURE 24.2 Alexander Fleming and His Culture of *Penicillium*. (**A**) Alexander Fleming reported the existence of penicillin in 1928 but was unable to purify it for use as an antimicrobial agent. (**B**) The actual photograph of Fleming's culture plate shows how staphylococci in the region of the *Penicillium* colony have been killed (they are "undergoing lysis") by some unknown substance produced by the mold. Fleming called the substance penicillin. »» How can you tell the bacteria have lysed near the *Penicillium* colony?

oped the techniques for the large-scale production of penicillin and made the drug available for commercial use.

CONCEPT AND REASONING CHECKS

24.2 Identify how penicillin was discovered and later developed into a useful drug.

Antimicrobial Agents Have a Number of Important Properties

KEY CONCEPT

3. Antimicrobial agents vary in their form and activity.

The preceding historical accounts illustrate the origins for the two groups of antimicrobial agents. Those drugs, like Salvarsan and Prontosil, which are made (synthesized) in the pharmaceutical laboratory, are called **synthetic drugs**. By contrast, drugs like penicillin, which are products of or derived from the metabolism of living microorganisms are called **antibiotics** (*anti* = "against;" *biosis* = "life"); that is, the agents work to kill or inhibit living organisms.

Today, many antibiotics are produced by a process that is partly of laboratory origin and partly of microbial origin. Such **semisynthetic drugs**, for simplicity, will be included with the discussions of the true antibiotics.

The synthetic and antibiotic drugs have certain important properties that need to be considered when prescribing a drug for an infection or disease.

Selective Toxicity. Ehrlich's idea of a magic bullet was based on **selective toxicity**, which says that an antimicrobial drug should harm the infectious agent but not the host. Today, two terms are used when considering the toxicity of a drug. The **toxic dose** refers to the concentration of the drug causing harm to the host. The **therapeutic dose** refers to the concentration of the drug that effectively destroys or eliminates the pathogen from the host. Together these can be used to formulate the **chemotherapeutic index**, which is the highest concentration (per kilogram of body weight) of the drug tolerated by the host divided by the lowest concentration (per kilogram body weight) of the drug that will eliminate the infection or disease agent.

The chemotherapeutic index of each antimicrobial drug must be considered, and the efficacy in eliminating disease and providing symptom relief must outweigh the associated toxicity and adverse events (**FIGURE 24.3**).

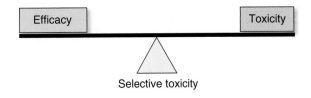

$$\text{Chemotherapeutic index} = \frac{\text{Toxic dose}}{\text{Therapeutic dose}}$$

FIGURE 24.3 **A Representation of the Chemotherapeutic Index.** Finding the correct drug means considering the efficacy of the drug and its toxicity to the pathogen being considered. »» Determine whether a drug with a high or low chemotherapeutic index would be the more desirable.

As we will see below, the best way to accomplish this is to develop drugs that target a specific component of microbial cells, such as the cell wall or a certain metabolic pathway, which is absent in human cells.

Antimicrobial Spectrum. Another important property in prescribing an appropriate drug is identifying the range of pathogens to which a particular drug will work. This range of antimicrobial action is the **antimicrobial spectrum**. Those drugs affecting many taxonomic groups are considered as having a **broad spectrum** of action, while those affecting few pathogens have a **narrow spectrum** of action (**FIGURE 24.4**). We will identify the drug spectrum with each antimicrobial drug discussed.

CONCEPT AND REASONING CHECKS

24.3 Assess the importance of the chemotherapeutic index and the antimicrobial drug spectrum in prescribing treatment for an infectious disease.

Antibiotics Are Agents of Natural Biological Warfare

KEY CONCEPT

4. Antibiotics can be toxic to sensitive bacterial competitors.

Before we proceed to examine specific antibiotics, it is worth noting the reasons why some bacterial and fungal species produce antibiotics. In the natural environment (soil, water), there can be fierce competition between microbes for limited nutrients. Therefore, if a bacterial or fungal species has the ability to secrete an antibiotic, it may kill or inhibit the growth of those competitors sensitive to the chemical. Antibiotic production thus gives the producer a selective advantage.

Efficacy:
The ability to produce the necessary or desired results.

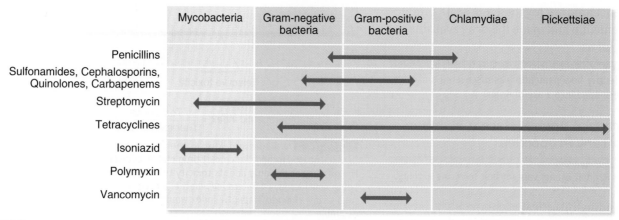

	Mycobacteria	Gram-negative bacteria	Gram-positive bacteria	Chlamydiae	Rickettsiae
Penicillins			◄───────────►		
Sulfonamides, Cephalosporins, Quinolones, Carbapenems		◄───────►			
Streptomycin	◄───────────►				
Tetracyclines		◄───────────────────────────────►			
Isoniazid	◄───►				
Polymyxin		◄───►			
Vancomycin			◄───►		

FIGURE 24.4 The Antimicrobial Spectrum of Activity. Antimicrobial drugs have a limited range of activity in the domain *Bacteria*. »» According to the figure, which drugs have a broad spectrum and which ones have a narrow spectrum?

In response to antibiotic secretion, some species may develop mutations or gain plasmids capable of making the bacterial cell resistant to one or more antibiotics. So, in a natural habitat, it has been a continuous evolutionary process for bacterial species to compete against one another, either through antibiotic production or by developing resistance. MicroFocus 24.1 describes one particular "bug battle."

CONCEPT AND REASONING CHECKS

24.4 What advantage is afforded by bacteria that secrete antibiotics in the soil?

24.2 The Synthetic Antibacterial Agents

The work of Gerhard Domagk in identifying the usefulness for Prontosil set the stage for the identification of many other synthetic antimicrobials. In 1935, a group at the Pasteur Institute, headed by Jacques and Therese Tréfouël, isolated the active form of Prontosil. They found it to be a chemical called sulfanilamide, which was highly active against gram-positive bacteria; it quickly became a mainstay for treating wound-related infections during World War II.

Sulfanilamide and Other Sulfonamides Target Specific Metabolic Reactions

KEY CONCEPT

5. Sulfonamides affect DNA synthesis.

Sulfanilamide was the first of a group of broad-spectrum synthetic agents known as **sulfonamides**. These so-called "sulfa drugs" interfere with the metabolism of bacterial cells without damaging body tissues. Here's how they work.

Bacterial cells synthesize an important vitamin called **folic acid** for use in nucleic acid synthesis.

These organisms possess the necessary enzyme to manufacture folic acid; in fact, until recently they were incapable of absorbing folic acid from the surrounding environment. Humans cannot synthesize folic acid and must consume it in foods or vitamin capsules.

To produce folic acid, a bacterial enzyme joins together three important components, one of which is **para-aminobenzoic acid (PABA)**. This molecule is similar to sulfanilamide in chemical structure (**FIGURE 24.5**). However, if the cell contains large amounts of sulfanilamide, the sulfanilamide competes with PABA for the active site in the bacterial enzyme. Such **competitive inhibition** (see Chapter 8) results in a reduction of folic acid, which stops nucleic acid synthesis and DNA replication. The bacterial cells eventually die.

Drug resistance is now quite common as mutations have arisen that allow the microbes to absorb folic acid from outside sources. Newer formulations of modern sulfonamides are typified by sulfamethoxazole, which is typically combined

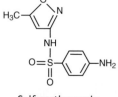

Sulfamethoxazole

MICROFOCUS 24.1: Evolution

Bug Battles

One of the natural roles of antibiotics is in chemical warfare. When two or more soil-dwelling bacterial species are competing for the same nutrients or growth space, the ability of one species to kill or inhibit its neighbors gives that species the advantage. Antibiotics can provide that advantage. But does this actually occur in nature?

Biologists at the Massachusetts Institute of Technology (MIT) have pitted one bacterial species against another in a battle for survival. In one corner is *Rhodococcus fascians* (see figure), an aerobic, non–spore-forming, gram-positive bacterial species typically found in the soil. When this plant pathogen was sequenced, the MIT biologists discovered the organism contained a number of genes coding for secondary metabolic products; that is, compounds like antibiotics, toxins, and pigments. Although *R. fascians* does not normally produce antibiotics, many bacteria species have genes for antibiotics that are only activated when the organism is threatened in some way. If pure cultures of *R. fascians* were stressed by placing them in higher or lower temperatures, or altering the growth medium, no antibiotics could be identified. So, physical or nutrient conditions in the environment do not seem to affect *R. fascians*, as far as antibiotic production is concerned.

In the other corner is another soil-dwelling bacterial species, *Streptomyces padanus* (see figure). Like many species of *Streptomyces*, *S. padanus* produces an antibiotic that normally kills other bacteria.

The definitive battle for survival occurred when the MIT biologists placed *R. fascians* in the same culture tube with *S. padanus*. What ensued was quite remarkable—*R. fascians* killed *S. padanus*! What happened was that on mixing the two species, *R. fascians* started producing its own antibiotic and wiped out *S. padanus*!

The biologists then isolated the antibiotic, which they called rhodostreptomycin, and tested it to see what other species of bacteria could be killed. The antibiotic proved effective against several bacterial species, but it was most effective against *Helicobacter pylori*, a nasty stomach pathogen responsible for stomach ulcers and, in some cases, stomach cancer (see Chapter 11). So, perhaps here is a new and potentially useful antibiotic to treat *H. pylori*.

But why did *R. fascians* start producing the antibiotic when pitted against *S. padanus*? Perhaps the most logical scenario is that the stressful environment triggered *R. fascians* to turn on new gene activity responsible for synthesizing rhodostreptomycin. But what caused the new gene activity? The MIT biologists discovered that in the battle *R. fascians* picked up a large plasmid from *S. padanus*. Perhaps that plasmid was the "stress trigger" for the new gene activity.

Whatever the case, the experiment shows that bug battles in the natural environmental can trigger offensive responses, supporting the idea of competitive advantage—and demonstrate there may be many other antibiotics "out there" that could be useful in fighting infectious diseases.

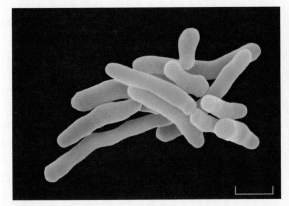 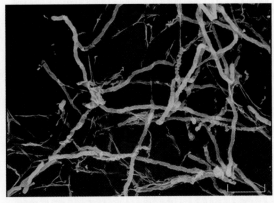

False-color scanning electron microscope images of *Rhodococcus fascians* (left; Bar = 10 μm) and a *Streptomyces* species (right; Bar = 10 μm).

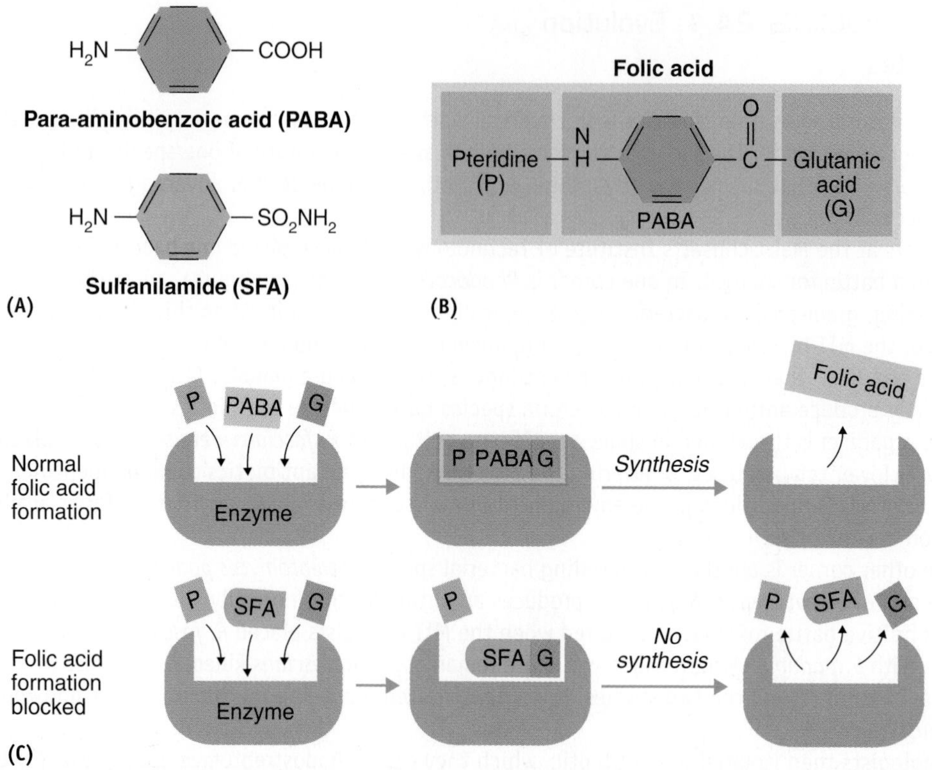

FIGURE 24.5 The Disruption of Folic Acid Synthesis by Competitive Inhibition. (**A**) The chemical structures of para-aminobenzoic acid (PABA) and sulfanilamide (SFA) are very similar. (**B**) Folic acid is made up of three components: pteridine (P), PABA, and glutamic acid (G). (**C**) In the normal synthesis of folic acid, a bacterial enzyme joins the three components to form folic acid. However, in competitive inhibition, SFA competes for the active site because of its great abundance. The SFA assumes the position normally reserved for PABA, and folic acid cannot form. »» Why does a bacterial cell die without folic acid?

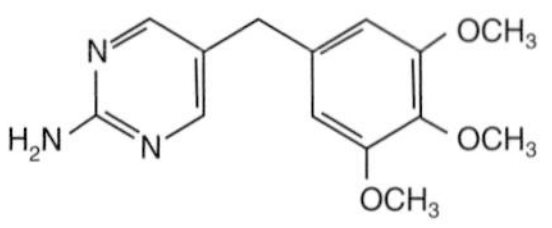

Trimethoprim

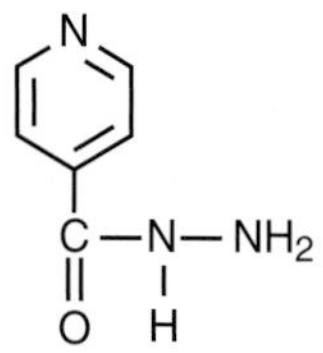

Isoniazid

with trimethoprim, another synthetic agent that inhibits a different step in folic acid synthesis. The "synergistic" drug, called co-trimoxazole (Bactrim) has the advantage of being effective in much lower doses than either drug alone and makes the generation of resistance less likely. The drug is prescribed for urinary tract infections due to gram-negative rods.

Two other important synthetic drugs blocking PABA metabolism in *Mycobacterium* species are *p*-aminosalicylic acid (PAS), which is used for treating tuberculosis, and dapsone (diaminodiphenylsulfone), which is effective against leprosy (see Chapter 13).

CONCEPT AND REASONING CHECKS

24.5 Explain how sulfanilamide interferes with DNA replication.

Other Synthetic Antimicrobials Have Additional Bacterial Cell Targets

KEY CONCEPT

6. Cell walls or DNA are the targets for other synthetic drugs.

Several synthetic agents that target other bacterial structures are currently in wide use.

Isoniazid (isonicotinic acid hydrazide, or INH) has a very narrow drug spectrum as the active form of the drug specifically interferes with cell wall synthesis in *Mycobacterium* species by inhibiting the production of mycolic acid, a component of the wall (MicroFocus 24.2). Isoniazid often is combined in therapy with such drugs as rifampin and ethambutol. **Ethambutol** is a synthetic, well-absorbed drug, but side effects include visual disturbances, thus limiting its use in the treatment of tuberculosis.

MICROFOCUS 24.2: Public Health
The Trojan Horse

In Greek mythology, the Trojan horse was a colossal statue of a horse in which Greek soldiers hid to gain access to the city of Troy (see figure). Now, scientists have uncovered a similar plot used by the drug isoniazid to kill *Mycobacterium tuberculosis,* the agent responsible for tuberculosis (TB).

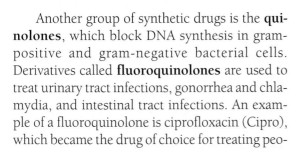

Isoniazid, a small, synthetic molecule, enters the bacterial cytoplasm as a benign, nontoxic chemical substance no different than any nutrient passing through the cell wall and membrane. Once in the cytoplasm, the drug reveals its true identity. A common cytoplasmic enzyme called catalase activates the drug and converts it to a toxic form. The activated, now toxic drug then attacks a protein used by the tubercle bacillus to synthesize mycolic acid, a key component of its cell wall. Without a strong cell wall, the organism is left vulnerable to osmotic lysis (see Chapter 4).

In the Greek tale, the city of Troy fell to invaders. However, bacterial cells don't read books, and unlike the literary Trojan horse, the mycobacteria fight back and resist the drug. What happens is the bacterial cell stops producing catalase by switching off the gene encoding the enzyme. Thus, the isoniazid remains inactive. However, the action also leaves the bacterial cell in a tenuous position because catalase breaks down hydrogen peroxide, a corrosive compound normally produced during bacterial metabolism.

To resolve this dilemma, a second bacterial gene encodes a second enzyme (alkyl hydroperoxidase), which takes over the job of catalase and destroys any hydrogen peroxide generated.

In mythology, the Greeks carried the day and won the city of Troy. And in the early halcyon days of antibiotic use, isoniazid was a prime weapon in the fight against TB. In 2009, 14% of TB cases were resistant to isoniazid and more than 50 million people worldwide may be infected with drug-resistant strains of *M. tuberculosis*. Tubercle bacilli are learning to resist the Trojan horse. Let's hope other antituberculoid drugs can find alternative ways to enter into the bacterial "gates of Troy" and reverse the trend.

Another group of synthetic drugs is the **quinolones**, which block DNA synthesis in gram-positive and gram-negative bacterial cells. Derivatives called **fluoroquinolones** are used to treat urinary tract infections, gonorrhea and chlamydia, and intestinal tract infections. An example of a fluoroquinolone is ciprofloxacin (Cipro), which became the drug of choice for treating people exposed to anthrax spores during the 2001 anthrax bioterror incidents. The fluoroquinolones are expected to be the number one selling class of antibiotics by 2011.

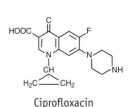

Ciprofloxacin

CONCEPT AND REASONING CHECKS

24.6 Describe the mode of action for (A) isoniazid and (B) the fluoroquinolones.

24.3 The Beta-Lactam Family of Antibiotics

One of the most common mechanisms of antibiotic action is blocking the synthesis of the bacterial cell wall. Besides the synthetic drugs isoniazid and ethambutol, which target mycobacterial species, there are a large number of antibiotics or semisynthetic ones targeting the assembly of the peptidoglycan component of bacterial cell walls.

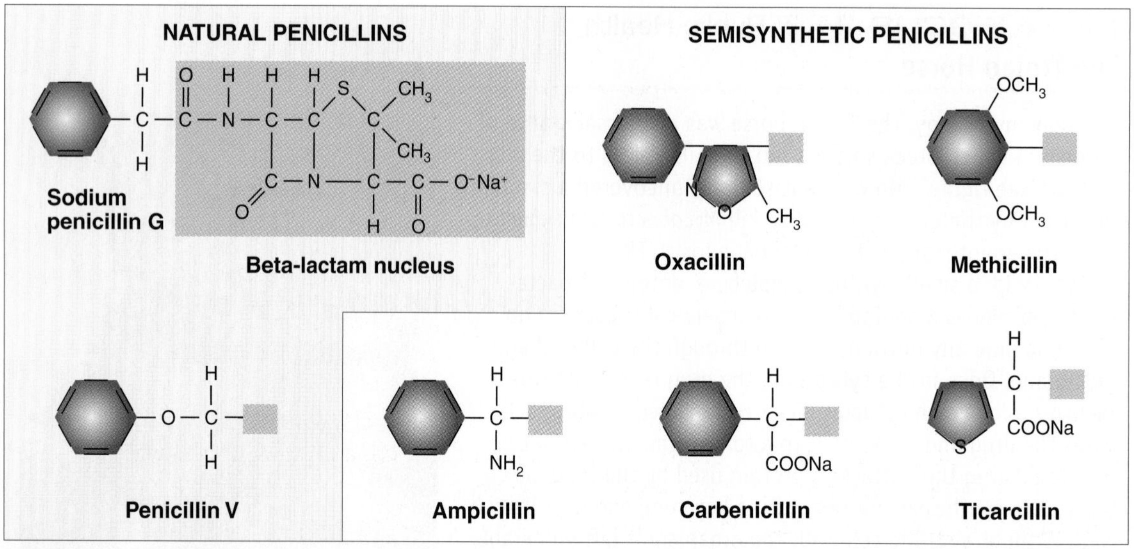

FIGURE 24.6 Some Members of the Penicillin Group of Antibiotics. The beta-lactam nucleus is common to all the penicillins. Different penicillins are formed by varying the side group on the molecule. »» Why is there a need for so many semisynthetic penicillins?

Penicillin Has Remained the Most Widely Used Antibiotic

KEY CONCEPT

7. Penicillin is a bactericidal agent effective against gram-positive cells.

Thanks to the purification and production work of Florey, Chain, and co-workers in the 1940s, **penicillin** has saved the lives of millions of individuals. Its high chemotherapeutic index has made it the drug of choice in eradicating many infections.

Penicillin G has been the most popular penicillin antibiotic and is usually the one intended when doctors prescribe "penicillin." It is sensitive to acid, so it is primarily given intravenously. Penicillin V is more acid resistant and can be given orally. Both forms of penicillin have a beta-lactam nucleus and differ in the side groups attached (**FIGURE 24.6**).

The penicillins are active against a variety of gram-positive bacteria, including staphylococci, streptococci, clostridia, and pneumococci. In higher concentrations, they also are inhibitory to the gram-negative diplococci causing gonorrhea and meningitis, and they are useful against syphilis spirochetes.

Penicillin functions during the synthesis of the bacterial cell wall. It inhibits the peptide cross-linking of carbohydrates between peptidoglycan layers during wall formation. This results in such a weak wall that internal osmotic pressure allows the cell to swell and burst. Penicillin is therefore bactericidal in rapidly multiplying bacteria (as in an infection). Where bacterial cells are multiplying slowly or are dormant, the drug may have only a bacteriostatic effect or no effect at all.

Over the years, two major drawbacks to the use of penicillin have surfaced. The first is the anaphylactic reaction occurring in allergic individuals (see Chapter 23). This allergy applies to all compounds related to penicillin. Swelling around the eyes or wrists, flushed or itchy skin, shortness of breath, and a series of hives are signals of a hypersensitivity; penicillin therapy should cease immediately if these symptoms occur.

The second disadvantage is the evolution of penicillin-resistant bacterial species. Many of these organisms produce **beta-lactamases**, which inactivate beta-lactam antibiotics. For example, **penicillinase** opens the beta-lactam ring, converting penicillin G into harmless penicilloic acid (**FIGURE 24.7**). The ability to produce penicillinase probably has existed in certain bacterial mutants, but the ability manifests itself when the organisms are confronted with the drug. Thus, a process of natural selection takes place, and the rapid multiplication of penicillinase-producing bacterial cells yields organisms over which penicillin has no effect.

In the late 1950s, the beta-lactam nucleus of the penicillin molecule was identified and synthesized, and scientists found they could attach

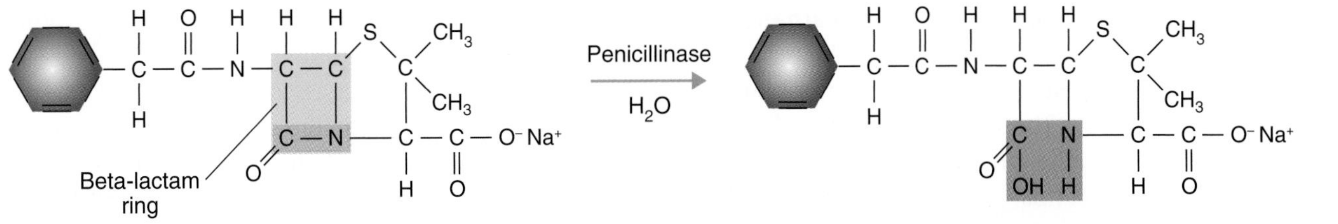

FIGURE 24.7 The Action of Penicillinase on Sodium Penicillin G. The enzyme penicillinase converts penicillin to harmless penicilloic acid by opening the beta-lactam ring and inserting a hydroxyl group to the carbon and a hydrogen to the nitrogen. »» What does the action of penicillinase tell you about the role of the beta-lactam ring in the structure of penicillin?

various groups to this nucleus and create new penicillins. In the following years, numerous "semisynthetic" penicillins emerged (Figure 24.6). Oxacillin and methicillin are penicillinase-resistant penicillins, while ampicillin is a broad spectrum drug of value against some gram-negative rods (*Escherichia, Proteus, Haemophilus*) as well as gonococci and meningococci. The drugs resist stomach acid and are absorbed from the intestine after oral consumption. Other broad spectrum drugs, such as carbenicillin and ticarcillin, are effective against even a broader range of gram-negative bacteria and can be used for infections of the urinary tract. Penicillins, such as amoxicillin, have been used in combination with chemicals like clavulanic acid (the combination is called Augmentin). The clavulanic acid inactivates penicillinase, allowing amoxicillin to affect cell wall synthesis.

CONCEPT AND REASONING CHECKS

24.7 Explain how penicillin works and how penicillinase affects penicillin.

Other Beta-Lactam Antibiotics Also Inhibit Cell Wall Synthesis

KEY CONCEPT

8. Cephalosporins and carbapenems also contain a beta-lactam ring and inhibit peptidoglycan cross-linking.

Besides the penicillins, there are other beta-lactam antibiotics affecting cell wall synthesis.

Cephalosporins. While evaluating seawater samples along the coast of Sardinia in 1945, an Italian microbiologist named Giuseppe Brotzu observed a striking difference in the amount of *Escherichia coli* in two adjoining areas. Subsequently he discovered that a fungus, *Cephalosporium acremonium,* was producing

an antibacterial substance in the water. The substance, named cephalosporin C, was later isolated and characterized by scientists as having a beta-lactam nucleus. Today, the cephalosporins and penicillins make up more than 50% of all antibiotics produced and used worldwide.

Cephalosporins resemble penicillins in chemical structure, except the beta-lactam nucleus has a slightly different composition. They are used as alternatives to penicillin where resistance is encountered, or in cases where penicillin allergy exists. Cephalosporins also have a broader antibacterial spectrum, are longer lasting in the body, and are resistant to many beta-lactamases (cephalosporinases).

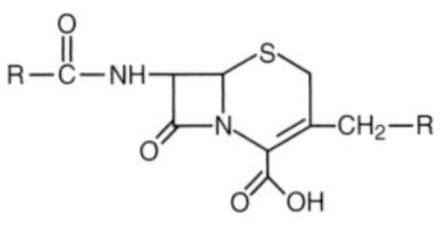

First-generation cephalosporins are variably absorbed from the intestines and are useful against gram-positive cocci, and certain gram-negative rods. They include cephalexin (Keflex) and cephalothin (Keflin). Modifications to the basic cephalosporin chemical structure have generated newer semisynthetic generation drugs that are more active and longer lasting.

Second-generation drugs, such as cefacor and cefuroxime, have an expanded activity against gram-positive cocci, as well as numerous gram-negative rods (e.g., *Haemophilus influenzae*). The third-generation cephalosporins, including cefotaxime and ceftriaxone, are used primarily against gram-negative rods (e.g., *Pseudomonas aeruginosa*) and for treating diseases of the central nervous system. The fourth generation cephalosporins (e.g., cefepime) have improved activity against the gram-negative bacteria involved with urinary tract infections.

Cephalosporin

CONCEPT AND REASONING CHECKS

24.8A List the advantages of the cephalosporins over the penicillins.

Monobactams and Carbapenems. A group of narrow-spectrum antibiotics first produced in the early 1980s is the **monobactams**. Isolated from the bacterium *Chromobacter violaceum,* a purple-pigmented bacterial species, monobactams, like moxalactam, are active against aerobic, gram-negative rods, especially those involved in nosocomial diseases and bacterial meningitis. Interference with platelet function and severe bleeding has limited their use.

Another set of beta-lactam drugs are the broad spectrum **carbapenems**, one of the most important groups of clinically useful antibiotics today. The representative of this group, imipenem, is derived from a compound produced by the bacterium *Streptomyces cattley*. Imipenem

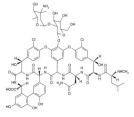

Imipenem

is effective against a variety of aerobic gram-positive bacteria and gram-negative rods, as well as anaerobes (e.g., *Bacteroides fragilis*). It is not active against methicillin-resistant *Staphylococcus aureus.* Imipenem is often prescribed in cases where resistances occur, and it appears to have minimal side effects, although people allergic to penicillin and other beta-lactam antibiotics should not take imipenem. The drug normally is degraded by the kidneys before it can effect its action in the body. Therefore, imipenem usually is prescribed in combination with cilastatin. Cilastatin prevents the kidneys from degrading the drug. The imipenem/cilastatin combination is known as Primaxin.

CONCEPT AND REASONING CHECKS

24.8B   List the advantages of the carbapenems.

24.4   Other Bacterially Produced Antibiotics

Besides antibiotics that target the bacterial cell wall, there are many others that affect the cell membrane, protein synthesis, or nucleic acid synthesis. However, we begin our discussion with some non-beta-lactam drugs affecting cell walls.

Vancomycin Also Inhibits Cell Wall Synthesis

KEY CONCEPT

9.   Vancomycin is a glycopeptide antibiotic and cell wall assembly inhibitor.

Vancomycin

Bacitracin

Vancomycin, a cell wall inhibitor of gram-positive bacteria, is a product of *Amycolatopsis* (formerly *Streptomyces*) *orientalis*. It is administered by intravenous injection against diseases caused by gram-positive bacteria, especially severe staphylococcal diseases where penicillin allergy or bacterial resistance is found. It also is used against *Clostridium* and *Enterococcus* species (enterococci).

As drug resistance has developed and spread among staphylococci, the choice of antibiotics has gradually diminished, and vancomycin has emerged as a key treatment in therapy; it often is referred to as the "drug of last resort." Unfortunately, resistance to vancomycin among bacterial groups such as the enterococci (*Enterococcus faecium* and *E. faecalis*) populations has also been observed. The enterococci account for over 10% of all hospital acquired infections in the United States.

CONCEPT AND REASONING CHECKS

24.9   What problems are associated with vancomycin use?

Polypeptide Antibiotics Affect the Cell Membrane

KEY CONCEPT

10.   Antibiotics targeting the cell membrane either interfere with cell wall synthesis or disrupt membrane structure.

Both bacitracin and polymyxin B are polypeptide antibiotics produced by *Bacillus* species. These antibiotics are quite toxic internally and can cause kidney damage. Therefore, they generally are restricted to topical use, such as on the skin.

Bacitracin is a cyclic polypeptide that interferes with the transport of cell wall precursors through the cell membrane. Because it is very toxic taken internally, bacitracin is only available in ointments for topical treatment of skin infections caused by gram-positive bacteria or for the prevention of wound infections. When combined with neomycin (see below) and polymyxin B, it is sold under the brand Neosporin.

Polymyxins are cyclic polypeptides that insert into the cell membrane. Acting like detergents, they increase permeability and lead to cell death. They are most active against gram-negative rods. Polymyxin B is valuable against *Pseudomonas aeruginosa* and other gram-negative bacilli, particularly those causing superficial infections in wounds, abrasions, and burns. The two antibiotics, bacitracin and polymyxin B, are combined with gramicidin (which increases permeability of the cell membrane) in Polysporin.

CONCEPT AND REASONING CHECKS

24.10 Why are the polypeptide antibiotics only used topically?

Many Antibiotics Affect Protein Synthesis

KEY CONCEPT

11. Many antibiotics target the ribosome's ability to translate mRNA.

There are several groups of bacterially produced antibiotics that target the protein synthesizing machinery in bacterial cells by binding to the 30S (small) or 50S (large) subunit (**FIGURE 24.8**).

Aminoglycosides. The **aminoglycosides** are a group of bactericidal antibiotic compounds that attach irreversibly to the 30S subunit of bacterial ribosomes, thereby blocking the reading of the genetic code on messenger RNA (mRNA) molecules. Because oral absorption is negligible, the antibiotics must be administered by intramuscular injection. Their use has declined in recent years with the introduction of second- and third-generation cephalosporins, and with the introduction and development of quinolone drugs, such as the fluoroquinolones.

In 1943, the first aminoglycoside, streptomycin, was discovered by Selman Waksman's group at Rutgers University. At the time, the discovery was sensational because streptomycin was useful against tuberculosis and numerous other diseases caused by gram-negative bacteria. Since then, it has been largely replaced by safer drugs, although streptomycin still is prescribed on occasion for tuberculosis. The major side effect is damage to the auditory branch of the nerve extending from the inner ear—deafness may result.

Gentamicin, a still-useful aminoglycoside, is administered for serious infections of the urinary tract caused by gram-negative bacteria. The antibiotic is produced by a species of *Micromonospora,* a bacterial species related to *Streptomyces.* As already mentioned, neomycin, which also was discovered in Waksman's lab, is available in combination with polymyxin B and bacitracin as Neosporin. Physicians use an aerosolized version of another aminoglycoside, tobramycin (Tobi®), to treat *Pseudomonas*-caused respiratory infections in patients with cystic fibrosis.

Chloramphenicol. An antibiotic with a broad spectrum is **chloramphenicol**. Its discovery from *Streptomyces venezuelae* was hailed as a milestone in microbiology because the drug is capable of inhibiting a wide variety of gram-positive and gram-negative bacteria, as well as several species of rickettsiae and fungi.

Chloramphenicol is a small molecule that passes into the tissues, where it interferes with peptide bond formation by the 50S ribosome subunit. It remains the drug of choice in the treatment of typhoid fever and is an alternative to tetracycline for epidemic typhus and Rocky Mountain spotted fever (see Chapter 12).

Streptomycin

Gentamicin

Cystic fibrosis: A hereditary disease starting in infancy and affecting various glands, resulting in secretion of thick mucus that blocks internal passages of the lungs and causes respiratory infections.

Chloramphenicol

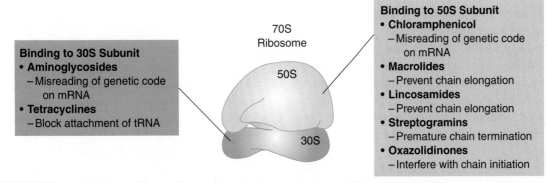

Binding to 30S Subunit
- **Aminoglycosides**
  - Misreading of genetic code on mRNA
- **Tetracyclines**
  - Block attachment of tRNA

70S Ribosome

50S

30S

Binding to 50S Subunit
- **Chloramphenicol**
  - Misreading of genetic code on mRNA
- **Macrolides**
  - Prevent chain elongation
- **Lincosamides**
  - Prevent chain elongation
- **Streptogramins**
  - Premature chain termination
- **Oxazolidinones**
  - Interfere with chain initiation

FIGURE 24.8 **Antibiotics and Their Affect on Protein Synthesis.** Many antibiotics interfere with a translations step on the 30S (small) or 50S (large) ribosome subunit. »» Why are there so many antibiotics that affect translation?

■ Aplastic anemia:
Refers to the inability of the bone marrow to produce new blood cells.

These drugs usually are reserved for treating serious and life-threatening infections because of their side effects. In the bone marrow, it prevents hemoglobin incorporation into the red blood cells, causing a condition called aplastic anemia. Chloramphenicol also accumulates in the blood of newborns, causing a toxic reaction and sudden breakdown of the cardiovascular system known as the **gray syndrome**. Still, it is used to treat endemic cholera in some parts of the world.

Tetracyclines. In 1948, scientists discovered chlortetracycline, the first of the **tetracycline** antibiotics. This finding completed the initial quartet of "wonder drugs": penicillin, streptomycin, chloramphenicol, and tetracycline.

The tetracyclines are a group of broad-spectrum bacteriostatic antibiotics that block attachment of the tRNA to the 30S subunit. There are naturally occurring chlortetracyclines, isolated from species of *Streptomyces*, and semisynthetic tetracyclines, such as minocycline and doxycycline (FIGURE 24.9A). All have the four benzene ring chemical structure.

Tetracycline antibiotics may be taken orally, a factor that led to their indiscriminate use in the 1950s and 1960s. The antibiotics were consumed in huge quantities by tens of millions of people, and in some people, the normal microbiota of the intestine was destroyed. With these natural controls eliminated, fungi such as *Candida albicans* flourished. Tetracyclines also cause a yellow-gray-brown discoloration of teeth and stunted bones in children (FIGURE 24.9B). These problems are minimized by restricting use of the antibiotic in pregnant women and children through the teen years.

Despite these side effects, tetracyclines remain the drugs of choice for most rickettsial and chlamydial diseases. They are used against a wide range of gram-negative bacteria, and they are valuable for treating primary atypical pneumonia, syphilis, gonorrhea, and pneumococcal pneumonia. Although resistance to these medications has occurred, newer tetracyclines such as minocycline and doxycycline appear to circumvent this. Evidence indicates that tetracycline may have been present in the food of ancient people (MicroFocus 24.3).

In 2005, the U.S. Food and Drug Administration (FDA) approved a new class of antibiotics

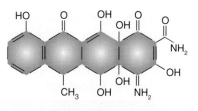

Doxycycline

(A)

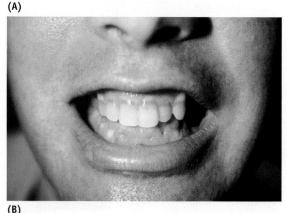

(B)

FIGURE 24.9 The Tetracyclines. (**A**) The chemical structure of doxycycline. (**B**) The staining of teeth associated with tetracycline use. »» What chemical feature characterizes all tetracycline antibiotics?

related to the tetracyclines, called the **glycylcyclines**. The new drug, tigecycline (Tygacil) is effective against methicillin-resistant *S. aureus* (MRSA) infections.

Macrolides. The **macrolides** are a group of bacteriostatic antibiotics consisting of large carbon rings attached to unusual carbohydrate molecules. They block protein synthesis by inhibiting chain elongation (see Chapter 8).

In 1949, Abelardo Aguilar, a Filipino scientist, sent some soil samples to his employer Eli Lilly. In the soil was the species *Streptomyces erythreus* (now called *Saccaropolyspora erythraea*), from which the Lilly scientists isolated a drug called erythromycin. In the 1970s, researchers discovered that erythromycin was effective for treating primary atypical pneumonia and Legionnaires' disease. It is recommended for use against gram-positive bacteria in patients with penicillin allergy and against both *Neisseria* and *Chlamydia* species affecting the eyes of newborns. Although it has few side effects, at higher doses it sometimes causes nausea, vomiting, and diarrhea.

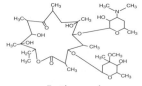

Erythromycin

MICROFOCUS 24.3: History
Wonder Bread

In September 1980, a chance observation led to the discovery that antibiotics were protecting humans from disease long before anyone had suspected.

The remarkable find was made by Debra L. Martin, a graduate student at Detroit's Henry Ford Hospital. After preparing thin bone sections for microscopic observation, Martin placed her slides under a fluorescence microscope because no other microscope was available for her use at the time. When illuminated with ultraviolet light, the sections glowed with a peculiar yellow-green color. Her colleagues attributed the glow to the antibiotic tetracycline.

These were no ordinary bone sections. Rather, they were from the mummified remains of Nubian people excavated along the floodplain of the Nile River. Anthropologists from the University of Massachusetts, led by George Armelagos, had previously established that the Nubian population was remarkably free of infectious disease, and now Martin's discovery gave a possible reason why.

Streptomyces species are very common in desert soil, and the anthropologists postulated that the bacterial cells may have contaminated the grain bins and deposited tetracycline. Bread made from the antibiotic-rich grain then conferred freedom from disease. The theory was strengthened when the amount of tetracycline in the ancient bone was shown to be equivalent to that in the therapeutic doses used in modern medicine.

A modern-day Nubian mother and child.

Another practice, reported in 1944, indicates that modern people may be more deliberate in their use of contaminated bread. A doctor traveling in Europe noted that a loaf of moldy bread hung in the kitchens of many homes. He inquired about it and was told that when a wound or abrasion was sustained, a sliver of the bread was mixed with water to form a paste; the paste was then applied to the skin. A wound so treated was less likely to become infected. Presumably the modern bread, like the ancient bread, contained a chemical that would be recognized today as an antibiotic.

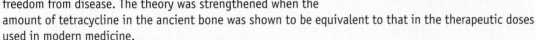

Other macrolides with a broader spectrum include clarithromycin, (Biaxin) and azithromycin (Zithromax), one of the world's best-selling antibiotics. Both macrolides are semisynthetic drugs associated with causing serious allergic and dermatologic reactions.

Lincosamides. The **lincosamides** bind to the 50S subunit and inhibit chain elongation. *Streptomyces lincolnensis* produces an antibiotic called lincomycin from which the semisynthetic drug **clindamycin** is derived. This bacteriostatic drug is an alternative in cases where penicillin resistance is encountered. Clindamycin is active against aerobic, gram-positive cocci and anaerobic, gram-negative bacilli (e.g., *Bacteroides* species). Use of the antibiotic is limited to serious infections because the drugs eliminate competing organisms from the intestine and permit *Clostridium difficile* to overgrow the area. The clostridial toxins then may induce a potentially lethal condition called **pseudomembranous colitis**, in which membranous lesions cover the intestinal wall (see Chapter 13).

Streptogramins. Another group of cyclic peptides are the **streptogramins**. Discovered in 1962 in yet another species of *Streptomyces*, these antibiotics are prescribed as a combination of two cyclic peptides, called quinupristin-dalfopristin (Synercid). Both components interfere with protein synthesis on the 50S subunit,

Clindamycin

Streptogramin A

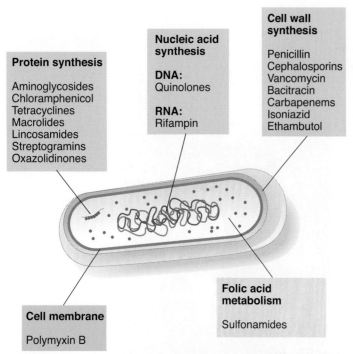

Protein synthesis

Aminoglycosides
Chloramphenicol
Tetracyclines
Macrolides
Lincosamides
Streptogramins
Oxazolidinones

Nucleic acid synthesis

DNA:
Quinolones

RNA:
Rifampin

Cell wall synthesis

Penicillin
Cephalosporins
Vancomycin
Bacitracin
Carbapenems
Isoniazid
Ethambutol

Folic acid metabolism

Sulfonamides

Cell membrane

Polymyxin B

FIGURE 24.10 **The Targets for Antibacterial Agents.** There are six major targets for antibacterial agents: the cell wall, cell membrane, ribosomes (protein synthesis), nucleic acid synthesis (RNA and DNA synthesis), and metabolic reactions. »» Hypothesize why there are so many antibiotics that affect protein synthesis.

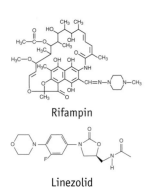

Rifampin

Linezolid

the interference being synergistic and bactericidal. The drugs are effective against a broad range of gram-positive bacteria, including *Staphylococcus aureus*, and respiratory pathogens.

Oxazolidinones. After the identification of the streptogramins in 1962, no new structural classes of antibiotics were discovered until 2000. These were the **oxazolidinones**, which interfere with chain initiation by the 50S subunit. Drugs such as linezolid (Zyvox), are effective in treating gram-positive bacteria, including MRSA. However, it can produce allergic reac-

tions and is toxic to mitochondria. Therefore, the oxazolidinones are drugs of last resort, being used only where every other antibiotic has failed.

CONCEPT AND REASONING CHECKS

24.11   Identify the bacterial source for the aminoglycosides, chloramphenicol, tetracyclines, macrolides, clindamycin, and the streptogramins.

Some Antibiotics Inhibit Nucleic Acid Synthesis

KEY CONCEPT

12.   Rifampin inhibits transcription by inhibiting RNA synthesis.

The synthetic quinolones that inhibit bacterial DNA synthesis have already been described.

Other drugs affect transcription. **Rifampin**, a semisynthetic bactericidal drug derived from *Streptomyces mediterranei*, interferes with RNA synthesis. Because mycobacterial resistance to rifampin can develop quickly, it is prescribed in combination with isoniazid and ethambutol for tuberculosis and leprosy patients. It also is administered to carriers of *Neisseria* and *Haemophilus* species that cause meningitis and as a prophylactic when exposure has occurred. Rifampin therapy may cause the urine, feces, tears, and other body secretions to assume an orange-red color and may cause a rash and liver damage. The drug is administered orally and is well absorbed.

FIGURE 24.10 summarizes the sites of activity for the antibacterial agents. **TABLE 24.1** reviews those antibacterial drugs currently in use.

CONCEPT AND REASONING CHECKS

24.12   To what enzyme must rifampin bind to inhibit transcription?

24.5 Antiviral, Antifungal, and Antiparasitic Drugs

If you look at the CDC's web site that lists diseases and conditions, about 36% of the diseases have a bacterial origin, 33% have a viral origin, 6% are fungal, and 25% are due to parasites (**FIGURE 24.11**). If we look at the developing nations, the percentage of viral and especially parasitic diseases would be much higher. Although these data are important for monitoring trends and for targeting research, prevention, and control efforts, they do not include non-notifiable diseases, such as

colds, forms of gastroenteritis, yeast infections, etc. Still, we can make the following generalization: we have dozens more antibiotics in our antimicrobial arsenal to fight bacterial infections than we have drugs to fight viral, fungal, and parasite diseases. In addition, the antiviral drugs we do have are useful against only a very limited number of viral diseases and most have been developed to fight HIV disease/AIDS. Importantly, unlike antibiotics, there are no antiviral drugs that will definitely cure

24.1 A Summary of Major Antibacterial Drugs (by Mode of Action)

Antibacterial Agent	Source	Antibacterial Spectrum	Side Effects
Competitive inhibitors of essential metabolic reactions			

Sulfonamides
  Sulfanilamide
  Sulfamethoxazole/
  Trimethoprim
	Synthetic	Broad	Kidney and liver damage
			Allergic reactions

Inhibitors of cell wall synthesis			
Isoniazid	Synthetic	Tubercle bacilli	Liver damage
Ethambutol	Synthetic	Tubercle bacilli	Visual disturbances

Penicillins
  Penicillin G
  Ampicillin
  Amoxicillin
  Nafcillin
  Oxacillin
	Penicillium notatum and *Penicillium chrysogenum*
	Some semisynthetic
	Some synthetic
		Broad, especially gram-positive bacteria	Allergic reactions
			Selection of penicillinase-producing strains

Cephalosporins
  Cephalothin
  Cephalexin
	Cephalosporium species
	Some semisynthetic
		Gram-positive bacteria
		Broad
			Occasional allergic reactions

Carbapenems Imipenem	*Chromobacter violaceum*	Broad	Few reported
Vancomycin	*Streptomyces orientalis*	Gram-positive bacteria, especially staphylococci	Ear and kidney damage
Bacitracin	*Bacillus subtilis*	Gram-positive bacteria, especially staphylococci	Kidney damage
Inhibitors of cell membrane function			
Polymyxin	*Bacillus polymyxa*	Gram-negative rods, especially in wounds	Kidney damage
Inhibitors of protein synthesis			

Aminoglycosides
  Streptomycin
  Gentamicin
  Neomycin
	Streptomyces species
	Micromonospora species
		Broad, especially gram-negative bacteria	Hearing defects
			Kidney damage

Chloramphenicol	*Streptomyces venezuelae*	Broad, especially typhoid bacilli	Aplastic anemia Gray syndrome

Tetracyclines
  Chlortetracycline
  Minocycline
  Doxycycline
	Streptomyces species
	Some semisynthetic
		Broad, rickettsiae, chlamydiae, gram-negative bacteria	Destruction of natural flora
			Discoloration of teeth
			Stunted bones

Macrolides
  Erythromycin
  Clarithromycin
  Azithromycin
	Streptomyces erythreus
	Some semisynthetic
		Broad, gram-positive bacteria, *Mycoplasma*	Gastrointestinal distress

Lincosamides Clindamycin	*Streptomyces lincolnesis*	Gram-positive bacteria	Pseudomembranous colitis

Streptogramins
  Quinupristin/
  Dalfopristin
	Streptomyces species	Resistant strains of *Staphylococcus aureus*	Muscle aches, rash, headache

Oxazolidinones Linezolid	Synthetic	Last resort antibiotics, gram-positive bacteria	Toxic to mitochondria
Inhibitors of DNA synthesis			
Fluoroquinolones Ciprofloxacin	Synthetic	Broad	Few reported
Inhibitors of RNA synthesis			
Rifampin	*Streptomyces mediterranei* Semisynthetic	Tubercle bacilli Gram-positive cocci	Liver damage

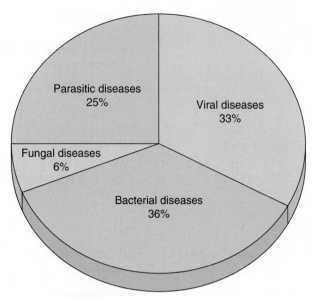

FIGURE 24.11 **Types of Human Infectious Disease.** The percentage of bacterial and viral disease is about equal.
»» Provide a hypothesis as to why there are fewer fungal diseases than there are diseases of the other groups.

a person of any viral disease. Therefore, whether it is an antiviral drug directed against HIV, herpes, or influenza, antiviral chemotherapy only lessens the effects of the infection.

The ability to use antimicrobial agents effectively against fungi presents another problem not encountered with bacterial pathogens. The major problem facing medical mycologists is finding drugs affecting the fungal pathogen but not the host tissue. Because fungi are members of the *Eukarya*, they possess much of the cellular machinery common to animals and humans; thus, many drugs targeting fungi will also target host tissues and potentially be quite toxic. Therefore, most antifungal agents are used only topically on the skin. The few drugs that are selectively toxic and systemic are targeted at unique fungal structures or metabolic processes.

Antiviral Drugs Can Be Used to Treat a Limited Number of Human Viral Diseases

KEY CONCEPT

13. Antiviral drugs interfere with specific stages of the viral replication cycle.

The body's normal immunological defenses can be summoned to actively fight disease either naturally

as a result of being infected (Chapter 21) or artificially as a result of a vaccination (Chapter 22). However, antibiotics will not work against viral diseases because viruses lack the structures and metabolic machinery with which antibiotics interfere. For example, penicillin is useless for inhibiting viruses because viruses have no cell wall.

A limited number of antiviral drugs are of value for treating diseases caused by picornaviruses, herpesviruses, hepatitis B and C viruses, HIV, and influenza viruses. These drugs affect viral penetration/uncoating, genome replication, or viral maturation/release (TABLE 24.2).

Some antivirals block the ability of viruses to penetrate or uncoat in the host cell. Antiviral drugs such as pleconaril integrate into the picornavirus capsid and prevent capsid uncoating. Amantadine and rimantadine prevent the influenza A viruses from escaping the endocytotic vacuole in which they are contained. HIV penetration is blocked by enfuvirtide, which blocks the fusion of the HIV envelope with the host cell plasma membrane.

Most of the antiviral drugs designed target the DNA polymerase of the herpesviruses or the reverse transcriptase of HIV and hepatitis B virus. Most of these drugs represent **base analogs** (see Chapter 8) and insert themselves into the replicating DNA strand. Insertion blocks the ability of the virus to continue replicating its genome. Azidothymidine (AZT) and several other drugs have been used to treat HIV infections and AIDS. Another drug called foscarnet is used in patients having retinal disease (retinitis) caused by the cytomegalovirus (CMV). **Reverse transcriptase inhibitors**, such as nevirapine, bind directly to reverse transcriptase and inhibit transcription, thereby preventing the synthesis of DNA in retroviruses.

Other antiviral drugs work by blocking the maturation process or release of the virions from the host cells. **Protease inhibitors**, such as indinavir, react with the HIV protease, the enzyme that trims viral proteins down to working size for the construction of the capsid. In the type A influenza viruses, neuraminidase is an enzyme in the spike that helps the viruses spread to other cells (Chapter 15). The **neuraminidase inhibi-**

TABLE

24.2 Examples of Antiviral Drugs, Their Mode of Action, and Targeted Viruses

Antiviral Drug	Mode of Action	Example
Blocking of Penetration/Uncoating		
Amantadine (Symmetrel)[1] Rimantadine (Flumadine)	Block viral uncoating	Influenza A viruses
Pleconaril (Picovir)	Blocks uncoating	Picornavirus infections
Enfuvirtide (Fuzeon)	Blocks penetration/uncoating	HIV disease/AIDS
Maraviroc (Selzentry)	Blocks attachment/penetration	HIV disease/AIDS
Inhibition of Genome Replication		
Acyclovir (Zovirax)	Base analog (mimics guanine) inhibiting DNA replication	Genital herpes, chickenpox, shingles
Ganciclovir (Cytovene)	Base analog (mimics guanine) inhibiting DNA polymerase	Herpesvirus infections (retinitis)
Ribavirin (Virazole)	Base analog (mimics guanine) inhibiting viral replication	Broad range of viruses, including hepatitis B and C
Azidothymidine (AZT; Retrovir)	Base analog (mimics adenine) terminating DNA chain elongation	HIV and other retrovirus infections
Vidarabine (Vira-A)	Terminates DNA chain elongation	Shingles and herpesvirus-caused encephalitis
Idoxuridine (Stoxil) Trifluridine (Viroptic)	Base analogs (mimic thymidine) causing replication errors (mutations)	Keratitis caused by herpesvirus infection
Dideoxycytidine (Hivid) Dideoxyinosine (Videx)	Base analogs (mimic cytosine and adenine, respectively) terminating DNA chain elongation	HIV disease/AIDS
Foscarnet (Foscavir)	Inhibits DNA polymerases	Retinitis caused by herpesvirus infection
Nevirapine (Viramune) Delavirdine (Rescriptor)	Reverse transcriptase inhibitors	Used primarily against HIV
Raltegravir	Block provirus integration	HIV disease/AIDS
Inhibition of Virion Maturation/Release		
Indinavir (Crixivan) Ritonavir (Norvir) Saquinavir (Invirase)	Inhibit HIV protease	HIV disease/AIDS
Oseltamivir (Tamiflu) Zanamivir (Relenza)	Inhibit viral release	Influenza A viruses

[1]Trade name

tors zanamivir and oseltamivir block the action of neuraminidase, preventing release of new virions—and thereby limiting disease spread in the body. The development of all these drugs is based on biochemical knowledge of virus function coupled with human ingenuity.

CONCEPT AND REASONING CHECKS

24.13 Why have most of the antiviral agents synthesized been targeted against viral genome replication?

Several Classes of Antifungal Drugs Cause Membrane Damage

KEY CONCEPT

14. Selective toxicity is important when treating fungal infections and diseases.

Polyenes. The **polyenes** are large, ring-shaped organic compounds containing alternating double and single carbon-carbon bonds

Flucytosine

Griseofulvin

Amphotericin B

Miconazole

Quinine

on one side of the ring and multiple hydroxyl (–OH) groups bonded to carbons on the other side of the ring. The polyenes bind to ergosterol, a sterol found in the fungal plasma membrane, causing the cell's contents to leak out and cause cell death. Human and animal cells lack this sterol.

For infections of the intestine, vagina, or oral cavity due to *Candida albicans,* physicians often prescribe nystatin. This product of *Streptomyces noursei* is commercially sold in ointment, cream, or suppository form. Often it is combined with antibacterial antibiotics to retard *Candida* overgrowth of the intestines during the treatment of bacterial diseases.

For serious systemic fungal infections, the drug of choice is amphotericin B. This broad spectrum antibiotic, extracted from *Streptomyces nodosus,* is used intravenously with immunocompromised patients, and for treating aspergillosis, cryptococcosis, and candidiasis. However, it causes a wide variety of side effects, including kidney damage, and therefore is used only in progressive and potentially fatal cases.

Imidazoles. Synthetic antifungal drugs, called **imidazoles,** inhibit an enzyme needed to form ergosterol in the fungal plasma membrane. Again, the lack of the sterol causes the cell's contents to leak out and, depending on the fungus and the drug, either brings about an inhibition of fungal growth (fungistatic) or cell death (fungicidal).

The imidazoles include clotrimazole, miconazole, and ketoconazole. Clotrimazole (Gyne-Lotrimin) is used topically for *Candida* skin infections, while the other drugs are used topically as well as internally for systemic diseases. Side effects are uncommon. Miconazole is commercially available as Micatin for athlete's foot and Monistat for yeast infections. Ketoconazole has been used to treat fungal infections in immunocompromised patients, including those with AIDS.

Echinocandins. Because fungi have cell walls, this structure represents another unique site for antifungal drug activity. The semisynthetic **echinocandins** inhibit the synthesis of the fungal cell wall. The drugs are fungistatic against *Aspergillus* and fungicidal against *Candida.* Currently, the only echinocandin approved by the FDA is caspofungin, which is used to treat invasive candidiasis or aspergillosis.

Flucytosine. The antimetabolite drug **flucytosine** is converted in fungal cells to an inhibitor that interrupts nucleic acid synthesis. The drug is active against some strains of *Candida* and *Cryptococcus.* The drug usually is used in combination with amphotericin B.

Griseofulvin. The drug **griseofulvin** is a product of *Penicillium griseofulvum* and is taken orally. Griseofulvin interferes with mitosis by binding to microtubules (see Chapter 3). It is used for fungal infections of the skin, hair, and nails, such as ringworm and athlete's foot.

CONCEPT AND REASONING CHECK

24.14 Assess the emphasis for developing antifungal drugs targeting the fungal plasma membrane or cell wall.

The Goal of Antiprotozoal Agents Is to Eradicate the Parasite

KEY CONCEPT

15. Antiprotozoal drugs target unique aspects of parasites.

Because protozoa also are members of the domain *Eukarya,* antiprotozoal agents attempt to target unique aspects of nucleic acid synthesis, protein synthesis, or metabolic pathways.

Aminoquinolines. The **aminoquinolines** are antimalarial drugs that accumulate in parasitized red blood cells. The aminoquinolines appear to be toxic to the malaria parasite, interfering with the parasite's ability to break down and digest hemoglobin (see Chapter 18). The parasite thus starves or the drug causes the accumulation of toxic products resulting from the degradation of hemoglobin in the parasite.

Quinine was one of the first natural antimicrobials and is derived from the bark of the South American cinchona tree (MICROFOCUS 24.4). It was the primary agent used to treat malaria until substantial resistance to quinine developed. More effective synthetic drugs, such as chloroquine, mefloquine, and primaquine, are now used. Chloroquine remains the drug of choice to treat all species of *Plasmodium* while mefloquine (Lariam) is primarily used for malaria caused by *P. falciparum.* Reports have been made that mefloquine can have rare but serious side-effects, including severe depression, anxiety, paranoia, nightmares, and insomnia. Besides its use in treating malaria, primaquine is

MICROFOCUS 24.4: History
The Fever Tree

Rarely had a tree caused such a stir in Europe. In the 1500s, Spaniards returning from the New World told of its magical powers on malaria patients, and before long, the tree was dubbed "the fever tree." The tall evergreen grew only on the eastern slopes of the Andes Mountains (see figure). According to legend, the Countess of Chinchón, wife of the Spanish ambassador to Peru, developed malaria in 1638 and agreed to be treated with its bark. When she recovered, she spread news of the tree throughout Europe, and a century later, Linnaeus named it *Cinchona* after her.

For the next two centuries, cinchona bark remained a staple for malaria treatment. Peruvian Indians called the bark quina-quina (bark of bark), and the term *quinine* gradually evolved. In 1820, two French chemists, Pierre

Leaves of a cinchona tree.

Pelletier and Joseph Caventou, extracted pure quinine from the bark and increased its availability still further. The ensuing rush to stockpile the chemical led to a rapid decline in the supply of cinchona trees from Peru, but Dutch farmers made new plantings in Indonesia, where the climate was similar. The island of Java eventually became the primary source of quinine for the world.

During World War II, Southeast Asia came under Japanese domination, and the supply of quinine to the West was drastically reduced. Scientists synthesized quinine shortly thereafter, but production costs were prohibitive. Finally, two useful substitutes were synthesized in chloroquine and primaquine. Today, as resistance to these drugs is increasingly observed in malarial parasites, scientists once again are looking for another "fever tree" to help control malaria.

---

used with clindamycin to treat *Pneumocystis* pneumonia (Chapter 17).

Sulfonamides. Similar to bacterial cells, protozoal cells require folic acid for the synthesis of nucleic acids, but are unable to absorb it from the environment. Therefore, the same sulfonamides that are used against bacterial pathogens will produce a similar result with the protozoa. Another drug, diaminopyrimidine, can be used with trimethoprim to achieve a synergistic effect. These two drugs in combination with sulfamethoxazole are effective in treating toxoplasmosis (see Chapter 18).

Nitroimidazoles. Two common drugs in the **nitroimidazole** family, metronidazole (Flagyl) and tinidazole, interfere with DNA synthesis. These drugs are effective in the treatment of amebiasis, giardiasis, and trichomoniasis (see Chapter 18). Metronidazole can be mutagenic to the host.

Heavy Metals. Arsenic and antimony derivatives have been used since ancient times. The arsenic derivative, melarsoprol (Mel B) is used to

treat African trypanosomiasis, while antimoniate is effective against leishmaniasis (see Chapter 18). Although the drugs are toxic in the nervous system and kidneys, they primarily affect the intense metabolism of protozoal cells.

Other Drugs. Several of the antibiotics that affect bacterial protein synthesis also inhibit protein synthesis in protozoa. Clindamycin and the tetracyclines, such as doxycycline, are used to treat malaria. Another protein synthesis inhibitor, paromomycin, can be used against cryptosporidiosis (see Chapter 18).

Pentamidine, a drug of unknown action, has been used against the parasites causing leishmaniasis. **Artemisinin** is used to treat multi-drug resistant strains of *P. falciparum*. The drug, isolated from the shrub *Artemisia annua* (sweet wormwood) has long been used by Chinese herbalists (**FIGURE 24.12**). In red blood cells, the drug releases free radicals that destroy the malarial parasites.

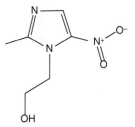

Metronidazole

Artemisinin

■ Free radicals: Highly reactive atoms or groups of atoms with unpaired electrons.

CONCEPT AND REASONING CHECKS

24.15 Identify the cellular structures targeted by the antiprotozoal agents.

Praziquantel

FIGURE 24.12 The Source of Artemisinin. *Artemisia* shrubs are grown in Africa, as indicated in this harvest by a Kenyan farmer. »» What is the mode of action of artemisinin?

Antihelminthic Agents Are Targeted at Nondividing Helminths

KEY CONCEPT

24.16   Antihelminthic drugs affect helminth neuromuscular coordination, carbohydrate metabolism, or microtubule structure.

The helminths are eukaryotic, multicellular parasites. Although most antimicrobial drugs are targeted at actively dividing cells of the pathogen, the mechanism of action of most antihelminthic drugs targets the nondividing organisms.

Praziquantel is thought to change the permeability of the parasite plasma membranes of cestodes and trematodes. The permeability change causes the inflow of calcium ions, which leads to muscle contraction and paralysis in the parasite. Parasites then cannot feed and eventually die. Praziquantel has been the drug of choice for mass therapy campaigns, including the treatment of schistosomiasis (see Chapter 18 and MicroFocus 16.1).

Mebendazole inhibits uptake of glucose and other nutrients by adult and larval worms from the host intestine where helminths are located. Without the ability to carry out ATP synthesis, the parasites will die. The drug has a wide spectrum, affecting many nematodes and cestodes. The drug also disrupts microtubules and cell division.

The **avermectins** are antihelminthic drugs derived from *Streptomyces avermitilis*. The drugs are effective in extremely low doses and against a wide variety of roundworms. The avermectins have an effect on the nematode nervous system such that muscle paralysis results. Ivermectin (Stromectol) is one drug currently used to treat onchocerciasis (river blindness) and is being investigated for treatment of a number of nematode infections. **TABLE 24.3** summarizes the characteristics of the antifungal and antiparasitic agents.

CONCEPT AND REASONING CHECKS

24.16   Provide several reasons why there are fewer effective antifungal and antiparasitic drugs to treat infections.

24.6 Antibiotic Assays and Resistance

The substantial variety of antibiotics and chemotherapeutic agents developed since the 1930s necessitates that medical professionals know which one is best for the patient under the circumstances of the infection. Accordingly, antibiotic sensitivity assays can be performed. The assay also helps determine whether a microorganism is resistant to a particular antibiotic, a problem that has developed into a major concern of modern medicine, as described in the introduction to this chapter.

There Are Several Antibiotic Susceptibility Assays

KEY CONCEPT

17.   Antibiotic susceptibility assays are used to study the inhibition of a test organism by one or more antimicrobial agents.

Two general methods are in common use to test the susceptibility of a bacterial pathogen to specific antibiotics. These are the tube dilution method and the agar disk diffusion method.

TABLE

24.3 A Summary of Antifungal and Antiparasitic Drugs

Antifungal Agent	Source	Antifungal Spectrum	Activity Impeded	Side Effects
Polyenes Nystatin	*Streptomyces noursei*	*Candida albicans*	Cell membrane function	Few reported
Amphotericin B	*Streptomyces nodosus*	Systemic infections	Cell membrane function	Fever, gastrointestinal distress
Imidazoles Clotrimazole Ketoconazole Miconazole	Synthetic	Superficial infections	Inhibit sterol synthesis	Few reported
Echinocandins Caspofungin	Semisynthetic	*Aspergillus, Candida*	Cell wall synthesis	Low incidence
Flucytosine	Synthetic	*Candida, Cryptococcus*	Nucleic acid synthesis	Adverse renal and liver effects
Griseofulvin	*Penicillium griseofulvum*	Superficial infection	Microtubules	Occasional allergic reactions

Antiparasitic Agent	Source	Antiparasite Spectrum	Activity Impeded	Side Effects
Aminoquinolines Quinine Chloroquine Mefloquine Primaquine	Synthetic	*Plasmodium* species	Parasite digestion of hemoglobin	Severe neurological and behavioral side effect possible with mefloquine
Sulfonamides Diaminopyramidine	Synthetic	*Toxoplasma, Plasmodium*	Folic acid metabolism	Kidney and liver damage; allergic reactions
Nitroimidazoles Metronidazole Tinidazole	Synthetic	*Trichomonas*	DNA synthesis	Tumors in mice
Artemisinin	*Artemisia annua*	*Plasmodium* species	Free radicals destroy parasites	Few reports of adverse effects
Praziquantel	Synthetic	*Schistosoma*	Membrane permeability	Result from killing of parasites
Mebendazole	Synthetic	Broad	Glucose uptake	Diarrhea, stomach pain
Ivermectin	*Streptomyces avermitilis*	*Ascaris, Enterobius*	Nervous system	Low incidence

The **tube dilution method** determines the lowest concentration of antibiotic that will prevent growth of the pathogen. This amount is known as the **minimum inhibitory concentration (MIC)**. To determine the MIC, the microbiologist prepares a set of tubes with different concentrations (dilutions) of a particular antibiotic (FIGURE 24.13). Each tube is inoculated with an identical number of cells, incubated, and examined for the growth of bacterial cells. The extent of growth diminishes as the concentration of antibiotic increases, and eventually an antibiotic concentration is observed at which growth no longer occurs. This is the MIC.

The second method is an antibiotic susceptibility test, which operates on the principle that an antibiotic will diffuse from a paper strip or disk into an agar medium containing a test organism (FIGURE 24.14A). In one version of the test, called the **E Test**, a paper strip impregnated with a marked gradient of antibiotic is placed on the plate. As the drug diffuses into the agar culture, the higher drug concentrations will inhibit growth of a susceptible bacterium. The inhibition of growth is observed as a clear, oval halo on the plate. By reading the number on the strip (antibiotic concentration) where growth intersects the strip, the MIC can be determined.

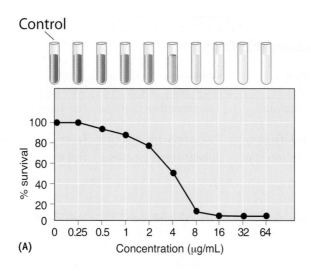

(A)

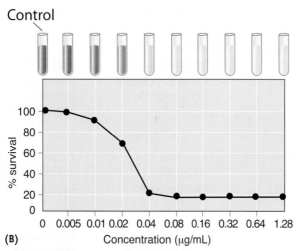

(B)

FIGURE 24.13 Determination of Minimal Inhibitory Concentration. The minimal inhibitory concentration (MIC) for ampicillin against (**A**) *Staphylococcus aureus* and (**B**) *Streptococcus pyogenes*. From the tube dilutions (top), the MIC can be plotted and determined. »» What are the MICs for *Staphylococcus aureus* and *Streptococcus pyogenes*? *Source:* Data modified from: Atwal, R. (2003). *JARVM* 1(3).

In another version of the test, called the **disk diffusion method**, inhibition of bacterial growth is again observed as a failure of a susceptible bacterium to grow, leaving a clear halo around the paper disk. A common application of the agar disk diffusion method is the **Kirby-Bauer test**, named after W. M. Kirby and A. W. Bauer, who developed it in the 1960s. This procedure determines the susceptibility of a microorganism to a series of antibiotics and is performed according to standards established by the FDA. A more advanced procedure is noted in FIGURE 24.14B, C .

CONCEPT AND REASONING CHECKS

24.17 Explain the significance of knowing a drug's minimal inhibitory concentration.

There Are Four Major Mechanisms of Antibiotic Resistance

KEY CONCEPT

18. Bacterial species have several ways to generate resistance to antibacterial agents.

Since the mid-1900s, an increasing number of bacterial species have become resistant to antibiotics (FIGURE 24.15); that is, the bacterial cells are not effectively inhibited or killed with the use of antibiotics, allowing the resistant cells to continue multiplying in the presence of therapeutic levels of an antibiotic. Resistance develops in the following two ways.

Horizontal Gene Transfer. Sometimes resistance comes from bacterial cells picking up resistance genes from other bacterial cells by

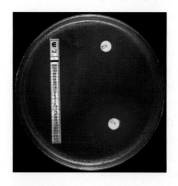

(A)

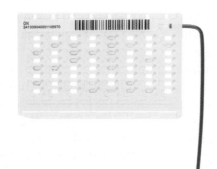

(B)

(C)

FIGURE 24.14 Antimicrobial Susceptibility Testing. (**A**) This photograph shows both the disk diffusion assay (right half) and the E Test (left half). In the disk diffusion assay, the bacterium is susceptible to erythromycin (E15 disk) but resistant to tetracycline (TE30 disk). In the E test, the antibiotic-containing strip can be used to determine the minimum inhibitory concentration (MIC). (**B**) In a modern automated system, bacterial cells are automatically inoculated into wells on a card, each well containing a different antibiotic. The card is then incubated, and the presence or absence of growth is assayed by a computer (**C**). A printout related the organism's susceptibility (no growth) to the antibiotic, or its resistance (growth). »» Using the E test, what is the MIC in A?

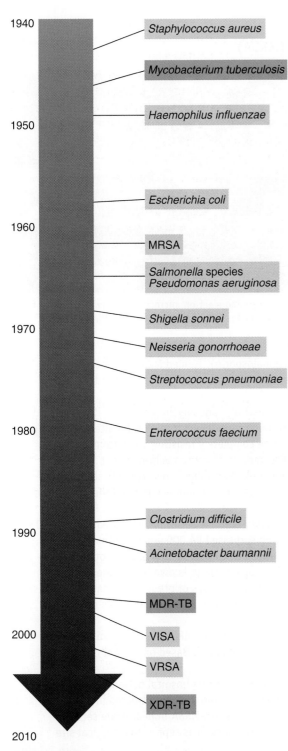

FIGURE 24.15 **Timeline for Appearance of Antibiotic Resistance.** This timeline shows the approximate dates when some notable bacterial pathogens (pink = gram-negative species; purple = gram-positive species; red = acid-fast species) were identified as being antibiotic resistant. MRSA (VISA, VRSA) = methicillin-resistant (vancomycin intermediary resistant; vancomycin-resistant) *Staphylococcus aureus*; MDR (XDR)-TB = multidrug-resistant (extensively drug resistant) tuberculosis.
»» What are the two ways the bacteria species gained resistance?

means of **horizontal gene transfer** (see Chapter 9). These transfer mechanisms can allow bacterial cells to acquire antibiotic resistance genes from donor bacterial cells by undergoing **conjugation** or **transformation**, or by **transformation**, where bacteriophages pass resistance genes between bacterial cells.

Mutations. Other times resistance is of its own making, such as through **mutations**, those rare, spontaneous changes in a bacterial cell's genetic material. For example, one strain of *Staphylococcus aureus* in an infected patient evolved resistance to vancomycin over a period of 30 months. During that time the *S. aureus* strain accumulated 18 sequential mutations, making it resistant not only to vancomycin but also to several other antibiotics. The scary part is that the patient was never treated with these other antibiotics! The mutations produced "spontaneous" resistance. Such resistant organisms are increasingly responsible for human diseases of the intestinal tract, lungs, skin, and urinary tract. Those in intensive care units and burn wards are particularly vulnerable, as are infants, the elderly, and the infirm. Common diseases like bacterial pneumonia, tuberculosis, streptococcal sore throat, and gonorrhea, which until recently succumbed to a single dose of antibiotics, are now among the most difficult to treat.

Any bacterial cells that acquire resistance genes, whether by gene transfer with other bacteria or mutation, have the ability to resist one or more antibiotics. Because bacteria can collect multiple resistance traits over time, they can become resistant to many different families of antibiotics. One of the major concerns to public health officials is the bacterial species *Staphylococcus aureus* (MicroFocus 24.5).

As already mentioned, microorganisms in the environment have been involved in chemical warfare with their neighbors for hundreds of millions of years. To counteract the production of antibiotics by some microbes, others developed self-protective (resistance) mechanisms to the natural antibiotics. Thus, it is not surprising that microbes quickly develop resistance to the natural antibiotics used to combat infectious diseases. Bacterial resistance to antibiotics can develop through four major mechanisms (**FIGURE 24.16**).

MICROFOCUS 24.5: Public Health

Methicillin-Resistant *Staphylococcus aureus* (MRSA)

Staphylococcus aureus, often referred to simply as "staph," is a gram-positive bacterial species that commonly colonizes the skin or the nose of some 25% to 30% of healthy people—without causing an infection (see figure). However, if it invades the body, the organism can cause septicemia, arthritis, pneumonia, endocarditis, and meningitis. In fact, *S. aureus* is involved in more than 250,000 infections in Americans each year, primarily involving patients in hospitals and nursing homes.

Of major concern to health officials is that over the years strains of *S. aureus* have developed resistance to penicillin and other broad-spectrum drugs. Then, several decades ago, *S. aureus* strains were identified that were resistant to the action of methicillin

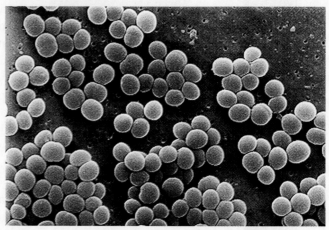

A false-color scanning electron microscope image of clusters of *Staphylococcus aureus* cells. Courtesy of Janice Haney Carr/CDC

and related beta-lactam antibiotics (e.g., penicillin and cephalosporin). Termed MRSA, for methicillin-resistant *S. aureus*, the strains could only be treated with more potent, expensive, and often toxic drugs, like vancomycin.

In 1997, a MRSA strain was identified with intermediate (partial) vancomycin resistance; it was named VISA, for vancomycin-intermediate resistant *S. aureus*. Although researchers found useful alternative drug combinations, they were worried that soon *S. aureus* would be completely resistant to vancomycin, often the drug of "last resort" for MRSA patients. Indeed, in 2001, reports of vancomycin-resistant *S. aureus* (VRSA) were reported in many countries, indicating global spread. Vancomycin resistance is still a rare occurrence. Unfortunately, VRSA may also be resistant to meropenem and imipenem, two other antibiotics used to treat MRSA patients.

Prevalence. In 2007, a report in the *Journal of the American Medical Association* estimated that in 2005 the number of Americans developing a serious MRSA infection surpassed 94,000 and more than 18,000 died. While most were invasive MRSA infections that were traced to a hospital stay or some other healthcare exposure, about 15% of invasive infections occurred in people with no known healthcare risk. Two-thirds of the 85% of MRSA infections that could be traced to hospital stays or other health care exposures occurred among people who were no longer hospitalized. People over age 65 were four times more likely than the general population to get a MRSA infection.

Today, MRSA is often sub-categorized as hospital-associated MRSA (HA-MRSA) or community-associated MRSA (CA-MRSA), depending upon how the strain was acquired. Based on current data, these are distinct strains of *S. aureus*.

HA-MRSA. The hospital-associated strain occurs most frequently among hospital patients who undergo invasive medical procedures, or who have weakened immune systems and are being treated in hospitals and healthcare facilities, such as nursing homes and dialysis centers. HA-MRSA in healthcare settings commonly causes serious and potentially life-threatening infections, such as septicemia, surgical site infections, or pneumonia.

Transmission commonly comes from HA-MRSA patients who already have an infection or who carry the bacterial cells on their bodies (colonized) but are asymptomatic. The main mode of transmission to other patients is through human hands, especially those of healthcare workers after contact with infected or colonized patients. If appropriate hand hygiene, such as washing with soap and water or using an alcohol-based hand sanitizer, is not performed, the bacterial cells can be spread when the healthcare worker touches other patients.

CA-MRSA. A CA-MRSA infection is defined as a MRSA infection occurring in otherwise healthy people who have not been recently (within the past year) hospitalized, or had a medical procedure (such as dialysis, surgery, catheters). Studies have shown that rates of CA-MRSA infection are growing fast. One study of children in south Texas found that cases of CA-MRSA had a 14-fold increase between 1999 and 2001. In 2007, at least three school-age children died from MRSA infections, causing the schools to close while extensive disinfection measures were carried out and control measures instituted.

The CA-MRSA strain, called USA 300, is very contagious. However, about 75% of CA-MRSA infections are usually skin infections, causing abscesses, boils, and other pus-filled lesions that can be treated effectively. Health experts worry though that USA 300 will soon display enhanced virulence and cause more severe illness to vital organs, leading to widespread infection (sepsis), toxic shock syndrome, and pneumonia. It is not known why some healthy people develop CA-MRSA skin infections that are treatable whereas others infected with the same strain develop severe, fatal infections. CA-MRSA skin infections have been identified among certain populations that share close quarters or experience more skin-to-skin contact. Examples are team athletes, military recruits, and prisoners. However, more and more CA-MRSA infections are being seen in the general community as well.

CA-MRSA also appears to infect younger people. In a Minnesota study published in the *Journal of the American Medical Association*, the authors reported that the average age of people with MRSA in a hospital or healthcare facility was 68; the average age of a person with CA-MRSA was only 23.

Protection. With the spread of antibiotic resistance, there are procedures that you can take to lessen the chances of a MRSA infection whether in the community or as a patient in a healthcare facility (see Table)—of course, hand washing is at the top of the list. Recently, scientists have been working on a *S. aureus* vaccine that is showing good success in mice. Hopefully, such results will also be seen when human studies are started.

TABLE

Measures to Protect Yourself in Community and Healthcare Settings

Community

Wash hands frequently.
Cover cuts and scrapes with a bandage.
Do not touch wounds (or bandages) on other people.
Do not share personal items such as towels and razors.

Healthcare

Request all hospital staff to wash their hands before touching you.
Ask what the hospital/healthcare facility is doing to prevent MRSA infections.
Make sure IV tubes and catheters are inserted and removed under sterile conditions.

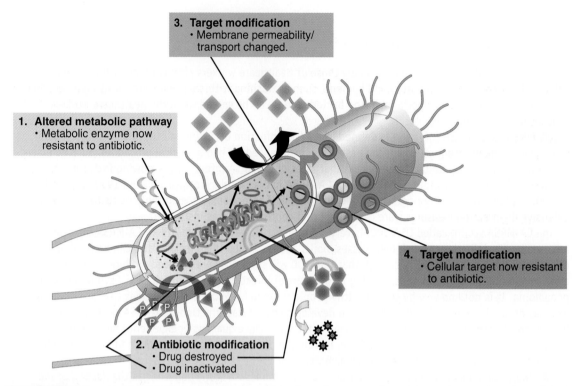

3. Target modification
• Membrane permeability/ transport changed.

1. Altered metabolic pathway
• Metabolic enzyme now resistant to antibiotic.

4. Target modification
• Cellular target now resistant to antibiotic.

2. Antibiotic modification
• Drug destroyed
• Drug inactivated

FIGURE 24.16 **Types of Antibiotic Resistance.** Through horizontal gene transfers and genetic mutations, bacterial cells can develop different mechanisms of antibiotic resistance. »» Which of the resistance mechanisms appears to be more "defensive" in action and "offensive" in action? Please explain why.

1. Altered Metabolic Pathway. Resistance to sulfonamides may develop when the drug fails to bind with enzymes that synthesize folic acid. This can come about because the enzyme's structure has changed. Moreover, drug resistance may be due to an altered metabolic pathway in the microorganism—a pathway that bypasses the reaction inhibited by the drug.

2. Antibiotic Modification. Resistance can arise from the microorganism's ability to enzymatically split apart (destroy) the antibiotic. The production of penicillinases (beta-lactamases) by penicillin-resistant gonococci is an example. By breaking the beta-lactam ring, penicillin- or cephalosporin-resistant bacterial cells have a mechanism to prevent blockage of cell wall synthesis.

The aminoglycosides normally block mRNA translation on bacterial ribosomes. Aminoglycoside-resistant bacterial cells have developed ways to enzymatically modify (inactivate) aminoglycosides so they cannot bind to ribosomes. The enzymatic modifications, including phosphorylation, acetylation, or adenylation of the antibiotic, prevent binding of the antibiotic to the ribosome.

3. Reduced Permeability/Active Export of Antibiotics. Another resistance mechanism involves the ability of microbes to prevent drug entry into the cytoplasm. For example, changes in membrane permeability in penicillin-resistant *Pseudomonas* prevent the antibiotic from entering the cytoplasm. Bacterial species such as *E. coli* and *S. aureus,* which are resistant to tetracyclines, actively export (pump out) the drug. Cytoplasmic and membrane proteins in these bacterial cells act as pumps to remove the antibiotic before it can affect the ribosomes in the cytoplasm.

4. Target Modification. A fourth route leading to microbial resistance involves altering the drug target. Some streptomycin-resistant bacterial species can modify the structure of their ribosomes so the antibiotic cannot bind to the ribosome and protein synthesis is not inhibited. Other targets include RNA polymerase and enzymes involved in DNA replication. TABLE 24.4 summarizes the resistance mechanisms in relation to the antibiotics mode of action and class of drugs.

CONCEPT AND REASONING CHECKS

24.18   How do the resistance mechanisms confer resistance to an antibiotic?

Phosphorylation, acetylation, or adenylation: The addition, respectively, of a phosphate group, an acetyl group, or an adenyl group to the structure of an antibiotic.

TABLE 24.4 A Summary of Bacterial Resistance Mechanisms

Antibiotic Target	Antibiotic Class	Resistance Mechanisms
Inhibition of cell wall synthesis	Penicillins, cephalosporins, carbapenems, vancomycin	Altered wall composition; drug destruction by beta-lactamases
Membrane permeability and transport	Polypeptide antibiotics	Altered membrane structure
Inhibition of DNA synthesis	Fluoroquinolones	Inactivation of drug; altered drug target
Inhibition of RNA synthesis	Rifampin	Altered drug target
Inhibition of protein synthesis	Aminoglycosides, chloramphenicol, tetracyclines, macrolides, clindamycin, streptogramins, oxazolidinones	Altered membrane permeability; drug pumping; antibiotic inactivation; altered drug target
Inhibition of folic acid synthesis	Sulfonamides	Alternate metabolic pathway; altered drug target

Antibiotic Resistance Is of Grave Concern in the Medical Community

KEY CONCEPT

19. Antibiotic misuse and abuse encourage the emergence of resistant forms.

Antibiotic resistance is primarily the result of improper use of antibiotics. For example, drug companies promote antibiotics heavily, patients pressure doctors for quick cures, and physicians sometimes write prescriptions to avoid ordering costly tests to pinpoint the patient's illness. In addition, people may diagnose their own illness and take leftover antibiotics from their medicine chests for ailments where antibiotics are useless. Moreover, many people fail to complete their prescription, and some organisms remain alive to evolve to resistant forms. The survivors proliferate well because they face reduced competition from susceptible organisms **FIGURE 24.17**.

Hospitals are another forcing ground for the emergence of resistant bacterial species. In many cases, physicians use unnecessarily large doses of antibiotics to prevent infection during and following surgery. This increases the possibility that resistant strains will overgrow susceptible strains and subsequently spread to other patients, causing a **superinfection**. Antibiotic-resistant *E. coli*, *P. aeruginosa*, *Serratia marcescens*, and *Proteus* species now are widely encountered causes of illness in hospital settings.

Antibiotics also are abused in developing countries where they often are available without prescription (**FIGURE 24.18**). Some countries permit the over-the-counter sale of potent antibiotics, and large doses encourage resistance to develop. Between 1968 and 1971, some 12,000 people died in Guatemala from shigellosis attributed to antibiotic-resistant *Shigella dysenteriae*.

Moreover, antibiotic use is widespread in livestock feeds. Each year, more than 500 metric tons of antibiotics are produced and used worldwide (**FIGURE 24.19**). An astonishing 40% or more of all the antibiotics used in the United States find their way into animal feeds. Unfortunately, these antibiotics are not being used because the animals are sick. Rather, they are used for "nontherapeutic" purposes, mainly to make the animals grow larger and faster, and to compensate for crowded, unsanitary production methods. By killing off less hardy bacterial species, chronic low doses of antibiotics fed to animals create ideal growth environments for resistant strains. If these strains can be transferred to humans through contaminated meat, the resistant organisms may cause intractable illnesses.

Allied to the problem of antibiotic resistance is the transfer of the resistance between bacterial organisms. Researchers have demonstrated that plasmids and transposons account for the movement of antibiotic-resistant genes among bacterial species (see Chapter 9). Thus, the resistance in a

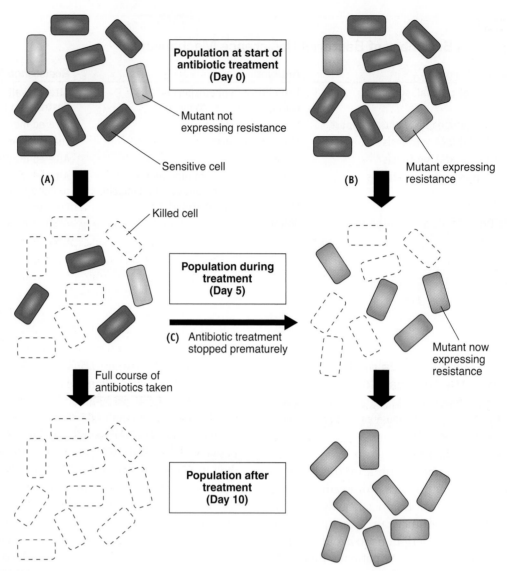

Population at start of antibiotic treatment (Day 0)

Mutant not expressing resistance

Sensitive cell

(A)

(B)

Mutant expressing resistance

Killed cell

Population during treatment (Day 5)

(C) Antibiotic treatment stopped prematurely

Mutant now expressing resistance

Full course of antibiotics taken

Population after treatment (Day 10)

FIGURE 24.17 **The Possible Outcomes of Antibiotic Treatment.** (**A**) Ideally, with a complete course of antibiotics, all pathogens will be destroyed. However, (**B**) if there are some resistant cells in the infecting population, they will survive and grow without any competition. If the person 'feels better' and prematurely stops taking the antibiotic (**C**), any mutant cells may have had the opportunity to express antibiotic resistance, survive, and again grow without any competition. »» What factor(s) would have triggered the mutant to start expressing antibiotic resistance?

relatively harmless bacterial species may be passed to a pathogenic one where the potential for disease is then increased by resistance to standardized antibiotic treatment.

Antibiotics have been known as miracle drugs, but they are becoming overworked miracles. Some researchers suggest that antibiotics should be controlled as strictly as narcotics. Curbing the overuse of antibiotics through public and physician education and govern-

mental regulation is a useful approach as well (although this approach will be particularly difficult, because the antibiotics market is currently worth more than $25 billion per year in sales). The antibiotic roulette that is currently taking place should be a matter of discussion to all individuals concerned about infectious disease, be they scientist or student.

CONCEPT AND REASONING CHECKS

24.19 Summarize the misuses and abuses of antibiotics.

FIGURE 24.18 Antibiotics On Sale. A sign in front of a foreign pharmacy selling antibiotics over the counters. »» What are the risks of buying such over-the-counter drugs?

New Approaches to Antibiotic Therapy Are Needed

KEY CONCEPT

20. The discovery of new targets and the development of new antibiotics may help fight antibiotic resistance.

Dealing with emerging antibiotic resistance is a major problem confronting contemporary researchers. As reports of antibiotic resistance and harmful side effects appear in the literature, scientists hasten their search for new approaches to antibiotic therapy as well as identifying new targets to which pathogens are susceptible. For example, it should be possible to prevent a bacterial cell from pumping an antibiotic like tetracycline out of its cytoplasm.

One targeted approach is to focus on the lipopolysaccharide (LPS) in the outer membrane of gram-negative bacteria (see Chapter 4). The LPS contains several unusual carbohydrates representing possible targets for new drugs.

Another approach is to interfere with an enzyme called DNA adenine methylase, which bacterial cells use to coat DNA with methyl groups and regulate DNA replication and repair. Researchers have significantly reduced the virulence of a *Salmonella* species by disabling the gene encoding the methylating enzyme. The researchers now

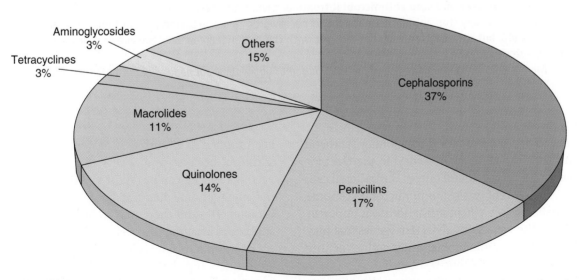

FIGURE 24.19 Annual Global Production of Antibiotics. More than half of the antibiotic production and use ends up in livestock. »» What bacterial structure is being targeted by more than 50% of the antibiotics produced?

MICROFOCUS 24.6: Public Health
Preventing Antibiotic Resistance: Steps You Can Take

The Centers for Disease Control and Prevention's (CDC) Campaign to Prevent Antimicrobial Resistance (http://www.cdc.gov/drugresistance/healthcare/) aims to prevent antimicrobial resistance in healthcare settings. However, the four main strategies—prevent infection, diagnose and treat infection, use antimicrobials wisely, and prevent transmission—can be slightly modified for the general public.

A. Prevent Infections
Prevention will decrease antibiotic use—if:
Step 1. You are vaccinated. Keeping up with all vaccinations will limit any infection by a pathogen. In addition, children should be immunized on schedule, people should get annual influenza vaccinations and keep up with other required booster shots, and those over 60 should receive a pneumococcal vaccination.
Step 2. Avoid use of indwelling instruments, if possible. Indwelling instruments are a major source for potential infection in a healthcare setting. Request that catheters and other indwelling instruments be used only when essential and for minimal duration. Catheters should be removed aseptically when no longer essential.
The bottom line: Maintaining high herd immunity and minimal use of indwelling instruments will eliminate or minimize any need for antibiotics.

B. Diagnose and Treat Infection Effectively
Proper diagnosis and infection treatment will decrease antibiotic use—if:
Step 3. The correct pathogen is identified. The clinical lab should attempt to identify the pathogen, if not known, and then prescribe the appropriate narrow-spectrum antibiotic.
Step 4. Reliable experts are involved. For complicated or serious infections, consult an infectious disease expert. Consider a second opinion if unsatisfied with the antibiotic treatment or recovery.
The bottom line: Appropriate diagnosis and antibiotic treatment will limit the chances of resistance development.

C. Use Antimicrobials Wisely
Proper use of antibiotics will decrease antibiotic use—if:
Step 5. You are an informed "drug user." Minimize use of broad-spectrum antibiotics as these can lead to more rapid antibiotic resistance. Also, avoid chronic or long-term antimicrobial use.
Step 6. You take and stop antimicrobial treatment when indicated. When cultures are negative and infection is unlikely stop antibiotic treatment. If positive, do not stop taking the antibiotic treatment prematurely, but once the infection has resolved with the prescribed course of antibiotic treatment, antibiotic use should stop.
The bottom line: Appropriate use of and completion of antibiotic treatment will limit potential resistance.

hope to develop a chemical compound (synthetic antibiotic?) to neutralize the enzyme, thereby pinpointing a protein found only in the bacterial cell.

The current explosion in microbial genome sequencing (see Chapter 9) affords an opportunity to discover essential bacterial genes that may be targets for antimicrobial drugs and to identify other forms of antibiotic resistance.

Because many of the newer antibiotics are simply minor modifications of existing drugs, there is a need to find new and unique antibiotics. The marine microbial community has barely been tapped as a source of new antibiotics. In addition, because only a very small portion of bacterial species can be cultured, 90% to 99% of the metagenome has not been sampled (see Chapter 9)—and there may be many beneficial antibiotics awaiting discovery.

Many microbiology labs around the world are sequencing the genomes of dozens of different pathogens thought to contain genes for undiscovered antibiotics. For example, the genus *Streptomyces* is the source of more than 65% of the antibiotics in current use. Almost certainly there

Prevent Transmission

Preventing transmission will decrease antibiotic use—if:

Step 7. The chain of contagion is broken. Stay home when sick and cover your mouth when you cough or sneeze. Educate family members and coworkers as to proper hygiene.

Step 8. You and others perform hand hygiene. Use alcohol-based handrubs or wash your hands often, especially when sick. Encourage the same for all family members and coworkers.

The bottom line: Proper hygiene will limit pathogen transmission and minimize the need for antibiotics and the generation of resistance.

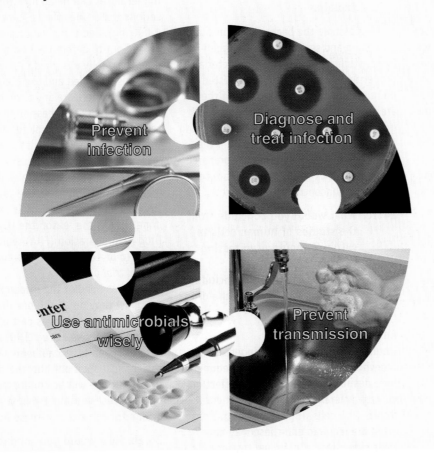

are more antibiotics produced by this genus, as the opener to Chapter 9 describes. Through genome sequencing, researchers have identified clusters of genes in *Streptomyces coelicolor* that may be sources for new antibiotic agents. What other, perhaps novel, natural products exist is an area of active research.

Although antimicrobial drugs probably are not the magic bullets once perceived by Ehrlich, new drug discovery may provide us with many new antimicrobial agents to fight pathogens and resistance. MicroInquiry 24 looks at what is involved in bringing a drug to the pharmacy shelf and medicine cabinet.

The public also needs to take responsibility in limiting the spread of antibiotic resistance. To this end, the CDC has promoted a multi-step program designed to prevent microbial resistance. MicroFocus 24.6 outlines these steps.

CONCEPT AND REASONING CHECKS

24.20 Assess the need for new and novel antibiotics and targets for those antibiotics.

MICROINQUIRY 24

Testing Drugs—Clinical Trials

Before any antibiotic or chemotherapeutic agent can be used in therapy, it must be tested to ensure its safety and efficacy. The process runs from basic biomedical laboratory studies to approval of the product to improve health care. On average it takes over eight years to study the drug in the laboratory (preclinical testing), test it in animals, and finally run human clinical trials.

In this MicroInquiry, we examine a condensed version of the steps through which a typical antibiotic would pass. Some questions asked are followed by the answer so you can progress further in your analysis. Please try to answer these questions before progressing through the scenario. Also, methods are used that you have studied in this and previous chapters. A real pharmaceutical company might have other, more sophisticated methods available. Answers can be found in **Appendix D**.

The Scenario

You are head of a drug research group with a pharmaceutical company that has isolated a chemical compound (let's call it FM04) that is believed to have antibacterial properties and thus potential chemotherapeutic benefit. As head researcher, you must move this drug through the development pipeline from testing to clinical trials, evaluating, at each step, whether to proceed with further testing.

Preclinical Testing

Many experiments need to be done before the chemical compound can be tested in humans. Because FM04 is "believed" to have antibacterial properties,

24.1a. What would be the first experimental tests to be carried out?

You would experimentally test the drug on bacterial cells in culture to ascertain its relative strength and potency as an antibacterial agent. This testing might include the agar disk diffusion method described in this chapter. The figure to the right shows three diffusion disk plates, each plated with a

different bacterial species (*Escherichia coli, Staphylococcus aureus,* and *Pseudomonas aeruginosa*) and disks with different concentrations of FM04 (0, 1, 10, 100 µg).

24.1b. What can you conclude from these disk diffusion studies? Should drug testing continue? Explain.

Let's assume the bacterial studies look promising and several months of additional studies confirm the antibacterial properties. FM04 now would be tested on human cells in culture to see if there are any toxic effects on metabolism and growth. **Table A** presents the results using the same drug concentrations from the bacterial studies. (The drug was prepared as a liquid stock solution to prepare the final drug concentration in culture.)

24.1c. What would you conclude from the studies of human cells in culture? Should drug testing continue? Explain.

24.1d. What was the point of including a solution with 0 µg of the drug?

Animal Testing

When drugs are injected or ingested into the whole body, concentrations required for a desired chemotherapeutic effect may be very different from the experimental results with cells. Higher concentrations often are required and these may have toxic side effects. In animal testing, efforts should be made to use as few animals as possible and they should be subjected to humane and proper care. Let's assume the research studies of FM04 on mice used oral drug concentrations of

1, 10, 100, and 1,000 mg. **Table B** presents the simplified results.

24.2a. What would you conclude from the mouse studies? Should drug testing continue? Explain.

Because the results look positive and higher doses can be administered without serious toxic side effects (except at 1,000 mg), additional studies would be carried out to determine how much of the drug is actually absorbed into the blood, how it is degraded and excreted in the animal, and if there are any toxic breakdown products produced. These studies could take a few years to complete. Again, for the sake of the exercise, let's say the studies with FM04 remain promising.

Although mice are quite similar to humans physiologically, they are not identical, so human clinical trials need to be done. This is the reason the U.S. Food and Drug Administration (FDA) requires all new drugs to be tested through a *three-phase clinical trial period*.

First, as head of the research team, you must provide all the preclinical and animal testing results as part of an FDA *Investigational New Drug* (IND) *Application* before human tests can begin. The application must spell out how the clinical investigations will be conducted. As head of the investigational team, you must design the protocol for these human trials.

24.2b. How would you conduct the human tests to see if FM04 has positive therapeutic effects as an antibiotic in humans?

Briefly, there are many factors to take into consideration in designing the protocol.

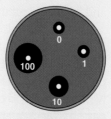

E. coli

S. aureus

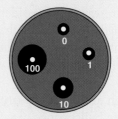

P. aeruginosa

(i) How will the drug be administered? Let's assume it will be oral in the form of pills.

(ii) You need to recruit groups of patients (volunteers) of similar age, weight, and health status (healthy patients as well as patients with the infectious condition to which the drug may be therapeutic). Note that the bacterial studies suggested FM04 had its most potent effect on *Staphylococcus aureus,* a gram-positive species. Perhaps patients with a staph infection would be recruited.

(iii) Eventually, volunteers (patients) need to be split into two groups, one that receives the drug and one that receives a placebo (an inactive compound that looks like the FM04 pill).

(iv) The clinical studies should be carried out as "double blind" studies. In these studies, neither the patients nor the investigators know which patients are getting the drug or placebo. This prevents patients and investigators from "wishful thinking" as to the outcome of the studies.

In actual clinical studies there are even more factors to consider, but we will assume that FM04 wins approval for clinical testing.

Clinical Trials

In *Phase 1 studies,* a small number (20 to 100) of healthy patients are treated with the test drug at different concentrations (doses) to see if there are any adverse side effects. These studies normally take at least several months. If unfavorable side effects are minimal, Phase 2 trials begin.

Phase 2 studies use a larger population of patients (several hundred) who have the disease the drug is designed to treat. Here, it is especially important to split the patients into the test and control groups because the major purpose of these trials is to study drug effectiveness. There needs to be a control group to accurately

contrast effectiveness. These trials can take anywhere from several months to two years to complete. Provided there are no serious side effects or toxic reactions, or a lack of effectiveness, Phase 3 trials begin. There is an ethical question that can arise at this stage in the clinical trials regarding making the drug available, especially if the drug shows signs of reversing an illness that might be life-threatening. Even medical experts disagree as to the answer.

24.3 Is it ethical to give ill patients placebos when effective treatment appears available?

Phase 3 studies include several hundred to a few thousand patients. The trials can take several years to complete because the purpose is to evaluate safety, dosage, and effectiveness of the drug.

If the clinical trials are positive, as head researcher you can recommend that the pharmaceutical manufacturer apply to the FDA for a New Drug Application (NDA). The application is reviewed by an internal FDA committee that examines all clinical data, proposed labeling, and manufacturing procedures. The most important questions that need to be addressed are:

(i) Were the clinical studies well controlled to provide "substantial evidence of effectiveness"?

(ii) Did the clinical study results demonstrate that the drug was safe; that is, do the benefits outweigh the risks?

Based on its findings, the committee then makes its recommendation to the FDA. If approved, an approval letter is given to the company to market the drug.

Note that very few drugs (perhaps 5 in 5,000 tested compounds) actually make it to human clinical trials. Then, only about 1 of those 5 are found to be safe and effective enough to reach the pharmacy.

TABLE **A**	Adverse Cellular Effects Based on FM04 Drug Concentration	
Final Drug Concentration (µg/culture)	**Adverse Cellular Effects**
0	None
1	None
10	None
100	Abnormal cells and cell death

TABLE **B**	Toxicity Results on 50 Mice Given Different Concentrations of FM04		
Drug Concentration (mg)	**Number of Mice**	**Adverse Reactions**
0	10	None
1	10	None
10	10	None
100	10	None
1,000	10	Tremors and seizures in 8 mice

SUMMARY OF KEY CONCEPTS

24.1 The History and Properties of Antimicrobial Agents

1. Ehrlich saw **antibiotics** as a chemotherapeutic approach to alleviating disease. One of his students isolated the first chemotherapeutic agent, **Salvarsan**. Domagk used **Prontosil** to treat bacterial infections.

2. Penicillin was discovered by Fleming, and purified and prepared for chemotherapy by Florey and Chain.

3. All antimicrobial agents represent **synthetic drugs** or **antibiotics**, demonstrate **selective toxicity**, and have a **broad** or **narrow drug spectrum**.

4. Antibiotics are natural compounds toxic to some competitive bacterial species.

24.2 The Synthetic Antibacterial Agents

5. The **sulfonamides** interfere with the production of **folic acid** through **competitive inhibition**. Modern sulfonamides are trimethoprim and sulfamethoxazole.

6. **Isoniazid** and **ethambutol** are antituberculosis drugs affecting wall synthesis. **Ciprofloxacin** is a fluoroquinolone that interacts with DNA to inhibit replication.

24.3 The Beta-Lactam Family of Antibiotics

7. **Penicillin** interferes with cell wall synthesis in gram-positive bacterial cells. It can cause an anaphylactic reaction in sensitive individuals. Numerous synthetic and semisynthetic forms of penicillin are more resistant to **beta-lactamase** activity.

8. Certain **cephalosporin** drugs are first-choice antibiotics for penicillin-resistant bacterial species and a wide variety of these drugs is currently in use. The **carbapenems** (imipenem) inhibit cell wall synthesis and are broad spectrum drugs.

24.4 Other Bacterially Produced Antibiotics

9. **Vancomycin** inhibits cell wall synthesis.

10. **Bacitracin** and **polymyxin B** affect the permeability of the cell membrane.

11. **Aminoglycosides** (gentamicin, neomycin) inhibit protein synthesis in gram-negative bacterial cells. **Chloramphenicol** is a broad-spectrum antibiotic used against gram-positive and gram-negative bacteria. Less severe side effects accompany tetracycline use, and this antibiotic is recommended against gram-negative bacteria as well as rickettsiae and chlamydiae. The **macrolides**, **lincosamides**, and **streptogramins** inhibit protein synthesis.

12. Besides the quinolones that inhibit DNA replication, **rifampin** interferes with transcription.

24.5 Antiviral, Antifungal, and Antiparasitic Drugs

13. Antiviral drugs interfere with viral entry, replication, or maturation.

14. The **polyenes** (nystatin, amphotericin B) and **imidazoles** are valuable against fungal infections. The **echinocandins** inhibit fungal cell wall synthesis, **flucytosine** interrupts nucleic acid synthesis, and **griseofulvin** interferes with mitosis.

15. Antiprotozoal agents include the **aminoquinolines** (quinine and chloroquine) that are used to treat malaria; **sulfonamides** interfere with folic acid synthesis; and **nitroimidazoles** inhibit nucleic acid synthesis.

16. Antihelminthic drugs include **praziquantel**, **mebendazole**, and **avermectins**.

24.6 Antibiotic Assays and Resistance

17. Antibiotic assays include the **tube dilution assay**, which measures the **minimal inhibitory concentration**, and the **agar disk diffusion method**, which determines the susceptibility of a bacterial species to a series of antibiotics.

18. Bacterial species have developed resistance to antimicrobial agents by altering metabolic pathways, inactivating antibiotics, reducing membrane permeability, or modifying the drug target.

19. Arising from any of several sources such as changes in microbial biochemistry, antibiotic resistance threatens to put an end to the cures of infectious disease that have come to be expected in contemporary medicine.

20. New approaches include the discovery of new drug targets and developing new antibiotics to fight infectious disease and antibiotic resistance.

LEARNING OBJECTIVES

After understanding the textbook reading, you should be capable of writing a paragraph that includes the appropriate terms and pertinent information to answer the objective.

1. Summarize the accomplishments of Paul Ehrlich and Gerhard Domagk to early advances in antimicrobial **chemotherapy**.

2. Describe the important advances made by Fleming, and Florey and Chain toward chemotherapy.

3. Identify the important properties of antimicrobial agents.

4. Explain why some microbes produce antibiotics in the environment.

5. Explain how **sulfonamide** drugs block the **folic acid** metabolic pathway in bacterial cells.

6. Compare and contrast the mechanism of action for **isoniazid** and the **quinolones**.

7. Summarize the mechanism of action of **penicillin** and its semisynthetic derivatives.

8. Discuss the other beta-lactam antibiotics that affect cell wall synthesis.

9. Indicate how **vancomycin** differs from the beta-lactam antibiotics.

10. Distinguish between the mechanisms of action for the polypeptide antibiotics.

11. Hypothesize why there is such a large number of natural antibiotics that target the ribosome and protein synthesis.

12. Identify the antibiotics affecting RNA synthesis.

13. Describe the targets for antiviral drugs.

14. Indicate the targets for antifungal drugs.

15. Identify the drug targets in the protozoal parasites.

16. Identify several antihelminthic drugs and their targets.

17. Contrast between the two antibiotic susceptibility assays.

18. Describe the four mechanisms used by bacterial species to generate resistance to antibiotics.

19. Justify the statement that antibiotic resistance has resulted from their misuse and abuse.

20. Identify ways researchers are attempting to discover new drug targets and antibiotics.

STEP A: SELF-TEST

Each of the following questions is designed to assess your ability to remember or recall factual or conceptual knowledge related to this chapter. Read each question carefully, then select the **one** answer that best fits the question or statement. Answers to even-numbered questions can be found in **Appendix C**.

1. Ehrlich and Hata discovered _____ that was used to treat _____.
   A. Salvarsan; syphilis
   B. penicillin; surgical wounds
   C. Salvarsan; malaria
   D. Prontosil; malaria

2. The re-isolation and purification of penicillin was carried out by
   A. Waksman.
   B. Florey and Chain.
   C. Domagk.
   D. Fleming.

3. The concentration of an antibiotic causing harm to the host is called the
   A. toxic dosage level.
   B. therapeutic dosage level.
   C. minimal inhibitory concentration.
   D. chemotherapeutic index.

4. In the soil, antibiotics are produced by some
   A. fungal species and viruses.
   B. bacterial species and viruses.
   C. viruses and helminthic species.
   D. fungal and bacterial species.

5. Trimethoprim and sulfamethoxazole are examples of _____ that block _____ synthesis.
   A. sulfonamides; PABA
   B. penicillins; cell wall
   C. sulfonamides; folic acid
   D. macrolides; protein

6. Isoniazid and ethambutol are used to treat
   A. cholera.
   B. influenza.
   C. MRSA.
   D. tuberculosis.

7. Penicillins are useful in treating
   A. gram-positive infections.
   B. leprosy.
   C. gram-negative infections.
   D. tuberculosis.

8. All the following are drugs or drug classes blocking cell wall synthesis *except*:
   A. cephalosporins.
   B. carbapenems.
   C. monobactams.
   D. tetracyclines.

9. Vancomycin inhibits _____ synthesis.
   A. protein
   B. DNA
   C. bacterial cell wall
   D. RNA

10. Two cyclic polypeptide antibiotics are
    A. vancomycin and streptomycin.
    B. penicillin and cephalosporin.
    C. bacitracin and polymyxins.
    D. gentamicin and chloramphenicol.

11. Which one of the following is NOT an inhibitor of protein synthesis?
    A. Clindamycin
    B. Macrolides
    C. Rifampin
    D. Chloramphenicol

12. Antibiotics inhibiting nucleic acid synthesis include
    A. rifampin and quinolones.
    B. aminoglycosides and tetracyclines.
    C. lincosamides and streptogramins.
    D. macrolides and aminoglycosides.

13. Antiviral drugs that are base analogs inhibit
    A. viral entry.
    B. genome replication.
    C. uncoating.
    D. naturation.

14. Antifungal drugs, such as _____, inhibit proper formation of _____.
    A. miconazole; a plasma membrane
    B. griseofulvin; DNA
    C. ketoconazole; a cell wall
    D. flucytosine; a microtubule

15. All of the following antiprotozoal drugs have been used to treat malaria *except*:
    A. melarsoprol.
    B. quinine.
    C. mefloquine.
    D. chloroquine.

16. This antihelminthic agent makes the membrane permeable to calcium ions.
    A. Mebendazole
    B. Pentamidine
    C. Ivermectin
    D. Praziquantel

17. The _____ is used to determine an antibiotic's minimal inhibitory concentration (MIC).
    A. Ames test
    B. tube dilution method
    C. agar disk diffusion method
    D. Kirby-Bauer test

18. The phosphorylation of an antibiotic is an example of which mechanism of resistance?
    A. Target modification
    B. Reduced permeability
    C. Antibiotic inactivation
    D. Altered metabolic pathway

19. A superinfection could arise from
    A. using antibiotics when not needed.
    B. using unnecessarily large doses of antibiotics.
    C. stopping antibiotic treatment prematurely.
    D. All of the above (**A–C**) are correct.

20. New approaches to antibiotic therapy include
    A. carbohydrate targets in the wall LPS.
    B. unregulating DNA replication.
    C. discovering new and unique antibiotics.
    D. All of the above (**A–C**) are correct.

STEP B: REVIEW—ANTIMICROBIAL DRUGS

On completing your study of these pages, test your understanding of their contents by deciding whether the following statements are true or false. If the statement is true, write "True" in the space. If false, substitute a word for the underlined word to make the statement true. The answers to even-numbered statements are listed in **Appendix C**.

_____ 21. Tetracycline has <u>four</u> benzene rings in the molecule.

_____ 22. Sulfonamides block folic acid formation in <u>fungi</u>.

_____ 23. Gentamicin is an antibiotic that interferes with <u>transcription</u>.

_____ 24. A side effect of <u>chloramphenicol</u> is a discoloration of the teeth.

_____ 25. Penicillinase also is known as an <u>alpha-lactamase</u>.

_____ 26. <u>Polymyxin B</u> affects membrane permeability.

_____ 27. Folic acid synthesis is inhibited by <u>rifampin</u>.

_____ 28. <u>Penicillin</u> has no effect on viruses

_____ 29. <u>Artemisinin</u> is a drug that releases free radicals.

_____ 30. Neosporin contains bacitracin, polymyxin B, and <u>tetracycline</u>.

STEP B: REVIEW—IDENTIFICATION

Identify the bacterial structure inhibited or affected by each of the antibiotics in the description below. Answers to even-numbered descriptions can be found in **Appendix C**.

Descriptions
31. Sulfonamides act here.
32. Polymyxin B affects this structure.
33. Rifampin blocks this process.
34. Quinolones act here.
35. Tetracyclines inhibit the function of this structure.
36. Cephalosporins target this structure.
37. The macrolides act here.
38. Vancomycin inhibits the assembly of this structure.
39. Aminoglycosides inhibit this structure.
40. Penicillin acts here.

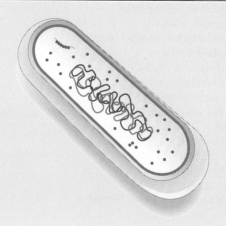

STEP C: APPLICATIONS

Answers to even-numbered questions can be found in **Appendix C**.

41. In 1877, Pasteur and his assistant Joubert observed that anthrax bacilli grew vigorously in sterile urine but failed to grow when the urine was contaminated with other bacilli. What was happening?

42. In May 1953, Edmund Hillary and Tenzing Norgay were the first to reach the summit of Mount Everest, the world's highest mountain. Since that time over 150 other mountaineers have reached the summit, and groups have gone to Nepal from all over the world on expeditions. The arrival of "civilization" has brought a drastic change to the lifestyle of Nepal's Sherpa mountain people. For example, half of all Sherpas used to die before the age of 20, but partly due to antibiotics available to fight disease, the population has grown from 9 million to more than 23 million. Medical enthusiasts are proud of this increase in the life expectancy, but population ecologists see a bleaker side. What do you suspect they foresee, and what does this tell you about the impact antibiotics have on a culture?

43. Why would a synthetically produced antibiotic be more advantageous than a naturally occurring antibiotic? Why would it be less advantageous?

STEP D: QUESTIONS FOR THOUGHT AND DISCUSSION

Answers to even-numbered questions can be found in **Appendix C**.

44. According to historians, 2,500 years ago the Chinese learned to treat superficial infections such as boils by applying moldy soybean curds to the skin. Can you suggest what this implies?

45. Most naturally occurring antibiotics appear to be products of the soil bacteria belonging to the genus *Streptomyces*. Can you draw any connection between the habitat of these organisms and their ability to produce antibiotics?

46. Of the thousands and thousands of types of organisms screened for antibiotics since 1940, only five bacterial genera appear capable of producing these chemicals. Does this strike you as unusual? What factors might eliminate potentially useful antibiotics?

47. Is an antibiotic that cannot be absorbed from the gastrointestinal tract necessarily useless? How about one that is rapidly expelled from the blood into the urine? Explain your answers?

48. The antibiotic resistance issue can be argued from two perspectives. Some people contend that because of side effects and microbial resistance, antibiotics will eventually be abandoned in medicine. Others see the future development of a superantibiotic, a type of "miracle drug." What arguments can you offer for either view? Which direction in medicine do you support?

49. History shows that over and over, creativity is a communal act, not an individual one. How does the 1945 Nobel Prize to Fleming, Florey, and Chain reflect this notion?

HTTP://MICROBIOLOGY.JBPUB.COM/9E

The site features eLearning, an online review area that provides quizzes and other tools to help you study for your class. You can also follow useful links for in-depth information, read more MicroFocus stories, or just find out the latest microbiology news.

Appendix A: Metric Measurement

	Fundamental Unit	Quantity	Symbol	Numerical Unit	Scientific Notation
Length	Meter (m)				
		kilometer	km	1,000 m	10^3
		centimeter	cm	0.01m	10^{-2}
		millimeter	mm	0.001 m	10^{-3}
		micrometer	μm	0.000001m	10^{-6}
		nanometer	nm	0.000000001 m	10^{-9}
Volume (liquids)	Liter (L)				
		milliliter	mL	0.001 L	10^{-3}
		microliter	μl	0.000001 L	10^{-6}
Mass	Gram (g)				
		kilogram	kg	1,000 g	10^3
		milligram	mg	0.001 g	10^{-3}
		microgram	μg	0.000001 g	10^{-6}

Appendix B: Temperature Conversion Chart

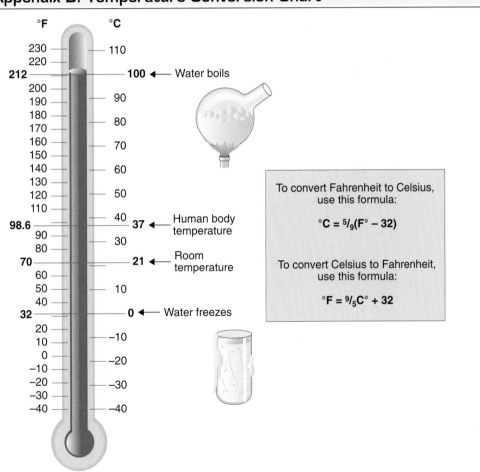

To convert Fahrenheit to Celsius, use this formula:

$$°C = \tfrac{5}{9}(F° - 32)$$

To convert Celsius to Fahrenheit, use this formula:

$$°F = \tfrac{9}{5}C° + 32$$

The answers to even-numbered end-of-chapter questions or statements can be found on the student companion Web site in **Appendix C**.

The answers to Textbook Case and MicroInquiry questions or statements can be found on the student companion Web site in **Appendix D**.

Glossary

This glossary contains concise definitions for microbiological terms and concepts only. **Please refer to the index for specific infectious agents, infectious diseases, anatomical terms, and immune disorders.**

A

abscess   A circumscribed pus-filled lesion characteristic of staphylococcal skin disease; also called a boil.

abyssal zone   The environment at the bottom of oceanic trenches.

acetyl CoA   One of the starting compounds for the Krebs cycle.

acid   A substance that releases hydrogen ions (H⁺) in solution; *see also* **base**.

acid-fast technique   A staining process in which mycobacteria resist decolorization with acid alcohol.

acidic dye   A negatively charged colored substance in solution that is used to stain an area around cells.

acidophile   A microorganism that grows at acidic pHs below 4.

acquired immunity   A response to a specific immune stimulus that involves immune defensive cells and frequently leads to the establishment of host immunity.

Actinobacteria   A phylum in the domain *Bacteria* that exhibits fungus-like properties when cultivated in the laboratory.

actinomycete   A soil bacterium that exhibits fungus-like properties when cultivated in the laboratory.

activated sludge   Aerated sewage containing microorganisms added to untreated sewage to purify it by accelerating its bacterial decomposition.

activation energy   The energy required for a chemical reaction to occur.

active immunity   The immune system responds to antigen by producing antibodies and specific lymphocytes.

active site   The region of an enzyme where the substrate binds.

active transport   An energy-requiring movement of substances from an area of lower concentration across a biological membrane to a region of higher concentration by means of a membrane-spanning carrier protein.

acute disease   A disease that develops rapidly, exhibits substantial symptoms, and lasts only a short time.

acute period   The phase of a disease during which specific symptoms occur and the disease is at its height.

acute phase protein   A defensive blood protein secreted by liver cells that elevates the inflammatory and complement responses to infection.

acute rejection   A transplant rejection response that takes 10 to 30 days.

acute wound   One that arises from cuts, lacerations, bites, or surgical procedures.

adenosine diphosphate (ADP)   A molecule in cells that is the product of ATP hydrolysis.

adenosine triphosphate (ATP)   A molecule in cells that provides most of the energy for metabolism.

adenylyl cyclase   An enzyme that catalyzes the conversion of ATP to cyclic adenosine monophosphate (cAMP).

adhesin   A protein in bacterial pili that assists in attachment to the surface molecules of cells.

adjuvant   An agent added to a vaccine to increase the vaccine's effectiveness.

ADP   *See* **adenosine diphosphate**.

aerobe   An organism that uses oxygen gas (O₂) for metabolism.

aerobic respiration   The process for transforming energy to ATP in which the final electron acceptor in the electron transport chain is oxygen gas (O₂).

aerosol   A small particle transmitted through the air.

aerotolerant   A bacterium not inhibited by oxygen gas (O₂).

aflatoxin   A toxin produced by *Aspergillus flavus* that is cancer causing in vertebrates.

agar   A polysaccharide derived from marine seaweed that is used as a solidifying agent in many microbiological culture media.

agglutination   A type of antigen-antibody reaction that results in visible clumps of organisms or other material.

agranulocyte   A white blood cell lacking visible granules; includes the lymphocytes and monocytes; *see also* **granulocyte**.

alcoholic fermentation   A catabolic process that forms ethyl alcohol during the reoxidation of NADH to NAD⁺ for reuse in glycolysis to generate ATP.

alga (pl. algae)   An organism in the kingdom Protista that performs photosynthesis.

algal bloom   An excessive growth of algae on or near the surface of water, often the result of an oversupply of nutrients from organic pollution.

alginate   A sticky substance used as a thickener in foods and beverages.

allergen   An antigenic substance that stimulates an allergic reaction in the body.

allergen immunotherapy   A procedure involving a series of allergen shots over several years to reduce the sensitivity to allergens.

allergic rhinitis   The technical name for hay fever.

allergy shot   *See* **allergen immunotherapy**.

allograft   A tissue graft between two members of the same species, such as between two humans (not identical twins).

alpha helix   The spiral structure of a polypeptide consisting of amino acids stabilized by hydrogen bonds.

alpha (α) hemolytic   Referring to those bacterial species that when plated on blood agar cause a partial destruction of red blood cells as seen by an olive green color in the agar around colonies.

amatoxin   A group of toxic compounds found in several genera of poisonous mushrooms.

Ames test   A diagnostic procedure used to detect potential cancer-causing agents in humans by the ability of the agent to cause mutations in bacterial cells.

amino acid   An organic acid containing one or more amino groups; the monomers that build proteins in all living cells.

aminoglycoside   An antibiotic that contains amino groups bonded to carbohydrate groups that inhibit protein synthesis; examples are gentamicin, streptomycin, and neomycin.

aminoquinoline   An antiprotozoal drug that is toxic to the malarial parasite.

amoeba   A protozoan that undergoes a crawling movement by forming cytoplasmic projections into the environment.

amoeboid motion   A crawling type of movement caused by the flow of cytoplasm into plasma membrane projections; typical of the amoebas.

Amoebozoa   One of the super groups in the kingdom Protista characterized by amoeboid movement and pseudopodia.

anabolism   An energy-requiring process involving the synthesis of larger organic compounds from smaller ones; *see also* **catabolism**.

anaerobe   An organism that does not require or cannot use oxygen gas (O_2) for metabolism.

anaerobic respiration   The production of ATP where the final electron acceptor is an inorganic molecule other than oxygen gas (O_2); examples include nitrate and sulfate.

animalcule   A microscopic organism observed by Leeuwenhoek.

anion   An ion with a negative charge; *see also* **cation**.

anoxygenic photosynthesis   A form of photosynthesis in which molecular oxyten (O_2) is not produced.

antibiotic   A substance naturally produced by a few bacterial or fungal species that inhibits or kills other microorganisms.

antibody   A highly specific protein produced by the body in response to a foreign substance, such as a bacterium or virus, and capable of binding to the substance.

anticodon   A three-base sequence on the tRNA molecule that binds to the codon on the mRNA molecule during translation.

antigen   A chemical substance that stimulates the production of antibodies by the body's immune system.

antigen-antibody complex   The interaction of an antigen with antibodies.

antigen binding site   The region on an antibody that binds to an antigen.

antigen presenting cell (APC)   A macrophage or dendritic cell that exposes antigen peptide fragments on its surface to T cells.

antigenic determinant   A section of an antigen molecule that stimulates antibody formation and to which the antibody binds; also called epitope.

antigenic drift   A minor variation over time in the antigenic composition of influenza viruses.

antigenic shift   A major change over time in the antigenic composition of influenza viruses.

antihistamine   A drug that blocks cell receptors for histamine, preventing allergic effects such as sneezing and itching.

antimicrobial agent (drug)   A chemical that inhibits or kills the growth of microorganisms.

antimicrobial spectrum   The range of antimicrobial drug action.

antisense molecule   RNA segments that complementarily bind to a messenger RNA and block translation.

antisepsis   The use of chemical methods for eliminating or reducing microorganisms on the skin.

antiseptic   A chemical used to reduce or kill pathogenic microorganisms on a living object, such as the surface of the human body.

antiserum (pl. antisera)   A blood-derived fluid containing antibodies and used to provide temporary immunity.

antitoxin   An antibody produced by the body that circulates in the bloodstream to provide protection against toxins by neutralizing them.

antiviral protein   A protein made in response to interferon and that blocks viral replication.

antiviral state   A cell capable of inhibiting viral protein synthesis due to interferon activation.

apicomplexan   A protozoan containing a number of organelles at one end of the cell that are used for host penetration; no motion is observed in adult forms.

APTIMA assay   A test used to amplify HIV RNA.

aqueous solution   One or more substances dissolved in water.

arbovirus   A virus transmitted by arthropods (e.g., insects).

Archaea   The domain of living organisms that excludes the *Bacteria* and *Eukarya*.

artemisinin   An anti-protozoal drug used to treat malaria.

arthrospore   An asexual fungal spore formed by fragmentation of a septate hypha.

Arthus phenomenon   An immune complex hypersensitivity when large amounts of IgG antibody for complexes with antigens in blood vessels or near the site of antigen entry.

artificially acquired active immunity   The production of antibodies by the body in response to antigens in a vaccination; or the passive transfer of antibodies formed in one individual or animal to another susceptible person.

artificially acquired   The active production of antibodies by the body in response to antigens in a vaccination; or the passive transfer of antibodies formed in one individual or animal to another susceptible person.

Ascomycota   A phylum of fungi whose members have septate hyphae and form ascospores within saclike asci, among other notable characteristics.

ascospore   A sexually produced fungal spore formed by members of the ascomycetes.

ascus (pl. asci)   A saclike structure containing ascospores; formed by the ascomycetes.

asepsis   The process or method of bringing about a condition in which no unwanted microbes are present.

asexual reproduction   The form of reproduction that maintains genetic constancy while increasing cell numbers.

asthma   A reversible airway constriction caused by the release of inflammatory mediators from mast cells upon an encounter with an allergen.

atom   The smallest portion into which an element can be divided and still enter into a chemical reaction.

atomic nucleus   The positively charged core of an atom, consisting of protons and neutrons that make up most of the mass.

atomic number   The number of protons in the nucleus of an atom.

atopic disease   A condition resulting from the body's response to certain allergens and producing a localized reaction in the body; examples include hay fever and food allergies.

ATP   *See* **adenosine triphosphate**.

ATP/ADP cycle   The cellular processes of synthesis and hydrolysis of ATP.

ATP synthase   The enzyme involved in forming ATP by using the energy in a proton gradient.

attenuate   Referring to the reduced ability of bacterial cells or viruses to do damage to the exposed individual.

attractant   A substance that attracts cells through motility.

autoclave   An instrument used to sterilize microbiological materials by means of high temperature using steam under pressure.

autograft   Tissue taken from one part of the body and grafted to another part of the same body.

autoimmune disorder   A reaction in which antibodies react with an individual's own chemical substances and cells.

autotroph   An organism that uses carbon dioxide (CO_2) as a carbon source; *see also* **chemoautotroph** *and* **photoautotroph**.

autotrophy   The process by which an organism makes its own food.

auxotroph   A mutant strain of an organism lacking the ability to synthesize a nutritional need; *see also* **prototroph**.

avermectin   An antihelminthic drug that causes muscle paralysis in nematodes.

avirulent   Referring to an organism that is not likely to cause disease.

B

bacillé Calmette Guérin (BCG)   A strain of attenuated *Mycobacterium bovis* used for immunization against tuberculosis and, on occasion, leprosy.

bacillus (pl. bacilli)   (1) Any rod-shaped bacterial or archaeal cell. (2) When referring to the genus *Bacillus*, it refers to an aerobic or facultatively anaerobic, rod-shaped, endospore-producing, gram-positive bacterial cell.

bacitracin   An antibiotic derived from a *Bacillus* species, effective against gram-positive bacteria when used topically.

bacteremia   The presence of live bacterial cells in the blood.

Bacteria   The domain of living things that includes all organisms not classified as *Archaea* or *Eukarya*.

bacterial growth curve   The events occurring over time within a population of growing and dividing prokaryotic cells.

bactericidal   Referring to any agent that kills bacterial cells.

bacteriochlorophyll   A pigment located in the membrane systems of purple sulfur bacteria that upon excitement by light, loses electrons and initiates photosynthetic reactions.

bacteriocin   One of a group of bacterial proteins toxic to other bacterial cells.

bacteriology   The scientific study of prokaryotes; originally used to describe the study of bacteria.

bacteriophage (phage)   A virus that infects and replicates within bacterial cells.

bacteriostatic   Referring to any substance that prevents the growth of bacteria.

bacterium (pl. **bacteria**)   A single-celled microorganism lacking a cell nucleus and membrane-enclosed compartments, and often having peptidoglycan in the cell wall.

barophile   A microorganism that lives under conditions of high atmospheric pressure.

base   A chemical compound that accepts hydrogen ions (H^+) in solution; *see also* **acid**.

base analog   A nitrogenous base with a similar structure to a natural base but differing slightly in composition.

basic dye   A positively charged colored substance in solution that is used to stain cells.

Basidiomycota   The phylum of fungi whose members have septate hyphae and form basidiospores on supportive basidia, among other notable characteristics.

basidiospore   A sexually produced fungal spore formed by members of the basidiomycetes.

basidium (pl. **basidia**)   A club-like structure containing basidiospores; formed by the basidiomycetes.

basophil   A type of white blood cell with granules that functions in allergic reactions.

B cell   *See* **B lymphocyte**.

benign   Referring to a tumor that usually is not life threatening or likely to spread to another part of the body.

benthic zone   The environment at the bottom of a deep river, lake, or sea.

benzoic acid   A chemical preservative used to protect beverages, catsup, and margarine.

beta (β) hemolytic   Referring to those bacterial species that when plated on blood agar completely destroy the red blood cells as seen by a clearing in the agar around the colonies.

beta-lactamase   The enzyme that converts the beta-lactam antibiotics (penicillins, cephalosporins, and carbapenems) into inactive forms.

beta oxidation   The breakdown of fatty acids during cellular metabolism through the successive removal from one end of two carbon units.

binary fission   An asexual process in bacterial and archaeal cells by which a cell divides to form two new cells while maintaining genetic constancy.

biocatalysis   The metabolic reactions of microorganisms carried out at an industrial scale.

biochemical oxygen demand (**BOD**)   A number referring to the amount of oxygen used by the microorganisms in a sample of water during a 5-day period of incubation.

biocrime   The intentional introduction of biological agents into food or water, or by injection, to harm or kill individuals.

biofilm   A complex community of microorganisms that form a protective and adhesive matrix that attaches to a surface, such as a catheter or industrial pipeline.

biological pollution   The presence of microorganisms from human waste in water.

biological safety cabinet   A cabinet or hood used to prevent contamination of biological materials.

biological vector   An infected arthropod, such as a mosquito or tick, that transmits disease-causing organisms between hosts; *see also* **mechanical vector**.

bioreactor   A large fermentation tank for growing microorganisms used in industrial production; also called a fermentor.

bioremediation   The use of microorganisms to degrade toxic wastes and other synthetic products of industrial pollution.

biotechnology   The commercial application of genetic engineering using living organisms.

bioterrorism   The intentional or threatened use of biological agents to cause fear in or actually inflict death or disease upon a large population.

bipolar staining   A characteristic of *Yersinia* and *Francisella* species in which stain gathers at the poles of the cells, yielding the appearance of safety pins.

bisphenol   A combination of two phenol molecules used in disinfection.

blanching   A process of putting food in boiling water for a few seconds to destroy enzymes.

blastospore   A fungal spore formed by budding.

B lymphocyte (B cell)   A white blood cell that matures into memory cells and plasma cells that secrete antibody.

bone marrow   A soft reddish substance inside some bones that is involved in the production of blood cells.

booster shot   A repeat dose of a vaccine given some years after the initial course to maintain a high level of immunity.

bright-field microscope   An instrument that magnifies an object by passing visible light directly through the lenses and object; *see also* **light microscope**.

broad spectrum   Referring to an antimicrobial drug useful for treating many groups of microorganisms, including gram-positive and gram-negative bacteria; *see also* **narrow-spectrum**.

bubo   A swelling of the lymph nodes due to inflammation.

budding   (1) An asexual process of reproduction in fungi, in which a new cell forms as a swelling at the border of the parent cell and then breaks free to live independently. (2) The controlled release of virus particles from an infected animal cell.

buffer   (1) A compound that minimizes pH changes in a solution by neutralizing added acids and bases. (2) Refers to a solution containing such a substance.

bull's eye rash   A circular lesion on the skin with a red border and central clearing; characteristic of Lyme disease.

Burkitt lymphoma   A tumor of the jaw triggered by the Epstein-Barr virus.

burst size   The number of virus particles released from an infected bacterial cell.

C

cancer   A disease characterized by the radiating spread of malignant cells that reproduce at an uncontrolled rate.

capnophilic   Referring to a prokaryotic cell requiring low oxygen gas (O_2) and a high concentration of carbon dioxide gas (CO_2) for metabolism.

capsid   The protein coat that encloses the genome of a virus.

capsomere   Any of the protein subunits of a capsid.

capsule   A layer of polysaccharides and small proteins covalently bound some prokaryotic cells; *see also* **slime layer** and **glycocalyx**.

carbapenem   An antibacterial drug derived from a species of *Streptomyces* that is effective against gram-positive and gram-negative bacterial cells by inhibiting cell wall synthesis.

carbohydrate   An organic compound consisting of carbon, hydrogen, and oxygen that is an important source of carbon and energy for all organisms; examples include simple sugars, starch, and cellulose.

carbon cycle   A series of interlinked processes involving carbon compound exchange between living organisms and the nonliving environment.

carbon-fixing reaction   A chemical reaction in the second part of photosynthesis in which carbohydrates are formed.

carbuncle   An enlarged abscess formed from the union of several smaller abscesses or boils.

carcinogen   Any physical or chemical substance that causes tumor formation.

carrier   An individual who has recovered from a disease but retains the infectious agents in the body and continues to shed them.

casein   The major protein in milk.

casing soil   A non-nutritious soil used to provide moisture for mushroom formation.

catabolism   An energy-liberating process in which larger organic compounds are broken down into smaller ones; *see also* **anabolism**.

cation   A positively charged ion; *see also* **anion**.

CD4 T cell   A lymphocyte expressing the CD4 receptor on the cell surface.

CD8 T cell   A lymphocyte expressing the CD8 receptor on the cell surface.

cell envelope   The cell wall and cell membrane of a bacterial or archaeal cell.

cell line   A group of identical cells in culture and derived from a single cell.

cell-mediated immune response   The body's ability to resist infection through the activity of T-lymphocyte recognition of antigen peptides presented on macrophages and dendritic cells and on infected cells.

cell membrane   A thin bilayer of phospholipids and proteins that surrounds the prokaryotic cell cytoplasm. *See also* **plasma membrane**.

cell theory   The tenet that all organisms are made of cells and arise from preexisting cells.

cellular chemistry   The chemical study of cells and cellular molecules.

cellular respiration   The process of converting chemical energy into cellular energy in the form of ATP.

cell wall   A carbohydrate-containing structure surrounding fungal, algal, and most bacterial and archaeal cells.

central dogma   The doctrine that DNA codes for RNA through transcription and RNA is converted to protein through translation.

centrosome   The microtubule-organizing center of a eukaryotic cell.

cephalosporin   An antibiotic derived from the mold *Cephalosporium* that inhibits cell wall synthesis in gram-positive bacteria and certain gram-negative bacteria.

cercaria (pl. **cercariae**)   A tadpole-like larva form in the life cycle of a trematode.

cervical cancer   Cancer of the cervix caused by several papillomaviruses.

cesspool   Concrete cylindrical rings with pores in the walls that is used to collect human waste.

cestode   A flatworm, commonly known as a tapeworm, that lives as a parasite in the gut of vertebrates.

CFU   *See* **colony forming unit**.

chain of transmission   How infectious diseases can be spread from human to human (or animal to human).

chancre   A painless, circular, purplish hard ulcer with a raised margin that occurs during primary syphilis.

chaperone   A protein that ensures a polypeptide folds into the proper shape.

chemical bond   A force between two or more atoms that tends to bind those atoms together.

chemical element   Any substance that cannot be broken down into a simpler one by a chemical reaction.

chemical pollution   The presence of inorganic and organic waste in water.

chemical reaction   A process that changes the molecular composition of a substance by redistributing atoms or groups of atoms without altering the number of atoms.

chemiosmosis   The use of a proton gradient across a membrane to generate cellular energy in the form of ATP.

chemoautotroph   An organism that derives energy from inorganic chemicals and uses the energy to synthesize nutrients from carbon dioxide gas (CO_2).

chemoheterotroph   An organism that derives energy from organic chemicals and uses the energy to synthesize nutrients from carbon compounds other than carbon dioxide gas (CO_2).

chemokine   A protein that prompts specific white blood cells to migrate to an infection site and carry out their immune system functions.

chemotaxis   A movement of a cell or organism toward a chemical or nutrient.

chemotherapeutic agent   A chemical compound used to treat diseases and infections in the body.

chemotherapeutic index   A number that represents the highest level of an antimicrobial drug tolerated by the host divided by the lowest level of the drug that eliminates the infectious agent.

chemotherapy   The process of using chemical agents to treat diseases and infections, or other disorders, such as cancer.

chitin   A polymer of acetylglucosamine units that provides rigidity in the cell walls of fungi.

Chlamydae   A phylum of extremely small bacteria that can be cultured only in living cells.

chloramphenicol   A broad-spectrum antibiotic derived from a *Streptomyces* species that interferes with protein synthesis.

chlorination   The process of treating water with chlorine to kill harmful organisms.

chlorine dioxide   A gas used to sterilize objects or instruments.

chlorophyll   A green or purple pigment in algae and some bacterial cells that functions in capturing light for photosynthesis.

chloroplast   A double membrane-enclosed compartment in algae that contains chlorophyll and other pigments for photosynthesis.

cholera toxin   An enterotoxin that triggers an unrelenting loss of fluid.

chromalveolata   One of the super groups in the kingdom Protista that includes the dinoflagellates, diatoms, ciliates, and the apicomplexans.

chromosome   A structure in the nucleoid or cell nucleus that carries hereditary information in the form of genes.

chronic disease   A disease that develops slowly, tends to linger for a long time, and requires a long convalescence.

chronic rejection   A form of transplant rejection that takes months to years to develop.

chronic wound   An open sore, such as a leg ulcer or bed sore.

chytrid   A fungus in the phylum Chytridiomycota.

Chytridiomycota   A phylum of predominantly aquatic fungi.

ciliate   A protozoan that moves with the aid of cilia.

cilium (pl. **cilia**)   A hair-like projection on some eukaryotic cells that along with many others assist in the motion of some protozoa and beat rhythmically to aid the movement of a fluid past the respiratory epithelial cells in humans.

cirrhosis   Extensive injury of cells of the liver.

citric acid cycle   A metabolic pathway in which acetyl groups are completely oxidized to carbon dioxide gas and some ATP molecules are formed. Also called Krebs cycle.

class   A category of related organisms consisting of one or more orders.

classification   The arrangement of organisms into hierarchal groups based on relatedness.

climax   *See* **acute period**.

clindamycin   An antibiotic that inhibits protein synthesis and is used as a penicillin substitute for certain anaerobic bacterial diseases.

clinical disease   A disease in which the symptoms are apparent.

clonal selection   The theory that certain lymphocytes are activated from the mixed population of B or T lymphocytes when stimulated by antigen or antigen peptide fragments.

clone   A population of genetically identical cells (or plasmids).

cloning vector   A plasmid used to introduce genes into a bacterial cell.

coagulase   An enzyme produced by some staphylococci that catalyzes the formation of a fibrin clot.

coccus (pl. **cocci**)   A spherical-shaped bacterial or archaeal cell.

codon   A three-base sequence on the mRNA molecule that specifies a particular amino acid insertion in a polypeptide.

coenocytic   Referring to a fungus containing no septa (cross-walls) and multinucleate hyphae.

coenzyme   A small, organic molecule that forms the nonprotein part of an enzyme molecule.

coenzyme A (**CoA**)   A small, organic molecule of cellular respiration that functions in release of carbon dioxide gas (CO_2) and the transfer of electrons and protons to another coenzyme.

cofactor   A inorganic substance that acts with and is essential to the activity of an enzyme; examples include metal ions and some vitamins.

coliform bacterium   A gram-negative, nonspore-forming, rodshaped cell that ferments lactose to acid and gas and usually is found in the human and animal intestine; high numbers in water is an indicator of contamination.

colony   A visible mass of microorganisms of one type.

colony forming unit (**CFU**)   A measure of the viable cells by counting the number of colonies on a plate; each colony presumably started from one viable cell.

colostrum   The yellowish fluid rich in antibodies secreted from the mammary glands of animals or humans prior to the production of true milk.

comedo (pl. **comedones**)   A plugged sebaceous gland.

commensalism   A close and permanent association between two species of organisms in which one species benefits and the other remains unharmed and unaffected.

commercial sterilization   A canning process to eliminate the most resistant bacterial spores.

communicable disease   A disease that is readily transmissible between hosts.

comparative genomics   The comparison of DNA sequences between organisms.

competence   Referring to the ability of a cell to take up naked DNA from the environment.

competitive inhibition   The prevention of a chemical reaction by a chemical that competes with the normal substrate for an enzyme's active site; *see also* **noncompetitive inhibition**.

complement   A group of blood proteins that functions in a cascading series of reactions with antibodies to recognize and help eliminate certain antigens or infectious agents.

complement fixation test   A serological procedure to detect antibodies to any of a variety of pathogens by identifying antibody-antigen-complement complexes.

complex   Referring to one form of symmetry found in some viral capsids.

complex medium   A chemically undefined medium in which the nature and quantity of each component has not been identified; *see also* **synthetic medium**.

compound   A substance made by the combination of two or more different chemical elements.

compound microscope   *See* **light microscope**.

condensation reaction   *See* **dehydration synthesis reaction**.

conidiophore   The supportive structure on which conidia form.

conidium (pl. **conidia**)   An asexually produced fungal spore formed on a supportive structure without an enclosing sac.

conjugate vaccine   An antigen preparation consisting of the antigen bound to a carrier protein.

conjugation   (1) In *Bacteria* and *Archaea*, a unidirectional transfer of genetic material from a live donor cell into a live recipient cell during a period of cell contact. (2) In the protozoan ciliates, a sexual process involving the reciprocal transfer of micronuclei between cells in contact.

conjugation pilus (pl. **pili**)   A protein filament essential for conjugation between donor and recipient bacterial cells.

constant domain   The invariable amino acids in the light and heavy chains of an antibody.

contact dermatitis   A delayed hypersensitivity reaction occurring after contact with an allergen.

contagious   Referring to a disease whose agent passes with particular ease among susceptible individuals.

contractile vacuole   A membrane-enclosed structure within a cell's cytoplasm that regulates the water content by absorbing water and then contracting to expel it.

control   That part of an experiment not exposed to or treated with the factor being tested.

contrast   In microscopy, to be able to see an object against the background.

convalescent serum   Antibody-rich serum obtained from a convalescing patient.

Coombs test   An antibody test used to detect Rh antibodies involved in hemolytic disease of the newborn.

corticosteroid   A synthetic drug used to control allergic disorders by blocking the release of chemical mediators.

covalent bond   A chemical linkage formed by the sharing of electrons between atoms or molecules.

Crenarchaeota   A group within the domain *Archaea* that tend to grow in hot or cold environments.

critical control point (CCP)   In the food processing industry, a place where contamination of the food product could occur.

cyanobacterium (pl. **cyanobacteria**)   An oxygen-producing, pigmented bacterial cell in a unicellular and filamentous form that carries out photosynthesis.

cyst   A dormant and very resistant form of a protozoan and multicellular parasite.

cystitis   An inflammation of the urinary bladder.

cytochrome   A compound containing protein and iron that plays a role as an electron carrier in cellular respiration and photosynthesis; *see also* **electron transport chain**.

cytokine   Small proteins released by immune defensive cells that affects other cells and the immune response to an infectious agent.

cytopathic effect (CPE)   Visible effect that can be seen in a virus-infected host cell.

cytoplasm   The complex of chemicals and structures within a cell; in plant and animal cells excluding the nucleus.

cytoskeleton   (1) The structural proteins in a prokaryotic cell that help control cell shape and cell division. (2) In a eukaryotic cell, the internal network of protein filaments and microtubules that control the cell's shape and movement.

cytosol   The fluid, ions, and compounds of a cell's cytoplasm excluding organelles and other structures.

cytotoxic hypersensitivity   A cell-damaging or cell-destroying hypersensitivity that develops when IgG reacts with antigens on the surfaces of cells.

cytotoxic T cell   The type of T lymphocyte that searches out and destroys infected cells.

cytotoxin   A chemical that is poisonous to cells.

D

dapsone   A chemotherapeutic agent used to treat leprosy patients.

dark-field microscopy   An optical system on the light microscope that scatters light such that the specimen appears white on a black background.

deamination   A biochemical process in which amino groups are enzymatically removed from amino acids or other organic compound.

decimal reduction time (D valve)   The time required to kill 90% of the viable organisms at a specified temperature.

decline phase   The final portion of a bacterial growth curve in which environmental factors overwhelm the population and induce death; also called death phase.

decomposer   An organism, such as a bacterium or fungus, that recycles dead or decaying matter.

defensin   An antimicrobial peptide present in white blood cells that plays a role in the prevention or elimination of infection.

definitive host   An organism that harbors the adult, sexually mature form of a parasite.

degerm   To mechanically remove organisms from a surface.

degranulation   The release of cell mediators from mast cells and basophils.

dehydration synthesis reaction   A process of bonding two molecules together by removing the products of water and joining the open bonds.

delayed hypersensitivity   An immune response to an allergen that takes 2 to 3 days to develop.

denaturation   A process caused by heat or pH in which proteins lose their function due to changes in their 3-D structure.

dendritic cell   A white blood cell having long finger-like extensions and found within all tissues; it engulfs and digests foreign material, such as bacterial cells and viruses, and presents antigen peptides on its surface.

dental plaque   A biofilm consisting of salivary proteins, food debris, and bacteria attached to the tooth surface.

deoxyribonucleic acid (DNA)   The genetic material of all cells and many viruses.

dermatophyte   A pathogenic fungus that affects the skin, hair, or nails.

desensitization   A process in which minute doses of antigens are used to remove antibodies from the body tissues to prevent a later allergic reaction.

detergent   A synthetic cleansing substance that dissolves dirt and oil.

diapedesis   A process by which phagocytes move out of the blood vessels by migrating between capillary cells.

diatom   One of a group of microscopic marine algae that performs photosynthesis.

dichotomous key   A method of deducing the correct species assignment of a living organism by offering two alternatives at each juncture, with the choice of one of those alternatives determining the next step.

differential medium   A growth medium in which different species of microorganisms can be distinguished visually.

differential staining procedure   A technique using two dyes to differentiate cells or cellular objects based on their staining; *see also* **simple stain technique**.

dikaryon   A fungal cell in which two genetically different haploid nuclei closely pair.

dimorphic   Referring to pathogenic fungi that take a yeast form in the human body and a filamentous form when cultivated in the laboratory.

dinoflagellate   A microscopic photosynthetic marine alga that forms one of the foundations of the food chain in the ocean.

dipicolinic acid   An organic substance that helps stabilize the proteins and DNA in a bacterial spore, thereby increasing spore resistance.

diplobacillus   A pair of rod-shaped bacterial or archaeal cells.

diplococcus (pl. diplococci)   A pair of spherical-shaped bacterial or archaeal cells.

diplomonad   A protozoan that contains four pair of flagella, two haploid nuclei, and live in low oxygen or anaerobic environments; most members are symbiotic in animals.

direct contact   The form of disease transmission involving close association between hosts; *see also* **indirect contact**.

direct microscopic count   Estimation of the number of cells by observation with the light microscope.

disaccharide   A sugar formed from two single sugar molecules; examples include sucrose and lactose.

disease   Any change from the general state of good health.

disinfectant   A chemical used to kill or inhibit pathogenic microorganisms on a lifeless object such as a tabletop.

disinfection   The process of killing or inhibiting the growth of pathogens.

disk diffusion method   A procedure for determining bacterial susceptibility to an antibiotic by determining if bacterial growth occurs around an antibiotic disk; also called the Kirby-Bauer test.

disulfide bridge   A covalent bond between sulfur-containing R groups in amino acids.

DNA   *See* **deoxyribonucleic acid**.

DNA chip   *See* **DNA microarray**.

DNA ligase   An enzyme that binds together DNA fragments.

DNA microarray   A series of DNA segments used to study changes in gene expression.

DNA polymerase III   An enzyme that catalyzes DNA replication by combining complementary nucleotides to an existing strand.

DNA probe   A short segment of single stranded DNA used to locate a complementary strand among many other DNA strands.

DNA replication   The process of copying the genetic material in a cell.

DNA vaccine   A preparation that consists of a DNA plasmid containing the gene for a pathogen protein.

domain   (1) The most inclusive taxonomic level of classification; consists of the *Archaea*, *Bacteria*, and *Eukarya*.

double diffusion assay   Another name for the diffusion of antibodies and antigens.

double helix   The structure of DNA, in which the two complementary strands are connected by hydrogen bonds between complementary nitrogenous bases and wound in opposing spirals.

Downey cell   A swollen lymphocyte with foamy cytoplasm and many vacuoles that develops as a result of infection with infectious mononucleosis viruses.

droplet transmission   The movement of an airborne particle of mucus and sputum from the respiratory tract.

E

echinocandin   An antifungal drug that inhibits cell wall synthesis.

eclipse period   The period of a viral infection when no viruses can be found inside the infected cell.

edema   A swelling of the tissues brought about by an accumulation of fluid.

effector cell   An activated immune cell targeting a pathogen.

electron   A negatively charged particle with a small mass that moves around the nucleus of an atom.

electron microscope   An instrument that uses electrons and a system of electromagnetic lenses to produce a greatly magnified image of an object; *see also* **transmission electron microscope** and **scanning electron microscope**.

electron shell   An energy level surrounding the atomic nucleus that contains one or more electrons.

electron transport chain   A series of proteins that transfer electrons in cellular respiration to generate ATP.

elementary body   An infectious form of *Chlamydia* in the early stage of reproduction.

ELISA   *See* **enzyme-linked immunosorbent assay**.

elongation   (1) The addition of complementary nucleotides to a parental DNA strand. (2) The addition of addition of amino acids onto the forming polypeptide during translation.

emerging infectious disease   A new disease or changing disease that is seen for the first time; *see also* **reemerging infectious disease**.

endemic   Referring to a disease that is constantly present in a specific area or region.

endergonic reaction   A chemical process that requires energy; *see also* **exergonic reaction**.

endocytosis   The process by which many eukaryotic cells take up substances, cells, or viruses from the environment.

endoflagellum   A microscopic fiber located along cell walls in certain species of spirochetes; contractions of the filaments yield undulating motion in the cell.

endogenous infection   A disorder that starts with a microbe or virus that already was in or on the body as part of the microbiota; *see also* **exogenous infection**.

endomembrane system   A cytoplasmic set of membranes that function in the transport, modification, and sorting of proteins and lipids in eukaryotic cells.

endophyte   A fungus that lives within plants and does not cause any known disease.

endoplasmic reticulum (ER)   A network of membranous plates and tubes in the eukaryotic cell cytoplasm responsible for the synthesis and transport of materials from the cell.

endospore   An extremely resistant dormant cell produced by some gram-positive bacterial species.

endosymbiotic model   The idea that mitochondria and chloroplasts originated from bacterial cells and cyanobacteria that took up residence in a primitive eukaryotic cell.

endotoxin   A metabolic poison, produced chiefly by gram-negative bacteria, that are part of the bacterial cell wall and consequently are released on cell disintegration; composed of lipid-polysaccharide-peptide complexes.

endotoxin shock   A drop in blood pressure due to an endotoxin.

energy-fixing reaction   A reaction in the first part of photosynthesis where light energy is converted into chemical energy in the form of ATP.

enriched medium   A growth medium in which special nutrients must be added to get an species to grow.

enterotoxin   A toxin that is active in the gastrointestinal tract of the host.

envelope   The flexible membrane of protein and lipid that surrounds many types of viruses.

enzyme   A reusable protein molecule that brings about a chemical change while itself remaining unchanged.

enzyme-linked immunosorbent assay (ELISA)   A serological test in which an enzyme system is used to detect an individual's exposure to a pathogen.

enzyme-substrate complex   The association of an enzyme with its substrate at the active site.

eosinophil   A type of white blood cell with granules that stains with the dye eosin and plays a role in allergic reactions and the body's response to parasitic infections.

epidemic   Referring to a disease that spreads more quickly and more extensively within a population than normally expected.

epidemiology   The scientific study of the source, cause, and transmission of disease within a population.

epitope   *See* **antigenic determinant**.

ergotism   A condition inducing convulsions and hallucinations from fungal toxins in rye.

erythema   A zone of redness in the skin due to a widening of blood vessels near the skin surface.

erythema migrans (EM)   An expanding circular red rash that occurs on the skin of patients with Lyme disease.

erythrogenic   Referring to a streptococcal poison that leads to the rash in scarlet fever.

E-test   An antibiotic sensitivity test using a paper strip containing a marked gradient of antibiotic.

Ethambutol   A synthetic antimicrobial drug used to treat tuberculosis.

ethylene oxide　A chemical gas that is used to sterilize many objects and instruments.

Eukarya　The taxonomic domain encompassing all eukaryotic organisms.

eukaryote　An organism whose cells contain a cell nucleus with multiple chromosomes, a nuclear envelope, and membrane-bound compartments; *see also* **prokaryote**.

eukaryotic　Referring to a cell or organism containing a cell nucleus with multiple chromosomes, a nuclear envelope, and membrane-bound compartments.

Euryarchaeota　A group within the domain *Archaea* that contains the methanogens and extreme halophiles.

exanthema　A maculopapular rash occurring on the skin surface.

Excavata　One of the super groups in the kingdom Protista, many lacking true mitochondria.

excision repair　A mechanism to correct improperly bonded bases in a DNA sequence.

exergonic reaction　A chemical process releasing energy; *see also* **endergonic reaction**.

exogenous infection　A disorder that starts with a microbe or virus that entered the body from the environment; *see also* **endogenous infection**.

exon　The coding sequence in a split gene; *see also* **intron**.

exotoxin　A bacterial metabolic poison composed of protein that is released to the environment; in the human body, it can affect various organs and systems.

experiment　A test or trial to verify or refute a hypothesis.

extreme halophile　An archaeal organism living at an extremely acidic pH.

extremophile　A microorganism that lives in extreme environments, such as high temperature, high acidity, or high salt.

F

Fab fragment　The branched portion of an antibody molecule that combines with an antigenic determinant of an antigen.

facilitated diffusion　The movement of substances from an area of higher concentration across a biological membrane to a region of lower concentration by means of a membrane-spanning channel or carrier protein.

facultative　Referring to an organism that grows in the presence or absence of oxygen gas (O_2).

family　A category of related organisms consisting of one or more genera.

Fc fragment　The stem portion of an antibody molecule that combines with phagocytes, mast cells, or complement.

feedback inhibition　The slowing down or prevention of a metabolic pathway when excess end product binds noncompetitively to an enzyme in the pathway.

fermentation　A metabolic pathway in which carbohydrates serve as electron donors, the final electron acceptor is not oxygen gas (O_2), and NADH is reoxidized to NAD^+ for reuse in glycolysis for generation of ATP.

fermentor　*See* **bioreactor**.

fever　An abnormally high body temperature that is usually caused by a bacterial or viral infection.

F factor　A plasmid containing genes for plasmid replication and conjugation pilus formation.

field　The circular area seen when looking in the light microscope.

filariform　Referring to the thread-like shape of the larvae of the hookworm.

filopodium (pl. **filopodia**)　A thin protrusion from a cell, such as a phagocyte.

filtration　A mechanical method to remove microorganisms by passing a liquid or air through a filter.

Firmicutes　A phylum in the domain *Bacteria* that contains many of the gram-positive species.

flaccid paralysis　A loss of voluntary movement in which the limbs have little tone and become flabby.

flagellum (pl. **flagella**)　A long, hair-like appendage composed of protein and responsible for motion in microorganisms; found in some bacterial, archaeal protozoal, algal, and fungal cells.

flare　A spreading zone of redness around a wheal that occurs during an allergic reaction; *also see* **wheal**.

flash pasteurization method　A treatment in which milk is heated at 71.6°C for 15 seconds and then cooled rapidly to eliminate harmful bacteria; also called HTST ("high temperature, short time") method.

flatworm　A multicellular parasite with a flattened body; examples include the tapeworms.

floc　A jelly-like mass that forms in a liquid and made up of coagulated particles.

flocculation　(1) A serological reaction in which particulate antigens react with antibodies to form visible aggregates of material. (2) The formation of jelly-like masses of coagulated material in the water-purification process.

flucytosine　An antifungal drug that interrupts nucleic acid synthesis.

fluid mosaic model   The representation for the cell (plasma) membrane where proteins "float" within or on a bilayer of phospholipid.

fluke   *See* **trematode**.

fluorescence microscopy   An optical system on the light microscope that uses ultraviolet light to excite dye-containing objects to fluoresce.

fluorescent antibody technique   A diagnostic tool that uses fluorescent antibodies with the fluorescence microscope to identify an unknown organism.

fluoroquinolone   A drug used to treat urinary and intestinal tract infections.

flush   A burst of mushroom growth.

folic acid   The organic compound in bacteria whose synthesis is blocked by sulfonamide drugs.

fomite   An inanimate object, such as clothing or a utensil, that carries disease organisms.

food vacuole   A membrane-enclosed compartment in some eukaryotic that results from the intake of large molecules, particles, or cells, for digestion.

foraminiferan   A shell-containing amoeboid protozoan having a chalky skeleton with window-like openings between sections of the shell.

formalin   A solution of formaldehyde used as embalming fluid, in the inactivation of viruses, and as a disinfectant.

F plasmid   A DNA plasmid in the cytoplasm of an F^+ bacterial cell that may be transferred to a recipient bacterial cell during conjugation.

F' plasmid   An F plasmid carrying a bacterial chromosome fragment.

fractional sterilization   A sterilization method in which materials are heated in free-flowing steam for 30 minutes on each of three successive days; also called tyndallization.

fruiting body   The general name for a reproductive structure of a fungus from which spores are produced.

functional genomics   The identification of gene function from a gene sequence.

functional group   A group of atoms on hydrocarbons that function in chemical reactions.

fungemia   The dissemination of fungi through the circulatory system.

Fungi   One of the four kingdoms in the domain *Eukarya*; composed of the molds and yeasts.

fungicidal   Referring to any agent that kills fungi.

fungistatic   Referring to any substance that inhibits the growth of fungi.

furuncle   An infection of a hair follicle.

G

gametocyte   The stage in the life cycle of the malaria parasite during which it reproduces sexually in the blood of a mosquito.

gamma globulin   A general term for antibody-rich serum.

gamma ray   An ionizing radiation that can be used to sterilize objects.

gangrene   A physiological process in which the enzymes from wounded tissue digest the surrounding layer of cells, inducing a spreading death to the tissue cells.

gastroenteritis   Infection of the stomach and intestinal tract often due to a virus.

gas vesicle   A cytoplasmic compartment in some bacterial and archaeal cells used to regulate buoyancy.

gene   A segment of a DNA molecule that provides the biochemical information for a polypeptide or for a functioning RNA molecule.

gene probe   A small, single-stranded DNA fragment labeled for identification of a specific DNA segment.

generalized transduction   A process by which a bacteriophage carries a bacterial chromosome fragment from one cell to another; *see also* **specialized transduction**.

generation time   The time interval for a cell population to double in number.

genetically modified organism (GMO)   An organism produced by genetic engineering.

genetic code   The specific order of nucleotide sequences in DNA or RNA that encode specific amino acids for protein synthesis.

genetic engineering   The use of bacterial and microbial genetics to isolate, manipulate, recombine, and express genes.

genetic recombination   The process of bring together different segments of DNA.

genome   The complete set of genes in a virus or an organism.

genome annotation   The identification of gene locations and function in a gene sequence.

genomic island   A series of up to 25 genes absent in other strains of the same prokaryotic species.

genomics   The study of an organism's gene structure and gene function in viruses and organisms.

genus (pl. **genera**)   A rank in the classification system of organisms composed of one or more species; a collection of genera constitute a family.

germicide   Any agent that kills microorganisms.

germ theory The principle formulated by Pasteur and proved by Koch that microorganisms are responsible for infectious diseases.

Glomeromycota A group of mycorrhizal fungi that exist within the roots of most land plants.

glucose A six-carbon sugar used as a major energy source for metabolism.

glutaraldehyde A liquid chemical used for sterilization.

glycocalyx A viscous polysaccharide material covering many prokaryotic cells to assist in attachment to a surface and impart resistance to desiccation; *see also* **capsule** *and* **slime layer**.

glycolysis A metabolic pathway in which glucose is broken down into two molecules of pyruvate with a net gain of two ATP molecules.

glycylcycline A drug that is effective against antibiotic resistant *Staphylococcus aureus*.

Golgi apparatus A stack of flattened, membrane-enclosed compartments in eukaryotic cells involved in the modification and sorting of lipids and proteins.

gonococcus A colloquial name for *Neisseria gonorrhoeae*.

graft-versus-host (GVH) reaction A phenomenon in which a tissue graft or transplant produces immune substances against the recipient.

gram-negative Referring to a bacterial cell that stains red after Gram staining.

gram-positive Referring to a bacterial cell that stains purple after Gram staining.

Gram stain technique A staining procedure used to identify bacterial cells as gram-positive or gram-negative.

granulocyte A white blood cell with visible granules in the cytoplasm; includes neutrophils, eosinophils, and basophils; *see also* **agranulocyte**.

granuloma A small lesion caused by an infection.

granzyme A cytotoxic enzyme that causes infected cells to undergo a programmed cell death.

gray syndrome A side effect of chloramphenicol therapy, characterized by a sudden breakdown of the cardiovascular system.

griseofulvin An antifungal drug used against infections of the skin, hair, and nails.

groundwater Water originating from deep wells and subterranean springs; *see also* **surface water**.

group A streptococci (GAS) Contains those streptococcal species that are β-hemolytic organisms.

Guillain-Barré syndrome (GBS) A complication of influenza and chickenpox, characterized by nerve damage and polio-like paralysis.

gumma A soft, granular lesion that forms in the cardiovascular and/or nervous systems during tertiary syphilis.

H

HAART *See* **highly active antiretroviral therapy**.

HACCP *See* **hazard analysis critical control point**.

halogen A chemical element whose atoms have seven electrons in their outer shell; examples include iodine and chlorine.

halophile An organism that lives in environments with high concentrations of salt.

halotolerant A microbe that grows best without salt but can tolerate low concentrations.

hapten A small molecule that combines with tissue proteins or polysaccharides to form an antigen.

Hazard Analysis Critical Control Point (HACCP) A set of federally enforced regulations to ensure the dietary safety of seafood, meat, and poultry.

healthcare-associated infection (HAI) An infection resulting from treatment for another condition while in a hospital or healthcare facility.

heat fixation The use of warm temperatures to prepare microorganisms for staining and viewing with the light microscope.

heavy (H) chain The larger polypeptide in an antibody.

heavy metal A chemical element often toxic to microorganisms; examples include mercury, copper, and silver.

helical One form of symmetry found in some viral capsids.

helminth A term referring to a multicellular parasite; includes roundworms and flatworms.

helminthic load The number of worms in the human body.

helper T (T_H) cell A type of lymphocyte involved in regulation of cell-mediated immunity and in helping B cells to produce antibody.

helper T1 (T_H1) cell A T lymphocyte that enhances the activity of B lymphocytes.

helper T2 (T_H2) cell A T lymphocyte that stimulates destruction of macrophages infected with bacterial cells.

hemagglutination The formation of clumps of red blood cells.

hemagglutination-inhibition (HAI) test A test using a patient's serum to detect the presence of antibodies against a specific infectious agent.

hemagglutinin   (1) An enzyme composing one type of surface spike on influenza viruses that enables the viruses to bind to the host cell. (2) An agent such as a virus or an antibody that causes red blood cells to clump together.

hemolysin   A bacterial toxin that destroys red blood cells.

hemolysis   The destruction of red blood cells.

hepatitis B core antigen (HBcAG)   An antigen located in the inner lipoprotein coat enclosing the DNA of a hepatitis B virus.

hepatitis B surface antigen (HBsAg)   An antigen located in the outer surface coat of a hepatitis B virus.

hepatocellular carcinoma (HCC)   Cancer of the liver tissue.

herd immunity   The proportion of a population that is immune to a disease; also called community immunity.

heterokaryon   A fungal cell that has two or more genetically different nuclei.

heterotroph   An organism that requires preformed organic matter for its energy and carbon needs; *see also* **photoheterotroph** *and* **chemoheterotroph**.

heterotrophy   The process by which an organism uses preformed organic compounds for metabolism.

high frequency of recombination (Hfr)   Referring to a bacterial cell containing an F factor incorporated into the bacterial chromosome.

highly active antiretroviral therapy (HAART) The combination of several (typically three or four) antiretroviral drugs for the treatment of infections caused by retroviruses, especially the human immunodeficiency virus.

histamine   A mediator in type I hypersensitivity reactions that is released from the granules in mast cells and basophils and causes the contraction of smooth muscles.

holding method   A pasteurization process that exposes a liquid to 62.9° for 30 min.

homeostasis   The tendency of an organism to maintain a steady state or equilibrium with respect to specific functions and processes.

hops   The dried petals of the vine *Humulus lupulus* that impart flavor, color, and stability to beer.

horizontal gene transfer (HGT)   The movement of genes from one organism to another within the same generation; also called lateral gene transfer.

horizontal transmission   The spread of disease from one person to another.

host   An organism on or in which a microorganism lives and grows, or a virus replicates.

host range   The variety of species that a disease-causing microorganism can infect.

human genome   The complete set a genetic information in a human cell.

human leukocyte antigen (HLA)   Cell surface antigens involved with tissue transplantation.

humoral immune response   The immune reaction of producing antibodies directed against antigens in the body fluids.

hyaluronidase   An enzyme that digests hyaluronic acid and thereby permits the penetration of pathogens through connective tissue.

hybridoma   The cell resulting from the fusion of a B cell and a cancer cell.

hydatid cyst   A thick-walled body formed in the human liver by *Echinococcus granulosus* and containing larvae.

hydrocarbon   An organic molecule containing only hydrogen and carbon atoms that are connected by a sharing of electrons.

hydrogen bond   A weak attraction between a positively charged hydrogen atom (covalently bonded to oxygen or nitrogen) and a covalently bonded, negatively charged oxygen or nitrogen atom in the same or separate molecules.

hydrogen peroxide   An unstable liquid that readily decomposes in water and oxygen gas (O_2).

hydrolysis reaction   A process in which a molecule is split into two parts through the interaction of H^+ and $(OH)^-$ of a water molecule.

hydrophilic   Referring to a substance that dissolves in or mixes easily with water; *see also* **hydrophobic**.

hydrophobia   An emotional condition ("fear of water") arising from the inability to swallow as a consequence of rabies.

hydrophobic   Referring to a substance that does not dissolve in or mix easily with water; *see also* **hydrophilic**.

hyperacute rejection   An immediate rejection response to a transplant.

hyperimmune serum   The fluid portion of the blood containing a higher than normal amount of a particular antibody.

hypersensitivity   An immunological response to exposure to an allergen or antigen.

hyperthermophile   A prokaryote that has an optimal growth temperature above 80°C.

hypha (pl. hyphae)   A microscopic filament of cells representing the vegetative portion of a fungus.

hypothesis   An educated guess or answer to a properly framed question.

I

icosahedral   Referring to a symmetrical figure composed of 20 triangular faces and 12 points; one of the forms of symmetry found in some viral capsids.

IgA   The class of antibodies found in respiratory and gastrointestinal secretions that help neutralize pathogens.

IgD   The class of antibodies found on the surface of B cells that act as receptors for binding antigen.

IgE   The class of antibodies responsible for type I hypersensitivities.

IgG   The class of antibodies abundant in serum that are major diseases fighters.

IgM   The first class of antibodies to appear in serum in helping fight pathogens.

imidazole   An antifungal drug that interferes with sterol synthesis in fungal cell membranes; examples are miconazole and ketoconazole.

immediate hypersensitivity   Immune reactions to an allergen that occur within minutes to a few hours.

immune complex   A combination of antibody and antigen capable of complement activation and characteristic of type III hypersensitivity reactions.

immune complex hypersensitivity   An immune response involving antibodies combining with antigens, the complex accumulating in blood vessels or on tissue surfaces.

immune deficiency   The lack of an adequate immune system response.

immunity   The body's ability to resist infectious disease through innate and acquired mechanisms.

immunization   The process of making an individual resistant to a particular disease by administering a vaccine; *see also* **vaccination**.

immunocompetent   The ability of the body to develop an immune response in the presence of a disease-causing agent.

immunocompromised   Referring to an inadequate immune response as a result of disease, exposure to radiation, or treatment with immunosuppressive drugs.

immunodeficiency disorder   The inability, either inborn or acquired, of the body to produce an adequate immune response to fight disease.

immunodiffusion   The movement of antigen and antibody toward one another through a gel or agar to produce a visible precipitate.

immunoelectrophoresis   A laboratory diagnostic procedure in which antigen molecules move through an electric field and then diffuse to meet antibody molecules to form a precipitation line; *see also* **electrophoresis**.

immunogenic   Capable of generating an immune response.

immunoglobulin (Ig)   The class of immunological proteins that react with an antigen; an alternate term for antibody.

immunological memory   The long-term ability of the immune system to remember past pathogen exposures.

immunology   The scientific study of how the immune system works and responds to non-self agents.

immunomodulation   The modification of some aspect of the immune system as part of a treatment, especially the suppression of an overactive inflammatory response.

immunosuppressive agent   A drug given to block transplant rejection by the immune system.

incidence   The number of reported disease cases in a given time frame.

incineration   To burn to ashes, or cause something to burn to ashes.

inclusion   (1) A granule-like storage structure found in the prokaryotic cell cytoplasm. (2) A virus in the cytoplasm or nucleus of an infected cell.

incubation period   The time that elapses between the entry of a pathogen into the host and the appearance of signs and symptoms.

index of refraction   A measure of the light bending ability of a medium through which light passes.

indicator organism   A microorganism whose presence signals fecal contamination of water.

indigenous microbiota   The microbial agents that are associated with an animal for long periods of time without causing disease; *see also* **transient microbiota**.

indirect contact   The mode of disease transmission involving nonliving objects; *see also* **direct contact**.

induced mutation   A change in the sequence of nucleotide bases in a DNA molecule arising from a mutagenic agent used under controlled laboratory conditions.

inducer   A substance that binds to a repressor protein and "turns on" new gene transcription.

induration   A thickening and drying of the skin tissue that occurs in type IV hypersensitivity reactions.

industrial fermentation   Any large scale industrial process, with or without oxygen gas (O_2), for growing microorganisms; *see also* **fermentation**.

infection   The relationship between two organisms and the competition for supremacy that takes place between them.

infection allergy   A type IV hypersensitivity reaction in which the immune system responds to the presence of certain microbial agents.

infectious disease   A disorder arising from a pathogen invading a susceptible host and inducing medically significant symptoms.

infectious dose   The number of microorganisms needed to bring about infection.

inflammation   A nonspecific defensive response to injury; usually characterized by redness, warmth, swelling, and pain.

initiation   (1) The unwinding and separating of DNA strands during replication. (2) The beginning of translation.

innate immunity   An inborn set of the preexisting defenses against infectious agents; includes the skin, mucous membranes, and secretions.

insertion sequence (IS)   A segment of DNA that forms a copy of itself, after which the copy moves into areas of gene activity to interrupt the genetic coding sequence.

interferon   An antiviral protein produced by body cells on exposure to viruses and which trigger the synthesis of antiviral proteins.

interleukin   A chemical cytokine produced by white blood cells that causes other white blood cells to divide.

intermediate host   The host in which the larval or asexual stage of a parasite is found.

intoxication   The presence of microbial toxins in the body.

intraepithelial lymphatocyte (IEL)   An intestinal epithelial cell that removes injured or infected cells.

intron   A non-coding sequence in a split gene; *see also* **exon**.

in use test   A procedure used to determine the value of a disinfectant or antiseptic.

invasiveness   The ability of a pathogen to spread from one point to adjacent areas in the host and cause structural damage to those tissues.

invasive wound infection   Microbial invasion of living tissue beneath a burn.

iodophor   A complex of iodine and detergents that is used as an antiseptic and disinfectant.

ion   An electrically charged atom.

ionic bond   The electrical attraction between oppositely charged ions.

ionizing radiation   A type of radiation such as gamma rays and X rays that causes the separation of atoms or a molecule into ions.

isograft   Tissue taken from one identical twin and grafted to the other twin.

isomer   A molecule with the same molecular formula but different structural formula.

isoniazid (INH)   An antimicrobial drug effective against the tubercle bacillus.

isotope   An atom of the same element in which the number of neutrons differs; *see also* **radioisotope**.

K

Kaposi sarcoma   A type of cancer in immunocompromised individuals, such as AIDS patients, where cancer cells and an abnormal growth of blood vessels form solid lesions in connective tissue.

keratin   The tough, fibrous protein produced by keratinocytes.

keratinocyte   The most common cell type of the epidermis.

kinetoplast   The mass of DNA found in the mitochondria of the kinetoplastids.

kinetoplastid   A protozoan with one flagellum; most are parasitic in aerobic or anaerobic environments.

kingdom   The most inclusive taxonomic rank below the domain level.

Kirby-Bauer test   *See* **disk diffusion method**.

Koch's postulates   A set of procedures by which a specific organism can be related to a specific disease.

Koplik spots   Red patches with white central lesions that form on the gums and walls of the pharynx during the early stages of measles.

Krebs cycle   *See* **citric acid cycle**.

L

lactic acid fermentation   A catabolic process that produces lactic acid during the reoxidation of NADH to NAD$^+$ for reuse in glycolysis to generate ATP.

lactone   A chemical compound used as a flavoring ingredient in foods and beverages.

lagering   The secondary aging of beer.

lagging strand   During DNA replication, the new strand that is synthesized discontinuously; *see also* **leading strand**.

lag phase   A portion of a bacterial growth curve encompassing the first few hours of the population's history when no growth occurs.

latency   A condition in which a virus integrates into a host chromosome without immediately causing a disease.

latent infection   A viral infection where the viral DNA remains "dormant" within a host chromosome.

latent phase   The period during a viral infection when virus particles cannot be found outside the infected cells.

late-phase anaphylaxis   After an acute IgE mediated reaction, a second response occurs many hours after the initial response, and appears to be based on the activity of eosinophils.

leading strand   During DNA replication, the new strand that is synthesized continuously; *see also* **lagging strand**.

leproma   A tumor-like growth on the skin associated with leprosy.

leukocidin   A bacterial enzyme that destroys phagocytes, thereby preventing phagocytosis.

leukocyte   Any of a number of types of white blood cells.

leukopenia   A condition characterized by an abnormal drop in the normal number of white blood cells.

leukotriene   A substance that acts as a mediator during type I hypersensitivity reactions; formed from arachidonic acid after the antigen-antibody reaction has taken place.

lichen   An association between a fungal mycelium and a cyanobacterium or alga.

light (L) chain   A smaller polypeptide in an antibody.

light microscope   An instrument that uses visible light and a system of glass lenses to produce a magnified image of an object; also called a compound microscope.

lincosamide   An antibiotic that blocks protein synthesis.

lipid   A nonpolar organic compound composed of carbon, hydrogen, and oxygen; examples include triglycerides and phospholipids.

lipid A   A component in the outer membrane of the gram-negative cell wall.

lipopolysaccharide (LPS)   A molecule composed of lipid and polysaccharide that is found in the outer membrane of the gram-negative cell wall of bacterial cells.

littoral zone   The environment along the shoreline of an ocean.

local disease   A disease restricted to a single area of the body.

lockjaw   *See* **trismus**.

logarithmic (log) phase   The portion of a bacterial growth curve during which active growth leads to a rapid rise in cell numbers.

looped domain structure   The term used to describe organization and packing of the prokaryotic chromosome.

lymph   The tissue fluid that contains white blood cells and drains tissue spaces through the lymphatic system.

lymphedema   An abnormal swelling due to the loss of normal lymph vessel drainage of the affected part; typical of elephantiasis.

lymph node   A bean-shaped organ located along lymph vessels that is involved in the immune response and contains phagocytes and lymphocytes.

lymphocyte   A type of white blood cell that functions in the immune system.

lymphoid progenitor   A bone marrow cell that gives rise to lymphocytes; *see also* **myeloid progenitor**.

lyophilization   A process in which food or other material is deep frozen, after which its liquid is drawn off by a vacuum; also called freeze-drying.

lysis   The rupture of a cell and the loss of cell contents.

lysogenic cycle   The events of a bacterial virus infection that result in the integration of its DNA into the bacterial chromosome.

lysosome   A membrane-enclosed compartment in many eukaryotic cells that contains enzymes to degrade or digest substances.

lysozyme   An enzyme found in tears and saliva that digests the peptidoglycan of gram-positive bacterial cell walls.

lytic cycle   A process by which a bacterial virus replicates within a host cell and ultimately destroys the host cell.

M

macrolide   An antibiotic that blocks protein synthesis.

macromolecule   A large, chemically-bonded substance built from smaller building blocks.

macronucleus   The larger of two nuclei in most ciliates that is involved in controlling metabolism; *see also* **micronucleus**.

macrophage   A large cell derived from monocytes that is found within various tissues and actively engulfs foreign material, including infecting bacterial cells and viruses.

macule   A pink-red skin spot associated with infectious disease.

magnetosome   A cytoplasmic inclusion body in some bacterial cells that assists orientation to the environment by aligning with the magnetic field.

magnification   The increase in the apparent size of an object observed with a microscope.

major histocompatibility complex (MHC)   A set of genes that controls the expression of MHC proteins; involved in transplant rejection; *see also* **MHC class I** *and* **MHC class II**.

malignant   Referring to a tumor that invades the tissue around it and may spread to other parts of the body.

malting   The process whereby barley grains are treated to digest starch to maltose.

Mantoux test   The infection of a tuberculosis purified protein derivative into the forearm.

mashing   A fermentable mixture of hot water and barley grain from which alcohol is distilled.

mass   The amount of matter in a sample.

mass number   The total number of protons and neutrons in an atom.

mast cell   A type of cell in connective tissue which releases histamine during allergic attacks.

matrix protein   A protein shell found in some viruses between the genome and capsid.

matter   Any substance that has mass and occupies space.

M cell   An intestinal epithelial cell that processes foreign antigens.

mebendazole   An antihelminthic drug that inhibits the uptake of nutrients.

mechanical vector   A living organism, or an object, that transmits disease agents on its surface; *see also* **biological vector**.

membrane attack complex   A set of complement proteins that forms holes in a bacterial cell membrane and leads to cell destruction.

membrane filter   A pad of cellulose acetate or polycarbonate.

memory cell   A cell derived from B lymphocytes or T lymphocytes that reacts rapidly upon re-exposure to antigen.

meningitis   A general term for inflammation of the covering layers of the brain and spinal cord due to any of several bacteria, fungi, viruses, or protozoa.

merozoite   A stage in the life cycle of the malaria parasite that invades red blood cells in the human host.

mesophile   An organism that grows in temperature ranges of 20°C to 40°C.

messenger RNA (mRNA)   An RNA transcript containing the information for synthesizing a specific polypeptide.

metabolic pathway   A sequence of linked enzyme-catalyzed reactions in a cell.

metabolism   The sum of all biochemical processes taking place in a living cell; *see also* **anabolism** *and* **catabolism**.

metachromatic granule   A polyphosphate-storing granule commonly found in *Corynebacterium diphtheriae* that stains deeply with methylene blue; also called volutin.

metagenome   The collective genomes from a population of organisms.

metagenomics   The study of genes isolated directly from environmental samples.

metastasize   Referring to a tumor that spreads from the site of origin to other tissues in the body.

methanogen   An archaeal organism that lives on simple compounds in anaerobic environments and produces methane during its metabolism.

MHC class I   A group of glycoproteins found on all nucleated cells to which cytotoxic T cells can bind.

MHC class II   A group of glycoproteins found on B cells, Macrophages, and dendritic cells that are recognized by helper T cells.

miasma   An ill-defined idea of the 1700s and 1800s that suggested diseases were caused by an altered chemical quality of the atmosphere.

microaerophile   An organism that grows best in an oxygen-reduced environment.

microbe   *See* **microorganism**.

microbial forensics   The discipline involved with the recognition, identification, and control of a pathogen.

microbial genomics   The discipline of sequencing, analyzing, and comparing microbial genomes.

microbicidal   Referring to any agent that kills microbes.

microbiology   The scientific study of microscopic organisms and viruses, and their roles in human disease as well as beneficial processes.

microbiome   The community of microorganisms in an environment or body location.

microbiostatic   Referring to any agent that inhibits microbial growth.

microbiota   The population of microorganisms that colonize various parts of the human body and do not cause disease in a healthy individual.

microcompartment   A region in some bacterial cells surrounded by a protein shell.

microenvironment   A cell's or organism's physical and chemical surroundings.

microfilaria (pl. microfilariae)   The larva stage of the parasitic nematode *Wuchereria bancrofti*.

micrometer (μm)   A unit of measurement equivalent to one millionth of a meter; commonly used in measuring the size of microorganisms.

micronucleus   The smaller of the two nuclei in most ciliates that contains genetic material and is involved in sexual reproduction; *see also* **macronucleus**.

microorganism (microbe)   A microscopic form of life including bacterial, archaeal, fungal, and protozoal cells.

mineralization   The conversion of organic compounds to inorganic compounds and ammonia.

minimum inhibitory concentration (MIC)   The lowest concentration of an antimicrobial agent that will inhibit its growth.

miracidium (pl. **miracidia**)   A ciliated larva representing an intermediary stage in the life cycle of a trematode.

mismatch repair   A mechanism to correct mismatched bases in the DNA; *see also* **excision repair**.

mitochondrion (pl. **mitochondria**)   A double membrane-enclosed compartment in eukaryotic cells that carries out aerobic respiration.

mitosporic fungus   A fungus without a known sexual stage of reproduction.

mold   A type of fungus that consists of chains of cells and appears as a fuzzy mass of thin filaments in culture.

molecular formula   The representation of the kinds and numbers of each atom in a molecule.

molecular taxonomy   The systematized arrangement of related organisms based on molecular characteristics, such as ribosomal DNA nucleotide sequences.

molecular weight   The sum of the atomic masses of all atoms in a molecule.

molecule   Two or more atoms held together by a sharing of electron pairs.

monobactam   A synthetic antibacterial drug used to inhibit cell wall synthesis in gram-negative bacteria.

monoclonal antibody   A type of antibody produced by a clone of hybridoma cells, consisting of antigen-stimulated B cells fused to myeloma cells.

monocyte   A circulating white blood cell with a large bean-shaped nucleus that is the precursor to a macrophage.

monolayer   A single layer of cultured cells.

monomer   A simple organic molecule that can join in long chains with other molecules to form a more complex molecule; *see also* **polymer**.

monophyletic group   A group of organisms descended from a common ancestor.

monosaccharide   A simple sugar that cannot be broken down into simpler sugars; examples include glucose and fructose.

monospot test   A method used to detect the presence of heterophile antibodies, which is indicative of infectious mononucleosis.

monotherapy   The use of a single drug to treat a disease.

most probable number (**MPN**)   A laboratory test in which a statistical evaluation is used to estimate the number of bacterial cells in a sample of fluid; often employed in determinations of coliform bacteria in water.

M protein   A protein that enhances the pathogenicity of streptococci by allowing organisms to resist phagocytosis and adhere firmly to tissue.

mucous membrane (**mucosa**)   A moist lining in the body passages of all mammals that contains mucus-secreting cells and is open directly or indirectly to the external environment.

mucus   A sticky secretion of glycoproteins.

mushroom   A spore-bearing fruiting body typical of many members of the basdiomycetes.

must   The juice resulting from crushing grapes.

mutagen   A chemical or physical agent that causes a mutation.

mutant   An organism carrying a mutation.

mutation   A permanent alteration of a DNA sequence.

mutualism   A close and permanent association between two populations of organisms in which both benefit from the association.

mycelium (pl. **mycelia**)   A mass of fungal filaments from which most fungi are built.

mycetism   Mushroom poisoning.

mycolic acid   A waxy lipid composing the cell wall of mycobacterial species.

mycology   The scientific study of fungi.

mycorrhiza (pl. **mycorrhizae**)   A close association between a fungus and the roots of many plants.

mycosis (pl. **mycoses**)   A disease caused by a fungus.

mycotoxin   A poison produce by a fungus that adversely affects other organisms.

myeloid progenitor   A bone marrow cell that gives rise to red blood cells and all white blood cells except lymphocytes; *see also* **lymphoid progenitor**.

myeloma   A malignant tumor that develops in the blood-cell–producing cells of the bone marrow; the cells are used in the procedure to produce monoclonal antibodies.

myxobacteria   A group of soil-dwelling bacterial species that exhibit multicellular behaviors.

N

nanometer (**nm**)   A unit of measurement equivalent to one billionth of a meter; the unit used in measuring viruses and the wavelength of energy forms.

narrow spectrum   Referring to an antimicrobial drug that is useful for a restricted group of microorganisms; *see also* **broad spectrum**.

nasopharyngeal carcinoma   A cancer originating in the nasopharynx and linked to infection by the Epstein-Barr virus.

natural killer (NK) cell   A type of defensive body cell that attacks and destroys cancer cells and infected cells without the involvement of the immune system.

naturally acquired active immunity   A host response resulting in antibody production as a result of experiencing the disease agent, or a passive response resulting from the passage of antibodies to the fetus via the placenta or the milk of a nursing mother.

necrosis   Cell or tissue death.

negative control   A form of gene regulation where a repressor protein binds to an operator and blocks transcription.

negative selection   A method for identifying mutations by selecting cells or colonies that do not grow when replica plated.

negative stain technique   A staining process that results in colorless bacterial cells on a stained background when viewed with the light microscope.

negative strand   Referring to those RNA viruses whose genome cannot be directly transcribed into protein.

Negri body   A cytoplasmic inclusion that occurs in brain cells infected with rabies viruses.

neuraminidase   An enzyme composing one type of surface spike of influenza viruses that facilitates viral release from the host cell.

neuraminidase inhibitor   A compound that inhibits neuraminidase, preventing the release of flu viruses from an infected cell.

neutralization   A type of antigen-antibody reaction in which the activity of a toxin is inactivated.

neutron   An uncharged particle in the atomic nucleus.

neutrophil   The most common type of white blood cell; functions chiefly to engulf and destroy foreign material, including bacterial cells and viruses that have entered the body.

neutrophil extracellular traps (NETS)   The nuclear DNA and antimicrobial proteins of dead neutrophils that form a fibrous mesh trapping pathogens.

nitrogen cycle   The processes that convert nitrogen gas (N_2) to nitrogen-containing substances in soil and living organisms, then reconverted to the gas.

nitrogen fixation   The chemical process by which microorganisms convert nitrogen gas (N_2) into ammonia.

nitrogenous base   Any of five nitrogen-containing compounds found in nucleic acids, including adenine, guanine, cytosine, thymine, and uracil.

nitroimidazole   An antiprotozoal drug that interferes with DNA synthesis.

noncommunicable disease   A disease whose causative agent is acquired from the environment and is not transmitted to another individual.

noncompetitive inhibition   The prevention of a chemical reaction by a chemical that binds elsewhere than to active site of an enzyme; *see also* **competitive inhibition**.

nonhalophile   A microorganism that cannot grow in the presence of added sodium chloride.

nonhemolytic   Referring to those bacterial species that when plated on blood agar cause no destruction of the red blood cells.

nonpolar molecule   A covalently bonded substance in which there is no electrical charge; *see also* **polar molecule**.

nosocomial infection   A disorder acquired during an individual's stay at a hospital or chronic care facility.

nucleic acid   A high-molecular-weight molecule consisting of nucleotide chains that convey genetic information and are found in all living cells and viruses; *see* **deoxyribonucleic acid** *and* **ribonucleic acid**.

nucleocapsid   The combination of genome and capsid of a virus.

nucleoid   The chromosomal region of a bacterial and archaeal cells.

nucleotide   A component of a nucleic acid consisting of a carbohydrate molecule, a phosphate group, and a nitrogenous base.

O

obligate aerobe   An organism that requires oxygen gas (O_2) for metabolism.

obligate anaerobe   An organism that cannot use oxygen gas (O_2) for metabolism.

observation   The use of the senses or instruments to gather information on which science inquiry is based.

Okazaki fragment   A segment of DNA resulting from discontinuous DNA replication.

oncogene   A segment of DNA that can induce uncontrolled growth of a cell if permitted to function.

oncogenic   Referring to any agent such as viruses that can cause tumors.

operator   A sequences of bases in the DNA to which a repressor protein can bind.

operon   The unit of bacterial DNA consisting of a promoter, operator, and a set of structural genes.

opisthotonus   An arching of the back that is characteristic of tetanus.

opportunist   A microorganism that invades the tissues when body defenses are suppressed.

opportunistic infection   A disorder caused by a microorganism that does not cause disease but that can become pathogenic or life-threatening if the host has a low level of immunity.

opsonin   An antibody or complement component that encourages phagocytosis.

opsonization   Enhanced phagocytosis due to the activity of antibodies or complement.

oral rehydration solution (ORS)   A mixture of blood salts and glucose in water.

order   A category of related organisms consisting of one or more families.

organelle   A specialized compartment in cells that has a particular function.

organic compound   A substance characterized by chains or rings of carbon atoms that are linked to atoms of hydrogen and sometimes oxygen, nitrogen, and other elements.

origin of transfer   The fixed point on an F plasmid (factor) where one strand is nicked and transferred to a recipient cell.

osmosis   The net movement of water molecules from where they are in a high concentration through a semipermeable membrane to a region where they are in a lower concentration.

osmotic pressure   The force that must be applied to a solution to inhibit the inward movement of water across a membrane.

outbreak   A small, localized epidemic.

outer membrane   A bilayer membrane forming part of the cell wall of gram-negative bacteria.

oxazolidinone   An antibiotic that blocks protein synthesis and is effective in treating gram-positive bacteria.

oxidation   A chemical change in which electrons are lost by an atom; *see also* **reduction**.

oxidation lagoon   A large pond in which sewage is allowed to remain undisturbed so that digestion of organic matter can occur.

oxidative phosphorylation   A series of sequential steps in which energy is released from electrons as they pass from coenzymes to cytochromes, and ultimately, to oxygen gas (O_2); the energy is used to combine phosphate ions with ADP molecules to form ATP molecules.

oxygenic photosynthesis   A form of photosynthesis in which molecular oxygen (O_2) is produced.

P

pandemic   A worldwide epidemic.

papule   A pink pimple on the skin.

para-aminobenzoic acid (PABA)   A precursor for folic acid synthesis.

parabasalid   A protozoan that contains numerous of flagella and lives in low oxygen or anaerobic environments; most members are symbiotic in animals.

parasite   A type of heterotrophic organism that feeds on live organic matter such as another organism.

parasitemia   The spread of protozoa and multicellular worms through the circulatory system.

parasitism   A close association between two organisms in which one (the parasite) feeds on the other (the host) and may cause injury to the host.

parasitology   The scientific study of parasites.

paroxysm   A sudden intensification of symptoms, such as a severe bout of coughing.

particle transmission   The movement of particles suspended in the air.

passive agglutination   An immunological procedure in which antigen molecules are adsorbed to the surface of latex spheres or other carriers that agglutinate when combined with antibodies.

passive immunity   The temporary immunity that comes from receiving antibodies from another source.

pasteurization   A heating process that destroys pathogenic bacteria in a fluid such as milk and lowers the overall number of bacterial cells in the fluid.

pasteurizing dose   The amount of irradiation used to eliminate pathogens.

pathogen   A microorganism or virus that causes disease.

pathogen-associated molecular pattern (PAMP)   A unique microbial molecular sequence recognized by innate immune system receptors.

pathogenicity   The ability of a disease-causing agent to gain entry to a host and bring about a physiological or anatomical change interpreted as disease.

pathogenicity island   A set of adjacent genes that encode virulence factors.

penicillin   Any of a group of antibiotics derived from *Penicillium* species or produced synthetically; effective against gram-positive bacteria and several gram-negative bacteria by interfering with cell wall synthesis.

penicillinase   An enzyme produced by certain microorganisms that converts penicillin to penicilloic acid and thereby confers resistance against penicillin.

peptide bond   A linkage between the amino group on one amino acid and the carboxyl group on another amino acid.

peptidoglycan   A complex molecule of the bacterial cell wall composed of alternating units of N-acetylglucosamine and N-acetylmuramic acid cross linked by short peptides.

perforin   A protein secreted by cytotoxic T lymphocytes and natural killer cells that forms holes in the plasma membrane of a targeted infected cell.

period of convalescence   The phase of a disease during which the body's systems return to normal.

period of decline   The phase of a disease during which symptoms subside.

periplasm   A metabolic region between the cell membrane and outer membrane of gram-negative cells.

peroxide   A compound containing an oxygen–oxygen single bond.

pH   An abbreviation for the power of the hydrogen ion concentration [H$^+$] of a solution.

phage   *See* **bacteriophage**.

phage typing   A procedure of using specific bacterial viruses to identify a particular strain of a bacterial species.

phagocyte   A white blood cell capable of engulfing and destroying foreign materials, including bacterial cells and viruses.

phagocytosis   A process by which foreign material or cells are taken into a white blood cell and destroyed.

phagolyososme   A membrane-enclosed compartment resulting from the fusion of a phagosome with lysosomes and in which the foreign material is digested.

phagosome   A membrane-enclosed compartment containing foreign material or infectious agents that the cell has engulfed.

phallotoxin   A group of chemical compounds present in the mushroom *Amanita phalloides*.

pharyngitis   An inflammation of the pharynx; commonly called a sore throat.

phase-contrast microscopy   An optical system on the light microscope that uses a special condenser and objective lenses to examine cell structure.

phenol   A chemical compound that has one or more hydroxyl groups attached to a benzene ring and derivatives are used as an antiseptic or disinfectant; also called carbolic acid.

phenol coefficient (**PC**)   A number that indicates the effectiveness of an antiseptic or disinfectant compared to phenol.

phenotype   The visible (physical) appearance of an organism resulting from the interaction between its genetic makeup and the environment.

phospholipid   A water-insoluble compound containing glycerol, two fatty acids, and a phosphate head group; forms part of the membrane in all cells.

phosphorylation   The addition of a phosphate group to a molecule.

photoautotroph   An organism that uses light energy to synthesize nutrients from carbon dioxide gas (CO_2).

photoheterotroph   An organism that uses light energy to synthesize nutrients from organic carbon compounds.

photophobia   Sensitivity to bright light.

photophosphorylatyion   The generation of ATP through the trapping of light.

photosynthesis   A biochemical process in which light energy is converted to chemical energy, which is then used for carbohydrate synthesis.

photosystem   A group of pigments that act as a light trapping system for photosynthesis.

pH scale   A range of values that extends from 0 to 14 and indicates the degree of acidity or alkalinity of a solution.

phycology   The scientific study of algae.

phylochip   *See* **DNA microarray**.

phylum   A category of organisms consisting of one or more classes.

physical pollution   The presence of particulate matter in water.

phytoplankton   Microscopic free-floating communities of cyanobacteria and unicellular algae.

pilus (pl. **pili**)   A short hair-like structure used by bacterial cells for attachment.

pinkeye   *See* **conjunctivitis**.

planktonic bacteria   Referring to bacterial cells that live as individual cells.

plaque   A clear area on a lawn of bacterial cells where viruses have destroyed the bacterial cells.

plasma   The fluid portion of blood remaining after the cells have been removed; *see also* **serum**.

plasma cell   An antibody-producing cell derived from B lymphocytes.

plasma membrane   The phospholipid bilayer with proteins that surrounds the eukaryotic cell cytoplasm; *see also* **cell membrane**.

plasmid   A small, closed-loop molecule of DNA apart from the chromosome that replicates independently and carries nonessential genetic information.

pleated sheet   The zig-zag secondary structure of a polypeptide in a flat plane.

pneumonia   An inflammation of the bronchial tubes and one or both lungs.

point mutation   The replacement of one base in a DNA strand with another base.

polar molecule   A substance with electrically-charged poles; *see also* **nonpolar molecule**.

polyclonal antibodies   Antibodies produced from different clones of B cell.

polyene An antifungal drug that destroys the plasma membrane; examples include nystatin.

polymer A substance formed by combining smaller molecules into larger ones; *see also* **monomer**.

polymerase chain reaction (PCR) A technique used to replicate a fragment of DNA many times.

polymicrobial disease A disorder caused by more than one infectious agent.

polymyxin An antibiotic derived from a *Bacillus* species that disrupts the cell membrane of gram-negative rods.

polynucleotide A chain of linked nucleotides.

polypeptide A chain of linked amino acids.

polysaccharide A complex carbohydrate made up of sugar molecules linked into a branched or chain structure; examples include starch and cellulose.

polysome A cluster of ribosomes linked by a strand of mRNA and all translating the mRNA.

porin A protein in the outer membrane of gram-negative bacteria that acts as a channel for the passage of small molecules.

portal of entry The site at which a pathogen enters the host.

portal of exit The site at which a pathogen leaves the host.

positive selection A method for selecting mutant cells by their growth as colonies on agar.

positive strand Referring to the RNA viruses whose genome consists of a mRNA molecule.

post-exposure immunization The receiving of a vaccine after contracting the pathogen.

post-polio syndrome A condition that affects polio survivors years after recovery from an initial acute attack by the polio virus.

potability Referring to water that is safe to drink because it contains no harmful material or microbes.

pour plate method A process by which a mixed culture can be separated into pure colonies and the colonies isolated; *see also* **streak plate method**.

pox Pitted scars remaining on the skin of individuals who have recovered from smallpox.

praziquantel An antihelminthic drug that alters the permeability of the plasma membrane.

precipitation A type of antigen-antibody reaction in which thousands of molecules of antigen and antibody cross-link to form visible aggregates.

prevalence The percentage of the population affected by a disease.

prevacuum autoclave An instrument that uses saturated steam at high temperatures and pressure for short time periods to sterilize materials.

preventative vaccine A vaccine that could be given to HIV-negative individuals to prevent infection to HIV.

primary antibody response The first contact between an antigen and the immune system, characterized by the synthesis of IgM and then IgG antibodies; *see also* **secondary antibody response**.

primary cell culture Animal cells separated from tissue and grown in cell culture.

primary infection A disease that develops in an otherwise healthy individual.

primary lymphoid tissue A site where immune cells form and mature; examples are the thymus and bone marrow; *see also* **secondary metabolite**.

primary metabolite A small molecule essential to the survival and growth of an organism; *see also* **secondary metabolite**.

primary structure The sequence of amino acids in a polypeptide.

prion An infectious, self-replicating protein involved in human and animal diseases of the brain.

prodromal phase The phase of a disease during which general symptoms occur in the body.

product A substance or substances resulting from a chemical reaction.

productive infection The active assembly and maturation of viruses in an animal cell.

proglottid One of a series of segments that make up the body of a tapeworm.

prokaryote A microorganism in the domain *Bacteria* or *Archaea* composed of single cells having a single chromosome but no cell nucleus or other membrane-bound compartments.

prokaryotic Referring to cells or organisms having a single chromosome but no cell nucleus or other membrane-bound compartments.

promoter The region of a template DNA strand or operon to which RNA polymerase binds.

Prontosil A red dye found by Domagk to have significant antimicrobial activity when tested in live animals, and from which sulfanilamide was later isolated.

prophage The viral DNA of a bacterial virus that is inserted into the bacterial DNA and is passed on from one generation to the next during binary fission.

propionic acid A chemical preservative used in cheese, breads, and other bakery products.

prostaglandin A hormone-like substance that acts as a mediator in type I hypersensitivity reactions.

protease inhibitor A compound that breaks down the enzyme protease, inhibiting the replication of some viruses, such as HIV.

protein A chain or chains of linked amino acids used as a structural material or enzyme in living cells.

protein-only hypothesis The idea that prions are composed solely of protein and contain no nucleic acid.

protein synthesis The process of forming a polypeptide or protein through a series of chemical reactions involving amino acids.

Proteobacteria A phylum of gram-negative, chemoheterotrophic species in the domain *Bacteria* that are defined primarily in terms of their ribosomal RNA (rRNA) sequences; examples include *Escherichia coli*, *Salmonella*, and the rickettsiae.

protist A member of the kingdom Protista.

Protista One of the four kingdoms in the domain Eukarya composed of the protozoa and unicellular algae.

proton A positively charged particle in the atomic nucleus.

proto-oncogene A region of DNA in the chromosome of human cells; they are altered by carcinogens into oncogenes that transform cells.

prototroph An organism that contains all its nutritional needs; *see also* **auxotroph**.

protozoan (pl. **protozoa**) A single-celled eukaryotic organism that lacks a cell wall and usually exhibits chemoheterotrophic metabolism.

protozoology The scientific study of protozoa.

provirus The viral DNA that has integrated into a eukaryotic host chromosome and is then passed on from one generation to the next through cell division.

pseudomembrane An accumulation of mucus, leukocytes, bacteria, and dead tissue in the respiratory passages of diphtheria patients.

pseudopeptidoglycan A complex molecule of some archaeal cell walls composed of alternating units of N-acetylglucosamine and N-acetyltalosamine uronic acid.

pseudopod A projection of the plasma membrane that allows movement in members of the amoebozoans.

psychrophile An organism that lives at cold temperature ranges of 0°C to 20°C.

psychrotolerant Referring to microorganisms that grow at 0°C but have a temperature optima of 20° to 40°C.

psychrotroph *See* **psychrotolerant**.

pure culture An accumulation or colony of microorganisms of one species.

pus A mixture of dead tissue cells, leukocytes, and bacteria that accumulates at the site of infection.

pyrogen A fever-producing substance.

pyruvate The end product of the glycolysis metabolic pathway.

Q

quat *See* **quaternary ammonium compound**.

quaternary ammonium compound (**quat**) A positively charged detergent with four organic groups attached to a central nitrogen atom; used as a disinfectant.

quaternary structure The association of two or more polypeptides in a protein.

question A statement written from an observation and used to formulate a hypothesis.

quinolone A synthetic antimicrobial drug that blocks DNA synthesis.

quorum sensing The ability of bacteria to chemically communicate and coordinate behavior via signaling molecules.

R

radioallergosorbent test (**RAST**) A type of radioimmunoassay in which antigens for the unknown antibody are attached to matrix particles.

radioimmunoassay (**RIA**) An immunological procedure that uses radioactive-tagged antigens to determine the identity and amount of antibodies in a sample.

radioisotope A unstable form of a chemical element that is radioactive; *see also* **isotope**.

radiolarian A single-celled marine organism with a round silica-containing shell that has radiating arms to catch prey.

reactant A substance that interacts with another in a chemical reaction.

recombinant DNA molecule A DNA molecule containing DNA from two different sources.

recombinant F⁻ cell that received a few chromosomal genes and partial F factor genes from a donor cell during conjugation.

recombinant subunit vaccine The synthesis of antigens in a microorganism using recombined genes for the purpose of producing a vaccine.

reduction The gain of electrons by a molecule; *see also* **oxydation**.

redundancy Referring to multiple codons coding for the same amino acid.

reemerging infectious disease A disease showing a resurgence in incidence or a spread in its geographical area; *see also* **emerging infectious disease**.

regulatory gene A DNA segment that codes for a repressor protein.

regulatory T cell A population of lymphocytes that prevent other T lymphocytes from attacking self.

replication factory   The location in a cell where DNA synthesis occurs.

replication fork   The point where complementary strands of DNA separate and new complementary strands are synthesized.

replication origin   The fixed point on a DNA molecule where copying of the molecule starts.

repressor protein   A protein that when bound to the operator blocks transcription.

reservoir   The location or organism where disease-causing agents exist and maintain their ability for infection.

resolving power   The numerical value of a lens system indicating the size of the smallest object that can be seen clearly when using that system.

respiratory droplet   Small liquid droplets expelled by sneezing or coughing.

restriction endonuclease   A type of enzyme that splits open a DNA molecule at a specific restricted point; important in genetic engineering techniques.

reticulate body   The replicating, intracellular, non-infectious stage of *Chlamydia trachomatis*.

reverse transcriptase   An enzyme that synthesizes a DNA molecule from the code supplied by an RNA molecule.

reverse transcriptase inhibitor   A compound that inhibits the action of reverse transcriptase, preventing the viral genome from being replicated.

revertant   Referring to a mutant organism or cell that has reacquired its original phenotype or metabolic ability.

Reye syndrome   A complication of influenza and chickenpox, characterized by vomiting and convulsions as well as liver and brain damage.

R group   The side chain on an amino acid.

rhabditiform   Referring to the elongated, rod-like shape of the larvae of the hookworm.

rhinitis   An inflammation of the nasal passages.

RhoGAM   Rh-positive antibodies.

ribonucleic acid (**RNA**)   The nucleic acid involved in protein synthesis and gene control; also the genetic information in some viruses.

ribosomal RNA (**rRNA**)   An RNA transcript that forms part of the ribosome's structure.

ribosome   A cellular structure made of RNA and protein that participates in protein synthesis.

ribozyme   An RNA molecule that can catalyze a chemical reaction in a cell.

rice-water stool   A colorless, watery diarrhea containing particles of intestinal tissue in cholera patients.

rifampin   An antibiotic prescribed for tuberculosis and leprosy patients and for carriers of *Neisseria* and *Haemophilus* species.

Rivers' postulates   A set of procedures by which a specific virus can be associated with a specific disease.

RNA   *See* **ribonucleic acid**.

RNA interference (**RNAi**)   A process whereby translation is silenced by the binding of double-stranded RNA to specific messenger RNA molecules.

RNA polymerase   The enzyme that synthesizes an RNA polynucleotide from a DNA template.

rolling circle mechanism   A type of DNA replication in which a strand of DNA "rolls off" the loop and serves as a template for the synthesis of a complementary strand of DNA.

rose spots   Bright red skin spots associated with diseases such as typhoid fever and relapsing fever.

roundworm   A multicellular parasite with a round body; examples include the nematodes.

R plasmid   A small, circular DNA molecule that occurs frequently in bacterial cells and carries genes for drug resistance.

S

salt   An ionic compound formed from a reaction between most positively and negatively charged atoms.

Salvarsan   The first modern synthetic antimicrobial agent.

sanitization   To remove microbes or reduce their populations to a safe level as determined by public health standards.

sanitize   Referring to the reduction of a microbial population to a safe level.

saprobe   A type of heterotrophic organism that feeds on dead organic matter, such as rotting wood or compost.

sarcina (pl. **sarcinae**)   (1) A packet of eight spherical-shaped prokaryotic cells. (2) A genus of gram-positive, anaerobic spheres.

saturated   Referring to a water-insoluble compound that cannot incorporate any additional hydrogen atoms; *see also* **unsaturated**.

scanning electron microscope (**SEM**)   The type of electron microscope that allows electrons to scan across an object, generating a three-dimensional image of the object.

Schick test   A skin test used to determine the effectiveness of diphtheria immunization.

schmutzdecke   In water purification, a slimy layer of microorganisms that develops in a slow sand filter.

scientific inquiry   The way a science problem is investigated by formulating a question, developing a hypothesis, collecting data about it through observation, and experiment, and interpreting the results; also called scientific method.

sclerotium (pl. **sclerotia**)   A hard purple body that forms in grains contaminated with *Claviceps purpurea*.

sebum   An oily substance produced by the sebaceous glands that keep the skin and hair soft and moist.

secondary antibody response   A second or ensuing response triggered by memory cells to an antigen and characterized by substantial production of IgG antibodies; *see also* **primary antibody response**.

secondary infection   A disorder caused by an opportunistic microbe as a result of a primary infection weakening the host.

secondary lymphoid tissue   A site where mature immune cells interact with pathogens; examples includes the spleen and lymph nodes; *see also* **primary lymphoid tissue**.

secondary metabolite   A small molecule not essential to the survival and growth of an organism; *see also* **primary metabolite**.

secondary structure   The region of a polypepetide folded into an alpha helix or pleated sheet.

secretory IgA (S-IgA)   The form of IgA antibodies secreted into the intestinal lumen.

sedimentation   The removal of soil particulates from water.

selective medium   A growth medium that contains ingredients to inhibit certain microorganisms while encouraging the growth of others.

selective toxicity   A property of many antimicrobial drugs that harm the infectious agent but not the host.

semiconservative replication   The DNA copying process where each parent (old) strand serves as a template for a new complementary strand.

semisynthetic drug   A chemical substance synthesized from natural and lab components used to treat disease.

senescence   Deteriorative changes in a cell or organism with aging.

sense codon   A nucleotide sequence that specifies an amino acid.

sensitizing dose   The first exposure to an allergy-causing antigen.

sepsis   The growth and spreading of bacteria or their toxins in the blood and tissues.

septate   Referring to the cross-walls formed in the filaments of many fungi.

septic shock   A collapse of the circulatory and respiratory systems caused by an overwhelming immune response.

septicemia   A growth and spreading of bacterial cells in the bloodstream.

septic tank   An enclosed concrete box that collects waste from the home.

septum (pl. septa)   A cross-wall in the hypha of a fungus.

seroconversion   The time when antibodies to a disease agent can be detected in the blood.

serological reaction   An antigen-antibody reaction studied under laboratory conditions and involving serum.

serology   A branch of immunology that studies serological reactions.

serum (pl. sera)   The fluid portion of the blood consisting of water, minerals, salts, proteins, and other organic substances, including antibodies; contains no clotting agents; *see also* **plasma**.

serum sickness   A type of hypersensitivity reaction in which the body responds to proteins contained in foreign serum.

sexually transmitted disease (STD)   A disease such as gonorrhea or chlamydia that is normally passed from one person to another through sexual activity.

sexually transmitted infection (STI)   *See* **sexually transmitted disease**.

Shiga toxin   A bacterial poison that inhibits protein synthesis in target cells.

sign   An indication of the presence of a disease, especially one observed by a doctor but not apparent to the patient; *see also* **symptom**.

simple stain technique   The use of a single cationic dye to contrast cells; *see also* **differential stain technique**.

single-dose vaccine   The combination of several vaccines into one measured quantity.

sinusitis   An inflammation of the membrane lining a sinus.

S-layer   The cell wall of most archaeal species consisting of protein or glycoprotein assembled in a crystalline lattice.

slide agglutination test   *See* **VDRL** test.

slime layer   A thin, loosely bound layer of polysaccharide covering some prokaryotic cells; *see also* **capsule** and **glycocalyx**.

sludge   The solids in sewage that separate out during sewage treatment.

sludge tank   The area in which secondary water treatment occurs.

somatic recombination   The reshuffling of antibody genetic segments in a B lymphocyte as it matures.

soredium (pl. soredia)   The disseminated group of fungal and photosynthetic cells formed by a lichen.

sour curd   The acidification of milk, causing a change in the structure of milk proteins.

spawn   A mushroom mycelium used to start a new culture of the fungus.

specialized transduction   The transfer of a few bacterial genes by a bacterial virus that carries the genes to another bacterial cell; *see also* **generalized transduction**.

species   The fundamental rank in the classification system of organisms.

specific epithet   The second of the two scientific names for a species.

spike   A protein projecting from the viral envelope or capsid that aids in attachment and penetration of a host cell.

spiral   A shape of many bacterial and archaeal cells.

spirillum (pl. **spirilla**)   (1) A bacterial cell shape characterized by twisted or curved rods. (2) A genus of aerobic, helical cells usually with many flagella.

Spirochaetes   A phylum in the domain *Bacteria* whose members possess a helical cell shape.

spirochete   A twisted bacterial rod with a flexible cell wall containing endoflagella for motility.

spontaneous generation   The doctrine that nonliving matter could spontaneously give rise to living things.

spontaneous mutation   A mutation that arises from natural phenomena in the environment.

sporangiospore   Asexual spore produced by many fungi.

sporangium (pl. **sporangia**)   The structures in fungi in which asexual spores are formed.

spore   (1) A reproductive structure formed by a fungus. (2) A highly resistant dormant structure formed from vegetative cells in several genera of bacteria, including *Bacillus* and *Clostridium*; *see also* **endospore**.

sporozoite   A stage in the life cycle of the malaria parasite that enters the human body.

sporulation   The process of spore formation.

sputum   Thick, expectorated matter from the lower respiratory tract.

stabilizing protein   A protein that keeps the DNA template strands separated during DNA replication.

standard plate count   A procedure to estimate the number of cells in a sample dilution spread on an agar plate.

standard precautions   Using those measures to avoid contact with a patient's bodily fluids; examples include wearing gloves, goggles, and proper disposal of used hypodermic needles.

staphylococcus (pl. **staphylococci**)   (1) An arrangement of bacterial cells characterized by spheres in a grapelike cluster. (2) A genus of facultatively anaerobic, nonmotile, nonsporeforming, gram-positive spheres in clusters.

start codon   The starting nucleotide sequence (AUG) in translation.

stationary phase   The portion of a bacterial growth curve in which the reproductive and death rates of cells are equal.

stem cell   An undifferentiated cell from which specialized cells arise.

sterile   Free from living microorganisms, spores, and viruses.

sterilization   The removal of all life forms, including bacterial spores.

sterol   A organic solid containing several carbon rings with side chains; examples include cholesterol.

stop codon   The nucleotide sequence that terminates translation.

streak plate method   A process by which a mixed culture can be streaked onto an agar plate and pure colonies isolated; *see also* **pour plate method**.

streptobacillus (pl. **streptobacilli**)   (1) A chain of bacterial rods. (2) A genus of facultatively anaerobic, nonmotile, gram-negative rods.

streptococcus (pl. **streptococci**)   (1) A chain of bacterial cocci. (2) A genus of facultatively anaerobic, nonmotile, nonsporeforming, gram-positive spheres in chains.

streptogramin   An antibiotic that block protein synthesis.

streptokinase   An enzyme that dissolves blood clots; produced by virulent streptococci.

stridor   A high-pitched wheezing sound during breathing.

stromatolite   A microbial mat consisting of many layers that can become fossilized.

structural formula   A chemical diagram representing the arrangement of atoms and bonds within a molecule.

structural gene   A segment of a DNA molecule that provides the biochemical information for a polypeptide.

subclinical disease   A disease in which there are few or inapparent symptoms.

substrate   The substance or substances upon which an enzyme acts.

substrate-level phosphorylation   The formation of ATP resulting from the transfer of phosphate from a substrate to ADP.

subunit vaccine   A vaccine that contains parts of microorganisms, such as capsular polysaccharides or purified fimbriae.

sulfonamide   A synthetic, sulfur-containing antibacterial agent; also called sulfa drug.

sulfur cycle   The processes by which sulfur moves through and is recycled in the environment.

sulfur dioxide   A chemical preservative used in dried fruits.

superantigen   An antigen that stimulates an immune response without any prior processing.

supercoiled domain   A loop of wound DNA consisting of 10,000 bases.

supercoiling   The process by which a chromosome is twisted and packed.

superinfection   The overgrowth of susceptible strains by antibiotic resistant ones.

suppressor T cell   A group of lymphocytes that regulate IgE antibody production.

surface water   The water in lakes, streams, and shallow wells.

surfactant   A synthetic chemical, such as a detergent, that emulsifies and solubilizes particles attached to surfaces by reducing the surface tension.

symbiosis   An interrelationship between two populations of organisms where there is a close and permanent association.

symptom   An indication of some disease or other disorder that is experienced by the patient; *see also* **sign**.

syncytium (pl. **syncytia**)   A giant tissue cell formed by the fusion of cells infected with respiratory syncytial viruses.

syndrome   A collection of signs or symptoms that together are characteristic of a disease.

synthetic biology   A field of study that attempts to "build" new living organisms by combining parts of other species.

synthetic drug (**agent**)   A substance made in the lab to prevent illness or treat disease.

synthetic medium   A chemically defined medium in which the nature and quantity of each component is identified; *see also* **complex medium**.

systemic anaphylaxis   The release of cell mediators throughout the body.

systemic disease   A disorder that disseminates to the deeper organs and systems of the body.

T

tapeworm   *See* **cestode**.

taxonomy   The science dealing with the systematized arrangements of related living things in categories.

T cell   *See* **T lymphocyte**.

T-dependent antigen   An antigen that requires the assistance of T_H2 lymphocytes to stimulate antibody-mediated immunity.

teichoic acid   A negatively charged polysaccharide in the cell wall of gram-positive bacteria.

temperate   Referring to a bacterial virus that enters a bacterial cell and then the viral DNA integrates into the bacterial cell's chromosome.

termination   (1) The completion of DNA synthesis during DNA replication. (2) The release of a polypeptide from a ribosome during translation.

termination factor   A protein that triggers the release of a polypeptide from a ribosome.

tertiary structure   The folding of a polypeptide back on itself.

tetanospasmin   An exotoxin produced by *Clostridium tetani* that acts at synapses, thereby stimulating muscle contractions.

tetracycline   An antibiotic characterized by four benzene rings with attached side groups that blocks protein synthesis in many gram-negative bacteria, rickettsiae, and chlamydiae.

tetrad   An arrangement of four bacterial cells in a cube shape.

theory   A scientific explanation supported by many experiments done by separate individuals.

therapeutic dose   The concentration of an antimicrobial drug that effectively destroys an infectious agent.

therapeutic serum   Antibody-rich serum used to treat a specified condition.

therapeutic vaccine   A vaccine that could be given to HIV-positive individuals to control HIV disease.

thermal death point (**TDP**)   The temperature required to kill a bacterial population in a given length of time.

thermal death time (**TDT**)   The length of time required to kill a bacterial population at a given temperature.

thermoduric   Referring to an organism that tolerates the heat of the pasteurization process.

thermophile   An organism that lives at high temperature ranges of 40°C to 90°C.

thimerosal   A stabilizer put in some vaccines as a preservative.

thioglycollate broth   A microbiological medium containing a chemical that binds oxygen from the atmosphere and creates an environment suitable for anaerobic growth.

three domain system   The classification scheme placing all living organisms into one of three groups based, in part, on ribosomal RNA sequences.

thymus   A flat, bilobed organ where T lymphocytes mature.

tissue tropism   Refers to the specific tissues within a host that a virus infects.

titer   A measurement of the amount of antibody in a sample of serum that is determined by the most dilute concentration of antibody that will yield a positive reaction with a specific antigen.

titration   A method of calculating the concentration of a dissolved substance, such as an antibody, by adding quantities of a reagent of known concentration to a known volume of test solution until a reaction occurs.

T lymphocyte (**T cell**)   A type of white blood cell that matures in the thymus gland and is associated with cell-mediated immunity.

toll-like receptor (TLR) A signaling molecule on immune cells that recognizes a unique molecular pattern on an infectious agent.

TORCH An acronym for four diseases that pass from the mother to the unborn child: toxoplasmosis, rubella, cytomegalovirus disease, and herpes simplex; the O stands for other diseases.

toxemia The presence of toxins in the blood.

toxic dose (1) The amount of toxin need to cause a disease. (2) The amount of an antimicrobial drug that causes harm to the host.

toxigenicity The ability of an organism to produce a toxin.

toxin A poisonous chemical substance produced by an organism.

toxoid A preparation of a microbial toxin that has been rendered harmless by chemical treatment but that is capable of stimulating antibodies; used as vaccines.

transcription The biochemical process in which RNA is synthesized according to a code supplied by the template strand of a gene in the DNA molecule.

transduction The transfer of a few bacterial genes from a donor cell to a recipient cell via a bacterial virus.

transfer RNA (tRNA) A molecule of RNA that unites with amino acids and transports them to the ribosome in protein synthesis.

transformation (1) The transfer and integration of DNA fragments from a dead and lysed donor cells to a recipient cell's chromosome. (2) The conversion of a normal cell into a malignant cell due to the action of a carcinogen or virus.

transient microbiota The microbial agents that are associated with an animal for short periods of time without causing disease; see also **indigenous microbiota**.

translation The biochemical process in which the code on the mRNA molecule is converted into a sequence of amino acids in a polypeptide.

transmissible spongiform encephalopathy (TSE) A group of progressive conditions that affect the brain and nervous system.

transmission electron microscope (TEM) The type of electron microscope that allows electrons to pass through a thin section of the object, resulting in a detailed view of the object's structure.

transposable genetic element Fragments of DNA called insertion sequences or transposons that can cause mutations.

transposon A segment of DNA that moves from one site on a DNA molecule to another site, carrying information for protein synthesis.

traumatic wound A deep cut, compound fracture, or thermal burn.

trematode A flatworm, commonly known as a fluke, that lives as a parasite in the liver, gut, lungs, or blood vessels of vertebrates.

triclosan A phenol derivative incorporated as an antimicrobial agent into a wide variety of household products.

trismus A sustained spasm of the jaw muscles, characteristic of the early stages of tetanus; also called lockjaw.

trivalent vaccine A vaccine consisting of three components, each of which stimulates immunity.

trophozoite The feeding form of a microorganism, such as a protozoan.

tube dilution method A procedure for determining bacterial susceptibility to an antibiotic by determining the minimal amount of the drug needed to inhibit growth of the pathogen; see **minimal inhibitory concentration**.

tubercle A hard nodule that develops in tissue infected with *Mycobacterium tuberculosis*.

tuberculin test A procedure performed by applying purified protein derivative from *Mycobacterium tuberculosis* to the skin and noting if a thickening of the skin with a raised vesicle appears within a few days; used to establish if someone has been exposed to the bacterium.

tumor An abnormal uncontrolled growth of cells that has no physiological function.

turbidity The cloudiness of a broth culture due to bacterial growth.

tyndallization See **fractional sterilization**.

U

ultra-high temperature (URT) method A treatment in which milk is heated at 82°C for 3 seconds to destroy pathogens.

ultrastructure The detailed structure of an cell, virus, or other object when viewed with the electron microscope.

ultraviolet (UV) light A type of electromagnetic radiation of short wavelengths that damages DNA.

uncoating Referring to the loss of the viral capsid inside an infected eukaryotic cell.

unsaturated Referring to a water-insoluble compound that can incorporate additional hydrogen atoms; see also **saturated fat**.

urethritis An inflammation of the urethra.

urinary tract infection (UTI) A common infection, particularly in young women, caused by a variety of bacterial, fungal, or protozoal species.

urticaria A hive-like rash of the skin.

V

vaccination Inoculation with weakened or dead microbes, or viruses, in order to generate immunity; *see also* **immunization**.

vaccine A preparation containing weakened or dead microorganisms or viruses, treated toxins, or parts of microorganisms or viruses to stimulate immune resistance.

vaccine adverse events reporting system (**VAERS**) A reporting system designed to identify any serious adverse reactions to a vaccination.

vancomycin An antibacterial drug that inhibits cell wall synthesis and is used in treating diseases caused by gram-positive bacteria, especially staphylococci.

variable That part of an experiment exposed to or treated with the factor being tested.

variable domain The different amino acids in different antibody light and heavy chains.

variolation A 14th to 18th century method to inoculate a susceptible person with material from a smallpox vesicle to render that person resistant to infection.

vasodilation A widening of the blood vessels, especially the arteries, leading to increased blood flow.

VBNC Referring to prokaryotes that are "viable but not culturable."

VDRL test A screening procedure used in the detection of syphilis antibodies.

vector (1) An arthropod that transmits the agents of disease from an infected host to a susceptible host. (2) A plasmid used in genetic engineering to carry a DNA segment into a bacterium or other cell.

vehicle transmission The spread of disease through contaminated food and water.

vertical gene transfer The passing of genes from one cell generation to the next; *see also* **horizontal gene transfer**.

vertical transmission The spread of disease from mother to fetus or newborn.

viable count The living cells identified from a standard plate count.

vibrio (1) A prokaryotic cell shape occurring as a curved rod. (2) A genus of facultatively anaerobic, gram-negative curved rods with flagella.

viral inhibition The prevention of a virus infection by antibodies binding to molecules on the viral surface.

viral load test A method used to detect the RNA genome of HIV.

viremia The presence and spread of viruses through the blood.

virion A completely assembled virus outside its host cell.

viroid An infectious RNA segment associated with certain plant diseases.

virology The scientific study of viruses.

virosphere Refers to all places where viruses are found or interact with their hosts.

virulence The degree to which a pathogen is capable of causing a disease.

virulence factor A structure or molecule possessed by a pathogen that increases its ability to invade or cause disease to a host.

virulent Referring to a virus or microorganism that can be extremely damaging when in the host.

virus An infectious agent consisting of DNA or RNA and surrounded by a protein sheath; in some cases, a membranous envelope surrounds the coat.

W

wheal An enlarged, hive-like zone of puffiness on the skin, often due to an allergic reaction; *see also* **flare**.

white blood cell *See* **leukocyte**.

wild type The form of an organism or gene isolated from nature.

wort A sugary liquid produced from crushed malted grain and water to which is added yeast and hops for the brewing of beer.

X

xenograft A tissue graft between members of different species, such as between a pig and a human.

X ray An ionizing radiation that can be used to sterilize objects.

Y

yeast (1) A type of unicellular, nonfilamentous fungus that resembles bacterial colonies when grown in culture. (2) A term sometimes used to denote the unicellular form of pathogenic fungi.

Z

zone of equivalence The region in a precipitation reaction where ideal concentrations of antigen and antibody occur.

zoonosis (pl. **zoonoses**) An animal disease that may be transmitted to humans.

Zygomycota A phylum of fungi whose members have coenocytic hyphae and form zygospores, among other notable characteristics.

zygospore A sexually produced spore formed by members of the Zygomycota.

Index

NOTE: page numbers followed by *b* indicate MicroInquiry or MicroFocus boxes; *f*, figures, and *t*, tables. Page numbers in color indicate chapters online only.

Photograph Acknowledgments

Title page © Eye of Science/Photo Researchers, Inc.

Chapter opener images Courtesy of C. S. Goldsmith and A. Balish/CDC

TABLE OF CONTENTS

Page vii © Geir Olav Lyngfjell/ShutterStock, Inc.; **page viii** © Aleix Ventayol Farrés/ShutterStock, Inc.; **page xiii** © Jaimie Duplass/ShutterStock, Inc.

Part Opener 1 © Dr. Gopal Murti/Photo Researchers, Inc.; **Part 1 Microbiology Pathways** Courtesy of CDC

CHAPTER 1

1.1A © Steve Gschmeissner/Photo Researchers, Inc.; **1.1B** © Reuters/Eliana Aponte/Landov; **MicroFocus 1** © William-Robinson/Alamy Images; **1.2A** © Library of Congress [LC-USZ62-95187]; **1.2B** © National Library of Medicine; **1.2C** Collection of the University of Michigan Health System, Gift of Pfizer, Inc. (UMHS.15); **1.2E** © Royal Society, London; **1.4** © National Library of Medicine; **MicroFocus 2A** © Dr. Dennis Kunkel/Visuals Unlimited; **MircoFocus 2B** Reproduced from V. Noireaux and A. Libchaber, *PNAS*, 101 (2004), 17669-17674; Copyright (2004) National Academy of Science, U.S.A. Photo courtesy of Vincent Noireaux and Albert Libchabe; **1.5A** Courtesy of Frerichs, R. R. John Snow website: http://www.ph.ucla.edu/epi/snow.html, 2006; **1.5B** Collection of the University of Michigan Health System, Gift of Pfizer, Inc. (UMHS.26); **1.5 insert** © National Library of Medicine; **MicroFocus 3** Courtesy of the Public Health Foundation/CDC; **1.6** Collection of the University of Michigan Health System, Gift of Pfizer, Inc. (UMHS.23); **1.7A** © National Library of Medicine; **1.8** © Mary Evans Picture Library/Alamy Images; **1.8 insert** © National Library of Medicine; **1.9A** © National Library of Medicine; **1.9B insert** Courtesy of CDC; **MicroFocus 4** © National Library of Medicine; **1.10** © Institut Pasteur, Paris; **1.12C** © Dr. Hans Gelderblom/Visuals Unlimited; **1.13A** © Professor P. Motta and T. Naguro/Photo Researchers, Inc.; **1.13B** © Dr. Dennis Kunkel/Visuals Unlimited; **1.14A** Collection of the University of Michigan Health System, Gift of Pfizer, Inc. (UMHS.44); **1.14B** © St. Mary's Hospital Medical School/Photo Researchers, Inc.; **1.14C** Used with permission from Pfizer, Inc.; **1.16A** Reproduced with permission of the New York State Department of Health; **1.16B** © Photodisc; **1.17B** Courtesy of the Exxon Valdez Oil Spill Trustee Council/NOAA

CHAPTER 2

2.1A Courtesy of Jim Peaco/Yellowstone National Park; **2.1B** © Geir Olav Lyngfjell/ShutterStock, Inc.; **MicroFocus 1** Courtesy of ESA-D.Ducros; **MicroFocus 4** © Dr. David Phillips/Visuals Unlimited; **Textbook Case** Courtesy of CDC

CHAPTER 3

3.1 Courtesy of GeoEye and NASA. Copyright 2008. All rights reserved; **3.2A** © SPL/Photo Researchers, Inc.; **3.2C** Modified from David G. Davies, Binghamton University, Binghamton NY; **3.3A** © Mona Lisa Production/Photo Researchers, Inc.; **3.3B** Courtesy of Rodney M. Donlan, Ph.D. and Janice Carr/CDC; **MicroFocus 1A-B** © Scientifica/Visuals Unlimited; **3.5** Reprinted with permission from the American Society for Microbiology (*Microbe*, January, 2006, p. 20–24). Photo courtesy of Dr. Thomas A. Bobik, Department of Biochemistry, Biophysics and Molecular, Iowa State University; **3.6** Courtesy of Biolog, Inc.; **3.10A** Courtesy of Carl Zeiss MicroImaging, Inc.; **MicroFocus 5** Courtesy of Heide Schulz, Max Planck Institute of Marine Microbiology, Germany; **Textbook Case** Courtesy of Dr. W. A. Clark/CDC; **3.14A-C** © David M. Phillips/Visuals Unlimited; **3.15** Courtesy of Dr. Peter Lewis; School of Environmental and Life Sciences, University of Newcastle; **3.16A** Courtesy of Carl Zeiss, NTS; **3.17A** © CNRI/Photo Researchers, Inc.; **3.17B** © Dr. Dennis Kunkel/Visuals Unlimited; **Application Question** Courtesy of Janice Haney Carr/CDC

CHAPTER 4

4.1A © Aleix Ventayol Farrés/ShutterStock, Inc.; **4.1B** © Corbis/age fotostock; **4.3C** © Don W. Fawcett/Photo Researchers, Inc.; **4.3D** Courtesy of Dr. David Berd/CDC; **4.4A** © Dr. Kari Lounatmaa/Photo Researchers, Inc.; **4.4B** © Aerial Archives/Alamy Images; **4.4C** © Eye of Science/Photo Researchers, Inc.; **4.6** © George Chapman/Visuals Unlimited; **4.8** © Dr. Dennis Kunkel/Visuals Unlimited; **4.12A** Courtesy of Elliot Juni, Department of Microbiology and Immunology, The University of Michigan; **4.12B** © George Musil/Visuals Unlimited; **4.13** © CNRI/Photo Researchers, Inc.; **Textbook Case** Courtesy of Dr. Rodney M. Donlan and Janice Carr/CDC; **4.18** © Dr. Dennis Kunkel/Visuals Unlimited; **4.19A** © Science Source/Photo Researchers, Inc.; **MicroFocus 4** © Dr. Dennis Kunkel/Visuals Unlimited; **MicroFocus 5** © Dr. David M. Phillips/Visuals Unlimited; **4.20A** Courtesy of Rut Carballido-López, University of Oxford; **4.20B** Reprinted from *Cell*, vol 115, Jacobs-Wagner, C., cover, copyright 2003, with permission from Elsevier. Photo courtesy of Christine Jacobs-Wagner, Yale University; **4.21** Courtesy of Dr. Peter Lewis; School of Environmental and Life Sciences, University of Newcastle; **MicroInquiry** © Carlos Arguelles/ShutterStock, Inc.

CHAPTER 5

5.1 © Photodisc/age fotostock; **5.2B** © Lee Simon/Photo Researchers, Inc.; **5.5A** Courtesy of Dr. J. J. Farmer/CDC; **5.5B** Courtesy of CDC; **5.6A** Courtesy of CDC; **5.6B** © Scott Camazine/Alamy Images; **5.6C** Courtesy of Janice Carr/CDC; **Textbook Case** Courtesy of Sheila Mitchell/CDC; **5.9C** © Scott Coutts/Alamy Images; **MicroFocus 4** © Doug Pearson/age fotostock; **MicroFocus 6** © CDC/Science Source/Photo Researchers, Inc.; **5.14E** Courtesy of James Gathany/CDC; **5.16** © R.A. Longuehaye/Photo Researchers, Inc.

CHAPTER 6

6.1 © Dr. Dennis Kunkel/Visuals Unlimited; **MicroFocus 4** Courtesy of Catherine Billick, www.flickr.com/photos/catbcorner/; **6.14B (diary)** © niderlander/ShutterStock, Inc.; **16.14B (alcohol)** © drKaczmar/ShutterStock, Inc.; **16.14B (bacteria)** © SPL/Photo Researchers, Inc.; **16.14B (cheese)** © Elena Elisseeva/ShutterStock, Inc.; **16.14B (solvents)** © Esteban De Armas/ShutterStock, Inc.; **6.15A** © Dr. Dennis Kunkel/Visuals Unlimited; **6.15B** © Roland Birke/Peter Arnold, Inc.; **MicroFocus 5** Courtesy of OAR/National Undersea Research Program (NURP)/NOAA

CHAPTER 7

7.1B © Flat Earth/FotoSearch; **7.2** Courtesy of Journalist 2nd Class Shane Tuc/U.S Navy; **7.4** © Simon Fraser/Photo Researchers,

Inc.; **MicroFocus 1** © Robert A. Levy Photography, LLC/ShutterStock, Inc.; **7.7** © Apply Pictures/Alamy Images; **7.8B** Courtesy of Pall Corporation; **7.8C** © Christine Case/Visuals Unlimited; **7.11B** Courtesy of US Army Natick Soldier Center; **MicroFocus 5** © Photodisc; **MicroFocus 7A** © Reuters/Department of Justice/Handout/Landov; **MicroFocus 7B** Courtesy of the US Census Bureau, Public Information Office

Part Opener 2 © Eye of Science/Photo Researchers, Inc.; **Part 2 Microbiology Pathways** © Jon Feingersh Photogr/Blend Images/age fotostock

CHAPTER 8

8.1 Courtesy of Abigail Allwood, Geologist at NASA Jet Propulsion Laboratory; **MicroFocus 1** © Vittorio Luzzati/Photo Researchers, Inc.; **8.4B** © G. Murti/Photo Researchers, Inc.; **8.13** Reproduced from O. L. Miller, B. A. Hamkalo, and C. A. Thomas, *Science* 169 (1977): 392. Reprinted with permission from AAAS; **8.15A-B** Courtesy of Dr. Peter Lewis; School of Environmental and Life Sciences, University of Newcastle; **MicroFocus 5** © Dennis Kunkel Microscopy, Inc./Phototake/Alamy Images; **MicroFocus 6** © National Library of Medicine/Photo Researchers, Inc.

CHAPTER 9

9.1 Courtesy of John Innes Centre www.jic.ac.uk; **MicroFocus 2** Courtesy of Chris Gotschalk, University of California, Santa Barbara; **9.5** © Dr. Dennis Kunkel/Visuals Unlimited; **Textbook Case** © Medical-on-Line/Alamy Images; **MicroInquiry B** © Martin Shields/Alamy Images; **MicroFocus 5** © Photos.com; **MicroFocus 6** © Alfred Pasieka/Photo Researchers, Inc.; **MicroFocus 7** Reproduced from Claire M. Fraser et al., *Science* 270 (1995): 397–404. Reprinted with permission from AAAS

Part Opener 3 © Kwangshin Kim/Photo Researchers, Inc.; **Part 3 Microbiology Pathways** © Viktor Pryymachuk/ShutterStock, Inc.

CHAPTER 10

10.1 © Dr. George J. Wilder/Visuals Unlimited; **10.4** © Juergen Berger/Photo Researchers, Inc.; **10.5A** © Medical-on-Line/Alamy Images; **10.5B** © imagebroker/Alamy Images; **10.6** © Medical-on-Line/Alamy Images; **MicroFocus 1** © Volker Steger/Peter Arnold, Inc.; **10.10** © CAMR, A.B. Dowsett/Photo Researchers, Inc.; **10.12** © NIBSC/Photo Researchers, Inc.; **10.15A** © James Cavallini/Photo Researchers, Inc.; **10.15B** © Manfred Kage/Peter Arnold, Inc.; **10.16** © Bart's Medical Library/Phototake/Alamy Images; **MicroFocus 3** © Allik Camazine/Alamy Images; **10.18** Courtesy of Dr. Mike Miller/CDC; **10.19A-B** © Michael Gabridge/Visuals Unlimited; **10.20A** © Collection CNRI/Phototake/Alamy Images; **10.20B** Courtesy of Don Howard/CDC; **Textbook Case** Courtesy of CDC; **10.21** Courtesy of Rocky Mountain Laboratories, NIAID, NIH

CHAPTER 11

11.1 © SPL/Photo Researchers, Inc.; **11.2** © Dr. Hans Ackermann/Visuals Unlimited; **MicroFocus 1** © Scott Camazine/Alamy Images; **11.6B** © BSIP/Photo Researchers, Inc.; **11.8A** © Dr. Tony Brain/Photo Researchers, Inc.; **11.8B** © Steve Gschmeissner/Photo Researchers, Inc.; **11.9A** © Medical-on-Line/Alamy Images; **11.9B** © SPL/Photo Researchers, Inc.; **11.11A** © Dr. Gary Gaugler/Visuals Unlimited; **11.B** Courtesy of Paul Gulig, Donna Duckworth and Julio Martin, University of Florida; **11.12** © CNRI/Photo Researchers, Inc.; **11.13** © Michael N. Paras/age fotostock; **11.14A** © SAS/Alamy Images; **11.15** © Stephanie

Schuller/Photo Researchers, Inc.; **11.16** © David M. Martin, M.D./Photo Researchers, Inc.; **11.17** Courtesy of Dr. Balasubr Swaminathan and Peggy Hayes/CDC; **11.19A** © Dr. John D. Cunningham/Visuals Unlimited; **11.19B** © Scimat/Photo Researchers, Inc.; **11.20** © Jeff Chiu/AP Photos; **Textbook Case** Courtesy of CDC; **11.21** © David Scharf/Peter Arnold, Inc.; **11.22** Reproduced from *ASM News*, January 2002, p. 20–24 and with permission from the American Society for Microbiology. Photo courtesy of Doctor Virginia L. Miller, University of North Carolina; **MicroInquiry A** © Adrienne Marshall 2009; **MicroInquiry B** © SPL/Photo Researchers, Inc.; **11.23 insert** © P. Hawtin/Photo Researchers, Inc.

CHAPTER 12

12.1 © Photos.com; **12.2A** © Michael Abbey/Visuals Unlimited; **12.2B** © Dennis Kunkel Microscopy, Inc./Phototake/Alamy Images; **12.3** Courtesy of James H. Steele/CDC; **MicroFocus 1** © Scotsman Publ Ltd; **12.4** Courtesy of Dr. Jack Poland/CDC; **MicroFocus 2** © Oscar Knott/FogStock/Alamy Images; **12.5** © Eye of Science/Photo Researchers, Inc.; **MicroFocus 3** © Andrea Seemann/ShutterStock, Inc.; **12.6A-B** Courtesy of CDC; **12.7** Courtesy of Dr. Brachman/CDC; **12.8A** © Microworks/Phototake/Alamy Images; **12.8B** Courtesy of Scott Bauer/USDA; **12.10** Courtesy of James Gathany/CDC; **Textbook Case** © Arthur Siegelman/Visuals Unlimited; **12.12A** Courtesy of James Gathany/CDC; **12.12B** Courtesy of W.H.O/CDC; **12.12C** Courtesy of World Health Organization/CDC; **12.13A** © Science VU/Visuals Unlimited; **12.13B** Courtesy of CDC; **MicroFocus 6** © National Library of Medicine; **12.14** Courtesy of Armed Forces Institute of Pathology

CHAPTER 13

13.4 © Dennis Kunkel Microscopy, Inc./Phototake/Alamy Images; **13.7** © Dr. R. Dourmashkin/Photo Researchers, Inc.; **Textbook Case** Courtesy of Dr. E. Arum/Dr. N. Jacobs/CDC; **MicroFocus 1** © CNRI/Photo Researchers, Inc.; **13.8** © Dr. David M. Phillips/Visuals Unlimited; **13.9** Courtesy of Joe Miller/CDC; **13.10A** Courtesy of M. Rein, VD/CDC; **13.10B** Courtesy of Dr. Gavin Hart/CDC; **13.10C** Courtesy of Susan Lindsley/CDC; **MicroFocus 2** © Mary Evans Picture Library/Alamy Images; **13.11** Courtesy of the CDC; **13.13A** © Dr. Fred Hossler/Visuals Unlimited; **13.13B** Courtesy of CHROMagar; **MicroFocus 3** © Dennis Kunkel Microscopy, Inc./Phototake/Alamy Images; **13.15A** © Medical-on-Line/Alamy Images; **13.15B** © Kwangshin Kim/Science Photo Library; **13.17B** Courtesy of the CDC; **13.18** Courtesy of CDC; **13.19A** © Medical-on-Line/Alamy Images; **13.19B** © Gladden Willis, M.D./Visuals Unlimited; **13.19C** Courtesy of Vincent A. Fischetti, Ph.D., Head of the Laboratory of Bacterial Pathogenesis at Rockefeller University; **13.20** Courtesy of Dr. Thomas F. Sellers, Emory University/CDC; **13.21** © Dr. M.A. Ansary/Photo Researchers, Inc.; **13.22** © Dr. M.A. Ansary/Photo Researchers, Inc.; **13.23A** Courtesy of Armed Forces Institute of Pathology; **13.23B** Image courtesy of Division of Pediatric Surgery, Brown Medical School; **13.24A-B** Courtesy of American Leprosy Mission, 1 ALM Way Greenville, SC 29601; **MicroFocus 5** Photo courtesy of, US Department of Health and Human Services, Health Resources Services Administration, Bureau of Primary Health Care, National Hansen's Disease Programs; **13.26** © Medical-on-Line/Alamy Images; **13.27** Courtesy of CDC; **13.29** © Adrian Arbib/Peter Arnold, Inc.

Part Opener 4 © Eye of Science/Photo Researchers, Inc.; **Part 4 Microbiology Pathways** Courtesy of James Gathany/CDC

CHAPTER 14

14.1 Courtesy of Ranco Los Amigos National Rehabilitation Center/LADHS; **14.3A** © Science Photo Library/Photo Researchers, Inc.; **14.3B** Courtesy of Clemson University—USDA Cooperative Extension Slide Series, Bugwood.org; **14.4** © Dennis Kunkel Microscopy, Inc./Phototake/Alamy Images; **MicroFocus 2** Courtesy of Didier Raoult, Rickettsia Laboratory, La Timone, Marseille, France; **14.7B** © Eye of Science/Photo Researchers, Inc.; **MicroFocus 3** © Lee D. Simon/Photo Researchers, Inc.; **14.13** © Dennis Kunkel Microscopy, Inc./Phototake/Alamy Images; **14.14A** Courtesy of Greg Knobloch/CDC; **14.14B** © James King-holmes/Photo Researchers, Inc.; **14.14C** Courtesy of Giles Scientific Inc, CA, www.biomic.com; **MicroFocus 5** Courtesy of Dr. G. William, Jr./CDC; **MicroFocus 6** Courtesy of Susy Mercado/CDC; **14.19A** Courtesy of APHIS photo by DR. Al Jenny/CDC

CHAPTER 15

15.1 © Science Photo Library/Photo Researchers, Inc.; **15.2** Human rhinovirus 16: Picornaviridae; Rhinovirus; Human rhinovirus A; strain (NA). Hadfield, A.T., Lee, W.M., Zhao, R., Oliveira, M.A., Minor, I., Rueckert, R.R. and Rossmann, M.G. (1997). The refined structure of human rhinovirus 16 at 2.15 A resolution: implications for the viral life cycle. *Structure*, 5, 427–441. (PDB-ID: 1AYM) Image by J.Y. Sgro, UW-Madison; **MicroFocus 1** © Tihis/ShutterStock, Inc.; **15.4** © Dr. Linda Stannard, UCT/Photo Researchers, Inc.; **MicroFocus 2** © Cristina Fumi/ShutterStock, Inc.; **15.7** © Gopal Murti/Phototake/Alamy Images; **15.8** © Dr. Gary D. Gaugler/Phototake/Alamy Images; **Textbook Case** Courtesy of Dr. Sherif Zaki/CDC; **15.9** © Chris Bjornberg/Photo Researchers, Inc.; **15.11A** © Phototake/Alamy Images; **15.11B** Courtesy of Dr. Hermann/CDC; **15.12A-D** Reproduced from *J. Clin. Microbiol.*, 1985, vol. 22, pp. 366–368, DOI and reproduced with permission from the American Society for Microbiology. Photo courtesy of Woody Spruance, M.D, Professor of Internal Medicine, at University of Utah; **15.13A** © SW Productions/age fotostock; **15.13B** © Science Photo Library/Photo Researchers, Inc.; **15.14** © Scott Camazine & Sue Trainor/Photo Researchers, Inc.; **MicroFocus 3** © Leah-Anne Thompson/ShutterStock, Inc.; **15.15** Courtesy of National Cancer Institute; **15.16A** © Medical-on-Line/Alamy Images; **15.16B** Courtesy of CDC; **15.17** Courtesy of NIP/Barbara Rice/CDC; **MicroFocus 4** © Martin Bowker/ShutterStock, Inc.; **15.18** © Dr. K. G. Murti/Visuals Unlimited; **15.19** © Dr. Ken Greer/Visuals Unlimited; **15.20** Courtesy Dr. Wiesner/CDC; **15.21** Courtesy CDC; **MicroFocus 5** © Medical-on-Line/Alamy Images; **15.22B** © Science VU/Visuals Unlimited; **15.22C** © Dr. F.A. Murphy/Visuals Unlimited; **15.23A** Courtesy of World Health Organization; Diagnosis of Smallpox Slide Series/CDC; **15.23B** Courtesy of James Hicks/CDC

CHAPTER 16

16.4 © Chris Bjornberg/Photo Researchers, Inc.; **16.5** Courtesy of Cynthia Goldsmith/CDC; **16.7** © James Cavallini/Photo Researchers, Inc.; **16.8** © Science Source/Photo Researchers, Inc.; **16.9** © Eye of Science/Photo Researchers, Inc.; **16.11** © Science VU/CDC/Visuals Unlimited; **MicroFocus 4** Courtesy of Chris Zahniser/CDC; **16.13** Courtesy of Cynthia Goldsmith/CDC; **MicroFocus 5** Courtesy of Robert S. Craig/CDC

CHAPTER 17

17.1A © Scott Threlkeld, The Times-Picayune/Landov; **17.1B** © Reuters/Charles W. Luzier/Landov; **17.2B** Courtesy of John Pitt of CSIRO Food Science, Australia; **17.3A** © Dr. Dennis Kunkel/Visuals Unlimited; **MicroFocus 1**© Izatul Lail bin Mohd Yasar/ShutterStock, Inc.; **17.5A** © Dr. Gerald Van Dyke/Visuals Unlimited; **17.5B** © Science VU/R.Roncadori/Visuals Unlimited; **MicroFocus 3** Courtesy of Dr. A. Elizabeth Arnold, Assistant Professor & Curator, Gilbertson Mycological Herbarium, Department of Plant Sciences, University of Arizona; **17.6A** © Andrew Syred/Photo Researchers, Inc.; **17.6B** © Dennis Kunkel Microscopy, Inc./Phototake/Alamy Images; **17.7** © Medical-on-Line/Alamy Images; **17.8** © Science Photo Library/Photo Researchers, Inc.; **MicroFocus 4** Courtesy of Brad Wilson, DVM; **17.12** © Carolina Biological Supply Company/Phototake/Alamy Images; **17.13A** © Dr. John D. Cunningham/Visuals Unlimited; **17.13B** © Liane Matrisch/Dreamstime.com; **17.14A** © Biodisc/Visuals Unlimited; **17.14B** © Dr. John D. Cunningham/Visuals Unlimited; **17.15B** © V. Ahmadjian/Visuals Unlimited; **17.16C** © Anthony Collins/Alamy Images; **17.17B** © Dr. Richard Kessel & Dr. Gene Shih/Visuals Unlimited; **17.17C** © Eye of Science/Photo Researchers, Inc.; **MicroFocus 5** © North Wind Picture Archives/Alamy Images; **17.19A** © Niels-DK/Alamy Images; **17.19B** Courtesy of Misterzin, www.flickr.com/photos/misterzin; **17.20A** Courtesy of Dr. Libero Ajello/CDC; **17.20B** Courtesy of CDC; **17.21A** Courtesy of Dr. Godon Roberstad/CDC; **17.21B** Courtesy of CDC; **17.22A** © E. Gueho—CNRI/Photo Researchers, Inc.; **17.22B** © Everett Beneke/Visuals Unlimited; **17.23A** Courtesy of Dr. Leanor Haley/CDC; **17.23B** © Dr. G. W. Willis/Visuals Unlimited; **17.24A** Courtesy of Dr. Libero Ajello/CDC; **17.24B** © Scott Camazine/Alamy Images; **Textbook Case** Courtesy of Courtesy of Dr. Errol Reiss/CDC, modified

CHAPTER 18

18.3A © M I (Spike) Walker/Alamy Images; **18.3B** © Manfred Kage/Peter Arnold, Inc.; **18.3C** © Mark Bond/ShutterStock, Inc.; **18.3D** Courtesy of Eduardo A. Morales, Ph.D., The Academy of Natural Sciences of Philadelphia; **MicroFocus 1** © Holt Studios International Ltd/Alamy Images; **18.4A** © Wim van Egmond/Visuals Unlimited; **18.4B** © Michael Abbey/Visuals Unlimited; **18.4C** Courtesy of the Laboratory of Parasitology, University of Pennsylvania School of Veterinary Medicine; **18.5A** © M I (Spike) Walker/Alamy Images; **18.5B** © Roland Birke/Phototake/Alamy Images; **18.8A** © Dennis Kunkel Microscopy, Inc./Phototake/Alamy Images; **18.8B** Courtesy of WHO/CDC; **18.8C** © Medical-on-Line/Alamy Images; **MicroFocus 2** © Leslie E. Kossoff/AP Photos; **MicroFocus 3** Courtesy of Christopher J. Rapuano, M.D.; Director, Cornea Service, Wills Eye. Professor, Jefferson Medical College of Thomas Jefferson University, Philadelphia, PA; **18.10** © Jerome Paulin/Visuals Unlimited; **18.11** Reproduced from Ma, P. and R. Soave, *J. Infect Dis.* 147:5 (1983): 824–828. With permission from University of Chicago Press; **18.12** © Dr. Dennis Kunkel/Visuals Unlimited; **MicroFocus 4** © Mark Pink/Alamy Images; **18.14A** Courtesy of Dr. Mae Melvin/CDC; **18.14B** Courtesy of Peggy Greb/USDA ARS; **18.15** © Dennis Kunkel Microscopy, Inc./Phototake/Alamy Images; **Textbook Case** Courtesy of CDC; **18.17** Courtesy of Dr. Govinda S. Visvesvara/CDC; **18.18A** © Sinclair Stammers/Photo Researchers, Inc.; **18.18B** © Medical-on-Line/Alamy Images; **18.18C** © Dennis Kunkel Microscopy, Inc./Phototake/Alamy Images; **18.19** © R.F. Ashley/Visuals Unlimited; **MicroFocus 5** © CNRI/Photo Researchers, Inc.; **MicroInquiry** © Eye of Science/Photo Researchers, Inc.; **18.21A** © Biodisc/Visuals Unlimited; **18.21B** © Alfred Pasieka/Photo Researchers, Inc.; **18.22 insert** © Pr. Bouree/Photo Researchers, Inc.; **18.23** Courtesy of CDC; **18.24** © Dennis Kunkel Microscopy, Inc./Phototake/Alamy Images; **18.26** © John Greim/Photo Researchers, Inc.

Part Opener 5 © David M. Phillips/Photo Researchers, Inc.; **Part 5 Microbiology Pathways** Courtesy of Ethleen Lloyd/CDC

CHAPTER 19

MicroFocus 1 © SPL/Photo Researchers, Inc.; **MicroFocus 3** © Comstock Images/Jupiterimages; **19.9** © Dennis Kunkel Microscopy, Inc./Phototake/Alamy Images; **19.10** © Dr. K. G. Murti/Visuals Unlimited; **19.11A** Reproduced from *Trends in Microbiology*, Vol. 1, Tilney, L.G. and Tilney, M.S., The wily ways of a parasite: induction of actin assembly by Listeria, pp. 25–31, © 1993, with permission from Elsevier. [http://www.sciencedirect.com/science/journal/0966842X.]; **19.16** © Matt Meadows/Peter Arnold, Inc.; **MicroFocus 4** Courtesy of Jay Herman/NASA; **MicroFocus 6A** Courtesy of CDC; **MicroFocus 6B** © CNRI/Photo Researchers, Inc.

CHAPTER 20

MicroFocus 1 © nazira_g/ShutterStock, Inc.; **20.4 insert** © Dennis Kunkel Microscopy, Inc./Phototake/Alamy Images; **MicroFocus 2** © Dr. Gopal Murti/Photo Researchers, Inc.; **MicroFocus 3** © Volker Steger/Photo Researchers, Inc.; **MicroFocus 4** © Blend Images/Jupiterimages

CHAPTER 21

21.1A © Dr. Klaus Boller/Photo Researchers, Inc.; **21.1B** © NIBSC/Photo Researchers, Inc.; **21.7B** Genetics Home Reference. [Internet] Bethesda (MD): National Library of Medicine (US); 2003- [updated 2004 Aug 4; cited 2006 Aug 08] Immunoglobulin G (IgG) Available from http://ghr.nlm.nih.gov/handbook/illustrations/igg; **21.14** © Steve Gschmeissner/Photo Researchers, Inc.

CHAPTER 22

MicroFocus 1 © James King-Holmes/Photo Researchers, Inc.; **MicroFocus 2** © Jaimie Duplass/ShutterStock, Inc.; **MicroFocus 5** © Olivier Asselin/Alamy Images; **22.9C** Courtesy of Gunnar Flåten, Bionor Laboratories AS; **22.9D** © Ed Reschke/Peter Arnold, Inc.; **22.11A-C** Reproduced from *Appli. Environ. Microbiol*, 1996, vol. 62(2), p. 347–352, DOI and reproduced with permission from the American Society for Microbiology. Photo courtesy of Doctor Bruce J. Paster, Senior Member at Department of Molecular Genetics, The Forsyth Institute; **MicroInquiry** Courtesy of Home Access Health Corporation; **22.15** © Dr Jeremy Burgess/Photo Researchers, Inc.; **22.16 insert** Image courtesy of Abbott Laboratories

CHAPTER 23

MicroFocus 1 Courtesy of USDA-ARS; **23.6** © Michael Donne/Photo Researchers, Inc.; **MicroInquiry** © Jean Schweitzer/Dreamstime.com; **MicroFocus 2** © Stockbyte Silver/Getty Images; **23.12** © Bart's Medical Library/Phototake/Alamy Images; **23.13B** © Bill Beatty/Visuals Unlimited; **23.15** © Scott Camazine/Alamy Images; **MicroFocus 3** © Photodisc; **23.17** © Peter Menzel/Photo Researchers, Inc.; **MicroFocus 4** © Dana Ward/ShutterStock, Inc.; **MicroFocus 5** Courtesy of OraSure Technologies, Inc.; **MicroFocus 6** © Anette Linnea Rasmussen/ShutterStock, Inc.

CHAPTER 24

24.2A Collection of the University of Michigan Health System, Gift of Pfizer, Inc. (UMHS.44); **24.2B** © National Library of Medicine; **MicroFocus 1A-B** © Dennis Kunkel Microscopy, Inc./Phototake/Alamy Images; **MicroFocus 2** © imagebroker/Alamy Images; **24.9B** © Kenneth E. Greer/Visuals Unlimited; **MicroFocus 3** © Gary Cook/Alamy Images; **MicroFocus 4** Photo by Forest & Kim Starr; **24.12** © Jack Barker/Alamy Images; **24.14A** Courtesy of Dr. Richard Facklam/CDC; **24.14B-C** Courtesy of bioMerieux, Inc.; **MicroFocus 5** Courtesy of Janice Haney Carr/CDC; **24.18** © Stephen Saks Photography/Alamy Images; **MicroFocus 6A** © Imagemore/age fotostock; **MicroFocus 6B** Courtesy of Dr. J. J. farmer/CDC; **MicroFocus 6C** © Anyka/ShutterStock, Inc.; **MicroFocus 6D** Courtesy of Kimberly Smith/CDC

(On-line only) Part Opener 6 © Scimat/Photo Researchers, Inc.; **Part 6 Microbiology Pathways** © Shout/Alamy Images

CHAPTER 25

25.4 © Dennis Kunkel Microscopy, Inc./Phototake/Alamy Images; **25.5** © Inga Spence/Visuals Unlimited; **25.6** © Brian Klimek, *The Laurinburg Exchange*/AP Photos; **25.8** © Scimat/Photo Researchers, Inc.; **25.9** © Robert Longuehaye, NIBSC/SPL/Photo Researchers, Inc.; **MicroFocus 6** © Stephen Coburn/ShutterStock, Inc**;** **25.10** Courtesy of USDA; **25.12** © Pacific Press Service/Alamy Images; **25.13** © David R. Frazier Photolibrary, Inc./Alamy Images; **25.14** © Scimat/Photo Researchers, Inc.

CHAPTER 26

26.2A © Eric Grave/Photo Researchers, Inc.; **26.2B** © EM Unit, VLA/Photo Researchers, Inc.; **26.4A** Courtesy of Professor Gordon T. Taylor, Stony Brook University/NSF Polar Programs/NOAA, **26.4B** Courtesy of GSFC/LaRC/JPL, MISR Team/NASA; **26.5A** Courtesy of CDC; **26.5B** Courtesy of Don Howard/CDC; **MicroFocus 1** Courtesy of Chau Nguyen, M.D., Memorial University of Newfoundland; **26.8** Courtesy of Dr. Rodney M. Donlan and Janice Carr/CDC; **26.9** Courtesy of Scott Bauer/USDA; **26.12A** Courtesy of Harold Evans; **26.12B** © Medical-on-Line/Alamy Images; **MicroFocus 4** Photo by Forest & Kim Starr

CHAPTER 27

27.1B © Maximilian Stock LTD/Phototake/Alamy Images; **27.5C** © LiquidLibrary; **27.6A** © Christine Case/Visuals Unlimited; **27.6B** © Scimat/Photo Researchers, Inc.; **27.7A** © George Chapman/Visuals Unlimited; **27.7B** © Holt Studios International Ltd/Alamy Images; **27.8B** © Mashkov Yuri, *Itar-Tass*/Landov; **27.9** Courtesy of the Exxon Valdez Oil Spill Trust Council/NOAA; **27.10** © Dr. Gopal Murti/Visuals Unlimited

Unless otherwise indicated, all photographs and illustrations are under copyright of Jones and Bartlett Publishers, LLC, or have been provided by the author.

Pronouncing Organism Names (*continued*)

Microsporum racemosum mī-krō-spô'rum ras-ē-mōs'um

Morchella esculentum môr-che'lä es-kyū-len'tum

Mucor mū'kôr

Mycobacterium avium mī-kō-bak-ti'rē-um ā'vē-um

M. bovis bō'vis

M. cheloni kē-lō'ē

M. haemophilum hē-mo'fil-um

M. kansasii kan-sä-sē'ī

M. leprae lep'rī

M. marinum mār'in-um

M. tuberculosis tū-bėr-kū-lō'sis

Mycoplasma genitalium mī-kō-plaz'mä jen'i-tä-lē-um

M. pneumoniae nu-mō'nē-ī

Myxococcus xanthus micks-ō-kok'kus zan'thus

Naegleria fowleri nī-gle'rē-ä fou'lėr-ē

Nannizzia nan'nė-zė-ä

Necator americanus ne-kā'tôr ä-me-ri-ka'nus

Neisseria gonorrhoeae nī-se'rē-ä go-nôr-rē'ī

N. meningitidis me-nin ji'ti-dis

Neurospora nū-ros'pōr-ä

Nitrobacter nī-trō-bak'tėr

Nitrosomonas nī-trō-sō-mō'näs

Nocardia asteroids nō-kär'dē-a as-tėr-oi'dēz

Nostoc nos'tok

Paramecium pär-ä-mē'sē-um

Pasteurella multocida pas-tyėr-el'lä mul-tō'si-dä

Pelagibacter ubique pel-aj'ē-bak-tėr ū-bēk

Penicillium camemberti pen-i-sil'lē-um kam-am-bėr'tē

P. chrysogenum krī-so'gen-um

P. griseofulvin gris-ē-ō-fül'vin

P. notatum nō-tä'tum

P. roqueforti rō-kō-fôr'tē

Peptostreptococcus pep-tō-strep'tō-kok'kus

Pfiesteria piscicida fes-ter'ē-ä pis-si-sē'dä

Photorhabdus luminescens fō-tō-rab'dus lü-mi-nes'senz

Phytophthora infestans fī-tof'thô-rä in-fes'tans

P. ramorum rä-môr'-um

Plasmodium falciparum plaz-mō'dē-um fal-sip'är-um

P. malariae mä-lā'rē-ī

P. ovale ō'va'lē

P. vivax vī'vaks

Pneumocystis jiroveci nü-mō-sis'tis jėr-ō-vek'ē

Porphyromonas gingivalis pôr'fī-rō-mō-näs jin-ji-val'is

Prochlorococcus prō-klôr-ō-kok'kus

Propionibacterium acnes prō-pē-on'ē-bak-ti-rē-um ak'nēz

Proteus mirabilis prō'tē-us mi-ra'bi-lis

Pseudomonas aeruginosa sū-dō-mō'näs ā-rü ji-nō'sä

P. cepacia se-pā'sē-ä

P. marginalis mär-gin-al'is

Rhizobium rī-zō'bē-um

Rhizopus stolonifer rī'zo-pus stō-lon-i-fėr

Rhodospirillum rubrum rō-dō-spī-ril'um rūb'rum

Rickettsia akari ri-ket'sē-ä ä-kär'ī

R. prowazekii prou-wa-ze'kē-ē

R. rickettsii ri-ket'sē-ē

R. tsutsugamushi tsü-tsü-gäm-ü'shē

R. typhi tī'fē

Saccharomyces carlsbergensis sak-ä-rō-mī'sēs kä-rls-bėr-gen'sis

S. cerevisiae se-ri-vis'ē-ī

S. ellipsoideus ē-lip-soi'dē-us

Saccharopolyspora erythraea sak-kär-ō-pol'ē-spo-rä ē-rith'rä-ē

Salmonella enterica säl-mōn-el'lä en-tėr-i'kä

S. enterica serotype Enteritidis en-tėr-i-tī'dis

S. enterica serotype Typhi tī'fē

S. enterica serotype Typhimurium tī-fi-mur'ē-um

Schistosoma haematobium shis-tō-sō'mä hē-mä-tō'bē-um

S. japonicum ja-pō'nē-kum

S. mansoni man'-sō-nē

Serratia marcescens ser-rä'tē-ä mär-ses'sens

Shigella dysenteriae shi-gel'lä dis-en-te'rē-ī

S. sonnei son'nē-ē

Spirillum minus spī'ril-lum mī'nus

Sporothrix schenkii spô-rō'thriks shen'kē-ē

Staphylococcus aureus staf-i-lō-kok'kus ô-rē-us

S. epidermidis e-pi-der'mi-dis

S. saprophyticus sa-prō-fi'ti-kus

Streptobacillus moniliformis strep-tō-bä-sil'lus mon-i-li-fôr'mis

Streptococcus agalactiae strep-tō-kok'kus a-gal-ac'tē-ī

S. cremoris kre-mōr'is

S. lactis lak'tis

S. mutans mū'tans

S. pneumoniae nü-mō'nē-ī

S. pyogenes pī-äj'en-ēz

S. sobrinus sō'bri-nus

S. thermophilus thėr-mo'fil-us

Streptomyces cattley strep-tō-mī'sēs kat-tel'-ē

S. coelicolor kō'lē-ku-lėr

S. erythraeus er-i-thrā-us

S. griseus gri'sē-us

S. lincolnensis lin-kōl-nen'sis